W9-CNS-865

ENCYCLOPEDIA OF

Microbiology

Volume 2 D–L

Editorial Advisory Board

Editor-in-Chief
Joshua Lederberg
Rockefeller University
New York, New York

ENCYCLOPEDIA OF Microbiology

Volume 2 D–L

ACADEMIC PRESS, INC.
A Division of Harcourt Brace & Company
San Diego New York Boston London Sydney Tokyo Toronto

This book is printed on acid-free paper. 

Academic Press, Inc.
1250 Sixth Avenue, San Diego, California 92101-4311

United Kingdom Edition published by
Academic Press Limited
24–28 Oval Road, London NW1 7DX

Library of Congress Cataloging-in-Publication Data

Encyclopedia of microbiology / edited by Joshua Lederberg
p. cm.
Includes bibliographical references and indexes.
ISBN 0-12-226891-1 (v. 1). -- ISBN 0-12-226892-X (v. 2). -- ISBN 0-12-226893-8 (v. 3). -- ISBN 0-12-226894-6 (v. 4)
1. Microbiology--Encyclopedias. I. Lederberg, Joshua.
QR9E53 1992
576'.03--dc20 92-4429
CIP

PRINTED IN THE UNITED STATES OF AMERICA
93 94 95 96 97 EB 9 8 7 6 5 4 3 2

Contents

Please refer to Volume 4 for the following information:

Preface

For the purposes of this encyclopedia, microbiology has been understood to embrace the study of "microorganisms," including the basic science and the roles of these organisms in practical arts (agriculture and technology) and in disease (public health and medicine). Microorganisms do not constitute a well-defined taxonomic group; they include the two kingdoms of Archaebacteria and Eubacteria, as well as protozoa and those fungi and algae that are predominantly unicellular in their habit. Viruses are also an important constituent, albeit they are not quite "organisms." Whether to include the mitochondria and chloroplasts of higher eukaryotes is a matter of choice, since these organelles are believed to be descended from free-living bacteria. Cell biology is practiced extensively with tissue cells in culture, where the cells are manipulated very much as though they were autonomous microbes; however, we shall exclude this branch of research. Microbiology also is enmeshed thoroughly with biotechnology, biochemistry, and genetics, since microbes are the canonical substrates for many investigations of genes, enzymes, and metabolic pathways, as well as the technical vehicles for discovery and manufacture of new biological products, for example, recombinant human insulin.

Within these arbitrarily designated limits, let us consider the overall volume of published literature in microbiology, where to find its core, and strategies for searching for current information on particular topics. Most of the data for this preface are derived from the 1988 Journal Citation Reports Current Contents (T) of the Institute for Scientific Information (ISI). Table I lists the 53 most consequential journals in microbiology, assessed by citation impact factor, the average number of literature citations per article published in a given journal. Table II presents that list sorted by the total number of articles printed in each journal in 1988. Table III shows the distribution of journals citing the *Journal of Bacteriology* and the distribution of journals cited in it.

Obviously, the publications of the American Society for Microbiology (indicated by AMS in the tables) play a commanding role. The society is now making its journals available in electronically searchable form (on optical disks), which will greatly facilitate locating and retrieving the most up-to-date information on any given subject. In addition, interdisciplinary journals such as *Nature (London), Science,* and the *Proceedings of the National Academy of Sciences, U.S.A.* are important sources of prompt news of scientific developments in microbiology. It is difficult to assess how much of their total publication addresses microbiology. As seen in Table III, the bibliographies in the *Journal of Bacteriology* cite half as many articles from the *Proceedings* (2348) as from the *Journal of Bacteriology* itself (5708). The 7038 articles indicated in Table II probably reach some 10,000 per year when these interdisciplinary and other dispersed sources are taken into account. An equal number might be added from overlapping aspects of molecular biology and genetics. To find and read all these titles would tax any scholar, although it could be done as a near full-time occupation with the help of the weekly Current Contents (T) of the ISI. To start afresh, with perhaps a decade's accumulation of timely background, would be beyond reasonable human competence. No one person would intelligently peruse more than a small fraction of the total texts.

The "Encyclopedia of Microbiology" is intended to survey the entire field coherently, complementing material that would be included in an advanced undergraduate and graduate major course of university study. Particular topics should be accessible to talented high school and college students, as well as graduates involved in teaching, research, and technical practice of microbiology.

Even these hefty volumes cannot embrace all current knowledge in the field. Each article does provide key references to the literature available at the time of writing. Acquisition of more detailed and up-to-date knowledge depends on (1) exploiting the review and monographic literature and (2) bibliographic retrieval of the preceding and current research literature. To make greatest use of review literature and monographs, the journals listed in Table II are invaluable. Titles such as *Annual Reviews* should not be misunderstood: these journals appear at annual intervals, but 5 or 10 years of accumulated research is necessary for the inclusion of a focused treatment of a given subject.

To access bibliographic materials in microbiology, the main retrieval resources are Medline, sponsored by the U.S. National Library of Medicine, and the Science Citation Index of the ISI. With governmental subsidy, Medline is widely available at modest cost: terminals are available at every medical school and at many other academic centers. Medline provides searches of the recent literature by author, title, and key word, and offers on-line displays of the relevant bibliographies and abstracts. Medical aspects of microbiology are covered exhaustively; general microbiology is covered in reasonable depth. The Science Citation Index must recover its costs from user fees, but is widely available at major research centers. It offers additional search capabilities, especially by citation linkage. Therefore, starting with the bibliography of a given encyclopedia article, one can quickly find (1) all articles more recently published that have cited those bibliographic reference starting points and (2) all other recent articles that share bibliographic information with the others. With luck, one of these articles may be identified as another comprehensive review that has digested more recent or broader primary material.

On a weekly basis, services such as Current Contents on Diskette (ISI) and Reference Update offer still more timely access to current literature as well as abstracts with a variety of useful features. Under the impetus of intense competition, these services are evolving rapidly, to the great benefit of a user community desperate for electronic assistance in coping with the rapidly growing and intertwined networks of discovery. The bibliographic services of Chemical Abstracts and Biological Abstracts would also be potentially invaluable; however, their coverage of microbiology is rather limited.

In addition, major monographs have appeared from time to time—"The Bacteria," "The Prokaryotes," and many others. Your local reference library should be consulted for these volumes.

Valuable collections of reviews also include *Critical Reviews for Microbiology, Symposia of the Society for General Microbiology, Monographs of the ASM,* and *Proceedings of the International Congresses of Microbiology.*

The articles in this encyclopedia are intended to be accessible to a broader audience, not to take the place of review articles with comprehensive bibliographies. Citations should be sufficient to give the reader access to the latter, as may be required. We do apologize to many individuals whose contributions to the growth of microbiology could not be adequately embraced by the secondary bibliographies included here.

The organization of encyclopedic knowledge is a daunting task in any discipline; it is all the more complex in such a diversified and rapidly moving domain as microbiology. The best way to anticipate the rapid further growth that we can expect in the near future is unclear. Perhaps more specialized series in subfields of microbiology would be more appropriate. The publishers and editors would welcome readers' comments on these points, as well as on any deficiencies that may be perceived in the current effort.

My personal thanks are extended to Kathryn Linenger at Academic Press for her diligent, patient, and professional work in overseeing this series; to my coeditors, Martin Alexander, David A. Hopwood, Barbara H. Iglewski, and Allan I. Laskin; above all, to the many very busy scientists who took time to draft and review each of these articles.

Joshua Lederberg

Table I The Top Journals in Microbiology Listed by Impact Factor

Citation impact rank	Journal title	Number of articles published in 1988	Citation impact rank	Journal title	Number of articles published in 1988
1	*Microbiol. Rev.*	28	28	*FEMS Microbiol. Lett.*	365
2	*Adv. Microb. Ecol.*	10	29	*Am. J. Reprod. Immunol.*	50
3	*Annu. Rev. Microbiol.*	29	30	*Infection*	103
4	*FEMS Microbiol. Rev.*	13	31	*Can. J. Microbiol.*	236
5	*Yeast*	NA	32	*Curr. Microbiol.*	87
6	*J. Bacteriol.*	915	33	*J. Appl. Bacteriol.*	125
7	*Mol. Microbiol.*	94	34	*J. Microbiol. Meth.*	34
8	*Antimicrob. Agents Ch.*	408	35	*B. I. Pasteur*	20
9	*Rev. Infect. Dis.*	213	36	*ZBL Bakt. Mikr. Hyg. A*	164
10	*CRC Crit. Rev. Microbiol.*	12	37	*Ann. Inst. Pasteur Mic.*	58
11	*Syst. Appl. Microbiol.*	52	38	*Vet. Microbiol.*	104
12	*Int. J. Syst. Bacteriol.*	83	39	*Acta Path. Micro. Im. B*	NA
13	*J. Antimicrob. Chemoth.*	352	40	*Protistologica*	NA
14	*Appl. Environ. Microb.*	588	41	*Med. Microbiol. Immun.*	37
15	*J. Clin. Microbiol.*	619	42	*Diagn. Micr. Infec. Dis.*	60
16	*Adv. Appl. Microbiol.*	8	43	*Int. J. Food Microbiol.*	66
17	*Curr. Top. Microbiol.*	53	44	*J. Gen. Appl. Microbiol.*	27
18	*Arch. Microbiol.*	173	45	*Microbiol. Immunol.*	122
19	*J. Gen. Microbiol.*	367	46	*Lett. Appl. Microbiol.*	81
20	*Enzyme Microb. Tech.*	108	47	*Gen. Physiol. Biophys.*	57
21	*Eur. J. Clin. Microbiol.*	161	48	*A. Van Leeuw. J. Microb.*	51
22	*FEMS Microbiol. Ecol.*	42	49	*Symbiosis*	14
23	*J. Med. Microbiol.*	124	50	*Comp. Immunol. Microb.*	27
24	*J. Infection*	68	51	*Microbios.*	61
25	*Eur. J. Protistol.*	37	52	*ZBL Bakt. Mikr. Hyg. B*	76
26	*Microbiol. Sci.*	70	53	*J. Basic Microb.*	69
27	*Appl. Microbiol. Biot.*	270			

NA, Not available.

Table II Microbiology Journals Listed by Total Number of Articles Published per Year (1988)

Journal title	Number of articles published in 1988	Journal title	Number of articles published in 1988
J. Bacteriol.	915	*Int. J. Food Microbiol.*	66
J. Clin. Microbiol.	619	*Microbios.*	61
Appl. Environ. Microb.	588	*Diagn. Micr. Infec. Dis.*	60
Antimicrob. Agents Ch.	408	*Ann. Inst. Pasteur Mic.*	58
J. Gen. Microbiol.	367	*Gen. Physiol. Biophys.*	57
FEMS Microbiol. Lett.	365	*Curr. Top. Microbiol.*	53
J. Antimicrob. Chemoth.	352	*Syst. Appl. Microbiol.*	52
Appl. Microbiol. Biot.	270	*A. Van Leeuw. J. Microb.*	51
ZBL Bakt. Mikr. Hyg. A	240	*Am. J. Reprod. Immunol.*	50
Can. J. Microbiol.	236	*FEMS Microbiol. Ecol.*	42
Rev. Infect. Dis.	213	*Med. Microbiol. Immun.*	37
Arch. Microbiol.	173	*Eur. J. Protistol.*	37
Eur. J. Clin. Microbiol.	161	*J. Microbiol. Meth.*	34
J. Appl. Bacteriol.	125	*Eur. J. Protistology*	29
J. Med. Microbiol.	124	*Annu. Rev. Microbiol.*	29
Microbiol Immunol.	122	*Microbiol. Rev.*	28
Enzyme Microb. Tech.	108	*J. Gen. Appl. Microbiol.*	27
Vet. Microbiol.	104	*Comp. Immunol. Microb.*	27
Infection	103	*B. I. Pasteur*	20
Mol. Microbiol.	94	*Acta Path. Micro. Im.*	18
Curr. Microbiol.	87	*Symbiosis*	14
Int. J. Syst. Bacteriol.	83	*FEMS Microbiol. Rev.*	13
Lett. Appl. Microbiol.	81	*CRC Crit. R. Microbiol.*	12
Microbiol. Sci.	70	*Adv. Microb. Ecol.*	10
J. Basic Microb.	69	*Adv. Appl. Microbiol.*	8
		Total	7038

Table III.A Distribution of Journals Cited in *Journal of Bacteriology*, 1979–1988

Journal cited	Number of citations	Journal cited	Number of citations
J. Bacteriol.	5708	*Genetics*	183
P. Natl. Acad. Sci. U.S.A.	2348	*Can. J. Microbiol.*	139
J. Biol. Chem.	1698	*Arch. Biochem. Biophys.*	127
Mol. Gen. Genet.	1157	*Virology*	123
J. Mol. Biol.	1148	*Bacteriol. Rev.*	118
Gene	902	*Cold Spring Harb. Sym.*	110
Nature (London)	820	*Antimicrob. Agents Ch.*	109
Nucleic Acids Res.	804	*Escherichia Coli Sal.*	95
Cell	802	*Plant Physiol.*	80
J. Gen. Microbiol.	701	*J. Biochem.-Tokyo*	78
Infect. Immun.	478	*J. Virol.*	78
Methods Enzymol.	434	*Mol. Cell. Biol.*	68
Anal. Biochem.	411	*J. Infect. Dis.*	67
Biochim. Biophys. Acta	401	*Bio-Technol.*	61
Eur. J. Biochem.	376	*Exp. Gene Fusions*	60
Mol. Cloning Laboratory	363	*Trends Biochem. Sci.*	60
Microbiol. Rev.	361	*Mutat. Res.*	59
Arch. Microbiol.	347	*Syst. Appl. Microbiol.*	55
Embo J.	327	*Phytopathology*	51
Biochemistry-U.S.	310	*Adv. Bacterial Genet.*	50
Science	301	*Photochem. Photobiol.*	50
Appl. Environ. Microb.	294	*Biochimie*	49
FEMS Microbiol. Lett.	257	*J. Exp. Med.*	48
Exp. Mol. Genetics	234	*Agr. Biol. Chem. Tokyo*	47
Plasmid	234	*Int. J. Syst. Bacteriol.*	44
Biochem. Bioph. Res. Commun.	224	*FEMS Microbiol. Rev.*	43
FEBS Lett.	213	*J. Clin. Microbiol.*	42
Biochem. J.	207	*Curr. Microbiol.*	41
Annu. Rev. Microbiol.	194	*J. Cell Biol.*	41
Annu. Rev. Biochem.	188		
Annu. Rev. Genet.	187	All other (1301)	4311

(continues)

Table III.B *(continued)* Distribution of Journals Citing *Journal of Bacteriology*, 1979–1988

Journal citing	Number of citations	Journal citing	Number of citations
J. Bacteriol.	5708	Curr. Genet.	117
J. Biol. Chem.	1119	FEMS Microbiol. Rev.	115
J. Gen. Microbiol.	963	J. Basic Microb.	115
Mol. Gen. Genet.	896	J. Antimicrob. Chemoth.	112
Appl. Environ. Microb.	890	Microb. Pathogenesis	110
Microbiol. Rev.	759	Science	104
Infect. Immun.	663	Ann. Inst. Pasteur Mic.	101
FEMS Microbiol. Lett.	648	Methods Enzymol.	99
Gene	599	ZBL Bakt. Mikr. Hyg. A	98
P. Natl. Acad. Sci. U.S.A.	588	A. Van Leeuw. J. Microb.	95
Can. J. Microbiol.	579	Annu. Rev. Biochem.	94
Arch. Microbiol.	484	Plant Physiol.	88
Mol. Microbiol.	452	J. Infect. Dis.	86
J. Mol. Biol.	434	J. Med. Microbiol.	85
Nucleic Acids Res.	431	Folia Microbiol.	79
Biochim. Biophys. Acta	378	Genetika	79
Eur. J. Biochem.	350	Gene Dev.	78
Antimicrob. Agents Ch.	340	Microbios.	77
Annu. Rev. Microbiol.	316	Arch. Biochem. Biophys.	75
Cell	246	Biotechnol. Bioeng.	73
Biochimie	238	Nature (London)	69
Biochemistry-U.S.	236	Syst. Appl. Microbiol.	69
Plasmid	236	Zh. Mikrob. Epid. Immun.	67
Embo J.	234	J. Antibiot.	66
J. Clin. Microbiol.	214	Annu. Rev. Genet.	65
Genetics	201	Microbiol. Immunol.	65
Adv. Microb. Physiol.	199	J. Biochem.-Tokyo	64
Agr. Biol. Chem. Tokyo	198	Microbial Ecol.	60
Mol. Cell. Biol.	197	Plant Soil	58
CRC Crit. R. Microbiol.	194	Anal. Biochem	56
Curr. Microbiol	193	Annu. Rev. Cell Biol.	55
Appl. Microbiol. Biot.	183	Biotechnol. Lett.	54
J. Appl. Bacteriol.	169	Adv. Microb. Ecol.	53
Mutat. Res.	160	Enzyme Microb. Tech.	53
Biochem. Bioph. Res. Commun.	152	Curr. Sci. India	52
Rev. Infect. Dis.	141	Eur. J. Clin. Microbiol.	51
Biochem. J.	137	J. Theor. Biol.	51
Microbiol. Sci.	135	Bot. Acta	50
Int. J. Syst. Bacteriol.	128	Photochem. Photobiol.	50
FEBS Lett	125		

How to Use the Encyclopedia

This encyclopedia is organized in a manner that we believe will be the most useful to you, and we would like to acquaint you with some of its features.

The volumes are organized alphabetically as you would expect to find them in, for example, magazine articles. Thus, "Foodborne Illness" is listed as such and would not be found under "Illness, Foodborne." If the first words in a title are not the primary subject matter contained in an article, the main subject of the title is listed first (e.g., "Heavy Metals, Bacterial Resistances," "Marine Habitats, Bacteria," "Method, Philosophy," "Transcription, Viral"). This is also true if the primary word of a title is too general (e.g.,"Bacteriocins, Molecular Biology"). Here, the word "bacteriocins" is listed first because "molecular biology" is a very broad topic. Titles are alphabetized letter-by-letter so that "Cell Membrane: Structure and Function" is followed by "Cellulases" and then by "Cell Walls of Bacteria."

Each article contains a brief introductory Glossary wherein terms that may be unfamiliar to you are defined *in the context of their use in the article*. Thus, a term may appear in another article defined in a slightly different manner or with a subtle pedagogic nuance that is specific to that particular article. For clarity, we have allowed these differences in definition to remain so that the terms are defined relative to the context of each article.

Articles about closely related subjects are identified in the Index of Related Titles at the end of the last volume (Volume 4). The article titles that are cross-referenced within each article may be found in this index, along with other articles on related topics.

The Subject Index contains specific, detailed information about any subject discussed in the *Encyclopedia*. Entries appear with the source volume number in boldface followed by a colon and the page number in that volume where the information occurs (e.g., "DNA repair by bacterial cells, **2**:9"). Each article is also indexed by its title (or a shortened version thereof), and the page ranges of the article appear in boldface (e.g. "Lyme disease, **2:639–646**" means that the primary coverage of the topic of Lyme disease occurs on pages 639–646 of Volume 2).

If a topic is covered primarily under one heading but additional related information may be found elsewhere, a cross-reference is given to the related material. For example, "Biodegradation" would contain all the page numbers where relevant information occurs, followed by "*See also* Bioremediation; Pesticide biodegradation" for different but related information. Similarly, a "*See*" reference refers the reader from a less-used synonym (or acronym) to a more specific or descriptive subject heading. For example, "Immunogens, synthetic. *See* Vaccines, synthetic." A *See under* cross-reference guides the reader to a specific subheading under a term. For example, "Mixis. *See under* Genome rearrangement."

An additional feature of the Subject Index is the identification of Glossary terms. These appear in the index where the word "defined" (or the words "definition of") follows an entry. As we noted earlier, there may be more than one definition for a particular term, and as when using a dictionary, you will be able to choose among several different usages to find the particular meaning that is specifically of interest to you.

D

Dairy Products

Mary Ellen Sanders
Consultant; Littleton, Colorado

Glossary

Bacteriophage Virus infecting a bacterium

Commercial sterility Processing (usually retort processing) of food to eliminate all pathogenic and spoilage microorganisms that can contribute to food spoilage under normal storage conditions. The only microbes, if any, remaining in a commercially sterile product are extremely heat-resistant bacterial spores, which can cause spoilage of product stored at unusually high storage temperatures

Fluid milk Milk that is prepared to be consumed as a natural liquid product, including raw or pasteurized milks with different milk fat contents or vitamin or milk solids fortifications

Lactic acid bacteria Name of a group of bacteria belonging to a diversity of genera used to effect food fermentations; its composed chiefly of bacteria whose primary metabolic end product from carbohydrate metabolism is lactic acid, although poor lactate producers (e.g., leuconostocs and propionibacteria) are sometimes included because of their association with food fermentations

Milk products Products manufactured from fluid milk, including natural cheeses, processed cheeses, fermented milks, yogurts, butter, ice cream, sour cream, whipped cream, canned milks, and dried milk

Starter culture Microbial strain or mixture of strains, species, or genera used to effect a fermentation and bring about functional changes in milk that lead to desirable characteristics in the fermented product

THE MICROBIOLOGY OF DAIRY PRODUCTS encompasses the safety and spoilage of fluid milk and milk products, as well as the fermentation of milk into a plethora of cheeses, yogurts, and fermented milks produced worldwide. These two facets of dairy microbiology are intricately associated, because undoubtedly, the first fermented milk products made some 8000 years ago provided a means for preserving milk as a safe and wholesome food. A diversity of microbes is associated with dairy products, including gram-positive and gram-negative bacteria, molds, yeasts, and bacteriophages. Spoilage and pathogenic microorganisms are predominantly controlled by pasteurization, refrigeration, fermentation, and limiting postprocesses contamination. Reduced water activity, high salt content, and heat sterilization also contribute to the preservation of some dairy products. Great effort is expended to control microbes responsible for spoilage or pathogenicity, but microbes are intentional additives to milk destined for fermentation. These microbes serve to preserve milk primarily through the production of organic acids. The dairy microbiologist must balance microbial populations and activities in milk so that positive effects are enhanced and spoilage and pathogenesis are discouraged or eliminated.

I. Natural Flora of Milk

Milk, as it is produced by the mammal, is sterile. Bacteria inhabiting the teat do, however, migrate up into the interior, causing even aseptically drawn milk to contain some bacteria, predominantly micrococci, streptococci, and *Corynebacterium bovis*. Milk taken from a mastitic animal (one with a teat or udder infection) will show dramatic levels of microbes, including streptococci, staphylococci, coliforms, *Pseudomonas aeruginosa*, and *Corynebacterium pyogenes*. Animals sick with other infections may also shed pathogenic microbes, including *Mycobacterium* species, *Brucella* species, mycoplasma, and *Coxiella burnetii*. Milk from a healthy animal does develop a complex flora on milking. Because milk is an animal product, microbes associated with mammals, farms, agricultural feedstuffs, and green plant material are often present in milk. Bacilli from the soil, clostridia from silage, coliforms from manure and bedding, and streptococci from green plant material commonly contaminate milk. In addition, the storage and processing environment and equipment, including milking machines, farm storage tanks, transportation equipment, cooling tanks, and milk processing equipment contribute greatly to the microbial flora of fluid milk.

II. Microbial Spoilage

A. Fluid Milk

The spoilage of fluid milk is dominated by the effects of pasteurization, which limits the presence of many microbes, and of refrigeration, which controls the growth of all but psychrotrophic microbes. Pasteurization processes for milk are designed to kill all microbes that are likely to contribute to disease from the consumption of milk. Along with the pathogens, yeasts, molds, and gram-negative and many gram-positive microbes are killed, greatly contributing not only to the safety, but the microbial stability of milk during storage. Because thermoduric and thermophilic microbes can survive pasteurization and to ensure processors that the minimum heating is achieved, processors often use temperatures and times somewhat beyond what is minimally required.

The greatest challenge to fluid milk processors is to limit the contamination that occurs after pasteurization during transport and bottling of milk. The extent of this postprocess contamination is directly related to the level of sanitation in the processing plant and suitable refrigeration.

By far, the most significant group of microbes in the spoilage of fluid milk is the psychrotrophs. Psychrotrophs are microorganisms that are capable of growing at refrigeration temperatures, although their *optimum* growth temperature may be much higher. Psychrotrophs include species from at least 27 genera of bacteria, four genera of yeast, and four genera of molds (Table I). Proper refrigeration is of the utmost importance to the control of microbial growth, because a small increase in storage temperature can result in a large decrease in bacterial generation times. [*See* REFRIGERATED FOODS.]

B. Cheese

The final composition of cheese relative to moisture content, salt content, fat content, and pH can vary tremendously in different varieties. Because all these factors contribute to the microbial stability of cheese, the only general statement about microbial stability that can be made is that cheeses are more stable than the milk from which they were made. Moisture content can range from 80% in cottage and cream cheeses down to 35% in hard, grating cheeses. Salt ranges from 1.5 to 5%, although a more significant effect on water activity than expected is seen because the salt is concentrated in the aqueous phase. Final titratable acidities during fermentation may also vary; however, a substandard fermentation risks both rapid spoilage and health hazards.

Molds, yeasts, and anaerobic sporeformers are involved most often in the spoilage of cheese, although a diversity of psychrotrophic microbes can spoil high-moisture cheeses, such as cottage cheese. Molds cause an unsightly appearance to cheese surfaces and can pose a health threat (see Section III). Anaerobic sporeformers such as clostridia, coliforms, and even unbalanced levels of gas-producing starter strains can cause abnormal gas formation. Spoilage during ripening of cheese can be controlled by maintaining low ripening temperatures and low humidity, factors that may inhibit microbial or enzymatic ripening processes. Sporeformers have been successfully controlled using nisin, an antibiotic produced by *Lactococcus lactis* subsp. *lactis*, or with the enzyme lysozyme.

Table I Genera of Bacteria, Yeasts, and Molds Containing Psychrotrophic Species

Bacteria	Yeasts	Molds
Acinetobacter	*Candida*	*Aspergillus*
Aeromonas	*Cryptococcus*	*Cladosporium*
Alcaligenes	*Rhodotorula*	*Penicillium*
Arthrobacter	*Torulopsis*	*Trichothecium*
Bacillus		
Chromobacterium		
Citrobacter		
Clostridium		
Corynebacterium		
Enterobacter		
Erwinia		
Escherichia		
Flavobacterium		
Klebsiella		
Lactobacillus		
Leuconostoc		
Listeria		
Microbacterium		
Micrococcus		
Moraxella		
Proteus		
Pseudomonas		
Serratia		
Streptomyces		
Streptococcus		
Vibrio		
Yersinia		

[Adapted from International Commission on Microbiological Specifications for Foods (1980). "*Microbial Ecology of Foods*," Vol. 1, p. 5. Academic Press, New York.]

C. Fermented Milks

Microbial growth is controlled in fermented milks by the low pH and high titratable acidity achieved during fermentation as well as dominance of the flora by starter bacteria. The lactic cultures can lower the pH to 4.8 and lower, depending on the product, and the pH may continue to drop during refrigerated storage. At this level of acidity, pathogens are effectively inhibited and even killed on storage. Spoilage is limited to yeasts and molds, which may accompany addition of flavorings and fruits that may be added to the product. As long as acid production is not inhibited during fermentation, a safe and long-shelf-life product results.

D. Dried Milk Products

Dried milks products, often prepared for use as ingredients in other foods, include whole milk, skim milk, whey, buttermilk, cheese, and cream. The drying process, although conducted at elevated temperatures, cannot be relied on as a method of microbial destruction. Proper pasteurization, sanitation, and product handling are the only consistent controls over the safety of these products. Once dried, these products are microbiologically stable. However, any remaining pathogens could be a threat in food subsequently formulated with the contaminated ingredient.

E. Canned Milks

Canned evaporated milk is heated to achieve commercial sterility. Therefore, all pathogens are destroyed, although some extremely heat-resistant spores such as *Bacillus stearothermophilus* might survive. This product is both shelf-stable under normal storage temperatures and pathogen-free. Sweetened condensed milks rely on pasteurization, low water activity, and high sugar content for preservation. These products are not heat processed after can closing and therefore can spoil from molds or yeast that enter during the filling operation or through can defects. Another class of commercially sterile milks has been developed. These undergo high-temperature–short-time sterilization and are aseptically packaged in special paper containers. These products are stable at room temperature.

F. Ice Cream Mixes

These mixes are formulated with milk, cream, sugar, emulsifiers and stabilizers, colors, fruits, nuts, and flavorings. Therefore, the microbial content of these mixes can be diverse, depending on the content of the respective ingredients. Although the final product from these mixes is frozen and as such does not support microbial growth, treatment and formulation of the product before freezing can result in spoilage and safety concerns. Procurement of ingredients of good microbial quality, proper pasteurization, careful postpasteurization handling, and refrigeration prevent microbial problems. Adherence to government regulations on bacterial counts for these products help assure Good Manufacturing Practices. [*See* FOODS, QUALITY CONTROL.]

III. Pathogens of Concern in Dairy Products

Milk was once the vehicle of transmission of typhoid fever, scarlet fever, septic sore throat, diphtheria, tuberculosis, and shigellosis. The frequency and severity of these diseases prompted large-scale adoption of milk pasteurization by the end of World War II. The dairy industry emerged from this concern about milkborne disease as the segment of the food industry having a model commitment to sanitation. Although raw milk is still consumed in some regions, this practice is questionable from a human health perspective. Milkborne infections of current importance are listeriosis, salmonellosis, campylobacterosis, brucellosis, and yersiniosis, although properly pasteurized milk that is free from postprocess contaminants is not a vehicle for transmission of these or other diseases.

Dairy products and milk accounted for 4% of the outbreaks and 14% of the cases of foodborne disease with a known vehicle reported between 1973 and 1987 in the United States. The pathogens responsible (number of outbreaks is in parentheses) were *Salmonella* (50), *Campylobacter* (25), *Staphylococcus aureus* (6), *Brucella* (3), *Yersinia entercolitica* (2), hepatitis A virus (2), *Bacillus cereus* (1), and *Clostridium botulinum* (1). Contributing factors to the outbreaks included improper holding temperatures, inadequate cooking, contaminated equipment, unsafe source of food, and poor personal hygiene. Postprocess contamination by any of these pathogens or consumption of raw milk or its products can result in a significant health threat. [*See* FOODBORNE ILLNESS.]

Not mentioned in the above list is a recently recognized threat of listeriosis. Cheeses (i.e., Brie and Camembert) are ideally suited as vehicles for listeriosis if made from raw milk. Although fermentation reduces the pH to an inhibitory level, ripening of these cheeses involves the surface growth of mold. These molds produce alkaline metabolites and raise the surface pH of the cheese. The lack of refrigeration during ripening further encourages the growth of *Listeria,* although *Listeria* can grow also at refrigeration temperatures. The application of mold suspensions to the surface of these cheeses can provide another mechanism of contamination and dissemination of undesired microbes if suspensions are not prepared and used in a sanitary fashion. These factors combine to make listeriosis a significant threat in these raw milk cheeses. These same conditions can also encourage growth of other pathogens, emphasizing the importance of using pasteurized milk for the manufacture of these cheeses.

Some microbial toxins are found in dairy products or milk. These include staphylococcal toxin, aflatoxins, other mycotoxins, and biogenic amines. *Staphylococcus aureus* can contaminate milk from mastitic cows (cows with an udder infection). If improperly pasteurized or used raw, this milk can support the growth of *S. aureus,* leading to toxin production. Pasteurization will not inactivate this heat-stable toxin. Milk contaminated with *S. aureus* is especially dangerous if involved in a substandard fermentation. *Staphylococcus aureus* can grow and produce toxin during the fermentation period if acid is not being generated. Some common *Aspergillus* species produce a family of toxic and carcinogenic aflatoxins during growth on damaged grains and other substrates. Dairy cow consumption of these grains leads to ingestion of these toxins. One type of aflatoxin, M_1, is excreted in milk and is retained in products made from contaminated milk. *Aspergillus* ssp. producing aflatoxin can also grow on the surface of cheese, providing another means of contamination. Penicillic acid, patulin, and ochratoxin A are other mycotoxins that have been found in moldy cheese trimmings. The incidence of mycotoxin contamination of cheese is low; storage temperatures below 7°C will greatly discourage toxin formation. Biogenic amines including tyramine and histamine have been found in cheese. These vasoactive compounds can be toxic at high levels or in people with compromised metabolic ability to deaminate these amines. These amines are formed by the decarboxylation of tyrosine and histidine catalyzed by decarboxylases found in microbes commonly associated with cheese.

IV. Fermentation of Dairy Products

A. Microbes Associated with Dairy Fermentations

Fermented dairy products derive their characteristic flavor and texture from the microbial action of starter cultures. These are as many as 2000 different varieties of cheeses and fermented milks produced worldwide. Although there is some difference in the source (e.g., cow, sheep, goat, buffalo) and composition of the milk used as a raw material, this great diversity of fermented foods stems primarily from the microbes used to execute the fermentation and the physical treatments of the fermented product.

Fermentation has been used to preserve milk for millennia. Before a technical understanding of the fermentation was available, microbes naturally present in the milk or in the containers used for milk storage executed the fermentation. As awareness of the role of microbes in the fermentation process increased, "back slopping" (a technique to carry over some of a previously fermented milk to bring about the fermentation of a new batch of milk) was used. In modern industrial fermentation environments, purified and characterized starter cultures are intentionally inoculated into the milk. This enables much more control over the fermentation and characteristics of the final product.

The microbes associated with fermented dairy products are listed in Table II. They are nonpathogenic and contribute specific attributes to the final product. They generally are at levels of 10^8/g in freshly fermented products. During ripening or storage, the types and levels of microbes change, depending on the product.

B. Function of Microbes in Dairy Fermentations

The fermentation of milk sugar, lactose, to lactic acid and the proteolytic degradation of milk protein, casein, to usable nitrogen are the primary requirements of a starter culture. However, these microbes contribute to flavor and texture development in many more ways. The production and degree of metabolic end products (e.g., lactic, propionic, and acetic acids, ethanol, CO_2, diacetyl and other flavor compounds, dimethyl sulfide, proteases, and lipases) help determine the unique attributes of a given fermented milk product. Lactic acid is important to all fermented dairy products. It provides the acidity necessary for a tart flavor and for changes in the structure of casein to achieve syneresis and de-

Table II Microbes Associated with Fermented Dairy Products

Microbes	Product association[a]	Function
Lactococcus lactis subsp. *cremoris* and *lactis*	American cheeses, buttermilk, cottage cheese, soft cheese, viili	Lactic acid production at <40°C; some strains provide ropiness in viili
Lactococcus lactis subsp. *lactis* var. *diacetylactis*	Buttermilk, soft cheese, sour cream, cottage cheese dressing	Lactic acid, CO_2, and diacetyl
Lactobacillus delbrueckii subsp. *bulgaricus, L. lactis, L. helveticus, L. casei*	Italian and Swiss cheeses, yogurt, fermented milks; often paired with *S. thermophilus*	Lactic acid production at <50°C; some ropy strains
Streptococcus salivarius subsp. *thermophilus*	Italian and Swiss cheeses, yogurt; often paired with the lactobacilli	Lactic acid production <50°C; some ropy strains
Leuconostoc	Buttermilk, Roquefort cheese, cottage cheese	Diacetyl, and CO_2, ethanol, and acetic acid
Propionibacterium ssp.	Swiss cheeses	CO_2 causing eye formation, propionic and acetic acids from lactate
Lactobacillus acidophilus	Therapeutic milks	Lactic acid and undefined factors that may aid in promotion of health
Penicillium ssp.	Soft cheeses, blue-veined cheese	White surface mold, increase pH, blue vein production
Geotrichum ssp.	Soft cheeses	White surface mold
Lactose-fermenting yeast	Kefir and other mixed fermentation beverages	Ethanol and CO_2
Bifidobacterium	Therapeutic milks	Organic acids and undefined factors that may aid in promotion of health

[a] American cheeses: Chedder, brick, Monterey Jack, Muenster; soft cheeses: Camembert, Brie; Italian cheeses: Parmesan, provolone, ricotta, mozzarella.

sired functional characteristics. Propionic acid gives Swiss cheeses their characteristic nutty flavor. Acetic acid and ethanol must be balanced in yogurt to promote proper flavor. Ethanol and lactic acids combine in mixed lactic bacteria/yeast fermented milks to provide the desired flavor. Carbon dioxide provides effervescence to fermented milks, eye formation in some cheeses, and an open texture in others. Diacetyl provides the buttery flavor important in buttermilk, soft cheeses, and cottage cheese dressing. Starters that produce extracellular polysaccharides can improve the mouth-feel of some fermented milks and yogurts or may cause a slime defect in products where ropiness is not desirable. Proteases degrade milk proteins to peptides, on which starter and nonstarter peptidases act further. Some flavor defects such as bitterness and brothiness stem from the presence of certain amino acids or peptides. The action and control of starter proteases and peptidases are critical to proper ripening of most aged cheeses. Attempts to accelerate the ripening of aged cheeses has led to the development of culture and/or enzyme additives that promote proteolysis important in characteristic aged flavor and body. Lipases are very important to the proper flavor of some Italian cheeses. Frequently, starter lipases are not sufficient, and animal-derived lipases are added for proper ripening and flavor development. Nonstarter bacteria can also contribute to flavor production during ripening (providing the justification for using raw milk in cheese manufacture). In addition to the contribution of microbial factors, numerous physical manipulations help determine characteristics of the final product. The timing and extent of stirring, mixing, stretching, cheddaring, temperature and moisture control, and product formulation are important to the flavor, texture, and body of fermented dairy products.

C. Genetics of Dairy Starter Cultures

For years, observant dairy technologists noticed that lactic acid bacteria often did not retain some desirable traits when the cultures were held for long periods of time or sequentially transferred. Lactic cultures often lost their ability to ferment milk rapidly, to produce diacetyl, or to resist bacteriophage infection. Since the mid 1970s, researchers have focused on the genetic basis for this occurrence. Genes encoding lactose metabolism, protease production, diacetyl formation, bacteriophage resistance, and bacteriocin production have been found to be linked to plasmid DNA. In some cases, these plasmid-encoded genes are also flanked by insertion sequences. The linkage of these traits to naturally occurring plasmids and the involvement of insertion sequences has provided a mechanism for rapid evolution and genetic shift in this group of bacteria. Genes for some of these and other traits from lactococci, *Streptococcus thermophilus,* and some lactobacilli have been cloned, and the DNA has been sequenced.

This knowledge of the genetics of lactic acid bacteria led to efforts to apply directed genetic techniques to the improvement of these industrially important bacteria. Cloning vectors, a prerequisite for some of this work, have been constructed. Early in this work, vectors from related gram-positive bacteria (e.g., streptococci, bacilli, enterococci, and staphylococci) were used in lactic cultures. Subsequently, small cryptic, naturally occurring plasmids were mapped, sequenced, and used to form cloning vectors specific for lactic acid bacteria. The development of genetic transfer systems including conjugation, transformation, and transduction has enabled the transfer of genes into new hosts to study genetic stability and gene expression. The directed conjugal transfer of characterized phage-resistant plasmids into phage-sensitive recipient strains has provided the first example of use of this technology to improve dairy strains genetically. This field of research represents a most exciting area and will continue to provide the tools for directed genetic improvement of starter cultures. [*See* PLASMIDS.]

V. Bacteriophages in Dairy Fermentations

The failure of a dairy lactic fermentation results in significant product loss and economic setback. Problems with starter culture performance can be caused by the presence of antibiotics in milk, the presence of natural inhibitors in milk, use of an inactive starter inoculum, use of temperatures during processing inhibitory to the culture, and the presence of bacteriophage. The presence of bacteriophage, or phage, is by far the most significant of these factors.

Bacteriophage infecting lactic cultures harbor double-stranded DNA of 20–60 kb surrounded by a protein coat. This phage "head" has a diameter of about 40–90 nm and can be isometric-sided or prolate in shape. Most lactic phages have tails, which aid in DNA injection, ranging in length from 20 to 400 nm. Morphologically, there are many different types of lactic phages, and phages have been isolated for all genera and many species of lactic cultures. They are parasitic in nature, requiring host cell functions to replicate and, during a lytic cycle, will lyse the host cell to release newly formed phage particles. The tremendous impact that a lytic phage infection can have on a fermentation is due to the potential of phage to infect rapidly, lyse, and release large numbers of progeny for subsequent infections. It may take less than 60 min for a complete replication cycle (latent period) and up to 200 viable phage progeny can be released (burst size). This form of replication allows phage to out-pace the slower binary fission replication of bacteria. Phage can exist in one of two life states: lytic or lysogenic. A lytic phage adsorbs to the surface of a cell and injects its DNA, using cellular replication machinery and sophisticated regulatory control; the phage DNA directs the expression of phage proteins, and phage DNA is replicated and packaged in new phage coat proteins. Finally the cell is lysed to release newly formed phage. In the lysogenic cycle, after injection of the phage DNA, the DNA combines with host DNA and remains latent, being silently replicated along with each cellular replication cycle. The phage switches to a lytic cycle only after some environmental stimulus signals DNA excision and replication. Many lactic cultures have been shown to carry lysogenic phage on their DNA. However, molecular differences between lysogenic and lytic phages suggest that lysogenic phage do not contribute greatly to the lytic phage problem. [*See* BACTERIOPHAGES.]

Dairy fermentations are plagued with phage difficulties. Contributing factors to the impact of phage include the nonaseptic nature of the fermentations; the modern constraints of time schedules and product consistencies that make fluctuations in make times and product quality unacceptable; the frequent use of large, open vats for cheesemaking; the fluid nature of milk and frequent mixing during cheesemaking, allowing rapid dissemination of newly formed phage particles; the regular microbial growth and replication, which provides susceptible hosts for the phage; the ability of phages to be transmitted as an aerosol; and the rapid replication rates of some lytic phages. Some cultures seem more susceptible to phage infection than others. The lactococci appear to be much more frequently attacked than the lactobacilli or *Streptococcus thermophilus,* although phages for all these cultures are known.

The dramatic influence of phage on dairy fermentations has led to a variety of measures designed to control the effect of phage. Some control measures are listed below:

1. *Effective Processing Plant Design and Sanitation* Improved control over bacteriophage can be designed into a plant. A separate starter preparation room, control over air, personnel and product flows in the plant, use of easily cleaned fermentation vessels, and regular cleaning and sanitation of equipment provide means to reduce phage levels in the dairy plant.
2. *Use of Phage Inhibitory Media* These media have been formulated to inhibit phage growth in the bulk starter vessel. This decreases the chance that any lytic phage contaminating the bulk starter vessel do not replicate before cheese vat inoculation. If phage levels are kept low, the fermentation may proceed before phage levels build to a destructive level. Formulations generally involve the chelation of Ca^{2+} ions with citrate or phosphate. Calcium ions are required for the phage to infect the host cells.
3. *Strain Selection and Use* Next to fermentative ability, the largest single factor for strain selection for use in a dairy fermentation is resistance to bacteriophage. Great efforts are made to select and even design strains that will resist phage. Once proper effective strains are identified, they are often used in rotation, so that phage do not have an opportunity to build to high levels on any given strain, and in combination with one or more other strains. Some cheesemakers prefer to use large numbers of undefined strains. If any strain is destroyed by phage, another strain should be present in the mixture to carry out the fermentation.
4. *Monitoring Whey for Phage* Conducting phage tests on whey from current fermentations informs the cheesemaker which strains in use are developing phage. Judgments can be made with this information on the best strains to keep as current production and backup strains.

VI. Healthful Attributes of Culture-Containing Milk Products

Milk and milk products have provided humans an excellent source of nutrition (i.e., protein, calories, calcium and other minerals) since the domestication of the mammal. In addition to nutritional benefits of milk consumption, health promotion has also been attributed to the cultures used in fermenting milk products. The most commonly cited cultures used for "therapeutic" purposes are *Lactobacillus acidophilus* and *Bifidobacterium* ssp. Healthful benefits attributed to ingestion of these cultures in fermented or unfermented milk products include alleviation of lactase deficiency symptoms, abatement of antibiotic-associated diarrhea, stimulation of the immune system, anticancer effects, and cholesterol assimilation. Different levels of scientific support exist for each of these claims. Overall, it appears that some cultures can help alleviate lactase deficiency and may be useful for shortening the duration of some diarrheal diseases. No convincing evidence for the other claims has been provided in humans. Some promising results have been generated with immune system stimulation in animal models and with decreasing levels of fecal enzymes involved in carcinogen formation in humans. This latter effect may influence the rate of colon cancer occurrence. However, this has not been substantiated clinically. A broad-based belief exists that these lactic cultures have therapeutic attributes. Unfortunately, in many cases, well-substantiated scientific fact does not support the testimonials. Perhaps continued research will determine the true role of lactic bacteria in promoting human health.

Bibliography

Hall, R. J., and Franks, P. A. (1985). Cheese starters. *In* "Comprehensive Biotechnology. The Principles, Applications and Regulations of Biotechnology in Industry, Agriculture and Medicine, " Vol. 3. "The Practice of Biotechnology: Current Commodity Products."(M. Moo-Young, ed.), pp. 507–522, Pergamon Press, Oxford.

International Commission on Microbiological Specifications for Foods. (1980). "Microbial Ecology of Foods," Vol. II. Academic Press, New York.

Irvine, D. M., and Hill, A. R. (1985). Cheese technology. *In* "Comprehensive Biotechnology. The Principles, Applications and Regulations of Biotechnology in Industry, Agriculture and Medicine," Vol. 3. "The Practice of Biotechnology: Current Commodity Products."(M. Moo-Young, ed.), pp. 523–565, Pergamon Press, Oxford.

DNA Repair by Bacterial Cells

Lawrence Grossman
Johns Hopkins University

Glossary

Endonuclease Nuclease that hydrolyzes internal phosphodiester bonds

Excision Removal of damaged nucleotides from incised nucleic acids

Exonuclease Nuclease that hydrolyzes terminal phosphodiester bonds

Glycosylase Enzymes that hydrolyze N-glycosyl bonds linking purines and pyrimidines to carbohydrate components of nucleic acids

Incision Endonucleolytic break in damaged nucleic acids

Ligation Phosphodiester bond formation as the final stage in repair

Nuclease Enzyme that hydrolyzes internucleotide phosphodiester bonds in nucleic acids

Resynthesis Polymerization of nucleotides into excised regions of damaged nucleic acids

THE ABILITY OF CELLS to survive hostile environments is due in part to surveillance systems that recognize damaged sites in DNA and are capable of either reversing the damage or removing damaged bases or nucleotides, generating sites that lead to a cascade of events restoring DNA to its original structural and biological integrity.

Both endogenous and exogenous environmental agents can damage DNA. A number of repair systems are regulated by the stressful effects of such damage, by affecting the levels of responsible enzymes or by modifying their specificity. Repair enzymes apparently are the most highly conserved proteins showing their important role throughout evolution.

The enzyme systems can either directly reverse the damage to form the normal purine or pyrimidine bases or the modified bases can be removed together with surrounding bases through a succession of events involving nucleases, DNA polymerizing enzymes, and polynucleotide ligases, which assist in restoring the biological and genetic integrity to DNA.

I. Damage

As a target for damage, DNA possesses a multitude of sites that differ in their receptiveness to modification. On a stereochemical level, nucleotides in the major groove are more receptive to modification than those in the minor groove, the termini of DNA chains expose reactive groups, and some atoms of a purine or pyrimidine are more susceptible than others. As a consequence, the structure of DNA represents a heterogenous target in which certain nucleotide sequences also contribute to the susceptibility of DNA to genotoxic agents.

A. Endogenous Damage

Even at physiological pHs and temperatures in the absence of extraneous agents the primary structure of DNA undergoes alterations. A number of specific reactions directly influence the informational content as well the integrity of DNA. Although the rate constants for many reactions are inherently low, the enormous size of DNA and its persistence in cellular life cycles cause the significant long-term effects of the accumulation of these changes.

1. Deamination

The hydrolytic conversion of adenine to hypoxanthine, guanine to xanthine, and cytosine to uracil-containing nucleotides is of sufficient magnitude to affect the informational content of DNA (Fig. 1).

This article was first published in the *Encyclopedia of Human Biology*, Volume 6.

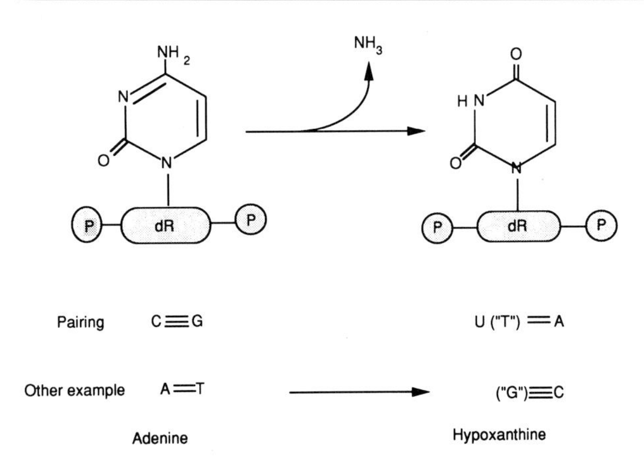

Figure 1 Deamination reactions have mutagenic consequences because the deaminated bases cause false recognition.

2. Depurination

The glycosylic bonds linking guanine in nucleotides are especially sensitive to hydrolysis—more than the adenine and pyrimidine glycosylic links. The end result is that apurinic [or apyrimidinic (AP) sites] are recognized by surveillance systems and, as a consequence, are repaired.

3. Mismatched Bases

During the course of DNA replication, noncomplementary nucleotides are incorrectly incorporated into DNA and manage to escape the editing functions of the DNA polymerases. The proper strand as well as the mismatched base is recognized and repaired.

4. Metabolic Damage

When thymine incorporation into DNA is limited either through restricted precursor deoxyuridine triphosphate (dUTP) availability or inhibition of the thymidylate synthetase system, dUTP is utilized as a substitute for thymidine triphosphate. The presence of uracil is identified as a damaged site and acted upon by repair processes.

5. Oxygen Damage

The production of oxygen, superoxide, or hydroxyl radicals as a metabolic consequence as well as at inflammatory sites causes sugar destruction, which eventually leads to strand breakage.

B. Exogenous Damage

The concept of DNA repair in biological systems arose from those studies by photobiologists and radiobiologists studying the viability and mutagenicity in biological systems exposed to either ionizing or ultraviolet irradiation. Target theories, derived from the random statistical nature of photon bombardment, led to the identification of DNA as the primary target for the cytotoxicity and mutagenicity of ultraviolet light. In addition, identification of most of the structural and regulatory genes controlling DNA repair in *Escherichia coli* were identified, facilitating the isolation and molecular characterization of the relevant enzymes and proteins.

1. Ionizing Radiation

The primary cellular effect of ionizing radiation is the radiolysis of water, which mainly generates hydroxyl radicals (HO·). The HO· is capable of abstracting protons from the C-4′ position of the deoxyribose moiety of DNA, thereby labilizing the phosphodiester bonds and generating single- and double-strand breaks. The pyrimidine bases are also subject to HO· addition reactions.

2. Ultraviolet Irradiation

Most of the ultraviolet photoproducts are chemically stable; their recognition provided direct biochemical evidence for DNA repair. The major photoproducts are 5,6-cyclobutane dimers of neighboring pyrimidines (intrastrand dimers), 6,4-pyrimidine–pyrimidone dimers (6-4 adducts), and 5,6-water addition products of cytosine (cytosine hydrates).

3. Alkylation Damage

Alkylation damage occurs on purine ring nitrogens (cytotoxic adducts), the O^6– position on guanine, the O^4 positions of the pyrimidines (mutagenic lesions), and the oxygen residues of the phosphodiester bonds of the DNA backbone (biologically silent). Alkylating agents are environmentally pervasive, arising indirectly from many foodstuffs and from automobile exhaust in which internal combustion of atmospheric nitrogen results in the formation of nitrate and nitrites.

4. Bulky Adducts

Large bulky polycyclic aromatic hydrocarbon modification occurs primarily on the N-2, N-7, and C-8 position of guanines, invariably from the metabolic activation of these large, hydrophobic, uncharged macromolecules to their epoxide analogs. The major source of these substances is from the combustion of tobacco, petroleum products, and foodstuffs.

II. Direct Removal Mechanisms

The simplest repair mechanisms involve the direct photoreversal of pyrimidine dimers to their normal homologs and the removal of O-alkyl groups from the O^{6-} methylguanine and from the phosphotriester backbone as a consequence of alkylation damage to DNA.

A. Photolyases (Photoreversal)

The direct reversal of pyrimidine dimers to the monomeric pyrimidines is the simplest mechanism (Fig. 2), and parenthetically it is chronologically the first mechanism described for the repair of photochemically damaged DNA. It is a unique mechanism characterized by a requirement for visible light as the sole source of energy for breaking two carbon–carbon bonds.

The enzyme protein has two associated light-absorbing molecules (chromophores), which can form an active light-dependent enzyme. One chromophore is reduced flavin adenine dinucleotide ($FADH_2$) and the other is either a pterin or a deazaflavin that can absorb the effective wavelengths of 365–400 nm required for photoreactivation of pyrimidine dimers. Photoreversal perhaps involves energy transfer from the pterin molecule to $FADH_2$, with electron transfer to the pyrimidine dimer resulting in nonsynchronous cleavage of the C5 and C6 cyclobutane bonds.

Enzymes that carry out photoreactivation have been identifed in both prokaryotes and eukaryotes.

B. Alkyl Group Removal (Methyltransferases)

Bacterial cells pretreated with less than cytotoxic or genotoxic levels of alkylating agents before lethal or mutagenic doses are more resistant. This is an **adaptive** phenomenon with antimutagenic and anticytotoxic significance. During this adaptive period, a 39-kDa Ada protein is synthesized and specifically removes a methyl group from a phosphotriester bond and from an O^6- methyl group of guanine (or from O^4- methylthymine). The O^6- methyl group of guanine is not liberated as free O^6- methylguanine during this process but, rather, is transferred directly from the alkylated DNA to this protein (Fig. 3); the Ada protein (methyltransferase) and an unmodified guanine are simultaneously generated. These alkyl groups specifically methylate cysteine 69 and cysteine 321, respectively, in the protein.

The methyltransferase is used stoichiometrically in the process (does not turn over) and is permanently inactivated in the process. Nascent enzyme is, however, generated because the mono- or dimethylated transferase activates transcription of its own "regulon," which includes, in addition to the *ada* gene, the *alkB* gene of undefined activity and the *alkA* gene, which sponsors a DNA glycosylase. The latter enzyme acts on 3-methyladenine, 3-methylguanine, O^2- methylcytosine, and O^2- methylthymine. The methylated Ada protein can specifically bind to the operator of the *ada* gene, acting as a positive regulator. Down-regulation may be controlled by proteases acting at two hinge sites in the Ada protein.

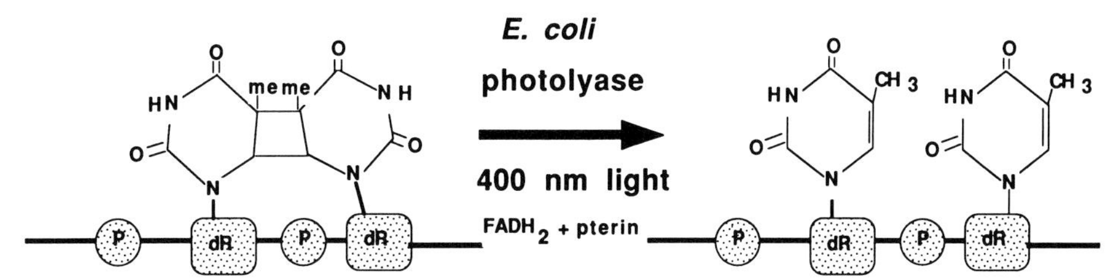

Figure 2 The enzymatic photoreactivation of cyclobutane pyrimidine dimers in the presence of visible light.

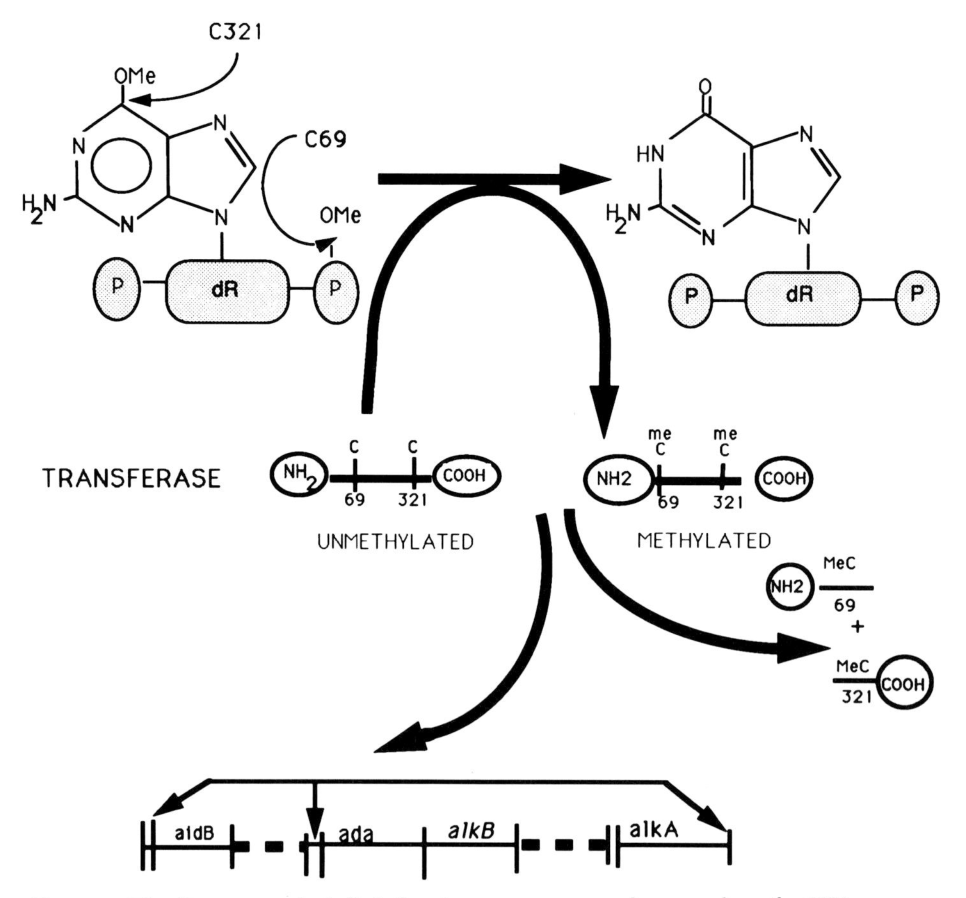

Figure 3 The direct reversal of alkylation damage removes such groups from the DNA backbone and the O^6- position of guanine. Such alkyl groups are transferred directly to specific cysteine residues on the transferase, the levels of which are influenced adaptively by levels of the alkylating agents. The methylation of the transferase inactivates the enzyme, which is used up stoichiometrically in the reaction. The alkylated transferase acts as a positive transcriptive signal turning on the synthesis of unique messenger RNA. Regulation of transferase levels may be influenced by a unique protease.

III. Base-Specific Responses

A. Base Excision Repair by Glycosylases and Apyrimidinic Endonucleases

Bases modified by deamination can be repaired by a group of enzymes called DNA glycosylases, which specifically hydrolyze the N-glycosyl bond of that base and the deoxyribose of the DNA backbone, generating an apyrimidinic, or an apurinic, site (AP site) (Fig. 4). These are rather small, highly specific enzymes that require no cofactor for functioning. They are the most highly conserved proteins attesting to the evolutionary unity both structurally and mechanistically from bacteria to humans.

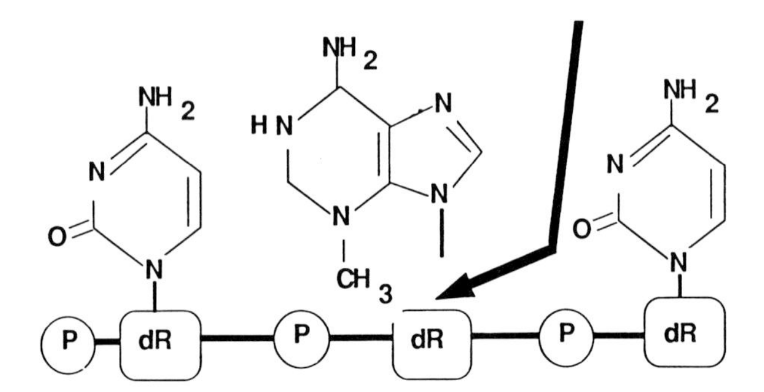

Figure 4 *Escherichia coli alkA* DNA glycosylase removal of 3-methyladenine and other lesions. DNA glycosylases hydrolyze the N-glycosyl bond between damaged bases and deoxyribose generating an apyrimidinic, or apurinic, site.

As a consequence of DNA glycosylase action, the

AP sites generated in the DNA are acted upon by a phosphodiesterase specific for such sites, which can nick the DNA either 5′ and/or 3′ to such damaged sites (Fig. 5). If there is a sequential action of a 5′-acting and a 3′-acting AP endonuclease, the AP site is excised, generating a gap in the DNA strand.

B. Glycosylase-Associated Endonucleases

An enzyme from bacteria and phage-infected bacteria, encoded in the latter case by a single gene (*denV*), hydrolyzes the N-glycosyl bond of the 5′-thymine moiety of a pyrimidine dimer followed by hydrolysis of the phosphodiester bond between the two thymine residues of the dimer (Fig. 6). This enzyme, referred to as the pyrimidine dimer DNA glycosylase, is found in *Micrococcus luteus* and phage T4-infected *Escherichia coli*. This small, uncomplicated enzyme does not require cofactors and is presumed to act by a series of linked β-elimination reactions.

An enzyme behaving in a similar glycosylase–endonuclease fashion but acting on the radiolysis product of thymine, thymine glycol has been isolated from *E. coli* and is referred to as endonuclease III.

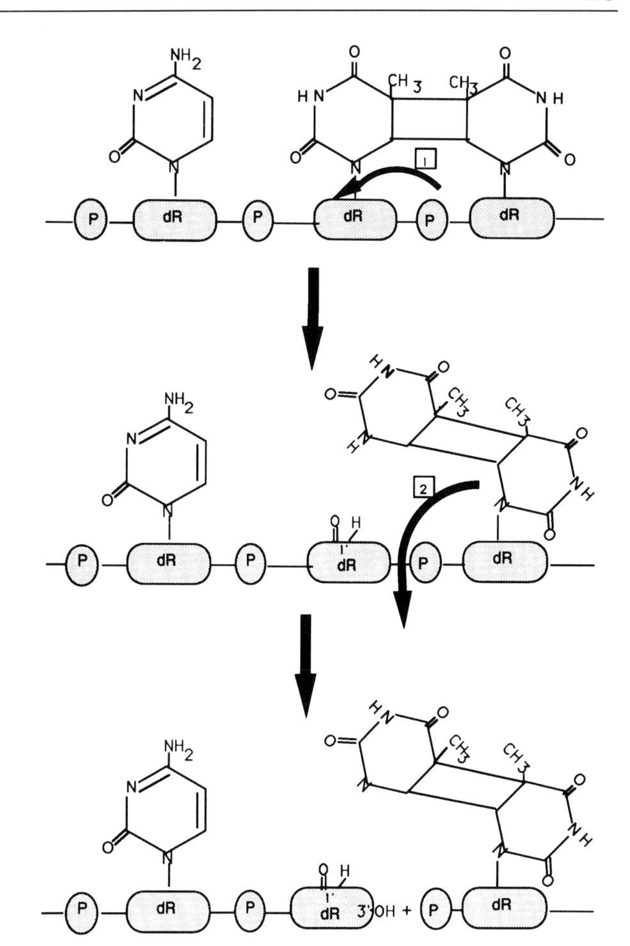

Figure 6 The same enzyme that can hydrolyze the N-glycosyl residue of a damaged nucleotide also hydrolyzes the phosphodiester bond linking the apyrimidinic site generated in the first N-glycosylase reaction.

IV. Nucleotide Excision Repair

The ideal repair system is one that is somewhat indiscriminate and that can respond to virtually any kind of damage. Such a repair system has been characterized in *E. coli*, where it consists of at least six gene products of the *uvr* system. This ensemble of proteins consists of the UvrA protein that binds as a dimer to DNA in the presence of adenosine triphosphate (ATP), followed by the UvrB protein, which cannot bind DNA by itself. Translocation of the $UvrA_2B$ complex from initial undamaged DNA sites to damaged sites is driven by a cryptic ATPase associated with UvrB, which is activated by the formation of the $UvrA_2B$-undamaged DNA complex. This complex is now poised for endonucleolytic activity catalyzed by the interaction of the $UvrA_2B$-damaged DNA complex with UvrC to generate two nicks in the DNA seven nucleotides 5′ to the damaged site and three to four nucleotides 3′ to the same site (Fig. 7). These sites of breakage are invariant regardless of the nature of the damage. In the presence of the UvrD (helicase III), DNA polymerase I, and substrate deoxynucleoside triphosphates, the damaged fragment is released ac-

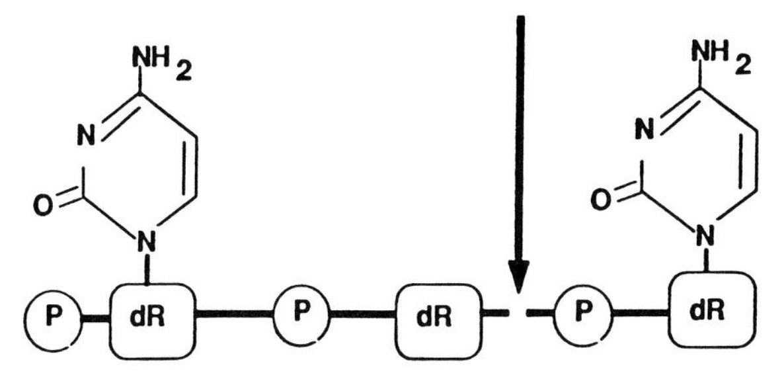

Figure 5 Endonucleases recognize AP sites and hydrolyze the phosphodiester bonds either 3′, 5′, or both sides of the deoxyribose moiety in damaged DNA.

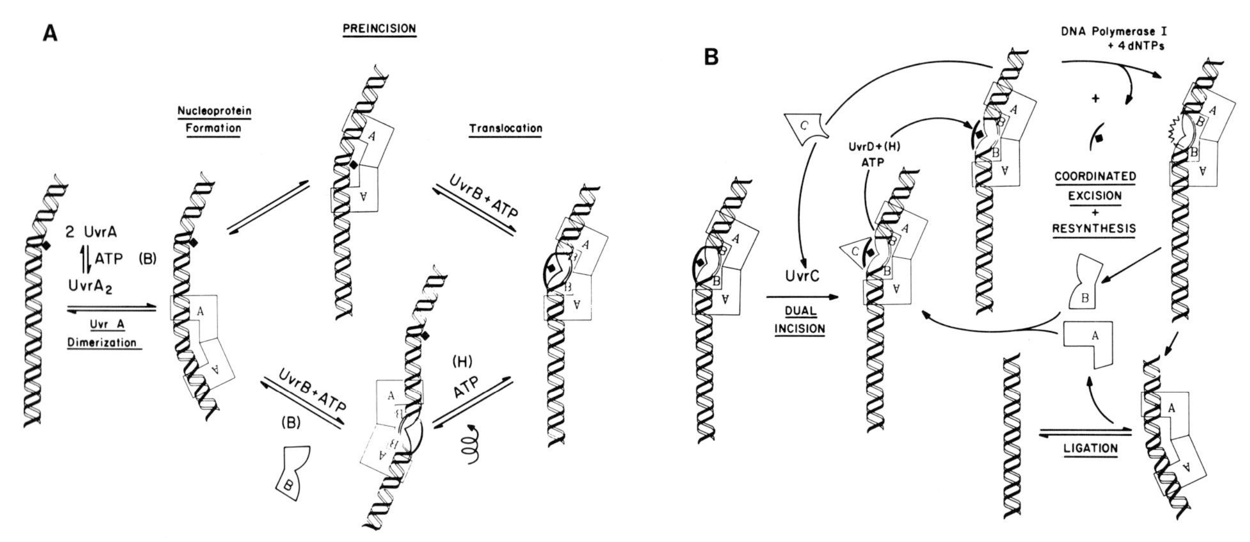

Figure 7 Nucleotide excision reactions. (A) Preincision reactions. (B) Postincision reactions. In this multiprotein enzyme system, the UvrABC proteins catalyze a dual incision reaction seven nucleotides 5′ and three to four nucleotides 3′ to a damaged site. The UvrA protein, as a dimer, binds to undamaged sites initially and in the presence of UvrB, whose cryptic adenosine triphosphatase (ATPase) is manifested in the presence of UvrA, providing the energy necessary for translocation to a damaged site. This preincision complex interacts with UvrC, leading to the dual incision reaction. The incised DNA–UvrABC does not turn over and requires the coordinated participation of the UvrD and DNA polymerase reactions for damaged fragment release and turnover of the UvrABC proteins. Ligation, the final reaction, restores integrity to the DNA stands. dNTP, deoxynucleoside triphosphate.

companied by the turnover of the UvrA, UvrB, and UvrC proteins. The continuity of the DNA helix is maintained based on the sequence of the opposite strand. The final integrity of the interrupted strands is restored by the action of DNA polymerase I, which copies the other strand, and by polynucleotide ligase, which seals the gap.

The levels of the Uvr proteins are regulated in *E. coli* by an SOS regulon monitoring a large number of genes, which include the *uvrA*, *uvrB* (possibly *uvrC*), and *uvrD* as part of the excision repair system; it also includes the regulators of the SOS system; the LexA and RecA proteins, cell division genes *sulA* and *sulB;* recombination genes *recA*, *recN*, *recQ*, *uvrD*, and *ruv;* mutagenic by-pass mechanisms (*umuDC*, *recA*); damage-inducible genes; and the lysogenic phage λ. The LexA protein negatively regulates these genes as a repressor by binding to unique operator regions. When the DNA is damaged (e.g., by ultraviolet light), a signal in the form of a DNA repair intermediate induces the synthesis of the RecA protein. When induced, the RecA protein acts as a protease assisting the LexA protein to degrade itself, activating its own synthesis and that of the RecA protein as well as some 20 other different genes. These genes permit the survival of

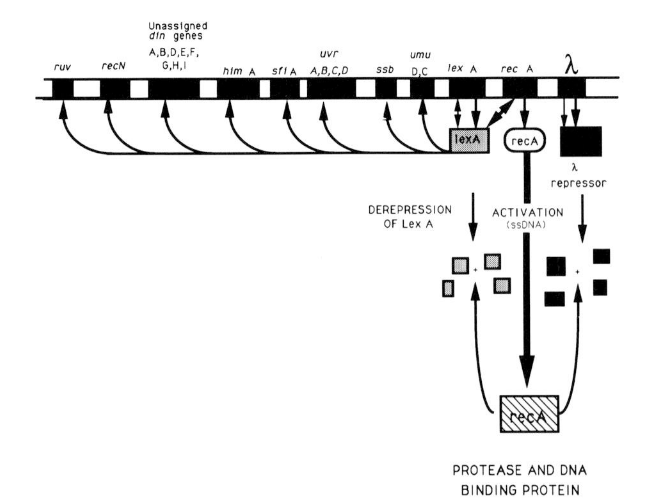

Figure 8 Regulation of the nucleotide excision pathway by the SOS system. The lexA and phage λ repressors negatively control a multitude of genes that are turned on when bacterial cells are damaged, leading to the overproduction of the RecA protein, which assists in the proteolysis of the LexA and phage λ proteins, thereby derepressing the controlled gene systems. When DNA is fully repaired, the level of RecA declines, restoring the SOS system to negative control. ssDNA, single-stranded DNA.

the cell in the face of life-threatening environmental damages. Upon repair of the damaged DNA, the level of the signal subsides, reducing the level of RecA and stabilizing the integrity of the intact LexA protein and its repressive properties on all the other genes (Fig. 8). Then the cell returns to its normal state. [*See* RECA.]

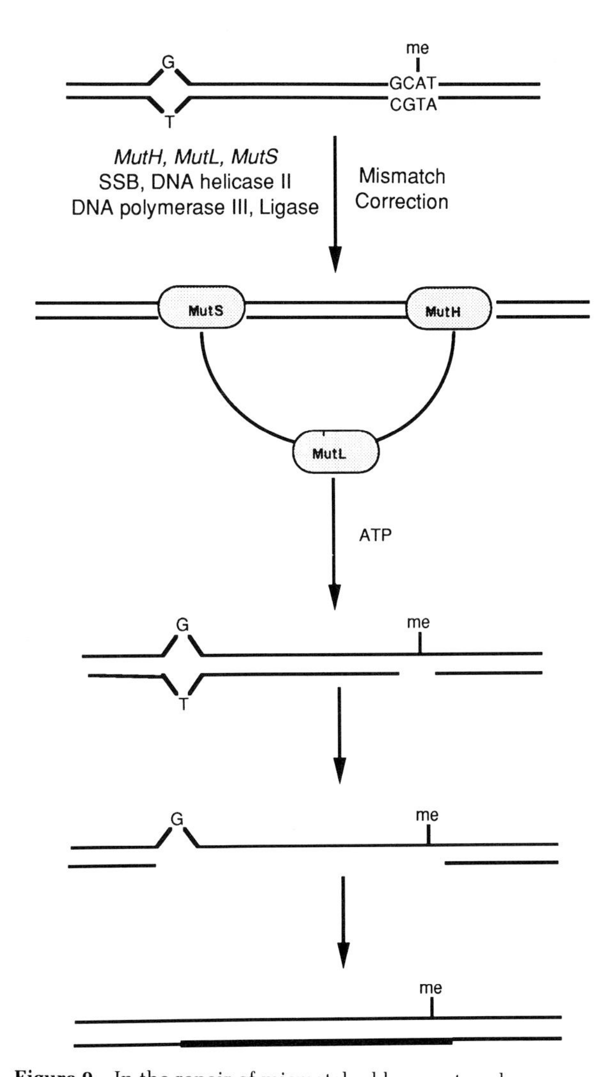

Figure 9 In the repair of mismatched bases, strand distinction can be achieved by the delay in adenine methylation during replication. The nascent unmethylated strand serves as a template for the incision reactions catalyzed by a number of proteins specifically engaged in mismatch repair processes.

The structural and biological specificity associated with transcriptional processes apparently limit DNA repair to those damaged regions of the chromosome undergoing transcription and damage in those quiescent regions is persistent. Within expressed genes, the repair process is selective for the transcribed DNA strand for damage such as pyrimidine dimers, and this "coupling" to transcription has been shown in *E. coli*.

V. Mismatch Correction

A number of mechanisms recognize not damage but, rather, mispairing errors that occur in all biological systems. In *E. coli*, mismatch correction is controlled by seven mutator genes: *dam* (methyl-directed), *mutD, mutH, mutL, mutS, mutU* (*uvrD*), and *mutY*. In mismatch correction, one of the two strands of the mismatches is corrected to conform with the other strand (Fig. 9). Strand selection is one of the intrinsic problems in mismatch repair, and the selection is operated in bacterial systems by the adenine methylation that takes place at d(GATC) sequences. Because such methylation occurs after DNA has replicated, only the template strand of the nascent duplex is methylated. In mismatch repair, only the unmethylated strand is repaired, thus retaining the original nucleotide sequence. The MutH, MutL, and MutS proteins apparently are involved in the incision reaction on this strand, with the remainder of the proteins plus DNA polymerase III and polynucleotide ligase participating in the excision–resynthesis reactions.

Bibliography

Friedberg, E. C. (1984). "DNA Repair." Freeman, New York.
Grossman, L., Caron, P. R., Mazur, S. J., and Oh, E. Y. (1988). *Faseb J.* **2,** 2696–2701.
Lindahl, T., Sedgwick, B., Sekiguchi, M., and Nakabeppu, Y. (1988). *Annu. Rev. Biochem.* **57,** 133–157.
Modrich, P. (1989). *J. Biol. Chem.* **264,** 6597–6600.
Sancar, A., and Sancar, G. B. (1988). *Annu. Rev. Biochem.* **57,** 29–67.
Walker, G. C. (1985). *Annu. Rev. Biochem.* **54,** 425–457.
Weiss, B., and Grossman, L. (1987). *Adv. Enzymol.* **60,** 1–34.

DNA Replication

Robb E. Moses
Oregon Health Sciences University

Glossary

Chromosome Package of genes representing part or all of the inherited information of the organism

Exonuclease Enzyme that degrades DNA from a terminus

Gyrase Enzyme that introduces coiling into the DNA duplex in an adenosine triphosphage-dependent reaction

Helicase Enzyme that unwinds duplex DNA and requires adenosine triphosphate

Polymerase Enzyme that synthesizes a nucleic acid polymer

Replication Act of duplicating the genome of a cell

Replicon Replicative unit, either part or the whole of the genome; in *Escherichia coli*, the entire genome is considered a replicon

DNA REPLICATION in *Escherichia coli* is a carefully regulated process involving multiple components representing more than 20 genes participating in duplication of the genome. The process is divided into distinct phases: initiation, elongation, and termination. The synthesis of a new chromosome involves an array of complex protein assemblies acting in sequential fashion in a carefully regulated and reiterated overall pattern. The scheme for DNA replication is under careful genetic control. The process is localized on the DNA structure by both DNA sequence and topology and requires specific protein–DNA interactions. DNA replication in *E. coli* is bidirectional and symmetrical.

I. Development of the Field

The development of our understanding of DNA replication in prokaryotes depends on a combination of biochemical and genetic approaches. Using several selection techniques, a number of laboratories isolated *E. coli* mutants that were conditionally defective (usually temperature-sensitive) in DNA replication. This method of identifying genes involved in DNA replication assumed that defects of such genes resulted in the death of the cell. When a large series of mutants was assembled, they fell clearly into two broad categories: those in which DNA replication ceased abruptly following a shift to restrictive conditions and those in which DNA replication ceased slowly. The former class is called fast-stop, and the latter slow-stop. The first category represents cells containing mutations in gene products that are required for the elongation phase of DNA replication, and the latter category contains cells with defects in gene products that are required for the initiation of new rounds of DNA replication.

The identification of *dnats* mutants represents one requirement for our understanding DNA replication. The second requirement was the development of systems that could be biochemically manipulated but that represented all or part of the authentic DNA replication process in *E. coli*. Several successive systems offered increasing advantages. Permeable cells that allow free access of small molecules to minimally disturbed chromosomes were earliest. These systems allowed definition of the energy and cofactor requirements for the elongation phase of DNA replication. The limitation was that they did not permit access by macromolecules to the replication apparatus and therefore did not allow complementation of defects in proteins required for DNA replication. Also, such systems did not allow the initiation of new rounds of replication.

The development of lysate replication systems rested upon the recognition that the failure to maintain the complex process of DNA replication in early studies was due to dilution of the components, resulting in disassembly of the replication structure and loss of functions required for DNA replication.

The concentrated lysate systems depended on the bacterial chromosome, but it was quickly recognized that small bacterial phage chromosomes could be utilized as exogenous templates for DNA replication. Lysate systems using circular phage chromosomes formed the basis for the biochemical definition of replication because they permitted the addition of proteins to allow complementation of defects in DNA replication. Lysates made from mutants defective in a specific step of DNA replication could be used to define the step at which the defect occurred and to identify the protein product complementing the defect. This allowed assignment of protein products to genes. Such systems, however, did not allow the study of initiation of new rounds of DNA synthesis on the host chromosome.

Two theoretical shortcomings of such systems are (1) that such systems might not define all of the proteins required for DNA replication by the host and (2) that such systems might require a protein for replication of the phage genome not ordinarily required by the host genome.

One general point arising from these studies is worth remembering: Although *E. coli* contains numerous proteins that have overlapping or similar enzymatic function, the participation of a protein in the replication process is carefully regulated, reflecting a specific role. No class of enzymes is a better example than the DNA polymerases. Each of the three recognized DNA polymerases of *E. coli* has similar enzymatic capabilities, but ordinarily only DNA polymerase III catalyzes replication. This restriction of activity can be partially explained on the basis of protein–protein interactions. The complete basis for the regulation is not understood and the reasons for its being advantageous to the cell are not clear (Table I).

Table I DNA Polymerases of *Escherichia coli*

	I	II	III
Molecular mass	103 kDa	88 kDa	140 kDa
Synthesis	$5' \rightarrow 3'$	$5' \rightarrow 3'$	$5' \rightarrow 3'$
Initiation	No	No	No
5′-Exonuclease	Yes	Yes	Yes
3′-Exonuclease	Yes	Yes	Yes[a]
Gene	*polA*	*polB*	*polC* (*DnaE*)

[a] In a separate protein.

II. Control of DNA Replication

Control of DNA replication rests on the regulation of new rounds of replication. In *E. coli*, there are two components of control: the DnaA protein and the structure of the origin of DNA replication (*oriC*). The region of the *E. coli* origin of DNA replication is at 83.5 min on the genetic map (basis of 100 min for the map). Thus, there is a fixed site on the *E. coli* genome that represents the appropriate place for the initiation of DNA replication. This DNA initiation is referred to as "macroinitiation," as opposed to the repetitive initiation that must occur multiple times during the "elongation" phase of DNA replication. The latter is referred to as "microinitiation." It seems that the requirements for macroinitiation at the *oriC* region include those needed for microinitiation plus additional ones. There is a region of approximately 250 bp that must be present for DNA replication to initiate.

A. Macroinitiation

Several features regarding this region are notable. There are multiple binding sites for the DnaA protein (the DNA boxes). This is a nine-nucleotide sequence that has been shown to bind the DnaA protein. There are also multiple promoter elements, suggesting the involvement of RNA polymerase in macroinitiation. Possible DNA gyrase binding sites are also present. Another notable feature is the presence of multiple Dam-methylase restriction sites (GATC sequence).

The definition of the *oriC* region rests on cloning of this region into plasmids constructed so that replication of the plasmid depends on function of the *oriC* sequence. This has allowed development of an *in vitro* assay system for macroinitiation of DNA replication. The cloning of *oriC* confirmed that the specificity of macroinitiation in *E. coli* resides in the origin.

The primary protein actor in macroinitiation is the DnaA protein. This protein binds at multiple sites within the *oriC* structure as noted. In addition to binding at the DnaA box consensus sequence, the DnaA protein displays a DNA-dependent adenosine triphosphatase (ATPase) activity and appears to display cooperative binding properties. This suggests that the possible role in initiation is a change of conformation of DNA by DnaA protein interactions. The DnaA protein also binds in the promoter region of the DnaA gene itself, suggesting autoregulation, which is supported by genetic studies. In any event, it appears that the DnaA protein must act positively to initiate DNA replication in *E. coli*.

Both protein and RNA synthesis are required for macroinitiation to occur in *E. coli*. The macroinitiation phase may be further subdivided into stages. The earliest step involves the binding of the DnaA protein to the *oriC* structure. This results in a conformational change of the origin. This complex then binds the DnaA to DnaB and DnaC proteins (which are required as well for the elongation phase of DNA replication). DnaC plays a unique role in the delivery of the DnaB protein to the replication structure. The resulting complex appears to unwind the DNA strands since the DnaB protein functions as a helicase, an ATP-dependent unwinding activity. This allows the binding of single-stranded binding (Ssb) protein, which allows priming like that which occurs in the elongation of DNA replication. This stage is followed by the propagation of microinitiation and elongation phases of replication.

Thus, the proteins required for the macroinitiation of *E. coli* DNA replication appear to include DnaA and DnaC in specific roles as well as DnaB, Ssb protein, gyrase, the DnaG primase protein, and the replicative apparatus of DNA polymerase III holoenzyme complex. Studies also suggest a direct role for RNA polymerase in the macroinitiation of DNA replication.

The DnaA protein offers important support of the replicon hypothesis. Mutations in the DnaA protein shows that all of the *E. coli* chromosome is under a unit control mechanism, defining it as a single replicon. Integration of certain low-copy number plasmids into the chromosome suppresses the phenotype in DnaA mutants that were defective in macroinitiation. This "integrative suppression" shows a general control of macroinitiation and supported the replicon hypothesis.

B. Microinitiation

Microinitiation is the hallmark of the elongation phase of DNA replication. During this phase, repeated initiation occurs along the DNA. The microinitiation step appears to be analogous to the initiation step studied in the *in vitro* lysate systems using small circular phage genomes. In the prokaryotic cell, the requirements for microinitiation appear to mimic those of the phage systems G4 and ΦX174, which do not display a requirement for the DnaA protein nor the features of the *oriC* region.

Cell proteins required for microinitiation include the DnaB protein. The DnaB protein contains a nucleoside triphosphate activity that is stimulated by single-stranded DNA. It also displays DNA helicase activity. In addition, it appears to undergo protein–protein interactions with the DnaC protein. The DnaB mutants are notable for a rapid cessation of DNA synthesis at restrictive conditions. It appears that the DnaB protein is one of the "motors" that moves the replication complex along the DNA (or moves the DNA through the replication complex). The DnaB protein is typical of the proteins involved in DNA replication in that it may have more than one role.

The DnaG protein of *E. coli* is the primase. This particular protein has the capability of synthesizing oligonucleotides utilizing nucleoside (or deoxynucleoside) triphosphates. It appears that physiologically its role is to synthesize RNA primers, which can be utilized by the synthesis apparatus of the cell. As displayed in Fig. 1, at least some portion of DNA replication in most organisms is discontinuous. That is, part of the DNA is synthesized in short pieces (termed Okazaki pieces). This is the result of the restriction for DNA synthesis in the $5' \rightarrow 3'$ direction. Since the replication fork requires apparent growth of the nascent strands in both the $5' \rightarrow 3'$ direction and the $3' \rightarrow 5'$ direction, effort was spent looking for precursors or enzymes that would allow growth in the $3' \rightarrow 5'$ direction. None were found. The hypothesis of discontinuous synthesis states that, on a microscopic scale, DNA is synthesized discontinuously in a $5' \rightarrow 3'$ direction in small pieces to allow an overall growth in the $3' \rightarrow 5'$ direction on one strand. This hypothesis predicts the existence of a relatively uniform class of small nascent DNA strands prior to joining, and it makes a second prediction of joining activity for such DNA strands. Both of these predictions are fulfilled.

It appears that in *E. coli* DNA is synthesized on one lagging strand in pieces of approximately 1000 nucleotides, which are then covalently linked via the action of DNA ligase following synthesis. Because none of the DNA polymerases in prokaryotes have been found to initiate synthesis *de novo*, this hypothesis further leads to the prediction that RNA synthesis, which *can* be demonstrated to initiate *de novo*, forms primers that are utilized for DNA strand synthesis. Identification of the DnaG primase activity satisfies this prediction.

The polarity restriction of DNA synthesis by DNA polymerases permits one strand to be made continuously, as indicated in the model. It appears

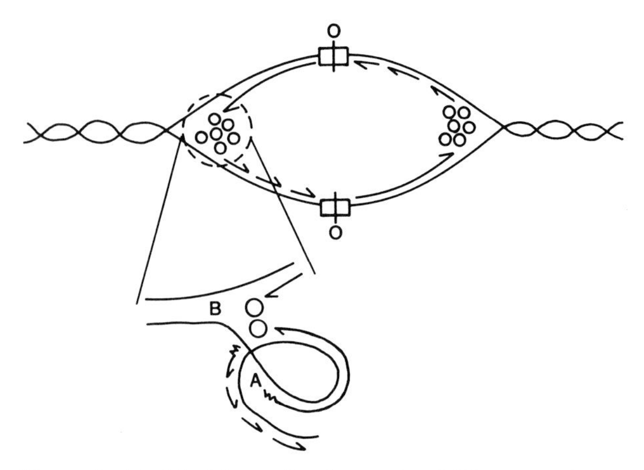

Figure 1 Model of the replication region. "O" represents the origin. In the expanded view, "A" represents a primed, elongating nascent strand; "B" represents the point at which the replicative enzyme will release from the "A" strand and re-initiate a new discontinuous strand.

that relatively few initiations are made in this (leading) strand and, in fact, that macroinitiation may serve to prime that strand.

Ssb protein has analogs found throughout nature. In *E. coli,* this protein is a relatively small protein (about 19 kDa) and functions as a tetramer. This protein is required for DNA replication. It appears to have several roles. In the single-strand phage systems *in vitro,* it confers specificity on the origin of DNA replication and there is not reason to doubt that it peforms a similar role in *E. coli.* It probably maintains DNA in a somewhat more open state under physiologic conditions during DNA replication in the cell. Some evidence suggests that the Ssb protein may participate in a nucleosomelike structure, perhaps with *E. coli* HU protein. It is possible that the coating of the DNA strands by Ssb protein protects against nucleolytic degradation during replication. Lastly, Ssb protein appears to stimulate the synthesis rate of DNA polymerases under particular conditions. Whether or not this is the case during DNA replication is not clear.

DNA ligase is required for joining the Okazaki pieces made during discontinuous DNA synthesis. This enzyme is also required for DNA replication because mutants conditionally defective in ligase are conditionally defective in DNa synthesis.

In addition to the previously mentioned proteins, other proteins such as DNA gyrase or DNA topoisomerase I (ω protein) may play a critical role during the microinitiation phase of DNA replication.

Additional genes *priA, B,* and *C* (for primosome) and their products can be shown to play a role in assembly of the primosome structure and are required for replication of at least some single-stranded phages and plasmids. However, the effect on the cell of a deficiency is modest.

III. Genetics of DNA Strand Synthesis

The genetics of DNA strand synthesis are reflected in the DNA polymerase. *Escherichia coli* is known to contain at least three distinct and separate DNA polymerases, all possessing similar enzymatic activities: synthesis exclusively in the $5' \rightarrow 3'$ direction, utilization of 5′-deoxynucleoside triphosphates for substrates, the copying of single-stranded DNA template, incorporation of base analogs or ribonucleoside triphosphates at low efficiency under altered conditions (such as in the presence of manganese), a 3′ editorial exonuclease, which preferentially removes mismatched 3′ termini, and a rate of synthesis that does not approach that of DNA replication in the cell (Table I). Despite these similarities, distinct physical differences exist, and the cell uses exclusively DNA polymerase III for DNA replication. The synthesis subunit for DNA polymerase III, the α-subunit, is encoded by the *polC* (*dnaE*) gene.

A. DNA Polymerases

DNA polymerase I, encoded by the *polA* gene, appears to be an auxiliary protein for DNA replication. Cells lacking this enzyme demonstrate viability, although those lacking the notable $5' \rightarrow 3'$ exonuclease activity of this enzyme are only partially viable unless grown in high salt. DNA polymerase I is strikingly important for survival of the cell following many types of DNA damage, and in its absence the cell has persistent single-stranded breaks that promote DNA recombination. DNA polymerase I appears to be a particularly potent effector with DNA ligase in sealing single-stranded nicks, perhaps because of its ability to catalyze nick-translation in which the 5′-exonucleolytic removal of bases is coupled to the synthesis activity. Neither of the other DNA polymerases appears to possess this property.

DNA polymerase II is an enzyme without a defined role in the cell. It has recently been cloned and overproduced and has been found to bear a closer relationship to T4 DNA polymerase and human polymerase alpha than to either of the other two *E. coli* DNA polymerases. Nevertheless, it is capable of interacting with the subunits of the DNA polymerase III complex. Cells that completely lack the structural gene for DNA polymerase II (the *polB* gene) show normal viability and normal repair after DNA damage under many circumstances.

DNA polymerase III is the required replicase of *E. coli*. The fact that it plays a significant role in DNA replication is demonstrated because *dnaE*ts mutants contain a temperature-sensitive DNA polymerase III. Despite having properties similar to those of DNA polymerase I and II, DNA polymerase III is specifically required for DNA replication. This is a reflection of its ability to interact with a set of subunits that confer particular properties upon the complex. In the complex (termed the holoenzyme) DNA polymerase III takes on the properties of a high rate of synthesis and great processivity. Intuitively, processivity may be thought of as the ability of an enzyme catalyzing the synthesis of DNA to remain tracking on one template for a long period of time before disassociating and initiating synthesis on another template. Highly processive enzymes are capable of synthesizing thousands of nucleotides at a single stretch before releasing the template. DNA polymerase III appears to be uniquely processive among the *E. coli* DNA polymerases.

B. Holoenzyme DNA Polymerase III

The constituents of the DNA polymerase III holoenzyme complex, in addition to the α-synthesis subunit, number at least six. Each of them appears to be the product of a required gene (with the exception of the θ-subunit, which is uncertain) as demonstrated by the fact that mutations in that gene produce conditional cessation of DNA replication. The β-subunit is the product of the *dnaN* gene. This subunit appears to confer specificity for primer utilization upon the complex and to increase the processivity. The γ-protein and the δ-protein appear to be products of the same *dnaZ* gene that produces the τ-protein. This is a case where frame shifting termination of protein synthesis plays a role in producing different proteins from the same gene. The ε protein of the holoenzyme complex is known to provide a powerful 3′-editorial exonuclease activity. This is manifest by the fact that in addition to lethal mutants in this gene (*mutD*), mutants that show increased error rates in DNA replication (mutators) can be isolated. Lastly, the θ-subunit of the holoenzyme remains to be assigned to a gene and, thus, demonstrated to be a required protein. It may affect the interaction of the protein subunits.

Physical studies as well as genetic studies argue that the DNA polymerase III holoenzyme complex exists in a dimer form. The stoichiometry of the various subunits suggests that the dimer is not exactly symmetrical, but it does appear to be symmetrical for the α-, β-, and ε-subunits. The physical and genetic evidence supporting dimerization of DNA polymerase III fits nicely with a structural model for replication. This is a so-called inch-worm, or trombone, model of DNA replication. As indicated in the model, a dimer at the growing fork would allow coupling of rates of synthesis on the leading and lagging strands, i.e., the strand made continuously and the strand made discontinuously. Because the strand made discontinuously may require frequent initiation, one might expect synthesis of this nascent DNA strand to be slower. To prevent a discrepancy in growth rate between the strands, dimerization of the synthesis units for the two strands would be a method for locking the rates in step. It is possible to do this by assuming that the DNA template for the lagging strand loops out in such a way as to provide permissible polarity for the nascent strand.

IV. Termination of Replication

The termination of DNA replication is not a passive process. The product requires resolution of a catenate (interlocked circles). We might anticipate that topoisomerase I (the ω protein) or DNA gyrase (topoisomerase II) will be required for separation of the products.

In addition, there is a specific region of the chromosome, approximately directly opposite the origin on the *E. coli* map at the 30–32 min region, which represents the termination region. This region is particularly sparce in genetic markers. The DnaT protein is required for the proper termination of

DNA replication. Mutants in this gene produce premature disassembly of the replication complex and poor genome segregation. Both *cis* and *trans* elements must act for termination. These include proteins binding to DNA and specific recognition signals in the DNA. DNA replication termination may also be a control for cell division.

Bibliography

Bramhill, D., and Kornberg, A. (1988). *Cell* **54,** 915–918.

Echols, H. (1986). *Science* **233,** 1050–1056.

McHenry, C. S. (1988). *Annu. Rev. Biochem.* **57,** 519–550.

McMaken, R., Silver, L., and Georgepoulos, C. (1987). DNA replication in *E. coli* and *S. typhimurium*. *In* "Cellular and Molecular Biology" (F. Niedhardt, ed.), pp. 564–612. American Society for Microbiology, Washington, D.C.

Dutch Elm Disease and Elm Yellows

Lawrence R. Schreiber
U.S. Department of Agriculture, Agricultural Research Service

Glossary

Disease cycle Events leading up to and through the end of the disease process
Incubation period Period of pathogen growth
Infectious Can serve as an infecting agent
Inoculum Portion of the pathogen that can cause infection
Mycoplasmalike organism Organism with characteristics of a mycoplasma, including undifferentiated cells that vary in shape and lack cell walls and are smaller than bacteria but larger than viruses
Tyloses Balloonlike growths into the xylem vessels from adjacent parenchyma cells
Vector Carrier of inoculum

ELMS are among the most frequently planted and valuable trees in the world. Their diversity in size, shape, environmental adaptability, and stress tolerance have made them the trees of choice in many landscape settings. The monetary value of elms, particularly in urban areas, where they contribute significantly to property value, is difficult to assess. Elms provide shade and beauty in the landscape; shelter for humans and animals as shelterbelts and windbreaks; lumber for building boats, tools, and furniture; and fodder for cattle. In North America, the American elm (*Ulmus americana*) embodied many of the desirable characteristics of landscape trees in rural, suburban, and especially urban sites. Thus, it was extensively planted in many cities and towns almost to the exclusion of other species. As a result of widespread planting of susceptible elms in Europe, Britain, and North America, Dutch elm disease and elm yellows epidemics have had a devastating worldwide impact both aesthetically and economically.

I. Dutch Elm Disease

A. Background

Dutch elm disease (DED) is the most widespread and destructive tree disease in the world. The fungal pathogen, *Ophiostoma ulmi* (Buism.) Nannf., was first associated with the disease and identified in The Netherlands in 1920. By that time, DED had already been known to occur widely in Belgium and France as well as in The Netherlands. How or when the disease was introduced into Europe is not certain. Some speculate that the pathogen originated and was introduced from Asia because of the high levels of disease resistance found among Asiatic elms. This resistance suggests that these species evolved in contact with the pathogen through natural selection. The disease spread widely through Europe and European Russia between 1920 and 1939 and was first reported in Britain in 1927.

It is generally accepted that the DED fungus was introduced into North America on elm logs from France sometime prior to identification in 1930 in Ohio. Burled elm logs were used for veneer as furniture in the Eastern and Midwestern United States. Thus, the disease began to appear at ports of entry on the East Coast and along railroad rights-of-way between there and the Midwest. DED spread from loci in the Midwest and East Coast into Canada and all states in the continental United States except Florida, Nevada, Utah, Arizona, and New Mexico.

B. Symptoms

DED produces foliar symptoms including wilting and inward rolling of the leaves followed by drying, browning, and defoliation. Affected branches die. Disease symptoms usually appear in one or several branches beginning with the outer foliage. Symptoms then spread to other parts of the crown. In some instances, symptoms develop simultaneously throughout the crown and trees may die rapidly, within a few weeks. Such a disease pattern usually occurs in highly susceptible trees when infection occurred late in the season prior to symptom expression. Usually trees that become infected in the spring or early summer die quickly, whereas those that become infected late in the summer may not die until the following year. Trees affected by DED develop brownish black streaking in the water-conducting vessels of the current year's wood. This may be seen by stripping the bark from, or making a horizontal cut through, a branch with foliar symptoms.

Blockage of the vascular tissue occurs in response to the presence of the fungus as a result of gelatinous materials secreted into the xylem vessels. Balloonlike tyloses develop from xylem parenchyma cells through pits in the xylem vessel walls. These occlusions in the vascular system plus toxin activity result in excessive water loss and other imbalances that lead to wilting.

DED is one of several diseases that have symptoms that might be mistaken for one another. These include *Verticillium* wilt, *Dothiorella* wilt, and elm yellows. The former two diseases are caused by fungal pathogens and the latter by mycoplasmalike organisms. Thus, unequivocal identification requires isolation and identification of the DED pathogen. The inability to isolate the pathogen, particularly late in the summer or fall, does not preclude the presence of the disease.

C. The Pathogen

The DED fungus, first isolated and identified by a Dutch plant pathologist, was named *Graphium ulmi* Schwartz after the asexual stage. The sexual or perfect stage of the fungus was described in 1932 as *Ceratostomella ulmi* Buism., in 1934 as *Ophiostoma ulmi* Nannf., and in 1952 as *Ceratocystis ulmi* (Buism.) C. Moreau. Currently, the name *O. ulmi* is again in favor.

Ophiostoma ulmi is a highly selective pathogen that infects only elms and close elm relatives such as *Zelkova serrata*. It is a saprophyte that can grow on a wide variety of substrates including plant debris and agars. The fungus synthesizes enzymes that enable it to penetrate cell walls and cause adjacent cells to expand tyloses into xylem vessels, thus producing blockages that contribute to foliar wilting. Other enzymes contribute to wilting by destroying cell membrane permeability. *Ophiostoma ulmi* produces the plant toxin cerato-ulmin that is significantly involved in symptom development.

The worldwide distribution and regional isolation of the fungus has led to the development of morphological and pathological strains. These have been termed aggressive and nonaggressive. Evidence indicates that the nonaggressive strains were prevalent in Europe and Britain in the 1920s and 1930s and caused little elm mortality. However, a new series of DED epidemics began in the 1960s in Britain with the introduction of the aggressive strain on elm logs from Canada. About 70% of the elms were killed between 1971 and 1978. The aggressive strain was also found in Europe at about the same time. Elms developed in the Dutch breeding program, resistant to the earlier strains, were lost during the 1970s. Genetic variability may occur in *O. ulmi* through somatic mutations or by recombination of compatible mating types.

D. Hosts

The genus *Ulmus* includes about 32 species, most of which are native to central Asia; 5 species are native to Europe and 8 to North America. Elms have adapted to a great diversity of geographic and environmental conditions ranging from sea level to >10,000 ft and in northern tundra to southern tropics. Elms are native to China and Japan, European and Asiatic Russia, North Africa, and North America. They have been extensively planted beyond their natural range in North America. Elm species are now also established in Australia and South America.

Although elm species and selections vary greatly in their susceptibility to DED, none is immune. Unfortunately, species native to North America are among the most susceptible. These include the American elm (*Ulma americana*), our most valuable and widely planted species, as well as winged elm (*U. alata*), red elm (*U. rubra*), and September elm (*U. serotina*). At the other end of the spectrum are the highly resistant Asiatic species including Siberian elm (*U. pumila*), Chinese elm (*U. parvafolia* and *U. wilsoniana*). For the most part, European elms such as smooth-leaved elm (*U. carpinifolia*), Scots elm (*U. glabra*), and European white elm (*U. laevis*) are also very susceptible but selections have been made within these species that have demonstrated moderate levels of resistance. In Europe, prior to the 1960s, disease-resistant elms were selected by inoculating trees with predominantly nonaggressive isolates. These selections were not resistant to the more aggressive strains introduced later. This was not a problem in the North American breeding program because of the predominance of the aggressive strains used to test new selections.

E. Insect Vectors

The DED fungus is transmitted overland on the bodies of a number of species of elm bark beetles. The primary vectors in Great Britain and Europe are *Scolytus multistriatus*, the smaller European elm bark beetle, and *S. scolytus*, the larger European elm bark beetle. In North America, the vectors are *S. multistriatus* and *Hylurgopinus rufipes*, the native elm bark beetle. *Scolytus* spp. are native to Europe and Asia and were introduced into North America before the 1900s the same way the pathogen was introduced, on elm logs used for veneer. *Scolytus multistriatus* became well established as the most important vector in North America, displacing *H. rufipes* in most areas.

F. Disease Cycle

The adult *S. multistriatus* feeds and breeds in elms. Feeding occurs primarily in crotches of 2–4-yr-old twigs of healthy elms. Beetles breed in dead or dying elms or large branches where the bark still adheres to the wood. Adult mating beetles bore through the bark of brood trees and lay their eggs between the wood and bark. Complete development from eggs to adults may occur in about 6 wk. Eggs laid later in the summer hatch and develop into larvae that emerge as adults about the time elms break dormancy and are the most susceptible. Emerging adults fly or are blown by the wind to healthy elms where they feed.

Breeding occurs in dead or dying elm wood from whatever cause, including DED. If brood wood has resulted from DED, the emerging beetles may carry the fungus on their bodies and mouth parts and infect healthy elms during feeding. Fungus spores introduced into the large, open spring wood vessels move readily throughout the tree. Beetle emergence occurs between April and October with both breeding and feeding continuing throughout the growing season. There are two to three generations a year, depending on the latitude.

Hylurgopinus rufipes inhabits the colder, more northern latitudes and overwinters as adults in bark at the base of healthy elms. They feed on the lower trunk in the fall and in the branches in the spring. *Hylurgopinus rufipes* produces one or more generations a year.

In addition to overland infection by means of feeding contaminated beetles, the pathogen may move between infected and healthy elms through root grafts. Root grafts form between adjacent elms when their roots come in contact. In some instances, functional unions occur between xylem vessels. In these cases, *O. ulmi* may grow from the points of infection in the tree crown, through the vascular system in the branches, trunk, and roots, through the root grafts, to the adjacent healthy tree. As a result of active fungus growth and by means of transpiration pull by the healthy tree, fungus propagules will move into and infect the healthy elm.

G. Control

Control of DED centers around two major strategies: prevention and therapy. While neither approach used separately or in combination can eliminate the disease, they can significantly reduce infection and subsequent losses of valuable trees.

Disease prevention is by far the most effective means of controlling the disease. The single most important part of a disease prevention program is sanitation. Sanitation is directed at reducing both the fungus inoculum and the insect vector and involves removing dead or dying elms or parts of elms. Removal of dead and dying wood reduces breeding sites for the elm bark beetle vectors. Reducing beetle populations reduces the likelihood of spread of the fungus. If the wood that is removed has been infected, then sanitation achieves a second valuable function of reducing the availability of the pathogen.

To be effective, sanitation must begin as early as possible, preferably before the disease is know to occur. To assure the earliest possible identification and removal of hazardous elms, surveys should be made of all elms to determine the presence of the disease and the presence of any elms or elm wood that could serve as a breeding site for beetles. Surveys should be made as frequently as possible throughout the growing season. Symptomatic trees, identified between April and September, should be removed within 30 days to prevent beetle emergence. Elm wood can be properly disposed of by debarking, burning, or burying in a landfill. Beetle feeding may be reduced through application of insecticides. To be effective, all tree parts to be protected must be covered because insecticides serve to kill feeding beetles before they can cause infection. Insecticides are generally applied during the dormant season because foliage would impede proper coverage of feeding sites in the tree's crown. [*See* INSECTICIDES, MICROBIAL.]

Fungicides that prevent or retard infection and disease development are available. For best results, these chemicals are injected directly into the tree's trunk and roots. While disease prevention is more effective than therapy, early application of fungicides to diseased trees may also be effective.

All of the preceding recommendations are designed to protect susceptible elms already in the landscape. Disease control through the development and release of disease-resistant elms aims to replace trees lost to the disease. A DED resistance program was initiated in The Netherlands 50 years ago and has provided useful trees for many areas in Europe. Many of the Dutch trees were not suitable for use in North America because they lacked climatic adaptability as well as being susceptible to strains of the fungus present in North America. While none are immune, elms suitable for planting in North America have resulted from crosses between more resistant European and Asiatic elms. Moderate resistance in American elms has been found in a few disease survivors. Because the levels of resistance in these American elms has not been fully established, but may be lower than that found in Asiatic or European selections and crosses, large-scale plantings should be avoided. In addition, none is known to be resistant to elm yellows.

II. Elm Yellows (Elm Phloem Necrosis)

A. Background

Elm yellows, previously known as elm phloem necrosis, was first reported in 1938 following investigations of dying elms in various parts of Ohio from 1918 to 1935. The causal agent was considered to be a virus. While it was not known where the disease originated, sightings and reports indicated that it was probably present in parts of Indiana, Kentucky, and Illinois for many years. Reports from Kentucky as early as 1893 and 1899 indicated dying elms from similar causes. The disease probably occurred in Illinois at least as early as 1882. The described symptoms of these early epidemics closely resemble those subsequently attributed to elm yellows. However, because the causes were never definitely determined and the symptoms were not adequately described for positive identification, we are not certain that it was elm yellows that was being described. Initially, epidemics were considered to be limited to the central and southern United States but were later found in New York and Pennsylvania in the 1960s and 1970s.

B. Symptoms

Symptoms frequently do not appear for 6 mo to 1 yr or more after infection. Foliar symptoms of elm yellows first appear in July–September. Symptoms vary but are usually characterized by a gradual and general decline throughout the crown with initial symptoms first appearing in a portion of the tree or even in a single branch. Leaves droop and the blades curl inward, giving the impression of a scarcity of foliage that starts at the top and periphery of the crown. Leaf color appears to change to a lighter green or a gray due to the upward curling of the leaves and exposure of their undersurface. Later, loss of foliage and increased yellowing gives the tree the appearance seen at the outset of fall dormancy or of symptoms associated with drought. These symptoms may be followed by drying and browning of the leaves and defoliation. In some cases, leaves are lost following yellowing of the foliage. The rate of onset of foliar symptoms varies among trees.

Elm yellows causes the death of tree roots starting with feeder roots. Tan discoloration and necrosis develop in the inner bark or phloem of larger roots and extends into the trunk and some branches, hence the name phloem necrosis originally given to the disease. These symptoms are more likely to occur in the upper branches of smaller trees and become apparent in trees of all sizes prior to the appearance of foliar symptoms. The discoloration is often described as butterscotch in color with scattered black or brown spots. It later becomes darker brown. In the earlier stages of discoloration, a slight wintergreen odor, caused by methyl salicylate, may be detected. This is most easily discerned by placing a piece of freshly cut discolored bark in a small sealed container for a few minutes to allow volatilization.

The two symptoms of discolored phloem and the wintergreen odor are specific symptoms for the identification of elm yellows. The foliar symptoms may not as clearly differentiate elm yellows from other elm diseases, especially DED. Because normal bark from healthy elms will become brown when exposed to the air, diagnosis of elm yellows should be made only with freshly cut bark. The absence of any of these symptoms does not preclude the presence of the disease, but observations at a later date may be necessary to determine its presence.

C. The Pathogen

The causal agent of elm yellows was incorrectly identified in the 1930s and 1940s. Work by scientists in the U.S. Department of Agriculture pointed to a virus as the disease agent. There were good reasons for this early misconception. Symptoms and other disease characteristics were those typically associated with a virus disease. At that time, viruses were too small to be seen through microscopes. The presence of viruses was postulated by their highly infectious nature in the absence of any other recognizable agent. Numerous attempts to infect trees with organisms isolated from diseased elms failed. In addition, the disease agent mimicked the properties and behavior associated with viruses. Healthy elms grafted with bark patches or other portions of diseased trees became diseased within 6–24 mo. In addition to graft transmission, the role of an insect vector was demonstrated.

In the 1970s, several scientists reported the presence of mycoplasmalike organisms (MLOs) associated with the phloem of elm yellows-infected American elms. These bodies were absent from similar cells of healthy elms. MLOs have been known to cause diseases in animals since 1809, but they have only been associated with plant diseases since 1967. Mycoplasmas are smaller than bacteria and larger than viruses. They are irregular in shape and surrounded by an elastic membrane but lack a cell wall. Nuclei are absent. [*See* MOLLICUTES (MYCOPLASMAS).]

Beside elm yellows, it is now recognized that MLOs are probably the cause of several plant diseases. These are collectively referred to as "yellows-type" diseases. Evidence for MLOs as the causal agent for elm yellows depends on the consistent association of the organism with diseased plants, its absence from healthy ones, and its transmission by leafhopper vectors. Tetracycline antibiotics, known to be active against animal mycoplasmas, have produced disease remission in plants with MLO diseases. Antibiotic treatment of insect vectors known to harbor MLOs has rendered them noninfectious. This good circumstantial evidence relates these bodies to yellows diseases.

D. Hosts

The earliest reported elm hosts of elm yellows were American elm and the winged elm. Elm species vary in their susceptibility to elm yellows. American elms are extremely susceptible, with most trees dying in 1 yr or less. Red or slippery elms are less susceptible and may die 2 or more yr after infection. Other native North American species that are susceptible include cedar (*U. crassifolia*) and September elms. Asiatic and European species are most resistant. Attempts at graft transmission from American and red elms to Siberian and Scotch elms resulted in no external symptoms while similar grafts to *U. carpinifolia* produced "witches broom" symptoms but was not fatal. Epidemics in New York killed American and red elms, but Scotch, European smooth-leaved, and Siberian elms were not affected.

Graft transmission of elm yellows, but not natural infection, has been successful in European white elm and Chinese elm as well as smooth-leaved elm. Infection in those species produces symptoms including yellowing of foliage, stunting, and witch's brooming of branches. In most cases, infected trees do not die.

The only nonelm host of the elm yellow MLOs is periwinkle (*Vinca rosea*). Transmission was accomplished experimentally through dodder (*Cuscuta epithymum*) trained simultaneously on yellows infected elm and healthy periwinkle.

E. Vectors

The primary vector of elm yellows is the white-banded elm leafhopper (*Scaphoideus lutelous* Van Duzee). This insect is found throughout the range of the disease and beyond in northern areas from Maine to Minnesota. The insect overwinters as eggs in bark crevices of small elm branches. Eggs start to hatch in the spring as buds break and leaves begin to expand. The insect develops to the adult stage by June and is present until frost in the fall. Nymphs are distinguished by a white band across the middle of their back. Adults and nymphs feed by sucking through mouthparts inserted into the phloem of leaves and green herbaceous shoots. Feeding occurs in all parts of the crown. Leafhoppers become infected with MLOs by feeding on diseased trees. The MLO multiplies in the insect and moves to the salivary glands. This process occurs in about 3 wk after infection, and the insect then remains infective until death.

F. Disease Cycle

Elm yellows can be transmitted through root grafts when elms are in close proximity or by feeding leafhoppers. Leafhoppers emerge from eggs as elms break dormancy in the spring. The nymphs and later the adults feed on elm foliage. If host trees are infected with elm yellows, subsequent feeding by the leafhoppers may result in infection of healthy elms. MLOs move from the phloem in the tree's crown downward into the roots. Inoculum build-up occurs in the roots and results in the death of feeder roots before foliar symptoms are present. As damage to larger roots occurs, foliar symptoms begin and early leaf drop occurs. Following foliar symptoms, discoloration begins in the phloem, which is associated with the onset of a detectable wintergreen odor in the inner bark. Continued decline of the tree is usually followed by death by the end of the second year.

G. Control

There are no satisfactory means of controlling elm yellows. Some studies have been conducted using insecticides to control the leafhopper vector, but satisfactory results have not been reported. Antibiotic injections with tetracycline and oxytetracycline have arrested foliar symptoms, but symptoms return unless treatments are continued. Thus, a chemical cure for the disease is not available. The problem of early chemical injection treatment is increased, in more susceptible species, by advanced stages of root deterioration that occur prior to the earliest foliar symptoms. To be successful, chemical treatments must be administered immediately prior to or following infection.

Genetic resistance to elm yellows has been investigated in American elms. Unfortunately, elms that showed yellows resistance proved to be susceptible to DED. Combining resistances to both diseases into a single American elm selection has not yet been possible.

Bibliography

Brasier, C. M. (1979). *Nature (London)* **281,** 78–80.

Braun, E. J., and Sinclair, W. A. (1976). *Phytopathology* **66,** 598–607.

MacHardy, W. E., and Beckman, C. H. (1973). *Phytopathology* **63,** 98–103.

Neely, D. (1975). *In* "Proceedings of the 1973 International Union for Research Organization Conference on Dutch Elm Disease" (D. A. Burdekin and H. M. Heybrook, eds.). U.S. Department of Agriculture Forest Service, Northeast Forest Experiment Station, Upper Darby, Pennsylvania.

Richards, W. C., and Takai, S. (1984). *Can. J. Plant Path.* **6,** 291–298.

Sinclair, W. A., Braun, E. J., and Larsen, A. O. (1976). *J. Arboric.* **2,** 106–113.

Wilson, C. J., Seliscar, C. E., and Krause, C. R. (1972). *Phytopathology* **62,** 140–143.

Yang, D., Jeng, R. S., and Hubbes, M. (1989). *Can. J. Bot.* **67,** 3490–3497.

Dysentery, Bacillary

Samuel B. Formal, David N. Taylor, and Jerry M. Buysse
Walter Reed Army Institute of Research

Glossary

Diarrhea Abnormal frequency and liquidity of fecal discharges

Dysentery Disease with inflammation of the large intestine, severe abdominal pains, constant desire to evacuate the bowels, and the discharge of blood, mucus, and inflammatory cells in the feces

Northern blot Procedure in which electrophoretically separated RNA molecules, immobilized on a nitrocellulose filter, are hybridized with homologous radiolabeled DNA molecules

Open reading frame Genetic unit containing all information encoding a particular protein

Plasmid Extrachromosomal, self-replicating DNA element in a cell

Transposon mutagenesis Procedure by which mutations are randomly generated in a bacterial cell via insertion of a transposable genetic element encoding antibiotic resistance

Western blot Procedure in which electrophoretically separated proteins, immobilized on a nitrocellulose filter, are reacted with antisera containing antibody directed against the proteins

BACILLARY DYSENTERY has been recognized since biblical times. Most early descriptions of dysentery were associated with war, where the crowding of men and poor sanitation contributed to the spread of this disease and made it an easily recognizable syndrome. Along with typhus, cholera, plague, typhoid fever, influenza, and malaria, bacillary dysentery has influenced the outcome of battles as much as those who planned, conducted, and waged them. From the wars between the Persians and the Greeks (431 B.C.) to the present time, few conflicts have escaped its disruptive impact. In the twentieth century, the defeat of the British at Gallipoli (World War I) and the Germans at El Alamein (World War II) were due in part to epidemics of dysentery.

Bacillary dysentery is caused by members of the genus *Shigella*. These organisms have a worldwide distribution and cause a disease of major importance in third world countries, where morbidity and mortality are high. The importance of epidemic shigellosis caused by *Shigella dysenteriae* 1 has been apparent since the late 1960s, when epidemics occurred in Central America, the Indian subcontinent, and Africa. Sporadic cases and small epidemics caused by *Shigella flexneri* or *Shigella sonnei* can be seen in developed nations, but disease is particularly common in closed populations, such as children in daycare centers and inmates of prisons and mental institutions.

The symptoms of the disease range from mild diarrhea to dysentery. Diarrhea is a watery stool, whereas dysentery is defined as the presence of blood, mucus, and inflammatory cells in the stool. The term shigellosis is often used instead of bacillary dysentery and covers the range of symptoms from diarrhea to classical dysentery.

I. Pathogen

Kiyoshi Shiga is generally given credit for isolating the first strain of dysentery bacilli in Japan in 1898, although Chantemesse and Widal described an organism similar to Shiga's isolate 10 years earlier. The organisms that cause bacillary dysentery are classified in the family Enterobacteriaceae, tribe Escherichieae, genus *Shigella*. Shigellae are gram-negative, nonmotile, asporogenous rods that are facultative anaerobes. They are lysine decarboxylase-negative, phenylalanine deaminase-negative, ornithine decarboxylase-negative (with the exception of *Shigella boydii* 13 and *S. sonnei*), lactose-negative (with the exception of *S. sonnei*, which ferments lactose slowly), and anaerogenic (with the exception of some strains of *S. flexneri* 6).

The genus *Shigella* is divided into four subgroups on the basis of biochemical and serological properties (Table I). Group A (*S. dysenteriae*) consists of shigellae that do not utilize mannitol nor serologically cross-react, to a great extent, with members of the other subgroups, or with each other. There are 11 serotypes in Group A. Group B (*S. flexneri*) ferments mannitol. All six serotypes in this group serologically cross-react. On the basis of these cross-reactions, the six serotypes are further subdivided into an additional nine subtypes. Group C (*S. boydii*) is biochemically, but not serologically, related to Group B. There are 18 serotypes of *S. boydii*, and, with a few exceptions, they do not cross-react. Group D (*S. sonnei*) ferments lactose slowly and decarboxylates ornithine. There is a single serotype of virulent *S. sonnei* called Form I. This mutates at a high frequency to a rough avirulent organism termed Form II. *Shigella sonnei* can be further subdivided on the basis of colicin typing, but this procedure is not used routinely.

Shigellae are closely related to organisms of the genus *Escherichia*.While typical *Escherichia coli* strains can easily be distinguished from shigellae on the basis of metabolic reactions, there are atypical strains of *E. coli* that have many of the biochemical characteristics of shigellae. Indeed, some *E. coli* are antigenically identical to some shigellae and may even possess the virulence factors that enable them to cause a disease similar to shigellosis. Furthermore, the DNA relatedness of *E. coli* and *Shigella* suggests that they are a single species (*S. boydii* 13 is an exception).

The somatic (O) antigens on the cell surface of the dysentery group are typical of those of the family Enterobacteriaceae in that they consist of a lipopolysaccharide core to which oligosaccharide side chains that confer the serologic specificity to the organism are attached (Fig. 1). The structures of the antigens of some of the more common *Shigella* types has been determined. *Shigella dysenteriae* 1 has an oligosaccharide made up of repeat units of galactopyranose–rhamnopyranose–rhamnopyranose–*N*-acetylglucosamine. The specific side chains of *S. flexneri* types consist of repeat units of rhamnose–rhamnose–rhamnose–*N*-acetylglucosamine. The specificity of the types and subgroups is determined by differences in the positions of α-glucosyl or O-acetylated α-glucosyl residues attached to the primary chain structure (Fig. 1). Some mutant strains of *S. flexneri* lack the moities that confer type specificity and possess only the group antigen(s). These are called *S. flexneri* X or Y depending on which group antigen is expressed. Two uncommon sugars form the basic structure of *S. sonnei* I antigen. The disaccharide components are 2-amino-2-deoxyl-L-altruonic acid and 2-acetamido-4-amino-2,4,6-trideoxy-D-galactose. The structures of *S. boydii* somatic antigens have not been determined.

Some shigellae express K-type surface antigens, which may interfere with their capacity to agglutinate in typing sera. In this case, agglutination in O-antiserum can be induced by heating cultures at 100°C for 2 hr. Some strains of *S. flexneri* types 1–5 produce filamentous structures (fimbriae) on their surfaces. Neither the K antigens nor the fimbria have, at present, a defined role in virulence or in diagnostics.

Shigella dysenteriae 1 produces a potent toxin (Shiga toxin) with three activities. When discovered in 1903, the toxin was thought to be a neurotoxin because when injected into mice or rabbits it caused paralylsis in the hindlegs or forelegs, respectively. However, later studies have indicated that the action of the toxin is not directly on neurons but, rather, on blood vessels in the central nervous system. A second action is the cytotoxicity of the toxin on some mammalian cells. The toxin also has enterotoxic activity causing fluid secretion in ligated loops of rabbit ileum.

The toxin is cell-associated and is released into the culture medium after the death of the bacterial cell. As with tetanus and diphtheria toxins, production of Shiga toxin is increased by limiting the concentration of iron in the medium. The purified toxin has a molecular weight (MW) of approximately 70,000 and consists of one A-subunit (MW approximately 32,000) and five copies of a B-subunit (MW approximately 7700). The B-subunit is responsible for the binding of the toxin to a glycolipid on the surface of the target cell. The A-subunit must be enzymatically nicked to be activated and internalized into the cell, most likely by endocytosis. The enzymatically active A fragment binds to and structurally alters 60S ribosomal subunits causing an irreversible inhibition of protein synthesis and cell death. This mechanism of action is like that of the

Table I Laboratory Characteristics of the Genus *Shigella*

Species	Serogroup	Serotypes (subtypes)	Mannitol utilization	Lactose utilization	Gas production	Ornithine decarboxylase	Lysine decarboxylase	H_2S production	Motility
S. dysenteriae	A	11	−	−	−	−	−	−	−
S. flexneri	B	6 (15)	+	−	−[a]	−	−	−	−
S. boydii	C	18	+	−	−	−	−	−	−
S. sonnei	D	1[b]	+	−	−	+	−	−	−
E. coli—noninvasive	NA[c]	NA	+	+	+	±	±	−	+
E. coli—invasive	NA	13[d]	+	±	−	±	±	−	±
Salmonella sp.	NA	NA	+	−	+	+	+	+	+

[a] A few *S. flexneri* 6 produce gas.

[b] Forms I and II are serotypically distinguishable, and multiple subtypes can be identified by colicin typing.

[c] NA, not applicable.

[d] The following *E. coli* serotypes are invasive: O28a,28c:NM; O112a,112c:NM; O124:NM; O124:H30; O124:H32; O136:NM; O143:NM; O144:NM; O152:NM; O159:H2; O164:NM; O167:H4; O167:H5.

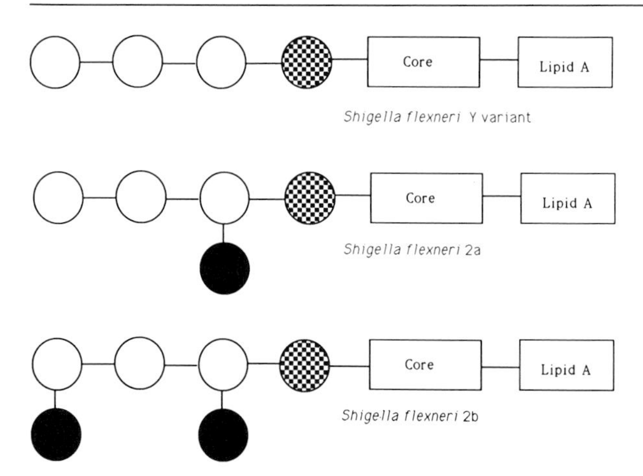

Figure 1 Schematic representation of the somatic antigens of *Shigella flexneri* Y variant, *S. flexneri* 2a, and *S. flexneri* 2b somatic antigens. The postions of the glucosyl residues determine the type specificity of the antigen. O, Rhamnose; ●, glucosyl residue; ⊖, N-acetyglucosamine.

plant lectin, ricin, and indeed the structures of the A-subunits of the two toxins are similar. All shigellae so far tested produce small amounts of a toxin with the biological properties of Shiga toxin, but this toxin is under the control of genes different from those of Shiga toxin. Some serotypes of enteropathogenic *E. coli* (O26:H−) and enterohemorrhagic *E. coli* (O157:H7) produce a Shiga-like toxin (SLT I) with all the biological properties of Shiga toxin and neutralizable by Shiga antitoxin. In addition, enterohemorrhagic *E. coli* strains produce a second SLT (SLT II) that is not neutralized by Shiga antitoxin, but that also has the biological properties of Shiga toxin. The SLTs are also called verotoxins because of their cytotoxic effect on Vero cells.

II. Clinical Disease

As with most infectious diseases, the severity of shigellosis varies from patient to patient. Following ingestion of the organism and an incubation period of usually <4 days, nonspecific signs of lethergy, anorexia, and fever appear, and within another day diarrhea and abdominal cramping occur. A day or so later, blood and inflammatory cells can be observed in stools. As the disease progresses, stool frequency increases to ≥4/hr, the intestine is completely evacuated, and stools consist of only blood, mucus, and pus. Cramps and bowel contractions persist, and the urge to evacuate with no intestinal contents to pass cause the very uncomfortable feeling known as tenesmus. Inflammation and ulceration can be seen throughout the large intestine but are most severe in the distal colon. The ulcer can reach into the lamina propria, and the inflammatory process may extend from the surface epithelium to the submucosa. The disease is usually self-limited, lasts for perhaps 4–7 days, and in otherwise healthy individuals resolves without treatment.

Several local and systemic complications can occur as a result of shigellosis. Rectal prolapse (protrusion of the rectal mucosa) caused by repeated attacks of tenesmus can be seen in children. Massive dilatation of the colon (toxic megacolon), such as that observed in severe ulcerative colitis, may also be seen and is thought to be due to colonic stasis. Perforation of the colon with ensuing peritonitis presents another possible lethal complication that may occur in cases of severe ulceration or toxic megacolon. In situations where extensive ulceration exists, a protein-losing enteropathy may cause nutritional wasting and protracted infection.

Serious systemic complications occur in some patients with shigellosis. Convulsions may be experienced early in the disease, especially in children, and are associated with high fever. Seizures also occur after other infections such as salmonellosis but less often than in shigellosis. The seizures leave no residual effects. Bacteremia is uncommon except in the young or the elderly who have a severe illness. It is associated with high mortality.

The hemolytic uremic syndrome (HUS) is seen following infection with *S. dysenteriae* 1 or with SLT-producing strains of *E. coli*. A week or so after onset of the disease, when the severity of the intestinal symptoms is subsiding, oliguria, which may gradually progress to anuria, occurs. Hemolysis, often severe and rapid, may require multiple transfusions, and hemorrhage or purpura, due to thrombocytopenia, is occasionally seen. Leukemoid reaction, defined as a white blood cell count of >50,000, may occur as part of the HUS or alone, and this reaction is associated with increased mortality.

Reiter's syndrome (arthritis, irititis, and urethritis) has been described in patients of the HLA-B27 histocompatibility complex as a sequellae of shigellosis. A reactive arthritis may occur alone and may progress to a debilitating ankylosing spondylitis.

III. Pathogenesis

Infection is initiated by ingestion of the pathogen. The organism is acid-resistant and survives in an environment of pH 2 for 4 hr. Thus, without neutralizing stomach acid, the dose causing infection in 50% of humans is <500 organisms. The primary step in the disease process is the invasion into the cells of the colonic epithelium by the pathogen. To reach the epithelial cells, the organism must traverse both the mucous and glycocalyx layers, which coat the epithelium. It is not known how this step is achieved. A general pattern of the events of invasion following contact of the pathogen with epithelial cells has been revealed from microscopic studies of tissues from infected animals or from infected monolayers of cultured cells. The first observed alteration is invasion of the colonic intestinal epithelium (Fig. 2). The pathogen is engulfed by an invagination of the cell membrane. Microscopic examination of contact points between the host and bacterial cells, using flourescent labels to selectively stain host actin and myosin, has shown the condensation of filamentous actin underneath the host plasma membrane in areas of close apposition with the invading bacterial cell; myosin also accumulates in these areas. The actin assembly appears to result from specific interaction with invasion-positive shigellae, because noninvasive mutants do not induce any cytoskeletal reaction. Clathrin also accumulates underneath regions of actin polymerization. Cells depleted of intracellular K^+, a process that normally arrests receptor-mediated endocytosis, cannot internalize shigellae, and accumulations of clathrin are not detected in these cells. These findings, however, do not imply that shigellae invade epithelial cells through an exclusive receptor-mediated endocytic event, but rather use directed phagocytosis involving various cytoskeletal and membrane components of the host cell, including clatharin, to gain entry.

After invasion, the pathogen is observed in a vacuole (consisting of the host cell membrane) within the epithelial cell. The integrity of the plasma membrane is not compromised during invasion. Lysis of the vacuole precedes multiplication and intercellular spread of the organism. The capacity to lyse the vacuole membrane is also a property of invasive shigellae, since noninvasive mutants do not rupture the endocytic vacuole when taken up by J774 macrophages; invasive shigellae are readily released from the vacuole. The ability to lyse the endocytic vacuole has been correlated with contact hemolysin activity, a property of invasive shigellae.

After the bacteria escape the endocytic vacuole and spread throughout the cytoplasm, via actin microfilaments, they invade an adjacent cell through the lateral membrane. The result of this process is the formation of ulcers through which blood and inflammatory cells reach the lumen of the intestine (Fig. 3). Dysentery bacilli that reach the lamina propria evoke an intense inflammatory reaction and rarely penetrate further to the submucosa.

Although the broad aspects of the pathogenesis of dysentery are understood, little is known about the process of fluid loss and diarrhea. Studies in Rhesus monkeys have demonstrated that diarrhea and dys-

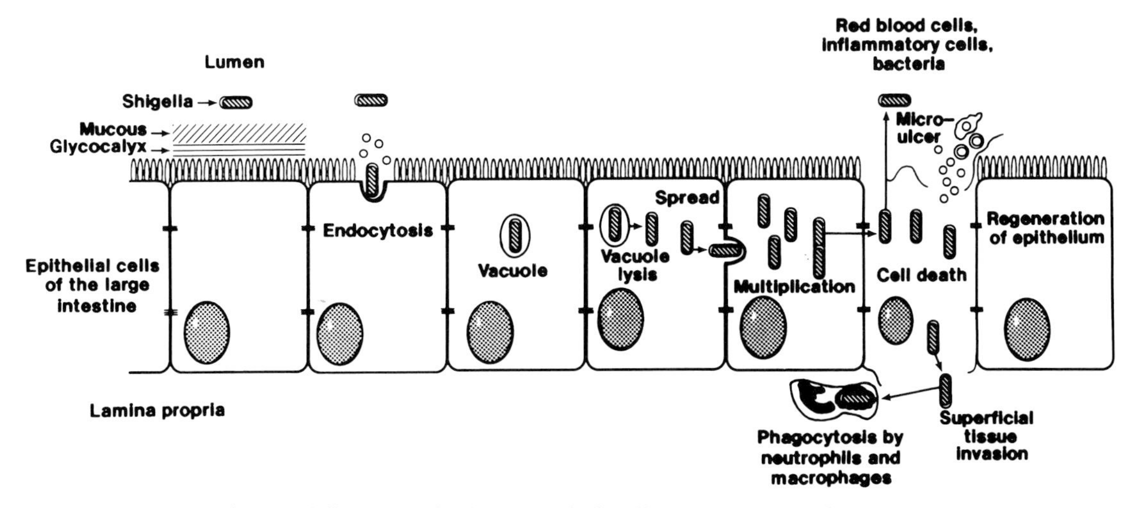

Figure 2 Sequence of events following oral infection with shigellae: penetration of the mucous–glycocalyx layers; destruction of the microvilli, and endocytosis of the pathogen; containment in and escape from the endocytic vacuole; multiplication in the enterocyte and spread to adjacent cells; formation of a micro-ulcer, inflammatory reaction in the lamina propria, and release of white cells, red cells, mucus from depleted goblet cells, and cellular debris into the intestinal lumen.

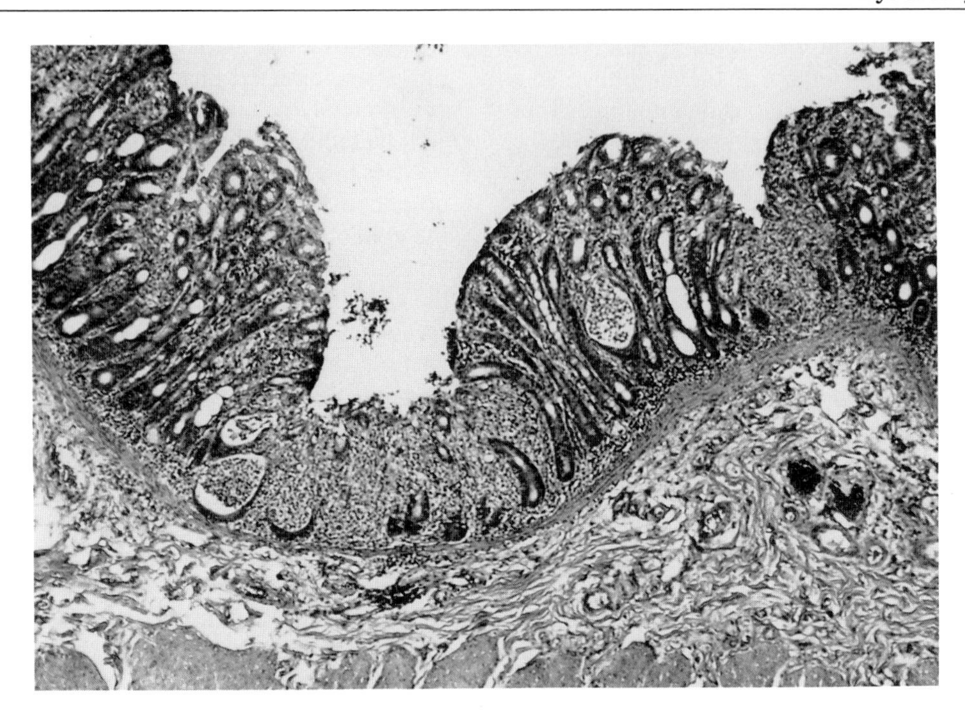

Figure 3 Colon of a monkey with shigellosis. Mucosal ulcerations and focal crypt abscesses containing large numbers of inflammatory cells are numerous. The tubular glands are elongated. Goblet cells are depleted of mucus.

entery occur in animals when the organisms are ingested and traverse the small intestine, causing secretion of fluid into the small bowel. The volume of the diarrheal stool depends on the amount of water entering the colon and on the extent to which water absorption is inhibited in this organ. However, only dysentery results following the inoculation of shigellae directly into the caecum. Either absorption of fluid is inhibited in the colon or a small net secretion occurs.

The role of Shiga toxin and SLTs in pathogenesis of shigellosis is still unclear. As the organisms pass through the small intestine, the toxins could cause the secretion of fluid that is observed in the diarrheal phase of the disease, or following invasion of the colonic epithelial cell, the toxin could cause cell death by inhibiting protein synthesis, thus contributing to the formation of colonic ulcers. There remain conflicting observations concerning the role of Shiga and SLTs in the disease. *Shigella sonnei* is the species that commonly causes diarrhea, yet it does not produce Shiga toxin. Furthermore, a mutant of *S. dysenteriae* 1 that does not produce Shiga toxin causes a disease similar to the Shiga toxin-producing parent strain. It is ironic that we do not yet understand the function of a toxin that is as lethal for laboratory animals as diphtheria toxin.

A. Laboratory Models

Several laboratory models are available to study various aspects of the disease process. Tissue culture monolayers of various epithelial cells can be used to study invasion, and when the cells are contiguous, bacterial spread from cell to cell can be demonstrated by the formation of plaques. Subhuman primates are naturally susceptible to infection and can be used to study the pathophysiology of infection and to assess the immune response and efficacy of vaccines. Oral infection of starved, opiated guinea pigs results in a fatal enteric infection with ulcerative lesions of the intestine. Similar lesions and fluid secretion follow the inoculation of virulent shigellae into ligated ileal loops of rabbits. The Sereny procedure is perhaps the most commonly used animal test to assess virulence. Keratoconjunctivitis with ulcerative lesions of the cornea is observed in rabbits and guinea pigs after depositing virulent shigellae in the conjunctival sac. Simple conjunctivitis occurs when the test is done in mice.

IV. Genetics of Virulence

The infectious phenotype of shigellae results from the coordinate expression of a number of virulence (*vir*) genes encoded on the chromosome and on a large extrachromosomal DNA element, commonly referred to as the invasion plasmid. Virulence genes on the invasion plasmid encode products that account for the major attributes of shigella virulence, namely adherence to and invasion of host cells and dissemination of the bacteria to contiguous cells. Chromosomal virulence genes generally direct the synthesis of products that may exacerbate these disease steps, an example being the synthesis of Shiga toxin, or provide functions that allow the organism to blunt host defenses, as demonstrated by the protection lipopolysaccharide side chains afford the bacteria against phagocytosis. Additionally, certain chromosomal determinants regulate the expression of plasmid virulence genes, ensuring that the synthesis of these products occurs as the bacterium passes from a free-living to host environment. [*See* PLASMIDS.]

Because *Shigella* and *E. coli* K12 are closely related in nucleotide sequence (89–90% homology) and chromosomal gene order, conventional genetic manipulations involving conjugation, transduction, and transformation between the species have been exploited to define chromosomal and plasmid virulence genes in *Shigella*. For instance, conjugal transfer of the 140–180 mDa invasion plasmid from a virulent *Shigella* donor to an avirulent, plasmidless *E. coli* K12 strain confers the invasive phenotype on the transconjugants; further genetic transfer of defined *Shigella* chromosomal DNA segments, via conjugation and P1 phage transduction, results in the reconstitution of the virulence phenotype in the *E. coli* host. With these genetic manipulations, and the aid of recombinant DNA technology, the multideterminant nature of the virulence phenotype in *Shigella* has been established.

A. Chromosomal Determinants of Virulence

Components of the virulence phenotype encoded by chromosomal genes may be classified into three groups, comprising (1) determinants that encode the synthesis of products associated with a particular virulence function, (2) determinants that regulate the expression of virulence genes, and (3) growth or metabolic genes that influence the survival of the bacteria within the host.

1. Chromosomal Determinants Encoding a Particular Virulence Function

a. Lipopolysaccharide Genes

A smooth phenotype is necessary for the virulence of dysentery bacilli. Rough mutants of shigellae that do not produce the O-repeat antigen but, rather, synthesize complete to incomplete core structures retain the ability to invade cultured epithelial cells; however, they are uniformly avirulent in animal models. The essential function of the oligosaccharide side chains (somatic antigen) is to render the organism more resistant to host defense mechanisms such as phagocytosis.

Shigella lipopolysaccharide (LPS) core oligosaccharide genes (*rfa* alleles) have been mapped at chromosomal minute (min.) 81 on the circular 100 min. genetic map, analogous to the map position of these genes in *E. coli* and *Salmonella* (Table II). The location of the genes that determine the expression of the somatic antigen varies according to the *Shigella* species examined. In *S. flexneri*, genes controlling synthesis of the serogroup B somatic antigen are positioned near the *his* operon at 44 min. and correspond to the *his*-linked *rfbABD* locus of *E. coli*. Genes encoding the Group 3,4 or Y variant antigen and the serologically distinct Group 7,8 or X variant antigen are unique; however, their location remains associated with the 44-min. region of the chromosome. *Shigella flexneri* serotype specificity is encoded by a distinct set of genes whose products enzymatically modify the O-antigen chemical backbone. Many of these *S. flexneri* serotype specificities are under the genetic control of temperate phage that use the Y-group antigen as a receptor for adsorption. These lysogenic phage integrate into the *S. flexneri* chromosome at the type or T-locus, embedded in the *pro-lac* region of the chromosome (6–8 min.). Conjugal matings of *E. coli* and *S. flexneri* have established the close genetic linkage of the type-specific antigen genes I, IV, and V with the *lac* (min. 8) marker and the type II genes with the *pro* locus (min. 6). Converting phages that encode the type III and IV specificities map at an undetermined region of the chromosome.

While *S. flexneri* LPS genes are chromosomally encoded, some of the genes for complete LPS antigen expression by *S. dysenteriae* 1 and *S. sonnei* are on plasmids. A small 9-kb plasmid of *S. dysenteriae*

Table II Chromosomal Determinants of Virulence

Locus	Map position (minute)	Function	Mutant phenotype
rfa (allele)	81	Synthesis of somatic antigen basal core	Negative Sereny test; decreased intercellular spread
rfb (allele)	44	Synthesis of group-specific somatic antigen	Negative Sereny test; decreased intercellular spread
T locus	7	Integration site for lysogenic phage encoding type-specific somatic antigen	Null or slight decrease in virulence
stx	28	Synthesis of Shiga toxin	Decreased vascular damage in infected colonic epithelium
virR (*osmZ*)	27	Synthesis of histonelike protein 1(H-NS); repression of plasmid invasion genes at low temperatures	Invasive in tissue culture at 30°C
ompB (*ompR-envZ*)	75	Two-component sensor of osmotic shifts; induction of plasmid invasion loci in response to high osmolarity	Negative Sereny test; decreased tissue culture invasion
kcpA	12	Positive regulation of plasmid gene *virG* (*icsA*)	Negative Sereny test; limited inter- and intracellular spread
iucABCD-iutA	83	Synthesis of aerobactin and aerobactin receptor protein	Delayed Sereny reaction; decreased histopathology and fluid accumulation in rabbit-ligated ileal loops
sod	Unknown	Synthesis of superoxide dismutase	Negative Sereny test; decreased histopathology in ligated rabbit ileal loops

1 carries a gene, designated *rfp*, which is essential for O-antigen production and virulence in *S. dysenteriae* 1 strains. When introduced into *E. coli*, the cloned *rfp* gene product modifies the *E. coli* LPS core by adding a galactose residue, the first sugar of the *S. dysenteriae* 1 O-antigen. In addition to this plasmid, *his*-linked *S. dysenteriae* 1 genes are necessary for core and repeat unit assembly in *E. coli*. The synthesis of four disaccharide amino sugar repeat units comprising the Form I O-antigen of *S. sonnei* is directed by genes located on the 180-kb invasion plasmid. Unique *his*-linked genes from the *S. sonnei* chromosome are not required for Form I antigen expression, because appropriate subclones of the invasion plasmid can produce the antigen when transformed into *E. coli*.

b. Shiga Toxin (*stx*)

The Shiga toxin protein consists of one A-subunit and five B-subunits. The corresponding genes for the subunits are designated *stxA* and *stxB*, and this locus (*stx*) is linked to the 28-min. *trp-pyrF* region of the *S. dysenteriae* 1 genetic map. The *stxA* and *stxB* genes are cotranscribed in a single operon, and β-galactosidase gene fusions constructed with each gene indicate that the subunits are transcribed at approximately a 1:1 ratio. The mechanism of the translational stoichiometry that results in a 1:5 ratio of subunits in the holotoxin is unknown. The *stx* operator region contains a binding site for the Fur protein, a negative transcriptional regulator that represses *stx* expression when the organisms are grown in a high iron environment. Other *Shigella* species produce low levels of a cytotoxin whose activity is neutralized by antiserum to Shiga toxin. However, neither the genetics of toxin production nor the biochemical nature of the toxin has been determined.

2. Chromosomal Genes that Regulate the Expression of Virulence Genes

a. Keratoconjunctivitis Provocation Locus (*kcpA*)

The chromosomal marker *kcpA* was originally identified in *S. flexneri* hybrids that had inherited the *purE* (12 min.) locus of *E. coli* K12. These hybrids could not evoke a keratoconjunctivitis reaction in the guinea pig corneal epithelium. *KcpA*− mutants retain the ability to invade epithelial cells but lack the capacity to spread from cell to cell. They do not form plaques in monolayers of contiguous HeLa cells. The *kcpA* locus encodes a trans-effector of *virG* expression. *VirG,* a gene on the large invasion plasmid, directs the synthesis of a 130-kDa protein that is associated with actin-dependent spread of bacteria within and between cells (see Section IV,B,3). *Escherichia coli* strains carry an inactive form of the *kcpA* gene that may reactivate at a low (one in 1×10^6 cells) frequency.

b. *ompB* (*ompR-envZ*)

As shigellae pass from the external to the host environment, the organisms are subject to environmental shifts in pH, osmolarity, iron concentration, and oxygen tension. Bacteria have evolved a variety of two-component regulatory systems that transmit these environmental stimuli into coordinated gene expression. An example of a two-component regulatory system that influences *vir* gene expression in *Shigella* is found in the *ompR-envZ* loci. These linked genes, which map at min. 75 next to the *malA* operon, were originally defined in *E. coli* as osmodependent transcriptional regulators of the *ompF* and *ompC* genes, which encode the major outer membrane proteins of *E. coli.* Molecules of EnvZ span the outer membrane, with the extracellular domain of the protein acting as a "sensor" of osmotic shifts and the cytpolasmic domain filling the role of a "transmitter," which transfers the signal by its kinase activity. EnvZ protein phosphorylates a histidine residue in the amino end of the OmpR protein (the "receiver" domain), and the phosphorylated OmpR protein then binds to specific gene promoter regions, activating transcription of those genes.

In *Shigella,* the *ompR-envZ* genes regulate the expression of selected *vir* genes. The introduction of specific *envZ* mutations into shigellae results in decreased expression of *vir* genes in both low- and high-osmolarity growth media. Introduction of a more extensive deletion that removes the entire *ompB* locus (i.e., both *ompR and envZ*) severely impaires virulence, leaving the cells invasion-positive but plaque assay and Sereny test-negative.

A spontaneous avirulent mutant of *S. flexneri* 5, in which the invasion-essential cell envelope proteins IpaB, IpaC, and IpaD (see later) were not produced, was restored to full virulence after conjugal transfer of the *malA-ompB* genetic region from *E. coli* K12. It is reasonable to assume that the products of the *ompB* locus may modulate the expression of invasion plasmid-encoded *vir* genes when shigellae pass from the environment and are exposed to the hypertonic colonic contents of the primate intestine.

c. *vir* R (*osmZ*)

The *virR* (*osmZ*) locus is another regulatory gene whose product coordinates the expression of *vir* genes in response to environmental signals. Shigellae that are grown at 30°C fail to invade epithelial cells. Genes in the virulence regulon (i.e., *vir* genes) are repressed at this temperature but are fully expressed at 37°C. Genetic mapping has localized the *virR* temperature regulator between *galU* and *trp* at 27 min. Mutations in this gene result in a cell that constitutively expresses the invasive phenotype at both 30° and 37°C.

Genetic transduction of a mutant *Shigella virR* allele (*virR::Tn10*) into *E. coli* produces transductants whose phenotype mimics that found in *osmZ* mutants of *E. coli.* Because the *osmZ* gene maps to the same genetic position as *virR,* the determinants are assumed to be allelic, a conclusion supported by DNA sequence analysis of the respective genes. *OsmZ* encodes the synthesis of a histonelike protein known as H-NS, which binds without sequence specificity to DNA. This binding affects chromatin structure such that, when H-NS interacts with DNA, the topology and supercoiling of the DNA target is altered, thereby influencing gene expression at the transcriptional level. DNA that is normally supercoiled in a $virR^+$ strain is present in a relaxed, nonsupercoiled configuration in $virR^-$-negative mutants, making the *vir* genes accessible to the transcriptional apparatus. Because DNA supercoiling varies in response to environmental signals such as osmolarity, temperature, and anaerobic growth, the product of the *osmZ/virR* gene may "signal" changes in the extracellular environment by altering the degree of DNA supercoiling.

3. Chromosomal Genes that Influence the Survival of Shigellae in the Host

a. Aerobactin (*iuc*), Enterobactin (*ent*), and Hemin-Binding Protein Genes

Multiplication of shigellae within the host requires that the organisms compete successfully for all essential nutrients. Iron acquisition has been shown to be a critical determinant in the ability of a particular pathogen to establish itself in the host. Although iron-binding siderophores play a major role in the virulence of some bacterial pathogens, these compounds appear to be less critical for the virulence of shigellae. Nonetheless, *Shigella* possess a redundant array of genes whose products either specifically bind iron or facilitate its transport into the bacterial cell (siderophores). Selective mutagenesis of these genes does influence the virulence phenotype when animal models are used to assess the effects of the mutations.

Siderophores are secreted in response to iron deprivation and, once complexed to iron, are transported back into the cell via specific receptors. No single siderpohore is common to all *Shigella,* but the genes for two siderophore types, enterobactin and aerobactin, have been found in the four *Shigella* species. The aerobactin gene cluster includes the structural genes (*iucABCD*) encoding this hydroxymate siderophore and the 76-kDa *iutA*-encoded receptor protein that transports the ferrisiderophore into the bacterial cell. This collection of genes is found in *S. boydii, S. flexneri,* and *S. sonnei.* In *S. flexneri,* the aerobactin gene cluster is located near min. 83, linked with the *tnaA* locus. Site-directed mutation of the *iuc* locus and reintroduction of the mutated allele into *S. flexneri* does not affect bacterial multiplication within cultured epithelial cells, indicating that the intracellular pool of reduced iron is adequate for growth. However, in animal models, a small inoculum (1×10^7) of the *iuc*$^-$ mutant produces a delayed reaction in the Sereny test and minimal mucosal inflammation and fluid accumulation in rabbit ileal loops. These aberrant effects are overcome by increasing the inoculum size 10–100-fold, indicating that attenuation by the loss of siderophore expression can be overcome by increases in infectious dose.

Shigella dysenteriae 1 and some strains of *S. sonnei* elaborate the enterobactin catechol siderophore instead of the aerobactin siderophore. Although most clinical isolates of *S. flexneri* do not produce enterobactin, all of the strains tested to date possess the enterobactin gene cluster on the chromosome. The *ent* genes are collected in a 20-kb segment of the chromosome, mapping, in *E. coli,* near 13 min. The role this siderophore may contribute in the virulence of shigellae remains to be established.

b. Catalase (*kat*) and Superoxide Dismutase (*sod*) Genes

Shigellae that breach the epithelial cell wall are faced with a pronounced change in oxygen tension when passing from the relatively anaerobic environment of the host intestine to the aerobic environment of colonic epithelial cells. This change produces damaging oxygen radicals; therefore, to survive, shigellae must produce catalase and superoxide dimutase to avoid the toxic effects of oxygen by-products. Mutations in either the catalase (*kat*) or the superoxide dismutase (*sod*) genes of *S. flexneri* render the cells extremely sensitive to killing by mouse peritoneal macrophages and human polymorphonuclear leukocytes as compared to the wild-type parent. The *kat*$^-$ mutant is 10 times less sensitive to killing by macrophages than the *sod*$^-$ mutant. However, when tested in rabbit ileal loops, the *kat*$^-$ mutant still damages the intestinal epithelium, while the *sod*$^-$ mutant is avirulent. These observations indicate that superoxide dismutase may play a more important role than catalase in defense of the organism from oxidative stress.

B. Invasion Plasmid Determinants of Virulence

Virulent *Shigella* and enteroinvasive *E. coli* (EIEC) strains carry a large (180–220 kb), nonconjugative plasmid of the IncFI or IncFII incompatibility groups. Loss of this plasmid is associated with avirulence because plasmid-cured segregants cannot invade epithelial cells. Genetic transfer of the 220-kb *S. flexneri* invasion plasmid confers the invasive phenotype on a wide range of plasmid-cured *Shigella* hosts and even provides an invasive capacity to noninvasive *E. coli* K12. The invasion plasmids of *Shigella* species are functionally interchangeable and, although their restriction endonuclease patterns vary considerablly from one serotype to another, DNA hybridization experiments indicate the conservation of all identified virulence-related genes (Table III; Fig. 4).

1. Invasion Plasmid Antigen (*ipaBCDA*) Genes

Specific invasion plasmid-encoded proteins have been identified in virulent shigellae and EIEC. Analysis of these proteins reveals a complement of seven polypeptides, designated a–g, that are unique to the invasion plasmid and are not produced by other *incFI* or *incFII* group plasmids. Polypeptides a–d and an additional plasmid-encoded 130-kDa protein are the principal immunogens recognized by sera from convalescent shigellosis patients. These proteins are serologically similar among the various serotypes of *Shigella* and EIEC.

Transposon mutagenesis has been used to define

Table III Plasmids Determinants of Virulence

Locus	Function	Mutant phenotype
ipaA	Synthesis of 70-kDa cell envelope protein	Unknown
ipaB	Synthesis of 62-kDa cell envelope protein	Invasion negative
ipaC	Synthesis of 43-kDa cell envelope protein	Invasion negative
ipaD	Synthesis of 37-kDa cell envelope protein	Invasion negative; loss of adherence to host cell membrane
ipaH	Synthesis of 60-kDa cell envelope protein	Unknown
virG (*icsA*)	Synthesis of 120-kDa outer membrane protein	Loss of intracellular and intercellular spreading phenotypes
virF	Synthesis of 30-kDa positive regulatory protein	Loss of *vir*G, *ipa*R expression
ipaR (*virB*, *invE*)	Synthesis of 34-kDa positive regulatory protein	Loss of *ipa*BCDA, Regions 3–5 and Pcr^{+} expression
mxi AB	Membrane expression/transport of invasion plasmid antigens	Loss of Ipa protein export

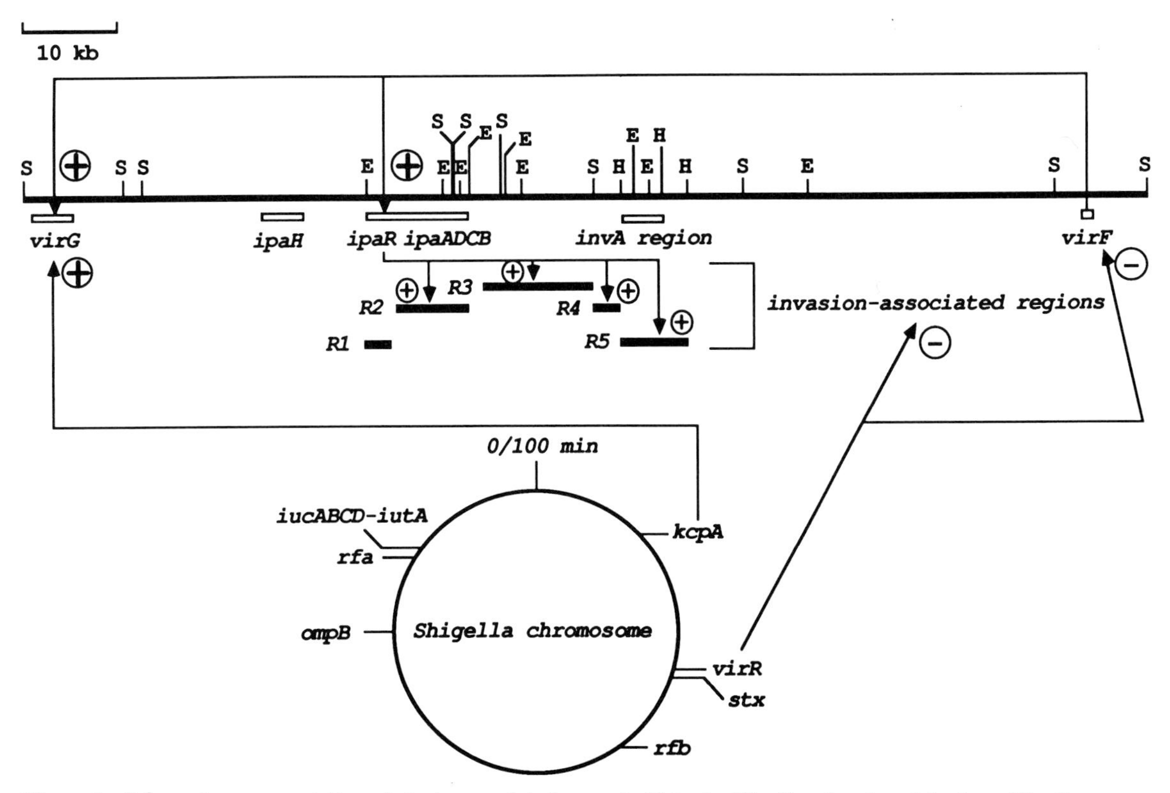

Figure 4 Schematic representation of virulence-related genes in *Shigella*. The line drawing at the top of the figure depicts the genetic organization of virulence genes found on the invasion plasmid of *S. flexneri* 5. The relative map positions of *virG*, *ipaH*, *ipaBCDAR*, *invA*, and *virF* are indicated with open rectangles; invasion-associated regions R1–R5 are indicated with shaded rectangles. A partial restriction map of the corresponding plasmid DNA is shown on the black line: E, *Eco*RI; S, *Sal*I; H, *Hin*dIII. The bottom circle represents the *Shigella* chromosome and virulence related genes positioned on the 100-minute circular map of the chromosome. Thin-lined arrows extending from the various regulatory genes to their respective targets indicate the mode of action for the regulator. + , positive effector; − , negative effector.

five invasion-associated genetic regions (R1–R5) on the *S. flexneri* invasion plasmid (Fig. 4). These contiguous regions span 35 kb of the plasmid and are conserved on the invasion plasmids of *Shigella* species and EIEC strains. Transposon mutations that abolish the expression of polypeptides b, c, and d, significantly decrease invasiveness, whereas insertions that block the expression of polypeptide a or the 130-kDa antigen have no effect on invasiveness, indicating that the b, c, and d antigens are primary invasion determinants. In addition, the invasion plasmid encodes a specific adherence function since plasmid-cured *S. flexneri* 5 cells are 10-fold less adherent when tested in a cultured HeLa cell model than their invasive isogenic parent.

The genetic organization of the a–d immunogen genes, designated invasion plasmid antigen (*ipa*) genes *ipaA*, *ipaB*, *ipaC*, and ipaD, respectively, has been established (Fig. 4). The genetic map reveals that the four *ipa* genes are clustered in a 6-kb segment of the invasion plasmid. Affinity-purified antibodies and monoclonal antibodies have been used to show that the Ipa proteins are immunologically distinct and that each molecule contains separable epitope units. Interestingly, the genetic map position of these epitopes exactly overlaps regions of strong hydrophilicity predicted by DNA sequence analysis of the genes (see later), in agreement with the postulated cell-surface exposure of the antigens.

The DNA sequence of the *ipa* gene region has been determined from two *S. flexneri* serotypes (5 and 2a) and a compilation of these data shows that the 8880 base pairs sequenced encode the synthesis of seven polypeptides. The genes for these proteins are transcribed from the same sense strand of DNA, using all three reading frames, and are organized from the 5′ end as follows: 24-kDa open reading frame (ORF), 18-kDa ORF (*ippl*), *ipaB* (62 kDa), *ipaC* (43 kDa), *ipaD* (37 kDa), *ipaA* (70 kDa), and *ipaR* (34 kDa) (Fig. 4). The G + C content of the *ipa* genes averages 37%, considerably lower than the 50% G + C content of the *Shigella* chromosome.

Significant regions of dyad symmetry are found at the boundaries of the *ipa* genes. Two overlapping inverted repeats are embedded in the junctions between the *ipaB*/*ipaC* and *ipaC*/ipaD termination and initiation codons, while three overlapping inverted repeats occur near the 3′ end of the *ipaD* gene. It is likely that the resultant "hairpin loop" structures serve to sequester ribosome or RNA polymerase binding sites and so may have an important role in the regulation of *ipa* gene expression; in fact, Shine-Delgarno sequences for both the *ipaC* and *ipaD* proteins constitute a portion of the repeats bracketing the 5′ ends of these genes. Transcriptional analysis of the *ipa* gene region reveals that a number of messenger RNA (mRNA) molecules are synthesized from the genes. Northern blot analysis of *S. flexneri* total mRNA prepared from virulent cells grown at 30° and 37°C has demonstrated that temperature regulation of *ipa* gene expression occurs at the level of transcription.

Although the invasion-essential IpaB, IpaC, and IpaD proteins have not been purified by standard protein biochemical techniques, the nucleotide sequence of the genes suggests these antigens are hydrophilic. Of the four antigens, the IpaD protein is the most hydrophilic, with the IpaB, IpaC, and IpaA proteins consisting of hydrophilic domains at the amino- and carboxyl-terminal ends with significant hydrophobic regions in the interiors of the molecules. Epitopes defined by monoclonal antibodies raised against IpaB and IpaC correspond in their genetic map position with hydrophilic regions on both molecules. The proteins are also distinguished by their lack of cysteine residues (one in IpaA, IpaD, and IpaB; none in IpaC) and by the absence of signal peptide sequences at their amino termini. The degree of Ipa protein intergration, if any, with the outer membrane has not been determined, although it is likely that the antigens are only loosely affiliated with the outer envelope since the four antigens are extracted readily by washing the cells in distilled water. Monoclonal antibodies that recognize amino-terminal epitopes of IpaB and IpaC bind to whole bacterial cells in an enzyme-linked immunosorbent assay. This observation indicates that the antigens are largely exposed on the cell surface and, in fact, may be excreted into the medium. Furthermore, an IpaB-specific Mab reduces the plaque-forming efficiency of shigellae in BHK cell monolayers, while an IpaC-specific Mab enhances plaque formation; both of these observations indicate that unencumbered IpaB and IpaC epitopes are needed for a wild-type level of invasion.

A series of genes located 10–15 kb upstream of the *ipa* regulon and overlapping Regions 3, 4, and 5 (*inv*A region; Fig. 4) are critical for the proper post-translational modification, transport, or presentation and excretion of the Ipa antigens on to the cell surface and into the growth medium. The associated phenotype has been designated Mxi$^+$ (membrane expression of Ipa antigens) or Spa$^+$ (surface presentation of Ipa antigens). Mutations in the *mxi*/*spa* genes do not block Ipa antigen synthesis but do prevent proper export of the antigens.

2. Invasion Plasmid Antigen H (*ipaH*)

In addition to the A–D invasion plasmid antigens, a fifth antigen, designated IpaH, is present in all virulent *Shigella*/EIEC strains. The molecular mass of this antigen (60 kDa) is similar to that of IpaB (62 kDa) but the protein is distinct as determined by DNA sequence homology measurements. The *ipaH* antigen gene is present in multiple copies (four to six) on the invasion plasmids of a number of *Shigella* isolates, in contrast to the unit copy representation found for the *ipaB, ipaC, ipaD,* and *ipaA* genes. IpaH is expressed during *Shigella* infections since antibody recognizing this protein is present in convalescent human and monkey antisera. Unlike other plasmid virulence genes, expression of *ipaH* is independent of temperature regulation or induction by *virF* or *ipaR* (see later). Although the conservation of *ipaH* in multiple copies suggests a powerful selection for this gene, the presence of these copies has precluded the construction of IpaH$^-$ mutants and so has hindered the identification of a virulence-associated IpaH phenotype.

3. Intercellular Spreading (*virG icsA*)

After lysis of the phagocytic vacuole two genetic loci influence the ability of the pathogen to spread throughout the epithelial cell layer. The previously described chromosomal *kcpA* locus regulates the production of the plasmid-encoded *virG* protein, an essential component of contiguous cell infection. Transposon mutagenesis of *virG* blocks infection of adjacent cells in tissue culture monolayers or within the corneal epithelium; the locus has also been designated *icsA* to denote the associated intercellular spreading phenotype. Genetic complementation of the *virG* region has established that the determinant consists of a single cistron and the nucleotide sequence indicates that the largest ORF codes for a protein of 1102 amino acid residues with a molecular mass of 116. Since a number of in-frame initiation codons are dispersed throughout the *virG* gene, it is conceivable that more than one protein product is produced by the locus, either through selective processing of a primary transcript, differential translation of the mRNA, or posttranslational modification of the 130-kDa product. The VirG protein is a major antigen, and immunoblots using convalescent antiserum detect only a single, large product in invasion plasmid-cured *Shigella* strains carrying the cloned *virG* gene.

Shigellae that express the VirG protein are distinguished by their ability to promote the deposition of F-actin trailing one pole of the bacterial cell and extending in a filament through the host epithelial cytoplasm. This F-actin polymerization often gives the appearance of a "comet tail," and it is postulated that the F-actin trail provides a motive force for the intracellular motion of the bacteria. Actin "tails" are often found associated with bacteria inside protrusions of the host cell plasma membrane, and these structures may represent the first stage in bacterial spread to adjacent cells.

4. Positive Transcriptional Regulators (*vir*F and *ipa*R)

The sequential expression of phenotypes that result in the dissemination of shigellae throughout the colonic mucosa requires carefully orchestrated synthesis of virulence gene products. The chromosomal *virR* regulatory locus controls *vir* gene expression in a general way by repressing the synthesis of a number of *vir* operons in response to low ambient temperature. More specific examples of *vir* gene regulation are found in the invasion plasmid-encoded activators *virF* and *ipaR*. Genetic analysis of mutations in *virF* and *ipaR* show that the products act in a coordinated cascade to initiate transcription of *vir* genes; thus, the VirF protein activates *virG* and *ipaR* transcription, while IpaR initiates expression of *vir* genes in regions 2–5 (Fig. 4).

The *virF* locus is situated 55 kb from the 5′ end of the *ipa* gene regulon. Mutations in *virF* render the cell invasion negative, and such mutants cannot bind the dye Congo red, a phenotype (Pcr$^+$) that correlates with virulence in *Shigella*. The expression of invasion plasmid antigens IpaA, IpaB, IpaC, IpaD, and VirG is decreased significantly in *virF* mutants and transcriptional analysis has indicated that the *virF* gene is a positive regulator of the *virG* locus. VirF also influences *ipaBCDA* expression by serving as a transcriptional activator of *ipaR*, whose product in turn potentiates mRNA synthesis from the *ipaBCDA* genes (see later). The DNA sequence of *virF* encodes a 30-kDa, 262-amino acid residue protein product that is classified as a member of the AraC family of positive regulators, by virtue of the amino acid homology the protein shares with other members of the group.

Spontaneously occurring mutations in the *virF* locus have pleiotropic effects not only on *vir* gene expression but also on colony morphology and cell metabolism. Opaque colonial variants of *S. flexneri* 2a and *S. dysenteriae* 1 have been shown to result from specific mutations in the *virF* gene. In the case of *S. flexneri* 2a, a specific IS*1* insertion into the amino end of the *virF* gene has been documented in all spontaneous opaque variants isolated; such variants are restored to full virulence by the transfer of a functional *virF* gene *in trans*.

Transposon insertions in Region 1 of the invasion plasmid (Fig. 4) eliminate the invasive phenotype and synthesis of the IpaA, IpaB, IpaC, and IpaD antigens. Molecular cloning of the Region 1 wild-type allele, variously designated as *ipaR, invE,* or *virB,* and complementation of the corresponding mutants has demonstrated that Ipa antigen synthesis is regulated by the *ipaR* product at the level of transcription. Additionally, *lacZ* gene fusions in determinants carried by Regions 3–5 (Fig. 4) are transcriptionally regulated by *ipaR,* as is the Pcr^+ phenotype. DNA sequence analysis of the *ipaR* locus revealed a 34-kDa, 309-residue hydrophilic protein with an isoelectric point of 9.7, a pI indicative of a DNA binding protein. The IpaR sequence exhibits marked homology (42.8% identity over 278 amino acid residues) with the ParB protein of phage P1 and the SopB protein of the F plasmid. The latter proteins are known to bind to specific palindromic sequences on their respective replicons, facilitating the interaction of the molecules with DNA partition sites on the cell membrane. It is conceivable that the IpaR protein is also a DNA binding protein that recognizes palindromic sequences such as those that bracket the various *ipa* genes. By analogy with the ParB and SopB proteins, IpaR may bind invasion plasmid DNA sequences and ensure the close association of the DNA with discrete sites in the inner membrane important for invasion plasmid antigen transport.

V. Diagnosis

The definitive diagnosis of shigellosis is made by isolating and identifying *Shigella* species from fecal specimens. Fresh stool samples, where mucous plugs can be selected for further processing, are preferred. If these cannot be obtained, rectal swabs, passed through the anal sphincter and rubbed against the rectal wall are satisfactory. If possible, the samples should be plated on differential media at the bedside; if a delay occurs, they should be plated in a holding medium such as buffered glycerol-saline until they can be processed. Differential media contain lactose and an indicator dye to identify lactose-negative colonies. In addition, they contain substances that inhibit the growth of gram-positive organisms. Some of these media also have inhibitors that depress the growth of coliforms. It is best to employ a medium such as MacConkey agar, which supports the growth of coliforms, shigellae, etc., and also more selective media such as xylose–lysine–deoxycholate agar or Hektoen enteric agar. After incubation on differential media, lactose-negative colonies are transferred to a medium such as Kliegler iron agar, which indicates glucose and lactose fermentation, gas, and hydrogen sulfide production. Glucose-positive, lactose-negative, hydrogen sulfide-negative, anaerogenic cultures are presumptive shigellae. Confirmation is achieved by standard biochemical tests to classify members of the family enterobacteriaceae and by serotyping. Well-equipped laboratories have polyvalent antisera to identify the four serogroups of shigellae, but serotyping is usually done by reference laboratories.

A presumptive identification and serogrouping of a dysentery bacillus from a stool requires 48 hr. Obviously, more rapid and effective procedures for the identification of shigellae are needed. Two promising procedures identify organisms harboring the virulence plasmid. One employs DNA probes to hybridize with common plasmid fragments; the other an enzyme-linked immunoabsorbent assay with antiserum specific for plasmid-encoded proteins. Both techniques are satisfactory for identifying isolated colonies, but neither have the sensitivity to detect organisms in stool specimens. A method employing the polymerase chain reaction (PCR) procedure has been used to detect shigellae and EIEC. This procedure amplifies a region of an invasion-associated locus and has been used with crude DNA extracts of feces. The PCR technique appears to be more sensitive than standard procedures in detecting these organisms and potentially offers an additional method to the diagnostic laboratory. [*See* Polymerase Chain Reaction (PCR).]

VI. Epidemiology

Shigellosis persists where there is poor sanitation, poor personal hygiene habits, inadequate water supplies, and crowding. While these conditions describe the situation in underdeveloped countries, they also can occur in inner city and impoverished rural areas of developed nations. Shigellosis occurs most frequently in young children but can affect all ages. In third world regions, the incidence may reach two cases per child per year. The incidence of disease is lower in breast-fed babies than in non-breast-fed babies, and weaning is associated with an increased incidence of disease.

For unexplained reasons, the predominant serotype in developed countries is *S. sonnei* (67% in the United States), whereas cases in the third world are due to *S. flexneri*. Cases of *S. dysenteriae* 1 in developed countries are usually imported by travelers from places where the infection is endemic or epidemic.

Because the infectious dose is small (as few as 10 organisms), person-to-person spread of shigellosis is considered to be the most important route of transmission. A day-care attendant may pass the infection from one child to another if proper precautions are not taken when diapers are changes; the child may then infect its mother who may infect other family members by contaminating their food. If feces are not processed properly, flies can become contaminated and infect the food of a neighboring family. In many places, it is difficult to break this chain of transmission. Point-source transmissions involving thousands of people also occur. Vehicles for these outbreaks include infected water, milk, ice cream, uncooked tofu, and salad.

VII. Control and Prevention

The most effective way to control shigellosis is by the good public health procedures practiced in developed countries. Safe water supplies, effective waste disposal, and good sanitary habits by the populace limit, to a large extent, the spread of shigellae. Handwashing with soap has been shown to be an effective means of reducing the spread of infection in developing countries; however, this simple and effective procedure requires education and is costly. Once the training has ceased, the population often returns to the habits that its ancestors have practiced for centuries, and the incidence of disease returns to its former level. Control of flies can also reduce the incidence of shigellosis, but it is not likely to be a practical procedure in developing countries. Vaccination is an attractive alternative. The fact that in highly endemic areas shigellosis is a disease of the young suggests that older individuals are immune because of prior infections. It is possible that this immunity could be induced by vaccination. Several live, attenuated oral vaccines have protected volunteers from experimental infection and populations from natural infection. Because of problems with stability or of unacceptable side effects, they are not in widespread use. With our present knowledge of pathogenesis and the genetics of virulence of dysentery bacilli, it seems likely that a safe and effective vaccine could be produced. The prophylactic use of antibiotics has been proposed for use in travelers, but widespread resistance to antimicrobial agents makes it difficult to predict the correct drug and precludes their general use for prophylaxis.

In those individuals who do become ill, the first concern is to correct and maintain fluid and electrolyte balance. In most cases of shigellosis, the diarrhea is not severe and rehydration can be accomplished with the oral rehydration solution developed by the World Health Organization. The decision to treat with antibiotics depends on the age and nutritional status of the patient, the severity of the illness, and the concern for potential spread of infection. Early treatment reduces the duration of illness and reduces the time the patient excretes the organism to 2 or 3 days. If the infecting organism is sensitive, ampicillin, trimethoprim-sulfamethoxazole, nalidixic acid, and ciprofloxacin are effective. Maintaining caloric intake is important in children, especially those who are malnourished.

Bibliography

Ewing, W. H. (1986). "Identification of Enterobacteriaceae," 4th ed. Elsevier, New York.

Felsen, J. (1945). "Bacillary Dysentery and Ulcerative Colitis." W. B. Saunders, Philadelphia.

Finlay, B. B., and Falkow, S. (1989). *Microbiol. Rev.* **53,** 210–230.

Formal, S. B., Hale, T. L., and Sansonetti, P. J. (1983). *Rev. Inf. Dis.* **5(Suppl. 4),** S702–707.

Hale, T. L. (1991). *Microbiol. Rev.* **55,** 206–224.

Keusch, G. T., Formal, S. B., and Bennish, M. L. (1989). Shigellosis. *In* "Tropical and Geographical Medicine," 2nd ed. (K. S. Warren and A. F. Mahmoud, eds.), pp. 762–775. McGraw Hill Information Services, New York.

Lindberg, A. A., Karnell, A., and Weintraaub, A. (1991). *Rev. Inf. Dis.* **13(Suppl. 4),** S279–284.

O'Brien, A., and Holmes, R. K. (1987). *Microbiol. Rev.* **51,** 206–220.

Sansonetti, P. (1991). *Rev. Infect. Dis.* **13(Suppl. 4),** S285–292.

Simmons, D. A. R., and Romanowska, E. (1987). *J. Med. Microbiol.* **23,** 289–302.

E

Ecology, Microbial

Michael J. Klug
Michigan State University

David A. Odelson
Central Michigan University

Glossary

Cell sorter Instrument that uses optical or mechanical technologies, which allow the separation of cells on the basis of size or cellular properties

Community Assemblage of species of microorganisms, which occur and interact within a given habitat

Confocal microscope Microscope that uses intense laser light beams and computer assisted image enhancement to provide a nearly three-dimensional image

Food web Interaction of communities of organisms with varying functional capabilities

Koch postulates Concept embodied in these postulates define the fundamental questions needed to address the function of a microbial population within a given habitat or role in a function

Microelectrode Micro version of pH, O_2, and specific ion electrodes that allow the exploration of microhabitats

Niche Location or habitat occupied by microorganisms and (indirectly) the properties of the organisms, which allows them to survive and grow within the habitat

Phylloplane Aboveground exposed surfaces of plants that are available for the colonization of microorganisms

Population Group of individuals of one species within a defined area or space

Syntrophic Interpopulation interaction that involves two or more populations that provide nutritional requirements for each other

MICROORGANISMS are often considered to be among the first life forms on earth. Fascination with this life form was initially predisposed to their small size and simple forms. Early observations of microorganisms (in the 17th century) by van Leeuwenhoek with the aid of the first microscope were followed by the demonstration of the role of microorganisms in the process of fermentation and spoilage by Pasteur and eventually of the development of a means to prevent the growth of microorganisms (i.e., pasteurization).

I. Evolution of Microbial Ecology as a Discipline

Isolation of causative microorganisms of disease, as well as pure culture techniques, evolved in the late 1800s; Koch postulates provided a solid basis for studying microorganisms and their roles. The era

Encyclopedia of Microbiology, Volume 2

that followed provided for the continued isolation of microorganisms, the definition of their metabolic capabilities, and in turn, their implied roles in important biogeochemical processes (i.e., in the nitrogen and sulfur cycles). Attention was given to organisms from specific habitats (i.e., soil, water, animals, and plants). The findings demonstrated the vast numbers of diverse microbiologic forms and functions, which were found in nearly every location that was sampled. Eventually microbiologic subdisciplines dealing with microbial associations of natural (i.e., soil, water) and human-made (i.e., food, industrial, and others) environments were established. [*See* History of Microbiology.]

Within the past 30 years it has been recognized that common microorganisms are observed in many habitats and common principles are involved in the mechanisms describing the reasons for the association of those microbes in these varying habitats. The observation that individual populations of microorganisms are rarely found alone suggests potential interactions between populations with each other and their surrounding physicochemical environment. Additionally, their abilities to associate themselves with specific "strata" (e.g., phyloplane, rhizosphere) or carry out specific metabolic functions (i.e., nitrogen fixation and cellulose hydrolysis) were recognized as competitive traits that allowed their colonization and growth in these diverse habitats.

The field of ecology is defined as a discipline of biology that deals with organisms' interactions with each other and surrounding environments. As one might expect, these aforementioned observations of microorganisms and their habitats led to the development of the subdiscipline known as microbial ecology. This development further emphasizes the need to establish a union between the examinations of the physicochemical nature of a habitat along with microbiologic investigations. An equally important emphasis in microbial ecology is to examine the structure and activities associated with microbial communities rather than the general emphasis of microbiology of individual populations.

II. Microbial Communities

Although observations of microorganisms and their associations have been made in numerous habitats, a few of these observations are felt to highlight the importance of the interactions between microorganisms, which leads to community versus population responses.

The rumen ecosystem has received considerable attention, which has been principally driven by economic consideration. Research in this system has, however, defined the mutualistic relationship between members of the microbial fermentative community and an animal's growth and survival. Technically, approaches to this ecologic niche presents somewhat greater difficulties than those of soil or natural waters. The ecosystem is internal to the animal, and the microbes are strict anaerobes. Mechanistically, however, rumen microbial ecology is relatively easier to discern, inasmuch as both input and output to and from the system are clearly defined, in much a similar relationship as an industrial bioreactor. It is interesting to note that syntrophic community level interactions are frequently illustrated with the rumen system, in regard to hydrogen transfer and methane production. [*See* Rumen Microbiology.]

The classic description by Hungate, and later by Wolfe, of the anaerobic food web that occurs in the rumen is illustrated in Fig. 1. Interpopulation interactions have been described between bacterial communities capable of plant polysaccharide hydrolysis (e.g., cellulose), monomer fermenting communities, fatty acid oxidizing communities, and finally, terminal communities (e.g., methanogenesis) that oxidize fatty acids and reduce CO_2 to CH_4 with hydrogen derived during the previous oxidative steps. Rates of and extent of metabolism is controlled by the interdependence of one community on each other. The rumen system has also served as an example of a strategy for, an approach to, and methods to conduct similar investigations of animal–microbe associations, be it the crop of tropical birds or the intestinal tract of termites. Similar interactions have been observed in other anaerobic systems (i.e., sewage sludge, lake and marine sediments), which suggests common controls and mechanisms associated with metabolism of complex organic compounds in all of these systems. It seems likely that future resolve will focus on the functional versus molecular diversity within populations associated within these systems. Ideally, standard molecular probes (see below) will discern whether stress (i.e., substrate manipulations) will change the microbial community at the functional or genetic level. Do populations adapt and respond phenotypically or do populations replace each other? [*See* Methanogenesis.]

Similarily, a large number of interactions occur

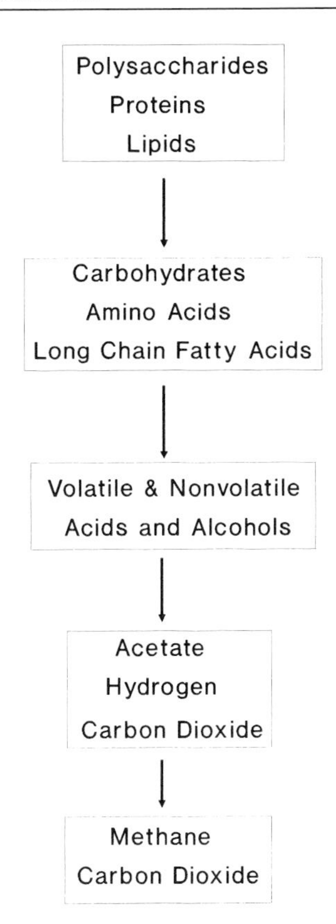

Figure 1 Anaerobic microbial food web involved in conversion of complex organic matter to carbon dioxide and methane. [Modified from Wolfe, R. S. (1971). *Adv. Microb. Physiol.* **6,** 107–145.]

between microorganisms and higher plants and animals in terrestrial and aquatic systems. Traditional food web descriptions in aquatic and terrestrial ecosystems have often failed to consider the role of microbial interactions as a contribution to carbon and nitrogen cycling in these systems. In the past decade there is a growing realization that microbial food webs or loops play a significant role in these systems. Early recognition of these contributions was blurred by the examination of individual populations of organisms and by inadequately examining the interactions and controls of population size and distribution in the surrounding communities. Figure 2 illustrates in its simplest form the relationship between primary producers in aquatic/marine or terrestrial environments, heterotrophic bacteria, phagotrophic protozoans, or zooplankton. [*See* NITROGEN CYCLE.]

Primary producers release soluble organic compounds (i.e., root exudates, algal metabolites), which are consumed by heterotrophic bacteria that

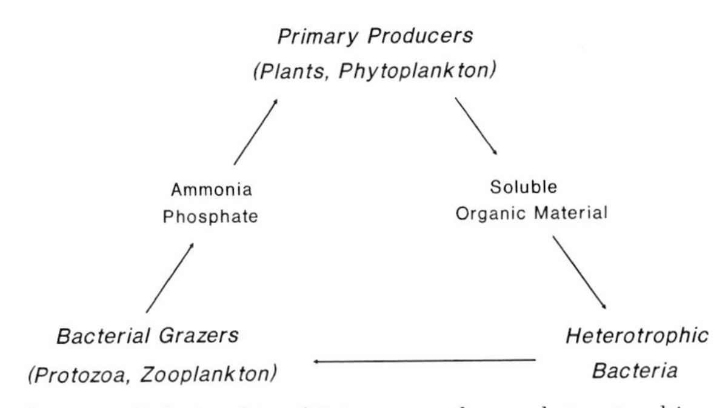

Figure 2 Relationship of Primary producers, heterotrophic bacteria, and bacterial consumers in aquatic–marine and terrestrial environments.

are subsequently grazed by phagotrophic protozoans or zooplankton. The grazers excrete nitrogen and phosphorus, which is used by the primary producers.

In total, these observations have strikingly modified the contemporary view of the structure and mechanistic controls and regulation of growth of higher plant and animal forms in aquatic and terrestrial systems. They also point to the importance of a thorough understanding of these interactions if one is to consider management of associations in applied applications (i.e., sustainable agriculture or aquaculture).

The above illustrations serve as examples of ecologic interactions that occur between microorganisms. Some of these microbial systems are expanded on in other articles. It is important to realize that approaches to microbial ecology must recognize not only isolated individual populations, but more directly the interactions of populations within a community and the resultant affect on the overall functions of the community.

III. Approaches to Microbial Ecology

The fundamental approach to higher plant and animal ecologic studies uses quantitative observations of specific populations in different environments. For more than 100 years plant and animal ecologists have observed and detailed the frequency of occurrence of specific plant and animal populations under various environmental conditions. These observations have led to the development of models for prediction of the occurrence as well as relationship between the organism and the associated environments in which they were observed. Recently these observations have been complemented with physiologic and genetic experiments that suggest there are specific mechanisms of selection that have led to the observed frequency or distribution of organisms. Unfortunately, microbial systems are still embryonic in both description and prediction.

Although the diversity of size and shape of microorganisms is easily discerned by use of a microscope, rapid quantification of microbial numbers and definition as specific populations are not readily possible by microbial ecologists. It is principally the size of the microorganisms and, equally important, the size of the local habitat that creates a considerable constraint on the observation of microorganisms in natural habitats. The microbial ecologist is further hampered by the shear number of microorganisms found in minute habitats, ranging from 1 million cells per milliliter of water to 10 billion cells per gram of human fecal material. This complexity is illustrated in Fig. 3, which depicts the associated bacterial and fungal community with a decomposing deciduous leaf in a woodland stream. Nondescript fungal hyphae and bacterial cells entwined in a complex organic matrix make quantitative enumeration and taxonomic speciation difficult at best. *In situ* observations are further complicated by our inability to observe this environment without disturbing the microorganisms' natural habitat. The various surfaces of the leaf provide varying degrees of carbon and nitrogen resources and physicochemical environments (e.g., gaseous exchange) for colonizing microorganisms. Additionally, other organisms, micro and macro alike, have a potential impact on naturally occurring organisms. Grazing of microorganisms by other microbes or faunal components of a habitat such as this have a significant impact on numbers and types of microorganisms present.

In some respects, microbial ecology evolved as a cross discipline of standard microbiology and environmental analytic analysis. The activity of a specific population or community of microorganisms was inferred by estimating their relative number by either direct microscope counts, viable colonial or turbidimetric determinations, or specific chemical analyses (e.g., chlorophyll for algae) in relation to the overall community structure. Unfortunately,

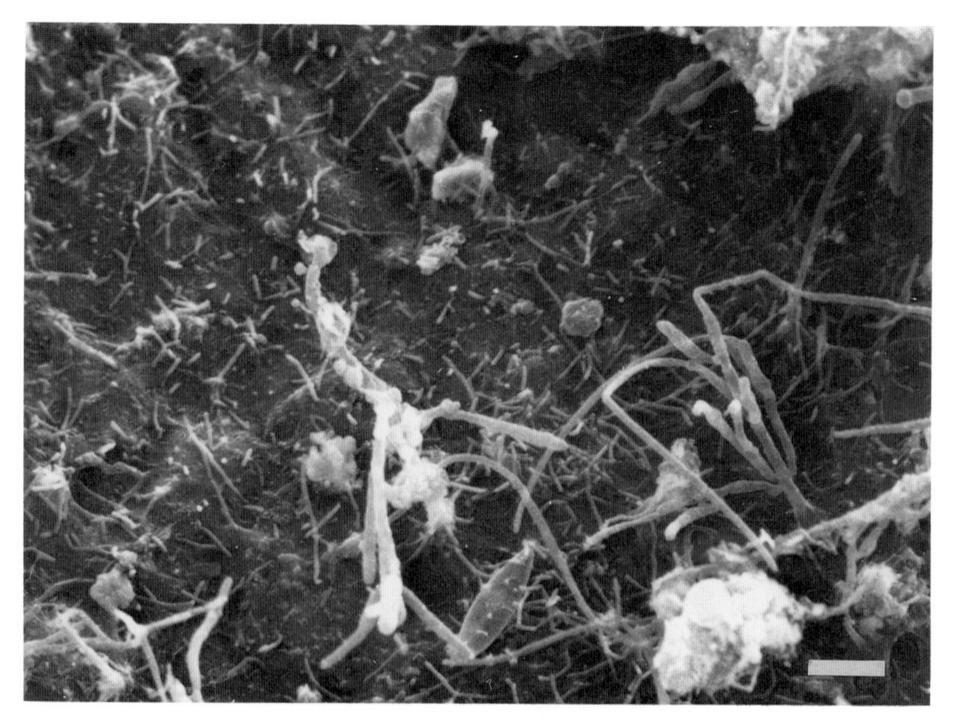

Figure 3 Scanning electron micrograph of microorganisms on deciduous leaf surfaces following incubation in a stream in South Western Michigan. Bar = 10 μm (1000×).

limitations by both the microbial and the environmental analyses have in most cases failed to accurately define the ecology of micobial habitats. Recently, novel analytic and microbiologic methodologies have provided tools for expanding our view of microbial habitats, in a fashion that is not constrained by the microscale of the environment nor limited by our ability to selectively cultivate members of a community.

In situ observation of microorganisms has been expanded by the advent of laser confocal microscopy, a recent advance that provides for a somewhat "x-ray" imaging of material with neither disturbance or fixation. Similarly, laser optical trapping allows the removal of individual cells from a habitat. These techniques provide new insights to our understanding of population components of communities. These increases in optical resolution have also provided a means of sorting communities by either size or chemical factors through the use of cell sorters.

Advancements have also been made in developing methodology to simulate the microhabitats and physiochemical gradients that occur in these habitats. It is interesting to note that one of the pioneering environmental microbiologists, Winogradsky, simulated gradients of light, water, sulfide, and oxygen to describe relationships between sulfide oxidizing photoautotrophic bacteria and sulfate reducing and chemoautotrophic sulfur oxidizing organisms in his Winogradsky column. Recent advances using gels and gradostats to stimulate diffusion barriers will allow us to expand our knowledge concerning relationships between the spatial heterogeneity of the physicochemical environment and the distribution of diverse groups of microorganism. Combined with microelectrode techniques, accurate analytic measurements will complement these microbial investigations. Although our abilities to observe organisms and to better understand the physical and chemical nature of their habits have improved, our abilities to isolate a higher percentage of observable organisms have not improved. Nevertheless within the past 30 years the number of previously undescribed microorganisms has significantly increased.

In an effort to forego the inherent problems of the microscopic and numerical diversity of the microbial world, recent method development has approached the microbial component of the unseen niches at the macromolecular level. The recognition that diversity is intrinsically related to genetics has provided a sound basis for separation and characterization of both microscopic and macroscopic life. In turn, recombinant DNA technology, which has allowed one to identify and determine the specific nucleotide sequence of genes, has provided ecologists with novel methods to pursue community structure. In practice, microbial community samples can be analyzed without microbial cultivation or microscopic observation. In brief, nucleic acid is isolated from the sample and used for hybridization studies with the corresponding gene of interest (i.e., probe). This probe can represent either a metabolic gene or a systematic determinant, such as the 16S rRNA gene sequence. In turn, as one may infer, the sample can be evaluated in terms of both population diversity at the species level and community diversity at the kingdom level (eukaryotic versus prokaryotic). Although intrinsically devoted to the gene level, similar methodology allows one to discern gene expression by hybridization with community mRNA. Most recently all these analyses have been expanded by inclusion of the polymerase chain reaction (PCR), a method allowing amplification and subsequent detection of as few as 10 microorganisms. Interestingly, although these methods have in fact relied on previous isolation and characterization of specific microbial populations, before isolation of a gene for use as a probe, recent investigations have illustrated the universal nature of certain sequences, such as from 16S rRNA genes, for random use in community analyses. [*See* MOLECULAR MICROBIAL ECOLOGY; POLYMERASE CHAIN REACTION (PCR).]

One area that still remains to be resolved in microbial ecology is the relationship of microbial diversity to the response of microbial communities to varying degrees of disturbance. In higher plant and animal systems, the relationship of diversity within communities to their resistance to change or recovery after disturbance (e.g., fire, tillage) has been discussed and debated for decades. Various indices have been used to calculate diversity within communities. In their simplest form these indices represent the number of species found within the community; therefore communities with many species are described as having high diversity or vice versa. Other indices relate the dominance of specific species to the total diversity such that even communities with high diversity can have a few populations that dominate the activity within the community. Our current inabilities to adequately isolate all microorganisms and the lack of distinctive morphologic characteristics fail to provide us with accurate methods for measuring indices of diversity. Improvements in our analytic skills to identify the macromolecule characteristics of microorganisms may provide a basis for estimating diversity and phylogeny. Profiles of cellular or phospholipid fatty acids from lipids extracted from communities or specific chemical or antigenic determinants may provide a snapshot image of changes within microbial communities, after disturbance.

IV. Future Directions in Microbial Ecology

It has been often said that advancements in microbial ecology are limited by the methods available to analyze microbial systems. Some of the advances made in the analytic and molecular methods over the past decade have been illustrated in this overview. Although continued advancements in the use of microelectrodes and optical methods will increase our abilities to measure and observe microorganisms, emphasis must increase in the development of techniques to discern changes in microbial community structure and the resultant impact of these changes on the function of the community. In this regard a significant area of contemporary concern is in the area of the remediation of habitats contaminated with anthropogenic sources of organic and inorganic compounds.

The applied area of bioremediation is, and will continue to be, a timely subject in the decades to come. Efforts to maintain and clean our environment will provide a stimulus for an applied aspect of microbial ecology. Although this area is yet to be accurately defined, in principle the directive is to promote microbial dissimilation of anthropogenic compounds in a controlled manner. In some respects this aspect is fairly ancient in use. Wastewater treatment facilities are, in fact, closed systems of controlled microbial metabolism. Interestingly, current highlights of the biostimulation of oil degradation in the Alaskan waterways and polychlorinated biphenyl degradation within the Hudson River sediment have provided an audience touched by microbes. It would seem prudent for microbial ecology to further promote this microbial usage, perhaps as a tool for examining whether basic principles of ecology can be applied to the practical use of environmental control. Indeed, preliminary information has suggested that our knowledge of microbial activity in soils is partially accurate. In general, bioremediation is often limited by nutrients such as nitrogen and not by microorganisms capable of metabolizing the compounds. Similarly, the community activity of a system has also been discerned, inasmuch as addition of populations capable of metabolizing the compounds of interest does not lead to remedial activity. Ideally, bioremedial activity will depend on evaluating the system as a community and in turn providing directives to manipulate the community. This, in fact, requires an understanding of the interaction of the environment and the microbes. This, in fact, is microbial ecology. [*See* BIOREMEDIATION.]

Bibliography

Amann, R. I., Binder, B. J., Olson, R. J., Chisholm, S. W., Devereux, R., and Stahl, D. A. (1990). Appl. Environ. Microbiol. **56**(6), 1919–1925.

Arndt-Jovin, D. J., Robert-Nicoud, M., Kaufman, S. J., and Jovin, T. M. (1985). *Science* **235,** 247–256.

Ashkin, A., Dziendzic, J. M., and Yamane, T. (1987). *Nature* **330,** 769–771.

Atlas, R. M. (1984). Diversity in microbial communities. *In* "Advances in Microbial Ecology." (K. C. Marshall, ed.), pp. 1–40. Plenum Press, New York.

Atlas, R. M., and Bartha, R. (1987). "Microbial Ecology." Benjamin and Cummings, Menlo Park, California.

Brown, J. F., Jr., Bedard, D. L., Brennan, M. J., Carnahan, J. C., Feng, H., and Wagner, R. E. (1987). *Science* **236,** 709–712.

Chaudhry, G. R., Toranzos, G. A. and Bhatti, A. R. (1989). *Appl. Environmental Microbiol.* **55**(5), 1301–1304.

Clarholm, M. (1984). Heterotrophic; Free-living protozoa: Neglected microorganisms with an important task in regulating bacterial populations. *In* "Current Perspectives in Microbial Ecology." (M. J. Klug and C. A. Reddy, eds.), pp. 321–326. ASM Press, Washington, D.C.

Hedrick, D. B., Richards, B., Jewell, W., Guckert, J. B., and White, D. C. (1991). J. Ind. Microbiol. **9,** 91–98.

Lindstrom, J. E., Prince, R. C., Clark, J. C., Grossman, M. J., Yeager, T. R., Braddock, J. F., and Brown, E. J. (1991). *Appl. Environ. Microbiol.* **57**(9), 2514–2522.

Pace, M. L., and Funke, E. (1991). *Ecology* **72,** 904–914.

Pichard, S. L., and Paul, J. H. (1991). *Appl. Environ. Microbiol.* **57**(6), 1721–1727.

Revsbech, N. P., and Jorgensen, B. B. (1986). Microelectrodes: Their use in microbial ecology. *In* "Advances in Microbial Ecology." (K. C. Marshall, ed.), pp. 293–352. Plenum Press, New York.

Sayler, G. S., and Layton, A. C. (1990). *Annu. Rev. Microbiol.* **44,** 625–648.

Smith, G. B., and Tiedje, J. M. (1992). *Appl. Environ. Microbiol.* **58**(1), 376–384.

Steffan, R. J., and Atlas, E. J. (1988). *Appl. Environ. Microbiol.* **54**(9), 2185–2191.

Torsvik, V., Salte, K., Sorheim, R., and Goksoyr, J. (1990). *Appl. Environ. Microbiol.* **56**(3), 776–781.

Wimpenny, J. W. T., Coombs, J. P., and Lovitt, R. W. (1984). Growth and interactions of microorganisms in spatially heterogeneous ecosystems. *In* "Current Perspectives in Microbial Ecology." (M. J. Klug and C. A. Reddy, eds.), pp. 291–299. ASM Press, Washington, D.C.

Wolfe, R. S. (1971). *Adv. Microb. Physiol.* **6,** 107–145.

Electron Microscopy, Microbial

William J. Todd and M. D. Socolofsky
Louisiana State University and Louisiana State University Agricultural Center

Glossary

Artifact Investigator-induced alteration of the native state, usually introduced by methods of specimen preparation

Contrast Prominence of shade or tone to distinguish detailed features

Fixation Preservation of structural detail by altering the native state, usually accomplished by chemical cross-linkers or ultralow temperatures

Magnification Increase in apparent size; real magnification is always accompanied by an increase in resolution

Resolution Ability to distinguish fine detail, such as to form separate images of two points; the limit of resolving two points is approximately one-half the wavelength of the imaging wave

THE ELECTRON MICROSCOPE, invented in the 1930s, has enabled scientists to perceive the detailed structure of microorganisms because of its high resolving power. With its use has developed a knowledge of the structure of viruses and the detailed architecture of bacteria and higher microorganisms. The implementation of the terms prokaryotic and eukaryotic has come from an appreciation of the differences in the architectural ground plan of these two groups of organisms as revealed by the electron microscope.

I. Electron Microscopes

A. Transmission Electron Microscope

Two types of electron microscopes are in general use (Fig. 1). The first to be developed, the transmission electron microscope (TEM), has many features in common with the light microscope (LM). Both the TEM and the LM involve the formation of a magnified image of the specimen after a beam of electromagnetic energy consisting of electrons or photons is passed through the specimen. Because electrons have a much shorter wavelength than the photons of white light, electron waves can read the specimen with superior resolution of structural detail, which in turn permits greater effective magnification or enlargement of the image.

Over the past several centuries, LMs have extended human powers of observation to see microorganisms alive and in color. This is not the case with electron microscopy. Because the electron beam in a TEM must be maintained in a high vacuum, water must be removed from the specimen, rendering the material lifeless. Furthermore, the image is made up of gradations of dark and light; color cannot be observed.

The TEM employs a beam of electrons projected from an electron gun, which is directed or focused by an electromagnetic condenser lens onto the thin specimen. The electrons of the beam are differentially scattered by the number and the mass of atoms in the specimen. Those that pass through the specimen are gathered and focused by an electromagnetic objective lens, which presents an image of the specimen to the projector lens system for further enlargement. The image, dependent on the distribution of electrons that pass through the specimen, is made visible by allowing it to impinge on a screen that fluoresces when struck with the electrons. The image can be recorded on photographic film.

B. Scanning Electron Microscope

The principles governing the construction of the second major instrument, the scanning electron microscope (SEM), were presented in the mid-1930s but, because of technological problems, the first commercial instrument was not produced until about 30 years later. The SEM, while generally yielding lower resolving power than the TEM, can produce a great

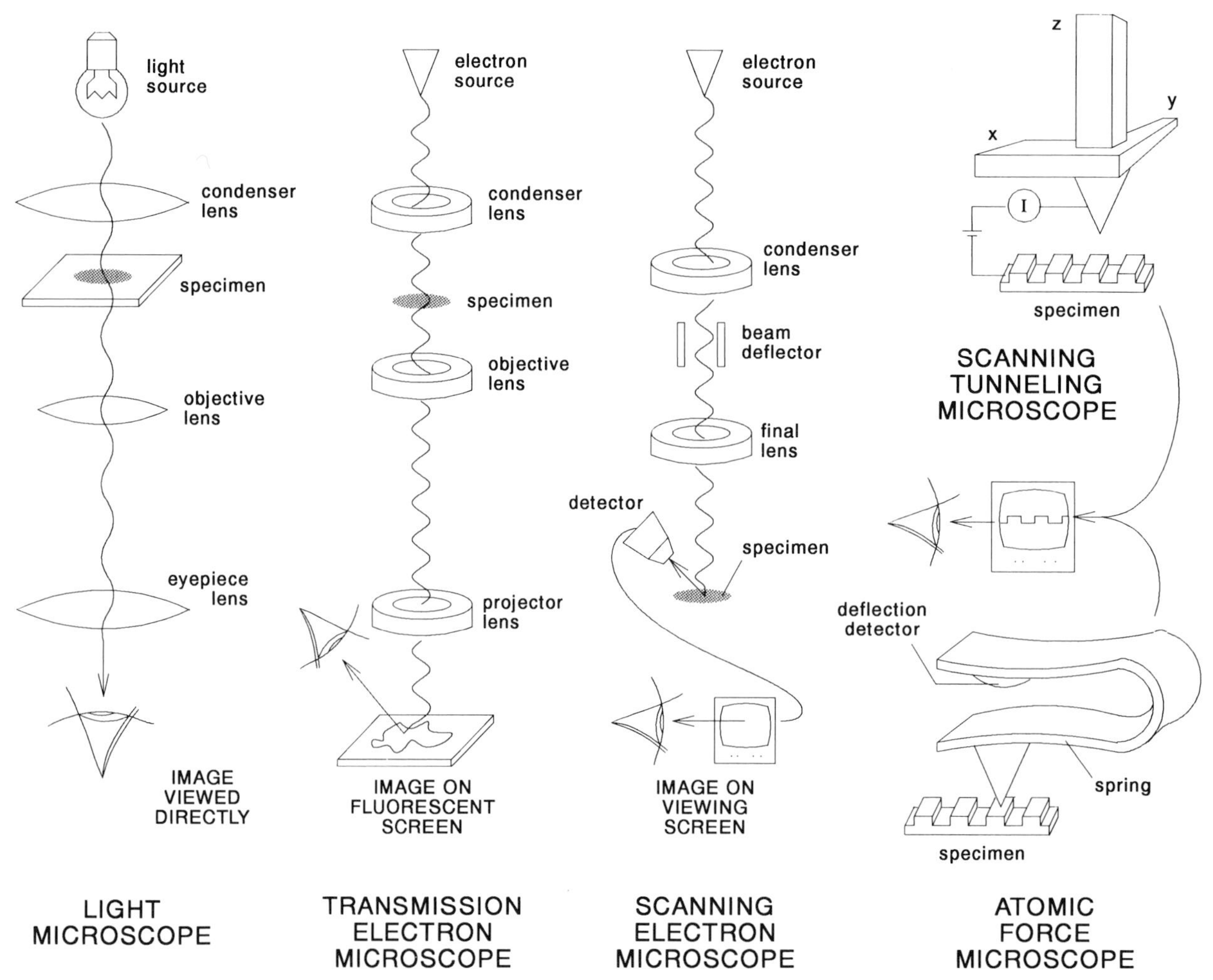

Figure 1 Schematic diagram displaying the primary features of microscope technology. Three groups of instruments are currently in use. The light microscope (LM) was developed hand-in-hand with concepts of magnification and wave function. By analogy, the principles of LM were applied to transmission electron microscopy; by using the shorter electron wavelengths in place of the longer wavelengths of light, greatly enhanced resolution is obtained. Scanning electron microscopy captures back-scattered electrons and energy released from the interaction between the electrons and atoms of the specimen. Images are constructed by examining multiple points across the specimen. The scanning tunneling microscope (STM) and its relative, the atomic force microscope, are conceptionally unique from the traditional microscopes as wave functions of light and electrons are not used. STM relies on the conductance of electrons between the probe and the specimen. The atomic force microscope (AFM) directly traces the molecular contour of the specimen. In both AFM and STM, images are formed from multiple points of analysis across the surface of the specimen.

deal of useful information. It is particularly effective in presenting a three-dimensional image of the surface structure of microorganisms (Fig. 2). In addition, it is capable of giving information on the elemental chemical composition of the specimen.

In reality, the SEM is not a true microscope but, rather, a hybrid of an electron microprobe and a television set. Electrons leave the electron gun (as in the TEM) and are focused by means of electron lenses into a very fine point or probe that interacts with the surface of the specimen. As the electrons interact with the specimen, they evoke the release of different forms of radiation from the surface of the material, each of which can be captured by an appropriate detector, amplified, and then imaged on a television screen. There is a one-to-one correspondence

Figure 2 The diverse microflora found on the skin of the bowhead whale (*Balaena mysticetus*) is revealed by scanning electron microscopy. The large saucer-shaped structures (P) are single-cell plants called diatoms; many different species of bacteria (B) are also attached to the surface (S) of the whale's skin. Magnification: 3900× [Courtesy of William G. Henk.]

between the location of the probe on the specimen and the placement of the resulting energy information on the TV screen. Thus, the beam sweeps the specimen in synchrony with the display presentation and in essence paints a picture of the images point-by-point. The most widely used information for the development of a surface image is derived by collecting secondary electrons emitted after beam interaction with atoms of the specimen. Magnification of the image is accomplished by having the display raster on the TV presented much larger than the size of the corresponding beam sweep of the specimen. Resolution consequently largely depends on the size of the scanning probe.

C. Scanning Tunneling Microscope

Brilliant conceptual advances by Ernst Ruska coupled with amazing progress in micromachinery have recently led to the development of new types of microscopes capable of analysis at the molecular and atomic levels. The scanning tunneling microscope (STM) functions by repeatedly scanning a micropoint across the surface of the specimen. Electrons jump or tunnel through the gap between the probe point and the conducting surface of the specimen. While maintaining either a constant signal or constant distance from the specimen surface, the vertical movement of the probe is recorded as a drawing of the specimen surface. By decreasing the gap distance between the probe and the specimen, the resolution can be improved down to the atomic level. To escape the limits imposed by the requirement for a conduction surface, the atomic force microscope was developed. This modification of the STM directly traces the molecular surface of the specimen to produce a topography formed by repeatedly scanning the probe back and forth across the specimen surface.

II. Specimen Preparation

Electron microscopy (EM) remains both the principal approach to obtain information about subcellular structure and the primary method to coordinate biochemical, immunological, and biophysical information to enhance our understanding of the relationship between subcellular structure and function. There are numerous methods and types of equipment used in electron microscopy, each designed to yield different information about high-resolution structure, chemical identity, and subcellular function. Each different method can yield surprisingly varied images of the same subject; thus, interpretations of electron microscopic images are often difficult and should be approached with caution. The best aid to interpret electron micrographs is to understand what happens to the specimen during the essential preparative techniques. Herein we briefly examine central features of the more commonly used methods. The methods are designed principally to prepare and preserve microbial features to permit bombardment by electrons in a vacuum. Without such preparation, most structures would be destroyed. A regrettable compromise is made between first altering the structure by the preparative methods and losing the structure entirely to the vacuum and electron beam. Despite the limitations, EM has arguably added more information to our understanding of the microbial world than any other method. EM data also have inherent artistic qualities, which adds to the beauty of research through ultrastructural approaches.

A. Negative Stains

Negative staining, the simplest of the EM preparative methods, is an effective approach to study small structures such as purified components, protruding surface features of bacteria, and viruses. Support grids are coated with either a thin film of carbon or a plastic such as parlodion or formvar, to provide an electron-transparent support to hold the specimen of interest. To add contrast to the specimen, staining solutions of heavy metal salts, usually phosphotungstic acid, or uranyl salts are added as a drop to the grid and excess stain wicked away with adsorbent paper. An alternative and more effective approach is to mix the specimen with the stain and spray the mixture onto the grid in the form of a mist or fine spray. By either method the electron-dense metal salts will flow into and fill the crevices and empty space surrounding the specimen, without actually staining the biological material itself. Hence the term negative stain as everything but the structure of interest is stained! When exposed to the electron beam, the heavy metal salts will scatter the electrons outside of the field of focus, eliminating them from the image. Electrons that pass through the specimen will be focused by the lenses to form the specimen image, revealing the morphology in great detail. Without the heavy metal salts, there would not be enough contrast to detect the details of the specimen. Disadvantages include limitations on the specimen thickness and surface tension artifacts due to air drying.

B. Shadowing

Specimen preparation for shadowing also requires placing the specimen upon a coated grid. To obtain the necessary specimen contrast, heavy metals such as platinum or platinum–palladium alloys are vaporized in a vacuum chamber at a low angle to the specimen. The coating of the specimen will be uneven with respect to the metal deposit depending on the molecular contours of the specimen. The resulting effect is to add contrast in the form of micro-shadows on the specimen itself and shadows cast on the specimen support. In recent years, fixed angle shadowing is most commonly used with freeze-fracture technology. During shadowing, the specimen may also be rotated, as is common in nucleic acid spreading methods, which can distinguish double strands from single strands, provide measurements of contour length, and reveal structures of interest such as replication forks, stem loops, and areas of hybridization.

C. Fixation and Embedding

Fixation is a prerequisite for all embedding methods. Because most specimens of interest are too thick to yield analyzable electron images, they are usually studied in thin sections of plastic-embedded specimens (Fig. 3). The plastic is necessary to support the specimen during sectioning and to hold the ultrathin tissue section together for bombardment by the electron beam. The thin-section technology permits individual cells to be cut into hundreds of sections. For the embedding process to be successful, the biological structures must first be stabilized to withstand the process of plastic infiltration. The most common fixation approach is to cross-link protein constituents by reaction with bifunctional chemicals such as glutaraldehyde and to stabilize lipids and other constituents by reaction with osmium tetroxide. After fixation, the water is gradually replaced by alcohol or other organic solvents as a transition to embedding with the usually hydrophobic embedding medium. After dehydration, the embedding medium is gradually infiltrated into the specimen and heat cured to form a solid plastic suitable for cutting into ultrathin sections. Although a great deal of informa-

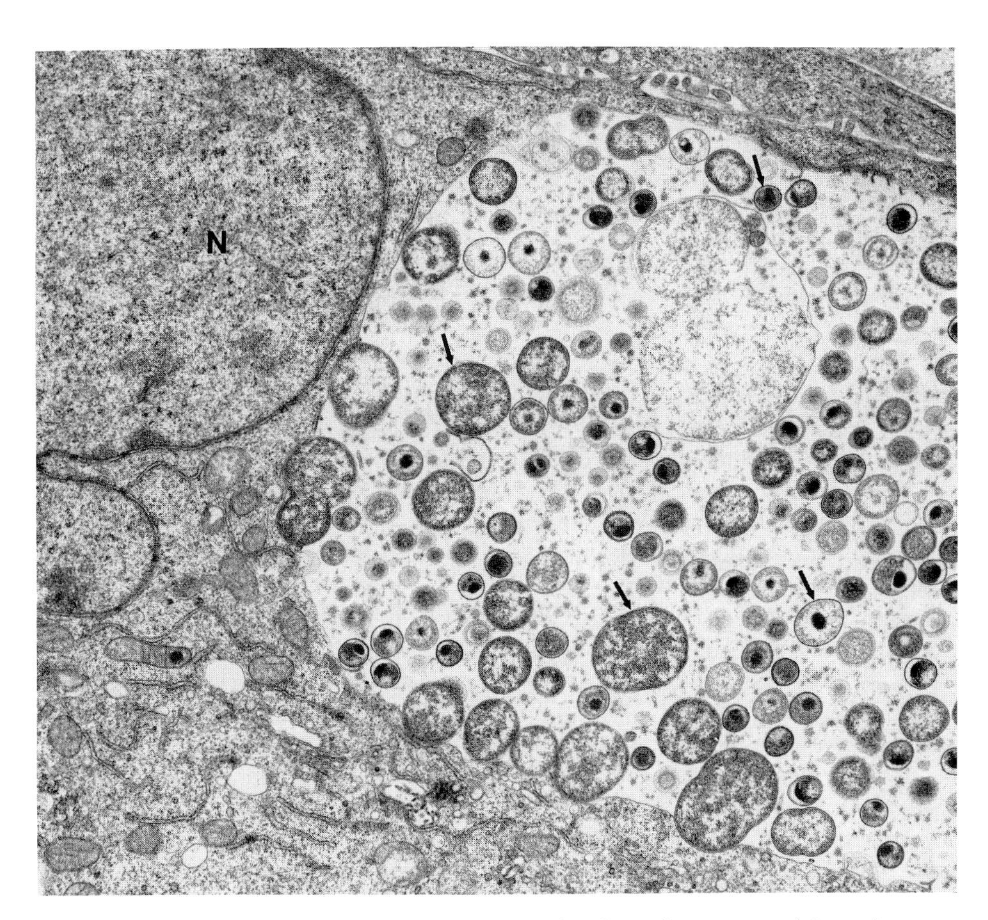

Figure 3 Transmission electron microscopy was used to detect the presence of the prokaryote *Chlamydia trachomatis*, an obligate intracellular pathogen, within a host cell. This thin section of the infected host cell was stained using uranyl acetate and lead citrate to reveal the complex inclusion containing the morphologically distinct forms of this pathogen (arrows). N marks the nucleus of the host cell. Magnification: 1020×.

tion about microorganisms and the infectious disease process has been obtained by studying thin sections, concerns remain about the artifacts and loss of structural information by this process. More recent trends in the fixation and embedding process include rapid low-temperature preservation followed by low-temperature infiltration of fixatives in a process called freeze substitution. Fewer distortions and better preservation are reported using this fixation method. For example, the low-temperature methods have led to a better understanding of the structure of the microbial envelope than is obtainable by more traditional methods.

Unfortunately, all fixation and embedding methods distort the native ultrastructure to some extent. Subcellular and molecular functions are lost, selective extraction of microbial constituents occurs, and shape distortion, primarily by shrinkage, happens during both fixation and embedding. Despite the limitations, a rather complete picture of microbial structure can be obtained by applying several different preparative methods to the same subject and combining the relevant data obtained from each to reach appropriate conclusions.

D. Stains

Stains are essential in most applications of EM to increase contrast of the specimen. For example, consider plastic-embedded specimens used in the thin-sectioning technology. Both the specimen and the embedding medium consist largely of carbon atoms. It would therefore be difficult for the electron beam to distinguish between the specimen and the supporting matrix. If osmium tetroxide were used during fixation, this heavy metal would impart some density to the specimen, but not enough to provide enough contrast to resolve effectively detailed ultrastructure. Further staining is required. During specimen preparation, other metal stains are commonly used such as dissolved uranyl salts. This step can also improve fixation as well as improve contrast. For example, the inner membranes of gram-negative bacteria are better preserved using uranyl salts than without. The use of other stains such as ruthenium red can give some indication of the chemical nature of the constituents; primarily carbohydrates are stained by ruthenium red. Although staining the specimens prior to embedding is useful, most embedded specimens are also stained postsectioning using uranyl acetate and lead citrate. This is accomplished simply by floating a thin section on the surface of a drop of the stain. The stain will wick into the specimen tissue but will not stain the plastic itself. In this manner, the contrast between the biological tissue and the background of the embedding media can be greatly improved to obtain sharp images of microbial structure. The beautiful and informative electron micrographs obtained from stained thin sections of embedded specimens are, in fact, not much more than heavy metal staining patterns; little of the biological material is directly detected. Constituents extracted during the fixation and embedding process, and those refractile to the stains, are not detected by this technology. The problem of refractile constituents can sometimes be solved by coating the ultrastructural constituents with a mordant such as tannic acid prior to embedding. The mordant will pick up the stains and impart sufficient contrast to detect the presence of the constituent.

E. Histochemistry

Although the heavy metal stains are effective, they lack specificity. It is becoming increasingly important to selectively identify subcellular microbial constituents in order to improve our understanding of microbial structure and function and to enhance our ability to comprehend the role of microorganisms in disease processes. There are several practical methods to precisely identify microbial constituents at the ultrastructural level.

1. Antibodies

Antibodies are the most versatile of the specific labels used in electron microscopy because they can be produced to recognize almost any specific microbial constituent. The same antibodies can also be used with Western blots and other biochemical methods to provide corroborating data. With the advent of monoclonal antibodies, applications of antibodies to EM have bloomed. The method is most commonly used with the technology of negative stains and thin sections but is also an effective technique with freeze fracture and SEM. To be visible, the antibodies are usually coupled either directly or indirectly to an electron-dense or morphologically distinct label to improve contrast. However, because of the resolving power of modern electron microscopes, antibodies can be detected unlabeled using negative stain technology. This is commonly accomplished in diagnostic virology, whereby antibodies can be used to aggregate and identify viruses. Nevertheless, for most experimental purposes, electron-dense labels are required, and the current labels of choice are microspheres of gold, which are easy to make in a desirable variety of sizes, are extremely electron dense, and can be coupled to antibodies either directly or indirectly via ligands such as protein A, protein G, or anti-antibodies. Because uniform particles of different sizes can be made, multiple labeling studies are possible.

2. Lectins

Lectins are proteins that recognize specific carbohydrate moieties. They are commonly found in the seeds of various plants. These lectins, of known carbohydrate specificity, can be coupled to electron-dense markers and used to determine the presence and location of the carbohydrate target. This approach has been most fruitful in studies of bacterial surfaces that are often rich in carbohydrates such as the glycocalyx. There are literally hundreds of lectins with different specificities that can be used for this purpose.

3. Enzymes

Two key features of enzymes, substrate specificity and production of enzyme-specific reaction products, are useful for ultrastructural histochemistry. The molecular specificity of the enzyme's reaction site for its substrate is analogous to the specificity of the antibody binding site for its antigen. Enzymes can be labeled with electron-dense markers such as colloidal gold and used to identify specific substrate sites in cells prepared for EM. Even in fixed specimens, the molecular features of the substrate are often sufficiently intact for enzyme recognition and binding, but generation of a reaction product usually cannot occur. Hence, the labeled enzymes accumulate at the substrate sites, identifying the chemistry of those sites. Furthermore, many of the enzymes present within specimens remain functional even after they are immobilized by specimen fixation. The location of the functional enzymes may be determined by infiltrating the appropriate substrates and capturing the reaction products in an electron-dense form; thus, the enzymes are located *in situ* through accumulation of their reaction products. This is most effective where the enzyme reaction products are anions that can easily be captured by precipitation with heavy metal salts at the reaction site.

III. Key Accomplishments

Many of the fascinating advancements in microbiology have depended on electron microscopes and associated methods of specimen preparation. Listed below are but a few of the significant contributions of EM that changed our understanding of the microbial world.

1. *Distinctions between Prokaryotes and Eukaryotes* It has been demonstrated that two major ground plans of cell structure exist in nature. They differ greatly in organization. One organizational unit, the prokaryotic cell, does not have a separate membrane surrounding its genetic material. The other design, the eukaryotic pattern, consists of cells in which the nuclear material is bounded by a separate nuclear membrane.

2. *Membrane Structure* Through TEM, it has been observed that the plane associated with the fracture of frozen cells often passes through the hydrophobic interior of the lipid bilayers of the cytoplasmic membrane, revealing the two interior faces of the membrane. These faces are studded with small particles consisting primarily of large proteins that are embedded in the bimolecular leaflets in an organized fashion. [*See* CELL MEMBRANE: STRUCTURE AND FUNCTION.]

3. *Flagellar Structure* The elucidation of the structure of the bacterial flagellum consisting of an external filament—a shoulderlike hook that joins the filament to a rotatable basal assembly that attaches the flagellar apparatus to the cell envelope—has been possible using TEM. A model of flagellar operation based on structure and function has been developed. [*See* FLAGELLA.]

4. *Fimbriae* The presence of protein projections from the surface of certain bacteria was observed with TEM. These structures—called fimbriae and pili—have been shown to be involved in such activities as attachment to surfaces or the transfer of genetic material from one cell to another.

5. *Gram-Positive and Gram-Negative Bacterial Wall Structure* Thin sections of bacteria revealed the existence of a rather thick layer of cell wall peptidoglycan external to a single inner membrane in gram-positive bacteria. In gram-negative bacteria, the presence of a second membrane, the outer membrane, was detected exterior to both a very thin layer of peptidoglycan and the cytoplasmic membrane. The existence of a periplasmic space was identified in gram-negative bacteria lying between

the inner and outer membranes, perhaps creating a more constant internal environment for cell growth. [*See* CELL WALLS OF BACTERIA.]

6. *Chromosome Organization* The existence of a large circular DNA molecule as the key organizational feature of the bacterial genetic material was identified by TEM; this double-stranded DNA chromosome was found within a ribosome-free region of the cell. Upon gentle release from the cell, extensive folding and twisting is apparent, indicating that the DNA is in a super coiled configuration. EM was used to establish the existence of "nicks," which could be imparted into the DNA molecule affecting its degree of relaxation and helping to define regions of topological constraint. [*See* CHROMOSOME, BACTERIAL.]

7. *Plasmids* TEM was pivitol to identify the presence of small, circular, and self-replicating double-stranded DNA molecules in bacteria. Plasmids commonly code for drug resistance; they have found wide use in genetic engineering. [*See* PLASMIDS.]

8. *Ribosome Structure* The development of a model for the shape of 70S ribosomes consisting of 30S and 50S subunits was accomplished using TEM electron micrographs. The organizational structure contains units such as a body, a head, a cleft, and a platform and has grooves that accommodate both a growing polypeptide chain and a messenger RNA molecule. [*See* RIBOSOMES.]

9. *Mitochondrial DNA* The existence of mitochondrial DNA in yeast and protozoa as well as other eukaryotic organisms was substantiated using both thin sections of fixed cells and EM of the DNA molecules after their release from purified mitochondria onto coated grids. This helped establish the evolutionary relatedness between certain eukaryotic organelles and bacteria.

10. *Cytoplasmic Filaments* The existence of cellular organelles composed of protein in eukaryotes known as microtubules, microfilaments, and intermediate filaments was observed using TEM. These fibrillar units function to both change and maintain the shape of a cell and to carry out other activities such as transport and chromosomal separation.

11. *Cellular Organelles* The existence of small (0.5 μm) particles in eukaryotic cells known as microbodies, which were surrounded by a single membrane and contained a granular matrix, was shown upon examining cells by EM procedures. Later, biochemical evidence revealed that these organelles house a diversity of enzymatic functions ranging from hydrolytic enzymes to peroxidases.

12. *Oncogene Products* The location of an oncogene product, in cells infected with the Rous sarcoma virus, was determined to be along the plasma membrane of the infected cells.

13. *Cytotoxic Killing* Both SEM and TEM procedures were used to establish the manner in which cytotoxic T cells (killer cells) destroy tumor cells by making contact and releasing a protein that causes the target cell to become leaky.

14. *Assembly of Complex Molecules* The construction of certain simple viruses consisting of a single strand of RNA packaged within a rod of protein may take place by self-assembly in a test tube to yield a product recognizable by TEM as a copy of the parent virus. This provided a model for the manner in which larger structures might be built within living cells.

15. *Virus Assembly* The method in which animal viruses utilize the normal biochemical machinery of the host cell to produce more virus was advanced largely through EM methods.

16. *Prions and Viroids* The existence of agents structurally simpler than viruses, but capable of infecting cells, has received support through EM. Prions seem to consist only of protein yet can cause disease, whereas viroids consist of single-stranded RNA but may infect certain plants. [*See* TRANSMISSIBLE SPONGIFORM ENCEPHALOPATHIES.]

Bibliography

Aldrich, H. C., and Todd, W. J. (eds.) (1986). "Ultrastructure Techniques for Microorganisms." Plenum Press, New York.

Bozzola, J. J., and Russell, L. D. (1992). "Electron Microscopy. Principles and Techniques for Biologists." Jones and Bartlett, Boston.

Glauert, A. M. (ed.) (1972–1991). "Practical Methods in Electron Microscopy," Vols. 1–13. North-Holland Publishing Company, Amsterdam.

Graham, L. L., Harris, R., Villiger, W., and Beveridge, T. J. (1991). *J. Bacteriol.* **173,** 1623–1633.

Hansma, P. K., Elings, V. B., Marti, O., and Bracker, C. E. (1988). *Science* **242,** 209–216.

Holt, S. C., and Beveridge, T. J. (1982). *Can. J. Microbiol.* **28,** 1–53.

ELISA Technology

Abdallah M. Isa
Tennessee State University

Glossary

Absorbance Optical density at specific wavelength
Enzyme conjugate Enzyme-immunoreactant conjugate (antigen or antibody)
Substrate Chemical degraded by enzyme activity to produce visible color

ENZYME-LINKED IMMUNOSORBENT ASSAY (ELISA), conveniently referred to as enzyme immunoassay (EIA), is a diagnostic test based on enzyme-labeled immunoreactants for use in the detection of analytes in clinical specimens. Prior to the advent of EIA, the diagnostic field was monopolized by radioimmunoassay (RIA) and immunofluorescence (IFA). Both of these assay systems have problems inherent to the particular assay. RIA suffers from regulatory inconveniences due to health hazards of the radioisotope, its disposal, and its short half-life. It is now confined primarily to endocrinological assays. On the other hand, IFA, which is still widely used for antibody and antigen detection, has its drawbacks, because of its subjectivity and the need for highly trained personnel to perform the assay. As opposed to RIA and IFA, EIA is not subject to legislation and its reagents have a long shelf-life. It is adaptable to several assays in small and large automated laboratories. It is unique in that results can be read visually or photometrically with the use of several colored, luminiscent, or fluorescent substrates.

I. Enzyme Immunoassay Applications

A. Assays for Antibody

1. Indirect Antibody Assays

The antigen is immobilized onto solid phase matrix such as tubes or microtiter plates by passively adsorbing the antigen onto such surfaces. Upon washing and drying, the coated solid phase is stable for a long time. Specific antibody to the antigen (serum) is added to the immobilized antigen and incubated to allow formation of a primary antigen–antibody complex. After washing, an enzyme–antispecies antibody conjugate is added and incubated further to permit formation of a secondary antigen–antibody–enzyme complex. After washing to remove unbound conjugate, an enzyme substrate is added and within a few minutes of incubation a visible color appears due to the degradation of the substrate by the enzyme in the secondary complex. The intensity of the color produced is proportional to the amount of antibody present in the specimen tested. The color can be read visually or quantitated spectrophotometrically.

immobilized antigen + antibody → primary Ag-Ab complex
Ag-Ab complex + E-Ab conjugate → secondary Ag-Ab-E complex
Ag-Ab-E complex + substrate → color

where Ab is antibody, Ag is antigen, and E is enzyme.

2. IgM Antibody Capture Assay

The indirect EIA mentioned above is suitable for the detection of IgG and total antibody. It is not suitable for the detection of specific IgM antibody because

1. false-positive reactions occur due to rheumatoid factor (IgM–anti-IgG antibodies)

and to competition with IgG antibodies present in the serum and
2. false-negative reactions may occur due to the presence of excess IgG antibodies in the specimen. The IgG antibodies compete with IgM for the limited antigen sites.

To reduce the possibility of false-positive and negative reactions, one of two approaches can be followed:

1. pretreatment of the specimen with anti-IgG and/or aggregated IgG or
2. pre-coat the solid phase with anti-IgM antibodies. Using this approach, all IgM molecules in the specimen will be immobilized on the solid phase by forming an IgM–anti-IgM complex. Upon addition of the antigen (unlabeled or labeled with the enzyme), a secondary antigen–antibody complex forms. If unlabeled antigen is used, it must be followed by the addition of enzyme-labeled–antibody conjugate (specific to the antigen). Color develops upon the addition of the enzyme substrate. If, on the other hand, enzyme-labeled antigen is added, color will develop upon the addition of the enzyme substrate. The intensity of color produced is proportional to the amount of IgM antibody present in the original specimen.

a. Use of Unlabeled Antigen

immobolized anti-IgM + IgM (serum) → IgM–anti-IgM complex
IgM–anti-IgM complex + antigen → IgM–anti-IgM–Ag complex
IgM–anti-IgM–Ag + E antibody conjugate → IgM–anti-IgM–Ag–Ab–E
IgM–anti-IgM–Ag–Ab–E complex + substrate → color

b. Use of Enzyme-Labeled Antigen

Immobilized anti-IgM + IgM (serum) → IgM–anti-IgM complex
IgM–anti-IgM + Ag-E conjugate → IgM–anti-IgM–Ag–E complex
IgM–anti-IgM–Ag–E complex + substrate → color

B. Antigen Capture: The Sandwich Method

Antibody specific to antigen is used to coat the solid phase. Antigen in the specimen to be assayed is captured by the specific antibody coating the solid phase after addition of the antigen and incubation. A primary Ag-Ab complex is formed. After washing to remove excess proteins, an enzyme-labeled antibody (specific to the antigen) is added and incubated to permit formation of a secondary Ab-Ag-Ab-E complex. After washing to remove excess conjugate, enzyme substrate is added and incubated. Color develops, the intensity of which is proportional to the amount of antigen present in the specimen.

The antibody coating the solid phase may be a monoclonal antibody specific to one epitope on the antigen and the enzyme conjugate is another monoclonal antibody specific to another epitope on the same antigen. If the coating antibody is polyclonal, the antibody in the enzyme conjugate must be produced in a species that is different from that used to produce the coating polyclonal antibody.

Sensitivity of EIA has been enhanced by the incorporation of the biotin–avidin system. The sandwich assay is performed in the usual manner up to and through the sample addition step. This is followed by the addition of biotin-labeled antibody and avidin-labeled enzyme, respectively.

immobilized Ab + Ag(sample) → Ab-Ag complex
Ab-Ag complex + biotin-Ab → Ab-Ag-biotin-Ab complex
Ab-Ag-biotin-Ab + avidin-enzyme → Ab-Ag-biotin-Ab-avidin-E
Ab-Ag-biotin-Ab-avidin-E + substrate → amplified color

II. Components of the Enzyme Immunoassay System

A. Solid Phase

The solid phase is plastic in nature and may consist of tubes, paddles, discs, or microtiter plate wells. The discussion in this paper will emphasize microtiter plate well strips, because they are the most popular carriers of the analyte. Microtiter plates are designed in removable strips consisting of 8 or 12 wells/strip and each plate consists of 8 12-well or 12 8-well strips. Microtiter plate wells are coated with antigen or antibody in carbonate–bicarbonate buffer, pH 9.6. The coated plates are incubated for a few hours at 37°C and overnight at 2–8°C. The efficiency of coating depends on the quality of the plastic, nature of the material to be coated, and also the pH of the coating buffer. The optimal concentration of the coating material must be predetermined prior to coating. After coating, plates are washed with wash buffer containing Tween-20, the remaining free sites on the wells are blocked with bovine serum albumin. Plates are washed, dried, and sealed. Plates treated in this manner are stable for long periods of time at 2–8°C.

B. Enzyme Conjugate

Most conjugates used in EIA are enzyme-labeled antibodies or antigens (Table I). Several conjugation methods are available with the glutaraldehyde method being the most popular. After conjugation, unbound enzyme is removed by gel filtration or by other separation techniques.

Most enzyme substrates are unstable and must be used within a few hours of preparation. Stop reagents in use consist of corrosive inorganic acids and alkali. One unique and ideal EIA system is the histidine-rich protein (HRP) enzyme and the 2,2′-azinubis (3-ethylbenz-thiazoline sulfunic acid (ABTS) substrate. The HRP–antibody conjugate and the ABTS substrate are offered in the ready-to-use format and were found to be stable for >1 yr at refrigeration temperatures. No stop reagent is needed and the color produced was stable for 4 hr at room temperature and for 48 hr at 2–8°C. In this system, test results can be read visually or photometrically. The absorbance values (optical density) at 405 nm of specimens, and controls in the absence of stop reagent do not change if test results are read within the time frame previously mentioned.

Table I Common Enzymes Used and Their Substrates

Enzyme	Substrate	Comment
Horseradish Peroxidase (HRP)	OPD, TMB, 5-ASA	Need stop reagent
	4-chloro α-naphthol, DAB	Precipitate
	ABTS	No stop reagent
Alkaline phosphatase	PNP, MUFP, Fast Red	Need stop reagent
β-Galacosidase	O-nitrophenyl β-D-galactosidease	Need stop reagent
	Methyl umbelliferyl galactoside	Need stop reagent
Glucose oxidase	OPD	Need stop reagent
	D-glucose–nitro-blue tetrazolium	Precipitate
	DAB	Precipitate
Urease	Urea	Sensitive to pH

Precision and accuracy are extremely important in any serologic assay. This is particularly important in the multicomponent EIA. To obtain sensitivity and specificity of any assay depends largely on the standardization of all components of the system.

III. Optimization of the System

A. Antigen Coat

The optimal concentration of antigen to be used for coating the microtiter plate wells is determined by the checkerboard titration. Antigen is diluted serially and 100-μl aliquots are applied to vertical rows of wells. Plates are incubated at 37°C for 4 hr and at 2–8°C overnight. Wells are washed and blocked. Serial dilutions of a strong positive serum, a weak positive serum, and negative serum are applied to horizontal rows of wells. Incubate for 30 min at room temperature (20–30°C) and wash. Enzyme–antibody conjugate is added to all wells and the wells are incubated again for 30 min at room temperature (20–30°C). Wells are washed, and enzyme substrate is added to all wells. Incubation is carried out for 30 min. Reactions (color) are read visually or photometrically. The highest dilution of antigen giving high color intensity or high absorbance value is the one chosen to coat the wells.

B. Enzyme–Antibody Conjugate

Prepare three serial dilutions of a strong positive serum, weak positive serum, and negative serum. Apply 100 μl of each serum dilution into separate wells in horizontal rows of antigen-coated wells, incubate, and wash. Prepare serial dilutions of the enzyme–antibody conjugate and apply 100 μl of each dilution to separate rows of vertical wells, incubate, and wash. Apply 100 μl of enzyme substrate to each well and incubate. Read color visually or spectrophotometrically. The enzyme–antibody conjugate dilution giving the strongest color (absorbance) is chosen as the working dilution of the conjugate.

Bibliography

Amico Laboratories (1991). Amizyme test system, product inserts. Amico Laboratories Inc., Nashville, Tennessee 37209.

Ishikawa, E., Imagawa, M., Hashida, S., Yoshitake, S., Hamaguchi, Y., and Ueno, T. (1983). *J. Immunoassay* **4,** 209.

Voller, A., and Bidwell, D. (1986). Enzyme-linked immunosorbent assay. *In* "Manual of Clinical Laboratory Immunology," 3rd ed. (N. R. Rose, H. Friedman, and J. L. Fahey eds.), pp. 99–109. Washington, D.C.

Enteropathogens

Herbert L. DuPont
The University of Texas Medical School

Glossary

Colitis Inflammation of the colon

Diarrhea Passage of stools of increased number and decreased form from usual state

Dysentery Passage of small volume stools that contain blood and mucus

Enteritis Inflammation of the small bowel

Enterotoxin(s) Exotoxin produced by bacterial pathogens, which produce losses of fluids and electrolytes

ACUTE DIARRHEA is a major medical problem in industrialized areas, especially among infants in day-care centers, those confined to custodial institutions for the mentally retarded, gay males, and travelers to developing tropical regions. It is a major cause of infant mortality in developing regions. Most acute diarrhea is caused by one or more virulent microorganisms infecting the intestines (enteropathogens): bacterial, viral, and parasites. This article will review the microbiology of the important enteropathogens and briefly review the epidemiology and clinical features of the diseases they produce.

I. Bacterial Agents

A. *Escherichia coli*

There are at least four well-defined types of diarrheagenic *E. coli* that produce human disease. While they each are biochemically classified as *E. coli*, they show important differences in terms of virulence characteristics and disease produced.

1. Enteropathogenic *E. coli*

Enteropathogenic *E. coli* (EPEC) serotypes were identified in the 1940s and 1950s in association with hospital nursery outbreaks among infants. During the 1950s, procedures for serotyping the strains were developed based on their characteristic somatic (O) and flagellar (H) antigens. Classical EPECs belong to approximately 12 of the recognized 164 O antigen groups and show characteristic O and H types. When an *E. coli* strain is suspected as being an EPEC, the serotype is determined by using polyvalent pools of EPEC antisera. In addition to belonging to certain O and H types, most EPECs show localized adherence to the intestine resulting in cupping of the enterocyte and destruction of microvilli. This attachment property can be identified in the laboratory by documenting mannose-resistant attachment to HEp-2 tissue culture cells. The genes encoding for attachment have been identified on a plasmid and adherent EPECs can be identified by hybridization with DNA probes derived from the plasmid. One family of EPEC-like organisms apparently do not belong to EPEC serotypes but show the adherence property. EPEC strains are important causes of hospital nursery and day-care center outbreaks, of endemic pediatric disease, and of persistent diarrhea in infants living in developing tropical areas.

2. Enterotoxigenic *E. coli*

Enterotoxigenic *E. coli* (ETEC) are endemic during the summer months in tropical and semitropical regions of Latin America, Africa, and Southern Asia. These strains characteristically belong to 1 of about 20 serotypes and produce one or both of well-defined enterotoxins differing in terms of heat susceptibility. A heat-labile choleralike enterotoxin (LT) possesses two important constituents: five binding (B) subunits that attach to membrane enterocyte receptors and an active (A) subunit, which then stimulates mucosal adenylate cyclase after entering the mucosa leading to fluid and electrolyte transport. The smaller molecular weight, poorly antigenic, heat-stable enterotoxin (ST) exerts its intestinal secretory effect after producing increases in mucosal guanylate cyclase. A third virulence factor is intestinal adherence mediated by one of a limited number of species-specific surface fimbrial antigens. Both toxin production and fimbrial antigens are plasmid-mediated, and strains are now identified by hybridization with DNA probes that detect LT and ST. ETEC, produce approximately one-third of infantile and pediatric diarrhea cases that occur in developing tropical areas. They cause approximately 40% of diarrhea among persons visiting these areas from industrialized areas.

3. Enteroinvasive *E. coli*

Enteroinvasive *E. coli* (EIEC) were first demonstrated as causes of feline dysenteric (passage of bloody mucoid stools) disease in Japan and Brazil. The strains possess *Shigella*-like intestinal invasive capability that can be shown either by the strain's ability to produce keratoconjunctivitis in a guinea pig eye (Sereny) model or by hybridization with a specific DNA probe to detect the presence of genes encoding for invasiveness. As with other diarrheagenic *E. coli,* relationship between EIEC and *E. coli* serotype is poorly understood. These strains are unusual causes of food-borne outbreaks of diarrheal disease in the United States. The outbreaks may be extensive.

4. Enterohemmorhagic *E. coli*

Strains of *E. coli* serotype 0157 : H7 were first documented as causes of diarrhea associated with consumption of undercooked hamburgers. The organism is characterized by its ability to ferment sorbitol, by inhibition of motility in media containing H7 antisera, and by production of Shiga-like toxin in a tissue culture assay. Certain strains of *E. coli* belonging to serotype 026 : H11 have also been shown to produce hemorrhagic colitis. Enterohemmorhagic *E. coli* (EHEC) have been incriminated in epidemic and endemic disease throughout the industrialized world with poorly understood predilection for certain geographic regions. Cattle and beef food products appear to serve as the primary reservoir. Hemorrhagic colitis is a distinct clinical syndrome. Patients pass bloody stools and an intense mucosal hemorrhagic process can be found by endoscopy; however, in contrast to EIEC disease, fever is not usually present or, if it is, it is low grade. Because of the production of Shiga-like toxin, the hemolytic uremic syndrome is a complication of EHEC enteric infection.

B. *Shigella*

Shigella spp. are small, nonmotile, gram-negative rods that are identified by characteristic biochemical reactions and classified into one of four species—(1) *S. dysenteriae,* (2) *S. flexneri,* (3) *S. boydii,* (4) *S. sonnei*—and approximately 40 serotypes by agglutination using group- or type-specific antisera. Shigellae produce disease with a low inoculum (100 organisms or less), which explains its potential for person-to-person spread. The organisms invade the colonic mucosa and spread laterally in the epithelium, producing ulcerations and inflammation. The capacity to invade is controlled by a large (140 Md) plasmid and chromosomal genes. The Shiga bacillus (*Shigella dysenteriae* 1) produces a cytotoxin that has been associated with production of hemolytic uremic syndrome. Shigellosis is associated with fever in one-third of cases and dysentery in half. Clinically, patients with shigellosis are more systemically ill when compared to other forms of enteric infection and the disease persists longer. In developing areas, *Shigella* infection, particularly that due to *S. dysenteriae,* produces high rates of death in poorly nourished infants and children. Complications of disease include dehydration, pneumonia, sepsis, and hemolytic uremic syndrome. Cyclic epidemics of type- or group-specific disease occur over periods of time of 20–40 years. The first epidemic during the first 25 years of this century was caused by *S. dysenteriae* 1 followed by *S. flexneri* serotypes in the late 1920s and early 1930s. We currently are seeing a predominance of *S. sonnei* in most areas of the world.

C. *Salmonella*

Salmonella spp. are flagellated gram-negative bacilli that are classified into *S. typhi, S. cholera-suis,* and *S. enteritidis* in the diagnostic microbiology laboratory based on biochemical properties. *Salmonella enteritidis* is further divided into more than 2000 serotypes by reference laboratories. Serotyping has epidemiologic importance in allowing the study of the occurrence of outbreaks and secondary transmission of an identified strain. An estimated 2–4 million cases of nontyphoid *Salmonella* infections occur annually in the United States. Animals represent the most important reservoirs of nontyphoid *Salmonella,* whereas humans serve as the sole source of typhoid salmonellosis. Nontyphoid *Salmonella* commonly contaminate raw foods of animal origin (poultry, pork, beef, milk, and eggs). Most outbreaks reflect inadequate food handling. The highest incidence of infection is in infants and young children <5 yr of age, with an unexplained high frequency in infants <1 yr of age. The clinical expressions of disease include enteric or typhoid fever, enterocolitis, bacteremia, extraintestinal focal infections, and asymptomatic carriage. In certain developing countries, typhoid fever is a major public health problem and an important cause of death. High fever, systemic toxicity, abdominal complaints (constipation, diarrhea, ileus), and bacteremia characterize the disease. After the first week of untreated illness, complications may occur: intestinal perforation or hemorrhage. Relapses following therapy are common due to the intracellular nature of the infection. [*See* FOODBORNE ILLNESS.]

In the United States, acute enterocolitis is the most common and most important form of the illness. Affected persons experience diarrhea and abdominal cramps and pain between 8 and 48 hr after eating a contaminated meal. About 50% of the patients will experience fever. Transient bacteremia occurs in between 5% and 8% of the cases.

Bacteremia is seen in *Salmonella*-infected patients with altered immunity (AIDS or advanced cancer). Osteomyelitis, central nervous system infection, endocarditis, and local abscesses may occur in normal or debilitated hosts. Osteomyelitis classically occurs in patients with sickle cell anemia.

D. *Campylobacter* and *Helicobacter*

Campylobacter species require the use of specialized methods for isolation, which include a microaerophilic atmosphere (5% oxygen) and the use of a selective antibiotic-containing media. *Campylobacter jejuni* is an important cause of acute diarrhea, particularly in young children. *Campylobacter fetus* is a cause of bacteremia and systemic infection in persons who are immunosuppressed. *Helicobacter pylori* is a related organism associated with the gastric mucosa, which has been implicated as a cause of gastritis. Like *S. enteritidis, C. jejuni* lives in the intestinal tract of mammalian and avian species. Poultry and animal products represent the major vehicles of transmission. The organism invades the intestinal mucosa like *Shigella* strains resulting in local inflammation and crypt abscesses. Although *C. jejuni* produce enterotoxins and cytotoxins, their biologic importance remains unestablished. While diarrhea is the major clinical finding in *C. jejuni* infection, fever and dysentery are common. *Helicobacter pylori,* a gastric urease-positive organism, appears to be part of gastric flora in certain persons, with rates of colonization increasing with age. The organism is acid-susceptible and survives due to the urease produced that generates ammonia which reduces local pH. Also, the mucus layer serves as an anatomic barrier to stomach acid. There is a close relationship between gastritis and nearly all forms of stomach pathology and *H. pylori* infection. The evidence is convincing that the organism is important in the pathogenesis of gastritis, perhaps peptic ulcer disease, and possibly gastric neoplasia.

E. *Aeromonas* Species

The motile, mesophilic aeromonads *A. hydrophila, A. sobria,* and *A. caviae* are human pathogens. The three important species have been further divided into 13 DNA hybridization groups. *Aeromonas* spp. grow on routinely employed, less selective gram-negative media or broth such as MacConkey agar, alkaline peptone water broth, and blood agar with 10 mcg/ml ampicillin. On blood agar, many of the strains are hemolytic, most do not ferment lactose, and the organisms are identified by properties of glucose fermentation, cytochrome oxidase positivity, and a resistance to vibrostatic compound 0/129. The three pathogenic species are further divided based on esculin hydrolysis, gas from glucose, lysine decarboxylase activity, growth at 42°C, arabinose fermentation, and salicin fermentation.

The three major human *Aeromonas* spp. produce numerous extracellular enzymes, hemagglutinins, and adherence fimbriae and invade intestinal cells or tissue culture cells. The exotoxins produces are hemolysins, cytotoxins, and enterotoxins. Two hemolysins are produced: alpha and beta, which hemolyze erythrocytes and produce cytotoxic changes in tissue culture cells. The exotoxin that causes fluid accumulation in suckling mice and rabbit intestinal loops is referred to as an enterotoxin.

The major reservoir of *Aeromonas* strains is environmental surface water, which becomes heavily colonized during summer months. The agents also have been found in processed chickens and grocery store produce. The virulence of environmental strains does not appear to differ from strains associated with human illness.

Infectious diseases of humans caused by these

strains include the following: septicemia in persons with underlying liver disease or malignancy; wound and soft tissue infections (often associated with exposure to environmental water sources); miscellaneous infections of the respiratory tract, meninges of the brain, bones, and intra-abdominal cavity or pelvic infections of women; and diarrheal disease. Approximately one-half of the isolated *Aeromonas* strains are associated with diarrhea. *Aeromonas* spp. produce between 1% and 10% of acute diarrhea, with the highest rates seen in infants <3 yr of age, during summer months, and in tropical areas. While watery diarrhea is the most common clinical presentation, approximately one-half of the cases will show one or more of the following: fever, dysentery (passage of bloody mucoid stools), and persistence (diarrhea lasting 14 days or longer).

F. *Plesiomonas shigelloides*

Environmental surface water and seafood during summer months serve as the reservoir for this member of the family Vibrionaceae. The organism produces acute diarrhea in previously healthy persons, and the two epidemiologic associations are travel to developing areas and consumption of seafood. Bacteremia has been reported particularly in those who are immunocompromised. The organism is recovered using routine gram-negative selective media. It is oxidase-positive.

G. *Yersinia enterocolitica*

Yersinia enterocolitica is a gram-negative, lactose-negative, urease-positive bacillus. It grows on conventional enteric media, but at 25°C. While there are more than 50 serotypes of the organism, most infections are caused by strains belonging to serotypes 03, 08, and 09. The agent is an important cause of diarrhea in Northern Europe and Canada and occurs in all areas of the world. Contaminated food or water represent the usual vehicle of transmission. The natural reservoir of the organism is a variety of mammals. The most common clinical manifestation of intestinal infection is enterocolitis, with patients presenting with diarrhea, fever, and abdominal pain. Other patients present with mesenteric adenitis, terminal ileitis, and reactive polyarthritis.

H. Vibrios

With the exception of *Vibrio cholerae,* the pathogenic vibrios require salt for growth. Thiosulfate citrate bile salt sucrose (TCBS) medium is used for recovery of *Vibro* species from a stool sample. *Vibrio cholerae* 01, the cause of cholera, can be divided into El tor or classical biotypes. An El tor biotype of *V. cholerae* 01 is currently causing a large epidemic in Latin America. *Vibrio cholerae* 01 produces a heat-labile enterotoxin that, like ETEC LT, possesses binding subunits and an active subunit that produces diarrhea via intestinal adenylate cyclase stimulation. Cholera is associated with profound fluid electrolyte losses with rapid development of dehydration. Non-O1 *V. cholerae* are common in estuarine water and in seafood, particularly oysters. These noncholera vibrios and *Vibrio parahemolyticus* each show the same sources (water and seafood) and produce diarrhea. *Vibrio parahemolyticus* isolated from cases of diarrhea are characteristically β-hemolytic on blood agar (Kanagawa-positive), whereas most environmental isolates are Kanagawa-negative.

I. *Clostridium perfringens*

Clostridium perfringens is commonly found in soil and the gastrointestinal tract of many mammals. The organism is relatively aerotolerent, grows rapidly under anaerobic conditions, and shows stormy fermentation in milk and a double zone of hemolysis on blood agar. *Clostridium perfringens* type A is a common cause of food-borne enteric infection. In this setting, it produces outbreaks of usually mild illness characterized by diarrhea and abdominal cramps, with few systemic symptoms. In this form of *C. perfringens* infection, an exotoxin with enterotoxic activity is responsible for the clinical manifestations of the disease. A *C. perfringens* food-borne outbreak is suspected when nonfebrile diarrhea occurs 6–16 hr after ingestion of a common food. Recovery of the organism from food and identification of 10^6 *C. perfringens* cells per gram of stool from illness cases represent diagnostic criteria.

J. *Clostridium difficile*

Clostridium difficile is a spore-forming gram-negative, oblique anaerobic bacillus. It is part of the intestinal flora of 3–4% of healthy adult persons. Following receipt of antimicrobial therapy to which the organism is resistant, it may grow to high concentrations in the gut. The organism produces sufficient quantities of toxins A and B, which induce secretory and inflammatory changes of gut mucosa. Watery diarrhea followed by passage of stools that contain microscopic leukocytes and blood (macroscopic or by occult blood detected by biochemical assay) associated with fever represents the classical clinical presentation. Pseudomembranous nodules can be seen on endoscopic examination of the colonic mucosa. The diagnosis is made in older children and adults with a characteristic clinical picture (acute diarrhea following initiation of an antimicrobial) who have *C. difficile* cytotoxin detected in stool.

II. Viral Agents

A. Norwalk Virus

Norwalk virus is a round virus, 27-nm in diameter, with a single RNA structural protein and appears to be related to calicivirus. It has remained refractory to *in vitro* cultivation. Recently, the viral genome has been cloned, which will help future characterization. The virus is endemic in developing regions, and most persons by 3 yr of age have serologic evidence of infection. In industrialized areas such as the United States, a large percentage of the population is susceptible. It is an important cause of water- and food-borne epidemics and also a cause of shellfish-associated gastroenteritis. Diagnostic reagents will soon be available for making the diagnosis of Norwalk disease.

B. *Rotavirus*

Rotaviruses are 70-nm double-shelled capsid, double-stranded, segmented RNA viruses. They represent the most common definable causes of gastroenteritis in young infants under the age of 2 yr in many regions of the world, including the United States. By the age of 3 yr, nearly all persons are resistant to disease, although symptomatic excretion of the virus occurs in adults. In temperate climates, rotavirus infection typically occurs in the winter, with low rates of illness in the summer. The striking seasonal pattern in temperate climates is obvious as one looks at infection patterns closer to the equator. Rotavirus is the major cause of death in cholera-nonendemic areas. It has been estimated that 1 million rotavirus gastroenteritis deaths occur each year. Rotaviruses produced nearly half of the hospitalizations for infectious diarrhea in infants and children. The coexistence of vomiting and diarrhea in a very young infant leads to rapid dehydration. A number of reliable and sensitive commercial test kits are available to detect rotavirus antigen in stool.

C. Enteric Adenoviruses

Types 40 and 41 enteric adenoviruses are important causes of gastroenteritis. The viruses do not grow efficiently in cell cultures and can be detected in stool by serologic or morphologic studies (electron microscopy).

D. Caliciviruses

Caliciviruses, like Norwalk virus, are small, round, single-stranded RNA viruses (27–38 nm diameter). They are an important cause of gastroenteritis in young children and presently can be detected by serologic procedures or by electron microscopy only in research laboratories. [*See* ENTEROVIRUSES.]

III. Parasites

While a number of parasites can cause gastrointestinal symptoms and diarrhea, the intestinal protozoa represent the most important parasitic etiologic agents and will be considered here.

A. *Entamoeba histolytica*

Entamoeba histolytica recovered from patients will illness show isoenzyme patterns that can be used to help differentiate virulent (invasive) from avirulent (noninvasive) strains. The organism has two forms: cysts and trophozoites. The infectious particle is the cyst. After ingestion, the cyst undergoes nuclear division and excyst to form eight trophozoites. The trophozoite is responsible for invasive disease. The disease produced, amebiasis, is primarily a problem of persons in developing regions of lower socioeconomic level. While most infected persons show few symptoms, abdominal pain can occur. The classic form of intestinal amebiasis is dysenteric disease with passage of bloody stools. Extraintestinal spread occurs particularly in men (women during reproductive years are protected from dissemination) in <5% intestinal infections. Liver abscess is the most common extraintestinal form of the disease. Diagnosis of amebiasis is made by finding the unidirectionally motile trophozoites containing ingested red blood cells in freshly passed stool of a patient with diarrhea, or in a patient with liver abscess, a high titer of *E. histolytica* serum antibodies is sufficient to make the diagnosis.

B. *Giardia lamblia*

Many animal species have been shown to be infected with morphologically similar organisms. The infection is produced when as few as 10 cysts are ingested. The cysts excyst and produce trophozoites. Infection characteristically occurs in certain settings: travel to mountainous areas of North America with exposure to surface water; living in rural, tropical, and developing areas of the world; attending a day-care center in urban regions; and among gay males. A number of community-wide waterborne outbreaks have been well documented. The protozoan can be cultured but the diagnosis is generally made by finding the organism by microscopy of *Giardia* antigen in stool of patients with acute or subacute diarrhea.

C. *Cryptosporidium*

Cryptosporidium is a sporozoan parasite that infects many animal species. The oocyst is the environmentally resistant and infectious form of the parasite. After ingestion, the oocyte excysts, producing infective sporozoites that attach to host mucosa. *Cryptosporidium* characteristically occurs in animal handlers, infants and staff in day-care centers, patients with AIDS, and persons who ingest water in Leningrad. The diagnosis is made by identifying oocysts in stool samples. The most commonly used staining procedure is the modified Kinyoun acid-fast stain.

Bibliography

DuPont H. L. (1990). Gastrointestinal infections. *In* "Internal Medicine," 3rd ed. Ed. (J. H. Stein ed.), pp. 13, 1–13. Little, Brown and Company, Boston.

DuPont, H. L., and Mathewson, J. J. (1991). *Escherichia coli* diarrhea. *In* "Bacterial Infections of Humans: Epidemiology and Control," 2nd ed (A.S. Evans, and P. S. Brachman, eds.), pp. 239–254. Plenum Medical Book Company, New York.

DuPont, H. L., and Pickering, L. K. (1980). Infections of the Gastrointestinal Tract. Microbiology, Pathophysiology, and Clinical Features." Plenum, New York.

Gorbach, S. L. (1986). "Infectious Diarrhea." Blackwell Scientific Publications, Boston.

Guerrant, R. L., Lohr, J. A., and Williams, E. K. (1986). *Pediatr. Infect. Dis. J.* **5,** 353–359.

Mandell, G. T., Douglas, R. G., Jr., and Bennett, J. E. (1990). "Principles and Practice of Infectious Diseases," 3rd ed. Churchill Livingstone, New York.

Enteroviruses

Joseph L. Melnick
Baylor College of Medicine

Glossary

Attenuated virus vaccine Immunizing agent composed of a selected or modified strain of an originally virulent virus, which does not cause clinical signs associated with the parent virus but still infects and multiplies in the host so as to induce immunity

Coxsackieviruses Viruses named after the town of Coxsackie, New York, the home of the patient from whom the first Group A virus was isolated and recognized in 1948; the first Group B virus, isolated from a patient in Connecticut, was described in 1949

Cytopathic effect Microscopic changes in virus-infected cultured cells, such as rounding of individual cells, fusion of large numbers of cells, or production of new intracellular structures

Echoviruses Viruses named from the acronym for enteric cytopathogenic human orphan viruses plus the suffix *-viruses;* due to their unknown relationship to human disease and because they failed to produce illness in laboratory animals, including infant mice, they were called orphan viruses

Incubation period Time between exposure to virus and first clinical evidence of disease

THE ENTEROVIRUS GROUP includes polioviruses, coxsackieviruses, and echoviruses. These agents, which share a number of clinical, epidemiological, and ecological characteristics as well as physical and biochemical properties, are classified the *Enterovirus* genus of the family *Picornaviridae*. Poliovirus type 1 is the type species of the genus.

Enteroviruses are transient inhabitants of the human alimentary tract and may be isolated from the throat of the infected individual (from just before disease onset to 1 wk or less after onset) or from the lower intestine (from before onset to several weeks after onset). Healthy carriers are common; they usually excrete virus for a period of several weeks. A number of animal species serve as hosts to enteroviruses specific for each species.

I. Taxonomy

Human enteroviruses include the following:

1. *Polioviruses:* types 1–3. Neurovirulent for monkeys.
2. *Coxsackieviruses A:* types A1–A24 and several variants. In infant mice, Group A viruses characteristically produce flaccid paralysis as a result of extensive myositis, the infection and destruction of striated muscle throughout the body.
3. *Coxsackieviruses B:* types B1–B6. The disease caused in infant mice by Group B viruses is characterized by a widespread infection of the brain and fat pads and a more limited infection of muscle and other tissue. Cultured primate cells are highly susceptible to infection.
4. *Echoviruses:* 31 types. Cytopathogenic for cells grown in culture but not for mice. Although types were numbered sequentially 1–34, three have been reclassified: echovirus 10 as a reovirus, echovirus 28 as a rhinovirus, and echovirus 34 as a variant of coxsackievirus A24. Echovirus 9 is antigenically similar to coxsackievirus A23 but lacks pathogenicity for mice.
5. *Enterovirus:* types 68–71. These later additions would formerly have been classified as either

coxsackievirus or echovirus types, but the distinction—based on host susceptibility—has since become blurred.
6. *Enterovirus type 72*. Hepatitis A virus was provisionally classified here but is now classified as a separate genus within the picornavirus family.

Intratypic differences exist within each enterovirus type. The intratypic variants within the types of poliovirus can be distinguished by use of the recently developed tools of molecular biology: highly strain-specific adsorbed or monoclonal sera, oligonucleotide mapping, and sequencing of the bases of the viral genome. Isolates are identified by type, country (or city), strain number, and year of isolation. Thus, P1/Houston/23/62 designates a type 1 poliovirus strain, number 23, isolated in Houston in 1962.

II. Properties

Among the enteroviruses, poliovirus has proven to be the one most often producing serious disease in humans. Consequently, poliovirus has been the enterovirus studied in most detail and, in this article, will serve as the model.

The typical enterovirus is approximately 27 nm in diameter and consists of a capsid shell of 60 subunits, each with four proteins (VP1–VP4) arranged in icosahedral symmetry around a genome made up of a single strand of positive-sense RNA (Fig. 1). These polypeptides/proteins are cleaved from a larger precursor polyprotein. X-ray diffraction studies of poliovirus have revealed the three-dimensional molecular structure of the virion. The three largest proteins (VP1–VP3) are similar in core structure; the peptide backbone of the protein loops back upon itself, forming a barrel of eight strands held together by hydrogen bonds (the β-barrel). Between the β-barrel and the amino- and carboxyl-terminal portions of the protein, the amino acid chain contains a series of loops, which include the chief antigenic sites found on the virion surface; these sites are involved in the neutralization of virus infection. The smallest (internal) protein VP4 is associated with viral RNA.

Enteroviruses are stable at acid pH for 1–3 hr and have a buoyant density in cesium chloride of about 1.34 g/ml. Because the virion has no lipid-containing envelope, it is not affected by lipid solvents such as

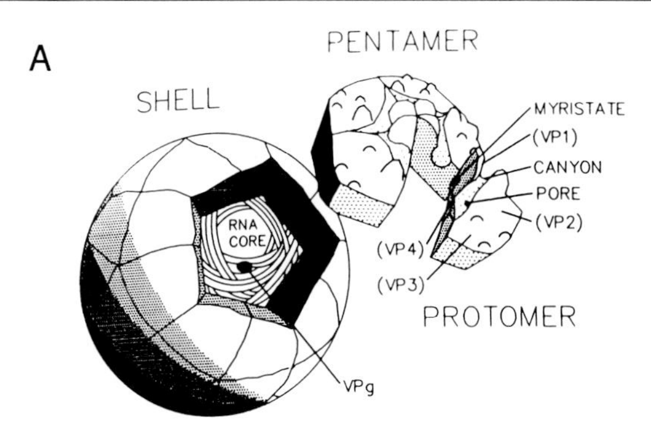

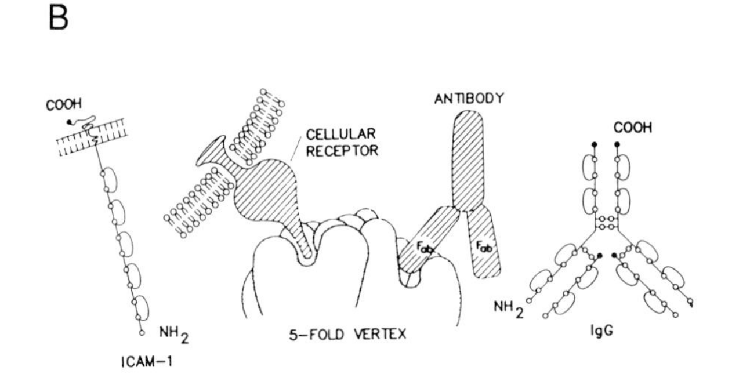

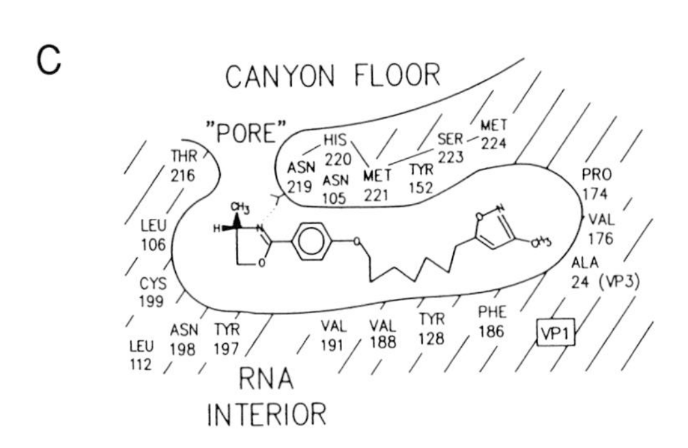

Figure 1 Structure of a typical picornavirus. (A) Exploded diagram showing internal location of the RNA genome surrounded by capsid composed of pentamers of proteins VP1, VP2, VP3, and VP4. Note the "canyon" depression surrounding the vertex of the pentamer. (B) Binding of cellular receptor to the floor of the canyon. The major rhinovirus receptor (ICAM-1 molecule) has a diameter roughly half that of an IgG antibody molecule. (C) Location of a drug-binding site in VP1 of a rhinovirus. The antiviral drug shown, WIN 52084, prevents viral attachment by deforming part of the canyon floor. [From Rueckert R. R. (1990). Picornaviridae and their replication. *In* "Virology," 2nd ed. (B. N. Fields, D. M. Knipe, *et al.*, eds.), pp. 507–548. Raven Press, New York.]

ether or sodium deoxycholate. The virus is completely inactivated when heated at 55°C for 30 min. In the presence of molar magnesium chloride, poliovirus is protected against thermal inactivation, a property that has been utilized in stabilizing oral poliovirus vaccine.

Enteroviruses replicate in the cytoplasm of the infected cell. Mature virus is released by lysis of the cell. In the diagnostic laboratory, virus is cultivated in primary or continuous-line cell cultures from various human or monkey tissues. The typical enterovirus infects only primate cells that contain a specific membrane receptor for the virus on the cell surface.

III. Replication

Understanding the mode of replication of an enterovirus provides useful insights into its host range and its transmission. Polioviruses adsorb to cells at specific cell receptor sites. This is evidenced by the fact that intact poliovirus infects only primate cells in culture, whereas the viral RNA isolated from its surrounding capsid also infects nonprimate cells (rabbit, guinea pig, chick), where only one cycle of multiplication occurs. Multiple cycles of infection are not observed in nonprimate cells because the resulting virus progeny possess protein coats that do not bind to the cells lacking the viral receptors.

After attachment, the virus particles are taken into the cell, and the viral RNA is uncoated. The single-stranded genomic RNA serves as its own messenger RNA (mRNA). This mRNA is translated, resulting in the formation of an enzyme, RNA polymerase. This enzyme is necessary for the formation of a replicative intermediate and then a replicative form, which is a double-stranded molecule consisting of a positive parental RNA strand and a complementary negative strand, and also for the formation of inhibitors that turn off the synthesis of cellular RNA and protein. Single-stranded viral RNA molecules—positive strands—are then synthesized from the replicative form. The newly synthesized positive-strand RNA molecules perform any of the following functions: (1) serve as mRNA for synthesis of structural proteins; (2) serve as template for continued RNA replication; or (3) become encapsidated, resulting in mature progeny virions. The synthesis of viral capsid proteins is initiated at about the same time as RNA synthesis.

The entire poliovirus genome, acting as its own mRNA, is translated to form a single large polypeptide that is subsequently cleaved to produce the various viral capsid polypeptides. Thus, the poliovirus genome serves as a polycistronic messenger molecule. Completion of encapsidation produces mature virus particles that are then released when the cell undergoes lysis.

Like all living things, polioviruses mutate, and this property presents both opportunities and problems in relation to human health. Pathogenic polioviruses, by propagation under manipulated conditions through many generations in cultured monkey kidney cells, were deliberately attentuated so as to be no longer neurovirulent for humans. By this means, attenuated live poliovirus vaccines were developed. However, the living vaccine viruses also may mutate, and their progeny may spread to persons not vaccinated—a property that has both advantages and disadvantages (see Section VIII).

IV. Pathogenesis

The portal of entry of the virus is the mouth. Primary multiplication takes place in the oropharynx or the intestine, and for a few days virus may appear in the blood. The virus can be isolated regularly from the throat just before and at the first signs of illness. The incubation period is usually between 7 and 14 days but may range from 3 to 35 days. By 1 wk after onset, there is little virus in the throat, but large amounts of virus continue to be excreted in the stools for several weeks, even though humoral antibodies usually develop during the same period.

When the infecting enterovirus is a poliovirus, virus enters the bloodstream early in the infection. Virus from the bloodstream sometimes invades the central nervous system, but this route can be blocked if neutralizing antibodies have been acquired previously. Within the central nervous system, the virus spreads along nerve fibers, and in the process of its intracellular multiplication it may damage or completely destroy the invaded nerve cells. Flaccid paralysis of the corresponding muscles is the result. The anterior horn cells of the spinal cord are the most prominently involved, but in severe cases the posterior horn and dorsal root ganglia may also be involved. In the brain, the reticular formation, vestibular nuclei, and deep cerebellar nuclei are most often affected. The cortex is almost completely spared, except for the motor cortex along the precentral gyrus.

Poliovirus does not multiply in muscle tissue *in vivo*. The malfunctions that occur in peripheral nerves and voluntary muscles follow upon the replication of virus in nerve cells. Changes occur rapidly in nerve cells, from mild chromatolysis to neuronophagia and complete destruction. Some nerve cells that lose their function may ultimately recover completely. Inflammation, chiefly by lymphocytes, is a secondary result of the attack on nerve cells.

By applying the methods of molecular biology, one laboratory has made remarkable progress in understanding the pathogenesis of poliovirus. The cell receptor has been shown to be a novel member of the immunoglobulin superfamily of proteins.

Introduction of human genomic DNA containing the poliovirus receptor gene, or cloned human poliovirus receptor complementary DNA (cDNA), into mouse L cells was followed by their acquiring susceptibility to multicycle viral infection. However, there may also be some binding of virus to tissues that are not sites of poliovirus replication. Poliovirus receptor RNA and protein are expressed naturally in several human tissues, including those that do not act as host cells *in vivo*. Thus, true tissue susceptibility to poliovirus is not governed solely by expression of the receptor in tissues.

To elucidate poliovirus host range and tissue tropism, transgenic mice containing the human poliovirus receptor gene in their germ line were developed. The transgenic mice express poliovirus receptor transcripts and poliovirus-binding sites in many different tissues. Inoculation of poliovirus receptor transgenic mice with wild poliovirus induces a paralytic disease that clinically and histopathologically is like human poliomyelitis. However, poliovirus receptor transgenic mice inoculated with an attenuated poliovirus (the Sabin vaccine strain) do not develop disease. These studies demonstrate that the poliovirus receptor gene is the major determinant of poliovirus host range. Transgenic mice expressing the poliovirus receptor gene offer a new and convenient model for studying poliovirus neurovirulence and attenuation.

V. Diseases

A. Poliomyelitis

When a susceptible individual is exposed to infection by a virulent poliovirus, the response may be (1) inapparent infection without symptoms, (2) mild illness, (3) aseptic meningitis, or (4) paralytic poliomyelitis. Only about 1% of poliovirus infections are recognized clinically. Shedding of virus from the throat and feces and resultant transmission of infection can take place without invasion of the central nervous system.

The mild illness may include fever, malaise, drowsiness, headache, nausea, vomiting, constipation, and sore throat, in various combinations. The patient recovers in a few days. Nonparalytic poliomyelitis (aseptic meningitis), in addition to some of the preceding signs and symptoms, includes stiffness and pain in the back and neck; the illness lasts 2–10 days, with rapid and complete recovery. It is important to emphasize that poliovirus is only one of the many enteroviruses and other viruses that produce aseptic meningitis.

In paralytic poliomyelitis, onset of the major disease may follow a minor illness such as those already described, or it may occur without any antecedent first phase. The predominating complaint is flaccid muscle paralysis resulting from lower motor neuron damage. The maximal recovery usually occurs within 6 mo, but residual paralysis usually lasts much longer.

Certain other enteroviruses, and perhaps other agents still unidentified, can, on rare occasions, cause a syndrome difficult or impossible to distinguish on clinical grounds from paralytic polio. As the worldwide drive to eradicate poliomyelitis and the polioviruses proceeds and surveillance for cases becomes crucial, recognition of these other paralytogenic agents will be important.

Mention should also be made of late-onset postpolio syndrome (PPS), consisting of muscle weakness, muscle pain, and unaccustomed fatigue, which has been reported with increasing frequency among former poliomyelitis patients. Studies have shown that the age of the patient with PPS is less important as a determinant of PPS onset than the length of the interval following the acute illness. PPS incidence peaks at an interval of about 30 years following the time of acute poliomyelitis. It seems that PPS results from a neuromuscular disease process initiated at the time of the acute illness in susceptible persons. PPS definitely is not caused by a reactivation of the original infecting poliovirus or by a reinfection with a current strain.

B. Coxsackievirus Diseases

Coxsackieviruses, a large subgroup of the enteroviruses, are divided into groups A and B based on different pathogenic potentials for mice. They produce a variety of illnesses in humans. Herpangina, hand, foot, and mouth disease, and acute hemorrhagic conjunctivitis are caused by certain Group A serotypes. Pleurodynia (devil's grip), myocarditis, pericarditis, and meningoencephalitis are caused by Group B viruses. In addition to these, a number of Group A and B serotypes can give rise to aseptic meningitis, respiratory and undifferentiated febrile illnesses, hepatitis, and paralysis. Generally, paralysis produced by nonpolio enteroviruses is incomplete and reversible. Coxsackie B viruses are the most commonly identified causative agents of viral heart disease in humans.

1. Herpangina

Herpangina is caused by certain Group A viruses (types 2, 4, 5, 6, 8, and 10). There is an abrupt onset of fever, sore throat, anorexia, dysphagia, vomiting, or abdominal pain. The pharynx may develop characteristic discrete vesicles. The illness is self-limited and is most frequent in small children.

2. Summer Minor Illnesses

Coxsackieviruses are often isolated from patients with acute febrile illnesses of short duration that occur during the summer or fall and are without distinctive features.

3. Pleurodynia

Pleurodynia (epidemic myalgia, Bornholm disease) is caused by Group B viruses. Fever and chest pain are usually abrupt in onset but are sometimes preceded by malaise, headache, and anorexia. The chest pain may be located on either side or substernally, is intensified by movement, and may last from 2 days to 2 wk. Abdominal pain occurs in approximately half of the cases, and in children this may be the chief complaint. The illness is self-limited and recovery is complete, although relapses are common.

4. Aseptic Meningitis and Mild Paresis

Aseptic meningitis and mild paresis are often caused by all Group B coxsackieviruses and by types A7, A9, and A24. Fever, malaise, headache, nausea, and abdominal pain are common early symptoms. Signs of meningeal irritation, stiff neck or back, and vomiting may appear 1–2 days later. The disease sometimes progresses to mild muscle weakness suggestive of paralytic poliomyelitis. Patients almost always recover completely.

5. Neonatal Disease

Neonatal disease may be caused by Group B viruses, with lethargy, feeding difficulty, and vomiting, with or without fever. Myocarditis or pericarditis may occur within the first 8 days of life. The clinical course may be rapidly fatal, or the patient may recover completely. Myocarditis has also been caused by some Group A viruses.

6. Colds

A number of the enteroviruses have been associated with common colds. Among these are types A10, A21, A24, and B3.

7. Hand, Foot, and Mouth Disease

Hand, foot, and mouth disease has been associated particularly with type A16, but A4, A5, A7, A9, and A10 have also been implicated. Virus may be recovered not only from the stool and pharyngeal secretions but also from vesicular fluid.

The syndrome is characterized by oral and pharyngeal ulcerations and a vesicular rash of the palms and soles, which may spread to the arms and legs. Vesicles heal without crusting, which clinically differentiates them from the vesicles of herpesviruses and poxviruses.

8. Myocardiopathy

Group B infections are increasingly recognized as a cause of primary myocardial disease in adults as well as children. Group A and echoviruses have been implicated but to a lesser degree. In experimental animals, the severity of acute viral myocardiopathy is greatly increased by vigorous exercise, hydrocortisone, alcohol consumption, pregnancy, and undernutrition and is greater in males than in females. In human illnesses, these factors may similarly increase the severity of the disease.

9. Postviral Fatigue Syndrome

Patients have a history of months to years of excessive muscle fatigue accompanied by myalgia, with or without an acute viral infection at onset. Chronic infections with group B viruses have been reported in some of these patients.

10. Acute Hemorrhagic Conjunctivitis

Type A24 is one of the agents that can cause this disease.

11. Diabetes Mellitus

Serological studies sugest an association of type 1 diabetes with past infection by coxsackievirus B4 and perhaps other members of the B group. In mice, another picornavirus, encephalomyocarditis virus, induces lesions in the pancreatic islets of Langerhans as well as an accompanying diabetes.

12. Swine Vesicular Disease

The agent of swine vesicular disease is an enterovirus antigenically related to coxsackievirus B5. Furthermore, the swine virus can also infect humans.

C. Echovirus Diseases

Echoviruses (enteric cytopathogenic human orphan viruses) are grouped together because they infect the human enteric tract and because they can be recovered from humans only by inoculation of certain tissue cultures. More than 30 serotypes are known, but not all cause human illness. Aseptic meningitis, febrile illnesses with or without rash, common colds, and acute hemorrhagic conjunctivitis are among the diseases caused by echoviruses.

To establish etiological association of echovirus with disease, the following criteria are used. (1) There is a much higher rate of recovery of virus from patients with the disease than from healthy individuals of the same age and socioeconomic level living in the same area at the same time. (2) Antibodies against the virus develop during the course of the disease. If the clinical syndrome can be caused by other known agents, then virological or serological evidence must be negative for concurrent infection with such agents. (3) The virus is isolated from body fluids or tissues manifesting lesions (e.g., from the cerebrospinal fluid in cases of aseptic meningitis).

Echoviruses 4, 6, 9, 11, 14, 16, 18, and others have been associated with aseptic meningitis. Rashes are common in infections with types 4, 9, 16 (Boston exanthem disease) and 18. Rashes are most common in young children. Occasionally, there is conjunctivitis, muscle weakness, and spasm (types 6, 9, and others). Infantile diarrhea may be associated with some types (e.g., 18, 20). Echovirus type 28 isolated from patients with upper respiratory illness causes "colds" in volunteers and has been reclassified as rhinovirus type 1. For many echoviruses (and some coxsackieviruses), no disease entities have been defined.

With the virtual elimination of poliomyelitis in developed countries, the central nervous system syndromes associated with echoviruses and coxsackieviruses have assumed greater prominence. The latter in children <1 yr old may lead to neurological sequelae and mental impairment. This does not appear to happen in older children.

D. Other Enterovirus Diseases

Four of the later enteroviruses (types 68–71) grow in monkey kidney cultures, and three of them cause human disease. Enterovirus 68 has been isolated from the respiratory tracts of children with bronchiolitis or pneumonia.

Enterovirus 70 is the chief cause of acute hemorrhagic conjunctivitis. It was isolated from the conjunctiva of patients with this striking eye disease, which occurred in pandemic form in 1969–1971 in Africa and Southeast Asia. It was not diagnosed in the United States until its importation into Florida in 1981. Acute hemorrhagic conjunctivitis has a sudden onset of subconjunctival hemorrhage ranging from small petechiae to large blotches covering the bulbar conjunctiva. There may also be epithelial keratitis and occasionally lumbar radiculomyelopathy. The disease is most common in adults, with an incubation period of 1 day and a duration of 8–10 days. Complete recovery is the rule. The virus is highly communicable and spreads rapidly under crowded or unhygienic conditions. There is no effective treatment.

Enterovirus 71 has been isolated from patients with meningitis, encephalitis, and paralysis resembling poliomyelitis. It continues to be one of the main causes of central nervous system disease, sometimes fatal, around the world. In some areas, particularly in Japan and Sweden, the virus has caused outbreaks of hand, foot, and mouth disease.

E. Diagnosis

In an individual case, diagnosing an enterovirus infection on clinical grounds is impossible. However, in the following epidemic situations, enteroviruses must be considered: (1) summer outbreaks of aseptic meningitis; (2) summer epidemics, especially in young children, of a febrile illness with rash; and (3) outbreaks of diarrheal disease in young infants from whom no pathogenic enterobacteria can be recovered.

The diagnosis depends on laboratory tests. The procedure of choice is isolation of virus from throat swabs, stools, rectal swabs, and, in aseptic meningitis, cerebrospinal fluid. Serological tests are impractical—because of the many different virus types—except when a virus has been isolated from a patient or during an outbreak of typical clinical illness. Neutralizing and hemagglutination-inhibiting antibodies are type-specific and may persist for years. Complement-fixing antibodies give many heterotypic responses.

If an agent is isolated in tissue culture, it is tested with specially designed combination pools of antisera against enteroviruses, known as the LBM pools. Determination of the type of virus present depends on neutralization by a single specific combination of antisera. At times, infection with two enteroviruses may occur simultaneously.

VI. Epidemiology

A. General

It is important to emphasize that by far the most common form of infection with poliovirus is a mild or silent episode and that severe manifestations are rare. Paralytic poliomyelitis remained an epidemiological enigma until this concept was understood.

In warm countries and in families living under unsanitary and poor socioeconomic conditions in temperate zones, children are infected very early in life, and 90% may have already experienced infections with all three serotypes of poliovirus, and with many other enteroviruses, before the age of 5 yr. In such settings, paralytic poliomyelitis is rarely recognized and epidemics generally do not occur. When infection is delayed to older childhood and young adult life, the reported incidence of paralytic poliomyelitis rises. The ratio of inapparent infections to paralytic cases is about 100:1; it may be higher among infants.

Poliomyelitis can be viewed as having three major epidemiological phases: endemic, epidemic, and vaccine-era. These phases have occurred sequentially in history, but they all coexist at the present time in different regions of the world. The first two can be termed prevaccine, whether in reference to the historical time before vaccines had been developed or to the situation in a region where vaccine coverage has been inadequate. [*See* EPIDEMIOLOGIC CONCEPTS.]

Sporadic cases of paralytic poliomyelitis apparently have occurred for as long as history has been recorded. From ancient times into the late 1800s, polioviruses became established throughout most of the world's populations and survived for many centuries in an endemic fashion, continuously infecting new susceptible infants born into the community. In this endemic pattern, with antibody to all three poliovirus types present almost universally in women of childbearing age, passive immunity was transferred from mother to offspring, and many infants subsequently experienced their first poliovirus infections during the first few months of life while maternal antibodies still provided protection. Furthermore, because such a large proportion of poliovirus infections take an inapparent or subclinical form, the paralytic cases that did occur could go unnoted in populations faced with very high infant and child mortality rates.

B. Epidemiological Patterns in Industrialized Countries

In the latter part of the nineteenth century and early in the twentieth century, the urban, industrialized parts of northern Europe and the United States began to suffer from epidemics of paralytic polio that became larger, more frequent, and more severe in more and more localities. The generally accepted explanation is that, with increased economic development and the corresponding resources for community and household hygiene, the opportunities for infection among infants and young children were reduced. Thus, increasing numbers of persons encountered poliovirus for the first time in later childhood or in adult life, at ages when poliovirus infections are more likely to take the paralytic form. Furthermore, the delay in exposure increased the pool of susceptible persons, opening the way for rapid and explosive spread of the viruses once they did enter the population, in contrast to the steady endemic transmission of the preceding phase. This transition from the endemic phase to the epidemic phase sometimes took place abruptly and sometimes after gradual increases in the annual case rates of "sporadic" poliomyelitis.

In many developed countries, the vaccine era in poliomyelitis began after 1955, when inactivated poliovirus vaccine (IPV) was introduced, and notably after 1960, when live attenuated poliovaccine became available on a large scale. Rarely has a serious disease been controlled so quickly and dramatically as was poliomyelitis in these countries. In 1955, the Soviet Union, 23 other European countries, the United States, Canada, Australia, and New Zealand experienced a total of >76,000 reported cases of poliomyelitis. Only 12 yr later, in 1967, 1013 cases were recorded in these same countries, a reduction of almost 99%. Numbers of poliomyelitis cases continued to be reduced until now, in industrialized countries, poliomyelitis is a rare disease.

In the United States, before the inactivated virus vaccine became available, rates of the paralytic form of the disease were between 5 and 10/100,000 population (10,000–21,000 paralytic cases per year). After the inactivated virus vaccine came into use, cases were far fewer, in some years reaching incidence rates as low as 0.5/100,000 population. But this still meant that significant numbers of cases were occurring; in 1960, there were >2500. Concomitant with the use of inactivated virus vaccine in the United States and the decreasing number of cases as compared with the 1955 totals was the finding that some cases were occurring among the fully vaccinated. In a study of several thousand paralytic cases, 17% were in children who had received three injections of IPV. Some of the disappointing results were due to potency problems, which have since been corrected.

After the introduction in the United States of oral poliomyelitis vaccine (OPV) made from live attenuated virus, the numbers of cases dropped precipitously, and from 1969 onward the largest number in any year was 32 in 1970. For the decade 1970–1979, the average annual number of cases was 17. For the period 1980–1990, the annual average decreased to about 5/yr. This has meant case rates as low as 0.001/100,000 population.

For many years, the United States has relied almost completely on OPV, and it appears that there is no longer any endogenous reservior of wild polioviruses within the country. Wild strains continue to be introduced, particularly from Mexico, but even such imported cases are sporadic and almost never result in secondary cases. The use of live poliovirus vaccine has achieved this result by establishing intestinal resistance extensively within the population, reducing the pool of susceptible individuals below the level required for perpetuation of wild polioviruses, and a true break in the chain of infection has been achieved. Many other countries with extensive and continuing live vaccine programs also are reporting virtually no cases and few, if any, poliovirus isolates other than vaccinelike strains.

As mentioned, good results also have been obtained with inactivated poliovirus vaccines, especially in Sweden, Finland, and The Netherlands, coutries of relatively small populations (a total of only 26 million persons) that are culturally homogeneous and socially advanced. They have excellent public health services, and inactivated virus vaccine has been administered in intensive and regularly maintained immunization programs, achieving vaccine coverage among children that approaches 100%. Within these countries, wild polioviruses no longer circulate endemically.

However, even in well-vaccinated countries that have achieved almost complete control of poliomyelitis, there may be important gaps in protection. This vulnerability was shown by the oubreaks of poliomyelitis among persons refusing vaccine on religious grounds in The Netherlands, Canada, and the United States in 1978 and 1979. The virulent epidemic strain was an imported one, believed to have come from Turkey, and no cases occurred beyond these interconnected unvaccinated groups in any of the countries involved. There were no cases in the vaccinated Dutch population, but in nursery and primary schools in some of the affected communities 24% of the vaccinated and 71% of the nonvaccinated children tested were excreting the wild poliovirus.

In the IPV-vaccinated population of Finland—after 20 yr of freedom from poliomyelitis—nine cases of paralytic polio and one nonparalytic case occurred between August 1984 and January 1985. On the basis of virus isolations from healthy individuals and from sewage, it is estimated that at least 100,000 persons in the general population were infected. OPV was widely used in a mass campaign covering about 95% of the entire population, and 1.5 million extra doses of IPV were administered to persons <18 yr old. The outbreak was halted in February 1985.

The epidemic strain of poliovirus (P3/Finland/84) was tested extensively by new molecular techniques in several laboratories and was found to differ in both immunological and molecular properties from the type 3 vaccine strains. Analysis of sera collected before the outbreak indicated that prevalence of neutralizing antibodies against most of the new type 3 isolates was much lower than that of antibodies to the Saukett strain (the type 3 strain used in the killed vaccine). The precipitating factor in the outbreak was the appearance of a strain of poliovirus type 3 that was antigenically aberrant enough to break through low-grade herd immunity that resulted from the low immunogenicity of the type 3 component of the Finnish IPV. After immunization with the new enhanced IPV (eIPV) or with OPV, high serum antibody titers are produced against the Finnish epidemic type 3 strain. Thus, although differences among strains from various parts of the world have indeed been demonstrated for many years, the major neutralizing antigen of each poliovirus serotype is remarkably stable.

Another well-studied outbreak, which occurred in Israel in 1988, has yielded important information on immunological control. Israel began administration of OPV in 1961, and the continued widespread use of OPV in infancy, without further doses, has led to effective control in that country. However, an outbreak of 15 cases of paralytic polio caused by type 1 poliovirus occurred in 1988. Most of the patients were $\geq$15 yr old. The focus of the outbreak (12 of the 15 cases) was an area where eIPV had been the only poliovirus vaccine used for infants since 1982. Surveys of neutralizing antibodies carried out in recent years in Israel provided evidence of relatively low immunity to type 1 poliovirus in teenagers who had been immunized in the first year of life with trivalent OPV, but who had received no subsequent doses. A number of healthy infants tested at the onset of the 1988 outbreak were found to be excreting wild poliovirus.

A combination of factors apparently were responsible for the outbreak: (1) lack of booster doses of OPV, which led to impaired immunity in the young adult population; (2) introduction of wild poliovirus from surrounding countries in which polio is endemic; and (3) transmission of wild poliovirus to susceptibles by eIPV-vaccinated infants, who were themselves protected from development of the paralytic disease, but who acquired intestinal infection and excreted the epidemic virus, thus spreading it to susceptibles in the community. The lessons from this outbreak are (i) that IPV alone—even of enhanced potency—does not interrupt the circulation of wild virus, which singles out susceptible contacts as targets; and (ii) that OPV alone—administered in infancy—is not completely effective for providing lifelong protection.

The outbreaks previously described were unfortunate reminders that virulent polioviruses do indeed still exist and can readily circulate if given the right opportunity. Thus, importation of virulent wild polioviruses remains a clear danger against which no nation can be secure except by maintaining high levels of immunity through vaccination.

C. Epidemiological Patterns in Developing Nations

A prevaccine phase of polio epidemiology is of more than historical interest, for it is also a current situation that many nations face as of this writing. Worldwide, according to the World Health Organization (WHO), in the 1980s well over 300,000 cases of poliomyelitis were occurring each year. In many instances, the reported cases represented $<10\%$ of the actual number of cases that occurred. Poliomyelitis clearly remains an urgent problem for many of the world's people; however, with properly conducted vaccination programs, polio can be controlled everywhere. Through such programs utilizing OPV, polio is now a rare disease in the entire Western Hemisphere.

VII. Immunity

Passive immunity to an enterovirus is transferred from mother to offspring. The maternal antibodies gradually disappear during the first 6 mo of life. Passively administered antibody lasts only 3–5 wk.

Virus-neutralizing antibody forms within a few days after exposure to the virus, often before the onset of illness, and persists, apparently, for life. Its formation early in the infection is a result of virus multiplication in the intestinal tract and deep lymphatic structures. With poliovirus infections, antibodies must be present in the blood to prevent the dissemination of virus to the brain. Immunization is of value in poliovirus infections only if it precedes the onset of symptoms referable to the nervous system.

Circulating serum antibody is not the only source of protection against infection. Local or cellular immunity is manifested by protection against intestinal reinfection after recovery from a natural infection or after immunization with the live polio vaccine. Local or secretory antibody is increasingly recognized as having an important role in defense against infections by poliovirus or by other enteroviruses.

Viral infections in general carry an increased risk for persons with various immunodeficiencies in either humoral or cell-mediated immunity. In such persons, poliovirus infection—either by wild virus or by vaccine strains—may develop in an atypical manner, with an incubation period longer than 28 days, a high mortality after a long chronic illness, and unusual distribution of lesions in the central nervous system. With other enteroviruses also, severe chronic infections have been recorded in immunodeficient individuals. [*See* IMMUNE SUPPRESSION.]

A decrease in resistance to poliovirus accompanies removal of tonsils and adenoids. Preexisting secretory antibody levels in the nasopharynx decrease sharply following operation, without any change in antibody levels in serum. Local antibody remains at low levels or is absent for as long as 7 mo. In seronegative children, nasopharyngeal antibody response to poliovirus vaccine develops significantly later and to lower titers in children previously tonsillectomized than in those with intact tonsils.

The VP1 surface protein of poliovirus, and presumably other enteroviruses, contains several virus-neutralizing epitopes, each of which may contain fewer than 10 amino acids. Each epitope is capable of inducing virus-neutralizing antibodies.

VIII. Prevention

Although improved sanitation and hygiene help to limit the spread of polioviruses, the only specific means of preventing paralytic polio is immunization with live, attenuated OPV or with formalin IPV. Both contain viruses of all three poliovirus serotypes.

IPV, also known as Salk vaccine, is prepared from virus grown in monkey kidney cultures—originally in primary cultures, now chiefly in continuous-passage Vero monkey cells. Repeated booster inoculations have been required to maintain immunity with the IPV that has been generally available up to the present writing. New formulations of IPV with higher concentrations of antigen (eIPV) have been developed. These new vaccines were developed in the hope of inducing satisfactory and lasting antibody responses in a large proportion of vaccines after only two doses. However, field experiences to date have not yielded unequivocally clear results as to their clinical efficacy and overall long-term effect. The merits and problems of IPV are listed in Table I.

Table I Inactivated Poliovirus Vaccine: Merits and Problems

Merits
• Confers humoral immunity in vaccinees if sufficient doses of potent vaccine are given.
• Can be incorporated into regular pediatric immunization, with other injectable vaccines DPT (diphtheria, pertussis, tetanus).
• Absence of living virus excludes potential for mutation and reversion to virulence.
• Absence of living virus permits its use in immunodeficient or immuosuppressed individuals and their households.
• Has greatly reduced the spread of polioviruses in small countries where it has been properly used (wide and frequent coverage).
• Beneficial in certain tropical areas where live vaccine has failed to "take" in some young infants.
Problems
• Early studies indicated a disappointing record in percentage of vaccinees developing antibodies after three doses, but more immunopotent antigens are now produced.
• Generally, with the vaccines that have been commercially available, repeated boosters have been required to maintain detectable antibody levels.
• Does not induce sufficient local intestinal immunity in the vaccinee to block transmission of wild polioviruses by the fecal–oral route.
• More expensive than live vaccine.
• Growing scarcity of monkeys for kidney tissue substrate was a problem but has been overcome by use of continuous-passage monkey cells (Vero) for vaccine production.
• Use of virulent polioviruses as vaccine seed creates potential for tragedy if a single failure in virus inactivation were to occur in a batch of released vaccine. This risk is somewhat increased since monkey neurovirulence tests are no longer required before release of inactivated vaccine. However, this problem could be overcome by use of attenuated strains for production.

[Modified from Melnick, J. L. (1978). *Bull. WHO* **56**, 21–38.]

OPV, also known as Sabin vaccine, contains virus grown in primary monkey cell cultures or in human diploid cell cultures. OPV can be stabilized by the addition of magnesium chloride. Stabilized OPV can be stored at 0° to 8°C for 1 yr without significant loss of titer. When the distribution and administration of the vaccine is not imminent, OPV may be stored at −20°C; at this temperature, it retains full potency for years. At elevated temperatures, stabilized OPV retains the minimum potency for shorter periods of time: 7–14 days at 26°C and 2 days at 31°C. At 37°C, the vaccine loses about 0.15 $\log_{10}$ every day. At this rate of degradation, a vaccine with an assumed total virus content of 6.15 $\log_{10}$ should not be used if it has been held at 37°C for longer than 3 days. Further efforts are needed to increase the heat stability of OPV. The merits and problems of OPV are listed in Table II.

Both IPV and OPV induce humoral antibodies that circulate in the blood and protect the central nervous system from subsequent invasion by wild virus. The immunity induced by IPV, however, has little effect on intestinal infection and carriage of virus, and thus IPV recipients remain potential vehi-

Table II Live Poliovirus Vaccine: Merits and Problems

Merits

- Confers both humoral and intestinal immunity, like the natural infection.
- Immunity, once adequately induced, is enduring.
- Oral administration is more acceptable to vaccinees than injection and is easier to accomplish.
- Induces antibody very quickly in a large proportion of vaccinees.
- Administration does not require use of highly trained personnel.
- When properly stabilized, it can retain potency under difficult field conditions with less refrigeration and no freezers.
- Under epidemic conditions, it not only induces antibody quickly but also rapidly infects the alimentary tract, blocking spread of the epidemic virus.
- Is relatively inexpensive, both to produce and to administer and does not require continuing booster doses.
- Can be prepared in pretested human or monkey cells, thus does not depend on monkeys except for neurovirulence testing.
- Use of continuous cell lines eliminates the theoretical risk of including monkey virus contaminants in the vaccine.

Problems

- Vaccine viruses may mutate and, in very rare instances, revert toward neurovirulence sufficient to cause paralytic polio in recipients or their contacts.
- Vaccine progeny virus spreads to household contacts.
- Vaccine virus also spreads to persons in the community. (Some people consider this spread into the household and the community to be an advantage, but the progeny virus excreted and spread by vaccinees often is a mutated virus and obviously cannot be a safety-tested vaccine, licensed for use in the general population.)
- In certain tropical countries, induction of antibodies in a satisfactorily high proportion of vaccinees has been difficult to accomplish.
- Contraindicated in those with immunodeficiency diseases and in their household associates as well as in persons undergoing immunosuppressive therapy, and their households.
- Requires monkeys for safety testing.

[Modified from Melnick, J. L. (1978). *Bull. WHO* **56**, 21–38.]

cles for spread of wild virus to susceptible persons with whom they come in contact.

In contrast, OPV infects, multiplies, and thus immunizes in a manner that parallels natural infection. OPV induces not only long-lasting IgG antibodies in the blood but also secretory IgA antibodies in the pharynx and intestine, which then become resistant to infection by wild virus.

The live attenuated polioviruses, especially type 3, mutate to some degree during their multiplication in vaccinated children, but not to full neurovirulence, and only extremely rare cases of paralytic poliomyelitis have occurred in recipients or their close contacts. The risks of paralytic polio associated with reversion of OPV are exceedingly small. On the basis of several large studies, including a 10-yr study conducted by WHO, one case of vaccine-associated paralysis was estimated to have occurred for every 2–4 million doses of trivalent OPV distributed.

In developing countries, the primary immunization schedule should begin early and be completed early in infancy. The immunization schedule recommended by WHO is a series of three doses of OPV. These should be administered at 6, 10, and 14 wk of age, or as soon as possible thereafter. Intervals >4 wk between doses do not require restarting the series.

Where children are born in hospitals (or other maternity institutions) or where children come in contact with health services early in life, these opportunities should be used to administer an extra dose of OPV. It should not replace a dose from the regular OPV series. Although the serological response to OPV in the first week of life is less than that observed with immunization of older infants, >70% of neonates benefit by developing local immunity in the intestinal tract. In addition, 30–50% of neonates develop serum antibodies to one or more poliovirus types. Many of the remaining infants become immunologically primed and therefore respond promptly to additional doses later in life. A dose of OPV at birth is particularly important in cities, in other areas with high population density, and where cases occur in the first year of life. In large cities in many parts of the world, 50% of polio cases occur in infants. This illustrates the need to complete the series of polio immunizations as early in life as possible.

In some, but not all, developing countries, children have demonstrated a lower-than-expected serological response to three or more doses of OPV, possibly because of difficulties in maintaining the cold chain during transit to target areas, interference with vaccine "take" by the presence of other enteroviruses, or interference by nonspecific inhibitors in the gut.

A combined regimen of IPV and OPV has proven

successful in overriding the factors interfering with the serological response (by use of IPV) and in increasing intestinal immunity (by use of OPV). The combined regimen also eliminates the rare risk of paralysis caused by a mutated type 3 poliovirus in the vaccinated host. To eradicate polio from the world, combined regimens of IPV and OPV may be necessary in some areas.

The application of recombinant DNA technology may permit the development of a live poliovirus that cannot mutate to increased neurovirulence. A key technological advance was the construction of infectious cDNA clones that allowed the manipulation of nucleotide sequences to generate poliovirus mutants with specific and desirable alterations in the genome. Recombinant viruses have been constructed from parental viruses belonging to different poliovirus serotypes and between virulent and attenuated strains of the same serotype. Sequences in the viral genome that are responsible for an attenuated phenotype have been identified.

The type 1 vaccine virus, which is extremely stable genetically, has been used as a vector for type 2 and type 3 nucleotide sequences encoding immunogenic regions of their VP1 proteins. The new "chimeric" strains have the desired biological characteristics of type 1 but the immunogenic properties of type 2 or type 3, respectively. A poliovirus chimera has been constructed that carries a short nucleotide sequence of type 3 inserted into the type 1 genome. The chimeric virus expressed a fusion protein containing eight amino acids of type 3 VP1 and induced antibodies to both type 1 and type 3 viruses.

Scientific advances such as those described earlier may lead to a more genetically stable type 3 vaccine. However, field-testing such a new vaccine candidate will be difficult, because it will be necessary to prove that the new vaccine produces fewer than one vaccine-associated case per million susceptible recipients.

Also, the success of the global application of OPV as currently constituted and readily available is leading to interruption in the transmission of wild poliovirus and to the eradication of poliomyelitis, before any newly developed vaccine strains can be properly field-tested. The key to the success of OPV in both industrialized and developing countries hinges not only on the vaccine itself but also on the energy and diligence of those concerned with vaccine administration and subsequent surveillance to ensure its effectiveness in blocking circulation of wild virus.

Bibliography

Burke, K. L., Almond, J. W., and Evans, D. J. (1991). *Prog. Med. Virol.* **38,** 56–68.

Melnick, J. L. (1988). Live attenuated poliovaccines. *In* "Vaccines" (S. A. Plotkin and E. A. Mortiner, Jr., eds.), pp. 115–157. W. B. Saunders, Philadelphia.

Melnick, J. L. (1989). Enteroviruses. *In* "Viral Infections of Humans: Epidemiology and Control," 3rd ed. (A. S. Evans, ed.), pp. 191–263. Plenum Medical Books, New York.

Melnick, J. L. (1990). Enteroviruses: Polioviruses, coxsackieviruses, echoviruses, and newer enteroviruses. *In* "Fields Virology," 2nd ed. (B. N. Fields, D. M. Knipe, *et al.*, eds.), pp. 549–605. Raven Press, New York.

Ren, R., Costantini, F., Gorgacz, E. J., Lee, J. J., and Racaniello, V. R. (1990). *Cell* **63,** 353–362.

Rueckert, R. R. (1990). Picornaviridae and their replication. *In* "Fields Virology," 2nd ed. (B. N. Fields, D. M. Knipe, *et al.*, eds.), pp. 507–548. Raven Press, New York.

Salk, J., and Drucker J. (1988). Noninfectious poliovirus vaccine. *In* "Vaccines" (S. A. Plotkin and E. A. Mortimer, Jr., eds.), pp. 158–181. W. B. Saunders, Philadelphia.

Enzymes, Extracellular

Fergus G. Priest
Heriot Watt University

Glossary

Chaperone Cytoplasmic protein that maintains a protein destined for secretion in a secretion-compatible configuration

Signal peptide Extension to a protein, normally N terminal and of 15–40 amino acids, that is responsible for secretion of the protein across the membrane

Temporal regulation Derepression of enzyme synthesis that occurs as a batch culture enters stationary phase

EXTRACELLULAR ENZYMES are generally secreted by bacteria and molds in order to hydrolyze high-molecular weight molecules in the environment that, in their native form, would be too large to enter the cell. The resultant products are then assimilated by the microorganism as sources of nutrients. Consequently, extracellular enzymes are commonly secreted by soil microorganisms such as filamentous fungi and bacteria of the genera *Bacillus* and *Streptomyces*. The enzymes are responsible for the hydrolysis of plant polysaccharides including starch, cellulose and pectin, proteins, lipids, and nucleic acids. The high yields of these enzymes in culture fluids and their robust properties, especially tolerance to extremes of pH and temperature, led to their application in various industries, particularly in starch and food-processing and the leather industry, where they replaced several unsatisfactory chemical processes. It is now apparent that the general process for the transfer of proteins across membranes using signal sequences, first demonstrated in eukaryotic cells, is also applicable to bacteria and that complex genetic controls, including two-component signal transduction systems, govern the regulation of extracellular enzyme synthesis in some bacteria.

I. Localization of Enzymes

Extracellular enzymes are generally defined as enzymes that have crossed the cytoplasmic membrane of the cell. This definition includes proteins attached to the outer surface of the membrane since such molecules have at least initiated the export process. The final destination of an extracellular enzyme therefore depends on the cell structure. Gram-positive bacteria and fungi are surrounded by a thick cell wall comprised of peptidoglycan or chitin–glucan, respectively. There is no apparent compartmentation in these walls, so enzymes either are released from the outer surface of the membrane and subsequently diffuse through the wall to accumulate in the surrounding environment (truly extracellular enzymes) or are anchored to the membrane (membrane-bound enzymes). Some enzymes may be localized in the wall itself, particularly those involved in cell wall synthesis and turnover.

The aerobic, endospore-forming bacteria of the genus *Bacillus* have enzymes that display these various locations. In *Bacillus licheniformis*, penicillinase (β-lactamase), a protective enzyme that degrades exogenous penicillin, is present as a membrane-bound enzyme in the form of a lipoprotein. This protein has a glyceride thioether moiety attached to the N-terminal cysteine that anchors the hydrophilic protein securely to the lipid membrane. Proteolytic removal of the lipid anchor releases the enzyme into the culture supernatant as a hydrophilic enzyme. The bound enzyme presumably offers improved protection against penicillin, although it may be an intermediate in the process of secretion.

Other enzymes such as alkaline phosphatase and β-glucosidase are distributed to the membrane or the culture fluid, depending on growth conditions. These proteins are not modified as lipoproteins.

Gram-negative bacteria are surrounded by two hydrophobic barriers: the cytoplasmic and outer membranes between which lies the periplasm. Enzymes may therefore be located on or in the cytoplasmic membrane, in the periplasm, or in the outer membrane of these bacteria. The periplasmic enzymes are commonly considered to be the equivalent of the membrane and wall-bound enzymes of the gram-positive cell, and, indeed, many similar enzymes such as alkaline phosphatase and maltodextrin-hydrolyzing enzymes are found in these locations. Although gram-negative bacteria are often considered to secrete enzymes seldomly, bacteria such as *Pseudomonas, Erwinia,* and *Serratia* secrete large amounts of proteases, amylases, and pectinases into the environment.

Recent studies have revealed some interesting variations on the localization of extracellular hydrolases. Many gram-positive (e.g., *Clostridium thermocellum*) and gram-negative (e.g., *Bacteroides* species) cellulolytic bacteria synthesize large protuberances on the cell surface called cellulosomes. Massive numbers of cellulosomes cover the surface of these bacteria when they are grown in the presence of cellulose. The cellulosome comprises at least 14 different enzymes (mostly endocellulases; no exohydrolases) assuming a total molecular weight of about 2 million daltons. An attachment factor in the cellulosome is responsible for attachment of the bacterium to the insoluble cellulose fiber and, in providing close contact, enables cellulose hydrolysis by the endocellulases.

II. Extracellular Enzymes: Their Characteristics and Applications

A. Starch-Hydrolyzing Enzymes

Starch comprises a combination of two polysaccharides: amylose, which is a linear chain of 1,4-α-linked glucose residues, and amylopectin, which is a branched molecule of amylose chains linked by 1,6-α-branch points. On average, amylose comprises chains of about 10^3 glucose residues, whereas amylopectin contains about 10^4–10^5 residues in chains of about 20–25 residues in length. The proportion of the two molecules in starches varies depending on the source but is generally in the region of 70% amylopectin. Starch is present in plant materials in the form of granules that vary in size and shape but are usually 25–100 μm in diameter. Related α-glucans are glycogen, which is a highly branched form of amylopectin, and pullulan, which is a linear molecule of maltotriose units joined by 1,6-α-linkages and is derived from the mold *Aureobasidium pullulans*.

1. α-Amylase

α-Amylase is the most common of microbial starch hydrolyzing enzymes. It is an endo-acting enzyme that hydrolyzes the internal 1,4-α-bonds in amylose, amylopectin, and glycogen. This results in the rapid reduction in viscosity and iodine staining power of starch and the slow release of reducing sugars. Most α-amylases can be grouped into two broad classes: the liquifying enzymes and the saccharifying enzymes. The former hydrolyze amylose to oligosaccharides of five or six glucose residues with small amounts of maltose and glucose. The best studied examples derive from *Bacillus,* in particular *Bacillus amyloliquefaciens* (a close relative of *B. subtilis*), *B. licheniformis,* and the thermophile *Bacillus stearothermophilus*. Saccharifying enzymes on the other hand conduct a more extensive depolymerization of amylose resulting in large amounts of glucose, maltose and maltotriose. *B. subtilis* secretes a saccharifying α-amylase but the industrial enzymes are produced from fungi in particular *Aspergillus oryzae* and some *Rhizopus* species.

α-Amylases are calcium metalloenzymes generally requiring at least one atom of this metal per molecule. In the presence of calcium, they are stable to extremes of pH with optimal activity occurring between pH 4.8 and 6.5 (exceptional enzymes from acidophilic or alkaliphilic bacteria have more extreme pH optima). The fungal enzymes usually have the lower pH optima. Most have molecular masses of about 50 kDa, the fungal (but not the bacterial) enzymes being glycoproteins. Although many α-amylases are rapidly denatured above 50°C in their pure state, the liquifying enzymes from some bacilli are exceptionally thermostable. For example, the optimum operating temperature of the α-amylase from *B. licheniformis* is 90°C, and it can be used industrially at temperatures in excess of

100°C. Interestingly this enzyme has a low calcium requirement.

2. Glucoamylase

Glucoamylase (also called amyloglucosidase) also hydrolyzes the 1.4-α-linkages in amylopectin, amylose, and glycogen, but it does so by attacking consecutive bonds starting at the nonreducing chain end. It is therefore an exo-acting enzyme, and it releases glucose residues in their α-anomeric form. Like other exo-attacking enzymes, it reduces the viscosity and iodine staining power of starch slowly but produces a rapid release of reducing power. Glucoamylase will also hydrolyze the 1,6-α-branch points in amylopectin, although it does this slowly. Nevertheless, it has the potential to hydrolyze starch totally to glucose (in practice, total conversion is impossible, but 96% conversion can be achieved). Although some bacteria secrete glucoamylases, the industrial enzymes are produced from *Aspergillus* and *Rhizopus* species. Like other fungal amylases, the enzymes are thermolabile (optimum temperature about 60°C), have low pH optima (about 4.5–5), and are glycoproteins containing about 5–20% carbohydrate. Molecular masses range from 27 to 112 kDa.

3. β-Amylase

For many years it was thought that bacteria secreted α-amylase exclusively, but extended searches revealed that organisms such as *Bacillus polymyxa, Bacillus cereus,* and *Bacillus megaterium* as well as some clostridia secreted β-amylase. Like glucoamylase, this enzyme attacks amylose in an exo-fashion, hydrolyzing alternate 1,4-α-linkages starting at the nonreducing end of the chain and releasing maltose in the β-configuration. It does not hydrolyze the 1,6-α-bond; therefore, the end-product from amylopectin hydrolysis is a large β-limit dextrin bounded by 1,6-α-linked residues. These enzymes are rather thermolabile and are not produced on an industrial scale. However, there is a requirement for maltose production, and this has prompted the search for alternative enzymes. This has resulted in the discovery of "maltogenic α-amylases", i.e., enzymes that are exo-acting but produce maltose in the α-configuration from starch. These are produced by some thermophilic bacilli.

Other exo-acting amylases that produce maltotetraose (from *Pseudomonas stutzeri*) and maltohexaose (from *Klebsiella pneumoniae*) have been described but are not produced commercially.

4. Debranching Enzymes

Enzymes that hydrolyze the 1,6-α-bonds in amylopectin, glycogen, and pullulan are called debranching enzymes. The two major classes are isoamylase, which hydrolyzes glycogen but not pullulan, and pullulanase, first recognized by its ability to hydrolyze the 1,6-α-linkages in pullulan as well as amylopectin but which has low activity on glycogen. Hydrolysis of amylopectin by pullulanase yields amylose, and hydrolysis of pullulan produces maltotriose. Pullulanases are common in bacteria and are produced commercially from *K. pneumoniae* and *Bacillus acidopullulyticus*. The pH optima occur between pH 5 and 7 and temperature optima between 45° and 50°C, although more temperature-stable versions from thermophilic bacteria are being discovered.

Several novel pullulan-hydrolyzing enzymes have been described recently. These include an α-amylase from a *Thermoactinomyces* species that hydrolyzes the 1,4-α-bond closest to the nonreducing chain end of pullulan to produce panose. A similar enzyme—called neopullulanase—has been isolated from some thermophilic bacilli. Finally, isopullulanase recovered from culture filtrates of *Aspergillus niger* hydrolyzes the other 1,4-α-bond in pullulan to yield isopanose.

5. Applications of Starch Hydrolyzing Enzymes

A major use of these enzymes is in the starch processing industry, where they have now replaced the traditional acid hydrolysis of starch. The basic process begins with "liquifaction." A starch–water slurry comprising insoluble starch granules is heated at 105–110°C to burst the granules and release the starch into solution. Thermostable α-amylase (from *B. licheniformis*) is added both before and after this gelatinization step to effect partial hydrolysis of the highly viscous starch solution to maltodextrins of about 40 glucose units. Liquifaction at the high temperatures used (about 85°C) is rapid (generally within 2 hr) and the degree of hydrolysis can be varied according to requirements by adjusting the period of hydrolysis or amount of enzyme used. Liquifaction reduces the viscosity of the starch solution prior to further treatment, although the dextrins may be dried and used at this stage.

Saccharyifying α-amylases and glucoamylase are used to produce syrups from the liquified starch in a process termed saccharification. These fungal enzymes are relatively thermolabile, and so the process is conducted at lower temperatures and for longer periods (e.g., 55°C for about 40–96 hr). Moreover, the pH has to be reduced to the more acid optima of these enzymes. "High maltose syrups" are produced using fungal α-amylase and contain about 50% maltose, the residue being some glucose, maltotriose, and α-limit dextrins. More recently, the highly maltogenic α-amylases have been introduced (in conjunction with a debranching enzyme) for the production of syrups comprising almost entirely maltose. These have application in the brewing industry because they resemble malt hydrolysates more closely than other syrups and their low hygroscopicity and resistance to crystallization makes them attractive to the confectionary and food industry.

"High conversion syrups" are produced from liquified starch by hydrolysis with fungal α-amylase and glucoamylase. These syrups comprise about 40% glucose, 45% maltose, and the remainder maltotriose. They are used extensively in the brewing, baking, confectionary, and soft drink industries. Varying the ratios of amylase and glucoamylase gives different proportions of end-products, but the maximum glucose concentration is about 43%—above this crystallization becomes a problem.

The major development in the starch processing industry in recent years has been the introduction of high fructose syrups. Because fructose tastes about twice as sweet as glucose, it can be used to replace sucrose in foods and beverages and provides the same sweetness and calorific value. High fructose corn syrups (HFCS) are made from liquified starch by exhaustive hydrolysis with glucoamylase. This yields about 96% glucose. The syrup is then isomerized using the enzyme glucose isomerase. This intracellular enzyme derived from various bacteria catalyzes the reversible isomerization of xylose to xyulose. It also converts glucose to fructose. The enzyme is used in immobilized form. First the pH of glucose syrup must be raised and the material deionized (calcium inhibits glucose isomerase). After isomerization, the syrup contains about 42% fructose and 54% glucose. Higher levels of fructose can only be obtained by nonenzymatic treatment because the reversible reaction reaches an equilibrium. In the United States, HFCS have replaced sucrose in many important applications, particularly sweetening of beverages, but in the European community strict production quotas have been imposed to protect the sugar beet farming industry.

Other applications for α-amylases are various. In the textile industry, cotton is soaked in starch "size" before weaving to provide tensile strength. After weaving, the size is removed with α-amylase before the fabric is dyed. The distilling and brewing industries use amylases extensively to aid the conversion of starch to fermentable sugars during mashing. Similarly, in bread production, the flour may contain insufficient endogenous amylase. Wheats raised in the hot dry climate of North America, for example, are often deficient in amylase. Thus, when the yeast begins to raise the dough, it is restrained by lack of fermentable sugar and the process halts. Supplementation of flours with fungal amylase ensures adequate maltose levels and consistent leavening of the dough. Moreover, crust color is brought about by chemical browning of free glucose. This can be enhanced by the addition of glucoamylase. There are numerous other minor applications of amylases which together make amylases an important product in the enzyme market.

B. Proteases

From an economic point of view, the proteases are the most important industrial enzymes and together comprise some 40% of the enzyme market. Proteases can be classified on the basis of their catalytic properties into four groups: the serine proteases, metalloproteases, cysteine proteases, and aspartic proteases. This correlates well with pH optima; thus, the serine proteases have alkaline optima, the metalloproteases are optimally active around neutrality, and the cysteine and aspartic proteases have acidic optima. All of these proteases are of commercial interest.

1. Serine Proteases

Serine proteases are secreted by many bacteria, the enzymes from *B. licheniformis* and *B. amyloliquefaciens,* and various alkaliphilic bacilli being produced for industrial usage. These small enzymes (25–30 kDa) exhibit maximal activity at pH 9–11, have no metal ion requirement, and resemble the animal enzyme trypsin. The most famous alkaline serine proteases are the subtilisins. *Subtilisin Carlsberg* was prepared and crystallized in 1952. It is secreted by *B. licheniformis* (then confused with *B. subtilis*), whereas *Subtilisin Novo*, sometimes called *Subtilisin BPN*, is produced by *B. amyloliquefaciens*. The enzymes are very similar, but *Subtilisin Carlsberg* is the enzyme made famous by its inclusion in washing powders.

The concept of improving laundering efficiency by inclusion of protease is not new and was developed in 1913 by Rohm, who used pancreatic enzyme preparations. These were not very effective, but the properties of the alkaline protease from *B. licheniformis*, notably high pH optimum (pH 8–9), reasonable resistance to oxidizing agents (now improved by protein engineering), lack of metal ion requirements (detergents contain metal chelators), and reasonable temperature tolerance made it ideal as a laundering aid. Initial problems with hypersensitive reactions among workers were solved by manufacturing dust-free preparations, and enzyme-containing detergents quickly gained, and consequently retained, large market volumes particularly in Europe. "Dirt" often adheres to fabrics by being bound by protein, so protease-containing detergents not only remove obvious proteinaceous stains such as sweat and blood but also help overall laundering efficiency.

The search for improved alkali stability encouraged the isolation and screening of alkaliphilic bacilli for proteases. These bacteria grow at high pH and not at neutrality, so it was considered likely that their extracellular enzymes would be well suited to an alkaline environment. Indeed, serine proteases from these bacteria have exceptionally high pH optima (up to about 12), and they are now preferred for use in high pH detergent formulations.

These highly alkali-tolerant serine proteases also opened up new markets in the leather industry. The traditional way to remove hair from cowhides is to use sodium sulfide and slaked lime. These chemicals dissolve the hair and open up the fiber structure. Enzyme-assisted dehairing uses alkaline proteases in combination with lime, thus reducing the toxic sulfide requirement. One of the oldest applications of industrially made enzymes is in the bating process, in which dehaired skins are soaked in enzyme to make them pliable. Originally this used pancreatic extracts, but alkaline proteases quickly replaced the animal extracts.

2. Metalloproteases

These enzymes are widely distributed in microorganisms. They contain an essential metal atom, usually zinc, have a molecular weight of 35–40 kDa, and have a pH optimum near neutrality. They are generally less thermostable than the serine proteases. They are produced commercially from *B. amyloliquefaciens* and *B. stearothermophilus* ("Thermolysin") and have application in the food industry. For example, flour for biscuits and cookies should ideally be low in gluten to produce the desired dough, and metalloproteases are well suited to this. They also have uses in brewing to increase the amount of available nitrogen to the fermentation and in fish meal processing.

3. Cysteine Proteases

The thiol proteases have a cysteine residue at their active site and are typified by the plant enzyme papain. There are few microbial sources of these enzymes, and the industrial products are still manufactured from plant sources. The enzymes have optimal activity just below neutrality and have application in the brewery for removal of protein hazes, which occur upon chilling the beer and for tenderizing meat. Both applications require an enzyme with a high specificity to avoid spoilage of the product.

4. Acid Proteases

These enzymes are widely distributed in molds and yeast but are seldom found in bacteria. They have pH optima at 3–4 and, like the animal enzymes pepsin and rennin, their active sites contain an aspartate residue. The most important usage of these enzymes is in the dairy industry as rennin substitutes. The coagulation of milk to produce cheese is traditionally brought about using extracts (rennet) from the stomachs of young calves. As total calf slaughter has declined, so cheese manufacture and consumption has increased, providing a requirement for alternative sources of milk-coagulating enzymes since the cost of the animal product has soared. It is necessary that the enzyme has the correct specificity for coagulation of milk and does not lead to proteolytic degradation of the casein. Moreover, contaminating lipases can give rise to bitterness in the product and rancidity. Rennets from *Mucor miehi* and *Mucor pusillus* contain acid proteases with pH optima around 4–4.5 and produce a similar pattern of hydrolysis products from casein as calf rennet. These microbial rennets are now used extensively for the production of hard "cheddar-style" cheeses, but for specialist cheeses animal rennets are still used. The cloning of the gene for calf rennin indicates that microbial calf rennin may be available for cheese manufacture in the near future.

C. Other Enzymes

Together, proteases and starch hydrolyzing enzymes constitute about 90% of the world market but, nevertheless, some other important extracellular enzymes are manufactured on a commercial scale.

1. Cellulase

The hydrolysis of crystalline cellulose to sugar is an attractive proposition but with current enzyme preparations is uneconomic. Cellulose comprises linear chains of 1,4-β-linked glucose residues held in a crystalline form by hydrogen bonds, and it is very resistant to enzymatic hydrolysis. Fungal enzymes such as those from *Trichoderma* are mixtures of exo-acting (cellobiohydrolases) and endo-acting (endocellulases) enzymes together with β-glucosidase. These are among the most effective at attacking crystalline cellulose and it is thought that attack by exo-enzymes, which begins the hydrolysis at amorphous regions in the cellulose fibril, followed by synergistic hydrolysis by endocellulolytic enzymes is responsible for the degradation of the polysaccharide. Some bacteria such as *Clostridium thermocellum* produce endoglucanases in the form of cellulosomes (see Section I) that are effective at hydrolyzing native cellulose apparently in the absence of exo-enzymes. Many bacilli produce endocellulases that hydrolyze soluble cellulose derivatives such as carboxymethyl cellulose but are inactive on the crystalline substrate. Some of these enzymes may have application in the detergent industry as color "brighteners." Tiny fibrils are generated by wear on cotton fabrics. These reflect light and give a worn, bleached appearance to the garment. Removal of these fibrils by presoaking in endocellulase restores the original color and softens the fabric. [*See* CELLULASES.]

2. Pectinase

Molds are primarily used for production of pectinases on a commercial scale. These enzymes from aspergilli are mixtures of pectin esterases and depolymerizing enzymes and are particularly well suited to their principal application, which is the clarification and reduction in viscosity of fruit juices by the removal of pectin. The application of pectinase mash enzymes in the processing of berries, grapes, apples, and pears is standard practice today, and these enzymes are also used in the citrus industry for juice processing.

3. β-Glucanase

Mixed linkage 1,3–1,4-β-glucans are present in barley cell walls and can cause viscosity and turbidity problems in brewing. These difficulties can be alleviated by adding β-glucanase from *Bacillus* species to the mash. Barley β-glucans can also cause problems in poultry feeds where the animal is fed on cereal-rich diets. Poultry cannot digest β-glucans, resulting in wet, sticky droppings and reduction in utilization of the carbohydrate content of the feed. Addition of β-glucanase to the feed not only cures this problem but also provides more metabollically useful sugars.

4. Lipases

The natural substrates of lipases are triglycerides of long-chain fatty acids. The substrate is therefore insoluble in water and the enzymes characteristically catalyze the hydrolysis of the ester bonds at the interface between the aqueous phase and the solid substrate. Lipases are included in washing detergents for the removal of fatty stains, but major developments are expected in the oils and fats industry, where specificity of microbial lipases can be harnessed for the modification of plant oils for inclusion in margarines and cocoa.

These major enzymes and some others of minor or developing importance are listed in Table I.

III. Regulation of Synthesis

Extracellular enzymes are secreted by microorganisms to provide assimilable sources of nutrients. Early genetic and physiological studies that suggested that some enzymes, in particular proteases, were in some way involved in differentiation processes such as sporulation in bacilli, have now been shown to have been interpreted incorrectly. Indeed, the recent construction by genetic engineering techniques of strains of *B. subtilis* completely deficient in extracellular protease activity, but able to sporulate, normally has totally separated proteases (and other extracellular enzymes) from a role in sporulation. Given that these are therefore "scavenger" enzymes, it remains important for the organism to exercise some control over their synthesis. This is particularly important when the enzyme is

Table I Some Common Extracellular Enzymes, their Sources, and Uses

Enzyme	Source	Principal uses
α-Amylase	*Aspergillus oryzae* *Bacillus amyloliquefaciens* *Bacillus licheniformis*	Starch hydrolysis for sugar syrups, brewing Textiles and paper processing
β-Glucanase	*Bacillus subtilis* *Aspergillus niger*	β-glucan hydrolysis in brewing
Cellulase	*Aspergillus* sp. *Trichoderma reesei* *Penicillium* sp.	Fruit and vegetable processing Detergents
Glucoamylase	*Aspergillus niger* *Rhizopus* sp.	Glucose syrup production from liquified starch
Glucose isomerase[a]	*Actinoplanes missouriensis* *Streptomyces* sp.	Isomerization of glucose into high fructose syrups
Lactase	*Saccharomyces* sp. *Kluyveromyces marxianus*	Hydrolysis of lactose in milk and whey
Lipase	*Aspergillus* sp. *Mucor* sp. *Rhizopus* sp.	Cheese and butter flavor modification; fat and oil processing
Pectinase	*Aspergillus niger*	Extraction and clarification of fruit juices
Penicillin acylase	*Bacillus megaterium* *Escherichia coli*	Synthesis of 6-aminopenicillanic acid for manufacture of semisynthetic antibiotics
Protease (acid)	*Endothea parasitica* *Mucor miehei*	Cheese manufacture
Protease (alkaline)	*Bacillus licheniformis* *Alkaliphilic bacilli*	Detergent and leather industries
Protease (neutral)	*Bacillus amyloliquefaciens*	Baking and brewing
Pullulanase	*Klebsiella penumoniae* *Bacillus acidopullulyticus*	Debranching starch in sugar syrup manufacture

[a] Glucose isomerase is an intracellular enzyme.

destined for the environment for two reasons. First, the external conditions may not be conducive to enzymatic activity due to extremes of pH or presence of inhibitory or inactivating ions. It would therefore be wasteful to continue to secrete enzyme. Second, even if the conditions are appropriate for the enzyme, the products from the enzymatic degradation of the macromolecular substrate might be rapidly assimilated by large numbers of competing microorganisms. It might therefore be expected that sophisticated regulatory systems will have evolved for the control of synthesis of these enzymes. Such systems do indeed exist and can be grouped into three classes.

A. Induction

Extracellular enzymes are commonly secreted at a basal level. If they encounter substrate, the material is hydrolyzed and the low-molecular weight products are released and diffuse. In the absence of competing microorganisms, product will accumulate to a threshold concentration and will be transported into the cell. Intracellular product then induces further synthesis of the enzyme. Thus, the microorganism can monitor the level of *product* in the environment and regulate synthesis of the enzyme accordingly. Many amylases, cellulases, β-glucanases, and other extracellular enzymes are regulated in this way. Inducing sugars generally range from disaccharides to tetrasaccharides and sometimes higher oligosaccharides. In the case of endocellulase (carboxymethyl cellulase) from several bacilli, glucose is often the inducer, which is unusual given its role in catabolite repression (see later). Not all extracellular enzymes are inducible in this way, and the high levels of amylase and protease synthesized by *B. amyloliquefaciens* and *B. licheniformis* are constitutive.

The mechanisms of induction are now being elucidated in several bacteria. One of the best-studied systems is pullulanase and amylase synthesis in *K. pneumoniae*, which are organized as part of the maltose catabolic system. The two enzymes reduce starch to short oligosaccharides that are transported via the periplasm (where further hydrolysis can occur) to the cytoplasm, where they are catabolized by maltodextrin phosphorylase and amylomaltase. All these enzymes and the transport proteins are governed by the mechanisms of classical operon control, in this case a positively controlled system. The genes of the maltose "regulon" are organized in three operons. The *malA* area comprises the enzymes of maltodextrin catabolism and contains the regulatory gene *malT*. The *malB* locus contains the genes for the transport proteins, and the *malS* (amylase) and *pulA* (pullulanase) genes are situated elsewhere. MalT protein is an activator of transcription that is required for efficient transcription of maltose genes by RNA polymerase. MalT is inactive in the absence of maltose (although there is presumably some active protein that gives rise to the low basal level of expression) and active in the presence of the inducer maltose. There is no element of negative control in this operon. All promoters for maltose-regulated genes contain a consensus sequence 34 or 35 bases downstream from the start point of transcription, the *malT* box. This is the binding site for MalT protein to exert its activation of transcription, but the situation is complicated by the multiplicity of these boxes. For example, the *malP* promoter has three copies of these boxes in various locations, while the *pulA* gene has only one. The precise way in which MalT enhances transcription is not clear at present.

Not all extracellular enzymes are controlled by standard operon systems. The synthesis of extracellular levansucrase in *B. subtilis* in reponse to sucrose is governed by an attenuation process. Levansucrase is a transferase that, in hydrolyzing sucrose, releases glucose for catabolism and synthesizes a high-molecular weight fructose polymer, levan. The *sacX* gene, which may be a minor enzyme II^{scr} of the phosphotransferase sugar transport system, acts as a sucrose sensor and is phosphorylated in the presence of relatively low sucrose concentrations. In its phosphorylated form, it inactivates/inhibits a second protein, the *sacY* gene product. This is an antiterminator. When inactivated, it allows premature termination of transcription of the levansucrase gene (*sacB*), and only a short transcript, which does not include the protein coding sequence, is produced. Levansucrase synthesis is therefore prevented. In the presence of high levels of sucrose, SacX is not phosphorylated and cannot interact with SacY. Thus, the antiterminator activity of SacY is functional, and it prevents the premature termination of transcription of the *sacB* gene. A complete transcript is produced and levansucrase synthesized. [*See* PEP: CARBOHYDRATE PHOSPHOTRANSFERASE SYSTEM.]

Induction of extracellular enzymes may therefore be brought about by different molecular mechanisms. It would seem likely that operon control would be more prevalent than the attenuation system, which may have evolved in response to the unusual demands of the levansucrase enzyme. This enzyme is induced largely independent of the sucrose catabolic pathway and only when sucrose is abundant in the environment. This fits with the probable function of the levan, which is as a storage polymer or antidessicant. Thus, it is appropriate to induce the enzyme only when the sugar is plentiful and the SacX protein would be a suitable sensor of the external sucrose concentration.

B. Catabolite Repression

Virtually all extracellular enzymes are controlled by catabolite repression. That is, in the presence of glucose or some other rapidly metabolized carbon source such as mannitol, there is repression of enzyme synthesis. It must be remembered that this control is not exclusively operated by sugars—in pseudomonads, for example, succinate is a strong catabolite repressor. Catabolite repression makes for energy efficiency, because it is wasteful for a microorganism to synthesize a range of enzymes for energy production in response to a range of inducers—better to use the carbon sources in turn, starting with the most easily metabolized. Catabolite repression provides this coordination of carbon metabolism and to exert its influence must override induction control. Although virtually all microorganisms display catabolite control of metabolism, the molecular mechanisms vary. Here we will consider two examples. [*See* CATABOLITE REPRESSION.]

Expression of the maltose regulon including the pullulanase and amylase genes of *K. pneumoniae* is very strongly inhibited by glucose. Like other enterobacteria, catabolite repression in *K. aerogenes* is effected by the intracellular concentration of cyclic adenosine monophosphate (cAMP). Thus catabolite responsive operons require cAMP together with the activator protein CAP for efficient transcription. In the absence of glucose, intracellular cAMP levels are high and such operons can be induced, but in the presence of glucose cAMP is lacking and, even in the presence of inducer, transcription is inefficient. Synthesis of the maltose operon activator protein MalT is subject to catabolite control and, in the absence of glucose expression of *malT*, is two- to four-fold greater. This activator of transcription will therefore be able to operate if maltose is present in the environment. The promoters for *malT* and the transport proteins all have CAP-binding sites and are therefore activated by cAMP/CAP in the absence of glucose and presence of maltose. However, the genes for pullulanase and the maltose catabolic enzymes (*malPQ*) lack the CAP-binding sites and do not therefore come directly under catabolite control. Regulation of these enzymes must be by inducer exclusion, in which the repression of the transport proteins by glucose will prevent uptake of maltose. As a consequence, there will be insufficient cytoplasmic maltose to effect induction of pullulanase or the catabolic enzymes. The modulation of the activator protein, MalT, will also play a part. This system illustrates well the complexities of glucose and catabolite repression of extracellular enzyme synthesis.

In *Bacillus* and *Streptomyces*, cAMP is not involved in catabolite repression. Because the synthesis of most extracellular enzymes in these bacteria is repressed by glucose, some other system must be operating. In both bacteria, control is transcriptional, and some progress is now being made in our understanding of amylase synthesis in *B. subtilis* with the isolation of promoter (*amyR*) mutations that are insensitive to glucose repression of amylase synthesis. It seems likely that catabolite repression in *B. subtilis* is a negative control system, but the details are not known.

C. Temporal Regulation

In bacilli and streptomycetes, synthesis of many inducible and constitutive extracellular enzymes is repressed during exponential growth and derepressed during early stationary phase. This temporal regulation is true of the proteases and some amylases of *B. subtilis* and *B. licheniformis* but it can be strain-dependent.

Sporulation in *B. subtilis* is a differentiation process in which sets of genes are expressed in a predetermined temporal sequence. This is brought about, at least in part, by changes to the specificity of RNA polymerase. RNA polymerase is directed to specific DNA sequences in promoters by the sigma (σ) factor. As sporulation ensues, new sigma factors are synthesized that replace existing sigma factors and, thus, direct RNA polymerase to bind and transcribe specific sets of sporulation genes. The vegetative cell sigma factor, σ^A, is replaced by sporulation-specific forms soon after the end of exponential growth. An attractive theory for the derepression of protease genes was that these were transcribed by a form of RNA polymerase containing a postexponential phase sigma factor. This is now known to be incorrect; all extracellular enzyme genes from *Bacillus* characterized to date have σ^A-specific promoters.

So some other system must be responsible for the delay in the appearance of extracellular enzyme synthesis until after the end of exponential growth. Bacteria constantly sense their environment and respond to it by regulating protein synthesis in a coordinated way. One mechanism for achieving this involves two-component signal transduction systems. Briefly, a sensor protein, often but not invariably located spanning the cytoplasmic membrane, responds to changes in a specific factor (such as pH, osmolarity, or oxygen) by phosphorylating an internal portion of the protein at a conserved histidine residue using adenosine triphosphate. This histidine protein kinase sensor then transfers the phosphate to a second protein in the chain, a response regulator. In its phosphorylated form, the response regulator usually activates (but may repress) transcription of certain operons under its control. In this way, sets of genes (regulons) respond to certain environmental stimuli. Extracellular enzymes in *Bacillus* (and other bacteria such as *Staphylococcus aureus* and *Pseudomonas aeruginosa*) are controlled by such two-component systems.

Early genetic studies revealed several mutations (*pap, amyB, sacU,* etc.) that caused hyperproduction of various extracellular enzymes including amylase, β-glucanase, levansucrase, and proteases as well as affecting motility and DNA-mediated transformation. These mutations have now been attributed to changes in the DegS/DegU two-component regulatory system. DegS refers to a histidine protein kinase sensor protein (although it is not located in the membrane) that, under certain conditions (perhaps associated with entry into stationary phase) phosphorylates DegU. The latter is a response regulator that, when phosphorylated, activates the operons under its control, including the various extracellular enzymes mentioned previously. It also represses flagella synthesis when phosphorylated, thus explaining the early mutations affected in motility.

A second regulatory system that is more closely involved with sporulation controls serine protease synthesis in *B. subtilis*. The Abr protein is a repressor of several genes including some sporulation genes and *aprE,* the gene for extracellular serine protease. During exponential growth, Abr ensures repression of these important genes. As the cell enters stationary phase, SpoOA protein, which is a response regulator in a signal transduction pathway associated with the initiation of sporulation, is phosphorylated and represses *abr* expression. This, in turn, alleviates the repression of *aprE,* and serine protease is synthesized and secreted.

Several other genes that affect extracellular enzyme synthesis in *B. subtilis* have been cloned and characterized, but their exact roles in the regulation of synthesis remain unclear.

IV. Secretion

The process of the translocation of proteins across membranes in both eukaryotic and prokaryotic cells is largely in accord with the signal model. In this scheme, proteins destined for secretion are synthesized with an N-terminal extension of some 15–40 amino acids, the signal peptide. In the original model, this peptide interacted with membrane proteins, forming a pore through which the protein was transported as it was translated. This cotranslational secretion resulted in a nascent protein on the outside of the membrane, which was processed by the removal of the signal peptide by signal peptidase. The protein then adopted its native configuration. A later refinement of the process was the discovery of the signal recognition particle (SRP), an RNA–protein complex that binds to the ribosome as the signal peptide emerges and halts further translation. This stable messenger mRNA–ribosome–SRP complex migrates to the inner surface of the membrane where it interacts with the "docking" protein. At this point, translation ensues and secretion through the membrane is effected.

The signal hypothesis was so influential that it was adopted as a universal scheme for the secretion of proteins in eukaryotes and prokaryotes. Indeed, the process has been conserved to a remarkable extent throughout evolution; however, as more systems are studied, exceptions to the original model are uncovered. Analysis of signal or leader peptide sequences from a variety of sources reveals limited homology of primary sequences but a conserved structure. Signal peptides comprise three characteristic regions. First, a positively charged N-terminal end. This seems to be important in targeting the peptide to the membrane. It is followed by a hydrophobic region of 10–15 amino acids, which is important for the translocation of the protein across the membrane. Inclusion of charged amino acid residues in this region interrupts secretion as do changes to its length. The C-terminal end of the signal peptide includes the cleavage site for signal peptidase. The universal structure of signal peptides would indicate that they could function in heterologous hosts, and this is indeed often the case, especially for closely related species. Indeed, randomly cloned fragments of DNA will often effect secretion of proteins if they code for tripartite peptides similar in structure to signal sequences.

Although the signal sequence is now firmly established as having a central role in secretion, the nature of its exact function is not so clear. A bacterial equivalent of the eukaryotic SRP has been indicated in *Escherichia coli,* but the major difference between bacteria and eukaryotes seems to be in the linking of secretion to translation and the nature of the passage through the membrane. Bacterial proteins may be secreted cotranslationally, posttranslationally, or commonly as a mixture of the two, whereas in eukaryotes cotranslational secretion seems to predominate. It is also unlikely that bacterial proteins are transported through pores in the membrane. In the loop hypothesis, the positively charged N-terminus of the signal peptide was envisaged to interact with the negatively charged inner surface of the membrane. The hydrophobic core of the peptide would then be inserted into the membrane as it was translated and would traverse the membrane, revealing the cleavage site on the external surface. The coding sequence is then passed through the membrane and the signal peptide removed by signal peptidase. Variations of this model have been proposed, but its general format has not been seriously challenged.

The trigger hypothesis was devised to explain posttranslational secretion and suggested that the signal peptide was not solely involved in targeting but encouraged the initial folding of the protein in a manner acceptable to the aqueous cytoplasm. This would involve burying apolar residues, and it is now known that this is largely achieved through the action of "chaperone" molecules such as "trigger factor" and SecB protein. At the membrane, interaction with lipids or membrane receptor proteins displaces the chaperones and encourages rapid intramolecular refolding of the protein concomitant with passage across the membrane. The protein then assumes its native configuration on the outside as the signal peptide is removed. In this system, export is posttranslational and coupled to correct folding of the molecule guided by signal peptide and chaperones. An important implication of the trigger hypothesis is that information for export is encoded not only in the signal peptide but in the folding pathway of the entire polypeptide chain. Thus, only proteins that had evolved in association with secretion could be transferred across membranes. Evidence for this had previously originated from abortive attempts to secrete cytoplasmic proteins by the fusion of signal sequences.

An export machinery in the *E. coli* membrane has been identified through the isolation of secretion-defective (Sec$^-$) mutants. *secA* and *secY* are parts of a membrane located "protein translocating ATPase," which may be responsible for mediating the actual translocation event, and *secB* is a chaperone molecule for the periplasmic maltose-binding molecule. The products of *secD* and *secE* have not been identified yet. A *secY* homolog has recently been cloned and characterized from *B. subtilis*, indicating the universal importance of this protein.

The final stage of secretion is processing to remove the signal peptide. In *E. coli*, there are two signal peptidases. Signal peptidase I is the general enzyme that processes several periplasmic and outer membrane proteins, whereas signal peptidase II, or prolipoprotein signal peptidase, seems to be specific for processing of prolipoproteins such as the outer membrane protein OmpA.

In gram-positive bacteria, the exported protein is generally thought to diffuse through the cell wall to become completely extracellular. It has been pointed out that the barrier imposed by the peptidoglycan should not be so readily dismissed and that some form of translocation process might operate to mobilize the proteins. In support of this, pulse-labeling studies have shown that some proteins are externalized by *B. subtilis* more quickly than others.

In gram-negative bacteria, the protein must pass through the outer membrane to become completely external. Some proteins such as cholera toxin and pectinases of *Erwinia chrysanthemi* pass through the periplasm and then cross the outer membrane. Other proteins such as exotoxin A of *P. aeruginosa* and hemolysin of *E. coli* do not enter the periplasm and appear to proceed through the lipid bridges ("Bayer bridges") that span the two membranes. Pullulanase is a secreted lipoprotein and is processed by lipoprotein signal peptidase. The initial stage of pullulanase secretion is via the general pathway, but passage from the periplasm through the outer membrane relies on specific pullulanase secretion genes.

Since multiple systems have evolved to cope with secretion of proteins or families of proteins, it might be expected that secretion of heterologous proteins by *E. coli* or *B. subtilis* will be impossible or, at least, inefficient. *Escherichia coli* will recognize most gram-positive signal peptides, but secretion to the external medium is rare and the proteins usually accumulate in the periplasm. *Bacillus subtilis* will process and secrete most extracellular proteins from other gram-positive bacteria such as other bacilli or staphylococci. Most gram-negative proteins and those from eukaryotes are not secreted however. In these instances, it is necessary to provide the gene with a native signal sequence to ensure secretion. For example, using a protease–gene-based secretion system, efficient secretion of human growth hormone and other products in *Bacillus* has been achieved.

V. Industrial Production

Manufacture of extracellular enzymes is done on an industrial scale, and so the processes involved must be kept relatively simple. For example, in most cases no attempt is made to purify the enzymes; they are sold as enzyme mixtures, and in many applications this is desirable. The "contaminating" amylase in a β-glucanase preparation destined for the mash tun of a brewery will catalyze useful hydrolysis of starch. Similarly, enzyme mixtures in a washing detergent will be beneficial. Where purity is necessary, additional steps are taken to purify the enzyme, and this will obviously add to the cost.

The main problems for production are to devise an economic process that complies with the strict codes for safety and production laid down by the regulatory authorities such as the Food and Drug Administration in the United States or the Health and Safety Inspectorate in the United Kingdom. In this respect, the producer organism is obviously of crucial importance. Some bacteria such as *B. amyloliquefaciens* or *B. licheniformis* automatically have clearance for production of food-grade enzymes, but other products from unlisted sources must be subject to extensive safety testing. Many enzymes are now manufactured from genetically engineered strains and this introduces additional legislative problems, although these can be fewer if the host already has clearance. Fermentation media are generally based on agricultural by-products such as corn and barley starches, soybean extracts, fish meal, corn steep liquor, and other similar materials depending on local availability and world markets. Media compositions are closely guarded secrets, but usually the fermentation is balanced empirically to provide for a rapid period of exponential growth (e.g., in an amylase fermentation, 10–20 hr) followed by a prolonged slowing of growth and extensive stationary phase, lasting perhaps 100 hr, during which the enzyme accumulates. Virtually all fermentations now use submerged culture, although some fungal enzymes are manufactured as solid substrate fermentations on moistened wheat bran in open trays. [*See* FOODS, QUALITY CONTROL.]

Asporogenous mutants of bacilli are usually employed for production purposes to give higher yields and to prevent spores of the producer strain being recovered from the finished product. At the end of the fermentation, the broth is cooled and centrifuged to remove the cells or mycelium, perhaps with some form of flocculating pretreatment. The enzyme is then concentrated by ultrafiltration or vacuum drying. The enzyme can then be filtered, preservatives added, and the material standardized prior to being packaged as a liquid concentrate. Alternatively, after filtration, extending agents are added, it is spray-dried, and it is packaged as a powder after standardization. In the case of proteases, the powder is pearled, or marumized, into a dust-free form by covering it in a waxy coating in the form of small beads.

VI. The Future

Recent developments in biotechnology have encouraged new enzyme technologies, but this is still an industry in which the available products are far more numerous than the market can bear. Nevertheless, new technologies based on enzymes are the subjects of intensive research in biotechnology companies and several trends are evident. [*See* FOOD BIOTECHNOLOGY.]

Most applications currently involve hydrolysis of macromolecules, but the industrial use of enzymes for biosynthetic purposes holds great promise. Enzyme applications in medicine include the production of several drugs—for example, penicillins and cortisone for arthritis treatment. The cost of the latter has decreased 200-fold to less than $1/g using an enzyme-based manufacturing process. A lipase for the biosynthesis of glycolipid surfactants from sugar derivatives and fatty acids has also been developed. Moreover, the high specificity of enzymes offers great advantage over chemical processes for the synthesis of speciality amino acids in specific steroismeric form.

Enzymes will probably play a key role in the introduction of environmentally friendly processes that form cleaner products in a much milder, efficient, and economical way than traditional chemical processes based on large amounts of noxious solvents, acids, alkali, or other corrosive agents. For example, chlorine is normally required to pretreat wood pulp before it is bleached with chlorine dioxide to make quality white paper. The pretreatment removes the lignin from the wood, which imparts the "brown bag" color to the paper, but the liquid waste from pulping mills contains highly toxic chlorinated organic compounds. An alkaline xylanase that solubilizes the lignin so that it can be drawn off in alkaline fluids prior to "brightening" has recently been developed. This reduces chlorine levels in effluent considerably, but a slightly different approach, which uses hemicellulase enzymes for pulp treatment, is claimed to dispense with chlorine altogether. Gene technology and protein engineering will play a major part in the development of these new processes, but screening programs aimed at finding new microorganisms will continue to reveal novel and interesting extracellular enzymes.

Bibliography

Beguin, P. (1990). *Annu. Rev. Microbiol.* **44,** 219–248.

Friedman, R. B. (ed.) (1991). "Biotechnology of Amylodextrin Oligosaccharides." American Chemical Society, Washington, DC.

Gross, R., Aric, B., and Rappuoli, R. (1989). *Mol. Microbiol.* **3,** 1661–1667.

Steinmetz, M., and Aymeritch, S. (1990). The *Bacillus subtilis sac-deg* constellation: how and why? *In* "Genetics and Biotechnology of Bacilli," Vol. 3 (M. M. Zukowski, A. T. Ganesan, and J. A. Hoch, eds.), pp. 303–312. Academic Press, San Diego.

Vihenen, M., and Mäntsälä, P. (1989). *CRC Crit. Rev. Biotechnol.* **4,** 329–417.

Ward, O. P. (1983). Proteases. *In* "Microbial Enzymes and Biotechnology" (W. M. Fogarty, ed.), pp. 251–317. Applied Science, London.

Wickner, W. (1989). *Trends Biochem. Sci.* **14,** 280–283.

Wandersman, C. (1989). *Mol. Microbiol.* **3,** 1825–1831.

Epidemiologic Concepts

Craig A. Molgaard
San Diego State University

Stephanie K. Brodine
Naval Health Research Center

Glossary[1]

Agent (of disease) Factor, such as a microorganism, chemical substance, or form of radiation, whose presence, excessive presence, or (in deficiency diseases) relative absence is essential for the occurrence of a disease; a disease may have a single agent, a number of independent alternative agents (at least one of which must be present), or a complex of two or more factors whose combined presence is essential for the development of the disease

Analytic study Study designed to examine associations, commonly putative or hypothesized causal relationships; usually concerned with identifying or measuring the effects of risk factors or with the health effects of specific exposure (s); contrast descriptive study, which does not test hypotheses; common types of analytic study are cross-sectional, cohort, and case control; in an analytic study, individuals in the study population may be classified according to absence or presence (or future development) of specific disease and according to "attributes" that may influence disease occurrence (attributes may include age, race, sex, other disease(s), genetic, biochemical, and physiological characteristics, economic status, occupation, residence, and various aspects of the environment or personal behavior)

Association [Syn.: correlation, (statistical) dependence, relationship] Statistical dependence between two or more events, characteristics, or other variables; present if the probability of occurrence of an event or characteristic, or the quantity of a variable, depends on the occurrence of one or more other events, the presence of one or more other characteristics, or the quantity of one or more other variables; the association between two variables is described as positive when the occurrence of higher values of a variable is associated with the occurrence of higher values of another variable; in a negative association, the occurrence of higher values of one variable is associated with lower values of the other variable; an association may be fortuitous or may be produced by various other circumstances; the presence of an association does not necessarily imply a causal relationship; if the use of the term "association" is confined to situations in which the relationship between two variables is statistically significant, the terms "statistical association" and "statistically significant association" become tautological; however, ordinary usage is seldom so precise as this; the terms "association" and "relationship" are often used interchangeably; associations can be broadly grouped under two headings, symmetrical or noncausal and asymmetrical or causal

Attack rate Also called case rate, Cumulative incidence rate often used for particular groups, observed for limited periods and under special circumstances, as in an epidemic; the secondary attack rate is the number of cases among contacts occurring within the accepted incubation period following exposure to a primary case, in relation to the total of exposed contacts; the de-

[1] All glossary entries are from "A Dictionary of Epidemiology," (1988). (John Last, ed.). 2nd Edition. Oxford University Press, New York. By permission.

nominator may be restricted to susceptible contacts when determinable; infection rate is the incidence of manifest plus inapparent infections, which can be identified (e.g., by seroepidemiology)

Bias Deviation of results or inferences from the truth, or processes leading to such deviation; any trend in the collection, analysis, interpretation, publication, or review of data that can lead to conclusions that are systematically different from the truth. Among the ways in which deviation from the truth can occur, are the following: (1) systematic (one-sided) variation of measurements from the true values (syno: systematic error), (2) variation of statistical summary measures (means, rates, measures of association, etc.) from their true values as a result of systematic variation of measurements, other flaws in data collection, or flaws in study design or analysis, (3) deviation of inferences from the truth as a result of flaws in study design, data collection, or the analysis or interpretation of results, (4) a tendency of procedures (in study design, data collecton, analysis, interpretation, review, or publication) to yield results or conclusions that depart from the truth, (5) prejudice leading to the conscious or unconscious selection of study procedures that depart from the truth in a particular direction, or to one-sidedness in the interpretation of results; the term "bias" does not necessarily carry an imputation of prejudice or other subjective factor, such as the experimenter's desire for a particular outcome; this differs from conventional usage in which bias refers to a partisan point of view; many varieties of bias have been described

Endemic disease Constant presence of a disease or infectious agent within a given geographic area or population group; may also refer to the usual prevalence of a given disease within such an area or group

Epidemic [Grk. *epi* (upon), *demos* (people)] Occurrence in a community or region of cases of an illness, specific health-related behavior, or other health-related events clearly in excess of normal expectancy; the community or region and the time period in which the cases occur are specified precisely; the number of cases indicating the presence of an epidemic will vary according to the agent, size and type of population exposed, previous experience or lack of exposure to the disease, and time and place of occurrence; epidemicity is thus relative to usual frequency of the disease in the same area, among the specified population, at the same season of the year; a single case of a communicable disease long absent from a population or first invasion by a disease not previously recognized in that area requires immediate reporting and full-field investigation; two cases of such a disease associated in time and place may be sufficient evidence to be considered an epidemic; the word may also be used to describe outbreaks of disease in animal or bird populations

Epidemiology Study of the distribution and determinants of health-related states and events in populations and the application of this study to control health problems

There have been many definitions of epidemiology. In the past 50 years or so, the definition has broadened from the concern with communicable disease epidemics to take in all phenomena related to health in populations.

The Oxford English Dictionary (OED) gives as a definition: "That branch of medical science which treats of epidemics" and cites Parkin (1873) as a source. However, there was a London Epidemiological Society in the 1850s. The identity of the scholar who first used the word at that time has been lost. Epidemiologica appears in the title of a Spanish history of epidemics, Epidemiologica espanola, Madrid, 1802.

Epidemic is much older. The word appears in Johnson's Dictionary (1755), and OED gives a citation dated 1603. The word was, of course, used by Hippocrates.

Incidence rate Rate at which new events occur in the population; the numerator is the number of new events that occur in a defined period, and the denominator is the population at risk of experiencing the event during this period, sometimes expressed as person-time; the incidence rate most often used in public health practice is calculated by the formula (*No. of new events in specified period*/No. of persons exposed to risk during this period) 10^n; in a dynamic population, the denominator is the average size of the population, often the estimated population at the mid-period; if the period is a year, this is the annual incidence rate; this rate is an estimate of the person-time incidence rate, i.e., the rate per 10^n person-years; if the rate is low, as with many chronic diseases, it is also a good estimate of the cumulative incidence rate; in follow-up studies with no censoring, the incidence rate is calculated by dividing the number of new cases in a specified period by the initial size of the cohort of persons being followed; this is equivalent to the cumulative incidence rate during the period; if the number of new cases during a specified period is divided by the sum of the person-time units at risk for all persons during the period, the result is the person-time incidence rate

Incubation period Time interval between invasion by an infectious agent and appearance of the first sign or symptom of the disease in question; in a vector, the period between entry of the infectious agent into the vector and the time at which the vector becomes infective; i.e., transmission of the infectious agent from the vector to a fresh final host is possible (extrinsic incubation period)

Infection (Syn.: colonization) Entry and envelopment or multiplication of an infectious agent in the body of humans or animals; not synonymous with infectious disease—the result may be inapparent or manifest; the presence of living infectious agents on exterior surfaces of the body is "infestation" (e.g., pediculosis, scabies); the presence of living infectious agents upon articles of apparel or soiled articles is not infection but represents contamination of such articles

Prevalence rate (ratio) Total number of all individuals who have an attribute or disease at a particular time (or during a particular period) divided by the population at risk of having the attribute or disease at this point in time or midway through the period; a problem may arise with calculating period prevalence rates because of the difficulty of defining the most appropriate denominator

Quarantine, The 14th edition of *Control of Cummunicable Disease in Man* gives the following:

Restriction of the activities of well persons or animals who have been exposed to a case of communicable disease during its period of communicability (i.e., contacts) to prevent disease transmission during the incubation period if infection should occur.

1. *Absolute or Complete Quarantine* The limitation of freedom of movement of those exposed to a communicable disease for a period of time not longer than the longest usual incubation period of that disease, in such manner as to prevent effective contact with those not so exposed.

2. *Modified Quarantine* A selective, partial limitation of freedom of movement of contacts, commonly on the basis of known or presumed differences in susceptibility and related to the danger of disease transmission. It may be designed to meet particular situations. Examples are exclusion of children from school, exemption of immune persons from provisions applicable to susceptible persons, or restriction of military populations to the post or to quarters. It includes: personal surveillance, the practice of close medical or other supervision of contacts in order to permit prompt recognition of infection or illness but without restricting their movements; and segregation, the separation of some part of a group of persons or domestic animals from the others for special consideration, control or observation-removal of susceptible children to homes of immune persons, or establishment of a sanitary boundary to protect uninfected from infected portions of a population.

Seroepidemiology Epidemiologic study or activity based on the detection on serological testing of characteristic change in the serum level of specific antibodies; latent, subclinical infections and carrier states can thus be detected in addition to clinically overt cases

Transmission of infection Any mechanism by which an infectious agent either directly or indirectly is spread through the environment or to another person

These mechanisms are defined in *Control of Communicable Diseases in Man* as follows:

1. *Direct Transmission* Direct and essentially immediate transfer of infectious agents (other than from an arthropod in which the organism has undergone essential multiplication or development) to a receptive portal of entry through which human infection may take place. This may be by direct contact as by touching, kissing, or sexual intercourse, or by the direct projection (droplet spread) of droplet spray onto the conjunctiva or onto the mucous membranes of the nose or mouth during sneezing, coughing, spitting, singing, or talking (usually limited to a distance of about 1 m or less). It may also be by

direct exposure of susceptible tissue to an agent in soil, compost, or decaying vegetable matter in which it normally leads a saprophytic existence, (e.g., the systemic mycoses), or by the bite of a rabid animal. Transplacental transmission is another form of direct transmission.

2. *Indirect Transmission*/(a) *Vehicle-borne* Contaminated materials or objects (fomites) such as toys, handkerchiefs, soiled clothes, bedding, cooking or eating utensils, and surgical instruments or dressings (indirect contact); water, food, milk, biological products (e.g., blood, serum, plasma, tissues, or organs); or any substance serving as an intermediate means by which an infectious agent is transported and introduced into a susceptible host through a suitable portal of entry. The agent may or may not have multiplied or developed in or on the vehicle before being transmitted. (b) *Vector-borne* (i) *Mechanical:* Includes simple mechanical carriage by a crawling or flying insect through soiling of its feet or proboscis, or by passage of organisms through its gastrointestinal tract. This does not require multiplication or development of the organism. (ii) *Biological:* Propagation (multiplication), cyclic development, or a combination of these (cyclopropagative) is required before the arthropod can transmit the infective form of the agent to man. An incubation period (extrinsic) is required following infection before the arthropod becomes infective. The infectious agent may be passed vertically to succeeding generations (transovarian transmission); transstadial transmission is its passage from one stage of life cycle to another, as nymph to adult. Transmission may be by saliva during biting or by regurgitation or deposition on the skin of feces or other material capable of penetrating subsequently through the bite wound or through an area of trauma from scratching or rubbing. This is transmission by an infected nonvertebrate host and must be differentiated for epidemiologic purposes from simple mechanical carriage by a vector in the role of a vehicle. An arthropod in either role is termed a "vector." (c) *Airborne* The dissemination of microbial aerosols to a suitable portal of entry, usually the respiratory tract. Microbial aerosols are suspensions in the air of particles consisting partially or wholly of microorganisms. Particles in the 1-5 range are easily drawn into the alveoli without deposition. They may remain suspended in the air for long periods of time, some retaining and others losing infectivity or virulence. Not considered as airborne are droplets and other large particles that promptly settle out.

The following are airborne and their mode of transmission is direct: (a) *Droplet nuclei:* Usually the small residues that result from evaporation of fluid from droplets emitted by an infected host (see earlier). Droplet nuclei also may be created purposely by a variety of atomizing devices, or accidentally as in microbiology laboratories or in abattoirs, rendering plants, or autopsy rooms. They usually remain suspended in the air for long periods of time. (b) *Dust:* The small particles of widely varying size that may arise from soil (e.g., fungus spores separated from dry soil by wind or mechanical agitation), clothers, bedding, or contaminated floors.

Vector-borne infection Several classes of infections, each with epidemiologic features that are determined by the interaction between the infectious agent and the human host on the one hand and the vector on the other; therefore, environmental factors such as climatic and seasonal variations influence the epidemiologic pattern by virtue of their effects on the vector and its habits; terms used to describe specific features of these infections are *biological transmission*, transmission of the infectious agent to susceptible host by bite of blood-feeding (arthropod) vector as in malaria, or by other inoculation, as in Schistosoma infection; *Extrinsic incubation period*, time necessary after acquisition of infection by the (arthropod) vector for the infectious agent to multiply or develop sufficiently so that it can be transmitted by the vector to a vertebrate host; *hibernation*, a possible mechanism by which the infected vector survives adverse cold weather by becoming dormant; *inapparent infection*, response to infection without developing overt signs of illness. If this is accompanied by viremia or bacteremia in a high proportion of infected animals or persons, the receptor species is well suited as an epidemiologically important host in the transmission cycle; *mechanical transmission*, transport of the infectious agent between hosts by arthropod vectors with contaminated mouthparts, antennae, or limbs. There is no multiplication of the infectious agent in the vector; *overwintering*, persistence of the infectious microorganism in the vector for extended periods, such as the cooler winter months, during which the vector has no opportunity to be reinfected or to infect a vertebrate host. Overwintering is an important concept in the epidemiology of vector-borne diseases since the annual recrudescence of viral activity after periods (winter, dry season) adverse to continual transmission depends upon a mechanism for local survival in the local winter reservoir. Because overwinter survival may in turn depend upon the level of activity of the microorganism during the preceding summer-fall, outbreaks sometimes occur for two or more successive years; and *transovarial infection (transmission)*, of the infectious microorganism from the affected female arthropod to her progeny.

THE RELATIONSHIP between microbiology and epidemiology is both venerable and modern. It is venerable in the sense that the key definition of epidemiology—the study of the distribution and determinants of disease at the population level—can clearly be traced to the Henle–Koch postulates. It is modern in the sense that current concern with controlling AIDS and other newly identified retroviral diseases is heavily dependent on the joint research efforts and knowledge bases of microbiologists and epidemiologists.

The Henle–Koch and Evans postulates have often been used to evaluate the causal relationship of a new infectious agent to a clinical disease with which it is apparently associated. The Henle-Koch postulates first appeared in 1840 and are usually attributed to a professor of anatomy in Zurich named Jakob Henle. The postulates were further elaborated by Robert Koch, a student of Henle.

The three basic concepts of the Henle–Koch postulates are summarized in Table I. When all three conditions were satisfied, the belief was that the occurrence of the parasite with the disease could no longer be accidental but, rather, must be the cause of the disease. At the time, diseases such as anthrax, tuberculosis, and tetanus were thought to be clear examples of the Henle–Koch postulates.

During the last century, various exceptions to the postulates were noted. The most difficult was the problem of producing the disease anew in an experimental host. In general, although seldom recommended as rigid criteria of causation, the Henle-Koch postulates served as convenient intellectual guideposts to microbiologically based epidemiologic reasoning and inference during early decades of the twentieth century. These have been further refined by Evans (Table II) in the face of continued expansion and elaboration of microbiological and biomedical knowledge.

Table I Henle–Koch Postulates

I. The parasite occurs in every case of the disease in question and under circumstances that can account for pathological changes
II. It occurs in no other disease as a fortuitous and nonpathogenic parasite.
III. After being fully isolated from the body and repeatedly grown in pure culture, it can induce the disease over.

I. Historical Antecedents to Modern Epidemiology

Modern epidemiology is preeminently a science of the twentieth century. The *American Journal of Hygiene* was first published in 1921. In 1938 the format was changed to specifically include a section on epidemiology, biostatistics, and general topics. The practitioners of modern epidemiology are varied, because the field itself is dynamic and multifaceted in its use of any and all weapons in the battle against disease. Physicians, sociologists, anthropologists, psychologists, economists, demographers, statisticians, computer scientists, nurses, microbiologists, sanitarians, and many others all combine their skills and knowledge in an effort to maintain human health at its optimal level.

However, many of the basic concepts relating to the study of disease at the population level emerged piecemeal through the centuries. It is generally believed that Hippocrates used the words "epidemeion" (epidemic) and "endemeion" (endemic) at the school of Cos over 2400 years before the present era. Other epidemiologic concepts attributed to Hippocrates (usually from his work *Airs,*

Table II Evans Postulates

1. Prevalence of the disease should be significantly higher in those exposed to the hypothesized cause than in controls not so exposed.
2. Exposure to the hypothesized cause should be more frequent among those with the disease than in controls without the disease—when all other risk factors are held constant.
3. Incidence of the disease should be significantly higher in those exposed to the hypothesized cause than in those not so exposed, as shown by prospective studies.
4. The disease should follow exposure to the hypothesized causative agent with a distribution of incubation periods on a bell-shaped curve.
5. A spectrum of host responses should follow exposure to the hypothesized agent along a logical biological gradient from mild to severe.
6. A measurable host response following exposure to the hypothesized cause should have a high probability of appearing in those lacking this before exposure (e.g., antibody, cancer cells) or should increase in magnitude if present before exposure. This response pattern should occur infrequently in persons not so exposed.
7. Experimental reproduction of the disease should occur more frequently in animals or humans appropriately exposed to the hypothesized cause than in those not so exposed; this exposure may be deliberate in volunteers, experimentally induced in the laboratory, or represent a regulation of natural exposure.
8. Elimination or modification of the hypothesized cause should decrease the incidence of the disease (e.g., attenuation of a virus, removal of tar from cigarettes).
9. Prevention or modification of the host's response on exposure to the hypothesized cause should decrease or eliminate the disease (e.g., immunization, drugs to lower cholesterol, specific lymphocyte transfer factor in cancer).
10. All of the relationships and findings should make biological and epidemiologic sense.

Waters, Places) are the following: the notion that disease is distributed in terms of time, space, and the people affected by it; distribution of disease by age classes, and the influence of climate, body build, and behavioral habits on the distribution of disease.

Numerous other examples illustrate what could be called "folk epidemiology" from essentially preliterate societies. Among the Celts of prehistoric Europe, a number of saga cycles concerned the Celtic folk hero Cu Chulainn. The most important of these is Tain Bo Cualnge (The Cattle-Raid of Cooley). Tain Bo Cualnge is the story of the theft of a prized bull owned by Conchobar mac Nessa, King of Ulster, by Medb, the queen of Connaught. When, during the course of the story, the Ulster men are stricken down by a mysterious illness they are defended by their champion Cu Chulainn, who being of a different race than those from Ulster is immune to the disease. This clearly shows that the epidemiologic concept of immunity is of ancient lineage.

This charming story from the dawn of recorded history also touches upon a major factor in the early development of epidemiology. This was the need for societies' armed forces, even those from tribal bands or clans from preliterate times, to remain free of disease. This concept is known as military epidemiology. In the United States, this discipline emerged with the formation of the Army Epidemiological Board in 1940, which evolved into the Armed Forces Epidemiological Board by 1948. This group of civilian infectious disease specialists was brought together to give advice to the military on a multitude of infectious disease threats, ranging from rheumatic fever from steptococcal infections in recruit camps to tropical diseases such as malaria and typhoid fever experienced by Allied forces.

Military epidemiology has served as an impetus to the evolution and integration of epidemiologic concepts into a formal scientific discipline based on the need of societies' leaders to maintain a standing, disease-free army and navy. Failure to do so in the past has often resulted in the most dire consequences for nations both geopolitically and demographically. An example is the "plague of Athens" during the Peloponnesian War against Sparta and her allies, which had a great deal to do with Athens defeat. Another example is Napoleon's decline and defeat during his bid to dominate Europe in the 1800s, which began with the destruction of his Grand Armee in Russia. Often attributed to the cold and snow of the Russian winter, it is now known that "General Winter" had far less to do with the destruction of France's military prowess than did a devastating epidemic of typhus. Similarly, in the 1300s a Tartar army besieged a Genoese trading colony on the Black Sea coast. The siege did not progress well, with plague eventually breaking out among the Tartars in their camps. With a fine sense of the epidemiologic concept of contagion, the Tartar commander then began using his siege machines to catapult the diseased bodies of his army casualties over the city wall in an attempt to spread the plague to the Italian garrison. The stratagem was successful, but far more deadly than anticipated, as the return of part of the infected garrison to Genoa was a major route by which the Black Death entered Italy and then Europe. Other examples also exist, but the point is made. The need for a healthy military has been a driving impetus to the development of modern epidemiology in Western nations during the last century. [*See* BIOLOGICAL WARFARE.]

II. Early Theoretical Controversies

It has been said that epidemiology is a method in search of a theory. Others have argued that it is an art in search of a method. Finally, George Comstock has suggested that the art of epidemiology is to draw sensible conclusions from imperfect data. Yet its early theoretical development, although based on anecdotes and simple observations in many ways, was closely related to two major historical events—the industrial and the French revolutions. The theoretical development per se revolved about the debate between the contagionists and the miasmatists.

Prior to 1874 and the development of the germ theory of disease, the theory of miasma was dominant. This held that when the air was of bad quality, having something to do with decaying organic matter, breathing it resulted in illness. Malaria (or "bad air" was a classic example of a disease long-attributed to miasmata. The theory held that once an individual was affected the disease could spread to other susceptibles. Another classic example of the use of the theory of miasma was William Farr's explanation of cholera. As an aside, a related theory, held by Virchow, Villermé, and Alison among others, was that poverty and social conditions in general somehow generated disease. This variant of the miasma theory, for better or worse, still has many proponents in the modern field of social and behavioral epidemiology. The problem with the "social conditions" variant of the miasma theory is that, like similar theories from other fields and disciplines, it is generally true. It therefore explains everything in general and nothing in specific, and it is very hard to introduce as a public health intervention in specific environments.

The impact of the French Revolution on the development of public health and epidemiology has received less notice than that of the Industrial Revolution. The impact of industrialization and exploitation of the working class first came to attention with the work of Villermé in the textile factories of 1820s France, while William Farr of England also described excesses of mortality in the poorer, laboring classes. But the precedent for this attention was the reorganization of the state incumbent in the French Revolution. The community per se was finally given consideration in terms of allocation of resources and professional expertise vis-à-vis the king and the nobility. A concomitant creation and expansion of a modern road network in Napoleonic Europe allowed a marked increase in communication and transportation of foodstuffs, which was further enhanced by the development of a railroad network. In summary, at approximately the same time as populist and population-based social and civil reforms of the type carried by the Napoleonic Code were being implemented in Europe, skilled public health workers in England (Farr), France (Villermé), Denmark (Panum), and Germany (Franck) were carrying out research, proposing reforms, and training students whose concern was improvement of the health of the general population. In addition, the frequency of famines was markedly reduced as the ability to transport foodstuffs within national boundaries was improved in the early half of the nineteenth century. When breakdown occurred in the transportation of food in the face of famine potential, as in Ireland during the potato famine, it was often the result of nationalistic machinations or specific programs rather than inefficiency in the system of dispersal.

Once again, however, we need to be clear on the driving mechanism of militarism on these changes. Napoleon built modern roads so that he could move his armies and couriers quickly through Europe to control it. Well-fed, disease-free, demographically expanding populations were a necessity for the large, standing armies in favor in Europe from 1815 to 1945. Yet the benefits of military epidemiology were not the sole monopoly of the armies. The work of Lind, Cook, and Blane on scurvy among the sailors of the English Navy was crucial in producing scurvy-free British naval crews (issued limes and other fruits as part of their daily rations to prevent the disease—hence, "limeys") who were more effective and efficient on long-distance cruises than had previously been the case. Similarly, the means for effective control of beriberi were discovered by Takiki at the Tokyo Naval Hospital in the 1880s, resulting in large-scale dietary reforms (e.g., more fresh meat and vegetables; barley substitued for rice at some meals) instituted by the Japanese Admiralty. The development of a modern effective Japanese fleet was partly the results of such activities.

III. The Modern Era

The transition to modern epidemiology can be viewed as the result of the growth and maturation of three separate fields: medicine, statistics, and computer science. In the flow and interchange of ideas and technology among the three areas, the modern approach to combatting disease at the population level solidified during the 1950s and 1960s. The global eradication of smallpox by the 1970s—perhaps an event of equal magnitude with human conquest of space—signaled the tremendous power of epidemiologic concepts developed by pioneers such as John Snow and Wade Hampton Frost when supported by international expertise from these three areas. [*See* SMALLPOX.]

The evolution of medicine to include an active specialty area of preventive and/or community medicine, which usually includes epidemiology as one emphasis, followed a path of internal reform and regulation. Acceptance of the doctrine of the germ theory of disease was followed by the Rockefeller Foundation's Flexner Report and the creation of the model medical school, with a new uniformity and standardization of training at the beginning of this century. As technological improvements continued in the first half of the twentieth century, medical knowledge and specialization increased. Preventive or community medicine departments emerged as part and parcel of this dual process of regulation and specialization of the medical profession. Simultaneously, continued and expanded involvement of the U.S. government as witnessed by the Centers for Disease Control and National Institutes of Health during the 1950s and 1960s resulted in increased emphasis on research on chronic diseases as the population of the United States and other industrialized countries aged. The classic example of this process was the series of unique studies coming out of Framingham, Massachusetts, devoted to cardiovascular epidemiology during the 1970s and 1980s. Such studies in turn led to the concept of the "public health burden" of a specific disease, that is, the economic cost of health care delivery and service for victims of specific chronic diseases. As the skyrocketing economic costs of diseases such as Alzheimer's disease and AIDS were projected and discussed, the importance of prevention in combatting disease began to garner attention from both policy-makers and the public. With prevention and preventive medicine both popular and economically feasible, epidemiological risk assessment and intervention at the population level reached its modern conception: a dual focus on biology and behavior. Perhaps the classic example of this orientation has been that of Pekka Puska and his Finnish collaborators involved in the North Karelia Heart Disease Project. The advent of retroviral epidemiology has continued to emphasize this dual approach in the battle against AIDS, and it has also been witnessed in the national behavioral epidemiology surveys mounted in the United States by the Centers for Disease Control in the late 1980s. [*See* ACQUIRED IMMUNODEFICIENCY SYNDROME (AIDS).]

Advances in statistical theory and methodology were less spectacular but just as profound. The key role of the Rothhamsted Experimental Station in England in the development of modern statistics needs to be stressed. There, beginning around 1918 in what was to be a "temporary" 6-month job, R. A. Fisher revolutionized the scientific application of mathematics while carrying out agricultural research on hybrid seeds. Concepts of randomization, trial, Latin square, split-plot design, incomplete blocks, and many others became firmly embedded into the methods of experimental design, while decision rules using the theories of statistical inference and hypothesis testing (created by Fisher, Karl and Egon Pearson, and Jerzy Neyman) were applied to the experimental environment. This history was recently celebrated at the centenary of R. A. Fisher's birth by the British Biometry Society at Rothhamsted in 1990. With a keynote address by Frank Yates, the immeasurable contributions of Fisher and his colleagues were noted at this meeting. This fine tradition of English biostatistics was carried on and expanded by Peter Armitage in the area of clinical trials and Sir David Cox in the area of survival analysis as well as in other areas by their numerous colleagues. On the "other side of the pond," a similar research environment in the area of agricultural experimentation, developed at Iowa State University, also expanded to meet other applied needs in medicine and public health. The American tradition of biostatistics was associated with George Snedecor at Iowa State and Jerzy Neyman when he continued his movement from Poland to London to the University of California at Berkeley. There Neyman influenced Chin Long Chiang, Elizabeth Scott, R. A. Yershalmy, Steve Selvin, and their students in the tradition of modern biostatistics. Other strong traditions of American biostatistics should also be noted: The University of North Carolina at Chapel Hill, Standford University, Iowa State University, and the University of Washington being especially important at this time. [*See* STATISTICAL METHODS FOR MICROBIOLOGY.]

The continuing interface between statistics and medicine can also be illustrated by a little-known anecdote. When R. A. Fisher retired from Cambridge, he moved to Melbourne, Australia, where he continued his research interests at the University. There in the late 1950s and early 1960s Fisher, who was both a geneticist and a statistician, contributed (along with Leonard Kurland) to the development of a slow virus explanation of a disease known to New Guinea tribesmen as Kuru. Kuru was eventually discovered to be transmitted by ritual endocannibalism and was the first "slow virus" with a very long latency period to be identified. Carleton Gadjusek received the Nobel Prize for his extensive and brilliant work on the epidemiology of this disease. Research in slow viral epidemiology has continued to be an important focus since that time, with important work being carried out on Jakob–Creutzfeldt disease and scrapie. Recent work on mad cow disease in England is also part of this tradition.

The final thread of the modern era of epidemiology is that of computer science. The computer revolution in the 1960s and 1970s allowed the collection, maintenance, and analysis of very large data sets from large human populations. However, this required a mainframe and support staff until two crucial events occurred: The first was the development and manufacture of inexpensive and relatively quick micro, or personal, computers by a host of manufacturers in the United States, Japan, and elsewhere; the second was the development of software packages that were available and inexpensive for the individual researcher. Such packages often would consist of a series of programs allowing the individual investigator to enter, clean, and analyze small to mid-size data sets in a convenient and fast fashion. Here the interaction between theoretical biostatistics and computer scientists was crucial in developing the potential of computer hardware with user-friendly software for analytic purposes. Respected software packages in this regard, often geared for both mainframe and personal computer applications, include BMDP, SPSSX, SAS, SYSTAT, EPISTAT, EPILOGUE, and many others that continue to appear on the research horizon with amazing rapidity. As an example of this expansion, *Epidemiology Monitor* now carries a section on computer software.

IV. Current Issues

A. Emerging Viral Diseases

Viruses are an ever-changing and continually evolving human parasite. As a result of micro- or macroenvironmental disruptions that increase or change the nature of human contact with animals or vectors of old viruses, "new" viral diseases occur. Other truly new viruses have emerged in animals as a result of single-point mutations or genetic recombination between viruses. [*See* VIRUSES, EMERGING.]

Some viruses, such as influenza A and B, are difficult to control because they continually evolve. These antigenically changed mutant viruses are new viruses in a sense, but they cause the same clinical disease known for decades. However, the continual evolution of such a virus means that individuals have no or little chance to develop immunity. [*See* INFLUENZA.]

Poliomyelitis can be considered an example of a previous extant virus causing only sporadic disease that caused an epidemic outbreak of disease as a result of changed environmental conditions. Following the great sanitary reforms of the last half of the nineteenth century and the first half of the twentieth century, infection with poliomyelitis was moved from infants, where it was often subclinical, to older children and adults with an associated paralysis. The development of polio vaccines and polio control in many countries was one of the greatest victories in the modern era for public health and epidemiology.

Often, however, new outbreaks occur when human-made environmental changes put humans in closer contact with infected animals or increase the population of viral vectors. The syndrome of hemorrhagic fever is an example, whether caused by arthropod-borne viruses (dengue and yellow fevers) or rodent-borne bunyaviruses (Korean hemorrhagic fever) and arenaviruses (Lassa, Argentine, and Bolivian hemorrhagic fevers). For dengue fever, the environmental change involved increased reproduction of the mosquito vector *Aedes aegypti* or, in the United States, importation of a new vector from Asia known as *Aedes albopictus*. Alteration of rodent ecology in corn and rice planting and harvesting by humans is thought to be involved with Korean hemorrhagic fever and Argentine and Bolivian hemorrhagic fevers.

In terms of the creation of new viruses, these are relatively uncommon. Point mutations of sites on viral proteins can both increase virulence and change transmissibility. Influenza viruses are especially known for this, as are parvoviruses. Genetic reassortment has been known to be responsible for the pandemic influenza viruses of 1957 and 1968 (involving human and avian influenza viruses). Intramolecular recombination has been noted for Western equine encephalitis, which apparently came about from the recombination of Eastern equine encephalitis virus and a Sindbis-like virus.

The extreme mutability of the HIV virus has been well documented. Within a single individual there are rapid and extensive genetic mutations, resulting in several coexistant "strains." The real danger, of course, is that highly mutable viruses such as HIV may prove to be like influenza—difficult to control through vaccination.

B. HTLV-I

Human T-cell leukemia virus (HTLV-I) has been identified as the etiologic agent for adult T-cell leukemia/lymphoma as well as a chronic demyelinating neurologic condition known as HTLV-I associated myelopathy or tropical spastic paraparesis. Epidemiologic research on such a retrovirus is a special priority because HTLV-I can be transmitted sexually, parenterally, and perinatally and is closely related to HIV. Whatever we can learn about HTLV-I in terms of its clinical outcomes and course, mode of transmission, pathology, and risk factors will assist in prevention of HTLV-I-associated disorders as well as advance our knowledge base for HIV.

Serological surveys have documented high seropositivity rates of HTLV-I in southwestern Japan, the Caribbean, parts of Central and South America, and Africa. Okinawa has one of the highest seroprevalence rates of HTLV-I in the world, with approximately 15% of the population infected with the virus.

Because Okinawa is the only identified hyperendemic area for HTLV-I in which U.S. military personnel are stationed for extended periods of time, the U.S. Navy carried out an epidemiologic study on Okinawa. In essence, this study was a form of natural experiment, in which a population that was largely negative in terms of exposure to a particular virus (U.S. Marines) was moved in large numbers to live with a population that was heavily afflicted with the virus (the native population of Okinawa). As such, the Okinawa study was crucial in terms of offering both an epidemiologic natural experiment and a strategic research focus for retroviral epidemiology.

Among 5,255 active duty U.S. Marines on permanent tour in Okinawa, Japan, screened for human T-cell leukemia/lymphoma virus type I (HTLV-I) seropositivity, 3 (0.06%) were confirmed by Western blot analysis to have core and envelope reactivity. All three seropositive individuals had a history of prolonged sexual contact with Okinawan women, and two of the three individuals were married to seropositive Okinawan wives. Two gave a prior history of gonorrhea, while all three were negative for syphilis (MHA-TP) and hepatis B. No other risk factors associated with HTLV-I seropositivity in the United States were identified. A banked sample from one individual, obtained 8 months after initial sexual relations with his HTLV-I seropositive Okinawan spouse and 20 months before being retested in the survey, showed a pattern suggesting seroconversion.

This study suggests that female to male heterosexual transmission of HTLV-I may occur. The low rate of seropositivity among U.S. Marines may be due to (1) relative inefficiency of female to male heterosexual transmission and/or (2) a lack of significant sexual contact between U.S. Marines and infected Okinawan women. [*See* SEXUALLY TRANSMITTED DISEASES.]

As usual with epidemiologic research, the attempt to answer a few simple questions with a well-designed study generates more questions, usually complicated. Epidemiologic concepts and methodology are powerful scientific tools for just this reason—The relationship among the original research question, the research finding, and the modified research question based on such findings is quick and challenging. It is also why the discipline of epidemiology is fascinating for its adherents.

Bibliography

Armitage, P. (1971). "Statistical Methods in Medical Research." Blackwell, Oxford.

Benenson, A. S. (ed.) (1990). "Control of Communicable Diseases in Man," 15th ed. American Public Health Association, Washington, D.C.

Bone, C. M., Molgaard, C. A., Helmkamp, J. C., and Golbeck, A. L. (1988). Are nuclear ships safer than conventionally powered ships: A comparison of health outcomes among occupational cohorts. *J. Environ. Health* **50,** 277–281.

Brodine, S. K., Oldfield, E. C., Corwin, A. L., Thomas, R. J., Ryan, A. B., Holmberg, J., Molgaard, C. A., Golbeck, A. L., Ryden, L. A., Benenson, A. S., and Blattner, W. A. (1989). Seroprevalence of HTLV-I among U.S. Marines stationed in a hyperendemic area. *Proceedings of the V International Conference on AIDS,* p. 145. The Scientific and Social Challenge, Montreal, Quebec, Canada (abstract). [Also in *J. AIDS* (1992). **5,** 158–162.

Buck, C., Llopis, A., Najera, E., and Terris, M. (1988). "The Challenge of Epidemiology: Issues and Selected Readings." Pan American Health Organization, Washington, D.C.

Chiang, C. L. (1980). "Introduction to Stochastic Processes in Biostatistics" Robert E. Krieger Publishing Company Incorporated, Huntington, New York.

Cox, D. R., and Hinkley, D. V. (1974). "Theoretical Statistics." Chapman and Hall, New York.

Evans, A. S. (1976). Causation and disease: The Henle–Koch postulates revisited. *Yale J. Biol. Med.* **49,** 175–195.

Fleiss, J. L. (1981). "Statistical Methods for Rates and Proportions," 2nd ed. Wiley, New York.

Kilbourne, E. A. (1990). New Viral Diseases: A Real and Potential Problem Without Boundaries. *JAMA* **264,** 68–70.

Kleinbaum, D. G., Kupper, L. L., and Morgenstern, H. (1982). "Epidemiology—Principles and Quantitative Methods." Lifetime Learning Publications, Belmont, California.

Kurland, L., and Molgaard, C. A. (1981). The patient record in epidemiology. *Sci. Am.* **245,** 54–63.

Last, J. M. (ed.) (1991). "Maxcy-Rosenau public health and preventive medicine," 12th ed. Appleton-Century-Crofts, New York.

Last, J. M. (ed.) (1988). "A Dictionary of Epidemiology" 2nd ed. Oxford University Press, Oxford, United Kingdom.

Mausner, J. S., and Kramer, S. (1984). "Epidemiology: An Introductory Text," 2nd ed. Saunders, Philadelphia.

Molgaard, C. A., Nakamura, C., Hovell, M., and Elder, J. P. (1988). Assessing alcoholism as a risk factor for acquired immunodeficiency syndrome (AIDS). *Soc. Sci. Med.* **17,** 1147–1152.

Molgaard, C. A., Poikolainen, K., Elder, J. P., Nissinen, A., Pekkanen, J., Golbeck, A. L., deMoor, C., Lahtela, K., and Puska, P. (1991). Depression late after combat: A follow-up of Finnish World War Two veterans from the Seven Countries East–West Cohort. *Mil. Med.* **156,** 219–222.

Molgaard, C. A., Stanford, E. P., Morton, D. J., Ryden, L. A., Golbeck, A. L., and Schubert, K. R. (1990). The epidemiology of head trauma and neurocognitive impairment in a multiethnic population. *Neuroepidemiology* **9,** 233–242.

Sackett, D. L. (1979). Bias in analytic research. *J. Chron. Dis.* **32,** 51–63.

Schlesselman, J. J. (1982). "Case-control studies: Design, conduct, analysis." Oxford University Press, New York.

Selvin, S. (1991). "Statistical Analysis of Epidemiologic Data." Oxford University Press, New York.

Zaltchuk, R., Jenkins, D. P., Bellamy, R. F., Ingram, V. M., and Quick, C. M. (1990). "The Armed Forces Epidemiological Board—Its First Fifty Years." Office of the Surgeon General, Department of the Army, Falls Church, Virginia.

Escherichia coli and *Salmonella typhimurium*, Mutagenesis

Patricia L. Foster
Boston University

Glossary

Allele One of the alternative states of a gene at a specific genetic locus; a gene with a mutation is a mutant allele of the wild type allele

Base substitution A mutation in which one DNA base pair has been changed for another

Frameshift A mutation that shifts the translational reading frame of the DNA

Genotype The specific sequence of an organisms' genomic DNA

Missense mutation A mutation that changes the codon for one amino acid into that of another

Mutant An individual organism that carries a mutation in its genome; an allele of a gene with a mutation

Mutation An heritable change in the sequence of genomic DNA

Nonsense mutation A mutation that changes the codon for an amino acid into a translational stop codon

Phenotype The characteristics of an individual organism, usually conferred by its genotype, that are observable or measurable

Wild type A nonmutant individual; the nonmutant allele of a gene

MUTATIONS are heritable changes in the sequence of genomic DNA. In general, point mutations, involving a small number of DNA base pairs, and large rearrangements, which can involve as much as a third of the genome, proceed by different mechanisms. Mutations can be further classified by their mode of induction. Spontaneous mutations result from the normal life processes of the cell, whereas induced mutations result from DNA damage produced by environmental agents. These can usually be distinguished by their chacteristics. However, because some metabolic intermediates can damage DNA, and some mutagens may interfere with normal cell processes, the distinction between spontaneous and induced mutations may not always be mechanistically meaningful.

I. What are Mutations

It is useful to differentiate between a mutation, which is a change in DNA sequence, and a mutant, which is an individual that carries a mutation in its genome. A mutant individual is usually distinguished from a nonmutant, or wild type, individual by the possession of a distinct phenotype. An individual's phenotype is what is externally observable or measurable, whereas its genotype is the specific sequence of its genomic DNA. While it is usually true that the phenotype of a individual is an expression of its genotype, phenotypic changes can also arise from other sources of variation, although such phenotypes may not be stable or heritable. It is also true that the acquisition of a mutation does not necessarily confer an observable phenotype; an individual containing such a "silent" mutation can only be identified as mutant by demonstrating that it has a change in its DNA sequence relative to the wild type.

Mutations are rare events and, in nature, usually result in deleterious consequences for the organism. To investigate mutagenic processes in the face of these limitations, scientist have used microorganisms, such as *Escherichia coli* and *Salmonella typhimurium* and their bacteriophages, which can be grown to large populations in short periods of time. To identify the rare mutations in a large population, the organisms are genetically manipulated so that mutant cells have a survival or growth advantage over wild type cells (i.e., the mutation results a "selectable" phenotype) and the original mutant gives rise to a clone of identical descendents. Inevitably, the requirement for a selectable phenotype has imposed biases on the type of mutations that can be studied. It is only recently that technology has been developed to allow unselected mutations to be investigated, and it remains to be seen if our understanding of mutagenic processes will be changed as a result.

II. Types of Mutational Events

A. Point Mutations

Point mutations involve the substitution of one base pair for another (base substitutions) or the addition or deletion of a small number of bases (Table I). Base substitutions are further classified by whether the purine-pyrimidine orientation between the DNA strands is maintained (transitions) or inverted (transversions) relative to the original sequence. When base substitutions occur within the coding sequence of a gene, they can change the amino acid sequence of a protein (missense mutations) or create aberrant stop codons (nonsense mutations) that result in a truncated protein. Frameshift mutations are the gain or loss of $3N \pm 1$ bases (N = an integer). As the name implies, such mutations alter the reading frame of the RNA transcript, inevitably resulting in a nonfunctional protein downstream of the site of the mutation. Gains or losses of small multiples of three bases, which will not shift the reading frame, are often silent (i.e., confer no detectable phenotype) unless the amino acids involved are essential to protein function.

Table I Types of Point Mutations

Type of mutation	Sequence change
Transition	G to A; C to T A to G; T to C
Transversion	G to T; C to A A to C; T to G A to T; T to A G to C; C to G
Missense	The codon for one amino acid to the codon for a different amino acid.
Nonsense	The codon for an amino acid to one of the three stop codons: TAG, TAA, or TGA.
Frameshift	$\pm(3N \pm 1)$ bases; N, an integer

B. Rearrangements

Larger genomic rearrangements, such as deletions, insertions, duplications, and inversions, which can involve more than one gene, can have much more drastic effects on the cell than point mutations. In addition, rearrangements often proceed by different mechanisms. Whereas point mutations are dependent on DNA replication either for their creation or for their fixation, genomic rearrangements can be due to a recombination event between regions of DNA homology. Thus, a small deletion might be distinguished from a large frameshift by the fact that the former may have been the result of a recombinational event between flanking homologous sequences, whereas the latter may have been the result of slippage of the replication machinery along a series of repeated bases.

Genomic rearrangements can also be mediated by mobile genetic elements (IS elements, transposons). These elements create mutations when they insert in the bacterial genome and can also induce deletions and inversions in their vicinity. In addition, multiple elements can provide regions of homology leading to large genomic rearrangements.

This article will focus on point mutations. [*See* Transposable Elements; Recombinant DNA, Basic Procedures.]

III. Spontaneous Mutation

A. DNA Replication Errors

Replication errors constitute a major category of spontaneous mutations. *E. coli* replicates its DNA with a fidelity of 10^{-10} to 10^{-9} mistakes per base pair. With a genome of 5×10^6 base pairs, this represents a mutation rate of roughly 10^{-3} per cell per generation. This extraordinary accuracy is achieved by a variety of error prevention and correction mechanisms.

During replication, DNA polymerase inserts the correctly paired base opposite the template base with an error rate of 10^{-4} to 10^{-5}. If a mistake is made, the mispaired terminus is degraded by an associated 3′ to 5′ exonuclease and a new base inserted. This process, called proofreading, decreases the frequency of mispairing an additional 10- to 200-fold. Another type of replication error is believed to be the source of frameshifts and small insertions and deletions. During synthesis past a run of repeated bases, the template and newly synthesized strand can become misaligned, throwing the replication complex out of register. Depending on which way the misalignment occurs, this mechanism can result in the addition or deletion of bases.

Mistakes made during replication, if uncorrected, will result in permanent sequence changes after the next round of DNA synthesis. A further 10- to 1000-fold increase in replication accuracy is achieved by enzyme systems that either prevent mispairing or recognize and correct mispairs after replication. To provide a net reduction in error rate, these error-correcting pathways must have some way to distinguish the correct base from the incorrect base. For example, the product of the *mutT* gene appears to act at the replication fork to prevent the mistaken insertion of guanines opposite template adenines, thereby preventing A to C transversions. A second enzyme, the product of the *mutY* gene, acts after replication by recognizing G:A mispairs and removing the adenine, preventing G to T transversions. MutY achieves a net reduction in mutations because, after the activity of MutT, the majority of G:A mispairs in the newly replicated DNA are those that were created by misinsertion of an A opposite the template G.

The most powerful postreplication error-correction pathway is the methyl-directed mismatch repair system encoded by the *mutH, mutL, mutS,* and *uvrD* genes. In *E. coli* and *S. typhimurium,* the N6 position of adenines in GATC sequences is methylated. Methylation, however, lags behind replication about 10% of a generation, resulting in a brief period during which GATC sequences are hemimethylated. Enzymes of the methyl-directed mismatch repair system recognize mispaired bases, cut the DNA at a nearby GATC sequence, and initiate degradation and resynthesis of the unmethylated strand. Because the methylated strand serves as template, mismatches are preferentially corrected in favor of the preexisting DNA, resulting in a net error-reduction. Misalignment errors are also recognized by this enzyme system, achieving a substantial reduction in frameshift mutations.

These various error-correcting mechanisms are not essential. In *E. coli,* each is defined by one or more "mutator alleles" that dramatically increase the rate of spontaneous mutation (Table II). It is noteworthy that no antimutators that increase the accuracy of replication have been identified in *E. coli* or *S. typhimurium,* although mutants that decrease at least some replication errors have been found in bacteriophage T4. This failure suggests that the accuracy of replication in these bacteria may be close to the maximum possible consistent with the energy available for the process. [*See* DNA REPLICATION.]

B. Endogenous DNA Lesions

Spontaneous mutations can also result from endogenously occurring DNA damage. For example, cytosine deaminates spontaneously to yield uracil. If not repaired, uracil residues code as thymines, giving rise to C to T transitions. A more insidious problem is created by the deamination of 5-methyl cytosine, which yields thymine, a normal DNA base. *Escherichia coli* and *S. typhimurium* methylate the second cytosine in CCAGG sequences, and these sequences are strong hot spots for C to T transitions. DNA bases, particularly purines, also are spontaneously loss from the DNA by breakage of the glycosy-

Table II Mutator Allels

Pathway involved	Allele	Increase in mutation frequency	Major type of mutation
Proofreading	*dnaQ49*	10^3–10^4	Transversions
Proofreading	*mutD5*	10^3–10^4	Transitions, transversions, frameshifts
Replication-directed mismatch repair	*mutT*	10^3–10^4	A to C transversions
Replication-directed mismatch repair	*mutY*	$>10^2$	G to T transversions
Methyl-directed mismatch repair	*mutH,S,L*	10^2–10^3	Transitions, frameshifts

[Adapted from Foster P. L. (1991). *Meth. Enzymol.* **204,** 114–125.]

lic bond between the base and the sugar. The rate of depurination under normal conditions *in vivo* is estimated to be about 10^{-7} per base per hour.

A variety of metabolic activities may yield DNA damaging intermediates. Oxygen metabolism produces intermediates, such as hydrogen peroxide, superoxide radical, and hydroxyl radical, that are highly reactive to DNA. Because certain DNA repair enzymes are induced by anaerobic conditions, it is likely that anaerobic metabolism also results in DNA-damaging intermediates. DNA bases also may be methylated by endogenous methylation agents, such as S-adenosyl-L-methionine, independently of DNA-sequence dependent methylases.

Endogenously produced DNA lesions may themselves be mutagenic or may give rise to mutagenic products. For example, a particularly reactive site for methylation is the N7 position of guanine; methylation of this position can cause weakening of the glycosylic bond, resulting in spontaneous loss of the base or opening of the imidazole ring, yielding a methylformamidopyrimidine residue. In addition, bacteria possess a variety of repair pathways that are initiated by the recognition and enzymatic removal of a damaged base by a specific glycosylase. The resulting abasic sites may be the actual mutagenic intermediates common to a number of spontaneous and enzymatic processes.

C. Directed Mutation

It is generally believed that spontaneous mutations occur at random and without regard to their ultimate utility to the cell. Evidence has suggested, however, that when *E. coli* is subjected to nonlethal selection, the frequency of useful mutations (i.e., those that are being selected for) increases relative to the frequency of neutral mutations. Thus, the mutational process appears to be "directed" by the selective agent.

Although it is not possible at this point to give a definitive explanation for this phenomenon, the data suggest that two factors may be at work. First, the production of genetic variation appears to increase in nutritionally deprived cells. Second, the variants appear to be transient, only immortalized as mutations (i.e., heritable changes in the DNA sequence) if the cell derives some benefit from them. Thus, for example, a DNA error may occur that would normally be corrected by one of the mechanisms already discussed; however, if that error-containing DNA is transcribed and produces a useful mutant protein, the DNA might be replicated before correction, creating a mutation. The result would be a mechanism for "trial and error" whereby each variant is tested for utility before being immortalized.

IV. Induced Mutation

A. Accurate DNA Repair Pathways

Mutations are induced by a variety of exogenous DNA-damaging agents. Some, such as UV-light and oxygen, are ubiquitous, and both *E. coli* and *S. typhimurium* have evolved extensive enzyme systems to repair the DNA damage created by these agents. Other repair pathways appear to be more specific for endogenously-produced lesions, but are also active against similar DNA damage produced by exogenous mutagens. Several accurate repair pathways, and mutants that inactivate them, are summarized in Table III. [*See* DNA REPAIR BY BACTERIAL CELLS.]

B. The SOS Response

In *E. coli* and *S. typhimurium*, certain types of DNA damage result in the increased transcription of approximately 20 genes, collectively known as the SOS response. Some of these genes code for accurate DNA repair enzymes, but a small subset code for proteins that are actively involved in producing

Table III DNA Repair Pathways

Repair pathway	Substrates	Mutations
Uracil glycosylase	Uracil residues	*ung*
Abasic site repair	Abasic sites	*xth, nfo*
UvrABC excision repair	Bulky lesions	*uvrA,B,C*
Photoreactivation	Pyrimidine dimers	*phr*
Adaptive response to alkylation	O^6-Alkylguanine, O^4-alkylthymine	*ada*
Adaptive response to alkylation	N3-Alkylpurines, O2-alkylpyrimidines	*alkA*
Adaptive response to oxidation	Abasic sites, oxygen damaged bases	*soxR, nfo*

[Adapted from Foster P. L. (1991). *Meth. Enzymol.* **204**, 114–125.]

mutations. This process, also called "error-prone repair," has been extensively studied in *E. coli* and has been shown to require the products of *recA* gene and the *umuDC* operon.

RecA protein is responsible for most homologous recombination in *E. coli*. It binds single-stranded DNA and promotes strand exchange while hydrolyzing ATP; however, in the presence of an as-yet unidentified inducing signal produced when the DNA is damaged, RecA is activated to a state, called RecA*, in which it facilitates the cleavage of a number of proteins. These include LexA, the common repressor of the SOS genes; cI, the bateriophage λ repressor; and UmuD. Cleavage of LexA inactivates it, resulting in the induction of the SOS genes. Both *recA* and *lexA* are repressed by LexA, a genetic refinement that allows the system to rapidly amplify when DNA damage is present and rapidly shut-down when the inducing signal decays. Like LexA, cI is inactivated by cleavage, resulting in the induction of λ prophage if present in the cell. In contrast, cleavage of UmuD results in the production of a fragment, called UmuD*, that is active in mutagenesis.

While the genetic regulation of the SOS response is well established, the actual molecular mechanism by which mutations are produced is as yet unknown. Evidence suggests that DNA polymerase III is the replicative enzyme involved, and that it must be modified in some way to achieve "translesion synthesis." The first step of translesion synthesis—insertion of a base opposite a DNA lesion—apparently does not require any of the known SOS proteins, although mutant RecA proteins can interfere with the process. The second step—extension of replication from the mispaired terminus—requires UmuD*, UmuC, and RecA. UmuD* and C may act as accessory proteins to promote extension. RecA may "target" the lesion and assist in the formation of the active UmuD*C complex. It is also possible that one or more of these proteins may inactivate the proofreader function of DNA polymerase to allow extension after misinsertion.

As elaborate as this process appears, there still remain several SOS-associated phenomena that are not explained by this scheme outline. These are currently active areas of research. It is also not clear how widespread the phenomenon of SOS-mutagenesis is, since some species of bacteria are known to lack the SOS system and are essentially nonmutable by SOS-dependent mutagens.

C. DNA Damaging Agents

The SOS mutagenic pathway is not required for the production of mutations induced by many mutagenic agents. Prominent among these SOS-independent mutagens are base analogs, such as 5-bromouracil and 2-aminopurine, that have inherently ambiguous coding properties. Thus, for example, 5-bromouracil may be incorporated into the DNA opposite adenine and, in a subsequent round of replication, pair with guanine. It is a property of base analogs that they produce only transition mutations.

Certain chemicals act on DNA bases *in situ* to change their coding properties. Nitrous acid, for example, induces oxidative deamination of cytosine, adenine, and guanine, to create uracil, hypoxanthine, and xanthine, respectively. All but xanthine mispair to produce transitions. Of more biological relevance are alkylating agents, such as the alkylnitrosoguanidines and alkylnitrosoureas. Although these compounds form a variety of DNA-alkyl lesions *in vivo*, they have a high affinity for alkylating the exocyclic oxygens of guanine and thymine. Both O^6-alkylguanine and O^4-alkylthymine mispair and, like other base analogs, induce transition mutations.

A separate class of SOS-independent mutagens are typified by the class of acridine-based dyes, particularly alkylated acridines. These planar molecules apparently intercalate into the DNA causing helix perturbations, resulting in frameshift mutations by misalignment during replication; however, the mechanism of action of these compounds is far from determined, and some classic frameshift mutagens have been shown to induce a variety of DNA lesions. For example, it has been proposed that aberrant processing of single strand breaks in the DNA is an important mechanism for producing frameshift mutations.

In contrast to many SOS-independent mutagens, SOS-dependent mutagens characteristically produce DNA lesions that are blocks to DNA replication. Examples of such blocking lesions are pyrimidine dimers induced by UV light; bulky purine adducts produced aflatoxin B1 and benzo [a]pyrene; apurinic sites induced by heat or methylation; inter-and intrastand cross-links produced by bifunctional agents such as mitomycin C and *cis*-diamminedichloroplatinum(II) (cisplatin). The DNA replication complex apparently cannot bypass such lesions unless the SOS response is active, and SOS-aided bypass results in mutations.

Although mutagens are classified as SOS-independent or SOS-dependent, the distinction is actually a reflection of the DNA lesions induced, not the mutagen itself. Alkylating agents, for example, produce a variety of DNA lesions, some of which, such as O^6-alkylguanine, are directly mispairing, and some of which, such as abasic sites, require SOS processing to be mutagenic.

D. Mutagenic Specificity

Determining the mutagenic specificity of a mutagen provides clues to the identity of the important DNA lesions that it may induce and points to the mechanism by which the lesions are processed into mutations. The mutagenic specificity of a mutagen is described by two parameters: (1) the types of mutational events that it induces and (2) the sites in the DNA at which mutations are induced.

As previously mentioned, many SOS-independent mutagens are themselves base analogs or create base analogs *in situ;* the mutational events that are induced are readily explained by mispairing. In contrast, the mutagenic specificity of SOS-dependent mutagens is not, as yet, predictable from the structure or chemistry of the DNA lesions produced. From examining a large number of SOS-dependent mutagens, it is clear that the great majority of the mutations induced by these agents are "targeted" (i.e., they occur at the sites of DNA damage, not as a result of "error-prone" replication of undamaged bases). It is also clear that different DNA lesions induce different mutations. For example, both 4-nitroquinoline oxide and N-2-aminofluorene bind to the C8 position of guanine, but 4-nitroquinoline oxide induces predominantly G to A transitions, whereas N-2-aminofluorene induces predominantly G to T transversions. To further complicate the analysis, the SOS system itself appears to have some specificity. The mutagenic characteristics of several SOS-dependent mutagens can be explained by assuming that they induce "noninformational" lesions opposite which the modified replication complex preferentially inserts an adenine. Exceptions to this "put in an A rule," however, suggest that some lesions may be "pseudo-informational," influencing to some extent which base is inserted opposite them. It may be that abasic sites are the only truly noninformational lesions.

The spatial distribution of the mutations induced by most mutagens is strikingly nonrandom. Within a given sequence of DNA, mutagens will have characteristic "hot spots" and "cold spots." While the distribution of mutations must, at some level, reflect the distribution of DNA lesions, the underlying physical and chemical rules determining site susceptibility have not been elucidated. Local DNA sequence, secondary and tertiary DNA structure, and availability to DNA repair enzymes, may all influence the mutagenicity of a given site.

V. Systems to Analyze Mutations

To investigate mutagenic processes, mutations must be quantified and identified. Because mutations are rare events, however, in most cases only those mutations that lead to selectable phenotypes can be studied. Systems that are currently used to study mutagenic processes in *E. coli* and *S. typhimurium* employ either reversion, in which a function is restored to an already defective gene, or forward mutation, in which a function is lost by a wild type gene. To be generally useful, these systems must also embody some way to easily identify what mutations have occurred.

A. Reversion Systems

Strains of microorganisms that have lost the ability to synthesize an essential nutrient are called "auxotrophs." Wild type, nutritionally competent, strains are "prototrophs." Many auxotrophic strains of *E. coli* and *S. typhimurium* were created and characterized by scientists investigating biosynthetic pathways. Reversion of these auxotrophies, particularly for amino acids, has been widely used to assay for mutagenesis. Revertible auxotrophies are usually caused by either missense or nonsense mutations in a gene coding for a biosynthetic enzyme. Both classes of mutation can be reverted by mutations that restore a functional amino acid (true reversions) or by mutations that compensate in some other way for the original mutation (suppressor mutations). True reversions are more properly called "on site" reversions because, in many cases, the original amino acid need not be restored by the reverting mutation. In general, the class of suppressor mutations can include a large variety of mutations in the same or other genes. Suppressors of nonsense mutations, however, are usually a mutation in the anticodon of a transfer RNA that allows it to read the nonsense codon and insert a functional amino acid.

Missense tRNA suppressors also exist, although they are less common. For reasons that are still a mystery, tRNAs are particularly good targets for mutations.

The most commonly used reversion system was developed by Bruce Ames and is known as the "Ames test," this test, which monitors the reversion of several histidine auxotrophies in *S. typhimurium* is widely used to screen chemicals for potential genotoxicity and has been optimized for convenience and sensitivity. The bacteria are exposed to the agent and then plated on minimum media containing a limiting amount of histidine. All cells undergo several generations of growth, allowing DNA lesions to be processed into mutations and the mutant phenotype to be expressed. After exhaustion of the histidine, His^+ revertants grow as distinct colonies, each representing an independent mutational event. In some Ames strains, the bacterial cell wall has been genetically modified to increase its permeability, and a major DNA repair pathway has been disabled to increase the bacteria's sensitivity to DNA damage. *S. typhimurium* is naturally less mutable than *E. coli;* to detect SOS-dependent mutagens, certain strains harbor a plasmid, pKM101, which carries hyperactive analogues of the *imuDC* genes. Finally, many compounds are genotoxic only after enzymatic conversion to a more active derivative; mammalian microsomes containing the necessary enzymes can be incorporated into the Ames assay.

Charles Yanofsky devised a similar system in *E. coli* using the reversion of tryptophan auxotrophies. Because revertants caused by different mutational events can be distinguished phenotypically, the Yanofsky system allows the mutagenic specificity of a mutagen to be determined. Similar phenotypic tests for various *his* alleles used in the Ames test have also been devised. In both cases, the basis of the phenotypic analysis is that, as has already been mentioned, not all reverted enzymes have the same amino acid sequence, although each must function in its biosynthetic pathway. By finding a set of conditions that distinguishes between the different possible enzymes and by knowing the sequence of the original mutant and every possible revertant, each revertant can be assigned a specific mutational event.

Determining mutagenic specificity by phenotypic screening is laborious. With the knowledge of the DNA sequence of a gene and a detailed understanding of the biochemical activity of the protein it encodes, it is possible to create mutant alleles than can only be reverted by one mutational event. Specificity is achieved by changing a codon for an essential amino acid in the active site of an enzyme so that only one type of mutation will restore the required wild type codon. Such "tester strains" have been made using the β-lactamase gene carried on plasmid pBR322 and the *lacZ* gene, which codes for β-galactosidase. In each case, only one specific mutation will give rise to ampicillin resistance or the ability to grow on lactose, respectively. An entire set of *lacZ* alleles has been created that will detect each class of base substitutions and several kinds of frameshifts.

B. Forward Mutational Systems

To determine the overall site and event preference of a mutagen or a mutational process, a large and unbiased collection of mutations are required. In general, the number of mutational events that can revert a given mutant allele is inherently limited. Thus, all reversion systems are biased for the detection of a few specific events at a few specific sites. Less limiting, in theory, would be a forward mutational system that monitored mutation of a wild type gene. Mutations that render bacteria resistant to antibiotics, such as streptomycin, nalidixic acid, and rifampicin, although widely used, are particularly unsuitable for this purpose. The genes involved are essential, and only a few specific mutations can maintain gene function and at the same time confer resistance to the antibiotic. The most unbiased collection of mutations would be those that result in the loss of function of a nonessential gene, but it is difficult to find instances in which loss of function of a gene results in a selectable phenotype.

Several forward mutational systems have been developed using genes that code for repressors, such as the cI repressor of bacteriophage λ, the *mnt* repressor/operator of bacteria phage P22, and the *lacI* repressor in *E. coli*. However, the *lacI* system, developed by J. Miller, has provided most of the information on mutational specificity to date. The *lacI* gene codes for the repressor of the genes required to metabolize lactose. $LacI^-$ mutants, which express the *lac* operon constitutively, are selected for by their ability to grow on phenyl-β-galactoside, a lactose analog that can be metabolized but that cannot induce the operon. The *lacI* system can be used to determine mutagenic specificity in two ways. There are over 60 sites in the gene that can give rise to one or more nonsense codons by a single base substitution. Each nonsense mutation can be identified phenotypically and genetically mapped to one of the sites. Because both the nonsense codon and the codon that gave rise to it are known, the mutational event is identified. Alternatively, the DNA from LacI$^-$ mutations can be isolated, cloned, and sequenced.

C. New Technological Advances

New technological advances may solve some of the technical problems that have limited investigations of mutational processes. When dealing with a defined mutational target, colony hybridization can be used to rapidly identify the specific mutations carried by a large number of mutants. Oligonucleotides, short pieces of single-stranded DNA, that carry each mutant sequence are synthesized and labeled with radioactivity. The DNA from colonies of mutant bacteria is isolated *in situ* on a filter and "hybridized" to each of the mutant oligonucleotides. With the proper temperature and conditions, hybridization of the correct mutant sequence is favored over all others, and the mutant colony can be detected on photographic film. This technique has been successfully applied to the Ames strains that revert to histidine prototrophy by a limited number of base changes.

The polymerase chain reaction (PCR) allows defined segments of genomic DNA to be amplified by repeated rounds of DNA synthesis initiated from artificial primers. The increase in DNA is exponential, and the amplified DNA can be sequenced after simple purification. The ability to directly sequence genomic DNA obviates the need to clone a large collection of mutant DNA's before sequencing. Although the procedure is subject to the high error frequency of the polymerase used for amplification, this problem can be circumvented, if necessary, by use of newer polymerases with improved accuracy or simply by repeating the process several times.

The need to isolate individual mutants prior to analysis can also be circumvented by combining PCR with denaturant-gradient gel electrophoresis (DGGE) to identify mutated segments of DNA directly. The migration of double-stranded DNA in an electric field is abruptly reduced when it is denatured (melted). Because the concentration of denaturant at which DNA melts is exquisitely dependent on its sequence, segments of DNA with mutant sequences can be readily distinguished from each other and from the wild type on such gels. With this technique, DNA from a mixed population of mutants can be amplified by PCR and mutant segments identified by their altered migration on the denaturant gel. The DNA is then excised from the gel and sequenced. This technique may even eventually allow mutants to be identified in a population of cells never exposed to selection, producing a truly unbiased collection of mutations. [*See* POLYMERASE CHAIN REACTION (PCR).]

Bibliography

Drake, J. W. (1991). *Ann. Rev. Genet.* **25,** 125–146.

Eisenstadt, E. (1987). Analysis of Mutagenesis. *In* "*Escherichia coli* and *Salmonella typhimurium,* Cellular and Molecular Biology." (F. Neidhardt, J. L. Ingraham, K. B. Low, B. Magasanik, M. Schaechter, and H. E. Umbarger, eds.), pp. 1016–1033. American Society for Microbiology, Washington, DC.

Foster, P. L. (1991). Directed Mutation in *Escherichia coli:* Theories and Mechanisms. *In* "Organism and the Origins of Self." (A. I. Tauber, ed), pp. 213–234. Kluwer Academic Publishers, The Netherlands.

Foster, P. L. (1991). *Meth. Enzymol.* **204,** 114–125.

Loeb, L. (1986). *Ann. Rev. Genet.* **20,** 201–230.

Radman, M., and Wagner, R. E. Jr. (1986). *Ann. Rev. Genet.* **20,** 523–538.

Ripley, L. S. (1990). *Ann. Rev. Genet.* **24,** 189–214.

Walker, G. W. (1987). The SOS Response of *Escherichia coli. In* "*Escherichia coli* and *Salmonella typhimurium,* Cellular and Molecular Biology." (F. Neidhardt, J. L. Ingraham, K. B. Low, B. Magasanik, M. Schaechter, and H. E. Umbarger, eds.), pp. 1346–1357. American Society for Microbiology, Washington, DC.

Escherichia coli, General Biology

Moselio Schaechter
Tufts University School of Medicine

Glossary

Enterohemorrhagic *Escherichia coli* (EHEC) Cause hemorrhagic colitis

Enteropathogenic *Escherichia coli* (EPEC) Causes dysentery after colonization of the middistal small intestine

Enterotoxigenic *Escherichia coli* (ETEC) Most common bacterial agents of diarrhea in the United States and Europe

Hemorrhagic colitis Nonfebrile bloody diarrhea caused by enterohemorrhagic (EHEC) *Escherichia coli*

Phase variation Phenomenon occurring in several strains of the *Escherichia coli* in which the bacteria alternate between the piliated and nonpiliated conditions

ESCHERICHIA COLI is a gram-negative, facultative anaerobic, non-spore-forming, motile rod. The species belongs to the Enterobacteriaceae and includes a large number of strains that differ in pathogenic potential. Certain strains are common, innocuous residents of the intestine of mammals; others cause human and animal infections of the digestive and urinary tracts, blood, and central nervous system. The structure, biochemical functions, and genetics of this organism are well studied, making it the best known of all cellular forms of life. This organism has occupied center stage in the development of molecular biology.

I. Taxonomy

Escherichia coli is one of five recognized species of the genus *Escherichia*. The genus is named after Theodor Escherich, who first isolated *E. coli* in 1884. The species is defined on the basis of certain readily measurable biochemical activities shared by most strains. Thus, *E. coli* generally ferments lactose, possesses lysine decarboxylase, produces indole, does not grow on citrate, does not produce H_2S, and is Voges-Proskauer-negative (does not produce acetoin). Its chromosomal DNA is 49–52% G + C.

The limits of the taxonomic definition are under scrutiny because DNA hybridization data suggest that, contrary to tradition, *Escherichia* and *Shigella* belong to the same genus. *Escherichia coli* comprises a number of strains that share the same basic taxonomic features, with about 70% DNA homology at the extremes.

Escherichia coli strains are defined mainly by their antigenic composition. Of taxonomic import are over 170 different serological types of lipopolysaccharide antigens (O antigens) and 80 types of capsular (K antigens). Other properties that are used to define individual strains are H antigens (flagellar proteins), F antigens (fimbrial proteins), and phage and colicin sensitivity.

Recent quantitative approaches to define taxonomic relationships are based on patterns of isozymes of metabolically important enzymes, restriction fragment-length polymorphism, and protein composition of the outer membrane. Although there is a huge number of combinations of these properties, the variety of strains that have been isolated, although large, is circumscribed (perhaps in the thousands). By the criteria used, most of the strains in today's world appear to be clonal descendants of relatively few ancestors. Genetic exchange leading to recombinational events thus seems to be an infrequent event in the environment. It has been estimated that, for this species, major episodes of selection have occurred once in 30,000 years.

II. Ecology

Escherichia coli is the most abundant facultative anaerobe in the feces, and therefore the colon, of normal humans and many mammals. It is commonly present in concentrations of 10^7–10^8 live organisms per gram of feces. Thus, the total number of individual *E. coli* cells present on earth at any one time exceeds 10^{20}. *Escherichia coli* is far from the most abundant organism in the colon and is outnumbered by strict anaerobes by 100-fold or more. Most fecal *E. coli* isolates are well adapted to colonizing the mammalian large intestine and seldom cause disease. When human strains are cultivated in the laboratory for a long time, they tend to lose the ability to colonize. Included among these is K12, the most widely used strain in the molecular microbiology.

Escherichia coli cells are periodically deposited from their intestinal residence into soils and waters. It has been thought that they do not survive for extended periods of time in such environments and could be cultured only for a few days (seldomly weeks) after their introduction. For this reason, their presence has been taken as a measure of recent fecal contamination, and the coliform count of the drinking water supply or of swimming facilities is still a common measure of microbiological water purity. The notion that *E. coli* has a short survival time in the environment has been challenged, and new work suggests that the presence of these organisms may not be a reliable indication of recent fecal pollution. [*See* WATERBORNE DISEASES.]

Mammals become colonized with *E. coli* within a few days of birth, possibly from the mother or other attendants. How the organisms are transmitted to neonate is not known with certainty; this may occur during passage through the birth canal or, shortly after birth, via the fecal–oral.

III. Structure and Function of Cell Parts

In both structure and function, *E. coli* serves as the prototype for members of the Enterobacteriaceae. This article is limited to the features that are typical of *E. coli*. For information regarding more general aspects of bacterial structure and function, see appropriate entries elsewhere in this work. An example of the overall composition of this organism in its growth phase is shown in Table I. [*See* ENTEROPATHOGENS.]

A. Pili (Fimbriae)

Different *E. coli* strains carry one or two kinds of pili: common and conjugative (sex pili). The common pili fall into two groups, depending on the *in vitro* effect of mannose on the adhesion of the organisms to animal cell receptors. Type 1 (mannose-sensitive) pili are found in numbers of 100–300 per cell and consist mainly of an acidic hydrophobic protein called pilin. Pili are highly antigenic, comprising many so-called F antigens. *Escherichia coli* strain K12 possesses only type 1 common pili. This strain and others alternate between the piliated and nonpiliated condition, a phenomenon known as phase variation. It is thought that the presence of common pili allows organisms in their first efforts to colonize their host by attaching to epithelial cells. Inside the body, turning off the synthesis of common pili may lessen the chances that the organisms will be phagocytized by white blood cells.

The sex pilus is encoded by plasmids such as F or R and is usually present in one or a few copies per cell. This structure causes donor and recipient bacteria to make contact, allowing the transfer of DNA during conjugation.

B. Flagella

Escherichia coli is not strongly flagellated and is usually endowed with only four to eight flagella per cell. As typical bacterial flagella, they are composed of a long filament, a hook, and a basal body. The principal component of *E. coli* flagella is an *N*-methyl-lysine-rich protein known as flagellin. Its size, usually around 55 kDa, varies among strains. Flagellin self-assembles *in vitro* into flagellalike filamentous cylindric lattices with hexagonal packing. [*See* FLAGELLA.]

Escherichia coli flagella are highly antigenic, comprising a large number of so-called H antigens. The N and C termini of various H antigens are highly conserved, the major antigenic divergence being found in the central region of the molecule.

C. Capsule and Outer Membrane

In some strains, the outer membrane of *E. coli* is covered by a polysaccharide capsule composed of so-called K antigens. Other polysaccharides, the M antigens (colanic acids, polymers of glucose, galactose, fucose, and galacturonic acid), are synthesized when the organisms are placed under conditions of high osmolarity, low temperature, and low humid-

Table I Overall Macromolecular Composition of an Average *E. coli* Strain B/r Cell[a]

Macromolecule	Percentage of total dry weight	Weight per cell (10^{15} × weight, g)		Molecular weight	Number of molecules per cell	Different kinds of molecule
Protein	55.0	155.0		4.0×10^4	2,360,000	1050
RNA	20.5	59.0				
23S ribosomal RNA			31.0	1.0×10^6	18,700	1
16S ribosomal RNA			16.0	5.0×10^5	18,700	1
5S ribosomal RNA			1.0	3.9×10^4	18,700	1
Transfer			8.6	2.5×10^4	205,000	60
Messenger			2.4	1.0×10^9	1380	400
DNA	3.1	9.0		2.5×10^9	2.13	1
Lipid	9.1	26.0		705	22,000,000	4
Lipopolysaccharide	3.4	10.0		4346	1,200,000	1
Murein	2.5	7.0			1	1
Glycogen	2.5	7.0		1.0×10^6	4360	1
Total macromolecules	96.1	273.0				
Soluble pool	2.9	8.0				
Building blocks			7.0			
Metabolites, vitamins			1.0			
Inorganic ions	1.0	3.0				
Total dry weight	100.0	284.00				
Total dry weight/cell		2.8×10^{-13} g				
Water (at 70% of cell)		6.7×10^{-13} g				

[a] These values are for *E. coli* K12 in balanced growth in a glucose-minimal medium at 37°C.
[From Neidhardt, F. C., Ingraham, J. L., and Schaechter, M. (1990). "Physiology of the Bacterial Cell." Sinauer Associates, Sunderland, Massachusetts.]

ity, suggesting that normally these compounds may be made in response to stressful conditions in the external environment.

The major outer membrane proteins include the pore-forming proteins, called porins Omp C, Omp F, and Pho E. Together, these porins are present in about 10^5 copies per cell. Their sizes vary from 36.7 to 38.3 kDa. The diameters of the pores are 1.16 nm for Omp F and Pho E and 1.08 nm for Omp C. The synthesis of these porins is also regulated by environmental conditions; thus, Omp F is repressed by high osmotic conditions or high temperature, Omp C is derepressed by high osmotic conditions, and Pho E is synthesized when cells are starved for phosphate. These findings suggest that the organisms utilize narrower porin channels in the animal host than in the outside environment, which invites speculation about the nature of chemical and metabolic challenges faced under the two conditions.

Certain compounds that are too large to diffuse through *E. coli* porin channels are carried across the outer membrane by special transport proteins. These compounds include maltose oligosaccharides, nucleosides, various iron chelates, and vitamin B_{12}. The proteins involved, as well as the porins, also act as receptors for the attachment of bacteriophages and colicins.

D. Periplasm and Cell Wall

By functional tests of solute partition and electron microscopy, the periplasm of *E. coli* makes up 20–40% of the cell volume. This compartment is osmotically active, in part because it contains large amounts of membrane-derived oligosaccharides (molecules of 8–10 linked glucose residues substituted with 1-phosphoglycerol and O-succinyl esters). Some evidence indicates that the contents of the periplasmic space form a gel. The *E. coli* periplasm contains about 40 known proteins, including (a) binding proteins for amino acids, sugars, vitamins, and ions, (b) degradative enzymes (phosphatases, proteases, endonucleases), and (c) antibiotic detoxifying enzymes (β-lactamases, alkyl sulfodehydrases, aminoglycoside phosphorylating enzymes).

As in most bacteria, the cell wall of *E. coli* consists of a peptidoglycan layer responsible for cell shape and rigidity. In this organism, peptidoglycan is one or, at most, a few molecules in thickness. It is anchored to the outer membrane at some 400,000 sites via covalent links to a major membrane lipoprotein and noncovalent links to porins.

Evidence indicates that the periplasm is spanned by 200–400 adhesion zones between the outer membrane and the cytoplasmic membrane. These appear to be the sites of attachment of certain bacteriophages and of export of outer membrane proteins and lipopolysaccharide. In addition to these apparently scattered junctions, the two membranes are joined at defined periseptal annuli, ring-shaped adhesion zones that are formed near the septum.

E. Cytoplasmic Membrane

The cytoplasmic membrane of *E. coli* is made up of about 200 different proteins and four kinds of phospholipids. Proteins comprise about 70% of the weight of the structure. Under aerobic conditions, the *E. coli* cytoplasmic membrane contains a number of dehydrogenases (e.g., NADH-, D- and L-lactate, succinate dehydrogenases), pyruvate oxidase, cytochromes (of the o and d complexes), and quinones (mainly 8-ubiquinone). Anaerobically grown *E. coli* may contain other dehydrogenases (e.g., formate- and glycerol-3-phosphate dehydrogenases) and enzymes involved in anaerobic respiration (nitrate and fumarate reductases). The cytoplasmic membrane is the site of adenosine triphosphate synthesis. The cytoplasmic membrane systems involved in the transport of solutes are highly efficient and permit this species to grow in relatively dilute nutrient solutions. An example of the variety of transport systems is shown for amino acids in Table II.

The cytoplasmic membrane of *E. coli* contains over 20 proteins involved in various aspects of peptidoglycan biosynthesis, cell wall elongation, and cell division. About 10 of these proteins have been identified by their ability to covalently bind β-lactam antibiotics. They are known as the penicillin-binding proteins, and some have been shown to be involved directly in cell wall synthesis.

Table II Transport Systems for Amino Acids in *E. coli*

1. Glycine–alanine
2. Threonine–serine
3. Leucine–isoleucine–valine
4. Phenylalanine–tyrosine–tryptophan
5. Methionine
6. Proline
7. Lysine–ornithine–arginine
8. Cystine
9. Asparagine
10. Glutamine
11. Aspartate
12. Glutamate
13. Histidine
14. Cysteine (probably)

[From Neidhardt, F. C., Ingraham, J. L., and Schaechter, M. (1990). "Physiology of the Bacterial Cell." Sinauer Associates, Sunderland, Massachusetts.]

F. Cytoplasm

Most of the 2000 or so biochemical reactions necessary for the growth of *E. coli* take place in the cytoplasm. These activities are divided into those concerned with metabolic fueling (production of energy, reducing power, and precursor metabolites), biosynthesis of building blocks, polymerization into macromolecules, and assembly of cell structures. For *E. coli*, each of these activities is generally the same as for all microorganisms but with certain species-specific characteristics.

In fast-growing *E. coli*, much of the cytoplasmic space is taken up by ribosomes. The number of ribosomes per cell is proportional to the growth rate, ranging from about 2000 in cells growing at 37°C at doubling rates of 0.2 hr^{-1} to >70,000 at a doubling rate of 2.5 hr^{-1}, where they make up about 40% of the cell mass. In *E. coli*, the genes for the four ribosomal RNAs (16S, 23S, 5S, transfer) are arranged in seven operons located at different sites on the chromosome. Most of these operons are found near the origin of chromosome replication and, thus, are replicated early. This arrangement ensures that ribosomal RNAs are made in large amounts during rapid growth. Each operon encodes one, two, or three transfer RNAs at sites between the 16S and 23S RNA genes and/or at the end of the operon. The 52 ribosomal proteins are encoded by 21 transcriptional units. Ribosomal RNAs and proteins assemble into particles via a precise sequence of reactions that has been well studied *in vitro*. [*See* RIBOSOMES.]

G. Nucleoid

The nucleoid of *E. coli* is a highly lobular intracytoplasmic region, generally located toward the center of the cell. Within this region, the DNA is found at a local concentration of 2–5% (w/v). The reason why this long molecule (4.4×10^6 bp) is folded and physically limited to the nucleoid region is not well understood. It is known that *in vivo* the DNA is negatively supercoiled into some 50 individual domains. Nucleoids of superhelicity and dimension similar to those seen intracellularly can be isolated by gentle breakage of cells in the presence of divalent cations.

Transcription takes place at the nucleoid–cytoplasm interface, as the nucleoid is thought to form a significant barrier to the diffusion of macromolecules. The reason for the highly irregular shape of the nucleoid may be to contribute to the availability of genes for transcription. At least four small molecular weight proteins that bind to DNA are known to play a role in transcription, recombination, and replication. These nucleoid-associated proteins range in molecular weight from 9200 to 15,400. Two, HU and IHF, are among the abundant *E. coli* proteins and are present in 20–50,000 monomers per cell.

Initiation of DNA replication takes place at a specific origin site, *oriC*, and is under the influence of a protein that is highly conserved among many bacteria, DnaA. Once initiated, DNA replication takes place at a nearly constant rate in moderately fast and fast-growing *E. coli*, until it reaches a terminus. Doubling of the chromosome takes 40 min at 37°C, which requires that in cultures growing faster than this time initiation of chromosome replication take places before the end of the previous round [*See* DNA REPLICATION.]

Little is known about the mode of segregation of the nucleoids. The process takes place with considerable fidelity and, thus, cannot result from partitioning into progeny cells by chance alone. The widespread view is that the chromosome is attached to the cell membrane and that movement of the membrane serves as a primitive mitotic apparatus. However, this view is based largely on cell fractionation studies and is not supported by functional tests. A specific affinity of recently replicated (hemimethylated) origin DNA for the membrane has been demonstrated *in vitro*.

IV. Metabolism and Growth

A. Biosynthetic and Fueling Reactions

The central metabolism of *E. coli* is carried out via the Embden–Meyerhof–Parnas pathway, the pentose pathway, and the tricarboxylic acid cycle plus, for the metabolism of gluconate, the Entner–Doudoroff pathway. As a facultative anaerobe, *E. coli* meets its energy needs either by respiratory or fermentative pathways. *Escherichia coli* carries out a mixed acid fermentation of glucose that results in the formation of a large number of products. Under anaerobic conditions, the main products (and the moles formed per 100 moles of glucose used) are formate (2.4), acetate (37), lactate (80), succinate (12), ethanol (50), 2,3-butanediol (0.3), CO_2 (88), and H_2 (75).

The need for biosynthetic building blocks is met in *E. coli* by the production of 12 precursor metabolites common to all bacteria. The number of molecules needed for the manufacture of the major building blocks as well as the requirements for energy and reducing power are shown in Tables III and IV.

The precursor metabolites do not contain nitrogen or sulfur, which must enter the metabolic circuit independently. *Escherichia coli* does not fix dinitrogen gas but can use a number of compounds as a source of nitrogen, including ammonium ions and various amino acids. It can utilize nitrate and nitrite as terminal electron acceptors during anaerobic respiration by activating nitrate or nitrite reductases. However, under anaerobic conditions, no energy is generated by this process, and a source of reduced nitrogen is necessary for the anaerobic growth of *E. coli*. The incorporation of ammonium ion into organic compounds is catalyzed either by L-glutamate dehydrogenase when ammonia is abundant or by glutamine synthetase and glutamate synthase, acting together, when ammonia is limiting. [*See* ANAEROBIC RESPIRATION.]

The common sources of sulfur in the *E. coli* environment are sulfate and sulfur-containing amino acids. Sulfate is transported into the cell after being reduced to H_2S by a sulfite reductase. Sulfur is then assimilated from H_2S using O-acetylserine sulfohydrolase to produce L-cysteine. *Escherichia coli* has a complex system of transporting and utilizing organic

Table III Energy Requirements for Polymerization of the Macromolecules in 1 g of Cells[a]

Macromolecule	Amount of energy required (μmol ~ P)
From activated building blocks	
DNA: from dNTPs	136
RNA: from NTPs	236
Protein: from aminoacyl–transfer RNAs	11,808
Murein: in part from activated building blocks	138
Phospholipids: in part from activated building blocks	258
Lipopolysaccharide	0
Polysaccharide (glycogen)	0
Total energy	12,576
From unactivated building blocks	
DNA: from dNMPs	336
RNA: from NMPs	1516
Protein: from amino acids	21,970
Murein: in part from activated building blocks	138
Phospholipids: in part from activated building blocks	258
Lipopolysaccharide	0
Polysaccharide (glycogen)	0
Total energy	24,218

[a] *1 g of cells contains 961 mg macromolecules.*
dNMP and NMP, deoriboxynucleotide monophosphates and ribonucleotide monophosphates; dNTP and NTP, deoxyribonucleotide triphosphates and ribonucleotide triphosphates.
[Data from Neidhardt, F. C., Ingraham, J. L., and Schaechter, M. (1990). "Physiology of the Bacterial Cell." Sinauer Associates, Sunderland, Massachusetts.]

phosphates, including an inducible alkaline phosphatase in its periplasm.

B. Nutrition and Growth

Escherichia coli is a chemoheterotroph capable of growing on any of a large number of sugars or amino acids provided individually or in mixtures. Some strains found in nature have single auxotrophic requirements, among which thiamine deficiency is common. Growth of many strains is inhibited by the presence of single amino acids, such as serine, valine, or cysteine. *Escherichia coli* grows faster with glucose than with any other single carbon and energy source and reaches a doubling time of 50 min under well oxygenated conditions at 37°C. Doubling times with less favored substrates may be hours in length. Slow rates of growth can also be achieved by using an externally controlled continuous culture device or by adding to the culture a metabolic analog and its antagonist at proper ratios. When the medium is supplemented with building blocks such as amino acids, nucleosides, sugars, and vitamin precursors, *E. coli* grows more rapidly, reaching doubling times of 20 min at 37°C in rich nutrient broths.

Escherichia coli can grow at temperatures between 8° and 48°C, depending on the strain and the nutrient medium. Its optimum growth temperature is 39°C. *Escherichia coli* does not grow in media containing a NaCl concentration greater than about 0.65 *M*. In response to changes in the osmotic pressure of the medium, *E. coli* increases its concentration of ions, especially K^+ and glutamate. The pH range for growth is between pH 6.0 and 8.0, although some growth is possible at values approximately 1 pH unit above and below this range.

V. Pathogenesis

Different strains of *E. coli* are responsible for a large number of clinical diseases. In their manifestation, some of these diseases overlap with those caused by other species (e.g., *Shigella, Salmonella*). The most common infections caused by *E. coli* involve the intestinal and urinary tracts of humans and other mammals, where they produce simple watery diarrhea or locally invasive forms of infection (e.g., dysentery). *Escherichia coli* infects deeper tissues, including the blood (septicemia) in patients with compromised defense mechanisms and, additionally, the meninges in infants. The organism also causes mastitis in cattle. Strains of *E. coli* that produce intestinal infections are divided into groups according to the clinical picture they produce and their known virulence factors (Table V). These strains are denoted by abbreviations (ETEC, EPEC, etc., where the terminal EC stands for *E. coli*), the number of which is proliferating. [*See* GASTROINTESTINAL MICROBIOLOGY; DYSENTERY, BACILLARY.]

A. Enterotoxigenic Strains

Enterotoxigenic (ETEC) strains acquired from food or water contaminated with human or animal feces

Table IV Building Blocks Needed to Produce 1 g of *E. coli* Protoplasm[a]

Building block	Amount present in *E. coli* B/r (μmol/g dried cells)	Cost of making 1 μmol of each of these building blocks (μmol/μmol)						
		Metabolites	ATP	NADH	NADPH	1-C	NH^+_4	S
Protein amino acids								
Alanine	488	1 pyr	0	0	1	0	1	0
Arginine	281	1 akg	7	−1	4	0	4	0
Asparagine	229	1 oaa	3	0	1	0	2	0
Aspartate	229	1 oaa	0	0	1	0	1	0
Cysteine	87	1 pga	4	−1	5	0	1	1
Glutamate	250	1 akg	0	0	1	0	1	0
Glutamine	250	1 akg	1	0	1	0	2	0
Glycine	582	1 pga	0	−1	1	−1	1	0
Histidine	90	1 penP	6	−3	1	1	3	0
Isoleucine	276	1 oaa, 1 pyr	2	0	5	0	1	0
Leucine	428	2 pyr, 1 acCoA	0	−1	2	0	1	0
Lysine	326	1 oaa, 1 pyr	2	0	4	0	2	0
Methionine	146	1 oaa	7	0	8	1	1	1
Phenylalanine	176	1 eryP, 2 pep	1	0	2	0	1	0
Proline	210	1 akg	1	0	3	0	1	0
Serine	205	1 pga	0	−1	1	0	1	0
Threonine	241	1 oaa	2	0	3	0	1	0
Tryptophan	54	1 penP, 1 eryP, 1 pep	5	−2	3	0	2	0
Tyrosine	131	1 eryP, 2 pep	1	−1	2	0	1	0
Valine	402	2 pyr	0	0	2	0	1	0
RNA nucleotides								
ATP	165	1 penP, 1 pga	11	−3	1	1	5	0
GTP	203	1 penP, 1 pga	13	−3	0	1	5	0
CTP	126	1 penP, 1 oaa	9	0	1	0	3	0
UTP	136	1 penP, 1 oaa	7	0	1	0	2	0
DNA nucleotides								
dATP	24.7	1 penP, 1 pga	11	−3	2	1	5	0
dGTP	25.4	1 penP, 1 pga	13	−3	1	1	5	0
dCTP	25.4	1 penP, 1 oaa	9	0	2	0	3	0
dTTP	24.7	1 penP, 1 oaa	10.5	0	3	1	2	0
Lipid components								
Glycerol phosphate	129	1 triosP	0	0	1	0	0	0
Serine	129	1 pga	0	−1	1	0	1	0
$C_{16:0}$ fatty acid (43%)		8 acCoA	7	0	14	0	0	0
$C_{16:1}$ fatty acid (33%)		8 acCoA	7	0	13	0	0	0
$C_{18:1}$ fatty acid (24%)		9 acCoA	8	0	15	0	0	0
Average fatty acid	258	8.2 acCoA	7.2	0	14	0	0	0
LPS components								
UDP-glucose	15.7	1 gluP	1	0	0	0	0	0

Continues

Table IV (*Continued*)

Building block	Amount present in *E. coli* B/r (μmol/g dried cells)	Cost of making 1 μmol of each of these building blocks (μmol/μmol)						
		Metabolites	ATP	NADH	NADPH	1-C	NH^+_4	S
(CDP)ethanolamine	23.5	1 pga	3	−1	1	0	1	0
OH-myristic acid	23.5	7 acCoA	6	0	11	0	0	0
$C_{14:0}$ fatty acid	23.5	7 acCoA	6	0	12	0	0	0
(CMP)KDO	23.5	1 penP, 1 pep	2	0	0	0	0	0
(NDP) heptose	23.5	1.5 gluP	1	0	−4	0	0	0
(TDP) glucosamine	15.7	1 fruP	2	0	0	0	1	0
Peptidoglycan monomers								
UDP-N-acetylglucosamine	27.6	1 fruP, 1 acCoA	3	0	0	0	1	0
UDP-N-acetylmuramic acid	27.6	1 fruP, 1 pep, 1 acCoA	4	0	1	0	1	0
Alanine	55.2	1 pyr	0	0	1	0	1	0
Diaminopimelate	27.6	1 oaa, 1 pyr	2	0	3	0	2	0
Glutamate	27.6	1 akg	0	0	1	0	1	0
Glycogen monomers								
Glucose	154	1 gluP	1	0	0	0	0	0
1-Carbon requirement								
Serine	48.5	1 pga	0	−1	1	0	0	0
Polyamines								
Ornithine equivalents	59.3	1 akg	2	0	3	0	2	0

[a] Other (small) molecules (<3% of cell dry weight). Coenzymes: nicotinamide adenine dinucleotide, nicotinamide adenine dinucleotide phosphate, CoA, CoQ, bacteroprenoid, tetrahydrofolate, cyanocobalamin, pyridoxal phosphate. Prosthetic groups: flavin mononucleotide, flavin adenine dinucleotide, biotin, cytochromes, lipoic acid, thiamine pyrophosphate. Pool of unpolymerized monomers: average approximately 1% of amount of macromolecules.

ATP, adenosine triphosphate; CDP, CMP, and CTP, cytidine di-, mono-, and triphosphate; d-, deoxy-; GTP, guanosine triphosphate; KDO, ketodeoxyoctonate; NADH, nicotinamide adenine dinucleotide, reduced; NADPH, nicotinamide adenine dinucleotide phosphate, reduced; NDP, nucleoside diphosphate; TDP and TTP, ribosylthymine di- and triphosphate; UDP and UTP, uridine di- and triphosphate; acCoA, acetyl CoA; akg, α-ketoglutarate; eryP, erythrose-4-phosphate; fruP, fructose-6-phosphate; gluP, glucose-6-phosphate; oaa, oxaloacetate; penP, ribose-5-phosphate; pep, phosphoenolypyruvate; pga, 3-phosphoglycerate; pyr, pyruvate; triosP, triose phosphate.

[From Neidhardt, F. C., Ingraham, J. L., and Schaechter, M. (1990). "Physiology of the Bacterial Cell." Sinauer Associates, Sunderland, Massachusetts.]

are the most common bacterial agents of diarrhea in the United States and Europe. These strains circulate among the local population, but the majority of people (especially adults) usually remain asymptomatic, most likely due to immunity afforded by previous exposure. ETEC strains are responsible for the "tourist's diarrhea" that frequently affects persons traveling to countries with a low level of sanitation. Watery diarrhea due to *E. coli* resembles that seen in mild cases of cholera.

ETEC strains colonize the small intestine, where they produce one or both of two enterotoxins called heat labile (LT) and heat stable (ST). Both toxins act by changing the net fluid transport activity in the gut from absorption to secretion. LT is structurally similar to cholera toxin and activates the adenylate cyclase–cyclic adenosine monophosphate system, whereas ST works on guanylate cyclase. The intestinal mucosa is not visibly damaged, the watery stool does not contain white or red blood cells, and no inflammatory process occurs in the gut wall. Gut cells activated by LT or cholera toxin remain in that state until they die, whereas the effects of ST on guanylate cyclase are turned off when the toxin is washed away from the cell.

Table V Examples of Serotypes of Intestinal Pathogenic *E. coli*

	Symptoms	Typical serotypes[a]	Number of serotypes
Enteropathogenic	"Traveler's diarrhea"	O26:H111	Many
Enterotoxigenic	Watery diarrhea	O6:H$^-$	A few more
Enteroinvasive	Bloody diarrhea and dysentery	O29:H$^-$	Many more
Enterohemorrhagic	Bloody diarrhea and dysentery	O157:H7	None or few others

[a] The letter O refers to "somatic" antigens, part of the bacterial lipopolysaccharide; the letter H to "flagellar" antigens, and the numbers to different antigenic serotypes.
[From Schaechter, M., Schlessinger, D., and Medoff, G. (1989). "Mechanisms of Microbial Disease," p. 259. Williams & Wilkins, Baltimore.]

B. Enteropathogenic Strains

Enteropathogenic (EPEC) strains cause dysentery after colonization of the mid-distal small intestine (ileum). They recognize their preferred hosts and tissues by means of surface adhesins specific for receptors on the intestinal brush-border membranes. A particular virulence factor of these strains is the EPEC adherence protein that mediates binding of the organism to target epithelial cells in a characteristic localized pattern on regions of the plasma membrane. Another protein, the "attaching and effacing factor," permits tight adherence of the organism and leads to cytoskeletal rearrangements that damage microvilli. The affected cells form a broad flat pedestal (effacement) beneath the attached microorganism, which, by damaging the absorptive surface, contributes to the diarrhea. [*See* ADHESION, BACTERIAL.]

C. Enterohemorrhagic Strains

The enterohemorrhagic (EHEC) *E. coli* comprise a limited number of serotypes that cause a characteristic nonfebrile bloody diarrhea known as hemorrhagic colitis. The most common of these serotypes in the United States is O157 : H7, whereas others, particularly O26, are found with greater frequency elsewhere in the world. *Escherichia coli* O157 : H7 causes both outbreaks and sporadic disease. The organism is commonly isolated from cattle and several outbreaks have been traced to undercooked hamburger meat.

EHEC strains have two special characteristics of pathogenic importance. First, they produce high levels of two related cytotoxins that resemble toxins of *Shigella* in both structure and function and that possess the same protein synthesis-inhibitory action and binding specificity. These toxins are therefore called Shiga-like toxins (SLT I or II—they are often called verotoxins in Europe and Canada). *In vitro* the SLT toxins are cytotoxic for endothelial cells in culture. Second, they possess a gene highly homologous to the EPEC attaching and effacing gene. In combination, the protein encoded by this gene and the Shiga-like toxins presumably damage the gut mucosa in a manner characteristic of hemorrhagic colitis.

EHEC strains cause systemic manifestations (hemolytic–uremic syndrome or thrombotic thrombocytopenic purpura) that are believed to be related to systemic absorption of Shiga-like toxin, possibly in combination with endotoxin. These syndromes represent the clinical response to endothelial damage of glomeruli and the central nervous system.

D. Other Strains That Cause Intestinal Infections

Enteroinvasive (EIEC) *E. coli* strains cause dysentery, resembling that due to *Shigella*. Unlike the strains described earlier, which are noninvasive, EIEC strains are selectively taken up into epithelial cells of the colon, requiring for this process a specific outer membrane protein. EIEC strains also make Shiga-like toxins. Cell damage by these strains triggers an intense inflammatory response.

Other strains, called EAggEC, are associated with diarrhea in infants under 6 mo of age, often persisting for weeks with marked nutritional consequences. EAggEC strains spontaneously agglutinate (aggregate) in tissue culture.

E. Strains That Infect the Genitourinary Tract

Escherichia coli strains are the most common cause of genitourinary tract infections of humans, including cystitis, pyelonephritis, and prostatitis. Many of the strains that cause pyelonephritis possess pili that bind specifically to a glycolipid constituent of kidney tissue. They are called P pili because the receptor is a complex galactose-containing molecule that is part of the P blood group antigen. About 1% of humans are P antigen-negative and, not carrying the P pilus receptor, are not susceptible to colonization by P pilus-carrying strains. These persons do not suffer from urinary tract infections mediated by the usual route (i.e., mucosal colonization followed by ascending invasion of the bacteria into the bladder). Such individuals may, however, become infected when the normal route is bypassed (e.g., by the use of an indwelling urinary catheter).

F. Other Invasive Strains

Strains that possess a sialic acid-containing capsular polysaccharide, called K1 antigen, cause invasive diseases, such as septicemia and meningitis, in young infants. *Escherichia coli* is also a common cause of septicemia in adults, especially in patients who are immunocompromised. Many patients with this manifestation acquire it as a consequence of infection of the urinary tract, often followed by manipulations such as urinary catheterization.

VI. Principles of Diagnosis Using Clinical Specimens

All naturally occurring strains of *E. coli* look alike, with respect to both the morphology of their colonies on agar plates and their shape under the microscope. They can be distinguished on the basis of biochemical and nutritional properties. *Escherichia coli* differs from some of the classical intestinal pathogens, such as *Salmonella* and *Shigella* in that it ferments lactose. For this reason, this sugar is included together with a pH indicator dye in agar media. Lactose-fermenting colonies (presumptively, those of *E. coli*) turn a distinctive color due to the production of acid.

With the help of an ingenious array of differential and selective media, it is usually simple to isolate *E. coli* from samples, such as feces, that contain a preponderance of many other bacteria. These media and other special tests permit the laboratory to narrow down the identification to the main genera of Enterobacteriaceae. Classifying *E. coli* into serological subgroups is not a task that most clinical laboratories are prepared to carry out. The serological reagents most readily available commercially are antisera directed against EPEC strains. [*See* ISOLATION.]

EHEC strains of serotype O157 : H7 are nearly unique in their inability to ferment sorbitol. Nonfermenting colonies are detected on sorbitol–MacConkey agar and confirmed with a sensitive and specific latex agglutination test.

Bibliography

Levi, Primo. (1989). An interview with *Escherichia coli*. (One of "Five Intimate Interviews.") *In* "The Mirror Maker." Schoecken Books Inc., New York.

Neidhardt, F. C., Ingraham, J. L., Low, K. B., Magasanik, B., Schaechter, M., and Umbarger, H. E. (1988). *Escherichia coli* and *Salmonella typhimiurium*. "Cellular and Molecular Biology." American Society for Microbiology, Washington, D.C.

Neidhardt, F. C., Ingraham, J. L., and Schaechter, M. (1990). "Physiology of the Bacterial Cell." Sinauer Associates, Sunderland, Massachusetts.

Evolution, Experimental

Richard E. Lenski
Michigan State University

Glossary

Adaptation Match between a particular feature of an organism and its environment, which results from natural selection

Evolution Change in the genetic properties of populations and species over generations, which requires the origin of variation (by mutation and/or mixis) as well as the subsequent spread or extinction of variants (by natural selection and/or genetic drift)

Fitness Average reproductive success of a genotype in a particular environment, usually expressed relative to another genotype

Genetic drift Changes in gene frequency caused by the random sampling of genes during transmission across generations, rather than by any detrimental or beneficial effects of those genes

Mixis Production of a new multilocus genotype by recombination of genes from two sources

Natural selection Changes in gene frequency caused by specific detrimental or beneficial effects of those genes

Population Group of individuals belonging to the same species and living in close proximity, so that individuals may potentially recombine, compete for limiting resources, or otherwise interact

EXPERIMENTAL EVOLUTION is the study, in the laboratory, of the fundamental processes of evolutionary change. These processes include spontaneous mutation and adaptation by natural selection, and they give rise to various patterns of genetic diversity within and between populations. Microorganisms have proven to be useful subjects for this research as a consequence of their large population sizes and short generation times, the ease with which their environments and genetic systems can be manipulated, and other desirable properties. Experimental studies of microbial evolution have generally confirmed the basic principles of modern evolutionary theory, while also providing new insights into the genetics, physiology, and ecology of microorganisms.

I. Review of Evolutionary Theory

Evolutionary theory seeks to explain the patterns of biological diversity in terms of a relatively few fundamental evolutionary processes. These processes are presumed not only to have operated in the past, but also to continue to operate today. Therefore, they can be studied by direct experimentation in the laboratory. Before discussing experiments that have used microorganisms to examine evolutionary processes, the major elements of evolutionary theory will be reviewed.

A. Evolutionary Patterns

Three of the most conspicuous products of organic evolution are (1) the wealth of genetic variation that exists within almost every species, (2) the apparent divergence of populations and species from one another and from their common ancestors, and (3) the manifest adaptation, or fit, of organisms to the environments in which they live.

1. Genetic Variation

The existence of extensive genetic variation within species has been demonstrated by a variety of means. Variation in certain traits, such as seed shape in pea plants and blood type in humans, can be shown to have a genetic basis by careful examination of pedigrees. For many other traits, such as milk production in cows or body weight in humans, quantitative genetic analyses are required to partition the phenotypic variation that is due to genetic versus environmental influences. Biochemical and molecular techniques have also revealed extensive variation in DNA sequences and the proteins they encode.

2. Divergence and Speciation

All biological species differ from one another in some respects. It is generally possible to arrange species hierarchically, depending on the extent and nature of their similarities and differences. This hierarchy is reflected in the taxonomic classification scheme of Linnaeus (species, genus, family, and so on). This hierarchical arrangement also suggests a sort of "tree of life," in which the degree of taxonomic relatedness reflects descent with modification from some common ancestor in the more or less distant past.

Investigating the origins of particular traits and the relationships of taxa requires an historical approach, which is not amenable to direct experimentation. Even so, historically based hypotheses can often be tested using phylogenetic and comparative methods, which utilize data on the distribution of character states across various taxa and environments, sometimes supplemented with information from the fossil record or biogeography.

The extent of evolutionary divergence that is necessary for two groups of organisms to be regarded as distinct species is embodied in the biological species concept, according to which "species are groups of actually or potentially interbreeding populations, which are reproductively isolated from other such groups" (E. Mayr, 1942, "Systematics and the Origin of Species," Columbia University Press). Speciation thus refers to the historical events by which groups of organisms have become so different from one another that they no longer can interbreed. However, many organisms (including most microorganisms) reproduce primarily or exclusively asexually, and the preceding species definition is not applicable. For such organisms, the extent of evolutionary divergence that corresponds to distinct species is somewhat arbitrary and often more a matter of convenience than of scientific principle. [*See* TAXONOMIC METHODS.]

3. Adaptation

The various features of organisms often exhibit an exquisite match to their environments. For example, the bacteria that live in hot springs have special physiological and biochemical properties that allow them to survive and grow at very high temperatures, which would kill most other bacteria; often these thermophiles cannot grow at all under the much more benign conditions where most other bacteria thrive. Nevertheless, organisms are by no means *perfectly* adapted to the environments in which they live. Evidence for the imperfection of adaptation can be seen when species go extinct, usually as a consequence of some change in the environment.

B. Evolutionary Processes

Biological evolution occurs whenever the genetic composition of a population or species changes over a period of generations. Four basic processes contribute to such change: mutation, mixis, natural selection, and genetic drift. Selection and drift cannot act unless genetic variation exists among individuals.

1. Sources of Genetic Variation

Genetic variation among individuals is generated by two distinct processes: mutation and mixis. In terms of evolutionary theory, these processes are usually distinguished as follows: Mutation refers to a change at a single gene locus from one allelic state to another (e.g., *abcd* → *Abcd*), whereas mixis refers to the production of some new multilocus genotype by the recombination of two different genotypes (e.g., *abcd* + *ABCD* → *aBcD*).

a. Mutation

There are many different types of mutations, including point mutations, rearrangements, and transposition of mobile genetic elements from one site in the genome to another. Some mutations cause major changes in an organism's phenotype; for example, a bacterium may become resistant to attack by a virus (bacteriophage) as the result of a mutation that alters a receptor on the cell surface. Other mutations have little or even no effect on an organism's phenotype: Many point mutations have absolutely no effect on amino acid sequence (and, hence, protein structure and function) because of the redundancy that exists in the genetic code. [*See* TRANSPOSABLE ELE-

MENTS; *ESCHERICHIA COLI* AND *SALMONELLA TYPHIMURIUM*, MUTAGENESIS.]

Any number of factors may affect mutation rates, including both environmental agents (e.g., intensity of ultraviolet irradiation) and the organism's own genetic constitution (e.g., presence or absence of transposons). Evolutionary theory makes almost no assumptions about the rates of mutations or their biophysical bases, with one exception: *Mutations are assumed to occur spontaneously, i.e., irrespective of their beneficial or harmful effects on the organism.*

Although particular mutations are assumed to occur without regard to their selective value for the organism, it is quite possible that organims have evolved characteristic mutation rates, which may reflect a balance between beneficial and harmful effects of mutations. I shall return to this point later.

b. Mixis

Recombination among genomes can occur by a number of different mechanisms. The most familiar mechanism is eukaryotic sex, which arises from Mendelian segregation (meiosis) and reassortment of chromosomes (fertilization). Many eukaryotic microorganisms, including fungi and protozoa, engage in sexual mixis. Bacteria generally reproduce asexually but may undergo mixis via conjugation (plasmid-mediated), transduction (virus-mediated), or transformation. Even viruses may recombine when two or more co-infect a single host cell. [*See* CONJUGATION, GENETICS; PLASMIDS.]

Unlike mutation, these various mechanisms do not necessarily produce organisms with new genes; instead, they may produce organisms that possess new *combinations* of genes. This can have very important consequences in evolutionary theory. In the absence of mixis, two or more mutations can be incorporated into an evolving population only if they occur *sequentially* in a single lineage (Fig. 1a). With mixis, however, mutations that occur in separate lineages can be incorporated *simultaneously* into an evolving population (Fig. 1b). Thus, mixis may accelerate the rate of adaptive evolution, at least in some circumstances, by bringing together favorable combinations of alleles.

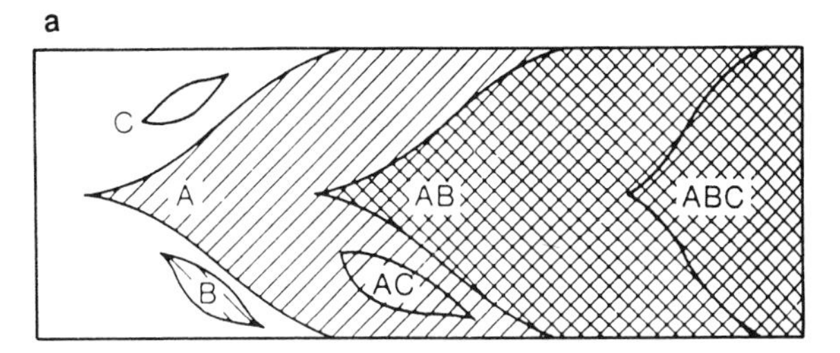

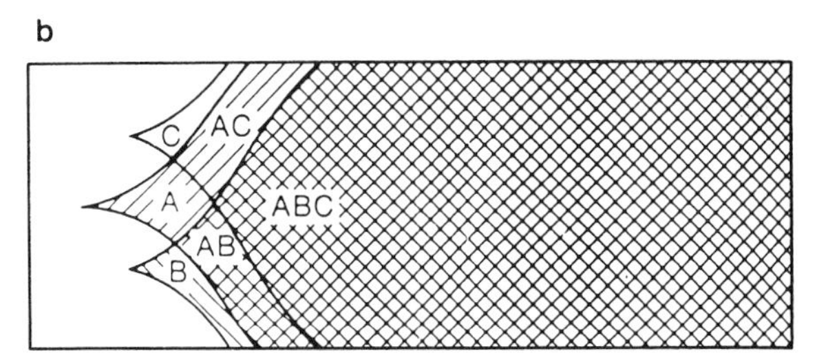

Figure 1 Substitution of advantageous mutations in large populations of asexual (a) and sexual (b) organisms. The variously hatched and shaded areas indicate the changing frequencies of mutant alleles at three loci with time. Recombination of genes from different individuals allows the favored alleles at these loci to be substituted simultaneously. In the absence of mixis, however, the favored alleles must be fixed sequentially in a single lineage. Sexuality and other forms of mixis may thereby accelerate the rate of adaptive evolution. [From Crow, J. F., and Kimura, M. (1965). *Am. Nat.* **99**, 439–450. The University of Chicago Press, Chicago.]

2. Natural Selection

One of the most conspicuous features of biological evolution is the evident "fit" (adaptation) of organisms to the environments in which they live. For many centuries, this match between organism and environment was taken as evidence for the design of a Creator. But in 1859, Charles Darwin published "The Origin of Species," in which he set forth the principle of adaptation by natural selection. This principle follows logically from three simple premises. First, variation among individuals exists for many phenotypic traits. Second, these phenotypic traits influence individual survival and reproductive success. Third, phenotypic variation in those characters that affect survival and reproductive success is heritable, at least in part. (Many phenotypic traits are subject to both genetic and environmental influences.) Hence, individuals in later generations will tend to be better adapted to their environment than were individuals in earlier generations, provided that the environment itself has not changed too much in the intervening time. (Environments do sometimes change, of course, and when this happens a population or species may go extinct if it cannot adapt to these changes.)

Darwin himself did not know about the material basis of heredity (DNA and chromosomes), nor did he even understand the precise causes of heritable variation among individuals (mutation and mixis). What he clearly understood, however, was that this heritable variation did exist and its causes (whatever they were) could be logically separated from its consequences for the reproductive success of individuals and the resulting adaptation of species to their environments.

Darwin's theories were influenced, in part, by his observations on the practices of breeders of domesticated animals and plants. These practices are now commonly referred to as artificial selection. It is useful to distinguish between artificial and natural selection and to relate this distinction to experimental evolution in the laboratory. Under artificial selection, individual organisms are chosen directly by a breeder, who allows some but not all individuals within a population to survive and reproduce. Individuals are thus selected on the basis of particular traits that are desirable to the breeder. By contrast, under natural selection, no one consciously chooses which individuals within a population will survive and reproduce and which will not. Instead, the match between organismal traits and environmental factors determines whether or not a particular individual will survive and reproduce.

At first glance, one might regard all laboratory studies of selection as studies of artificial selection, because they are necessarily performed under unnatural environmental conditions. Such usage, however, would not reflect the critical distinction between artificial and natural selection that I have outlined earlier, i.e., whether a breeder or the environment determines which individuals will survive and reproduce. In experimental evolution, an investigator typically manipulates environmental factors, such as temperature and resource concentration, but he or she does not directly choose which individuals within an experimental population will survive and reproduce. Instead, *natural selection in the laboratory, like natural selection in the wild, depends on the match between organismal traits and environmental factors.*

3. Genetic Drift

The process of adaptation by natural selection has sometimes been criticized because a "just-so story" can be offered to explain the value of almost any phenotypic trait. In fact, the frequency of genes within populations, and hence also the distribution of phenotypic traits, may change not only as the result of natural selection, but also as a consequence of the random sampling of genes during transmission across generations.

In practice, it can be difficult to distinguish between natural selection and genetic drift. This difficulty is especially evident when the only available data consist of static distributions of gene frequencies or phenotypic traits. What is usually needed to resolve this problem is some independent method for directly assessing the effects of particular genes or phenotypic traits on survival and reproductive success.

By using microorganisms to study evolution experimentally, it is possible to compare the survival and reproductive success of different genotypes that are placed in direct competition with one another. With proper replication of such experiments, it becomes possible to distinguish systematic differences in survival and reproductive success from chance deviations that are due to random genetic sampling.

II. Experimental Tests of Fundamental Principles

Two of the most important principles of modern evolutionary theory are the spontaneity of mutation and adaptation by natural selection. According to the former, mutations occur irrespective of any beneficial or harmful effects they may have on the individual. According to the latter, individuals in later generations will tend to be better adapted to their environment than were individuals in earlier generations, provided that the necessary genetic variation exists and the environment itself does not change.

A. Spontaneous Mutation

For many years, it was known that bacteria could adapt to various environmental challenges. For example, the introduction of bacteriophage into a population of susceptible bacteria often caused the bacterial population to become resistant to further viral infection. It was unclear, however, whether the mutations that were responsible for bacterial adaptation were caused directly by exposure to the selective agent, or this adaptation was the result of spontaneous mutation and subsequent natural selection. Two elegant experiments were performed during the 1940s and 1950s, which demonstrated that mutations existed *prior to* exposure to the selective agent, so that these mutations could not logically have been *caused by* that exposure.

1. Fluctuation Test

The first of these experiments was published by Salvador Luria and Max Delbrück in 1943 and relied on subtle mathematical reasoning. Imagine a set of bacterial populations, each of which is allowed to grow from a single cell to some larger number of cells (N); the founding cells are identical in all of the populations. If exposure to the selective agent causes a bacterial cell to mutate with some low probability (p), then the number of mutants in a population is expected to be, on average, pN. Although this probability is the same for each of the replicate populations, the exact number of mutants in each population may vary somewhat due to chance (just as the number of heads and tails in 20 flips of a fair coin will not always equal exactly 10). If the hypothesis that exposure to the selective agent causes these mutations is correct, then mathematical theory shows that the expected *variance* in the number of mutants among the set of replicate populations is equal to the average number of mutants. A typical outcome expected under this hypothesis is shown in Fig. 2a.

Now imagine this same set of bacterial populations, but assume that mutations occur spontaneously, i.e., independent of exposure to the selective agent. During each generation of binary fission, there is a certain probability that one of the two daughter cells is a mutant. A mutant cell's progeny are themselves also mutants, and so on. According to mathematical theory, under this hypothesis the expected variance in the number of mutants among the set of replicate populations is much greater than the average number of mutants. This large variance comes about because mutations will, by chance, occur earlier in some replicate populations than in others, and these early ("jackpot") mutations will leave numerous mutant progeny owing to the geometric growth of the population. Figure 2b shows a typical outcome expected under the hypothesis of spontaneous mutation.

Luria and Delbrück designed experiments that allowed them to compute both the average and the variance of the number of mutants in a set of bacterial populations. In these experiments, the observed variances were much greater than expected under the hypothesis that exposure to the selective agent caused the mutations. Hence, Luria and Delbrück's results provided strong evidence in support of the hypothesis of spontaneous mutation.

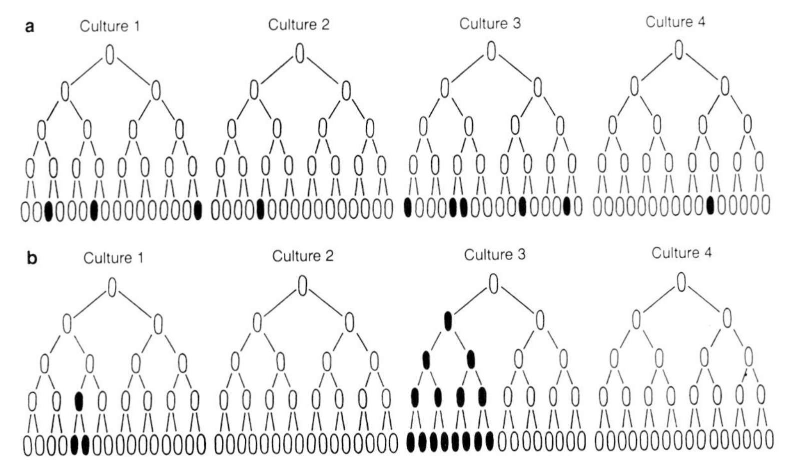

Figure 2 Schematic illustration of the origin of mutants in a set of four cultures of bacteria. Each culture is initiated from a single cell, and cells are spread on a selective medium after the four generations of binary fission that correspond to the bottom row. (a) Typical distribution expected under the hypothesis that exposure to the selective agent causes bacterial cells to mutate with some low probability. (b) Typical distribution expected under the hypothesis that mutations occur spontaneously, i.e., independent of exposure of the cells to the selective agent. The two distributions are not distinguished by the mean number of mutants in the replicated cultures but, rather, by the ratio of the variance to the mean. [From Molecular Genetics: An Introductory Narrative by Gunther S. Stent and Richard Calendar. Copyright (c) 1971, 1978, by W. H. Freeman and Company, New York. Reprinted by permission.]

2. Replica-Plating Experiment

Joshua and Esther Lederberg devised a more direct demonstration of the spontaneous origin of mutations, which they published in 1952. In their experiment, cells are spread on an agar plate that does not contain the selective agent, so that each cell grows until it produces a discrete colony (master plate). Cells from each of these colonies are then transferred onto several other agar plates that contain the selective agent, which prevents the growth of colonies *except* by those cells that have the appropriate mutation (replica plates). If mutations are caused by exposure to the selective agent, then there should be no tendency for mutant colonies detected on the replica plates to be derived from a restricted subset of the colonies on the master plate. But if mutations occur during the growth of the colony on the master plate (i.e., prior to the cells' exposure to the selective agent), then those master colonies that give rise to mutant colonies on one replica plate should also give rise to mutant colonies on the other replica plates. Indeed, Lederberg and Lederberg observed that master colonies giving rise to mutants on one replica plate gave rise to mutants on the other replica plates, thus demonstrating that the mutations had occurred spontaneously during the growth of the colony on the master plate.

B. Adaptation by Natural Selection

In addition to demonstrating the spontaneous occurrence of mutations, both the fluctuation test and the replica-plating experiment demonstrate adaptation by natural selection. Two other types of experiments also demonstrate adaptation by natural selection.

1. "Periodic" Selection

When a population is propagated in a constant environment, classes of mutant genotypes that are selectively neutral will tend to accumulate owing to recurring mutations. Early studies of bacterial populations in chemostats documented the expected accumulation with time of certain mutants that were readily scored by the investigator. However, these studies, as well as other more recent studies, also showed that the frequency of mutants did not increase continuously but, instead, exhibited a sawtooth trajectory of steady increases punctuated by sudden declines (Fig. 3). What causes these unusual dynamics? [*See* PERIODIC SELECTION.]

The mutation that is readily scored by the investigator can be designated as a to A. Now consider a mutation at another locus, which is not scored by the investigator but which is highly advantageous to the organism, designated as b to B. If the frequency of

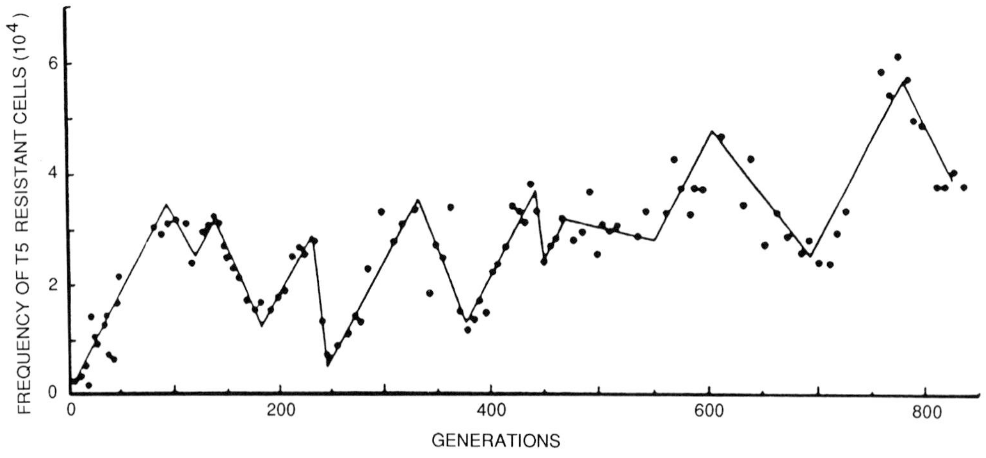

Figure 3 Changes in the frequency of T5-resistant mutants in a population of *E. coli* propagated for 800 generations in a glucose-limited chemostat. T5-resistant mutants are selectively neutral in this environment, but they increase in frequency owing to recurring mutations. Periods of steady increase are interrupted, however, by sudden declines in the frequency of these mutants. These declines result from selection for favorable mutants at other loci, which tend to occur in the numerically dominant T5-sensitive (or wild-type) portion of the population. As these favorable mutants increase in frequency, they competitively exclude their progenitors, including also the T5-resistant derivatives thereof. With time, T5-resistant mutants accumulate anew in the now dominant genetic background that contains the favorable mutant allele. This cycle may occur repeatedly and is commonly referred to as periodic selection. [From Helling, R. B., Vargas, C. N., and Adams, J. (1987). *Genetics* **116**, 349–358. Genetics Society of America,Chapel Hill, North Carolina.]

the *a* allele is much greater than the frequency of the *A* allele (as is the case before too much time has elapsed in such an experiment), then a mutation from *ab* to *aB* is much more common than a mutation from *Ab* to *AB*. Because the *aB* genotype is more fit than either *ab* or *Ab* (the first locus being selectively neutral), the *aB* genotype will outcompete the others. As this happens, cells containing the mutant *A* alleles, which had been accumulating in the *b* background, will be purged. With time, however, new mutations from *a* to *A* will begin accumulating in the *B* background. Subsequent favorable mutations at other loci will give rise to additional "periodic" reversals in the accumulation of *A*. (Note that this explanation depends on the asexuality of bacterial reproduction, which causes linkage disequilibrium between the alleles at the two loci.) In other words, each saw-tooth corresponds to the substitution of a favorable allele by natural selection.

One can test this explanation further by isolating clones from both before and after sharp down-turns in the frequency of *A* and then placing these clones in direct competition with one another under the same culture conditions from which they were isolated. Such tests have been performed repeatedly and confirm adaptation by natural selection.

2. Direct Estimation of Fitness Relative to an Ancestor

It is also possible to demonstrate adaptation by natural selection without tracking the dynamics at any particular locus. To do this, a population is founded using an *ancestral* clone, which is also stored in a nongrowing state (usually at a very low temperature). The population is then propagated under defined environmental conditions, and *derived* clones are isolated from it at arbitrary intervals. A derived clone is placed in direct competition with the ancestral clone under these same defined environmental conditions, after each clone has been allowed to acclimate physiologically to these conditions. If, in competition, the derived clone's population density increases relative to the ancestral clone's density in a systematic and statistically reproducible fashion, then the derived clone has become more fit than its ancestor, in the particular experimental conditions, as the result of mutation and natural selection (Fig. 4).

To distinguish the derived and ancestral clones from one another in a competition experiment, it is usually necessary to introduce a genetic marker that can be scored into one (or both) of the clones. This

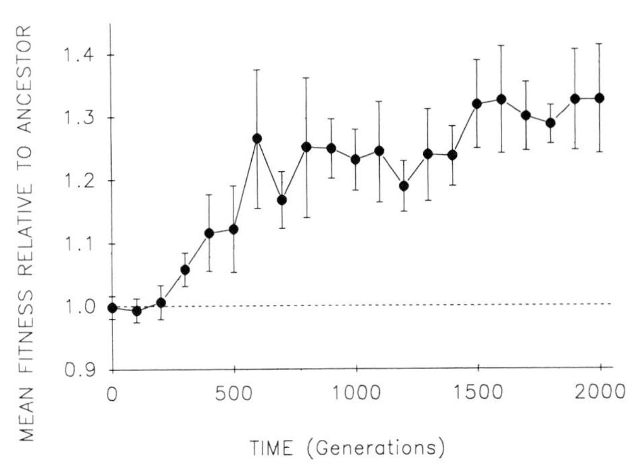

Figure 4 Changes in the mean fitness of 12 populations of *E. coli* during serial propagation for 2000 generations in a glucose-limited minimal medium. Fitness is expressed relative to the ancestral clone, which has been stored at −80°C; relative fitnesses were estimated by competing ancestral and derived strains in this same medium. Error bars indicate 95% confidence intervals about the sample means. The mean fitness of the derived strains relative to their ancestor increased by more than 30% during these 2000 generations of experimental evolution. [From Lenski, R. E., Rose, M. R., Simpson, S. C., and Tadler, S. C. (1991). *Am. Nat.* **138**, 1315–1341. The University of Chicago Press, Chicago.]

genetic manipulation necessitates an appropriate control experiment to estimate the effect of the genetic marker on fitness.

III. Genetic and Physiological Bases of Fitness

The fact that one clone may be more fit than another in a particular environment usually tells us little about the causes of that difference. It is interesting to know why one clone is more fit than another in terms of their genotypes and their physiological properties. There are two distinct approaches that have been employed in trying to elucidate the genetic and physiological bases of differences in fitness. The "bottom-up" approach uses clones that are well characterized genetically and seeks to determine the consequence of their genetic differences for physiological performance and for relative fitness. By contrast, the "top-down" approach uses clones that have been derived from some ancestor by propagation in a defined environment and seeks to elucidate the physiological and even genetic changes that have occurred as the result of this experimental evolution. Each approach has its strengths and limitations. The bottom-up approach permits more direct causal inferences to be drawn with respect to the effects of particular genetic differences, provided that proper care is taken to ensure that clones are otherwise isogenic (genetically identical). The bottom-up approach is limited, however, in that it cannot easily address whether or not these defined genetic differences are representative of the genetic changes that are available to an evolving population. The top-down approach uses precisely those genetic changes that have been important during the evolution of a particular population. However, many of these genetic changes are very difficult to analyze using standard microbial genetic methods because they cause phenotypic changes that cannot be scored in a simple "either–or" manner.

A. Fitness Effects due to Possession of Unused Functions

A number of studies have used well-characterized bacterial genotypes to examine the effects on fitness caused by the carriage and expression of superfluous gene functions. These studies have measured the relative fitnesses of (1) bacteria with constitutive (high-level) and repressed (low-level) expression of enzymes for catabolism of carbon sources in media where those resources are not available; (2) prototrophic bacteria (which produce an amino acid or other required nutrient) and auxotrophic mutants (which cannot produce the required nutrient) in media where the required nutrients are supplied; (3) phage-sensitive bacteria and phage-resistant mutants in environments where phages are absent; and (4) bacteria with plasmid-encoded resistance to antibiotics and isogenic plasmid-free bacteria in media that contain no antibiotics.

These studies have often, but not always, demonstrated substantial fitness disadvantages associated with possession of unnecessary gene functions. In many of the cases where such disadvantages have been detected, they are much greater than can be explained on the basis of the energetic costs associated with the synthesis of unneeded proteins and other metabolites. For example, one study found that the fitness disadvantage associated with synthesis of the amino acid tryptophan, when it was supplied in the medium, was 1000-fold greater than could be explained on the basis of energetic costs. Evidently, the expression of superfluous functions can sometimes have strong indirect effects, which may arise through the disruption of other physiological processes.

B. Effects due to Variation in Essential Metabolic Activities

It is clear that the expression of unnecessary metabolic functions is often disadvantageous to a microorganism. An equally important issue concerns the relationship between fitness and the level of expression of metabolic functions that are *required* for growth in a particular environment. This latter issue is generally much more difficult to address experimentally, because it necessitates detailed analyses of subtle differences between strains in biochemical activities rather than the mere manipulation of the presence or absence of some function.

Daniel Dykhuizen, Anthony Dean, and Daniel Hartl performed a pioneering study to examine the relationships among genotype, biochemical activities in a required metabolic pathway, and fitness. Their study examined growth on lactose by genotypes of *Escherichia coli* that varied in their levels of expression of the permease that is required for active transport of lactose into the cell and the β-galactosidase that is required for hydrolysis of the lactose. Given that both enzymes are necessary for growth on lactose, how do the activities at each step affect the net flux through this metabolic pathway? And how does net flux affect fitness?

Using metabolic control theory, Dykhuizen and

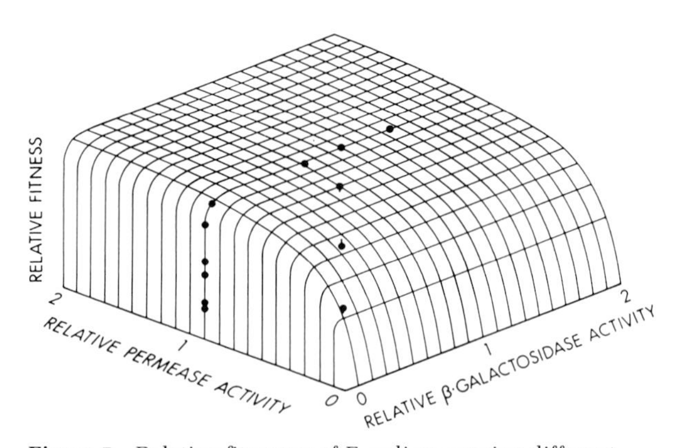

Figure 5 Relative fitnesses of *E. coli* expressing different levels of permease and β-galactosidase activities in the lactose operon. The fitness surface is predicted from metabolic control theory using estimates of the biochemical activities of the two enzymes. Estimates of relative fitnesses are shown as points above or below the fitness surface; these were obtained by competing strains with different enzyme activities in chemostats in which lactose was the sole source of energy. [From Dykhuizen, D. E., Dean, A. M., and Hartl, D. L. (1987). *Genetics* **115**, 25–31. Genetics Society of America, Chapel Hill, North Carolina.]

his co-workers could predict how the equilibrium flux through this pathway would depend jointly on the activities of the permease and β-galactosidase enzymes. They estimated these activities for their genotypes using appropriate biochemical methods. They then predicted that the relative fitness of any two strains would be directly proportional to their relative fluxes whenever lactose provided the sole energy source.

To test the model and its predictions, Dykhuizen and his colleagues estimated the relative fitnesses of the various genotypes in a medium in which lactose provided the sole source of energy for growth. The observed relative fitnesses were extremely close to those predicted from the model, as shown in Fig. 5. Interestingly, their results also show that a small change in permease activity from the wild-type level (=1, after standardization) has a much greater effect on fitness than does a comparable change in β-galactosidase activity. This difference may indicate that the permease activity is suboptimal for lactose transport, perhaps owing to some opposing selective pressure on the permease.

C. Effects of Genetic Background

It is obvious that the fitness effects caused by particular genetic differences depend strongly on the environment into which an organsim is placed. For example, the same antibiotic-resistance gene function that is essential for survival and replication of a bacterium in the presence of antibiotic may hinder growth in an antibiotic-free environment. Similarly, the fitness effects that are due to particular gene functions may often depend on the genetic background in which those genes are found.

For example, one study showed that several different alleles at the 6-phosphogluconate dehydrogenase locus in *E. coli* had very similar fitnesses in a gluconate-limited medium, provided that these alleles were introduced into a genetic background that also encoded an alternative metabolic pathway for 6-phosphogluconate utilization. In a genetic background where this alternative pathway was defective, however, these several alleles had quite variable fitnesses in the gluconate-limited medium.

In another study with *E. coli*, it was observed that the selective disadvantage associated with resistance mutations in a bacteriophage-free environment was reduced by about 50% over the course of several hundred generations of experimental evolution. This improvement resulted from secondary mutations in the genetic background, which compensated for the maladaptive side effects of the resistance mutations but had no effect on the expression of resistance itself.

IV. Genetic Variation within and between Populations

In nature, genetic variation is abundant in most species, including microorganisms. Some of this variation exists within local populations, while other variation may distinguish one population from another. In this section, we will consider studies that have addressed the various processes that influence the maintenance of genetic variation within a population as well as those that may contribute to the divergence of populations.

A. Transient Polymorphisms

A population can be said to be polymorphic whenever two or more genotypes are present in the population. A polymorphism exists, for example, while an advantageous mutation is increasing in frequency relative to the ancestral allele. This type of polymorphism is said to be transient, because eventually the favored genotype will exclude the ancestral genotype. Transient polymorphisms must necessarily exist during any substitution of one allele for another by natural selection.

B. Selective Neutrality

At the other extreme, some polymorphisms may exist almost indefinitely precisely because the alleles that are involved have little or no differential effect on fitness. Such selectively neutral alleles are subject only to genetic drift. Experimental studies have sought to determine whether some polymorphic loci in natural populations of *E. coli* might exist because of selective neutrality, or other explanations are needed. To that end, naturally occurring alleles at particular loci were transferred into a common genetic background, and the fitness effects associated with the various alleles were determined. Even when the bacteria were grown under conditions where growth was directly dependent on the particular enzymatic steps encoded by these loci, there were in many cases no discernible effects on fitness due to the different alleles. These studies thus support the hypothesis that random mutation and genetic drift may be responsible for a substantial amount of the genetic variation that is present within natural populatons. (It should be noted, however, that this does not imply that most *substitutions* of one allele for another are due to genetic drift.)

C. Frequency-Dependent Selection

To this point, we have implicitly assumed that the fitness of one genotype relative to another is independent of their relative abundances. But this may not always be the case. In the course of growth and competition in a particular environment, microorganisms modify their environment through the depletion of resources, the secretion of metabolites, and so on. When this happens, the relative fitnesses of genotypes may depend on the frequency with which they are represented in a population, and the process of selection is said to be frequency-dependent. Frequency-dependent selection can give rise to several different patterns of genetic variation within and between populations, as will now be discussed.

1. Stable Equilibria

Two (or more) genotypes can coexist indefinitely when each genotype has some competitive advantage that disappears as that genotype becomes more common. In that case, each genotype can invade a population consisting largely of the other genotype but cannot exclude that other genotype, so that a stable equilibrium results.

A number of different ecological interactions between genotypes can promote these stable equilibria. For example, an environment may contain two different carbon sources. If one genotype is better at exploiting one resource and the other genotype is superior in competition for the other resource, then whichever genotype is rarer will tend to have more resource available to it, thereby promoting their stable coexistence. In some cases, a resource that is essential for one genotype may actually be produced as a metabolic by-product of growth by another genotype; such interactions are termed cross-feeding. Stable coexistence of genotypes in one population can also occur when the environment contains a population of predators (or parasites); predator-mediated coexistence requires that one of the prey genotypes be better at exploiting the limiting resource while the other prey genotype is more resistant to being exploited by the predator. The evolution of stably coexisting bacterial genotypes from a single ancestral genotype has been demonstrated in several experiments involving both cross-feeding and predator–prey interactions.

2. Unstable Equilibria

Those ecological interactions that promote the stable coexistence of two or more genotypes contribute to the maintenance of genetic variation in populations. However, certain frequency-dependent ecological interactions can actually give rise to *un*stable equilibria. An unstable equilibrium exists when each of two genotypes can prevent the other genotype from increasing in number.

One type of ecological interaction that can give rise to an unstable equilibrium is interference competition. Interference competition occurs when one genotype produces an allelopathic (toxic) substance, which inhibits the growth of competing genotypes; it is distinguished from exploitative competition, which occurs simply by the depletion of resources. Many microorganisms secrete allelopathic compounds, including fungi, which produce antibiotics. Certain strains of *E. coli* produce colicins, which kill competing strains of *E. coli* but to which the colicinogenic (colicin-producing) strain is immune. Colicinogenic genotypes, when common, produce so much toxin that they can exclude a colicin-sensitive genotype that is otherwise more efficient in exploitative competition for a limiting resource. When the colicinogenic genotype is rare, however, the cost of colicin synthesis is greater than the benefit of the resource that is made available to it by the killing of colicin-sensitive cells, and so the colicinogenic genotype loses out to the more efficient colicin-sensitive competitor. (The outcome of competition between colicinogenic and colicin-sensitive genotypes also depends on the physical structure of the environment. In particular, the advantage shifts to the colicinogenic bacteria on surfaces, even when they are rare, because the resources made available by the killing action of the colicins accrue locally, rather than being dispersed as in liquid.)

Ecological interactions that give rise to unstable equilibria do not promote genetic polymorphisms within a particular population. However, they may contribute to the maintenance of genetic differences between populations, because neither type can invade a resident population of the other type.

3. Nontransitive Interactions

In some cases, frequency-dependent selection may give rise to nontransitive competitive interactions. Nontransitivity exists, for example, when genotype *A* out-competes genotype *B* and genotype *B* out-competes genotype *C*, but genotype *C* out-competes genotype *A*. Nontransitive interactions among genotypes were demonstrated in one study with populations of the yeast *Saccharomyces cerevisiae* evolving in a chemostat, in which glucose was supplied as the sole carbon source. Mathematical models of competition for a single growth rate-limiting resource in a spatially and temporally homogeneous environment (such as a chemostat) predict strictly transitive interactions among genotypes, so that the demonstration of nontransitivity apparently indicates the involvement of other limiting factors, such as the accumulation of allelopathic metabolites in the culture medium.

Nontransitive interactions can give rise to situations in which the mean fitness of an evolving population relative to some distant ancestor may actually decline with time, even though each genotype has increased fitness relative to its immediate predecessor. Nontransitive interactions may also maintain genetic diversity within populations over time by recycling genotypes that would otherwise be lost.

D. Divergence of Populations

One very interesting issue is the extent to which experimental evolution in the laboratory is reproducible. If identical genotypes are introduced into identical environments, will the replicate populations adapt to their environment in similar or different ways? One might evaluate similarities and differences in adaptation by several criteria: the extent of improvement in ecological performance (e.g., fitness relative to a common ancestor), the physiological bases of enhanced performance (e.g., increased transport of some limiting nutrient into the cell), or the genetic changes underlying adaptation (e.g., the particular genes in which the beneficial mutations occur). Few studies have addressed this issue directly, and none so far has systematically examined divergence at all of these different levels. Nonetheless, several studies suggest that the process of adaptation by natural selection can be remarkably reproducible, given its dependence on the generation of variation by random mutation. In one study, it was observed that 12 replicate populations of *E. coli* increased in mean fitness *relative to their common ancestor* by >30% during 2000 generations of experimental evolution, and yet during this time the evolving populations diverged in mean fitness *relative to one another* by only 3% or so. Several other studies provide evidence for the similarity of the physiological and genetic changes underlying adaptations in replicate populations. For example, replicated populations of *E. coli* placed in chemostats in which lactose provides the sole energy source almost invariably evolve constitutive expression of the enzymes encoded by the lactose operon.

V. Coevolution of Interacting Genomes

Microorganisms in nature rarely, if ever, exist as single species, as they are usually studied in the laboratory. Rather, they exist in complex natural communities that contain many interacting populations. Some interactions are exploitative: One population makes its living by parasitizing or preying upon another population. Other interactions are mutualistic, so that each population obtains some benefit from its association with the other. In many cases, these interactions can be quite plastic both genetically and ecologically. For example, a single mutation in a bacterium may render it resistant to lethal infection by a bacteriophage. And a plasmid that confers antibiotic resistance may be beneficial to its bacterial host in an antibiotic-laden environment but detrimental in an antibiotic-free environment.

As a consequence of this variability, microorganisms have proven very useful for investigating experimentally the ecological and genetic factors that shape the coevolution of interacting populations. Are there evolutionary "arms races" between host defenses and parasite counterdefenses? Why are some parasites so virulent to their hosts, whereas others are relatively benign? How can mutualistic interactions evolve, if natural selection favors inherently "selfish" genes?

A. Exploitative Interactions

A number of studies have demonstrated the stable coexistence of virulent bacteriophage (lytic viruses) and bacteria in continuous culture. In these studies, the virus population may hold the bacterial population in check at a density that is several orders of magnitude below the density that would be permitted by the available resource if viruses were not present. Typically, however, bacterial mutants eventually appear that are resistant to the virus, and these mutants have a pronounced selective advantage over their virus-sensitive progenitors. The proliferation of bacteria that are resistant to infection by the original virus provides a selective advantage to host-range viral mutants, which are capable of infecting these resistant bacteria. Thus, one can imagine, in principle, an endless "arms race" between resistant bacteria and extended host-range viruses.

In fact, however, there appear to be important constraints that prevent the realization of this outcome. Bacterial mutants may appear, sooner or later, against which it is difficult or impossible to isolate corresponding host-range viral mutants. This asymmetry may arise because bacterial resistance can occur via mutations that cause either the alteration or the complete loss of certain receptors on the bacterial surface, whereas viral host-range mutations can counter only the former. Despite this asymmetry, the virus population persists if the virus-resistant bacterial mutants are less efficient than their sensitive progenitors in competing for limiting resources. In such cases, a dynamic equilibrium is obtained in which the growth-rate advantage of the sensitive bacterium relative to the resistant bacterium is offset by death due to viral infection. Such trade-offs between competitiveness and resistance commonly occur, in fact, because the receptors that are used by viruses to adsorb to the cell surface often serve also to transport nutrients into the cell or to maintain the structural integrity of the cell envelope.

A commonly held belief is that a predator or parasite that is too efficient or virulent will drive its prey or host population extinct, thereby causing its own demise. We have just seen, however, that virulent bacteriophage can stably coexist with bacteria, despite the fact that successful reproduction of the viral genome is necessarily lethal to the infected bacterium. Moreover, the process of natural selection neither requires nor permits foresight, so that the mere *prospect* of extinction cannot deter the evolution of more efficient predators or more virulent parasites. Nevertheless, there do indeed exist many viruses (lysogenic and filamentous bacteriophages) that are replicated alongside the host genome and whose infections, although deleterious, are not necessarily lethal. These viruses, as well as conjugative plasmids, have life cycles that include both horizontal (infectious) and vertical (intergenerational) transmission.

At present, the evolutionary factors that favor these alternative modes of transmission are not fully understood. One factor that is likely to be important, however, is the density of hosts. If susceptible hosts are abundant, then the opportunity for horizontal transmission is correspondingly great. In such circumstances, selection may favor those parasites that replicate and infectiously transmit themselves most rapidly, regardless of the consequences of these activities for the host's fitness. But if susceptible hosts are scarce, then the potential for horizontal transmission is limited. Vertical transmission, by contrast, does not depend on the parasite or its progeny "finding" another host. Instead, the success of a vertically transmitted parasite is inexorably linked to the success of its infected host. The greater the burden that such a parasite imposes on its host, the slower that host will be able to reproduce its own genome and that of the parasite. Hence, when the density of susceptible hosts is low, selection may favor those parasites that minimize their replicative and infectious activities, thereby minimizing their deleterious effects on the host. It is hoped that more studies will address this interesting problem in the future.

B. Mutualistic Interactions

It has often been suggested that many mutualisms have evolved from formerly exploitative interactions. Indeed, the hypothesis advanced above implies that, at sufficiently low host densities, genetic elements such as plasmids and phage can persist *only* if they are actually beneficial to the host. Many plasmids do encode functions that are useful to their bacterial hosts, including resistance to various antibiotics, restriction immunity to certain phages, production of bacteriocins, and so on. And some of these plasmids are incapable of conjugation, instead relying exclusively on vertical transmission. Moreover, several studies have demonstrated unexpected competitive advantages for bacteria that are infected by plasmids, transposons, and even temperate phage, relative to cells that are not infected but are otherwise genetically identical.

Two studies have even demonstrated the evolution of mutualistic interactions from formerly antagonistic associations. In one study, the growth rate of a strain of *Amoeba proteus* was shown initially to be severely reduced by a virulent bacterial infection.

The harmful effects of the bacteria were diminished by propagation of the infected amebae for several years, and the amebae eventually became dependent on the bacterial infection for their viability. In the other study, a plasmid initially reduced the fitness of its bacterial host in antibiotic-free medium; after 500 generations had elapsed, however, the plasmid enhanced the fitness of its host in this same medium. Interestingly, the genetic change responsible for the newly evolved mutualistic interaction was in the host chromosome, not in the plasmid genome. Both of these studies demonstrate that hosts can become dependent on, or otherwise benefit from, formerly parasitic genomes, thus giving rise to mutualistic interactions.

VI. Evolution of New Metabolic Functions

Microorganisms exhibit a tremendous diversity of metabolic activities, some of which function in biodegradative pathways (catabolism) while others work in biosynthetic pathways (anabolism). How has this diversity evolved? One major avenue of research in the field of experimental evolution seeks to elucidate the various processes by which microorganisms can acquire new metabolic functions. This research is particularly timely as humans seek to harness microorganisms that can be used, for example, to degrade toxic pollutants in the environment. [*See* BIOREMEDIATION.]

A. Acquisition by Genetic Exchange

Perhaps the simplest way in which a microorganism can acquire some new metabolic function is by genetic exchange with another microorganism that already possesses that function. For example, antibiotic resistance functions are frequently encoded by plasmids, which are transmitted from donors to recipients by conjugation. Acquisition by genetic exchange is not always so simple a solution, however. Effective biodegradation of certain recalcitrant compounds may require complex coordination of several steps in a biochemical pathway, which are encoded by complementary genes from two (or more) different microorganisms. The acquisition of activities that depend on such pathways may typically require not only genetic exchange, but also subsequent refinement of the new function by mutation and natural selection.

B. Changes in Regulatory and Structural Genes

In several cases, microorganisms have been shown to acquire new metabolic activities without any genetic exchange. Instead, the acquisition of a new metabolic function may occur by selection for one or more mutations in existing regulatory or structural genes, which normally have some other function. For example, the bacterium *Klebsiella aerogenes* cannot normally grow on the sugar D-arabinose, although it does possess an enzyme, isomerase, that is capable of catalyzing the conversion of D-arabinose into an intermediate, D-ribulose, which can be further degraded to provide energy to the cell. This isomerase is normally expressed at a very low level, however, which does not permit growth on D-arabinose. Mutations that increase the level of expression of this isomerase are sufficient to enable growth by *K. aerogenes* on D-arabinose. The ability of this bacterium to grow on D-arabinose may be further improved by certain mutations in the structural gene, which change the amino acid sequence of the isomerase in such a way as to improve the efficiency of the catalytic conversion of D-arabinose to D-ribulose.

In essence, the evolution of new metabolic activities may depend on the microorganism "borrowing" gene products that were previously used for other metabolic activities. It is perhaps not surprising that this process may sometimes also encroach upon those gene products' previous metabolic activities. Such encroachment could, in turn, favor gene duplication, a type of mutation whereby a single copy of an ancestral gene gives rise to two homologous copies, each of which may subsequently evolve toward different metabolic capabilities.

C. Reactivation of Cryptic Genes

Selection for novel metabolic activities has occasionally revealed the existence of "cryptic" genes, which are apparently nonfunctional but can be made functional by one or a few mutations. Cryptic genes are presumably derived from once-active genes, which have been silenced by mutations that destroyed their functions. In the course of experimental evolutionary studies, the existence of such cryptic genes has been revealed by selection for new mutations that reverse or suppress these earlier mutations, thus restoring the lost metabolic activities.

VII. Evolution of Genetic Systems

The process of adaptation by natural selection requires genetic variation in those characters that influence the survival and reproduction of organisms. The two sources of genetic variation are mutation and mixis. In general, rates of mutation and mixis depend not only on environmental factors (e.g., the intensity of ultraviolet irradiation), but also on the properties of the "genetic system" intrinsic to the organism in question. Here, genetic system is taken to mean all those aspects of the physiology, biochemistry, and reproductive biology of an organism that influence rates of mutation and mixis. For example, organisms have mechanisms of varying efficacy to promote the accurate replication and repair of their DNA. And while sex is an integral part of reproduction for some organisms, many others (including numerous microorganisms) reproduce asexually, so that the resulting progeny are usually genetically identical to their parent and to one another.

Among the most interesting questions in evolutionary biology are those that concern the adaptive significance and evolutionary consequences of alternative genetic systems. Why do some organisms reproduce sexually, whereas others reproduce asexually? If mutation generates variation that is necessary for adaptation by the species, but most mutations have deleterious effects on the individual organism, then what mutation rate is optimal? Might organisms somehow be able to choose only those mutations that are beneficial to them, given their present ecological circumstances?

A. Sexuality and Mixis

The hypothesized advantages for sexuality depend, in one way or another, on the genetic variation that is produced by mixis. Efforts to address these hypotheses have been based primarily on mathematical models and on phylogenetic and ecological patterns of the distribution of sexual versus asexual life cycles. Only a few studies have examined experimentally the evolutionary consequences of mixis, and most of these have used microorganisms, for which it is often possible to manipulate the extent of intergenomic recombination. For example, mixis in bacterial viruses can be manipulated by varying the multiplicity of infection (MOI) of host cells, since recombination of viral genotypes can occur only if two or more viruses infect the same host cell. One study compared the rate of adaptive evolution of a bacterial virus at high and low MOI; the total size of the viral population was standardized for both treatments. The average fitness increased more rapidly under the high MOI (=high recombination) treatment than under the low MOI (=low recombination) treatment. This result is consistent with the hypothesis that sexual populations can adapt more rapidly than asexual populations because two or more advantageous mutations can be incorporated simultaneously in the former, but only sequentially in the latter (see Fig. 1).

Some experiments have suggested that another advantage of mixis may arise when the overall rate of deleterious mutation is high and the effective population size is very small. Such conditions may apply to microorganisms with high error rates during replication (e.g., RNA viruses) or those with relatively large genomes (e.g., protozoa), if their populations are also subject to periodic "bottlenecks." In these cases, deleterious mutations tend to accumulate indefinitely in asexual lineages, a process called "Muller's ratchet" (after the geneticist H. J. Muller, who first described this phenomenon). However, even occasional mixis can purge lineages of their accumulated load of deleterious mutations. This effect occurs because two recombining genomes may each complement the deleterious mutations that are present in the other, thereby producing some progeny with a reduced load of deleterious mutations (as well as other progeny with an increased load, which will tend to be removed by natural selection).

In still other cases, mixis appears to be less an adaptation to recombine genes per se than a coincidental consequence of the movement between cells of parasitic entities. In many bacteria, for example, recombination of chromosomal genes occurs only when cells are infected by viruses (transduction) or plasmids (conjugation). The new combinations of chromosomal genes that may result from such infections will occasionally be advantageous. One need not regard phages and plasmids as benevolent agents of bacterial carnal pleasure, however, because their effects are more often deleterious to the host.

B. Evolutionary Effects of Mutator Genes

"Mutator" genes increase the rate of mutation elsewhere in the genome by disrupting aspects of DNA replication and repair. Mobile genetic elements may also behave like mutator genes, as their physical transposition in the genome can alter the expression of other genes. Several studies have investigated the effects of mutator genes, including transposons, on the evolution of bacteria. These studies have revealed a pattern that seems, at first glance, rather curious (Fig. 6). When a mutator gene is introduced into a population above a certain initial frequency (e.g., 0.01%), it tends to increase in frequency over the long-term. But if that mutator gene is introduced at a frequency below that threshold, then it tends to be out-competed and go extinct over the long-term.

What causes this threshold phenomenon? In a sense, there is an evolutionary race between two clones, one with and one without the mutator gene, to see which one gets the next advantageous mutation. The rate of appearance of advantageous mutations for each clone depends on the product of its population size, N, and its corresponding mutation rate, u. So when the ratio of the mutation rates of the mutator and nonmutator clones, u'/u, is greater than the inverse ratio of their population sizes, N/N', then the mutator clone is more likely to have the next favorable mutation. But when u'/u, is less than N/N', the nonmutator clone, by virtue of its greater numbers, is likely to produce the next beneficial mutation.

This explanation, while almost certainly the correct one for these laboratory experiments, presents two difficulties for understanding the possible evolution of mutator genes in nature. First, if mutator genes are advantageous only when they are common, then how do they *become* common? Second, for how long can this process continue before a mutator clone "uses up" its advantageous mutations? The answer to this second question almost certainly depends on the extent of environmental variability. In particular, it has been hypothesized that the advantage of a mutator clone will progressively deteriorate in a constant environment, as the mutations that produce further improvement in fitness in that environment are exhausted. As a consequence, the ratio of beneficial to harmful mutations caused by the mutator gene will decline, and its effect will become progressively more deleterious with time.

Thus, we see that aspects of genetic systems that increase variation—whether by mutation or mixis—may accelerate adaptive evolution. But mutation and mixis can also break down genotypes that are already well adapted to particular environments. The evolution of genetic systems may therefore represent a balance between these opposing pressures.

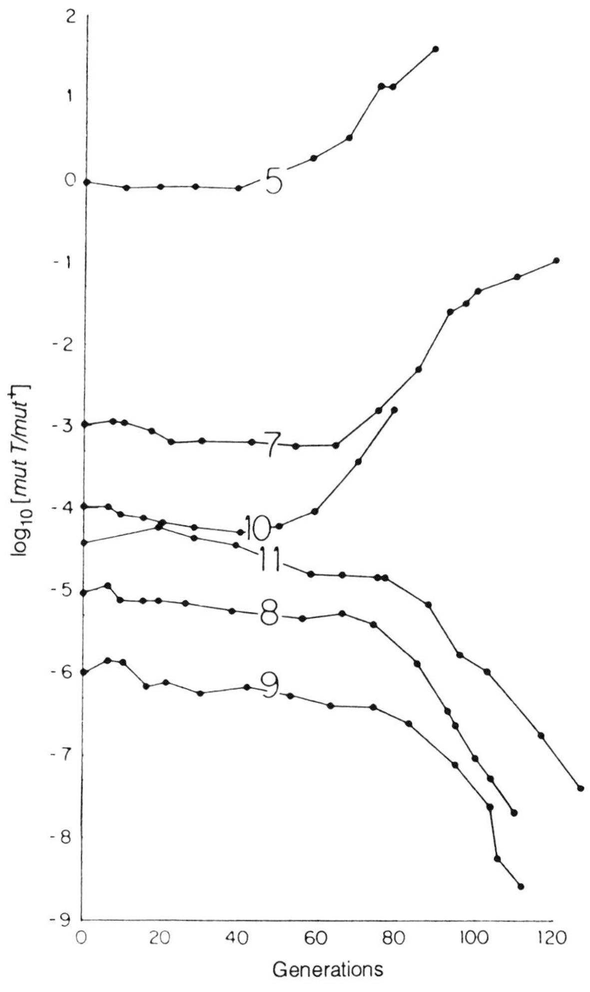

Figure 6 Changes in the frequency of an *E. coli* clone expressing a mutator gene during competition with a nonmutator (or wild-type) clone. The several numbered lines indicate separate populations in which the initial frequency of the clone with the mutator gene was varied over several orders of magnitude. When the clone expressing the mutator gene is initially present above a certain threshold frequency, it is likely to get a highly advantageous mutation before the nonmutator clone gets such a mutation. When the initial frequency of the mutator gene is below this threshold, however, the greater numbers of the nonmutator clone are more than sufficient to offset the higher mutation rate of the mutator clone. [From Chao, L., and Cox, E. C. (1983). *Evolution* **37**, 125–134. Society for the Study of Evolution, Santa Barbara, California.]

C. Directed Mutation: A Controversy

We have seen already how the fluctuation test of Luria and Delbrück and the replica-plating experiment of the Lederbergs were used to demonstrate that bacterial mutations arose prior to the cells' exposure to the selective agent and, hence, could not have been caused by the organism's response to that agent. Recently, the generality of spontaneous mutation has been called into question, however, as several studies have purported to show that certain bacterial mutations occur only (or more often) when the mutants are favored. Two lines of evidence may suggest the existence of these so-called "directed" mutations. (1) The distribution of certain mutants among replicate cultures has a variance that is lower than that expected under the Luria–Delbrück hypothesis of spontaneous mutation. (2) Certain mutants appear on selective media after delays that are inconsistent with the mutations having occurred prior to this exposure, contrary to the Lederbergs' test.

At this time, however, it is unclear whether or not these claims of directed mutation will withstand further scrutiny. With respect to the first type of evidence, it has been shown that many processes other than directed mutation can produce similar deviations from the Luria–Delbrück hypothesis. For example, if mutants grow more slowly than nonmutants prior to selective plating, then this also reduces the expected variance of the distribution of mutants among replicate cultures. Similarly, the fact that certain mutations occur after plating on selective medium does not necessarily imply that the cell perceives the selective agent and mutates accordingly. For example, the rates of certain classes of mutation may increase sharply as cells starve, irrespective of the presence or absence of the selective agent that permits growth of the mutants. General acceptance of the directed mutation hypothesis will require more careful experiments to exclude these alternative hypotheses, as well as demonstration of a physiological mechanism that permits directed mutation.

Acknowledgments

I wish to dedicate this article to my parents, Jean and Gerry Lenski. I thank Al Bennett and Dan Dykhuizen for valuable comments and discussion. Preparation of this article was supported by research grants from the Whitaker Foundation and the National Science Foundation.

Bibliography

Bell, G. (1988). "Sex and Death in Protozoa." Cambridge University Press, Cambridge.

Dykhuizen, D. E. (1990). *Annu. Rev. Ecol. Syst.* **21,** 373–398.

Dykhuizen, D. E., and Dean, A. M. (1990). *Trends Ecol. Evol.* **5,** 257–262.

Futuyma, D. J. (1986). "Evolutionary Biology," 2nd ed. Sinauer Associates, Sunderland, Massachusetts.

Hartl, D. L., and Clark, A. G. (1989). "Principles of Population Genetics," 2nd ed. Sinauer Associates, Sunderland, Massachusetts.

Hartl, D. L., and Dykhuizen, D. E. (1985). The neutral theory and the molecular basis of preadaptation. *In* "Population Genetics and Molecular Evolution" (T. Ohta and K. Aoki, eds.) pp. 107–124. Japan Scientific Societies, Tokyo.

Lenski, R. E. (1988). *Adv. Microb. Ecol.* **10,** 1–44.

Lenski, R. E. (1989). *Trends Ecol. Evol.* **4,** 148–150.

Levin, B. R. (1988). *Phil. Trans. R. Soc. Lond. B* **319,** 459–472.

Levin, B. R., and Lenski, R. E. (1983). Coevolution in bacteria and their viruses and plasmids. *In* "Coevolution" (D. J. Futuyma and M. Slatkin, eds.) pp. 99–127. Sinauer Associates, Sunderland, Massachusetts.

Michod, R. E., and Levin, B. R. (eds.) (1988). "The Evolution of Sex." Sinauer Associates, Sunderland, Massachusetts.

Mortlock, R. P. (ed.) (1984) "Microorganisms as Model Systems for Studying Evolution." Plenum Press, New York.

Evolution, Viral

Stephen S. Morse
The Rockefeller University

Glossary

Evolution Dynamic process defined (in its biological sense) by Darwin as "descent with modification"

Genome Genetic information content of an organism; normally made of DNA, except for RNA viruses, whose genomic material is RNA

Homology Similarity due to common origin or genetic relatedness; also used loosely (if, strictly speaking, incorrectly) to refer to similarities in genetic (DNA or RNA) sequences being compared

Mutation Alteration in a genomic sequence

Negative-strand virus RNA virus whose genomic RNA has the opposite polarity (i.e., is the complementary sequence) to messenger RNA

Phylogeny Evolutionary history of a group

Positive-strand virus RNA virus whose genome has the same polarity as messenger RNA

Retroelement Genetic element (DNA or RNA sequence) that encodes a reverse transcriptase

Reverse transcriptase RNA-dependent DNA polymerase, an enzyme that uses an RNA template to make a DNA copy

Retrovirus Family of RNA viruses possessing the enzyme reverse transcriptase

BIOLOGICAL, OR ORGANIC, EVOLUTION is the concept that living species are not static archetypes but undergo change, and the process by which this occurs. Charles Darwin's formulation, "descent with modification," is widely accepted as a definition of evolution. Evolutionary biology since Darwin has generally encompassed two major questions: the study of the evolutionary history of life (lines of descent) and the study of the causes and mechanisms of evolution. Virologists generally accept the assumption that viruses have evolved and continue to do so. The assumption that this occurs in nature is supported by evidence such as geographic differences between viral isolates (for many viruses that have been studied) and the periodic emergence of new viral variants, of which influenza A virus provides the best-known examples.

I. Background and Methods

A. General Background

Specifically, viral evolution today largely signifies molecular mechanisms of viral variation and genetic relationships of various viruses. In the past, the term viral evolution often referred to the origins of viruses.

However, the scientific study of viral evolution is still young. Until recently, many of the major tools for studying evolution in animals and plants were not applicable to viruses because of their small size and other limitations. The recent advent of techniques for studying molecular evolution by genetic sequence analysis revolutionized the study of evolution for all organisms, and many evolutionary questions about viruses are beginning to be addressed.

Viral evolution has both scientific and practical interest. People are naturally curious about origins and wish to understand how life developed. The apparent simplicity of viruses as well as their great diversity makes them seem especially interesting objects for evolutionary inquiry. On the practical side, current taxonomic classification systems in biology are theoretically based on grouping organisms by relatedness to one another, which requires making assumptions about evolutionary lineages. Classifying an organism is therefore influenced by what we know, or what we believe, about its evolutionary history. This type of classification is still in the promising early stages of being adapted to the taxonomy of viruses.

Perhaps most important, viral variation must be taken into account in the development of vaccines and antiviral agents. Many feared diseases are caused by viruses, so the ways in which viruses evolve, and whether or not viral evolution can lead to the development of new diseases, has serious practical consequences in human and animal health and in evaluating strategies for vaccine development. [*See* VIRUSES, EMERGING.]

This article will review in broad outline some of our present knowledge of viral evolution, attempting to place the information available within a general evolutionary framework. The vast, and rapidly expanding, amount of information now available on viral genetic sequences and the large literature on evolutionary theory in other organisms makes it impracticable to cover all aspects of viral evolution in a article of this scope.

B. Historical Background

Charles Darwin did not actually use the word "evolution" but, rather, referred to his work as the "theory of descent with modification through natural selection." It is generally agreed that the idea of natural selection in evolution was a major innovation, offering as it did a mechanism through which evolution could act. In the century since Darwin (and, independently, Alfred Russel Wallace) first made evolutionary biology a viable scientific discipline by introducing the concept of natural selection, interpretation of evolutionary events has sparked some of the most enduring controversies in biology. Some debates that are continuing today include whether natural selection is the actual cause of evolution or merely a passive agent acting after the fact, at what level natural selection and significant evolutionary change are carried out (whether the targets of selection are genome, individual, species, or ecosystem), and what determines the tempo of evolutionary change. Such controversies are likely to continue and, in fact, are healthy as a stimulus to further analysis and new research. The complexity of biological systems and of their interactions with their environments as well as the difficulty of studying evolution at work have made evolutionary theory necessarily more complex than the "Central Dogma" of molecular biology, the other great unifying concept in twentieth-century biology.

Despite this complexity, the concept of evolution has had a profound influence on the development of biological thought, and its value is unquestionable. The concept has been applied to all of the living world. As J. Maynard Smith stated, "the main unifying idea in biology is Darwin's theory of evolution through natural selection." Dobzhansky's famous statement that "nothing in biology makes sense except in the light of evolution" is also often quoted; Dobzhansky himself liked the notion so much that he made it the title of an article. The broad implication of biological evolution—that living things are not static and that populations undergo change, that change is a fact of life—is now taken for granted in virtually every field of biology. While evolutionists still debate the exact role of natural selection as a cause of evolution, its importance in evolution is also widely accepted. [*See* PERIODIC SELECTION; EVOLUTION, EXPERIMENTAL.]

C. Methods for Studying Viral Evolution

Many viral diseases have been known since antiquity. These include, among others, herpes simplex, which was described by Hippocrates, and zoster, which may be described in the same sources; paralytic polio, possibly depicted in an Egyptian frieze; probably measles; possibly influenza; smallpox, probably known to ancients and apparently described in Indian, Chinese, and Persian sources as far back as the fourth century; and Korean hemorrhagic fever (Hantaan), which some believe was described in Chinese medical texts 1000 years ago. However, the actual identification and study of viruses was possible only recently. While virologists early in the twentieth century were aware of Darwin's work, work on viral evolution was difficult because only a limited number of characteristics could be studied.

Viral evolution has always presented the problem of what to study. Classical evolutionists could study organisms in an ecological setting using phenotypic characteristics such as morphology. The nature of viruses makes such classical approaches difficult. Phenotype is difficult to study with viruses, a problem even greater before tissue culture methods were widely available. The number, range, and resolution of phenotypic characters that could be studied were limited. Despite these limitations, some very interesting early work was done, such as Macfarlane Burnet's work with influenza using characteristics such as neutralization by certain antisera and heat stability of certain functions. Antigenic relatedness has also been widely used with other viruses, notably with "arboviruses" (arthropod-borne viruses), and still finds useful practical application for classification.

The development of molecular technologies, first for protein identification and then for nucleic acids, profoundly affected the study of viral evolution, which became almost entirely a field of molecular comparisons. A similar development occurred in the field of evolutionary biology as a whole, where molecular methods revolutionized the study of evolution in all organisms. The various molecular biological methods for estimating genetic relatedness, initially by nucleic acid hybridization, later by directly determining protein or nucleic acid sequences (and most recently using the polymerase chain reaction (PCR) for detection and to generate samples), have been employed.

The development of computer methods using molecular sequence comparisons to construct phylogenetic, or family, trees from molecular sequence data is one of the most powerful and valuable applications of molecular evolutionary theory and has been most useful for elucidating relationships among viruses. Part of the rationale was derived by Kimura as a valuable corollary of his neutral theory (see Section V.C), but, as Joseph Felsenstein pointed out, for the purpose of constructing phylogenies "it does not matter whether nucleotide substitutions are neutral or selective." Successes include valuable insights into the evolution of influenza viruses and of retroviruses, including HIV. [*See* TAXONOMIC METHODS.]

II. Viral Origins

A. Hypotheses Concerning Origins of Viruses

Even phylogenetic methods have not quite resolved the question of viral origins. Some authors have pointed out that the notion of "viral origins" is itself ambiguous and depends on what is considered an origin. Despite this objection, the question is often asked. Historically, three main hypotheses have been proposed to explain how viruses originated: viruses as "degenerate" bacteria; viruses as vestiges of precellular life; and viruses as genetic sequences that originated, or escaped, from cells. The last theory is the most widely accepted view at present.

When viruses were first identified, bacteria had recently been recognized as major causes of disease. It was already known that viruses are generally smaller than bacteria. Reasoning by analogy, it was easy to envision viruses as "degenerate" bacteria. This view of viral origins held for a number of years. The great Sir Macfarlane Burnet as recently as 1952 spoke of "the likelihood that a virus like influenza is the diminished descendant of some ancient bacterial form." At that time, the *Rickettsia* and *Chlamydia* were also classified with the viruses. These organisms have clear bacterial affinities, and it would be reasonable to argue today that they were descended from bacterial ancestors. This is not true of viruses, however, and the regressive evolution of viruses was gradually abandoned as more detailed molecular characterization began to elucidate major differences between viruses and cellular organisms including bacteria.

The view that viruses were prebiotic nucleic acid, vestiges of earlier (and possibly otherwise largely extinct) life, has been proposed at various times. Many viruses have unique features: The diversity of viral genome types, particularly the uniquely viral use of RNA as primary genetic material and especially the existence of negative-strand genomes, were major arguments in favor of this hypothesis. It was also once thought that most viruses showed only limited genetic homology with cells, further emphasizing a separate origin of these viruses. As opinion changed on this matter, the prebiotic theory has fallen out of favor. The exquisite dependence of viruses on host cells was also a strong argument against the prebiotic view for most viruses; it was hard to understand how viruses, dependent as they are on host cells for their replication, could predate the evolution of this indispensable synthetic machinery. However, after being in abeyance for several years, the prebiotic theory has been revitalized by recent findings that some RNA molecules possess enzymatic activity and can undergo self-splicing. RNA molecules with enzymelike catalytic activities are termed ribozymes. Thomas Cech and Sidney Altman received the 1989 Nobel Prize for Chemistry for this discovery. In 1986, Walter Gilbert suggested that this function of RNA might stem from an ancient origin of RNA-based "life" predating the advent of DNA, a stage he called the RNA world. Although the notion of an RNA world remains controversial, it can be invoked for the origin of at least some RNA viruses and would at least partially explain why there are so many negative-strand viruses, despite the fact that their genomes are in a form that cannot be read by the cell. It could also be used to explain the great variety of genome structures in RNA viruses, representing almost every conceivable variation on expression of RNA as genetic material. Some still favor a prebiotic origin for the small RNA agents known as viroids.

Since roughly the 1970s, an increasing consensus has favored the endogenous theory—that viruses originated from cellular genetic sequences. This is now the most widely accepted view as well as probably the best supported by evidence. Howard Temin has been the most prominent current proponent and elaborator of this view for animal viruses. The original formal expression by Temin was his protovirus theory, which "suggests that leukemia viruses do not preexist but arise from other elements, protoviruses, by genetic change." Temin later refined and generalized the hypothesis. For retroviruses, the supporting data are extensive. As suggested by the protovirus theory, such genetic elements can be seen as past or potential (prospective) viruses.

Although Temin developed his original theory with retroviruses, oncogenes, and cellular genes controlling growth and differentiation, apparent similarities between viral genetic sequences and other types of mobile elements lend additional support to the view that this may be generalizable to most, and perhaps all, viruses. While an origin in cellular DNA seems straightforward for retroviruses and DNA viruses, other RNA viruses, which do not undergo a DNA phase in the cell, could be more problematic. Explaining the origins of specialized viral polymerases is therefore essential to any unified theory of viral origins; this may be possible, at least in principle (Section II.B.1). Various suggestions have also been made for the acquisition of an origin of replication, which would be required for independent existence of a genetic element. On the other hand, at the moment, origins of viral capsid proteins are harder to explain, as fewer cellular homologies are apparent and a suitable mechanism for co-opting both protein (in some cases, perhaps from nucleic acid binding proteins) and the gene coding for it has not yet been demonstrated. However, some putative homologies have been suggested, and a number of possible reasons could be invoked to explain failure to identify cellular homologs for other viral genes, such as divergent evolution, acquisition of ancestral viral genes from host genes that have not yet been extensively sequenced, or origins of ancestral viruses in unstudied organisms that are phylogenetically distant from well-characterized hosts.

On balance, most of the present evidence would seem to favor a cellular origin for many viruses, presumably from mobile genetic elements, although other origins cannot definitively be ruled out. It is also conceivable, if less likely, that different types of viruses might have had different origins.

A question an evolutionist might ask is how many separate lineages of viruses exist. Are all existing viruses descended from a single ancestor, or did individual ancestors arising at different times in evolutionary history give rise to different virus families? Stated formally, are viruses a true monophyletic group (all descended from a single common ancestor), or are they polyphyletic? The evidence is insufficient to decide, but the great diversity of known viruses might seem to argue against a monophyletic origin. Although the question seems abstract, determination of ancestry is important for classification, among other uses.

The most conservative scheme might suggest separate origins for DNA viruses, perhaps with retroviruses, and for several groups of RNA viruses. Baltimore is representative of the majority of informed opinion when he suggested in a 1980 article that "within any one class of viruses, it is reasonable to suggest that they evolved one from another. It is more difficult to see how negative strand viruses and positive strand viruses could be related. . . . [F]rom many points of view the RNA polymerases of negative and positive strand viruses seem very different and must have either arisen separately or be only very distant relatives." Speaking of his four classes of RNA viruses (positive strand, negative strand, double-stranded RNA, and retroviruses), Baltimore concluded that they were likely to be "four independent evolutionary lines" and further that "[t]hree of the viral systems may well have shared genes but the double-stranded RNA viruses appear to represent a very different evolutionary line." He speculates that the latter may have arisen in fungi since they are common there. A recent phylogenetic analysis of viroids (small pathogenic RNA agents of plants, which are smaller than viruses and lack protein coats) hypothesizes a monophyletic origin for these agents.

More detailed recent molecular genetic analysis has allowed greater efforts to bridge the gaps between viral families (Section IV.C), in superfamily classifications. While such arrangements might seem to indicate a separate origin for each superfamily, Ellen and James Strauss advance the suggestion that the high evolutionary rate of RNA viruses could have allowed all of the present families to arise by divergence from a small number of common ancestors, possibly even a single common ancestor. The evidence is insufficient to prove this conjecture, but apparent homologies of replicases and other functions may be consistent with this suggestion.

B. Host Gene Homologies and Origins of Viral Functions

1. Nucleic Acid Replicases

Recent work defining putative phylogeny of nucleic acid polymerases might also help to resolve issues of viral origins. The negative-strand RNA viruses require specialized transcriptases to allow their RNA to function in the host cell, but most viruses, including many of the DNA viruses whose genomes could be read and replicated by host cell enzymes, possess their own specialized polymerases. Some homologies have been suggested both between cellular and viral polymerases and among polymerases of different viral families. While there is still no clear phylogeny that encompasses all the known polymerases, remarkable strides have recently been made in identifying motifs (short regions of similar nucleic acid or amino acid sequence) common to many of them. Gregory Kamer and Patrick Argos, Olivier Poch and his colleagues, and a group including Teresa Wang, Scott Wong, and David Korn, have identified common motifs in many RNA and DNA polymerases (Fig. 1). It is not possible at this point to ascertain whether these similarities are the result of a common ancestry or of convergent evolution (unrelated proteins that share necessary similarities imposed by their similar functions), but it is tantalizing to speculate about possible common origins.

2. Other Cellular Homologies

It should not be surprising that there are many homologies between cellular and viral genes. After all, in the words of Joshua Lederberg, "the very essence of the virus is its fundamental entanglement with the genetic and metabolic machinery of the host."

Many examples are known, and molecular comparisons will discover more similarities both between virus and host and between different viruses.

Additional recent examples have involved promoter sequences and other transcriptional regulatory elements, which are ubiquitous in both cells and viruses. Indeed, many apparent homologies involving such elements can be found in a variety of otherwise apparently unrelated viruses, such as HIV and SV40. It is tempting to think that this may reflect common origin from cells, at least of the sequence if not of the viruses themselves, but this is another situation where one may favor homology (common origin) as an explanation but cannot rule out convergence.

Retroviruses afford many examples of clear homologies. The best known are the numerous homologies between cellular oncogenes, which control growth and differentiation, and retroviral oncogenes, which were discovered because they caused transformation when inappropriately expressed in the infected cell.

A remarkable probable homology involves superantigens in mice. Superantigens, some of which are coded by host *Mls* (minor lymphocyte stimulatory) genes, interact with products of the major histocompatibility complex (MHC) and play an important role during thymic T-cell development by inducing selective deletion of specific clones. (By contrast, these products induce cell division in mature lymphocytes.) This process is thought to be important in regulating T-cell repertoire and eliminating self-reactive clones. In 1990 and 1991, several groups reported evidence that a retrovirus, mouse mammary tumor virus (MMTV), encoded a superantigen (coded in its 3′ long terminal repeat), and that T-cell clonal deletions in some mouse colonies could be attributed to infection with MMTV. These findings also supported long-standing suggestions, as well as some recent data, implicating retroviruses in autoimmune disease. There is recent evidence that HIV may also encode a superantigen.

All the host *Mls* genes have not been definitively identified or sequenced, but recent work suggests that several specific mouse *Mls* genes are encoded by specific endogenous MMTV or other retroviral proviruses, or their cellular homologs. Even more remarkably, a comparison between the MMTV sequences and a portion of the immediate-early gene

```
              Sequence               DIR PPR REF    Species
              --------               --- --- ---    -------
1 2 3 4 5 6 7 8 9101112131415

N L E V I Y G D T D S I M I N  DNA DNA  2  human alpha gene
N L L V V Y G D T D S V M I D  DNA DNA  2  S.cerevisiae pol I
S M R I I Y G D T D S I F V L  DNA DNA  2  herpes simplex virus
E A R V I Y G D T D S V F V R  DNA DNA  2  cytomegalovirus
Q L R V I Y G D T D S L F I E  DNA DNA  2  Epstein-Barr virus
R F R S V Y G D T D S V F T E  DNA DNA  2  vaccinia virus
A E R P L Y C D T D S I I C R  DNA DNA 30  bacteriophage PRD1
E D F I A A G D T D S V Y V C  DNA DNA  2  bacteriophage T4
Y D R I I Y C D T D S I H L T  DNA DNA  2  bacteriophage phi-29
P L K S V Y G D T D S L F V T  DNA DNA  2  adenovirus 2
E V K V I Y G D T D S V F I R  DNA DNA 31  varicella-zoster virus

M Y I I H Y M D - D I L I A G  RNA DNA  7  simian retrovirus(GNLJMP)
L I V I H Y M D - D I L I C H  RNA DNA  7  hamster A-particle(GNHYIH)
C T I L Q Y M D - D I L L A S  RNA DNA  1  virus HTLV-I
S T I V Q Y M D - D I L L A S  RNA DNA  7  virus HTLV-II(GNLJH2)
V I I I Q Y M D - D I L I A S  RNA DNA  7  AIDS virus HIV-II ROD(GNLJG2)
I V I Y Q Y M D - D L Y V G S  RNA DNA  3  AIDS virus HTLV-III
C L A F S Y M D - D V V L G A  RNA DNA  1  human hepatitis B virus
C L A F A Y M D - D L V L G A  RNA DNA  7  squirrel hepatitis B virus(JDVLS)
C V V F A Y M D - D L V L G A  RNA DNA  7  woodchuck hepatitis B virus(JDVLV)
V W T F T Y M D - D F L L C H  RNA DNA  7  Duck hepatitis B virus(JDVLC)
L C M L H Y M D - D L L L A A  RNA DNA  1  Rous sarcoma virus
S L L V S Y M D - D I L I A S  RNA DNA  3  bovine leukemia virus
L I L L Q Y V D - D L L L A A  RNA DNA  1  maloney murine leukemia virus
I Q F G I Y M D - D I Y I G S  RNA DNA  7  visna lentivirus(GNLJVS)
V Q L Y Q Y M D - D L F V G S  RNA DNA  7  equine infectious anemia virus(GNLJEV)
K F C C V Y V D - D I L V F S  RNA DNA  1  cauliflower mosaic virus

I Y V L L Y V D - D V V I A T  RNA DNA  7  D.melanogaster copia transposon(OFFFCP)
K H C L V Y L D - D I I V F S  RNA DNA  7  D.melanogaster 17.6 transposon(GNFF17)
V T I C L F V D - D M V L F S  RNA DNA  7  S.cerevisiae Ty912 transposon(B22671)
V S V I A Y L D - D L L I V G  RNA DNA  7  Dictyostelium DIRS-2 transposon(C24785)

F R M I A Y G D - D V I A S Y  RNA RNA 24  coxsackievirus B3
D R L L F S G D - D S L A F S  RNA RNA 24  cucumber mosaic virus
C K F F A N G D - D L I I A I  RNA RNA 24  tobacco vien mottling virus
S R L I N N G D - D C V L I C  RNA RNA 24  carnation mottle virus
L I G P K C G D - D G L S R A  RNA RNA 24  black beetle virus
V M V T Y G G D - D S L I A F  RNA RNA 24  tobacco ringspot virus
I V Y Y V N G D - D L L I A I  RNA RNA 24  tobacco etch virus
G S L G I Y G D - D I I V P V  RNA RNA  7  bacteriophage GA beta chain(RRBPBG)
G T I G I Y G D - D I I C P S  RNA RNA  1  bacteriophage MS2 beta chain
L R I L C Y G D - D V L I V F  RNA RNA  7  hepatitis A virus(GNNYHR)
N A S C A A M D - D F Q L I P  RNA RNA  1  influenza P2 polypeptide
L K M I A Y G D - D V I A S Y  RNA RNA  1  polio virus
I G L V T Y G D - D N L I S V  RNA RNA  1  cowpea mosaic virus
L K I I A Y G D - D V I F S Y  RNA RNA  7  Rhinovirus 2(GNNYH2)
L K I L A Y G D - D L I V S Y  RNA RNA 24  Rhinovirus 14
V K V L S Y G D - D D L L V A  RNA RNA  1  encephalomyocarditis virus
Y T M I S Y G D - D I V V A S  RNA RNA  1  foot-and-mouth disease virus
K C A A F I G D - D N I V H G  RNA RNA  7  middleburg virus(MNWVM)
R C A A F I G D - D N I I H G  RNA RNA  1  sindbis virus
D C A I F S G D - D S L I I S  RNA RNA  1  brome mosaic virus
N F V V A S G D - D S L I G T  RNA RNA  1  alfalfa mosaic virus
S R M A V S G D - D C V V K P  RNA RNA  7  West Nile virus(GNWVWV)
K R M A V S G D - D C V V R P  RNA RNA  7  yellow fever virus(GNWVY)
I K G A F C G D - D S L L Y F  RNA RNA  1  tobacco mosaic virus
A A Q V Y A G D - D M S I D Y  RNA RNA 26  white clover mosaic virus
P W C I A M G D - D S V E G F  RNA RNA 25  southern bean mosaic virus

- - H H - Y G D - D - H H - -  CONSENSUS(Hy=hydrophobic)
    y y     M         y y
```

Figure 1 Comparison of a conserved amino acid motif and consensus sequence of DNA and RNA polymerases from a variety of cellular and viral sources. Amino acid sequences are in single-letter codes. Note consensus sequence with highly conserved YG/MD-D motif (Y = tyrosine, G = glycine, M = methionine, D = aspartate). DIR, template ("directed by"); PPR, "polymerized product." See original source for key to reference numbers in "REF" column. Letters in parentheses after species names are code names of the relevant files from the Protein Identification Resource (PIR), a molecular sequence databank. [From Argos, P. (1988). *Nucleic Acids Res.* **16,** 9909–9914 (illustration from p. 9910). Reprinted with permission, from, Oxford University Press.]

of *Herpesvirus saimiri* (a lymphotropic herpesvirus of monkeys) showed 25% identity and 46% overall similarity, with one 57-amino acid region showing 43% identity and 60% similarity. As a possible parallel, another herpesvirus, human cytomegalovirus, was reported in 1988 to encode a glycoprotein homologous to an MHC Class I antigen.

Finally, as a last example of virus–host homologies, human DNA apparently contains numerous copies of a DNA sequence homologous with the X gene of hepatitis B virus.

III. Retroviral Phylogeny

Retroviruses have recently provided their share of examples in viral evolution. Of all viral phylogenies, that of retroviruses is probably the most explored and best understood at present. Temin and Baltimore have independently suggested a retroviral origin from moveable genetic elements (transposons) in cells. Temin has pointed out the striking similarities both among various types of cellular sequences that encode a reverse transcriptase (retroelements), including several in yeast and *Drosophila,* and between retroelements and transposons. In addition to cellular and retroviral homologs of oncogenes, numerous endogenous retroviruses (retrovirallike DNA sequences integrated into chromosomal DNA in the host germline) are known in mice, humans, and other mammalian species. Transposition of a human retroelement was recently shown to cause a genetic disease, hemophilia A, which occurred when the element inserted into the gene for clotting factor VIII, disrupting the gene. Other retrovirallike mobile genetic elements have been identified, including a class of hybrid dysgenesis elements in *Drosophila* and "retroelements" in a wide variety of other species, including bacteria. This last is also supported by the recent identification in bacteria of reverse transcriptase and of a defective bacteriophage in *Escherichia coli,* retronphage ϕR73, that contains a retroelement and integrates into a bacterial gene.

The retroviral phylogeny devised by Russell Doolittle and his colleagues unites all genetic elements that possess reverse transcriptase as the offspring of a common ancestor of comparative antiquity. Sharing a common descent would be reverse transcriptase-encoding genetic elements in yeast, *Drosophila,* humans, and many other eukaryotes. These lineages include not only retroviruses but also the hepadnaviruses (hepatitis B virus and related viruses infecting other mammals and birds) and the caulimoviruses of plants, which contain DNA but also encode a reverse transcriptase, a feature that makes these viruses "retroviruses in disguise."

In computing phylogenetic distances from molecular sequence data, it should be noted that, in general, different genes can diverge at different rates due to differing mutation rates, presumably as a result of selective pressure. It has long been known that sequences coding for essential functions show a slower rate of change than do pseudogenes, for example. Doolittle and his colleagues looked at the evolutionary rates of change of 10 genes from retroviruses. Overall, the reverse transcriptase showed the slowest rate of change, and the outer portion of the envelope protein the most rapid, evolving three times faster. The core portion of the *gag* protein changed about 1.6 times as fast as the transcriptase, the protease 1.8 times as fast, and the 140 amino acids at the amino terminal of *gag* 2.5 times as fast. Doolittle has published phylogenetic trees for mammalian retroviruses based on these calculations (Fig. 2). One surprise was an apparent recombination event between two murine retroviruses, Moloney murine leukemia virus and MMTV.

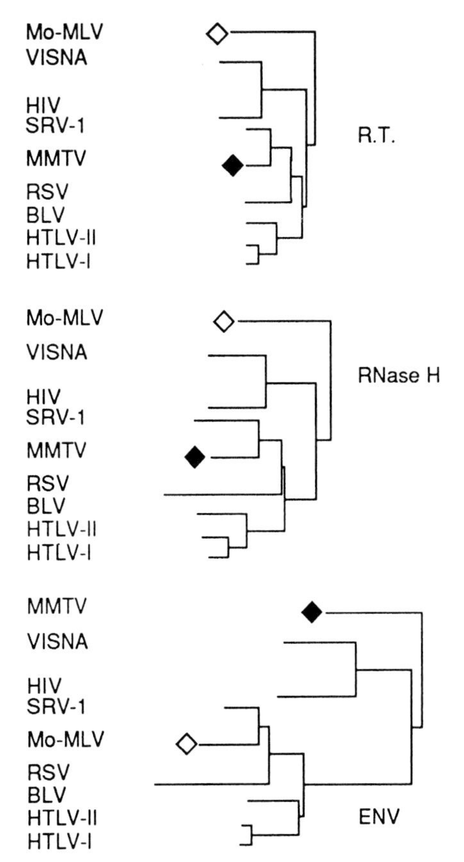

Figure 2 Phylogenetic tree analysis of retroviruses by Doolittle, based on analysis of genes for reverse transcriptase (R. T.), ribonuclease H (RNase H), and envelope (ENV). Viruses included in the analysis are Mo-MLV, Moloney murine leukemia virus; VISNA, Visna (lentivirus); HIV, human immunodeficiency virus; SRV-I, Simian retrovirus I; MMTV, mouse mammary tumor virus; RSV, Rous sarcoma virus; BLV, bovine leukemia virus; HTLV, human T-lymphotropic virus. Diamonds indicate a past recombination event between Mo-MLV and MMTV involving the *env* (envelope) gene. Open diamond, Mo-MLV; closed diamond, present MMTV sequence. [From McClure, M. A., Johnson, M. S., Feng, D.-F., and Doolittle, R. F. (1988). *Proc. Natl. Acad. Sci. USA* **85,** 2469–2473.]

From phylogenetic analysis, Russell Doolittle

suggested that retrovirus evolution is sporadic, with retroviruses evolving at different rates in different situations. For instance, the human endogenous retroviral element HERC is shared with chimpanzees, indicating no change in over 8 million years, whereas strains of HIV have diverged in mere decades. Doolittle suggested that retroviruses became infectious relatively recently, certainly after the emergence of vertebrates, arising from endogenous retroviruses that escaped one species and infected another by horizontal transmission. Murine retroviruses have demonstrated this property in the past. A retrovirus then may integrate into germline cells as a new endogenous retrovirus, and Doolittle suggests that the fate of most endogenous retroviruses is eventually to become degenerate sequences accumulating in the germline. [*See* RETROVIRUSES.]

IV. Viral Taxonomy and Diversity of Genome Types in Viruses

A. Criteria Conventionally Used in Viral Taxonomy

To appreciate some of the difficulties of sorting out viral taxonomy and evolution, let us consider genomic types in viruses and their classification. As an increasing number of viruses were examined by electron microscopy and chemical analysis, the morphologic and especially genetic diversity of viruses became apparent. Viruses can be divided into major groups defined by morphology (size, general shape, whether or not surrounded by a membranelike envelope) and by the type of nucleic acid present in the viral particle. These criteria, along with some host-range criteria, serve as the major basis for the official taxonomy of viruses, which is the responsibility of the International Committee on the Taxonomy of Viruses. In 1991, the committee executive body agreed to accept a definition of viral species as a "class of viruses that constitutes a replicating lineage and occupies a particular ecological niche."

Viral morphology is largely constrained by limited coding capacity and the ability of subunits to self-assemble, so viruses generally fall into one of several possible shapes, usually icosahedral or helical. On the other hand, there is great diversity in the nucleic acid. Some viruses contain DNA genomes, while others use RNA as their genetic material. After the universality of DNA genomes in cellular organisms became generally appreciated, it was realized that the RNA viruses were unique in their use of RNA as primary genetic information. It also became evident that the RNA viruses were not a monolithic group but, rather, different types of RNA viruses used different forms of RNA as genetic information.

B. Classification of Genome Type

In a seminal 1971 article in *Bacteriological Reviews*, David Baltimore categorized viral genome types into broad classes depending on how messenger RNA (mRNA) is coded and produced during infection. Baltimore pointed out that the virus has two requirements: synthesis of viral products, which requires mRNA-directed protein synthesis, and replication of the viral genome.

DNA viruses contain either double-stranded or single-stranded DNA. Families of double-stranded DNA viruses of vertebrates include the Adenoviridae, Herpesviridae, and Poxviridae, and the Hepadnaviridae (hepatitis B virus and relatives; the Hepadnaviridae are actually somewhat unlike the other families of DNA viruses of animals). Single-stranded DNA viruses include the parvoviruses and the DNA bacteriophages ϕX174 and fd. In DNA viruses, the infected cell produces mRNA from the viral genome in exactly the same way it ordinarily does when it transcribes its own cellular DNA. For viral replication, although the cell can carry out DNA replication, most DNA viruses use their own specialized DNA polymerases, presumably for greater efficiency.

It is in the RNA viruses that the uniqueness and diversity of viral genetic expression becomes clearly apparent. RNA viruses represent a great diversity of viruses and replication strategies. One group, viruses with double-stranded RNA, such as the reoviruses, are in a formal sense analogous to the DNA viruses: One strand of the RNA has the same polarity as the mRNA for a particular gene and the other is complementary. However, double-stranded RNA is unusual even among viruses; most RNA viruses contain only single-stranded RNA. The single-stranded RNA viruses can be subdivided into two major classes depending on whether or not the RNA in the viral particle (virion) is functionally equivalent to mRNA (i.e., if introduced into a host cell, the virion RNA would at least formally code for viral proteins just like a mRNA). In Baltimore's convention, mRNA is termed "positive," and viruses whose virion RNA is essentially like viral mRNA are therefore termed "positive-strand viruses." Many small RNA viruses are positive strand, including the picornaviruses (examples include poliovirus and hepatitis A virus), the togaviruses (rubella), the coronaviruses, all known RNA bacteriophages, and others.

By analogy, RNA of opposite polarity (i.e., RNA whose sequence is complementary to viral mRNA and therefore cannot code directly for viral proteins) is termed "negative," and the viruses possessing this type of RNA in the virion are called "negative-strand viruses." The distinction appears fundamental, as positive and negative strand viruses require different replication strategies. The negative-strand viruses comprise a large and diverse group with many important disease agents of humans and domestic animals. Some families of negative-strand viruses (and representative members) are the Orthomyxoviridae (influenza), the Paramyxoviridae (measles, mumps, Newcastle disease of poultry, and some respiratory viruses), and the Rhabdoviridae (rabies). All exhibit helical symmetry (ropelike or rodlike core; the Rhabdoviruses are bullet-shaped). Another negative-strand family, the Bunyaviridae, is possibly the largest viral family in number of known members, including many arthropod-borne diseases as well as others. Finally, the Arenaviridae family includes several viruses of rodents. All of these are enveloped (enclosed in an "envelope" derived from modified host cell membrane). The viruses now known as the Retroviridae are placed in a separate group. These viruses contain RNA (positive sense) in their virion, but replication in the cell requires a DNA intermediate that is made from the virion RNA by a specialized viral enzyme, reverse transcriptase. The hepadnaviruses (hepatitis B) contain a DNA genome, but considerable evidence suggests that the hepadnaviruses share a common origin with retroviruses, including an obligatory function for reverse transcriptase in their replication.

Unlike RNA viruses, cells do not use RNA as primary genetic material; therefore, replication of RNA must be carried out by virus-coded enzymes that can perform a task for which the host cell is generally unequipped. Because the genome of a negative-strand virus cannot code for products when introduced into the cell, a specialized enzyme to convert the viral genome into mRNA sequences is indispensable. This enzyme is normally carried in the virion and introduced during initial infection. For this reason, negative-strand and positive-strand viruses have somewhat different replication strategies after infection. The origins of these specialized viral enzymes must also be accounted for in any theory of viral origins.

Table I Superfamilies of RNA Viruses (Strauss' System)[a]

Superfamily	Virus groups included[b]	Common features
Sindbis-like	Togaviridae (I/V), Bromovirus (P), Cucumovirus (P), Tobamovirus (P) Ilavirus (P), Tobravirus (P), Potexvirus (P)	RNA genome of plus (positive) polarity; 5′ caps on viral RNA; Messenger RNA is subgenomic; no overlapping ORFs; read-through (in most members)
Picornavirus-like	Picornaviridae (I/V), Caliciviridae (V), Comovirus (P), Nepovirus (I/P) Potyvirus (P)	Genome is positive polarity; a viral protein ("5′-VPg") is attached to 5′ end of viral genome; viral genomic RNA has 3′ poly-A tail; no subgenomic mRNA; viral structural proteins are made as a polyprotein and cleaved; no overlapping ORFs
Minus-stranded	Paramyxoviridae (V), Rhabdoviridae (I, P, V), Orthomyxoviridae (V), Bunyaviridae (I, V), Arenaviridae (V)	Genome RNA is minus (negative) polarity; self-complementary ends; helical capsid; overlapping ORFs
Double-stranded	Reoviridae (I, P, V); fungal viruses (Birnaviridae)	Double-stranded RNA; segmented genome; genomic RNA has 5′ cap and free OH on 3′ end; single-stranded RNA intermediates
Flavivirus-like	Flaviviridae (I, V), Pestivirus (V)	Capsid proteins coded at 5′ end of viral genome; one ORF; 3′ end lacks a poly-A tail
Coronaviridae	Coronaviridae (V)	5′ caps, 3′ poly-A (RNA is of plus polarity)

[a] Retroviruses and hepadnaviruses (hepatitis B and related viruses) are not included; can be assigned to a separate superfamily. Unassigned: Nodaviridae (I). Within a given superfamily, morphology and hosts are divergent characteristics (differ among superfamily members).

[b] Endings in *-idae* are viral families; others are genera.

I; invertebrates only (usually insects); P, plant viruses; V, vertebrates. ORF, open reading frame.

[Modified from Strauss, E. G., Strauss, J. H., and Levine, A. J. (1990). Virus evolution. *In* "Virology" (B. N. Fields, D. M. Knipe *et al.*, eds.), pp. 167–190. Raven Press, New York.]

C. Superfamily Arrangements

Recent superfamily arrangements of RNA viruses can be viewed as an extension of this approach. Such systems use limited nucleic acid homologies, generally in the RNA replicases, and similarities in genome organization and form to classify many of the RNA viruses into major "superfamilies" that unite some conventional viral families into a more comprehensive phylogenetic scheme. The most extensive superfamily scheme so far is that proposed by James and Ellen Strauss (Table I). Other schemes for some viruses based on similar principles have also been devised by Rob Goldbach and Edward Rybicki. A considerable advantage of a superfamily arrangement is that it attempts to emphasize common features, making possible relationships and origins more apparent, especially when combined with formal phylogenetic analysis.

V. Molecular Evolution of Viral Genomes

A. Mechanisms of Molecular Evolution of Viruses

Phylogenetic analysis is based on the comparison of molecular sequences, with the assumption that genes diverge in sequence as they evolve from a common ancestor. Genomes evolve through such genetic mechanisms as mutation, recombination, and gene duplication (Table II). The mechanisms for DNA viruses presumably resemble those available in a host cell, and those DNA viruses that integrate into host DNA are afforded similar opportunities for recombination. Some examples are given in the preceding section on gene homologies with host (Section II.B). For retroviruses, as Temin points out, the existence of the DNA proviral step in viral replication allows opportunities for recombination with host cellular genes in order to introduce new sequences into the retrovirus genome. Many such events are known—for example, with oncogenes; presumably the recombination involving an exchange of envelope genes between Moloney murine leukemia virus and MMTV also occurred during this step.

With other RNA viruses, these possibilities may seem somewhat more limited, and it was once thought that RNA recombination did not occur. However, RNA recombination has long since been demonstrated in the laboratory and has been demonstrated *in vivo* as well. Strauss showed that Western equine encephalomyelitis was an apparent recombinant virus of which one parent was Eastern equine encephalomyelitis; other examples likely exist also. Such findings suggest that both gene homologies and genome organization in viruses may be more fluid than previously believed. The recently identified—and surprising—homology between the esterase protein of influenza C (a negative-strand RNA virus) and coronaviruses (positive-strand) may be additional evidence of such fluidity. This notion of interchangeable functions, swapped among viruses, has been stated more formally as the "modular" scheme of viral evolution.

B. Heterogeneity and Mutation Rates of Viruses

Whatever the significance of recombination events in viral evolution, it is well established that viruses tend to have very high rates of mutation. This is theoretically understandable. J. W. Drake noted

Table II Examples of Mechanisms in Genomic Evolution of Viruses

Mechanism	Example
Base changes (point mutations)	Genetic drift in influenza
Biased hypermutation	Measles virus (U $\rightarrow$ C)
Gene reassortment	Origin of pandemic influenza viruses of 1957, 1968 [surface protein gene(s) from avian virus]
Deletions	Defective retroviruses
Nonsense mutation	Precore mutants of hepatitis B (associated with fulminant hepatitis infection)
Intramolecular recombination	Insertion of gene cassette; Western equine encephalomyelitis as recombinant of Eastern equine encephalomyelitis and Sindbis-like parents
Recombination with host gene	Cellular protooncogenes and retroviral oncogenes; X gene of hepatitis B
Recombination between deletion mutants	Regeneration of functional plant virus genome

[Adapted, with modification, from Kilbourne, E. D. (1991). *Curr. Opinion Immunol.* **3**, 518–524. ©Current Biology Ltd.]

that the spontaneous mutation rates of all living organisms tend to be inversely proportional to the sizes of their genomes. As Manfred Eigen pointed out, because of their small genomes, viruses should therefore mutate more rapidly than organisms with larger genomes. Eigen developed a theoretical treatment of how such error-prone replication would soon result in an extremely heterogeneous collection of viral genomes in the progeny of even a single replicating virus particle. In other words, most RNA viruses actually consist of a population of genomes showing considerable variation around a "master" sequence. This theoretical work on what Eigen terms "quasispecies" populations (H. M. Temin uses the term "swarm" synonymously) has been confirmed with many viruses. John Holland has been particularly influential in establishing and extending these results for animal viruses. More recently, similar "swarms" or heterogeneous populations have been demonstrated in strains of HIV isolated from individual patients.

While this result would theoretically hold true irrespective of whether the genome is DNA or RNA, RNA viruses are generally thought to have higher mutation rates than DNA viruses of the same genome size. This difference has been attributed to the error-prone nature of RNA replication, presumably due to the lack of a "proofreading" function in this process, while most DNA polymerases can correct their errors. John Holland suggests that many viral RNA polymerases may have evolved to be almost as error-prone as is consistent with maintaining the information content of the genome. Although the error frequencies of the RNA polymerases are not known in all cases, the attribution of high mutation rates to error-prone polymerases is reasonable based on estimates that have been made. Several studies have estimated error frequencies of $\geq 10^{-4}$ per site for avian myeloblastosis virus (AMV) reverse transcriptase and comparable misincorporation rates for purified poliovirus RNA polymerase.

Different classes of viruses show different error rates. Temin, Doolittle, and others have pointed out that retroviruses span two worlds in this respect. Reverse transcriptase is error-prone, so infectious retroviruses show a high mutation rate in replication. On the other hand, retroviruses integrated in the host DNA, such as endogenous retroviruses carried in the germline, are copied by relatively accurate host DNA replicative machinery and, hence, evolve very slowly compared to infective retroviruses.

Some estimates have been made of mutation rates during viral replication. While RNA viruses, which have been the most studied, tend to show higher mutation rates overall than DNA viruses, all viruses that have been carefully examined appear to show mutation rates greatly in excess of those found in cellular DNA replication, bearing out Eigen's suggestion. Exact mutation rates vary greatly depending on the method used to detect the mutations as well as on the gene examined. Typical rates reported are on the order of 10^{-5} mutations per base site per replication cycle for influenza, and 10^{-3}–10^{-4} for vesicular stomatitis virus (a rhabdovirus). Reported mutation rates were lower for polio (roughly 10^{-6}, although higher mutation rates have been reported), and higher for retroviruses (10^{-4} to 2×10^{-5} substitutions per base pair per replication cycle for retroviruses like Rous sarcoma virus). Using a system that allowed measuring mutation rates in a single cycle of retroviral replication, Temin obtained a rate for base pair substitution of 7×10^{-6} (per replication cycle per base pair) and approximately 10^{-6} for frameshifts and deletions. Most estimates suggest that these values are at least slightly higher for HIV. There are fewer estimates for DNA viruses, but mutation rates comparable to the RNA viruses can be found cited in the literature. Many DNA viruses use their own DNA polymerases, and these might possibly show less fidelity than cellular polymerases. For comparison, estimated mutation rate for cellular DNA replication in eukaryotes is approximately 1–7×10^{-9} substitutions per site per year.

C. Neutral Mutations and the Neutral Theory

The difference in mutation rates between viruses and cellular organisms is considerable, but this difference may conceptually be more quantitative than qualitative. Beginning in the 1960s, when it became possible to determine amino acid and nucleotide sequences for organisms, it was noted that (in Kimura's words) "calculating the rate of evolution in terms of nucleotide substitutions seems to give a value so high that many of the mutations involved must be neutral ones." Kimura developed the neutral theory as the formal expression of this effect in a population. The neutral theory gives an important role to genetic drift—random mutations—as a mechanism of evolution. In Kimura's view, mutations that are selectively neutral or near neutral can become randomly "fixed" in a population. In the past few years, both Kimura himself and Howard Temin have been interested in applying the neutral theory to viruses. Kimura's group has adduced evidence for neutral mutations in influenza and in HIV, while Temin has extended Kimura's neutral theory to both oncogenes and viruses. Temin has been particularly interested in the idea of mutation-driven evolution, that mutation can drive the evolution of viruses. In Temin's words, "[T]he high rate of virus genetic variation allows mutation-driven evolution. . . . A consequence of the high rate of mutation and recombination in retrovirus replication is that many variant viruses will be present in any retrovirus population. . . . This phenomenon allows retroviruses to undergo multiple mutations and recombinations before they are subject to selection (analogous to the effect of recessive mutations in diploid organisms)."

VI. Viral Evolution in Nature

A. Development of Viral Variants during Infection

Several examples of variants with altered properties arising during natural infection have also been described. Variants were reported in mice infected at birth with lymphocytic choriomeningitis virus (an arenavirus). In this infection, lymphotropic and neurotropic variants of the virus can be identified. Each has a different phenotype, the neurotropic form causing acute, and the lymphotropic chronic, infections. A single amino acid change in the viral glycoprotein (from phenyalanine to leucine at residue 260) was correlated with this altered tropism; both variants could be isolated from many of the same carrier mice that had initially been infected with an acute strain of the virus. Thus, some variants with differing tissue tropisms can be internally generated during infection. The work previously referred to on populations of HIV variants in infected individuals suggests the possibility that HIV variants with differing tropisms could similarly be produced during infection and that this could occur with other infections as well. In what could be an analogous situation in humans, it was recently suggested that fulminant hepatitis B infection in humans was associated with a variant virus containing a mutation in the viral precore region (preceding the gene for core protein). Vaccine escape mutants of hepatitis B were also recently described in infected people. Finally, a group in Japan determined the rate of evolution of the delta antigen gene in two patients infected with hepatitis delta (hepatitis D, an RNA agent that requires hepatitis B as a helper virus and uses its coat protein). They estimated a mutation rate of 0.57×10^{-3} (per nucleotide site per year) in one patient and 0.64×10^{-3} in another.

B. Evolution of Viruses in Natural Infection

While a considerable amount of information on viral mutation and evolution is available from *in vitro* systems, studying viruses in their natural settings has proved considerably more difficult. The evolution of influenza virus in nature was examined by Peter Palese and colleagues. Using phylogenetic analysis to compare genetic sequences from influenza A strains isolated from successive epidemics, they estimated an evolution rate *in vivo* of 2×10^{-3} changes per RNA base site per year. In contrast to the mutation rate already described, this rate describes the actual evolution rate of the virus in human infection. Their tree analysis gave two parallel lines: the main trunk, representing early H1 isolates (virus strains with hemagglutinin subtype 1) and continuing to later isolates with H2 and H3, and a branch, representing the H1 viruses isolated after 1977. Despite having diverged, the evolutionary rate was approximately the same in both lines. This and subsequent analyses also indicate that, with influenza A in humans, the virus exhibits rapid evolution, in which a parent sheds off variants, only one of which may be successful. Rapid replacement appears to be the rule for any variant in the human population.

Although the question is controversial, some evidence suggests that this may be driven by immune selection. Additional supporting evidence is work demonstrating different rates of evolution of influenza nucleoprotein genes in avian and human hosts, which may depend on differing degrees of selection in each host species (Fig. 3). [*See* INFLUENZA.]

The classic study bearing on viral evolution in a natural habitat involves myxoma virus (rabbitpox) in Australia. The virus was introduced to control the rabbit population. Rabbits had been introduced by European settlers some years earlier and their population increase had become a major nuisance. Fenner and colleagues followed the coevolution of both the virus and the rabbits. In terms of the evolution of the virus, a highly lethal strain of virus evolved toward intermediate virulence (Table III). As pointed out by Robert May, the level of virulence eventually attained by the virus (i.e., by the strains or variants that predominated in the population) represented a trade-off between virulence and transmissibility.

VII. Viruses and the Evolutionary Synthesis

A. The Evolutionary Synthesis

The modern theory of evolution now known as the "evolutionary synthesis" was largely forged in the 1940s. The nature of this synthesis, in the words of Ernst Mayr, one of its chief architects, "was not one of great innovations but rather of mutual education. Naturalists who had not known it before learned from the geneticists that . . . [t]here can be no . . . inheritance of acquired characters. . . . Another finding of genetics, its Mendelian (particulate) character, was also finally universally adopted." Mayr writes that the education of naturalists and geneticists was mutual: "The claim, frequently made, that the evolutionary synthesis was nothing but the application of Mendelian inheritance to evolutionary biology overlooks how much the geneticists had to learn from the naturalists about the importance of population thinking, of the geographical dimension, and of the individual as the unit of selection . . . ."

Two major points were to accept that small mutations could lead to major changes and thus eventually to macroevolution and to reaffirm, again in Mayr's words, "the Darwinian formulation that all adaptive evolutionary change is due to the directing force of natural selection on abundantly available variation."

There have been several challenges to the "synthetic theory" since then, including Kimura's neutral theory of mutation (discussed earlier) and punctuated equilibrium. But, in general, most evolutionists emphasize their general agreement with the broad outlines of the "synthesis," especially natural selection. Most recently, the "synthetic theory" has made gradual adjustments to accommodate successive challenges; each challenge was eventually fit into the synthesis by slightly modifying the emphasis of the "orthodoxy."

B. Natural Selection in Viruses

Is natural selection a factor in viral evolution? Finding suitable evidence to answer the question of whether or not and how selection acts in nature is often difficult. As Felsenstein notes, "The controversies between neutralists and selectionists have continued for 20 years with no clear resolution, primarily due to the low resolving power of the data—natural selection many orders of magnitude weaker than we can detect in the laboratory can be effective in nature." For viruses, this is further complicated by the limitations of studying viruses under natural conditions. However, based on such evidence as whether the host influences stability of viral genotype or phenotype, it is reasonable to believe that natural selection does operate in viruses and would appear to be as important as in more familiar organisms.

Some of the work discussed in previous sections of this article, such as the evolution of myxoma virus in Australia and differential rates of variation of influenza genes in avian and human hosts, strongly suggests that natural selection is stabilizing viral variation. Similarly suggestive is a recent report of repeated independent isolation from pigs of the identical pig-adapted influenza variant after inoculation of a different variant that was not pig-adapted. A further indication of strong constraint not predictable from genetic composition of the virus comes from work on the ability of various avian influenza A recombinants to replicate in primates. In these experiments from Brian Murphy's laboratory, several gene combinations that appear as if they ought to replicate adequately in primates are unable to do so.

Despite the great variability and high mutation rates of virus populations, virus strains in nature can also show remarkable relative stability over years or decades at least possibly comparable to the "punctuated equilibrium" proposed by Eldredge and Gould for other organisms. Taken together, such evidence indicates that there are factors strongly stabilizing viral phenotype and even the viral genome. The best candidate for a stabilizing force is natural selection. One can of course debate the precise role of neutral mutation, as two groups recently have, with Palese and Fitch suggesting that the appearance of influenza A variants is driven by immune selection in a Darwinian manner and Kimura's group favoring random fixation of neutral or near neutral mutations, but this does not obviate a central role for natural selection. The framework of evolutionary biology has itself been undergoing adjustment, and this debate is still ongoing, although it would appear that a resolution is possible within the overall framework of the evolutionary synthesis. For example, Kimura differs in such particulars as the level at which selection acts in populations, but

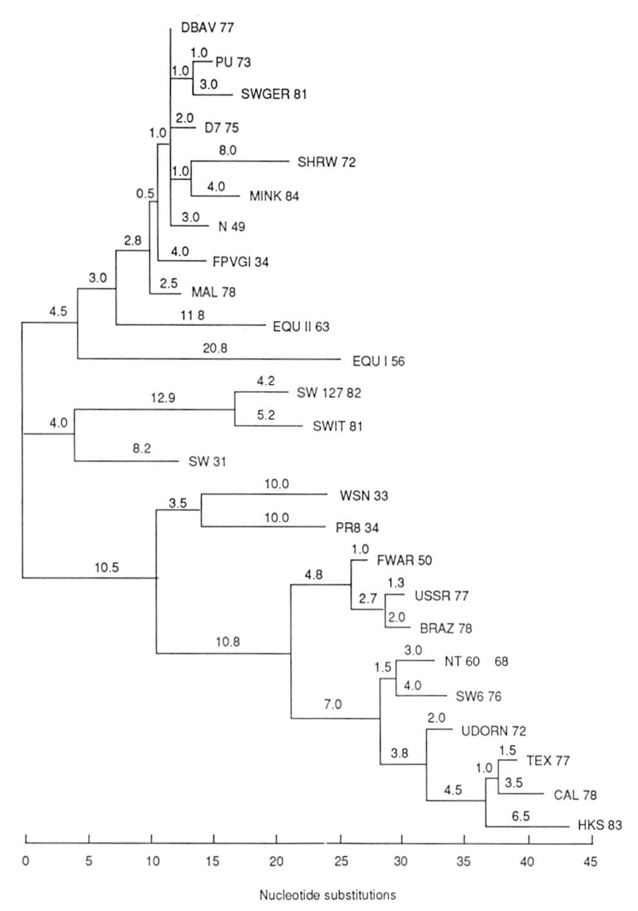

Figure 3 Phylogenetic tree of influenza A virus NS (nonstructural protein) amino acid sequences, comparing influenza A strains isolated from a number of species. Divergence increases to the right. Note that the strains at the top of the diagram are mostly avian strains, and those at the bottom are human isolates. Isolates from other mammals (notably horses, EQU, and pigs, SW strains) tend to fall in the center. Strain designations are as follows: WSN = A/WSN/33, H1N1; PR8 = A/PR/8/34, H1N1; FWAR = A/Fort Warren/1/50, H1N1; USSR = A/USSR/90/77, H1N1; BRAZ = A/Brazil/11/78, H1N1; NT60 = A/NT/60/68, H3N2; UDORN = A/Udorn/307/72, H3N2; TEX = A/Texas/1/77, H3N2; Cal = A/California/10/78, H1N1; HK5 = A/Hong Kong/5/83, H3N2; SW6 = A/Swine/Hong Kong/6/76, H3N2; SWIT = A/Swine/Italy/147/81, H1N1; SW127 = A/Swine/Hong Kong/127/82, H3N2; SW = A/Swine/1976/31, H1N1; SWGER = A/Swine/Germany/2/81, H1N1; DBAV = A/Duck/Bavaria/2/77, H1N1; D7 = A/Duck/Hong Kong/7/75, H3N2; PU = A/Parrot/Ulster/73, H7N1; SHRW = A/Shearwater/Australia/72, H6N5; N = A/Chicken/Germany "N"/49, H10N7; FPVGi = A/FPV/Rostock/34 Giessen-isolate, H7N1; MINK = A/Mink/Sweden/84, H10N4; MAL = A/Mallard/NY/6750/78, H2N2; EQUII = A/Equi/Miami/1/63, H3N8; EQUI = A/Equi/Prague/56, H7N7. The year of isolation is included with the strain. [From Gammelin, M., Altmüller, A., Reinhardt, U., Mandler, J., Harley, V. R., Hudson, P. J., Fitch, W. M., and Scholtissek, C. (1990). *Mol. Biol. Evol.* **7**, 194–200. © 1990 by the University of Chicago Press.

he does not reject a key role for selection; he notes that stabilizing selection is the most prevalent type of selection in nature. In 1990, Kimura proposed a model for evolution, the four-step model, in which a key initial step is the relaxation of an existing constraint such as stabilizing selection.

C. Viruses and the Evolutionary Paradigm

Until the development of molecular methods for evolutionary comparisons, there had been considerable thought about viral and microbial evolution but little data. All the data then possible were based on rather gross phenotypic characters, and influence of genotype could not readily be studied. Since the development of molecular evolutionary methods, a large amount of comparative molecular data have been assembled. More recent studies have consequently emphasized genotype. With the current availability of techniques for studying molecular evolution, and greater quantities of comparative data, it would now seem an opportune time to reconcile genotype and phenotype in viral evolution. Although it is still not generally possible to predict biological characteristics from genotype, increased understanding is gradually emerging. Approaches for defining evolutionary constraints in nature, particularly the role of natural selection, and for defining the molecular bases of host range, virulence, pathogenesis, and host interactions should also be emphasized.

Table III Virulence of Myxoma (Rabbitpox) Virus Strains Recovered from the Field in Australia

Years	Percentage of isolates in virulence grade[a]				
	I	II	III	IV	V
1950–1951	100				
1952–1955	13.3	20.0	53.3	13.3	0
1955–1958	0.7	5.3	54.6	24.1	15.5
1959–1963	1.7	11.1	60.6	21.8	4.7
1964–1966	0.7	0.3	63.7	34.0	1.3
1967–1969	0	0	62.4	35.8	1.7
1970–1974	0.6	4.6	74.1	20.7	0
1975–1981	1.9	3.3	67.0	27.8	0

[a] Grades represent decreasing severity in a standardized rabbit stock used for testing: I is most severe, with case fatality rate >99%; II, case fatality rate 95–99%; III, 70–95%; IV, 50–70%; and V, <50%.
[Data from Fenner, F. (1983). *Proc. R. Soc. (London) B* **218**, 259–285.]

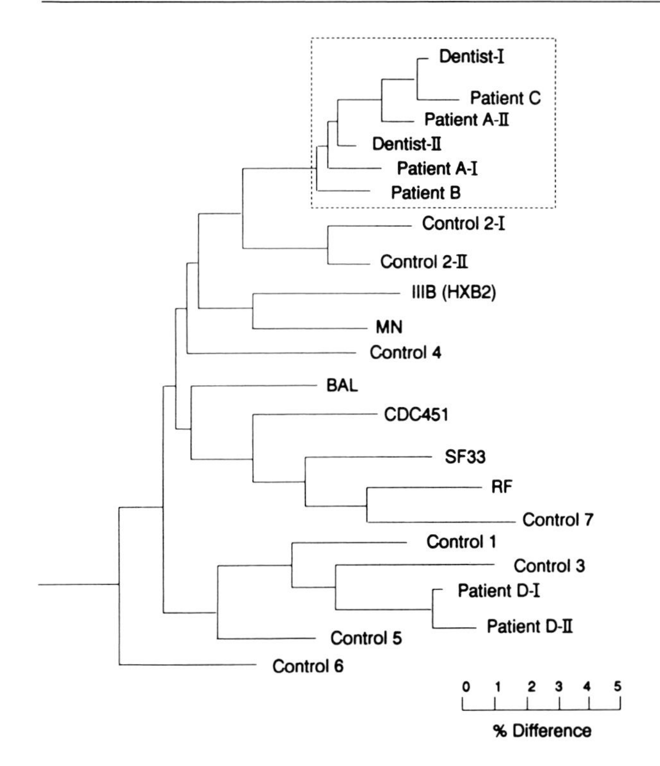

Figure 4 Application of molecular epidemiology to tracking putative source of an HIV infection. Tree analysis of HIV strains from a Florida dentist and from four HIV-infected patients of the dentist (patients A, B, C, and D), compared with local HIV strains (controls 1–7) and standard North American HIV isolates [IIIB (HXB2), MN, BAL, CDC451, SF33, and RF]. [From Centers for Disease Control (1991). *Morbid. Mortal. W. Rep.* **40,** 21–27, 33 (illustration from p. 24).]

Another outgrowth of molecular evolution, the development of phylogenetic analyses for viruses, has shed considerable light and is likely to remain highly productive. A very promising line of molecular epidemiology is developing from comparison of strains over time and from different geographic regions, using the methods already discussed for constructing phylogenetic trees. One recent application was to track an incident of HIV transmission (Fig. 4). Further understanding of geographic variants from the standpoint of evolutionary theory (potentially analogous to geographic variants of animal species developing in isolation) has begun and should be developed further. These discoveries will make the study of viral evolution an exciting challenge for some time to come. As it begins to place viruses in the broader framework of evolutionary biology, virology may also help shed some light on evolutionary processes in general.

Acknowledgments

I thank Joshua Lederberg, Mirko Grmek, Howard Temin, Merrill W. Chase, John Holland, Frank Fenner, Robert May, Walter Fitch, Gerald Myers, Ernst Mayr, David Baltimore, Edwin D. Kilbourne, Robert E. Shope, Mitchell Feigenbaum, Jan Geliebter, Niles Eldridge, Lily Kay, Peter Palese, Hugh Robertson, Baruch Blumberg, Seymour Cohen, Paul Ewald, and Bruce Levin for invaluable comments and discussions. I am supported by the National Institutes of Health (NIH), U.S. Department of Health and Human Services (RR 03121 and RR 01180). Work relating to emerging viruses was additionally supported by the Division of Microbiology and Infectious Diseases (DMID), National Institute of Allergy and Infectious Diseases, NIH, and by the Fogarty International Center of NIH. I am grateful especially to John R. La Montagne, Director, DMID, and to Ann Schluederberg, Virology Branch Chief, for support and encouragement. Portions of this article were adapted from my article, "Evolving Views of Viral Evolution," *in History and Philosophy of the Life Sciences* (1992), in press.

Bibliography

Doolittle, R. F., Feng, D.-F., Johnson, M. S., and McClure, M. A. (1989). *Q. Rev. Biol.* **64,** 1–30.

Felsenstein, J. (1988). *Annu. Rev. Genet.* **22,** 521–565.

Futuyma, D. (1986). "Evolutionary Biology," 2nd ed. Sinauer Associates, Sunderland, Massachusetts.

Gibbs, Adrian (1987). *J. Cell Sci.* **7**(suppl.), 319–337.

Li, W.-H., and Graur, D. (1991). "Fundamentals of Molecular Evolution." Sinauer Associates, Sunderland, Massachusetts.

May, R. M., and Anderson, R. M. (1983). *Proc. R. Soc. (London) B* **219,** 281–313.

Mayr, E. (1988). "Toward a New Philosophy of Biology." Harvard University Press, Cambridge, Massachusetts.

Morse, S. S., and Schluederberg, A. (1990). *J. Infect. Dis.* **162,** 1–7.

Nei, Masatoshi (1987). "Molecular Evolutionary Genetics." Columbia University Press, New York.

Smith, D. B., and Inglis, S. C. (1987). *J. Gen. Virol.* **68,** 2729–2740.

Steinhauer, D., and Holland, J. J. (1987). *Annu. Rev. Microbiol.* **41,** 409–433.

Strauss, J. H., and Strauss, E. (1988). *Annu. Rev. Microbiol.* **42,** 657–683.

Strauss, E. G., Strauss, J. H., and Levine, A. J. (1990). Virus evolution. *In* "Virology" (B. N. Field, D. M. Knipe *et al.*, eds.), pp. 167–190. Raven Press, New York.

Temin, H. M. (1980). *Cell* **21,** 599–600.

Fire Blight, Potato Blight, and Walnut Blight

Dorothy McMeekin
Michigan State University

Glossary

bp Base pairs in a nucleic acid

Cosmid Molecule used in inserting DNA that combines the useful features of virus and plasmid vectors

Eukaryote Cell with parts enclosed in membranes: nucleus, mitochondria

Gametangium Cell in which gametes or haploid reproductive cells are formed; they fuse to form diploid zygotes (see meiosis below)

Gram-negative Bacteria that do not retain a crystal violet-iodine stain when alcohol is applied

Hypha Single filament of a fungus

kb, or kbp Kilobase, unit of length for nucleic acids consisting of 1000 nucleotides

Meiosis Cell division in which the two sets of chromosomes (2N, diploid) per nucleus are reduced to one set (1N, haploid) per nucleus

Peritrichous Having flagellae all over the surface

Phenotype Physical appearance of the organism

THE WORD BLIGHT is used for many plant diseases. The definition varies, but always includes the death of leaf tissue (Fig. 1). This necrosis may extend to the stem and floral parts.

All blights in commercial crops are, of some economic importance, but potato blight has had a very significant impact on human history. The potato originated into the Western Hemisphere and was introduced in Europe during the sixteenth and seventeenth centuries. The potato eventually became the staple food in Ireland, because its underground tubers could not be destroyed during the frequent political conflicts. In the nineteenth century, the arrival of the fungus *Phytophthora infestans* (Mont.) deBary, the cause of the potato blight, led not only to the destruction of the potato, but also to either the starvation or emigration of millions of people. The losses caused by the blights on pear and walnut, associated with the bacterial parasites *Erwinia amylovora* (Burrill) Winslow *et al.* and *Xanthomonas campestris* pv. *juglandis* (Pierce) Dye, respectively, are not as conspicuous in history, but they reduce yearly fruit yield.

The history of the study of plant disease is associated with the development of scientific theory and technology: the light microscope, the cell theory, chemistry, the electron microscope, genetics, and, most recently, recombinant DNA techniques. Through research into plant disease, plant pathologists seek to understand the mechanisms involved in the pathogenicity of parasites and the resistance of hosts in order to reduce crop losses.

Figure 1 Symptoms of fire blight, late blight, and walnut blight. (A) *Erwinia amylovora* on pear leaves. (B) *Phytophthora infestans* on a potato leaf. (C, D) *Xanthomonas campestris* pv. *juglandis* on walnut leaves and fruit. H, healthy tissue; N, necrotic tissue.

I. Blight Symptoms

Initially with potato blight, the leaf margins appear darker green and water-soaked. As these areas enlarge, the center turns dark brown or black, and the lesion has a yellow (chloronemic) border where, if humidity is high, sporangiophores (Fig. 2) of *P. infestans* form. In wet weather, the infection spreads rapidly to involve the whole leaf, the petiole, and the stem. When the fungus reaches the tuber, the skin becomes discolored at first and then water-soaked. In storage, the potato tissue shows a reddish brown firm rot, which makes it susceptible to secondary soft (wet) rot organisms. *Phytophthora infestans* is limited to solanaceous plants: potato, tomato, tobacco, eggplant, and some weeds and ornamentals.

Fire blight begins in the flower after petal fall. The remaining blossom parts and developing fruit, become water-soaked, shrivel, and turn black. The pathogen *E. amylovora* frequently spreads from the flower to the stem and leaves, which turn brown or black. Cankers (dead, slightly sunken oval areas in the bark) appear on the stem. Liquid exudes from these affected parts. The blight organism is restricted to the Rosaceae family and is most common in the Pomoideae: pears, apples, quinces, and hawthorns.

The symptoms of walnut blight (*X. campestris* pv. *juglandis*) appear first on the new leaves that were outermost in the overwintering bud and move inward as other leaves become infected. Water-soaked spots develop on the leaves and young fruit in wet weather. These spots enlarge, become sunken, and turn dark brown. Cankers form on the stem. An exudate may seep out of the affected parts. The bacterium infects black walnut, Japanese walnut, and is particularly severe in Persian walnut, which is the most susceptible.

II. Etiology or Cause

A. The Microorganisms: Structure, Classification, and Genetics

Phytophthora infestans is a filamentous fungus of the class Oomycetes and family Peronosporaceae. Although it is a eukaryote with organized nuclei, the vegetative hyphae are coenocytic (not partitioned by cross walls). It reproduces asexually by means of sporangia that may germinate either with a germ tube or by producing motile zoospores (Fig. 2). In sexual reproduction, gametangia (containing many nuclei produced by meiosis), the oogonium or female and the antheridium or male, come into contact and one of the haploid (1N) nuclei from the antheridium passes through a fertilization tube to fuse with one haploid (1N) nucleus in the oogonium to form the oospore with a diploid (2N) nucleus (Fig. 2). The other nuclei in the gametangia disintegrate. The mature oospore has an impermeable outer wall, an inner wall of carbohydrates, and a vacuole containing lipids. When the oospore germinates, the inner wall thins, and the diploid nucleus undergoes numerous mitotic divisions. The germinating oospore may produce a germ tube that branches to form hyphae if nutrients are abundant or a sporangium (Fig. 2) if nutrients are limited.

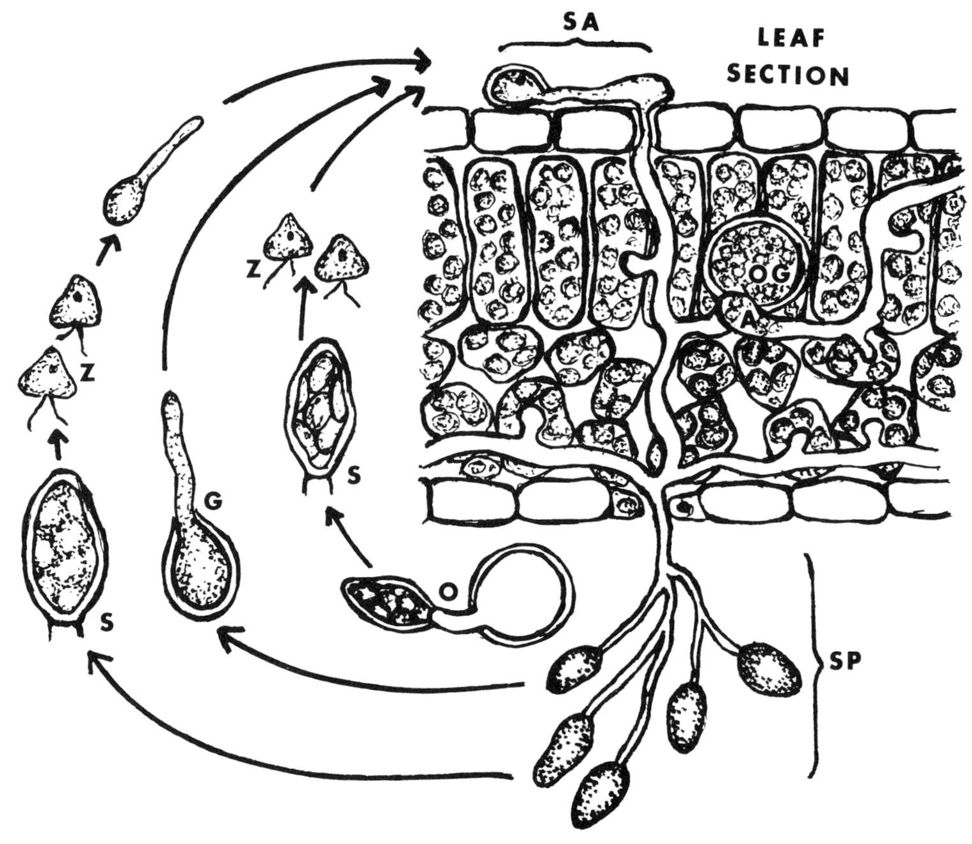

Figure 2 Life cycle of *Phytophthora infestans*. A, antheridium; G, sporangiospore with germ tube; H, hyphae in leaf intercellular spaces sending haustoria into cells; O, oospore germinating and producing a sporangium; OG, oogonium; S, sporangium; SA, germinating sporangiospore forming appresorium and infection peg; SP, sporangiophore with sporangia; Z, zoospore.

There are difficulties in conducting genetic studies with *P. infestans*. They require the presence of two strains (heterothallic, called A1 and A2) for oospore formation, but occasionally they produce oospores in single isolate culture. Oospore germination is low and mating behavior in the field is not fully understood. When field isolates of A1 strain are self-crossed the offspring are all A1. When the A2 strain is self-crossed, it yields both A1 and A2 zoospores from the same sporangium. This indicates that A2 is genetically mixed (heterozygous) whereas A1 is not (homozygous). The size and organization of the genome is not clearly established. The DNA content of the zoospores of some isolates indicates that they are diploid, but other isolates may be triploid, tetraploid, or aneuploid (nuclei with three, four or more sets of chromosomes). [*See* Hypha, Fungal.]

Erwinia amylovora is a gram-negative, rod-shaped bacterium with peritrichous flagella that can be seen with microscopic examination of the exudate from infected plants. It produces white, dome-shaped colonies in culture. It is a facultative aerobe and, except for its ability to infect plants, has many qualities similar to those of intestinal bacteria such as *Escherichia coli*. It is classified in the family Enterobacteriaceae. *Escherichia coli* and most gram-negative bacteria have an inner cytoplasmic membrane, a central peptidoglycan layer, and an outer membrane. The wall polysaccharides are in two layers: lipopolysaccharide and capsular polysaccharide, which may form a capsule that adheres to the cell or, as in *E. amylovora*, is excreted as an extracellular polysaccharide (EPS) slime.

Xanthomonas campestris pv. *juglandis*, present in infected walnut exudate, is also a gram-negative, rod-shaped bacterium with its motility controlled by a polar flagellum. It is an obligate aerobe and usually forms yellow colonies in culture. It is classified in the family Pseudomonadaceae. There are many strains of *X. campestris*, but pathovar (pv.) *juglan-*

dis is distinguished from them primarily by its pathogenicity on walnut (*Juglans* sp.).

B. Transmission, Penetration, and Infection

Sporangia of *Phytophthora infestans* are carried by wind or water to the host where a film of water is necessary for germination. Below 18°C, each sporangium forms two to eight zoospores, which in 3 hr swim about, lose their flagella, and produce a germ tube. At higher temperatures, up to 32°C, the sporangium produces a germ tube directly without forming zoospores. A swelling or appresorium forms on the end of the germ tube above the place where the infection peg of the fungus penetrates the cuticle and enters the epidermal cells of the leaf (Fig. 2). The hyphae grow between the leaf cells and obtain their nutrients by sending short branches (haustoria) into the host cells. Cool wet weather favors the growth of the parasite. If the humidity is >90%, the fungus forms more sporangia on sporangiophores that emerge from the host leaf openings or stomates.

Erwinia amylovora, present in the ooze from cankers produced by the previous season's infections, is transferred by splashing water to the spring buds. Bees pollinating the flowers spread the bacteria. *Erwinia amylovora* may enter the host through the nectarthodes, hydathodes, lenticels, unwounded leaf and stem surfaces, and wounded places. The almost 100% relative humidity in the host intercellular spaces favors the pathogen when it has entered through a natural opening. Wounded places allow for systemic infection or long-distance movement in the water-conducting tubes, the xylem. The bacterium may also move with the flow of liquids and nutrients in the phloem. Within the host, *E. amylovora* does not move by using its flagella, which are absent there, but *in vitro* after 10 hr the virulent strains have all of their normal size flagella and the avirulent have relatively fewer shorter flagella, which suggests that bacterial motility is significant in initiating an infection: The virulent forms move more rapidly. However, even if flagella and motility are absent, *E. amylovora* still infects.

The EPSs of *E. amylovora* may produce the water-soaked symptoms and the blocked xylem vessels that lead to wilting and plasmolysis of the xylem parenchyma cells. EPS may protect the pathogen from hosts defenses and prevent recognition of the pathogen by the host.

Xanthomonas campestris pv. *juglandis* overwinters primarily in infected walnut buds and is splashed to other parts of the tree in the spring. The wind-borne pollen from diseased catkins may also spread the pathogen. Detailed studies of what happens on the microscopic level to *X. campestris* pv. *juglandis,* at the point where it interfaces with the host cells, that are comparable to those conducted with either *P. infestans, E. amylovora,* or the other pathovars of *X. campestris,* have not been conducted.

C. Host Resistance: Hypersensitive Reaction and Phytoalexins

There are differences in susceptibility among the varieties of potato (*Solanum tuberosum*) to different isolates of *P. infestans,* which makes it possible to distinguish several races of this parasite. Some host variety–pathogen race combinations show more susceptibility or resistance than others. Resistance is seen as the limitation of the fungus to a very small necrotic spot on the host, and this is called the hypersensitive reaction (HR). A cell-free extract of any race of *P. infestans* is nonspecific and will produce the HR on the potato. The nonspecific substances involved are called elicitors and have been identified as two polyunsaturated fatty acids: arachidonic acid (AA) and eicosapentaenoic acid (EPA). These acids do not occur naturally in potato and are known in plants only in the oomycetes and mosses. AA and EPA accumulate in solanaceous plants only when they are infected with *P. infestans. In vitro,* AA and EPA are released by lipases, and thus the lipases in potato may release them from the fungus. Antimicrobial low-molecular weight metabolic products formed in response to parasitic attack are called phytoalexins. The HR and phytoalexin production often coincide, so it has been said that the resistance of potato to *P. infestans* is due to the phytoalexins AA and EPA.

The cell death that characterizes the HR is accompanied by activation of the phenylpropanoid pathway and by the accumulation of ligninlike compounds and sesquiterpenes. Glucans increase the accumulation of sesquiterpenes elicited by AA; therefore, the glucans are called enhancers. Their mode of action is not known.

If potato tuber slices are inoculated with a compatible *P. infestans* race, the HR to inoculation later with an incompatible *P. infestans* race does not occur. Therefore, it is proposed that the compatible race produced a suppressor that prevented the HR. A glucan extracted from a compatible *P. infestans* race suppresses browning as well as sesquiterpene formation induced in potato by *P. infestans* elicitor preparations (AA and EPA). These suppressors are race-specific. How this glucan, which nonspecifically enhances the HR, specifically suppresses it has not been explained.

At least superficially, the HR produced by *E. amylovora* and *X. campestris* pv. *juglandis* are the same, but the details of their effects have not been studied. Fire blight (*E. amylovora*) will be discussed further in Section III.B.

III. Control

A. Cultural Practices and Plant Breeding

Early infection by *P. infestans* can be avoided by planting healthy tubers and eliminating volunteer or discarded plants at the margins of the field, because they may harbor the pathogen. Killing the potato vines 2–3 wk before harvest and having dry tubers before storage will prevent tuber infection. Infection during the growing season can be reduced with protective and/or systemic sprays. Fungicides (Zineb, Maneb, and Daconil) are applied weekly on the leaf surface. If cool wet weather that favors the disease continues for several weeks, then fixed copper sprays should be used. The systemic fungicide spray (Metalaxyl) moves into the plant and, thus, does not have to be applied so frequently because it cannot be washed off. It stops established infections and prevents new ones. Data on temperature and rainfall is used to forecast disease prevalence and indicate the most effective spraying time.

Moderately resistant varieties of the potato (*Solanum tuberosum* cultivar Sebago) have been developed by crossing with *Solanum demissum*, a wild potato relative that is immune to *P. infestans*. The immunity introduced into the domestic potato is only temporary, since new successful races of the pathogen continually arise due to mutation and genetic recombination during sexual reproduction.

Blighted pear and apple twigs and cankers infected with *E. amylovora* can be removed by pruning at least 2 in. below the dead twig tissue and $\frac{1}{4}$ inch around the necrotic part of the canker. Very young developing leaves susceptible to fire blight should be sprayed with Bordeaux mixture (copper sulfate–lime–water, 8 : 8 : 10). The blossoms can also be protected with antibiotics (streptomycin, terramycin) or with a lower concentration of Bordeaux mixture (2 : 6 : 10).

The cankers caused by *X. campestris* pv. *juglandis* on walnut can also be removed surgically, but the effectiveness of this is questionable, because the primary overwintering site and source of inoculum in the spring is located internally in the buds. Some control can be obtained with bacteriocides applied when new leaves appear, especially after pollination, when it is wet and humid. Sites exposed to full sunshine should be used for new plantings.

B. Recombinant DNA Technology

Recombinant DNA technology involves the following questions: What is the genetic material that is transferred, how is it transferred, and what effect does this transfer have on the virulence of the pathogen? Together these topics are referred to as genetic recombination. [*See* RECOMBINANT DNA, BASIC PROCEDURES.]

What may be transferred? The bacterial cell has three arrangements of nucleic acids that may determine its ability to parasitize a plant: its chromosome, a closed circular DNA molecule; its plasmids (p), DNA molecules that are inherited but not linked to the chromosome; and its transposable elements, small mobile (800–5700 bp) DNA molecules. The chromosome is coded for the processes needed for the survival of the bacteria. The plasmids provide for variability: They may code for toxin production which increases the virulence of the pathogen or antibiotic resistance. Variability in resistance to antibiotics will determine whether the pathogen can be controlled by antibiotics. The transposable elements may migrate to or from the chromosome or the plasmids and influence the expression of the plasmid and chromosome genes. Some transposable elements that are <2 kb long function only to assist in inserting other DNA and are labeled IS 1, IS 2, IS 3, etc., while others, called transposons (Tn), are longer than 2 kb and carry genes controlling phenotype. [*See* CHROMOSOME, BACTERIAL; PLASMIDS; TRANSPOSABLE ELEMENTS.]

How is genetic material transferred? There are three general processes: conjugation, transduction, and transformation. In conjugation or mating, DNA may be transferred by direct contact between whole cells. If DNA is carried from one cell to another by a virus, it is called transduction. Transformation is the introduction of genetic material from extracts of other cells. The DNA fragments transferred may be plasmids, or DNA molecules may be cut into 4–6-nucleotide-long pieces by very specific restriction enzymes. In the latter case, either a virus, plasmid, or transposon is used to carry the fragment into a host cell and then a ligase enzyme may be used to attach the DNA fragment to the DNA of the host.

What has been discovered with these approaches about the pathogens responsible for fire blight, potato blight, and walnut blight? Some plasmids introduced into the chromosome by conjugation lead to a strain of bacteria that has a high frequency of recombination (Hfr). Chromosomes can be mapped by interrupting the process of Hfr incorporation at different time intervals to see what genes have been transferred. The genes for the synthesis of eight amino acids, the utilization of two sugars, and virulence have been mapped in *E. amylovora* with this technique. It has been suggested that because there are Hfr strains of *E. amylovora* (like *E. coli*), their chromosomes must have an insertion sequencer (like *E. coli*) recognized by the F genome or plastid, which is a circular DNA molecule with a molecular weight of 63×10^6. In *E. coli,* F and F-like plasmids code for a sex pilus or hollow tube that usually extends about 2 μm from the surface of the bacterial cell and is involved in conjugation.

R plasmids, common in gram-negative bacteria, contain transposons and through conjugation transfer the genes for drug resistance. R plasmids are not associated with F plasmids. Streptomycin resistance can be transferred by conjugation with streptomycin-sensitive strains. A 33-kb plasmid was present in all streptomycin-resistant field strains of *E. amylovora,* but not in streptomycin-sensitive strains.

Because extracellular polysaccharide (EPS) is associated with the pathogenicity of *E. amylovora,* the genetic control of EPS has been investigated. The reports are somewhat conflicting. A bacteriophage (virus) carrying the code for a depolymerase for the EPS of *E. amylovora* was cloned and placed in *E. coli,* and then this cloned DNA was put into *E. amylovora,* which in turn produced less EPS. These transformed *E. amylovora* cells produced necrotic lesions but not ooze on pear fruit, which indicates that the ooze is not responsible for the death of the tissue. In another experiment, an *E. amylovora* mutant was made by Tn5 mutagenesis with the cloned gene rcsA (controls capsule production in *E. coli*), which in turn was recombined into the chromosome of wild-type *E. amylovora*. This reduced the synthesis of EPS in the wild type by 90% and indicates that rcsA controls EPS production. This mutated wild type was still virulent on pear, but less so. The HR reaction was lost, suggesting that the genes for HR are needed for pathogenicity. *Erwinia amylovora* may mutate to a type that is nonmucoid in culture but mucoid on pear, which suggests that the host can determine the production of EPS. Nonpathogenic Tn5 mutants of *E. amylovora* that did not produce the HR reaction on tobacco leaves were partially restored to pathogenicity and the HR reaction on tobacco when they received the wild-type DNA.

Analysis of the mutations of *E. amylovora* has led to the identification of a cluster of hypersensitive reaction genes (hrp) that elicit the HR reaction: They span about 40 kb of the chromosome. Eighteen distinct transposon-induced mutants that were restored by a single cosmid (pCPP430) with a 45-kb insert have been described. These hrp genes may or may not be functional in *E. amylovora;* they are also found in *E. coli,* which is not a plant pathogen.

The genetics of *X. campestris* is being studied on other pathovars (pv. *oryzae* on rice), and it can be assumed that these investigations eventually will include the pathovar on walnut.

The actual genetic engineering of plants that are resistant to parasites has just begun. When chestnut trees are infected with a hypovirulent (less virulent) strain of the blight fungus *Cryphonectria parasitica,* the tree is resistant to the virulent strain of this fungus. Hypovirulence in *C. parasitica* is associated with double-stranded RNA (dsRNA). These genes are in the cytoplasm and are transmitted when hyphae from two different strains fuse. The fusion is controlled by nuclear genes. But, in *P. infestans* the most virulent strains from Mexico have the most dsRNA, whereas the less virulent strains of the parasite from the United States and Europe have little or no dsRNA. Jonathan apple can be protected against *E. amylovora* by avirulent strains of *E. amylovora, Pseudomonas tabaci,* and *Erwinia herbicola,* but this protection is influenced by environmental factors and the time of inoculation.

In general, the specific genes of a parasite control the formation of unique enzymes that direct particular cellular reactions in the parasite, which in turn interact with the distinctive make-up of the host cells. A comparison of the molecular research conducted on the disease complexes called fire blight and late blight indicates that the emphasis has differed. More is known about the genetic components and their control in *E. amylovora* than in *P. infestans*, for which we know more about the response in the potato host (HR). The similarity of *Erwinia amylovora* to the nonpathogenic *E. coli* has made it possible to use the techniques and information from many years of research on *E. coli* in the study of *E. amylovora*. Research on the molecular genetics of *P. infestans* is less advanced, but more is known about its effect in the host because solanaceous plants (potato, tomato, tobacco) and their tissues are easier to work with in the laboratory than are those of pear and walnut trees. Discoveries made on other pathovars of *X. campestris* will eventually be used in the study of pathovar *juglandis* on walnut, which has hardly begun. In the future, these deficiencies will be addressed: More information will be obtained on the links between the reactions produced in the hosts by the genetically controlled metabolites of *E. amylovora*, and the well-documented responses of the hosts to *P. infestans* will be correlated with the latter's genetically controlled metabolism. Presumably this knowledge will lead to new ways to control these diseases.

Bibliography

Bennett, J. W., and Lasure, L. (1985). "Gene Manipulations in Fungi." Academic Press, Orlando, Florida.

Bernhard, F., Poetter, K., Geider, K., and Coplin, D. (1990). *Mol. Plant–Microbe Interact.* **3**, 429–437.

Chatterjee, A., Chun, W., and Chatterjee, A. K. (1990). *Mol. Plant–Microbe Interact.* **3**, 144–147.

Chiou, C. S., and Jones, A. L. (1991). *Phytopathology* **81**, 710–714.

Coplin, D. L. (1989). *Annu. Rev. Phytopathol.* **27**, 187–212.

Daniels, M. J., Dow, J. M., and Osbourn, A. E. (1988). *Annu. Rev. Phytopathol.* **26**, 285–312.

Forde, H. I. (1979). Persian walnuts in the western United States. *In* "Nut Tree Culture in North America" (R. A. Jaynes, ed.). pp. 84–97 The Northern Nut Growers Association, Inc., Hamden, Connecticut.

Goodman, R. N., Kiraly, Z., and Wood, K. R. (1986). "The Biochemistry and Physiology of Plant Disease." University of Missouri Press, Columbia.

Hardy, K. (1986). "Bacterial Plasmids." American Society for Microbiology, Washington, D.C.

Hartung, J. A., Fulbright, D. W., and Klos, E. J. (1988). *Mol. Plant–Microbe Interact.* **1**, 87–93.

Keen, N. T. (1990). *Annu. Rev. Genetics* **24**, 447–463.

Ko, W. (1988). *Annu. Rev. Phytopathol.* **26**, 57–73.

Krieg, N. R. (ed.) (1984). "Bergey's Manual of Systematic Bacteriology," Vol. 1. Williams and Wilkins, Baltimore, Maryland.

Kuc, J., and Preisig, C. (1984). *Mycologia* **76**, 767–784.

Laurent, J., Barny, M. A., Kotovjansky, A., Dufriche, P., and Vanneste, J. L. (1989). *Mol. Plant–Microbe Interact.* **2**, 160–164.

Lazo, G. R., and Gabriel, D. W. (1987). *Phytopathology* **77**, 448–453.

Leong, S. A. (1988). *Adv. Plant Pathol.* **6**, 1–26.

Matton, D. P., and Brisson, N. (1989). *Mol. Plant–Microbe Interact.* **2**, 325–331.

Mulrean, E. N., and Schroth, M. N. (1982). *Phytopathology* **72**, 434–438.

Nuss, D. L., and Koltin, Y. (1990). *Annu. Rev. Phytopathol.* **28**, 37–58.

Panopoulos, N. J. and Peet, R. C. (1985). *Annu. Rev. Phytopathol.* **23**, 381–419.

Pegg, G. F., and Ayres, P. G. (1987). "Fungal Infection of Plants." Cambridge University Press, New York.

Roberts, D. A., and Boothroyd, C. W. (1984). "Fundamentals of Plant Pathology." W. H. Freeman and Co., New York.

Shaw, D. S. (1988). *Adv. Plant Pathol.* **6**, 27–51.

Steinberger, E. M., and Beer, S. V. (1988). *Mol. Plant–Microbe Interact.* **1**, 135–144.

Willis, D. K., Rich, J. J., and Hrabak, E. M. (1991). *Mol. Plant–Microbe Interact.* **4**, 132–138.

Flagella

Robert M. Macnab
Yale University

Glossary

Basal body Portion of the flagellum that is embedded in the cell surface

Filament Thin helical structure that when rotated propels the cell

Flagellin Protein from which the flagellar filament is constructed

Switch Substructure that determines whether the motor is rotating counterclockwise or clockwise

Taxis Behavioral response that results in selective migration through the environment

FLAGELLA are organelles that exist in many bacterial species and are used to propel the cell. They are not to be confused with eukaryotic flagella, which are quite distinct in terms of their structure, energy source, and mechanism. Bacterial flagella rotate, driven by transmembrane ion potentials. They switch back and forth between clockwise and counterclockwise rotation, the two directions having quite different consequences in terms of the motion of the cell. (Some species employ a more rudimentary mechanism and simply switch the motor on or off.) Sensory information controls the switching process and enables the cell to display migratory behavior, or taxis. The flagellar gene system is complex and displays a hierarchical pattern of expression. The process of flagellar assembly is also complex and includes a unique pathway for export of external flagellar protein subunits. *Escherichia coli* and *Salmonella typhimurium* will be used as reference points because they are the most highly studied species with respect to flagellar structure and genetics.

I. Occurrence of Flagellation among Bacterial Species

Movement is a widely distributed characteristic among bacteria. It is not by any means universal, however, and one can make some generalizations on taxonomic grounds concerning its presence or absence. One of these is that cocci (species with spherically shaped cells) are seldom motile, whereas rod-shaped species are commonly so. In most cases, motility is driven by an organelle called the flagellum; the principal exception is a type of movement called gliding, which occurs at interfaces (e.g., the surface of an agar plate) and is not well understood. [*See* MOTILITY.]

II. Flagellar Morphology

When a flagellated cell is viewed in the electron microscope, one normally sees only the external part of the flagellum, consisting of a filament that is exceedingly thin (typically 20 nm diameter) and long (typically 10 μm, or about 10 times the length of the cell body) (Fig. 1A). The other striking feature of the filament is that it is neither straight nor randomly bent but, rather, has a regular waveform that is a perfect helix. This, plus the fact that it is rotated by the motor at its base, is the key to flagellar motility: Rotation of a helix causes its waveform to travel along the helical axis and, if this is resisted, then the rotation develops both torque and thrust. (The term "flagellum" is something of a misnomer for the bacterial organelle, because it is the Latin word for "whip," thus implying propagation of a bending wave rather than rotation of a helix. It *is* an appro-

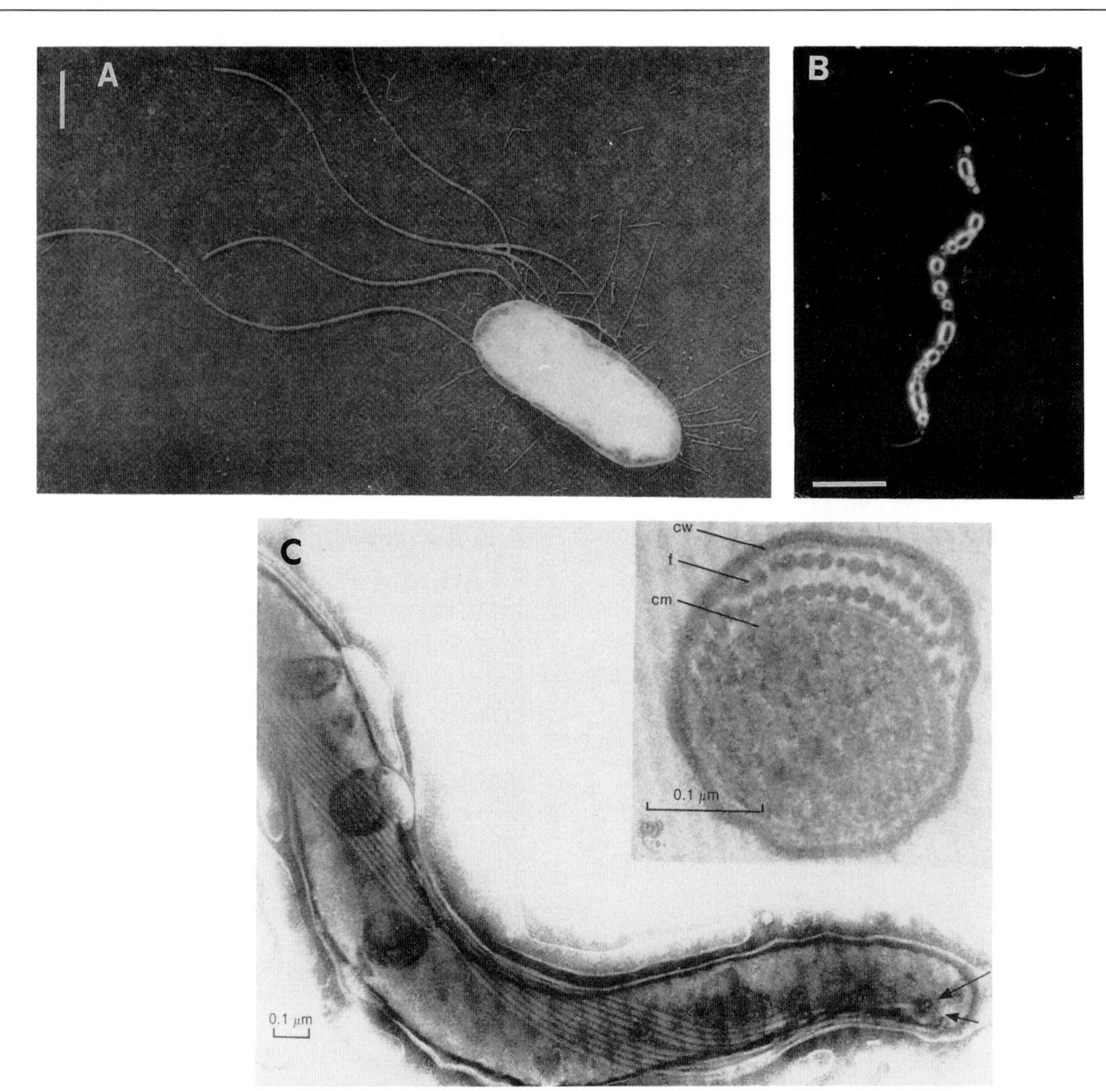

Figure 1 (A) A peritrichously flagellated bacterium, *Escherichia coli*, with several flagella randomly located around the cell. Bar = 0.5 μm. [From Berg, H. C. (1975). *Sci. Am.* **233,** 36–44.] (B) A spirillum, *Spirillum volutans*, with polar flagella and a helical body shape. Bar = 10 μm. [From Macnab, R. M. (1979). "Physiology of Movements." Springer-Verlag.] (C) A large oral spirochete, with flagella (f) contained between the cell membrane (cm) and the outer cell wall (cw). Bars = 0.1 μm. [From Listgarten, M. A., and Socransky, S. S. (1964). *J. Bacteriol.* **88,** 1087–1103.]

priate term for the eukaryotic organelle, which does actively bend).

Unless the cell is disrupted, little else can be discerned in the way of flagellar structure. By dissolving away the cell surface with enzymes and detergents, however, it is possible to isolate a structure called a filament hook–basal body, which consists of the filament, joined to a short curved section called the hook, joined in turn to the basal body (Fig. 2). The basal body has a central rod that passes through several rings (the number of rings varies among species but is typically two for gram-positive bacteria and four for gram-negative bacteria). Together, the rod, hook, and filament form a continuous thin axial structure that extends from the cell surface for a distance that is large compared with the size of the cell. The innermost pair of rings is associated with the cell membrane, and the outer rings (where present) with the outer membrane and the periplasmic compartment between the two membranes.

Other flagellar structures have been detected by

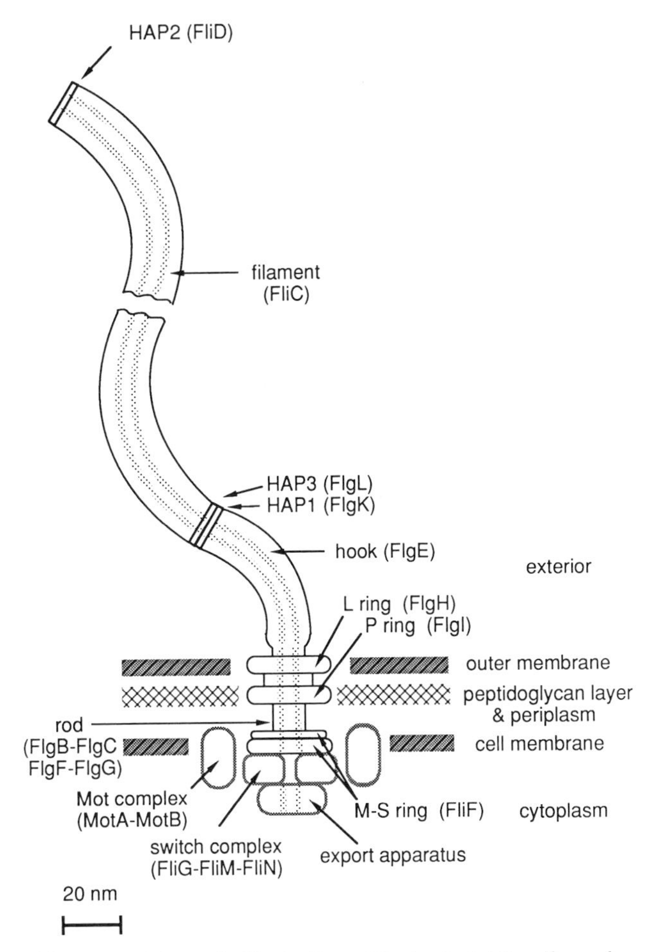

Figure 2 Schematic illustration of the bacterial flagellum of *Salmonella typhimurium*. Morphological features such as rings and rod are indicated, together with the gene products from which they are constructed. The filament is much longer than shown here. The locations of the MotA and MotB proteins, the three switch proteins, and the proteins that comprise the flagellum-specific export apparatus are based on a variety of lines of indirect evidence. The channel through which the axial protein subunits, such as flagellin, travel to reach their site of assembly is shown in dotted outline. HAP, hook-associated protein. [Modified from Macnab, R. M. (1990). Genetics, structure and assembly of the bacterial flagellum. *In* "Biology of the Chemotactic Response." Soc. Gen. *Microbiol. Symp.* **46**, 77–106. Cambridge University Press.]

techniques such as freeze-fracture of the cell membrane. They include large disks in the outer membrane and circlets of studs in the inner membrane. Structures extending inward from the M ring of the basal body have also been seen. The significance of these will be considered in Sections V and VI. [*See* CELL MEMBRANE: STRUCTURE AND FUNCTION.]

III. Organization of Flagella

A. Cells with External Helical Flagella

The most common type of flagellation involves external helical flagella that rotate and propel the cell. Bacteria with this type of flagellation can be classified according to how many flagella they have and where on the cell surface they originate (Table I). The simplest case is where the cell has just one flagellum. Most often, this is located either at or close to the pole (polar or subpolar monoflagellate). In some examples, it originates at the midpoint of the long axis of the cell (medial monoflagellate). Monoflagellate cells generally have a simple type of motile behavior, consisting of forward and backward swimming.

At the next level of complexity, there are polar multiflagellate species with a tuft of several flagella at one pole. Further elaboration comes when both poles are flagellated, either singly (bipolar monoflagellate) or multiply (bipolar multiflagellate). The distinction between polar and bipolar can become blurred depending on the time relationship between cell division and flagellar assembly. Thus, *Halobacterium halobium* is bipolar multiflagellate shortly before division but polar multiflagellate after division.

One of the most common patterns, called peritrichous flagellation, involves multiple flagella (often around 5–10, but many more than that in some species such as *Proteus mirabilis*) located at essentially

Table I Classification of Flagellated Bacteria According to Flagellar Number and Location

Classification	Examples
Polar monoflagellate	Pseudomonads, *Caulobacter crescentus*
Medial monoflagellate	*Rhodobacter sphaeroides*
Polar multiflagellate	*Chromatium* spp.
Bipolar monoflagellate	*Wolinella succinogenes*
Bipolar multiflagellate	*Halobacterium halobium*
Peritrichous multiflagellate	*Escherichia coli*, *Salmonella typhimurium*, *Bacillus subtilis*
Spirillar	*Rhodospirillum rubrum*, *Aquaspirillum serpens*
Spirochetal	*Treponema pallidum*, *Spirochaeta aurantia*, *Leptospira interrogans*

random positions on the cell surface. This results in a more complex type of motility, where one direction of flagellar rotation causes forward swimming with the flagella formed into a bundle, whereas the other direction causes a chaotic tumbling behavior with little net forward or backward motion. The physical basis for tumbling is rather esoteric, involving intricate structural changes in the filament itself.

Some species have a lateral row of flagella along their long body axis and use them to swarm on solid surfaces. For example, *Vibrio parahaemolyticus* has such a pattern (but has a single polar flagellum when free swimming in aqueous medium).

B. Flagellated Cells with Helical Bodies (Spirilla)

In the species already considered, the helicity of the filaments is responsible for converting torque into thrust. In some organisms, notably the spirilla, the cell body is helical and its counter-rotation in reaction to flagellar rotation produces thrust (Fig. 1B).

C. Cells with Internalized Flagella (Spirochetes)

A group of bacteria, including some that are clinically very important such as *Treponema pallidum* (the causative agent of syphilis), have flagella that are internalized between the inner and outer membranes of the cell. They are collectively called spirochetes (Fig. 1C). Two different mechanisms apparently can cause these internalized flagella to transmit force to the medium. In one, exemplified by *Spirochaeta aurantia,* researchers believe that the outer membrane is free to roll with respect to the cell membrane and that the rotation of the flagella in the periplasmic space causes this to occur; because the outer membrane conforms to the helical shape of the cell body, this rolling develops both torque and thrust from the resistance of the medium. In the second class of mechanism, employed by the leptospira, the flagellum at one pole drives a precessional bending of the cell body that results in body rotation and the generation of thrust by virtue of the helical shape of the body at the other end. [*See* SPIROCHETES.]

IV. Flagellar Genetics

A. Chromosomal Organization and Control of Gene Expression

Judging by the number of genes involved, the assembly and function of flagella is one of the most complex tasks that a bacterium undertakes. In *E. coli,* for example >40 genes are involved directly plus another 6 are involved in signal transduction and perhaps 10 or so in signal reception; comparable numbers have been found for other species such as *Caulobacter crescentus* and *Bacillus subtilis.* This places the flagellar system second only to the ribosomal gene system in complexity. (Many other genes are also needed for flagellar synthesis and energization, such as the genes for the electron transport chain; however, these are used by the cell generally and cannot really be considered as part of the flagellar gene system.)

Flagellar genes are highly localized in just a few clusters on the chromosome (Fig. 3). Within each

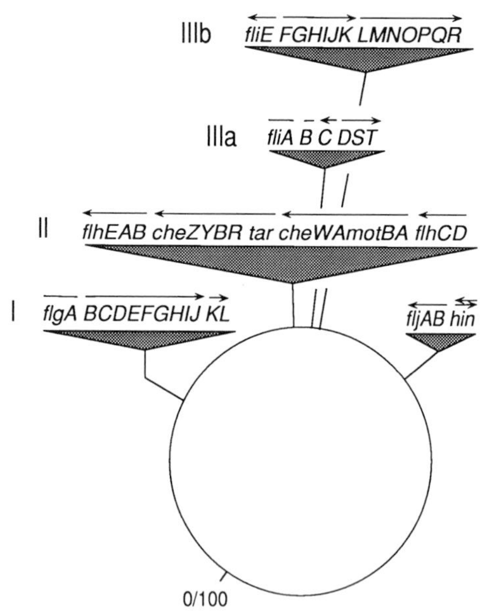

Figure 3 Flagellar and related genes of *Salmonella typhimurium*. The genes are highly clustered into four main regions. There is also a small region containing a second allele of the flagellin gene (*fljB*) and control genes for alternate expression of *fljB* and the other flagellin allele, *fliC*. Operons are shown by overlining, with an arrowhead where the polarity is known. The map has been rotated to place the flagellar genes at the top; the origin of the map is at 0/100. [From Macnab, R. M. (1990). Genetics, structure, and assembly of the bacterial flagellum. *In* "Biology of the Chemotactic Response." *Soc. Gen. Microbiol. Symp.* **46**, 77–106. Cambridge University Press.]

cluster, the flagellar-related genes are contiguous; i.e., there is no interspersal of unrelated genes. The genes commonly occur in multicistronic operons, with the largest in *E. coli* containing nine genes. Recent work indicates that in the gram-positive bacterium *B. subtilis* one extremely large operon may account for a majority of flagellar genes. Flagellar genes constitute a unit of genetic expression, or regulon, in that one master operon controls the expression of all flagellar-related genes. In *E. coli,* this operon is known to be positively regulated by cyclic adenosine monophosphate in association with a protein known as catabolite activator protein. Thus, under rich environmental conditions where migration is not needed (notably high glucose), the synthesis of flagella, which is biosynthetically costly, is repressed.

All flagellar operons other than the master operon have specialized promoter consensus sequences that are recognized by flagellum-specific sigma factors, so that transcription of flagellar genes is under separate control from transcription of other genes. This control occurs at two levels, with most of the structural genes existing at the first level and those genes that would be useful only after the basic flagellar structure was in place existing at the second level. The latter level includes the structural gene for flagellin, which is present at about 20,000 subunits per filament (or about 200,000 subunits per cell), and, thus, is an expensive item to synthesize. Also at this level are the genes for the motor and sensory systems.

B. Flagellar Phase Variation

Salmonella typhimurium has two distinct genes for flagellin and exhibits a remarkable form of regulation that results in their alternate expression. The phenomenon, called flagellar phase variation, involves the occasional stochastic inversion of a particular segment of DNA, such that in one orientation it creates a proper promoter for the phase-2 flagellin gene and a repressor for the phase-1 flagellin gene (so that phase-2 flagellin is synthesized but phase-1 flagellin is not), whereas in the other orientation it destroys the promoter for phase-2 flagellin and phase-1 repressor (so that phase-2 flagellin is not synthesized but phase-1 flagellin is). The gene for the enzyme that catalyzes the DNA inversion lies within the inverted segment and is expressed in both orientations. Although functionally equivalent in propulsion, filaments constructed from the different flagellins are antigenically distinct; the phase variation phenomenon may have evolved to take advantage of this fact.

C. Roles of the Flagellar Genes

Many of the flagellar genes are structural, encoding the components of features such as the filament or the rings of the basal body. Others encode presumptive structures, such as the switch or the export apparatus, which have not yet been seen in the electron microscope. Others are regulatory, either in the genetic sense (e.g., encoding a flagellum-specific sigma factor) or in the sense that they control the process of flagellar assembly (e.g., determining the length of the hook). The most complete assignment of flagellar genes to specific functions has been made for *E. coli* and *S. typhimurium* and is summarized in Table II.

Functionally related genes are often close to each other, in many cases on the same operon. On an even larger scale, the different flagellar clusters correspond roughly to different functional groupings: Region IIIb (Fig. 3) contains genes that are required for the structure and function of the earliest flagellar components in the assembly pathway (see later), region I to middle components including the basal body and hook, region IIIa to the filament, and region II to overall regulation, motor rotation, and the sensory system. However, the precise organization cannot be all that important, because it varies appreciably among the organisms where it has been established.

V. Molecular Structure of Flagella

The principal morphological features of the flagellum have already been outlined; helical filament, hook, basal body, and, in some cases, circlets of studs or other annular features in the cell surface. These will now be considered in more detail at the molecular level. Figure 2 depicts the known and presumed structures of the flagellum of *E. coli/S. typhimurium* and also indicates the proteins from which they are constructed.

A. The Flagellar Filament

In many species, the flagellar filament is built of subunits of a single protein. In other species (e.g., *C.*

Table II Flagellar and Motility Genes and Gene Products in *Salmonella typhimurium*[a,b]

Gene	Location/structure/function
Genes encoding known structural features	
fliF	Basal body: M–S ring
fliE	Basal body: unknown
flgB	Basal body: proximal rod
flgC	Basal body: proximal rod
flgF	Basal body: proximal rod
flgG	Basal body: distal rod
flgI	Basal body: P ring
flgH	Basal body: L ring
flgE	Hook
flgK	Hook–filament junction
flgL	Hook–filament junction
fliC	Flagellar filament (flagellin, phase 1)
fljB	Flagellar filament (flagellin, phase 2)
fliD	Filament cap
Genes encoding presumed structures	
motA	Motor rotation: energy transduction?
motB	Motor rotation: anchoring to cell surface?
fliG	Motor rotation and switching
fliM	Motor rotation and switching
fliN	Motor rotation and switching
flhA	Flagellum-specific export?
fliH	Flagellum-specific export?
fliI	Flagellum-specific export? Putative adenosine triphosphatase
Genes involved in control of assembly or modification of structure	
flgD	Basal-body rod modification
fliK	Hook length control
fliB	Methylation of lysine residues on flagellin
Genes involved in regulation of expression	
flhC	Sigma factor for middle genes?
flhD	Sigma factor for middle genes?
fliA	Sigma factor for late genes
fliS	Negative regulation of late genes?
fliT	Negative regulation of late genes?
hin	Site-specific inversion of promoter of phase-2 flagellin gene
fljA	Repressor of phase-1 flagellin gene
Genes of unknown function	
flgA, flgJ, flhB, flhE, fliJ, fliL, fliO, fliP, fliQ, fliR	

[a] Genes involved in chemotaxis are not included.
[b] With the exception of the genes for phase-2 flagellin and phase inversion, and the gene for the flagellin methylase, all genes and gene products also apply to *Escherichia coli*. A number of homologs to these genes have been established in *Caulobacter crescentus* and *Bacillus subtilis*.

crescentus, Rhizobium meliloti), more than one kind of subunit is used, although it is not obvious what advantage this confers. In yet other species (e.g., *Vibrio* spp.), a sheath structure surrounds the filament core.

Recall that the filament is helical and typically is long compared with the pitch (wavelength) of the helix, so that it contains several helical turns. The organization of the subunits within it can be described as follows (Fig. 4). Imagine constructing a cylinder from spherical subunits laid down according to the following rules. Each sphere is to be placed slightly higher than the previous one and at a position rotated at an angle about the axis of the cylinder. The values of the rotation are to be such that it takes $5\frac{1}{2}$ spheres to complete one turn; after 11 spheres, two turns will have been completed, and the next sphere will lie exactly above the first one. The values of the rise are to be such that the spheres nestle into each other and form essentially a hexagonal lattice around the surface of the cylinder. One of the three directions of the lattice is parallel to the cylinder axis and produces 11 "fibrils" or longitudinal strands of subunits. However, the structure generated in this way is a straight cylinder, whereas in reality the filament is not straight but helical. The simple description used therefore must be modified in two ways. (1) Even though the subunits are identical proteins, we must assume that they stack slightly differently, depending on which fibril they belong to. Specifically, the distance between subunits in a fibril on one side of the cylinder must be less than that for

Figure 4 Packing of flagellin subunits in the flagellar filament that results in its helical waveform. Subunits 1, 2, etc., are successive elements of the basic stacking pattern, which is shown here with exactly $5\frac{1}{2}$ subunits per turn. Subunits 6 and 7 are therefore above and to the left and right, respectively, of subunit 1, whereas subunit 12 is exactly above subunit 1. This produces a lattice that is almost hexagonal and wrapped around into a cylinder: subunits (shown in black, diagonal cross-hatching, and horizontal cross-hatching) are in close contact in the three directions (arrows at top left) of the lattice. In reality, there are not quite exactly $5\frac{1}{2}$ subunits per turn, and, thus, the column of subunits shown in black would be at a slight angle to the cylinder axis (arrows at top right). Also, a column of subunits on one side of the cylinder actually has a slightly shorter intersubunit distance than a column on the opposite side, causing the cylinder to be distorted from a straight into a helical one.

subunits on the other side, thus forcing a bending of the cylinder into an arc of a circle. (2) The number of subunits per turn must be made slightly different from $5\frac{1}{2}$ (say 5.52) so that now the fibrils are no longer exactly parallel to the cylinder axis but slowly wind around it. Now the different intersubunit distance in fibrils on opposite sides of the cylinder will cause it to be distorted into a helical shape (rather than an arc of a circle), just as is observed for the flagellar filament.

In terms of elastic properties, the filament is quite rigid, consonant with its use as a propeller; however, it undergoes discrete (inelastic) structural rearrangement when subjected to sufficient stress. This causes a dramatic change in helical shape, from a left-handed form to a right-handed form, with important effects on motility.

A remarkable feature of flagellin is that only its N- and C-terminal regions are important in terms of assembly into filament. These regions are also the most highly conserved among different bacterial species, presumably because of similar structural constraints. Much of the central part can be deleted or replaced by foreign sequence without affecting the ability of the protein to assemble into filament and propel the cell. In the three-dimensional structure of the protein, the central part of the primary sequence forms a knob that faces outward from the filament and does not interact with the equivalent knobs of adjacent subunits. This knob is, however, responsible for most of the epitopes that constitute the potent H antigen in many virulent species such as *S. typhi* and, thus, presumably is important to the cell in that regard.

B. The Hook

The hook is constructed in a fashion quite similar to the filament, but with some important differences. The lattice is qualitatively the same, but the 11 fibrils are at a sharper tilt to the cylindrical axis, and, thus, the pitch of the helix is shorter. However, because the hook is quite short (about 60 nm vs. 10 μm for the filament), only about one-quarter of a turn is present despite the short pitch, resulting in the characteristic "hooklike" appearance.

Unlike the filament, the hook is quite flexible, consistent with a role as a universal joint to allow rotation of the motor about an axis perpendicular to the cell surface to be transmitted to the filament, which is in general oriented differently. The flexibility may not be an elastic property but, rather, one deriving from the fundamental design of the structure and involving cycling of which fibrils are in the long and short state.

C. Junction Proteins

The filament and hook are not joined together directly. Intervening between the two structures is a short junction zone consisting of two kinds of proteins, called hook-associated proteins. It is thought that, although the filament and hook proteins are on similar lattices, they are not sufficiently similar to form a stable direct junction and that therefore the hook-associated proteins are needed to act as adaptors. At the tip of the filament, there are subunits of another protein that acts as a cap to prevent flagellin subunits from being lost during the assembly process.

D. The Basal-Body Rod

The rod, which is presumed to act as a transmission shaft between the motor and the hook-filament, consists of at least four different kinds of subunits. This may be because at different points along its length it must interact with different structures such as the M–S ring, the P ring, the L ring, and the hook.

E. The Axial Family of Flagellar Proteins

Together rod proteins, hook proteins, hook-associated proteins, flagellin, and filament capping protein constitute a family that shares certain primary sequence similarities and structural features that are believed to be important in the quaternary interactions among subunits. Specifically, the termini contain amphipathic α-helices that are thought to interact in a coiled-coil fashion and lock the subunits together.

F. The Inner Rings of the Basal Body

The basal-body rings that encircle the rod can be divided into two classes: inner and outer. In all species that have been examined, a pair of inner rings exist in or near the plane of the cell membrane. These have been termed the M (membrane) and S (supramembrane) rings, respectively. For a long time, they were believed to mutually rotate. The protein responsible for the M ring in *S. typhimurium* and *E. coli* was identified a number of years ago; attempts to identify that responsible for the S ring were unsuccessful. Recently it has been shown that both morpholigical features derive from the same protein; therefore, the idea of these rings as a rotor–stator pair must be abandoned. The M–S pair apparently is joined to the rod and presumably rotates with it.

Current evidence suggests that the role of the M–S ring in motor rotation is passive, because mutations causing paralyzed phenotype have never been found within the M–S ring gene. The best guess for its function is that it acts as a mounting plate for the active components.

G. The Outer Rings of the Basal Body

In gram-negative bacteria, there is usually a pair of outer rings called the P (periplasmic) and L (lipopolysaccharide) rings, which together constitute a structure called the outer cylinder comprising the ring features and a connecting wall. The P-ring feature and wall–L-ring features are constructed from subunits of different proteins.

The outer rings are presumed *not* to rotate but, rather, to act as a bushing to stabilize the rod against lateral shear forces. This idea is supported by the fact that in gram-positive species, where a thick strong wall exists, the outer rings are not needed.

H. Large-Diameter Disks

In several species with polar flagella—for example, the polar monoflagellate *Wolinella succinogenes* and the polar multiflagellate *Aquaspirillum serpens*—a very large-diameter disk (130 nm vs. 30 nm for the basal-body rings) associated with the flagellum exists; whether or not all polarly flagellated organisms possess these disks is not known. The precise role of this disk is not clear, but it may be present for mechanical reasons, to ensure that the thrust of the flagellum is stable and appropriately directed. In multiflagellate species, it may also function as a spacer between the individual flagella in a polar tuft. Because the organisms in which these structures have been seen are not well characterized genetically, it is not possible to comment on the genes involved or on the phenotype associated with disk defects.

I. Mot Proteins

Mot proteins are defined as proteins that are necessary for motor rotation to occur. Five such proteins are known to exist in the enterics, and homologs of several of these have been found in other species. Of the five, three are also involved in switching the direction of rotation and will be described in the next section.

MotA and MotB are the two proteins that are only involved in rotation and not in switching. They do not need to be present for the rest of the flagellar apparatus to assemble, and if their synthesis is artificially controlled they can be added to a preassembled flagellum and enable it to rotate. Both are membrane proteins. MotB has a very simple organization within the membrane, with just a single spanning segment and with the bulk of the structure constituting a domain in the periplasm. It is suggested that its role is in clamping MotA rigidly to the cell surface rather than in energy transduction per se. MotA is predicted to span the membrane several times, and analysis of mutants suggests that it does contribute to the energy transduction process and constitutes at least part of the conductance pathway for the protons that drive the motor. The circlets of studs that have been seen in freeze-fracture images of the membrane of flagellated cells are thought to consist of MotA–MotB complexes.

J. The Switch

The switch is the part of the flagellum that determines whether the motor is generating counterclockwise or clockwise rotation. However, the switch seems to do more than that and to contribute, along with the Mot proteins, to the rotation process itself. The evidence for this is that some mutations in switch proteins cause paralysis (no rotation), whereas others cause abnormal switch bias (too much time in one direction of rotation). They also differ from the Mot proteins in that they are necessary for the integrity of the rest of the flagellar structure. Thus, a severe mutation in the switch genes causes nonflagellate phenotype.

There are three different switch proteins in the enterics and, based on genetic data, they are thought to form a multisubunit complex at the flagellar base and facing the cytoplasm. The final output of the sensory system is binding of a protein, CheY, to the switch and shifting its bias toward clockwise rotation. It is not yet known which of the proteins within the switch contribute to the CheY-binding site. [*See* CHEMOTAXIS.]

The flagellar switch, although identified genetically and biochemically, has not yet been detected as a physical structure. Basal bodies prepared using slightly modified protocols are beginning to give evidence for additional structures at the cytoplasmic face of the M ring, but the proteins responsible for these additional structures have yet to be established.

K. The Export Apparatus

The process of flagellar assembly, to be described later, requires the export of many components. There is good reason to believe that a flagellum-specific export apparatus must exist distinctly from the primary cellular export apparatus and that this apparatus must be physically part of the flagellar structure. Several potential candidates for components of the flagellum-specific export apparatus have been identified. Interestingly, one of these is homologous to the adenosine triphosphate (ATP)-hydrolyzing subunit of one of the cell's major power stations, the F_0F_1 ATPase, which interconverts ATP/adenosine diphosphate energy and protonmotive force. The implications of this remain to be established, but one possibility is that the flagellar protein is an ATP-hydrolyzing subunit pumping flagellar proteins across the membrane.

As with the switch, physical identification of the export apparatus has not yet been achieved.

VI. The Process of Flagellar Assembly

A. Overall Assembly Scheme

The process of flagellar assembly must be viewed in terms of both the intrinsic structural complexity of the organelle and its complex relationship to the cell surface. Broadly, one can think in terms of components that are located (1) in the cytoplasm (e.g., regulatory components), (2) at the inner face of the cell membrane (e.g., the switch), (3) within the cell membrane (e.g., the M ring), (4) in the periplasm (e.g., the rod), (5) in the outer membrane (e.g., the L ring), and (6) entirely external to the cell (e.g., the filament). These categories have different constraints in terms of assembly.

The assembly process in *S. typhimurium* and *E. coli,* elucidated by a variety of techniques, largely occurs in the order just described (Fig. 5). One of the first events seems to be the insertion of the M ring into the cell membrane. This is probably followed by mounting of peripheral structures such as the switch and the export apparatus onto the M ring. Next follows export of rod proteins through the cell membrane and assembly onto the M ring. Then exported P- and L-ring proteins nucleate onto the rod to form the outer cylinder. After a mysterious "modification" process, the rod becomes competent to accept assembly of exported hook proteins at its tip. When the hook is complete (and this involves another mysterious process by which hook length is controlled), the subunits of the two junction proteins and the capping protein are exported and assembled at the hook tip in a triple layer. Finally, flagellin is exported and assembled by insertion at the far end, just under the cap.

The Mot proteins, which are integral membrane proteins, apparently can assemble at any point after the M ring and other early components are in place,

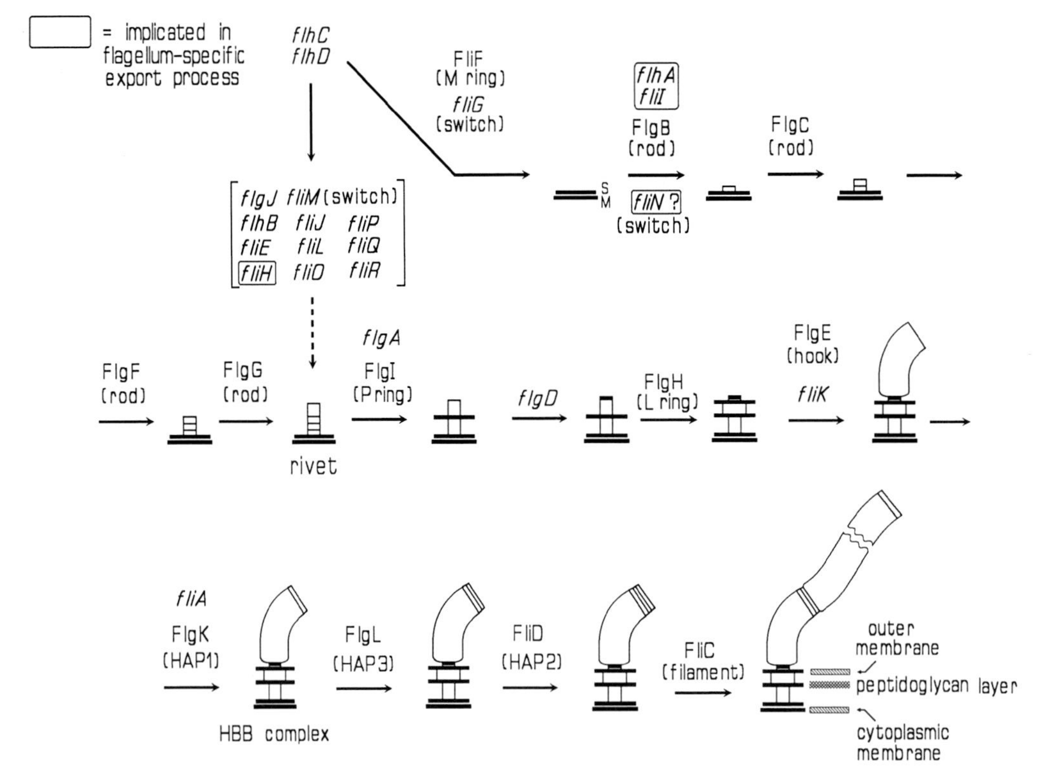

Figure 5 Assembly pathway for the flagellum of *Salmonella typhimurium*. Note that structure develops from components proximal to the cell to components progressively farther away from it. Where a protein has been identified biochemically in the structure and its gene has been established, the protein is indicated (e.g., FlgE); otherwise, the gene is indicated (e.g., *flgD*). The genes in brackets are needed prior to the structure called *rivet*, but where prior to that structure they participate is not known. HAP, hook-associated protein; HBB, hook–basal body. [Modified from Jones, C. J., and Macnab, R. M. (1990). *J. Bacteriol.* **172,** 1327–1339.]

in a process that is independent of the state of the periplasmic and external components.

B. Flagellar Assembly and the Cell Cycle

Given the intimate relationship between the flagellum and the cell surface, one might expect that the processes of flagellar assembly and cell growth and division would interact with each other. The organism in which this question has been examined most carefully is *C. crescentus*, which undergoes a clearly defined pattern of alternating assembly of a polar flagellum and a polar stalk that is used for attachment to solid surfaces. Here it is clear that both temporal information (to synthesize flagellar proteins at the appropriate time) and spatial information (to place the flagellum at the pole of the appropriate predivisional cell) must exist. [*See* CAULOBACTER DIFFERENTIATION.]

Although the event initiating flagellar gene expression has not yet been determined, it has been shown that such initiation does occur at a particular point in the cycle and that operons at different levels within the hierarchy of control are successively expressed in an order that corresponds to the assembly order previously described. However, the situation is more subtle, because it has also been shown that newly synthesized proteins segregate to the appropriate half of a predivisional cell even though no obvious barrier exists between it and the other half. It has been suggested that there may be differences in cell surface that result in sequestration of protein at the appropriate location, where it can be stored until needed in the assembly process. A more likely idea perhaps is that the two copies of the chromosome are differentiated in some way so that expression of flagellar proteins can only proceed from the copy in the half of the predivisional cell that is about to develop a flagellum and that cell division gets completed before the protein can equilibrate between the two halves.

With organisms such as *S. typhimurium,* which are flagellate throughout their cycle and where the flagella do not have well-defined locations, demonstrating linkage between flagellar assembly and cell cycle is much more difficult, but some evidence suggests that such linkage exists.

C. Pathways for Export of Flagellar Proteins

Bacterial proteins residing in the cell membrane insert spontaneously and require no export apparatus. Thus, assembly of the M ring, for example, seems to be a straightforward process that can occur unaided.

The P- and L-ring proteins are exported to the periplasm and outer membrane by the same process as most other bacterial proteins located in these compartments. The pathway, known as Sec (for secretion), is energy-requiring, involves several different components near and in the cell membrane, and employs recognition and cleavage of an N-terminal signal peptide from the protein that is being exported.

All of the other external flagellar proteins, namely, those that constitute the axial family of rod, hook, etc., employ a flagellum-specific export pathway that does not involve signal peptide cleavage. How these particular proteins are recognized for export by this pathway is not yet known. At the level of amino acid sequence, they do show some common features, but at least some of these (such as the occurrence of hydrophobic residues with a periodicity of seven in the primary sequence) are likely to be less related to the question of recognition for export than to the fact that these proteins form related higher-order structures (e.g., hook vs. filament).

As already noted, the overall process of flagellar assembly proceeds from proximal to distal structures, and, in the case of the filament (and probably also the hook and rod), the individual subunits of the exported proteins add to the far end of the growing structure. For a filament of average length, this means reaching an assembly point that is a great distance from the cell. It is inconceivable that this could be accomplished by diffusion through the medium in which the cell is living, and the evidence strongly indicates that the actual path is through a channel throughout the entire length of the axial structure of rod, hook, and filament (Fig. 6). The entrance to this channel obviously cannot be open to any molecule that encounters it; otherwise, the cellular content of the cell would be free to leak out. This again raises the question as to how flagellar proteins are recognized as being in a different category from all other proteins and molecules in the cell. The answer is not known, but a specific device must be responsible for the recognition, and perhaps active (energy-requiring) transport of the proteins across the membrane may also be needed. From then on, the process could be passive, because each subunit would be getting "pushed" by the one behind in what is essentially a single-file diffusion process. Possible candidates for the apparatus responsible for this process, including a putative ATPase, were mentioned earlier.

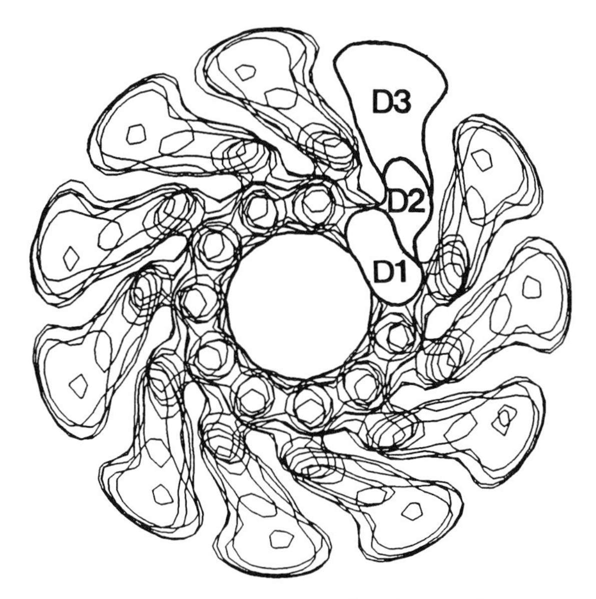

Figure 6 Cross-section of the flagellar filament, illustrating that it has a substantial channel down its center. This channel is presumed to be the conduit through which flagellin subunits pass to reach the tip and assemble into the growing filament. The cross-section is one view of a three-dimensional reconstruction of flagellar filament obtained by a combination of X-ray fiber diffraction analysis and electron microscopy. The knoblike feature (D3) at the outer radius derives from the central part of the primary sequence of flagellin; unlike domains D1 and D2, it is not involved in intersubunit interactions within the filament but is important antigenically. [From Namba, K., Yamashita, I., and Vonderviszt, F. (1989). *Nature (London)* **342,** 648–654.]

VII. Concluding Comments

As perhaps can be appreciated from the foregoing, the bacterial flagellum represents one of the richest topics in prokaryotic biology. It involves many diverse aspects: a complex genetic basis, in terms of both the number of genes and their regulation; a complex structure whose assembly involves a remarkable export process; a motor whose reversible rotary mechanism makes it unique; and, finally, a migrational behavior that represents one of the most basic examples of sensory physiology.

Bibliography

Jones, C. J., and Aizawa, S.-I. (1991). *Adv. Microb. Physiol.*, **32,** 109–172.

Jones, C. J., and Macnab, R. M. (1990). *J. Bacteriol.* **172,** 1327–1339.

Macnab, R. M. (1990). *Soc. General Microbiol. Symp.* **46,** 77–106.

Namba, K., Yamashita, I., and Vonderviszt, F. (1989). *Nature (London)* **342,** 648–654.

Vogler, A. P., Homma, M., Irikura, V. M., and Macnab, R. M. (1991). *J. Bacteriol.* **173,** 3564–3572.

Flow Cytometry

Howard M. Shapiro
West Newton, Pennsylvania

Glossary

Flow cytometry Process that measures physical or chemical characteristics of cells passing single file through the measuring apparatus in a fluid stream

Flow sorting Technique for physical isolation of cells with selected characteristics from the sample stream

Gated analysis Technique for restricting data analysis to cells with selected characteristics without physically isolating the cells

Parameter Cellular characteristic, such as cell size, DNA content, or membrane potential, measurable by flow cytometry

Probe Reagent used to measure a parameter

FLOW CYTOMETRY is a process that measures physical or chemical characteristics of cells or other biological particles as the cells or particles pass in single file through the measuring apparatus in a fluid stream. Simple flow cytometers have long been used for cell counting; more complex instruments now measure a wide range of cellular characteristics, or parameters.

I. What is Flow Cytometry?

A. Electronic and Optical Methods for Measurements of Single Cells

The most common electronic measurement determines cell volume by the Coulter principle. If cells, which are relatively nonconductive, are suspended in saline and passed through a small orifice, the electrical impedance of the orifice is higher when it contains a cell(s) than when it contains only saline. Each cell's passage produces an impedance change proportional to cell volume. Impedance measurements can also be made using alternating current; this provides some information about the internal structure of cells.

Optical flow cytometers are basically specialized light microscopes; cells are measured while passing through a sharply focused, high-intensity light beam derived from an arc lamp or laser. Measurements of cellular characteristics such as size and refractility can be determined in unstained samples from the amounts of light scattered at various angles to the beam by cells; most measurements, however, require that a light-absorbing or fluorescent reagent, or probe, be applied to the cells. The broadest range of cellular parameters is accessible to fluorescence measurements.

Parameters are classified as intrinsic or extrinsic, according to whether or not their measurement requires the use of reagents, and also as structural or functional. A list of parameters measurable by flow cytometry appears in Table I; Table II lists fluorescent probes usable for measurement of the most commonly studied extrinsic parameters.

B. History

Applications to cell counting motivated the first steps in the development of flow cytometry. Modern flow cytometers descend from aerosol particle counters used by the U.S. Army during World War II in experiments on detection of airborne bacteria and spores. The original apparatus subjected a flowing air sample to dark-field illumination. The intensity of light scattered by particles in this sample was used both to indicate their presence and to measure their size.

In the 1950s, instruments based on the same optical principles, but accommodating samples in saline solution, were used clinically for blood cell counting. Shortly thereafter, Wallace Coulter developed his electronic blood cell counter. Instruments based

Table I Parameters Measurable by Flow Cytometry

Parameter	Measurement method and probe if used
Intrinsic structural parameters (no probe)	
Cell size	Extinction or small angle (0–2°) (forward) light scattering, electronic impedance (Coulter principle)
Cell shape	Pulse shape analysis
Cytoplasmic granularity	Orthogonal (large angle, right angle, or 90°) light scattering
Pigment content (e.g., hemoglobin, photosynthetic pigments, porphyrins)	Absorption, fluorescence, multiangle scattering
Protein fluorescence (tryptophan, etc.)	Fluorescence
Intrinsic functional parameter (no probe)	
Redox state	Fluorescence (endogeneous pyridine–flavin nucleotides)
Extrinsic structural parameters	
DNA content	Fluorescence (e.g., propidium, DAPI)
DNA base ratio	Fluorescence (A-T and G-C preference dyes; e.g., Hoechst 33342 and chromomycin A_3)
Nucliec acid sequence	Fluorescence (labeled oligonucleotides)
Chromatin structure	Fluorescence (fluorochromes after DNA denaturation)
RNA content (single- and double-stranded)	Fluorescence (e.g., acridine orange)
Total protein	Fluorescence (covalent or ionic bonded acid dyes)
Basic protein	Fluorescence (acid dyes at high pH)
Sulfhydryl groups/glutathione	Fluorescence (e.g., bromobimanes)
Antigens	Fluorescence (labeled antibodies)
Lipids	Fluorescence (Nile red)
Surface sugars (lectin-binding sites)	Fluorescence (labeled lectins)
Extrinsic functional parameters	
Membrane integrity ("viability")	Absorption or scattering (e.g., trypan blue)
	Fluorescence [e.g., propidium, fluorescein diacetate (FDA)]
Surface receptors	Fluorescence (labeled ligands)
Intracellular receptors	Fluorescence (labeled ligands)
Surface charge	Fluorescence (labeled polyionic molecules)
Cytoskeletal organization	Fluorescence (e.g., labeled phallacidin)
Endocytosis	Fluorescence (e.g., labeled beads or bacteria)
Membrane permeability (dye–drug uptake–efflux)	Fluorescence (various probes)
Enzyme activity	Absorption or scattering (chromogenic substrates)
	Fluorescence (fluorogenic substrates)
DNA synthesis	Fluorescence (anti-BrdUrd antibodies; dye mixtures)
Membrane fluidity or microviscosity	Fluorescence polarization (diphenylhexatriene)
"Structuredness of cytoplasmic matrix"	Fluorescence polarization (FDA)
Cytoplasmic–mitochondrial membrane potential	Fluorescence (e.g., cyanine dyes, rhodamine 123)
"Membrane-bound" $[Ca^{++}]$	Fluorescence (chlorotetracycline)
Cytoplasmic $[Ca^{++}]$	Fluorescence ratio (indo-1)
Intracellular pH	Fluorescence ration (e.g., ADB, BCECF)
Oxidative metabolism	Fluorescence (dichlorofluorescein)

on this principle are now widely used in hematology laboratories.

The analytical capacity of flow cytometry was extended during the early 1960s by a researcher who was attempting to automate cancer cytology screening by adapting optical absorption measurement techniques, first used for quantitative microspectrophotometry of nucleic acids and protein in cells in the 1930s. These first optical flow cytometers measured light scattering, which was used for cell sizing, and absorption; fluorescence measurements were later added.

Flow sorters (i.e., devices that separate cells with selected characteristics from the sample stream) were developed in the late 1960s. Cell sorting was further refined by using argon laser illumination. Researchers were able to separate cells on the basis of their fluorescent antibody-binding characteristics; this has remained an important application of flow cytometry.

Table II Fluorescent Probes for Measurement of Extrinsic Parameters

Excitation	UV	UV	Violet to blue	Blue-green	UV to yellow	Blue-green	Yellow	Red
Emission	Violet to blue	Blue-green to green	Green	Green	Orange-red	Orange-red	Red	Deep red
Parameter Measured	Dye used							
DNA content	Hoechst dyes	DAPI	mithramycin, chromomycin A_3, olivomycin	acridine orange	ethidium, propidium			rhodamine 800
RNA content			thioflavin T	thiazole orange, auramine O	ethidium	acridine orange, pyronin Y		oxazine 1, thiazole blue
Total protein	SITS, coumarins	DANS	FITC	FITC	SR101, Texas red, XRITC	TRITC	SR101, Texas red, XRITC	
Surface structures	SITS, coumarins	coumarins, DANS	FITC	FITC		PE, TRITC; PE tandem conjugates, PerCP	XRITC, Texas red, PC	APC, PC, Cy5
Membrane integrity	primulin	FDA, COFDA	FDA, COFDA	FDA, COFDA	propidium			
Membrane potential	oxacyanines		oxacarbocyanines	oxacarbocyanines, rhodamine 123		indocarbo-cyanines, *bis*-oxonol	thiacarbocyanines	indo- and thiadicarbo-cyanines
Intracellular pH	ADB, 4-MU	ADB, 4-MU	COFDA, BCECF	COFDA, BCECF, SNARF		SNARF		
Enzyme activity	coumarin-based substrates	naphthol-based substrates	fluorescein-based substrates	fluorescein-based substrates			resorufin-based substrates	Vita blue-based substrates
Cytoplasmic Mic [Ca^{++}]	indo-1	indo-1		fluo-3				

ADB, 1,4 diacetoxy-2,3-dicyanobenzene; APC, allophycocyanin; BCECF, 2′,7′-bis(carboxyethyl)-5,6-carboxyfluorescein; COFDA, carboxyfluorescein diacetate; Cy5, reactive indodicarbocyanine; DANS, dansylchloride; DAPI, 4-6-diamidino-2-phenylinadole; FDA, fluorescein diacetate; FITC, fluorescein isothiocyanate; 4-MU, 4-methylumbelliferone; PC, phycocyanin; PE, phycoerythrin; PerCP, peridinin chlorophyll protein; SITS, stilbene disulfonic acid isotinocyanate; TRITC, tetramethylrhodamine isothiocyanate; UV, ultraviolet; XRITC, rhodamine 101 isothiocyanate.

In the 1970s, the commercial production of flow cytometers capable of light scattering, extinction, and fluorescence measurements began, a commercial version of the original cell sorter was introduced, and cell sorters derived from the 1960's designs were commercially produced. All of these instruments used argon lasers for illumination. A flow cytometric system for automated differential white blood cell counting was produced; this used chromogenic enzyme substrates to stain cells and made absorption and light-scattering measurements using an incandescent lamp as a light source. A commercial version of a fluorescence flow cytometer developed in the late 1960's, using the arc lamp illuminator and high-resolution, oil-immersion optics of a fluorescence microscope, was also produced; this apparatus was used primarily for precise measurements of cellular DNA content, which represents another major application of flow cytometry.

Since the late 1970s, significant improvements have been made in both laboratory-built and commercial flow cytometers. The introduction of microcomputer-based systems for data acquisition and analysis and the development of monoclonal antibodies and of new fluorescent labels have made it feasible to measure more characteristics of each cell analyzed than was previously possible. This sometimes requires the incorporation of two or more light sources emitting at different wavelengths into the apparatus. The precision and sensitivity of measurements have also increased. Instruments now in commercial production can detect a few hundred molecules of fluorescein, and experimental systems can measure light-scattering signals from individual virus particles.

C. Applications

Flow cytometry provides a rapid and precise means for detection, counting, and characterization of cells in mixed populations. Biological particles that have been subjected to flow cytometric analysis range from individual virus particles to multicellular organisms such as nematodes. Most applications require that cells be distinguished from particulate debris; it is also necessary to discriminate among different cell types within heterogeneous samples. Multiparameter flow cytometry, in which correlated measurements are made of several cellular characteristics, is particularly useful in studies of mixed populations. Patterns of correlation between measured variables can be used to identify cells as members of a particular subpopulation. Such identifications are used to control hardware and/or software that permits so-called gated analysis, which allows biochemical or physiologic characteristics of each subpopulation of interest to be defined separately. For example, thymic lymphocyte precursors can be identified by light scattering and by the presence of specific surface antigens; their proliferative status can then be assessed from the distribution of DNA and RNA content.

Gated analysis is also useful for measurements of weak signals. Immunofluorescence measurements, for example, are made only when the presence of a cell in the measurement system is indicated by a stronger signal, such as a light-scattering signal. Apart from their use in gated analysis, correlated multiparameter measurements are valuable because they can provide information about the interrelationships of cellular characteristics that would not be available if only a single parameter were measured.

In practice, flow cytometry has been employed predominantly for measurements of cell-surface and intracellular antigens, on the one hand, and of cellular nucleic acid (DNA and sometimes RNA) content, on the other. Qualitative and quantitative changes in these cellular parameters have been used to define and characterize normal and abnormal cellular differentiation and function and to determine the effects of drugs and biologicals on cells. Although now commonplace in clinical and experimental hematology, immunology, and pathology, flow cytometry has not been widely used in microbiology. However, the technique, in addition to offering researchers in this field a useful tool for quantitative analysis of microbial populations, may provide rapid, sensitive, and cost-effective means for bacterial detection and characterization in clinical laboratories.

II. Principles of Flow Cytometry

A. Instrumentation

Figure 1 shows a highly schematic diagram of a typical optical flow cytometer. The apparatus is configured to measure two parameters. The first parameter is forward scattered light, also frequently described as forward scatter, low-angle light scatter, or small-angle light scatter; as cells pass through the focused laser beam used for illumination, light scat-

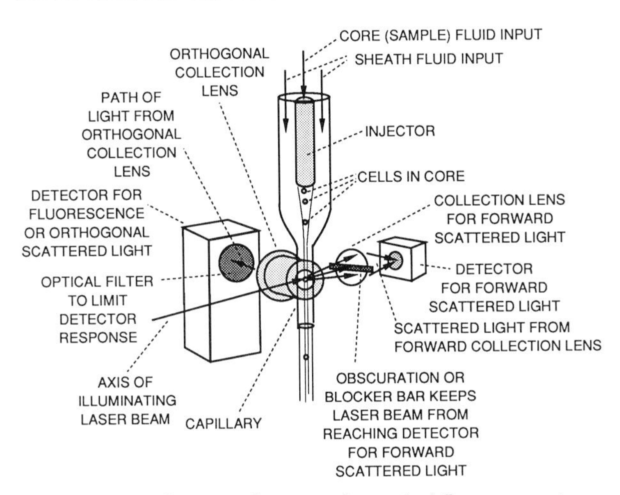

Figure 1 Schematic diagram of a typical flow cytometer.

tered by the cells at small angles to the beam is collected by a lens and directed to a photodetector, usually a photodiode. The laser beam itself is prevented from reaching the detector by an opaque obscuration or blocker bar; this and an iris (not shown) limit the angles at which scattered light can be collected.

The second parameter could be either fluorescence or light scattered at approximately a right angle to the incident beam (orthogonal, right angle, 90°, side, or large-angle light scatter). If the optical filter between the orthogonal collecting lens and the detector blocks the incident wavelength and passes longer wavelengths, the detector will respond to fluorescence emission at these longer wavelengths. If the filter passes the illuminating wavelength, the detector will respond to scattered light.

1. Flow Systems

In the system shown in Fig. 1, the single-file flow of cells is produced by introducing the sample or core fluid containing the cells through a narrow tube or injector, coaxial with a wider tube through which a cell-free sheath fluid is flowing. The sample and sheath fluids may be driven by air or nitrogen under pressure or by positive displacement pumps. Flow velocities are adjusted so that flow is laminar, confining the core fluid to the central region of the stream. Downstream from the end of the injector, the diameter of the stream is reduced by a gradual tapering of the tube, to prevent any disturbance of the laminar flow; at the observation or interrogation point, at which the illuminating beam intersects the cell stream and at which the cells are measured, the tube has been reduced to a capillary with inside diameter between 75 and 200 μm.

2. Light Sources and Optics

Like most laser-source flow cytometers, the apparatus illustrated uses what is called an orthogonal geometry, in which the direction of flow, the illuminating laser beam, and the optical axis of the fluorescence and/or orthogonal scatter collection lens are mutually perpendicular. The axis of the focused laser beam used for illumination is shown schematically as a dashed line. The usual practice is to focus laser beams to small round or elliptical spots so that the narrowest portion of the beam coincides with the observation or interrogation point. Typical spot sizes range from 5 to 50 μm in the direction of flow and from 50 to 130 μm perpendicular to the direction of flow. A small spot dimension in the direction of flow decreases the time taken for cells to pass through the beam, increasing throughput. The spot must be larger in the direction perpendicular to flow to assure even illumination, because cells are not always in the exact center of the stream. Optical designs for forward scatter collection vary widely from instrument to instrument. The obscuration bar must block transmission of the axial beam but permit light scattered by cells at small angles from the axis (0.5–10°, with both limits generally adjustable) to reach the forward scatter collection lens, which converges the scattered light toward the forward scatter detector. This lens need not collect light over a large solid angle.

The orthogonal collection lens is typically a microscope objective. This lens is used to collect fluorescence and/or orthogonal scatter signals. Because these are typically weaker signals than are forward scatter signals, it is advantageous to collect light over the greatest possible solid angle; thus, high numerical aperture lenses are preferred. The orthogonal lens will collect stray light as well as fluorescence emitted and light scattered by cells; to measure fluorescence, it is necessary to prevent most of the scattered and stray light from reaching the detector. This is done by using a combination of light shielding, field stops, and optical filters. In most flow cytometers, dichroic mirrors, which reflect and transmit different portions of the optical spectrum, are used to separate three or four different wavelength regions, diverting each to a separate detector. This allows the apparatus to use a single orthogonal collection lens for simultaneous measurements of orthogonal light scatter and of fluorescence in two or three distinct spectral regions.

Flow cytometers using arc lamp sources, like laser source instruments, usually have the axis of the fluorescence collection lens orthogonal to the direction of sample flow. The collection lens is used for (epi)illumination as well, following the same scheme used for incident light fluorescence microscopy.

3. Detectors and Signal Processing

The forward scatter detector is usually a photodiode, which generates a current proportional to the intensity of incident light. The detectors used for orthogonal scatter and fluorescence signals are usually photomultiplier tubes. These are advantageous for detection of weak signals because they have an internal amplification mechanism. Photomultipliers are also used to detect weak forward scatter signals from small particles such as bacteria and viruses.

The preamplifiers used to amplify the detector signals incorporate circuitry to subtract the baseline signal due to stray light reaching the detectors and, thus, produce pulses at their outputs whenever cells pass through. If the illuminated region is large enough to contain the entire cell, both the peak value or pulse height and the area or integrated value of the pulse are proportional to the total amount of fluorescent or scattering material in the cell. If the cell diameter is larger than the beam height (i.e., the dimension in the direction of flow), only the pulse area is proportional to the total amount of fluorescent or scattering material. The pulse height provides a measure of fluorescence or scatter per unit volume. In this circumstance, the duration or width of the pulse may be used as a measure of cell size.

The electronic circuitry needed to derive measures of pulse area, height, and width must be reset whenever a cell passes the observation point; to insure that cells are detected, a relatively strong, noise-free signal is used to synchronize, or trigger, the electronics for other, weaker signals. When the trigger signal level rises above a settable threshold value, a new measurement cycle is initiated.

Analog signal-processing electronics that permit the derivation of sums and differences of signal intensities are used to compensate for spectral overlap between different fluorescent dyes, and circuits that derive ratios of signal intensities are used for measurements involving dyes that respond to phenomena such as changes in pH or calcium ion concentration by changing their spectral characteristics. Analog logarithmic amplifiers are frequently used to transform signals from a linear to a logarithmic scale, to accommodate a large dynamic range. Most immunofluorescence measurements are recorded on logarithmic scales.

B. Data Analysis: Uni- and Multivariate Distributions

The data acquisition process in flow cytometry comprises all those operations that are required to make measurements of physical characteristics of cells in the sample. The data analysis phase of flow cytometry includes all of the subsequent operations used to derive information about the biological characteristics of some or all of the cells in the sample from the measured values of their physical characteristics. The first step in analysis of data obtained from flow cytometers usually involves examination of one- and two-dimensional frequency distributions of measured values of cellular parameters. In older instruments, which had only pulse height analyzers available for data analysis, one-dimensional frequency distributions were the only data recorded by the apparatus. In modern flow cytometers, which utilize mini- or microcomputers for data analysis, frequency distributions are generated from list mode data, i.e., raw data stored in computer memory and/or on magnetic storage media.

A typical one-dimensional distribution or histogram appears in Fig. 2, which shows the distribution of fluorescence signals from the DNA-specific fluorescent dye DAPI (4′-6-diamidino-2-phenylindole) applied to CCRF-CEM lymphoblasts. The fluorescence intensity of each cell is proportional to its DNA content. The distinctions among cells with diploid (G_0/G_1 phases of the cell cycle, represented by the major peak of the distribution), hyperdiploid

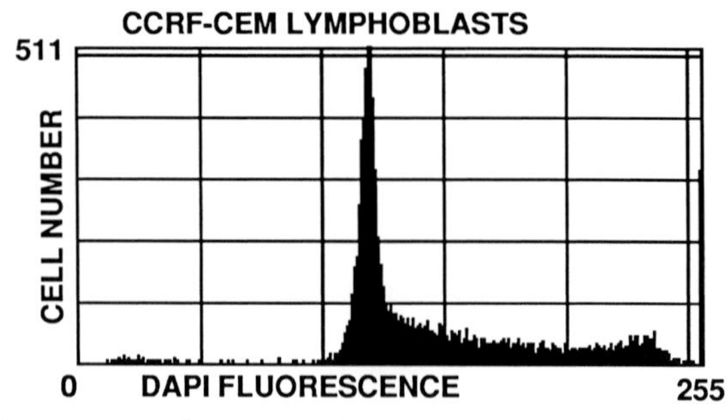

Figure 2 One-dimensional histogram of DAPI (4′-6-diamidino-2-phenylindole) fluorescence (DNA content) in a cultured lymphoblast line.

(S phase, represented by the portion of the distribution between the major and minor peaks), and tetraploid (G_2/M phases, the minor peak at the right) DNA content are fairly clear from the DAPI fluorescence histogram. A fair estimate of the percentages of cells in the G_0/G_1, S, and G_2/M phases of the cell cycle could be made simply by partitioning the histogram. An improved estimate might be obtained using more formal mathematical methods.

The diploid peaks of DNA content histograms are among the sharpest of distributions obtained from flow cytometry. Frequency distributions of any other single cellular parameter are broader and usually show considerable overlap from one cell type to another. It is almost always possible, however, to distinguish different types of cells by looking at multivariate frequency distributions of two or more parameters in spaces of two or more dimensions. Two-parameter dot plots or cytograms, and two-dimensional frequency distributions, shown as three-dimensional projections (peak-and-valley or "isometric" plots), as color- or gray-scale chromatic plots, or as contour plots, are the most commonly used forms of graphical presentation of multiparameter flow cytometric data. Figures 3–5 show two-color immunofluorescence data for monoclonal antibody-stained human peripheral lymphocytes; a gated analysis was based on a two-dimensional selection region defined by forward and orthogonal scatter values. Cells from whole blood were simultaneously stained, following lysis of erythrocytes, with fluorescein-labeled CD8, which defines the cytotoxic–suppressor T-cell population, and with phycoerythrin-labeled CD4, which defines the helper or inducer T-cell population. Forward scatter was used as the trigger parameter.

The dot plot of Fig. 3 shows several clusters of cells. It is partitioned into quadrants in which the counts of $CD4^-CD8^-$, $CD4^+CD8^-$ $CD4^-CD8^+$, and $CD4^+CD8^+$ cells are indicated. In a dot plot, each point or dot represents the measurement values obtained from a single cell. If the number of cells in the plot is relatively small, clusters of cells become visible and the intensity in different areas of the plot gives some indication of the frequency with which the corresponding values of the variables appear in the sample. However, dot plots give only a crude indication of a two-dimensional frequency distribution because the presence of a dot at given coordinates indicates only that at least one cell with the corresponding parameter values is present in the population.

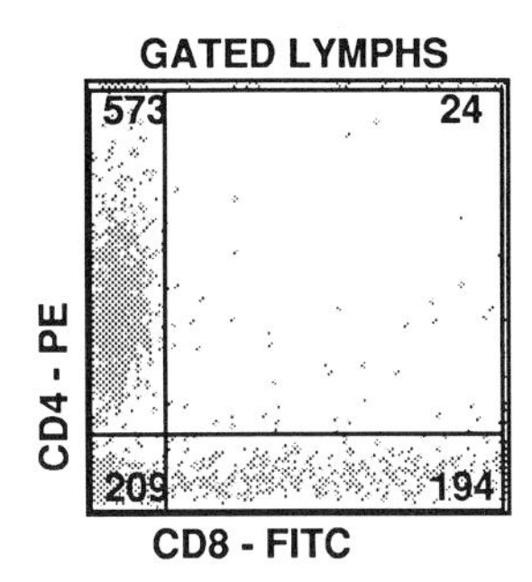

Figure 3 Dot plot showing fluorescence signal values from human peripheral lymphocytes stained to identify helper or inducer ($CD4^+$)and cytotoxic–suppressor ($CD8^+$) cells. The display is partitioned into quadrants and shows counts of cells in each. FITC, fluorescein; isothiocyanate; PE, phycoerythrin.

More information is present in the three-dimensional projection shown in Fig. 4, which represents the two-dimensional distribution of fluorescein and phycoerythrin fluorescence. In this type of plot, a simulated surface is created. The values of observed variables are plotted in the X-Y plane. The apparent height, or *Z*-value, at any point (X, Y) is made proportional to the frequency of occurrence of the paired data values (X, Y) in the sample. The relative volumes or masses (i.e., integrals, not heights) of "peaks" denote the relative numbers of cells in corresponding clusters or modes of the bivariate distribution.

In Fig. 5, the same bivariate distribution shown in Fig. 4 is represented as a gray-scale chromatic plot. Such plots are somewhat easier to generate than either three-dimensional projections or contour plots. In contour and chromatic plots, the values of the two observed variables are plotted in the X-Y plane as they are in a three-dimensional projection.

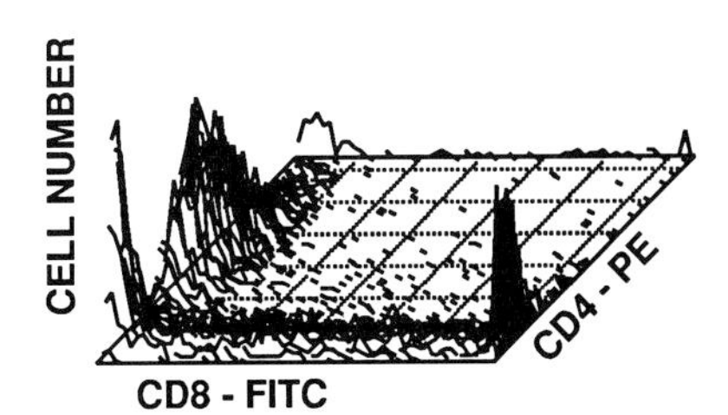

Figure 4 Two-dimensional distribution of fluorescence from human peripheral lymphocytes, displayed as a three-dimensional projection.

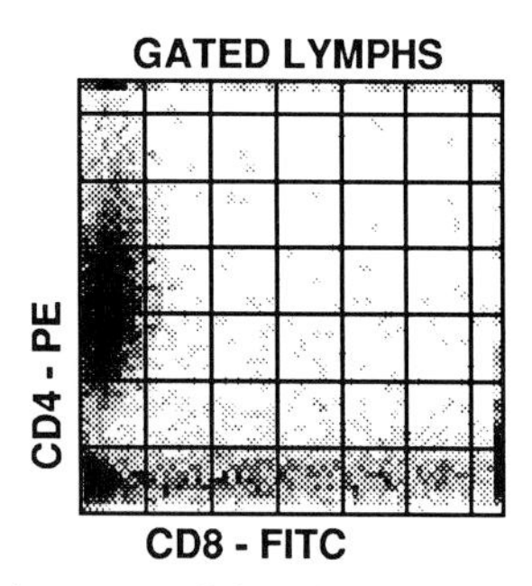

Figure 5 Two-dimensional distribution of fluorescence from human peripheral lymphocytes, displayed as a gray-scale chromatic plot.

In a chromatic plot, frequencies of occurrence are indicated for each point in the X-Y plane. Different frequencies of occurrence are represented as different colors, if a color display is used, or as different intensities or gray-scale levels, if a monochrome display is used. In a contour plot, a direct indication of frequency of occurrence is not given for each point in the X-Y plane. Instead, a series of contour lines are drawn connecting points for which data values occur with equal frequency. The information contents of different types of plots of the same histogram at the same resolution are, or should be, equal.

C. Cell Sorting

The addition of sorting capability to a flow cytometer makes it possible to isolate highly purified populations of cells with precisely defined values of any parameter(s) measurable in the cytometer, allowing further functional and/or biochemical characterization of the isolated subpopulations. In droplet sorting, the method most commonly used, an acoustic transducer stabilizes the point at which droplets break off from the main stream at a fixed distance downstream from the observation point. Between the time a cell traverses the observation point and the time the cell reaches this droplet breakoff point, the values of measured selection parameters are fed into electronic logic, which, if the cell meets the criteria set for sorting, applies a voltage to the stream, electrostatically charging the droplet containing the selected cell. The droplet stream, containing charged and uncharged droplets, passes through the electric field between two deflecting plates to which high voltages of opposite polarities have been applied. Charged droplets are deflected out of the main stream and collected, while the uncharged main stream passes into the waste reservoir. Commercially available sorters can sort cells at rates up to 5000/sec. Faster devices have been developed.

D. Precision, Sensitivity, and Accuracy

The performance of flow cytometers is generally discussed in terms of precision, sensitivity, and accuracy. Precision describes the extent to which identical values are obtained from measurements of identical particles. The measure of precision most commonly used is the coefficient of variation, or CV (i.e., the standard deviation divided by the mean) of a distribution of fluorescence or light-scattering intensities from particles. In fluorescence measurements, sensitivity refers to the number of molecules of a label such as fluorescein, which are detectable on a bead or cell. Beads bearing very small amounts of fluroescent material are commercially available for sensitivity determination and quality control. Most commercial flow cytometers can detect 2000 molecules of fluorescein. Detection of single molecules of phycoerythrin has been achieved using single photon counting techniques on an experimental system. The accuracy of flow cytometers, i.e., the degree to which the measured values approach the true values of the variables measured, is less often discussed than the precision or sensitivity. Accuracy is most often estimated by adding internal standards to specimens.

E. Special Considerations in Measurements of Microorganisms

Only a few dozen of thousands of publications on flow cytometry deal with microorganisms, reflecting the fact that most flow cytometers are designed for analysis of eukaryotic cells. Optimization for this purpose often compromises the instruments' sensitivity for work with smaller particles such as bacteria. Although, in principle, a flow cytometer can measure the same parameters in bacteria or even viruses as are commonly measured in eukaryotic cells, in practice, because the volume, mass, nucleic acid, protein content, etc., of bacteria are approximately $\frac{1}{1000}$th those of mammalian cells, apparatus must be designed and/or modified for increased sensitivity to make the measurements.

III. Parameters Measurable by Flow Cytometry

A. Intrinsic Parameters

1. Cell Size: Optical and Electronic Measurements

Cell size is estimated optically from the amplitude of forward light scattering. This can yield information about the size of particles below the resolution limit of light microscopes. Individual virus particles have, thus, been measured by flow cytometry. However, forward scatter intensity, while an indicator of particle size, strongly depends on the difference in refractive index between particles and the suspending medium. The range of variation in chemical composition of bacterial cell surfaces results in a wide range of variation in refractive index. A small, highly refractile organism will produce a larger light-scattering signal than that produced by a larger, less refractile one. This complication is largely eliminated when size is estimated using electronic (Coulter) volume sensors.

2. Multiangle Light-Scattering Measurements of Cell Shape and Texture

The amplitude of light scattered orthogonal to the illuminating beam is increased by particle asymmetry, by surface roughness, and by the presence of granular internal structures. Orthogonal scatter signals may therefore distinguish similarly sized particles that differ in any of these characteristics.

3. Measurements of Intrinsic Fluorescence

Bacterial chlorophylls and the highly fluorescent phycobiliprotein photosynthetic pigments found in algae and cyanobacteria are often measured *in situ* by marine biologists. The same materials are also extracted from their sources and used as fluorescent labels for antibodies.

B. Extrinsic Parameters

1. Fixation and Staining Mechanisms

While many relatively small cationic dye molecules readily enter intact cells, it is often necessary to permeabilize the cell membrane to permit staining. This can be accomplished using classical fixatives such as ethanol and aldehydes or by treatment with nonionic detergents. In gram-negative bacteria, the outer membrane presents a barrier to entry of many hydrophobic molecules which readily stain intact gram-positive organisms and eukaryotic cells. Staining may be facilitated by exposure of the organisms to EDTA or other calcium chelators.

2. DNA Content

DNA content is measured with fluorescent dyes such as propidium iodide, Hoechst 33258, DAPI, mithramycin, and acridine orange, which are also used for detection of small numbers of bacteria by fluorescence microscopy. Under well-controlled conditions, flow cytometry of bacterial DNA content can allow detection of drug effects on cell growth within a few generation times.

3. DNA Base Composition

DNA base composition, i.e., the ratio of adenine-thymine (A-T) to guanine-cytosine (G-C) base pairs in the genome, varies from species to species. Combinations of fluorescent dyes that preferentially stain A-T and G-C base pairs are used for flow cytometric determination of base composition. The ultraviolet-excited, blue fluorescent dyes Hoechst 33258 and DAPI are selective for A-T, whereas the blue-excited, yellow fluorescing chromomycin A_3 and mithramycin are specific for G-C. A mixture of Hoechst 33258 and chromomycin A_3, used with dual-beam excitation, can discriminate *Staphylococcus aureus* (31% G-C), *Escherichia coli* (50% G-C), and *Pseudomonas aeruginosa* (67% G-C). The same technique can also demonstrate bacteriophage multiplication in *E. coli*.

4. RNA Content

RNA content determination using fluorescent dyes such as acridine orange and pyronin Y can discriminate among mammalian cell populations in different phases of growth and development. RNA content is likelyl to be harder to measure in bacteria, because dyes used for RNA content determination will also stain other cellular constituents such as DNA, polysaccharides, and lipids.

5. Total and Basic Protein Content

Total protein content may be estimated using covalently or ionically bound acid dyes. Dyes such as fluorescein isothiocyanate (FITC), which bind covalently to protein amino groups, are preferable to other acid dyes for protein content determination in bacteria because washing or dilution can be used to lower background fluorescence which might otherwise interfere with measurements. FITC has been combined with propidium iodide for simultaneous flow cytometric analysis of DNA and protein content in bacteria. If the pH of the medium is altered so that only basic proteins retain reactive amino groups, the same reagents used for total protein determination can be used to measure basic protein content.

6. Cellular Antigens

The sensitivity attainable in flow cytometric measurements of immunofluorescence in bacteria is limited because only a few tens of thousands of molecules of antigen at most are present on or in an individual microorganism. While detection of specific antigens is possible, only small amounts of light can be collected from each cell, even when a relatively large laser is used as a light source.

Fluorescein, applied as FITC, remains the most popular fluorescent label for antibodies and other ligands. Its green fluorescence is optimally excited at the 488-nm operating wavelength of the lasers used in most flow cytometers. Phycoerythrin, which is also well excited at 488 nm, exhibits yellow-orange emission, which contrasts well with that of fluorescein. When excited at higher wavelengths (e.g., 530 or 546 nm), phycoerythrin provides increased sensitivity because cellular autofluorescence is diminished.

7. Specific Nucleic Acid Sequence Detection

Flow cytometric procedures using fluorescently labeled oligonucleotide probes have been developed for determination of specific nucleic acid sequences in eukaryotic cells and bacteria. Such analyses, like immunofluorescence analyses, are relatively difficult to perform because of the relatively small amount of fluorescent label that must be detected.

8. Fluorescent Probes of Functional Parameters

a. Enzyme Content and Activity

The presence or absence of specific enzymes may be useful for differentiating among cell types or species that are similar in other respects; indeed, many conventional procedures for bacterial identification rely on organisms' capacity or inability to ferment particular substrates. Flow cytometric detection of intracellular enzyme activity employs fluorogenic substrates, which are nonfluorescent derivatives of compounds such as fluorescein, methylumbelliferone, and resorufin. The substrates are taken up by cells. If the appropriate intracellular enzyme is present, substrate molecules are cleaved to release the highly fluorescent parent compound, which is retained in the cell for at least a short time, rendering the cells fluorescent. Cells lacking the required enzyme do not fluoresce. Although only a few enzyme molecules may be present in an individual microorganism, their activity can generate hundreds of thousands of molecules of fluorescent reaction product within a few minutes, providing a strong fluorescence signal.

b. Intracellular pH

Fluorogenic esters of fluorescein derivatives and of certain other compounds exhibit changes in their fluorescence excitation and/or emission spectra in response to changes in pH. These dyes can thus be utilized for measurements of intracellular pH and may be useful for demonstration of pH gradients between the bacterial cytoplasm and the medium, indicating metabolic activity.

c. Intracellular Calcium Ion Concentration

A variety of fluorescent indicator dyes, loaded into cells as fluorogenic esters, show differences in fluorescence on binding calcium ions; these have been used to study cellular calcium influxes and shifts mediated by ionophores, chemoattractants, peptide hormones, and neurotransmitters.

d. Membrane Potential

An electrical potential gradient, typically >100 mV, with the interior negative, is maintained across the plasma membranes of actively metabolizing microorganisms. A similar gradient is established across the mitochondrial membranes of eukaryotic cells, which also maintain a smaller potential difference across their cytoplasmic membranes. Bacterial membrane potential is decreased or lost within a few minutes following removal of energy sources. Membrane potential is also abolished by chemical or physical agents, which rupture the membrane (e.g., penicillin treatment in susceptible organisms, heat injury), and by ionophores that eliminate ion gradients across the membrane, including gramicidin and carbonyl cyanide chlorophenylhydrazone. [*See* CELL MEMBRANE: STRUCTURE AND FUNCTION.]

Bacterial membrane potentials can be estimated from the distribution of fluorescent dyes across membranes. According to the Nernst equation, at 37°C, every 60 mV of potential difference (interior negative) across the membrane contributes a factor of 10 to the interior–exterior concentration gradient of a monovalent cationic indicator. The 100–200-mV potential difference commonly observed across bacterial membranes corresponds to concentration gradients ranging from 50 : 1 to 1000 : 1. The concentration gradients actually observed are somewhat higher because the lipophilicity or hydrophobicity necessary to allow an indicator molecule to pass freely through the membrane makes it bind preferentially to cellular constituents even in the absence of a potential difference. A variety of cyanine and rhodamine dyes have been used for flow cytometric measurement of bacterial membrane potentials.

IV. Applications of Flow Cytometry in Microbiology

A. A Basic Problem: Identification of Cells in Mixed Populations

Flow cytometric measurements of forward and orthogonal light scattering, DNA and protein content, DNA base composition, immunofluorescence, nucleic acid sequences, and membrane potentials of bacteria, fungi, protozoa, and algae have been used with some success to discriminate among different species. Gating measurements on a relatively strong scatter or DNA fluorescence signal is essential for sensitive detection of the weak signals obtained from fluorescently labeled antibodies or nucleic acid probes. Measurement of additional parameters improves identification, especially when bacteria are present at low concentrations and when samples contain considerable particulate debris. A particle demonstrated to contain DNA or an enzyme or to exhibit a pH or membrane potential gradient is likely to be a microorganism. The identification becomes more reliable if two or more of these characteristics can be demonstrated in the same particle. *Escherichia coli* seeded into human blood samples at concentrations of 10–100 organisms/ml has been detected by measuring forward scatter and ethidium fluorescence following red cell lysis. Measurements of forward scatter, fluorescein immunofluorescence, and DNA content determined from propidium fluorescence have been used to detect *Listeria monocytogenes* in milk and to discriminate it from several other species. Thus, rapid, specific detection and identification of many clinically significant microorganisms probably could be accomplished using a multiple-illumination beam, multiparameter flow cytometer, which could measure samples stained with a mixture of several fluorescent dyes, labeled antibodies, and perhaps gene probes as well. However, such an apparatus might not be practical for most clinical microbiology laboratories. In designing a more practical, cost-effective flow cytometric system for clinical use, it is preferable to measure as few parameters as possible and desirable to use probes compatible with inexpensive light sources such as helium-neon or diode lasers.

B. Marine Microbiology and Microbial Ecology

Marine microorganisms, including bacteria, phytoplankton, and zooplankton, have been subjected to flow cytometry, often using apparatus carried on board ships. Intracellular toxin has been determined in the dinoflagellate *Gonyaulax;* quantification of the autofluorescence of luciferin in this organism was used to study the circadian cycle of its bioluminescence. [*See* BIOLUMINESCENCE, BACTERIAL.]

Multiangle light scattering, autofluorescence, DNA content, and genetic sequence have been examined as identifying parameters for phytoplankton, with a projected application to environmental quality assessment by monitoring population dynamics.

C. Industrial and Process Microbiology

Flow cytometry of bacteria has been advocated for fermentation process monitoring in industrial microbiology. Preparative sorting has been used to purify yeast basidiospores and might be used to isolate microbial subpopulations with desired metabolic characteristics following gene transfer. Monitoring of the metabolism of cultures over time may be facilitated by encapsulation of organisms in gel microdroplets containing indicator dyes.

D. Clinical Microbiology

Clinical microbiology requires detection, counting, characterization, and determination of drug effects on microbial cells. Existing conventional methods analyze many organisms or their metabolic products, requiring incubation periods for growth of the microbial population to produce enough material to be detected. Under these circumstances, nonviable bacteria in samples fail to produce progeny in the first stage of culture and are excluded from analysis, while viable organisms grow to outnumber organic and inorganic particulates originally present in the sample.

Flow cytometric approaches to rapid microbiological analysis attempt to detect and characterize organisms either in the absence of growth or after a minimal period in culture. For results to be reconcilable with those obtained using classical culture methods, the cytometric technique must not only detect and identify individual organisms, but also provide an indicator of their viability. Microorganisms must be distinguished from particulate debris as well as from one another.

1. Bacterial Detection and Identification

The ideal clinical specimen for flow cytometric analysis is one in which relatively few particles and few bacteria are found in the absence of an infection and in which relatively large numbers of bacteria are found when an infection is present. Urine meets these criteria. Between one-third and one-half of specimens submitted for culture are urine specimens. In uninfected individuals, urine is sterile and contains few particles, whereas in patients with urinary tract infections, urine typically contains at least tens and, more often, hundreds of thousands of bacteria per milliliter.

A first approximation to bacterial detection in urine can be obtained using an electronic (Coulter) counter to count particles in the appropriate (0.25–3.0 μm) size range. Several studies have shown this is effective in detecting bacteriuria; however, the technique can neither discriminate between bacteria and other small particles nor determine bacterial viability. The use of an optical flow cytometer, measuring DNA content and base composition or detecting bacterial enzymes, might improve detection and identification but would require a relatively expensive apparatus, as would strategies involving the use of fluorescently labeled antibodies or genetic probes.

A method that now appears promising identifies particles as viable microorganisms based on their membrane potentials; organic and inorganic particles of similar size, which may also be present in a sample, do not have membrane potentials. Viable bacteria can be distinguished by their size from fungi and from other cells and by their membrane potentials from particulates. Additional reagents permit discrimination between gram-positive and gram-negative bacteria on the basis of membrane permeability. The technique can be implemented using a small, inexpensive flow cytometer with a low-power red helium-neon laser or laser diode source, which simultaneously measures light scattering by bacteria and their uptake of a fluorescent membrane potential indicator dye. A 2-min analysis of two or three aliquots can detect as few as 10^4 organisms/ml in raw samples. Neither filtration nor centrifugation is required during sample preparation, although either concentration technique could be used to permit detection of lower concentrations of organisms.

The membrane potential assay can also be used to implement an analog of conventional classification procedures in which identification is accomplished by determining which of several media support or inhibit growth of the organisms. Bacteria will not grow in media that do not contain suitable energy sources. Conventional methods use the presence or absence of growth after an incubation period ranging from hours to days to determine this characteristic. However, bacteria become unable to maintain normal membrane potentials within a few minutes following transfer to media that do not contain appropriate energy sources and respond to inhibitors in a similarly short time. This allows growth and inhibition characteristics to be predicted by membrane potential assays in minutes rather than hours.

2. Antibiotic Sensitivity Testing

Effects of antibiotics, as well as those of other specific lethal agents such as antibiotics, antibody plus complement, bacteriophages, or colicins on microorganisms are detectable from changes in membrane potential and from changes in total bacterial counts over time. However, practical sensitivity testing, like thorough identification, requires the analysis of numerous aliquots of sample. Many existing automated microbiology instruments can examine multiple aliquots in parallel, while the complexity and cost of a flow cytometer would seem to dictate that a cytometric instrument perform analyses in a time-consuming serial fashion. Because membrane potential probes that could be used with extremely low-cost diode lasers are available, and because the construction of flow cytometers may be simplified considerably and the cost further decreased by the use of integrated optics, it may eventually be possible to speed up operation by incorporating several cytometers into a single microbiology system.

E. Flow Cytometry in Parasitology and Virology

Intracellular parasites such as plasmodia may be analyzed and quantified by flow cytometry. Nucleic acid stains have been used to detect the organisms inside erythrocytes and free in lysed preparations. However, the relatively high cost of even simple flow cytometers makes it unlikely that the technology will soon be widely applied in clinical parasitology.

Although experimental flow cytometers have demonstrated sufficient sensitivity in light scatter and fluorescence measurement to be able to detect signals from single virus particles, more practical applications of flow cytometry in virology are based on detection of viral or virus-related antigens and/or nucleic acid in infected cells. These techniques are presently being used in attempts to better understand the dynamics of human immunodeficiency virus infection.

Bibliography

Boye, E., and Løbner-Olesen, A. (1990). *New Biol.* **2,** 119–125.

Melamed, M. R., Lindmo, T., and Mendelsohn, M. L. (eds.) (1990). "Flow Cytometry and Sorting," 2nd ed. Wiley-Liss, New York.

Shapiro, H. M. (1988). "Practical Flow Cytometry," 2nd ed. Alan R. Liss, New York.

Yentsch, C. M., and Horan, P. K. (eds.) (1989). *Cytometry* **10,** 497–669.

Food Biotechnology

Susan Harlander
University of Minnesota

Glossary

Antibody Protein produced by humans and higher animals in response to the presence of a specific antigen

Biotechnology Collection of technologies that employ living systems, or compounds derived from these sytems, for the production of industrial goods and services

DNA probe Single-stranded DNA molecule that has been radioactively labeled and is used to locate a particular nucleotide sequence or gene on a DNA molecule

Enzyme engineering Use of site-specific mutagenesis to introduce single base substitutions in DNA that result in substitutions of amino acids at specific locations in an enzyme

Genetic engineering Technology used to alter the genetic material of living cells to make them capable of producing new substances or performing new functions

Monoclonal antibody Highly specific purified antibody that is derived from the fusion of an antibody-producing cell with a cancer cell; recognizes only one antigen

FOOD BIOTECHNOLOGY is the application of traditional and modern technologies that employ living systems of microbial, plant, or animal origin or components derived from these systems to enhance the production, processing, and distribution of safe, nutritious, palatable, and affordable food. The food chain can be viewed as a continuum from the planted seed, through the processing, distribution, and marketing of products, to the consumer. The food processing industry serves as the vital link between the farmer and the supermarket and includes everything that happens to raw agricultural products once they leave the farm gate and are transformed into forms acceptable and available for human consumption.

I. What Is Food Biotechnology?

The word "biotechnology" is derived from "bio," meaning life or living systems, and "technology," defined as scientific methods for achieving a practical purpose. Biotechnology is a collection of technologies that employ living systems, or compounds derived from these systems, for the production of industrial goods and services. The living systems can be of plant, animal, or microbial origin. Biotechnology is not new to the agricultural and food sector, since humans have been exploiting living systems for the production, processing, and preservation of food for centuries. Microorganisms including bacteria, yeast, and mold have been used since the beginning of recorded history for the production of fermented dairy, meat, and vegetable products, as well as for the fermenting of beverages such as wine and beer. Many ingredients used in foods as vitamins, stabilizers, flavors and flavor enhancers, colors, and preservatives are produced by microbes. Additionally, microbes have been used to degrade the waste products generated during the processing of food.

What distinguishes "modern" biotechnology from the more traditional examples just cited is the emergence within the last 20 years of genetic engineering and many other techniques that allow for genetic improvement of microorganisms with precision, predictability, and speed that was not attainable with traditional procedures. Genetic engineering provides the ability to cross species barriers, thus greatly expanding the available gene pool for improving plants, animals, and microorganisms.

To date, the major focus of genetic engineering has been its use in the construction of transgenic microorganisms capable of producing high-value human pharmaceuticals and health-care products including insulin, growth hormone, interferon, tissue plasminogen activator, and erythropoietin. Other examples of commercially viable biotechnology products include diagnostic kits for rapid detection of plant, animal, and human diseases and human and animal vaccines. Although it has long been predicted that genetic engineering would have a major impact on production agriculture and food processing, the first engineered crops, processing aids, and ingredients are just beginning to gain regulatory approval and move into the marketplace. The following discussion will focus on how biotechnology can be used to improve the processing of food.

II. Traditional Uses of Microorganisms in Foods

A. Food Fermentations

Bacteria, yeasts, and molds have been used for the production of fermented foods for centuries. Fermented foods are defined as those foods that have been subjected to the action of microorganisms (bacteria, filamentous fungi, or yeasts) or enzymes to produce desirable biochemical changes. The microorganisms may be the microflora indigenously present on vegetable or animal products that serve as the substrates for fermentation or they may be added starter cultures. Microbial metabolism is responsible for the production of preservative agents such as acids, carbon dioxide, and alcohol, as well as for chemical and physical changes that alter the flavor, texture, shelf life, safety, digestibility, and nutritional quality of fermented foods. The microorganisms involved are multifunctional and form an integral part of the end product. Fermentation is a relatively simple, natural, efficient, inexpensive, and low-energy food preservation process that reduces the need for refrigeration.

The end products of fermentation influence the character of the final product and depend on the particular microorganisms involved in the fermentation. Lactic acid bacteria belonging to the genera *Lactobacillus, Lactococcus, Streptococcus, Pediococcus,* and *Leuconostoc* are used for the production of fermented dairy, meat, and vegetable products, and produce lactic acid as the primary end product of fermentation. Fermentative yeasts from the genus *Saccharomyces,* used for the production of wine, beer, and bread, produce alcohol and carbon dioxide as primary end products of metabolism. Filamentous fungi such as *Aspergillus, Penicillium, Mucor,* and *Rhizopus* are equipped with a powerful arsenal of enzymes that contributes to degradation of substrates during fermentation. Fermented foods make a major contribution to the diet in all parts of the world; the classes of fermented foods produced in different regions of the world reflect the diet in each region. Some examples of fermented foods and the microorganisms responsible for the fermentation are provided in Table I.

B. Single-Cell Protein

Single-cell proteins are the dried cells of microorganisms such as algae, certain bacteria, yeasts, and molds that are grown in large-scale culture systems for use as protein for human or animal consumption. The products also contain other nutrients, including carbohydrates, fats, vitamins, and minerals. The ancient Aztecs of Mexico harvested algae of the genus *Spirulina* and consumed them as part of their diet. In Germany during both world wars, yeast was produced for consumption as a protein supplement. In more recent years, advances in scientific knowledge about the physiology, nutrition, and genetics of microorganisms have led to significant improvements in single-cell protein production from a wide range of microorganisms and raw materials.

The largest continuous microbial culture process for single-cell protein was designed by Imperial Chemical Industries in the United Kingdom. The fermenter used in this operation contained 1200 cubic meters of culture consuming 14 metric tons/hr of methanol and produced 50,000 metric tons per year of single-cell protein for animal feed. Although the production of single-cell protein is not a high priority at the present time in the United States due to the abundance of alternative protein sources, this technology will become increasingly important in the

Table I Examples of Fermented Foods

Class	Food product	Microorganism(s)
Beverages	Wine, beer	*Saccharomyces cerevisiae*
	Coffee, cocoa	*Erwinia, Bacillus, Streptococcus, Saccharomyces, Lactobacillus, Leuconostoc* spp.
	Tea	Indigenous enzymes
Cereal products		
Wheat	Bread, crackers	*Saccharomyces cerevisiae*
	Lavash, nan (flatbread)	*Saccharomyces cerevisiae, Lactobacillus, Streptococcus* spp.
	Sourdough bread	*Saccharomyces, Lactobacillus*
Cereal/legume	Hopper, idli, dosa, adai, vada, papadam	*Streptococcus, Pediococcus, Leuconostoc*
Rice	Lao-chao	Various species of molds
Dairy products	Cheese	*Lactococcus cremoris/lactis*
	Yogurt	*Streptococcus, Lactobacillus*
	Sweet acidophilus milk	*Lactobacillus*
Fish products	Rokorret, tarama, paak, fish sauce, momoni	*Micrococcus, Staphylococcus, Bacillus, Pediococcus*
Fruit products	Citron	Indigenous microorganisms
	Vanilla	*Leuconostoc, Lactobacillus, Streptococcus*
Vegetable products	Olives, sauerkraut, kimchi	*Leuconostoc, Streptococcus, Pediococcus, Lactobacillus*
Legume products	Tempe	*Rhizopus oligosporus*
	Soy sauce	Bacteria, yeasts, and molds
	Natto	*Bacillus subtilis* var. *natto*
	Miso	*Aspergillus oryzae*
Meat products	Salami, thuringer, pepperoni	*Pediococcus, Lactobacillus*
Starch products	Cassava	*Lactobacillus, Streptococcus*
	Taro (poi)	*Lactobacillus delbrueckii, Lactococcus lactis*

[Adapted from Campbell-Platt, G. (1987). "Fermented Foods of the World: A Dictionary and Guide." Butterworths, Boston.]

future as the need for inexpensive high quality protein increases. [*See* SINGLE-CELL PROTEINS.]

C. Probiotic Uses of Microorganisms

Microorganisms have been reported to play a key role in maintaining the health of humans and animals by controlling intestinal microorganisms capable of producing toxic effects in the host. For example, it has been suggested that lactobacilli assist in the digestion of lactose, provide important digestive enzymes, inactivate toxins, bind cancer-causing chemicals, modulate the gut flora, deconjugate bile acids, reduce cholesterol absorption in the gut, supply B vitamins, and even reduce bad breath. Probiotic effects have been studied extensively in animals; it is not uncommon to add certain organisms directly to animal feed to enhance digestibility of the feed and to protect the gastrointestinal tract from microbial invasion. Probiotic effects of microorganisms in humans are just beginning to be explored.

D. Microbial Production of Food Ingredients

Microorganisms produce a variety of secondary metabolites via fermentation that can be purified for use as food ingredients. Microorganisms are metabolically diverse, small in size, and easy to grow in large quantities on diverse substrates, making them ideal candidates for production of secondary metabolites. The types of chemicals produced by microbial fermentation include acidulants, amino acids, vitamins, flavors and flavor enhancers, pigments, stabilizers, thickeners, surfactants, sweeteners, antioxidants, and antimicrobial agents. Table II provides several examples of food ingredients, their functional uses in foods, and the producing microorganisms.

1. Processing Aids

Enzymes are protein catalysts that carry out all the synthetic and degradative reactions of living organisms. Enzymes are used extensively in the food

Table II Microbial Production of Food Ingredients

Ingredient	Function in foods	Microorganisms
Acetic acid	Acidulant	*Acetobacter pasteurianus*
D-Arabitol	Sugar	*Candida diddensii*
β-carotene	Pigment	*Blakeslea trispora*
Citric acid	Acidulant	*Ceratocystis* spp.
Diacetyl	Buttery flavor	*Leuconostoc cremoris*
Dextran	Thickener	*Leuconostoc mesenteroides*
Emulsifier	Emulsification	*Candida lipolytica*
Glycerol	Humectant	*Bacillus licheniformis*
Glutamic acid	Flavor enhancer	*Corynebacterium glutamicum*
Leucine	Amino acid	*Brevibacterium lactofermentum*
Mannitol	Sugar	*Torulopsis mannitofaciens*
Methylbutanol	Malt flavor	*Lactococcus lactis* subsp. *maltigenes*
Monascin	Pigment	*Monascus purpureus*
Nisin	Antimicrobial	*Lactococcus lactis*
Monosodium glutamate	Flavor enhancer	*Corynebacterium glutamicum*
L-Phenylalanine	Aspartame precursor	*Bacillus polymyxa*
Surfactant	Wettability	*Bacillus licheniformis*
Vitamin B_{12}	Vitamin	*Propionibacterium*
Xanthan gum	Thickener	*Xanthomonas campestris*
Xylitol	Sweetner	*Torulopsis candida*

[Reproduced from Wasserman, B. P., Montville, T. J., and Korwek, E. L. (1988). *Food Technol.* **42(1)**, 133–146.]

processing industry to control texture, appearance, and nutritive value and to generate desirable flavors and aromas. Although enzymes are produced by animals and plants as well as by microorganisms, the enzymes from microbial sources are generally most suitable for commercial applications. Microbial products can be mass-produced without the limitations that might be imposed by the season of the year or geographic locations, as could be the case for a plant-derived enzyme. In addition, microorganisms grow quickly and production costs are relatively low. In view of the metabolic diversity of microorganisms, nature has provided a vast reservoir of enzymes that act on all major biological molecules. Some examples of microbial enzymes used in the food processing industry are provided in Table III.

Enzymes are frequently used in batch food processing systems; however, enzymes can be immobilized and used in continuous processing systems, where applicable. For example, the enzymes used to convert starch in corn to high fructose corn syrup and the enzyme rennet used in cheese manufacture are immobilized and used continuously for weeks and sometimes months or years without substantial loss of activity.

2. Amino Acids and Vitamins

Amino acids are the building blocks of protein and can be obtained by hydrolysis of plant or animal protein. Alternatively, certain microorganisms are capable of synthesizing and excreting single amino acids that can be purified from fermentation broths and used as dietary supplements, antioxidants, and flavors and flavor enhances, as well as for manufacture of aspartame, a sweet dipeptide used in reduced-calorie foods. Table IV provides some examples of microbially produced amino acids and vitamins and their uses in foods.

3. Flavors and Pigments

Certain of the lactic acid bacteria used for the production of fermented dairy products are capable of producing lactic acid and the volatile compounds diacetyl and acetaldehyde. These compounds are produced during the fermentation as a result of microbial metabolism and are responsible for the characteristic buttery flavor and aroma of certain fermented dairy products. Diacetyl gives cottage cheese and buttermilk their buttery aroma; acetal-

Table III Applications of Selected Enzymes in Food Processing

Enzyme	Microorganisms[a]	Substrate	Function
α-Amylase	*Bacillus amyloliquifaciens*	Starch	Liquefaction to dextrins; brewing and baking
Cellulase	*Trichoderma reesei*	Cellulose	Juice clarification
D-Glucose isomerase	*Bacillus coagulans*	Glucose	High fructose corn syrup
Glucose oxidase	*Aspergillus niger*	Glucose	Flavor and color preservation in eggs and juices
Lactase	*Aspergillus niger*	Lactose	Improves milk digestibility; glucose production from cheese whey
Lipase	*Candida cylindracae*	Lipid	Cheese ripening
Pectinase	*Aspergillus niger*	Pectin	Wine/juice clarification
Proteinase	*Mucor miehei*	Protein	Meat tenderizer; sausage curing; dough conditioning; beer haze removal
Pullulanase	*Aerobacter aerogenes*	Amylopectin	Beer production; improves glucose and maltose release

[a] Only one of several producing microorganisms is included in this table.
[Adapted from Neidleman, S. (1986). Enzymology and food processing. *In* "Biotechnology in Food Processing" (S. K. Harlander and T. P. Labuza, eds.), pp. 37–56. Noyes Publications, Park Ridge, New Jersey. and Neidleman, S. (1989). The microbial production of biochemicals.] *In* "A Revolution in Biotechnology" (J. L. Marx, ed.), pp. 56–70. Cambridge University Press, New York.]

dehyde is the important volatile chemical of yogurt aroma.

Because of their size, metabolic diversity, and relatively simple growth requirements, bacteria, yeasts, and molds have been used commercially for the production of a vast array of flavors and aromas. Chemical synthesis is not possible for many of these compounds and microbial cells possess the relatively complex metabolic pathways essential for their biosynthesis. Some examples of microbially derived flavors are provided in Table V.

Microorganisms are also capable of producing pigments that could be used as alternatives to chemical dyes. These include polyketides produced by the mold *Monascus purpureus*, astaxanthin produced by the yeast *Phaffia rhodozyma*, and the carotenoids produced by the algae *Dunaliella bardarwil*.

Table IV Production of Amino Acids by Microbial Fermentation

Amino acids	Functional use	Microorganism
D, L-Alanine	Flavor	*Brevibacterium flavum*
L-Arginine	Dietary supplement	*Brevibacterium flavum*
L-Glutamic acid	Flavor enhancer	*Brevibacterium flavum*
L-Histidine	Dietary supplement	*Corynebacterium glutamicum*
L-Isoleucine	Dietary supplement	*Brevibacterium flavum*
L-Leucine	Dietary supplement	*Brevibacterium lactofermentum*
L-Lysine	Dietary supplement	*Corynebacterium glutamicum*
L-Methionine	Dietary supplement	*Brevibacterium flavum*
L-Phenylalanine	Manufacture of aspartame	*Brevibacterium lactofermentum*
L-Proline	Dietary supplement	*Cornyebacterium glutamicum*
L-Serine	Dietary supplement	*Corynebacterium hydrocarboclastus*
L-Threonine	Dietary supplement	*Corynebacterium glutamicum*
L-Tryptophan	Dietary supplement, sleep aid	*Brevibacterium flavum*
L-Tyrosine	Dietary supplement	*Corynebacterium glutamicum*

[Reproduced from Neidleman, S. (1989). The microbial production of biochemicals. *In* "Biotechnology in Food Processing" (S. K. Harlander and T. P. Labuza, eds.), p. 37–56. Noyes Publications, Park Ridge, New Jersey.]

Table V Microbiological Production of Flavor Chemicals

Compounds	Aroma/flavor	Microorganism
Anisaldehyde	Anise-like	*Trametes sauvolens*
Benzaldehyde	Almond-like	*Trametes sauvolens*
Benzyl alcohol	Fruity	*Phellinus igniarius*
Cinnamic acid methyl ester	Fruity, jasmine	*Inocybe corydalina*
Citronellol	Rose-like	*Ceratocystis variospora, Trametes odorata*
Citronellyl acetate	Fruity, rose-like	*Ceratocystis variospora*
γ-Decalactone	Peach	*moniliformis, Sporobolomyces odorus*
Diacetyl	Buttery	*Lactococcus lactis* subsp. *diacetylactis*
Ethyl benzoate	Fruity	*Phellinus igniarius*
Ethyl butyrate	Fruity	*Lactobacillus, Pseudomonas fragi*
Geranial	Rose-like	*Ceratocystis variospora*
Linalool	Floral	*C. variospora*
Methyl benzoate	Fruity	*Phellinus tremulus*
Methylphenylacetate	Honey-like	*Trametes odorata*
Methyl salicylate	Wintergreen	*Phellinus igniarius*
6-Pentyl-α-pyrone	Coconut-like	*Trichoderma viride*
Tetramethylpyrazines	Nutty	*Corynebacterium glutamicum*
p-Tolualdehyde	Almond-like	*Mycoacia uda*

[Reproduced from Neidleman, S. (1989). The microbial production of biochemicals. *In* "A Revolution in Biotechnology" (J. L. Marx, ed.), pp. 56–70. Cambridge University Press, New York.]

4. Polymers

Microorganisms synthesize a number of nonprotein extracellular polymers, primarily polysaccharides composed of chains of simple sugars such as glucose or fructose. Many of these compounds find application in the food industry as emulsifying agents and as substances that add bulk or texture to food products. They are also used as insoluble matrices for immobilizing enzymes and for encapsulating flavors. Microbial polymers and their producing microorganisms are listed in Table VI. One polymer of particular interest to the food industry is poly-β-hydroxybutyrate, which is synthesized from the simple 4-carbon fatty acid β-hydroxybutyrate. This compound can be polymerized into a film that is biodegradable and might be used for manufacture of food-packaging material.

E. Traditional Strain-Improvement Strategies

Despite their multitude of uses in food systems or for the production of ingredients, microorganisms are, in many cases, not ideally suited for the tasks at hand. Occasionally a strain can be isolated that functions more optimally, presumably due to a spontaneous mutational event in the affected metabolic pathway. Mutational events can have a dramatic effect on microbial metabolism; however, with the low rate of spontaneous generation of mutants (approximately 1 mutational event in every 10^6–10^7 cells per generation), it is impractical and time-consuming to effectively screen and select for such a rare event.

Strain-improvement programs that involve the purposeful generation of mutants that perform certain functions better or overproduce specific compounds have met with enormous success. Mutagenesis can be induced with short-wave ultraviolet light, ionizing radiation (e.g., X-rays, γ- and β-rays), or chemical agents (e.g., nitrous acid, hydroxylamine, alkylating agents) followed by screening and selection to identify improved strains. Yields greater than 100-fold have been attained with such traditional strain-improvement programs. Unfortunately, there are significant limitations to traditional strain-improvement strategies. For example, random and potentially deleterious mutations can occur during mutagenesis, and selection systems designed for the mutation of interest will not detect these other muta-

Table VI Microbial Production of Polymers

Polymer	Function/use	Microorganism
Alginate	Encapsulating agent	*Azotobacter vinelandii, Pseudomonas aeruginosa*
Cellulose	Anticaking agent	*Acetobacter* sp.
Curdlan	Thickening agent	*Agrobacterium* sp.
Cyclosophorans	Gelling agents	*Rhizobium, Agrobacterium, Xanthomonas* spp.
Dextran	Thickening agent	*Acetobacter* sp.
D-Fructan	Thickening agent	*Zymomonas mobilis*
Gellan	Gelling agent	*Auromonas elodea*
Levan	Thickening agent	*Bacillus* sp., *Leuconostoc mesenteroides, Pseudomonas* sp.
Phosphomannan	Gelling agent	*Hensenula capsulata, Rhizobium meliloti*
Poly-β-hydroxybutyrate	Packaging film	*Alcaligenes eutrophus, Methylobacterium organophilum*
Xanthan	Thickening agent	*Xanthomonas campestris*

[Adapted from Morris, V. J. (1990). *Food Biotechnol.* **4(1)**, 45–57, and Neidleman, S. (1989). The microbial production of biochemicals. *In* "A Revolution in Biotechnology" (J. L. Marx, ed.), pp. 56–70. Cambridge University Press, New York.]

tions. Mutation and selection procedures are time-consuming, laborious, and expensive, and offer no guarantee of improved performance. Finally, there is no opportunity to expand the genetic potential of microorganisms by adding specific genes.

III. Genetic Engineering of Foodgrade Microorganisms

Genetic engineering provides an alternative to classical mutation and selection for improving microbial starter cultures. Figure 1 illustrates the general method for gene cloning in microbial systems. The discovery in the early 1970s of restriction endonucleases, microbial enzymes that cleave DNA in specific places, heralded a new era in biology with a central focus on the molecular basis of living systems. Deoxyribonucleic acid (DNA), the universal code of life, is structurally and functionally identical in all living organisms; thus, it can be transferred among related and unrelated living organisms using gene cloning techniques. The essential elements for successful gene cloning include a transformable host, a vector that is capable of replicating in the host, high frequency gene transfer systems for introducing DNA into hosts, and an understanding of the structure, function, regulation, expression, and metabolic compatibility of the new genetic information. Genetic engineering has the potential to be more predictable, controllable, and precise than classical breeding and selection. In addition, genetic improvements can proceed at a much faster pace and the ability to cross species barriers greatly expands the available gene pool.

All the pioneering work in genetic engineering was conducted with the gram-negative bacterium, *Escherichia coli*. Much is known about the genetics and biochemistry of this organism: metabolic pathways are well understood, numerous mutations have been mapped on the chromosome, sophisticated vectors have been constructed, and high frequency gene transfer systems are readily available. Unfortunately, much of this information is not known for the yeasts, molds, and gram-positive organisms most frequently used in food fermentations. The techniques developed for *E. coli* are not readily

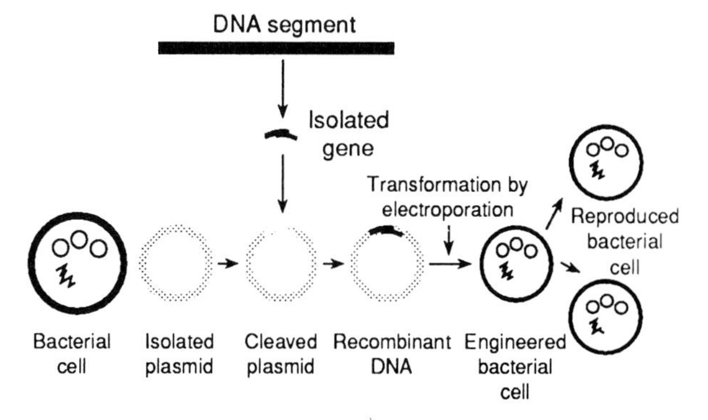

Figure 1 Schematic of genetic engineering. DNA can be isolated from any living organism (plant, animal, or microorganism), digested with restriction enzymes, and inserted into a plasmid vector. Bacterial cells are transformed with the recombinant DNA molecules and are grown to produce large quantities of the cloned gene product. [Reproduced from Harlander, S. K. (1989). *Food Technol.* **43(9)**, 196–206.]

transferable to these organisms, necessitating the development of parallel systems for foodgrade microorganisms.

A. Genetic Improvement of Starter Cultures

Much of the genetic engineering in *E. coli* has been done with the purpose of using the engineered strain as a factory for producing valuable single proteins that are purified from fermentation broths. The engineered *E. coli* cells are contained in fermentation vessels and destroyed during the downstream processing and purification of the gene products of interest. Genetic engineering of starter cultures, on the other hand, is done to improve fermentation properties and viable organisms remain as an integral part of the food that will be consumed by humans. Although single gene products may be of value, frequently transfer of whole metabolic pathways or several enzymes in a pathway may be required to improve fermentation parameters; in many cases, the basic biochemistry and genetics involved in the pathway are not well understood.

The fact that viable organisms will be released into the food supply and consumed by humans imposes some constraints that are not applicable in genetic engineering experiments with *E. coli*. For example, antibiotic resistance genes are used extensively as selectable markers on vectors to facilitate identification of transformed cells. Antibiotic resistance markers, particularly genes that confer resistance to those antibiotics used therapeutically, would be unacceptable in starter cultures because of the potential transfer of DNA to indigenous microorganisms, already present in the gastrointestinal tract, after consumption of the fermented food. Transfer of resistance genes to potentially pathogenic organisms would render these strains resistant to antibiotic therapy, which could result in life threatening situations. It is well known that antibiotic resistance genes can be readily transferred between microorganisms in natural environments, particularly under selective conditions. Thus, a tool that is readily available to *E. coli* geneticists cannot be used by the food biotechnologists, so alternative selection strategies must be developed. [*See* ANTIBIOTIC RESISTANCE.]

Genetic engineering strategies to date have involved the transfer of genetic information from one foodgrade organism to another. The products of these genes are already in the food supply and have been safely consumed by humans for hundreds of years. The introduction of gene(s) that encode proteins or metabolic end products not previously consumed in the human diet would need to be evaluated for digestibility and/or potentially negative effects on humans.

The production of fermented food is frequently more art than science for, although starter cultures have been used for centuries, their genetics and biochemistry are just beginning to be explored. Strain-improvement strategies have focused on traditional mutation and selection techniques to improve the metabolic properties of the dairy lactococci. Although application of genetic engineering offers exciting possibilities, the basic tools are just beginning to be developed. Current research is focused on the construction of foodgrade cloning vectors (multifunctional plasmids or integrative vectors derived solely from DNA of food-approved organisms); the development of high frequency gene transfer systems, particularly electroporation; and the identification and characterization of the structural and functional properties of desirable traits.

Table VII provides examples of how genetic engineering can be used to improve microorganisms used to produce fermented dairy, meat, and cereal products. Genetic engineering could be used to enhance the processing, nutritional value, microbiological safety, and shelf life of fermented foods. Prior to their use in foods, genetically engineered starter cultures will need to be approved by the Food and Drug Administration (FDA). The FDA and other federal agencies with oversight in the food biotechnology area are in the process of developing regulatory policies; until these guidelines are in place, the United States food processing industry will be reluctant to invest in the construction of improved organisms by novel techniques.

The first genetically engineered foodgrade starter culture was approved for use in March 1990. The British Ministry of Agriculture, Fisheries, and Food granted permission for broad-scale use in Great Britain of a baker's yeast strain, *Saccharomyces cerevisiae* 352Ng. This strain produces enhanced levels of carbon dioxide due to the production of elevated levels of two enzymes involved in starch utilization: maltose permease and maltase. The strain can be used in doughs containing widely different sugar concentrations and assists the baker in the control of product quality and consistency. As the first approved genetically engineered organism, this strain has historical significance and may serve

Table VII Genetic Improvement of Food Grade Microorganisms

Type of fermentation	Nature of improvement	Implications
Dairy		
Cheese	Bacteriophage (virus) resistance	Eliminates economic losses due to destruction of culture by viruses
	Increased levels of specific proteases involved in cheese ripening	Accelerated ripening and decreased storage costs
	Bacteriocin production	Improved safety due to inhibition of pathogens and spoilage organisms
Yogurt	Higher levels of the enzyme β-galactosidase	More digestible product for lactose-maldigesting individuals
	Thaumatin (sweet protein) production	Naturally sweet product with reduced calories
Meat		
Sausage	Bacteriocin production	Inhibition of pathogens and spoilage organisms
Cereal		
Beer	α-Amylase production	Production of "lite" or low-calorie beer
Bread	Higher levels of maltose permease and maltase	More consistent and improved leavening

[Reproduced from Harlander, S. K. (1990). *Cereal Foods World* **35(11)**, 1106–1109.]

as a prototype for the development and commercialization of other engineered foodgrade microorganisms.

B. Enzymes and Ingredients from Genetically Engineered Organisms

The use of genetically engineered microorganisms for the production of enzymes or food ingredients is analogous to cloning in *E. coli*, since the fermentation products will be purified and the producing organism will not remain in the product. An historic event for food biotechnology was the affirmation of GRAS (generally recognized as safe) status by the FDA for the first enzyme derived from a genetically engineered organism to be used directly in food. Recombinant chymosin or rennet is a proteolytic enzyme used to accelerate curd formation during the manufacture of cheese. Rennet is traditionally obtained as an extract of the forestomach of calves. To engineer the product, the structural gene for rennet was synthesized and inserted into an *E. coli* vector encoding antibiotic resistance; the gene product was produced by fermentation. The approval of this enzyme in March, 1990, was a significant milestone for the food industry since it established a critical regulatory precedent and serves as a model for other biotechnology-derived enzymes and ingredients. Other enzymes produced by genetically engineered microorganisms for which regulatory approval has been sought include forms of rennet produced by other organisms, as well as enzymes used in the production of high fructose corn syrup. It is predicted that genetic engineering will be used to improve many of the microorganisms discussed earlier that are used to produce food ingredients, processing aids, and nutritive additives of interest to the food-processing industry. [*See* GENETICALLY MODIFIED ORGANISMS: GUIDELINES AND REGULATIONS FOR RESEARCH.]

Because the regulatory status of the producing organism is of obvious concern to the ingredient manufacturer, much of the research in this area has focused on the use of foodgrade microorganisms as producing organisms. As genetic systems develop in foodgrade microorganisms, it will be possible to engineer these strains to produce valuable ingredients derived from virtually any living organism.

IV. Protein and Enzyme Engineering

A. Chemical Modification of Enzymes

The food processing industry is the largest single user of enzymes, accounting for, on average, more than 50% of enzyme sales. Proteases, lipases, pectinases, cellulases, amylases, and isomerases are used extensively to control the texture, appearance, flavor, and nutritional value of processed foods; however, these enzymes frequently do not function optimally under the conditions of temperature and pH used in food processing operations. Chemical modification of isolated enzymes can have a dramatic impact on enzymatic activity, specificity, and stability. Some examples are provided in Table VIII. Chemical modification of natural molecules has provided useful derivatives, but the general lack of specificity in the reagents and the requirement for difficult and tedious purification and characterization to insure homogeneity severely limits the power of the method when it is rigorously applied.

B. Site-Directed Mutagenesis

Genetic engineering has been used to introduce minor changes in the structure of enzymes that have dramatic effects on substrate specificity, pH and thermal stability, and resistance of the enzyme to proteolytic degradation. Using techniques of site-specific mutagenesis, base substitutions in the primary structure of DNA can result in substitutions of amino acids at specific locations in the protein molecule. For example, this technology has been used to substitute every amino acid at specific key locations in the active site of the enzyme subtilisin; properties of the enzyme could be dramatically altered, both positively and negatively, in comparison with the native enzyme. Some examples of positive effects of genetic engineering on enzyme activites are provided in Table IX.

Site-specific mutagenesis could improve the versatility of enzymes in food systems and decrease the cost of processing food. Examples of how site-directed mutagenesis could be used to improve enzymes are provided in Table X. This technology could also be used to modify other proteins of interest to the food-processing industry, with the possibility of altering functional properties or nutritional value.

V. Biosensors

The highly specific actions of certain biological molecules are potentially exploitable for the development of biosensors that can measure the concentration of specific components in a complex mixture. As illustrated in Fig. 2, enzymes, antibodies, or microbial cells can be immobilized onto a solid surface and the specific reactions they mediate can be detected electrochemically, thermometrically, mechanically, or photometrically.

Advances in several areas have contributed to the successful development of biosensors. Techniques for stabilizing enzymes, antibodies, and cells onto surfaces while retaining their biological activity are essential. Membranes are used to separate sensor elements and protect them from the external environment. Development of tailor-made membranes capable of separating solutes based on molecular size, charge, or solubility has greatly advanced biosensor construction. Advances in the semiconductor industry make it possible to combine chemical components and integrated circuits in a single miniaturized system.

A partial list of commercially available biosensors for food analysis is provided in Table XI. Biosensors could have broad applications in the food processing industry for the continuous monitoring of fermentation processes or of the concentrations of nutrients during processing. Ultimately, it will be possible to insert biosensors directly into food-

Table VIII Effects of Chemical Modification on Enzyme Activity

Enzyme (Source)	Chemical modification	Effect
α-Amylase (*Bacillus subtilis*)	Acetylation with p-nitrophenyl acetate	Increased thermostability above 70°C; reduced thermostability below 67°C
Carboxypeptidase A (mammals)	Acetylation or iodination of active-site tyrosine	Increased esterase and eliminated peptidase activity
Rennet (*Mucor pusillus*)	Acylation with anhydrides	Up to 2-fold increase in milk coagulating activity
Rennet (*Mucor* sp.)	Methionine oxidation as with H_2O_2	Decreased thermostability for easier inactivation during pasteurization in cheese making
Thermolysin (*Bacillus thermoproteolyticus*)	Acylation with amino acid *N*-hydroxysuccinimide esters	Increase in activity up to 70-fold

[Reproduced from Neidleman, S. (1986). Enzymology and food processing. *In* "Biotechnology in Food Processing" (S. K. Harlander and T. P. Labuza, eds.), pp. 37–56. Noyes Publications, Park Ridge, New Jersey.]

Table IX Positive Effects of Genetic Engineering on Enzyme Activities

Enzyme	Modification	New property
Subtilisin	Methionine^{222} → alanine Glycine^{166} → aspartic, glutamic acids	Greater bleach stability Altered substrate specificity
T4 lysozyme	Isoleucine^{3} → cysteine, then chemical cross-linking	Increased thermostability
Trypsin	Glycine^{226} → alanine	Altered substrate specificity
Tyrosyl-tRNA synthetase	Cysteine^{35} → serine	K_m for ATP lowered; increased enzyme activity
Amidase	Serine → Phenylalanine and others	Change in substrate range
Xanthine dehydrogenase or purine hydroxylase 1	Alteration in relative positions of catalytic or orienting sites	Change in substrate range

[Reproduced from Neidleman, S. (1986). Enzymology and food processing. *In* "Biotechnology in Food Processing" (S. K. Harlander and T. P. Labuza, eds.), pp. 37–56. Noyes Publications, Park Ridge, New Jersey.]

processing streams to obtain on-line, real-time measurements of important food-processing parameters. Biosensors could even be incorporated into food packages to monitor temperature abuse, microbial contamination, or loss of shelf life and to provide a visual indicator to consumers of the state of the product at the time of purchase.

VI. Food Safety

Although it is generally recognized that the United States enjoys the safest food supply in the world, pathogens not previously associated with food have emerged as a threat to particularly vulnerable segments of the population, especially infants and children, pregnant women, the aged, the chronically ill, and immunocompromised individuals. The dairy and meat industries have had to cope with the emergence of the pathogenic organism *Listeria monocytogenes,* which is capable of causing spontaneous abortion in pregnant women and meningitis in infants, the elderly, and the immunocompromised. A strain of *Salmonella enteritidis* has been isolated from intact eggs; entire flocks of poultry appear to be infected endemically with the organism. Other pathogens of growing concern to the food industry include *Yersinia enterocolitica, Campylobacter jejuni, Vibrio cholerae,* and *E. coli* 0157:H7. There is also increasing concern about microbial toxins, aflatoxin, chemical residues (e.g., herbicides, pesticides, fertilizers, and fungicides), antibiotics, and other animal drug residues in raw and processed foods.

Historically, food microbiologists have been concerned with the isolation and identification of spe-

Table X Suggestions for Improved Enzymatic Activity through Enzyme Engineering

Enzyme	Application	Useful improvement
α-Amylase	Starch liquefaction	Acid-tolerant and thermostable
Amyloglucosidase	High fructose corn syrup production	Immobilized with higher productivity
Esterases, lipases, proteases	Flavor development	Improved substrate specificity
Glucose isomerase	High fructose corn syrup production	Increased thermostability
Limoninase	Debittering of fruit juices	More complete limonin degradation
Protease	Beer chill-proofing	Improved substrate specificity
Pullulanase	High fructose corn syrup production	Increased thermostability

[Reproduced from Neidleman, S. (1986). Enzymology and food processing. *In* "Biotechnology in Food Processing" (S. K. Harlander and T. P. Labuza, eds.), pp. 37–56. Noyes Publications, Park Ridge, New Jersey.]

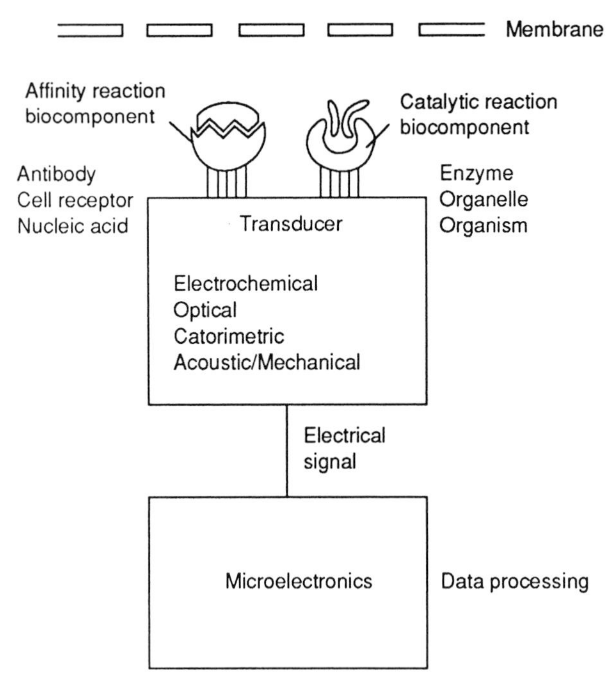

Figure 2 Generalized diagram of a biosensor. [Reproduced from Owen, V. M., and Turner, A. P. F. (1987). *Endeavour, New Series* **11(2)**, 100–104.]

cific organisms from foods. The traditional methods involve isolating the suspect organism(s) in pure culture and performing biochemical and serological tests to identify them. Often, nonselective and/or selective enrichments are required to increase the sensitivity of the culture methods. Standard microbiological tests based on culturing of organisms are laborious, time consuming (e.g., days or weeks for some organisms), and in many cases inaccurate.

The detection of foodborne pathogens is complicated because foods may be contaminated with low numbers of pathogenic organisms within a complex and varying nonpathogenic microbial flora also present in the same food. Additionally, food varies tremendously in physical and chemical composition and time is a major factor when analyzing highly perishable foods before release to the public. There is a need for the development of rapid tests that can be completed within minutes and can be used to assess the quality of raw products and ingredients before they are processed and during processing. Rapid, sensitive, and highly specific methods based on DNA probes and monoclonal antibodies are among the first successful applications of biotechnology to the food industry and will become increasingly important tools for ensuring the safety of the food supply. [*See* FOODS, QUALITY CONTROL.]

Table XI Commercially Available Biosensors for Food Analysis

Analyte	Biocomponent	Application
Glucose	Whole bacterial cells	Molases production
	Glucose oxidase	Brewing; various fermentations; fruit juice and soft drink manufacture; banana maturation
Lactose	β-Galactosidase	Raw milk
Sucrose	Invertase	Instant cocoa manufacture
Lactate	Lactate dehydrogenase	Dairy products; yogurt; whey
Ethanol	Alcohol dehydrogenase	Alcoholic beverage fermentations
Peptides	Amino peptidase	Casein hydrolysis
Amino acids	Amino acid dehydrogenase	Many foods
Glutamate	L-Glutamate oxidase	Soy sauce manufacture
Aspartame	L-Aspartase or alcohol oxidase	Level of sweetner in many foods including soft drinks
Ascorbic acid	Ascorbate oxidase	Fruit juices
Sulfite	Sulfite oxidase	Dry fruit; wine; vinegar; juices; potato flakes
Penicillin	Antibody–enzyme conjugate	Milk
PHB ester	*p*-Hydroxybenzoate hydroxylase	Fruit juices and drinks

[Reproduced from Wagner, G., and Schmid, R. D. (1990). *Food Biotechnol.* **4(1)**, 215–240.]

A. DNA Probe-Based Detection Systems

A DNA probe is a short (14–40-base) single-stranded sequence of nucleotide bases that will bind to specific regions of single-stranded target DNA sequences; the homology between the target and the DNA probe results in stable hybridization. Hybridization is monitored by labeling probes, by attachment to or incorporation into the probe, with compounds that can be detected visually or chemically. Isotopes such as ^{32}P can be incorporated into the structure of the probe. Enzymes such as alkaline phosphatase or horseradish peroxidase (HRP) are often linked to the probe via a chemical linkage; hybridization can be detected visually following addition of substrates. Fluorescently labeled compounds such as fluorescein isothiocyanate can also be attached directly to probes. Table XII provides several examples of DNA probes used for the detection of foodborne pathogens of concern to the food-processing industry; Fig. 3 illustrates the basic steps in a DNA–probe hybridization assay.

Sensitivity is one of the major concerns for food microbiologists, since the presence of a single organism can be significant in certain situations. DNA probe-based assay systems are capable of detecting in the range of 1000 to 100,000 microorganisms in a food sample; therefore, enrichment of organisms to detectable levels is frequently necessary. Probe-based systems do not provide certain information that culturing methods provide, such as strain or biotype identification or serotype. Culturing of samples is often still required because probes are not available for all pathogens. Additionally, probes detect both living and dead organisms; therefore, an infectious agent may be detected even if no viable cells are present. Probes do not detect preformed toxins; therefore, if toxin-producing cells are no longer present, a method of detection of the toxin must still be employed.

Different methods have been used to increase the sensitivity of DNA probe assays, but perhaps the most significant procedure involves the use of the polymerase chain reaction (PCR). Using PCR, the target DNA sequence can be selectively amplified, thus increasing the amount of target available for detection. The PCR technique is based on the reiteration of a three-step process: denaturation of dsDNA into single strands, annealing extension primers homologous to the target sequence to be amplified to the single-stranded DNA, and enzymatic extension of primers using a thermostable enzyme such as *Taq* DNA polymerase, as illustrated in Fig. 4. A typical amplification is 20–40 cycles and results in a million-fold amplification of the original target DNA. The DNA from a single cell would be detectable following PCR amplification. Increases in the amount of target DNA using PCR amplification

Table XII DNA Probes for Foodborne Pathogens

Foodborne pathogen	Component detected	Labeling system
Salmonella spp.	Chromosomal DNA fragment	Isotopic
	23S rRNA	FITC/HRP
Listeria spp.	β-Hemolysin (listeriolysin O gene)	Isotopic ^{32}P
	Species-specific 16S rRNA	FITC/HRP
Listeria monocytogenes	Plasmids	Hydrophobic grid-membrane filter
Escherichia coli	16S rRNA	FITC/HRP
	Heat-labile toxin (LT) gene	Isotopic ^{32}P
	Shiga-like toxin	Isotopic ^{32}P
	β-glucuronidase (GUD) gene	MUG
Yersinia enterocolitica	rRNA	FITC/HRP
Vibrio vulnificus	Hemolysin	Colony blot
Campylobacter spp.	rRNA	FITC/HRP
Staphylococcus aureus	Enterotoxins A, B, C1, and E	FITC/HRP
Shigella spp.	Virulence plasmids	^{32}P/colony-blot assay format

Abbreviations: FITC/HRP, fluorescein isothiocyanate/horseradish peroxidase; MUG, 4-methylumbelliferyl β-D-glucuronide (appearance of bluish fluorescence under long-wave ultraviolet light when 4-methylumbelliferyl is released by hydrolysis of the MUG with GUD).

[Reproduced from Harlander, S. K. (1989). *Food Technol.* **43(9),** 196–206, and Wolcott, M. J. (1991). *J. Food Protect.* **54(5),** 387–401.]

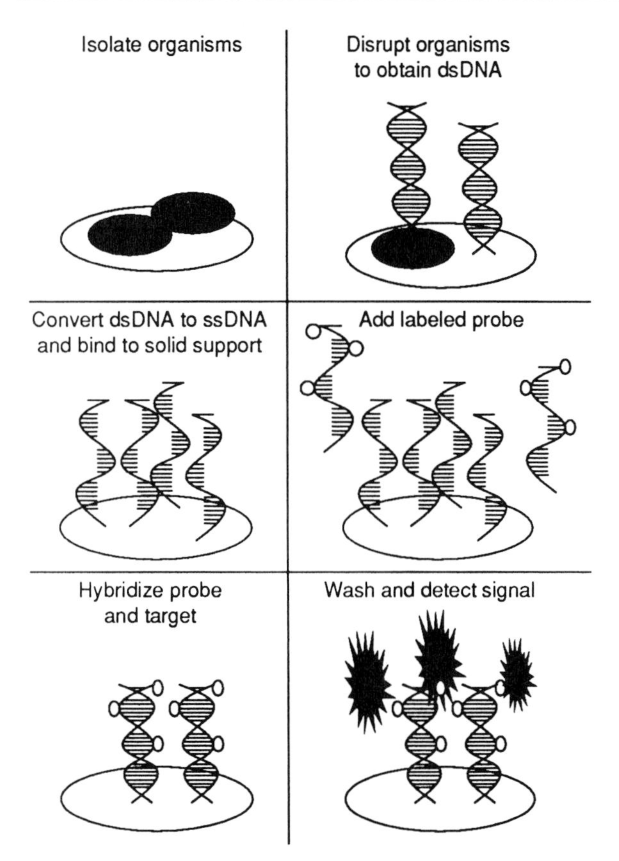

Figure 3 Basic steps in a DNA-probe hybridization assay. Organisms present in a food product are trapped on filters and disrupted to obtain double-stranded DNA. Following denaturation of the DNA to single strands, the labeled probe is allowed to hybridize with target DNA. Hybridization can be detected by a number of methods outlined in the text. [Reproduced from Wolcott, M. J. (1991). *J. Food Protect.* **54(5)**, 387–401.]

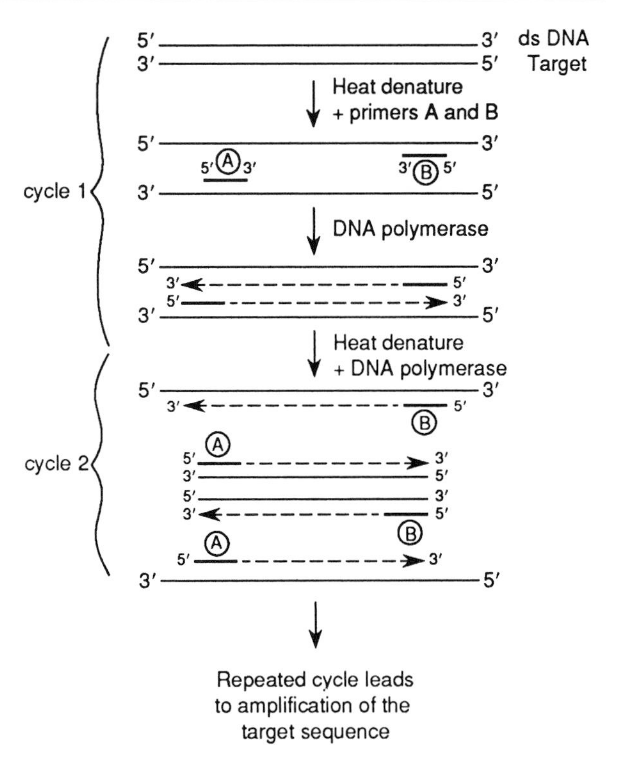

Figure 4 Polymerase chain reaction. A double-stranded DNA target is thermally denatured in the presence of primers A and B, which are complementary to opposite strands of the target DNA. After the primers are allowed to anneal to the denatured DNA strands, a DNA polymerase extends the primers using the target DNA strands as templates. In the second cycle, the newly synthesized DNA is thermally denatured from the target DNA strands. Primers A and B are again annealed to the appropriate DNA strands and the DNA polymerase extends the primers. Repeated cycles of denaturation and DNA synthesis produce an exponential replication of a segment of the target DNA molecule. [Reproduced from Kwoh, D. Y., and Kwoh, T. J. (1990). *Amer. Biotech. Lab.* **October**, 14–25.]

limit the usefulness of DNA probes for quantitative determinations unless the amplification process follows a defined kinetic reaction. [*See* POLYMERASE CHAIN REACTION (PCR).]

B. Polyclonal and Monoclonal Antibody-Based Detection Systems

When animals are exposed to foreign antigens, the body mounts an immune response and stimulates certain cells (B lymphocytes) to produce specific antibodies against the antigen. A vast array of compounds—including proteins, polysaccharides, glycoproteins, lipids, toxins, viruses, and chemical agents—functions as antigens. Antibodies are Y-shaped glycoproteins that circulate in the blood stream; they specifically bind to and inactivate their homologous antigen. Each B cell produces only one type of antibody in response to an antigen; however, each antigen has multiple epitopes and a mouse can make about 1000 different antibodies in response to a single antigenic determinant. The blood contains a mixture of antibodies, each recognizing a different epitope of the antigen; the antibodies in this mixture are referred to as polyclonal antibodies, the production of which is illustrated in Fig. 5A. Antibodies can be purified from the blood of an immunized animal and used for the development of assays to specifically detect known antigens.

Fusion of an antibody-producing cell, isolated from the spleen of an animal exposed to a specific antigen, with a cancer cell results in the formation of a continuous cell line that produces antibodies of a single specificity. These are referred to as monoclonal antibodies; the procedure for obtaining these antibodies is illustrated in Fig. 5B. Monoclonal antibodies are generally characterized by high-affinity, homogeneous composition, and molecular specificity that can also be used for the development of immunoassays to detect known antigens.

Immunoassays using either polyclonal or monoclonal antibodies have been developed for a number of molecules of interest to the food industry. One of the primary advantages of immunoassay-based systems is that they are capable of detecting both protein and non-protein target antigens. In addition to detection of spoilage organisms and foodborne pathogens, immunoassays can be used to detect microbial toxins, aflatoxins, viruses, agricultural chemicals (pesticides, herbicides, fungicides, and fertilizers), heavy metals, antibiotic residues, hormones, animal drug residues, and enzymes, making immunoassays far more versatile than DNA probes for many purposes.

For antibodies to be used in detection systems there must be some way of detecting the binding of antibodies to target antigens. This can be accomplished in much the same way and with many of the same reagents as used with the DNA probes discussed earlier. Many immunoassay formats have been developed, but the most popular is the enzyme-linked immunosorbant assay (ELISA).

The recent successful cloning of genes that encode the constant and variable regions of heavy and light chains of antibodies now makes it possible to produce monoclonal antibodies in *E. coli*. This will revolutionize immunoassay development since it totally bypasses the need for animal production of antibodies. Further, it will be possible to obtain antibodies to highly toxic compounds that normally kill the animal before an immune response can be mounted. Because this technology allows for an infinite combination of heavy and light chain variable regions, it is possible to construct a greater variety of high-affinity antibodies with even greater specificities than those achievable in animal hosts.

VII. Waste Management

Future approaches to food production and processing will have to be more concerned with waste management and conservation. Environmental concerns and economic issues necessitate better use of raw materials and further reduction of food processing waste. Innovative methods will need to be developed for more efficient use of cellulosic material (skins, peels, leaves, stalks, vines, shells, and pits) generated during fruit and vegetable processing; the fat, blood, collagen, and bone from meat processing; and the whey generated during cheese manufacture. To date, most approaches to the problems of food waste involve physical and chemical processing. With the advent of biotechnology, biological approaches to waste use employing intact microorganisms or cell-free enzyme systems have emerged.

A. Whole-Cell Fermentation

Food processing waste streams are frequently rich in carbohydrates, lipids, and/or proteins that serve as a good source of nutrients for microbial fermentation. For example, the whey generated during the manufacture of chedder cheese contains whey proteins and the disaccharide lactose, and serves as an excellent fermentation substrate for lactose-fermenting microorganisms. Over 40 billion pounds of whey are generated each year by the cheese manufacturing industry. Innovative methods of whey use have been developed and include the use of ultrafiltered whey as a feedstock for the production of ascorbic acid, a high-value food additive. Whey can also be used for the production of biofuels, such as ethanol and methane, and for the production of single-cell protein and baker's yeast.

The United States currently has a surplus of raw agricultural commodities such as wheat, corn, and milk. There is interest in application of biotechnology for nonfood uses of these renewable resources for the production of biofuel and hydrocarbons; biomaterials such as fibers, polymers, plastics, and ceramics; and specialty chemicals such as surfactants, fats and oils, flocculants, and proteins. Polyhydroxybutyrate is a compound produced by certain microorganisms that can be extracted and polymerized into a thermoplastic and biodegradable packaging film. Similarly, the over 10 billion pounds of potato processing wastes could be fermented to polylactic acid, which can also be polymerized into a biodegradable packaging film. In the future, it may be possible to use waste streams from food processing for the production of vaccines, therapeutic drugs, and other high-value pharmaceutical products.

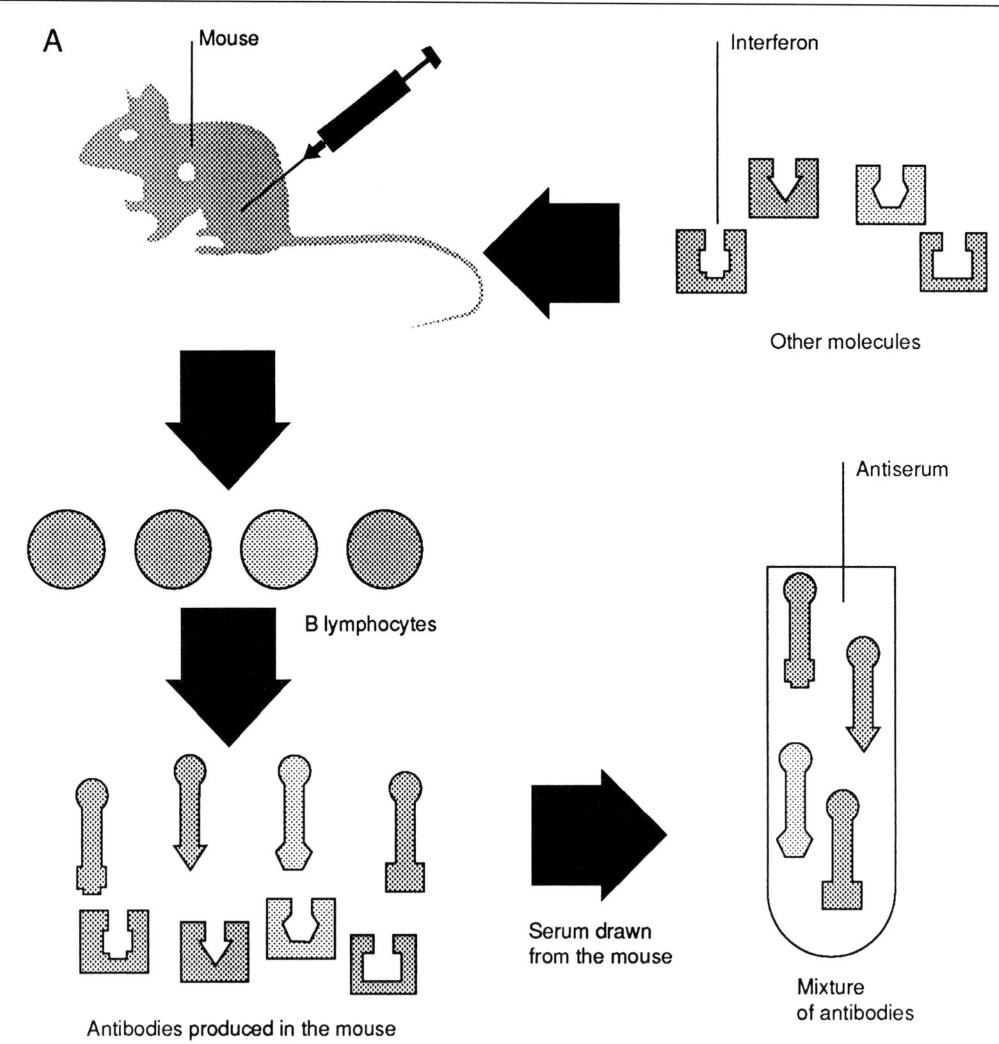

Figure 5 Polyclonal and monoclonal antibody production. (A) When an antigen (interferon, for instance) is injected into a mouse, several B lymphocytes (B cells) respond to different parts of the antigen and produce specific antibodies. These polyclonal antibodies can be harvested from the blood and contain a mixture of antibodies with different specificities for the antigen. (B) Following injection of the antigen (interferon) into a mouse, B cells isolated from the spleen can be fused with cancer cells to form a hybridoma. Each hybridoma produces a single type of antibody referred to as a monoclonal antibody. [Reproduced from Antebi, E., and Fishlock, D. (eds.) (1986). "Biotechnology: Strategies for Life." MIT Press, Cambridge.]

B. Cell-Free Enzyme Systems

Cell-free enzyme systems offer highly selective, controlled, and efficient processes for waste management in the food industry. For example, amylases can be used for the treatment of starch-containing waste waters generated during rice and potato processing; proteases can be used to upgrade normally undesirable fish to meal for animal feed; and ligninases can decolorize pulp mill effluents. Lactase specifically cleaves lactose present in whey into glucose and galactose, readily fermentable sugars that can be used as feedstock chemicals for subsequent fermentations. These enzymes can be used in batch treatments or be immobilized onto surfaces for continuous processing. Enzymes that do not function optimally could be improved for this purpose using the techniques of enzyme engineering described previously.

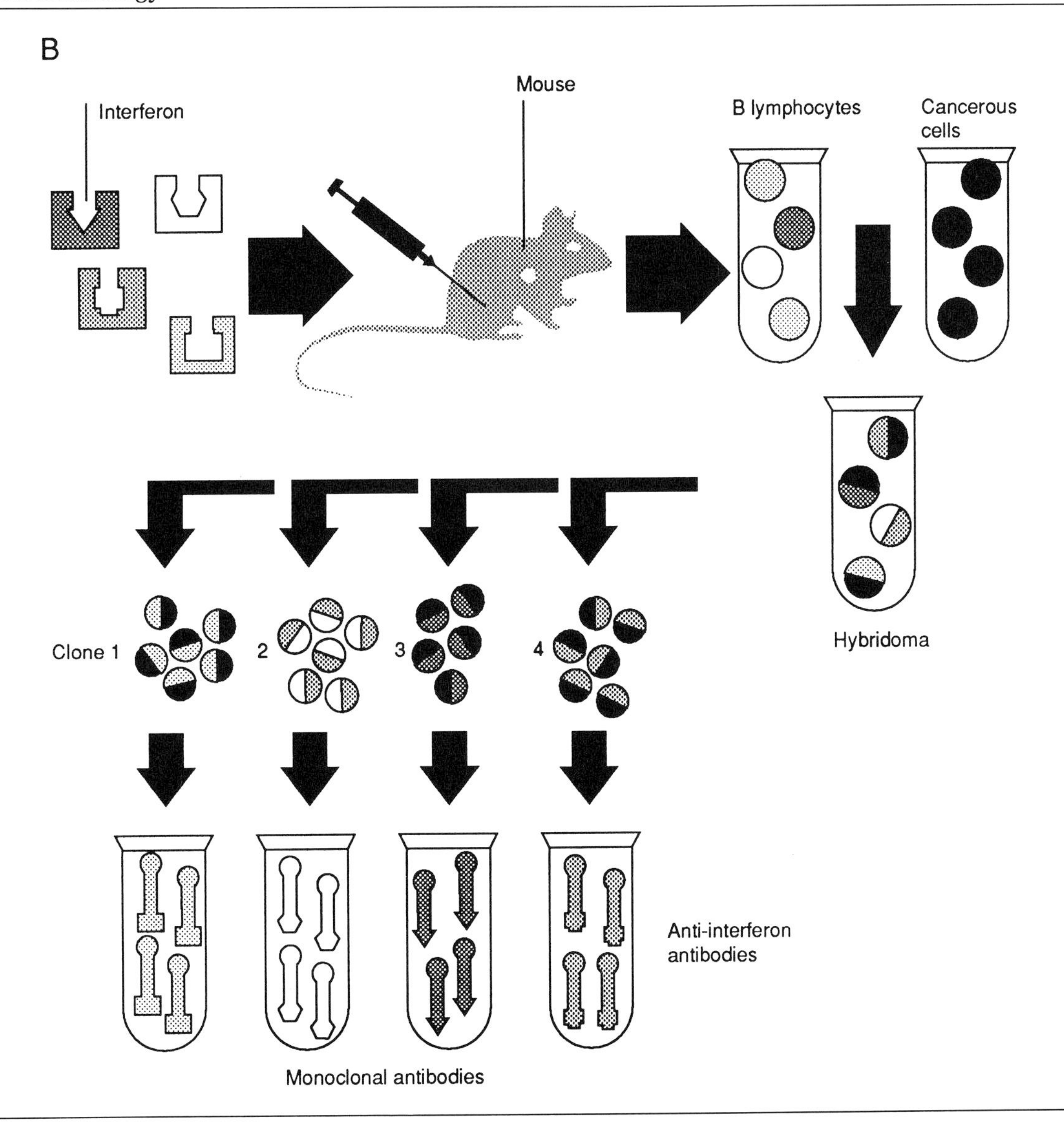

Figure 5 Continued

Bibliography

Antebi, E., and Fishlock, D. (eds.) (1986). "Biotechnology: Strategies for Life." MIT Press, Cambridge.

Campbell-Platt, G. (ed.) (1987). "Fermented Foods of the World: A Dictionary and Guide." Butterworths, Boston.

Harlander, S. K. (1989). *Food Technol.* **43(9),** 196–206.

Harlander, S. K. (1990). *Cereal Foods World* **35(11),** 1106–1109.

Harlander, S. K. (1990). *Food Biotechnol.* **4(1),** 515–526.

Harlander, S. K. (1991). *Food Technol.* **45(4),** 84–95.

Harlander, S. K., BeMiller, J. N., and Steenson, L. (1991). *In* "Agricultural Biotechnology: Issues and Choices" (B. R. Baumgardt and M. A. Martin, eds.), pp. 41–52. Purdue Research Foundation, West Lafayette, Indiana.

Kwoh, D. Y., and Kwoh, T. J. (1990). *Amer. Biotech. Lab.* **October,** 14–25.

Morris, V. J. (1990). *Food Biotechnol.* **4(1),** 45–57.

Neidleman, S. (1989). *In* "A Revolution in Biotechnology" (J. L. Marx, ed.), pp. 56–70. Cambridge University Press, New York.

Neidleman, S. (1986). *In* "Biotechnology in Food Processing" (S. K. Harlander and T. P. Labuza, eds.), pp. 37–56. Noyes Publications, Park Ridge, New Jersey.

Owen, V. M., and Turner, A. P. F. (1987). *Endeavour, New Series* **11(2),** 100–104.

Schleifer, K. H. (1990). *Food Biotechnol.* **4(1),** 585–598.

Wagner, G., and Schmid, R. D. (1990). *Food Biotechnol.* **4(1),** 215–240.

Wasserman, B. P., Montville, T. J., and Korwek, E. L. (1988). *Food Technol.* **42(1),** 133–146.

Wolcott, M. J. (1991). *J. Food Protect.* **54(5),** 387–401.

Foodborne Illness

Daniel Y. C. Fung
Kansas State University

Glossary

Endemic "Normal" cases of a particular sickness in a community

Epidemic "Abnormally" large number of cases of a particular sickness from a single source in a community

Epidemiology Study of chronic and sporadic diseases in population; uses clinical, laboratory, statistical methods; epidemiologists study disease patterns and causative agents in populations; physicians treat individuals

Etiological agent Chemical or microbiological agent identified as causing a specific disease

Foodborne disease case Contaminated food consumption by one susceptible individual who later becomes sick

Foodborne disease outbreak Contaminated food consumption from one source by one or many people who later become sick; can have 1 to 100,000 cases

Foodborne infection Ingestion of large numbers of viable, pathogenic microorganisms in foods by susceptible individuals who later become sick

Foodborne intoxication Ingestion of chemical toxins or preformed microbial toxic compounds in foods by susceptible individuals who later become sick

Food poisoning Ingestion of contaminated food containing either chemical preformed toxins or live microbes by susceptible individuals who later become sick

Pandemic Occurrence of a particular disease, usually contagious, affecting the entire world

CONSUMPTION OF FOOD is essential for human survival, but along with gaining good nutrition and satisfaction from eating food, occasionally human beings consume undesirable chemical and biological agents and toxins. When one is susceptible to these foodborne agents one might become sick; thus the term foodborne illness. The disease can be relatively mild (discomfort, diarrhea, vomit, fever for a few hours), severe (incapacitation for a few days), or even cause death (in the case of some chemical toxins or botulism). No one knows for certain the number of cases of foodborne intoxications and infections occurring annually in the world. In the United States, the rate is estimated at about 24 million cases per year, which means about 1 in 10 U.S. citizens is affected by foodborne disease per year. The economic lost because of foodborne illness runs into billions of dollars per year in the United States. In countries with poor sanitation, one can only guess that the number of foodborne disease cases is much higher. In more developed countries, much effort has been made to improve sanitation and nutrition for people; however, many people around the world are still suffering from poor nutrition, waterborne and foodborne diseases, and famine. The current concern in public health is "food safety." Consumers are much more aware of the potential for large-scale foodborne outbreaks because of mishandling or improper processing of foods, and they are demanding a safer food supply. Food microbiologists and public health professionals are charged with the responsibility of studying the occurrence, enumeration, isolation, detection, characterization, prevention, and control of foodborne microorganisms from food, water, and the environment.

I. Chemical Intoxication

All food are made of "chemicals." The proper composition of chemicals in foods—protein, carbohydrates, lipids, water, minerals, micro and macro nutrients—will provide us with good nutrition; however, excess consumption of any chemicals for a long time may lead to poor health. Chemical intoxication can occur over a long period of time (e.g., lead poisoning) or suddenly (e.g., arsenic poisoning), depending on the compound. Chemical intoxications are usually accidental (e.g., mistaking cyanide for table salt). People have been poisoned by inorganic compounds such as cyanide, arsenic, cadmium, mercury, selenium, antimony, lead, nickel, etc. The symptoms usually occur rapidly (a few minutes to hours), reactions are usually violent, and death can occur in cases of ingestion of large doses of toxic compounds. Immediate medical assistance is essential for victims in such cases.

There is another class of chemical toxicants that may create foodborne intoxication, and they are the naturally occurring food toxicants. Included in this group are toxic amino acids, glucosinolates (the source of organic nitriles, isothiocyanate, and thiocyanate (SCN ion), lectins (or hemagglutinins), carbohydrate inhibitors, cycasin, favism producing agents, estrogens, goitrogens, nitrosamines, polycyclic aromatic hydrocarbons, poisonous mushrooms, naturally occurring toxic plant compounds, and naturally occurring toxic animal compounds.

Through the ages, humans learned to avoid these plant or animal products that might be toxic and concentrated on consuming those that are palatable and safe. Occasionally, however, people still get sick or even die because of eating toxic plant and animal products.

II. Bacterial Intoxication

Clostridium botulinum causes botulism. The first recorded outbreak of botulism was in 1793 and involved sausages (botulus) in Germany. Since that time, many outbreaks all over the world have been reported. Between 1899 and 1977, there were 766 outbreaks recorded in the United States, with 1961 cases and 999 deaths. More recent data showed that between 1971 and 1985, 210 outbreaks were recorded with 485 cases and 55 deaths. The organism is a gram-positive, anaerobic, spore-forming rod and can grow at temperatures from 3.5°C to as high as 50°C. Most strains will grow well at 30°C. The spores formed by this organism can be found in soil, water, and the environment and can be transmitted to foods through harvesting, transportation, processing, and handling. Foods involved in botulism cases usually are improperly home-canned, medium or low acid foods. Many types of food were implicated in the transmission of botulism: Vegetables, fish and fish products, fruits, condiments, beef, milk and milk products, pork, and poultry. Information since 1899 indicates that about 70% of the outbreaks can be traced to improperly processed home-canned foods and 9% to commercially processed food, with the other outbreaks from unknown sources. Symptoms develop 18 to 96 hr after ingestion of foods containing the toxins and include vomiting, nausea, fatigue, dizziness, vertigo, headache, dryness of mouth, muscle paralysis, and death by asphyxiation. There are several types of botulin toxins (types A, B, C_1 , C_2 , D, E, F, and G). These are large molecular weight proteins (about 1 million), and the most important ones affecting human beings are toxins A, B, and E. Cultures producing type A are proteolytic, whereas both proteolytic and nonproteolytic strains exist for type B producers. Toxin E producers are nonproteolytic and are mainly involved with fish products. Food harboring proteolytic strains will develop an "off" odor, whereas food containing nonproteolytic strains will not. These toxins are among the most toxic materials produced by a biological system. It was estimated that 1 pure oz of botulin toxin can kill 200 million people. Treatment is by administration of monovalent E, bivalent AB, trivalent ABE, or polyvalent ABCDEF antisera, and early administration will reduce chances of paralysis and death. Fortunately, the toxins are heat sensitive: Boiling of the toxin for 10 min will destroy it. The toxins can be detected by animal tests using mice as well as immunological tests using specific antibodies (gel diffusion tests, ELISA, RIA tests, etc.). The key to preventing botulism is to know the composition (pH, A_w , oxidation-reduction potential, presence of inhibitory compounds, etc.) of the food and use proper time and temperature for processing as well as correct packaging and storage of the processed food. Much research has been done in the past 50 years to study the heating time and temperature to kill spores of *C. botulinum*. All high-moisture foods processed, then stored under anaerobic conditions should be subject to close scrutiny to avoid the possibility of *C. botulinum* surviving and later germinating and producing the toxins. When in doubt, always boil the suspected food for 10 min before discarding it. Fortunately, the number of cases of botulism is not high; however, once an outbreak occurs, it causes widespread alarm and fear. [*See* TETANUS AND BOTULISM.]

Staphylococcus aureus is a gram-positive, facultative anaerobic coccus occurring in clusters that look like bunches of grapes under the microscope. On agar medium, it produces golden colonies. The organism is ubiquitous and can be found in human skin, nose, hair, and many type of foods. When allowed to grow in food (especially protein foods), this organism may produce a class of low molecular weight (ca. 30,000) protein toxins called staphylococcal enterotoxins (A, B, C_1 , C_2 , C_3 , D, and E). These toxins, when ingested by a susceptible person, will cause severe nausea, vomiting, abdominal cramps, diarrhea, and prostration about 4 to 6 hr after consumption. Recovery from the disease takes about 24 to 72 hr. Victims will not die but may wish they had because the reactions are very violent. A special concern about these toxins is their heat stability. Once the toxins are formed in food, boiling will not destroy the preformed toxins. It will take autoclaving (pressure cooking) at 121°C for 30 min before the toxin is destroyed. The organism is not a good natural competitor compared with other spoilage organisms such as *Pseudomonas;* however, in the absence of competitors, such as in salty or processed foods, the organism can grow and produce the heat stable toxins. In this case, reheating of food before consumption will not solve the problem; therefore, it is essential to prevent *S. aureus* from growing in the food by proper refrigeration of cooked foods or keeping hot food hot. The organism can produce enough toxin in 4 hr at room temperature to cause a problem. People do not develop immunity against the toxins; thus, one can have several attacks in a life time. This organism is one of the three leading causes of foodborne disease in the United States.

Aspergillus flavus and *A. parasiticus* are molds that can produce the carcinogenic toxins called aflatoxins. In 1960 in England, 100,000 turkeys died of unknown causes, and the disease was called turkey x disease. After much research work, the contaminant was found to have originated from peanut meals from Brazil. The organisms responsible for producing the toxic compounds were isolated and identified as *A. flavus*. Later, *A. parasiticus* was also found to be able to produce the toxin. The mold can grow between 7.5° and 40°C, with optimal temperature at 24° to 28°C, and in 1 to 3 days of growth, the organism can produce the toxins, especially if the moisture of food is 16% or higher. The primary toxins are B_1 , B_2 , G_1 , and G_2 . "B" and "G" indicate that the toxins fluoresce blue or green under ultraviolet light. When cows consume B_1 and B_2 toxins, they can modify the toxins and excrete the toxins as M_1 and M_2 in milk. [*See* MYCOTOXICOSES.]

Spores of these molds are ubiquitous, and the organism has been found to grow in rice, sorghum, peanut, corn, wheat, and soybean corps, as well as in animal feed. In the field when the crops are healthy, invasion plants by mold is minimum; however, with an undesirable condition such as a drought, the plants are under stress, and this mold will invade the tissues and produce aflatoxin in the field, a highly undesirable situation. In experimental conditions, the toxin can be produced in cheese, meat, bread, and sausages and because the toxins are carcinogenic, they are under strict government scrutiny. Currently, the allowed limit is 20 ppb for foods and feed. Although no food-related aflatoxin cases have been reported in the United States, there were cases reported in Southeast Asia when people consumed heavily contaminated food. The toxins can be detected by animal tests using ducklings or chick embryo. Thin layer chromatography and high performance liquid chromatography can also be used to detect these toxins. Recently, monoclonal antibodies have been employed to detect these toxins with great rapidity (10 to 30 min) and sensitivity (1 ppb and lower).

The best way to control aflatoxin is to prevent the mold from growing in the food products. Mold inhibitor such as sorbate, benzone, and BHA can help prevent *A. flavus* and *A. parasiticus* from growing in foods, but detoxification of aflatoxins in foods and feeds is very complex and difficult.

Besides aflatoxins, a variety of toxins are produced by fungi and are collectively called mycotoxins. Among the more important ones are described in the following paragraphs.

Trichothecenes is a family of toxins characterized by their tetracyclic sesquiterpenoid structure and the presence of an epoxy group at the 12 and 13 carbon (e.g., T-2-toxin) produced by *Fusarium, Trichoderma, Cephalosporium,* etc.

Zearalenore is a class of resorcylic acid lactones and is produced by *Fusarium* and *Gibberella.*

Ochratoxins is a family of dihydroisocamarins

that is produced by *Aspergillus ochraceous* and *Penicillium*.

Sterigmatocystin is similar to aflatoxin and is produced by *Aspergillus* and *Bipolaris*.

Patulin are unsaturated lactone derivatives produced by *Pencillium*.

Alternaria metabolites are produced by *Alternaria*.

III. Exotoxins versus Endotoxins

It is necessary to differentiate these two major types of bacterial toxins before discussing foodborne infections and comparing them to foodborne intoxications. *Exotoxins* are toxins produced by an organism and later released into the environment with the cell intact. Ingestion of these preformed toxins along with food and drink causes foodborne intoxication. These toxins are protein toxins mainly produced by gram-positive organisms, and, because they are proteins, they can be neutralized by corresponding antibodies and detected by a variety of immunological methods. These toxins are relatively heat sensitive (except the staphylococcal enterotoxins, which are heat stable as described earlier). These toxins also have distinct pharmacology; some affect the intestines, whereas others affect nerve cells. Examples of exotoxins are staphylococcal enterotoxins and botulinum neurotoxins.

Endotoxins are part of the cell wall material of gram-negative cells. Every gram-negative bacterium examined has endotoxins, which are complex molecules containing protein, carbohydrate, and lipid (lipopolysaccharides). The protein moiety determines antigenicity, the carbohydrate moiety determines immunological specificity, and the lipid moiety causes toxicity. Unlike exotoxins, antibodies will not neutralize toxicity, because the toxic part is the lipid. All endotoxins have the same action and are released when the gram-negative bacterium lyses. These endotoxins cause fever by acting as *exogenous pyrogens*, which, when absorbed into the bloodstream, causes injury to the leukocytes, which, in turn, release an endogenous pyrogen. This endogenous pyrogen stimulates the thermoregulatory center of the brain at the hypothalamus and causes fever. Therefore, fever in a patient is indicative of a foodborne infection case. Endotoxins can be detected by *Limulus* amebocyte lysate (LAL) test. In the presence of endotoxins, the LAL will form a gel. The reaction takes about 1 hr. Currently, hospital materials should be pyrogen-free, and LAL is the standard test for pyrogens in the hospital environment.

IV. Bacterial Infection

One of the characteristics of bacterial infection is that large numbers of live pathogenic cells (1×10^6) must be ingested by a susceptible subject to cause a disease. Also, the onset time for disease is longer than that for foodborne intoxication, usually 24 to 48 hr. *Clostridium perfringens* occupies an interesting position as being both a foodborne infection agent as well as a foodborne intoxication agent. On the one hand, the susceptible subject has to ingest large numbers of viable (1×10^6) *C. perfringens* before coming down with a food poisoning case, while on the other hand, the organism produces an enterotoxin in the intestine to cause the illness. *Clostridium perfringens* is a gram-positive, anaerobic, spore-forming rod. Its generation time (doubling in number) in ideal conditions is as short as 9 min, making it the fastest growing organism known.

Spores of the organism distribute widely in nature and can easily contaminate foods through dust, water, equipment, and human handling of foods. Most of the incidences of *C. perfringens* food poisoning involve meats prepared in large quantities 1 day and consumed the next, after the food is held at lukewarm temperatures (e.g., 45° to 50°C). In such conditions, most vegetative cells of competitors die off during cooking, but spores of *C. perfringens* can survive. When food is held at a luke-warm temperature, the spores of *C. perfringens* have a chance to germinate, and the vegetative cells grow into large numbers with no competitors. The organisms, when ingested by a susceptible person, will start to sporulate because of the favorable anaerobic conditions in the small intestine.

The gene codes for sporulation also control the release of an enterotoxin, which is responsible for the mild diarrhea characteristic of *C. perfringens* food poisoning. Symptoms occur 8 to 20 hr after ingestion of a large number of viable *C. perfringens* and include acute abdominal pain, diarrhea, and nausea with rare vomiting. The symptoms are milder than those caused by *Salmonella*. Detection of this organism is by anaerobic cultivation of food using differential anaerobic agar, such as tryptose sulfite cycloserine agar. *Clostridium perfringens* forms black colonies in this agar medium. In 1991, the author developed a simple double tube system (Fung's double tube method), which can conveniently detect, isolate, and enumerate *C. perfringens* from foods. *Clostridium perfringens* causes about one-fifth of all foodborne cases in the United States annually.

Salmonella is the classic example of foodborne infection. This organism belongs to the family Enterobacteriaceae, which includes bacteria found in the intestine of humans and animals. *Salmonella enteritidis* was isolated in 1884 and still is an important foodborne organism. The organism is a gram-negative, facultative anaerobic, nonspore-forming rod, motile by peritrichous flagella. It does not ferment lactose and sucrose but ferments dulcitol, mannitol, and glucose. These sugars are useful in differentiating members of the Enterobacteriaceae. The organism is heat sensitive but can tolerate a variety of chemicals such as brilliant green, sodium lauryl sulfite, selenite, and tetrathionate. These compounds have been used in solid agar or liquid medium for the selective isolation of this organism from food and water. A variety of biochemical tests can be used to characterize the culture, but to confirm the isolate as a *Salmonella,* one must perform serological tests. Currently, there are more than 2000 serotypes of *Salmonella,* and each type is potentially pathogenic. *Salmonella* has been found in water, ice, milk, dairy products, shellfish, poultry and poultry meat products, eggs and egg products, animal feed, pets, etc. The organism is commonly found in poultry flocks and birds. Human beings can be healthy carriers of this organism, and it has been estimated that 4% of the general public carries this organism. There are actually three types of diseases caused by *Salmonella:* enteric fever (typhoid fever) caused by *S. typhosa,* in which the organism, ingested with food, finds its way into the blood stream and disseminates to the kidney and is excreted in the stools; septicemia caused by *S. cholera-suis,* in which the organism causes blood poisoning; and gastroenteritis caused by *S. typhimurium* and *S. enteritidis,* a true foodborne infection in which the organism is localized in the intestine. In the last case, large numbers of live *Salmonella* are ingested with food and, in 1 to 3 days, liberate the endotoxins that cause localized violent irritation of the mucous membrane with no invasion of blood stream and no distribution to other organs. Symptoms of salmonellosis occur 12 to 24 hr after ingestion of food containing 1 to 10 million *Salmonella* per gram and include nausea, vomiting, headache, chills, diarrhea, and fever. The illness lasts for 2 to 3 days. Most patients will recover; however, death can occur in the very old, the very young, and those with compromised immune systems. Patients can excrete *Salmonella* in their fecal materials and may become healthy carriers of the organism.

Because processed food is not allowed to have *Salmonella* present, it became necessary for the food industry to closely monitor the occurrence of this organism in foods. The conventional method of detection of *Salmonella* includes preenrichment, enrichment, selective enrichment, physiological tests, and, finally, serological tests. The entire sequence takes about 7 days. Currently, efforts are being made to shorten the detection time by such methods as ELISA test and DNA probes. It is possible now to provide a negative screen test in about 48 hr; However, once a sample is deemed positive, the conventional procedure is needed to confirm the presence of *Salmonella*. Because *Salmonella* is heat sensitive, proper cooking will destroy the organism, and proper chilling, refrigeration, and good sanitation will minimize the problem. *Salmonella* remains one of the most important food pathogens in our food supply and is ranked one of the top three foodborne pathogens annually.

Shigella is a gram-negative, facultative anaerobic, nonspore-forming rod often confused with *Salmonella* in the bacteriological diagnostic process. It is nonmotile and does not produce H_2S. On agar medium, the colonies are smaller than those of *Salmonella*. This organism is very important in waterborne diseases, especially in tropical countries where sanitation conditions are poor. The organism is transmitted by water, food, humans, and animals. The "4 F's" involved in the transmission of *Shigella* are food, fingers, feces, and flies. One to four days after ingestion of the organisms, there will be an inflammation of walls of the large intestines and ileum. The organism seldom invades the blood, but bloody stool will occur because of superficial ulceration. The cell wall of *Shigella,* when lysed, will release endotoxins. In addition, *S. dysenteriae* produces an exotoxin that is a highly toxic neurotoxin that can be neutralized by a specific antibody. The mortality rate of shigellosis is higher than that of salmonellosis. Prevention of shigellosis can be achieved by good sanitation, good hygiene, treatment of water, prevention of contamination, detection of carriers, and isolation of patients from the general public. [*See* DYSENTERY, BACILLARY.]

Vibrio cholerae, or *V. comma* as it used to be

named, was a very important disease-causing organism in the late nineteenth century and early twentieth century. It is still important in places with poor sanitary conditions, such as many developing countries because the organism can cause pandemic infection. It is a gram-negative, curved rod that looks like a comma under the microscope; thus, the origin of the name *V. comma*. No spore is formed by this organism. *Vibrio cholerae* grows well in an alkaline medium and is actively motile by a single polar flagellum. The organism is endemic in India and Southeast Asia and is spread by person-to-person contact, water, milk, food, and insects. The organism produces enterotoxins and endotoxins in the intestines and causes severe irritation to the mucous membranes, with resultant out-flow of fluid and salts that impairs the "sodium pump" of mammalian cells, thus causing severe diarrhea, dehydration, acidosis, shock, and even death. The mortality rate may be as high as 25 to 50%. The most effective therapy is replacement of water and electrolytes to correct severe dehydration and salt depletion. Recovery by this treatment can be very dramatic; a patient near death can be revived to health in a matter of days. *Vibrio cholerae* remains a dreaded communicable disease in many parts of the world and much education and public health work need to be done to reduce human suffering from this organism.

Vibrio parahemolyticus is an organism that has caused many cases of foodborne disease in Japan for many years. One of the reasons is that Japanese like to eat raw seafood, which harbors this pathogen. Most of the original reports and research works on this organism were in Japanese and not readily understandable or available to microbiologists in the West, but U.S. scientists started working on the organism around 1969. In 1971, three outbreaks of this organism occurred in the United States. The organism is a gram-negative, curved rod and is halophilic (salt loving); it grows best in 3 to 4% salt medium and can grow in media containing 8% salt, also. Growth temperature ranges from 15° to 40°C, and pH range is from 5 to 9.6. The organism is sensitive to streptomycin, tetracycline, chloramphenicol, and novobiocin but is resistant to polymyxin and colistin. The Kanagawa positive strains of this organism hemolyze human blood; environmental strains are negative for this test. The organism is distributed in marine food from seawater as well as from freshwater. Most of the outbreaks are recorded in the summer months when the water is warm in the northern hemisphere. Symptoms of the disease occur about 12 hr after ingestion of a large number of viable cells (10^5/g) and include abdominal pain, diarrhea, vomiting, mild chills, and headache. The symptoms are similar to those of salmonellosis but are more severe. It has been said that salmonellosis affects the "abdomen" of the patient where *V. parahemolyticus* infection affects the "stomach" of the patient. Detection of the organism is best achieved by a good selective medium, such as BTB-salt-Teepol agar. Prevention of infection by this organism is by adequate cooking of seafood and refraining from eating raw seafood from polluted water.

Bacillus cereus and other *Bacillus* species have been implicated in associated foodborne disease cases only in recent years, although these organisms have been suspected as agents of foodborne illness for a long time. These are gram-positive, aerobic, spore-forming rods occurring widely in nature and contaminating foods easily. Because of the general resistance (to heat, UV, chemical, sanitizers, etc.) of spores of these organisms and the prolific biochemical activity of the vegetative cells, *Bacillus* can be considered one of the most important environmental bacterial contaminants of foods. There are two distinct clinical syndromes caused by *B. cereus*. One, the "diarrheal syndrome," occurs 12 to 24 hr after ingestion of large numbers of viable *B. cereus* and includes abdominal pain, watery diarrhea, rectal tenesmus, and nausea without vomiting. The "diarrheal syndrome" is the result of consuming proteinaceous foods, pudding, sauces, and vegetables. The other, the "emetic syndrome," causes illness almost exclusively associated with cooked rice and is characterized by a rapid onset (1 to 5 hr) of nausea with vomiting and malaise. The two syndromes are related to two separate toxins, diarrheal enterotoxins and emetic toxins, produced by *B. cereus*.

Other *Bacillus* suspected of causing food-borne diseases when ingested in large numbers (10^6–10^7 organisms/g of food) by susceptible persons include *B. licheniformis* and *B. subtilis*. Control of *Bacillus* food poisoning is complicated by the ubiquitous nature of this organism. The best measures are to prevent the spore from germinating and to prevent multiplication of vegetative cells in cooked and ready-to-eat foods. Freshly cooked food eaten hot immediately after cooking should not be a problem; however, slow reheating of previously cooked rice products should be treated with caution. Rapid chilling of "left-over" cooked rice products is highly recommended as a preventive measure.

Campylobacter jejuni has been recognized as an "emerging" pathogen in the past 10 years and has been reported as the most common bacterial cause of gastrointestinal infection in humans requiring hospitalization, even surpassing rates of illness caused by *Salmonella* and *Shigella*. *Campylobacter* was originally called *Vibrio fetus*, because it was first recognized as an agent of infertility and abortion in sheep and cattle. The organism is included in the family *Spirillaceae* because of the physiological and morphological similarities to *Spirillum*. The organism is a gram-negative, slender, curved bacterium and motile by a single polar flagellum. It neither ferments nor oxidizes carbohydrates, is oxidase-positive, reduces nitrates, does not hydrolyze gelatin or urea, and is methyl red and Voges-Proskauer reaction negative. It will grow between 25° and 43°C. The organism is an obligate microaerophile and grows optimally in 5% oxygen. This attribute has been used for the isolation of the organism from animals, human subjects, and foods. The incubation time of *C. jejuni* food poisoning ranges from 2 to 5 days, and the duration of the sickness may be up to 10 days. The patient will exhibit enteritis, fever, malaise, abdominal pain, and headache. The stools become liquid and foul smelling with blood, bile, and mucus discharge occurring in serious cases. The organism has a worldwide distribution with outbreaks reported to be related to milk, poultry, eggs, red meat, pork, and water. Control and prevention of this organism are by proper food processing techniques (heating, cooling, chemical treatment of foods, etc.) because it is a fragile organism. Its prevalence can be attributed to postprocessing contaminations of food. Again, good sanitation and hygiene should reduce the incidence of this organism in our food supplies.

Escherichia coli is one of the most common bacteria in our environment. It is used extensively in genetic engineering and biotechnology, and it is probably the most studied microbe in human history. Most people do not think of *E. coli* as a food pathogen; however, recent research and information indicate that some strains of *E. coli* can indeed cause severe foodborne disease. *Escherichia coli* is a gram-negative, facultative anaerobic, nonspore-forming rod that occurs widely in nature, as well as in intestines of humans and animals. It is glucose and lactose positive, indole and methyl red positive, but Voges-Proskauer and citrate negative. The most useful way to classify the species is by serotyping, using antibodies against somatic (O), flagella (H), or capsular (K) antigens of various strains of *E. coli*. Most *E. coli* isolated from the environment are not pathogenic; however, a group of *E. coli* has been defined as EPEC or enteropathogenic *E. coli*—diarrheagenic *E. coli* belonging to serogroups epidemiologically incriminated as pathogens. Pathogenic mechanisms have not been proven to be related either to heat-liable enterotoxins (LT) or heat-stable enterotoxins (ST) or to *Shigella*-like invasiveness. The serotypes included in EPEC are 018ab, 018ac, 026, 044, 055, 086, 0111, 0114, 0119, 0125, 0126, 0127, 0128ab, 0142, and 0158. A newly recognized pathogenic *E. coli* is the vero cytotoxin-producing *E. coli* 0157:H7, which causes hemorrhage colitis, hemolytic uremic syndrome, and thrombotic thrombocytopenic purpiera. Another class of pathogenic *E. coli* is called enteroinvasive *E. coli* (EIEC) and resembles *Shigella* by producing an invasive, dysenteric form of diarrheal illness in humans. The serogroups associated with EIEC are 028ac, 029, 0124, 0136, 0143, 0144, 0152, 0164, and 0167. The last pathogenic group of *E. coli* is called the enterotoxigenic *E. coli* (ETEC). This group of organisms produces one or both of two well-established enterotoxins: heat-labile enterotoxin (LT) and a nonantigenic, heat-stable enterotoxin (ST). Serogroups include 06, 08, 015, 020, 027, and others.

Detection of EPEC, EIEC, and ETEC in foods can be accomplished by common procedures used in coliform isolation at 44° to 45°C (for fecal coliform). In the case of *E. coli* 0157:H7, however, 44° to 45°C will not allow this pathogenic *E. coli* to grow. A variety of methods are available for the isolation of these pathogenic *E. coli*.

Prevention and control of pathogenic *E. coli* are best done by education of food-handlers who should adhere to strict hygienic practices. Fecal and other waste materials from humans and animals should be decontaminated and not allowed in contact with water and food supplies.

Yersinia enterocolitica, a close relative of *Y. pestis* the dreaded plague causing bacterium, is a gram-negative, facultative anaerobic, nonspore-forming bacterium that is sucrose-positive, rhamnose-negative, indole-positive, motile at 20°C but not at 37°C, and highly virulent to mice. Serotyping is very important in separating this organism from other closely related gram-negative bacteria. Although *Y. enterocolitica* has an optimal growth temperature at around 32° to 34°C, it is often isolated on enteric agars at 22° to 25°C. It grows slowly in simple glucose-salts medium, but it grows much better with supplements, such as methionine or cysteine and thiamine. One important aspect of this organism is that it can grow in vacuum-packaged meat under refrigeration because it is a facultative anaerobe and a psychrotroph (an organism that can grow at refrigeration temperature). Thus, the two effective methods to control microbial growth, namely anaerobic conditions and cold temperature storage, do not control *Y. enterocolitica*. After ingestion of large numbers of this organism, the susceptible person can develop fever, abdominal pain, and diarrhea, with nausea and vomiting occurring less frequently. More serious intestinal disorders include enteritis, terminal ileitis, and mesenteric lymphadenitis. Extraintestinal infections of *Y. enterocolitica* have been reported, including septicemia, arthritis, erythema nodosum, sarcoidosis, skin infection, eye infection. Foods suspected of being a source of yersiniosis in the United States include chocolate milk, milk powder, chow mein, tofu, and pasteurized milk. Pork products have also been suspected.

Isolation of this organism typically goes through an enrichment step using nutrient broth or Rappaport broth, then through a plating medium using an enteric agar (SS, XLD, DCL, etc.). An excellent agar for this purpose is the CIN agar (cefsulodin-irgasan-novobiocin agar). Control of yersiniosis involves proper handling of raw and cooked food of all types, especially pork products, and proper handling of water for food processing.

Listeria monocytogenes has developed into a very important food pathogen in the past 10 years from the standpoint of its economic and public health impacts. The organism is a small, short, gram-positive rod, and nonspore forming. It is motile by a characteristic tumbling or slightly rotating motion. In 1991, a motility enrichment procedure has been developed to rapidly isolate, detect, and enumerate this organism by a new Fung-Yu motility chamber system developed in the author's laboratory. The organism grows on simple laboratory media in the pH range between 5 and 9. On solid agar, the colonies are translucent, dewdrop-like, and bluish when viewed by 45°C incident transmitted light (Henry's illumination step). Recently, some agar medium (such as MOX) have been developed that will allow *Listeria* to form black colonies for ease of recognition. Biochemically, this organism can be confused with such organisms as *Lactobacillus, Brochothrix, Erysipelohrix,* and *Kurthia*. A variety of biochemical tests have been devised to separate *L. monocytogenes* from other *Listeria* species such as *L. innocua, L. welshimeri,* and *L. murrayi*. Serotyping is also important in the identification of this organism, the most important ones being 1/2a, 1/2b, 1/2c, 3a, 3b, 3c, and 4b. *Listeria* is a psychotroph capable of growing at temperatures as low as 2.5°C and as high as 44°C. Because dairy products have been implicated in outbreaks of listeriosis, much research has been directed toward cheese and milk products, and the organism has been found to survive the processing of cottage cheese, cheddar cheese, and colby cheese. A question of great concern is whether or not *L. monocytogenes* can survive the current pasteurization temperature of milk (i.e., 63°C for 30 min or 72°C for 15 sec). Data on this issue are still inconclusive and research on this topic is still ongoing in many food-related laboratories. It is important to note that in 1991, the time and temperature regulation for pasteurization of milk has not been affected by the possible heat resistance of *L. monocytogenes*. The disease starts with infection of the intestine—the infective dose (the number of pathogens to cause a disease) is not known at this point. Patients may develop transitory flulike symptoms such as malaise, diarrhea, and mild fever. In severe cases, virulent strains are capable of multiplying in macrophages and later producing septicemia. When this occurs, the bacteria can affect the central nervous system, the heart, and the eyes and may invade the fetus of a pregnant woman and result in abortion, stillbirth, or neonatal sepsis. Because of the severity of listeriosis, the current regulation for cooked products is "zero" in the United States. In other countries, proposals are being made to tolerate up to 100 *Listeria* in some foods.

In recent years, several well-documented cases of listeriosis have been reported: Nova Scotia (1981), Massachusetts (1983), and the most well-known one involving Mexican-style soft cheese in southern California (1985), which involved 142 people and more than 50 deaths.

L. monocytogenes has been isolated in a variety of commodities such as poultry carcasses, meat and chopped beef, dry sausages, milk and milk products, cheese, vegetables, and surface water. Control measures include controlling the occurrence of the organism in the raw food materials, transporting vehicles, food processing plants (especially in controlling cross contamination of raw and finished products), practicing good general sanitation of the entire food processing environment, and regular monitoring of the occurrence of this organism in the food processing facilities. Because the organism is killed by heat, proper cooking of food will also help reduce risks.

Aeromonas hydrophilia has been associated with foodborne infection, although the evidence is not conclusive. The organism belongs to the family Vibrionaceae and is a facultative anaerobic, gram-negative, motile rod. Biochemically, it is similar to *E. coli* and *Klebsiella*. The optimal temperature for growth is 28°C and the maximum is 42°C. Many strains can grow at 5°C, which is a temperature usually considered adequate to prevent growth of foodborne pathogens. Diseases caused by *A. hydrophilia* include gastroenteritis (''choleralike'' illness and ''dysenterylike'' illness) and extraintestinal infections such as septicemia and meningitis. This organism has been isolated from fish, shrimp, crabs, scallops, oysters, red meats, poultry, raw milk, vacuum-packaged pork and beef, and even bottled mineral water. Because the organism is a psychrotroph, cold storage is not an adequate preventive measure, but proper heating of food offers sufficient protection against this organism. Consumption of undercooked food or raw food, such as raw shellfish, especially from polluted water, is highly undesirable.

Plesiomonas shigelloides has been a suspect in foodborne disease cases. The organism is gram-negative, facultative anaerobic, catalase negative, and fermentative. The organism is oxidase-positive and can be differentiated from bacteria in the family Enterobacteriaceae by this test, because the latter is oxidase negative. The organism also resembles *Shigella* but can be differentiated by being motile. It is capable of producing many diseases ranging from enteritis to meningitis. Gastroenteritis caused by *P. shigellosides* is characterized by diarrhea, abdominal pain, nausea, chills, fever, headache, and vomiting after an incubation time of 1 to 2 days. Symptoms last for a week or longer. All reported food involved with cases of gastroenteritis are from aquatic origin (salted fish, crabs, and oysters). The organism can be isolated from a variety of sources, including humans, birds, fish, reptiles, crustaceans. The true nature of this organism as a foodborne agent is not fully known because it has not been well studied to date. This organism may be an ''emerging'' pathogen in the near future.

A. Miscellaneous Bacterial Foodborne Pathogens

There are many other microbes suspected of being foodborne pathogens. They are not currently being labelled as true foodborne pathogens because of a lack of reports, as well as a lack of isolation methods and research on these organisms. Many of these organisms may very well be identified as foodborne pathogens in the future. A listing of these organisms follows:

Gram-negative bacteria. *Citrobacter, Edwardsiella, Enterobacter, Klebsiella, Hafnia, Kluyvera, Proteus, Providencia, Morganella, Serratia, Vibrios,* and *Pseudomonas.*

Gram-positive bacteria. Other species of *Bacillus* and *Clostridium, Corynebacterium, Streptococcus.*

Miscellaneous. *Brucella, Mycobacterium* (TB), *Coxiella burnetii* (Q-fever), *Leptospirosis, Erysipelas,* and *Tularemia.*

V. Foodborne Viruses

Viruses are much less studied by food microbiologists compared with bacteria, yeast, and fungi because of the difficulty of cultivating these ''entities''; conventional bacteriological media will not allow these particles to grow. There are very few laboratories capable of studying food virology at this moment. No doubt many foodborne outbreaks and cases are caused by a variety of viruses, but scientists in many cases are not able to identify the sources of the infection. Viruses that have been incriminated in foodborne diseases include hepatitis A virus (oysters, clams, donuts, sandwiches, and salad); Norwalk virus (oysters); polio virus (milk and oysters); ECHO virus (oysters); enteroviruses (oysters); and Coxsackie virus (oysters). Much more research needs to be done in the field of food virology to help reduce the incidences of food-borne diseases caused by viruses. [*See* ENTEROVIRUSES.]

VI. Nonmicrobial Foodborne Disease Agents

Consumption of food containing other living organisms can directly and indirectly cause foodborne diseases as well. The following is a listing of these agents:

Scombroid poisoning (associated with high level of histamine in scombroid fish).
Cestodes (flatworms such as *Taenia saginata, T. solium,* and *Diphyllobothrium latum*).
Nematodes (hookworm such as *Trichinella spiralis*).
Trematodes (fluke such as *Clonorchis sinensis*).
Protozoa (i.e., *Toxoplasma gondii*).
Shellfish (indirectly by toxin from the diinoflagellate *Gonyaulax catenella*).
Ciguatera (eating fish such as barracudas, groupers, sea basses, etc., that feed on toxic algae).
Other poisonous fishes (such as puffer fish and moray eel).

VII. Summary

Food safety is everybody's responsibility. Scientists are charged with identifying the agents causing foodborne infections and intoxications and with studying the mechanisms of the intoxication and infection, as well as working on the isolation, enumeration, and characterization of the causative agents. One of the most exciting developments of food microbiology is in the area of rapid methods and automation in microbiology. In this field, scientists are combining microbiological, chemical, physical, biochemical, biophysical, immunological, and semiautomated and automated methods to rapidly identify and enumerate microbes and their metabolites in foods and the environment. These methods should help scientists to control microbes in our food even better. The food industry uses this basic knowledge and applies it to good manufacturing practices to produce wholesome, nutritious, and safe foods. The consumer must also be educated in the handling of raw and cooked food at the point of purchase, as well as preparation of the food and final consumption. All three parties are responsible for the food safety of all involved. [*See* FOODS, QUALITY CONTROL.]

Acknowledgment

This material is based on work supported by the Cooperative State Research Service, U.S. Department of Agriculture under Agreement No. 89-34187-4511. Contribution No. 92-29-B, Kansas Agricultural Experiment Station, Manhattan, Kansas 66506.

Bibliography

Ali, M. S., D. Y. C. Fung, and Kastner, C. L. (1991). *J. Food Sci.* 56, 367–370.
Doyle, M. P. (1989). "Food Borne Bacterial Pathogens." Marcel Dekker, Inc., New York.
Fung, D. Y. C. (1987). Types of microorganisms. *In* "Microbiology of Poultry Meat Products." (F. E. Cunningham and N. A. Cox, ed.) Academic Press, New York.
Fung, D. Y. C., and Matthews, R. F. (1991). "Instrumental Methods for Quality Assurance in Foods." Marcel Dekker, Inc., New York.
Jay, J. M. (1986). "Modern Food Microbiology," 3rd ed., Van Nostrand Reinhold, New York.
Pierson, M. D., and Stearn, N. J. (1985). "Food-borne Microorganisms and Their Toxins: Developing Methodology." Marcel Dekker, Inc., New York.
Taylor, S. L., and Scanlan, R. A. (1989). "Food Toxicology." Marcel Dekker, Inc., New York.
Vanderzant, C., and Splittstoesser, D. (1992). "Compendium of Methods for the Microbiological Examination of Foods." Am. Public Health Assoc., Washington, DC.
Yu, L. S. L., and Fung, D. Y. C. (1991). *J. Food Safety*. 11, 149–162.

Foods, Quality Control

Richard B. Smittle
Silliker Laboratories

Glossary

American Public Health Association Group of scientists, physicians, health professionals, and health administrators interested in improving the health of humankind

Association of Official Analytical Chemists Association of scientists who strive to evaluate, standardize, and recommend methods of proven accuracy and reliability

Food and Drug Administration U.S. government agency responsible for regulating food, drugs, and cosmetics

Internal Commission on Microbiological Specifications for Foods Chosen group of microbiologists under the aegis of the International Union of Microbiological Societies who attempt to improve the microbiological safety and quality of foods in international trade

National Academy of Science Nongovernmental body of scientists and engineers who advise the executive branch of the government; the National Research Council is the working arm of the National Academy of Sciences

Sanitizer Chemical or physical agent used to eliminate or control microorganisms

U.S. Department of Agriculture Government agency responsible for regulating the meat and poultry industry

MICROBIOLOGICAL QUALITY CONTROL of foods is essential for a safe, wholesome, consistent food supply. Food quality is defined by microbiological criteria that are developed by manufacturers and government agencies. To have a full appreciation of food quality, a thorough understanding of the chemical, physical, and biological aspects of microbial ecology is necessary. The principles of statistical quality control is essential for properly evaluating ingredients and products from a production system with appropriate sampling plans. Adequate evaluation of production systems, ingredients, finished products, and environments requires methods that are accurate, reliable, and convenient to use. The key to producing safe, wholesome food is using properly engineered food equipment, employing comprehensive personal hygiene, cleaning, sanitation, and pest control programs. To maintain the highest possible quality, it is essential to control hazardous microorganisms and detect indicator microorganisms of sanitation and spoilage agents. The Hazard Analysis and Critical Control Point (HACCP) system is a recently developed systematic approach to food safety. It is the foundation for producing safe and stable food.

I. Introduction

With the advent of microbiology, it became clear that microorganisms, especially bacteria, play an important role in food quality. Some of the earliest studies by Louis Pasteur, one of the fathers of microbiology, dealt with the spoilage of wine and milk. Since these first studies, microbiologists have made great advances in food safety and wholesomeness. With increased knowledge of disease-producing microorganisms, microbial ecology, and physiology, it became possible to produce large quantities of safe wholesome food. Various modern schemes for en-

Encyclopedia of Microbiology, Volume 2

suring a safe food supply have been proposed. The traditional quality control relied on finished product testing. However, one of the earliest proponents for controlling the quality of food during production was Sir Graham Wilson, who proposed "intervention for prevention," which involved control of a process at designated critical steps in the manufacture and distribution. This has been recently refined in Europe, where risk assessment and preventative control are emphasized. In the United States, a scheme based on the intervention for prevention was developed by the Pillsbury Co. and the National Aeronautics and Space Administration, or NASA, to prevent health hazards in food for space flights—HACCP, which was first introduced in 1971. An extensive description of HACCP was presented by the International Commission on Microbiological Specifications for Foods (ICMSF) of the International Union of Microbiological Societies in 1988. More recently, this has been altered by the National Advisory Committee (NAC) of the National Academy of Science for a rational approach to regulatory agencies in the United States.

II. Microbiological Criteria

Microbiological criteria are defined by the microorganisms of concern, the method, the sampling plan, and the decision criteria. They are determined by municipal, state, and federal government regulatory agencies, companies, trade associations, and scientific societies. Owing to the importance of microorganisms to food stability and safety, establishing food microbilogical criteria for quality and hazards is necessary. The three types of criteria are as follows:

1. *Standard* is a criterion that is a mandatory government requirement.
2. *Specification* is a mandatory criterion where the acceptance of a food is dictated by the buyer.
3. *Guideline* is a criterion used by a manufacturer or regulatory agency to evaluate a process.

Infectious agents and toxins are not permitted in processed foods and are standards. Generally, infectious agents have a zero tolerance as defined by their absence in some specific quality of food. As a rule, they are not permitted in any quantity in food that will not undergo a final heating step for their destruction.

Specifications are useful in determining the microbiological state of ingredients sold by one company to another. In some cases, they may be eventually agreed upon in contracts or by reference to trade, scientific, or government publications. Guidelines are mainly used by manufacturers to steer production; however, some are published by government agencies and used to evaluate food in trade. Table I contains a list of references for published microbiological criteria.

III. Principles of Food Microbial Ecology

To establish a quality control program for foods, a firm knowledge of the ecological aspects must be understood. The chemical, physical, and biological properties of a food, the processing it requires, and the conditions under which it is stored, distributed,

Table I Reference Organization and Specific Foods with Microbiological Criteria (United States)

Product	Reference organization
All types	International Commission on Microbiological Specifications for Foods
Starch and sugar	National Food Processors
Granulate sugar	American Bottlers of Carbonated Beverages
Liquid sugar	American Bottlers of Carbonated Beverages
Dairy products	U.S. Public Health Service
Certified milk	American Association of Medical Milk Comm., Inc.
Milk for manufacturing and processing	USDA
Dry milk	U.S. Public Health Service
Dry milk	USDA
Dry milk	American Dry Milk Institute, Inc.
Frozen desserts	U.S. Public Health Ordinance Code
Tomato juice and products	FDA
All processed food pathogens	FDA and USDA
All foods in international trade	Codex Alimentarius Commission FAO/WHO

and handled before consumption dictates the type and number of microorganisms and their response to the food.

The important intrinsic characters of the food to consider are (1) pH, (2) type acid, (3) available water (a_w), (4) moisture content, (5) nutrient content, (6) oxidation-reduction potential (E_h), (7) biological structures, (8) natural inhibitors, and (9) added inhibitors. The processing of the food can alter many of these intrinsic characteristics but most importantly the microflora that may be added or altered to include pathogens, indicators, or spoilage agents. The extrinsic parameters of paramount importance are (1) temperature of storage and processing, (2) relative humidity, and (3) gaseous environment.

IV. Sampling and Sampling Plans

A. Statistical Sampling

Probability is the long-term proportion of positive samples to the total number of samples, assuming uniform distribution. For example, 100 positive samples out of 1000 total samples is 0.100 or 10%. Samples taken from a population can only approximate the true probability of their frequency. Because it is impractical to sample the whole food lot, a microbiologist must depend on an estimate of the true probability. Samples must be taken to estimate the character of the population. The larger the number of sample units tested, the larger number frequency distribution units should be obtained. To eliminate bias in determining the true population of a food, random choices must be made. Where possible, random number tables must be used to assign randomness to the food to be sampled. Random choice will reduce the risk of accepting or rejecting a good or bad lot.

Operating characteristic (OC) curves are used to determine the risk of acceptance or rejection of a good or bad unit of food. These curves are generated by calculations of binomial distribution or data taken from Poisson distribution tables. OC curves are used to determine the frequencies of numbers with various levels of defective units. Figure 1 is an operating characteristic curve for 10 sampled units (n) with 2 defective units (c). The probability of accepting a lot of food is 0.68 under these conditions. A lot of food with a 20% defective level will be accepted 68% of the time or, conversely, rejected 32% of the time.

Before sampling, what is to be sampled must be

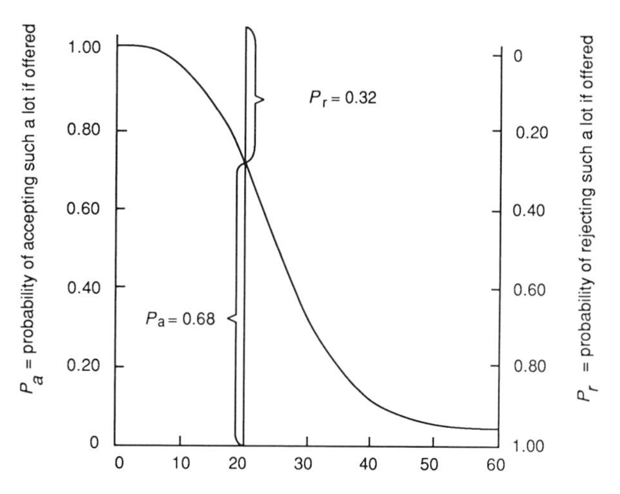

Figure 1 The operating characteristic curve for $n = 10$, $c = 2$, i.e., the probability of accepting lots, in relation to the proportion defective among the sample units comprising the lots. [From International Commission on Microbiological Specifications for Foods (1986). "Microorganisms in Foods. 2. Sampling for Microbiological Analyses: Principles and Specific Applications." 2nd. ed. University of Toronto Press, Toronto.]

determined. Generally, the food produced is defined in terms of a lot. In simplest terms, a lot is "the total amount of food produced and handled under uniform conditions." However, this definition does not always fit. Firstly, the microbial population in nonliquid foods is notoriously heterogeneous. Secondly, continuous systems are difficult to define, given the conditions. Continuous systems would be better defined as per length of time of production. Alternatively, a lot could be defined as any uniform batch of food produced over a short period of time with identifiable assigned code numbers. A lot should be uniform and with as little variation as possible. Any food produced over a prolonged period where there is a loss of quality—microbiological, chemical, sensory, color, etc.—is an unacceptable practice. Therefore, a lot can be defined in many ways depending on the food itself, process, storage, who is to consume it, sensitivity to production, etc.

What is a representative sample? Samples must be drawn to represent the product being investigated. Sampling at random is universally recognized as the means for eliminating bias where random numbers are used to take samples. In some circumstances, random sampling is undesirable. Is the food uniformly mixed? Has stratification occurred? Are there records for consistency? What is the value of the product versus the testing cost? What are the consequences of a wrong decision?

B. Sampling Plan

Traditional food manufacturing has resulted in Quality Control acting as police. This adversarial conflict between production and quality was common. Repeated sampling of out-of-specification lots continued until a desirable result was obtained with little design to sampling. Consequently, better sampling plans were needed to address these issues. The recommended sampling plan for resolving these conflicts is that proposed by the ICMSF. It is a two-class Attribute Plan and a three-class Attribute Plan. The two-class plan is for decision-making on good and bad lots ($c = o$). The three-class plan is for decision-making on acceptable, marginal, and unacceptable products ($n \geq c > o$). Table II contains the definition and symbols for this plan. Two-class plans are generally used for infectious pathogens where an accept/reject decision is required.

To choose a sampling plan, a knowledge of how a food is to be handled and the hazard involved is essential. Table III presents these concerns for choosing a plan, and they are listed as case 1–15, where case 15 is the most stringent. After having chosen the case, the table contains suggested sampling plans for each. These sampling plans are for normal foods. Where problems have been encountered before, suspicions of abuse, new and/or unknown sources, or food to be consumed by susceptible population, more stringent sampling is dictated. For class 2 illustration purposes, *Salmonella* concerns in roast beef would fall into case 12 and *Listeria (L.) monocytogenes* into case 15. *Salmonella* is considered a direct, moderate health hazard with potentially extensive spread and is in a food that when handled improperly in refrigerated storage would grow and increase the hazard (Table IV).

Table II ICMSF Definitions for Sampling and Testing

n =	number of samples taken per lot
c =	maximum number of allowable defects per lot
m =	lower specification
M =	maximum allowable specifications

Pasta is a good example for a class 3 and class 2 plan (Table IV). The hazards are from *Salmonella* (case 10, class plan 2) and *Staphylococcus (Staph.) aureus (case 8, class plan 3)*. *Staphylococcus aureus* in excess of 100/g but less than 10,000/g in one sample would be a marginal lot but acceptable. When any one sample >10,000/g or two or more samples >100/g are encountered, the product is rejected.

In the United States, *Salmonella* and *L. monocytogenes* sampling programs are outlined by the Food and Drug Administration (FDA) and the U.S. Department of Agriculture (USDA). Briefly, the FDA *Salmonella* plan categorizes foods as Category I foods, which would normally be in Category II, except that they are intended for consumption by the aged, the infirm, and infants; Category II foods, which would not normally be subjected to a process lethal to *Salmonella* between the time of sampling and consumption; and Category III foods, which would normally be subjected to a process lethal to *Salmonella* between the time of sampling and consumption. The sampling plan is as follows:

Category I:	60–25 g units
Category II:	30–25 g units
Category III:	15–25 g units

The maximum composite is 15 sample units × 25 g or 375 g/test.

The USDA *L. monocytogenes* and *Salmonella* sampling plan for cooked ready-to-eat meat or poultry products requires five samples from each lot and 5 g from each to be composited for one analytical test.

V. Methods

In determining microbiological criteria, choosing a standard method is essential. Standard and/or officially recognized methods provide (1) better reproducibility between laboratories for comparison purposes, (2) proven sensitivity, (3) security for legal disputes, and (4) standard reference for historical comparisons. The method must be validated to ensure its adequacy and must be sensitive enough to detect the levels of microorganisms required. Where possible, the method should be rapid and easily performed. Table V contains a list of references for the detection and enumeration of microorganisms in foods.

Table III Suggested Sampling Plans for Combinations of Degrees of Health and Conditions of Use (i.e., the 15 "Cases")

	Conditions in which food is expected to be handled and consumed after sampling, in the usual course of events[a]		
Degree of concern relative to utility and health hazard	Conditions reduce degree of concern	Conditions cause no change in concern	Conditions may increase concern
No direct health hazard Utility, (e.g., shelf-life and spoilage)	Increase shelf-life Case 1 3-class $n = 5, c = 3$	No change Case 2 3-class $n = 5, c = 2$	Reduce shelf-life Case 3 3-class $n = 5, c = 1$
Health hazard Low, indirect (indicator)	Reduce hazard Case 4 3-class $n = 5, c = 3$	No change Case 5 3-class $n = 5, c = 2$	Increase hazard Case 6 3-class $n = 5, c = 1$
Moderate, direct, limited spread[b]	Case 7 3-class $n = 5, c = 2$	Case 8 3-class $n = 5, c = 1$	Case 9 3-class $n = 10, c = 1$
Moderate, direct, potentially extensive spread[b]	Case 10 2-class $n = 5, c = 0$	Case 11 2-class $n = 10, c = 0$	Case 12 2-class $n = 20, c = 0$
Severe, direct	Case 13 2-class $n = 15, c = 0$	Case 14 2-class $n = 30, c = 0$	Case 15 2-class $n = 60, c = 0$

[a] More stringent sampling plans would generally be used for sensitive foods destined for susceptible populations.
[b] See 'Conclusions,' p. 46, for explanation of extensive and limited spread. [From International Commission on Microbiological Specifications for Foods (1974). "Microorganisms in Foods 2. Sampling for Microbiological Analyses: Principles and Specific Applications." University of Toronto Press, Toronto.]

VI. Sanitation

Sanitation is the total effort made to control the contamination and growth of undesirable microorganisms and the prevention of adulteration with filth from insects, rodents, birds, pets, other animals, and humans. It is the scientific application of microbiology, chemistry, engineering, physics, biology, and management to the hygienic control of adulteration. Commitment to an effective sanitation program is a responsibility of management that must be communicated to all employees. The keys to its success are proper training and education along with adequate supervision to ensure effectiveness.

The code by which foods are produced in the United States is outlined in the Current Good Manufacturing Practice in Manufacturing, Packing, or Holding Human Food. These are commonly called Good Manufacturing Practices (GMPs). The definition and the enforcement are described in the Food, Drug and Cosmetic Act. Basically, they describe when a food is considered adulterated during manufacture or held under unsanitary conditions that may lead to a contaminated product. Although in recent years the general GMPs have not had the full force of the law. They are still useful as good outlines for sanitation. However, the more specific GMPs for food such as low-acid canned foods are enforceable and very effective in preventing food-borne illness. The GMPs are broken down into (1) definitions, personnel, and personal hygiene, (2) sanitary considerations of building and facilities, (3) sanitary design of

Table IV Sampling Plans and Recommended Microbiological Limits

		Plan				Limit/g	
Product	Test	Case	Class	n	c	m	M
Roast beef	*Salmonella*	12	2	20	0	0	—
	Listeria monocytogenes	15	2	60	0	0	—
Dried Pasta	*Salmonella*	10	2	5	0	0	—
	Staphylococcus aureus	8	3	5	1	10^2	10^4

Table V Some References for the Detection and Enumeration of Microorganisms in Foods

Association of Official Analytical Chemists (AOAC) (1980). "Official Methods of Analysis." AOAC, Washington, D.C.

International Dairy Federation (IDF). (1981–1982). "International Dairy Federation Catalogue 1981–1982." IDF, Brussels, Belgium.

International Organization for Standardization (ISO) (1983). "International Standards Organization Catalogue. 1983." ISO Geneva, Switzerland.

International Commission on Microbiological Specifications for Food (1978). "Microorganisms in Foods, 1. Their Significance and Methods of Enumeration," 2nd ed. University of Toronto Press, Toronto.

Marth, E. H. (ed.) (1978). "Standard Methods for the Examination of Dairy Products," 14th ed. American Public Health Association, Washington, D.C.

National Food Processors (1968). "Laboratory Manual for Food Canners and Processors." Vol. 1, "Microbiology and Processing." National Canners Association, Washington, D.C.

Speck, M. S. (ed.) (1984). "Compendium of Methods for the Microbiological Examination of Food," 2nd ed. American Public Health Association, Washington, D.C.

U.S. Department of Agriculture (1977). "Microbiology Laboratory Guidebook." Food Safety Inspection Service, Washington, D.C.

U.S. Food and Drug Administration (1984). "Bacteriological Analytical Manual," 6th ed. AOAC, Arlington, Virginia.

equipment and utensils, and (4) production and process controls.

Personal hygiene and food handling are pivotal in sanitation. Humans who are involved in food handling and manufacture must be clean and healthy. Human contamination of foods with infectious agents and toxin-producing microorganisms is common. The microorganisms may be transmitted from the skin, hands, hair, eyes, nose, mouth, respiratory tract, and excretal organs. *Staph. aureus* can be transmitted from the skin and nasal passages; *Streptococcus pyogenes* from the throat and respiratory tract; and *Salmonella, Shigella, Vibrio cholera,* enteropathogenic *Escherchia coli,* protozoans, hepatitis, and Norwalk virus from excretal organs. Unhealthy employees must be segregated from direct handling and manufacture of food. Proper systems and facilities must be in place and available to prevent disease transmission, especially for asymptomatic and convalescing carriers. [*See* FOODBORNE ILLNESS.]

The food production area and equipment must be chosen and designed to prevent disease and adulteration. It must be easily cleaned, sanitized, and made from materials that are not toxic. Equipment for storing and conveying must be designed to exclude or minimize the growth of microorganisms.

Food held in the temperature danger zone of between 45° to 140°F for 2 hr or longer should be avoided. This also includes food buildup on contact surfaces. [*See* TEMPERATURE CONTROL.]

Proper cleaning followed by sanitizing is essential to control microbial hazards, spoilage organisms, and indicator bacteria. Chemical sanitation has traditionally been applied to food contact surfaces, especially in small operations and food service establishments. However, modern food plants employ Clean-in-Place systems where applicable. They are particularly effective in liquid foods such as milk and beverages. Some common chemical sanitizers and their recommended concentrations are found in Table VI.

Adequately designed sanitary facilities are crucial to a food manufacturing facility. The elements of well-designed sanitary facilities are (1) adequate and potable water supply, (2) proper plumbing for waste removal and cleaning, (3) convenient toilet facilities, (4) convenient hand washing and sanitization stations, and (5) a segregated efficient rubbish and offal disposal system. The facilities must be designed to eliminate or minimize the presence of insects, birds, rodents, and pets that might lead to

Table VI Some Common Sanitizers and Recommended Concentrations for Some Specific Food Processing Areas

Processing area	Sanitizer	Concentration
Hand dips	Quaternary compounds	25 ppm
	Iodophor	25 ppm
Rubber belts	Quaternary compounds	25–50 ppm
	Iodophor	25–50 ppm
Stainless Steel Equipment	Quaternary compounds	200 ppm
	Active chlorine	200 ppm
	Iodophor	200 ppm
Clean-in Place system	Acid sanitizer	100–150 ppm
	Active chlorine	200 ppm
Floors (concrete)	Active chlorine	1000 ppm
	Quaternary compounds	500 ppm
Walls	Active chlorine	25 ppm
	Quaternary compounds	25 ppm
	Iodophor	25 ppm
Water treatment	Active chlorine	20 ppm

food adulteration from filth and/or microbial contamination.

VII. Microbiological Quality

A. Indicators

Indicators of lack of sanitation in foods have been used in assessing the quality of water and foods since the turn of the century. Table VII contains a list of tests commonly used for this purpose. Coliforms and *E. coli* are the most frequently used to detect problems with sanitation. Coliforms are gram-negative, facultative anaerobic nonspore-forming rods that ferment lactose to acid and gas in 24–48 hr on liquid or solid media at 32–35°C. They are in the genera *Escherichia, Enterobacter, Klebsiella,* and *Citrobacter*. Coliforms are commonly found in feces of warm-blooded animals, vegetable material, and soil. However, *E. coli* is more specifically used as a direct indicator of fecal pollution in foods because of its closer association with feces. Coliforms capable of growing at 44.5°C in the official Association of Official Analytical Chemists procedures EC broth are fecal coliforms. Most are *E. coli* but not all. Being somewhat more specific than the standard coliform test, they are frequently used.

The fecal streptococci or enterococci are Lancefield's serologic Group D streptococci, which are *Streptococcus (S.) faecalis, Streptococcus faecium, Streptococcus bovis,* and *Streptococcus equinus*. These are intimately associated with the intestinal tracts of animals. They are more fastidious than the coliform group, with *S. faecalis* being the most common to the intestinal tract of humans but less specific than *E. coli*.

The Enterobacteriaceae family may be used as an indicator group, although it is infrequently used. On the other hand, total counts (aerobic plate count, standard plate count) are frequently used to indicate the handling of certain foods, especially those that have an extensive history. They are particularly useful in evaluating frozen, refrigerated, and dried products.

Table VII Commonly Used Indicator Microorganisms

Coliforms
Escherichia coli
Mesophilic aerobes
Fecal streptococci enterococci
Fecal coliforms
Enterobacteriaceae count

B. Spoilage

Table VIII contains a list of the more commonly used tests for spoilage organisms. These would be employed in products where the chemical, physical, and biological factors, both intrinsic and extrinsic, determine the spoilage flora. For example, salad dressing, with a low pH due to acetic acid, would be tested for yeast and mold, lactobacilli, lactic acid bacteria, and/or acetophiles.

C. Health Hazards

A variety of microorganisms are associated with food that cause disease (Table IX). They are broken down into groups, with the bacteria being the most prevalent, particularly in developed countries. The bacteria are further divided into infection producers, toxin producers, and enterotoxigenic types. All are classified as to their relative hazard, which is determined according to the organism pathogenicity, potential spread, and population at risk.

Table VIII Commonly Used Tests for Spoilage Microorganisms

Aerobic plate count
Standard plate count
Psychrophilic plate count
Anaerobic mesophilic plate count
Lactic acid bacteria count
Yeasts and molds
Lactobacillus
Osmophilic yeasts and molds
Lipolytic counts
Pectinolytic counts
Acetophilic counts
Thermophilic counts
Canned food enrichment procedures
Spore counts

Table IX Disease Microorganisms Associated with Foods

Microorganisms	Hazard
Infections Bacterial	
Salmonella typhi, paratyphi A and B	*
Salmonella	**
Shigella dysenteriae	*
Shigella sp.	**
Listeria monocytogenes	*
Brucella abortus	*
Brucella suis	*
Brucella melitensis	*
Mycobacterium bovis	*
Enteropathogenic *Escherichia coli*	**
Aeromonas hydrophila	***
Streptococcus pyogenes	***
Vibrio parahaemolyticus	***
Coxiella burnettii	***
Yersinia enterocolitica	***
Yersinia pseudotuberculosis	***
Campylobacter jejuni	**
Vibrio cholerae	*
Vibrio vulnificus	***
Plesiomonas shigelloides	***
Toxins Bacterial	
Clostridium botulinum	*
Staphylococcus aureus	***
Enterotoxigenic bacterial	
Clostridium perfringens	***
Bacillus cereus	***
Infections protozoan and parasitic	
Trichinella spiralis	***
Helminthic *parasites*	***
Taenia saginata	***
Taenia solium	***
Isospora (*Toxoplasma*) *gondii*	***
Entamoeba histolytica	***
Giardia lamblia	***
Cryptosporidium parvum	***
Echinococcus sp.	***
Diphyllobothrium pacifiam	***
Capillaria phillippinensis	***
Anisakiasis (nematodes)	***
Viral	
Rotavirus	**
Norwalk virus	**
Echovirus	**
Hepatitis A virus	*
Polio	*
Food–Microorganism interaction	
Fish and shellfish toxins	*
Vasoactive amines	***
Mold	
Aflatoxins	*
Mycotoxins (predominately produced by the genera *Aspergillus*, *Penicillium*, *Fusarium*, and *Mucor*)	***

*, Severe hazard; **, moderate hazard with potentially extensive spread; ***, moderate hazard with limited spread.

VIII. Hazard Analysis and Critical Control Point

HACCP is a rational systematic approach to food safety where hazards are identified with risks assigned and control measures implemented. It is a preventative system for controlling microbiological safety. Although designed primarily for safety, it can also be applied to spoilage control. Traditional microbiological quality control programs rely heavily on inspection and finished product testing for health and quality hazards. This approach results in information of little significance to safety, and test results are available long after the product is produced, consequently providing no flexibility in adjusting production for prevention. Obviously, a system such as HACCP, where control is placed on hazardous situations during production, is superior. Intervention at critical control points (CCPs) during production makes it a proactive program with great flexibility.

Since the introduction of HACCP in 1971, it has steadily gained in application. It has been gradually adopted by food manufacturers as a rational approach to food safety. Recently, it was outlined and adopted by the NAC on Microbiological Criteria for Foods, part of the NAC/National Research Council, as an effective quality assurance program. The most successful application of HACCP has been with the low-acid canned foods, where the critical control points are easily identified and controlled. To be effective, HACCP planning must include total commitment of the corporation from management to the factory worker. The HACCPs plan must be a team approach including at least a microbiologist, an engineer, a production supervisor, and quality control specialist as well as a sanitation supervisor. Depending on the food product, other areas of a business may be involved, such as sales and distribution. As proposed by the NAC, the HACCP system consists of seven principles.

1. *Assess Hazards Associated with Growing, Harvesting, Raw Materials and Ingredients, Processing, Manufacturing, Distribution, Marketing, Preparation, and Consumption of Foods* The type hazard will vary according to the food being produced. (Refer to Table IX for a list of potential hazards associated with foods.) For the purposes of demonstrating the application of HACCP, the manufacture of roast beef is chosen (Fig. 2). The hazards of primary concern are *Salmonella* and *L. monocytogenes*.

For spoilage microorganisms, the intrinsic and extrinsic factors would greatly influence the type of organisms causing the spoilage. For example, roast beef is highly nutritious, with a pH 5.6 and A_w 0.99, and has a low competing microflora because of roasting. It may or may not contain NO_3 and has only 1 or 2% salt. After roasting, it is rebagged and placed in an oxygen impermeable barrier under vacuum and then refrigerated. The extrinsic factors and intrinsic characteristics dictate the spoilage flora, which are primarily lactic acid-producing bacteria and heterofermentative psychotrophic lactobacilli.

The NAC assessed risk by ranking the hazards into categories A–F.

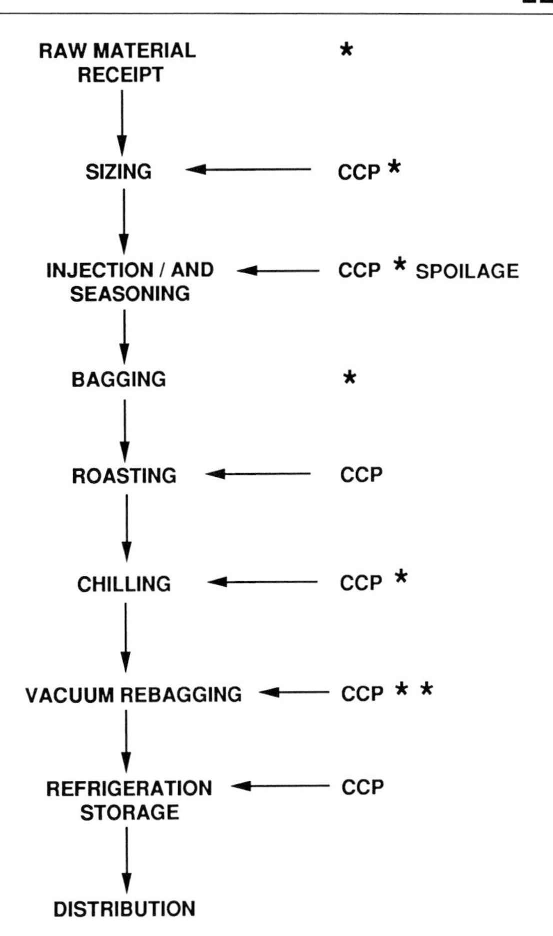

Figure 2 Flow diagram for the production of roast beef. CCP, critical control point; **, site of major contamination; *, site of minor contamination.

A. A special class that applies to nonsterile products designated and intended for consumption by at-risk populations (e.g., infants, the aged, the infirm, immunocompromised individuals).
B. The product contains "sensitive ingredients" in terms of microbiological hazards.
C. The process does not contain a controlled processing step that effectively destroys harmful microorganisms.
D. The product is subject to recontamination after processing before packaging.
E. There is substantial potential for abusive handling in distribution or in consumer handling that could render the product harmful when consumed.
F. There is not terminal heat process after packaging or when cooked in the home.

Risk is assigned according to the number of hazard categories identified for the food. There is one "special category," which is defined in hazard A, and all six hazard categories must be considered.

An alternative to this procedure has been outlined by the ICMSF where the severity of types of contaminants are considered. In Table IX, the severity of the organisms are ranked. To determine the ranking as to the hazard involved, refer to Table III. Considering roast beef as a model, *Salmonella* hazard is ranked as case 11 because *Salmonella* is "moderate, direct, potentially extensive spread" and under normal refrigeration (<45°F) the hazard is unlikely to change. In the case of *L. monocytogenes*, the type hazard is "severe, direct" and, because *L. monocytogenes* grows at refrigeration temperatures, it will increase the hazard to the highest case 15. This is in comparison to the NAC assessment for a Risk Category IV fulfilling categories B, D, E, and F. [*See* REFRIGERATED FOODS.]

2. *Determine CCP Required to Control the Identified Hazards* A CCP is any location, practice, manufacturing, handling, and storage procedure, or manufacturing process where a hazard can be controlled. Conversely, a CCP is any point, if failure occurs, that will result in an unacceptable health risk or a strong possibility for spoilage. An integral part of the HACCP plan is the development of a schematic diagram of the process flow identifying the CCPs and areas of significant contamination. For illustration purposes, Fig. 2 contains a schematic for the production of roast beef. The two most critical points are the roasting and refrigeration steps. Sizing is critical because roasting is calculated on a standard size and weight. Injection of brine and seasonings is important because lack of control on the brine would lead to contamination by heterofermentative lactobacilli, which results in gas pockets and/or sourness as the beef is being brought up to cooking temperatures. Rebagging into oxygen impermeable vacuum bags is critical because of recontamination by *Salmonella, L. monocytogenes,* and lactic acid bacteria.

3. *Establish the Critical Limits That Must be Met at each Identified CCP* Limits are set on physical, chemical, sensory, or biological attributes of a food in process. They can be temperature, pH, moisture, viscosity, visual appearance, available water, or any other parameter used to determine whether or not a CCP is under control. When these limits are not met, the CCP is considered to result in an unacceptable health or spoilage risk.

4. *Establish Procedures To Monitor CCP* Monitoring procedures must measure and quantify the control necessary in a CCP to ensure that it does not exceed the limits established for adequate control. This measurement should be continuous to be effective; however, spot checks can be useful where continuous monitoring is not practical. Monitoring can be chemical, physical, visual, sensory, and/or microbiological testing. The most rapid and useful are the chemical or physical tests such as pH, time, temperature, total acidity, salt, a_w , humidity, moisture, and viscosity. Microbiological tests may also be applicable when rapid enough to provide control, especially with regard to pathogens in critical environmental areas of continuous processes. In roast beef production, monitoring procedures are visual and odor inspection of incoming beef, weight control of sized beef cuts, visual inspection of injection equipment and microbiological testing of start-up brines, time/temperature records of roasting, chilling and refrigerated storage, sanitation records of cleaning and sanitation with visual inspection of contact surfaces during rebagging, and *Salmonella* and *L. monocytogenes* environmental samples of the chilling and rebagging areas.

5. *Establish Corrective Action to be Taken when a Deviation is Identified by Monitoring of a CCP* When a product that is out of the established limits is produced, it must be placed on hold. The HACCP plan must contain provisions for out-of-limit food products. It must contain the action plans to be taken to include reprocessing, reexamination, or disposal. Action must be taken to correct the out-of-contol CCP.

6. *Establish Effective Record-Keeping Systems That Document the HACCP Plan* One of the key elements of this plan is that the government will have access to the HACCP plan and all records pertinent to it. The records must be kept for ingredients, research in safety and stability, processing (CCP), packaging, storage and distribution, deviations, and modifications to the HACCP plan. Each CCP identified must have records kept to substantiate an observation. All must be validated as to appropriateness or efficacy.

7. *Establish Procedures for Verification That the HACCP System is Working Correctly* A procedure must be in place to determine if the HACCP plan is performing as intended. When the system is first introduced, extensive testing may be involved to verify the plan's efficacy. Auditing of the finished product will be a part of the HACCP. Changes in processing technology, suspicion of food-borne illness, and organisms or new food safety issues will require revalidation. The HACCP plan must be constantly verified through records review, plant inspections, and by random testing, all of which must be written to confirm compliance to the HACCP plan.

Simply testing final roast beef for *Salmonella* and *L. monocytogenes* would verify the system.

Bibliography

Food and Drug Administration (1986). "Current Good Manufacturing Practice in Manufacturing, Packaging, or Holding Human Food; Revised Current Good Manufacturing Practices." Federal Register 21 CFR Parts 20 and 110.

ICMSF (1980). "Microbial Ecology of Foods I. Factors Affecting Life and Death of Microorganisms." Academic Press, New York.

ICMSF (1980). "Microbial Ecology of Foods II. Food Commodities." Academic Press, New York.

ICMSF (1986). "Microorganisms in Foods 2. Sampling for Microbiological Analysis: Principles and Specific Applications," 2nd ed. University of Toronto Press, Toronto.

ICMSF (1988). "Microorganisms in Foods 4. Application of the Hazard Analysis Critical Control Point (HACCP) System to Ensure Microbiological Safety and Quality." Blackwell Scientific Publications, Oxford.

Jay, J. M. (1986). "Modern Food Microbiology," 3rd ed. Van Nostrand Reinhold, New York.

Marriott, N. G. (1989). "Principles of Food Sanitation," 2nd ed. Van Nostrand Reinhold, New York.

Mossell, D. A. A., van der Zee, H., Corry, J. E. L., and van Netten, P. (1984). Microbiological quality control. *In* "Quality Control in the Food Industry," Vol. 1 (S. M. Herschdoerfer, ed.), pp. 79–168. Academic Press, New York.

National Advisory Committee on Microbiological Criteria for Foods (1989). "HACCP Principles for Food Protection." USDA, FSIS.

Freeze-Drying of Microorganisms

Hiroshi Souzu
Hokkaido University

Glossary

Bound water Nearly 10% of the cellular water remains in unfrozen state in usual freeze-drying temperature range and water molecules are considered to be combined firmly to the cellular materials and exerting their force to keep the tissue structure

Ice inoculation Seeding of ice crystal to the suspension fluid for the prevention of super cooling and following rapid freezing of the materials

Metabolical injury Partial impairment of the activity of metabolic reactions brought about by freeze-thawing or freeze-drying operation. The cells viability varies depending upon the thawing or rehydration procedures.

Protective substance Smaller or larger molecular substances that are dispersed into the suspension to protect the materials from the injury arising from freezing or drying of living materials.

WATER MOLECULES are basically concerned with the functions of living organisms, either in close participation to their metabolic reactions or in the contribution to the structural stability of the cellular constituents and organelles in the cells. Freezing and freeze-drying are processes that are frequently utilized for the preservation of the biological substances by the restriction of the water activity in them. When these procedures on microorganisms are appropriately brought about, the cellular function is suspended temporarily by being cut off from the metabolic activities, which are mediated by the action of water molecules. As a result, the organisms are subjected to preservation for a given period of time. However, these operations involve many critical situations in each manipulation, and, even if a step of the procedures is not appropriately applied, the organisms cannot survive any longer.

I. Introduction

Originally designed for long-term preservation of biological materials, pharmaceuticals, and other delicate solvent-impregnated materials, freeze-drying is a conjugated stage operation of careful drying from the frozen state. Due to its reduced water activity, the dried products can be stored at a room-temperature range for a prolonged period if cut off from oxygen, moisture, and light. In addition, due to the highly porous structure of the products, they can easily reabsorb the water molecules and be restored to their original state.

The freeze-drying procedure consists of separate processes: solidification (freezing of materials to be dried) and drying of the frozen materials under reduced pressure (vacuum). This stage can be divided into two successive stages: (1) the stage of ice crystal sublimation and (2) the stage of isothermal desorption of the remaining liquid phase in the cellular material (referred to as the secondary drying stage). Storage and reconstitution (rehydration) of the dried product follow.

The freezing process is a very critical step to cell viability and drying efficiency and, furthermore, to the product stability during the storage period. Thus, the freezing program that is most adequate for the materials should be selected for each individual case. During the stage of sublimation, the specimen stays in a frozen state due to the latent heat of sublimation. Even after the drying is started from the liquid phase, the specimen freezes immediately as pressure reduction is started. However, to achieve the best results, the drying of the specimen proceeds after the freezing. The selection and incorporation of cryoprotectants are also introduced in this step.

The sublimation stage, as well as the subsequently accompanying secondary drying stage, is the most fundamental stage in the processes of freeze-drying. This step is generally carried out under reduced pressure, to promote the vapor flow through the systems containing the product itself and the equipment. At this stage, delicate care for the heat balance is required, for feeding the energy to promote the sublimation and to diminish the deleterious effect of the heating on still-frozen parts or already dried matrix, both of which are very sensitive to denaturation and resulting death of the microorganisms.

The reconstitution of the dried materials to restore them to their original state is also critical, even to the surviving cells. The viability of the cells in the products is usually not so high; thus, treatment of the cells in this stage should be done with special care to avoid damage and contact with undesirable contaminants.

As understood by the preceding description, freeze-drying is a multistage operation that contains each independent but mutually interacting step. In the following sections, detailed procedures and the care that should be taken to reduce damage in the freeze-drying of microorganisms will be described.

II. The Freeze-Drying Operation

A. Preliminary Preparation

Freeze-drying of microorganisms is usually carried out in aqueous suspensions. The preparation of appropriate suspending medium for each individual microorganism is the most important work in the procedure. This preparation includes adjustment of pH and ionic strength using a suitable buffering reagent and salts or sugars, because the medium and material properties combined in the suspending media sometimes cause a significant effect on cellular or cell material activity throughout the freeze-drying processes such as freezing, drying, storage, and restoration. The prepared medium should then be carefully processed ahead of time in much the same manner as sterilization, filtration, and degasification.

Cultivated cells are first collected by centrifugation, washed, and then finally resuspended in an appropriate concentration into a suspension fluid. The determination of the cell concentration in the suspension will also have a serious effect on the subsequent processes. When a highly concentrated cell suspension is used, it prevents an effective sublimation of ice crystals by the formation of a thick interstitial network. The thick interstitial network also prevents the desorption of unfreezable water molecules, which are universally distributed cohesively to the cellular materials. In addition, highly concentrated cell suspension causes difficulty in spreading out the specimen over the whole inside wall of the specimen container, which is essential to make a large specimen surface to obtain an economical drying efficiency. In contrast, when the cell concentration is too low, the cells, because they lack effective networks joining each of the cells, will disperse, led by the stream of water vapors during the ice crystal sublimation. Thus, an unaccountable amount of the cells will be lost in the process. When an appropriate specimen concentration is not obtainable, as when a very small amount of the cells are produced, the utilization of a bulking or connecting substance will be recommended. In the instance where very sensitive organisms are handled, an appropriate protective substance should also be incorporated at this stage.

B. Cooling Procedure

1. Temperature Range and Cooling Rates

The numerous cooling procedures include contact with a cold surface, immersion in a cold bath, direct spraying into liquid nitrogen, and utilization of liquified gas. The temperature range is also widely distributed from several degrees below zero to liquid nitrogen temperature (−196°C), and, in rare cases, liquid helium is also used. The temperature range, which usually has an injurious effect on the living materials, is considered to be approximately −80°C or higher.

In these steps, cooling-induced crystallization of the fluid varies in pattern, shape, and size and orientation of ice, etc., depending on the composition of the medium, its vessel configuration, the velocity and type of cooling, etc. These phenomena are of great importance in the overall freeze-drying processes, because the crystalline patterns significantly affect the subsequent velocity of drying as well as the structure, stability, and solubility of the final products. In general, slow freezing brings a rapid sublimation but slow secondary drying rates, and rapid freezing gives slow sublimation and rapid secondary drying rates (Fig. 1).

Cooling rates that are slower than 1°C/min are usually called slow freezing, and rates faster than 100°C/min are called rapid freezing; however, no accurate definition exists for slow and rapid freezing.

2. Intra- and Extracellular Freezing

In the cooling of an aqueous suspension of microorganisms, freezing of the suspension fluid commences at a temperature slightly <0°C; however, in such temperature regions, the cellular water still remains in a supercooled state. A higher vapor pressure of the supercooled water compared to that of ice in the same temperature expels the cellular water outside of the cells, and the water molecules freeze in an external site of the cells by being inoculated through contact with already frozen ice crystals. When the cooling rates of the specimens are low enough to equilibrate with the expulsion of the intracellular water, the water molecules successively move out of the cells and the cells shrink by reduction of the cellular water content, in which the concentration of the solutes increases accordingly. When the cooling reaches the eutectic temperature of the concentrated cellular fluid, the freezing terminates with the deposit of eutectic mixture intracellularly and pure ice crystals extracellularly (a small amount of the eutectic mixture of incorporated solutes in the suspending fluid is also present extracellularly). This mode of freezing is called extracellular freezing.

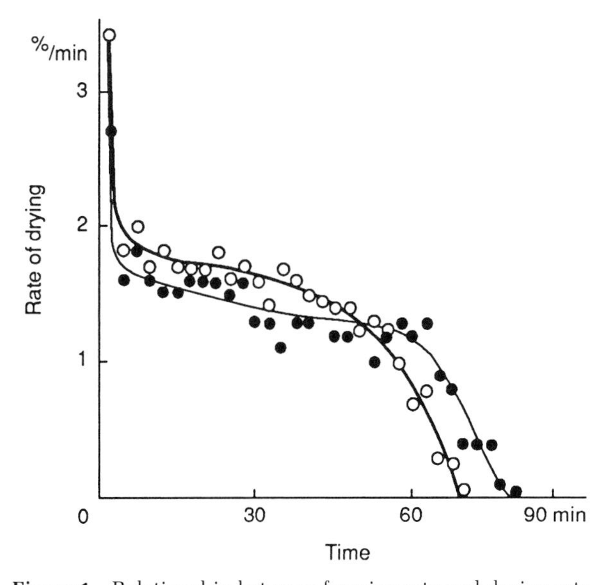

Figure 1 Relationship between freezing rate and drying rate in freeze-drying of rabbit serum. ●, Rapid freezing; ○, slow freezing. [From Nei, T., Souzu, H., and Hanafusa, N. (1964). *Contrib. Inst. Low Temp. Sci.* B-13, 7–13.]

When the rate of cooling is higher than the equilibration to the expelled water molecules from the cells, supercooling of the intracellular water increases, and ice crystallization takes place spontaneously in the cells. The result is called intracellular freezing.

The equilibration of the vapor pressure between intracellularly remaining water and the extracellular ice crystal in a given temperature is achieved by the shift of the water molecules out of the cells. Thus, the freezing rates that cause the intra- or extracellular freezing are determined by the rate of expulsion of intracellular water to the outside of the cells. The movability of the cellular water depends on the ratio of surface area to the volume of the cells (i.e., the largeness of the cells) and to the inherent permeability coefficient of the cell membrane of individual species. In other words, a smaller cell volume and higher permeability of the cell membrane of the organisms results in extracellular freezing in comparatively higher freezing rates, and vice versa.

The relationship between the amount of intracellularly remaining water and the cooling rates in yeast and red blood cells are theoretically calculated. In yeast, a significant amount of aqueous phase remains in the cells cooled at the rate of 10°C/min. However, in red blood cells, a similar amount of the water remains in the cells when they are cooled at the rate of 1000°C/min (Fig. 2). If it is hypothesized that the cellular water can be inoculated at −10°C, the yeast cells that are cooled at the rate of 100°C/min will result in almost 100% intracellular freezing. In contrast, when red blood cells are cooled at the same rate, only a small percentage of cellular water will freeze intracellularly. As a result, the same freezing rate applied to the yeast cells and to red blood cells gives significantly different influences on their viabilities.

In a comparison of four different kinds of cells, the freezing rates that give rise to most higher viabilities apparently differ by approximately 5000 times (Fig. 3). In general, intracellular freezing is considered to have a more damaging effect than extracellular freezing. The adoption of appropriate freezing procedures for each individual species is required, especially when freezing sensitive organisms. The

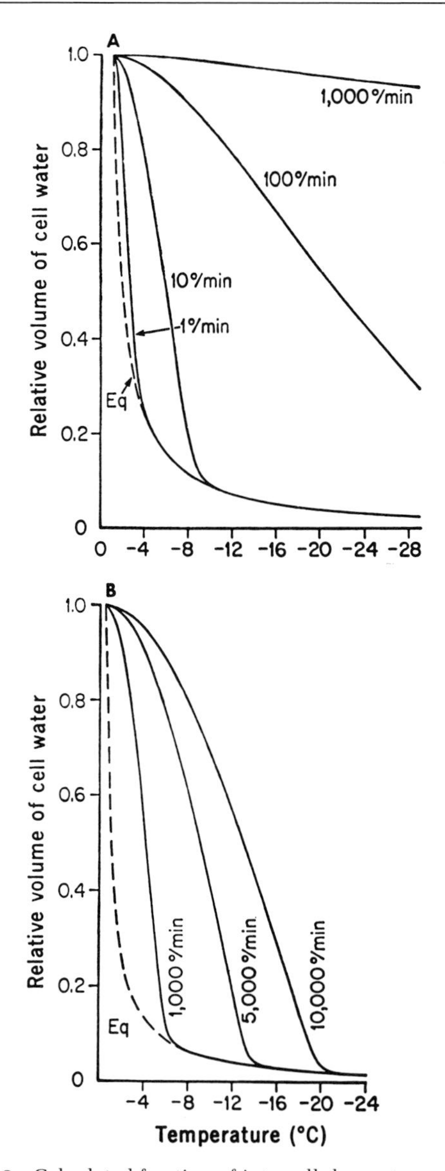

Figure 2 Calculated fraction of intracellular water remaining in (A) yeast cells and (B) human red cells as they cool to various temperatures at the indicated rates. Curve Eq represents the equilibrium water content. [From Mazur, P. (1970). *Science* **168**, 939–949. Reproduced with the permission of *Science*.]

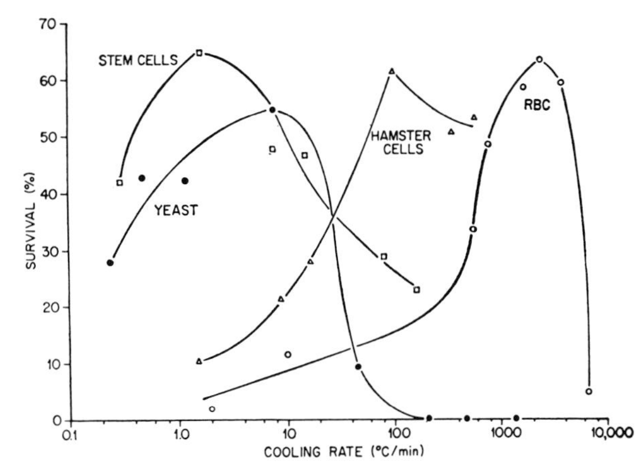

Figure 3 Comparative effects of cooling velocity on the survival of various cells cooled to −196°C and thawed rapidly. The yeast and human red cells (RBC) were frozen in distilled water and blood, respectively. The marrow stem cells and hamster cells were suspended in balanced salt solutions containing 1.25 *M* glycerol. [From Mazur, P. (1970). *Science* **168**, 939–949. Reproduced with the permission of *Science*.]

identification of the intra- or extracellular freezing can be attained with electron microscopic observation.

In any freezing procecure, a substantial amount of cellular water remains unfreezed in the cells. This water is called unfreezable water, or bound water, because it is considered to be firmly bound to the cellular materials and to be stabilizing their structure. A considerable amount of this water is known to be removed from the cells in the period of secondary drying.

3. Procedure of Slow Freezing

When specimens must be slowly frozen, they are precooled in a bath that was cooled to approximately 1° or 2°C lower than the freezing point of the suspension fluids. After a few minutes of equilibration at that temperature, the suspensions are inoculated by being touched with the tip of a Pasteur pipette containing ice, which also has been equilibrated to the same temperature. In another instance, a Pasteur pipette is replaced by a thin metal wire, the tip of which had been cooled by liquid nitrogen and then frosted on its surface by being kept briefly in the air. The inoculated specimens are held for several more minutes at that temperature to allow equilibration with respect to crystallization. Then the cooling bath temperature is lowered at the rate that is designed for the specimen.

This primary step of slow freezing is a very delicate stage of the procedure, because, in these lengthened periods, a substantial amount of the intracellular and interstitial fluid are separated out as pure ice crystal and, accordingly, the sensitive organisms are exposed to increasingly higher solute concentrations and associated hypertonic stress. For most sensitive microorganisms, utilization of appropriate cryoprotectants is absolutely required at this stage.

Slow freezing will often result in a concentrated cellular mass in the middle portion of the specimen as well as with a large interstitial network of concentrated fluids, both of which work deleteriously in the drying process. To overcome this phenomena, a device to cool the specimen evenly from the bottom of the vessel or other manipulations will be required.

If desired, the specimen that has been cooled to approximately −40°C can then be cooled rapidly to lower temperatures without any significant effect on the cell viability or other characters of the dried products.

4. Procedures for Rapid Freezing

In the treatment of freezing-resistant organisms, the specimen is usually frozen by immersion of the vessel into the cooling bath, which consists of either liquid nitrogen, dry ice–acetone, dry ice–alcohol, or any other refrigerated cooling media. In another case, the specimen is frozen in a deep-freezer. It is worthy to note that the spontaneous freezing in a deep-freezer results in rather rapid freezing, because the specimen is first supercooled and then subsequently frozen in low temperatures.

The most simple and more frequently utilized technique is as follows: An appropriate amount of the cell suspension dispensed in the container is cooled by manual rotation and dipped at a tilt angle into the freezing media. This procedure is recommended for obtaining a large surface area of the specimen spread around the whole inner surface of the sample vessel. In this freezing method, naturally, specimens freeze rapidly, but an accurately designed freezing rate is not obtainable.

When a higher-speed freezing of the specimen is required, the specimen vessel can be cooled by immersion in the liquified gas rather than direct immersion in liquid nitrogen, because the gas phase generated by boiling of liquid nitrogen significantly reduces the thermal conductivity. The gases can easily be liquified by direct spraying from the container into metal cups that have been cooled to the ranges between their liquifying and solidifying temperatures. These temperatures can be obtained by adjusting the distance of the cups from liquid nitrogen surface. The gases propane and butane and freons are commonly used.

In any event, the freezing step is considered to be most critical in the multistep processes, and it should be carried out according to well-defined methods specifically designed for each individual species.

C. Drying: Sublimation and Isothermal Desorption

Drying, the most substantial stage in the freeze-drying procedure, consists of two successive steps, namely sublimation of deposited ice crystals and evaporation of water molecules that remain in the cells absorbing to the cellular constituents. In practice, these two obviously different steps progress in tandem in the same specimen and cannot be handled separately; therefore, these two steps will be described in the same section.

Freeze-dryers consist mainly of three parts: a drying chamber (drying manifolds), a vacuum system, and a cold trap, which contains a coolant or refrigerant. The operation generally progresses under reduced pressure to boost the velocity of vapor flow throughout the system. The vapors released from the products can be eliminated directly by a pumping system, but this is not considered to be a practical solution because its capacity is usually not sufficient to handle a large volume of vapors in low pressure fields. Then, the cold trap is utilized to capture and condense the vapor on its surface. The cold trap also works to prevent the mixing of aqueous vapors into the rotary pump oil. The load on the vacuum pump itself is thus confined to the initial step of evacuation through the dryer system and a very minor one in the course of drying to eliminate the noncondensable gases evolving from the products or to handle small leaks in the apparatus.

Because the motivating force of the drying in the procedure arises from the difference of vapor pressures between the specimen surface and the cold trap surface, a coolant of a lower temperature is utilized in the cold trap, thus achieving a higher efficiency of drying. Dry ice–alcohol or dry ice–acetone or liquid nitrogen can be used efficiently. In long or large operations such as an industrial job, the utilization of a cold trap equipped with a refrigerator is recommended. Aqueous vapor pressures of ice and water in subzero temperature regions are listed in Table I.

In this sublimation stage, the delicate balance between the heating and its transfer to a drying surface must be maintained. Because a large amount of energy is required for sublimation, in contrast, a considerable amount of energy may bring about melting in the part where the bulk of ice crystals still

Table I Vapor Pressure of Water and Ice below 0°C[a]

Temp. (°C)	Water	Ice
0	4.579	4.579
−1	4.258	4.217
−2	3.956	3.880
−3	3.673	3.568
−4	3.410	3.280
−5	3.163	3.013
−10	2.149	1.950
−15	1.436	1.241
−20	—	0.776
−25	—	0.476
−30	—	0.286
−40	—	0.097
−50	—	0.030
−60	—	0.008
−70	—	0.002
−80	—	0.0004
−90	—	0.00007

[a] Pressure of aqueous vapor over water and ice in mmHg.
[From "CRC Handbook of Chemistry and Physics," 55th ed. Reproduced with the permission of CRC Press, Boca Raton, Florida.]

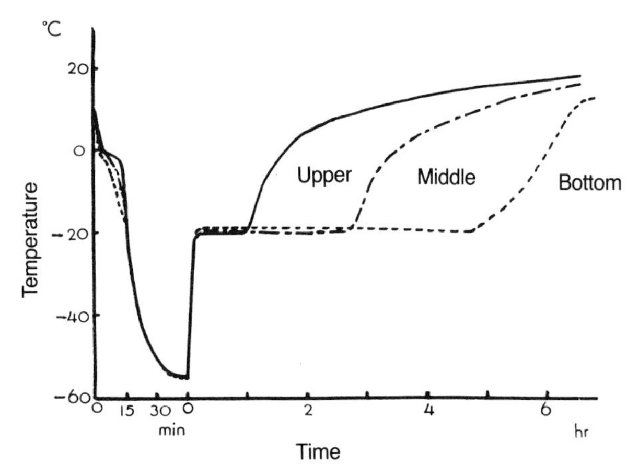

Figure 4 Progress of the specimen temperature during the drying of *Saccharomyces cerevisiae* cells at room temperature (20°C). [From Araki, T., and Nei, T. (1961). *Low Temp. Sci.* **B-19,** 43–47.]

remains and a heating denaturation of the products in the part already dried. Practically, throughout the sublimation stage, a nearly constant subzero temperature is held in the portion of the specimen in which the bulk of the ice crystal remains, but in the portions where drying surfaces passed through, the specimen temperature goes up quickly in accordance with ambient temperatures (Fig. 4). In a manifold-type dryer, in which the products' ampoules are exposed to moist air, an aqueous vapor freezes at the part of the vessel surface where the ice crystal still remains. Deposited frost layer significantly retards the sublimation process, with the inhibition of heat flow from atmosphere to drying products. This results in a longer term exposure of the portions of the product, where ice crystals disappeared, to atmospheric temperature. To promote the sublimation rate while avoiding such disadvantageous circumstances, a water bath of appropriate temperature is recommended. In the case of chamber drying, specimens are usually heated from the bottom by placing the vessels on shelves that are equipped with electric heating systems.

Changing the ambient temperature by using a water bath or some other device affects the specimen temperature as follows. Warming of the specimen gives rise to a slight increase of the specimen temperature. The difference between ambient temperature and specimen temperature clearly increases, indicating that the sublimation rate of ice is promoted by heating. In contrast, cooling the specimen to below −40°C decreases the specimen temperature according to the lowering of the ambient temperature. The difference between the ambient temperature and specimen temperature is reduced accordingly in these lower temperature regions, indicating that the sublimation rate is repressed in these temperature regions (Fig. 5). This result is also suggested by a steeper decrease of aqueous vapor pressure below −40°C (Table I).

To achieve an effective warming of the specimen without overheating, which brings about a deleterious effect on the nature of the products, monitoring of the specimen during the sublimation period has always been a point of great concern. Many different systems that have a relation to temperature measurements have been proposed. Considering that in many cases the temperature alone does not indicate much, one researcher introduced a special monitoring procedure. In this device, heat and vapor pressure in the system are automatically controlled by the electric characteristics of the products. When heating the products brings interstitial softening, the electric resistance drops sharply. It triggers the reduction of heating and leads to better vacuum. When

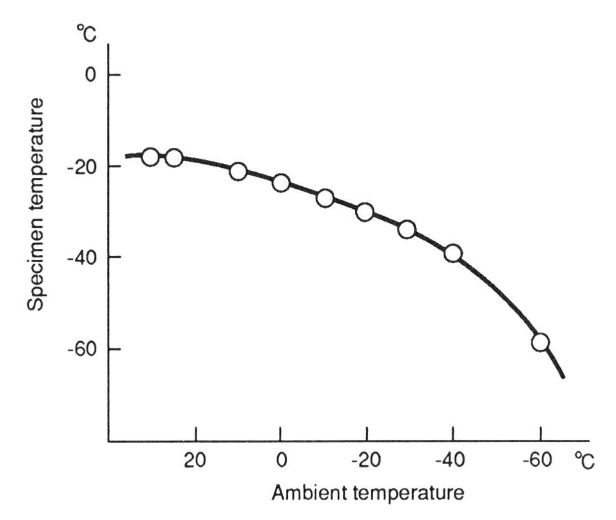

Figure 5 Relationship between specimen temperature and ambient temperature during freeze-drying of aqueous suspension of *Escherichia coli* cells. [From Nei, T., Souzu, H., and Araki, T. (1964). *Contrib. Inst. Low Temp. Sci.* **B-13**, 14–26.]

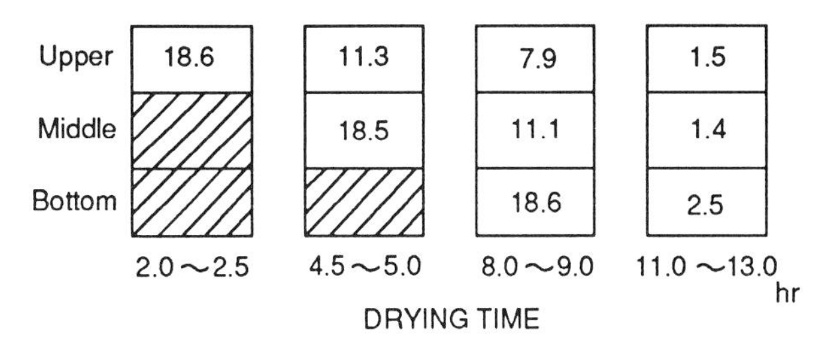

Figure 6 Variation of the residual moisture content at the different parts of the specimen during the drying of aqueous suspension of *Saccharomyces cerevisiae* cells. The residual moisture content is obtained in the equation $A - B/B \times 100$, where A = the weight of the specimen immediately after drying, and B = the weight of the specimen, which is obtained after drying of an additional 3 hr at 60°C. Hatched part shows the bulk of ice. [From Nei, T., Souzu, H., Hanafusa, N., and Araki, T. (1961). *Low Temp. Sci.* **B-19**, 59–72.]

the electric resistance climbs again, heating is increased to stimulate the sublimation rate again.

After some period of the primary drying process, in the part of the product where the bulk of ice crystals have already sublimated, the energy requirement drops steeply and the specimen temperature rises in accordance with the ambient temperature. In this situation, the residual moisture in the part just behind the sublimation surface might represent a substantial share of the initial content (approximately 15%). In the process of the so-called secondary drying stage, the residual moisture level decreases gradually from the portions farthest from the remaining ice crystal. However, still higher moisture remains in the specimen just after the ice crystals disappear completely (Fig. 6). The presence of these moisture levels does not allow a long-term preservation of the material, and they should be extracted by successive drying.

The secondary drying stage is also endothermic, and specifically devised heating procedures must be applied to the product, which is only partially dried and is extremely sensitive and should not be overheated. The energy supply at this stage to the required portions through the porous system under vacuum is very difficult to manage, because both have excellent insulating properties. Merely heating from the outside tends to overheat its surroundings, and the energy will not reach the portions where a true energy supply is required. Microwaves or infrared beams with a wavelength adequate to penetrate the products might render some improvements for reaching and hitting their specific targets, namely the adsorbed water molecules.

Another difficulty arises as to how to determine the end-point of the process. The higher moisture contents in the products will give damaging effects during storage, but overdrying of living cells will result in a serious disruption of the structure of the cellular materials with accompanying loss of their activity and, finally, to the reduction of the viability of the cells.

For the adequate determination of the end-point of this process, many procedures have been devised including pressure rise measurement, the Karl Fischer method, weight curves, vapor tension, moisture equilibrium, nuclear magnetic resonance, and dielectric measurement.

D. Storing the Products

The storage of dried products should be performed under ideal conditions, in such a way that the efforts accomplished in the previous processes can be prolonged over a long time. This step involves serious hazards, and the processing must be carried out under extremely cautious manipulations to avoid the undesirable contamination by such items as oxygen and moisture. The products usually are packed or sealed under vacuum or under an atmosphere of the dry inert gas—for instance, nitrogen or argon. These gases are introduced directly into the drying chamber or into ampoules at the end of the drying process to break the vacuum.

Although freeze-dried products can theoretically withstand storage over a wide temperature range, storage in a moderate temperature range will bring some additional amendments to the cells viability over a longer storage period. Light is also highly detrimental, and storage in a dark area or in an opaque or dark-colored container is recommended.

E. Reconstitution

In most cases, rehydration is performed by the mere addition of an exact amount of aqueous phase that was previously extracted from the system. This rapid rehydration has an advantage in that the products can avoid long-term exposure to concentrated solution. However, in the case where such rapid rehydration gives rise to deleterious results, the specimens are first rehydrated by a hypertonic solution, and then the solution is diluted by dialysis, changing the tonicity of the outer solutions gradually in the utilization of the appropriate buffering solutions, which contain a reducing amount of the solutes. In another instance, the specimen is first left standing to absorb airborne moisture at room temperature for an appropriate period, and then the aqueous phase is added to regain approximately the same concentration as in its original state.

The rehydration temperature is also a matter of consideration. Generally, rehydration is carried out by the addition of distilled water at room temperature to the dried products. In some instances, rehydration may be carried out in low temperatures to reduce the activities of the deleterious chemical and/or enzymatic reactions that will inevitably be introduced to the products following rehydration. However, this procedure cannot be used in specimens that are sensitive to chilling injury.

Another important point in which care should be taken is the differentiation of pH in the rehydrated suspension from the original ones. Because the solutes in the specimen remain exactly the same as those in dried materials, the reconstituted products will theoretically be restored to their original pH value. However, it is noteworthy that, in many instances, the materials leaking out of dead or injured cells considerably affect the pH of the reconstituted products.

When the reconstituted specimens have different compositions from the original ones, in the change of pH value or by an addition of enzyme inhibitor(s), or the reconstitution is accomplished with different amounts of fluid to obtain more concentrated or more diluted cell suspensions, the different osmolarity of the suspension media also causes a serious effect, especially on injured cells.

III. Cellular Damages Occurring in the Processes of the Operation

The previously described cellular damages that result from each individual processing of the operation are, of course, not restrictively brought about by these steps alone. The damage that is considered to have resulted from dehydration might have been initiated in the freezing stage. The occurrence of the chemical reactions, which are induced in the state of lower water activity, will progress throughout the period from the early drying stage to the storage stage. The effect of the rate of freezing appears just in the stage of rehydration. However, the damages are also mainly brought about in a characteristic fashion of each process. In this section, the damages that are considered to be mainly related to each operational step will be described in turn in the processes of the operation.

A. Damage Resulting from Freezing

Because the freezing stage is successively followed by a drying process, the extent of the damage cannot be clearly ascribed to this stage. However, a very serious portion of cellular damages are considered to be brought about in the freezing stage. The factor that mostly effects cellular damages in this stage is the rate of freezing. For instance, in the freeze-thawing of aerobically or anaerobically cultivated *Escherichia coli* cells, the viability reduction in both of the cells increases when the cooling rate exceeds 10°C min. The viability reduction of nonaerated cells increases more rapidly according to the increase of the freezing rates, compared to that of the aerated cells (Fig. 7). Electron microscopic observation has demonstrated that in the rapid freezing of aerobically cultivated specimens a number of cells shrunk but only a few cells were frozen intracellularly. In specimens that are nonaerobically cultivated, on the contrary, most of the cells are frozen intracellularly. The results indicate that nonaerobically cultivated cells, which have a low permeability of membrane to the water molecules, are more sensitive to rapid freezing. On the other hand, aerobically cultivated cells are tolerable to higher freezing rates, certainly

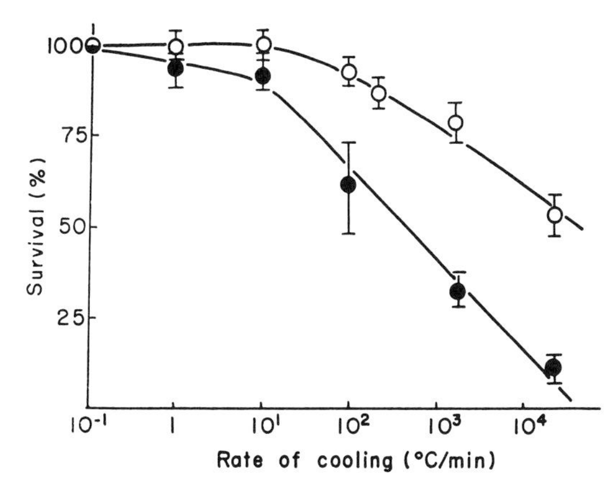

Figure 7 Survival rate of aerated (open circle) and nonaerated (solid circle) cultures of *Escherichia coli* cells frozen at various rates. [From Nei, T., Araki, T., and Matsusaka, T. (1969). Freezing injury to aerated and non-aerated cultures of *Escherichia coli*. *In* "Freezing and Drying of Microorganisms" (T. Nei, ed.), pp. 3–15. University of Tokyo Press, Tokyo.]

by the higher permeability of their membrane to the water molecules.

The composition of suspension medium is also a potentially effective factor on the freezing damage of *E. coli* cells. Among the many chemical compounds, alkaline metal salts are known to bring about an extremely lethal effect to the cells during the freezing and storing periods. It has been demonstrated, however, that Na-glutamate gives a remarkable protective action against the damage even in the presence of alkaline metal salts. Specimens just after freezing contain a significant portion of metabolically injured but survived cells among the dead cells. The percentage of the injured survivors are larger in the early stages of storage and decrease as the storage period is prolonged. Although numerous factors are considered to cause the death of the injured cells during the storage period, the deleterious metabolic and/or enzymatic reactions that progress even at storage temperatures might be targeted as one of the factors.

The deleterious enzymatic reaction in which activity appears after freezing have also been known in yeast cells. In cells just after thawing, the extractability of cellular phospholipids increases considerably. More than 30% of the cells still survived in this stage. During the following incubation of the thawed cell suspension at pH 4.4 for 30 min, however, a rapid phospholipid degradation progresses in the cells, and concomitantly the viability of the cells is reduced significantly. On the other hand, in the incubation of a similar specimen at pH 6.0 for the same period, a slight phospholipid degradation is observed and the cell viability is held nearly at the same level of that just after thawing. The results indicate that the enzymatic degradation of phospholipid, which takes place in the freeze-thawed cells, enhances the reduction of cell viability, and cellular death can be prevented by the repression of enzyme activity. Similar results are also obtained in the freeze-dried and rapidly rehydrated cells of the same species.

The concentration of the cells in the suspension also significantly affects the extent of freezing damage of *E. coli* cells. In a comparison of viability obtained by the same freezing protocol but in the different concentrations of the cells (which are approximately 0.5×10^7 cells/ml and approximately 5×10^9 cells/ml), the specimens with higher cell concentration showed a significantly advantageous result. The results suggest that the activity, either protective or injurious, of substances released from the cells might affect the extent of metabolic injury at higher cellular concentration.

B. Viability Reduction during the Drying Stage

The viability reduction of the cells in the freeze-drying operation is considered to be concentrated mostly on the secondary drying stage. The evaporation of unfreezable water molecules, which adsorb firmly to the cellular constituents and work for the stabilization of their structure, corresponds to this stage. The secondary drying process commences immediately after in the portion of the specimen where a drying surface has passed through. And, indeed, the reduction of the cell viability in these parts is known to increase considerably while the sublimation of ice is still progressing actively.

The numerous experimental results have suggested that the removal of unfreezable water portions of the cells are lethal to the cells. In an attempt to freeze-dry, the lowest residual water content in the specimens is required to achieve the best results for long preservation. However, a minimal amount of water is necessary to keep up the structural stability of cellular constituents or macromolecules and, accordingly, the cell viability. Thus, the viability reduction of the cells in this stage should be mainly brought about by overdrying. The experimental results have suggested that certain portions of the cells have already become overdried before the bulk of the ice crystal has completely disappeared (Fig. 6). To overcome these circumstances, care should be taken to prepare the specimen that can be dried homogeneously. Specimens with the thinnest layer and larger surface area may give some improvements. Another advantageous precaution is drying the specimens in a moderate vacuum and higher temperatures within the limits of possibility to avoid melting. In these conditions, the sublimation of ice crystals is quicker compared to the results obtained with lower pressure and lower temperature. Hence, an overdrying rate will be reduced considerably.

Overdrying will not only result in an impairment of cellular constituent construction but also trigger the causes of the oxidation reaction of lipids or proteins, which are considered to have been protected by the presence of hydrated water layers. There are injuries due to some specific chemical reactions that are brought about in the cells of lower water activity: These include aminocarbonyl reaction, free-radical generation, and unbalanced metabolism.

C. Damage Due to Rehydration

In general, the dried products are rehydrated by the addition of distilled water at room temperature. However, by such rapid reabsorption of water molecules, dried materials expand abruptly, and the disruption of their membrane structure may arise. The rapid rehydration also might result in inaccurate rearrangement of the membrane constituents. In experiments using *Saccharomyces cerevisiae* cells, it was demonstrated that the rapid rehydration of the dried cells with the addition of distilled water may bring about a degradation of phosphorus compounds including phospholipids. The extent of phosphate degradation is quite similar to the results that occur in a rapid freeze-thawing of the specimen. In contrast, slow rehydration of the specimen by absorption of the airborne moisture before the addition of water shows no phosphate compound degradation. The specimen that is rapidly frozen previous to drying shows a similar phosphate compound degradation even in a slow rehydration. The rate of the phosphate compound degradation is repressed significantly by incubation of the specimens in a pH range in which the phosphate compound degradation enzymes are inactive. Rapid cooling and rewarming of slowly rehydrated specimens shows similar phosphate compound degradation, denying the possibility that the enzyme activity is dismissed during the absorption of the moisture.

The hydration temperature is another important factor. One research group found that the maximal recovery of freeze-dried *Lactobacillus bulgaricus* cells was facilitated by rapid rehydration at 20°–25°C, and the maximum level of recovery varied depending on the presence of protective additives. Another research group (1985) showed a somewhat higher temperature range in which the maximum recovery of freeze-dried microorganisms of several species is attained (between 30° and 40°C). The recovery rates were lower in both higher and lower temperature regions. The rehydration temperature range that gives rise to maximum recovery might be related to the rigidity or morphism of cellular materials, especially in the phospholipid-containing systems.

IV. Methods to Overcome the Damages

A. Utilization of Protective Substances

When freeze-drying microorganisms, appropriate protective substances are indispensable for achieving sufficient results throughout the operation. The chemical characters of the protective substances widely utilized in this operation have something in common with those of the substances utilized in the freeze-storing of similar materials. The protective substances applied to freeze-storing are classified into three groups. The first group consists of glycerol, dimethylsulfoxide, etc., which can pass the cellular membrane freely. These substances have a strong hydrogen-bonding character and have been considered to exert their activity by modifying the ice crystal-forming fashion in the systems. Substances in the second group are sugars, containing glucose, sucrose, lactose, etc., and they also have a strong hydrogen-bonding character. These substances can also permeate the cellular membrane of the microorganisms in some cases; however, the permeability of the cellular membrane of the organisms are generally not high, compared with the substances belonging to the first group. Substances in the third group are proteins containing serum, gelatin, mutin, albumin, etc. These substances cannot pass the cellular membrane but strongly prevent freezing injury to the cells.

The protective substances commonly used in

freeze-drying are classified roughly into low-molecular weight compounds such as amino acids, organic acids, sugars, and sugar alcohol and high-molecular weight substances such as proteins, polysugars, polyvinylpyrrolidone, and synthetic polymers. Glycerol or other alcohols known as good protective agents for freezing injury cannot be used in the freeze-drying procedure because of their hygroscopic property.

It has been suggested that the low-molecular weight protective substances utilized in the freeze-drying operation have a common character in which each molecule contains a total of more than three of either hydrogen-bonding functions (-OH, $-NH_2$, =NH, =O) or ionic function groups (either acidic or basic) in a specifically organized structure. The physicochemical properties, characteristic of the high-molecular weight protective substances, are still not clarified. The action of these substances are conjectured to protect specifically the cell-surface structure from damage.

The utilization of these high-molecular weight substances can protect the cells somehow from the damage in the freeze-drying process; however, the restrictive use of these high-molecular weight substances alone is not considered to be sufficient for the protection of the cells from the damage that occurs in the storage period. Combined use of low- and high-molecular weight substances is indicated both in the promotion of the rate of drying and in the maintaining of higher stability of the cells in drying and storage periods (Table II).

It is suggested that the viability reduction of the cells during the drying process is mainly brought about by overdrying. The action of protective substances might be conjectured to be the fact that these substances exert a force to hold the minimal amount of the cellular water that is essential for the maintenance of the structural stability of the cellular constituents; however, no difference in the extent of hydration (mol H_2O/mol, 20°C) has been observed between the amino acids that have a higher protective effect and those that are noneffective. Besides, in an experiment using egg albumin as a model, a similar or rather reduced amount of residual water in the molecules, dried in the presence of protective substances (sugars and amino acids), was seen, compared to the results obtained without the use of these substances. It is suggested from the results that the protective action of these substances is by no means owed to holding the water, which is necessary to maintain the structure of the cellular constituents, but these substances are substituted for the water molecules and work to stabilize the structure of the cellular constituents at the absorbing sites. The site that is most positively protected will be the

Table II Protective Effect of Na-Glutamate and High-Molecular-Weight Substances in Freeze-Drying of *Lactobacillus Bifidus* Cells

	Survival of the cells ($\log_{10}$)			
	Freeze-dried		Stored at 45°C	
Protectants	Before	After	1 Mo	3 Mo
3% Na-glutamate	10.5	9.9	4.9	1.2
3% Soluble starch	10.5	8.5	0	—
3% Soluble starch + 3% Na-glutamate	10.5	9.9	7.8	7.4
3% Polyvinylpyrrolidone (PVP)	10.3	9.5	0	—
3% PVP + 3% Na-glutamate	10.5	9.9	8.5	8.0
3% Na-alginate	10.2	8.9	0	—
3% Na-alginate + 3% Na-glutamate	10.3	9.5	6.3	6.3
3% Dextran	10.5	9.4	0	—
3% Dextran + 3% Na-glutamate	10.4	9.9	8.0	5.2
1% Na-carboxymethylcellulose (CMC)	9.6	8.6	0	—
1% CMC + 3% Na-glutamate	9.8	9.0	7.0	5.5
3% Gelatin + 3% Na-glutamate	10.0	9.2	3.0	0

[From Obayashi, Y., Ota, S., and Arai, S. (1961). *J. Hyg., Cambridge* **59,** 77–91.]

cellular membrane, because many substances cannot pass the cellular membrane but, rather, exert a protective action to the cells. Indeed, the cellular membrane is considered to be the site that suffers from the damage most in the freezing and drying of microorganisms. [*See* CELL MEMBRANE: STRUCTURE AND FUNCTION.]

It is also known that, in some instances, the oxidation reaction, which proceeds in the product during the secondary drying and a subsequent storage period, leads to cellular death due to the generation of active per-oxi radicals. The viability reduction in dried bacterial products equally progresses with free-radical generation. The rate of free-radical generation depends on the presence of oxygen molecules, and it is believed to inhibit a flavin-linked enzyme activity. In such a case incorporation, in the beginning of the procedures, of an appropriate free-radical scavenger or antioxidants such as monosodium glutamate is recommended to protect the products from the oxidation reaction that inevitably would occur in the products of a dried state.

Table III Viability of Freeze-Thawed or Freeze-Dried *Escherichia coli* Cells Cultivated at Different Temperatures to Differing Growth Phases

Cultivations for:		Survival of the cells (%)		
		Freeze-thawed		Freeze-dried
°C	Hr[a]	−5°C	−30°C	
	2*	4.0	3.8	4.0
37	3**	52.3	50.4	24.9
	5***	93.9	92.5	38.6
	8*	6.2	5.7	0.2
17	20**	52.2	35.5	8.2
	48***	82.8	51.4	47.8

[a] Cells growing in logarithmic (*), early stationary (**), and stationary phase (***).

[From Souzu, H. (1985). The structural stability of *Escherichia coli* cell membranes related to the resistance of the cells to freeze-thawing and freeze-drying. *In* "Fundamentals and Applications of Freeze-Drying to Biological Materials, Drugs and Foodstuffs," pp. 247–253. International Institute of Refrigeration, Commission C1, Tokyo.]

B. Availability of the Growth Environments to the Tolerance of the Cell to Freezing and Drying

Cells growth phases and growth environments have a close relation to cell tolerance to freezing and also to drying. In general, logarithmically growing cells are very susceptible to freezing and freeze-drying. The resistivity of the cells increases as cell growth approaches its stationary phase. For instance, *E. coli* cells grown in two different temperature regions showed a similar resistivity increase to freeze-thawing and to freeze-drying, as cell growth phases progressed from logarithmic to early stationary and to the stationary phase, although the cultivation length necessary to approach growth phases differs considerably between these temperatures (Table III). The membrane structural stability is also known to increase according to the progress of cell growth phases.

A shift-up of the growth temperature of *E. coli* logarithmic phase cells from 30° to 42°C and the subsequent incubation at 42°C for 40 min increased the resistivity of the cells to drying significantly, compared to those of nonshift-up cultures. The cause of these results have been conjectured to be the heat-shock proteins, which are induced by a shift-up of growth temperature and might confer the stability of DNA and, subsequently, the viability increase of the cells in freeze-drying.

The growth circumstances of the cell also affect the resistivity of the cells to different rates of rehydration. The cells grown under aeration show a good recovery to both rapid and slow rehydration after slow dehydration. In contrast, the cells grown nonaerobically show a similar recovery in slow rehydration but a significantly reduced recovery in rapid rehydration. The result will affect the rehydration rates after freeze-drying of the cells grown in these two different environments.

V. Concluding Remarks

In the field of microbiology, the freeze-drying operation is mostly utilized for the long preservation, and subsequently for reproduction, of the species. To attain this objective, the products should be restored to their original state. In many instances, cell recovery decreases significantly by freeze-drying procedures. For obtaining higher recovery of the cells, the utilization and the selection of appropriate protective substances are inevitable. In general, the selection of the freeze-drying equipment itself is not so easy. However, many opportunities to choose the assortment of protective substances exist. The effort to find the effective protective substances for each individual case, and to establish a generalized definition to the protective action of these substances is of great importance.

Bibliography

DeLuca, P. P. (1985). Fundamentals of freeze-drying pharmaceuticals. *In* "Fundamentals and Applications of Freeze-Drying to Biological Materials, Drugs, and Foodstuffs," pp. 79–85. International Institute of Refrigeration, Commission C1, Tokyo.

Hanafusa, N. (1985). The interaction of hydration water and protein with cryoprotectant. *In* "Fundamentals and Applications of Freeze-Drying to Biological Materials, Drugs and Foodstuffs," pp. 59–64. International Institute of Refrigeration, Commission C1, Tokyo.

Mackenzie, A. P. (1985). A current understanding of the freeze-drying of representative aqueous solutions. *In* "Refrigeration Science and Technology: Fundamentals and Applications of Freeze-drying to Biological Materials, Drugs and Foodstuffs" (International Institute of Refrigeration, ed.), pp. 21–34. International Institute of Refrigeration, Paris.

Mazur, P. (1970). *Science* **168,** 939–949.

Obayashi, Y., Ota, S., and Arai, S. (1961). *J. Hyg., Cambridge* **59,** 77–91.

Rey, L. R. (1963). "Princips generaux de la liophilisation et l'humidite residulle des produits lyophilises." Publications de la Sociedad Espanola de Farmacotecnica, Barcelona.

Souzu, H. (1985). The structural stability of *Escherichia coli* cell membranes related to the resistance of the cells to freeze-thawing and freeze-drying. *In* "Fundamentals and Applications of Freeze-Drying to Biological Materials, Drugs and Foodstuffs," pp. 247–253. International Institute of Refrigeration, Commission C1, Tokyo.

Takano, M., Takemura, H., and Tsuchido, T. (1985). Freeze-drying tolerance of *Escherichia coli* recA mutants caused by growth temperature shift. *In* "Fundamentals and Applications of Freeze-Drying to Biological Materials, Drugs and Foodstuffs," pp. 279–284. International Institute of Refrigeration, Commission C1, Tokyo.

Willemer, H. (1985). Freeze-drying and advanced technology. *In* "Fundamentals and Applications of Freeze-Drying to Biological Materials, Drugs and Foodstuffs," pp. 201–207. International Institute of Refrigeration, Commission C1, Tokyo.

Willemer, H. (1987). Additional independent process control by process sampling for sensitive biomedical products. *In* "International du Froid," Vol. C, pp. 146–152. International Institute of Refrigeration, XV-II Congress, Wien.

Gastrointestinal Microbiology

Julie Parsonnet
Stanford University

Glossary

Cytotoxin Substance that inhibits or prevents cell function or destroys the cell

Dysentery Diarrheal illness with bloody stools, abdominal pain, and fever

Enterotoxin Cytotoxin that specifically attacks intestinal epithelial cells causing altered transport of ions and excessive secretion of fluids out of cells

Gastric Pertaining to the stomach

Infection Host damage or illness caused by the presence of a microorganism

Intestinal epithelium Thin layer of cells covering the luminal surface of intestinal organs

Normal flora Microorganisms that customarily reside in a host without causing tissue damage or illness

Pathogen Microorganism that causes tissue damage in the host that may result in illness

Peristalsis Wavelike motion of intestines that propels food down the gastrointestinal tract

Virulence Power of an organism to cause disease

THE GASTROINTESTINAL OR DIGESTIVE TRACT (Fig. 1) is a long, convoluted tube, open at both ends to the environment, that serves both as a barricade and a conduit to materials from the outside world. The tube begins in the mouth and continues through the esophagus, stomach, small intestine, and large intestine before it terminates at the anus. The gastrointestinal organs comprise distinct microbial ecosystems within which thrive an enormous diversity of organisms. Many of these organisms could be termed *normal flora,* flora that live symbiotically in the digestive tract, preventing colonization with more harmful infectious agents and occasionally supporting the host with metabolic functions. Gastrointestinal pathogens, however, ranging from viruses to worms, are all too common and plague each and every person with variable frequency throughout his or her lifetime. Some of these pathogens (e.g., the bacterium *Helicobacter pylori* and many intestinal parasites) are so ubiquitous and benign that they could almost be termed *normal flora*. Furthermore, as our understanding of the molecular aspects of gastrointestinal microbiology advances, it becomes apparent that the organisms we once called normal nonpathogens (e.g., *Escherichia coli* and yeasts) may not be innocuous under certain conditions. Through these changeable host–parasite interactions, we gain an understanding of both macro- and microorganisms.

I. Esophagus

A. Esophageal Ecosystem

The esophagus is a muscular tube that connects the throat (pharynx) with the stomach (Table I). Everything swallowed, whether it be food or saliva, passes through the esophagus, exposing the internal sur-

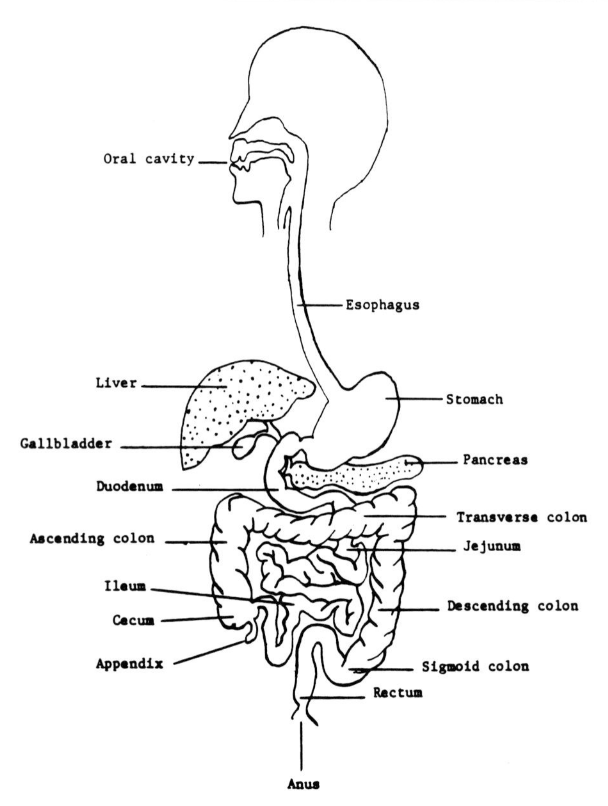

Figure 1 Human gastrointestinal tract.

face to organisms that survive the antimicrobial systems within the mouth. Most microorganisms that strive to cling to the relatively smooth esophageal epithelium are washed away in the relentless waves of foods, liquids, and oral secretions flowing into the stomach. Yet, at low levels, the esophagus can host many different anaerobic and aerobic bacterial species. Normal flora include aerobic and anaerobic streptococci, hemophilus, *Neisseria,* Enterobacteriaceae, bacteroides, and yeasts.

Table I Esophageal Ecosystem

Normal environment	
Neutral or slightly acid pH	
Smooth epithelium	
Vigorous peristalsis with frequent rapid passage of food and saliva	
Normal flora	
Quantity:	$<10^3$ organisms/ml
Variety:	similar to oropharyngeal flora
Common infections	
Viruses:	cytomegalovirus[a]
	herpes simplex[a]
Fungi:	yeasts[a]

[a] Seen predominantly in immunosuppressed or debilitated persons.

Infections of the esophagus rarely occur in the normal host. In immunosuppressed patients, however (e.g., in acquired immunodeficiency syndrome, cancer, or transplant patients), viral and fungal pathogens can cause serious and debilitating disease.

B. Yeasts

The most common infection of the esophagus are caused by yeasts of the *Candida* family. These organisms are frequent colonizers of the gastrointestinal tract and could be considered part of the normal flora. In certain debilitated or immunosuppressed persons, however, yeasts can cause extensive disease with invasion of tissue and marked inflammation.

Patients with candidal esophagitis may be asymptomatic; those with symptoms have burning pain on swallowing. Fiberoptic examination of the esophagus reveals characteristic white plaques surrounded by red inflamed tissue. Diagnosis of candidiasis can be made by 10% KOH preparation of material from the plaque or, more specifically, by culture of the plaque. Numerous topical and systemic antifungal agents are used to treat candidal esophagitis.

C. Viral Infections

Cytomegalovirus (CMV) and herpes simplex virus (HSV) are the two most common viral infections of the esophagus but only cause esophageal disease in immunosuppressed hosts. These large ubiquitous DNA viruses, both members of the herpesvirus family, usually infect children asymptomatically and thereafter remain latent within cells for a lifetime. Esophageal infections that occur in immunosuppressed hosts result from reactivation of latent virus. [*See* HERPESVIRUSES.]

Both CMV and HSV cause inflammation of the esophagus (esophagitis) that can range from mild to severe with ulcer formation. Although many patients have no symptoms, some develop intense, burning pain on swallowing. Definitive diagnosis is made by viral culture of tissue obtained by biopsy. Treatment with dihydroxypropoxymethyl guanine

(CMV) or acyclovir (HSV) may be advised in some patients.

II. Stomach

A. Gastric Ecosystem

The stomach is a hostile environment for microorganisms and acts as a barricade to entry of infectious agents into the lower gastrointestinal tract (Table II). In the normal adult, the stomach is extremely acid with a pH of less than 3, a pH that kills most microbial agents. Gastric enzymes further digest microorganisms while peristalsis propels survivors down the gastrointestinal tract. With these formidable antimicrobial defenses, the stomach only transiently contains low colony counts ($<10^3$) of relatively acid-resistant flora (yeasts, streptococci, lactobacilli, and oral anaerobes). If gastric acidity is diminished, however (e.g., with pernicious anemia, surgical excision of the acid-producing cells, or normal aging), the stomach can be colonized by more permanent bacterial flora including Enterobacteriaceae and anaerobes more typically found in the lower parts of the gastrointestinal tract.

Table II Gastric Ecosystem

Normal environment	
Highly acid pH	
Glandular epithelium	
Contains enzymes that digest proteins	
Normal flora	
Quantity:	few ($<10^3$)
Variety:	lactobacilli
	streptococci
	yeasts
	oral anaerobes
Common infections	
Viruses:	cytomegalovirus[a]
Bacteria:	*Helicobacter pylori*
Fungi:	yeasts[a]
Parasites:	Anisakidae

[a] Seen predominantly in immunosuppressed or debilitated persons.

B. *Helicobacter pylori*

One pathogen that has learned to adapt to the acidic ecosystem of the stomach is *Helicobacter pylori*, a biflagellate gram-negative rod that lives between the mucous layer that protects the gastric epithelium and the epithelium itself. One salient feature of the organism is its ability to produce large quantities of urease, an enzyme that splits urea into ammonia, CO_2, and water. The organism thus surrounds itself with a basic ammonia cloud, protecting it from gastric acid until it reaches the relatively acid-neutral haven underneath the mucous layer. Although *H. pylori* never invades the epithelial mucosal cells, infection is strongly associated with submucosal inflammation (gastritis) presumably mediated by *H. pylori* products that traverse or are transported across the epithelial cell layer.

Helicobacter pylori is extraordinarily common, infecting approximately 40% of adults in industrialized countries and 70% in developing countries. Transmission is felt to be by person-to-person spread, and no environmental source is known. Once infection occurs, it does not spontaneously resolve; untreated patients probably remain infected for their lifetimes. Most persons chronically infected with *H. pylori*, however, suffer no symptoms, although in a small proportion of infected hosts, chronic infection has been linked to the development of duodenal ulcers, gastric ulcers, and certain types of stomach cancer. Diagnosis of *H. pylori* is made by serologic testing, by histologic examination of biopsy material, or by culture of gastric tissue under microaerophilic conditions. *Helicobacter pylori* can be successfully eradicated with a combination of bismuth compounds and antibiotics.

C. Anisakidae

Parasitic infestation of the stomach is rare but is known to occur with ascarids of the worm family Anisakidae. The third-stage larval phase of these worms are found in raw, mildly salted, or pickled fish or squid. Human infestation can be acquired by eating dishes such as ceviche, sashimi, or gravlax. The ingested larvae tunnel into the stomach (less commonly into the small intestine or colon) and rapidly causes allergic, inflammatory swellings. Although a proportion of infected hosts will be asymptomatic, some persons will develop severe stomach pain, nausea, and vomiting between hours to weeks after ingestion. Diagnosis is made by patient history and, more definitively, by identification of larvae in the tissue. Symptoms normally subside without treatment, although on occasion the worm needs to be surgically excised.

D. Other Gastric Infections

In immunocompromised hosts, the stomach can become infected with CMV and *Candida* species in much the same manner as the esophagus. Infection of the stomach by either of these two agents usually reflects extensive disease throughout the gastrointestinal tract. Viral or yeast infections of only the stomach are rare.

III. Small Intestine

A. Ecosystem of the Small Intestine

The 12–20-foot-long small intestine (Table III) connects the stomach to the large intestine and contains three sequential sections; the part nearest the stomach is called the *duodenum,* followed by the *jejunum,* and finally the *ileum.* The primary functions of the small intestine are the digestion of food (predominantly in the duodenum and jejunum) and the absorption of dietary nutrients (all sections). To this end, the small intestine receives digestive enzymes and bile from the pancreas, liver, and gallbladder. The bicarbonate in these secretions counters gastric acid and creates a neutral environment in which digestion and absorption occur.

Because of the hostile barrier created by the stomach, as well as the peristaltic and digestive nature of the intestine itself, the first part of the small intestine contains relatively few ($<10^3$/ml) streptococci, lactobacilli, oral anaerobes, and yeasts. This amount gradually increases (10^7 to 10^8/ml) as the small bowel approaches the junction with the large bowel; fecal anaerobes (*Bacteroides, Bifidobacterium*) and Enterobacteriaceae begin to predominate. Although normal bacterial flora may enhance the digestive process by production of metabolic enzymes, bacterial overgrowth, which may occur in the absence of good intestinal motility, can result in poor absorption of essential nutrients. However, complete eradication of normal small bowel flora can eliminate a major source of vitamin K and several B vitamins in malnourished hosts. Because the bulk of gut normal flora live in the large intestine, its origins and metabolic activities will be discussed in Section IV.A.

Table III Small Intestinal Ecosystem

Normal environment	
Moderately basic pH	
Glandular epithelium	
Contains enzymes that digest proteins and sugars	
Vigorous peristalsis	
Normal flora	
Quantity:	few in duodenum ($<10^3$) with progressively more organisms distally to the ileum ($\sim10^8$)
Variety:	duodenum and jejunum, similar to stomach normal flora; ileum, similar to large intestinal normal flora
Infections	
Viruses:	rotavirus
	Norwalk agent and other small round viruses
	astrovirus
	adenovirus
	coronovirus
	enteroviruses
	cytomegalovirus[a]
Bacteria:	*Salmonellae*
	enterotoxigenic *E. coli*
	enteropathogenic *E. coli*
	Camplyobacter sp[b]
	Yersinia entercolitica
	Vibrio cholera
	Vibrio parahaemolyticus
Parasites:	*Giardia lamblia*
	Cryptosporidia[c]
	Isospora belli[c]
	ascaris
	hookworm
	Strongyloides stercoralis
	Fasciolopsis buski
	Taenia sp
	Hymenolepsis nana

[a] Predominantly in immunosuppressed hosts.
[b] Also infects large intestine.
[c] In the United States, most commonly seen in immunosuppressed hosts, particularly those with acquired immunodeficiency syndrome.

B. Viral Infections

1. Rotavirus

Rotavirus, a double-stranded RNA virus in the family Reoviridae, is the most common cause of diarrhea in infants and young children. The virus is well suited for small bowel infections; it is relatively resistant to destruction by gastric acid, and its infectivity is enhanced by intestinal enzymes such as trypsin, elastase, and pancreatine. Rotavirus infects epithelial cells of the duodenum and jejunum, causing inflammation, decreased intestinal surface area, and decreased absorptive capacity. Rotavirus infection also results in insufficient production of disaccharidases, enzymes that break down two-component sugars. Undigested disaccharides accumulate in the bowel, precipitate osmotic secretion of water, and thereby cause profuse, watery diarrhea.

Rotavirus diarrhea is largely a disease of infancy; almost all children have been infected at least once by the age of 5 years. Although adults may occasionally become ill, naturally acquired immunity frequently prevents clinical manifestations of disease. Most rotavirus infections occur in colder months, hence the nickname *winter diarrhea*. Transmission is felt to be person-to-person by an oral–fecal route, although there is increasing evidence for a component of respiratory transmission.

The hallmarks of rotavirus infection are vomiting, watery diarrhea, and low-grade fever lasting 2–8 days. Young infants can become dehydrated quickly; in developing countries with high rates of malnutrition and low access to medical care, rotavirus infection is the leading cause of death from diarrheal disease. The organism is not easily cultured, and diagnosis is usually made by direct immunoassay of stool. There are no specific antiviral therapies. Oral or intravenous rehydrating solutions containing water, glucose, and salt can be life-saving. Currently, rotavirus vaccines are being developed.

2. Caliciviruses and Norwalk Agent

Caliciviruses are small round viruses that cause outbreaks of vomiting and watery diarrhea. These single-stranded RNA viruses have never been successfully cultured, and their mechanism of pathogenesis is not well understood. The best known of the caliciviruses, the Norwalk agent, has been implicated in numerous diarrhea outbreaks worldwide. Biopsies from the small bowel of infected patients show structural changes in the epithelium, and physiologically, there is evidence of decreased absorption.

Norwalk diarrhea outbreaks are nonseasonal and result from fecal contamination of water or food; secondary transmission from person-to-person can also occur. Norwalk agent affects older children and adults, and previous infection does not impart protective immunity. The 1–5-day illness is characterized by vomiting, watery diarrhea, headache, and low-grade fever. Diagnosis is made serologically or by electron microscopic identification of the virus in stool. There is no specific therapy, but particularly ill patients may require oral or intravenous rehydration therapy.

3. Enteroviruses

Although enteroviruses take their name from the intestine (*entero* meaning intestine) and infect intestinal tissue, this generum of more than 70 picornaviruses does not usually cause disease referable to the gastrointestinal tract. These single-stranded RNA viruses, which are resistant to gastric acid and to digestive enzymes, infect mucosal cells of the pharynx and small intestinal as well as lymphocytes in the ileum. Within these cells, the viruses multiply, causing cell lysis and allowing virions to enter the blood to infect other body sites (see Table III). Virus can persist in the gut for 1 month before the body's immune response eradicates the infection. The most renowned of the enteroviruses is poliovirus, which predominantly causes paralytic disease of the central nervous system. Other enteroviruses include coxsackievirus and echovirus. [*See* ENTEROVIRUSES.]

4. Other Viral Infections of the Small Intestine

Adenoviruses, large double-stranded DNA viruses, are best known as respiratory pathogens. Recent evidence indicates, however, that serotypes 40 and 41 have an affinity for gastrointestinal tissue and may cause diarrhea in infants and in immunosuppressed patients. Astroviruses and coronoviruses have also been implicated as causes of diarrhea.

C. Bacterial Infections

Adhesion to intestinal cells, invasion of cells, and production of toxin are the traits that enable bacteria to cause intestinal disease. Adhesion is generally mediated by pili, hair-like bacterial projections that attach the organism to host cells. Invasion, the entry of bacteria into host cells, is controlled by a number of genes, some chromosomal and others plasmid-related. In some instances, invasion destroys epithelial cells (*Plesiomonas, Shigella*); in others (*Salmonella*), the cells are left intact. All gram-negative organisms contain at least one toxin, endotoxin, a lipopolysaccharide contained in the cell wall that can cause shock if the bacteria enter the bloodstream. Some strains also release enterotoxins and cytotoxins into the surrounding environment that augment bacterial virulence.

1. Salmonellae

Salmonellae are common human pathogens and cause three types of illness: enteric (or typhoid) fever, acute gastroenteritis, and septicemia (or infection of the blood). Salmonellae are classified into antigenic types based on the organisms' somatic (O) antigens and flagellar (H) antigens. Biochemical speciation further divides the antigenic types into three groups: *Salmonella typhi, S. choleraesuis,* and the *S. enteritidis* group. The *S. enteritidis* group comprises more than 2000 different antigenic types. All the organisms are members of the Enterobacteriaceae family and are motile, gram-negative rods that do not ferment lactose.

Salmonellae attach to the epithelial cell surface of the small bowel and are subsequently taken into the cell via vacuolization. The organisms then pass through the epithelial layer into the underlying lamina propria where they multiply; some may enter into the bloodstream to be carried to other sites in the body. The mechanisms by which salmonellae cause illness are not clear, although both endotoxin and exotoxins probably play a part.

Salmonella typhi is unique among the salmonellae in that (1) it causes the syndrome of enteric fever, (2) it only infects humans, and (3) a small subset of infected patients may asymptomatically carry the organism in the biliary system and shed it in their stools for a lifetime. In the absence of animal and environmental reservoirs, the human carrier state allows the organism to continue cycling through populations, particularly through populations with poor hygiene practices and inadequate sewage disposal. Infection is acquired by ingesting food or water contaminated with human waste. *Salmonella typhi* normally enters the bloodstream via the small intestine and disseminates widely throughout the body. A polysaccharide surface antigen unique to *S. typhi,* the Vi antigen, may help the organism evade the host's immune response.

The disease caused by *S. typhi,* enteric fever, is a potentially life-threatening illness characterized by high fever, headache, cough, and generalized aches; diarrhea is an uncommon complaint. Complications, such as perforation of the intestine and pneumonia, contribute to a mortality rate of up to 10% in untreated patients. Diagnosis can be made by culture of blood, bone marrow, urine, or stool. Enteric fever can be readily treated with antimicrobial agents, although bacterial resistance may become increasingly problematic in some regions of the world. Newer antimicrobial agents (e.g., the quinolones) hold promise in eradicating the chronic carrier state. Oral and intramuscular typhoid vaccines are moderately effective in preventing disease.

The nontyphoidal salmonellae infect mammals, birds, and reptiles. Human infections result from eating foods of animal origin that are contaminated with salmonella bacteria. Nontyphoidal salmonellae are a very common cause of gastroenteritis, and illness is characterized by 2–5 days of nonbloody diarrhea, fever, and abdominal cramps. Diagnosis is made by stool culture. Salmonella gastroenteritis requires no therapy. In fact, antimicrobial treatment may prolong stool carriage of the organism.

In immunosuppressed patients, infants, the elderly, and occasionally in normal hosts, nontyphoidal strains may enter the bloodstream causing sepsis as well as focal infections distant from the small intestine (e.g., in joints or bone). *Salmonella choleraesuis* is particularly notorious for causing bloodstream infections. Diagnosis is made by culture of the blood or of a focal site of infection. Septicemia and extraintestinal salmonellosis require aggressive management with antimicrobial agents.

2. Vibrionaceae

Vibrionaceae are curved gram-negative rods that commonly live in water. They include three genera pathogenic to humans: *Vibrio, Plesiomonas,* and *Aeromonas. Vibrio cholerae,* the agent responsible for the disease called *cholera,* is one of the best studied human bacterial pathogens. Knowledge of its mechanisms of virulence has fostered broader understanding of microbial pathogenesis and gastrointestinal physiology.

Vibrio cholerae is an acid-sensitive organism. As such, a large inoculum (approximately 10^{11} organisms) is needed to ensure that a few organisms evade gastric acid, enter the small intestine, and cause disease; persons with abnormally low gastric acid levels are more easily infected. If organisms do survive the stomach, they adhere to small intestinal epithelial cells where they secrete a powerful polypeptide enterotoxin called *cholera toxin.* The B subunit of cholera toxin binds to ganglioside receptors on the intestinal cell surface, allowing the A subunit to penetrate the epithelial cell membrane. This activates an intestinal enzyme, adenylate cyclase, which in turn precipitates an outpouring of salt and water into the intestinal lumen. To cause disease, *V. cholerae* requires both the ability to adhere and toxin production.

During the past decade, cholera epidemics have swept through Asia, Africa, and South America causing millions of illnesses and thousands of deaths. The principal sources of infection are food and water contaminated with human waste. Populations with inadequate sewage disposal and hygiene are therefore at particularly high risk. Shellfish are also reservoirs for *V. cholerae,* and seafood contamination is responsible for the few endemic cholera cases that have occurred along the U.S. coast of the Gulf of Mexico.

Cholera is among the most dramatic of human diseases. Patients have cramping, vomiting, and explosive, watery diarrhea with clear, odorless stools containing flecks of mucus (rice-water stools); patients may lose up to 1 liter of fluid per hour. In the absence of treatment, dehydration and metabolic abnormalities will cause death in 60% of cases. With adequate oral or intravenous fluids, the case-fatality rate is less than 1%. Antimicrobial therapy is unnecessary but may help eliminate carriage of the organism. The currently available vaccine is ineffective, and it is not recommended by the World Health Organization or the United States Centers for Disease Control.

Vibrio parahaemolyticus lives freely in coastal waters where it contaminates fish and shellfish. It causes similar but milder illness than *V. cholerae.* Infection is rare in the United States but is common in Japan, where raw fish is a common part of the diet.

Aeromonas hydrophila and *Plesiomonas shigelloides* are uncommon causes of diarrhea in the United States. On culture, they may be confused with the Enterobacteriaceae but, unlike Enterobacteriaceae, they are oxidase-positive. Both *A. hydrophila* and *P. shigelloides* cause acute self-limited diarrhea. *Aeromonas* may also cause chronic diarrhea in adults.

3. Escherichia coli

Escherichia coli is an ubiquitous member of the family Enterobacteriaceae and makes up a significant portion of the normal flora of the large intestine. Gram-negative, aerobic, lactose-fermenting rods, it is categorized into serotypes based on somatic (O), polysaccharide (K), and flagellar (H) antigens. Although most strains are harmless gut inhabitants, some acquire pathogenic traits that enable them to cause disease in the host. Two pathogenic *E. coli* types, enterotoxigenic (ETEC) and enteropathogenic (EPEC), affect the small intestine.

ETECs are among the most common causes of traveler's diarrhea. They can be divided into two types: strains that produce heat-labile toxin (LT) and strains that produce heat-stable toxin (ST). The genes for ST and LT reside in plasmids and virtually any *E. coli* serotype can contain them. LT is related structurally to cholera toxin and similarly activates adenylate cyclase, whereas ST activates guanylate cyclase. Both cause secretion of salt and water into the gut. Pili called colonization factor antigens enhance attachment to cells and are important for virulence. ETEC infection results in 2–4 days of watery diarrhea with vomiting. Infection is transmitted by food or water contaminated with human waste. Diagnosis can be made by LT- or ST-specific nucleic acid probe of stool or of *E. coli* isolates cultured from stool. Antibiotic therapy is not usually recommended for this self-limited disease.

EPEC causes diarrhea in infants living in developing countries. Classical EPEC strains belong to specific O serotypes that tightly stick to the surface of small intestinal mucosal cells and cause destruction of cellular microvilli. The mechanisms of adherence and pathogenesis are unknown but are mediated by a large plasmid. Illness is characterized by diarrhea and vomiting that may become prolonged, dehydrating, and life-threatening. Diagnosis can be made in research laboratories by nucleic acid probe of stool for *E. coli* adherence-related genes or by identification of the typical adherence pattern of the organisms in tissue culture.

A newly identified pathogenic type of *E. coli,* enteroadherent *E. coli,* appears to be associated with chronic diarrhea, particularly in children of developing countries. Like EPEC, enteroadherent *E. coli* can be identified by their adherence pattern in tissue culture.

4. Yersinia enterocolitica

Yersinia enterocolitica, like *E. coli* and salmonella, is a member of the family Enterobacteriaceae. It is a gram-negative, coccobacillary organism that has enhanced metabolic activity at lower temperatures (e.g., 25°C). Certain virulence factors of *Y. enterocolitica,* however (e.g., tissue invasiveness and creation of an antiphagocytic antigen complex), appear only at human body temperature (37°C). *Y. entero-*

colitica produces an exotoxin quite similar to *E. coli* ST that may play a part in disease pathogenesis.

Yersinia enterocolitica is particularly common in Scandinavia but occurs both sporadically and epidemically in the United States as well. The organism is known to be transmitted by ingestion or handling of contaminated foods, in particular pork or dairy products. Although *Y. enterocolitica* usually causes 1–2 weeks of diarrhea associated with fever and cramps, it can also cause an illness virtually indistinguishable from appendicitis. Diagnosis is made by culture of stools at both 25°C and 37°C. In the convalescent stage of disease, low levels of bacteria in the stools may necessitate special media and "cold enrichment" at 4°C.

D. Fungal Infections

Yeasts, in particular *Candida* species, are common normal flora of the small intestine. In immunosuppressed patients and patients receiving chemotherapy, however, *Candida* can overgrow the intestine and cause diarrhea. More importantly, in immunosuppressed hosts, the intestinal tract can serve as a primary source for deadly *Candida* infections that disseminate throughout the body. *Histoplasma capsulatum* may also infect the gastrointestinal tract of immunosuppressed patients causing diarrhea, malabsorption, and protein loss. Both *Candida* and *Histoplasma* infections can be treated with amphotericin B.

E. Parasite Infections

Although many in industrialized countries have come to regard parasitic infections as a thing of the past, this is far from the truth. At any given time, most of the world's people are infected with parasites. Rates of infection may be lower among persons in temperate regions with good hygiene than in impoverished persons of the tropics, but even under optimal conditions, parasitic infections, if not common, are also not rare. Parasites of the small intestine are of two types: with single-celled protozoa and with multicellular helminths or worms.

1. Protozoa

a. Flagellates

Giardia lamblia is a common flagellate protozoan that has a reservoir in woodland animals. Cysts excreted in the animals' stools contaminate water supplies; people who then drink the water unfiltered and untreated can become infected. Once in the human stomach, acid destroys the cyst wall (excystation) and trophozoites enter the small intestine. Here they attach themselves to the intestinal villi with a suction disk, causing malabsorption and diarrhea.

Giardia lamblia is the most common intestinal parasite in the United States and is found in 3% of stool samples. Although many infected persons are asymptomatic, characteristic symptoms include foul-smelling diarrhea, abdominal pain, and nausea; illness can become chronic. Diagnosis of *G. lamblia* is made by microscopic observation of cysts and/or trophozoites in specimens of stool or doudenal contents. Shedding of *Giardia* may be erratic, so several stool samples need to be examined. Boiling water before drinking it kills the cysts and prevents infection. *Giardia lamblia* infection can be treated with quinacrine or metronidazole.

b. Coccidia

Cryptosporidium and *Isospora belli* species are zoonotic coccidial parasites that can infect humans. Mature oocysts from contaminated food or water pass through the stomach and enter the small bowel where they release sporozoites. These then attach to and enter the intestinal epithelial cells where the sporozoites progress to trophozoites and then merozoites. Infection causes blunting of the intestinal villi and inflammation of the underlying tissue. Although *I. belli* attacks only the small intestine, *Cryptosporidium* may infect both small and large intestine.

Cryptosporidium and *I. belli* have entered the limelight in recent years because they are not infrequent causes of diarrhea in AIDS patients. *Cryptosporidium,* which commonly colonize farm animals and house pets, also has been implicated in large waterborne diarrhea outbreaks in normal hosts and frequently cause diarrhea in infants of developing countries.

Normal hosts with symptomatic cryptosporidiosis or *I. belli* infection will have watery, foul-smelling diarrhea with weight loss and cramps lasting 1–2 weeks. Immunosuppressed patients may have chronic, severe diarrhea lasting months. Diagnosis is made by acid-fast stain of stool specimens or by intestinal biopsy. Although some antiparasitic agents have activity against *Cryptosporidium,* none has been effective in curing disease. Trimethoprim-sulfamethoxazole can eradicate *I. belli.*

2. Helminths or Worms

Helminths are multicellular parasites often called by their common name, *worms*. There are three orders of helminths: nematodes (roundworms), trematodes (flukes), and cestodes (tapeworms).

a. Nematodes

Nematodes infections of the small intestine (ascaris, hookworm, and strongyloides) are among the most common infections of humans. Although the life cycles of the various nematodes vary somewhat, they have much in common. Immature forms enter the body from contaminated soil. Ascaris eggs need to be ingested; hookworm and strongyloides larvae penetrate the skin. Larval forms then migrate through the bloodstream, enter the lungs, are coughed up, swallowed, and grow to maturity in the small intestine. For all three nematode infections, multiple worms are common, and the degree of illness is related to the host's worm burden. Diagnosis can be made by examination of stools for eggs (ascaris and hookworm) or larvae (strongyloides). All three infections are readily treatable.

Ascaris lumbricoides, the largest of the small intestinal nematodes, may affect 75%–100% of the population in endemic tropical areas. Adult worms reside in the duodenum and jejunum and can attain lengths of 20–30 cm. Females lay abundant quantities of eggs (up to 200,000/day) that are excreted in stools, allowing the life cycle to repeat itself. Light ascaris infections are asymptomatic, but heavy worm burdens cause abdominal pain, vomiting, loss of appetite, impaired nutrition, and occasionally, gastrointestinal obstruction.

The two human hookworms, *Ancylostoma duodenale* and *Necator americanus,* are extremely common in tropical and subtropical zones where people without shoes come in contact with contaminated soil. Adult worms latch onto the intestinal mucosa with their teeth (buccal cavities) and feed on blood and other tissue fluids. Female worms lay 10,000–20,000 eggs/day, which are passed in the stool. Heavy worm burdens cause anemia, fatigue, abdominal discomfort, weight loss, and malnutrition.

Strongyloides stercoralis infection has some unique characteristics. Larvae, rather than eggs, are shed in the stool. These larvae can directly reenter the body through the skin or intestinal epithelium without passing part of the life cycle in the soil. This repetitive "autoinfection" allows this nematode infection to survive for decades in the host. Also unique to *Strongyloides* is its ability to cause "hyperinfection"; in immunosuppressed hosts (e.g., patients taking corticosteroids), larvae can disseminate widely throughout the body, killing the host. Most infected persons, however, are asymptomatic. Symptomatic persons have diarrhea, stomach pains, and itching at the site where the larvae enter the skin.

Rarer nematodes that cause human infection include *Capillaria philippinensis, Trichostrongylus,* and *Gnathostoma.*

b. Trematodes

Trematodes are typically zoonoses that only incidentally infect humans. They are considerably less common than nematode infections and are more localized in their geographic distribution. *Fasciolopsis buski,* the most important of the human intestinal trematodes, occurs in southeast Asia and results from eating freshwater plants (bamboo shoots or water chestnuts) contaminated with the infectious metacercarial stage. After ingestion, the metacercaria mature to adult worms in the small intestine, where they attach themselves to the bowel wall. Persons with heavy worm burdens experience abdominal pain and diarrhea with decreased absorption of nutrients. Obstruction of the bowel can also occur. Diagnosis is made by detection of eggs in the stool.

Rarer intestinal flukes include *Echinostomes* and *Heterophyids.*

c. Cestodes

Cestodes of the human small intestine include *Diphyllobothrium latum, Taenia solium, Taenia saginata,* and *Hymenolepis nana.* With the exception of *H. nana,* life cycles of cestodes are complicated and rely on nonhuman, intermediate hosts for worm maturation. Once in the human host, worms make a home in the small intestine where they attach themselves to the mucosa with a mouth-like organ called a *scolex.* They then grow by adding box-like structures called *proglottids* to their length. As these proglottids mature and fill with eggs, they break off and are shed in the stools. Although tape worms can grow to extraordinary lengths in the human gastrointestinal tract (up to 25 m), they usually cause no symptoms unless the worm burden is extraordinarily high. Heavy worm loads can cause vague abdominal pain or indigestion. *Diphyllobothrium latum* also decreases absorption of vitamin B12, causing pernicious anemia.

Cestode infection is usually acquired by eating contaminated food items: *D. latum* from uncooked freshwater fish, *T. solium* from undercooked pork, and *T. saginata* from undercooked beef. Unlike the other cestodes, *H. nana* is not acquired from food but is transferred from person to person.

IV. Large Intestine

A. General Microbiology

The large intestine is approximately 5 feet long and connects the small intestine to the anus. It comprises three parts: cecum, colon, and rectum. The cecum, with its attached appendix, is the large intestine's anteroom through which small intestinal contents pass from the ileum into the remainder of the colon. The colon, including the ascending, traverse, descending, and sigmoid colon, follows. Finally, the rectum connects the intestine to the anus and the outside world.

The large intestine's primary function is absorption of water from feces. Unlike the small intestine, the large intestine absorbs few nutrients and is unnecessary for human survival.

Both the small and the large intestine are sterile at birth. During the first two weeks of life, *E. coli* and *Streptococci* begin to colonize the mucosal surface. The consequent chemical reduction creates an appropriate environment for colonization by anaerobes including *Bifidobacteria, Clostridia,* and *Bacteroides*. *Bifidobacteria* predominate in the acidic intestines of breast-fed infants and cause frequent yellowish, loose stools. In formula-fed infants, *Bifidobacteria* coexist with *Clostridia, Bacteroides, E. coli,* and *Streptococci;* stools are firmer and less frequent. As solid foods are introduced, the levels of these latter organisms increase and the prior distinctions between formula-fed and breast-fed children disappear. When weaning is complete; the flora are similar to that of the adult: *Bacteroides,* anaerobic gram-positive cocci, and *Bifidobacteria* predominate with *Enterobactericeae* and *Clostridia* playing a lesser role (Table IV).

The bowel flora are in a steady state, varying little with age, diet, or location of residence. While bacteria multiply, others lose their adherence to the epithelial wall and are shed in the stool. Competition for mucosal space and for nutrients limits growth of individual species and prevents *de novo* colonization by strains alien to the host. Furthermore, the human immunolgic system tolerates "indigenous" flora, whereas antibodies combat nonnative residents (Table IV). Overall, 20% of fecal mass can be attributed to bacteria. The stability and magnitude of organisms may help to explain why fewer gastrointestinal pathogens can compete successfully in the large bowel than in the less heavily colonized small intestine.

The relationship between host and normal flora is truly symbiotic. While bacteria utilize ingested food and intestinal secretions to their advantage, they also provide beneficial functions to the host. These include vitamin synthesis, protein synthesis, production of energy sources for the colonic mucosa,

Table IV Large Intestinal Ecosystem

Normal environment	
Basic to normal pH	
Glandular epithelium	
Moderate peristalsis	
Normal flora	
Quantity:	large (10^{11})
Variety:	bacteroides
	Eubacterium
	anaerobic streptococci
	Bifidobacterium
	enterococci
	Enterobacteriaeae
	lactobacilli
	clostridia
	yeasts
Infections	
Viruses:	cytomegalovirus[a]
	herpes simplex
Bacteria:	*Shigella* sp
	Campylobacter sp[b]
	Salmonella sp[c]
	enterohemorrhagic *E. coli*
	enteroinvasive *E. coli*
Parasites:	*Entamoeba histolytica*
	Blastocystis hominis
	Balantidium coli
	Trichuris trichiura
	Enterobius vermicularis

[a] Predominantly in immunosuppressed hosts.
[b] Also infect small intestine.
[c] Predominantly infect small intestine.

and assistance with absorption. Balance is essential; both overgrowth of bacteria in a person with an inadequate diet can cause deficiencies of vitamins C, B_1, and A due to decomposition of the nutrients by bacteria. Furthermore, overgrowing bacteria break down intrinsic factor, a protein necessary for B_{12} absorption. On the other hand, low levels of intestinal flora (usually secondary to antibiotic use) in a malnourished host may result in folic acid and vitamin K deficiencies. While a good diet normally provides adequate quantities of these nutrients, malnourished persons may rely on bacterial production of folate and vitamin K to meet minimal requirements.

Normal flora are not required for digestion of food yet intestinal enzymes and sloughed cells can be digested by bacteria and the resultant amino acids and carbohydrates can be absorbed and used by the host. Bacterial fermentation of unabsorbed carbohydrates produces short-chain fatty acids favored by colonic epithelial cells as an energy source. Bacteria also assist in absorption of fats by degrading bile acids in the colon and allowing recycling of the metabolites through enterohepatic circulation. They metabolize steroids and cholesterol and allow reabsorption and reuse of their metabolites as well.

Yet despite these benefits, germfree animals with nutritionally complete diets live longer than colonized animals. Furthermore, antibiotics that destroy intestinal bacteria actually promote growth in livestock. It is hypothesized that normal flora, particularly *Enterobactericeae,* metabolize ingested substances into carcinogens (e.g., nitrosoamines) and toxins (e.g., ammonia) that reduce longevity and growth. Epidemiologists and microbiologists are currently studying whether dietary adaptations (decreasing dietary fats, proteins, and caloric intake) can foster the salutary effects of normal flora while eliminating their detrimental repercussions.

B. Bacterial Infections

As in the same intestine, pathogenicity of alien bacteria depends on ability to adhere, to invade, and to produce toxin. But whereas toxin production and secretion of large quantities of salt and fluid is the common *modus operandi* of bacteria in the small intestine, invasion with intestinal cell destruction is characteristic of bacterial infections of the colon.

1. Shigella

Shigella species, very close relatives of *E. coli,* are oxidase-negative, gram-negative, nonmotile Enterobacteriaceae. They can be categorized into four species based on the somatic O antigen: *S. dysenteriae* (group A), *S. flexneri* (group B), *S. boydii* (group C), and *S. sonnei* (group D). These species are further classified into multiple subtypes. *S. dysenteriae* type 1, the most virulent of the *Shigella* strains, is also known as the Shiga bacillus.

During the first 12 hours in the human host, *Shigella* multiply in the small intestine where they secrete a toxin that causes watery diarrhea. In the colon, they then enter epithelial cells via endocytosis and multiply, resulting in tissue destruction, mucosal ulceration, and marked inflammatory changes. A large "virulence plasmid" is necessary for this invasion. Some *Shigella* strains contain a chromosomal gene encoding for Shiga toxin, a toxin that inhibits protein synthesis and causes cell death. *S. dysenteriae* type 1 produces large quantities of Shiga toxin, which may in part account for its prodigious virulence. Few organisms are necessary for infection to the hold; less than 100 organisms can cause clinical illness.

Shigella infections occur worldwide. In the United States, *S. sonnei* is the most common species; *S. flexneri* predominates in developing countries. *Shigella dysenteriae* often occurs in epidemics, some of which have affected hundreds of thousands of people causing thousands of deaths. Humans are the only host for *Shigella,* and disease is usually transmitted from person to person. Children younger than 5 years old are most commonly infected. Poor sanitation and close quarters (e.g., daycare centers) foster disease transmission. Foodborne outbreaks also occur.

Clinically, *Shigella* causes a spectrum of illness ranging from mild diarrhea to bacillary dysentery, a potentially life-threatening disease. Dysentery is characterized by fever, severe abdominal cramps, and bloody diarrhea with mucus. *S. dysenteriae* and other Shiga toxin–producing strains may also cause kidney failure, anemia, abnormal bleeding, and seizures (hemolytic uremic syndrome). Diagnosis is made by stool culture. Increasing resistance of *Shigella* to available antibiotics is complicating disease therapy. In patients with mild disease, antimicrobial therapy is not warranted. With severe disease, therapy should be directed by antimicrobial resistance patterns. [*See* DYSENTERY, BACILLARY.]

2. Campylobacter

Campylobacter is a flagellated, oxidase-positive, catalase-positive, curved, gram-negative rod that grows best under microaerophilic conditions at 42°C. Two species, *C. jejuni* (hippurate-positive) and *C. coli* (hippurate-negative) account for most human disease. Other species, *C. laridis, C. cinaedi, C. fetus,* and *C. fennelliae,* are occasional pathogens.

Campylobacter invades small as well as large intestinal epithelium, causing mucosal destruction, ulceration, and inflammation. A recently described tissue cytotoxin is felt to play a role in disease pathogenesis. An enterotoxin similar to that of *E. coli* has also been postulated.

Campylobacter is common in domesticated animals, and infection is transmitted by eating contaminated foods of animal origin (typically poultry or unpasteurized milk) or by direct contact with animal carriers. Waterborne outbreaks also occur. There is a wide spectrum of clinical illness associated with *Campylobacter*. In developing countries, primary infection occurs in infancy, and immune adults tolerate infection asymptomatically. In industrialized nations, symptomatic infection strikes all age groups, and illness is characterized by self-limited diarrhea, fever, and cramps. Some patients will have a dysentery-like picture with bloody stools and marked gastrointestinal ulceration. Diagnosis is made by culture of stools. Antimicrobial treatment may be recommended in severe cases.

3. *Escherichia coli*

Escherichia comprise a significant proportion of the normal large intestinal flora. Although the overwhelming majority of strains are benign, two types of *E. coli* are large intestinal pathogens: enteroinvasive *E. coli* (EIEC) and enterohemorrhagic *E. coli* (EHEC). As mentioned above, *E coli* and *Shigella* are closely related and exchange genetic information. Both EIEC and EHEC are *E. coli* types that have adopted pathogenic mechanisms of *Shigella* strains.

EIEC contains the virulence plasmid of *Shigella*. Consequently, EIEC-related disease is virtually identical to bacillary dysentery associated with *Shigella*. EIEC infections are rare but can occur in outbreak settings.

EHECs cause bloody diarrhea without marked inflammation. Pathogenesis of disease appears to be linked to phage-mediated production of a cytotoxin that affects tissue-cultured Vero cells. This "verotoxin" is so similar to Shiga toxin that some investigators call it *Shiga-like toxin*. EHEC strains belong to specific serotypes with the cattle-borne *E. coli* O157:H7 serotype predominating. EHEC outbreaks have often been linked to ingestion of undercooked beef. EHEC, like Shiga toxin–producing *Shigella* strains, can cause deadly neurologic, kidney, and blood complications. Diagnosis can be made by identifying sorbital-negative *E. coli* strains from stools of infected patients.

4. *Clostridium difficile*

Clostridium difficile is associated with antibiotic-associated diarrhea. This bacterium is a strict anaerobe that colonizes the large intestine of a small proportion (less than 5%) of normal hosts. If an antimicrobial agent eliminates normal flora of the large intestine, *C. difficile* can proliferate and cause diarrheal disease. Diarrhea, which is probably mediated by two toxins, an enterotoxin (toxin A) and a cytotoxin (toxin B), can range from mild to life-threatening. The role of the cytotoxin is controversial. The most severe form of infection, pseudomembranous colitis, causes characteristic "membranes" of mucus and pus that overlie the large intestinal mucosa; illness is characterized by fever, sever cramps, and bloody stools. Diagnosis can be made by detection of the toxin in the stool (using a tissue culture cytotoxicity assay for toxin B or an ELISA for toxin A), by latex agglutination assay for a *C. difficile*–related protein or by anaerobic culture of stools. Patients with mild disease need no therapy except discontinuation of antibiotics. In patients with severe disease, *C. difficile* can be eradicated with vancomycin or metronidazole.

C. Parasites

1. Protozoa

Three protozoa infect the large intestine: *Entamoeba histolytica, Balantidium coli,* and *Blastocystis hominis*. The most important of these is *E. histolytica,* the agent of amebic dysentery and an important cause of death in developing countries.

a. *Entamoeba histolytica*

Ten percent of the world's population is felt to be infected with *E. histolytica* at any given time. Infection is acquired by ingesting food or water contaminated by cysts. In the small bowel, the cysts undergo excystation and the trophozoites pass into the large intestine. There, they adhere to invade, and destroy epithelial cells, causing ulcer formation, inflammation, and bleeding. Protein- and carbohydrate-dissolving enzymes (proteinases and glycosylases) elaborated by the protozoa accelerate tissue destruction.

In industrialized countries, amebiasis is found in the male homosexual population, in persons with AIDS, and in institutionalized populations. In highly endemic areas in the devloping world, the organism strikes people in all walks of life. Most persons infected with *E. histolytica* are asymptomatic. Why some people will get sick and others will not is unclear; it is hypothesized that illness is determined by both host factors (e.g., prior immunity) and virulence of the specific *E. histolytica* strains (zymodemes). The unfortunate minority who do have symptomatic disease may present with illness ranging from mild acute diarrhea to chronic diarrhea to dysentery (severe stomach cramps and bloody diarrhea). Fulminant disease with perforation of the colon and death occurs in previously debilitated or immunosuppressed persons. Ameba can also infect the liver, lung, and brain.

Diagnosis is made by microscopic observation of trophozoites or cysts in the stool. Distinguishing *E. histolytica* cysts from other cells that may be in stool is extremely difficult and only experienced technicians can be relied on. In the symptomatic host, several antiparasitic agents can be used to eradicate infection.

b. *Balantidium coli*

Balantidium coli is the only ciliated protozoa that infects humans. The cilia allow the organism to swim into the mucosa large intestine and ileum where they cause inflammation and ulceration. Primarily a disease of pigs and other animals, *B. coli* rarely causes human disease. As with *E. histolytica* infection, most *B. coli*–infected individuals are asymptomatic; those who are ill can have symptoms similar to those of *E. histolytica*. Diagnosis can be made by observation of trophozoites in stool or in diseased tissue. Tetracycline eradicates disease.

c. *Blastocystis hominis*

The role of *Blastocystis hominis* in human disease has been controversial. An anaerobic protozoan, *B. hominis* is commonly found in stool samples from asymptomatic persons. The most recent concensus, however, it that *B. hominis* should be considered diarrhea pathogen if (1) the organism is observed in large number on microscopic examination of stool and (2) no other stool pathogen is evident. Diarrhea attributed to *B. hominis* is often chronic and associated with bloating and fatigue. Metronidazole has been used successfully to eradicate infection.

2. Helminths

Two nematodes that principally infect the large intestine are *Enterobius vermicularis* (pinworm) and *Trichuris trichiura* (whipworm).

Humans are the only host for the tiny, white worm *E. vermicularis*. Ingested *Enterobius* eggs mature into larvae in the small intestine and into 5–10-mm-long adult worms in the cecum. Female adults then migrate to the anus where they lay their eggs. Transmission of infection results from close contact with infected persons who are shedding eggs.

Enterobius vermicularis is the most common worm in temperate, industrialized countries of North America. Children in crowded living conditions with poor sanitation are at particularly high risk for infection. Typically, infected children have anal itching without diarrhea or other gastrointestinal symptoms. Scratching the anus fosters hand contamination and spread of the worm to other children. Diagnosis is made by applying cellophane tape to the anus and microscopically examining the tape for eggs; eggs are not typically shed in stools. Several antiparasitic drugs are effective in eliminating the worm. If the living conditions that fostered transmission remain unchanged, however, reinfection rates are high.

Trichuris trichiura has a similar life cycle to *E. vermicularis*, with larvae maturing in the small intestine and adults in the cecum. The 4–5-cm adults burrow their heads into the cecal and large intestinal epithelium where the females lay eggs that are excreted in stools. *Trichuris trichiura* is found worldwide and infects more than 75% of persons in some highly endemic tropical regions with poor sanitation; children acquire infection by playing in dirt contaminated with human waste. Travelers to highly endemic areas are also at risk.

The degree of symptoms with *T. trichiura* is wholly dependent on worm burden. With light infections, children remain asymptomatic; with heavy infections, a frank dysenterylike syndrome can occur, often associated with anemia and malnutrition. diagosis is made by identification of eggs in the stools, and treatment with antiparasitic agents is highly effective.

D. Viruses

Viral infections of the large intestine are unusual. Cytomegalovirus can infect and cause diarrhea in immunosuppressed hosts. Herpes simplex infection of the rectum and anus causes painful, ulcerating disease (HSV proctitis). Typically, this is a sexually transmitted disease of homosexual men. [*See* HERPESVIRUSES.]

V. Conclusions

The human gastrointestinal tract harbors a panoply of organisms that often thrive despite the host's natural defenses. Both organism and host determine how the complex relationship proceeds. As the organisms affect the human body, so the human body affects the organsims within it, altering their metabolism and physical structure. This interaction is astonishingly complicated and poorly understood. Although it is evident that much research has been devoted to describing the clinical and pathologic consequences of organisms on a host, there is little understanding of reverse (i.e., how the host effects the microbial agents themselves). The future rests in focusing on host and parasite life forms as a unit, integrating molecular biology and physiology of the organisms with immunology of physiology of their host.

Bibliography

Bottone, E. F., Janda, J. M., Motyl, M. R., et al. (1986). Gastrointestinal specimens. *In* "Interpretive Medical Microbiology" (H. P. Dalton, ed.), pp. 427–522. Churchill Livingstone, New York.

Gutierrez, Y. (1990). "Diagnostic Pathology of Parasitic Infections with clinical Correlations." Lea & Febiger, Philadelphia.

Hentges, D. J. (ed.) (1983). "Human Intestinal Microflora in Health and Disease." Academic Press, New York.

Ryan, K. J. (1990). Enterobacteriaceae. *In* "Medical Microbiology: An Introduction to Infectious Diseases" (J. C. Sherris, ed.), pp. 357–383. Elsevier, New York.

Ryan, K. J. (1990). *Vibrio* and *Campylobacter*. *In* "Medical Microbiology: An Introduction to Infectious Disease" (J. C. Sherris, ed.), pp. 385–391. Elsevier, New York.

Genetically Engineered Microorganisms, Environmental Introduction

Martina McGloughlin and Roy H. Doi
University of California, Davis

Glossary

Chromosome Linkage structure consisting of a chromosome-specific linear sequence of genes

Gene Basic unit of Mendelian inheritance that represents a contiguous region of DNA (or RNA in some viruses) corresponding to one (less often two or more) transcription units

Genetic engineering Use of *in vitro* techniques (recombinant DNA technology) for the deliberate manipulation of genes within or between species for the purpose of gene analysis and product improvement

Genetic transformation Unidirectional transfer and incorporation of foreign DNA by prokaryotic and eukaryotic cells

Genotype Sum total of the genetic information contained in the chromosomes of the prokaryotes and eukaryotes, as distinguished from their phenotype

Phenotype Observable properties (structural and functional) of an organism, produced by the interaction of the organism's genotype and the environment within which it develops

Pleiotropic effect Genes or mutations that result in the production of apparently unrelated multiple effects at the phenotypic level

Proteases Enzymes that degrade proteins

Recombinant DNA technology Set of techniques that permit the formation of novel DNA sequences by *in vitro* combination of two nonhomologous DNA molecules

Recombinase Any of the enzymes that recognize specific DNA sequences, introduce single strand breaks, and are involved in general genetic recombination

Taxon Taxonomic group of individuals (population) recognized as a formal unit at any level of a hierarchial classification

Transformant Any cell that has either stably integrated transferred DNA into its chromosomal DNA or harbors the transferred DNA transiently

Transposable genetic elements Any of a class of diverse DNA segments that can insert into nonhomologous DNA, exit, and relocate in a reaction, which is independent of the general recombination function of the host

Transposition Change of position of transposable genetic elements from one site to another in the genetic material of prokaryotes and eukaryotes

Wild type Gene, genotype, or phenotype which predominates in ecological populations or in standard laboratory strains

BIOTECHNOLOGY, in the simplest and broadest sense, is a series of enabling technologies that involve the manipulation of living organisms or their subcellular components to make or modify products, to improve plants or animals, or to develop microorganisms for specific uses. In essence, this definition illustrates the breadth and implies the positive potential of biotechnology. The many and varied tools of this technology hold much promise for increasing the efficiency and sustainability of production agriculture and for assuring the abundance, variety, quality, and safety of food. One of the principle tools of biotechnology is recombinant DNA technology. This term covers a set of techniques that permit the formation of novel DNA sequences by *in vitro* combination of two nonhomologous DNA molecules. The use of organisms modified by these methods to improve agriculture and help the environment necessitates extensive and timely field testing within an environment similar to that in which they will be eventually used. Federal, state, and local oversight must be such as to insure that benefits are maximized while potential hazards to life and the environment are minimized. A scientific-based, flexible, and reasonable oversight system must be applied to insure that field testing of genetically modified organisms can proceed with minimum risk and maximum benefit.

I. Historical Context

What has made humans unique in the animal kingdom is the ability to manipulate our world. Many millennia B.C. people discovered that microorganisms could be used in fermentation processes, to bake bread, make alcohol, and produce cheese. Through mutation and selection processes, use of microorganisms as process tools became more and more sophisticated as time went by and this use took on another dimension with the advent of recombinant DNA technology in 1973. The molecular manipulation of living organisms is one of the principal tools of modern biotechnology. While biotechnology in the broadest sense is not new; what is new, however, is the level of complexity and precision involved in scientists' current ability to manipulate living things, making such manipulation predictable, precise, and controlled. This level of control is a tremendous asset in the quest to improve the quality of life.

While exploitation of the fundamental biological phenomenon of genetic change has been the principal well spring of agrarian evolution and, in no small way, has assisted the advancement of civilization, it did not emerge as an issue of social concern until scientists attained the means to manipulate life at the molecular level. Apart from the philosophical concerns with the technology, which are beyond the scope of this chapter, many other issues apply equally to traditional methods of modification, selection, and breeding but have taken on a different complexion in the context of current technological capabilities. These capabilities offer tremendous potential for addressing many of the pressing human and environmental needs, including increasing the efficiency and sustainability of production agriculture; assuring the abundance, variety, quality, and safety of food; providing means to monitor and reduce pollution; and offering versatile methods to combat infectious agents. Implicit in the effective utilization of these biotechnology applications is the requirement for the introduction of genetically engineered organisms (GEOs) into the environment.

To many, this fact is tantamount to opening a Pandora's box of ecological hazards within the environment. However, in essence, environmental introductions of GEOs is just another facet in a long history of human intervention in environmental processes. And, as such, the same parameters of risk-based assessment should apply. For many years, researchers and regulators have used the medium of "field tests" to assess both the efficacy and the safety of new biological products and systems. Such experimental terminology has historically not attained wide usage outside of the scientific community; however, of late, this research term often carries connotations of novel risks for many sectors of the community. This should not be the case: Any introduction into the environment carries an element of risk, but these risks must be weighed against the potential benefits and also the possible risks that may result from delay or nonimplementation. Environmental introduction of all types of organisms must be undertaken within a regulatory framework that ensures adequate protection of the environment and of the human and animal life that may come in contact with the introduced organism.

The gravity with which researchers hold these tenets is demonstrated by the fact that, following the development of the initial capabilities of recombinant DNA technology, Paul Berg and his distinguished colleagues, Baltimore, Boyer, and Cohen, published the findings of the National Academy of Science (NAS) Committee on Recombinant DNA Molecules in *Nature* on July 19, 1974, which effectively called for a moratorium on genetic engineering research. This publication spurred the convening of scientists at Asilomar in 1975, when every facet and implication of recombinant DNA research was explored. The product of this milestone conference was a set of guidelines that outlined strict procedures for ensuring the safety of genetic engineering experiments. These guidelines became the basis for the establishment of the National Institutes of Health (NIH) Recombinant DNA Advisory Committee (RAC) and the development and publication of the well-known RAC Guidelines in 1976. It is interesting to note that masquerading under the moniker of guidelines these directives carried the weight of regulatory oversight because NIH-funded researchers had to comply to avoid jeopardizing funding for their institutions. Over time, all federal and state funding agencies adopted the guidelines for recipients of their grants, thus covering all federally funded molecular genetic research. Although the NIH guidelines were adopted to exclusively cover the latter type of research, other institutions and biotechnology companies voluntarily complied with the guidelines. [*See* GENETICALLY MODIFIED ORGANISMS: GUIDELINES AND REGULATIONS FOR RESEARCH.]

In their initial form, any type of environmental release was absolutely prohibited by the Guidelines. Their primary goal was to minimize the chances of any recombinant organism ever leaving the research laboratory. Indeed, elaborate containment facilities, including negative pressure rooms, were required for some applications. To ensure compliance with the directives at the local level, Institutional Biosafety Committees (IBCs) were set up to assist the RAC by reviewing projects at each institution. The IBCs' function has evolved over the years, and they now represent the primary watchdogs at the institutional level for all biotechnology research requiring review, and the NIH have placed most decision-making at the level of the local IBCs. The guidelines themselves have been relaxed with time as a plethora of safe experiments have provided sufficient theoretical and practical information to allay concerns about risk.

As the focus for molecular biology research shifted from the basic pursuit of knowledge to the pursuit of lucrative applications, once again the specter of risk arose as the potential of new products and applications had to be evaluated outside the confines of a laboratory. However, the specter now became far more global as the implications of commercial applications brought not just worker safety into the loop but also the environment, agricultural and industrial products, and the safety and well-being of all living things. Beyond deliberate release, the RAC Guidelines were not designed to address these issues, so the issue moved into the realm of the federal agencies who had regulatory authority that could be interpreted to oversee biotechnology issues. This adaptation of oversight is very much a dynamic process as the various agencies wrestle with the task of applying existing regulations and developing new ones for oversight of this technology in transition.

The move from the laboratory to the field is indeed providing a challenge for the relevant agencies. In the short time that has elapsed since the initial development of the capabilities of recombinant DNA technology, the controversy over this research has shifted focus from the presumed risks associated with the possible escape of GEOs from research laboratories to the nature of the long-term environmental impact of GEOs that are intentionally released. The agencies must base their analysis of oversight not principally on any presumed theoretical "exotic" risks but, rather, exploit the vast cache of accumulated knowledge that has been amassed over years of research using highly developed scientific procedures for assessing field tests and planned introductions. Using such a scientific approach deduced from first principles is probably the most effective mechanism regulatory bodies can take in crafting effective regulations.

II. Regulations

A. Coordinated Framework for the Regulation of Biotechnology

The Office of Science and Technology Policy (OSTP) has established overall policy regarding how biotechnology will be regulated and coordinated in the United States. This policy has been outlined in formal notices published in the *Federal Register*. In 1986, the Coordinated Framework for the Regulation of Biotechnology policy statement was based on two premises: (1) that existing laws were sufficient to regulate GEOs and (2) that, to the extent possible, jurisdiction and regulations would be based on the use of the product, not the processes by which it was made. The framework described the comprehensive federal regulatory policy for ensuring the safety of biotechnology research and products built on experience with products developed by traditional genetic modification techniques. It determined that existing statutes provided a basic network of agency jurisdiction over both research and products assuring reasonable safeguards for the public. It did not deal extensively with the issue of environmental introduction of modified organisms; however, the Biotechnology Science Coordinating Committee (BSCC) did propose that the planned introduction of organisms meeting two different sets of criteria should be subject to oversight. The first were "intergeneric organisms" formed by deliberate combination of genetic material from sources in different genera. The second were organisms that belong to a pathogenic species or that contain genetic material from organisms that are pathogenic. Based on accumulated experience in the intervening years and the demand for a coordinated policy from researchers and regulators alike, the 1990 policy statement dealt with this issue in depth. The principle focus for this policy statement was to establish a standard scope definition of organisms that would be subject to oversight by the federal agencies, notwithstanding the fact that the statutory basis for regulation differ among the involved agencies. The scope definition was again revised in 1992. We will expand on the scope definition later, but first we will present an overview of the agencies and statutory authorities that presently are involved in oversight of the environmental introduction of GEOs.

Three agencies share primary responsibility for regulating the organisms, products, and processes of recombinant DNA technology, whether they be designed for closed systems or for environmental release. They are the Food and Drug Administration (FDA), the United States Department of Agriculture (USDA), and the Environmental Protection Agency (EPA). Regulatory authority under the Occupational Safety and Health Administration (OSHA) covers those working with recombinant DNA (rDNA) and the Department of Health and Human Services (DHHS) oversees the health of the general public. In addition, many states and local authorities were not satisfied with existing oversight, and some 14 states and several municipalities enacted their own biotechnology legislation.

Each federal agency is directed in its decision-making process by its own specific mandate from Congress when drawing up oversight guidelines and regulations. The EPA's primary concern is the environment, the FDA's is food and human safety, and the USDA's is crop protection and animal health. Agencies must also adhere to the umbrella National Environmental Policy Act (NEPA), which is binding on all federal agencies.

While all legislation includes safety as an important criteria in the review of field test applications, no existing oversight law specifically addresses the safety of biotechnology *per se;* they each deal with the use of products or the release of "material" into the environment. The NEPA ensures that actions taken by the different agencies are environmentally sound. Each must thoroughly evaluate the impact of the action under review on the environment and weigh the potential benefits with the possible adverse effects. In addition to examining all issues relevant to the case and exploring alternatives, the agencies are obliged to seek public comment. They are no longer required to consider worst-case scenarios.

B. Biotechnology Science Coordinating Committee

Because biotechnology cuts across so many government agencies, interagency coordination is vital for the development of consistent policies and resolving jurisdictional matters. A certain degree of consensual oversight has been achieved through the auspices of the BSCC, a committee of the Federal Coordinating Council for Science, Engineering and Technology, which reports to the OSTP, Executive Office of the President. The BSCC was formed as a result of the first OSTP publication of a Coordinated Framework for Biotechnology Regulation. The BSCC is composed of senior policy officials from the USDA, EPA, FDA, NIH, and the National Science Foundation (NSF). The BSCC is an extremely important forum for the many agencies to coordinate science policy and to reconcile review procedures within and between agencies.

The BSCC has helped to resolve some of the earlier difficulties with jurisdiction. A question of duplication of oversight may occur within many areas including food, food additives, pesticides, and animal health care products. For example, a veterinary science investigator may ask: Is the product a new animal drug within the jurisdiction of the FDA or a veterinary biologic falling under the aegis of the Animal and Plant Health Inspection Service (APHIS)? The BSCC helps to resolve those issues that are not clearly demarcated within the statutes of any one agency. When an obvious overlap occurs between agency jurisdiction, as is often the case with field tests of GEOs, procedures are now in place to determine which agency should take the lead and to ensure optimum coordination of review, with minimum duplication and burden on the applicant. [*See* FOOD BIOTECHNOLOGY.]

The BSCC is also a useful task force providing a forum for identifying relevant issues and fact finding through interagency funding of studies. A prime example of this was the report compiled by the National Research Council (NRC) of the NAS on

"Field Testing Genetically Modified Organisms: A Framework for Decision Making."

The rapid pace of biotechnology development and the differing missions of the many agencies who oversee its many facets makes the BSCC's task less than straightforward. Some of the many bones of contention include the development of consensual policies for risk assessment and risk management, the definition for organisms that should be subject to different levels of oversight, defining the terms of environmental release, and establishing coherent standards. The Coordinated Framework for the Regulation of Biotechnology, outlined earlier, is presently under review by the BSCC, and it was this organ that published that request for public comment in the *Federal Register* of July 31, 1990, on the "Principles for Federal Oversight of Biotechnology: Planned Introduction into the Environment of Organisms with Modified Heredity Traits." The principles for scope of oversight outlined in this document were lauded by the scientific community in that they were based on a risk-focused, scientifically deduced analysis process that took into account the traits of the organism and its environment and historical experience, and not its methods of modification. However, the exemption category for the most part stood in sharp contrast to this stated premise as all six exemption categories, except for the last, were defined by the method of their creation such as microorganisms modified solely through physical or chemical mutagenesis, vascular plants regenerated from tissue culture, and organisms resulting from deletions, rearrangements, and amplifications within a single genome. Such exemption categorizations implied the inherent safety of an organism without a requirement to examine more cogent criteria. The last exemption category, which covered organisms with new phenotypic traits conferring no greater risk to the target environment than a parental strain that is considered to be safe, was felt by many to be the only category that should have been defined within the list of exemptions. In response to public comment, the scope document was again published in the *Federal Register* of February 27, 1992; the principle change being the deletion of the six exemption categories.

C. President's Council on Competitiveness

The most recent policy statement on the issue of regulatory oversight was published in the President's Council on Competitiveness Report on National Biotechnology Policy. The Council recognized the fact that regulations may create substantial barriers to product development and that inconsistency in federal oversight may lead to the enactment of conflicting, duplicative, or burdensome state and local regulations. To address an uncertain regulatory environment and establish universal standards for oversight, the Council recommended four principles for regulatory review. They are as follows: (1) federal government regulatory oversight should focus on the characteristics and risks of the biotechnology product and not the process by which it is created; (2) for biotechnology products that require review, regulatory review should be designed to minimize regulatory burden while ensuring protection of public health and welfare; (3) regulatory programs should be designed to accommodate the rapid advances in biotechnology (performance standards are, therefore, generally preferred over design standards); and (4) to create opportunities for the application of innovative new biotechnology products, all regulation in environmental and health areas, whether or not they address biotechnology, should use performance standards rather than specifying rigid controls or specific designs for compliance.

The Council stated that the goal of these principles is to ensure that regulations and guidelines affecting biotechnology are based solely on the potential risks and are carefully constructed and monitored to avoid excessive restrictions that curtail the benefits of biotechnology to society.

The paths and processes by which the oversight regulations for environmental introduction of GEOs within the various agencies evolved under the guidance of the three policy statements of the OSTP is outlined in the following sections.

D. Environmental Protection Agency Oversight

Of the three agencies, the EPA and the USDA have the most involvement with environmental introduction of GEOs. And of those two, the former has the greater authority over genetically modified microorganisms. Before the age of recombinant DNA technology, the introduction of microbes such as viruses, bacteria, fungi, and protozoans into the environment were of little concern to the EPA. Until the 1980s, review and registration was only required of microorganisms that were used as pesticides. Even when these organisms were randomly mutated or genetically manipulated using "traditional" methods, there was little review involved. In fact, there was an attempt to minimize the review process for those products known as "biorational pesticides," because they were considered less hazardous than chemical pesticides. Following the development of recombinant DNA technology, the perceived limitless potential to create exotic constructs in the laboratory and let them loose on the world made the EPA sit up and take notice.

In 1984, the Office of Pesticides and Toxic Substances (OPTS) instituted the EPA Biotechnology Advisory Committee (BAC). They looked to their statute books and determined that they could and should regulate certain constructs of GEOs and their products under preexisting legislation. Of its seven environmental statutes, the EPA BAC decided that two of them were appropriate for oversight of biotechnology. They are the Toxic Substances Control Act (TSCA), administered by the Office of Toxic Substances (OTS), and the Federal Insecticide, Fungicide, and Rodenticide Act (FIFRA), administered by the Office of Pesticide Programs (OPP). These two statutes are termed "gateway" legislation because they are preventative rather than abatement-oriented (as is the case of most EPA statutes such as the Superfund law). They are invoked prior to the applications of a new product to the environment.

From the inception of FIFRA in 1947, any chemicals used as pesticides were reviewed and registered under this law. In 1948, the first microbial pesticides, *Bacillus popilliae* and *Bacillus lentimorbus*, were registered by the USDA. By 1970, the diverse authorities for pesticide registration were transferred to the newly formed EPA. In 1982, the agency included GEOs in its policy of regulating microbial pest-control agents (MPCAs; for the control of pests and weeds) as distinctive entities from chemicals. This was followed by the recombinant DNA testing guidelines, which were published in 1984.

FIFRA mandates the registration of pest-control products and "economic poisons" prior to production and sale. To test a product, an applicant must submit complete data as provided in the statute (7 USC 136c). During the course of its product evaluation, the OPP conducts the equivalent of an environmental assessment or environmental impact statements (EISs). Prior to 1984, MPCAs were not reviewed at the research level, notification was not needed for tests under 10 acres, and no experimental use permits (EUPs) were required. Starting in 1984, notification of intent to field test GEOs and nonengineered nonindigenous pathogenic and nonpathogenic microorganisms was required, regardless of testing site size.

Procedures are initiated when applicants notify OPP that they are ready to field test a potential pesticide. The EPA first set forth their review system in a Coordinated Framework document. It was to be a two-tiered review system for risk assessment of microbial pesticides with special consideration given to nonindigenous and GEOs. According to OPTS at that time, genetically altered or engineered was defined as "Any human intervention beyond removal from the environment and selection for the desired variant populations . . . should be considered to result in an engineered organism." At present, the agency has suspended the small-scale exemption for GEOs and requires EUPs for all such tests. In 1989, the EPA published its proposed amendments of EUPs. The issues it covered were the scope of genetically modified microbial pesticides subject to notification, the review of nonindigenous microbial pesticides at small-scale levels, and the establishment of expert review groups similar to the NIH IBCs. As the scope definition drawn up by the BSCC is in a state of flux, so too is that of the EPA, as discussed later.

The Office of Toxic Substances (OTS) administers the TSCA, which is often referred to as gap-filling legislation because chemicals not covered by other laws are covered by the TSCA. The TSCA is invoked through the concept of "newness." GEOs not used in pesticides, foods, food additives, cosmetics, drugs, and medical devices qualify for regulation under the TSCA. The TSCA basically covers new chemicals or new uses for existing chemicals and the EPA General Counsel felt in 1983 that because the TSCA covers all chemicals and DNA is a chemical(!), DNA is covered. They further concluded that a DNA molecule does not have any use except in the life form of which it is part. The inference was that it did not make sense to regulate new forms of DNA without regulating the life forms that carried them. However, prior to the publication of the "Coordinated Framework for Regulation of Biotechnology," the EPA paid little attention to the release of nonpesticide microorganisms, regardless of their progenitors or the method used for their production. However, the BSCC proposal stated that the planned introduction of organisms meeting the criteria of "intergeneric organisms" formed by deliberate combination of genetic material from sources in different genera be subject to oversight. Therefore, this first scope definition, which captured all recombinant DNA constructs under the rubric of "new" triggers a TSCA review and the possibility of notification requirement and regulation of microorganisms. This obviously processed-based all-encompassing definition of regulatable organisms met with much comment from the scientific community and the 1990 policy statement, while covering all eventualities, attempted to switch the emphasis to a more risk-based, product-oriented scope of oversight. In response to this, the EPA is now in the process of introducing a biotechnology rule under the TSCA. However, the agency has some way to go in reconciling the Administration's stated policy of focusing scrutiny on the phenotypic characteristics and related risks of the organism rather than on the method of genotypic modification with its (EPA's) clear inclusion of the latter criterium within the rule by virtue of its exclusion from the exemption category. This issue will be discussed in greater depth in Section III.A.

The TSCA requires manufacturers to notify the OTS through a formal Premanufacturing Notice (PMN) 90 days before they are going to manufacture and distribute new chemical substances for commercial purposes. When nucleic acids from different microbial genera are hybridized, the EPA defines the resultant recombinant as "new" for regulatory purposes. The same is true for microorganisms constructed by cell fusion. When manufacturers construct those "new" microbes, they must be reported to the OTS even if they are to be used in closed systems. At present, only recombinant GEOs constructed using well-characterized intergeneric nucleic acid segments of noncoding regulatory regions are exempt from this policy and do not require a PMN. But, as stated earlier, exemption categories and scope definitions are presently under review.

In addition to the newness of the organism, if a microbe is to be employed in a different application than originally submitted, it may become subject to the Significant New Use (SNU) regulation, based on changing populations subject to risk. For example, if Snomax™ (see Section III.B) was to be used to seed clouds rather than ski slopes, this would constitute a SNU.

In general, the TSCA will not be triggered if the microbes are used to produce foods, additives, drugs, vaccines, cosmetics, or medical devices if they are regulated by the FDA or USDA. Other than those listed exemptions, all other new or new use chemical end-products from microorganisms, which themselves may be considered new, are subject to the TSCA, and PMNs are required even within closed systems. The OTS uses consent orders and negotiated agreements as the mechanisms to cover environmental introductions.

In addition to the preceding regulations, the EPA regulates wastes resulting from the biotechnology industry under its Resource Conservation and Recovery Act (RCRA) for hazardous and industrial solid wastes and the Clean Water Act (CWA) for wastewater discharges. It has not made any special provisions for biotechnology under these acts. The application of GEOs in bioremediation at toxic waste sites has been included in new policies under the comprehensive Environmental Response, Compensation, and Liability Act (CERCLA/Superfund), but these additions under the act have not been biotechnology-specific.

The agency also created a Risk Assessment research program in its office of Research and Development in 1983 to develop methodology for biotechnology risk assessment. The program was intended to develop methods required for regulatory decision-making. It further expanded its biotechnology personnel by creating the position of special assistant (SA) for biotechnology in 1987 to aid and coordinate biotechnology activities of the assistant administrator of the OPTS. Its function is consensus development and policy review. The SA serves as a facilitator for the establishment of mutually acceptable terms for review. For example, the SA was instrumental in drawing up a consensual consent order between the OPTS and Monsanto, who was collaborating with Ellis Kline of Clemson University for field testing a strain of *Pseudomonas fluorescences* containing the *Escherichia coli lac*ZY gene (see Section III.B). The SA arranged intergovernmental cooperation among the OPTS, the USDA APHIS, the FDA, and RCRA and CWA experts.

The EPA has an 11-member Biotechnology Science Advisory Committee (BSCA) that was formed in 1986 to advise the administrator of the EPA. The BSCA creates subcommittees that review risk assessments produced by the OTS and the OPP and attempts to coordinate the actions of both offices. The committee also forms expert panels to investigate and review science and policy issues.

Between 1983 and 1991, the OPP and the OTS reviewed over 60 biotechnology products. Over half of those reviews involved just two bacterial species,

Rhizobium meliloti and *Bacillus thuringiensis*, either as the host or donor (Tables I and II). *Rhizobium meliloti* has been reviewed by the OTS under the TSCA as a new chemical, and *B. thuringiensis* (*Bt*) by the OPP under the FIFRA as a pesticide. EUPs were not required for small-scale testing of certain classes of microorganisms. Again one of Monsanto's applications is a good example. They were not required to obtain an EUP for testing a killed genetically engineered pseudomonad containing the *Bt* delta endotoxin. In fact, this was counted as an inanimate pesticide. Strains with transgenic markers being tested purely for tracing purposes such as the *Pseudomonas aureofaciens* containing the *E. coli lac*ZY submitted by Monsanto go through the TSCA analysis. The EPA claims to review each application on a case-by-case basis based on the product and the risk and not the means by which the organism was created. Yet it is interesting to note that no EUPs have been required for undirected mutagenesis, most transconjugants, and plasmid-cured strains. Yet EUPs were required for all live recombinant DNA GEOs irrespective of product or risk. Of all the field tests conducted to date, not one has resulted in any negative environmental impact or proved hazardous to human or animal life (if one discounts vandalism!).

Despite the recent efforts of the EPA under the FIFRA and TSCA, it is clear that considerable effort is still required in streamlining and developing a consistent, effective, science-based regulatory process within this agency. Issues such as scope definition, exemption categories, oversight of small- versus large-scale tests, distinctions between commercial and academic research and development, prescreening and procedures for developing criteria for decision-making, and data development all need to be analyzed and an effective system developed.

E. U.S. Department of Agriculture Oversight

The USDA has a responsibility both to fund basic research and to oversee the products of that research. Its scope for coverage is food and fiber products. It has a broad regulatory authority to protect against the adulteration of foods products made from livestock and poultry, to protect agriculture against threats to animal health, and to prevent the introduction and dissemination of plant pests. While its authority is as equally applicable to genetically modified animals, plants, and microorganisms, it is in one sense outside the circle of ecological safety, in that its primary concerns are the safety of crop plant and food animals and the safety and wholesomeness of food products.

The application of biotechnology to agriculture is of course not the sole premise of the USDA: It falls within the regulatory net of the EPA as outlined earlier and also the FDA under its mandate to oversee food safety and the regulation of new animal drugs, vaccines, and feed and the NIH RAC at the level of basic research.

The USDA has nine divisions that deal with biotechnology: the Agricultural Research Service (ARS), the Food Safety and Inspection Service (FSIS), APHIS, the Agricultural Marketing Service, the Cooperative State Research Service Extension Service, the National Agricultural Library (NAL), the Forest Service, and the Economic Research Service.

The Committee on Biotechnology in Agriculture is comprised of administrators of the USDA agencies with major activities involving biotechnology. The CBA operates on principles similar to those of the BSCC on a department level in that it provides coordinated interdivisionary review and recommendations on departmental policies involving research and regulatory matters. It seeks to develop policies that ensure adequate oversight but not to the detriment of innovative research in biotechnology.

Of the listed divisions, three play the most significant role in regulation of biotechnology. The ARS is research-centered and has formed a group similar to the NIH RAC called the Agricultural Biotechnology Recombinant DNA Advisory Committee (ABRAC) to review proposals and provide guidance on matters of biosafety in the development and use of biotechnology in agriculture. ARS has also incorporated a biotechnology information service as part of its National Biological Impact Assessment Program, and it provides procedures for field test applications on computer disks. The FSIS's function is to ensure the safety and wholesomeness of food products. APHIS is the watchdog that guards the licensing of veterinary biological material and issues permits for transport of biological material and field tests of genetically engineered plants and microorganisms. APHIS has formed the Biotechnology, Biologics and Environmental Protection Division with responsibility for all biotechnology products. Under the NEPA, the USDA has a responsibility for ensuring ecological safe utilization of crops, livestock, and veterinary products produced from both traditional

Table I Examples of Applications to the Environmental Protection Agency to Field Test Genetically Modified Organisms under the Federal Insecticide, Fungicide, and Rodenticide Act[a]

Organism	Trait	Source of genetic material	Issues
Pseudomonas syringe *Pseudomonas fluorescens*	Frost protection	Recombinant DNA deletion	Potential for adverse effects on weather may displace ice$^+$ strains
	Bioinsecticide	*Bacillus thuringiensis* delta endotoxin	Environmental problem due to toxin; effect on nontarget species
	Marker gene for tracking genetically engineered microorganisms in environment	*E. coli lacZY*	None
Bacillus thuringiensis	Transconjugant and plasmid-cured strains	Other *B. thuringiensis* strains	Effect on nontarget species
Trichoderna harzianum	Control of fungal pests	Hybrid strains with other *Trichoderna*, undirected mutagenesis	None identified
Clavibacter xyli	Control of corn borer	*B. thuringiensis* delta endotoxin	Mechanism of plant to plant transmission gene loss, competition
Colletotrichum gloesporides	Pesticide on rice and soybeans	Chemical mutagenesis	Host range changes
Sclerotinia sclerotiorum	Control plant pests	Ultraviolet mutagensis	Relative risk compared to parental strain
Trichoderna viride	Control of fungal pests	Undirected mutagensis	Level of risk
Autographa californica (nuclear polyhedrosis virus)	Nonpersistent viral insecticide	Recombinant deletion in polyhedrin gene	Persistence in environment
Tobacco, tomato, cotton	Insect resistance	*B. thuringiensis* delta endotoxin	Transfer of *Bt* genes to other plants

[a] Not all organisms are the products of rDNA research.
[Information courtesy of the Environmental Protection Agency, Office of Pesticide Programs.]

Table II Examples of Applications to the Environmental Protection Agency to Field Test Genetically Engineered Organisms under the Toxic Substances Control Act

Organism	Trait	Source of genetic material	Issues
Rhizobium meliloti	Nitrogen fixation	Various rhizobia	Genetic stability, host range, survival
	Marker gene	Various bacterial sources	
Bradyrhizobium japonicum	Nitrogen fixation	Bacteria	Antibiotic resistance in field plot
Pseudomonas aureofaciens	Tracking system for genetically engineered microorganisms	*E. coli lacZY*	None

[Information courtesy of the Environmental Protection Agency, Office of Toxic Substances.]

and recombinant DNA methods. [*See* RECOMBINANT DNA, BASIC PROCEDURES.]

In 1987, the USDA established a dedicated office called the Office of Agricultural Biotechnology under the Deputy Secretary of Agriculture to facilitate and ensure the coordination of all USDA biotechnology activities. It has taken the lead in the development of guidelines for field testing GEOs and has published a handbook to assist researchers applying for field tests.

The USDA policy on the regulation of biotechnology, consistent with the overall federal policy, does not view GEOs as fundamentally different from those produced using traditional methods. The USDA considered that the products of the new techniques of biotechnology were in principle covered by regulations that had been implemented for existing technologies. They did, however, consider that the assessment of the products of the new technologies in some instances required specific information that necessitated the introduction of some new regulations and the updating of some existing ones.

As federal policy now dictates, the USDA regulations focus on the product and any risks posed by the specific use of that product, rather than on the process used in its production. Thus, the USDA regulates biotechnology products, not because they are products of biotechnological processes per se, but rather because the products themselves are subject to regulation irrespective of the technology involved. Having said this, however, the guidelines that the USDA published for public comment in 1990 fell into the trap of basing their exemption category in large part on process-based assessments.

The principle organ that has been involved in implementing the regulation of field testing GEOs within the USDA is APHIS. Under the Plant Quarantine Act (PQA) and the Federal Plant Pest Act (FPPA), APHIS regulates the movement into and through the United States of plants, plant products, plant pests, and any product or article that may contain a regulated article at the time of movement. A plant pest is effectively any (nonvertebrate) organism that can damage a plant, or products thereof, including insects, snails, nematodes, bacteria, fungi, and viruses.

Regulations governing the introduction of conventional plant pests have been in place and permits have been issued since 1959. In 1987, APHIS introduced specific regulations to cover the introduction of organisms and products produced through genetic engineering where a plant pest is used in the construction. Because the Ti plasmid of the "pest" *Agribacterium tumifaciens* is the principle vector used in the transformation of plants and control regions from mosaic viruses such as the cauliflower mosaic virus 35S promoter are frequently used in constructs, field trials of recombinant plants will, in general, fall under APHIS's jurisdiction. Permits are required for the introduction of any regulated article, where introduction not only means environmental release, but also any movement into or within the states, and regulated article means any organism that has been altered or produced through genetic engineering, if the donor organism, recipient organism, or vector agents belongs to any genus or taxon designated on a specified list of organisms and that meets the definition of plant pest, or is an organism whose classification is unknown, or any product that contains an unclassified organism and may be a plant pest.

As the knowledge base evolves, specific exemptions of organisms that fall under the preceding definition have been made. For example, a movement permit is not required for genetic material from a plant pest contained in *E. coli* K12, sterile strains of *Saccharomyces cerevisiae*, and asporogenic *B. subtilis*. Over time, this list will no doubt grow.

Under the NEPA, an environmental assessment is prepared by the reviewing scientist in APHIS and is a key component in the permit review process. It is a comprehensive account of the analytical process the agency undertook in its decision to issue a permit. In addition to the regulations, it includes the conditions under which the permit is granted (or denied), precautions against environmental risk, the background biology of the organisms, and the possible environmental consequences of the field test. The test environment is described and the precautions to be undertaken to protect that environment, including plot design, inspection, monitoring, security, and disposal plans. All possible eventualities are examined taking into consideration the biology of the donor, recipient, and vector; potential for containment; risk to native flora and fauna; and impact on human health.

A particulary innovative feature of APHIS regulations is that within the regulations is a petition for their amendment. The petition is used to amend the list of regulated organisms and exemptions. A petition that meets the supporting requirements is published in the *Federal Register* for comment and, if approved, the changes to the regulation are published.

APHIS coordinates its oversight reviews with the FDA and the EPA and works with the EPA in reviews of applications for field tests where jurisdiction is shared—for example, when *Bt* endotoxin genes are involved. As of December 1991, APHIS has granted 190 permits to field test GEOs under the PQA and FPPA (Table III). The USDA and EPA have collaborated on the review of many of the engineered organisms, most specifically tobacco and tomato plants containing *Bt* delta endotoxin. And of all the field trials conducted under APHIS, not one has had any negative impact at either the ecological or societal level.

Veterinary biologic products are covered by APHIS under the Virus–Serum–Toxin Act (VSTA). This act covers all production, movement, and use of such products within the United States. The term biological products is very broad and covers all viruses, sera, toxins, and analogous products of a natural or synthetic origin, such as diagnostics, antitoxins, vaccines, live microorganisms, and antigenic or immunizing components of microorganisms intended for use in the diagnosis, treatment, or prevention of disease in animals.

APHIS issues licenses and permits for production, importation, sale, and experimental use of various types of biological products. Requests to use unlicensed products for experimental field studies must include information to demonstrate that the experimental conditions will be adequate to prevent the spread of the disease. Specific information that must be submitted for GEOs and products includes detailed information on stability, genetic constructs and vectors, and the effects of any insertions and deletions on the organism. GEOs for the purposes of the VSTA are classified into three groups: (1) inactivated products such as recombinant DNA-derived vaccines, bacterial toxins, viral and bacterial subunits, and monoclonal antibodies; (2) live products, such as those containing live or infective organisms modified by insertion or deletion of one or more genes; (3) live vectored products, such as products using live vectors to carry recombinant-derived foreign genes that code for immunizing antigens or other immune stimulants.

Where the introduction and field testing of GEOs is concerned, the Animal Quarantine Statutes (AQS) also apply. Permits are required for GEOs that, in their construction, use material from infectious, contagious, pathogenic, or oncogenic organisms. The submission requirements are similar to those described previously for biologics. The principle applications for field tests under this authority have been for recombinant animal vaccines such as pseudorabies and rinderpest (see Section III.B).

F. Food and Drug Administration Oversight

The FDA regulates biotechnology under the authority of the Food, Drug, and Cosmetic Act (FDCA) and the Public Health Services Act (PHSA). The agency has a mandate to ensure efficacy and safety of food and pharmaceutical products. Although the agency has a major responsibility in biotechnology in that 60% of the current market share of biotechnology products passes through the agency for review and it has already reviewed thousands of biotechnology products, it is only mentioned here in passing because its responsibilities to environmental integrity only exists insofar as it falls within the NEPA. The aspects of human safety with respect to field release of GEOs is covered by the EPA and USDA.

Table III Examples of Applications to the U.S. Department of Agriculture to Field Test Genetically Engineered Organisms (Containing Microbial Genetic Material) under the Animal and Plant Health Inspection Service

Organism	Trait	Source of genetic material	Issues
Under the Virus–Serum–Toxin Act and the Animal Quarantine Statutes			
Vaccinia virus	Immunization	Rabies virus	Persistence in environment, spread
		Pseudorabies virus Rinderpest virus	Persistence in environment, spread
Under the Plant Quarantine Act and the Federal Plant Pest Act			
Clavibacter xyli/corn	Control of corn borer	*B. thuringiensis* delta endotoxin	Mechanism of plant to plant transmission gene loss, competition
	Marker gene	Various bacterial sources	
Tomato	Insecticide resistance	*B. thuringiensis* delta endotoxin	Plant to plant transmission
	Virus resistance (TMV)	Mosaic virus	
Tobacco	Insect resistance	*B. thuringiensis* delta endotoxin	Mechanism of plant to plant transmission resistance
	TMV resistance	TMV coat protein gene	Resistance
	Marker gene	*E. coli* chloramphenicol acetyltransferase	Antibiotic resistance in environment
TMV/tobacco	Pharmaceutical production	Human melanin gene in TMV	
Cotton	Insect resistance	*B. thuringiensis* delta endotoxin	Plant to plant transmission
Potato	Potato virus X,Y resistance	Potato virus X,Y coat protein gene	
Walnut	Insect resistance	*B. thuringiensis* delta endotoxin	Mechanism of plant to plant transmission
	Marker gene	GUS gene	Pleiotrophic effects
Alfalfa	Resistance to AMV	AMV coat protein gene	
Cucumber	Resistance to CaMV	CaMV coat protein gene	
Cantaloupe	Resistance to CaMV	CaMV coat protein gene	

AMV, alfalfa mosaic virus; CaMV, cauliflower mosaic virus; TMV, tobacco mosaic virus.
[Information courtesy of the U.S. Department of Agriculture, Animal and Plant Health Inspection Service.]

G. Omnibus Bill

In addition to the above, Congress has been discussing a bill that is designed to rescue the federal government from the current delays and confusion that have accompanied attempts to establish regulatory policies over deliberate releases of genetically engineered organisms. The bill would establish (1) a Biotechnology Science Research Program within the OSTP; (2) permit-issuing processes for EPA under TSCA and for the USDA generally; (3) an application management board for certain proposals for releases; (4) provisions for state review; and (5) a 7 yr sunset provision. It may act as an impetus for the agencies to streamline their regulations and guidelines.

III. Assessment of Environmental Introduction of Genetically Modified Microorganisms

A. Risk Assessment

The issue of environmental introduction of GEDs, specifically microorganisms, has been the subject of much conjecture and polarizing rhetoric since the early 1980s when a researcher first applied to the RAC for permission to "go outside" with his "ice-minus" genetically engineered *Pseudomonas syringae* and *P. fluorescens* in 1982. The potential environmental impact has been debated by everyone from the most informed scientists to the least informed self-professed guardians of the public good. Over time there has been an evolution in the format of the debate by the scientifically enlightened over the introduction of GEOs into the environment. In its initial stages, the representatives from the two principle fields of expertise (molecular biology and ecology) were firmly entrenched in their view that each had sufficient knowledge to make proclamations on impact without recourse to the knowledge of the other and each had a plethora of historical precedent to support their views. By the end of the decade, the more enlightened molecular biologists and ecologists had progressed sufficiently in their thinking to realize that dialog, not conflict, was the most effective mechanism to assess impact and risk. They used their complementing abilities to determine what considerations are important to risk, how to develop effective overarching risk assessment frameworks, and how to scale regulatory scrutiny on the basis of risk.

The properties of the introduced organism and its target environment are the key features in the assessment of risk. Such factors include the demographic characterization of the introduced organisms; self-sustainability; dispersal ability; competitive ability; genetic stability, including the potential for horizontal transfer or outcrossing with pests or weeds; potential evolution; and the fit of the species to the physical and biological environment. The scale and frequency of the introductions are also important considerations. All factors apply equally to both modified and unmodified organisms and independently of the means used to effect any modification—the organism and its products are the important issues and not how they were constructed.

In 1982, the EPA Office of Strategic Assessments and Special Studies decided to initiate a fact-finding effort to anticipate the trends in biotechnology as it related to the environment. They selected the American Association for the Advancement of Science to collaborate on a seminar series to examine the issues relevant to this area. From a risk assessment point of view, they concluded that because information about the fate of GEOs in the environment is not directly available, risk assessments need to be based on analogy and extrapolation, and, although the historic record can be used to develop components of a recombinant DNA risk analysis scheme, more experience with genetic engineering is a prerequisite to better prediction.

The first noteworthy salvo in the debate on release was struck by a respected ecologist when he predicted an inevitable environmental disaster from environmental release of genetically engineered microorganisms (GEMs). The opening salvo on the part of molecular biologists was made by an equally respected microbiologist when he argued that plants and animals have been altered through traditional breeding by humans for centuries without any serious problems. Even more to the point, he felt that bacteria and fungi, including pathogenic organisms, have been added to soils and plants in an attempt to determine if beneficial uses could be found for these organisms. He concluded that these observations alone were sufficient on which to build a data base from which a risk assessment framework could be constructed even for GEOs.

Ecologists took umbrage at the nonquantitative nature of the preceding arguments. They observed that risk assessment based on quantitative factors is both appropriate and necessary. Any alteration to the genotype that may increase the flexibility and survival abilities of an organism in a given environment is not a "small" effect from an ecological point of view they argued. They also stated that, especially from the point of view of ecology and population dynamics, the phenotype of an organism is not predictable from the genotype. Extrapolation from laboratory experiences is not appropriate. Each release must be analyzed on a case-by-case basis. One researcher continued this argument when she stated that the absence of any ill effects to those who had worked with recombinant organisms for years in the laboratory was irrelevant in the assessment of whether or not such would be the case in an uncontained environment. She argued that shifts in environmental contexts can be as important as genetic modifications when determining the ecological effectives of an engineered organism relative to the parental strain. Molecular biologists struck back by claiming that there was nothing really new about using modified microbes—It was just an extension of the old process of selecting for useful organisms, in fermentation processes, in pest control, and in vaccine and antibiotic production. One researcher also somewhat contemptuously stated that ecologists' experience with transplanted higher organisms was less pertinent than the insights of fields closer to the specific properties of GEMs such as population genetics, bacterial physiology, epidemiology, and the study of pathogenesis.

The debate raged on with opponents citing examples of disastrous introductions such as Dutch elm disease, gypsy moth, and chestnut blight and proponents equally quoting beneficial introductions such as *Puccinia chondrillina* controlling skeletonweed in Australia, *Bacillus popilliae* controlling Japanese beetles in the United States, and *Rhizobia* inocula for nitrogen fixation.

In 1985, the NIH RAC (Recombinant DNA Technical Bulletin) had outlined some points to consider for environmental testing of microorganisms. They included genetic considerations of the modified organism to be tested including characteristics of the parent and molecular biology of the modified organism; habitat and geographical distribution; physical and chemical factors that can effect survival, reproduction, and dispersal; biological interactions; survival, replication, and dispersal of the modified organism; numbers and methods of application; plot design; containment; monitoring; risk analysis based on the nature of the organism; and the nature of the test.

The international community got in on the act through a subcommittee of the International Council of Scientific Unions (ICSU) called Scientific Committee on Problems of the Environment (SCOPE). The mandate of SCOPE is to assemble, review, and assess the information available on man-made environmental changes and the effects of these changes on humans, to assess and evaluate the methodologies of measurement of environmental parameters, and to establish itself as a corpus of informed advice on all matters dealing with the environment and environmental research. Another subcommittee of the ICSU, the Committee on Genetic Experimentation (COGENE), had been active since the early 1970s in evaluating the impact of recombinant DNA technology and providing international scientific evaluations to public discussions. A Working Group of this committee articulated the three most common concerns of recombinant DNA technology in the 1970s. GEOs may spread into the environment and disrupt ecological equilibria; they may produce some noxious substance or otherwise cause disease; and scientists may be crossing some hypothetical barrier to DNA exchange, thus affecting the pathogenicity or dispersion of pathological agents. After lengthy deliberation and consideration of all available evidence, the Working Group concluded that there were no scientific findings to justify those concerns. This international report became the basis for the relaxation of the NIH guidelines in the early 1980s. SCOPE and COGENE working together brought a global perspective to the issue of environmental introduction of GEOs. In their report on the introduction of GEOs into the environment, they stated that public discussion of the safety issues related to GEOs has frequently been clouded by decisions of regulatory bodies that were dictated more by politics than by logic. They continued that if political considerations are discounted, there are no convincing scientific grounds for distinguishing engineered organisms from natural ones. Because organisms of either type could pose unforeseen hazards, they cautioned that some safety testing is desirable before large-scale propagation. They see the need for gathering more knowledge, for the objective consideration of information that already exists, and for the formulation of both short-term and long-term solutions to the problems of the biosphere. They also consider that there is an immediate need to provide guidance on how to proceed prudently at the present time without stifling initiative and progress.

The NAS stepped into the debate in 1987 when they published a pamphlet on the key issues involved in environmental introductions of GEOs. Their conclusions on risk were the following: (1) no evidence indicates that unique hazards exist either from the use of recombinant DNA techniques or from the movement of genes between unrelated species; (2) the risks associated with the introduction of GEOs carrying recombinant DNA are the same in kind as those associated with the introduction of unmodified organisms and organisms modified by other methods; and (3) assessment of the risks of introducing GEOs carrying recombinant DNA into the environment should be based on the nature of the organism and the environment into which it is produced and independent of the method of production. Following protests on the lack of supporting documentation, the NRC of the NAS undertook an extensive review of the issue of risk assessment as it pertained to environmental introductions for GEOs. The Committee on Scientific Evaluation of the Introduction of Genetically Modified Microorganisms and Plants into the Environment concluded that no conceptual distinction exists between genetic modification of plants and microorganisms by classical methods or by molecular techniques that modify DNA and transfer genes, whether in the laboratory, in the field, or in large-scale environmental introductions. This formulation has three implications: (1) the product, not the process, of genetic modification and selection constitutes the primary basis for decisions about environmental introductions, (2) knowledge of the process provides information about the product that is useful for risk assessment but is not in itself a useful criterion to determine the appropriate level of oversight, and (3) organisms modified by modern molecular techniques are subject to the same laws of nature as are those produced using classical methods for which there is a wealth of experiential data. The NRC outlined the extent of the informational requirements for risk assessment in the form of three questions: (1) Are we familiar with the properties of the organism and the environment into which it may be introduced? (2) Can we confine or control the organism effectively? (3) What are the probable effects on the environment should the introduced organisms or a genetic trait persist longer than intended or spread to nontarget environments?

Interestingly enough, one of the members of the microorganism subcommittee of the NRC committee was the lead author of a report by the Ecological Society of America that appeared around the same time as the NRC report. Not surprisingly, it focused more on the ecological and evolutionary aspects of introductions and suggested four categories to be considered for oversight. The categories were (1) genetic alteration, (2) wild-type progenitor, (3) phenotypic attributes in comparison with the parental organism, and (4) environment. The authors provided a scale for each attribute and suggested that oversight should be commensurate with scale. They did, however, argue that most engineered organisms will pose a minimal environmental risk, but they did outline potential undesirable effects (which indeed could apply to any type of organism), creation of new pests, enhancements of effects of existing pests, harm to nontarget species, disruptive effects on biotic communities, adverse effects on ecosystem processes, incomplete degradation of hazardous chemicals, and squandering of valuable biological resources. The worst possible ecological impact a planned GEM introduction could have would be to disrupt a fundamental ecosystem process such as the cycling of a mineral or a nutrient, or the flow of energy in an ecosystem. However, the high degree of redundancy among microbes involved with such processes and the resilience and buffering in natural ecosystems mitigate against such an occurrence. One researcher has pointed out that the preceding categories are speculative because of the limited experience with the introduction of GEMs.

Conventional wisdom on the safety of introduced GEMs is based on four criterion: (1) almost all genetic change reduces fitness; (2) wild-type genotypes are almost always superior to introduced genotypes of the same species; (3) there is no "free lunch" in evolution—all advantages accruing from the engineered genes must be paid for, and this will lead to reduced fitness in the wild where, in the absence of selection in favor of the trait, the engineered gene is more likely to be a liability than a benefit; and (4) however, some engineered traits (increased tolerance to severe conditions) may be advantageous under field conditions and these traits may prosper away from the introduced environment.

There is a quantum difference between the perception of introducing genetically engineered plants and of introducing GEMs into the environment, the former having had a relatively noncontroversial right of passage, whereas the latter have been the subject of outright sabotage. The historical equation of microbes with disease in the public mind is almost as equally responsible for this effect as is the ecological assessment of relative risks.

While ecology literature is strewn with examples of the dynamics of species interaction and impacts of introduction among higher organisms, little has been devoted to microorganisms. One scientist points out that microorganisms live in ecosystems just as higher organisms do; they compete, prey on one another, and modify each others' environment chemically and physically. Microbial ecologists speak of "stable climax communities" of microbes that evolve by selection and successional processes. Microbial communities are normally resistant to penetration by new species but perturbations may result in the dominance of new species. Another researcher maintains that small size and high growth rate make prokaryotes adapt more quickly and dramatically to environmental changes assuming large numbers of genomes are correlated to diversity. He also maintains that they are easily dispersed and optimally adapted to their environments. However, in contrast to this niche saturation theory, two others maintain that there are unsaturated niches and an introduced organism in a microbial system is more likely to survive where competition is limited and that novel organisms will likely survive in unsaturated niches with limited competition. [*See* ECOLOGY, MICROBIAL.]

Of course, in addition to chance, undirected mutagenesis and selection, bacteria have another method for genotypic adapatation in natural habitats—

genetic recombination. Questions that this poses for introduction of GEMs include the following: What are the conditions that encourage transfer or maintenance of the inserted gene, and how likely is it that the genes could be transferred beyond the target organism? If transferred, will the new genetic material be expressed? If transferred and expressed, will there be any environmentally significant consequences positive or negative? There has been, however, no convincing evidence that gene transfer and subsequent fitness of the recipient have a higher frequency in soil and natural habitats. Two scientists studied the transfer of genetic material in nonsterile soil and freshwater *in situ,* and they found the frequencies to be very low, 1×10^{-9} and 3.3×10^{-8}, respectively, with the promiscuous plasmid pRD1. Despite the observation that in culture gene flow is continuous and rapid, especially for gram-negative bacteria, no experimental evidence indicates that genetic transfer occurs routinely and successfully in soil. Studies on the effects of adding various strains of *E. coli, Enterobacter cloacae, P. putida,* and *P. aeruginosa,* with and without plasmids carrying antibiotic resistance genes to soil have not shown any consistent and lasting effects on the gross metabolic activity (CO_2 emmission), the transformation of fixed nitrogen, the activity of soil enzymes (phosphatases, arylsulfatases, dehydrogenases), or the species diversity of the soil microbiota. However, the introduction into soil of a strain of *Streptomyces lividans* that contained a plasmid carrying a lignin peroxidase gene from *Streptomycos viridosporous* enhanced the rate of mineralization of soil carbon during the 30-day incubation period, especially when lignocellulose was added.

From the latter instance, it can be determined that in assessing risk, consideration of intrinsic factors such as the whole construct of the introduced gene, new host, and vector material may be helpful. The intracellular configuration of the new gene in the host organism may in fact determine its potential for movement to new hosts. Potential extrinsic factors affecting gene transfer include the receptivity of the habitat, selection pressure as in the case of the lignin peroxidase, and density of the introduced organisms and potential target organisms.

In any field release, it is of course of paramount importance to be able to monitor effectively the movement of GEMs within the environment. Useful tracking systems for the movement of GEMS within the environment include selective markers such as antibiotics, nutritional markers such as Monsanto's lactose-metabolizing insert and biochemical markers such as DNA probes. Researchers can also minimize gene transfer by using a gene that makes the bacterium viable only in the presence of a specific metabolite such as a pollutant or by immobilizing the vector. These will be demonstrated for the test cases outlined in the following section, where some examples of determined environmental impact with both classical organisms and GEOs are outlined.

B. Test Cases

Some recent proposals on oversight both here and in Europe are worrying in that, while purporting to be risk-based, they are clearly, through their exemption category, singling out recombinant organisms for special treatment. The BSCC's process-determined definition of "familiarity" inappropriately equates this with safety. It exempts from the oversight net experiments with organisms that are "familiar," defined solely by the test organisms being natural or having been created by classical genetic manipulation techniques.

The lack of wisdom in exempting an organism because of its genus, because it is not modified, or because of "familiarity" is equally as unscientific as the converse action. This is easily demonstrated by some historical examples. The house sparrow (*Passer domesticus*) and the tree sparrow (*Passer montanus*) were both introduced into the United States in the nineteenth century; the former species rapidly spread and ultimately occupied almost the entire North American continent and is now one of our most common birds. The tree sparrow has generally remained restricted to the vicinity of St. Louis, Missouri, where it was first released. The former causes much agricultural damage and has displaced many native insectivores, while there are no reports that the tree sparrow affects the resident community. There is no explanation for this striking difference. Both use similar foods and occupy similar niches. Even more remarkably, in some parts of the world (southern Asia) where both species were introduced, the tree sparrow spread widely and became more numerous than the house sparrow. The same scenario is true of two closely related subspecies of fire ant, one resident, one introduced in the southern United States, four species of mongoose introduced in various parts of the world, and numerous plant species.

Researchers did an analysis of the establishment of introduced species for biological control of arthropods between 1890 and 1968. The overall rate of establishment was a rather low 0.34 (N = 2295) and the rate of success in controlling the target pest was just 0.16 (N = 602). A researcher has interpreted these findings insofar as they portend to the fate of the introduction of GEOs for biological control. He concludes that planned introductions may not result in establishment, and that those that do get established may not provide an adequate solution to the target problem. The fate with weed control was analyzed for data collected between 1900 and 1980. With respect to insect introductions for weed control, the rate of establishment was 0.63 (N = 488). The rate of successful control was 0.36 (N = 151). The higher rate for biocontrol of weeds is most probably due to the requirement for more thorough preintroductory studies to meet regulatory requirements(!).

The world's first commercial biopesticide based on a live, genetically engineered microorganism debuted with the approval for sale of "NoGall" in Australia in February 1989. The agent is a genetically-modified strain of *Agrobacterium tumifaciens,* which is based on a biocontrol agent that has been used to combat crown gall disease of temperate fruit trees since the mid-1970s. The strain is missing the gene that permits transcongugation to occur, thus reducing the possibility of out-crossing with wild-type virulent strains. Ecological studies on the spread and persistance of the genetically modified inoculum are ongoing.

The first release of GEMs in the United States took place in a strawberry bed in Brentwood, California, and a potato patch in Tulelake, California, in April 1987. One researcher from the University of California, Berkeley, and one from Advanced Genetic Systems had first applied to the NIH RAC in September 1982 to conduct the test of Ice$^-$ *P. syringae* strains. After a 5-yr baptism by fire, where not just molecular biologists, ecologists, and regulators were involved but the halls of justice, Earth First!, county supervisors, and the media had their say, the tests finally went ahead. The ironic part of the hullabalu was that this was not the introduction of a novel organism: Ice$^-$ *P. syringae* strains exist in nature—It was merely an enrichment exercise. Indeed, enrichment with the converse Ice$^+$ strain saved the Calgary Winter Olympics when Snomax™ was sprayed on the slopes. Yet, because the latter was not the product of genetic manipulation, it did not even warrent a mention by the media.

The tests with Ice$^-$ *P. syringae* strains proved that the treated plots showed significantly less damage than untreated plants with a 50–85% decrease in freezing injury. Ice$^-$ strains represented about 20–90% of total bacteria on treated plants in the first month. The deleted mutants had demonstrated no competitive advantage over Ice$^+$ strains. Even when the latter were applied at much lower concentrations and 2 days later than Ice$^-$ strains, the Ice$^+$ strains were not eliminated. Ice$^-$ strains were dispersed all around the test plot but decreased almost logarithmically with distance and were found in vanishingly small numbers beyond 30 m. Indigenous Ice$^+$ strains were found in high numbers in adjacent wheat fields and recombinant Ice$^-$ strains were never found on those plants, presumably due to high competition pressure. Similarly, Ice$^-$ strains did not survive for more than 2 wk in soil in the field similar to their low survival rate in greenhouse soil. After the removal of all plant tissue, no recombinant Ice$^-$ strains were detectable in the plot site.

One of the most convincing experiments to track GEMs in soil was performed by Ellis Kline of Clemson University and Monsanto. In 1987, the collaborators initiated the first GEM tracking system in a wheat field in South Carolina. It was also the first field release of live bacteria carrying genes from two different strains. The tracking system consisted of *P. fluorescens* transformed with *lacZY* genes from *E. coli* K12 using a disarmed Tn7 transposition vector. The 18-mo study extended over three crops and more than 11,000 individual samples of plant and soil material. The test confirmed the effectiveness and the practicality of the lacZY marker to monitor the survival and location of GEMs under field conditions and displayed a pattern of bacterial survival with high populations on plant roots in the inoculated rows, but remarkably limited dissemination of the marked bacteria even centimeters away. Equally comforting was the fact that genetic exchange of the marker with native field strains was not detected by selection/hybridization capable of detecting such a transfer at a frequency as low as 9.8×10^{-8} events per gram of soil. They also found minimal carryover to noninoculated crops and no significant impact on crop yield caused by the marked strains. The test provided not only a valuable scientific data base and proved an effective monitoring system but also helped assuage the fear that genetically engineered microbial pesticides might not be easily monitored and safely applied.

Where the use of GEOs in biological control is concerned, *B. thuringiensis* as the source genetic material has been the subject of the greatest majority of field trials, not just of transgenic plants but also GEMs. The insecticidal crystal proteins (ICPs) of *B. thuringiensis* are toxic (but strain-specific) to many lepidoptera, diptera, and coleoptera larvae and those of *B. sphaericus* are limited to mosquito larvae. The genes for combinations of ICPs might be introduced into a variety of atypical systems to improve the total efficiency of this biocide in the field. For example, the gene has been introduced into bacteria commensurable with plants and has been field tested by coating seeds. Both Monsanto and Mycogen have cloned the *Bt* subsp. *kurstaki* ICP gene into pseudomonads. Monsanto used the host strain described earlier (with the inactivated transposase) to colonize cornroots for control of the black cut worm. Mycogen created a biological package, or microcapsule, for the toxin to protect it from the environment. The latter company also killed the cells, thus reducing the requirements for EPA and APHIS oversight. Crop Genetics International (CGI) inserted the ICP gene from the same *Bt* subspecies into the endophytic bacterium *Clavibacter xyli* for control of the European cornborer, one of the largest uncontrolled pests of corn. As an endophyte, *C. xyli* is capable of colonizing corn internally. CGI truncated the transacting recombinase gene segment in their gene cassette to reduce the possibility of gene transfer. Field tests of the preceding systems were conducted to determine recombinant levels in inoculated plants, to monitor mechanical and natural spread of the GEMs to corn and other plants, to monitor the GEM in plant residues, to monitor the GEMs in runoff water and soil, and to compare yields of colonized and control corn plants. The results were found to be the same as for the Monsanto and Kline study, where the GEMs colonized only the inoculated plants; there was no detectable gene transfer to resident species, nor were the GEMs detected more than centimeters away from the inoculated site. The CGI case provided some firsts for field testing GEMs. It was the first release by an agency of the U.S. government (ARS in Beltsville, Maryland), it was endorsed (with qualifications) by three leading U.S. environmental groups, it was assisted by a Washington, D.C.-formed Social Responsibility Committee to help win approval for release, and it was the first GEM endophyte and the first GEM developed to combat a major pest of the worldwide staple, corn.

On the scale of risk perception, viruses, because of their insidious nature, fall even lower on the acceptability scale than do bacteria. Yet the first release of a GEM in the United Kingdom was a genetically engineered baculovirus insecticide. Bob Possee's test marked the world's first release of a genetically engineered virus when, in 1986, he received approval to field test a recombinant *Autograpa californica* nuclear polyhedrosis virus (AcNPV) that contained an 80-bp synthetic noncoding oligonucleotide insert as a marker. This virus attacks insects and may provide the basis for an effective insecticide. The virus was tested with respect to host range in insects, genetic stability, and persistance in soil. In all respects, it was found to behave identically to the unmodified virus. Infected *Spodoptera exigua* larvae were placed in sugar beet plants in a netted enclosure. The larvae perished after 1 wk but the virus persisted in the soil for 6 mo. A recombinant, with the coat protein gene deleted and replaced with a unique oligonucleotide marker, was found to have the same host range but was susceptible to ultraviolet light and inactivation in the soil and did not persist in the environment. This scientist continues to scale up his tests and has not encountered negative impacts on any level.

The first U.S. direct recombinant virus release was covered by APHIS's VSTA and AQS, as it involved a rabies vaccine incorporated in meat traps for feral animals on Parramore Island off the coast of Virginia. This vaccine is based on the vaccinia virus and carries the gene for a key glycoprotein found on the surface of the rabies virus. After much to-and-fro with the various environmental organizations, the test was allowed to proceed in August 1990.

An interesting field test with a recombinant virus was performed by a company in California. They inserted the gene for melanin into a tobacco mosaic virus (TMV) vector. They wished to have the gene expressed in tobacco, but rather than using molecular techniques to transform plants in tissue culture they merely sprayed the recombinant TMV onto the plants to transfect them with the virus. Because TMV is a plant pathogen, the test was covered by APHIS's PQA and FPPA regulations. The commercial application for this strange construct is sunscreen products.

A case where the regulatory coordination broke down was demonstrated by the review of a University of Wisconsin researcher's field test application. This researcher pursued a labyrinthine path in his efforts to get the go-ahead to evaluate the ecological effects of the release of bacteriocin-producing strains of *Rhizobium*. These bacteria form symbiotic nitrogen-fixing nodules along the roots of leguminous plants. Bacteriocins are peptides made by one strain of a bacterium that interfere with the growth of other, ordinarily competing strains. This researcher wanted to know whether bacteriocins can help specific *Rhizobium* strains compete successfully for nitrogen-fixing nodulation space on the roots of commercially important plants. Field data are vital to ascertain the commercial viability of this scheme. However, despite what promised to be an easy passage through the federal approval system in June 1989, by late April 1990 the National Wildlife Federation (NWF) and the Environmental Defense Fund (EDF) insisted to USDA officials that, to meet requirements under the NEPA, the agency needed to prepare an environmental assessment of the proposal. By June, officials in the USDA Office of Agricultural Biotechnology were expected to rush through a hasty environmental assessment. The right to proceed had, however, already been granted by APHIS, the EPA, and the NIH RAC. Moreover, state and local officials, responding to briefings provided by Wisconsin scientists and administrators, appeared to be favorably disposed to this proposal. Nevertheless, the NWF and the EDF assert that his proposal should have gone to the USDA OAB. However, because the USDA still had not published its guidelines for research on GMOs, the assessment was carried out by APHIS, and indeed *Rhizobium* is on the APHIS list of regulated organisms. [*See* BACTERIOCINS: ACTIVITIES AND APPLICATIONS.]

IV. The Future for Genetically Engineered Microorganisms

While many of the bases have been covered in field introductions of GEMs, many more tests need to be undertaken from the environmental impact point of view (as opposed to testing the efficacy of the system being field tested per se). Studies that are needed include the evaluation of GEMs that have been engineered to perform specific enzymatic functions in the soil; the potential ecological effects of the accumulation in soil of the products of genes that have been introduced into new hosts to produce a toxin; the epidemiology in soil of biological control agents that are engineered to cause disease in pests; and, perhaps most importantly, the potential for pleiotrophic effects. Several examples of the latter effect have been observed on passage of GEMs through soil. When the yolk protein of *Drosophila grimshawi* is inserted into the *amp* gene of pBR322, extreme mucoidy in the *E. coli* host was found when it was reisolated from the soil (perhaps through read-through overproduction saturating the activity of the *lon* gene protease). Some researchers report unexpected biochemical and morpohological changes such as filament formation and increased cell fragility on reisolating an inoculated bacteria from the soil. Others found unpredicted phenotypic changes on passage through soil in phytopathogenic bacteria that contained plasmid pRD1 carrying genes for nitrogen fixation and antibiotic resistance, including increased virulence, altered utilization of amino and organic acids, and enhanced resistance to antibiotics for which genes were not present in the inoculum.

As advances in nucleotide chemistry make other techniques possible, new ways of immobilizing vectors and creating restricted and escape-retarded hosts are being explored. Research is also being directed toward increasing understanding of the ecology of different traits. Meanwhile, the implications of introducing any particular organism into the environment must be treated on its own merits with due regard for the possibility of unpredicted and unanticipated effects, but this does not mean that each introduction must be considered *de novo*. As experience accumulates with particular kinds of introductions in particular environments, more generic approaches to these classes of introductions can be developed, as was the case with the NIH RAC guidelines. The bases for action should be reviewed and updated continuously with precedents and cumulative experience smoothing the way for proven innocuous introductions and more caution be afforded to those where knowledge and predictions are less complete. Generalizations developed for particular groups of organisms cannot be extended automatically to other groups, which may have very different genetic and demographic characteristics, dispersal and reproductive mechanisms, and trophic positions. The conditions for containment, monitoring, and mitigation are very important as the possibility of recalling introduced microorganisms approaches zero.

Having said that, however, it should be reiterated that in almost two decades of research with recombinant DNA organisms, both within and outside the laboratory, not one single uncontrollable hazard has been created. No convincing evidence indicates that this situation will be changed by future developments.

Some scientists have devised an excellent risk-based scheme for oversight of field testing GEOs. It is based on the nature of the organism and of the test site into which the organism is to be introduced. It accommodates any organism, wild-type, natural, and modified, selected by any technique, or genetically manipulated by classical or recombinant mechanisms. To determine the degree of oversight of any organism, the investigator first determines from a table the overall level of concern, i.e., the overall perceived risk based on scientific knowledge and experience as determined by experts. That level of safety concern then is entered into a table that also considers the confinement level, i.e., the experimental site for the proposed experiment, and enables the investigator to read off the level of oversight required. The relationship among levels of safety concern, confinement, and oversight can be varied widely to reflect various considerations such as the amount of scrutiny judged to be appropriate, the regulations extant for both the investigator and the agency, etc. This scheme is scientifically defensible and has the flexibility to be refined and applied to any proposed scenario.

Ultimately, if the fruits of this technology are to achieve their true potential in the development of ecologically and economically sound approaches to agriculture and in providing scientists with new approaches to develop higher yielding, more nutritious, tastier, and longer-lasting crop varieties; increasing resistance to diseases and adverse conditions; improving our ability to monitor and reduce environmental contaminants; reducing or eliminating the need for pesticides, fertilizers, and other expensive and potentially hazardous agricultural chemicals; reducing dependency on nonrenewable resources; and increasing sustainability, then an adequate, workable, and effective regulatory environment must be in place for overseeing environmental release and the products of agricultural biotechnology. However, regulation must be product-oriented and risk-based and not be so punitively restrictive that it stymies innovative research and scares away scientists in both academia and the private sector from pursuing projects that will ultimately benefit society and the biosphere.

Bibliography

Animal and Plant Health Inspection Service (1990). A user's guide: Biotechnology permits, regulatory requirements for introduction. U.S. Department of Agriculture (draft copy).

Berg, P., Baltimore, D., Brenner, S., Robin, R. O., III, and Singer, M. F. (1975). *Science* **188,** 991–994.

Colwell, R. K., Norse, E. A., Pimentel, D., Sharples, F. E., and Simberloff, D. (1985). *Science* **229,** 111–112.

Cordle, M. K., Payne, J. H., and Young, A. L. (1991). Regulation and oversight of biotechnological applications for agriculture and forestry. *In* "Assessing Ecological Risks of Biotechnology" (L. R. Ginzburg, ed.). Butterworth Heinemann, Boston.

Crawley, M. J. (1991). The ecology of genetically engineered organisms: Assessing the environmental risks. *In* "Assessing Ecological Risks of Biotechnology" (L. R. Ginzburg, ed.). Butterworth Heinemann, Boston.

Davis, B. (1987). *Science* **235,** 1329–1335.

Devanas, M. A., and Stotzky, G. (1988). *In* "Developments in Industrial Microbiology," Vol. 29. (G. E. Pierce, ed.), pp. 287–296. *J. Indust. Microbiol.* (suppl. no. 3). Elsevier Science Publishers, Amsterdam.

Ehler, L. E. (1991). Planned introductions in biological control. *In* "Assessing Ecological Risks of Biotechnology" (L. R. Ginzburg, ed.). Butterworth Heinemann, Boston.

Environmental Protection Agency (1978). *Fed. Regist.* **43**(No. 52), March 16, pp. 3–103.

Faust, R. M., and Jayaraman, K. (1990). Current trends in the evaluation of the impact of deliberate release of microorganism in the environment: A case study with a bioinsecticidal bacterium. Introduction of genetically modified organisms into the environment (H. A. Mooney and G. Bernardi, eds.). *Scope* 44, Wiley, New York.

Levin, M., and Strauss, H. S. (1991). Overview of risk assessment and regulation of environmental biotechnology. *In* "Risk Assessment in Genetic Engineering." McGraw-Hill, New York.

Lindow, S. E., and Panopoulos, N. J. (1988). Field tests of recombinant ice$^-$ *Pseudomonas* syringae for biological frost control in potato. *In* "The Release of Genetically Engineered Micro-Organisms" (M. Sussman, G. H. Collins, F. A. Skinner, and D. E. Stewart-Tull, eds.), pp. 121–148. Academic Press, London.

Marois, J. J., and Bruening, G. (eds.) (1991). Risk Assessment in Agricultural Biotechnology: Proceedings of the International Conference. University of California Division of Agriculture and Natural Resources, Publication No. 1928, Oakland.

Miller, H. I., Burris, R. H., Vidaver, A. K., and Wivel, N. A. (1990). *Science* **250,** 490–491.

National Academy of Science (1987). "Introduction of Recombinant DNA-Engineered Organisms into the Environment: Key Issues. National Academy Press, Washington, D.C.

National Research Council (1989). "Field Testing Genetically Modified Organisms: Framework for Decisions." National Academy Press, Washington, D.C.

Organization for Economic Cooperation and Development (1986). Recombinant DNA safety considerations. OECD, Paris.

President's Council on Competitiveness (1991). *Fed. Regist.* **55**(147), 31118–31121.

President's Council on Competitiveness (1991). *Fed. Regist.* **56**(22), 4134–4151.

Purchase, H. G., and MacKenzie, D. R. (1990). Agricultural biotechnology: Introduction to field testing. Office of Agricultural Biotechnology, U.S. Department of Agriculture.

Sayler, G., and Stacey, G. (1985). Methods for evaluation of microorganism properties. *In* "The Suitability and Applicability of Risk Assessment Methods for Environmental Applications of Biotechnology" (V. T. Covello and J. R. Fiksel, eds.). National Science Foundation, Washington, D.C.

Sharples, F. E. (1987). *Science* **235,** 1329–1332.

Tiedje, J. M., Colwell, R. K., Grossman, Y. L., Hodson, R. E., Lenski, R. E., Mack, R. N., and Regal, P. J. (1989). *Ecology* **70**(2), 298–315.

Genetically Modified Organisms: Guidelines and Regulations for Research

Anne Vidaver
University of Nebraska

Sue Tolin
Virginia Polytechnic Institute and State University

Glossary

Confinement Procedures to keep genetically modified organisms within bounds or limits; usually in the environment with the result of preventing widespread dissemination

Containment Conditions or procedures that limit dissemination and exposure of humans and the environment to genetically modified organisms in laboratories, greenhouses, and some animal-holding facilities

Genetically modified organism (GMO) Any organism that acquires heritable traits not found in the parent organism; while traditional scientific techniques such as mutation can result in a GMO, the term is most frequently used to refer to modified plants, animals, and microorganisms that result from deliberate insertion, deletion, or other manipulation of DNA; also referred to as genetically engineered organisms or as organisms with modified hereditary traits

Oversight Application of appropriate laws, regulations, guidelines, or accepted standards of practice to control the use of an organism based on the degree of risk or uncertainty associated with that organism

Recombinant DNA Broad-range of techniques in which DNA, usually from different sources is combined *in vitro* and then transferred to a living organism to assess its properties.

GENETICALLY MODIFIED ORGANISM (GMO), as a term, is most frequently used to refer to an organism that has been changed genetically by recombinant DNA techniques. Historically, research with GMOs has been subject to special oversight that, to this day, differs depending on the location of the research with the organism, whether inside (contained) or outside (so-called field research), the type of organism (e.g., plant, animal, microorganism) or use (e.g., medical, agricultural, environmental), and country in which one works. The oversight mechanism for contained research is through guidelines developed by scientists and endorsed by the private and public sector. Outside research is currently overseen by a number of federal agencies. In some countries, such as the United States, there can be overlapping jurisdictions, differing interpretations of legal statutes, and different requirements or standards for compliance by scientists who do research with GMOs in the outside environment. Scientific issues deal with differences in perception of the risks of introductions of GMOs into the environment, the types of data required prior to introduction to conclude the experiment is of low risk, and the types of monitoring and mitigation practices, if necessary, to assure that the experiment is of low risk. Nonscientific issues are also considered and include those dealing with legal and social concerns. These differences in interpretation have resulted in few introductions into the environment of microorganisms.

I. Concern over Genetically Modified Organisms

A. The Concern over Safety

The new biology, dating to the 1970s and usually encompassing recombinant DNA techniques, enabled scientists to perform modifications of organisms with great precision and to combine DNA of organisms that can not, in current time, combine; yet these combinations are derived from components from naturally occurring organisms. The scientific community raised hypothetical questions about the safety of their genetically engineered organisms, and the public questioned the potential adverse effects of organisms with the new combinations of genetic information on humans and the environment. It was argued that, as such, the organisms have not been subjected to evolutionary pressures, including dissemination and selection, and may pose a risk to humans or the environment. However, it was recognized that genetic modifications can arise by classical or by molecular methods, ranging from selection of desirable combinations by farmers or bakers since antiquity to nucleotide insertion or substitution by molecular biologists.

The new biology, often called biotechnology, has generated fear, particularly fear of transfer of the modified trait to nontarget organisms and unpredictable survival and dissemination of these organisms. However, gene transfer occurs whether or not humans intervene. Such gene transfers are expected to have minimal adverse consequences unless selection is imposed. Increasing evidence supports the conclusion that microorganisms, particularly bacteria, usually maintain their fundamental characteristics and their essential identities and moderate the amount of change that can be absorbed by known and unknown mechanisms. Deleterious changes can occur and will, whether or not microorganisms are manipulated. The preliminary testing under contained conditions that is requisite and standard practice in science should, however, identify such gross changes. Most scientists agree that if one begins with a beneficial organism and imparts a neutral or beneficial trait, then the probability of harm of transfer of genetic information is small. Some scientists are more concerned about the widespread adoption of a beneficial organism in commerce rather than about any small-scale field trials, which are the basis of research.

The concern over GMOs, because of a perceived increase in ability to survive or persist, has been so great among some that they argue that oversight of GMOs should be as stringent as for toxic chemicals, physical disruptions, such as water control projects or exotic organism introductions. The most serious of these concerns is voiced by persons who compare GMOs with the introduction of exotic organisms. This analogy is inappropriate, however, because of the vast difference that exists between the whole organism and a modification that affects a small portion of the genome (Table I).

There is also a perception by some that, should there be a problem with survival or dissemination of a microorganism, nothing can be done. Essentially, the assumption is that once the gene(s) is out, it cannot be recalled. However, both orderly and inadvertent movement and dissemination of microorganisms occur repeatedly because of their presence on humans, plants, and animals that are moving throughout the world at increasing rates. There are also long-standing and environmental practices that are in use to decontaminate or mitigate unwanted effects of microorganisms. Such practices are known for microorganisms associated with plants and animals as well as for free-living microorganisms. Immediate decontamination methods include, among others, burning, chemical control, and sanitation by various means. Short-term and long-term methods are abundant for plant- and animal-associated microorganisms, since a great deal of research on developing mitigation methods is conducted by scientists in the disciplines of plant pathology, veterinary medicine, and human medicine. Many of these deal with management practices and the use of genetic resistance and application of biological control organisms. Immunization of humans and animals is another type of long-term management practice to mitigate the effects of microorganisms.

B. Concerns over Genetically Modified Domesticated Organisms in Agriculture and the Environment

Virtually all domesticated organisms used in the production of food and fiber have been genetically modified over long periods of time, including certain

Table I Comparison of Exotic Species and Genetically Engineered Organisms

	Exotic organism[a]	Engineered organism[b]
No. of genes introduced	4000 to >20,000	1–10
Evolutionary tuning	All genes have evolved to work together in a single package	Organism has several genes it may never have had before. These genes will often impose a cost or burden that will make the organism less able to compete with those not carrying the new genes.
Relationship of organism to receiving environment	Foreign	Familiar, with possible exception of new genes

[a] "Exotic organism" is used here to mean one not previously found in the habitat.
[b] "Engineered organism" is used here to mean a slightly modified (usually, but not always, by recombinant DNA techniques) form of an organism already present in the habitat.
[From U.S. Congress, Office of Technology Assessment (1988). "New Developments in Biotechnology: Field Testing Engineered Organisms: Genetic and Ecological Issues," Washington, D.C.]

live domesticated microorganisms used in making bread, beer, wine, various types of cheese, yoghurt, and other foods. Selected microorganisms that have been shown to be beneficial are also widely used in the environment. These uses include, among others, microorganisms that fix nitrogen and provide nutrients for trees (mycorrhizae) as well as used in sewage treatment plants and oil drilling. Also, naturally occurring pathogenic microorganisms are used in the testing of domesticated plants to ascertain their disease resistance. In such critical tests with known deleterious organisms, there has been no documented case of untoward effects, such as a plant disease epidemic, arising from such standard field trials.

It is widely accepted that the first step in risk assessment, whether in containment or in confined field trials, is identifying the risk by determining how much is known about the parental organism. It is also recognized that the risk can be minimized by the preferential selection of parental organisms that are generally recognized as safe because of their long history of use. In the oversight of food, such foods are categorized as GRAS, or generally recognized as safe. A similar category can be considered for microorganisms that would be introduced into the environment: GRACE, or those microorganisms that are generally recognized as compatible with the environment.

Examination of the food safety issues associated with genetic modifications has led to the conclusion that potential health risks are not expected to be any different in kind than those with traditional genetic modifications. All such evaluations rest on knowledge of the food, the genetic modification, the composition, and relevant toxicological data. A recent international body concluded that rarely, if ever, would it be necessary to pursue all such evaluations exhaustively. There is reasonable agreement that a threshold should exist for regulation, or even of concern below which further evaluations on a genetically modified food product or its individual components need not be conducted. The International Food Biotechnology Council recommends flexible, voluntary procedures between food producers and processors and a regulatory agency. [*See* FOOD BIOTECHNOLOGY.]

II. History of Guidelines and Regulations

The concern over the potential risks of GMOs led to the initiation of various mechanisms for the oversight of research conducted throughout the world. This oversight was in the form of both guidelines, a set of principles and practices for scientists to follow, and regulation by laws applicable to certain processes or organisms used for certain purposes. The legal profession claims that this is the only case in which hypothetical or speculative risks are the basis for regulation.

Codified guidelines date back to the 1970s, when the previously described concerns were raised. This led the United States National Institutes of Health (NIH) under the Department of Health and Human Services to develop guidelines for containment of research involving recombinant DNA molecules. The first guidelines were published in 1976 and have been most recently updated as of 1986 in the Federal Register (May 7). A revised version is due by early 1992. A public-meeting body of peers and nonscientists was assembled by the Office of Recombinant DNA Activities (ORDA) as the NIH's Recombinant DNA Advisory Committee (RAC) and was given the task to review all recombinant DNA experiments within the United States. Other countries soon followed suit.

The RAC was to assess the risk of the experiment and recommend containment conditions under which they thought the risk would be minimized for the laboratory worker and the environment. The resulting guidelines, which are available from NIH ORDA, spelled out recommended procedures for safety to individuals and to the environment. They included such specifics as the type of pipetting one should undertake, sterilization procedures, air filtration procedures, and decontamination and mitigation procedures.

Within a short time, most of the microorganisms and experiments had been assigned a containment level, and the responsibility for overseeing such experiments was decentralized and delegated to local institutional biosafety committees (IBCs) or other local institutions in other countries. Many experiments to modify common laboratory strains of bacteria and yeast were judged to pose no risk and were exempted from the guidelines or any containment requirements. Research with other microorganisms, including viruses, required containment no greater than one would use for research with the microorganism that did not involve genetic modification experiments. These assignments were generally consistent with the recommendations of the biomedical authorities, such as the Centers for Disease Control in the United States.

Experiments involving introduction of GMOs into the environment were begun in 1986 and have been constantly overseen by centralized authorities rather than through guidelines that describe principles and practices for confinement of the GMO to minimize risk. In the United States, regulatory oversight, which currently includes research conducted by any party, is under the jurisdiction of either the United States Department of Agriculture's Animal and Plant Health Inspection Service or the Environmental Protection Agency. The former generally oversees microorganisms that are or might be considered to be plant pests, while the latter oversees research with so-called pesticidal organisms and microorganisms for other uses. Where there is research with microorganisms and plants, both agencies may be involved, as well as the Food and Drug Administration. More detailed descriptions of legal and jurisdictional issues can be found in publications included in the bibliography.

At the present time, there is no decentralized body in any nation for oversight of field research that is comparable to the IBC for contained research (Table II). The United States Department of Agriculture (USDA) has recently published (Federal Register, Feb. 1, 1991) a draft of guidelines for conducting research under confinement in the open environment with a variety of organisms. However, it is not clear at present how these guidelines will be implemented and used. The draft guidelines provide generalized principles for evaluating risk and risk management of microorganisms, as well as plants and animals, that have been genetically modified, particularly by recombinant DNA. These guidelines are sufficiently generic that they should be applicable throughout the world.

In many countries, the oversight of GMOs is essentially the same as for unmodified organisms, except for the contentious issue of planned introduction into the environment. In countries such as the United States and Canada, a sizable bureaucracy has built up to oversee both the research and product development. Even though the risks remain

Table II Types of Oversight of Genetically Modified Organisms[a]

Stage of development	Oversight
Laboratory, greenhouse, animal pen[b]	Decentralized oversight from federal research or regulatory agency in the form of guidelines
Small-scale field research[c]	Guidelines and regulations: combination of decentralized and federal oversight
Scale-up or large-scale testing	Federal regulations
Commercial products	Federal, state regulations

[a] Reflects current practices: The degree of oversight differs in each country and among different funding and regulatory agencies.

[b] For unmodified organisms (naturally occurring, chemically altered, spontaneous, or selected mutants), standards of practice apply in research, whether conducted by the public or private sector.

[c] Includes tests on land and in enclosed waters.

speculative, the fear of litigation and unknown hazards has served to minimize the actual number of introductions, particularly of microorganisms. Of the approximately 200 tests of plants, animals, and microorganisms introduced into the environment, only about 10% have been microorganisms. Most of these microorganisms were modified to have marker genes that enabled them to be monitored in the environment. The functional genes added at the present time are those encoding an insecticidal toxin from *Bacillus thuringiensis* added to both a pseudomonad and a coryneform bacterium. In the former case, the modified organism is killed before it is marketed. In the latter case, field trials are still under evaluation for insecticidal activity, which is aimed against the corn borer.

III. Descriptions of Oversight Mechanisms

A. Standards of Practice

All trained professionals learn and adhere to certain accepted procedures and practices with respect to safety and to the appropriate scientific method for the profession. Most persons trained in microbiology and related disciplines using microorganisms, are exposed to the same standards of practice that are used in training medically oriented professionals. These include safe preparation, use, and disposal of inocula and inoculated organisms, whether plants or animals, and appropriate decontamination/mitigation procedures such as sterilization in contained facilities or incineration of animals or burial of plant material.

Standards of practice for all scientists include for example, procedures appropriate for conduct of experiments; the use of data, including statistical analysis techniques; publication ethics and standards; and sharing of biological materials after publication of results. Standards of practice in the open environment are particularly evident in agricultural and forestry research; the sites are evaluated, plots are designed to enable statistical analysis of results, criteria for evaluation of results are widely distributed and agreed upon through peer review, and results are disseminated through various publications. Practices to preclude significant risk to the environment and to mitigate possible untoward effects are routinely considered and used by researchers.

B. Guidelines and Directives

Guidelines may be considered a statement of policy by a group having authority over that policy. Sometimes such guidelines are also published as "points to consider." These guidelines offer assistance to investigators and do not have legal authority, except as adopted by a funding or regulatory authority. Guidelines are generally considered far more flexible than are directives or regulations. Guidelines that have been adopted worldwide for contained research with GMOs are those originating from the U.S. NIH. The NIH RAC guidelines have been considered *de facto* regulations, as commercial concerns have also adopted them as standards of practice.

Directives, particularly those issued by governments, are orders or instructions. Directives are currently an overarching method of oversight being implemented by the European Economic Community (EEC). Countries are being urged to implement the directives within the next few years that would ideally harmonize oversight within the EEC. However, individual countries would still have the prerogative of overseeing the details of such directives or even making them more stringent.

C. Regulations

Regulations are laws or rules to control or govern procedures or acts. In the case of GMOs, some countries have implemented new laws for oversight of field tests, especially in Europe. In other countries, no laws are currently applicable. In the United States, legal interpretations of current statutes have resulted in extensive oversight of research involving field trials. The laws have legal penalties for noncompliance, whether in the public or private sector. This type of oversight in the United States has resulted in an elaborate permitting, evaluation, interagency coordination and reporting system that many view as cumbersome and costly. New legislation may well occur, both to simplify and clarify oversight, and to close gaps regarding the oversight of aquatic species, for example. In addition to federal and national laws, other governmental entities such as states or cities have enacted their own regulations, compounding the difficulty and complexity of conducting field research.

Confidential business information is protected in all cases, including oversight by guidelines, by authorities, unless such information relates to safety issues.

IV. Appropriateness of Guidelines and Regulations for Research

A. Contained Conditions

The conditions described in NIH guidelines have served as a codification of practices for conducting research with both wild-type and modified microorganisms within a traditional laboratory setting and for large-scale fermentations with microorganisms. There is no indication that exempting certain organisms from containment requirements, or conducting most other experiments at the lowest containment conditions, has caused any problem to individual workers or the environment. The guidelines have thus served both national and world interests well for over 15 years.

Conditions for conducting research with organisms that require other conditions for optimum growth, such as plants, animals, and microorganisms associated with them, are not described in the guidelines. The next published version of the NIH guidelines, however, should include a complete description of appropriate practices for plants in greenhouses, other contained facilities, and for animals. The proposed practices are based on those developed as standards of practices in research.

B. Conditions for Planned Introduction into the Environment

The United States raised questions about the safety of planned introductions and examined such introductions in multiple ways. In 1987, the prestigious National Academy of Sciences made major conclusions regarding risk, one of which stated that there is no evidence that unique hazards exist, either from the use of recombinant DNA techniques or from the movement of genes between unrelated organisms. Furthermore, the risks of the introduction of GMOs carrying recombinant DNA are the same in kind as those associated with the introduction of unmodified organisms and organisms modified by other methods. The final conclusion was assessment of risks of introducing GMOs carrying recombinant DNA into the environment should be based on the phenotype of the organism and the environment into which it is introduced, not on the method by which the organism was produced. [*See* GENETICALLY ENGINEERED MICROORGANISMS, ENVIRONMENTAL INTRODUCTION.]

In going further with the assessment of risk, another study by the National Academy of Sciences in 1989 posed three fundamental questions to assess risk. (1) Are we familiar with the properties of the organism and the environment into which it may be introduced? (2) Can we confine or control the organism effectively? (3) What are the probable effects on the environment should the introduced organisms or genetic traits persist longer than intended or spread to nontarget environments? Similar questions for focusing on risk were elaborated by the Ecological Society of America. Questions still remain on how to scale regulatory oversight on the basis of risk.

The principles espoused in the preceding types of publications have played a major role in the risk assessment by the ECC and the Organization for Economic Cooperation and Development (OECD) as well as other countries. No regulatory mechanisms exist for conducting field trials in Latin America and the Caribbean, but guidelines based heavily on the OECD publications have recently been published by the Organization of American States. Various proposals have been put forth on the responsible oversight policies for recombinant DNA research and field trials in countries in which oversight mechanisms are lacking.

C. In Principle: Needs and Options

The objectives of a sound oversight policy are to develop a sensible, scientifically based policy that is consistent with a reasonable and accepted degree of safety (not absolute or "ensured" safety). Regulatory agencies also speak of procedures that do not pose an unreasonable risk to humans or the environment, i.e., one cannot be absolutely certain that no deleterious effects can occur. This is the case with wild-type organisms and items that are used in everyday life. Oversight ideally should balance risks with expected benefits. Many issues that deal with GMOs, particularly introduction into the environment, remain unresolved at this writing. These include the following:

1. The scope of oversight: What organisms should be subject to oversight?

2. By whom and at what level (e.g., professional organization, standards of practice, institutional biosafety committees, local regulatory bodies, federal agencies, or some combination) should oversight be conducted? Should all experiments be examined at a federal or country level or can some be decided upon at a local level? Decentralization of authority to make decisions on field releases has not yet occurred.
3. Can consistent definitions be developed? Definitions differ among agencies and countries and can lead to problems in legal interpretations. The meaning of the phrase "release or planned introduction into the environment," and even "pathogen," are yet to be agreed upon. This affects, for example, whether different considerations are given to plant pathogens that are applied as beneficial biological control agents.
4. Can a consistent or uniform policy for oversight of exploratory or discovery research be implemented? Ideally, there should be policies that apply regardless of whether or not the research is conducted in educational or commercial facilities.
5. To what degree should the manner of open and peer review be part of policy-making?
6. What appeal procedures, if any, should there be for scientists disagreeing with the oversight authorities?
7. Decision-making: Who should decide whether approval of field testing is needed? Is it the public, the scientists or the courts? What role is there for common sense? Who should analyze the risks and benefits?

There has been general agreement that a centralized data base for field trials would be desirable in order to compare information, including negative results that are not always published. The beginnings of such a data base have been proposed by the USDA through its National Biological Impact Assessment Program and by the OECD through its Biotrack monitoring data base. Whether or not these activities will serve the purpose of the scientific and commercial community and allay the concerns of the public remains to be seen. Over 200 tests worldwide have been conducted up to the present time and no unpredictable effects have been detected. However, it can be argued that such effects may occur only in later years, and, hence, monitoring will be necessary to assess any problems that might arise. Another question that remains unanswered is how long monitoring should occur, compared with wild-type organisms.

Given the different views of different countries and applicable laws, global agreements on oversight probably will not be forthcoming. However, there is reasonable general agreement on standards of practice through the scientific and professional societies of the world as well as guidelines for conducting such research. There are also areas of reasonable agreement in principle, although these are, as yet, few. These include acknowledging that the process by which a genetic modification is made is not as significant as the effects of that modification, i.e., the phenotype. The same degree of oversight is not now applicable to unmodified organisms and to organisms modified by traditional approaches. Hence, the process of genetic modification is still the "trigger" for oversight. A second area of agreement is that familiarity or knowledge of the organism and its modification are likely to be good predictors of the modified organism. A third is that knowledge of the ability to confine an organism or mitigate its effect, if need be, offers a reasonable indicator of expected risk.

V. Conclusion

Differing perspectives remain on the safety to humans and the environment of GMOs, particularly those that have been modified by recombinant DNA techniques. The concerns are particularly high with respect to the use of microorganisms in the environment. This discussion is likely to continue for several more years. It is too early to predict whether or not such tests will go forward with reasonable ease, given the stringency of the requirements to conduct the tests. In most cases, the scientific concerns do not warrant the expenditure of time, effort, and money to conduct field research, since no risks unique to GMOs have yet been identified.

Several potential oversight processes would be commensurate with risk assessment and risk management. These could include (1) categorical exclusions, (2) only notification requirements, (3) review and approval by a local organization (e.g., institutional biosafety committees), (4) review and approval by a federal agency with an advisory group consisting of members familiar with the relevant research area, or (5) review and approval by an international agency, in cooperation with a member country.

Relatively rapid change in oversight is occurring and will likely continue to occur. A reasonable policy of oversight will encourage research with GMOs, especially those modified by recombinant DNA techniques: A stringent policy will stifle it. Competing perspectives may differ in different countries and within a country and may not be reconciled. There is no perfect oversight mechanism for any human activity, including environmental releases. There is also the recognition that various viewpoints or perspectives cannot always be accommodated or reconciled. Hopefully this will not be the case in this area. Thus, persons of reason and broad vision will be needed to resolve some of the contentious issues dealing with planned introduction of GMOs into the environment.

Bibliography

Baumgardt, B. R., and Martens, M. A. (eds.) (1991). Agricultural biotechnology: Issues and choices. Purdue University Agricultural Experimental Station USA. 181 pp.

Cordle, M. K., Payne, J. H., and Young, A. L. (1991). Regulation and oversight of biotechnological applications for agriculture and forestry. *In* "Assessing Ecological Risks of Biotechnology" (L. R. Ginzberg, ed.), pp. 289–311. Butterworth-Heinemann, Boston.

International Food Biotechnology Council (1990). Biotechnologies and food: Assuring the safety of foods produced by genetic modification. *Regul. Toxicol. Pharmacol.* **12,** 196 pp.

Levin, M., and Strauss, H. (eds.) (1991). "Risk Assessment in Genetic Engineering." McGraw-Hill, New York. 403 pp.

MacKenzie, D. R., and Henry, S. C. (eds.) (1991). Biological monitoring of genetically engineered plants and microbes. Agricultural Research Institute, Maryland. 303 pp.

Miller, H. I., Burris, R. H., Vidaver, A. K., and Wivel, H. A. (1990). *Science* **250,** 490–491.

National Academy of Sciences (1987). Introduction of Recombinant DNA-Engineered Organisms into the Environment: Key Issues." Committee on the Introduction of Genetically-Engineered Organisms into the Environment. National Academy Press, Washington, D.C. 24 pp.

National Research Council (1989). "Field Testing Genetically-Modified Organisms: Framework for Decisions." Committee on Scientific Evaluation of the Introduction of Genetically-Modified Microorganisms of Plants into the Environment. National Academy Press, Washington, D.C. 170 pp.

Organization for Economic Cooperation and Development (1986). Recombinant DNA safety considerations. Safety considerations for industrial, agricultural and environmental applications of organisms derived by recombinant DNA techniques. OECD, Paris. 69 pp.

Organization for Economic Cooperation and Development (1990). Good development practices for small-scale field research with genetically modified plants and micro-organisms. A discussion document. OECD, Paris 36 pp.

Riley, M. (1989). Constancy and change in bacterial genomes. *In* "Bacteria in Nature," Vol. 3 (J. S. Poindexter and E. R. Ledbetter, eds.), pp. 359–388. Plenum, New York.

Tiedje, J. M., Colwell, R. K., Grossman, Y. L., Hodson, R. E., Lenski, R. E., Mack, R. N., and Regal, P. J. (1989). *Ecology* **70,** 298–315.

Tolin, S. A., and Vidaver, A. K. (1989). *Annu. Rev. Phytopathol.* **27,** 551–581.

U.S. Congress, Office of Technology Assessment (1988). New developments in biotechnology—Field-testing engineered organisms: Genetic and ecological issues. U.S. Government Printing Office, Washington, D.C. 152 pp.

Genetic Transformation, Evolution

Rick E. Hudson and Richard E. Michod
University of Arizona

Glossary

DNA damage Process and product of a physical alteration of DNA to an illegitimate form (e.g., cross-links between two DNA strands, breaks in one or both strands of a DNA molecule, demethylation of nucleotides, and thymine dimers), damaged DNA molecule is no longer a regular sequence of the standard four nucleotides; therefore, the damages themselves cannot be replicated (they are not standard nucleotides), but they can be recognized and repaired directly

Haploid State of having only one copy of a chromosome or gene

Homologous DNA DNA molecules or genes that have nucleotide sequences that are very similar to one another; this similarity allows the sequences to pair and recombine with one another

Linkage disequilibrium Population statistic measuring the degree of statistical dependency of the frequency of alleles at one locus with the frequency of alleles at another

Mixis Process and product of the mixing of genomes, or portions of genomes, to produce a new combination of genes

Mutation Process and product of a change in the nucleotide sequence of DNA (e.g., deletions, additions, and rearrangements of the standard four nucleotides present in a DNA molecule); important point is that a mutated DNA molecule is still a regular sequence of the four standard nucleotides, therefore mutations can be replicated but cannot usually be recognized and repaired by repair enzymes

Natural selection Process of genetic change in a population, which results from differences in viability and fertility of different genotypes

Outcrossing Process and product of bringing genomes, or portions of genomes, from different individuals into a common cell; during meiosis, this occurs as a result of fertilization; during transformation, this occurs as a result of the binding and uptake of DNA released by another cell

Phylogeny Evolutionary history of a group of organisms

Recombination Used in two different senses: (1) a synonym for mixis; (2) molecular process by which two homologous DNA molecules pair, break, and rejoin with one another; first meaning sometimes referred to as "allelic recombination"; second meaning referred to as "physical recombination" or "molecular recombination"; this article uses the term recombination in the molecular sense and uses the term mixis to refer to new combinations of genes

Sex There are three definitions in common use: (1) any process resulting in mixis; (2) any process involving outcrossing that results in mixis; and (3) any process involving homologous molecular recombination and outcrossing

NATURAL GENETIC TRANSFORMATION is a sexual process by which a bacterium brings external DNA into the cell and integrates the foreign DNA into its genome by homologous recombination. Why has this energetically costly process evolved? Several hypotheses exist to explain how transformation might benefit the recipient cell; none of these hypotheses are mutually exclusive. The transforming DNA might nourish the recipient cell, might serve to repair and replace damaged genes, or might benefit the cell by producing a novel combination of genes. Whether or not transformation evolved for the purpose of generating genetic variation, generate variation it does. This variability fuels the process of natural selection. In addition, transformation has important implications for the species concept as it applies to bacteria.

I. Transformation is a Sexual Process

The process of transformation involves several steps. Donor DNA is produced by cells either by active extrusion of their DNA or by passive release of their DNA after cell lysis. Before the recipient cell can be transformed, it must develop the physiological state known as competence, during which it can bind, fragment, take up, and recombine donor DNA. [*See* GENETIC TRANSFORMATION, MECHANISMS.]

Transformation is a sexual process. The word "sex" has a multitude of meanings. Sex often refers to gender, or to copulation; but here we will use it to refer to different meanings common in evolutionary biology. The process of sex has been variously defined by evolutionists (see Glossary) as mixis, mixis with outcrossing, and recombination with outcrossing. Genetic transformation satisfies all of these definitions as do two other prokaryotic processes of genetic exchange: conjugation and transduction. [*See* CONJUGATION, GENETICS.]

Transformation differs from meiotic sex in at least four important ways (the same is true for conjugation and transduction). First, sex and reproduction are temporally uncoupled. Bacteria have sex without reproducing; when they reproduce, they do so by directly replicating their genome without involving a sexual cycle. Second, the contribution of genes from the partners is unequal: most of the genes come from the recipient cell and only a few genes come from the donor cell. Third, most or all new combinations of genes are produced by the molecular aspects of recombination or by the transfer of extrachromosomal elements, such as plasmids. In eukaryotes, new combinations of genes are also produced by the reassortment of chromosomes, a process not available to the bacterium with its single chromosome. A fourth difference is that bacterial sex alters the genome only locally, where the recombination event occurred. The local nature of bacterial recombination has been demonstrated recently by sequencing specific regions of the chromosome in different individuals of the same species. Therefore, the impact of a recombination event on the linkage relationship of genes is much less severe in a bacterial genome than in a eukaryote. In population genetical terms, bacteria-style recombination is less efficient at reducing linkage disequilibrium than eukaryotic-style recombination.

II. Evolution of Transformation

A. Costs of Transformation

Transformation is a costly process. The first of these costs is that of maintaining genes that encode the transformation process. Some of this cost may be shared with other abilities of the cell, because some gene products may be used in more than one pathway. For example, in the competent bacterium *Bacillus subtilis,* the recA enzyme responsible for homologous integration is also involved in the general induction of the SOS repair system and in postreplication recombinational repair. However, other gene products are likely to be used exclusively by the transformation process, for example, the genes encoding DNA binding and uptake. A second cost is the cost of producing the proteins necessary for the transformation process. A third cost is the energy needed during the transformation process. A fourth cost is the time needed to complete transformation that could be used for other activities. In some bacterial species, cell growth and division are even arrested in cells competent for transformation. Therefore these individuals cannot replicate as quickly as their noncompetent (asexual) competitors. A fifth cost is that competent cells may be more likely to pick up genomic parasites than noncompetent cells, because viral and plasmid DNA can be taken up by the same process as homologous DNA. It also appears that viruses integrated into the chromosome are more likely to be induced to kill a bacterium when that bacterium becomes competent. Finally, the new genes brought in by transformation might disrupt the harmonious functioning of the genome and thereby lower fitness. Since these costs are all borne by individual bacterial cells, a theory for the evolution of transformation must show an advantage to the individual cells that at least compensates for these costs.

B. DNA as Nutrient

One possible advantage is that the transforming DNA serves as a nutrient for the recipient cell. According to the nutrient hypothesis, the DNA is taken up for its value as a raw material, that is, as a source of carbon, nitrogen, nucleotides, or energy. According to this perspective, the integration of the DNA and the subsequent creation and destruction of gene combinations are mere coincidental by-products. These byproducts occur because the cell maintains the integration machinery for another purpose, perhaps postreplication recombinational repair. This hypothesis has two attractive aspects. First, the need for nucleotides is ubiquitous. Second, nonhomologous DNA as well as homologous DNA would prove useful as nutrient. Several species of transformable bacteria take up nonhomologous DNA.

The problem with the "DNA as nutrient" hypothesis is that the bacteria seem to choose only specific states and types of DNA to take up. Such selectivity would be unnecessary if they were just using the DNA for food. For example, some competent species take up only one strand of the DNA duplex that they initially bind. Why discard one of the DNA strands if the cell needs nutrition? In addition, the entering DNA typically is protected from degradation after uptake. Competent *Haemophilus* species and *Neisseria gonorrhoeae* preferentially take up single strands that meet rigorous admission qualifications. In these bacteria, DNA molecules that contain certain sequences that are common among members of the same species but otherwise rare are much more likely to be bound to the recipient cell. A more fundamental objection to the nutrient hypothesis is that, once the cell integrates the DNA, the integrated DNA is unavailable for nutritional uses. Nevertheless, usually some of the DNA is not integrated.

C. Parasitic DNA

The parasitic DNA hypothesis contends that transformation evolved for the purpose of transferring infectious parasitic elements from cell to cell. According to this view, transformation is not advantageous to the recipient cell but advantageous to the individual gene sequences that enter a cell during transformation. To evaluate this hypothesis, it is helpful to contrast transformation with the two other prokaryotic processes of gene exchange: conjugation and transduction. Transformation is governed by genes found within the chromosome of the recipient rather than by extrachromosomal genes in the donor organism, as in transduction and conjugation. In addition, in transformation the recipient cell loses genetic information rather than spreads its genetic information. Furthermore, the genes in the recipient that encode transformation are often not the same genes that are spread by the process. In contrast, the genes directing conjugal transfer can be spread to individuals that lack these transfer genes. More generally, it is difficult to imagine that the bacterium would undergo such a costly process as transformation without some compensating benefit to the cell. For these reasons, we do not consider the parasitic DNA hypothesis a satisfactory explanation for transformation, although such a concept may be relevant to other processes of exchange such as conjugation and transduction.

D. DNA Repair Hypothesis

According to the repair hypothesis, transformation functions as a DNA repair system. When DNA is damaged by agents such as UV light and byproducts of oxidative metabolism, it is critical for the cell to repair this damage. Unrepaired damage interferes with DNA replication, leading either to cell death or to mutation-prone DNA synthesis. Both fates are undesirable. If only one of the two DNA strands is damaged, the damaged area can be removed and the information on the remaining strand can serve as a template to replace the removed area. This process occurs during excision repair. However, if both DNA strands are damaged, a different repair strategy must be used. To recover the information lost because of damage to both strands, that information can be obtained from a second homologous DNA molecule by recombinational repair. If there are two copies of a gene in the cell, recombinational repair can occur without the uptake of exogenous DNA. Such information recovery happens after replication in the well-studied process of postreplication recombinational repair. Otherwise, the cell must obtain the homologous DNA from another source. Competent cells can do this, and therefore may have an advantage over noncompetent cells in repairing DNA damage.

The plausibility of the repair hypothesis rests on the validity of its assumptions. The first assumption is that double-stranded damage is both common and lethal. It is further assumed that the bacterium has no internal source of homologous DNA, that is, that the cell is effectively haploid. Finally, the hypothesis requires that competent cells can find and take up DNA molecules appropriate for repairing the damaged genes. There is strong evidence for the validity of the first assumption. DNA damage is a ubiquitous problem for all life forms, as indicated by the variety of repair systems found even in simpler organisms such as *Escherichia coli,* bacteriophage T4, and *B.*

subtilis. The debilitating effects of repair-system mutations in many organisms and the high levels of repair-process waste products in rat and human urine provide experimental confirmation of this assumption. A comparison of the frequencies of different DNA damages produced by hydrogen peroxide demonstrates that double-stranded damage forms a significant class of all DNA damages. The second assumption, that competent individuals are effectively haploid, has been shown to hold for *B. subtilis*. *Bacillus subtilis,* like many other naturally transformable bacteria, expresses competence predominantly during the stationary phase of population growth. Bacterial cells are more likely to be haploid during the stationary phase than during the exponential phase of population growth. However, *Streptococcus pneumoniae* is competent during the rapidly dividing exponential stage of culture development.

The third assumption requires that a competent bacteria has a strong preference for integrating DNA homologous to its damaged genes. Competent bacteria take up an amount of DNA equivalent to only a small fraction of their genome. Therefore, the discrimination between DNA useful for repairing damage and other homologous DNA must take place before integration during the binding and uptake stages of transformation. Thus, if competent cells bring in DNA to repair damages, the damaged sites of the genome should be available at the DNA binding site to aid in choosing the appropriate DNA molecules.

Other circumstantial evidence lends support to the repair hypothesis. In *B. subtilis,* the recA enzyme has two functions: to regulate the SOS DNA repair systems and to promote homologous recombination. The levels of this protein increase under two circumstances: when the cell becomes competent or when the cell detects DNA damage. Thus, the same enzyme bears the responsibility for initiating both DNA repair and homologous recombination. In mammals, nondividing cells are more likely to accumulate DNA damage whereas dividing cells are more likely to accumulate mutations. If the accumulation of damage and mutations in bacterial cells follows the same pattern, then different hypotheses for transformation predict different timing for the development of competence. If transformation exists for the sake of repairing DNA damage, then it should occur at the end of the life stages that are more likely to accumulate DNA damage, that is, at the end of dormant stages such as spore stages. In *B. subtilis,* it appears that when spores germinate to become growing cells they both extrude DNA and become competent for transformation. In addition, cells of many species are more likely to be competent during later stages of culture grown when rates of cell division slow down and when inhibitors of cell division are added to the cultures. Among the important exceptions is the constitutively competent species *Neisseria gonorrhoea*.

The repair hypothesis has been investigated experimentally using competent cultures of *B. subtilis*. Competent cultures consist of a majority of noncompetent (asexual) cells and a minority of competent (sexual) cells. Therefore, the noncompetent and competent subpopulations of a culture may be studied under similar conditions. These conditions involve two different agents: UV light and added transforming DNA. The UV light serves to damage the recipient bacterial chromosomes and the added DNA contains a selectable marker to label competent cells. In these experiments, the DNA treatment took place both before the UV treatments (DNA–UV experiments) and after the UV treatments (UV–DNA experiments). The DNA added was either damaged or undamaged homologous *B. subtilis* DNA, nonhomologous plasmid DNA, or plasmid DNA containing a short homologous sequence. The homologous DNA could be used to repair the damaged recipient DNA, but only when it was added after the UV treatment. As the UV dose increased, the ratio of marked competent cells over unmarked total cells increased, but only in UV–DNA experiments. One interpretation of these results is that the marked cells were less affected by the increasing UV dose in the UV–DNA experiments because they had the additional advantage of transformational repair, which was not available in the DNA–UV experiments. A second interpretation is that UV light induced one or more of the stages of transformation, for example, the development of competence or the binding, uptake, or integration of the homologous donor DNA. This interpretation is also consistent with the repair hypothesis because, if the transformation system is designed to repair damage, it should be induced in the presence of damage. Further experiments using a nonhomologous plasmid as donor DNA indicated that the relative increase in density of transformed cells did not depend on the induction of competence, DNA binding, or DNA uptake. Other experiments ruled out the possibility that the phenomenon was solely due to the selective induction of other repair systems in competent cells. Experiments with plasmids that contain short homologous sequences and experiments with damaged donor DNA suggested that damage may directly stimulate recombination during transformation, as would be expected if the function of transformation was DNA repair. [*See* DNA Repair by Bacterial Cells.]

E. Variation Hypotheses

A common hypothesis for the evolution of sex asserts that the advantage of sex stems from genetic variation, which in the case of transformable bacteria, means new gene combinations created by transformation. According to the variation hypothesis, sexual individuals are taking a calculated risk that the payoff for creating a better combination of genes outweighs the penalty of creating a worse combination. Random changes to the genome are likely to be deleterious. However, if the payoff for a good combination is large enough, the gamble is worthwhile even if the odds of winning are low and most bets lose. The recipient cell undergoes transformation to attempt to create a more favorable gene combination that results in increased survival of the cell and increased numbers of descendants. The alternative outcome is the creation of a less favorable gene combination that may decrease the chance of survival or the number of descendants. Several different forms of the variation hypothesis can be distinguished according to whether they focus on beneficial or deleterious mutations in constant or varying environments.

1. Linkage Disequilibrium and Sex

Sex cannot create new traits; only mutation can do that. Sex reassorts and recombines traits that already exist in the population. Let us consider just two traits, e.g., responses to temperature and moisture, and suppose that organisms are adapted to hot or cold temperatures and to moist or dry environments. Given these assumptions, sex between two parents—one of which is adapted to hot, dry environments and the other to cold, wet environments—could produce an offspring that was unlike either parent (e.g. one adapted to hot wet environments). In this way, sex may create a new combination of traits and genes.

In order to rigorously discuss whether it is adaptive to create new combinations of genes, population geneticists need a measure of the association between the frequencies of the two kinds of traits and their underlying genes. Let "A" refer to one of the features (e.g., temperature response) and "B" refer to the other (e.g., moisture response). Let us consider just two alleles at each of the two loci controlling these responses, say A_1, and A_2 at the first locus and B_1 and B_2 at the second. $P(A_1, B_1)$, the probability that and genes A_1 and B_1 are found together in the same individual, measures the association between these two genes; similarly $P(A_1, B_2)$, $P(A_2, B_1)$, and $P(A_2, B_2)$ measure the probabilities of the other combinations. But we also need a statistic to measure the overall association between two kinds of traits. One such measure, extensively used in population genetics, is the coefficient of linkage disequilibrium, $D = P(A_1, B_1)P(A_2, B_2) - P(A_1, B_2)P(A_2, B_1)$. D is zero only if the genes for the traits combine randomly. In this case, there is no association between the genes for the two traits; in other words the genes at one locus are independent of the genes at the other.

Mathematical models have shown that sex can affect the distribution of traits, and hence the evolution of the population, only if D is non-zero (i.e., the traits and their genes are associated nonrandomly). Furthermore, if D is non-zero, sex has the tendency of reducing D to zero by making the distribution of traits more random. Decreasing D is the fundamental consequence of sex, indeed it is the only consequence of sex, on genetic variability.

For sex to have an effect on the evolution of a population, D must be non-zero. However, sex reduces D to zero. Consequently, for sex to be continuously advantageous, there must be some source that continuously restores D to a non-zero value. Otherwise sex would change the population until D becomes zero, and then cease to affect the population.

To intuitively see the reasons behind this last conclusion, recall that sex simply mixes traits from different individuals. If the population is already completely mixed ($D = 0$), further mixing can have no effect (just like stirring a bucket of paint in which the pigment has settled). The first few stirs have a big effect, but as the pigment and solvent become mixed, stirring has less and less of an effect. Once the pigment and solvent are completely mixed, further stirring can have no effect. For stirring to have an effect, the pigment must settle again. For similar reasons, sex can continue to have an effect on the mixing of genes only is there is some antagonistic process which keeps the genes unmixed. Among the antagonistic forces that may keep D non-zero in a population are finite population size, selection, and mutation.

2. Beneficial Mutations

Progress in evolution depends ultimately on the occurrence of mutations that are beneficial to the individuals that carry them. Such beneficial mutations are rare but are nonetheless of fundamental importance to the evolutionary process. In an asexual population, beneficial mutations occur in isolated clones and are thereby forced to compete with one another. To increase in frequency concurrently, two beneficial mutations would have to occur in the same individual cell. Beneficial mutations appear rarely. For this reason, it is very unlikely that two (or more) beneficial mutations appear initially in the same cell. Consequently, in an asexual population beneficial mutations usually exist in different individuals (clones) and must compete with one another. Only when the most beneficial mutation in a population replaces all other genotypes can another begin increasing in frequency.

In a sexual population, however, advantageous mutations can increase in frequency in parallel. Two beneficial mutations that initially appeared in separate cells may be combined in the same cell through sexual exchange of information. If this occurs, both mutations may increase in frequency at the same time.

Mathematical analysis has shown that a sexual population will evolve faster, that is, will more rapidly establish beneficial mutations, if two related assumptions are made. The first assumption is that beneficial mutations are rare and the second is that the population is not large. Put mathematically, the beneficial mutation rate should be much less than $1/N$, where N is the population size. As a consequence of these assumptions, it is unlikely that the two beneficial mutations initially appear in the same individual, except after the first mutation has reached a high frequency in the population.

A problem with this theory is that sex (transformation) can destroy the newly formed favorable combinations even as it creates them. Nonetheless, mathematical models have shown that, under certain conditions, combinations of beneficial mutations are established more quickly in sexual populations. A more significant problem with beneficial variation as an explanation for sex is that this advantage may be too weak to compensate for the large individual costs incurred by competent cells.

3. Deleterious Mutations

Although establishing beneficial mutations is important in evolution, it is equally, if not more, important to cope with the far larger class of deleterious mutations. Sex or transformation may be advantageous because it breaks up combinations of deleterious mutations (again creating beneficial combinations of, in this case, wild-type genes). In asexual organisms, combinations of deleterious mutations are not broken up, so they accumulate in a clone in a process known as Muller's ratchet, named after its discoverer, J. H. Muller. The occurrence of the ratchet process has been supported by mathematical models and by experiments that show that small asexual populations of ciliated protozoans and RNA bacteriophage lose fitness (i.e., viability) over time. The ratchet relies on many of the same assumptions as the beneficial mutation theory: favorable mutation (in this case, back mutation) rates should be small and populations cannot be large. In addition, the ratchet operates more quickly when mutations are not too deleterious, because they then accumulate more easily. These assumptions can be summarized by the relationship $\ln(N) << U/s$, where N is the population size, U is the rate of deleterious mutations per genome per generation, and s measures the decline in viability caused by one mutation.

It is easy to see how, in principle, the ratchet operates. Consider the class of cells with the lowest number of mutations. Initially, there may even be some cells with no mutations. Although cells with no mutations should have an advantage, they can be lost from the population by chance. When the conditions mentioned earlier hold, there is a significant chance that spontaneous loss will occur. If all the mutation-free cells are lost, Muller's "ratchet" has clicked one stop. At this point, all cells contain at least one deleterious mutation. The only way in which an asexual population can regenerate a mutation-free cell is by back mutation, a very rare event. However, the possibilities for a sexual population are different. Although all cells contain at least one mutation, these mutations are likely to exist in different genes. A competent cell might take up and integrate DNA that contains wild-type sequences at a site at which the cell carries a mutation. Thus, a transforming population has the potential to regenerate the mutation-free class of cells and avoid Muller's ratchet. However, there is also the potential of increasing the number of mutations by sex.

Just as a cell might lose mutations by transformation, it might gain mutations by bringing in DNA fragments with new mutations.

Like the previous scenario, the ratchet assumes that transforming cells have a better chance than noncompetent cells of producing a favorable variant. The risk of sex is that the transforming fragment may contain new mutations not present in the recipient. Nevertheless, there may still be a net benefit to the sexual cell, if the conditions of low back mutation rate and small population size exist.

Even in the absence of these two conditions, sexuality may still be advantageous if the benefit to fitness of losing mutations is greater than the cost to fitness of gaining mutations. This will occur if the mutations interact "synergistically," that is, if the detrimental effect of two deleterious mutations residing in one individual is worse than the additive effects of the individual mutations acting alone. In such a case, the cost of additional mutations is greater than the cost of the first mutation; put mathematically, if the probability of survival of an individual with one mutation is $(1-s)$, the probability of survival for an individual with two mutations will be less than $(1-s)^2$. If mutations are synergistic, transformation is a good risk for cells because the payoff for having fewer mutations is enhanced. There is evidence for the contention that mutations have synergistic effects on fitness in *Drosophila melanogaster* and in two RNA viruses, the influenza virus and vesicular stomatitis virus. However, in *E. coli* there is evidence that mutations in the *lac* genes do not act synergistically.

Mathematical models have demonstrated this advantage of transformation because of synergistically interacting mutations. It has been shown that, if the transforming fragments are large, the deleterious mutation rate is high, and the synergism of mutations is strong, transformation is advantageous even if the donor DNA contains many deleterious mutations. In other words, the mathematical models demonstrate that, even if the donor DNA is released from cells that died from too many mutations, the advantage of transformation can still persist under rather stringent conditions.

4. Varying Environments

Yet another version of the variation hypothesis is that transformation is advantageous in an environment that undergoes temporal (and perhaps spatial) fluctuations. The assumption is that different environments favor different combinations of alleles. When a different environment arises, a transformable individual, by recombination, should be more likely than an asexual individual, by mutation, to produce the new favorable combination. This view does not require that the population is finite or that mutations are rare.

In the past 10 or so years, a number of mathematical models based on the principles of population biology have been developed to test the logical rigor and consistency of this intuitively appealing idea. These models have shown sex to be advantageous in a varying environment only if the association between two states in the environment alternates with each generation. Consider two aspects of the environment, for example, temperature and moisture, that can each be present in either of two states, hot or cold and wet or dry. For sex to evolve, the correlation between temperature and moisture must change for each generation. This means that if for one generation hot environments tend to be moist and cold environments tend to be dry, then for the next generation the association must alternate so hot environments tend to be dry and cold environments wet. According to mathematical models, this alternation must go on continually if a sex gene is to increase in frequency.

According to these models, sex is not advantageous if the environment simply changes, if it is unpredictable, or if it is random. Sex is only advantageous if the association between two relevant states of the environment continually alternates. As the eminent biologist John Maynard Smith said, "It is difficult to believe that God is as bloody-minded as that!" Recently, several biologists, among them W. D. Hamilton of Oxford, have proposed that the source of flip-flopping environmental variation is biological rather than physical. Hamilton contends that the frequencies of various parasites change often enough to create an advantage for sex in their hosts. This idea could apply to transformation; bacteria are parasitized by bacteriophages and plasmids.

As discussed earlier, the costs of sex are paid by each individual cell every time it transforms. The benefit of generating or recreating beneficial variants may take generations to realize. A rare nontransforming cell would not have to pay the costs of transformation and should be able to devote more time and energy to survival and reproduction (ignoring for the time being any other benefits such as

DNA repair or nutrition) than a transforming cell. Although these nontransforming cells may not be able to produce favorable variants as quickly as the sexual cells, any long-term benefit of transformation would probably not prevent a nontransforming cell type from increasing in the short term and driving the transforming population to extinction.

Two aspects of meiotic sex that enhance the generation of variation (by reducing linkage disequilibrium) are missing from the transformation process. Unlike meiotic sex, transformation does not generate variation globally by the reciprocal exchange of gene loci that flank the integration site. In addition, transformation generates less variation than meiotic sex because the donor in transformation makes a much smaller contribution to the recombinant genotype.

Transformation may or may not have a significant impact on the amount of variation within a species. The evidence for greater variability in competent bacterial species than in noncompetent species conflicts. *Bacillus subtilis* strains isolated from desert soils have shown a large amount of genotype variation (i.e., a low amount of linkage disequilibrium) in protein electrophoretic types. However, when electrophoretic variants were investigated in studies of the naturally tranformable species *Haemophilus influenzae* and *Neisseria meningitidis* and the nontransforming bacteria *E. coli,* the amount of genotype variation detected was low and the amount of linkage disequilibrium was high. In addition, in *E. coli,* descendants of one clone usually dominate one habitat. If this is also true for transforming bacteria, then transformation would most often take place among genetically identical cells; therefore, the amount of variation that could be created would be restricted.

5. Group Selection

Many of the variation-based hypotheses just discussed rely on group or population level benefits to explain the evolution of sex. For example, it was argued that a transforming population might evolve faster than a nontransforming population. Such explanations are termed group-selection explanations; there has been extensive investigation during the last 20 years of the validity of such explanations. The basic question addressed is whether group selection can explain the evolution of complex and costly traits such as sex. Most evolutionary biologists feel that, to explain the evolution of a trait such as sex that is costly to individuals, one must understand why the benefits to individual cells outweigh the costs. Recent work has shown that some of the population advantages described earlier of producing new variations may also benefit the individuals involved.

F. Bacterial Species and Phylogeny

Transformation can and probably does generate new combinations of genes, though whether it is as important as mutation and plasmid transfer in generating new genotypes is debatable. Even if the production of variation is not the reason that transformation evolved, it is still an important consequence because the operation of natural selection requires genetic variability. After all, natural selection cannot occur if there are no alternatives from which to select.

Since members of a transforming population exchange genes, it may be considered a reproductive community analogous to a eukaryotic sexual species. In *B. subtilis, Pseudomonas stutzeri, H. influenzae,* and *N. gonorrhoeae,* there is evidence that the donation of DNA is an active process under the control of the cell. In the recipient cells, different mechanisms of processing incoming DNA give rise to different barriers to gene transfer. In *Haemophilus* and *N. gonorrhoeae,* donor DNA molecules lacking a certain base sequence characteristic of the species are usually rejected. On the other hand, species such as *B. subtilis* do not appear to discriminate among DNA molecules during DNA binding and uptake. There is evidence that restriction modification systems may act as barriers to transformation in *B. subtilis*. In experiments with *B. subtilis* isolates, it was also shown that the degree of similarity between the homologous DNA molecules influenced the transformation frequency, although even homologous molecules from other species can transform *B. subtilis*. The implication is that gene flow barriers between these bacterial species may be less rigid than the reproductive barriers between eukaryotic species. One consequence of these leaky gene flow barriers is that selection, rather than these barriers, may be the primary force maintaining homogeneity within a bacterial species.

The evolutionary history of bacteria, like that of other living things, has been represented with a branching tree-like diagram. For a branching process to be an appropriate description of historical relationships, the units of study (the species) must acquire their characteristics from their immediate ancestor in the tree. In different species of *Neisseria* and in different strains of *S. pneumoniae,* different regions of DNA give different patterns of similarity, that is, different trees. In these cases, evolutionary histories deduced from different genes are different. This result suggests that genes have been exchanged between members of these different groups. Gene flow between different units makes evolution more reticulate and less treelike.

III. Summary

Natural genetic transformation possesses the defining characteristics of sex: recombination and outcrossing. Unlike meiotic sex, sex during transformation is temporally uncoupled with reproduction, does not involve chromosomal reassortment or the reciprocal exchange (crossing over) of outside markers, and represents unequal genetic contributions from the parent cells. Unlike other sexual processes in bacteria (conjugation and transduction), it is directed by chromosomal genes in the recipient organism rather than by extrachromosomal genes in the donor. For this reason, transformation is unlikely to be a consequence of the spreading of parasitic genes or infectious elements. It is also unlikely that the main advantage of transformation is the use of donor DNA for food, although this may have been one of the original advantages of the DNA binding and uptake machinery. Transformation is more likely to confer an advantage on its possessors by repairing damaged genes or creating more favorable genetic variants. Even if the variation is not advantageous, it still facilitates the operation of natural selection and has other evolutionary consequences. One consequence is that variation may justify the classification of groups of bacteria as species, although transforming bacterial species are more likely than animal species to hybridize with other species. As a consequence of this gene flow, transformation and other means of gene exchange may obscure geneological relationships in bacteria.

Bibliography

Bernstein, C., and Bernstein, H. (1991). "Aging, Sex, and DNA Repair." Academic Press, San Diego.

Levin, B. R. (1988). *In* "The Evolution of Sex" (R. E. Michod and B. R. Levin, eds.), pp. 194–211. Sinauer, Sunderland, Massachusetts.

Michod, R. E., Wojciechowski, M. F., and Hoelzer, M. A. (1988). *Genetics* **121,** 411–422.

Michod, R. E., Wojciechowski, M. F., and Hoelzer, M. A. (1990). *In* "Molecular Evolution" (M. T. Clegg and S. J. O'Brien, eds.), pp. 135–144. Wiley-Liss, New York.

Redfield, R. J. (1988). *Genetics* **119,** 213–221.

Smith, J. M. (1990). *Ann. Rev. Ecol. Syst.* **21,** 1–12.

Stewart, G. J., and Carlson, C. A. (1986). *Ann. Rev. Microbiol.* **40,** 211–235.

Genetic Transformation, Mechanisms

Martin F. Wojciechowski
University of Arizona

Glossary

Competence Physiological state of cells characterized by the ability to bind and take up exogenous DNA

Heterologous DNA DNA having little, if any, nucleotide sequence similarity (i.e., DNA from an unrelated species)

Homologous DNA DNA having considerable nucleotide sequence similarity (i.e., DNA from the same or closely related species)

Recombination Any of a number of pathways (general or homologous, site-specific, transpositional) in which DNA molecules interact to bring about a change in the combination of genes or parts thereof; in the molecular sense, the physical breakage and rejoining or replacement of DNA molecules catalyzed by specific enzymes

GENETIC TRANSFORMATION is the process by which bacteria and a few eukaryotic microorganisms become competent to take up naked DNA from the environment and integrate it into their genomes. This process is one of three fundamental processes in bacteria that can mediate the exchange of genetic information, and has been found to occur naturally in a number of ecologically and phylogenetically diverse bacterial species. Genetic transformation in naturally transformable bacteria is the result of the development of physiological competence, a genetically regulated and distinct property of cells that is characterized by the ability to recognize, bind, and take up exogenous DNA, followed by the recombination and expression of this DNA in the recipient cell.

I. Introduction

Bacteria are haploid organisms that reproduce by binary fission, producing genetically identical or clonal progeny. Changes in the genetic material (mutations) of a cell brought about by replicational errors, or by the repair of damage induced by various chemical and physical agents, usually remain restricted to that particular cell and its clonal derivatives unless redistributed by processes (such as genetic exchange), which foster variation. Three fundamentally different mechanisms have evolved that facilitate the directional transfer or exchange of genetic information in bacterial systems. In each case, the process is mediated by the introduction of DNA into the recipient cell. These processes, which have been traditionally distinguished by the actual mode of DNA transmission, are conjugation, transduction, and transformation. Conjugation is the cell-contact-dependent transfer of a single-stranded copy of a conjugal plasmid and part or all of the chromosome of the donor cell to a recipient cell. In transduction, the transmission of small segments of the bacterial chromosome from one cell to another is ostensibly the incidental consequence of bacteriophage-mediated transfer of its own genome. Transformation results from the recognition and uptake of macromolecular DNA from the environment and its integration into the genome of the cell. It is important to emphasize that natural transformation is a function of normal cell physiology and is the only means for genetic exchange in bacteria with functions that are encoded entirely on the host-cell chromosome. Since each of these mechanisms involves the horizontal transmission of genetic information in a process that is often catalyzed by recombination, they are increasingly viewed as sexual processes. However, unlike genetic transformation, plasmid-mediated conjugation and phage-mediated transduction are mechanisms specialized for the transfer of plasmid and phage genomes into new host cells and apparently transfer chromosomal genes only accidentally. [*See* CONJUGATION, GENETICS.]

This article will focus on the general mechanisms of natural genetic transformation in bacteria, using the well-characterized systems of *Bacillus subtilis* and *Haemophilus influenzae* as paradigms for a comparative approach. The so-called artificially inducible transformation in other microbial systems, such as *Escherichia coli,* will also be discussed because of its practical applicability to the study of DNA transfer and its widespread use in molecular genetics research today. For a detailed discussion of the gene products involved in competence development, DNA uptake, or the mechanics and different pathways of recombination in bacteria, the reader is referred to the many reviews listed in the Bibliography.

II. History

The discovery of genetic transformation in bacteria is one of the most significant events in biology in this century. The first evidence for genetic recombination in bacteria was obtained in the late 1920s by the English physician Frederick Griffith who worked with the bacterium *Streptococcus pneumoniae* (known as pneumococcus), which causes a form of pneumonia. Griffith showed that if heat-killed cells of a virulent strain were injected into mice along with living cells of a nonvirulent mutant strain, only living virulent bacteria were recovered. Not until the early 1940s was a molecular explanation for the so-called pneumococcus "transforming principle" obtained. This was provided by Oswald Avery, Colin MacLeod, and Maclyn McCarty, who demonstrated that a cell-free extract of heat-killed virulent cells could induce transformation *in vitro,* and that the transforming activity was only present in purified preparations of DNA. In retrospect, this discovery marked the beginnings of molecular biology and molecular genetics, since it ultimately led to experiments proving without a doubt that DNA is the genetic material (RNA bacteriophages and viruses excepted), despite the widely-held belief at the time that genetic information was encoded by proteins. Subsequently, the elucidation of the three-dimensional structure of DNA by James Watson and Francis Crick in 1953 provided the theoretical framework necessary for understanding the replication of DNA and the mechanism by which it functions as the genetic material.

Since that time, genetic transformation has been found to occur among many phylogenetically divergent and ecologically diverse bacteria, including members of the traditional "gram-negative" and "gram-positive" groups of eubacteria, cyanobacteria (blue-green algae), archaebacteria, and possibly certain eukaryotic microorganisms (Table I). The list in Table I is based on a review of citations in both the primary and secondary literature and is

Table I Presence of Genetic Transformation in Microorganisms

Eubacteria	Cyanobacteria
Acinetobacter[a]	*Anabaena*
Achromobacter[a]	*Synechococcus*[a]
Agrobacterium	*Synechocystis*[a]
Azotobacter[a]	
Bacillus[a]	
Brevibacterium	Archaebacteria/Halophiles
Campylobacter[a]	*Halobacterium*
Clostridium	*Methanobacterium*
Corynebacterium	*Methanococcus*[a]
Erwinia	
Escherichia	
Haemophilus[a]	Mycoplasmas and Rickettsias
Lactococcus	*Acholeplasma*
Leuconostoc	*Myocoplasma*
Methylobacterium[a]	*Rochalimaea*
Micrococcus[a]	*Spiroplasma*[a]
Moraxella[a]	
Mycobacterium[a]	
Neisseria[a]	
Pasteurella[a]	Fungi
Pseudomonas[a]	*Aspergillus*
Psychrobacter	*Candida*
Rhizobium	*Gliocladium*
Rhodopseudomonas	*Hansenula*
Salmonella	*Pichia*
Staphylococcus	*Saccharomyces*
Streptococcus[a]	*Schizosaccharomyces*
Streptomyces[a]	*Trichosporon*
Thermoactinomyces[a]	*Ustilago*[a]

[a] Denotes those microorganisms known to possess a natural occurring physiological competence and efficient mechanism for uptake and integration of exogenous DNA.

intended to provide the reader with an appreciation for the diversity of microbial systems for which the ability to undergo transformation with purified DNA or the presence of a cellular mechanism for genetic exchange resembling transformation has been reported. These systems vary widely in the "naturalness" of the transformation process, perhaps reflecting the extent of our knowledge regarding the environmental and/or culture conditions that promote the ability to take up DNA in nature and also, perhaps, the strength of selection for transformation versus other forms of genetic exchange such as transduction.

Natural transformation in bacteria is a genetically determined, highly efficient, and evolved property of cells, and is dependent on the ability of cells to develop the physiological competence for DNA binding and uptake. This process makes evident the presence of an efficient membrane transport system for high-molecular-weight DNA molecules and, as such, poses an interesting problem in macromolecular transport. For some organisms, transformation is a principal, or perhaps the only, means of mediating the transfer of chromosomal DNA between individuals or populations of a species in nature, be they in the soil or in the gut of mammals. Experimentally, the acquisition of competence generally results from growth of a culture under defined physiologically realistic conditions, and is maximally expressed at a specific stage of growth (for example, stationary phase). In some species, competence is expressed constitutively during all phases of the growth cycle. Thus, transformation in naturally competent bacteria such as *B. subtilis* has usually been distinguished from the experimentally induced "competence" and manipulated transformation of microorganisms such as *E. coli* and yeast. Transformation in these systems is achieved by the treatment of otherwise growing but noncompetent cultures of these organisms with chemical or physical agents that render the cells capable of interacting with transforming DNA so as to promote its uptake. Transformation of this kind is now used as a common laboratory procedure for the introduction of DNA molecules, often engineered by *in vitro* recombinant methods, into a desired host organism for expression and further analysis. However, there is some evidence that suggests *E. coli* can be induced to a physiological state that resembles competence of naturally transformable bacteria (i.e., *Bacillus* and *Haemophilus*) and that transformation in this organism is not simply the result of poking large holes in the membrane through which DNA can pass. Whether the experimentally induced competence in these microorganisms reflects the presence of a less efficient, less highly evolved system for genetic transformation, or is truly artificial and only a consequence of the physical modification of the cellular membrane so as to make it less of a barrier to the passive entry of macromolecules like DNA is an intriguing but unresolved problem.

Historically, a great deal of study on natural transformation in bacteria has been devoted to only a few organisms. This situation has changed significantly in recent years as a result of the remarkable advances in both molecular and genetic methods for the detailed study and engineering of bacterial species not well suited to characterization by classical methods of genetic analysis. The best studied systems for natural transformation are those of *Streptococcus pneumoniae*, *S. sanguis*, *Haemophilus* (including *H. influenzae* and *H. parainfluenzae*), *Neisseria gonorrhoeae*, and *Bacillus*, *B. subtilis* being the most thoroughly characterized species. The results of work on these organisms during the 1960s and 1970s led Hamilton Smith to propose that natural transformation systems fall into two mechanistically distinct groups, one typical of the traditionally defined gram-negative bacteria such as *Haemophilus* and gonococcal species, and the other typical of gram-positive bacteria such as *B. subtilis* and *Streptococcus* species. However, this division has become less definitive in recent years as a result of more detailed studies of transformation in an ever-widening variety of organisms.

III. Development of Competence for Transformation

Competence may be defined as a physiological state of bacterial cells that is characterized by the ability to bind and take up exogenous DNA. Transformation then proceeds by the acquisition, transport, integration, and ultimate expression of this DNA by the recipient cell. Much of our present knowledge about the development of competence and the mechanism of DNA uptake and integration by naturally transformable bacteria has been obtained experimentally through the use of genetically characterized bacterial strains and chemically purified DNA in solution, thereby permitting both qualitative and quantitative aspects of the process to be investigated under rigorously controlled conditions. In contrast, the amount and sources of transforming DNA as well as the ecological significance of the transformation process itself in natural bacteria populations have received relatively little attention. In nature, transforming or "donor" DNA may be made available to competent cells either by the spontaneous autolysis of other cells or by active transport of DNA out of living cells.

In most naturally transformable bacteria studied in detail so far, the development of competence is an inducible, genetically determined (i.e., regulated), and physiologically discrete property of particular cells that is exhibited at a particular time during growth in culture. It has been likened to the process of differentiation in multicellular organisms. The development of competence and, thus, transformability is a variable trait even within a species; physiological differences between individual strains are known to exist. Further, it is important to note that competence is not a terminal state of most cells and that the physiological changes that cause or merely accompany competence are temporary and can be reversed, with a subsequent "loss" of competence.

The development of competence is accompanied by a number of spatial as well as temporal alterations in cellular physiology and structure that vary among the different genera of bacteria. For example, competent cells of *B. subtilis* are in so-called state of "biosynthetic latency," characterized by decreased rates of nucleic acid synthesis following the completion of a round of chromosomal replication and cell division. In *S. sanguis,* the rate of RNA synthesis declines with the onset of competence but the rate of DNA synthesis does not. Other specific changes have been documented as associated with or directly related to competence development and transformation. Among these are changes in the cell envelope, including the appearance and increased synthesis of a set of competent-cell-specific proteins that are located and presumably active at the cell membrane or outer surface, differences in the composition of the lipopolysaccharide layer, and changes in the density and location of membrane-bound structures. In some bacteria, there is also a marked increase in the capacity for recombination (and inducible DNA repair), as evidenced by increased frequencies of phage recombination and plasmid and chromosomal DNA-mediated transformation. The presumed role in the transformation process for many of these newly synthesized proteins (DNA-binding proteins and nucleases) and structural alterations is supported by their absence or reduction in various mutant strains that are deficient in competence and/or the ability to be genetically transformed. In contrast, relatively little is known about the role of the chromosomes of competent cells during transformation, although the appearance of sometimes extensive single-stranded gaps in the chromosomes of both *B. subtilis* and *H. influenzae* has been reported.

In *B. subtilis,* competence is transitory, as it is in most bacteria, and develops maximally after the exponential growth of a culture (in a glucose–minimal salts-based medium) has ceased. Competence is presumably induced during the transition to the slower stationary phase of growth. Competence in *B. subtilis* is just one of several stationary-phase-specific phenomena that are characteristically exhibited at this time in the growth cycle, perhaps the most familiar of which is the production of resistant endospores (sporulation). Indeed, the postexponential regulation of competence development in this bacterium strongly suggests the existence of a common pathway of metabolic induction and temporal regulation for these stationary phase responses, probably triggered initially by the exhaustion of some essential nutrient(s). Although sporulation seems to be regulated in part by a diffusible extracellular factor(s), it is not yet clear whether similar factors are even involved in the induction of competence in *B. subtilis.*

Perhaps more intriguing, the development of competence in *B. subtilis* is also subject to a cell-specific type of regulation, since only about 10% of the cells in a competent culture actually become competent for transformation. Whether this is merely a trivial consequence of the limited expression of some key metabolic component or the representation of a highly regulated developmental process is unknown. The heterogeneity of competence in *B. subtilis* populations was first suggested by the higher than expected (for independent transformation events) frequencies for cotransformation of "unlinked" or distant genetic markers. This heterogeneity was later confirmed directly by autoradiography and the resolution of transformed competent cultures into two distinct fractions or subpopulations of cells on the basis of their buoyant densities (by centrifugation), one a minority composed predominantly of competent cells and the other a majority composed predominantly of noncompetent cells. This heterogeneity of competence in *B. subtilis* may in fact represent a primitive sexual dimorphism, as recently suggested by David Dubnau: the noncompetent cells are specialized to extrude DNA and the competent cells are destined to take it up.

In comparison, competence in species of *Streptococcus* usually develops in all cells of a culture during the exponential phase of growth, apparently triggered by the accumulation of a soluble (extracellular) competence factor to a threshold level necessary for induction, and usually only when a culture reaches a density of about 10^8 cells per milliliter. Although this competence factor in *S. pneumoniae* has been partially characterized and appears to be a small basic protein, its role in competence induction remains obscure.

The traditional model for competence development in gram-negative bacteria is based almost exclusively on studies with *Haemophilus* species. As in *B. subtilis*, competence in cultures of *H. influenzae* and *H. parainfluenzae* is minimal during exponential growth and is induced only by conditions that slow or limit sustained growth, that is, conditions of nutritional downshift such as those that occur in early stationary phase cells. However, unlike *B. subtilis*, once induced virtually 100% of *Haemophilus* cells in a population can become competent for transformation. This condition can be maintained for some time in media that do not support growth. In *H. influenzae*, a less than maximal level of competence can also be induced in exponentially growing cells by the addition of adenosine-3′,5′-monophosphate (cyclic AMP), suggesting that a relief from the inhibitory effects of catabolite repression may be involved in induction.

Similarly, development of competence occurs at the onset of stationary phase in other naturally transformable gram-negative bacteria, including *Acinetobacter calcoaceticus* and *Pseudomonas stutzeri* (and related species). In contrast, competence in *Neisseria* and other gonococci is constitutive and apparently not regulated by the metabolic and temporal controls that operate in other bacteria. Although nearly all clinically isolated gonococcal strains are transformable, there is no convincing evidence either for a growth-stage dependence for competence development or for involvement of soluble competence factors. Interestingly, though, competence in gonococci is expressed constitutively only in piliated (P^+) cells; some 30–100% of piliated cells are competent for transformation. There has also been a report that suggests that pili may be necessary for genetic transformation in *Moraxella*. The role of pili in mediating the early stages of (pathogenic) genital infections by gonococci is believed to be the primary selective pressure for maintenance of the piliated, and thus competent, phase. The loss of pili (P^-) in gonococci during growth *in vitro* results in a dramatic decrease in transformability and the ability to take up DNA. Recent studies have shown that the P^+ to P^- phase transition in *N. gonorrhoeae* is associated with changes in a number of cellular proteins, including pilin. Thus, although presence of pili is strongly correlated with the ability to bind and take up exogenous transforming DNA in *Neisseria*, possibly by functioning to provide receptor sites or to act as channels through the outer membrane, there is no prevailing evidence for a direct role for pili in mediating the binding or promoting the uptake of DNA.

IV. Mechanism of DNA Binding and Uptake

Once cells have become competent, the process of binding donor DNA at their outer surface for subsequent uptake and integration into the genome is initiated. A considerable amount of progress has been made already toward understanding the various stages of genetic transformation at the molecular level, particularly the fate of transforming DNA during its binding, uptake, and recombination, at least in *B. subtilis*, *Streptococcus* spp., and *Haemophilus* spp. However, only now are investigators that work with these systems beginning to identify and characterize the many gene products involved and the role of the chromosome of the recipient cell in the process of transformation.

For most naturally transformable organisms, the overall process of transformation may be subdivided into several discernible stages—binding, fragmentation, uptake, and integration—related to the actual processing of the transforming donor DNA (see Fig. 1). Although transformation mechanisms in both gram-positive and gram-negative bacteria share the same basic steps, somewhat different mechanisms are used for binding and uptake, possibly as a consequence of the considerable differences in the structure of the bacterial cell envelopes. However, in both gram-positive and gram-negative bacteria, the transforming DNA enters the cytoplasm as a single-stranded molecule, a form that promotes the rapid and efficient integration of the molecule into the recipient chromosome. The ultimate fate of the donor DNA, as well as the efficiency of the transformation process itself, depends largely on the nature of the donor DNA molecules (i.e., linear versus circular; chromosomal, plasmid, or bacteriophage; homologous versus heterologous), although the binding and uptake of all DNAs may proceed via a common pathway in a given organism.

Binding by definition is the first stable attachment

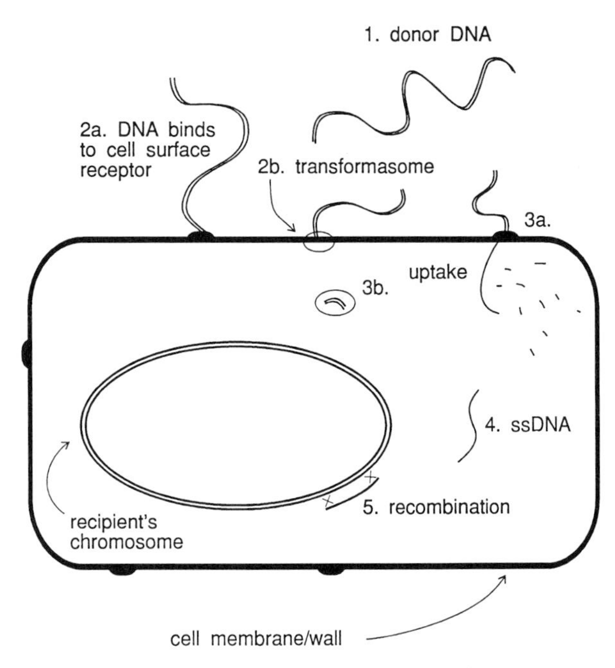

Figure 1 The mechanism of transformation in bacteria. Double-stranded donor DNA (1) in the environment becomes bound to specific receptors in gram-positive bacteria (2a) or to transformasomes in gram-negative bacteria (2b), present on the surface of competent cells. Uptake of bound transforming DNA in gram-positive bacteria is initiated by the action of a membrane-bound deoxyribonuclease, which degrades one strand of duplex DNA as the complementary strand of DNA is brought into the cell (3a). In gram-negative bacteria, uptake is mediated by the movement of the transformasome–DNA complex to the cytoplasm (3b) where it is followed by the transport of single-stranded DNA molecules out of the transformasome. In both groups of bacteria, the single-stranded transforming DNA molecules (4) are then rapidly integrated into the chromosome of the recipient cell, at the site of DNA sequence homology, by the enzyme-catalyzed process of recombination.

of transforming DNA to the surface of competent cells. The binding of DNA by cells may be loose so the DNA is susceptible to mechanical shearing, removal by washing, or displacement by competing DNA. DNA may be more tightly associated with the cell surface. Results from numerous studies have led to the following general conclusions regarding the binding process.

The attachment of transforming DNA to the surface of competent cells occurs at a limited number of binding sites; it is specific, irreversible, rapid, and proceeds to saturation according to the number of DNA molecules rather than the total mass of DNA. The substrate for binding and uptake in natural transformation systems is usually double-stranded or duplex DNA, although transformation by single-stranded DNA, albeit at lower levels, has been reported for *Haemophilus*, *B. subtilis*, and *E. coli*. The precise nature and location of DNA binding sites on the surface of competent cells are not known with certainty, although competence-specific DNA binding proteins or outer membrane structures have been described for most bacteria. Binding sites may be localized on the cell wall or membrane, or may represent specialized structures that span both layers. Since bound DNA is at least initially sensitive to mechanical shear and the action of exogenous deoxyribonuclease (DNase), it is probable that the entire molecule is physically extended into the aqueous environment and remains attached to the cell surface at relatively few points per molecule. Once bound in tight association, the DNA is taken up remarkably quickly (measured in minutes) and internalized.

Although tight binding of transforming DNA to competent cells has been demonstrated in all systems, the specificity and mechanism of binding appear to be significantly different for gram-positive and gram-negative bacteria, in part because of differences in the structures of their respective cell envelopes. Most gram-negative bacteria possess an envelope consisting of a cell membrane surrounded by a thin peptidoglycan layer and an outer membrane of unique biochemical composition. Some strains have, in addition, a thick external polysaccharide capsule. In contrast, gram-positive bacteria typically possess a cell membrane surrounded only be a thick peptidoglycan-containing wall.

In *B. subtilis* and *Streptococcus* species, a limited number of binding sites (20–50 for *B. subtilis;* 30–80 for *S. pneumoniae*) is present on the surface of a competent cell. These bacteria bind double-stranded DNA with little if any sequence specificity for homologous or "self" DNA, as evidenced by their ability to bind (and internalize) DNA from *E. coli* and a variety of phages and plasmids. However, double-stranded RNA, glucosylated DNA, and RNA–DNA hybrid molecules bind poorly or not at all to competent cells. The binding of DNA by competent cells is rapidly followed by single-strand nicking and/or double-strand cleavage. In *B. subtilis*, the fragmentation of DNA by double-strand cleavage produces a distribution of bound fragments ranging in size from 7.5 kb to 30 kb (kilobase pairs). In *S. pneumoniae* and *S. sanguis*, the binding of double-stranded DNA is accompanied by nicking of the DNA at intervals of 6–8 kb prior to double-strand cleavage. Although the cleaved molecules become relatively resistant to shearing, they are nonetheless bound in a form that is completely accessible to exogenous DNase and, thus, probably still exposed on the cell surface.

In *B. subtilis* and *Streptococcus,* uptake is then initiated by a membrane-associated exonuclease enzyme that requires divalent cations such as Mg^{2+} or Mn^{2+} as co-factors and a single-strand DNA binding protein; this results in the penetration of single-stranded DNA through the cell membrane. Detailed kinetic studies with *B. subtilis* have shown that uptake is linear with respect to the length of the donor DNA and that single-stranded DNA fragments are derived from surface-bound double-stranded fragments of similar size. These single-strand molecules are the precursors of the integrated DNA. When radioactively labeled donor DNA was used in such uptake studies, the appearance of single-stranded DNA was found to coincide with the release into the cell medium of acid-soluble materials: 5′ nucleotides, free bases and oligonucleotides (DNA degradation products), and about half the radiolabel. This result strongly suggests that single-strand penetration is concomitant with the complete degradation of the complementary strand. Furthermore, the appearance of the single-stranded molecules in the cell is also coincident with the loss of sensitivity of donor DNA to the action of DNase.

The energetics of DNA uptake and its transport into the competent cell have been examined so far only in *B. subtilis* and *S. pneumoniae*. According to a current model proposed by Sanford Lacks and his associates, transport occurs through a channel in the membrane that is associated with a nuclease asymmetrically located on the membrane. The entry of DNase-resistant single strands into the cell is apparently driven by the action of this membrane-bound nuclease. In *B. subtilis,* DNA uptake requires the proton motive force, especially the ΔpH component, and apparently is not driven directly by ATP (adenosine triphosphate) hydrolysis. In contrast, DNA uptake in *S. pneumoniae* is driven directly by ATP, although the overall process appears to be somewhat slower than in *B. subtilis*.

Several studies have shown that the binding and uptake of transforming DNA in *Streptococcus* is remarkably similar to the process in *B. subtilis*. Transport in *S. pneumoniae* also involves a membrane-localized nuclease that is required for the uptake of transforming DNA molecules and their concomitant conversion to a single-stranded form. Further, the mechanism of DNA uptake in *S. sanguis* appears to be very similar with the exception that the conversion of donor DNA molecules to a single-stranded form is not necessarily followed immediately by transport into the cell. In both *B. subtilis* and *Streptococcus,* the single-stranded donor DNA is protected from nonspecific degradation by nucleases during transport by the formation of a DNA–protein complex with a specific binding protein that is synthesized during competence development. As a consequence of the formation of this complex, both systems exhibit a phenomenon called the "eclipse" phase at which time the single-stranded donor DNA no longer possesses transforming activity when isolated just prior to integration. This phenomenon continues until the single-stranded DNA is actively integrated into the recipient chromosome.

The current model for transformation in *Haemophilus* (and the gonococci) suggests that a more elaborate mechanism for DNA binding and transport exists than previously described for bacteria such as *Bacillus* and *Streptococcus*. First, competent cells of both *Haemophilus* spp. and *N. gonorrhoeae* exhibit a novel preference for binding, uptake, and transformation by their own ("homospecific") DNA, in contrast to the *Bacillus* and *Streptococcus* systems. Second, specialized membranous vesicles present on the surface of the outer membrane of competent cells of *Haemophilus* spp. function to take up and protect the donor DNA during its entry into the cell.

This remarkable specificity for the uptake of homologous DNA by both *Haemophilus* and *N. gonorrhoeae* has been shown recently to depend on the presence of short repeated nucleotide sequences in their DNA and the interaction of these sequences with surface receptor sites (proteins). In these bacteria, the requirement for such signal or recognition sequences in donor DNA, however, is far from absolute, since DNA molecules lacking these specific sequences can still be taken up, and in similar amounts, but at a slower rate. *Haemophilus influenzae* and *N. gonorrhoeae* do not utilize the same recognition sequence since *H. influenzae* DNA does not compete with *N. gonorrhoeae* DNA for uptake. Both *H. influenzae* and *H. parainfluenzae* preferentially bind and take up DNA possessing the 11-bp sequence 5′-AAGTGCGGTCA-3′. The number of copies of this particular recognition sequence in the *Haemophilus* genome has been estimated at approximately 600 and corresponds to an average of one uptake sequence per 4 kb of DNA. This number was later substantiated by Southern hybridization analysis of *Haemophilus* spp. DNA using a radioactively labeled oligonucleotide of this specific 11-bp sequence as probe. Further, the particular sequence is not present to any significant degree in DNAs from other bacterial genera. In *N. gonorrhoeae,* the species-specific uptake of genetic transformation by DNA is mediated by the 10-bp sequence 5′-GCCGTCTGAA-3′. Indeed, insertion of various forms of this 10-bp sequence into an otherwise non-transforming plasmid results in an increase in both the uptake of double-stranded plasmid DNA into a DNase-resistant state and the genetic transformation of piliated gonococci strains by this foreign plasmid. Interestingly, this recognition sequence for transforming DNA in *Neisseria* is frequently found near sequences that presumably function as termination sites for transcription.

In *Haemophilus* spp., binding and uptake is apparently mediated by a small number (10–12 per cell) of receptor sites on membranous vesicles present only on the surface of competent cells. The association of transforming DNA with these vesicular structures renders the DNA insensitive to exogenous DNase. The evidence for the primary role of these vesicles, referred to as transformasomes, in DNA binding, comes from the analysis of competence-deficient mutants of both *H. influenzae* and *H. parainfluenzae* that shed vesicles during competence induction rather than accumulate them on the surface. Transformasomes from mutants of both species have a similar size and morphology and, when isolated from cell-free supernatants, possess far more DNA-binding ability than do intact cells. The addition of double-stranded donor DNA to competent *H. parainfluenzae* cells causes the rapid (within 3 min) internalization of these vesicles from the surface to the cytoplasmic side so donor DNA is protected from digestion by both exogenous DNase and intracellular restriction enzymes. Entry of DNA into the cytoplasm is then followed immediately by integration of the DNA into the genome, suggesting that recombination with the resident chromosome takes place in very close proximity to the vesicle. However, a different mechanism of transport apparently operates in *H. influenzae*. In this species, the bound double-stranded donor DNA is held and protected temporarily in transformasomes located on the surface, and is subsequently transported out of the transformasome directly into the cytoplasm, where integration occurs.

V. Integration and Resolution of Transforming DNA

The final stage of homologous DNA-mediated transformation is the enzyme-catalyzed integration of donor transforming DNA into the chromosome of the recipient cell via genetic recombination. [For a detailed discussion of the mechanism and genetics of recombination, see Smith (1988).] The homology-dependent (i.e., rec^{+}-dependent) nature of this stage of the transformation process in most bacteria has been confirmed through the isolation and extensive characterization of various recombination-deficient (rec^{-}) mutants. [*See* RECA.]

In *Bacillus, Haemophilus,* and *Streptococcus,* the end product of recombination is a heteroduplex molecule in which a single-strand donor molecule becomes stably but noncovalently paired to a complementary recipient strand to form a donor–recipient DNA complex. This pairing is made possible by the physical displacement and replacement of one of the homologous strand in the recipient duplex. Single-strands displacement is accompanied by the simultaneous degradation of the displaced recipient strand and other nonintegrated molecules of transforming DNA. When no sequence homology with the chromosome exists, as is the case with heterologous donor DNA, complete degradation ultimately occurs. The second unused strand of homologous donor DNA in gram-negative bacteria is similarly degraded and can be quickly reutilized as nutrients.

During the eclipse phase in *B. subtilis,* radiolabeled single-stranded donor molecules ([^{3}H] DNA) can be recovered from transformed cells. However, within minutes the radioactive label becomes associated with intact double-stranded DNA. The average size of the heteroduplex formed corresponds closely with the average size of the single-stranded donor fragments recovered from cells just moments before, demonstrating that the bulk of the single-stranded DNA mass (approximately 70%) is converted to an integrated form.

In *Haemophilus,* genetic experiments and experiments with radioactively labeled donor DNA give similar results concerning the integration step. Recombination can be detected as early as 5 min after the addition of duplex donor DNA and reaches a maximum level by 60 min. Although donor DNA fragments begin to exit the transformasome and become integrated almost immediately after uptake, their movement into the cell is accompanied by considerable degradation. In fact, recent studies indicate that only about 15% of the labeled donor DNA can be localized in donor–recipient duplex molecules that are indicative of recombination. Considerable experimental evidence suggests that one strand undergoes complete degradation during exit from the transformasome; the other strand is integrated but, at the same time, is subject to continuous degradation. As a consequence, the amount of degradation of the donor DNA molecules may, in fact, reflect the time it takes for the donor fragments to find homologous regions in the chromosome and become paired successfully.

The heteroduplex molecules that result from the integration of single-stranded donor DNA can then be resolved either by semi-conservative replication of the DNA or by mismatch repair followed by replication to produce a homoduplex (in terms of DNA sequence) DNA molecule. Resolution by replication alone yields genetically mixed clones of cells, that is, cells that contain both parental and transformant DNA, whereas mismatch repair followed by replication produces pure clones of either the parental or transformed genotype. The actual structure of the "joint" donor–recipient duplex molecules, as well as the mechanics of the recombination process, is largely unknown in any of the systems described.

VI. Characterization of Competence and the Transformation Process

Competence and the various stages in transformation have been characterized using a number of methods. The kinetics of binding and uptake of transforming DNA have been measured directly using radioisotopically labeled donor DNA ([^{3}H] DNA), alone or in competition experiments with unlabeled DNA, or using sedimentation analysis of labeled DNA–cell complexes. Autoradiography and electron microscopy, in combination with labeled DNAs, have been used to determine the location and number of DNA binding sites on cells and to monitor changes in membrane structures during competence and incubation with transforming DNA. At least for the *B. subtilis* system, specialized centrifugation methods permit the separation of competent from noncompetent cells based on differences in their bouyant densities in Renografin® (diatrizoate meglumine) or their sedimentation velocities in sucrose gradients, yielding two populations: a lighter buoyant density fraction composed of competent cells and a heavier noncompetent cell fraction. This method has been used successfully in several studies to investigate the inducible expression of specific genes in the competent and noncompetent subpopulations.

Competence can be measured experimentally, both qualitatively and quantitatively, in biologically relevant ways by the production of genetically transformed cells. The transformation process is dependent on both the amount and the kind of DNA used as donor. For example, the donor may be chromosomal, plasmid, or phage DNA; in each case the outcome of the transformation process will differ although the binding and uptake of these DNAs in a given bacterial species are thought to proceed by a common pathway.

Most studies of genetic transformation have employed homologous chromosomal DNA isolated from wild-type strains (for specific markers) or plasmids carrying antibiotic resistance genes, scoring transformation by the appearance of prototrophic or drug-resistant colonies under appropriate selection conditions, respectively. In chromosomal DNA-mediated transformation, the donor DNA is ultimately recombined into the homologous region of the chromosome in the recipient cell, effecting a heritable change in the genetic make-up of the recipient. In plasmid DNA-mediated transformation, an antibiotic-resistant transformant results from the *rec*-independent establishment of a stable autonomously replicating DNA element. When using phage DNA as donor (i.e., transfection), transformation is scored simply as the production of infectious centers ("plaques") on a lawn of cells of a sensitive "indicator" strain. The successful transformation of cells by plasmid and phage DNA molecules generally is not dependent on the presence of any sequence homology to the recipient genome; thus it is usually independent of the capacity of the cell for homologous recombination. In each type of transformation, the number of transformant colonies (colony-forming units) or infectious centers (plaque-forming units) is determined by counting and is expressed as a frequency of the number of total viable cells in a given transformed culture. Under conditions in which the concentration of donor DNA is saturating (i.e., not limiting), maximal transformation frequencies in bacteria range from 1×10^{-4} to 1×10^{-1} for a single chromosomal or plasmid-borne marker, or approximately 1 transformant per

10,000 to 1 transformant per 10 total viable cells. Here, genetic marker is defined as a recognizable phenotype that is specifically correlated with the presence of a certain allele. As an example, cells of an auxotrophic (mutant) strain of *B. subtilis* carrying a mutation at the *metB* locus (e.g., *metB5*) are able to grow on minimal glucose medium only when it is supplemented with the amino acid methionine (phenotypically Met$^-$), are transformed with DNA purified from a wild-type strain that is prototrophic for the same marker locus (Met$^+$). Met$^+$ transformants of the mutant strain are selected by plating aliquots of the transformed culture on minimal medium lacking methionine.

The efficiency of transformation also depends on the nature of the transforming DNA. Transformation with homologous chromosomal DNA is much more "efficient" than with either plasmid DNA-mediated transformation or phage DNA-mediated transfection, since there is nearly a one to one correspondence between the number of donor DNA molecules taken up and the number of genetic transformants. However, such comparisons do not reflect the fact that, although many more DNA molecules (10^3 to 10^4) are often taken up in plasmid and phage DNA-mediated transformation events, a very minor proportion of that DNA actually participates in the production of each transformed colony or plaque.

The chromosome of a single bacterial cell is a circular DNA molecule of approximately 2000–5000 kb with a molecular mass of 1.4–3.3 × 10^9 daltons. Assuming that an average protein contains about 330 amino acids (which represents 1 kb of DNA coding sequence), a bacterium such as *B. subtilis* would contain about 5000 genes. The average size of the DNA fragments integrated into *B. subtilis* cells during the transformation has been estimated to be 8–10 kb in length and each cell may integrate 10 molecules of this size. This amount of DNA corresponds to about 100 average-sized genes. Although transformation is an efficient process, it is impossible to transform all the competent cells in a culture for any one genetic marker (i.e., transformation at a single chromosomal locus) since only about 1 in 100 to 1 in 1000 DNA molecules in a purified DNA preparation contain the specific genetic marker or interest.

The minimum amount of highly purified DNA that will produce detectable numbers of transformant colonies (i.e., greater than about 100 transformant colonies per 10^8 total cells) under standard laboratory conditions using *B. subtilis* is about 0.00001 μg DNA. This amount of DNA, 10 pg (10 × 10^{-12}gm), is so small that it is virtually undetectable by standard physiochemical methods such as ultraviolet light spectroscopy, measured by absorbance at 260 nm, yet it is still biologically detectable. Just how much DNA is contained in 10 pg? Linear duplex DNA weighing 1 pg contains 9.1 × 10^8 bp and has a length of 30.9 cm! In standard laboratory experiments, a concentration of 0.1–5.0 μg of transforming DNA per ml of a competent culture (approximately 1 × 10^8 cells/ml) is typically used to produce maximal transformation frequencies for any given selectable marker.

VII. Artificial Competence and Transformation

Transformation of the bacterium *E. coli* and the yeast *Saccharomyces cerevisiae* exemplifies a second kind of transformation system in which "competence" for DNA binding and uptake is artificially induced by chemical or physical treatment of growing cells. The original method for introducing naked DNA into cells that are not naturally transformable is based on the observations of Mandel and Higa, who in 1970 demonstrated that incubation of a suspension of *E. coli* cells with bacteriophage λ DNA in an ice-cold solution of calcium chloride ($CaCl_2$), followed by a brief heat pulse at 42°C, resulted in the subsequent production of plaques. Bacteria treated accordingly yielded 10^5–10^6 transformed colonies per μg of supercoiled plasmid DNA.

Since then, many improvements on this basic method have been described, all directed toward optimizing the efficiency of transformation of bacterial strains by plasmids. The most notable include the construction of specific *E. coli* strains and treatment of cells with dimethyl sulfoxide, reducing agents, and high Mg^{2+} concentrations. As a consequence, transformation frequencies with standard *E. coli* strains have been increased some 100- to 1000-fold, a fact that has made the molecular cloning of entire eukaryotic genomes now practical. The logical application of this methodology to the study of DNA transfer in *E. coli* was later demonstrated by the genetic transformation of auxotrophic strains of *E. coli* with linear *E. coli* chromosomal DNA carrying prototrophic markers as well as by the transfection of a number of bacteriophages. The same basic techniques have proven successful for the transformation of *Staphylococcus aureus*, *Salmonella typhimurium*, and *Pseudomonas putida*.

Characterization of the transformation process in *E. coli* suggests that the preparation of cells for DNA uptake involves an alteration of the cell envelope to render the envelope accessible to interaction with transforming DNA. Although the molecular basis of competence in this bacterium is unknown, some general features have emerged that are reminiscent of the process as it occurs in naturally transformable bacteria. First, the process can be divided into two stages: the ability to establish competence and DNA uptake. Second, only a minor fraction (maximum of 10%) of cells in a population becomes "competent" and can be transformed under conditions of DNA excess. Third, surface–DNA interactions during binding and uptake do not appear to involve specific recognition sequences.

In *E. coli*, the outer membrane and inner membrane are fused to each other through holes in the peptidoglycan-containing cell wall called "zones of adhesion," through which macromolecules are believed to be transported. Considerable evidence suggests that transforming DNA enters the cell through these channels in the cell envelope. Since both DNA and the cell surface, especially the lipopolysaccharide component of the outer membrane, are essentially phosphate-rich polyanions, Hanahan has suggested that the combination of low temperatures, dimethyl sulfoxide, and cations acts to shield the charged phosphate groups on both the cell surface and the DNA, in effect creating favorable conditions for the association of these two structures.

Transformation in *S. cerevisiae* was first reported in 1960, but not until 1978 did Hinnen and associates succeed in reproducibly transforming yeast cells (spheroplasts) using an *E. coli* plasmid that contained genetically marked yeast DNA. This method gave transformation efficiencies of 3000–50,000 transformed cells per μg of plasmid DNA (2μ vectors), and frequencies of approximately 1 transformant per 10,000 viable cells. More recent transformation protocols use whole cells, rather than spheroplasts, made competent for DNA uptake by treatment with monovalent or divalent cations (such as lithium) in the presence of polyethylene glycol.

The general applicability of these and newer methods now has made possible the introduction of an extraordinary variety of circular and linear DNA molecules into many heretofore genetically intractable prokaryotic and eukaryotic systems. Indeed, a recent innovation, "electroporation," has become a valuable alternative technique for the transfer of nucleic acids into a variety of cell types, including mammalian cells, plant protoplasts, yeast, and *Dictyostelium*, that cannot be transformed or transfected by the standard chemical methods described previously. Electroporation, which uses an electrical field to reversibly permeabilize cells, was developed originally to introduce DNA into eukaryotic cells but has been used to "transform" *E. coli* and many other bacteria at efficiencies approaching 10^{10} transformants per μg DNA. Further, electroporation is increasingly being used to transfer other macromolecules, such as RNA and proteins, and even smaller molecules, such as nucleotides and fluorescent dyes, into cells.

VIII. Evolutionary Function of Natural Genetic Transformation

Natural transformation in bacteria is the result of a set of related biological processes that presumably has evolved in response to natural selection and confers an evolutionary advantage on cells. Several hypotheses have been advanced to explain the evolutionary function of transformation in bacteria. It has been suggested that competence evolved as a mechanism to acquire exogenous DNA as a source of nucleotides for DNA synthesis or to provide homologous single-strands as substrates for genetic variation (recombination) and DNA repair. A third possibility is that transformation evolved as the cellular counterpart of a transfer mechanism originally evolved and encoded by a "parasitic" element such as a plasmid or phage. (These hypotheses, which are not mutually exclusive, are explored more fully in the articles listed in the bibliography.) [*See* GENETIC TRANSFORMATION, EVOLUTION.]

Why study competence and transformation in microbial systems? The study of transformation has proven to be an extremely invaluable and powerful tool in the genetic analysis of many bacteria, and has been essential to the development and current use of microorganisms for recombinant DNA technology research in academic institutions and in the biotechnology industry. Further, an understanding of the mechanisms of genetic transformation is relevant to the larger issues of the evolutionary functions of competence itself, its role in the evolution of genetic exchange (recombination) and DNA repair systems in bacteria, and the evolution of prokaryotes in general.

Bibliography

Biswas, G. D., Thompson, S. A., and Sparling, P. F. (1989). *Clin. Microb. Rev.* **2,** S24–S28.

Bruschi, C. V., Comer, A. R., and Howe, G. A. (1987). *Yeast* **3,** 131–137.

Dubnau, D. (1991). *Microb. Rev.* **55,** 395–424.

Dubnau, D. (1991). *Mol. Microb.* **5,** 11–18.

Hanahan, D. (1987). *In "Escherichia coli* and *Salmonella typhimurium,* Cellular and Molecular Biology" (F. C. Neidhardt, ed.), pp. 1177–1183. American Society of Microbiology, Washington, D. C.

Kahn, M. E., and Smith, H. O. (1984). *Membrane Biol.* **81,** 89–103.

Shikegawa, K. and Dower, W. J. (1988). *BioTechniques* **6,** 742–751.

Smith, G. R. (1988). *Microb. Rev.* **52,** 1–28.

Stewart, G. J., and Carlson, C. A. (1986). *Ann. Rev. Microb.* **40,** 211–235.

Glycocalyx, Bacterial

J. William Costerton
University of Calgary

Hilary M. Lappin-Scott
University of Exeter

K. -J. Cheng
Agriculture Canada Research Station

Glossary

Biofilm bacteria Bacteria adherent to surfaces and growing in glycocalyx-enclosed populations

Genotypic Pertaining to the genome; genotypic changes are mutants in DNA structure

Glycocalyx Polysaccharide coat surrounding bacterial cells

Gram-negative Technically negative in the Gram strain; functionally having two cell membranes (inner and outer) and an extra measure of protection from antibacterial agents

Lipopolysaccharide Main component of the outer cell wall of gram-negative bacteria

Peptidoglycan Rigid component of bacterial cell wall

Phenotypic Pertaining to the manner in which the genome is expressed; phenotypic changes are changes based on gene expression not on genetic mutation

Planktonic Free-floating bacteria

GLYCOCALYX was originally defined as "those polysaccharide-containing structures, of bacterial origin, lying outside the integral elements of the outer membrane of gram-negative cells and the peptidoglycan of gram-positive cells." (Costerton *et al.*, 1981). In the interests of economy, the term *glycocalyx* was borrowed from botany and zoology, where it is used to describe very similar cell surface structures. The term glycocalyx can be further modified to describe these surface structures more accurately, and S-layers are clearly understood to be paracrystalline arrays of glycoprotein molecules at the cell surface, whereas capsules describe more amorphous fibrous exopolysaccharide layers of various thicknesses. The term *capsule* can be further modified to describe these structures more accurately:

1. *Rigid* A capsule sufficiently structurally coherent to exclude particles (e.g., India ink or nigrosin).
2. *Flexible* A capsule sufficiently deformable that it does not exclude particles.
3. *Integral* A capsule that is normally intimately associated with the cell surface.
4. *Peripheral* A capsule that may remain associated with the cell in some circumstances and may be shed into the menstruum (as an amorphous "slime") in others.

These descriptions are open-ended in that a species (e.g., *Pseudomonas aeruginosa*) can be said to produce a flexible peripheral capsule but another (e.g., *Streptococcus pneumoniae*) can be said to produce a rigid integral capsule.

All these glycocalyx structures, by definition, exist outside of the well-known cell wall structures of bacteria, which include the peripheral "O" antigen portion of the lipopolysaccharide (LPS) molecules of gram-negative bacteria and the outermost integral teichoic acid molecules of the gram-positive cell wall. Recently, very elegant freeze substitution work has shown that the distal portions of the LPS molecules of the gram-negative bacterial cell wall form a highly structured carpet-like "nap." Ion-related release data suggest that the oligosaccharide fibers that comprise the bacterial glycocalyx are probably anchored to this outermost cell wall component by cation-mediated ionic associations.

I. Discovery of the Glycocalyx

The physical nature of the bacterial glycocalyx delayed its discovery by microbiologists. This largely carbohydrate structure is highly hydrated (>99% water), and it was difficult to visualize by light microscopy, although some ephemeral sightings were recorded when antibodies were used to show it directly (the quellung reaction) and when inert particles (e.g., India ink) were used to show it by exclusion. The advent of electron microscopy produced only further confusion concerning the glycocalyx because the stains initially used to produce contrast in the specimen did not react strongly with carbohydrates. Furthermore, all early embodiments of this arcane discipline involved dehydration of the specimen, and the glycocalyx was only seen as the tangled wreckage of its real hydrated structure, even when we began to use a carbohydrate stain (ruthenium red) in specimen preparation. Electron micrographs sometimes showed wide spaces where we knew that cells were tightly apposed to a surface and sometimes showed wide clear zones surrounding cells within apparently continuous matrices, but it was not until the quellung reaction was extended to electron microscopy by Bayer and Thurow (1977) that we finally saw the glycocalyx directly and unequivocally. Specific antibodies react with the fibers of the glycocalyx, and their bivalent association stabilizes this structure and prevents its collapse during dehydration so that this extracellular structure is seen in something approaching its original dimensions (Fig. 1). This stabilization of the glycocalyx allowed us to examine the association of surface structures on a particular bacterial cell—the enterotoxigenic *Escherichia coli* (ETEC) strains that cause diarrhea in newborn calves. We used monoclonal antibodies to thicken the specific adhesion pili (K99 pili) of these adherent pathogens sufficiently so that they could be resolved in transmission electron microscopy (TEM) of sectioned material from an infected animal (Fig. 2), and we saw linear pili among

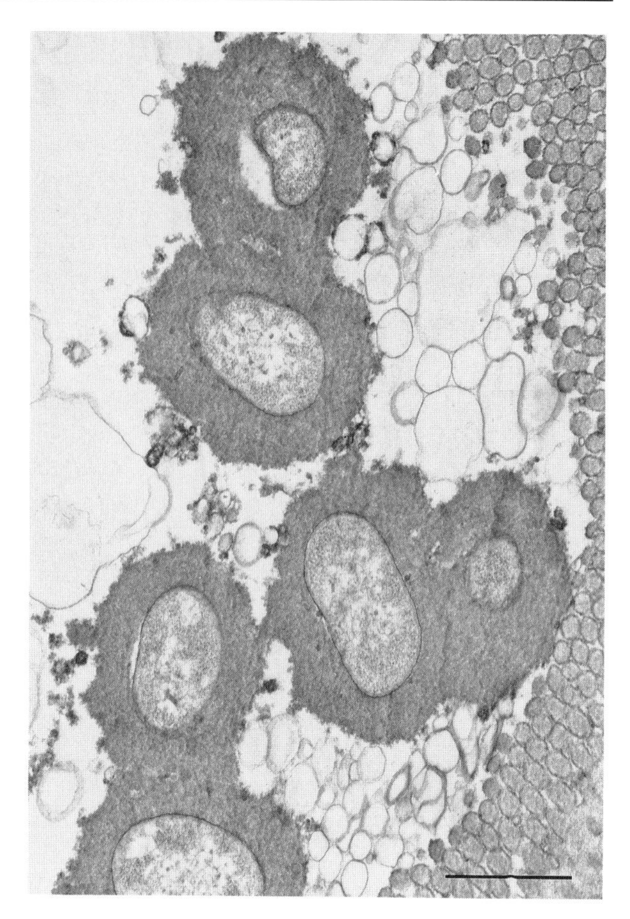

Figure 1 TEM of a preparation of calf intestine infected with enterotoxigenic *Escherichia coli* that had been stabilized with antiglycocalyx (anti-K30) antibody and stained with ruthenium red before fixation. Bivalent antibodies stabilize fibers of the exopolysaccharide glycocalyx and prevent its collapse during dehydration. These glycocalyces cover cells completely and comprise a volume greater than that of bacterial cells themselves. Bar = 1.0 μm in all figures.

the dehydrated and collapsed ruthenium red–stained wreckage of the glycocalyx. When the glycocalyx is stabilized, by the use of antibodies or lectins, the pili cannot be resolved (see Fig. 1). However, we can now understand that when the pili are not thickened and the glycocalyx is not stabilized but the glycocalyx residue are stained with ruthenium red, the exopolysaccharides collapse onto all available structures, including the rigid but invisible pili to produce an elaborate strutted condensed tangle (Fig. 3) during dehydration of the specimen. Stabilization with specific antibodies and lectins is a complex process that is very laborious and is only rarely feasible in natural mixed population ecosystems, but ruthe-

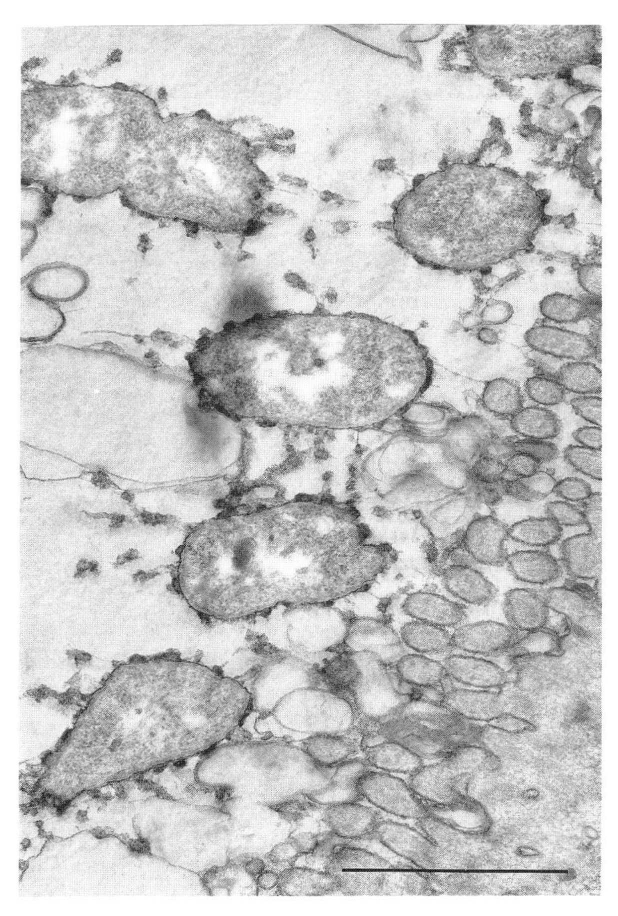

Figure 2 TEM of a preparation of calf intestine infected with ETEC that had been treated with monoclonal anti-K99 pilus antibody and stained with ruthenium red before fixation. Anti-K99 antibodies have thickened K99 pili so that they are visible in this sectioned material. The ruthenium red–stained glycocalyx has not been stabilized and has collapsed during dehydration to produce electron-dense accretions on the bacterial cell surface and on pili.

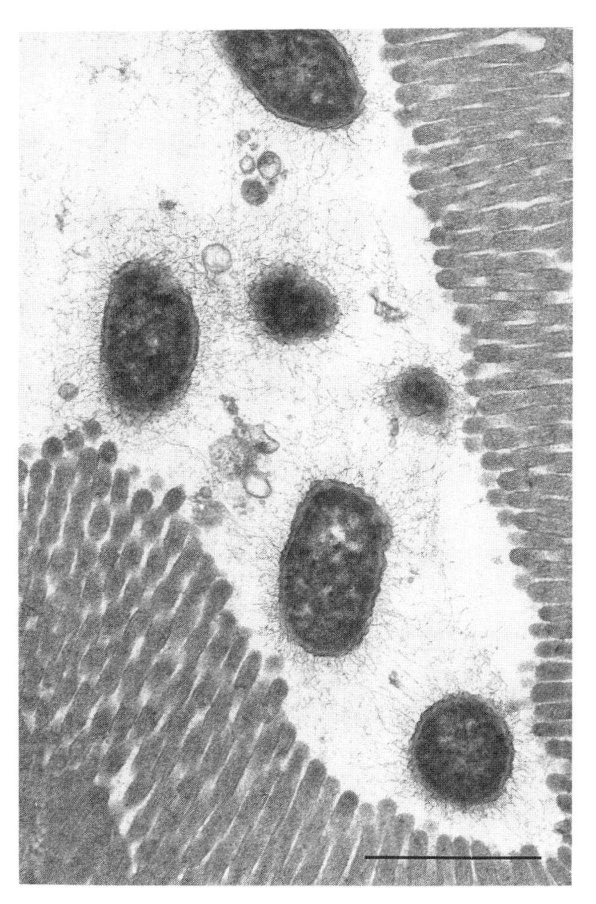

Figure 3 TEM preparation of intestine of an ETEC-infected calf that had not been treated with antibodies before ruthenium red staining and fixation. Glycocalyx material has condensed on bacterial cell surfaces and on pili in a fine electron-dense pattern of accretion.

nium red staining is widely applicable. However, the demonstration of the true extent of the glycocalyx in the ETEC and in several organisms examined by Bayer's group raised the awareness of morphologists concerning condensed fibrous residues in TEMs (Fig. 3) and similar condensed fibers (see Fig. 8) seen after dehydration for scanning electron microscopy (SEM). Soon, condensed glycocalyces were being reported in virtually all ecosystems. However, these demonstrations were not really convincing outside the morphological cabal because they involved tenuous extrapolations from structures badly distorted by dehydration.

The real structure of the bacterial glycocalyx was only recently defined by confocal scanning laser microscopy (CSLM). This technique uses a laser beam to image fully hydrated living bacterial populations, some of which are attached to surfaces, and the resultant signals can be analyzed to yield data on the relative densities of the glycocalyx matrix at various locations in a biofilm. This method has confirmed and reinforced our original impressions, based on light and electron microscopy, that almost all bacterial cells are surrounded by an extracellular layer of exopolysaccharide material that form a highly structured, highly hydrated matrix when the cells adhere to a surface and proceed to produce a biofilm (Fig. 4). [*See* Biofilms and Biofouling.]

The discovery of the ubiquity and the impressive dimensions of the bacterial glycocalyx had to await refinements of microscopy that allowed the resolution and stabilization of these very ephemeral hydrated exopolysaccharide structures. This impor-

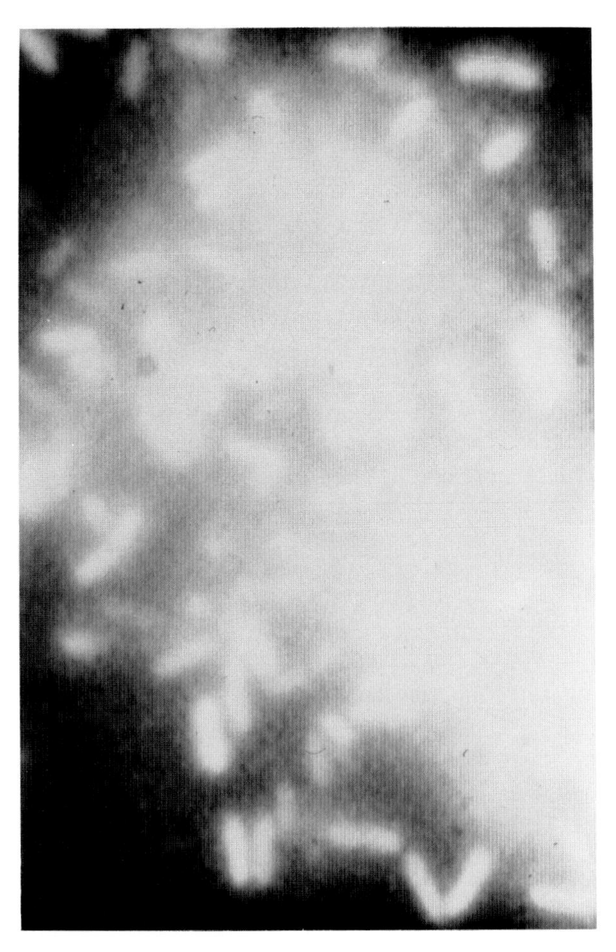

Figure 4 Computer-enhanced light micrograph of a developing biofilm of cells of *P. aeruginosa* showing the adherent bacteria and their enveloping glycocalyces. Photo credit: G. James and D. E. Caldwell.

tant discovery, which changes our concept of the effective outer surface of bacteria, was further delayed because of a common artifact of the cultivation of bacteria. Since the time of Pasteur, microbiologists have systematically removed bacteria from multispecies populations in complex, often nutrient-limited, natural environments and have grown these organisms as single species in nutrient excess throughout thousands of transfers and millions of generations. Recent comparisons of cells in laboratory cultures with cells of the same species growing in their natural environment have shown that the single species batch culture grown in nutrient excess is, in itself, an artifact. One of the first bacterial structures lost during repeated culture *in vitro* is the glycocalyx, whose protective functions are not necessary in single species culture. Our collective failure to appreciate the extent of this culture artifact further delayed the discovery of the ubiquity and importance of the glycocalyx in the reaction and adaptation of bacteria to their many environments.

However, despite the loss of the glycocalyx during laboratory cultivation of bacteria and despite the initial difficulties in the visualization of this structure in many types of microscopy, this surface structure is now established as being real, substantial in dimensions, and pivotal in function, based on a large, recently generated body of direct observations of the structure and function of bacteria.

II. Role of the Glycocalyx in Bacterial Cell Function

Because the glycocalyx comprises the outermost component of the bacterial cell, through which only pili and flagella are exposed, this surface layer mediates virtually all bacterial associations with surfaces and with other cells in any given environment. Glycocalyx components are involved in the nonspecific adhesion of bacteria to inert surfaces in aquatic ecosystems, and these anionic exopolysaccharide molecules (e.g., uronic acids) are typical of many fibrous molecules that initiate and cement cell–surface interactions in many areas of aquatic biology (e.g., diatom and mussel adhesion). In many instances of lectin-mediated specific adhesion of bacteria to plant and animal surfaces (e.g., the trifolin-mediated association of *Rhizobium* with the roots of clover plants), the microbial ligand in the adhesion complex is a component of the bacterial glycocalyx. Because bacteria can really only interact via their surface components, their glylcocalyces are pivotally important in both their associations with inert surfaces and with other bacterial cells. In the rumen, amylolytic bacteria use their glycocalyces to adhere to their specific substrate (starch) and cellulolytic bacteria use their glycocalyces (Fig. 5) to adhere to cellulose and to initiate the formation of metabolically cooperative microbial consortia with cells of other bacterial species (Fig. 5). Thus, in a real and important way, the physical location of a bacterial cell within a given ecosystem and the juxtaposition of that cell with cells of its own or of physiologically cooperative species both depend on the composition and chemical affinities of its glycocalyx. Microbial ecosystems are highly structured and complex and the chemical code that dictates location, juxtapo-

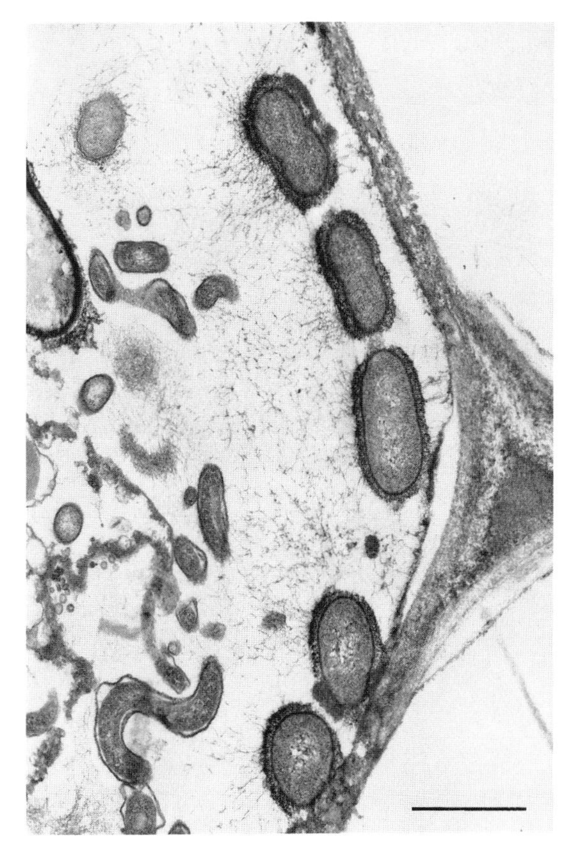

Figure 5 TEM of a ruthenium red–stained preparation of partly digested forage material from the bovine rumen. A row of cellulolytic bacteria have used their well-defined glycocalyces to adhere to their specific substrate (cellulose) and to mediate an association with cooperative spiral-shaped bacteria (*Tremponema* sp) with which they have a well-documented physiological cooperation within a microbial consortium.

sition, and, eventually, success is simply contained in the exposed chemical ligands of the bacterial glycocalyx. Bacterial glycocalyces show a wide variety of chemical structures, ranging from polypeptides to oligosaccharides, and a single species may produce as many as 100 chemically distinct capsule types. Generally, however, glycocalyx chemistry is fairly stable in most species, and most of these surface structures are composed of long polymers of repeating saccharide units—typically highly anionic uronic acids.

Because they lack the advantages of the homeostasis found in higher life forms, bacteria must enclose their fragile membrane-bound protoplasts within a whole series of structured layers that act to condition the microenvironment of the cell and cushion the effects of physical and chemical antibacterial shocks. In gram-negative bacteria, the peptidoglycan and the periplasmic space constitute the microenvironment within which the cytoplasmic membrane actually operates, and this zone is highly conditioned by the selective permeability of the outer membrane and the ion exchange behavior of the glycocalyx. We have actually determined, from enzyme denaturation data, that the effective pH in the periplasmic space of a gram-negative bacterial cell may be two full pH units higher (2 logs higher in proton concentration) than that of the bulk fluid in which the planktonic cell was floating. Cells in bacterial biofilms are inherently protected against the effects of a wide variety of antibacterial agents, but even planktonic cells with well-developed glycocalyces enjoy a large measure of this same protection.

III. A Scientist's Garden of Amazing Glycocalyces

When we use direct electron microscopy methods to examine the actual bacterial cells within a natural ecosystem like the bovine rumen (Figs. 6–8), we see an amazing variety of glycocalyx structures that can never be seen when we examine cells from large numbers of pure cultures derived from the ecosystem in question. We can only conclude that these elaborate external structures, which may be composed of complex radial and concentric elements, serve a purpose in the natural ecosystem (e.g., protection) that has no positive selection value in pure monospecies culture. These structures have a basic aesthetic fascination, but they also raise intriguing biochemical questions of cell organization because, outside of the active synthesis and the structural actin-myosin framework of the membrane-enclosed bacterial cell, elaborate scaffoldings of fibrous elements must be assembled into patterns of bewildering complexity using information that was imparted before they left the membrane itself. This, surely, is directed self-assembly on a scale that exceeds that of the virus particle or the bacterial flagellum and is all the more remarkable because its basic building blocks appear to be simple carbohydrates instead of information-rich protein molecules. It is clear that, decades from now, morphologists and biochemists will still be unraveling the details of synthesis and assembly of the hundreds of different bacterial glycocalyces that can be seen in a single natural ecosystem (e.g., bovine rumen).

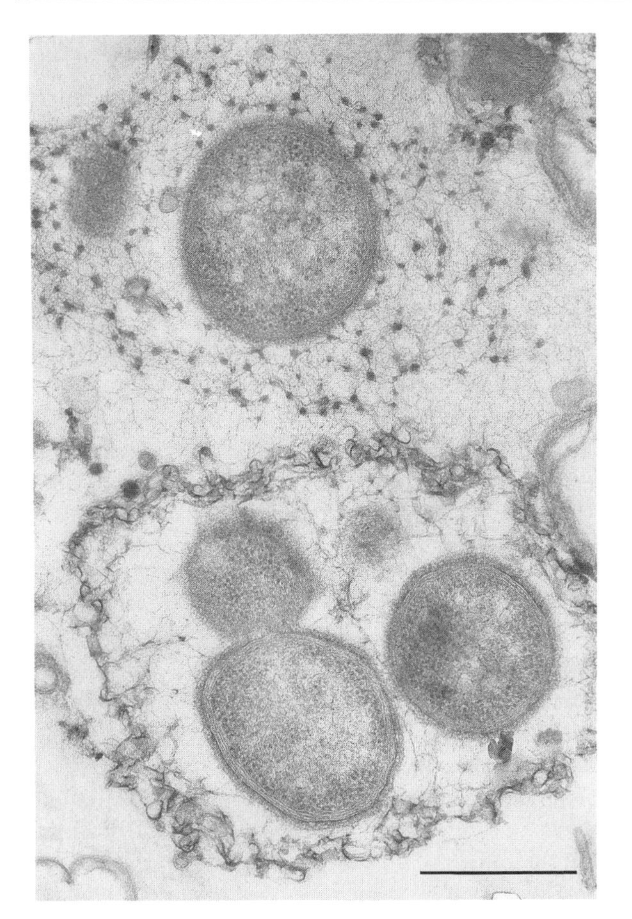

Figure 6 TEM of a ruthenium red–stained preparation of material from the bovine rumen showing the remarkable radial and concentric structures of glycocalyces that surround most of the bacteria in this competitive ecosystem. These elaborate glycocalyces have not been protected from dehydration during preparation for electron microscopy, and we can only guess at their complexity in their native, fully hydrated state.

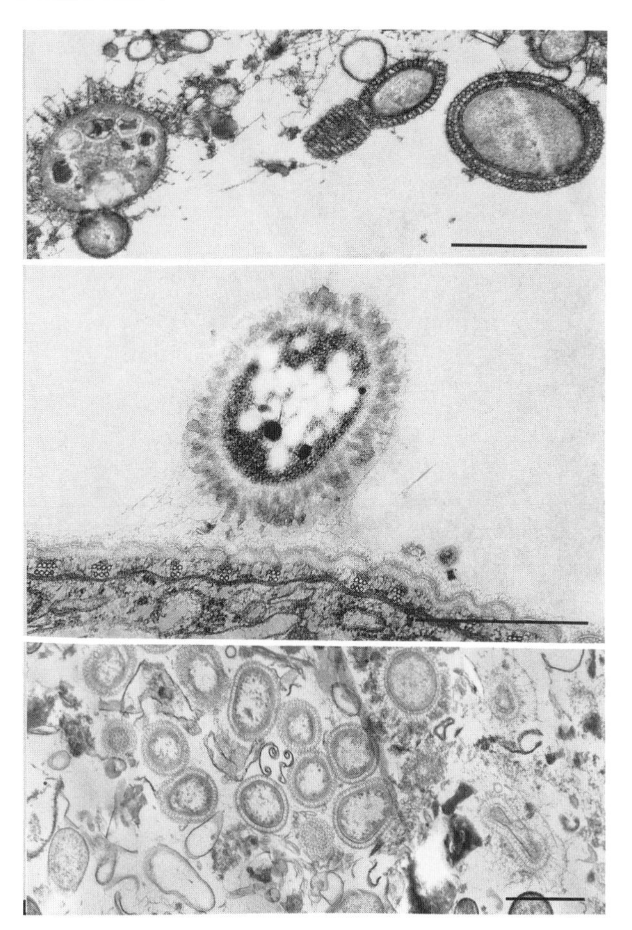

Figure 7 Further examples of the remarkable glycocalyces shown to surround bacterial cells in the bovine rumen. Isolates derived from these rumen bacteria have never been seen to retain these elaborate glycocalyces in pure cultures in laboratory media.

IV. Concept of Plasticity and the Bacterial Glycocalyx

In a landmark publication, Mike Brown's group (Aston University, United Kingdom) promulgated the theory of phenotypic plasticity. This theory states that, without genomic mutations, bacteria can respond to alterations in their microenvironment by profound changes in their structure and their metabolic processes. The production of elaborate glycocalyces in a competitive multispecies environment, and their subsequent loss during culture in noncompetitive single species culture, is a case-in-point. The glycocalyx represents a significant expense of metabolic energy, and if its function is partly nutrient acquisition and partly protection, it is not really required during growth in nutrient-rich single species cultures. This extreme morphological plasticity may be only the most visible of hundreds of phenotypic changes undergone by bacteria removed from their natural environments and cultured in the laboratory, and the extent to which this pure culture artifact has been incorporated into dogma in various fields of microbiology urgently needs examination.

Phenotypic plasticity also operates in the opposite direction, and cells from *in vitro* cultures (Fig. 9A) that find themselves in a hostile tissue environment as a result of an animal experiment are seen to alter their cell wall structure and to synthesize fibrous exopolysaccharides so that they produce a function-

Figure 8 SEM of bovine rumen contents showing condensed fibrils of dehydration-collapsed bacterial glycocalyces and illustrating that many of these organisms are enclosed by elaborate surface structures.

ally protective "glycocalyx" (Fig. 9B) in response to their altered microenvironment. The simplest of these ersatz "glycocalyces" appears to be synthesized by a simple overproduction of cell wall materials (Fig. 9B) and is not a true glycocalyx, but the cells often adapt further and combine cell wall overproduction with exopolysacchride synthesis. Because bacteria live partly as free-living single cells that are likely to be swept abruptly from one environment to another, they have succeeded by using their phenomenal phenotypic plasticity to adapt to these changes much more rapidly than higher organisms can adapt by the selection of mutants. Bacteria can also use selection of mutants to colonize new ecosystems, and this double capabilty goes a long way toward explaining the phenomenal success of these fragile membrane-enclosed, single-celled creatures in even the most hostile of ecosystems.

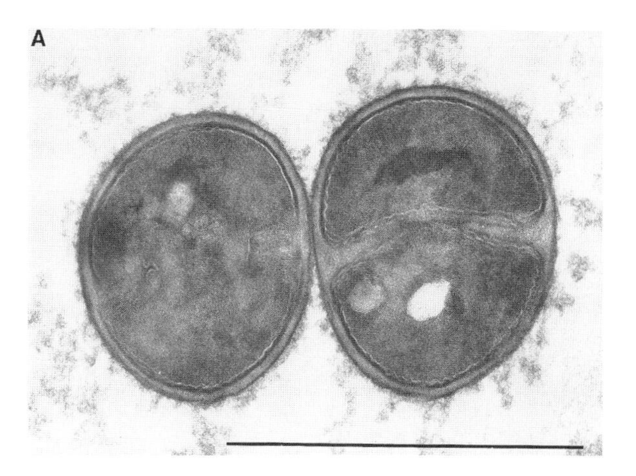

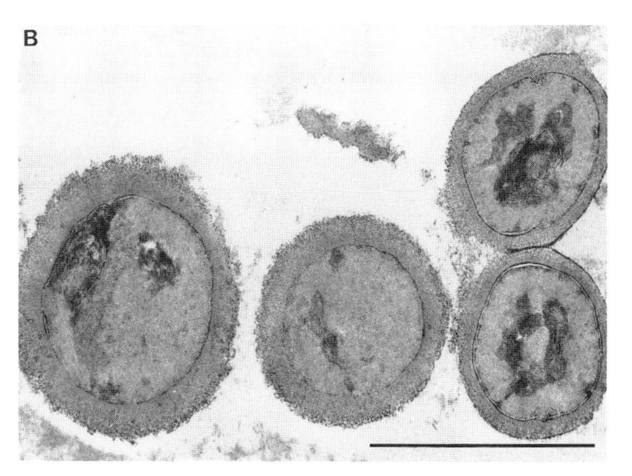

Figure 9 TEM of ruthenium red–stained preparations of cells of *Staphylococcus aureus* from a fast-growing laboratory culture (A) and from infected tissue in experimental osteomyelitis (B). In the hostile environment of infected tissue, cell walls of these gram-positive organisms are often seen to thicken and to produce a fibrous external layer that somewhat resembles a glycocalyx and may serve the same function.

Bibliography

Bayer, M. E., and Thurow, H. (1977). *J. Bacteriol.* **130,** 911–936.

Brown, M. R. W., and Williams, P. (1985). *Annu. Rev. Microbiol.* **39,** 527–556.

Costerton, J. W., Geesey, G. G., and Cheng, K.-J. (1978). *Sci. Am.* **238,** 86–95.

Costerton, J. W., Irvin, R. T., and Cheng, K.-J. (1981). *CRC Crit. Rev. Microbiol.* **8,** 303–338.

Costerton, J. W., Cheng, K.-J., Geesey, G. G., Ladd, T. I., Nickel, J. C., Dasgupta, M., and Marrie, T. J. (1987). *Annu. Rev. Microbiol.* **41,** 435–464.

Shotton, D. M. (1989). *J. Cell Sci.* **94,** 175–206.

Gram-Positive Cocci

Sybil Wellstood
Food and Drug Administration

Glossary

Catalase Enzyme that decomposes hydrogen peroxide into oxygen and water

Coagulase Enzyme that causes clotting of plasma; may be "free" (extracellular) or bound (component of cell wall)

C-substance Serologically distinct, group-specific cell wall antigen of streptococci; may be carbohydrate or teichoic acid

G+C value Relative amount of guanine and cytosine in the chromosomal DNA of an organism; expressed in molar percent (mol%)

Teichoic acid Polymer of ribitol or glycerol phosphates covalently linked to the peptidoglycan; regulates cell functions and serves as phage receptor sites

GRAM-POSITIVE COCCI (GPC) are spherical bacteria ranging in diameter from 0.5 to 3.5 μm. Organisms occur singly or in pairs, chains, clusters, tetrads, or packets, depending on the plane of cellular division. Cells dividing in one plane and remaining attached form pairs or chains. Chains vary in length from 2 to 4 cells or more than 20 cells. Organisms dividing in two planes form tetrads. Others, dividing in three planes, form irregular bunches or sheets of cells or produce cuboidal packets of four to eight cells.

Most GPC have typical, thick, gram-positive cell walls containing large amounts of peptidoglycan. When gram-stained, the walls retain the crystal-violet complex and cells appear purple. Teichoic acids are also present in the cell walls of some genera. Polysaccharide capsules or slime layers surround the cell walls of some species and contribute to the virulence of these organisms.

As chemoorganotrophs, the GPC use carbohydrates and/or amino acids as carbon and energy sources. All grow on laboratory media. The GPC are mesophiles and grow best at temperatures between 30° and 44°C. Oxygen requirements for growth vary from strictly aerobic genera that require oxygen for growth to obligate anaerobes that grow only in the absence of oxygen. Others are facultative anaerobes that grow with or without oxygen or microaerophiles that grow in reduced oxygen atmospheres.

Although the GPC are non-spore-formers, many are resistant to drying and survive in air, soil, and other adverse environments. Survival is also enhanced in some species by carotenoid pigments. Carotenoids are photoprotective agents and absorb harmful light in the atmosphere. By quenching toxic forms of oxygen produced by phagocytes, carotenoids also enable pathogenic species such as *Staphylococcus aureus* to resist phagocytosis. Organisms containing carotenoids form colonies ranging in color from yellow, orange, and pink to bright red.

Current molecular approaches to taxonomy, including determination of the G+C content of the DNA, indicate little relationship and great diversity among the 17 genera of GPC. Many taxonomic revisions have been made during the past 10 years and work is ongoing to clarify the taxonomy of the GPC. Presently, the 17 genera are classified in two families—the Micrococcaceae and Deinococcaceae—and two groups—the facultative anaerobic and microaerophilic catalase-negative cocci and the strict anaerobes.

I. Micrococcaceae

Usually catalase-positive, the Micrococcaceae are a family of GPC arranged in clusters, tetrads, and cuboidal packets. Phylogenetically, the four genera currently included in the family appear unrelated.

The mol% G+C content of their DNA ranges from a low of 30–40 for *Staphylococcus* to a high of 64–75 for *Micrococcus*. *Planococcus* and *Stomatococcus* fall in-between with 39–52 and 56–60.4, respectively. Table I lists some of the differential features of the four genera.

A. *Micrococcus*

Micrococcus sp. are aerobes that oxidatively produce acid from glucose. Because their cell walls lack glycine, the micrococci are resistant to lysostaphin, an enzyme that attacks glycyl–glycine linkages. Lysostaphin resistance is an important diagnostic feature of the micrococci.

Most species produce yellow, orange, pink, or red colonies. Except for *Micrococcus agilis*, organisms are nonmotile. In rare cases, strains may be opportunistic pathogens. Otherwise, the micrococci are nonpathogenic and live primarily on the surfaces of mammalian skin, on meats or dairy products, in soil, and in water.

B. *Stomatococcus*

Stomatococcus is the newest member of the Micrococcaceae. Described in 1982, the genus contains a single species, *Stomatococcus mucilaginosus*. Cells appear mainly in clusters and occasionally in tetrads. Although sometimes mistaken for staphylococci, stomatococci differ in catalase reactivity. Most are weakly catalase-positive, and almost half lack catalase activity. Colonies are mucoid, adhere to the agar, and lack pigments. A large polysaccharide capsule surrounds a cell wall lacking teichoic acids.

Stomatococci usually reside as commensals in the human mouth and upper respiratory tract. Recently, however, strains have been implicated in septicemias and endocarditis, particularly in drug abusers and compromised hosts. The treatment of choice is penicillin.

C. *Planococcus*

Members of the genus *Planococcus* are free-living saprophytes found in marine environments such as seawater, clams, shrimp, and prawns. Based on their G+C base composition, isolates are divided into two groups—those containing 39–42 mol% and the remainder with 48–52 mol%. Both species, *Planococcus citreus* and *Planococcus halophilus*, are motile (usually by one or two flagella) and form yellow or orange colonies. They oxidatively produce acid from glucose, hydrolyze gelatin, and grow well on nutrient agar containing 12% NaCl.

D. *Staphylococcus*

Staphylococci are ubiquitous organisms living on the skin or mucous membranes of humans, other mammals, and birds. Dividing in more than one plane, cells form irregular, "grapelike" clusters. Species are nonmotile and produce golden-yellow or white colonies. Among the Micrococcaceae, only staphylococcal cell walls contain teichoic acids. Organisms are hardy and survive exposure to adverse conditions such as drying, high salt and sucrose concentrations, some disinfectants, and numerous antimicrobial agents. Of the 27 species of staphylococci, 6 are medically significant.

1. *Staphylococcus aureus*

Staphylococcus aureus, the most important and virulent species of staphylococci, is part of the indigenous human flora. Infections result when the protective barriers of the skin or mucous membranes are breached. Burns, traumatic wounds, surgical procedures, and intravenous drug abuse are some predisposing factors. Skin infections are not only the most common staphylococcal infection; they are also the most common human bacterial infection. Infections range from localized boils, abscesses, and lesions to deep wounds. Serious or life-threatening disseminated diseases including pneumonia (up to 50% mortality rate), endocarditis, bacteremia, meningitis, osteomyelitis, food poisoning, and toxic shock syndrome (TSS) also occur.

Many people are asymptomatic carriers of *S. aureus*, often serving as resevoirs and sources of infection. The most frequent carrier sites are the anterior nares and perineum. Organisms are transmitted directly through contact with carriers and those shedding the organisms or, indirectly, through contact with contaminated objects.

Several factors contribute to the pathogenicity of *S. aureus*. Structurally, the cell wall contains a tightly cross-linked peptidoglycan layer that protects organisms in host tissues. A unique protein, protein A, also covalently linked to the cell wall, inhibits phagocytosis. A polysaccharide capsule produced by some strains affords additional protection from phagocytosis.

Staphylococcus aureus elaborates a variety of extracellular enzymes and toxins that contribute to

Table I Differentiating Characteristics of the Micrococcaceae

Characteristic	Genus			
	Micrococcus	*Staphylococcus*	*Stomatococcus*	*Planococcus*
Typical cellular morphology	Irregular clusters, tetrads	Irregular clusters, tetrads	Irregular clusters, tetrads	Pairs, tetrads
Oxygen tolerance	Strict aerobes	Facultative anaerobes	Facultative anaerobes	Strict aerobes
Motile species	–[a]	–	–	+
Capsule production	–	+	+	–
Pathogenic species	–	+	+	–
Tolerance to 15% NaCl	–	+	–	–
Pigment production	+	+	–	+
Resistance to lysostaphin (200 μg/ml)	+	–	+	+
Resistance to bacitracin (0.04 U)	–	+	–	NT
Cell walls contain teichoic acids	–	+	–	–
G+C content of DNA (mol%)	64–75	30–40	56–60.4	39–52
No. species	9	27	1	2

[a] Strains of *Micrococcus agilis* are motile;
NT, not tested.

their virulence. Enzymes allow organisms to establish themselves on skin and mucous membranes and counteract host defenses. Four types of hemolytic toxins—alpha, beta, gamma, and delta—are produced. All damage erythrocytes and produce local or systemic infections. Leucocidin, a nonhemolytic toxin, attacks white blood cells. [*See* ENZYMES, EXTRACELLULAR.]

Heat-stable enterotoxins, Types A–F, are responsible for acute vomiting and diarrhea associated with food-poisoning, with Type A being the most frequently implicated in foodborne outbreaks. About 50% of *S. aureus* strains produce enterotoxins, with some strains producing more than one type.

The exotoxin, TSST-1, is the major toxin involved in TSS. Others, unknown at this time, may also play a role. TSST-1 appears similar to enterotoxin F. The toxin causes fever, shock, and skin rash. Frequently, but not exclusively, the syndrome is associated with use of tampons. Males and menopausal women also have been victims of TSS with toxin-producing strains isolated from nongenital sources.

Staphylococcus aureus infections are treated with semisynthetic penicillins (cloxacillin/dicloxacillin), erythromycin, or methicillin. Most isolates (85–90%) have developed resistance to penicillin and ampicillin. Methicillin resistance is also emerging as a serious problem in hospitals and nursing homes. Methicillin-resistant *S. aureus* are resistant to other commonly used antimicrobials. Serious infections require therapy with vancomycin. Local infections are treated by incision and drainage.

Laboratory identification of *S. aureus* is accomplished by performing a coagulase test. Using rabbit plasma, two simple tests detect "free" or bound coagulase. Free coagulase production is determined by the tube method using a small amount of plasma and a loopful of growth from a suspected colony. Coagulase-positive organisms clot the plasma in 1–4 hr. Although less accurate than the tube method, a screening test for bound coagulase is performed by emulsifying growth in a drop of water on a slide, adding a drop of plasma, mixing, and looking for the formation of clumps in 5 sec. In addition to *S. aureus, Staphylococcus lugdunensis* and *Staphylococcus schleiferi* are positive by the slide method, while only *S. aureus* is positive for free coagulase production.

During outbreaks, identification of *S. aureus* strains may be important to trace the source of the offending organism. Phage typing has been a useful epidemiologic tool for this purpose. Each strain of *S. aureus* is susceptible to lysis by one or more specific phages. The particular phage type is a stable, genetically determined characteristic. Not all strains are typeable by the current scheme.

2. Coagulase-Negative Staphylococci

Formerly considered commensals and normal skin flora, some species of coagulase-negative staphylococci (CNS) are now recognized as opportunistic pathogens. Infections are usually related to the use of invasive medical procedures or indwelling devices. In the past, inadequate methods for identification and lack of appreciation for the disease potential of the CNS precluded species identification. All were called CNS, *Staphylococcus epidermidis,* or micrococci.

a. *Staphylococcus epidermidis*

The most important species of CNS, *S. epidermidis,* causes a variety of hosptial-acquired infections. Although organisms usually have low virulence, serious, life-threatening infections may occur when host defenses are compromised. In particular, *S. epidermidis* is the most common isolate from infected cardiac valves, total hip replacements, and central nervous system shunts and accounts for 74–94% of all CNS bacteremias. Pacemaker, vascular graft, prosthetic joint, and intravenous catheter infections also occur.

The production of a viscous, extracellular substance enables *S. epidermidis* to adhere to smooth surfaces such as artificial heart values, indwelling devices, and hip prostheses. Organisms grow in the extracellular slime on the surface of the foreign body and produce a biofilm. Devices may be contaminated during implantation or later by bacteremias. The slime layer also protects organisms from antibiotics and the hosts' cell defenses.

Many strains of *S. epidermidis* are multiply resistant to antibiotics including methicillin. Vancomycin plus rifampin or an aminoglycoside is the treatment of choice for serious infections.

b. Other Pathogenic Species

Staphylococcus haemolyticus, S. lugdunensis, S. schleiferi, and *S. saprophyticus* are other CNS implicated in human infections similar to those caused by *S. epidermidis.* Species identification is best achieved by using commercially available miniaturized biochemical test kits. Most of the numerous parameters required for speciation are available in these systems. Results are often available more rapidly than conventional test results.

II. Deinococcaceae

The family Deinococcaceae contains one genus, *Deinococcus,* and four species. They form pairs or tetrads of cells that often appear larger than other GPC. The mol% G+C content of their DNA ranges from 62 to 70. All species contain a variety of carotenoid pigments and produce pink, orange-red, or bright red colonies. Strains grow on tryptone–yeast-extract media at 30°C and require 2–7 days for complete growth.

Deinococci are unique among the GPC in their ability to resist gamma and ultraviolet irradiation. Some strains are more resistant than bacterial endospores. Radiation resistance is used as a selective procedure for isolating organisms from meat, fish, or sawdust. Samples are exposed to 1–2 Mrad of gamma radiation. Resistance is mutable, however, and susceptible strains have been isolated. Although organisms are highly resistant to dessication and mutagenic chemicals, they are not heat-resistant. Growth does not occur >44°C.

The structural complexity and chemical composition of deinococcal cell walls is also distinctive among the GPC. Their cell walls consist of three or more layers including a unique outer membrane layer. Outer membranes are usually found only in gram-negative bacterial cell walls. Deinococcal membranes contain proteins and lipids. The fatty acids contain large amounts of even-numbered, straight chain, saturated and unsaturated acids resembling those found in gram-negative bacteria. Palmitoliate, representing 25% of the fatty acid content, predominates.

The peptidoglycan layer of the wall is also thick and contains ornithine instead of lysine as the diamino acid. The interpeptide linkage contains glycine. No teichoic acids have been detected.

The natural habitat of Deinococcus is uncertain. However, strains have been isolated near atomic reactors and other potentially lethal radiation sources and from ground meat, soil, dust, and filtered air.

III. Facultative Anaerobic and Microaerophilic Catalase-Negative Gram-Positive Cocci

The streptococci and related genera of catalase-negative GPC ferment glucose producing either lactic acid (homofermenters) or lactic acid, ethanol, and CO_2 (heterofermenters). The seven genera in the group and their distinguishing features are presented in Table II.

A. *Streptococcus*

Streptococcus is a large, diverse genus containing many medically and commercially important species. Organisms divide in one plane resulting in pairs or chains of cells. All are homofermentative.

Several classification schemes have evolved to group the streptococci into major categories. Hemolytic activity on blood agar provides a preliminary identification. Some organisms produce hemolysins that completely disrupted red blood cells resulting in a clear zone of hemolysis around colonies called β-hemolysis. Others produce hemolysins that only partially hemolyze red blood cells. These α-hemolytic colonies appear green. Nonhemolytic species producing no hemolysins (gamma) also occur.

Rebecca Lancefield developed a classification scheme to differentiate streptococci serologically based on cell wall antigens called the C-substance. The C-substances can be extracted from cells and tested against specific antisera for groups A–U. Group specificity is conferred by the amino sugar present. In group A, for example, the C-substance contains rhamnose-*N*-acetylglucosamine. The Lancefield grouping system is mainly useful for classifying the β-hemolytic streptococci.

Classification systems based on biochemical and physiological parameters are also used for classification, particularly when species cannot be serologically grouped. The α- and nonhemolytic streptococci are classified in this manner.

1. β-Hemolytic Streptococci

a. Group A

Among the β-hemolytic streptococci, several groups are pathogenic for humans and animals with group A, *Streptococcus pyogenes,* implicated most frequently in human infections. Pharnygitis is the

Table II Typical Characteristics of the Catalase-Negative Cocci

Characteristic	Genus						
	Streptococcus	*Enterococcus*	*Lactococcus*	*Leuconostoc*	*Pediococcus*	*Aerococcus*	*Gemella*
Cellular morphology	Pairs, chains	Pairs, chains	Pairs, chains	Pairs, chains	Pairs, tetrads	Tetrads	Single, pairs
Oxygen tolerance	Facultative, strict anaerobes	Facultative anaerobes	Facultative anaerobes	Facultative anaerobes	Microaerophiles	Microaerophiles	Facultative anaerobes
Habitat	Mouth, respiratory tract	Gastrointestines	Dairy products	Fruits, vegetables, milk	Spoiled beer	Air, vegetables	Bronchial secretions
Growth at 45°C	V	+	—	—	+	—	—
Growth in 6.5% NaCl	—	+	V	V	V	+	—
Motility	V	V	—	—	—	—	—
Hemolysis[a]	α, β, −	α, β, −	—	—	—	—	β, −
Capsule/slime layer	+	—	—	—	+	—	+
Teichoic acids in cell wall	+	+	+	—	—	—	—
Pathogenic species	+	+	—	—	—	—	—
L-Pyrrolidonyl B-Napthlyamide hydrolysis	V	+	V	—	—	+	+
G+C content (mol%)	34–46	38–45	34–43	38–44	32–42	35–40	33.5 ± 1.6

[a] −, Nonhemolytic (gamma).
V, variable reaction.

most common infection. Untreated, sore throats, particularly in children, may progress to rheumatic fever and acute glomerulonephritis. Impetigo, cellulitis, otitis media, tonsillitis, pneumonia, endocarditis, and septicemia are other infections caused by group A streptococci. Penicillin is the treatment of choice, with erthromycin and clindamycin as alternatives.

Streptococcus pyogenes elaborates numerous extracellular substances and toxins that play a significant role in their virulence. Most strains produce two types of β-hemolysins. Streptolysin O, inactivated by oxygen, and streptolysin S, oxygen-stable, lyse red and white blood cells.

The most important virulence factor is probably the M protein. Originating on the cell membrane and protruding through the cell wall as hairlike projections called fimbriae, the M protein promotes adherence of streptococcal cells to the hosts' epithelial cells. The M protein also prevents phagocytosis. Strains lacking M proteins are avirulent. There are more than 60 types of group A streptococci based on the type-specific M protein.

Streptococcus pyogenes is usually isolated in the laboratory on blood-supplemented media. Colonies are translucent and surrounded by a large zone of hemolysis. Organisms are presumptively identified by susceptibility to bacitracin (0.04 U). Biochemically, *S. pyogenes* hydrolyzes L-pyrrolidonyl B-napthlylamide (PYR). The PYR test is more sensitive and rapid than bacitracin susceptibility testing. Several commercial methods are available to perform PYR tests. Serological typing provides definitive identifications. New methods permit rapid extraction of the group-specific cell wall substance followed by typing using a slide agglutination method.

b. Group B

Group B streptococci (GBS; *Streptococcus agalactiae*) have also emerged as significant human and animal pathogens. They are the major cause of neonatal meningitis and sepsis. As part of the normal human genitourinary tract flora, GBS are probably transmitted to neonates during delivery. The mortality rate is frequently high. Other human infections include bacteremia, endocarditis, pneumonia, osteomyelitis, arthritis, and bactiuria. In animals, the GBS cause bovine mastidis. The treatment of choice is penicillin G.

Colonies of GBS on blood agar are usually large and mucoid surrounded by a narrow zone of β-hemolysis. Some strains (5–15%) are nonhemolytic. On appropriate media, most strains produce carotenoid pigments. Identification methods include hippurate hydrolysis and a positive CAMP test reaction. The CAMP reaction is demonstrated by streaking growth from a colony of the test strain perpendicular to a β-lysin producing strain of *S. aureus*. The interaction between the CAMP factor produced by GBS and the β-lysin produces an enhanced zone of hemolysis in the shape of an arrow.

c. Group C

Among the group C streptococci, *Streptococcus equisimilis* has been implicated in pharnygitis, endocarditis, bacteremia, osteomyelitis, pneumonia, and wound infections. *Streptococcus equi* causes diseases in horses. Treatment is similar to that for group A streptococcal infections.

Morphologically, group C colonies resemble those of a group A streptococci. Identification is usually performed serologically.

d. Group G

The group G streptococci cause cellulitis, bone or joint infections, and septicemias. Infections frequently require treatment with penicillin plus an aminoglycoside.

Unlike other streptococci, the group G strains do not have species designations. Two types of colonies are formed: the large type with a wide zone of hemolysis and a minute type. Serological methods identify the group G antigen.

B. α-Hemolytic Streptococci

1. Group D

Until recently, the group D streptococci included the enterococci and nonenterococci. Differentiation was based on the former's ability to grow in the presence of 6.5% NaCl. The enterococci, currently classified in a separate genus, *Enterococcus*, will be discussed later.

The group D antigen, unlike other group-specific antigens, is a glycerol teichoic acid containing glucose and alanine rather than a carbohydrate. The antigen is also associated with the membrane and not the cell wall.

The group D species, *S. equinus* and *S. bovis*, are α- or γ-hemolytic on blood agar and are distinguished from other streptococci by their ability to hydrolyze esculin, grow in the presence of 40% bile, and grow at 45°C.

They are associated with bacteremias and endocarditis. There is also a relationship between gastrointestinal tumors and *S. bovis* bacteremia. Isolates are often resistant to penicillin, and a combination of penicillin plus an aminoglycoside is required.

2. Oral Group

Several species of α-hemolytic streptococci have been grouped together and called the oral streptococci. Formerly, these organisms were called the "viridans" group. Many are part of the normal oral and respiratory flora of humans and animals. However, the oral streptococci are opportunists and the major cause of bacterial endocarditis. Predisposing factors include dental treatments, previously damaged heart valves, and neutropenia. *Streptococcus mutans* is associated with dental caries.

Virulence is related to the organisms' ability to adhere to epithelial and endothelial cells. Some species (*S. mutans*) form dextrans (extracellular polysaccharides) that help them attach to heart valves or tooth surfaces. Others produce enzymes and toxins.

These organisms do not hydrolyze esculin or grow at 10° or 45°C. All are susceptible to vancomycin. Miniaturized biochemical systems similar to those used for speciating staphylococci are available and may be used to speciate the members of the oral streptococci.

3. Nutritionally Variant Streptococci

Sometimes known as pyridoxal-dependent, satelliting, thiol-requiring, or nutritionally deficient, the nutritionally variant streptotocci are significant etiologic agents of endocarditis. Organisms require vitamin B_6 (pyridoxal) for growth and do not grow on standard blood agar used in laboratories. Satellite colonies, however, will develop on blood agar cross-streaked with *S. aureus. Staphylococcus aureus* provides the necessary growth factor.

4. *Streptococcus pneumoniae*

Streptococcus pneumoniae is the major cause of community-acquired pneumonia. Infections are related to predisposing factors (alcoholism, malnutrition, viral or other respiratory infections, circulatory problems) that interfere with natural defense mechanisms. Bacteremia, meningitis, sinusitis, and ear infections may follow respiratory involvement.

A type-specific polysaccharide capsule confers virulence to organisms by protecting them from phagocytosis. There are currently 83 capsular types. The particular type can be determined by the Quellung reaction performed by mixing growth from colonies or fresh sputum with type-specific antiserum on a slide. The swollen capsule can be seen using a microscope.

Pneumococci often appear as lancet-shaped diplococci on gram-stained smears of sputum or other specimens. On blood agar, they form α-hemolytic colonies that are, at first, dome-shaped. Later, the center sinks and colonies appear "checker"-shaped. With age, autolytic enzymes dissolve colonies. Resistance to optochin, a quinine derivative, is a common test to differentiate pneumococci from other α-hemolytic streptococci. Disks impregnated with 5 μg optochin are placed on blood agar plates streaked with suspect colonies. A zone of inhibited growth around disks indicates resistance to optochin.

C. *Enterococcus*

Established as a separate genus in 1984, *Enterococcus* includes 12 species formerly referred to as enterococci. Their natural habitat is the human gastrointestinal tract. They commonly cause urinary tract and wound infections, intra-abdominal abscesses, endocarditis, and bacteremia. Serious infections are treated with a penicillin plus an aminoglycoside.

Previously, enterococci were identified by their ability to hydrolyze bile-esculin and grow in 6.5% NaCl. However, the PYR test, described for group A streptococci, is more definitive and distinguishes *Entercoccus* sp. from look-alike species of *Leuconostoc* and *Pediococcus*.

D. *Lactococcus*

In 1985, former species of lactic streptococci belonging to Lancefield group N were transferred to a new genus, *Lactococcus. Lactococcus lactis* and *lactococcus cremoris* are important in the dairy industry to produce flavoring agents. *Lactococcus lactis* also produces an antibiotic, nisin, valuable as a natural inhibitor of pathogenic gram-positive organisms such as *Listeria*.

Although several strains of lactococci have been isolated from human sources, the organisms are regarded as nonpathogens at this time. Often confused with enterococci, the lactococci differ by their inability to grow at 45°C.

E. *Leuconostoc*

Leuconostoc may appear as elongated or lenticular chains of cells and are mistaken for gram-positive rods. Their optimal growth temperature is only 20–30°C. The four species are nonpathogenic for humans, animals, or plants. Some strains produce large amounts of dextrans used as plasma extenders and others are used as cheese and butter starters. Because they tolerate high sugar concentrations, *Leuconostoc* grows well in food products and causes spoilage.

F. *Pediococcus*

Pediococcus, a tetrad forming GPC, is microaerophilic, and several species tolerate low pH environments. Habitats include fermented products (sauerkraut, olives, wines, silage) and fresh or cured meat, raw sausage, poultry, and cheese. None of the eight species are pathogenic for humans or plants. However, some secrete large volumes of a slimy material responsible for spoiling beer. Pediococci are also used as starter cultures for sausage and bacon production.

G. *Aerococcus*

Aerococcus viridans, the only species in the genus *Aerococcus*, produces small colonies on blood agar surrounded by a zone of greening. Microscopically, organisms appear in tetrads. As a free-living saprophyte, *Aerococcus* is isolated from the air and marine environments. Growth occurs in the presence of 40% bile and 10% NaCl and at pH 9.6. Several carbohydrates are fermented with acid production. Otherwise, strains are biochemically inactive. *Aerococcus viridans* has occasionally been implicated in endocarditis, meningitis, and nosocomial infections in immunocompromised patients. They also cause disease in lobsters.

H. *Gemella*

The two species in the genus *Gemella*, *Gemella haemolysans* and *Gemella morbillorum*, have been isolated from the upper respiratory tract and bronchial secretions of humans. Both species are associated with endocarditis. Cells are small and those of *G. haemolysans* are flattened on adjacent sides. *Gemella haemolysans* also forms β-hemolytic colonies on horse or rabbit blood agar.

IV. Anaerobic Cocci

The anaerobic GPC are part of the normal flora in the human oral cavity, respiratory and urogenital tracts, colon, and skin. Some genera are also isolated from the rumens of animals. Clinical specimens frequently contain organisms in association with other bacteria, particularly gram-negative bacilli. The infections caused by the anaerobic GPC reflect their habitats as normal flora. Most infections are treated with penicillin or metronidozole.

Isolates are usually nonmotile. On vitamin K and hemin-supplemented blood agar, organisms form minute to small colonies within 48–72 hr of incubation in an oxygen-free environment. Table III presents distinguishing features of the five genera of anaerobic GPC. It is important to note that the genera and species are sometimes difficult to differentiate based solely on microscopic or colonial morphology and biochemical parameters. Gas–liquid chromatographic analyses of fatty acid metabolic end-products may be necessary for definitive identification. Fatty acid profiles are determined by growing isolates in peptone–yeast broth supplemented with glucose.

A. *Peptococcus*

Peptococcus is a monospecific genus, *Peptococcus niger*, presumptively identified by the formation of small, shiny black colonies. Unlike other anaerobic GPC, peptococci may be weakly catalase-positive. Organisms are rarely pathogenic but are frequently isolated the human vagina.

B. *Peptostreptococcus*

Peptostreptococci are involved in a variety of head and neck, intra-abdominal, obstetrical or gynecological infections, bacteremia, brain abscesses, and osteomyelitis. The most frequently isolated species are *P. magnus*, *P. assacharolyticus*, and *P. anaerobius*. *Peptostreptococcus anaerobius* is presumptively identified by susceptibility to sodium polyanethanosulfonate, determined by a simple disk

Table III Identifying Characteristics of the Anaerobic Cocci

Characteristic	Genus				
	Peptococcus	*Peptostreptococcus*	*Ruminococcus*	*Coprococcus*	*Sarcina*
Cellular arrangement	Pairs, tetrads, clusters	Pairs, short chains	Pairs, short chains	Pairs, chains of pairs	Cuboidal packets
Carbohydrates fermented	–	–[a]	+	+	+
Habitat	Female genital tract	Upper respiratory, intestinal, female genital tract	Rumen, colon	Rumen, colon	Soil, plants, feces
Pathogenic	–[b]	+	–	–	–
Major gas–liquid chromatography products[c]	B, C	A, B, IC		F, L, B	A, B
G+C content (mol%)	50–51	28–35	39–46	39–42	28–31
No. species	1	9	2	3	2

[a] Occasional strains are fermentative.
[b] Rarely pathogenic.
[c] Fatty acids: A, acetic acid; B, butyric acid; C, caproic acid; F, formic acid; IC, isocaproic acid; L, lactic acid.

diffusion test. *Peptostreptococcus magnus* forms large cells arranged in pairs or clusters and is sometimes mistaken for *Staphylococcus*.

C. *Ruminococcus, Coprococcus,* and *Sarcina*

These genera are isolated from human feces or colons and cattle rumens. *Sarcina* are also found in soil and on cereal grains. None are medically significant. *Ruminococcus* may form white, tan, yellow, or orange colonies. Some species are motile by one to three flagella, and some digest cellulose or hemicellulose in animals but not humans. *Sarcina* are distinguished microscopically by their large cells arranged in packets of eight or more.

V. Conclusion

In summary, the 17 genera of GPC described in this article are a diverse group of organisms with few characteristics in common. Species are widespread in nature and occupy a variety of ecological niches. Many are part of the normal human flora and are opportunistic pathogens. The taxonomy of the group is in a state of flux and many changes are anticipated in the near future.

Bibliography

Balows, A., Hausler, W., Herrmann, K., Isenberg, H., and Shadomy, H. (eds.) (1991). "Manual of Clinical Microbiology." American Society for Microbiology, Washington, D.C.

Kloos, W. (1980). *Annu. Rev. Microbiol.* **34,** 559–592.

Schleifer, K., and Balz, R. (1987). *Syst. Appl. Microbiol.* **10,** 1–19.

Sneath, P., Mair, A., Sharpe, M., and Holt, G. (eds.) (1986). "Bergey's Manual of Systematic Bacteriology," Vol. 2. The Williams and Wilkins Co., Baltimore, Maryland.

Starr, M., Stolp, P., Truper, H., Balows, A., and Schlegel, H. (eds.) (1981). "The Prokaryotes, A Handbook on the Habitats, Isolation, and Identification of Bacteria," Vol. 2. Springer-Verlag, New York.

Gram-Positive Rods

Jill E. Clarridge III
Baylor College of Medicine

Glossary

DNA homology Measure of the relatedness of organism based on the degree single strands of the DNA from two species anneal, reflecting sequence similarity

Gram-positive Property of retaining the gentian violet stain after fixation by tannic acid and destaining with alcohol; it correlates with a fundemental difference in the cell envelope

rRNA homology A measure of relatedness based on the ribosomal sequence similarity of two organisms

GRAM-POSITIVE BACTERIA possess a relatively uniform, thick cell wall composed primarily of peptidoglycan, teichoic acids, and polysaccharides. Although genetic relatedness studies based on DNA, 5S rRNA, and 16S rRNA homology studies have shown that the gram-positive cocci (generally round bacteria) of the genera *Staphylococcus* and *Streptococcus* are more closely related to the gram-positive rods (generally elongated bacteria) of the genera *Bacillus* and *Lactobacillus*, respectively, than to each other, the traditional morphologic separation of rods and cocci is a significant factor in initial identification. The gram-positive rods show a variety of distinctive shapes from the delicate pleomorphic *Arcanobacteria* to the extensively branching *Nocardia* to the large regular *Bacillus species*. (Fig. 1). Much studied organisms causing disease in humans include *Corynebacterium diphtheriae, Erysipelothrix rhusiopathiae, Listeria monocytogenes, Bacillus anthracis, Clostridium tetani, Mycobacterium tuberculosis*, and *Nocardia asteroides*.

I. Taxonomy

There are only two major taxonomic branches of gram-positive organisms compared to a much wider diversity among the gram-negative bacteria and other prokaryotes such as those shown in Fig. 1. The relationships and the names of the genera within the actinomycetes line, which is characterized by a high guanine-cytosine (G-C) content of the DNA and a pleomorphic cell shape, are shown in Fig. 2 as a dentogram. The major groups of the *Clostridium-Bacillus* line are shown in Table I. Relatedness is determined by biochemical characteristics such as cell wall constituents and fatty acid composition as well as DNA and rRNA homologies and morphologic similarities. [*See* GRAM-POSITIVE COCCI.]

II. Identification

Biochemical characteristics that are important in distinguishing the gram-positive rods include the production of catalase, nitrate reductase, urease, hemolysin, lecithinase, gelatinase, and hydrogen sulfide, the ability to use sugar substrates and to hydrolyze esculin, motility, atmospheric conditions for growth, colony morphology, and cellular morphology. Some of these characteristics are shown in Table II.

Structural characteristics of importance are whole cell fatty acid and whole cell wall composition. The polymerase chain reaction can also be used to detect specific organisms (i.e., Tuberculosis).

III. Gram-Positive Rods of Clinical Importance

The gram-positive rods exhibit a wide range of pathogenicity. As shown in Table II, some are normal flora and essentially never pathogenic (e.g., *Rothia, Brevibacterium, Lactobacillus*), while in

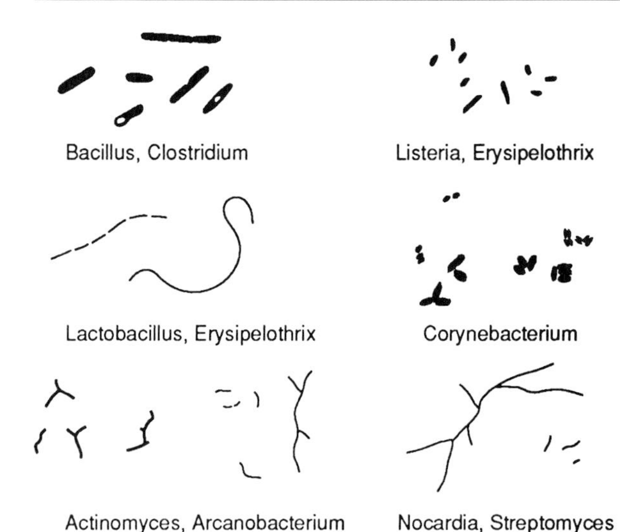

Figure 1 Representative morphologies of the gram-positive rods.

other groups most species exist commensally but there is at least one highly virulent strain (e.g., *Clostridium, Bacillus, Corynebacterium,* and *Actinomyces*). In still other groups most species are commonly found in the environment with only a few species being pathogenic whenever associated with humans (e.g., *Mycobacterium tuberculosis, Nocardia*); some genera are environmental and cause disease in humans only rarely and than usually in individuals who are immunologically compromised (Oerskovia, Streptomyces, Kurthia).

There are diseases associated with only a single species. Anthrax, a zoonosis which can be spread directly from animals, particularly sheep and goats, or from their products, is caused by *Bacillus anthracis* and has two major manifestations. The more common form is a cutaneous pustule, which developes after contact with infected material. The rarer form, inhalation anthrax (a severe pneumonia with septicemia) is almost always fatal. Erysipeloid, caused by *Erysipelothrix rhusiopathiae,* is usually seen as a skin infection but also can cause septicemia, arthritis or endocarditis. Leprosy and tuberculosis are caused by *Mycobacterium leprae* and *tuberculosis,* respectively. Although tuberculosis has been declining in the United States, the increasing number of cases in patients with AIDS, and the spread of organisms that are resistant to presently available antibiotics, has made tuberculosis an important public health problem. Diptheria, with the characteristic membrane formation in the throat and the systemic manifestations evoked by the

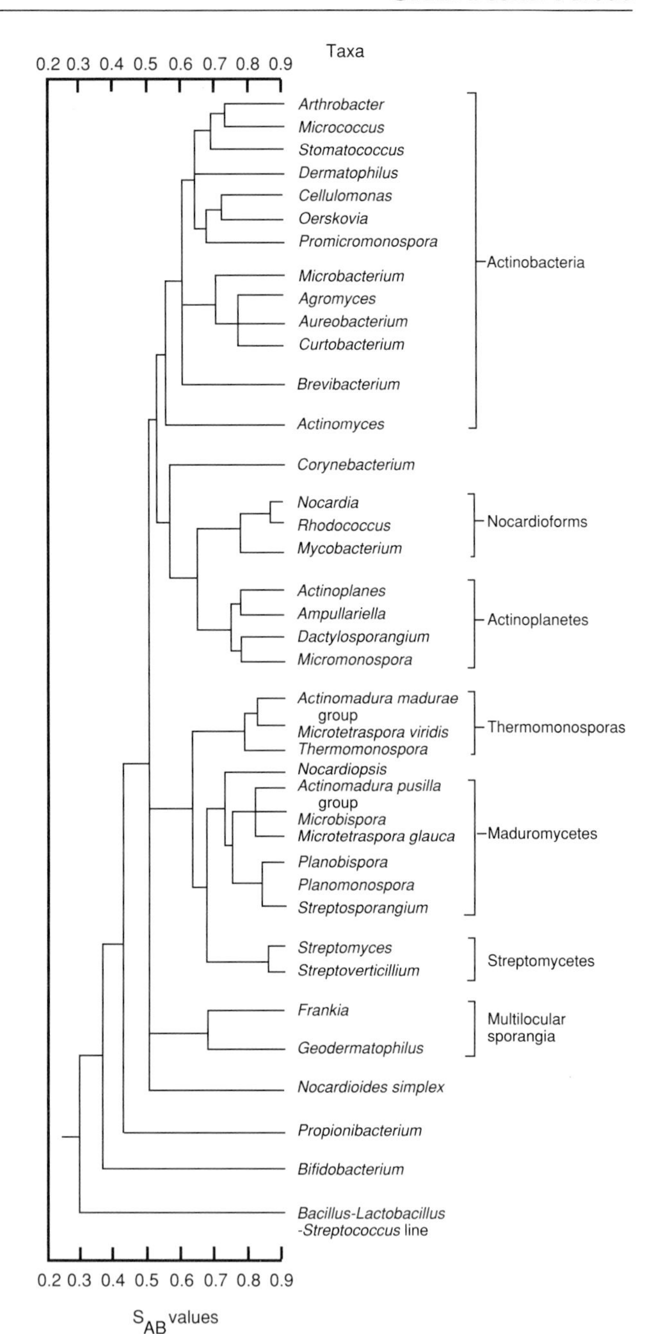

Figure 2 Relationships in the actinomycetes line. [From Goodfellow, M. (1989). Suprageneric classification of actinomycetes. *In*"Bergey's Manual of Systematic Bacteriology," Vol. 4. (S. T. Williams (eds.). Williams & Wilkins, Baltimore.

toxin, is caused by *Corynebacterium diphtheriae.* [*See* LEPROSY; ACQUIRED IMMUNODEFICIENCY SYNDROME (AIDS).]

However, purulent infections of skin and other organs can be caused by a great variety of gram-

Table I Groupings of the *Clostridium-Bacillus* Line

Order 1: Clostridiales
- Family 1: Clostridiaceae I
  - Genera: *Clostridium butyricum* and relatives, *Sarcina*
- Family 2: Clostridiacea II
  - Genera: *Clostridium lituseburense* and relatives, *Peptostreptococcus anaerobius, Eubacterium tenue*
- Family 3: Clostridiaceae III
  - Genera: *Clostridium sphenoides* and relatives

Order 2: Desulfotomaculales
- Family 1: Desulfotomaculaceae
  - Genera: *Desulfotomaculum, Ruminococcus*
- Of uncertain affiliation: *Peptococcus aerogenes*

Order 3: Mycoplasmatales
- Family 1: Mycoplasmataceae (true mycoplasma)
  - Genera: *Mycoplasma, Spiroplasma, Acholeplasma, Anaeroplasma, Ureaplasma*
- Family 2: Clostridiaceae IV
  - Genera: *Clostridium innocuum, Clostridium ramosum, Lactobacillus catenaforme, Erysipelothrix*

Order 4: Acetobacteriales
- Family 1: Acetobacteriaceae
  - Genera: *Acetobacterium, Eubacterium*
- Family 2: Thermoanaerobiaceae
  - Genera: *Thermoanaerobium, Acetogenium, Clostridium, Thermoaceticum* and relatives

Order 5: Bacillales
- Family 1: Bacillaceae
  - Genera: *Bacillus, Staphylococcus, Brochothrix, Listeria, Gemella*
- Family 2: Streptoccacceae
  - Genera: *Streptococcus, Enterococcus*
- Family 3: Lactobacillaceae
  - Genera: *Lactobacillus, Pediococcus, Leuconostoc*
- Of uncertain affiliation: *Aerococcus, Kurthia*
- Of uncertain affiliation: *Lactobacillus minutus*

[From Fox, G. E., and Stackebrundt, E. (1987). The application of 165 rRNA cataloguing and 5S rRNA sequencing in bacterial systematics. *In* "Methods Microbiology," Vol. 19. (R. Colwell and R. Grigorova (eds.), pp. 405–458. Academic Press, London.]

Table II Gram-Positive Rods Associated with Humans: Description and Diseases

Genus Name	Growth Atmosphere[a]	Morphology[b]	Catalase	Motility	Other	Habitat	Disease
Clostridium	An	S, Lg, Reg, Sp, Ch	−	V	Many produce toxins	Environmental, feces	Tetanus (*C. tetani*), gas gangrene (*C. perfringens*)
Eubacterium	An	S, Lg, Pl, Ch, CB	−	V		Feces	Rarely pathogenic alone
Bacillus	A	Lg, Reg, Sp, Ch	+	V	Many produce toxins	Environment	Anthrax (*B. anthracis*), food poisoning, ophtha
Listeria	F	S, Reg	+	+	Some hemolytic	Soil, food, vertebrae	Meningitis, abortion (*L. monocytogenes*)
Lactobacillus	F, An	S, Lg, Reg, Ch	−	−	Lactic acid produced	Vagina, feces, foods	Rarely pathogenic
Erysipelothrix	F	S, Reg, Fil	−	−	H_2S production	Soil, animals	Erysipeloid
Kurthia	A	R, Ch	+	+		Feces of farm animals	Nonpathogenic

(continued)

Table II (*Continued*)

Genus name	Growth atmosphere[a]	Morphology[b]	Catalase	Motility	Other	Habitat	Disease
Bifidobacterium	An	Pl, D	−	−		Feces, dental plaque	Rarely pathogenic
Propionibacterium	F, An	B, Fil, D	+	−	Propionic acid produced	Skin	Rarely pathogenic
Streptomyces	A	B, Fil	+	−		Soil	Rarely pathogenic
Rhodococcus	A	B, Fil, D, CB	+	−	Rod-coccus cycle	Soil	Rarely pathogenic, pulmonary disease
Nocardia	A	B, Fil, CB	+	−	Partly acid-fast	Soil	Abscess formation
Mycobacterium	A	S, Pl, C, rare B	+	−	Acid-fast	Environment	Tuberculosis, leprosy
Corynebacterium	A, F	Pl, D, CB	+	−		Skin	Diphtheria (*C. diphtheriae*)
Actinomyces	F, An	Pl, C, D, B	V	−		Mouth, skin	Abscess formation, actinomycosis
Arcanobacterium	F	S, Pl, C, D, B	−	−		Mouth, skin	Wound infection, pharyngitis
Arachnia	F, An	Pl, Ch, D, B, CB	+	−		Mouth	Actinomycosis
Rothia	A	D, B, CB	+	−		Mouth	Endocarditis, rarely pathogenic
Brevibacterium	A	Pl, D, B, CB	+	−	Rod-coccus cycle	Skin, dairy products	Nonpathogenic

[a] An, anaerobic; F, facultative growth (can grow with or without air); A, aerobic.
[b] Lg, large (0.7–1.0 = μm diameter, 2–4 μm long); Reg, regular or uniform; Sp, spores can be produced; Ch, sometimes chaining; Pl, pleomorphic; S, slender; C, curved; D, diphtheroid or club-shaped; B, branching; Fil, sometimes filaments; CB, coccobacillus.
[c] +, positive; −, negative; V, varies with species.

positive rods, including *Actinomyces sp.*, *Arcanobacterium haemolyticum, Corynebacterium pseudotuberculosis, Rhodoccoccus equi, Mycobacterium spp.*, such as *M. fortuitum, and Nocardia spp* as well as other organisms. Similarly, endocarditis, an infection of the heart valves, can be caused by organisms usually considered of low virulence. In this group are *Corynebacterium* spp. and rarely, *Rothia*, *Kurthia* and *Oerskovia*.

Bibliography

Clarridge, J. E. (1987). Gram positive bacilli in "Clinical and pathogenic microbiology." *In* (B. Howard, ed.). C. V. Mosby Co. pp. 417–434. St. Louis, Missouri.

Fox, G. E., and Stackebrandt, E. (1987). The application of 16S rRNA cataloguing and 5S rRNA sequencing in bacterial systematics. *In* "Methods Microbiology," Vol. 19. (R. Colwell and R. Grigorova (eds.), pp. 405–458. Academic Press, London.

Goodfellow, M. (1989). Suprageneric classification of actinomycetes. *In* "Bergey's Manual of Systematic Bacteriology," Vol 4. (S. T. Williams, ed.). Williams & Wilkins, Baltimore.

Sneath, P. H. A. (ed.) (1986). "Bergey's Manual of Systematic Bacteriology," Vol. 2. Williams & Wilkins, Baltimore.

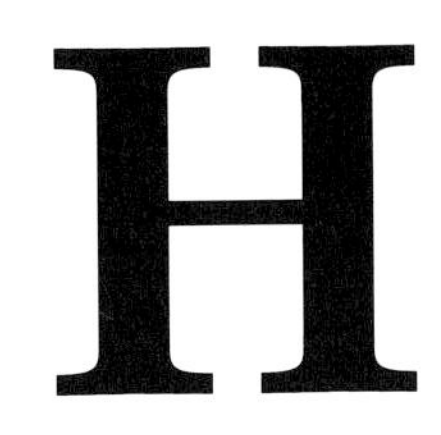

Hazardous Waste Treatment, Microbial Technologies

Brendlyn D. Faison
Oak Ridge National Laboratory

Glossary

Biodegradation Microbially catalyzed breakdown of an organic substrate, yielding carbon dioxide or carbon dioxide plus methane, water, and, in some cases, inorganic salts when carried to completion

Biosorption Binding of organic or inorganic substances into or onto microbial biomass via adsorption, ion-exchange, chelation, complexation, precipitation, particle entrapment, or uptake

Biotransformation Microbially catalyzed modification (incomplete degradation) of an organic substrate; microbially catalyzed modification (oxidation or reduction) or an inorganic substrate

THE PARAMOUNT ROLE of microorganisms in the global cycling of carbon and other elements has long been recognized. The diverse metabolic activities of microorganisms, and in particular their ability to interact with complex organic and inorganic substrates, are now being exploited in the treatment of hazardous wastes. "Hazardous wastes," for the purposes of this article, refers to these materials contaminated with substances deemed unusually dangerous to humans or to the environment because of certain chemical or physical attributes. (The legal definitions of hazardous materials, substances, or wastes will not be used here nor will the issue of infectious medical waste be addressed.) Novel technologies, including those based on microbial activities, are being evaluated for their ability to reduce the danger posed by the release of these wastes to the environment. Technology is applied science. The scientific basis for the development of microbial technologies for the treatment of these wastes will therefore be presented here, as will practical considerations relevant to the implementation of these technologies. Specific examples of microbial processes for the treatment of hazardous wastes before or after their introduction to the environment will be described.

I. Hazardous Waste: Politics and Science

The decades of the 1970s and 1980s saw a major increase in the amount of attention paid to environmental and health issues. A watershed event in the United States was the passage by Congress of the

National Environmental Policy Act in 1970. This act, the nation's first environmental legislation, requires that federal agencies address the environmental consequences of their actions. Although it neither addresses hazardous waste *per se* nor prohibits dangerous activities, it forms the precedent for subsequent policy.

A. U.S. Regulations Concerning Hazardous Waste

The Resource Conservation and Recovery Act of 1976 (RCRA) was promulgated to regulate the disposal of chemically hazardous wastes. The Environmental Protection Agency (EPA), which had been created in 1971, was given the authority and responsibility to define "hazardous waste" and to devise a system for the control of its management and disposal. The EPA definition of "hazardous waste" is too complex to be fully explored here. It is sufficient to describe these wastes as refuse containing materials that are toxic, carcinogenic, mutagenic, or teratogenic or that are ignitable, corrosive, highly reactive, or likely to leach out toxic chemicals. RCRA has created a "cradle-to-grave" liability for a waste (i.e., responsibility for its management from the time it is created until it is destroyed). Other permitting, recordkeeping, and training provisions have simultaneously increased the cost of waste management. Finally, RCRA includes enforcement provisions that may mandate fines of up to $250,000 and/or 5 years' imprisonment for individuals, and fines of up to $1 million for corporations.

RCRA is a major factor in the management of chemically hazardous wastes, but many of these wastes are exempt from regulation under RCRA. Industrial and municipal wastewater discharges are regulated by EPA under the Clean Water Act; gaseous emissions from industry and other sources are regulated by EPA under the Clean Air Act. There are several other types of environmental legislation—including but not limited to the Toxic Substances Control Act, the Safe Drinking Water Act, and the Occupational Safety and Health Act—that affect the use, storage, transportation, distribution, and disposal of chemically hazardous waste. Federal agencies other than EPA may participate in enforcement of these regulations.

Radioactive wastes generated during the handling of nuclear materials usually fall within the purview of the Nuclear Regulatory Commission (NRC), although other agencies, including EPA, may also be involved in the regulation of these wastes. Radioactive wastes are considered to be physically hazardous because of the deleterious biological effects of ionizing radiation. Some of these wastes are produced commercially. But many of these wastes are generated instead by the U.S. Departments of Energy (DOE) and Defense (DoD), whose activities encompass nuclear research and weapons production.

The regulations described above were all created at the federal level. Individual states and localities, in some cases as a consequence of grass-roots environmental activism, have also put forth environmental laws that are in some cases stricter than federal standards. Environmental awareness in other countries has led to analogous (or in some cases, even more rigorous) laws on the local, state, and national level. These international trends are encouraging the creation of a global environmental policy to which the United States would presumably also subscribe.

The increasingly stringent and complex regulatory requirements that have been set forth—and the intense congressional and public scrutiny that accompany them—have led to heightened concern about environmental compliance among chemically hazardous and/or radioactive waste generators. The threat of civil and criminal penalties for breaches of the law has become very real. Failure to adhere to environmental regulations has also been found to have a public relations cost: a poor reputation with respect to environmental protection has led in some cases to unfavorable notoriety, boycotts, and civil lawsuits. (As Abraham Lincoln said, "What kills a skunk is the publicity it gives itself.") Such a reputation, whether fairly earned or not, may also take a toll on worker morale. Thus, there has been clear business incentive for the development of good waste management practices.

In some respects, this new environmental awareness came too late. Past practices have resulted in environmental contamination, in some cases severe. The Comprehensive Environmental Response, Compensation, and Liability Act (CERCLA, or Superfund) and its amendments were promulgated to address the problem of environmental contamination, specifically hazardous waste spills and abandoned or inactive hazardous waste sites. CERCLA, which is enforced by EPA, provides for site restoration at government expense. However, CERCLA also establishes liability for the contamination problem and requires that responsible parties reimburse the government for costs incurred. Efforts at environmental restoration mandated by CERCLA will continue indefinitely.

It is evident that technologies are needed for the treatment of hazardous wastes in either an ongoing fashion for waste management, or retrospectively, for environmental restoration. The magnitude of this need derives in part from the remarkable variety of wastes involved.

B. Origin and Composition of Hazardous Wastes

Every activity of humans or other organisms generates waste. This maxim is especially true with respect to industrial processes (Table I). Industrial (nonmedical) wastes may be roughly ranked with respect to degree of hazard (Fig. 1). Definitions or examples of each follow.

Table I Examples of Hazardous Wastes

Origin (industry)	Waste components
Chemicals	Arenes, phenols, carbon disulfide, carbon tetrachloride, cyanides, Pb, As, Hg, Cd
Pulp/paper	Lignosulfonates, chlorolignins, formaldehyde, mercaptans, phenols
Leather/tanning	Phenols, Cr, Fe
Metal finishing	Cyanides, Cu, Ni, Zn, Cr, Fe
Rubber/plastic manufacture	Trichloroethylene, anilines, arenes
Explosives	Trinitrotoluene, nitroguanidine, RDX, picric acid, Cu
Pharmaceuticals	Anilines, phenols, formaldehyde
Petroleum/petrochemicals	Hydrocarbons (including polycyclic aromatics), mercaptans, H_2S
Agriculture	Pesticides (herbicides, insecticides, fungicides), phenols, As, Pb
Coal	Hydrocarbons (inc. polycyclic aromatics), pyridines
Combustion (general)	Sulfur oxides, nitrogen oxides, carbon monoxide
Nuclear	Radionuclides (U, Sr, Cs, Ra, Pu, Ru, Co, Ce, Y, Ru)
Textiles	Anilines, azo compounds, nitriles, ketones

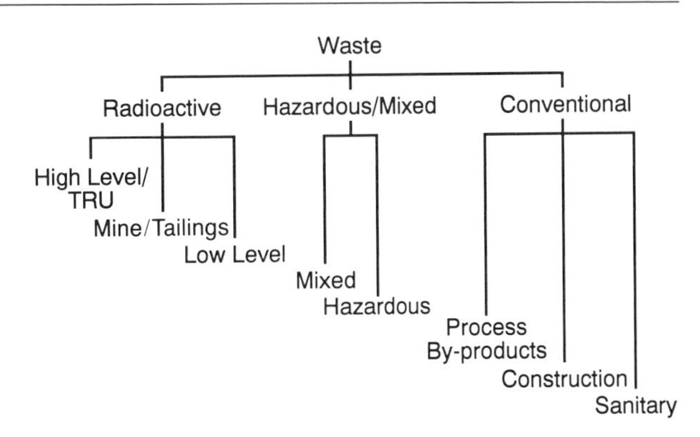

Figure 1 Waste classification scheme. A hierarchy of wastes is reflected here, ranging from the most dangerous wastes [high-level radioactive–transuranic (TRU)] to simple sanitary waste.

1. Radioactive Wastes

a. High-Level and Transuranic Wastes

The nuclear fuel cycle is initiated by the mining of ores containing fissile material, refining of the ore, its enrichment in fissile material, and its conversion to the proper form for use in reactors (i.e., fabrication). However, nuclear fuels, unlike fossil fuels, are not completely consumed within reactors. "Spent" fuel is reprocessed to recover its fuel value. High-level wastes are composites of the waste streams arising from reprocessing activities. Transuranic wastes are those reactor wastes that are enriched in elements with atomic numbers greater than that of uranium. High-level and transuranic wastes may be intensely radioactive and commonly contain uranium, plutonium, and their fission products. These radionuclides tend to be extremely long-lived (e.g., possessing half-lives of 10^4 to $>10^9$ years).

b. Mining Waste and Mill Tailings

Radioactive mining wastes are produced by the mining of uranium or thorium ore for the production of nuclear fuels. The subsequent crushing, washing, and concentration of the ores, carried out in processing plants adjacent to the mines, generate liquid and solid wastes that are referred to as mill tailings. Mining wastes and mill tailings contain residual uranium (which may be >95% of the amount originally contained in the ore) plus its decay products, including thorium, polonium, and lead. These radionuclides have a wide range of half-lives.

c. Low-Level Waste

Low-level waste is a catch-all term developed by NRC that basically covers all radioactive wastes that do not fall into one of the other two classifications. Most of these wastes are generated in the course of nuclear power production, nuclear energy research, or nuclear weapons production. However, coal and petroleum contain a small amount of natural radioactive material that may be incorporated in the waste generated by processing or combustion of these fuels. Radionuclides are used as tracers in industry, agriculture, medicine, and academic research, providing yet another source for radioactive waste. Nearly any radionuclide may be present in low-level waste, although it tends to contain primarily short-lived (i.e., well under 10^5 years) isotopes, such as those of carbon, hydrogen, and phosphorus.

2. Mixed and Hazardous Wastes

a. Mixed Waste

"Mixed waste" is a term with specific meaning to the nuclear industry. These wastes consist of mixtures of radioactive materials and nonradioactive, chemically hazardous material. These materials are primarily regulated by NRC and EPA, respectively. The frequently conflicting requirements of these agencies make the management of mixed waste particularly difficult. Mixed wastes are generated deliberately in the course of cleaning or decommissioning equipment contaminated with radionuclides, in certain industrial procedures, and in research. Typical hazardous components of mixed wastes include extractants such as tributyl phosphate or phenyl chloroform-based solvents; chelating agents such as ethylenediamine tetraacetate; and organic scintillation fluids used for radioactivity measurement in industrial and academic research. Mixed wastes may also be generated through the inadvertent mixing of waste streams, in which case any combination of radionuclide(s) and hazardous material(s) may be involved.

b. Hazardous Waste

EPA's definition of hazardous waste was mentioned above. Obviously, a multitude of compounds falls within this classification. Chemically hazardous wastes are generated throughout industry and agriculture and also from household sources (Table I). These wastes may be organic or inorganic.

3. Conventional Wastes

Conventional wastes, which are generated by "normal" industrial or household activities, are quite heterogeneous. Examples include whey from cheesemaking (a process by-product), certain wood products (construction waste), and sanitary waste. These wastes will not be considered further, except to note that hazardous (RCRA or radioactive) and conventional wastes may originate from the same facility. A prime example derives from the use of RCRA-regulated halogenated organic solvents to clean home septic systems.

C. Magnitude of the Hazardous Waste Problem

It is very difficult to assess the size of the hazardous waste problem in the United States because of the complex legal definition of these wastes, the number and variety of sources for these wastes, and the requirements of national security. Perhaps the most complete data are those compiled in 1962 through 1966 and reported in 1974. At that time, industry (exclusive of agricultural and food processing activities) produced a total of 10 billion kg of solid refuse, 50 billion liters (50 million m^3) of liquid effluents, and 39 billion kg of gaseous emissions, on an annual basis. There were 73,000 Ci of low- to intermediate-level radioactive waste released annually from federal facilities, and 59,000 Ci of radioactive waste (mainly in the form of gaseous emissions) were released each year from the three nuclear power plants that provided this information. No data on the generation of high-level or commercial radioactive waste were reported. Granted, these numbers are all overestimates, in that not all the waste released could be defined as hazardous. But this decades-old information certainly represents an underestimate of current waste generation rates. Additional preliminary data were compiled in 1991 by the National Science Foundation. These data suggested a then current hazardous (nonradioactive) waste generation rate of 1 billion kilograms per day. At that time, there were thought to be *25,000* hazardous waste sites in the United States that could be classified as dangerous. These sites accounted for an estimated environmental debt of *2 trillion* dollars.

II. Current Waste Management Technologies

The idealized industrial waste management strategy is to avoid generating waste. A less drastic approach is to minimize the amount of waste produced through source reduction and/or recycling. However, some waste will inevitably be produced. Very rarely can waste be legally discharged directly to the environment. The option of disposal, storage, or treatment must be exercised (Fig. 2). Each approach has the potential for environmental impact.

A. Disposal

Industrial waste disposal may be achieved through deep well injection, landfilling, burial in deep mines or salt deposits, discharge to surface impoundments (lagoons), geological isolation, seabed implacement, or ocean dumping. These practices result in the direct or ultimate introduction of intact hazardous wastes into the environment. Perhaps the most deleterious practices are those that affect groundwater quality:

Surface impoundments
Landfills
Home septic systems
Municipal sewage systems
Land disposal of sludge
Mine tailings
Accidental spills
Road runoff
Waste injection wells
Fertilizer/pesticide application
Leaks from underground storage tanks
Irrigation return flows
(Midnight dumpers?)

Groundwater provides 96% of all freshwater use in the United States, including 95% of rural water supplies, 40% of agricultural irrigation supplies, and 35% of municipal water supplies. Groundwater is, moreover, contiguous with surface water and indeed ultimately emerges to the surface. Groundwater contamination thus represents a vehicle for widespread exposure of humans, domestic animals, and wildlife to hazardous materials. Protection of groundwater has been the major factor underlying restrictions on many of these disposal practices. Similar considerations are limiting the landfilling of solid household wastes.

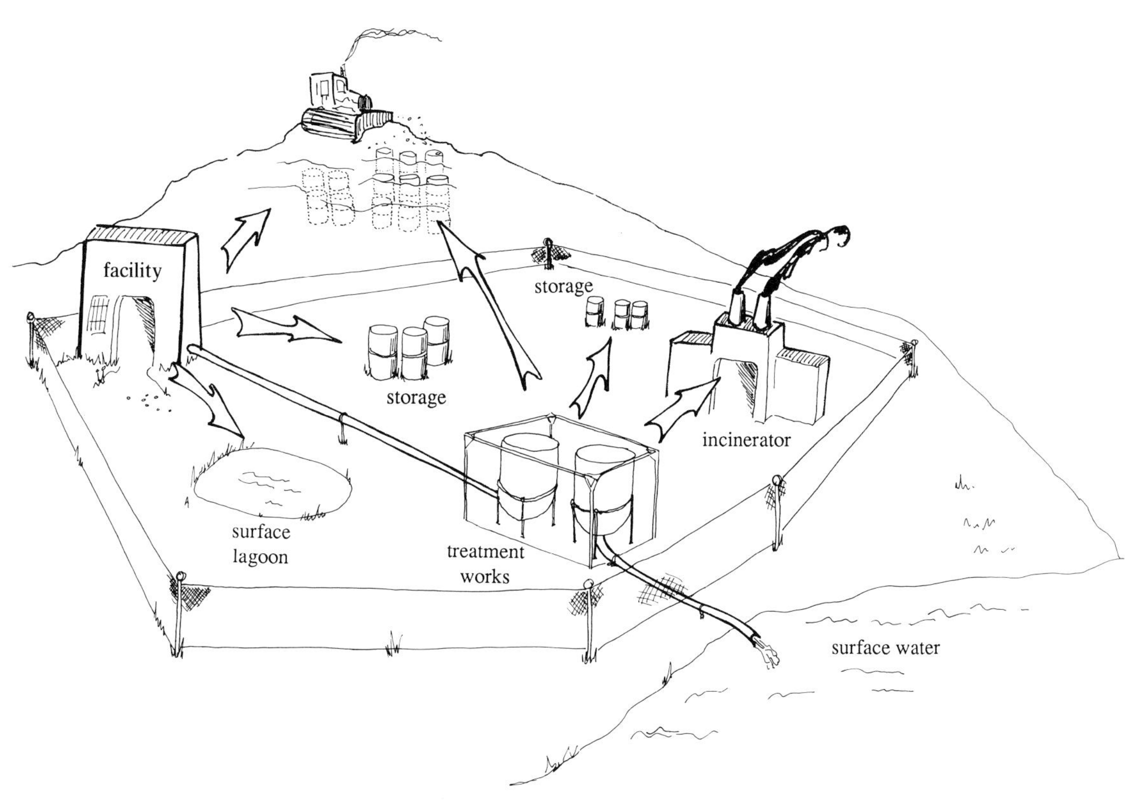

Figure 2 Industrial waste management. This schematic illustrates the three waste management options (treatment, storage, and disposal) as well as pathways for environmental impact.

B. Storage

Wastes may be stored aboveground or underground pending identification of a suitable technology for disposal or treatment. Storage containers may leak, overflow, corrode, or be ruptured accidentally, with escape of their contents. There is thus a clear potential for environmental damage associated with storage of hazardous waste. Storage is not considered an acceptable long-term approach to hazardous waste management.

C. Treatment

1. Overview of Waste Treatment Technologies

Hazardous wastes are treated to reduce their risk to humans and the environment. Treatment goals include destruction, detoxification, immobilization or containment, and volume reduction. A number of physical (nonthermal), chemical, and thermal treatment technologies have been developed:

- *Physical*
  Adsorption
  Air, steam, or vapor stripping
  Distillation
  Electrodialysis
  Encapsulation
  Flotation
  Ultrafiltration
- *Chemical*
  Hydrolysis
  Ion exchange
  Neutralization
  Oxidation/reduction
  Precipitation
  Solidification
  Solvent extraction
- *Thermal*
  Evaporation
  Incineration

However, these technologies are in some cases deficient. Some approaches (e.g., vapor stripping) simply transfer the hazardous material to another environmental compartment (itself a regulated activity). Other approaches, such as extraction with RCRA-regulated solvents, themselves generate unacceptable wastes. Still others, such as incineration, are capital- and/or energy-intensive and thus costly. Biological technologies, specifically those based on the activities of microorganisms, have been viewed with increasing favor as a method for hazardous waste treatment. These technologies have a number of advantages. First, microorganisms tend to be specific both with respect to the types of compounds attacked and to the reactions involved, allowing a process based on their activity to be fairly predictable. They are selective, allowing treatment of a specified component within a heterogeneous waste. They are flexible in that they will carry out their attack despite fluctuations in the waste composition. Microbial processes tend to be functional under mild operating conditions of near-ambient temperature and pressure and pH near neutrality. Microbial processes are generally environmentally compatible in that they generate a fairly clean waste stream and can usually be disposed of by conventional means. Microorganisms tend to be self-regenerating, meaning that they usually replicate themselves within the processing system. These properties combine to make microbially based processes relatively inexpensive. Most important, microbial technologies for waste treatment are proven, for they form the historical basis for treatment of sanitary waste within septic systems and sewage treatment plants alike. Finally, microbial treatment processes are integrable, meaning that they may be combined with other nonbiological unit processes (such as incineration or precipitation) as necessary.

2. Specific Microbial Waste Treatment Technologies

The microbial treatment technologies currently in use within industry are roughly modeled on those used in the treatment of household waste. The scale and physical appearance of many of these treatment processes may be realized by visiting a municipal wastewater treatment facility.

a. Trickling Filters

Trickling filter (biological bed) processes rely on a consortium of microorganisms (fungi, algae, and bacteria), protozoans, and macroorganisms (e.g., worms and flies, which occupy upper reaches of this food chain) grown in a biofilm on the surface of rocks or other solid support materials housed within an open-air cylinder. Liquid waste streams are sprayed over the flat upper surface of the filter and trickle down through its height. Properly operated systems are not allowed to become saturated with liquid. Trickling filter systems are highly aerobic.

b. Rotating Biological Contactors

Rotating contactor (biodisk) processes are very similar to those using trickling filters, except that the biofilm is supported on the surface of plastic or metal disks. These disks are arranged within a horizontally placed cylindrical apparatus by passing a rod through their centers. The wastewater is introduced into the cylinder to fill it only partially. As the disks rotate around the axis formed by the rod, they are intermittently submerged in the wastewater. The process is usually carried out in the open air. Because the biofilm on the disks is exposed to the atmosphere at least half of the time, the process is highly aerobic.

c. Activated Sludge

Aerobic decomposition may be carried out within large activated sludge units, which consist of bubbler tanks in which microorganisms are suspended within the wastewater of interest. The microbial population within these units is mainly composed of bacterial and fungi. After an appropriate incubation period (residence time), the sludge is sent to a clarifier in which microbial biomass is removed. This material is usually recycled into the reactor. This activated sludge process, like the biofilm processes described above, is highly aerobic.

d. Aerated Lagoons

Aerated ponds or lagoons are similar to activated sludge units except that they are mechanically agitated, not aerated. Biomass, consisting of bacteria and algae, is removed within a separate processing step by settling. These processes are carried out within open-air surface impoundments. The absence of a forced supply of oxygen results in an aerobic to microaerophilic process.

e. Waste Stabilization Ponds

A modified version of the aerated lagoon consists of open-air ponds in which the waste is incubated without agitation. Although these waste stabilization ponds are shallow, the absence of agitation or aeration results the settling of biomass *in situ* and the formation of anaerobic zones. Bacteria predominate within these systems, but some algae may also be present.

f. Anaerobic Digestors

Anaerobic digestors consist of closed vessels containing microbial biomass suspended in wastewater and thus are the anaerobic counterpart of activated sludge units. The digestors generate large amounts of methane that may be used as fuel sources for other industrial processes. The microbial biomass therein is thought to consist nearly exclusively of bacteria.

g. Anaerobic Filters

Anaerobic filters are similar to aerobic trickling filters, except that both the microbial biomass and support medium are wholly submerged in the wastewater. Microorganisms (probably bacteria) live both within biofilms on the surface of the support material and in the bulk fluid. "Stationary-type" anaerobic filter reactors contain a packed bed of support material onto which the wastewater is sprayed from the top. "Suspended-type" reactors receive wastewater through the bottom and represent fluidized beds operated in upflow mode. Both are completely anaerobic.

Most of the biological treatment processes described above were designed for the treatment of wastes contained as dissolved or suspended matter within aqueous effluents. However, industrial wastes may also include gaseous effluents, sludges, and solid materials (e.g., oily rags, or chemical or radiation workers' protective clothing that has become contaminated). These processes may be modified in a straightforward fashion to accommodate gaseous waste streams. Treatment of solid or semisolid materials, however, usually requires other technologies.

h. Composting

Composting is a process that facilitates the aerobic decomposition of solid materials. The refuse is mixed in piles, which may be supplemented with moisture and/or nutrients. The height of the piles, or windrows, is kept relatively low to facilitate oxygen penetration to all parts of the pile, although oxygen may also be supplied by forced-air systems. the piles may also be turned occasionally to encourage oxygen transfer, and they are not allowed to become saturated with moisture. The process is exothermic, and thermophilic fungi and actinomycetes predominate. A variation on the process is carried out in reactors.

i. Land Treatment

Land treatment is carried out by spreading a waste onto a land surface, where it is exposed to the metabolic activities of organisms present within the soil. Although sometimes referred to as land "farming," that term is a misnomer. Because hazardous materials are applied, the land used for this treatment is not suitable for agricultural activity. Land treatment is suitable for use with liquid, semisolid, and in some cases solid wastes. This technology is sometimes used for treatment of sludge formed as a residue of another treatment process. The waste is usually plowed into the land to maximize its availability to the microbial biomass. The process has both aerobic and anaerobic character. Because fungi and actinomycetes account for most of the biomass in soil, they may be assumed to play a major role in this treatment process; however, nonfilamentous bacteria and algae may also be involved. It is important to note that although land treatment does resemble disposal, it does not represent an indiscriminate dumping of waste.

j. Landfilling

It usually surprises the layperson to learn that landfilling does not constitute a biological treatment process. It is true that landfills achieve some contact between waste components and soil microorganisms. However, industrial and municipal landfills are generally designed in such a way as to prevent the infiltration or escape of moisture. This goal is achieved by containing the waste (and indeed, the entire landfill) within a liner. This approach is very different from that taken in land treatment, which does not involve a containment system. Water is essential to microbial activity. As a result, there is very little biodegradative activity within "well-designed" landfills.

III. Biochemical Basis for Microbial Technologies for Hazardous Waste Treatment

The goal of microbial waste treatment processes, like that of any waste treatment process, is either destruction, detoxification, immobilization/containment, or volume reduction. These microbial processes are based on one of the three mechanisms: biodegradation, biotransformation, or biosorption. Predictably, at least one of these mechanisms would be suitable for management of any hazardous organic or inorganic material.

A. Biodegradation

1. Hazardous Organics

Biodegradation achieves the complete destruction of hazardous organic compounds. These compounds may be assimilated directly as a primary source of carbon and energy for microbial growth via amphibolic pathways. Energy-yielding metabolism via strictly catabolic pathways may result in the complete mineralization of the target compound to CH_4 and/or CO_2. Alternatively, a hazardous organic may be cometabolized along with another more readily available organic nutrient. In the latter case, degradation of the target compound would occur only in the presence of a primary substrate. The so-called law of microbial infallibility dictates that for every naturally occurring organic compound, there exists at least one microbial species capable of its degradation, either directly or through cometabolism. The maxim can, in some cases, be extended to encompass man-made compounds. This category includes the halogenated organic compounds, which comprise 69 of the 129 priority pollutants designated by EPA. Halogenated organics are widely used as solvents (e.g., chloroform, methylene chloride, trichloroethylene), pesticides (e.g., pentachlorophenol, lindane, dieldrin), and in other applications, or occur as byproducts of industrial processes (e.g., chlorolignins, chloroanilines). These compounds are toxic, mutagenic, teratogenic, and/or carcinogenic. The complete degradation of these compounds yields salts in addition to the products mentioned. [*See* BIODEGRADATION.]

The ability of microorganisms to degrade petroleum hydrocarbons has received considerable attention as an approach to the bioremediation of crude oil spills. A wide variety of organisms have been shown to use the aliphatic components of crude oil, including *Achromobacter, Acinetobacter, Arthrobacter, Bacillus, Flavobacterium, Nocardia, Pseudomonas, Vibrio,* and the coryneforms. *Cladosporium,* a fungus commonly isolated from fuel tanks, is also capable of degrading aliphatic compounds of the type found in oil and thus may be useful in a process for removal of aliphatic and possibly aromatic hydrocarbons. Degradation of the toxic and/or mutagenic polycyclic aromatic hydrocarbons (PAHs) is carried out by various organisms. Naphthalene, the simplest PAH, is degraded by cyanobacteria and algae. The metabolic versatility of *Mycobacterium,* which has been shown to mineralize naphthalene as well as pyrene, benzo[a]pyrene, and fluoranthene, may be applied to the degradation of mixtures of PAHs. Other monoaromatic compounds, including phenols, benzoic acids, and their halogenated, nitrated, or otherwise substituted derivatives, are also susceptible to microbial degradation. Aerobic mineralization may be carried out by *Pseudomonas, Candida,* and *Trichosporon,* among other organisms. Anaerobic photodegradation by purple nonsulfur bacteria such as *Rhodopseudomonas* may occur. Anaerobic degradation may also be carried out chemotrophically by *Pseudomonas* or *Moraxella* under conditions that support nitrate respiration; by a coculture of *Pseudomonas* and *Desulfovibrio* in the presence of sulfate; or fermentatively, by methanogenic consortia. [*See* PETROLEUM MICROBIOLOGY.]

Various organisms capable of degrading nonhalogenated compounds of the type just described also attack halogenated derivatives. The *Mycobacterium* species is also known to attack chlorinated hydrocarbons of anthropogenic origin. This organism degrades trichloroethylene, which is toxic, as do various methanotrophic bacteria and toluene-degrading pseudomonads. Trichloroethylene degradation by each of these organisms occurs via cometabolism. However, significant mineralization usually occurs only in consortia containing these organisms. Pentachlorophenol, another toxic material, is degraded cometabolically by both bacteria (*Flavobacterium, Arthrobacter, Rhodococcus,* and *Pseudomonas*) and fungi (*Phanerochaete* and *Trametes*). Products formed from either chlorinated substrate include CO_2 and Cl^-. *Phanerochaete* has been reported to mineralize a remarkable range of substrates, including pesticides (e.g., DDT, chlordane, lindane), chlorinated dioxins, chloroanilines, and polychlorinated biphenyls (PCBs), as well as PAHs. In all cases, degradation of the hazardous material occurs cometabolically.

2. Hazardous Inorganics

It should not be necessary to point out that it is impossible to destroy a mineral through biochemical means. However, certain carbonaceous compounds, such as cyanide salts and carbon monoxide, are defined as inorganic compounds. The hazardous properties of both are well known. Various fungi such as *Stemphylium* are able to effect the mineralization of cyanide to CO_2, as are certain pseudomonads. Also, bacteria from several genera, known collectively as the carboxydotrophs, are able to convert CO to CO_2. Whether these reactions are more properly categorized as biodegradations or biotransformations is unclear. However, in either case, destruction and/or detoxification of the hazardous material is achieved.

B. Biotransformation

1. Hazardous Organics

In some cases, microbial attack on an organic substrate does not lead to mineralization. Intermediates in the degradative pathway may accumulate. Partial degradation is acceptable as long as these intermediates are less noxious than the original compound. In other words, biotransformation may be very useful in the detoxification of a hazardous waste.

Carbon tetrachloride, a toxic and carcinogenic compound, is partially dechlorinated by *Clostridium* to yield methylene chloride. The latter compound, although toxic, is significantly less so than is the original compound. (*Acetobacter,* an aerobe, is able to mineralize carbon tetrachloride to CO_2.) The fungus *Cunninghamella* attacks PAHs to form incompletely oxidized products. (Mineralization of PAHs was described previously.) These products, which are toxic, are then conjugated with glycosyl residues that are significantly less toxic than the intermediates or the original substrate. It is of particular interest to note that the fungal transformation reactions result in a detoxification, whereas those occurring in mammalian systems cause a deleterious activation of the hazardous compound.

Unfortunately, even in microbial systems, biotransformation can sometimes cause an increase in toxicity. The complete, aerobic degradation of trichloroethylene was previously described. Under anaerobiosis, trichloroethylene is transformed to vinyl chloride, a significantly more toxic compound. Vinyl chloride is stable under those conditions and accumulates. Trinitrotoluene, a toxic, explosive compound, is cometabolized by certain aerobic bacteria, yielding amines via the corresponding nitroso and hydroxylamino compounds. The amines are themselves toxic and carcinogenic. Other aerobic cultures can degrade the substrate only as far as the hydroxylamino compounds, which undergo spontaneous dimerization to form toxic, mutagenic compounds.

2. Hazardous Inorganics

Microbial treatment of hazardous organic compounds is an indirect reflection of the participation of microorganisms in the global carbon cycle. The activities of microorganisms in the biogeochemical cycling of nitrogen, sulfur, manganese, and iron may be similarly exploited. These elements provide energy through dissimilatory oxidation/reduction reactions, which for the purposes of this article represent biotransformations. Other elements may also undergo biotransformation, although energy may not be so derived. The physiological significance of these latter reactions is generally not well understood. However, these reactions may result in detoxification and/or immobilization of the hazardous inorganic.

There are at least two excellent examples of metal biotransformation activities that could form the basis of waste treatment processes. Chromium can exist in the 0, 2^+, 3^+, and 6^+ oxidation states. Hexavalent chromium is soluble, highly toxic, carcinogenic, and corrosive. Trivalent chromium is insoluble and significantly less hazardous. Members of several bacterial genera, including *Pseudomonas, Enterobacter, Bacillus, Flavobacterium,* and *Achromobacter,* have been shown to catalyze the reduction of Cr^{6+} to Cr^{3+}, resulting in a detoxification. Moreover, Cr^{3+} salts precipitate, causing their immobilization *in situ*. Uranium, all of whose isotopes are radioactive, can exist in several oxidation states. However, it is U^{6+} that is most environmentally significant. *Alteromonas,* which is known to catalyze the dissimilatory reduction of Fe^{3+} to Fe^{2+}, can also reduce U^{6+} to U^{4+}. Salts of tetravalent uranium are insoluble and precipitate from solution. *Alteromonas* and other organisms capable of this reaction may be used in a process for the indirect immobilization of this radioactive metal.

The potential for increasing the toxicity of hazardous organic compounds by biotransformation was described previously. There are examples of this phenomenon in the case of hazardous inorganics as well. Inorganic mercury can be methylated anaerobically by sulfate-reducing bacteria such as *Desulfovibrio* and methanogens such as *Methanobacterium*. Certain aerobic organisms, including both bacteria and fungi, are known to catalyze this reactions as well. But this activity is most significant in the context of anaerobic sediments that have been exposed to mercury over long periods. Both mercury salts and methyl mercury are extremely toxic; however, the methylated form, which is water-soluble, accumulates in living tissue and is poorly excreted. This characteristic leads to mercury's bioconcentration within the food chain. Thus, the microbial biotransformation results in a toxification through an increase in bioavailability. (Methylation of arsenic is carried out by the same organisms, to similar effect.) Mercury metal is volatile. Microbial demethylation of the organomercurial and subsequent reduction to mercury metal would result in detoxification of the liquid waste stream via release of mercury to the gaseous phase. Sequential reactions of this type are carried out by *Pseudomonas*. A process based on this activity may be useful in the restoration of anaerobic environments contaminated with mercury, at the expense of discharging mercury to the atmosphere. A similar process could be based on organisms such as *Bacillus* or the coryneform bacteria that are able to reduce mercuric ion to mercury metal. Finally, some organisms capable of methylating mercury are also able to methylate methyl mercury. The product of that activity is dimethyl mercury, which is volatile and thus would escape from the aqueous system. [*See* MERCURY CYCLE.]

C. Biosorption

1. Hazardous Organics

Municipal sewage treatment processes hinge on the concept of treatability (i.e., the effectiveness with which pollutants are removed from the wastewater). This concept incorporates both degradation and physical removal processes. The sorption of pollutants onto biomass is usually overlooked but may make a major contribution toward treatment.

A number of hazardous organic compounds sorb to microbial biomass. Indeed, sorption is a prerequisite for certain biodegradation and biotransformation processes. The fungi *Phanerochaete* and *Trametes* have been shown to sorb toxic, mutagenic chlorolignins as well as various toxic dye compounds. These activities cause a decolorization of waste streams containing these contaminants. The toxic, explosive compound RDX (hexahydro-1,3,5-trinitro-1,3,5-triazine) is also removed by biosorption, although indirectly. RDX is attacked by aerobic bacteria to form recalcitrant amino derivatives that undergo spontaneous conjugation. It is this conjugate that is sorptive to biomass. (A separate set of biotransformations is also possible, yielding noncyclic products that are mutagenic and carcinogenic.) Azo dyes have been shown to undergo a similar fate. It is entirely possible that biosorption is less a property of the organism tested than of the compound itself. As an example, toxic quaternary ammonium compounds are sorbed quite efficiently by biomass from a number of sources. Indeed, even activated charcoal can be biosorbed (by *Mucor, Aspergillus, Fusarium,* and *Penicillium*).

2. Hazardous Inorganics

A great number of metals have been shown to bind to microbial biomass:

Aluminum	Copper	Nickel
Beryllium	Gallium	Plutonium
Cadmium	Germanium	Radium
Calcium	Gold	Silver
Californium	Iron	Strontium
Cerium	Lanthanum	Thorium
Cesium	Lead	Titanium
Chromium	Manganese	Uranium
Cobalt	Neptunium	Zinc

The ability to bind metals is widespread among microorganisms and has been demonstrated in both gram-positive and gram-negative bacteria, fungi, and particularly algae. Indeed, the ability of municipal sludge to bind metal is as well known as its above-mentioned ability to bind organics. Biosorption results in the immobilization of the target metal, as described previously. In some cases, the metal may be removed from the microbial biomass by simple elution with dilute mineral acid, a salt solution, or a chelator. This treatment regenerates the biosorbent while concentrating the hazardous material in a small volume of liquid.

Biosorption may be a particularly useful approach to the management of radioactive waste. The destruction of radionuclides can be accomplished only by natural physical decay processes, which in the case of high-level or transuranic waste occur over trillions of years. (The rule of thumb is that a radioactive waste should be stored for 10 half-lives.) Current plans for management of this waste involve disposal in geological formations. Few acceptable disposal sites have been identified. Biosorptive treatment of contaminated liquid effluents may be useful in the immobilization of the radionuclides and in the reduction of the volume of contaminated material. This approach would facilitate containment of the waste and decrease the amount of space required for their disposal.

IV. Development of Microbial Processes for Treatment of Hazardous Waste

The realization of a microbial treatment technology requires the development of specific processes, based on specific microbial activities, for treatment of a specific waste in a specific context. Process development thus requires extrapolation from the laboratory to the field. These processes may be designed either as "end-of-the-pipe" treatments for waste before its release (i.e., waste management) or for remediation of sites already contaminated with hazardous waste components (i.e., environmental restoration).

A. Identification of Potentially Useful Microorganisms

1. Naturally Occurring Organisms

One of the most sensible approaches to the development of microbially based processes for treatment of hazardous wastes begins in the library. There is a huge amount of information detailing the metabolic capabilities of various microorganisms, particularly in the context of biodegradative activities. Microbial characterization generally includes an assessment of substrate range. An organism capable of degrading a naturally occurring compound that is structurally similar to a particular hazardous organic compound of anthropogenic origin may be an excellent candidate for a treatment process.

Mineralization of an organic compound requires the existence of enzymes that can catalyze a primary attack on the substrate, plus entry into the cell where ultimate degradation occurs. These activities may occur separately or may be temporally linked. Among the bacteria, *Pseudomonas* tends to be the most effective degrader of xenobiotic compounds yet identified. *Pseudomonas* is quite nutritionally adaptable; at least 100 different compounds capable of supporting its growth are known. This versatility implies either a great diversity of catabolic pathways or that the enzymes within these pathways are relatively nonspecific. Similarly, this characteristic indicates a relatively nonspecific transport system for uptake of these compounds. Among the fungi, the white-rot wood-decaying basidiomycetes are the most active degraders of xenobiotics known. The white-rotters, such as *Phanerochaete*, differ from *Pseudomonas* in that they tend to cometabolize organic substrates. This fungus is known to excrete highly nonspecific oxidative enzymes that attack organic substrates, leading in some cases to the assimilation and subsequent mineralization of the enzymatic degradation products. These enzymes are produced only in the presence of a primary substrate. The production of extracellular enzymes circumvents the need for uptake of the target compound before degradative attack.

The metabolic versatility exhibited by *Pseudomonas* and *Phanerochaete* does not and should not imply that these particular organisms are ideally suited for all waste treatment applications requiring biodegradative activity. A ready example focuses on *Stemphylium*, cited previously as a cyanide degrader. This organism is a plant pathogenic fungus. Many plants produce cyanide as a defense against pathogens, some of which have evolved in turn the ability to synthesize formamide hydratase as a detoxification mechanism. Clearly, organisms of this type would be obvious candidates in development of a process for cyanide removal. By extension, plant pathogens able to degrade wyerone, a phytoalexin incorporating triple-bonded carbons, would be reasonable candidates for a process to degrade acetylene or other anthropogenic alkynes.

A similar argument may be used to focus attempts to identify organisms capable of specific biotransformations. Both microbial reduction of mercuric ion and the decomposition of methyl mercury are accomplished by mercury-resistant strains of *Staphylococcus* as well as *Pseudomonas*. Thus, a survey may be based on the simple assumption that organisms able to survive in the presence of certain toxic metals would be those most likely to possess useful transformation activities. The ability to de/methylate mercury serves as a screen for the corresponding arsenic transformations. Such general observations of metal-transforming activity may be useful in the designation of organisms for a specific application. Finally, the ability to reduce ferric iron has served as a screen for the analogous uranium transformation. Examination of organisms that are dependent on dissimilatory metal transformation activities for energy-yielding metabolism (i.e., the chemolithotrophs) may thus be particularly fruitful.

The identification of organisms with putative ability to sorb organics or inorganics should perhaps focus on those species with certain cell surface characteristics, such as a thick cell wall furnishing many ion-exchangeable carboxylates within peptidoglycan (e.g., *Micrococcus*); the presence of chitin, whose nitrogen atom may coordinate certain metals (e.g., most fungi); or copious production of mucilaginous exopolymer (e.g., *Zoogloea*). Chemolithotrophs may be of specific interest, because their viability depends on their ability to bind metals, which is in turn a requisite for uptake and entry into central metabolism. Microbial resistance to metal toxicity is sometimes accomplished by immobilization of the metal on either extracellular or intracellular binding sites. As mentioned previously, metal-resistant organisms may be of great interest, here for the development of biosorption processes.

The preceding examples illustrate the benefit of exploiting organisms acclimated to the destruction, detoxification, or immobilization of a particular hazardous waste component or its surrogate. Such organisms may be obtained either through enrichment culture or by environmental sampling.

Sources in the United States for cultures to be enriched in the laboratory include the American Type Culture Collection, the U.S. Department of Agriculture's (USDA) National Center for Agricultural Utilization Research (formerly the Northern Regional Research Laboratory), the USDA Center for Forest Mycology, the U.S. Department of Health and Human Services' National Center for Toxicological Research, and university or industrial culture collections. Municipal sludge should not be overlooked as a potential source for organisms for hazardous waste treatment, not only because of household use of hazardous materials but also because of the great diversity and versatility of organisms therein. [*See* CULTURE COLLECTION, FUNCTIONS.]

Environmental sampling is an excellent—and perhaps the best—approach to obtaining useful cultures. Sampling should be carried out in areas adjacent to active and former industrial, agricultural, or mining sites whenever practicable and legal. Industrial landfills and land treatment locations may also yield useful organisms. In the context of environmental restoration, indigenous microflora may be those best suited for the desired application.

Process design should incorporate the use of mixed cultures (i.e., consortia) where necessary to achieve the desired performance goal (extent of cleanup). Indeed, this approach is frequently required to achieve complete mineralization of hazardous organic compounds. Similarly, thorough treatment of organometallic complexes such as those cyanide-containing wastes released by electroplating processes [e.g., tetracyanonickelate, $Ni(CN)_4^{2-}$] would require a culture capable of assimilating the cyanide and binding the nickel. Two or more members of a microbial consortium may be required to achieve this goal. Process design should also consider the waste's physical and chemical characteristics. For example, treatment of solid wastes may be best accomplished by mycelial organisms able to colonize the substrate. Treatment of acidic waste streams may suggest the use of acidophiles to eliminate a pretreatment step (chemical neutralization of the waste).

2. Genetically Engineered Organisms

The aforementioned metabolic diversity of *Pseudomonas* may reflect, at least in part, a remarkable genetic plasticity. Exchange of chromosomal elements among these organisms is well known. For this and other reasons, *Pseudomonas* is a favored organism for genetic manipulation, including the development of recombinants with enhanced biodegradative, biotransformative, or biosorptive activity. Other organisms, including but not limited to *Neurospora*, are being similarly engineered. Amplification of desirable genes present within in naturally occurring organisms is also being considered. Genetically engineered organisms may be extremely useful in processes for hazardous waste treatment. However, there are currently regulatory, legal, and sociopolitical barriers to their use. Also, a practical bioprocess (particularly one for biorestoration) would be carried out without benefit of sterilization. The robustness of the recombinants, meaning their ability to compete with organisms already present in the waste matrix, would be critical. The general ability of recombinants to survive in mixed cultures of this type is under assessment. [*See* GENETICALLY ENGINEERED MICROORGANISMS, ENVIRONMENTAL INTRODUCTION; GENETICALLY MODIFIED ORGANISMS: GUIDELINES AND REGULATIONS FOR RESEARCH.]

In light of these potential drawbacks to the use of recombinant-based processes, it is fortunate that the desired biochemical activities may often be enhanced by physiological manipulation. These enhancements would by necessity be based on a thorough understanding of culture's metabolic activities and ecological relations.

B. Application of Candidate Organisms

Process design incorporates a basic decision as to whether the microbial culture will be used within bioreactors or *in situ*. Waste management processes would almost certainly be carried out in bioreactors. Environmental restoration technologies based on processes carried out *in situ* are described elsewhere. However, bioreactors may be relevant to biorestoration in the form of pump-and-treat processes for remediation of groundwater or surface water or in soil washing processes. Ideally, the bioreactor would be portable for use at different sites or in different areas of a particular site. A rule of thumb defines portability in this context as follows: A complete bioreactor system for use in environmental restoration must measure $10 \times 12 \times 60$ ft, the size of a railroad boxcar, if it is to be used at multiple sites. [*See* BIOREACTORS.]

Microbial processes for hazardous waste treatment may be carried out in any of the suspended-growth, fixed-film, or slurry bioreactor systems described in Section II.C. A newer technology, cell immobilization, is now being applied to process development. Immobilization is similar to fixed-film processes in that it involves the attachment of cells to solid support materials. Fixed-film processes use only the surface of the support. However, immobilization of cells within hydrous gel beads permits maximal use of the support. There are several advantages to use of immobilized cells:

- Achieves high cell concentrations within bioreactor
- Establishes a controlled, favorable microenvironment for cells
- Prevents loss of cell from bioreactor
- Facilitates handling of cells
- Results in uniformity of particles
- Permits use of continuous or semicontinuous systems

It is likely that this approach will increase in popularity.

C. Real-World Considerations

1. Technical Issues

The waste stream that must be treated is rarely homogeneous or benevolent. The composition of industrial waste streams usually varies as a function of time, resulting in varying concentrations of the targeted waste component. Moreover, there is considerable potential for chemical interactions between components of the waste stream (e.g., the formation of organometallic complexes) that may alter the fundamental character of the targeted component. Finally, the waste components may themselves be toxic to the desired microbial activity.

Biodegradation processes require the use of living organisms. This requirement necessitates in turn that nutrients be supplied and that the process be operated under physiological conditions of temperature, pH, illumination, oxidation/reduction potential, etc. The general absence of biodegradative activity in landfills illustrates problems arising when these factors are not considered. Conflicting growth requirements may exist for processes employing more than one organism. Aerobic processes introduce the need for good oxygen mass transfer and all its accompanying problems. If nutrients are added, unused nutrients and metabolic products must be removed from the waste stream. Metabolic products may also interfere with the desired activity. Certain biosorption or (less likely) biotransformation processes may not require living cells. However, systems using nonliving biomass are not self-replenishing, and biodegradation of hazardous compounds also present in the waste would be impossible under these conditions.

The working life span of the bioprocess (i.e., its period of reliable operation) is a major determinant of process feasibility. It may be difficult to maintain living biomass for long periods, and autolysis may be an issue even in the case of nonliving biomass. The presence of competing microflora was mentioned earlier in the context of genetically engineered organisms but is equally relevant in the case of natural isolates. Even if stable coexistence was achieved, the contaminating organisms might excrete metabolic products that inhibit growth or activity of the desired culture.

Developmental work may result in the design of a microbial process that is effective under all these deleterious conditions. However, it is a maxim of applied microbiology that processes demonstrated in a test tube rarely perform exactly the same way when scaled up to 100-liter reactors. The scale-up issue may be expected to be a problem in the development of workable processes for hazardous waste treatment.

2. Competitiveness with Other Treatment Technologies

The list of nonbiological treatment technologies presented earlier represents the competition faced by new microbially based processes. This competition is based on cost, effectiveness, safety, and speed. The bioprocess must be based on cultures that are readily obtained and inexpensively produced, and it must be designed to minimize nutrient and co-factor requirements, simplify operating conditions, maximize life span, and eliminate complexity of operation. The bioprocess must be designed to achieve the stated treatment goals, or performance criteria. The bioprocess must use nonpathogenic organisms and not generate an environmentally unacceptable effluent (cf. examples of toxification through biotransformation). Finally, it must require a reasonably short residence time with the waste to be treated. In some cases, particularly where a treatment technology is lacking, bioprocesses may be quickly and gratefully adopted. But some specific biological processes will never supplant the role of conventional nonbiological treatment processes. However, as performance criteria become stricter, these latter bioprocesses may be used in conjunction with other unit processes, either as polishing steps to be implemented at the very end of the pipe, or as a separate process to treat the waste generated by nonbiological treatment.

3. Psychological Issues

There is inherent psychological resistance to any new technology. Fortunately, microbial processes for hazardous waste treatment may be viewed as "natural." It is critical, however, that those processes use nonpathogenic organisms. The use of nongenetically engineered organisms may also eliminate potential barriers (including regulatory ones). The greatest barrier may be in the mind of the potential user (i.e., the operator of the plant where the bioprocess would be introduced). Processes requiring minimal operator attention would probably be favored. Also, processes similar to those currently in use would be viewed as presenting minimal risk. Thus, it would be useful to develop a psychological link between the proposed bioprocess and those for treatment of household waste, or between the proposed process and nonbiological treatments such as precipitation.

V. Summary and Future Directions

The implementation of microbial technologies for treatment of hazardous waste requires research [i.e., the pursuit of basic (fundamental) or applied (specific) knowledge through scientific investigation]; development (i.e., small-scale experimentation and testing of equipment and systems designed to embody that new knowledge); and demonstration (i.e., engineered proof-of-principle in a real-world setting to show that the technology works as designed). Testing and evaluation of the technologies then follow. Technology implementation will be achieved only through the coupling of applied microbiological expertise with chemical engineering and/or environmental engineering skills.

The changing regulatory, legal, and sociopolitical environment suggests that microbial technologies for hazardous waste treatment will become more significant. New challenges, such as a demand for processes for treatment of gaseous oxides, will be presented. New, remote environments, such as deep thermal vents, will be sampled for unusual organisms, and new symbiotic associations, such as protozoan bacterial symbionts, will be discovered. As these cultures are characterized, potential applications of their novel metabolic capabilities will become apparent. Sophisticated methods for physiological manipulation of microorganisms will become possible. Advances in molecular genetics will permit further strain improvement. In short, there will be a multitude of opportunities for microbial ecologists, microbial physiologists, microbial biochemists, and molecular biologists to contribute to the development of microbial technologies for hazardous waste treatment.

Bibliography

Bond, R. G., and Straub, C. P. (1974). "Handbook of Environmental Control," 4 vols. CRC Press, Boca Raton, Florida.

Hobson, P. N., and Poole, N. J. (1988). Water pollution and its prevention. *In* "Micro-organisms in Action: Concepts and Applications in Microbial Ecology" (J. M. Lynch and J. E. Hobbie, eds.), pp. 302–321. Blackwell Scientific Publications, Oxford.

Leahy, J. G., and Colwell, R. R. (1990). *Microbiol. Rev.* **54,** 305–315.

Lewandowski, G., Armenante, P., and Baltzis, B., eds. (1988). "Biotechnology Applications in Hazardous Waste Treat ment." United Engineering Trustees, Inc., New York. 423 pp.

Macaskie, L. E., and A. C. R. Dean. (1989). *Adv. Biotechnol. Proc.* **12,** 159–202.

Neilson, A. H. (1990). *J. Appl. Bacteriol.* **69,** 445–470.

Sayler, G. S., Fox, R., and Blackburn, J. W., eds. (1991). "Environmental Biotechnology for Waste Treatment." Plenum Press, New York. 288 pp.

Schumacher, A. (1988). "A Guide to Hazardous Materials Management." Quorum Books, New York. 288 pp.

Weightman, A. J., and J. H. Slater. (1991). The problem of xenobiotics and recalcitrance. *In* "Micro-organisms in Action: Concepts and Applications in Microbial Ecology" (J. M. Lynch and J. E. Hobbie, eds.), pp. 322–348. Blackwell Scientific Publications, Oxford.

Wise, D. L. (1988). "Biotreatment Systems." CRC Press, Inc. Boca Raton. 3 vols. Hazardous Waste Treatment, Microbial Technologies.

Heavy Metal Pollutants: Environmental and Biotechnological Aspects

Geoffrey M. Gadd
University of Dundee

Glossary

Biosorption Removal of metal or metalloid species, compounds and particulates, radionuclides, and organometal(loid) compounds from solution by physicochemical interactions with biological material

Desorption Nondestructive recovery of, for example, metal or metalloid species, radionuclides, and organometal(loid) compounds from loaded biological material by physicochemical treatment(s)

Heavy metals Ill-defined group of biologically essential and inessential metallic elements, generally of density >5, exhibiting diverse physical, chemical, and biological properties with the potential to exert toxic effects on microorganisms and other life forms

Metal resistance Ability of a microorganism to survive toxic effects of heavy metal exposure by means of a detoxification mechanism usually produced in response to the metal species concerned

Metal tolerance Ability of a microorganism to survive toxic effects of heavy metal exposure because of intrinsic properties and/or environmental modification of toxicity

Metallothioneins Low-molecular-weight cysteine-rich proteins capable of binding essential metals (e.g., Cu and Zn), as well as inessential metals (e.g., Cd)

Organometallic compound Compound containing at least one metal-carbon bond, often exhibiting enhanced microbial toxicity. When such compounds contain "metalloid" elements (e.g., Ge, As, Se, and Te), the term "organometalloid" may be used

Phytochelatins Metal-binding cysteine-containing γ-glutamyl peptides of formula $(\gamma\text{Glu-Cys})_n$-Gly (n generally 2–5)

Siderophores Low-molecular-weight Fe^{3+} coordination compounds excreted by microorganisms to enable accumulation of iron from the external environment

HEAVY METALS comprise an ill-defined group of approximately 65 metallic elements, of density greater than 5, with diverse physical, chemical, and biological properties but generally having the ability to exert toxic effects toward microorganisms. Metals may be chemically classified according to their behavior as "hard" or "soft" acids, as well as other properties (e.g., ionic radii and complexing power). Hard acids (e.g., Na^+, K^+, Mg^{2+}) are generally small, exhibit high electronegativity and low polarizability, and tend to participate in electrostatic bonding with ligands. Soft acids (e.g., Ag^+, Cd^{2+}, Hg^{2+}) have low electronegativity and high polarizability and tend to participate in covalent bonding to ligands. Such definitions, as well as those based on equilibrium constants for metal ion-ligand complexes, may be used in certain contexts to describe metal–microbe interactions. Many essential metals (e.g., Fe^{3+}, Ca^{2+}, Mg^{2+}, K^+) are hard acids, whereas inessential toxic metals (e.g., Ag^+, Cd^{2+}, Hg^{2+}, Sn^{2+}) are soft acids. However, biologically inessential metals are also found among hard acids (e.g., Rb^+, Sr^{2+}, Al^{3+}), whereas a borderline metal category includes important essential metal ions (e.g., Cu^{2+}, Ni^{2+}, Zn^{2+}, Co^{2+}). Other definitions are based on toxicity and environmental impact, although toxicity is highly dependent on the particular element, speciation, concentration, and environmental parameters; environmental impact may further vary with anthropogenic activities. Many metals are essential for microbial growth and metabolism at low concentrations (e.g., Cu, Fe, Zn, Co, Mn), yet are toxic in excess, whereas both essential and nonessential metal ions may be accumulated by microbial cells by physicochemical and biological mechanisms. In this article, the term "heavy metal" will be used in a broad context and, as well as common environmental metal pollutants, will include actinides and organometal(loid) compounds. All these substances have a common potential for microbial toxicity and bioaccumulation and are of environmental significance either as waste pollutants or because of deliberate or accidental introduction as biocides, preservatives, or other products.

I. Environmental Aspects of Heavy Metal Pollution

A. Heavy Metals in the Environment

Although elevated levels of toxic heavy metals occur in some natural locations (e.g., volcanic soils, hot springs, and deep-sea vents), average environmental abundances are generally low, with most of that immobilized in sediments and ores being biologically unavailable. However, ore mining and processing, as well as a multiplicity of other industrial activities, have disrupted natural biogeochemical cycles, and there is increased atmospheric release as well as deposition into aquatic and terrestrial environments. Major sources of pollution include the combustion of fossil fuels, mineral mining and processing, nuclear and other industrial effluents and sludges, brewery and distillery wastes, biocides, and preservatives including organometallic compounds. It should be stressed that almost every industrial activity can lead to altered mobilization and distribution of heavy metals in the environment, and at local sites of pollution, effects on the biota may be severe. Microorganisms may need to respond using a variety of mechanisms that ensure survival.

B. Effects of Heavy Metals on Microbial Populations

For toxicity to occur, heavy metals must directly interact with microbial cells and/or indirectly affect growth and metabolism by interfering with uptake of an essential nutrient or by altering the physicochemical environment of the cell. A variety of nonspecific and specific mechanisms (e.g., cell wall composition and transport, respectively) determine metal entry into cells, and unless mechanism(s) for detoxification are possessed, cell death will ultimately result. Because of the wide range of metal-binding ligands found in the structural and biochemical components of living cells, many toxic interactions are possible. Almost every index of microbial metabolism and activity can be adversely affected by elevated concentrations of heavy metals. These include primary productivity, methanogenesis, nitrogen fixation, respiration, motility, biogeochemical cycling of C, N, S, P, and other elements, organic matter decomposition, enzyme synthesis and activity in soils, sediments, and waters. Because of the fundamental microbial involvement in such ecological processes, as well as in plant growth and symbiotic associations, heavy metal pollution can have severe short- and long-term effects and ultimately pose a threat to higher organisms, including humans, by bioaccumulation and transfer through food chains.

Despite this, the ability of microorganisms to survive and grow in the presence of heavy metals is a frequent phenomenon, and microbes from all major groups can be isolated from polluted habitats. However, generalizations about heavy metal effects on natural populations are difficult to make because of the diversity in chemical behavior, and toxicity of given metal species, the complex interactions that can occur between metal species and environmental components, and the morphological and metabolic diversity encountered in microorganisms. Furthermore, other environmental perturbations frequently associated with industrial heavy metal pollution (e.g., extremes of pH, salinity, nutrient limitations) may also affect microbial populations and activity *in situ* independent of effects caused by the metals present. Nevertheless, it is commonly stated that heavy metals may affect natural microbial populations by reducing numbers and species diversity and selecting for a "resistant" or "tolerant" population. Resistance and tolerance are arbitrarily defined terms, frequently interchangeable, and often based on whether given isolates can grow in the presence of selected heavy metal concentrations in laboratory media. It is probably more appropriate to use "resistance" to describe a direct mechanism resulting from heavy metal exposure (e.g., bacterial reduction of Hg^{2+} to Hg^0 and the synthesis of metallothioneins and/or γ-glutamyl peptides by yeasts). "Tolerance" may rely on intrinsic biochemical and structural properties of the host such as possession of impermeable cell walls, extracellular slime layers or polysaccharide, and metabolite excretion, as well as environmental modification of toxicity. However, distinctions are difficult in many cases because several direct and indirect mechanisms, both physicochemical and biological, may contribute to microbial survival and either or both these terms may be used.

Isolation of resistant bacteria, algae, and fungi from polluted environments is frequently described, although there may be differential responses between microbial groups and species. Such changes can affect interpretation of data based solely on numerical estimates. In general, gram-negative bacteria appear less sensitive to heavy metal pollution than gram-positive bacteria, contributory factors being cell structure as well as physiological and genetical mechanisms. On phylloplanes, bacteria may be more sensitive to aerially deposited heavy metals than fungi (e.g., *Aureobasidium pullulans*), although such observations may be complicated by the effects of other atmospheric pollutants (e.g., SO_2). In soils, fungal populations may not be significantly different at control or polluted sites and resistant/tolerant species can be obtained from both, although this may depend on the nature and extent of metal pollution. In severely polluted conditions, fungal species diversity may dramatically decrease with some species exhibiting extreme levels of tolerance (e.g., *Penicllium waksmanii, Penicillium ochro-chloron,* and *Paecilomyces* sp.). However, such tolerance may be dependent on highly acidic conditions, and sensitivity may be displayed at pH values approaching neutrality. In conclusion, although heavy metal pollution can qualitatively and quantitatively affect microbial populations in the environment, it may be difficult to distinguish heavy metal effects from those of environmental components, environmental influence on metal toxicity, and the nature of the microbial resistance/tolerance mechanisms involved. [*See* HEAVY METALS, BACTERIAL RESISTANCES.]

C. Environmental Modification of Heavy Metal Toxicity

The physicochemical characteristics of a given environment determine the chemical speciation and biological availability of heavy metals. Because most major mechanisms of heavy metal toxicity are a consequence of their strong coordinating properties, a reduction in bioavailability may reduce toxicity and enhance microbial survival. Such parameters as pH, temperature, aeration, soluble and particulate organic matter, clay minerals, and salinity can all influence heavy metal speciation, mobility, and toxicity.

The effects of pH are complex, and depending on the system examined, toxicity may be increased or decreased or be unaffected. Acidic conditions may increase metal availability, although H^+ may successfully compete with and reduce or prevent binding and transport into microbial cells. With increasing pH, there may be enhanced entry of metal cations into cells as well as the formation of hydroxides, oxides, and carbonates of varying solubility and toxicity. Some hydroxylated species may associate more efficiently with microbial cells than the corresponding metal cations, and increased accumulation and/or toxicity may therefore result at elevated pH values with certain metals. In other cases, a reduction in availability leads to a reduction in toxicity. Environmental pH also affects the rate and extent of metal complexation with organic components and inorganic anions (e.g., Cl^-). A reduction in toxicity in the presence of elevated concentrations of anions such as Cl^-, CO_3^{2-} and PO_4^{3-} is frequently observed, whereas the formation of insoluble nontoxic metal sulfides occurs in reducing environments (negative E_h) as a result of microbial H_2S production.

Mono-, di-, and multivalent cations may affect heavy metal toxicity by competing with binding sites on the cells as well as transport systems for intracellular accumulation. Clay minerals can bind heavy metal cations and reduce bioavailability, a high cation exchange capacity (CEC) being most effective in the amelioration of toxicity. Synthetic and naturally produced soluble and particulate organic substances may influence heavy metal toxicity by binding and complexation. Some organic materials may be produced by microorganisms as a result of "normal" metabolism or directly/indirectly in response to the presence of toxic heavy metals. Such behavior may enhance microbial survival. The removal of heavy metal species by intact living and dead microbial biomass, as a result of physicochemical and/or metabolic interactions, may also be a significant process in some locations. In a more general sense, microbial growth and metabolism is influenced by environmental parameters, including nutrient availability, and this may affect physiological, biochemical, and genetical responses to potentially toxic heavy metals.

D. Mechanisms of Microbial Heavy Metal Accumulation

Heavy metals, radionuclides, and organometal(loid) compounds can be accumulated by microbial cells by a variety of processes, both physicochemical and biological (Table I). Metabolism-independent binding or adsorption ("biosorption") to living or dead cells, extracellular polysaccharides, capsules, and slime layers is frequently rapid. Bacterial cell walls and envelopes, and the walls of algae, fungi, and yeasts, are efficient metal biosorbents with binding to charged groups frequently being followed by inorganic deposition of increased amounts of metal. These processes are influenced by environmental parameters, particularly pH and ion competition, as well as certain aspects of metabolism that may alter the microenvironment in the vicinity of the cell. Microbial cells can act as nuclei for the formation of crystalline metal deposits including phosphates, sulfides, and oxides. Crystallization of elemental gold and silver may occur as a result of reduction, whereas for uranium and thorium, formation of hydrolysis products can enhance precipitation in and around cell walls. These mechanisms can result in high levels of metal accumulation. Variations in the chemical behavior of metal species as well as the composition of microbial cell walls and extracellular materials can lead to wide differences in biosorptive capacities between different species.

Metabolism-dependent intracellular uptake is inhibited by low temperatures, metabolic inhibitors, or the absence of an energy source. Microbial transport systems are of varying specificity and essential and nonessential metal(loid) species may be taken up. Rates of uptake can depend on the physiological state of cells, as well as the nature and composition of the environment or growth medium. Integral to the transport of metal ions into cells are electrochemical gradients of, for example, H^+ and K^+, across membranes resulting from the operation of enzymatic pumps (ATPases) that transform the chemical energy of ATP into this form of biological energy. ATPases are also involved in ion efflux in a variety of organisms and organellar ion compartmentation in eukaryotes via operation across vacuolar membranes. With toxic heavy metals, permeabilization of cell membranes can result in further exposure of intracellular metal-binding sites and increase passive accumulation. Intracellular uptake may ultimately result in death of sensitive organisms unless a means of detoxification is induced or already possessed. [*See* ATPases and Ion Currents.]

Other mechanisms of microbial metal accumulation include iron-binding siderophores and the cotransport of metals with organic substrates (e.g.,

Table I Selected Examples of Microbial Heavy Metal and Actinide Accumulation to Industrially Significant Levels

Organism	Element	Uptake (% dry weight)
Bacteria		
Streptomyces sp.	Uranium	2–14
S. viridochromogenes[a]	Uranium	30
Thiobacillus ferrooxidans	Silver	25
Bacillus cereus	Cadmium	4–9
Zoogloea sp.	Cobalt	25
	Copper	34
	Nickel	13
Citrobacter sp.	Lead[b]	34–40
	Cadmium[b]	170
	Uranium[c]	900
Pseudomonas aeruginosa	Uranium	15
Mixed culture	Silver	32
Algae		
Chlorella vulgaris[a]	Gold	10
Chlorella regularis[a]	Uranium	15
Phoma sp.	Silver	2
Penicillium sp.	Uranium	8–17
Rhizopus arrhizus	Cadmium	3
	Lead	10
	Uranium	20
	Thorium	19
	Silver	5
	Mercury	6
Aspergillus niger	Thorium	19
	Uranium	22
Yeasts		
Saccharomyces cerevisiae	Uranium	10–15
	Thorium	12

[a] Immobilized cells.
[b] Phosphatase-mediated metal removal.
[c] Immobilized cells: phosphatase-mediated uranium precipitation as UO_2HPO_4.

germanium transport into *Pseudomonas putida* via a catechol transport system as a Ge-catechol complex).

E. Mechanisms of Microbial Heavy Metal Detoxification

Extracellular complexation, precipitation, and crystallization of heavy metals can result in detoxification. Polysaccharides, organic acids, pigments, proteins, and other metabolites can all remove metal ions from solution and/or convert them to less toxic species. Iron-chelating siderophores may chelate other metals and radionuclides and possibly reduce toxic effects. The production of H_2S by microorganisms can result in the formation of insoluble metal sulfides, which are deposited in and around cell surfaces in granular form. Anaerobic H_2S production (e.g., by *Desulphovibrio*) may also result in disproportionation of organometallic compounds to volatile products as well as insoluble sulfides:

$$2\,CH_3Hg^+ + H_2S \rightarrow (CH_3)_2Hg + HgS$$

$$2\,(CH_3)_3Pb^+ + H_2S \rightarrow (CH_3)_4Pb + (CH_3)_2PbS$$

Many other examples of metal crystallization and precipitation around microbial cells are known, mediated by processes dependent and independent of metabolism. Some of these are of great importance in biogeochemical cycles (e.g., microfossil formation, iron and manganese deposition, and silver and uranium mineralization).

Decreased accumulation, sometimes the result of efflux, and impermeability may be important survival mechanisms. Impermeability may be a consequence of cell wall structure and composition, changes in membrane proteins, lack of a transport mechanism, or increased turgor pressure. Plasmid-mediated bacterial resistance to Cd and As depends on efflux. Decreased heavy metal uptake has been observed in many resistant microbes, including bacteria, algae, fungi, and yeasts, although this is also dependent on environmental factors including pH and ion competition. Some resistant strains accumulate more metal than sensitive parental strains because of more efficient internal detoxification.

Inside cells, metal ions may be detoxified by chemical components, which include metal-binding proteins, or compartmentalized into specific organelles. Metal-sequestering granular material, including cyanophycin granules (in cyanobacteria) and polyphosphate, has been implicated in bacteria, cyanobacteria, algae, and fungi. Metal-binding proteins, including metallothioneins and metal γ-glutamyl peptides (phytochelatins), have been detected in all microbial groups examined. Metallothioneins are small, cysteine-rich polypeptides that can bind essential metals (e.g., Cu and Zn), in addition to nonessential metals (e.g., Cd). Metal γ-glutamyl peptides are short, cysteine-containing peptides of general formula $(\gamma\text{Glu-Cys})_n$-Gly. Peptides of $n = 2$ to 5 are most common, and these are important detoxification mechanisms in algae as well as several fungi and yeasts. In eukaryotic microorganisms, metal ions (e.g., Co, Zn, Mn) may preferentially accumulate in vacuoles, the vacuolar membrane (tonoplast) possessing transport systems for their accumulation from the cytosol.

Chemical transformations of heavy metal species by microorganisms may also constitute detoxification mechanisms. Plasmid-mediated bacterial Hg^{2+} reduction to Hg^0 is dealt with elsewhere. In addition to this, other examples of reduction are carried out by bacteria, algae, and fungi (e.g., Au^{3+} to Au^0, Ag^+ to Ag^0, and Cr^{6+} to Cr^{3+}). Methylation may be viewed as a detoxification mechanism for mercury and certain other metals and metalloids, because methylated species are usually volatile and lost from a given environment. Methylation of Hg^{2+}, by direct and indirect microbial action, involves methylcobalamin (CH_3CoB_{12}; vitamin B_{12}) with the formation of two products, CH_3Hg^+ and $(CH_3)_2Hg$. Methylcobalamin may also be involved in methylation of lead, tin, palladium, platinum, gold, and thallium, whereas S-adenosylmethionine (SAM) is the methylating agent, by transfer of carbonium ions (CH_3^+), for arsenic and selenium. Arsenic and selenium methylation appear to be significant processes in polluted environments.

Organometallic compounds may be detoxified by sequential removal of alkyl or aryl groups. Organomercurials can be degraded by organomercurial lyase, whereas organotin detoxification involves sequential removal of organic groups from the tin atom. [*See* HEAVY METALS, BACTERIAL RESISTANCES.]

$$R_4Sn \rightarrow R_3SnX \rightarrow R_2SnX_2 \rightarrow RSnX_3 \rightarrow SnX_4$$

It should be stressed that abiotic mechanisms of metal methylation and organometal(loid) degradation also contribute to their transformation and redistribution in aquatic, terrestrial, and aerial environments. The relative importance of biotic and abiotic mechanisms is often difficult to discern.

II. Biotechnological Aspects of Heavy Metal Pollution

A. Biotechnology of Microbial Heavy Metal Accumulation

The microbial removal of potentially toxic and/or valuable metal and metalloid species, radionuclides, organometal(loid)s, and metal particulates from aqueous solution can result in detoxification and safe environmental discharge. Subsequent treatment of loaded biomass can enable recovery of accumulated elements and/or containment of highly toxic/radioactive species. Biotechnological development of microbial systems may provide an alternative or adjunct to conventional physicochemical treatment methods for contaminated effluents and wastewaters. Growing evidence suggests that some biomass-related processes are economically competitive with existing treatments in mining and metallurgy.

1. Living Organisms

Living microorganisms may be subject to heavy metal toxicity unless growth is separated from the metal-contacting phase or resistant strains are used. However, some resistant strains exhibit reduced accumulation and/or may transform given metal species into others of altered volatility and toxicity by, for example, oxidation, reduction, or methylation. Intracellular uptake may result in irreversible metal sequestration by, for example, metal-binding proteins or organellar localization, which may necessitate destructive recovery. Most living cell systems exploited to date have been used for decontamination of effluents containing metals at sublethal concentrations. These may employ a mixture of microorganisms, as well as higher plants.

Blooms of algae and cyanobacteria, which may be encouraged by addition of sewage effluent, can reduce levels of Cu, Cd, Pb, Ni, Zn, Hg, and Fe in contaminated waters. ''Meander'' systems allow effluents to pass through engineered channels that contain cyanobacteria, algae, and higher plants including *Potomogeton* and *Typha*. Metals are removed by entrapment of particulates and general biosorption with high efficiency. Eventual decomposition of dead biomass in sediments can eventually result in H_2S production by sulfate-reducing bacteria (e.g., *Desulphovibrio* and/or *Desulphotomaculum* species). This precipitates heavy metals as sulfides (e.g., ZnS, CdS, CuS, and FeS), which is virtually an irreversible process if reducing conditions are maintained.

Naturally occurring and artificial bogs have potential for the treatment of acid mine drainage. These rely on a complex array of microorganisms and plants including mosses and sedges. Several removal mechanisms are involved, which range from cation exchange by wall constituents of mosses to precipitation by sulfide released from sulfate-reducing bacteria under anaerobiosis or bacterial oxidation and precipitation in aerobic conditions.

The most widely used living cell system capable of heavy metal removal from discharged effluents is sewage treatment. In most treatments, the primary sedimentation step removes up to 40% to 60% of total metals present; the remainder may pass into the biological treatment system, usually an activated sludge or a trickling filter system. Metals are predominantly bound by bacterial extracellular polysaccharides.

2. Immobilized Microbial Biomass

Immobilized biomass particles appear of greater potential in packed- or fluidized-bed reactors than freely suspended microbial cells, with advantages including better capability of regeneration and recirculation, easy separation of biomass and effluent, and minimal clogging under continuous flow. Living and dead cells, as well as derived products, can be used in immobilized forms, including biofilms, and many options have been explored for decontamination of effluents.

a. Immobilized Living Cell Biofilms

Although metal toxicity and/or other adverse properties of waste effluents may be problematical, living cell systems have potential for a long-term process if there is continual biomass replenishment. Furthermore, mechanisms only expressed by living cells may be exploited.

The method most often applied to metal removal is the use of cells grown as a biofilm on inert supports. Ideal supports are of large surface area but sufficiently porous to enable high effluent flow rates and reduce blocking. Support materials include those with planar surfaces (e.g., glass, metal sheets, plastics), uneven surfaces (e.g., wood shavings, clays, sand, crushed rock, coke), and porous materials (e.g., foams, sponges). Many of these materials have been used in a variety of reactor types including rotating biological contactors, trickle filters, fixed and fluidized beds, and air-lift bioreactors.

Living cell biofilms may provide a further capacity for the removal or degradation of other pollutants including hydrocarbons, pesticides, and nitrates. [*See* BIOFILMS AND BIOFOULING.]

b. Immobilized Particulate Biomass

Living and dead microoganisms from all the major groups have been used successfully as particulate forms for heavy metal removal. Immobilized biosorbent particles should ideally contain a minimal amount of the required binding or encapsulating agents and be of the size similar to that of other commercial adsorbents (i.e., 0.5 to 1.5 mm). Important properties are particle strength, high porosity, hydrophilicity, and resistance to harmful chemicals. A variety of support materials can be used to immobilize microbial cells including agar, cellulose, alginates, polyacrylamide, toluene diisocyanate, glutaraldehyde (cross-linking reagent), and silica gel. Biomass may be used in its "natural state" or modified by physical or chemical treatment (e.g., alkali treatment) to improve biosorption efficiency. Reactor systems employing granulated *Bacillus* are used in the AMT-Bioclaim process (Advanced Mineral Technologies, Inc., now VistaTech Partnership Ltd, Salt Lake City, Utah). A fixed-bed reactor containing approximately 20 kg of biosorbent is used for small flows ($<15\ l\ min^{-1}$), whereas a larger fluidized, pulsed-bed system containing approximately 80 to 90 kg of biomass is used for larger flows ($>35\ l\ min^{-1}$). In the latter, loaded dense granules sink to the bed bottom enabling addition of fresh biosorbent granules. Metals are removed from the biomass using, for example, sulfuric acid, sodium hydroxide, or complexing agents and recovered using, for example, electrowinning. Regeneration of granules may be achieved by alkali treatment.

Biosorption of uranium and other elements by immobilized fungi and algae has also received attention. Immobilized particles of *Rhizopus arrhizus* (~0.7–1.3-mm diameter), containing about 12% to 23% added polymer, can achieve complete uranium removal from dilute uranium ore bioleaching solutions (<300 mg $liter^{-1}$) with eluate concentrations after desorption being greater than 5000 mg $liter^{-1}$. Full loading capacity (~50 mg U g^{-1}) can be maintained over multiple biosorption–desorption cycles. Fluidized beds of alginate- and polyacrylamide-immobilized microalgae (e.g., *Chlorella* sp.) have been used to remove a variety of metals, including Cu^{2+}, Pb^{2+}, Zn^{2+}, and Au^{3+}, from mixtures, and several selective recovery schemes have been devised. AlgaSORB (Bio-Recovery Systems, Inc., Las Cruces, New Mexico) contains algal biomass immobilized in a silica matrix (40- to 100-mesh size) and can be used in batch or column systems for metal biosorption. After metal recovery, the regenerated biomass retains approximately 90% of the original metal uptake efficiency even after prolonged use (>18 months).

c. Immobilized Metal-Binding Compounds

Immobilization of low-molecular-weight metal-binding compounds requires association with a carrier material to provide the size and integrity appropriate for practical use. Biosorbent molecules are covalently attached to functional groups on the carrier materials, either directly or by means of an intermediate linkage or spacer group. The types of carriers available for immobilization of metal-binding ligands include natural and synthetic organic substances (e.g., dextran, cellulose, agar, alginate, polyacrylamide, polypropylene) and inorganic substances [e.g., controlled pore glass, silica gel, natural and synthetic zeolites (aluminosilicates)]. Selection for use depends on such factors as surface area, effective pore size and volume, particle size, mechanical strength, density, ease of preparation, and cost. A limited number of microbial metal-binding molecules have been immobilized to functionalized carriers [e.g., the siderophore desferrioxamine (Desferal)].

3. Growth-Decoupled Enzymatic Removal of Metals

Accumulation of heavy metals by growth-decoupled "resting cells" of a *Citrobacter* sp. is mediated by a surface-located acid-type phosphatase enzyme that releases HPO_4^{2-} from a supplied substrate (e.g., glycerol 2-phosphate) and precipitates divalent cations (M^{2+}) as $MHPO_4$ at cell surfaces. The process is nonspecific and depends on the insolubility of the particular metal phosphate involved; Cd^{2+}, Pb^{2+}, Cu^{2+}, and UO_2^{2+} can be precipitated singly or in combination. The *Citrobacter* process has potential as a long-term biofilm-containing filter for achievement of a high-metal load but with added potential for recovery and biomass regeneration if desired.

4. Microbial Transformations of Heavy Metals

Microbial transformations of heavy-metal and metalloid species include oxidation, reduction, methylation, and demethylation. Hg^{2+}-resistant bacteria and other microorganisms can reduce Hg^{2+} to Hg^0 with mercuric reductase. Hg^0 is volatile and may be removed from contaminated sewage at a rate of 2.5 mg $liter^{-1}$ hr^{-1}. Organomercurials can be detoxified by organomercurial lyase, the resulting Hg^{2+} then being reduced to Hg^0. Many microorganisms are capable of reducing Au^{3+} to elemental Au^0 and Ag^+ to elemental Ag^0. Treatment of arsenic-loaded sewage with arsenite oxidase-producing bacteria (which catalyze As^{3+} to As^{5+}) can improve certain arsenic removal methods because arsenate, As^{5+}, is more easily precipitated from wastewater by Fe^{3+} than is arsenite, As^{3+}. Chromate ($CrO_4{}^{2-}$)-reducing bacteria (e.g., *Enterobacter cloacae*) are resistant to chromate (10 mM) and anaerobically reduce Cr(VI) to Cr(III), which is precipitated and thus detoxified. Biomethylated metal derivatives are often volatile and may be lost from a given system although amounts liberated are usually low. Microbial dealkylation of organometallic compounds, which may constitute a detoxification mechanism, results in the eventual formation of ionic species that could possibly be removed using a biosorptive process.

5. Removal of Heavy Metals by Derived, Induced, or Excreted Microbial Products

Some microbial metal-binding biomolecules may be induced by the presence of appropriate heavy metals (e.g., metallothioneins and phytochelatins) or by the deprivation of an essential metal ion (e.g., Fe^{3+}), leading to siderophore production. Other metal-binding molecules may be overproduced as a result of exposure to heavy metals and interference with "normal" metabolism (e.g., fungal melanins). However, most microbial biomolecules with significant metal-binding abilities are synthesized as a result of "normal" growth and/or are important structural components. Relative efficiencies of metal binding may largely depend on the metal species present and the chemical nature and reactivity of the metal-binding ligands.

a. Siderophores

Siderophores are low-molecular-weight Fe^{3+} coordination compounds that are excreted under iron-limiting conditions by iron-dependent microorganisms to enable accumulation of iron. Although highly specific for Fe^{3+}, siderophores can complex other metals and actinides including Pu(IV), Ga(III), Cr(III), scandium (Sc), indium (In), nickel, uranium, and thorium.

b. Extracellular Microbial Polymers

Many microbial extracellular polymers possess significant metal-binding properties. These are mainly composed of polysaccharide, which may be associated with protein and certain other excretory products. Binding of metal ions to polymeric material may involve ion exchange, with cationic groups including carboxyl, organic phosphate, organic sulfate, and phenolic hydroxyl, complex formation, and/or metal deposition in a chemically altered form. In activated sludge, *Zoogloea ramigera* is important in flocculation and metal removal because of extensive exopolysaccharide production. Other bacterial polymers of importance include those from capsulate *Klebsiella (Enterobacter) aerogenes*, *Pseudomonas* sp., and *Arthrobacter viscosus*. In algae, not all polysaccharides exhibit metal binding, and a correlation generally exists between high anionic charge and complexing capacity. The bacterial emulsifying agent, emulsan, produced by *Arthrobacter*, *Pseudomonas*, and *Acinetobacter* sp., has the ability to bind high amounts of uranium. If emulsan is dispersed in water/hexadecane, the "emulsanosol" product can remove more than 800 mg uranium/g.

c. Metallothioneins and Phytochelatins

Metal-binding proteins have been recorded in all microbial groups examined, including cyanobacteria, bacteria, microalgae, and filamentous fungi, although most detailed work has been carried out with yeasts.

Copper resistance in *Saccharomyces cerevisiae* is mediated by the induction of a 6573-dalton cysteine-rich protein, copper metallothionein (Cu-MT). Yeast Cu-MT may be of potential in metal recovery because it can also bind other metals (e.g., Cd, Zn,Ag, Co, and Au), although these do not generally induce MT synthesis. The gene for copper resistance has been cloned in *Escherichia coli*, and this resulted in expression of a Cu-, Cd-, and Zn-binding protein product in the bacterium. Two other approaches for the exploitation of yeast metallothionein have been suggested. The first is the genetic engineering of yeast strains with constitutive expression of MT genes, which may then accumulate elevated amounts of metals that do not induce yeast MT transcription (e.g., gold), while the second relies on extracellular MT release by attaching the appropriate secretory signal sequence to MT genes. An ultimate aim may be the production of several metallothioneins specific for different metal ions.

Another group of metal-binding molecules are short, cysteine-containing γ-glutamyl peptides, commonly termed "phytochelatins." These peptides function in heavy-metal detoxification in algae as well as certain fungi and yeasts. *Schizosaccharomyces pombe* can make at least seven different peptides in response to, for example, Cd, Cu, Pb, Zn, and Ag, although most peptide synthesis occurs with Cd. Cd-induced γ-glutamyl peptides contain labile S, a major proportion of which may arise from Cd-stimulated H_2S production by, for example, *S. pombe* and *Candida glabrata*. $S^{2-}:Cd^{2+}$ complexes with a molar ratio of 0.7 contain a 20-Å diameter crystallite coated with γ-glutamyl peptides. These crystallites act as quantum particles and possess properties analogous to those of semiconductor clusters.

d. Heavy Metal Removal by Compounds Derived from Fungal Biomass

Waste fungal biomass arises from several industrial fermentations, and this may provide an economical source of biosorptive materials. Many species have a high-cell wall chitin content, and this polymer of *N*-acetyl glucosamine is an effective metal biosorbent. Actinide accumulation by intact biomass mainly comprises metabolism-independent biosorption, the main site of uptake being the cell wall. Chitosan, other chitin derivatives, and glucans also have a significant biosorptive capacity. Melanins are located in and/or exterior to fungal cell walls, where they may appear as electron-dense deposits and granules. Melanins and fungal phenolic polymers contain phenolic units, peptides, carbohydrates, aliphatic hydrocarbons, and fatty acids. Oxygen-containing groups in these substances, including carboxyl, phenolic and alcoholic hydroxyl, carbonyl, and methoxyl groups may be particularly significant in metal binding. Melanized cell forms (e.g., chlamydospores) can have high metal uptake capacities with virtually all bound metal being located in the cell wall.

6. Metal Recovery

Biotechnological exploitation of microbial metal accumulation may depend on the ease of biosorbent regeneration for metal recovery. The method used can depend on the element involved and the mechanism of accumulation. Metabolism-independent processes are frequently reversible by nondestructive methods and, in many cases, can be considered analogous to conventional ion exchange. Metabolism-dependent intracellular accumulation is often irreversible, requiring destructive recovery by incineration or dissolution in acids or alkalis. Most work has concentrated on nondestructive desorption, which, for maximum benefit, should be highly efficient and cheap and result in minimal damage to the biosorbent. Dilute mineral acids (~0.1 M) can be effective for metal removal, although more concentrated acids or lengthy exposure times may result in biomass damage. It may be possible to apply selective desorption of metal(loid) species from a loaded biosorbent using an appropriate elution scheme. For example, metal cations (e.g., Cu^{2+}, Cr^{3+}, Ni^{2+}, Pb^{2+}, Zn^{2+}, Cd^{2+}, and Co^{2+}) can be released from algal biomass using eluant at pH 2, whereas at higher pH values, anionic metal species (e.g., SeO_4^{2-}, CrO_4^{2-}, and MoO_4^{2-}) were removed. Au^{3+}, Ag^{+}, and Hg^{2+}, however, remained strongly bound at pH 2, and these were removed by addition of ligands that formed stable complexes with these metal ions. Carbonates and/or bicarbonates are efficient desorption agents with the potential for cheap, nondestructive metal recovery. Operating pH values for bicarbonates cause little damage to the biomass, which may retain at least 90% of the original uptake capacity.

7. Industrial and Economic Aspects

A variety of batch, semi-continuous, or continuous flow reactors can be used for biosorption, in series or in parallel, the latter often increasing the capacity and facilitating biosorbent regeneration. For batch and continuous flow reactors, separation of loaded biosorbent from the effluent may be achieved by, for example, settling, flotation, filtration, or centrifugation. The biosorbent may then be regenerated, ashed, or discarded. Several described biosorbent systems are based on packed- or fixed-bed reactor configurations with effluent upflow or downflow, although the latter may be particularly prone to clogging by suspended solids in the influent. The continuously operated pulsating-bed contactor employs intermittent withdrawal of metal-saturated biosorbent, for separate regeneration, and replacement by fresh biosorbent. Other systems include fluidized- or expanded-bed reactors, which generally use effluent upflow; metal-saturated biosorbent is removed from the column base for regeneration. Air-lift reactors are another useful system, and these may be used in series for large flow rates and adequate contact times. There is current interest in the application of high-gradient magnetic separation (HGMS) for the separation of freely suspended metal-accumulating microbes.

Several biosorptive processes possess advantages over existing physical and chemical treatments, which include high efficiency at low metal/radionuclide concentrations, low affinities for Ca^{2+} and Mg^{2+}, successful operation over wide ranges of pH and temperature (dead biomass or derived products) in multiple biosorption–desorption cycles, and compatibility with conventional ion exchange technology. In addition, microbial biomass may be available as a fermentation waste product or specifically grown using cheap substrates. Some physicochemical processes (e.g., evaporation and reverse osmosis) are economically feasible if metal concentrations are high. However, biosorptive treatments need not necessarily replace existing methodologies but may be used as "polishing" systems for processes that are not completely efficient. Selectivity may be achieved depending on the relative concentrations of different metals present in an effluent, by appropriate biosorbent selection and/or choice of appropriate selective recovery methods. To date, immobilized or pelleted preparations appear best for commercial use, with recovery using a cheap desorbing agent. Several well-publicized commercial processes appear competitive in cost and operation with traditional treatments and can treat effluent volumes greater than 100 m^3 day^{-1}. Significantly higher flow rates may, however, be encountered for many industrial effluents, including those arising from the nuclear industry, and these would require large amounts of biosorbent, which may adversely affect operation and cost. Nevertheless, scale-up may be possible, and it is advantageous that the design and operation of most biosorption equipment is similar to the technology already used for ion exchange or activated carbon adsorption.

There is now a great awareness of the potential dangers of environmental pollution by heavy metal compounds, metalloids, radionuclides, organometal(loid)s, and related substances. Removal of such pollutants from contaminated solutions by living or dead microbial biomass and derived or excreted products may provide an economically feasible and technically efficient means for element recovery and environmental protection. Development of this area of biotechnology is desirable on both environmental and economic grounds.

Bibliography

Belliveau, B. H., Starodub, M. E., Cotter, C., and Trevors, J. T. (1987). *Biotech. Adv.* **5,** 101–127.

Collins, Y. E., and Stotzky, G. (1989). Factors affecting the toxicity of heavy metals to microbes. *In* "Metal Ions and Bacteria" (T. J. Beveridge and R. J. Doyle, eds.), pp. 31–90. Wiley, New York.

Gadd, G. M. (1988). Accumulation of metals by microorganisms and algae. *In* "Biotechnology—A Comprehensive Treatise, Vol. 6b, Special Microbial Processes" (H.-J. Rehm, ed.), pp. 401–433. VCH Verlagsgesellschaft, Weinheim.

Gadd, G. M. (1991). Molecular biology and biotechnology of microbial interactions with organic and inorganic heavy metal compounds. *In* "Molecular Biology and Biotechnology of Extremophiles" (R. A. Herbert and R. J. Sharp, eds.), pp. 225–257. Blackie, Glasgow.

Gadd, G. M., and White, C. (1989). Heavy metal and radionuclide accumulation and toxicity in fungi and yeasts. *In* "Metal–Microbe Interactions" (R. K. Poole and G. M. Gadd, eds.), pp. 19–38. IRL Press, Oxford.

Hughes, M. N., and Poole, R. K. (1989). "Metals and Microorganisms." Chapman and Hall, London.

Macaskie, L. E., and Dean, A. C. R. (1989). Microbial metabolism, desolubilization, and deposition of heavy metals: Metal uptake by immobilized cells and application to the treatment of liquid wastes. *In* "Biological Waste Treatment" (A. Mizrahi, ed.), pp. 159–201. Alan R. Liss, New York.

Macaskie, L. E., and Dean, A. C. R. (1990). Metal-sequestering biochemicals. *In* "Biosorption of Heavy Metals" (B. Volesky, ed.), pp. 199–248. CRC Press, Boca Raton, Florida.

Rosen, B. P., and Silver, S. (1987). "Ion Transport in Prokaryotes." Academic Press, San Diego.

Tisa, L. S., and Rosen, B. P. (1990). *J. Bioenerg. Biomembr.* **22,** 493–507.

Tsezos, M. (1990). Engineering aspects of metal binding by biomass. *In* "Microbial Mineral Recovery" (H. L. Ehrlich and C. L. Brierley, eds.), pp. 325–339. McGraw-Hill, New York.

Winge, D. R., Reese, R. N., Mehra, R. K., Tarbet, E. B., Hughes, A. K., and Dameron, C. T. (1989). Structural aspects of metal-γ-glutamyl peptides. *In* "Metal Ion Homeostasis: Molecular Biology and Chemistry" (D. H. Hamer and D. R. Winge, eds.), pp. 301–311. Alan R. Liss, New York.

Heavy Metals, Bacterial Resistances

Tapan K. Misra
University of Illinois at Chicago

Glossary

Chemiosmosis Electron transport process that pumps protons across the inner membrane to the outer aqueous phase, generating a H^+ gradient across the inner membrane; the osmotic energy in this gradient supplies the energy for adenosine triphosphate synthesis

Constitutive gene Transcription of the gene depends on RNA polymerase only, and no other regulatory factor

Efflux Pumping out

Operon Segment of DNA constituting genes that are cotranscribed and the DNA element(s) that is recognized by the regulatory gene product

Promoter Specific segment of DNA recognized by RNA polymerase for initiation of transcription

"-10" Sequence RNA polymerase recognition sequence that is located approximately 10 bp upstream of the messenger RNA start site; the consensus sequence in bacterial DNA is TATAAT

"-35" Sequence RNA polymerase recognition sequence that is located approximately 35 bp upstream of messenger RNA start site; the consensus sequence in bacteria is TTGACA

Trans-acting element Element that can function on a physically separate expression system

Transcribe Synthesize RNA using the coding strand of a DNA

Transcription Synthesis of RNA using DNA as a template

Transposon Mobile genetic element that can replicate and then transfer a copy at new locations in the genome

Uncouplers Chemicals that can disrupt linkage between phosphorylation of adenosine diphosphate and electron transport

MOST HEAVY METALS are toxic to cells. Generally, the cations and anions formed from the metals, and not the reduced metallic material (which is inert), are toxic. Some of the heavy metals are components for many cellular enzymes and, therefore, are needed by cells in trace amounts. Examples of this group of metals are cobalt, manganese, nickel, and zinc. Another group of metals—antimony, arsenic, bismuth, cadmium, chromium, lead, mercury, and tellurium—are not needed by bacterial cells; they are very toxic. Resistant bacteria are prevalent in environments enriched with toxic compounds (soil, water). Some heavy metal compounds are used in industry as catalysts, some as preservatives (on seeds and wood products) to prevent microbial growth, and some as insecticides and herbicides, and some have been used as disinfectants in clinics, hospitals, and homes.

Apparently, separate genetically determined mechanisms of resistances exist against essentially each toxic metal ion. Due to their structural similarities, some metal ions are transported into the cells through specific transport systems designed for normally required metal ions. A few examples can be cited: arsenate (AsO_4^{3-}) uses the phosphate transport system(s), cadmium (Cd^{2+}) enters bacterial cells via manganese (Mn^{2+}) or zinc (Zn^{2+}) transport systems, and chromate (CrO_4^{2-}) uses the sulfate (SO_4^{2-}) transport system. One known exception is the mercuric ion (Hg^{2+}), which is transported into bacterial cells by an active transport process mediated by the protein products of genes that are cotranscribed with the other genes involved in the detoxification of Hg^{2+}.

Two basic mechanisms of resistance by cells against toxic ions can be envisaged: (1) specific alterations of ion transport (inward, preventing entry into the cell, or outward, pumping out of the cell) of the toxic ion, and (2) by chemical modification or by binding to cellular factor(s) resulting in a form that is no longer toxic to the cell. These mechanisms will be discussed with specific examples later. [*See* ION TRANSPORT.]

Genes conferring resistances to heavy metals are often found on mobile genetic elements such as plasmids and transposons. Coexistence of different metal ion resistance genes and antibiotic resistance genes on one plasmid is not rare. This is advantageous from an evolutionary point of view. Genetic exchange between different strains, species, or genera is facilitated by relatively small extrachromosomal elements rather than by complex rearrangement within the chromosome. The expression of many known metal ion resistance genes are inducible by the metal ion itself. When bacterial cells sense the metal ion, increased expression of genes conferring resistances to the metal ion takes place. Thus, these processes are efficiently controlled: Enzymes and other proteins involved in chemical transformations and transport of the toxic ions are produced only when they are required by cells (in a contaminated environment). The toxic ion "induces" the expression of ion-specific resistant gene(s). [*See* CATABOLIC PLASMIDS IN THE ENVIRONMENT; PLASMIDS.]

I. Antimony and Arsenic Resistances

Arsenicals have widely been used in cattle feed supplements, as herbicides, as insecticides, and in printing fabrics. Bacteria have developed distinct mechanisms of resistance against the oxyanions of arsenic (arsenite and arsenate) and antimony (antimonite). Arsenite resistance has been found in *Escherichia coli, Staphylococcus aureus, Pseudomonas pseudomalleli, Synechoccous,* and some strains of *Alcaligenes*. The biochemical and molecular basis of arsenite (AsO_2^-), arsenate (AsO_4^{3-}), and antimonite (SbO_2^-) resistances are most thoroughly known in *E. coli* and *S. aureus*. In both organisms, arsenite, arsenate, and antimonite resistances are determined by a single gene complex; therefore, they are coinduced by any of those ions. The resistance determinants are present on plasmids. [*See* HEAVY METAL POLLUTANTS: ENVIRONMENTAL AND BIOTECHNOLOGICAL ASPECTS.]

A. Biochemistry of Antimony and Arsenic Resistances

The biochemical mechanism of plasmid-determined arsenate, arsenite, and antimonite resistances are similar in *E. coli* and *S. aureus*. Resistances result from reduced net accumulation of these oxyanions in induced resistant cells. Arsenate is accumulated by virtue of its structural similarity to phosphate. Phosphate is essential for cellular growth. Many bacteria appear to have two separate chromosomally encoded phosphate transport systems. Arsenate enters the cell via both phosphate transport systems. Resistant cells quickly export arsenate but not phosphate. If the cells pumped out arsenate and phosphate by the same mechanism, then the cells would become phosphate-starved.

Arsenite, arsenate, and antimonite resistances were reported in *E. coli* and in *S. aureus* cells harboring resistance plasmids. Cells preloaded with radioactive arsenate ($^{74}AsO_4^{3-}$) efficiently pumped out $^{74}AsO_4^{3-}$ when the cells were induced with arsenate (or arsenite or antimonite) prior to loading with $^{74}AsO_4^{3-}$, and the uninduced cells failed to extrude $^{74}AsO_4^{3-}$. The efflux is energy-dependent and adenosine triphosphate (ATP) is the driving force for expulsion of arsenicals. Dicyclohexylcarbodiimide, an inhibitor of the oxidative phosphorylation ATPase, inhibited arsenate efflux. *Escherichia coli* strains lacking the ability to synthesize ATP from respiratory substrates could pump arsenate out in media containing glucose (allowing glycolytic ATP synthesis) but not in media containing succinate as the sole carbon source. Genetic and molecular biological studies advanced the biochemical understanding of the mechanism of arsenite, arsenate, and antimonite efflux. [*See* ATPASES AND ION CURRENTS.]

B. Molecular Genetics of Arsenic and Antimony Resistances

The best-characterized system is that of the plasmid R773 of *E. coli*. The arsenite, arsenate, and antimonite resistance operon (*ars*) consists of four genes. The DNA sequence of the *ars* operon is known, and the organization of the four genes is shown in Fig. 1. All four genes (*arsR*, *arsA*, *arsB*, and *arsC*) are cotranscribed from a single DNA site—the promoter. The product of the regulatory gene, ArsR, is a trans-acting regulatory protein that represses its own synthesis as well as the expression of the other downstream *ars* structural genes (in the absence of arsenicals and antimonite). ArsR, a 13-kDa protein, induces gene expression from the *ars* promoter in the presence of arsenite, arsenate, antimonite, and bismuth ion. The *ars* operon does not confer resistance to Bi^{3+}, but Bi^{3+} functions as an inducer; hence, it is known as a gratuitous inducer. Following the *arsR* gene, there are three other genes (Fig. 1): *arsA*, *arsB*, and *arsC*. *arsA* and *arsB* are necessary and sufficient for arsenite and antimonite resistance; *arsC* is also needed for arsenate resistance.

The 5′ and 3′ halves of the *arsA* gene have extensive sequence similarity, suggesting evolution of the gene through a gene duplication step and fusion. Thus, the ArsA protein is basically a fusion dimer with two glycine-rich adenylate binding sites, one near the N terminus of the polypeptide and the other near the middle. The purified ArsA protein functions as an arsenite- or antimonite-stimulated soluble ATPase. The protein exists in an equilibrium between monomeric and dimeric forms in solution, with equilibrium favoring dimerization upon binding with the anionic inducers.

How does the soluble ArsA ATPase export oxyanions of antimony and arsenic? The product of the *arsB* gene is a 45.6-kDa hydrophobic protein localized in the inner membrane of the cell. When the ArsA and ArsB proteins form a membrane-bound complex, this functions as an antimonite–arsenite-specific efflux pump, thus exporting these otherwise toxic anions as soon as they enter into the cell, resulting in reduced net accumulation of these ions inside the cell. The ArsC protein is needed for arsenate resistance. It is postulated that the ArsC protein brings about stereospecific modification of the ArsA–ArsB complex, rendering it active in pumping out arsenate from the cell.

P	arsR		arsA	arsB	arsC
	117 aa		583 aa	429 aa	141 aa
	13,213 Da		63,169 Da	45,577 Da	15,811 Da

Figure 1 Schematic representation of the arsenic resistance operon. P denotes promoter. Approximately 1-kb messenger RNA between the *arsR* and *arsA* transcript is not translated.

II. Cadmium Resistance

Cadmium salts are commercially used as lubricants, as ice nucleating agents, in dry cell batteries, in photography, in dyeing and calico printing, in electroplating and engraving, and in the manufacture of special mirrors. Cadmium resistance has been shown in certain strains of *Staphylococcus*, *Bacillus*, *Alcaligenes*, and *Pseudomonas*. The best-studied cadmium (Cd^{2+}) resistance determinants are those in *S. aureus*. There are at least two distinct Cd^{2+} resistance determinants in *S. aureus*, and they are both found on plasmids. One of them, known as the *cadA* system, confers about 100-fold greater tolerance to resistance cells over that for sensitive cells (without the cadmium resistance determinant).

A. Biochemistry of *cadA* Cadmium Resistance

Cadmium (Cd^{2+}) is transported into the cell by a chromosomally encoded manganese (Mn^{2+}) transport system (in gram-positive bacteria). The uptake of these ions is membrane potential-dependent, and accumulation of cadmium inside the cells results in respiratory arrest. The resistant cells rapidly pump Cd^{2+} out. The efflux was proposed as a chemiosmotic electroneutral process with the exchange of one Cd^{2+} for two protons. The evidence supporting this conclusion was that Cd^{2+} efflux is sensitive to agents that block or accelerate proton movement but not to agents that disrupt membrane potential. However, recent DNA sequence analysis of the *cadA* cadmium resistance determinant suggested strongly that an ATP is involved in pumping Cd^{2+} out from resistant cells. A currently accepted model is that both proton exchange and ATP hydrolysis are needed for Cd^{2+} efflux in *S. aureus* (Fig. 2).

B. Molecular Genetics of *cadA* Cadmium Resistance

The two structural genes that constitute the *cadA* determinant, *cadC* and *cadA*, are cotranscribed from a single promoter. The products of these two genes are necessary and sufficient for the maximum level of resistance to Cd^{2+} and to Zn^{2+}. [*Staphylococcus aureus* cells containing the *cadA* resistance determinant also confer resistance to zinc. The expression of the Cd^{2+}/Zn^{2+} resistance genes are inducible by both Cd^{2+} and Zn^{2+} as well as by Bi^{3+},

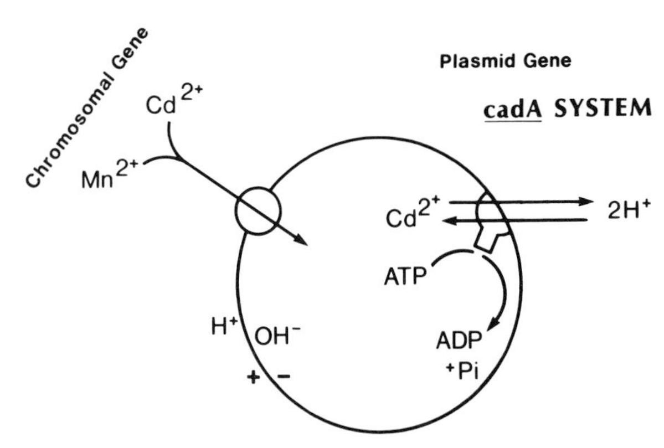

Figure 2 Current model for Cd^{2+} uptake and efflux in gram-positive bacteria. Both the chemiosmotic antiporter and the adenosine triphosphatase (ATPase) alternative for the efflux system are shown. ADP, adenosine diphosphate; Pi, inorganic phosphate. [Reproduced, with permission, from the Annual Reviews, Inc. From Silver, S., and Misra, T. K. (1988). *Annu. Rev. Microbiol.* **42,** 717–743.]

Co^{2+}, or Pb^{2+} (gratuitous inducers).] The location and nature of a hypothesized additional regulatory gene ("*cadR*") are unknown. The *cadC* gene product is a small, highly charged, soluble polypeptide, of 122 amino acids. The *cadA* gene product is a 727-amino acid membrane polypeptide with recognizable sequence similarity with the E1E2 class of cation-translocating ATPases of bacteria and all eukaryotes, including the well-known K^+ ATPases of *E. coli* and *Staphylococcus faecalis*, H^+ ATPases of yeast and plant cells, Na^+/K^+ ATPases of humans and electric eel, and Ca^{2+} ATPase of rabbit muscle.

III. Chromate Resistance

Chromium is used in industry primarily in the manufacture of alloys and in tanning. Chromium (VI) is the most toxic state of chromium salts and Cr(III) is less toxic. Chromate (CrO_4^{2-}) resistance has been observed among certain strains of *Pseudomonas, Alcaligenes, Salmonella, Streptococcus, Aeromonas,* and *Enterobacter.*

A. Biochemistry of Chromate Resistance

Chromate enters bacterial cells by the sulfate uptake system(s). There are at least two potential mechanisms of chromate resistance: (1) by reduction of Cr(VI) to Cr(III), which subsequently precipitates as $Cr(OH)_3$ (extracellular) and (2) by reduced uptake of chromate governed by a plasmid system in resistant cells. Is the reduced net uptake of CrO_4^{2-} due to inhibition of CrO_4^{2-} uptake directly or due to rapid efflux of CrO_4^{2-} by a mechanism parallel to those previously described for arsenic and cadmium? At present, no evidence distinguishes between the two models.

Chromate reductase activity has been studied in *Pseudomonas fluorescens, Pseudomonas aeruginosa,* and *Enterobacter cloacae.* In *Pseudomonas,* chromate reductase is a soluble enzyme. On the other hand, *E. cloacae* chromate reductase is located in the membrane. Interestingly, *E. cloacae* is resistant to chromate under both aerobic and anaerobic growth conditons, but chromate reduction occurs only under anaerobic conditions. Chromate reductase activity from *E. cloacae* is inhibited by cyanide and membrane uncouplers.

B. Molecular Genetics of Chromate Resistance

The genes conferring resistance to chromate in *P. aeruginosa* and *Alcaligenes eutrophus* have been cloned and sequenced. These genes were originally found on plasmids. Plasmid-determined chromate resistance is inducible in *A. eutrophus,* whereas chromate resistance is expressed constitutively in *P. aeruginosa* after cloning. In both cases, a single gene, *chrA,* is responsible for chromate resistance. The gene product is a 401- or 416-amino acid hydrophobic polypeptide. The ChrA proteins from the two sources are structurally related (primary amino acid sequence). It is hypothesized that the ChrA protein either limits the transport of CrO_4^{2-} in cells or is responsible for chromate efflux.

IV. Mercury and Organomercurial Resistance

Mercury compounds are used in industry as catalysts in oxidation of organic compounds. Other uses are in the extraction of gold and silver (currently not used in developed countries), in dry cell batteries, in electric rectifiers, etc. Mercury forms amalgams with other metals and silver-mercury amalgams are still used today in dental restorations. Phenylmercury, merthiolate, and mercurochrome had been used as household and hospital disinfectants. Phenylmercury and methylmercury were used in agriculture as fungicides on seeds and to keep golf course grass green. The largest source of environmental release of mercury compounds from human activities is that from burning coal and petroleum products, which are naturally enriched with mercury compounds. Mercury compounds are also leached from (weathering) natural deposits on rocks and soil. Mercury resistance is known in a wide number of bacterial genera. A few examples of gram-negative mercury-resistant bacteria are *Escherichia, Pseudomonas, Shigella, Serratia, Thiobacillus, Yersinia, Acinetobacter*, and *Alcaligenes*. Some examples of resistant gram-positive bacteria are *Staphylococcus, Bacillus, Mycobacterium, Streptococcus*, and *Streptomyces*. Of all the heavy metal resistances studied to date, mercury is certainly the most thoroughly studied system at the biochemical and molecular level. [*See* MERCURY CYCLE.]

A. Narrow- and Broad-Spectrum Mercury Resistance

Bacteria that are resistant primarily to inorganic mercury salts are called narrow-spectrum mercury-resistant, and those that are resistant to organomercurial compounds in addition to inorganic mercury salts are called broad-spectrum mercury-resistant.

B. Biochemistry of Mercurial Detoxification by Resistant Bacteria

All or essentially all of the mercury-resistant bacteria isolated from mercury-polluted soil or water have the same basic biochemical mechanism of detoxification of mercury compounds involving soluble enzymes. All produce mercuric reductase that reduces Hg^{2+} to Hg^0. Hg^0 is virtually insoluble in water and is relatively volatile. It is also less toxic. Mercuric reductase is a flavin adenine dinucleotide (FAD)-containing enzyme and reduced nicotinamide adenine dinucleotide phosphate ($NADPH^+{}_2$)-dependent. $NADPH^+{}_2$ functions as an electron donor, initially reducing the FAD. Broad-spectrum mercury-resistant bacteria produce mercuric reductase and a second enzyme, organomercurial lyase, which cleaves the carbon–mercury bond in compounds such as CH_3Hg^+ and $C_6H_5Hg^+$; Hg^{2+} is released and subsequently reduced by the mercuric reductase. These reactions are illustrated in Fig. 3.

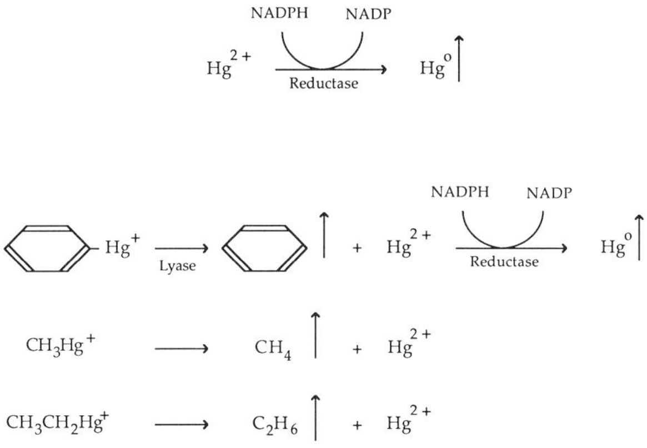

Both mercuric reductase and organomercurial lyase are soluble intracellular enzymes. Mercuric reductase is, as far as we know, a dimer consisting of two identical monomers of approximately 60-kDa polypeptide monomers, which has three pairs of conserved cysteine residues. Approximately 15 amino acid residues in the active site region are highly conserved in all the reductases from different bacteria. This region contains a pair of redox-active cysteine residues. Electrons donated by $NADPH^+{}_2$ are transferred via FAD to reduce the active site cystine, converting it to two cysteine residues. These cysteine residues then reduce enzyme-bound Hg^{2+} to Hg^0. Both inorganic mercury salts and organomercurial salts are highly toxic because they readily react with protein sulfhydryl groups. Hg^0 is lipophilic; it is also relatively biologically inert and volatile. Organomercurial lyase isolated from broad-spectrum mercury-resistant bacteria is a smaller monomeric polypeptide protein of approximately 23 kDa.

An alternative mechanism of mercury resistance by precipitation of insoluble HgS ($Hg^{2+} + H_2S = HgS + 2\ H^+$) has occasionally been proposed but never studied. If this occurs, then it is most likely a fortuitous by-product of a metabolic process.

$$Hg^{2+} \xrightarrow[\text{Reductase}]{\text{NADPH} \rightarrow \text{NADP}} Hg^0 \uparrow$$

$$C_6H_5Hg^+ \xrightarrow{\text{Lyase}} C_6H_6 \uparrow + Hg^{2+} \xrightarrow[\text{Reductase}]{\text{NADPH} \rightarrow \text{NADP}} Hg^0 \uparrow$$

$$CH_3Hg^+ \longrightarrow CH_4 \uparrow + Hg^{2+}$$

$$CH_3CH_2Hg^+ \longrightarrow C_2H_6 \uparrow + Hg^{2+}$$

Figure 3 Enzymatic detoxification of mercury compounds. NADP, nicotinamide adenine dinucleotide phosphate; NADPH, reduced NADP.

C. Molecular Genetics of Mercury Resistance

The genes conferring resistance to mercury compounds are physically clustered and functionally form a single transcriptional unit, or an operon. Mercury resistance operons from different bacteria have been cloned and their DNA sequences have been determined. The organization of the genes in gram-positive and gram-negative bacteria is somewhat different. These systems are discussed separately here.

1. Organization of the Mercury Resistance Genes in Gram-Negative Bacteria

The DNA sequence of the mercury resistance genes is 80–90% identical when compared in several different systems (with one known exception in *Thiobacillus ferrooxidans*). The physical organization of the broad-spectrum mercury resistance genes from the *Serratia* plasmid pDU1358 is shown in Fig. 4. The regulatory gene *MerR* is separated from the other *mer* genes by an operator–promoter region. The operator sequence is the one with which the product of the *merR* regulatory gene interacts, regulating expression of the remaining (structural) genes in the operon. There are two divergently oriented initiation sites for messenger RNA (mRNA) synthesis, i.e., promoters in this region; one is for the synthesis of the *merR* mRNA, and the other is for the expression of the other genes. The first two structural genes are *merT* and *merP,* whose products transport mercury ion across the cell membrane into the cell, and then follows the *merA, merB,* and *merD* genes. *merA* encodes mercuric reductase; *merB* encodes organomercurial lyase; and *merD* encodes a coregulatory protein, whose influence on the rate of operon expression is much smaller than the effects of MerR. The mercury resistance determinant of the plasmid R100 (from *Shigella flexneri*) contains the additional transport gene *merC,* located between *merP* and *merA*. In a chromosomally encoded mercury resistance determinant from *T. ferroxidans,* only the *merC* gene encodes the mercury transport protein. The *merC* gene as well as the *merT* and *merP* genes encode proteins for independent transport pathways.

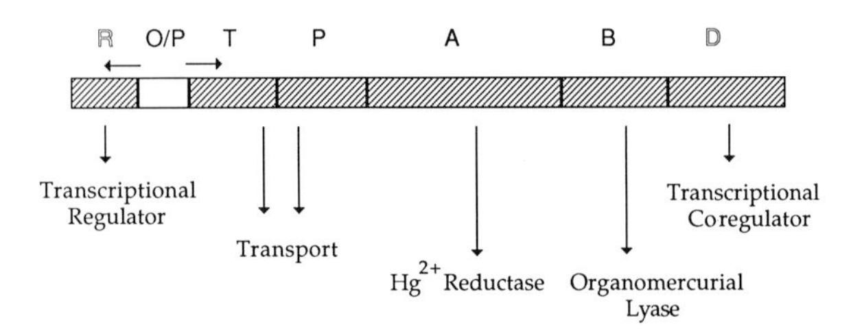

Figure 4 Schematic representation of the broad-spectrum mercury resistance genes of the *Serratia* plasmid pDU1358. Horizontal arrows show the direction of transcription. O/P, operator–promoter.

2. Organization of Mercury Resistance Genes in Gram-Positive Bacteria

Chromosomally encoded mercury resistance genes from a *Bacillus* species or those from *S. aureus* plasmids are organized in a single operon and are expressed from a single promoter. The regulatory gene is the first gene after the mRNA start site, and it is followed by three or four genes that are thought to encode transport proteins. Then there are genes for mercuric reductase and organomercurial lyase. No gene equivalent of *merD* of gram-negative bacteria has been identified in the mercury resistance operons of gram-positive bacteria.

Gene by gene, much is known about the transport proteins, detoxifying enzymes, and the regulation of the mercury resistance operons of different bacteria.

3. Transport of Mercury Ion into the Cell

The transport function was first defined in plasmid R100 by cloning and mutation studies. Insertion of a transposon disrupted the expression of the *merA* gene where it was inserted and of all other genes downstream from the site of insertion of the transposon in the operon. Such transposon-generated mutation in the mercuric reductase gene *merA* resulted in hypersensitivity to mercury salts. In other words, cells containing the *mer* operon with a mutation in *merA* are sensitive to lower concentrations of mercury salt compared to cells containing no *mer* operon. When incubated with $^{203}Hg^{2+}$, the hypersensitive cells accumulated higher concentrations of $^{203}Hg^{2+}$ as compared to cells containing no *mer* operon or those containing the intact *mer* operon. In these mutant cells, Hg^{2+} was transported inside but was not reduced to volatile Hg^0 because of the absence of the reductase enzyme. This accounts for the increased intracellular accumulation of Hg^{2+} and the hypersensitivity to mercury salts.

MerT is a hydrophobic, integral membrane protein. MerP is a periplasmic protein and has a leader sequence of 19 amino acids. Signal peptidase cuts the leader peptide, leaving the 72-amino acid periplasmic Hg^{2+}-binding protein. Mature MerP then delivers the bound Hg^{2+} to the MerT protein, and subsequently MerT delivers the Hg^{2+} to the mercuric reductase enzyme. MerC is a hydrophobic protein and is associated with the inner membrane of the cell. How organomercurial compounds are transported inside the cell is not known.

In gram-negative bacterial cells, the mercury

transport protein(s) are essential for resistance, and in their absence the cells are Hg^{2+}-sensitive. In gram-positive bacterial cells, the role of the transport proteins (there appear to be three membrane proteins in both *S. aureus* and *Bacillus*) is less distinct: With the transport proteins present but mercuric reductase missing, the cells are somewhat hypersensitive to Hg^{2+}, indicating that a transport system is functioning; however, with the mercuric reductase present but the transport proteins absent, the cells are still relatively resistant to Hg^{2+}, indicating that Hg^{2+} can cross the membrane into the cell by some alternative pathway.

4. Mercury-Detoxifying Enzymes: Conclusions from DNA Sequence Analysis

The biochemical properties of mercuric reductase and organomercurial lyase have already been discussed. What did we learn about these enzymes from DNA sequence analysis? The amino acid sequences (predicted from DNA sequence) of the mercuric reductase enzymes from both gram-positive and gram-negative bacteria are significantly similar (about 25% identical amino acids) with flavin-containing disulfide oxido-reductases such as glutathione reductase and lipoamide dehydrogenase (of both animal and bacterial origins). The active site region of human (red blood cell) glutathione reductase is known from biochemical studies. Amino acid sequence comparison between the glutathione reductase and mercuric reductase allowed the prediction of the active site region of mercuric reductase. This was confirmed subsequently by biochemical studies.

The N-terminal 80 or so amino acids of mercuric reductase are the only part of the enzyme that bears no relationhip to glutathione reductase, and apparently these represent an additional Hg^{2+}-related domain. The sequence of these amino acids of mercuric reductase and the sequence of the matured MerP protein are 35% identical. This homology strongly infers a common evolutionary origin of MerP and the N terminus of the mercuric reductase. The remainder of the mercuric reductase sequence and the sequences of glutathione reductase and lipoamide dehydrogenase share an evolutionary ancestry.

The amino acid sequences of four organomercurial lyases from broad-spectrum mercury resistance operons of the gram-negative and gram-positive bacteria are available and very similar. The homologies between the lyase structures fall mostly in the central half of 212- (gram-negative) to 216- (gram-positive) amino acid sequences with less homologies toward both ends. Unfortunately, no polypeptides in the available protein sequence libraries are related, so the sequences tell us little about enzyme function.

5. Regulation of Expression of the Mercury Resistance Operon

The mercury resistance gene cluster has become one of the best-understood metalloregulatory operons. The expression of this operon, as measured by mercuric reductase activity (^{203}Hg volatilization activity), is increased about 200-fold when cells are grown (induced) with subinhibitory concentrations of mercury salts. Regulation of the gram-positive and gram-negative *mer* operons are somewhat different. The regulation of the gram-negative mercury resistance operon is described first.

Mutation in the *merR* gene leads to constitutive expression of the operon, but at a very reduced level of approximately 2% of the fully induced rate. This suggests that MerR is needed as a repressor to suppress activity from the 2% level down to the 1/200 repressed level. Mutations in *merR* are complemented *in trans* (when intact *merR* gene is cloned in a different plasmid vector and introduced in cells containing the *mer* operon with a mutant *merR* gene). Full (100%) expression of the intact operon is induced in the presence of inorganic mercury salts. Thus, MerR protein also functions as a transcriptional activator in the presence of Hg^{2+}. The MerR protein has been purified, and its interaction with the operator–promoter region of DNA has been determined. Note that the *merR* gene and the other genes are expressed from two divergently oriented promoters (Fig. 4). MerR protein binds to the promoter region, which contains a 7-bp inverted repeat sequence. The site of binding to the DNA is not influenced by the presence or absence of Hg^{2+}. MerR occupies the *mer* promoter (P*mer*; promoter for the transcription of the structural genes) as a homodimer and represses the synthesis of the *merR* mRNA as well as the *mer* operon mRNA. The "-10" and "-35" canonical sequences (Fig. 5) of P*mer* are separated by 19 bases, which is longer than the 17 bases ideal for proper binding of RNA polymerase and efficient transcription. Binding of Hg^{2+} to the MerR protein results in a conformational change of the MerR protein itself, which, in turn, bends the

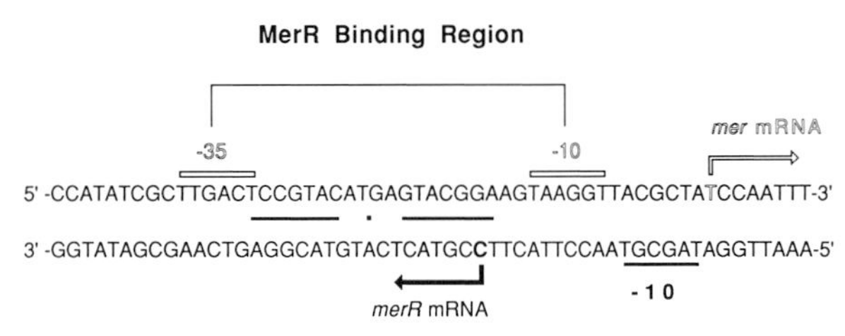

Figure 5 Nucleotide (DNA) sequence of the promoter region of a gram-negative mercury resistance operon. Directions of messenger RNA (mRNA) synthesis are shown by arrows. [Adapted from Walsh, C. T., Destefano, M. D., Moore, M. J., Shewchuk, L. M., and Verdine, G. L. (1988). *FASEB J.* **2**, 124–130.]

P*mer* DNA. The "-10" and "-35" sequences are brought closer together, thereby facilitating interaction of the promoter sequence with the RNA polymerase and allowing efficient transcription.

The narrow-spectrum *mer* operon is inducible with subinhibitory concentrations of inorganic mercury salts, but not by organomercurial compounds. On the other hand, the broad-spectrum *mer* operon of the *Serratia* plasmid pDU1358 is inducible by both inorganic and organic mercury salts. The amino acid sequences of the narrow- and broad-spectrum MerR are very similar except for the C-terminal 9-amino acid residues (Fig. 6). Deletion of 17 amino acids from the C-terminal end of the broad-spectrum MerR results in complete loss of inducibility with organomercurial compounds, but not with inorganic mercury salts. This suggests that the amino acid sequence at the C terminus of the broad-spectrum MerR is involved in the recognition of organomercurial inducers. This is a good example of how nature cleverly modifies gene structure to respond to changing environmental stimuli. Note in Fig. 6 that there are three conserved cysteine residues (bold face print), which are essential for Hg^{2+}-binding and transcriptional activation of the *mer* operon. These are conserved in all nine currently available sequences for MerR proteins.

The *merD* gene (see Fig. 4) is present in at least four gram-negative mercury resistance operons. Its product acts *in trans* and down-regulates transcription from the *mer* promoter (P*mer*). The MerD protein binds to the same site as the MerR in experiments with purified DNA and purified protein. The level of MerD in the cell is low (about 15-fold less than the mercuric reductase). The *in vivo* role of MerD is unclear. It appears to fine-tune *mer* operon expression as a coregulatory protein.

In gram-positive bacteria, the *merR* gene and the other *mer* genes are cotranscribed, unlike the gram-negative *mer* operon, where the *merR* gene and all the other *mer* structural genes are transcribed as two separate units. The gram-positive MerR and gram-negative MerR amino acid sequences are 35% identical, and there are three conserved cysteine residues. The operator sequence, where MerR binds the DNA (see Fig. 5) is very highly conserved in both gram-positive and gram-negative *mer* operons. Gram-positive MerR represses *mer* operon transcription in the absence of Hg^{2+} and activates transcription in the presence of Hg^{2+}.

V. Concluding Remarks

The foregoing examples of heavy metal resistance mechanisms are those that are best known. Genes governing resistances against many additional heavy metal ions, including those of tellurium, copper, cobalt, nickel, and lead, also have been identified. The mechanisms of resistance of these ions are not well understood, if at all, at this time. The processes by which bacteria convert more toxic forms of heavy metal ions to less toxic forms have the potential for application in bioremediation. We

```
Broad-spec    MEKNLENLTIGVFAKAAGVNVETIRFYQRKGLLPEPDKPYGSIRRYGEAD   50
Narrow-spec   **********************************R***************

Broad-spec    VTRVRFVKSAQRLGFSLDEIAELLRLDDGTHCEEASSLAEHKLQDVREKM  100
Narrow-spec   *V**K*****************************************K******

Broad-spec    TDLARMETVLSELVFACHARQGNVSCPLIASLQGEKEPRGADAV  144
Narrow-spec   A************C*****K***************AGLARSAMP
```

Figure 6 Comparison of amino acid sequences of the broad-spectrum (pDU1358) and narrow-spectrum (R100) MerR protein. Asterisks indicate identical amino acids. One letter code for amino acids used. Cysteine residues are bold.

need to study further to exploit such possibilities. [*See* BIOREMEDIATION.]

Bibliography

Cervantes, C., Ohtake, H., Chu, L., Misra, T. K., and Silver, S. (1990). *J. Bacteriol.* **172,** 287–291.

Foster, T. J. (1987). *CRC Crit. Rev. Microbiol.* **15,** 117–140.

Heltzel, A., Lee, I. W., Totis, P. A., and Summers, A. O. (1990). *Biochemistry* **29,** 9572–9584.

Nucifora, G., Chu, L., Silver, S., and Misra, T. K. (1989). *J. Bacteriol.* **171,** 4241–4247.

O'Halloran, T. V., Frantz, B., Shin, M. K., Ralston, D. M., and Wright, J. G. (1989). *Cell* **56,** 119–129.

Parkhill, J., and Brown, N. L. (1990). *Nucleic Acids Res.* **18,** 5157–5162.

Silver, S., and Laddaga, R. A. (1990). Molecular genetics of heavy metal resistances in *Staphylococcus* plasmids. *In* "Molecular Biology of Staphylococci" (R. P. Novick and R. A. Skurray, eds.), pp. 15–21. VCH Publishers, New York.

Silver, S., and Misra, T. K. (1988). *Annu. Rev. Microbiol.* **42,** 717–743.

Silver, S., Nucifora, G., and Misra, T. K. (1989). *Trends Biochem. Sci.* **14,** 76–80.

Tisa, L. S., and Rosen, B. P. (1990). *J. Bioenerg. Biomembr.* **22,** 493–507.

Walsh, C. T., Destefano, M. D., Moore, M. J., Shewchuk, L. M., and Verdine, G. L. (1988). *FASEB J.* **2,** 124–130.

Wang, P., Mori, T., Toda, K., and Ohtake, H. (1990). *J. Bacteriol.* **172,** 1670–1672.

Hepatitis

Gordon R. Dreesman
BioTech Resources, Inc.

Gregory R. Reyes
Genelabs, Inc.

Glossary

Chronic carrier Individual chronically infected with a given hepatitis virus

Endemic Present in a region (community) on a continuous basis

Fecal–oral transmission Virus excreted in feces contaminates food and/or water leading to ingestion and infection via the gastrointestinal tract

Hepatitis B core antigen Antigen associated with the core or nucleocapsid of the hepatitis B virus

Hepatitis B e antigen Antigen associated with a truncated polypeptide of the hepatitis B virus nucleocapsids

Hepatitis B surface antigen Complex antigen associated with surface or envelope material associated with the hepatitis B virus

IgM antibody High-molecular-weight antibody to distinct virus antigens elicited on initial exposure to virus

Point-source outbreaks Localization of an outbreak/epidemic to a specific event such as the fecal contamination of the drinking water for a community

Transaminases Enzyme elevations indicating hepatocellular damage; alanine aminotransferase (ALT or SGPT) and aspartate aminotransferase (AST or SGOT)

Zoonotic Infection naturally shared by humans and animal

ENTERICALLY AND PARENTERALLY transmitted forms of viral hepatitis have been historically recognized as major causes of morbidity and mortality. Viral hepatitis is grouped as a single classification based on generally indistinguishable histopathological lesions observed in the liver. However, each virological agent associated with this human disease is distinct. The viral agents associated with the two epidemiologically different forms include the hepatitis A virus (HAV) and hepatitis E virus (HEV) for enterically transmitted diseases and the hepatitis B virus (HBV), hepatitis C virus (HCV), and hepatitis Delta virus (HDV) for parenterally transmitted diseases. Until 1989 and 1990, respectively, the HCV was referred to as parenterally transmitted non-A, non-B hepatitis (PT-NANBH) and the HEV was coined as the enterically transmitted form, or ET-NANBH. In rare instances, other viruses associated with viral hepatitis include cytomegalovirus, Epstein–Barr virus, rubella virus, yellow fever virus, herpes simplex virus, adenoviruses, and selected enteroviruses. Due to their limited involvement, these latter viruses will not be discussed in this article.

I. Introduction

The general biophysical properties of the five known hepatitis viruses are summarized in Table I. As already pointed out, each of these five major agents is quite unique in that each is associated with distinctly different virus families. The two agents associated with enteric hepatitis are nonenveloped, whereas the three viruses identified with parenteral hepatitis are enveloped. The HDV differs in that it has properties similar to those in higher plant satellite

Table I Biophysical and Biochemical Properties of Major Human Hepatitis Viruses

Hepatitis virus	Virus family	Diameter (nm)	Envelope	Genome[a]
Enteric				
Hepatitis A	Picornaviridae	27–32	–	RNA, ss, 7.48 kb
Hepatitis E	Unclassified[b]	27–34	–	RNA, ss, 7.5 kb
Parenteral				
Hepatitis B	Hepadnaviridae	42–47	+	DNA, ds[c] 3.2 kb
Hepatitis C	Unclassified[d]	39–60	+	RNA, ss, 9.4 kb
Hepatitis D	Unclassified	28–39	+[e]	RNA, ss, 1.7 kb

[a] Genome-defined as RNA or DNA, single stranded (ss) or double stranded (ds) and number of nucleotides expressed in kilobases (kb).
[b] Unclassified but has biophysical properties similar to those of the Caliciviridae family.
[c] 15–60% of circle length as a single-stranded (ss) gap.
[d] Unclassified but has biophysical properties similar to those of the Flavi/Pesti-virus families.
[e] A poorly defined nucleocapsid coated with an hepatitis B surface antigen envelope (appears related to plant viroids or satellite RNAs).

viruses. The HDV RNA genome contains a coding region for a structural HDV specific polypeptide, but the virus is encapsulated by the surface antigen of its helper virus, the HBV. Therefore, the infectious HDV can only be produced by a cell that is co-infected with the HBV.

II. Epidemiology of Enteric Viruses

In the field of hepatitis agents, the enteric viruses are those spread by the fecal–oral route. The implication of these epidemiological findings is that viral infection results in primary pathology in the liver but ultimately leads to virus excreted via the bile into the gut and finally with the feces into the environment. Any predisposing environmental condition that leads to the contamination of foodstuffs or water supplies with excreta could lead to the development of a hepatitis outbreak. In some parts of the world, this can occur through crop fertilization, whereby untreated sewage can lead to crop contamination and the ingestion of sufficient numbers of virus particles to cause disease. Another documented, and not uncommon, source of contaminated foodstuffs are certain bivalve mollusks that upon filtering contaminated water concentrate virus particles. The common features of fecal contamination of food and drinking water can lead to the development of massive point-source outbreaks and epidemics. [*See* ENTEROVIRUSES; FOODBORNE ILLNESS.]

A. Hepatitis A Virus

The HAV was first identified epidemiologically in 1967. The occurrence of "epidemic jaundice" was, however, noted by the Greeks and Romans and also reported in Chinese medical literature. Two distinct forms of viral hepatitis with fecal–oral transmission are now recognized, and it does raise the possibility that the HEV (see later) was a possible cause of the epidemic jaundice. Historically, the HAV was first distinguished by its fecal–oral transmission. It was suggested then that the disease be referred to as infectious (or epidemic) hepatitis to distinguish it from "serum" hepatitis, which later became known as hepatitis B. [*See* EPIDEMIOLOGIC CONCEPTS.]

The HAV belongs to the family of small (pico), round RNA (rna) viruses known as the Picornaviridae. These viruses have various tissue tropisms (see, e.g., poliovirus and encephalomyocarditis) and infect various vertebrate species other than humans (see, e.g., foot and mouth disease virus). Putting tissue tropism aside, however, it is now clear that this diverse family of viruses share various features in common including their general genomic organization and expression strategy as well as their biophysical and biochemical properties.

1. Laboratory Diagnosis

The virus was first observed using electron microscopy to visualize virus particles that had been aggregated by sera from individuals in the convalescent phase of illness (immunoelectron microscopy IEM). The visualization of virus particles by IEM was an important benchmark in the establishment of the HAV as a distinct virologic entity. The course of assay development has led to the introduction of enzyme-linked immunosorbent assays (ELISAs) principally for the detection of antibody to the infecting agent. The basis for the assay involves the capture of the virus-specific antibody using anti-immunoglobulin (Ig)-specific antibodies that have been bound to a solid support (e.g., polystyrene plate or bead). The subsequent detection of the bound antibody is by reaction with HAV antigen, which itself is detected by an anti-HAV-specific IgG that has been conjugated with an enzyme (e.g., horseradish peroxidase). The assay is completed by the addition of the appropriate substrate to the reaction. The enzymatic conversion of substrate indicates the presence of anti-viral antibodies in the original test specimen. These so-called solid-phase "sandwich" techniques have, in general, formed the basis for a wide range of different assays because they are widely applicable if specific anti-virus antibody and/or a source of viral antigens are available. [*See* BIOASSAYS IN MICROBIOLOGY; ELISA TECHNOLOGY.]

Modifications to the assay make it possible to detect acute hepatitis A infections if the presence of IgM antibody is specifically sought. The presence of IgM is indicative of a primary exposure to an infecting agent and seroconversion as part of the initial human humoral immune response (Fig. 1). The later antibody shift to the IgG subclass indicates the development of protective immunity and protection from future infection with the HAV.

2. Principal Reservoir

Infections with the HAV occur in humans and by experimental transmission into nonhuman primates (e.g., marmoset, owl monkey, chimpanzee). A very early report has been noted as indicating that the transmission of the virus from chimpanzee back to humans can also occur. It is generally accepted, however, that the only principal reservoir for the HAV is humans. As noted earlier, transmission is via the fecal–oral route. As such, infections can be common in situations where the standards of hygiene and sanitation might predispose to fecal contamination of the water and/or food supply.

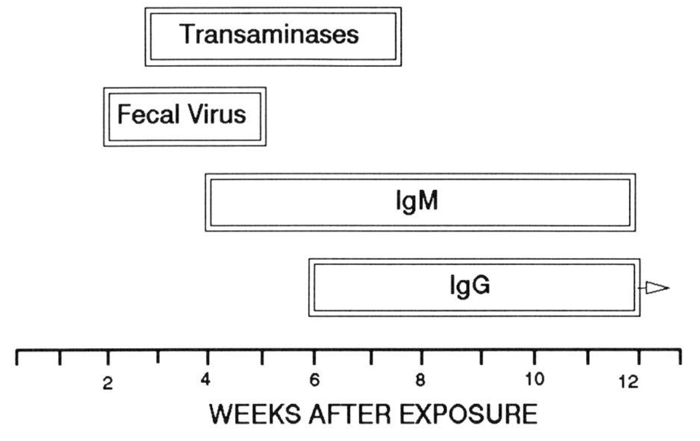

Figure 1 Hepatitis A virus (HAV). Typical immunological and biological events associated with an HAV infection.

3. Endemic Regions

HAV infections occur worldwide. Individuals of all ages can be infected, and the age distribution is generally determined by the particular sanitation conditions that predominate within any given area. A rise in the affluence of any particular region has generally been associated with improvements in the hygienic conditions of the populace. This has led to the recognition that there are two predominant seroprevalence patterns. One is typified by the extremely high seroprevalance rates that are associated with infections at an early age, thereby indicating the early and frequent environmental exposure to virus that is associated with depressed socioeconomic status or substandard living conditions that too often occur in the developing world. Close personal contact in crowded conditions such as prisons, schools, or daycare centers is also a factor associated with the occurrence of hepatitis A outbreaks.

The second incidence pattern is that seen in developed countries with a sufficient sanitation engineering infrastructure such that the rates of virus exposure during the early years are minimized. In these regions of the world, the pattern of infection increases with age into late adulthood. It should be noted, however, that in those areas of the world emerging into industrialized economies, epidemics can occur due to breakdowns in public sanitation leading to disease on a massive scale. Such a situation occurred in the Shanghai epidemic of 1988, when more than 310,000 cases of hepatitis A were reported.

4. Incidence Cycles

There does not appear to be any particular association or seasonal distribution of virus infection. Those cycles that have been observed appear artificial and based on migration patterns of humans from endemic to nonendemic areas. One example of this is the holiday period for German school children, which correlates with the peak of observed infections in late autumn and early winter.

5. Mode of Transmission

As already noted, the documented mode of transmission is fecal–oral. In any fecally–orally transmitted disease, person-to-person contact will also play an important role in viral transmission.

6. Development of Persistent Infection

HAV infections that occur later in life tend to be more serious. A truly persistent state of HAV infection does not exist. There are reports, however, of "smoldering" HAV infections, which can be severely debilitating for extended periods of time before the virus is eventually cleared.

B. Hepatitis E Virus

Hepatitis E has recently been recognized as a distinct disease with fecal–oral transmission similar to that seen for hepatitis A. This similar mode of transmission led to some early confusion as to whether or not there was a distinct etiologic agent that could be distinguished from the HAV. It was clearly differentiated from hepatitis A once diagnostic kits became available for both hepatitis A and hepatitis B. The term non-A, non-B hepatitis (NANBH) was, therefore, coined in order to group these forms of uncharacterized hepatitis from the etiologically recognized forms. This specific form of fecally–orally transmitted hepatitis was referred to as the enterically transmitted NANBH. Since that time, the application of molecular cloning techniques has permitted the cloning of the causative agent. The virus, now referred to as the HEV, is a small, round, nonenveloped virus of 27–34 nm in diameter. Although not yet classified, the virus has certain biochemical and biophysical properties that might place it in the family Caliciviridae.

1. Laboratory Diagnosis

The early work in this field relied on the use of IEM to confirm the diagnosis of hepatitis presumably due to the ET-NANBH agent. The IEM was critical to the confirmation of the cynomolgus macaque (cyno) monkey as an animal model for HEV infection. It was possible to aggregate viral particles from the feces of cynos using sera from humans infected in the course of ET-NANBH epidemics and outbreaks. The morphology of the aggregated particles was identical to that of particles aggregated from the feces of humans infected in the course of ET-NANBH epidemics. These same types of experiments, when performed using antisera from bona fide specimens of acute and convalescent cases of HAV, failed to aggregate virus particles, thereby conclusively demonstrating that the ET-NANBH epidemics resulted from a new form of fecally–orally transmitted agent different from that of the HAV.

The development of the animal model was also critical to the isolation of the first molecular clones in that cloning was facilitated by the availability of infected bile containing a relatively homogeneous population of virus particles. The subsequent characterization of HEV showed conclusively that this was a novel positive-sense, single-stranded polyadenylated RNA virus with a genomic organization and expression strategy different from other previously recognized agents of viral hepatitis.

The molecular cloning of the HEV has provided a source of recombinant proteins that have formed the basis for a diagnostic test. Subgenomic segments of the viral genome were identified and found to encode highly immunogenic epitopes. These epitopes have been incorporated into a prototype diagnostic ELISA for the direct detection of antibodies against the infecting HEV. Although not yet commercially available, the application of this prototype ELISA has modified the prevailing concepts of the epidemiology of the HEV (see later).

2. Principal Reservoir

It has been presumed that, as with the HAV, the principal reservoir for the virus is humans, with circulation into the environment and the maintenance of the virus, as with the HAV, by sporadic infections in the absence of overt contamination of common-source water supplies. There have been interesting reports, however, that the virus can be transmitted to domesticated farmstock (pigs). This raises the possibility that the virus has a zoonotic vector that might maintain the virus in the environment and the cycling of the virus between animals and humans.

3. Endemic Regions

The disease is clearly endemic in developing countries where sporadic infections occur in both adults and children. It was previously believed that the viral infections predominated in the young adult population. It is now clear, however, using the prototype recombinant ELISA, that the virus can lead to both acute and biochemical hepatitis (inapparent infections) in children. The endemicity of virus infections forms the background for widespread epidemics. These epidemics have all occurred in the developing regions of the world, where the status of publicly available sanitation might predispose to large point-source epidemics numbering into the tens of thousands of afflicted individuals. The question remains as to how an epidemic occurs in a population that had been previously exposed as children. We can speculate that a more virulent variant of the HEV may evolve from the normally endemic passage of the virus, or the reason could lie in the host immune response to the infecting HEV, i.e., the absence of a lasting protective immunity after infection. The antibody detected against the HEV does appear to be transient because in the majority of infected individuals detectable antibody disappears by 9–12 mo. These questions are the focus of current and future investigations.

Although developing countries are endemic for HEV-induced disease, evidence of past infection is growing in the developed countries of the world. Using the prototype diagnostic ELISA kit for the HEV, antibody reactivity in normal blood donors has been detected. The basis for this reactivity is unclear because there is no reported history of overt hepatitis in these individuals. Before these findings, the only reported cases of hepatitis in developed countries was that due to travelers returning from endemic areas. The presence of virus-specific antibodies to HEV recombinant proteins was confirmed using synthesized oligopeptides representing these same epitopes. The absence of any past history of infection or outbreaks of hepatitis E indicates that there might be a less pathogenic variant present in industrialized countries or that a related virus is circulating in the community. The rates of seropositivity noted in these studies (1–2%) would indicate that infection with this postulated HEV-like variant is relatively common.

4. Incidence Cycles

It is well documented that the rainy season has been associated with epidemic outbreaks of hepatitis E. In one of the best-studied outbreaks, there was a clear association with the seasonal monsoon rains leading to contamination of the village water supply. An epidemiological investigation showed that the incidence of disease increased with proximity to the stream and the utilization of the stream for cooking and washing. These epidemics obviously required that the fecal contamination of the water supply involved virus excretion by an individual currently infected. This same epidemiologic study also indicated that person-to-person contact was important in the subsequent outbreak that occurred in a nearby community.

5. Mode of Transmission

The HEV is transmitted by the fecal–oral route; however, as already noted, close personal contact forms the basis for much of this type of transmission.

6. Persistent Infection

Persistence, or the development of chronic HEV infection, has not been noted. HEV infections are resolved without any long-term sequelae. It has been noted, however, in many epidemiologic studies that the virus has an extremely high mortality rate in pregnant women (10–20%). The pathobiologic basis for this mortality is unknown, but death is reportedly due to fulminant hepatic failure. Studies are currently underway to determine if the HEV is responsible for other cases of hepatic failure observed outside of the epidemic setting.

III. Epidemiology of Parenteral Viruses

The parenterally transmitted group of hepatitis viruses are those that are spread via parenteral transfer of contaminated blood or blood products. This form of hepatitis was recognized as early as 1885, when it was documented that more than 1000 shipyard workers developed jaundice 2–8 mo following inoculation with a smallpox vaccine derived from human-derived materials. This early observation was documented by several studies carried out in the late 1930s. The seroepidemiologic distinction between hepatitis A (enteric hepatitis) and hepatitis B (parenteral hepatitis) was first defined in the 1960s with transmission studies performed in the Willowbrook State School for the Mentally Handicapped. An agent designated MS-2 (hepatitis B) was transmitted by the parenteral route and had an incubation period of approximately 3 mo.

The parenteral form of hepatitis is spread principally by transfusion of blood and/or blood products. In this context, inoculation with contaminated instruments, such as shared needles by intravenous (IV) drug users and tattoo parlors, or shared bath items (toothbrushes or razors), has also been implicated in disease transmission.

A. Hepatitis B Virus

In 1963, the specific discovery of the HBV was facilitated by the observation that serum derived from an Australian aborigine contained an antigen that reacted with serum antibody from an American hemophilia patient. Approximately 4 years later, this reactivity, called Australia antigen, was found in serum of patients diagnosed with hepatitis B. Subsequent studies showed that Australia antigen, now referred to as HBsAg, represented the envelope material of the DNA-containing virion (HBV) (originally referred to as the Dane particle). The HBV, along with related hepatitis viruses that infect a number of animal and/or avian species, has been placed in a new virus family known as Hepadnaviruses. Their biophysical properties are summarized in Table I.

1. Laboratory Diagnosis

An episode of acute hepatitis is clinically diagnosed by biochemical evaluation of liver function. The most important indicators of hepatocellular damage are noted by an elevation of one or both of two transaminases, serum alanine (ALT or SGPT) and aspartate (AST or SGOT) amino-transferase. In uncomplicated acute viral hepatitis, ALT levels are higher than AST levels, resulting in a low AST : ALT ratio (<0.7). High ALT values and low AST : ALT ratios are highly indicative of viral hepatitis. A number of serologic assays are available to distinguish HAV from HBV infections. The most common acute hepatitis profile includes assays for IgM anti-HAV (see Fig. 1), hepatitis B surface antigen (HBsAg), and IgM anti-hepatitis B core (anti-HBc) (see Fig. 2A). The majority of the commercial assays presently performed are in an ELISA format. The time sequence of these serologic markers, along with ALT rises, are noted in Fig. 2. It should be pointed out that the HBV consists of an electron-dense core (hepatitis B core antigen) that is antigenically distinct from the envelope or surface of the virion, which is designated HBsAg. A third antigen specificity is expressed when the major 22,000-dalton core polypeptide (P22) is secreted as a truncated 16-kDa polypeptide, which then expresses a new antigen specificity [hepatitis B e antigen (HBeAg)]. This antigen provides a key marker to assess the relative infectivity of the serum. HBsAg, which is expressed during the early phase of the disease (2–4 wk), gradually declines within 4–6 mo, at which time anti-HBs becomes detectable (Fig. 2A). Both anti-HBs and anti-HBc persist for many years.

Chronic liver disease develops in approximately 2–10% of adults infected with the HBV. Patients who test positive for HBsAg should be tested for IgM anti-HBc. If IgM anti-HBc activity is low or negative, development of chronic infection is indicated (see Fig. 2B). A follow-up test for HBeAg and anti-HBe is carried out to judge relative levels of infectivity. Patients positive for HBeAg are much more infectious than patients who are anti-HBe-positive.

2. Principal Reservoir

Infected blood derived from acutely infected individuals or chronic carriers represents the only known reservoir for this agent.

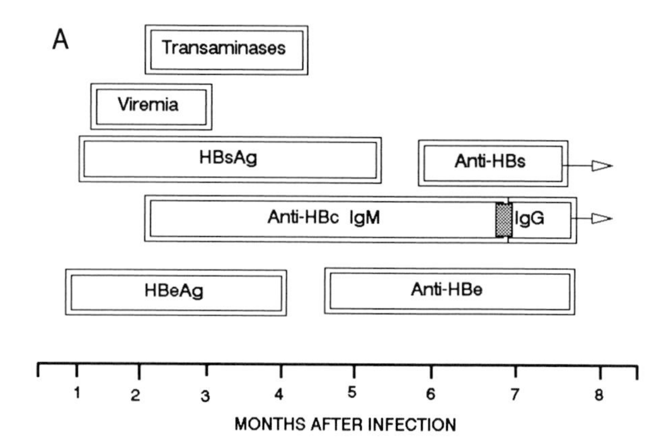

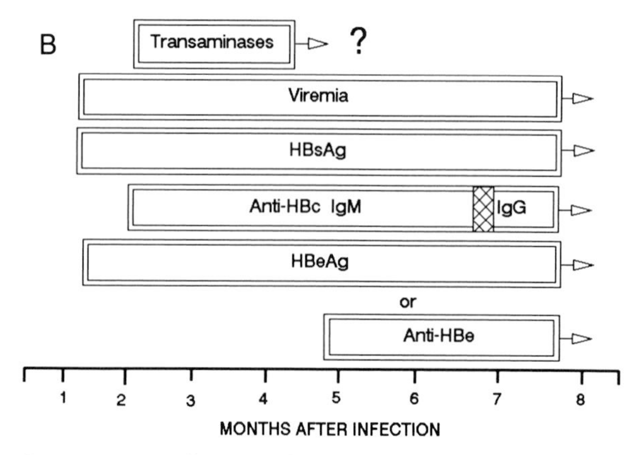

Figure 2 Typical immunological and biological events noted during acute hepatitis B virus (HBV) infection (A) and chronic HBV infection (B).

3. Endemic Regions

Worldwide there are approximately 176 million carriers. A prevalence for HBsAg of 5–20% is observed in Africa, Asia, and Alaska. Rates of 1–5% have been noted in the Middle East, North Africa, southern and eastern Europe, Russia, India, and South and Central America. Lower rates of 0.1–1% are observed in North America, western Europe, the Scandinavian countries, Australia, and New Zealand. Unlike the enteric hepatitis viruses, there does not appear to be a relationship between incidence of infection and any particular time of the year.

4. Mode of Transmission

As noted earlier, HBV-infected individuals, either acute or chronic, represent the reservoir for this virus. Parenteral transmission is the principal mode of transmission; therefore, transfusion of blood or blood products represents a major source of disease. Screening of donor blood units for HBsAg has significantly reduced the risk of HBV infections in transfusion recipients. IV drug users, healthcare workers, and family members of chronically infected patients sharing razors or toothbrushes are at risk. It is also spread by homosexual and heterosexual contact. Lastly, perinatal transmission from infected mother to child is especially high if the mother is acutely infected during the third trimester. Virtually all carrier women transmit the virus to their newborn. Therefore, HBV carrier women represent a large reservoir in developing countries.

5. Development of Persistent Infections

As already stated, 2–10% of adults infected with the HBV develop a life-long chronic infectious state. The course of disease in these carriers varies greatly. Persistent infections are most commonly noted following maternal–neonatal transmission with an incidence of 70–90% noted in neonates born of mothers who are HBeAg-positive (see Fig. 2B). In addition, it has been noted that nearly twice as many men develop chronic infections. The infection may remain quiescent for years or it may progress rapidly, resulting in cirrhosis and death within a short time frame. Histologically, these patients are classified as chronic persistent hepatitis, chronic lobular hepatitis, and chronic active hepatitis, with or without cirrhosis. People with a chronic persistent disease are usually asymptomatic. Some of these people develop symptoms similar to those observed during acute hepatitis. Periods of relapse and remissions are noted in people with infections classified as chronic lobular hepatitis. People with chronic active hepatitis usually have advanced liver histology, as indicated by observations of "ground glass" hepatocytes and involvement of the portal–periportal areas of the liver. As many as 40% of these patients develop sudden onset of acute hepatitis symptoms with biochemical abnormalities.

6. Association with Hepatocellular Carcinoma

A number of investigations have demonstrated a relationship between past or persistent HBV infections and development of primary hepatocellular carcinoma. This relationship is most significant in parts of the world where the prevalence of HBV is high. For example, one study showed that 70% of African hepatoma patients were persistently infected with the HBV. Usually the development of hepatocellular carcinoma follows the natural infection by a period of 20–40 years. Studies with a group of animal viruses related to the HBV have provided strong supportive evidence that the HBV has a causal relationship with hepatoma. The final proof to establish the HBV as a human oncogenic virus awaits the long-term influence that HBV vaccine trials have on reducing the prevalence of hepatoma in high-risk populations.

B. Hepatitis Delta Virus

HDV infections are observed only in acute or chronic HBV-infected individuals. Simultaneous infection with the HBV and HDV results in a clinical course of disease that is indistinguishable from that noted in an acute HBV-infected individual. Infection of an HBV carrier with HDV results in a clinical exacerbation after a short incubation period.

1. Laboratory Diagnosis

The most sensitive assay for detection of the HDV in infected blood derived from chronically infected patients is the polymerase chain reaction (PCR), designed for detection of HDV RNA. The only commercially available assays for serologic detection of the HDV are both a total anti-HDV and IgM anti-HDV test. In a co-infection with HBV, IgM anti-HDV is detected during the acute stage of infection, which disappears after several months. Superinfection usually results in the development of a chronic infection with the HDV. Generally, IgM anti-HDV is the marker of choice for diagnosing superinfection with the HDV. [*See* POLYMERASE CHAIN REACTION (PCR).]

2. Principal Reservoir

The principal reservoir is similar to that noted earlier for the HBV; humans are the only known reservoir for the HDV.

3. Endemic Regions

The virus is highly endemic in southern Italy and the South American countries of Venezuela, Colombia, and Brazil, In addition, a high incidence of anti-HDV has been noted among hemophiliacs and drug users in many parts of the world.

4. Mode of Transmission

The mode of transmission is parenteral. Therefore, the same high-risk populations are infected as those noted for the HBV. The unique characteristic of HDV transmission is that the host must be simultaneously infected with the HBV or an individual must be an HBV carrier to be superinfected with the HDV. People at highest risk for infection with the HDV include IV drug users, hemophiliacs, and HBV carriers receiving blood or blood products. Transmission through sexual practices does not appear to present a major risk. Perinatal transmission has been noted in Italy.

5. Development of Persistent Infections

The outcome of a simultaneous co-infection with the HBV and HDV leads to an acute phase illness where only 1–3% become chronically infected. In contrast, superinfection of HBV carriers leads to an HDV chronic infection in >70% of infected individuals. Fulminant hepatitis is almost 10 times greater in individuals infected with both viruses as compared to those infected with the HBV only. The mortality of fulminant hepatitis is roughly 80%.

6. Association with Hepatocellular Carcinoma

Hepatocellular carcinoma has been noted in chronic HDV-infected individuals. However, a direct carcinogenic effect for this virus has not been established.

C. Hepatitis C Virus

1. Laboratory Diagnosis

The possible existence of additional parenteral hepatitis viruses was first postulated in 1974, when it was established via the absence of serologic markers for the HBV and HAV that a significant number of posttransfusion hepatitis cases continued to occur. This failure to implicate either the HAV or HBV as the causative agent led to the term parenteral non-A, non-B hepatitis (PT-NANBH). The recent scientific findings of two different research groups facilitated the identification of an infectious agent for this disease. The first group constructed recombinant complementary DNA libraries from chimpanzee plasma with high titers of infectivity. Screening of these libraries with human antibody led to the discovery of a positive-sense, single-stranded RNA genome associated with the virus, hereafter referred to as HCV. A second research team reported replication of infectious HCV in normal primary chimpanzee hepatocytes with the morphology of viruses residing within the Flavivirus or Pestivirus family of viruses, including agents such as yellow fever within Dengue fever and hog cholera viruses. The properties of the viral genome corroborate the association with these related virus families.

Recombinant proteins derived from selected regions of the HCV genome have been used to develop an anti-HCV antibody assay to screen human sera for this HCV serologic marker. Based on studies reported around the world, it has been suggested that HCV is a major cause of transfusion-associated NANBH. It is of interest that positive sera, as monitored by the current antibody assay, are potentially infectious. This is confirmed in that HCV genomic RNA can be demonstrated by PCR in these chronic-stage antibody positive sera. This is in keeping with the observation that as many as 75% of individuals infected with HCV develop chronic infections (see Fig. 3B). However, studies with both HCV-infected humans and experimentally infected chimpanzees clearly show that a limited number of individuals develop a stage of convalescence. As shown in Fig. 3A, these individuals appear to develop protective antibody because animals cannot be reinfected when inoculated with homologous acute-phase plasma. Therefore, it is imperative that a commercial assay be made available for monitoring the development of the convalescence stage, i.e., detection of antibody that reacts specifically with the envelope subunit(s) of the virus.

2. Principal Reservoir

Contaminated blood from acute or chronically HCV-infected humans is the only known reservoir of this virus. However, up to 10% of transfusion-associated hepatitis cannot be HCV implicated with

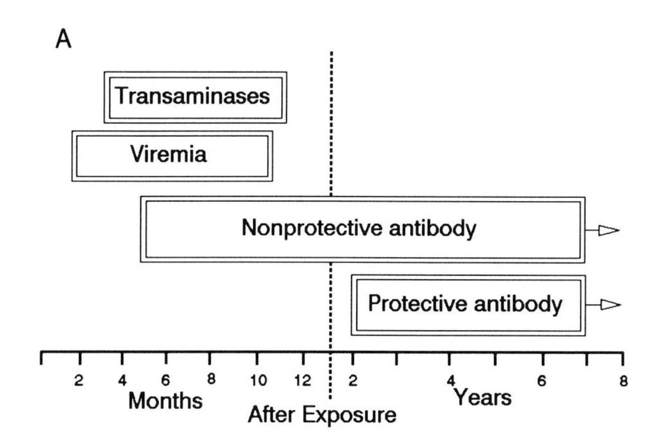

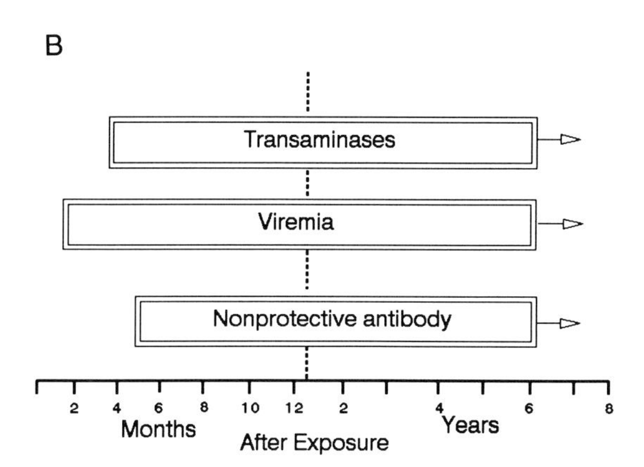

Figure 3 Typical immunological and biological events observed during acute hepatitis C virus (HCV) infection (A) and chronic HCV infection (B).

use of the present serologic assays. This indicates that there may be yet another unidentified PT-NANBH virus.

3. Endemic Regions

HCV-associated NANBH is found in all parts of the world. Up to 1% of the normal blood donor population in developed countries is anti-HCV-positive. Recent studies have shown that the HCV may be more prevalent in Japan. The practice in certain parts of the world of "therapeutic" phlebotomy can substantially increase the rates of infection in localized communities.

4. Mode of Transmission

Blood and blood products have been implicated in the transmission of the HCV. In addition, IV drug use and heterosexual sex have been implicated. These observations suggest that chronic HCV carriers may represent a major risk for transmission of the virus. The role of HCV infections in acute, sporadic, and resolving hepatitis is not clearly defined but may account for <50% of these cases.

5. Development of Persistent Infections

As already pointed out, persistent or chronic infections are noted in up to 75% of humans infected with the HCV. A proportion of these infected individuals remain chronically infected with this virus for life.

6. Association with Hepatocellular Carcinoma

It has been reported that 40–50% of individuals chronically infected with the HCV develop cirrhosis. It appears that hepatocellular carcinoma may complicate cirrhosis in chronic HCV-infected patients. Association of HCV infection with hepatocellular carcinoma appears to be more common in Japan than in other parts of the world. However, the causative role of the HCV for hepatocellular carcinoma has not been fully established.

Bibliography

Abb, J., and Deinhardt, F. (eds.) "Hepatitis A: Infection, Laboratory Diagnosis, and Immune Prophylaxis." Sorin Biomedica, Vercelli, Italy.

Blumberg, B. S., Gerstley, B. J. S., Hungerford, D. A., London, W. T., and Sutnick, A. I. (1967). *Ann. Intern. Med.* **66,** 924–931.

Choo, Q.-L., Kuo, G., Weiner, A. J., Overby, L. R., Bradley, D. W., and Houghton, M. (1989). *Science* **244,** 359–362.

Dawson, G. J., Chau, K. H., Cabal, C. M., Yarbough, P. O., Reyes, G. R., and Mushahwar, I. (1992). *J. Viral Methods* (in press).

Hollinger, F. B. (1990). Hepatitis B virus. *In* "Virology," Vol. 2 (B. N. Fields and D. M. Knipe, eds.), pp. 2171–2236. Raven Press, New York.

Hollinger, F. B., and Dreesman, G. R. (1986). Immunobiology of hepatitis viruses. *In* "Manual of Clinical Laboratory Immunology," 3rd ed. (N. R. Rose, H. Friedman, and J. L. Fahey, eds.), pp. 558–572. American Society for Microbiology, Washington, D.C.

Jacob, J. R., Burk, K. H., Eichberg, J. W., Dreesman, G. R., and Lanford, R. E. (1990). *J. Infect. Dis.* **161,** 1121–1127.

Jacob, J. R., Sureau, C., Burk, K. H., Eichberg, J. W., Dreesman, G. R., and Lanford, R. E. (1991). *In vitro* replication of non-A, non-B hepatitis virus. *In* "Viral Hepatitis and Liver Disease" (F. B. Hollinger, S. M. Lemon, and H. S. Margolis, eds.), pp. 387–392. Williams and Wilkens, Baltimore, Maryland.

Krugman, S., Giles, J. P., and Hammond, J. (1962). *Am. J. Med.* **32,** 717–728.

Purcell, R. H., and Gerin, J. L. (1990). Hepatitis delta virus. *In* "Virology," Vol. 2 (B. N. Fields and D. M. Knipe, eds.), pp. 2275–2287. Raven Press, New York.

Reyes, G. R., and Baroudy, B. M. (1991). Molecular biology of non-A, non-B hepatitis agents: Hepatitis C and hepatitis E viruses. *In* "Advances in Virus Research," Vol. 40 (K. Maramorosch, F. A. Murphy, and A. J. Shatkin, eds.), pp. 57–102. Academic Press, San Diego.

Reyes, G. R., Purdy, M. A., Kim, J. P., Luk, K. C., Young, L. A., Fry, K. E., and Bradley, D. W. (1990). *Science* **247,** 1335–1338.

Velasquez, O., Stelter, H. C., Avila, C., Ornelas, G., Alvarez, C., Hadler, S. C., Bradley, D. W., and Sepulveda, J. (1990). *JAMA* **263,** 3281–3285.

Zuckerman, A. J. (1983). The history of viral hepatitis from antiquity to the present. *In* "Viral Hepatitis: Laboratory and Clinical Science" (F. Deinhardt and J. Deinhardt, eds.), pp. 3–32. Marcel Dekker, New York.

Herpesviruses

E. Littler and K. L. Powell
Wellcome Research Laboratories

Glossary

α-, β-, and γ-Herpesviruses Subdivision of herpesvirus family into subclasses based on modes of replication and, more recently, DNA sequence of genomes—for example: α herpesviruses, lytic viruses including herpes simplex and varicella-zoster virus; β herpesviruses, slow-growing cell-associated viruses including cytomegaloviruses of various hosts; and γ herpesviruses, lytic infections in epithelia but latent in lymphocytes including Epstein–Barr Virus and herpesvirus saimiri

α-, β-, and γ-Proteins Herpesvirus proteins (particularly herpes simplex virus) are subdivided by their temporal classes: α, prior to *de novo* virus gene expression; β, postexpression of α-proteins and not requiring virus DNA synthesis; and γ, postexpression of α and β proteins and more or less dependent on virus DNA synthesis

HERPESVIRUSES are a large group of double-stranded DNA viruses whose genome, usually of >100 million molecular weight, is encapsidated within an icosahedral capsid, which itself is within a lipid envelope containing multiple virus glycoproteins. The viruses are characterized by an ability to become latent within their respective hosts, reappearing sometimes decades after primary infections. The group includes important pathogens of humans (Table I) and of domestic animals. Chemotherapy of herpesvirus disease is becoming commonplace and will undoubtedly increase. The complexity of the viruses and their ability to replicate under a wide variety of conditions makes them models for a range of eukaryotic cell functions including DNA replication, protein synthesis, glycosylation, and transport. Finally, because the viruses can transform cells in culture, they are an excellent system in which to study this important phenomenon.

I. Evolution

The herpesviruses are defined as "large DNA viruses with an icosahedral nucleocapsid, of 100nm in diameter, consisting of 162 capsomers surrounded by a membranous envelope." Some members of the group have been intensively studied, whereas for others little is known beyond this morphologic description. They replicate in the nuclei of eukaryotic cells and contain sufficient genetic information to code for many of the proteins necessary for control of transcription of the viral genes and for replication of the virus genome and its assembly and encapsulation. The group has been divided into three subgroups, namely, α-, β-, and γ-herpesviruses, initially largely based on their pathogenic properties, including site of latency, but more recently by comparison of the information encoded within the virus genome.

The evolutionary relationships among the members of the herpesvirus group have been studied for many years using a variety of approaches. Initial analyses of the size, composition, and organization of the virus genomes led to the conclusion that there was limited homology within members of the group. For example, individual genomes could be shown to vary in size between 100 and 240 kbp and in GC content between 32 and 75%. Liquid DNA hybrid-

Table I The Human Herpesviruses

Virus	Abbreviation	Common disease association
Herpes simplex virus type 1	HSV-1	Cold sores
Herpes simplex virus type 2	HSV-2	Genital herpes
Cytomegalovirus	CMV	Cytomegalic inclusion disease CMV retinitis (AIDS patients) CMV pneumonia (immuno-suppressed)
Varicella zoster Virus	VZV	Chicken pox and shingles
Epstein-Barr Virus	EBV	Infectious mononucleosis (glandular fever)
Human herpes virus 6	HHV-6	Roseola
Human herpes viruses 7 and 8	HHV-7/8	As yet unclear

ization subsequently showed that, with the exception of viruses closely related by host and pathology [i.e., herpes simplex virus type 1 and type 2 (HSV-1 and HSV-2)], the genomes had little sequence homology and varied in the size, location, and nature of terminal and internal repeats. However, the production of specific antisera (initially polyvalent and subsequently monoclonal) to individual herpesvirus proteins showed that a number of virus proteins (e.g., the major capsid protein) did have epitopes common to many members of the group.

The consensus of the evolutionary relationships of the herpesvirus has undergone a revolution with the advent of DNA sequencing technology. Indeed, the derivation of the entire genomic DNA sequence of Epstein–Barr virus (EBV), varicella zoster virus (VZV), HSV-1, and human cytomegalovirus (HCMV) represent some of the most substantial sequencing projects to date. A comparison of the amino acid sequence of predicted open reading frames (ORFs) of human herpesviruses, and other herpesviruses of animals, has led to major advances in our understanding of herpesviruses evolution. For example, it is now evident that a limited number of relatively well-conserved ORFs are coding for functions such as DNA polymerase, major capsid, major DNA-binding protein (MDBP), and glycoproteins B and H. In addition, the relative location of these genes is conserved in blocks or cassettes with each block varying in order and/or direction on the virus genome.

Such analyses shows that some pairs of viruses, such as HSV-1 and HSV-2 or EBV and herpesvirus saimiri (HVS) are relatively well conserved, whereas other viruses such as HCMV are quite distinct. Although the taxanomic relationships generated by DNA sequence comparisons are generally in accordance with those established previously based on the pathogenic properties of the viruses, there are exceptions. For example, Marek's disease virus (a pathogen of chickens) has been classified as a γ-herpesvirus based on its tropism for lymphoid cells; however, its DNA sequence is more similar to that of the α-herpesviruses (such as HSV-1) than to that of the γ-herpesviruses (such as EBV). Such variation is not surprising given that the family comprises such a large number of viruses from a wide selection of hosts, causing markedly different pathologic effects.

Recently, the sequence of channel catfish virus has been obtained and has shown that this virus, while maintaining a small number of partially conserved ORFs, is extremely divergent from other members of the herpesvirus group. No doubt other herpesviruses from such evolutionary primitive hosts will be found to be equally divergent.

II. Structure of the Herpesvirus Particle

The herpes simplex virus particle structure was described in some detail in a seminal paper in 1960. Using the negative-staining technique, researchers examined infected cell extracts and observed naked particles or capsids, some of which were penetrated by stain. The outer dimension of these structures was 105 ± 1 nm and the dimension of the core (the area penetrated by stain) was 77.5 ± 0.5 nm. The mature enveloped particles varied from 145 to 200 nm in diameter. Occasionally, two or more capsids were seen in one envelope. The capsid was icosahedral in shape and was composed of 162 capsomeres, which appeared as hollow tubes 10 nm in diameter and with a 4-nm hole. They were arranged on the capsid according to a T = 16 icosahedral symmetry, with five capsomeres along the edge of the triangular facets of the icosahedron. Recent evidence using three-dimensional reconstruction techniques from cryoelectron microscopy have confirmed these early studies and further confirmed that the capsomeres are either pentons (on the vertices of the icosahedron with fivefold symmetry) or hexons (on the faces with sixfold symmetry), with 150 hexons and 12 pentons in each particle. These more recent studies have also shown that the hexons are arranged in groups of three—"triplexes"—on the capsid surface.

The use of thin-section techniques showed that within the capsid was an electron-dense core consisting of a "toroid" of DNA spooled around a cylinder of protein. In mature virions, this structure was surrounded by the capsid, a "tegument" or inner envelope, and the envelope itself. The existence of the toroid has recently been confirmed by *in vitro* reconstruction studies.

The HSV particle is composed of proteins encoded by the virus genome. The exact number of polypeptide components is still unknown but probably exceeds 30. No host proteins are seen in virus particles. The envelope, like any other eukaryotic membrane, contains an array of glycoproteins labeled gB → gJ (what happened to gA is an interesting snippet of historical herpesvirology that is beyond the scope of this article). The tegument contains the majority of the nonglycosylated virion proteins, but their function is largely unknown. Some important identified functions, however, include that of the α-transinduction factor (α-TIF, or VMW65), a protein involved in shut-off of host functions, and a very large virion protein, VP1. The capsid in mature virions probably consists of seven proteins, including the major capsid protein VP1, VP19C, VP21–VP24, and a smaller 12,000-molecular weight protein. The particles of other herpesviruses are structurally similar but obviously vary widely in their exact composition.

The assembly of virus capsids is still a poorly understood process. Viral DNA is packaged into the toroid within preformed capsids in the cell nucleus. Packaging is a controlled phenomenon using concatomeric DNA, which is packaged into the capsid until signals are given that a "headful" of DNA has been achieved and that a second terminus of the DNA has been packaged. This is a fairly sketchy picture, and many of the principal steps (e.g., cleavage of the DNA as it enters the capsid) occur by unknown mechanisms. The completed capsid attaches to the internal surface of altered patches of the nuclear membrane. These areas of the nuclear membrane are modified by the addition of virus glycoproteins on the outer nuclear surface and virus tegument proteins on the inner surface. The capsid acquires both sets of proteins by budding through this structure. The next "black box" is how the apparently fully matured particle reaches the surface of the cell with its structure intact. Favored mechanisms include transport within cytoplasmic vacuoles and egress through the cisternae of the endoplasmic reticulum.

III. Organization of the Herpesvirus Genomes

For many years, the genomes of herpesviruses have been known to contain fixtures such as internal repeats and inverted terminal repeats. Individual members of the herpesvirus group have different patterns of organization of these repeat structures, as shown in Fig. 1. Thus, two unique regions of coding sequences are commonly surrounded by repeats. One interesting feature is the accumulation of ORFs in the short unique sequence (US) of the herpesviruses, which appear, circumstantially, to play a role in the pathogenesis of virus infection. For example, researchers have shown that all of the ORFs contained in the HSV US region may be deleted without a detrimental effect on virus replication *in vitro*. In addition, with the exception of HSV and VZV, the herpesviruses differ by a large extent in the content of the respective US region. For example, the EBV US contains genes coding only for two small noncoding transcripts, which may play a role in regulation of interferon expression within the infected cell and the promoters for several of the EBV latent genes. The terminal repeats have been shown to contain sequences that play a role in the nucleolytic scission of progeny DNA, a process necessary to form unit length genomes. Although this mechanism would imply the presence of a viral-coded nuclease capable of recognizing sequences in this region, whether or not such an enzyme does exist is not clear; however, one possible candidate would be the virus-coded alkaline nuclease or a complex of this enzyme and other factors involved in sequence recognition.

Within the herpesvirus genomes, a number of origins of DNA replication are found. In the case of HSV, two origins lie within the terminal repeats and one lies between the genes coding for the DNA poly-

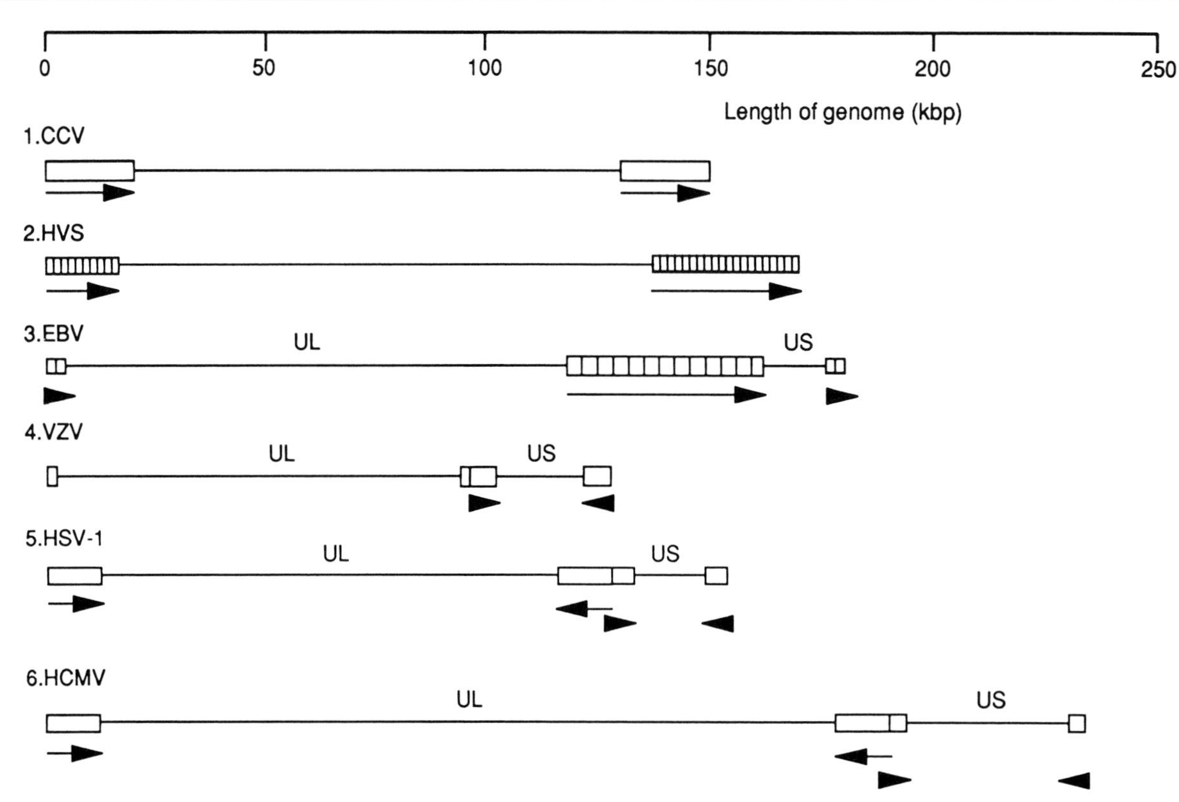

Figure 1 Five known genomic arrangements of herpesviruses. The genome structure of (1) channel catfish virus (CCV); (2) herpesvirus saimiri (HVS); (3) Epstein–Barr virus (EBV); (4) varicella zoster virus (VZV); (5) herpes simplex virus type 1 (HSV-1), and (6) human cytomegalovirus (HCMV) are presented as models for herpesvirus genomes (HSV and HCMV share the same structure). The repeat sequences are shown by rectangles, and their relative orientations are shown by arrows. UL and US denote the long and short unique sequences, respectively, in VZV, HSV-1, and HCMV. Each of the unique sequences is flanked by a pair of inverted repeats. The EBV terminal repeats and internal repeats are tandem in nature and variable in number.

merase and MDBP in the center of the long unique sequence (UL). This location may imply a coordinate role in HSV DNA replication from this origin and control of gene expression of the two flanking genes. In the case of EBV, one origin lies in the US region and one in the UL. The US origin has been shown to bind Epstein–Barr nuclear antigen-1 (EBNA-1) and to be thus recognized by the host DNA replication apparatus. It has been defined as the latent origin. The EBV ori lyt lies in the UL region of the EBV genome but not in the region of the DNA polymerase or MDBP.

The arrangement of other virus genes is best exemplified by recent DNA sequence analyses, which show that virus genes are maintained in conserved blocks that may change in their respective order or orientation in the genome.

IV. Control of Gene Expression

Herpesvirus gene expression has been shown to be coordinated in a highly regulated manner. Experiments with inhibitors of protein, RNA, and DNA synthesis showed that at least three kinetic classes of genes existed, namely, α (those genes transcribed in the absence of prior virus gene expression); β (those expressed before DNA replication), and γ (those expressed after DNA synthesis). It was always recognized that this simple model of cascade regulation of virus gene expression overshadowed more subtle control mechanisms. Recent work with HSV has shown that α-genes are themselves induced by a virus structural protein known as (α-TIF, or VMW65). A host protein that binds to a cis-acting sequence on upstream regulatory regions of α-genes has been identified and it is this complex between host protein and α genes that is in turn recognized by α-TIF.

The two HSV-1 immediate genes ICP0 (infected cell protein) and ICP4 have been shown to be transactivators of virus early promoters. However, although ICP0 has been shown to be a transactivator, its activity compared to that of ICP4 is weak. ICP4 is the major transactivator of HSV genes and is essential for virus replication. ICP4 can be shown to shut off its own expression by binding to sequences at or near the cap site of the transcript. For β- and γ-genes, ICP4 has been shown to act as a transactivator and a repressor, respectively. It has also been shown to act as a nonspecific transactivator of adenovirus and β-globin genes. Indeed, recent work has shown that the DNA sequences to which ICP4 can bind in a specific manner *in vitro* are not essential for the effect of ICP4 when the same sequences are examined upon introduction into the environment of the virus genome. This observation suggests that ICP4 DNA recognition is not tightly specific and that a number of these sites in close proximity to an individual β-promoter are not in themselves essential for the transactivation of ICP4 on that promoter.

Recent work on EBV gene regulation has identified that the protein product of the BZLFI ORF (sometimes known as ZEBRA protein) is an important protein in transactivation of other EBV promoters whose genes are involved in a cascade mechanism of gene expression. Gel shift and DNA footprint analysis would suggest that BZLFI binds to an AP-1 binding site and may function in a similar manner to the *fos–jun* complex. Other EBV ORFs involved in transactivation of virus promoters include BRRF3. It is important to note that the current status of this area of research is similar to that of HSV several years ago and that the *in vitro* findings must be substantiated in the context of the DNA sequences in the virus genome; however, this approach may prove practically difficult due to the lack of a true *in vitro* replicative system for EBV.

HCMV has been shown to code for at least two immediate–early (IE) genes whose transcripts constitute 0.6% of total infected cell RNA. The IE promoter of HCMV is the strongest eukaryotic promoter known. The region responsible for the strongest level of transcription contains a number of regions previously identified as reacting with eukaryotic transcription factors (Sp1, NFκB, AP-1, etc., and a cyclic adenosine monophosphate receptor). In several systems, the IE2 gene can be shown to increase transcription, whereas IE1 has no intrinsic activity but augments that of IE2.

V. Virus–Host Interactions

A virus is defined as being an obligate intracellular parasite, and as such it must interact with its host cells at many different levels. The initial interaction is between the virus particle and a receptor on its target cell. In the case of several herpesviruses, the interaction between virus and cell appears to be through a component of the immunoglobulin superfamily. For example, the receptor for EBV has been shown to be the CRD2 complement receptor, whereas that of HCMV is B–microglobulin. The HSV receptor remains undefined, but the fibroblast growth factor receptor and proteoglycans have been implicated. After entry into the host cell, the next point of interaction is with host factors involved in transcription. A structural component of the HSV particle has been shown to modulate the host cell transcriptional complex to stimulate the expression of viral IE genes (α-TIF; see Section II).

In the case of EBV, expression of virus genes can be shown to be correlated with the induction of cell-surface markers involved in processes such as cell adhesion. It is thought that these phenotypic changes in the EBV-infected B cell manipulate the behavior of the cell to make it more suitable to support viral latency. For example, they may be markers for a change in the infected cells' life span within the individual.

HCMV has been shown to induce a variety of host cell enzymes such as DNA polymerase and thymidine kinase, which may play a role in the manipulation of the environment of the infected cell, making it more amenable to virus replication.

Some evidence indicates that herpesviruses may affect the ability of an infected cell to play a role in the immune response. For example, HCMV has been shown to code for a protein with high similarity to the Class I histocompatability antigen. This protein may prevent the infected cell from presenting HCMV antigens in the correct manner to stimulate a cytotoxic T-cell response. More recently, EBV has been shown to code for a protein with high similarity to the cellular interleukin (IL)-10 gene responsible for the induction of IL-2 and γ-interferon in T-helper 1 cells. It is thought that IL-2 and γ-interferon levels may be important in the establishment of immortalized B cells by EBV *in vitro*.

In addition, some evidence indicates that two small RNA polymerase III transcripts from EBV (referred to as EBERS or JRNAs) modulate the interferon response of an infected cell.

VI. Functions of Herpesvirus Proteins

Herpesviruses vary widely in their genome coding capacity. HSV can encode about 72 proteins and, thus, represents an "average" member of the group. These proteins can be broadly divided into temporal classes and to structural or nonstructural roles.

A. Structural Proteins

Defining the components of the HSV virion can be a difficult exercise because obtaining highly purified preparations of these structures is not easy. It is generally agreed that the virus particle contains about 20 proteins, of which 4–9 are associated with the capsid. It has proven difficult to define roles for most of these polypeptides, so the following account is by no means complete. Functional capsids are largely composed of the major capsid protein of molecular mass 150 kDa, with five additional proteins, VP19, VP22a, VP23, VP24, and VP26, of molecular masses 53, 40, 33, 25, and 12 kDa, respectively. Both the pentons and the hexons of the virus particle are made up of the major capsid protein, whereas VP22a forms the toruslike structure in the virus core. The protein VP22a probably has a scaffoldinglike function for the virions, allowing virus DNA to be packaged. The role of the other capsid proteins in the virus are far from clear.

Proteins found in the envelope and accessible to labeling reagents include eight virus glycoproteins, some of which form "spikes" visible in electron micrographs of the purified virus. These proteins are important as targets for virus-specific neutralizing antibodies and, thus, as components of potential vaccines. The proteins located within the envelope include only one with a clearly defined function: This is the α-TIF, or VMW65, a protein that enhances the production of IE (α) virus proteins during the first stages of infection by the virus.

B. Nonstructural Proteins

1. Immediate–Early α-Proteins

The five α-proteins are made prior to virus DNA synthesis. All probably have regulatory functions, with the best characterized being ICP4, which is required for synthesis of β- and γ-proteins.

2. Early β-Proteins

The β-proteins are not expressed until after the synthesis of α-proteins, but their expression does not require prior DNA synthesis. The β-group (which may be further subdivided) includes the following functions: DNA polymerase, MDBP, DNA helicase, and DNA synthesis origin binding protein, all of which have been shown to be required to replicate plasmid replicons in an *in vitro* cell culture system. Other DNA-related metabolic functions include alkaline nuclease, thymidine kinase, ribonucleotide reductase, uracil-DNA glycosylase, and deoxyuridine triphosphatase (dUTPase). The virus is also known to encode a protein kinase. The detailed description of each of these functions, which have been identified by classical biochemical approaches or sequence homology comparisons, is beyond the scope of this article but we should note that about one-half of the virus genes still have no function allocated to them. (Of these, some will undoubtedly have replicative functions.)

3. Late γ-Proteins

The γ-proteins require virus DNA synthesis before they reach their highest rate of synthesis. Their level of dependence on this event varies, leading to further subdivision of the group. The majority of γ-proteins have a structural function; i.e., they form the virus particle including the virus glycoproteins and the major capsid protein. The aforementioned α-TIF (VMW65) protein also falls into the γ category.

4. Proteins Encoded by Other Herpesvirus Genomes

Although the outline of herpesvirus functions already given reflects the position for the HSV, it can be taken as a generic description of herpesvirus functions. Nevertheless, some interesting differences exist among the various members of the herpesvirus family—for example, VZV and HVS encode a thymidylate synthase; HCMV, a protein with homology to major histocompatibility complex I;

and human herpesvirus 6, a homolog of the AAV-2"rep" gene. Thus, herpesviruses may acquire functions that are intimately involved in their unique replication situations.

VII. Virus DNA Replication

The replication of virus DNA within infected cells is seen as an excellent model system for the more complex eukaryotic cell. Thus, this aspect of herpesvirus replication has had much attention. The major components of the virus DNA replication apparatus are identified as virus-encoded. These include origins of DNA replication [there are three in the HSV genome (see Section III)] and the replication proteins. It is important to remember that certain members of the family not only undergo lytic cycle DNA replication but are also replicated by the host cell machinery during alternative life-styles (e.g., EBV episomal replication). [*See* DNA REPLICATION.]

A. Mode of Replication

The virus DNA is present in infected cells as endless DNA (i.e., it is present in circular or concatomeric forms). It is assumed that circular DNA forms shortly after the release of the DNA from the nucleocapsid and that this DNA is rapidly replicated into branched forms. Late in infection, a complex, branched structure is formed. Some evidence indicates that herpesviruses replicate through a rolling circle mechanism.

B. The DNA Replication Apparatus

The central protein in the DNA replication apparatus is the MDBP. This is made in large amounts in infected cells and clearly has a stoichiometric function in replication. Lack of active DNA-binding protein (e.g., in ts mutant-infected cells) leads to loss of DNA synthesis and redistribution of the other DNA replication proteins. Associated with this protein is the DNA polymerase, which is composed of two subunits. The large polymerase subunit of about 140K molecular weight is associated with the $3' \rightarrow 5'$ exonuclease activity. The carboxy terminus of this protein is required for interaction with the small subunit of 65K molecular weight. This subunit is associated with increased progressivity of DNA replication of a primed template *in vitro*. Further proteins associated with the replication complex are the origin-binding protein, which is presumably required for accurate initiation of DNA synthesis, and three polypeptides, which are associated with the DNA helicase–primase complex.

C. Provision of Nucleotides for DNA Synthesis

As well as the proteins involved directly in DNA synthesis, the herpesviruses encode a variety of proteins whose whole role is to provide an appropriate level of precursors required for the synthesis of virus DNA. These enzymes include the inappropriately named thymidine kinase, which has a wide substrate range and has proven the "achilles heel" of the virus in that it will also activate a variety of antiviral drugs including acyclovir. The HSV also encodes a ribonucleotide reductase and a dUTPase, which presumably function to ensure an appropriate pool of nucleotides in the infected cell.

Other herpesviruses differ in their complement of enzymes; HCMV, for example, lacks a thymidine kinase (at least a recognizable enzyme) and HVS and herpesvirus aeteles encode both thymidylate synthase *and* dihydrofolate reductase enzymes. Late in infection (when virus DNA synthesis reaches its maximum rate), the host cell DNA is degraded and some of its constituents are utilized to make virus DNA. The significance of this recycling process and the role of the virus alkaline nuclease in it are at present unclear.

D. DNA Repair and Proofreading

Apart from the proofreading functions associated with the virus DNA polymerase, the virus encodes a uracil-DNA glycosylase. This enzyme functions to recognize the misincorporation of dUTP into the growing DNA strand and to repair DNA damaged by the deamination of cytosine residues. The virus protein encoded by the UL2 gene of 39K molecular weight is associated with this activity.

E. Isomerization of Herpesvirus DNA

One of the many unsolved questions of HSV replication is how cells infected by a single virus particle containing one isomer of the virus genome results in a cell containing all four potential isomers of the genome in equimolar proportions in 12–18 hr. The mechanism and even the function of this event remain obscure. Certainly a virus frozen (by deletion) in one isomeric form is viable.

VIII. Chemotherapy of Herpesvirus Infections

Practical antiviral chemotherapy began with the use of solutions of iododeoxyuridine (IUdR) for the treatment of occular herpes infections. This drug was too toxic for oral therapy but proved to be useful for this external site and some other topical applications. Trifluorothymidine, a drug with a similar lack of selectivity, has since superseded IUdR in this role, particularly in the United States. After this initial success, some years followed before a truly selective herpesvirus drug was discovered—acyclovir (ACV). This drug resembles guanosine with the 2′ and 3′ carbon atoms missing from the structure, forming a so-called acyclic sugar. Thus, ACV and related compounds are called acyclic nucleosides. It is selectively activated to the monophosphate by the viral thymidine kinase (tk) (Fig. 2), cell enzymes being incapable of efficient phosphorylation, and then to the triphosphate by cell enzymes. The triphosphate is a selective inhibitor of the viral DNA polymerase, giving a double selective effect. The drug is very effective and is thus widely used in the treatment and suppression of HSV infections. It is also used at higher doses for treatment of VZV infection but has little activity against acute HCMV infections, it may have sufficient activity for use in prophylaxis.

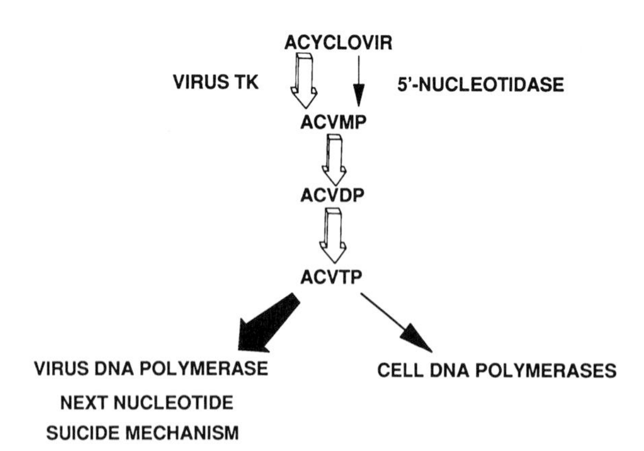

Figure 2 Acyclovir (ACV) has proven to be a prototype for most potential antiherpesvirus drugs. The drug is activated specifically by the virus thymidine kinase (TK) to the monophosphate and then by cell enzymes to the triphosphate. This active species then specifically inhibits the virus DNA polymerase by a well-understood mechanism.

HCMV is a ubiquitous infection but can cause severe disease, e.g., retinitis, pneumonia) in the immunocompromised, especially in AIDS patients or in those patients deliberately immunocompromised (e.g., for transplant). Currently, two drugs are used for the therapy of HCMV disease, but neither is truly selective. Ganciclovir (GCV), an ACV analog, is a reasonably potent HCMV inhibitor with a similar mechanism of action to ACV. GCV is effective in therapy but (unlike ACV) causes fairly severe side effects. Phosphonoformate is a small molecule inhibitor of herpesvirus DNA polymerases. Like GCV, it has poor oral bioavailability, and is effective against HCMV but has a poor side-effect profile.

Thus, although HSV infections are well controlled by ACV, new drugs are needed, especially for the treatment of HCMV and VZV infections. The need for such new agents is particularly urgent for treatment of the small percentage of immunocompromised patients whose virus has become resistant to either ACV or GCV. The mechanism of resistance to ACV has been extensively investigated: It can involve loss of the virus *tk* gene, alteration of the virus *tk* gene leading to changes in substrate recognition or changes in the DNA polymerase gene. Such mutated viruses usually have a reduced capacity to replicate in a healthy host, hence, they cause problems only in the immunocompromised group. Because GCV, due to its toxicity, can only be used in this population, the proportion of patients receiving the drug who have resistant virus is, as expected, much higher. Another complication of GCV use is that its toxicity is similar to that of the antihuman immunodeficiency virus drug zidovudine, or AZT, and this normally precludes the use of both drugs together in the AIDS patient.

New drugs in development for herpesvirus therapy include penciclovir, an analog of ACV, which is somewhat less potent *in vitro* than ACV, and the pyrimidine drug bromovinyl-ara-U (brovavir) with remarkable anti-VZV activity but with some carcinogenic potential. Both these agents rely on specific phosphorylation by the virus tk for their selective antiviral effect, and this emphasizes the importance of the wide substrate specificity of this virus enzyme as a target for chemotherapy.

The treatment of the other human herpesviruses is still in its infancy, and we know too little of the pathophysiology of EBV or the newer human herpesviruses to have developed the rational use of antiviral drugs for these infections.

IX. Immunology and Prophylaxis of Herpesvirus Infections

Most information concerning the immune response to herpesviruses is concerned with the production of antibody that recognizes virus-coded proteins. More interesting may be the role of the cell-mediated immune response, including the cytotoxic T-cell response. It has been shown that recovery from infections by herpesviruses involves the participation of cytotoxic T lymphocytes (CTL). In murine CMV, the principle protein recognized by BALB/C CTL is the nonstructural IE protein. In the case of HSV, the identity of target antigens for CTL is not clear, and each of the seven major glycoproteins have been shown to be suitable. Recently, it has been shown that ICP4, a nonstructural IE protein, contained epitopes recognized by the CTL response in mice; however, ICP0, another IE protein, did not. Vaccinia virus recombinants expressing ICP4 did not protect mice from challenge with HSV, suggesting that, whereas the CTL response may be important for recovery from an established virus infection, the antibody response may have a more protective role. Recently, several herpesviruses have become candidates for protective vaccination. To construct vaccines to protect an individual from a herpesvirus infection, a variety of approaches may be considered. For example, the benign herpesvirus of turkeys has been used as a vaccine to prevent Marek's disease virus from establishing infection in chickens; however, Marek's disease virus has recently been shown to be escaping from this cross-protection and efforts are in progress to develop alternative vaccines. [*See* T LYMPHOCYTES.]

Simple extracts of virus-infected cells that have been biologically inactivated have been used for vaccines against pseudorabies virus. However, the use of such vaccines to prevent human disease is limited by concerns of potential transforming properties of any residual herpesvirus DNA. One approach has been to develop attenuated virus strains by deleting the virus-coded thymidine kinase gene. The Ooka strain of VZV is now used in Japan as a vaccine, and similar genetically crippled mutants of HSV have also been derived. However, such an approach would not be useful to prevent the malignancies associated with EBV infection, because it is thought that the genes involved in transformation and immortalization are largely latent and, thus, expressed before virus DNA replication. Thus, work on development of a vaccine to prevent EBV-associated diseases and malignancies has recently concentrated on subunit or recombinant systems. Purified EBV glycoprotein gp350/220 (MA) has been shown in animal models to protect against challenge from virus. This ORF coding for MA has been expressed using bovine papilloma virus vectors, and protein obtained from this source will be used in phase I clinical trials in the United Kingdom in the near future. MA has also been cloned into recombinant vaccinia virus as a potential live vaccine vector. Although such recombinants have been shown to offer protection to challenge from EBV, whether or not this approach will achieve its exciting potential is not clear at this time.

X. Induction of Neoplasia

Clear, irrefutable evidence indicates that several herpesviruses that infect animals are the cause of neoplasia in their natural hosts. For example, Marek's disease virus has been shown to cause a malignant lymphoma in chickens (the ultimate proof of this statement is that the lymphoma may be prevented by prior vaccination).

A number of human cancers have been associated with herpesvirus infections. In the 1960s and 1970s, seroepidemiological evidence suggested an association between HSV-2 and cervical carcinoma. This association was supported by *in vitro* transformation studies in which HSV-2 DNA, or subgenomic fragments, could be shown to transform mammalian cells morphologically. Further evidence was provided by a variety of reports of HSV-2 DNA, RNA, or protein in cervical carcinoma biopsies. This evidence of a major causative role for HSV-2 in cervical carcinoma was largely refuted by a prospective epidemiological survey, which showed that prior infection by HSV-2 was not a prerequisite for cervical carcinoma. At the same time, it became clear that a wide variation existed in the efficiency of *in vitro* transformation and in the identity of HSV-2 DNA sequences responsible and in the DNA, RNA transcripts, and proteins that were found to be present in tumors. Recently, cervical carcinoma was shown to have a strong association with some types of human papilloma viruses; however, some evidence suggests that HSV-2 may be acting as a cofactor in tumor induction, perhaps using a "hit-and-run" mechanism.

The best evidence for an association between a herpesvirus and human cancer is for the association between EBV and Burkitt's lymphoma (a B-cell lymphoma of Equatorial Africa and New Guinea), nasopharyngeal carcinoma (NPC; a carcinoma of the nasopharynx found in populations in Southeast Asia), and B-cell lymphomas associated with immunocompromised individuals (such as AIDS patients). More recently, EBV has been shown to be associated with Hodgkin's lymphoma.

In each case, EBV DNA, RNA, and proteins have been detected in a very high proportion of biopsy samples. This, taken in conjunction with more indirect evidence such as strong seroepidemiological linkage, the ability of EBV to immortalize B lymphocytes *in vitro,* and animal model systems for EBV-induced malignancy, provides a body of proof of a role for EBV in the etiology of these tumors. The mechanism by which EBV transforms the two distinct tissue types (B lymphocytes and epithelia) is not clear; however, a number of EBV latent proteins have been shown to be present in the malignant cells. For example, EBNA-1 has been found in a large proportion of NPC biopsies with the EBV-coded latent membrane protein in approximately 50%, whereas in Burkitt's lymphoma only EBNA-1 has been shown to be expressed. A number of EBV genes have been shown to have *in vitro* properties that may have a role in neoplasia *in vivo*. EBNA-2 has been shown to be essential for the *in vitro* immortalizing properties of EBV, whereas the latent membrane protein has been shown to transform some rodent cells. The role of EBV genes in the establishment and maintenance of the viral-associated tumors is still not clear; however, further characterization of the molecular properties of EBV genes and their effects on epithelial cell culture systems should clarify the relationship. Ultimate proof of a direct causative role may have to wait for vaccination against EBV infection.

XI. Modes of Virus Latency

The ability of herpesviruses to remain latent in their hosts is one of the definitive characteristics of the group. Two fundamental questions are raised by the latency phenomena. First, what is the site or tissue in which the virus is latent? Second, what is the relationship between the virus and the latently infected cell? (The latter question includes more specific points such as which virus genes are expressed in the cell and what are their roles in the establishment and maintenance of latency.)

The site of virus latency varies from one virus to the next, but several preferred sites may suggest common pathological evolution of the viruses. Several viruses, such as HSV-1, HSV-2, VZV, and PRV, form latent infections in neural tissue. Transport within neurons occurs by neuronal transport mechanisms leading to the rapid appearance of the virus genome in an episomal form in the nucleus. In such neurotropic viruses, no evidence of extensive methylation of the virus DNA exists. A second common site of latency is the hemopoetic system, either T- or B-lymphoid cells or macrophages. For example, EBV is thought to establish a latent infection in B cells (or perhaps a particular subtype of B cells), whereas HVS forms a latent infection in cells of the T-cell lineage. Once again, latent virus exists as episomal DNA in the nuclei of infected cells.

There are a number of herpesviruses, including HCMV, for which no definitive evidence indicates the site of latency; instead, a number of sites have been suggested. This paucity of information is due to the inherent problem of defining what is a latent infection and defining which site is undergoing a latent infection. For example, HCMV has been shown to be not only species-specific but also tissue-specific. A number of sites (the CNS, lymphoid cells) may be infected with HCMV and only express a limited number of virus genes, but it is impossible without animal model systems, such as those that exist for HSV-1, to define whether or not an infection is truly latent.

The role of virus gene expression in latency is still not clear. For example, EBV is thought to have one of the best *in vitro* models for virus latency. B-cell lines that express a limited number of virus gene products exist. Although there are clear differences in gene expression between individual cell lines, it is apparent that only EBNA-1 (required as a DNA origin binding protein to maintain the copy number of episomal DNA) is truly required for latency *in vitro*. However, this system reflects the fact that EBV is latent in a rapidly dividing cell and (hence must replicate its genome), whereas HSV is latent in the genetically dormant site of the neuron. *In vivo* the role of other EBV latent genes (EBNA-2, latent membrane protein) must be important.

Transcription of HSV-1 during latency is limited to one region of the genome located in the terminal repeats. Three poly A-transcripts [latency-associated transcripts (LATs)] of 2.0, 1.5, and 1.45 kb are present and have been shown to partially overlap with the α-gene ICP0 arising from the opposite strand of the genome. Some deletion mutants of HSV-1, which do not express LATs, show impairment in their ability to reactivate; however, recent evidence would suggest that only part of the LATs are essential for this function. Similar transcripts are found in bovine mamillitis virus (LRT) and PRV. In each case, LAT overlaps an IE gene on the complementary strand. The mechanism by which LATs control latency of herpesviruses within a neuron is unknown. However, work with bovine mamillitis virus would suggest that down-regulation of latency-associated transcription correlates with reactivation. It is of interest to see if other herpesviruses that form latent infections in nonneuronal cells have a similar transcript or if this mechanism is specific for those viruses that establish latency in neurons.

Bibliography

Honess, R. W. (1984). *J. Gen. Virol.* **65,** 2077–2107.

Newcomb, W. W., and Brown, J. C. (1991). *J. Virol.* **65,** 613–620.

Roizman, B., and Sears, A. E. (1990). Herpes simplex viruses and their replication. *In* "Virology," 2nd ed. (B. N. Fields and D. M. Knipe, eds.) pp. 1795–1841. Raven Press, New York.

Wildy, P., Russell, W. C., and Horne, R. W. (1960). *Virology* **12,** 204–222.

Heterotrophic Microorganisms

James T. Staley
University of Washington

Glossary

Autotroph Organism that uses inorganic carbon (carbon dioxide) as a carbon source for growth

Chemoheterotroph (chemoorganotroph) Organism that used organic carbon as a source of both carbon and energy (most heterotrophs are chemoheterotrophs)

Eutroph (copiotroph) Organism that grows well when the concentration of organic substances is high

Heterotroph Organism that uses organic material as a source of carbon for growth

Oligotroph Organism that grows well on low concentrations of organic carbon

Photoheterotroph Organism that uses organic material as a source of carbon for growth and light as an energy source

HETEROTROPHIC microorganisms use organic carbon sources as their source of carbon for growth and synthesis of cell material. In contrast, autotrophic microorganisms derive their carbon from an inorganic source, carbon dioxide, from which they synthesize all their cellular material [i.e., proteins, fats, nucleic acids (i.e., DNA and RNA), and polysaccharides].

Heterotrophic organisms depend on autotrophic organisms, the primary producers, for the production of organic materials for their use. Therefore, they occupy higher trophic levels in the carbon cycle. The protozoa are microbial herbivores that may ingest microbial primary producers (algae) or bacteria. Other heterotrophic microbes are mineralizers (decomposers), which degrade the organic products or remains of other organisms. Many heterotrophic microorganisms are saprophytes, which derive their organic carbon nutrients by the decompositon of nonliving organic material. Some heterotrophic microorganisms live on or in close spatial association with other microorganisms, plants, or animals in various types of symbioses. An extreme of these associations occurs when the heterotrophic microbe is a parasite, predator, or pathogen and adversely affects the health or viability of its partner or host. [*See* ORGANIC MATTER, DECOMPOSITION.]

The overall role of heterotrophic organisms in the carbon cycle is to recycle the organic material into inorganic material. In this process, the organic carbon substrate is ultimately converted back into carbon dioxide, the substrate for autotrophic organisms. This process, called mineralization or decomposition, is accomplished by the concerted action of many different heterotrophic organisms including animals as well as heterotrophic microorganisms. But, because of their ubiquitous distribution, abundance, and diverse metabolic capabilities, it is the microbial heterotrophs that are primarily responsible for mineralization.

I. Nutritional and Physiological Groups of Microorganisms

A. Chemoheterotrophic versus Photoheterotrophic

Heterotrophic microorganisms are placed in subcategories based on their energy source (Table I). Chemoheterotrophy is very common in the microbial world, so common in fact that the term heterotrophy is normally meant to refer to this group of organisms. *Most bacteria are chemoheterotrophic; all fungi and protozoa are chemoheterotrophic.* Both the bacteria and the fungi have rigid cell walls and therefore obtain their organic carbon in soluble

Table I Nutritional Types of Microorganisms

Nutritional category	Microbial group(s)
I. Heterotrophic	
Derive carbon from organic sources	
A. Chemoheterotrophic	
Obtain energy from degradation of organic carbon sources	Protozoa, fungi, most bacteria
B. Photoheterotrophic	
Obtain energy from sunlight	Some bacteria
II. Autotrophic	
Derive carbon from inorganic sources	
A. Chemoautotrophic	
Obtain energy from oxidation of reduced inorganic compounds such as ammonia, nitrate, sulfide, sulfur, ferrous ion, and hydrogen gas	Some bacteria
B. Photoautotrophic	
Obtain energy from sunlight	Algae, some bacteria

forms from the environments in which they live. This type of heterotrophy is referred to as osmotrophic. In contrast, protozoa are small animals that lack rigid cell walls and feed by engulfing particulate materials. This type of feeding is referred to as phagotrophic. The protozoa serve as microbial grazers that can ingest bacteria, algae, and other microorganisms as well as small particles of organic material in the habitats in which they live.

Although bacteria and fungi have rigid cell walls that restrict them from ingesting particulate materials, some of them produce extracellular enzymes that allow them to degrade insoluble and refractory substances such as cellulose, lignin, chitin, and hydrocarbons (see later). The soluble by-products of degradation can then be transported into the cell as carbon and energy sources. [*See* ENZYMES, EXTRACELLULAR.]

Photoheterotrophy is confined to three bacterial groups: the green gliding bacteria such as *Chloroflexus*, the heliobacteria, and the purple bacteria. The purple bacteria can use sunlight as their energy source but can use certain organic carbon sources as carbon for growth while growing anaerobically in the presence of light. There are several genera of the so-called purple "nonsulfur" bacteria including *Rhodopseudomonas*, *Rhodospirillum*, *Rhodobacter*, *Rhodomicrobium*, and *Rhodocyclus* that use organic acids such as acetate or succinate or ethanol as a source of carbon and sunlight as a source of energy. Some of the purple "sulfur" bacteria, which normally grow as anaerobic photoautotrophs, can also use organic compounds.

B. Oxygen Requirements for Growth

Many microorganisms (and virtually all higher organisms) require oxygen for growth. Such organisms grow by aerobic respiration (see later) and are referred to as obligate aerobes. At the other extreme, the growth of some microbes is inhibited by oxygen. These are referred to as obligate anaerobes. Many of the obligate anaerobes such as members of the bacterial genus *Clostridium* obtain energy from fermentation of various sugars, amino acids, polysaccharides, and proteins. These obligate anaerobes live in the intestinal tracts of animals or in sediments and muds or in waterlogged soils.

Facultative anaerobes may grow either in the presence of oxygen or in its absence. Included in this group are organisms such as *Escherichia coli,* a common bacterium found in the intestinal tract of all humans as well as many other warm-blooded animals. *Saccharomyces cerevisiae,* the common baker's and brewer's yeast, is another example of a facultative anaerobe. Facultative anaerobes may grow by fermentation when they grow anaerobically or they may use alternate terminal electron acceptors to oxygen in anaerobic respiration (thus, nitrate or sulfate can be used by some bacteria) (see Section IV).

Another group of microorganisms require oxygen, but they cannot grow if the environment is saturated with oxygen at normal atmospheric pressures. These bacteria are called microaerophilic.

C. Water Activity Requirements

Life and living organisms require water, which must be in the liquid state. Microorganisms are no exception. Most bacteria cannot grow unless the environment has a relative humidity of 95% (this is equivalent to a water activity of 0.95). On the other hand, many microorganisms can be readily freeze-dried and stored for years in that state. Thus, under the proper conditons, organisms can survive in conditions of aridity they could not possibly tolerate for growth. Nonetheless, some microorganisms can actually grow under conditions of low relative humidity. For example, the extreme halophilic (i.e., salt-loving) bacteria, such as *Halobacterium* species can grow in saturated salt brines at a water activity of 0.75. *Staphylococcus aureus*, a bacterium that grows on the skin of humans, can grow at a water activity of 0.90. However, some of the fungi can grow under even more arid conditions. For example, the filamentous fungus *Xeromyces bisporus* and the yeast *Saccharomyces rouxii* grow at a relative humidity of 60%. Microorganisms that grow in environments with low relative humidity are called osmophiles.

II. Microbial Groups and Classification

A. Prokaryotic versus Eukaryotic

Bacteria are prokaryotic, whereas fungi and protozoa are eukaryotic. Prokaryotic organisms differ from eukaryotic organisms in several ways. First, and most distinctively, the cells of prokaryotic organisms are structurally more simple than eukaryotic organisms. The nucleus of prokayrotic cells contains only one chromosome (a double-stranded DNA molecule that carries the heridtary information of the organism in its genes), and this is not bound by a nuclear membrane (Fig. 1). In contrast, the nucleus of eukaryotic organisms contains more than one chromosome and is bound by a nuclear membrane (Fig. 2). Because eukaryotes have more chromosomes, their process of cell division, called mitosis, is more complex. Mitosis involves replication and separation of the chromosomes from one another prior to cell division to ensure that each daughter cell receives a complete set. Mitosis does not occur in prokaryotic organisms.

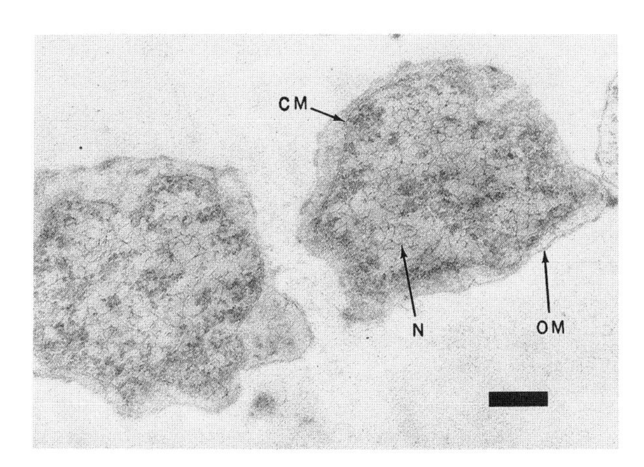

Figure 1 Thin section through the bacterium *Prosthecomicrobium enhydrum*. Note that the nuclear material (N) is not bound by a membrane but is within the cytoplasm. This is a gram-negative bacterium and the outer cell-wall membrane (OM) appears as a double-track membrane (it consists of lipid, protein, and polysaccharide). The peptidoglycan layer, although not discernible here, is located between the outer cell-wall layer and the cell membrane (CM). Bar = 0.2 μm.

A second major feature that distinguishes between these two groups is that eukaryotes have membrane-bound organelles within thier cells whereas prokaryotes do not. One example of this is the mitochondrion (Fig. 2). This organelle is actually about the size of a bacterial cell and is bound by its own protein–lipid cell membrane and even contains its own DNA and ribosomes. It is of interest to note that the DNA of the mitochondrion resembles that of bacteria in that it has no nuclear membrane around it and, furthermore, its ribosomes are small, like the ribosomes of prokaryotes, which are smaller than eukaryotic ribosomes. Evidence such as this has led some scientists to propose that the mitochondrion and the chloroplast (a similar membrane-bound organelle found in phototrophic eukaryotes that contains the chlorophyll pigments for photosynthesis) might have originated from prokaryotic organisms and become incorporated into eukaryotes through a process called endosymbiotic evolution.

Additional differences exist between prokaryotic and eukaryotic microorganisms (Table II). For example, the cells of eukaryotes are generally much larger compared to those of prokaryotes, and most eukaryotes are multicellular, a less common feature among prokaryotes.

Moreover, sexuality in bacteria is quite different from that of eukaryotic organisms. Eukaryotes have male and female mating types, which produce special reproductive cells called sperm and egg (or + and − cells). These cells fuse during mating (sexual reproduction) to form, either immediately or eventually, a fertilized diploid cell called a zygote, which receives one set of chromosomes from each parent. Prokaryotic organisms are more primitive. They are never diploid, but are always haploid, meaning that they do not have a second chromosome set with different genetic features (the closest to this in bacteria are plasmids, which are circular DNA pieces that contain genes and are found in the cytoplasm of some bacteria) Some types of genetic exchange can occur betweeen bacteria, but cell fusion, zygote formation, and diploidy are unknown in this primitive group of prokaryotes.

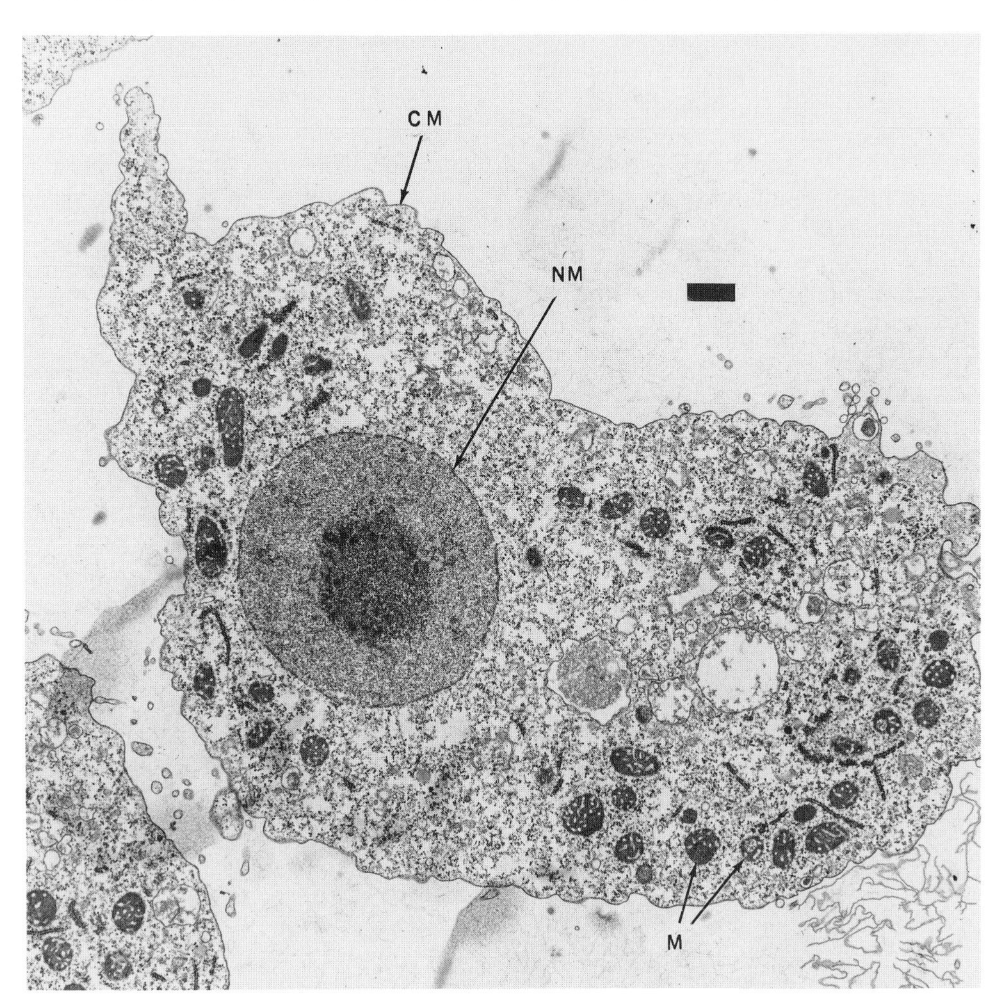

Figure 2 Thin section through a protozoan, an *Acanthamoeba* spp., which shows the nuclear membrane (NM) around the nucleus as well as many mitochondria (M). Unlike fungi and most bacteria, there is no cell wall in this member of the Protista, only a bounding cell membrane (CM). Also note how much larger this organism is than the bacterium in Fig. 1. Bar = 1.0 μm. [Courtesy of Thomas Fritsche.]

Table II Differences between Prokaryotic and Eukaryotic Cells

Characteristic	Prokaryotic	Eukaryotic
Nuclear features		
Nuclear membrane	–	+
Chromosome	1	>1
Mitosis	–	+
Cellular organelles		
Mitochondria	–	+
Chloroplasts	–	+ (if photosynthetic)
Ribosomes	small	large
Cell diameter (typical)	0.5–2 μm	>5 μm

B. Classification of Microorganisms

There are several groups of microorganisms. The bacteria and cyanobacteria are prokaryotic, and the fungi, algae, and protozoa are eukaryotic. Higher organisms, the plants and animals, are also eukaryotic. One popular classification system for organisms is the five-kingdom system proposed by R. Whittaker (Fig. 3). In this system, the prokaryotic organisms are placed in the kingdom Monera. The protozoa are placed in the kingdom Protista and the Fungi are placed in their own kingdom. The algae are classified in the Plant kingdom and the remaining kingdom is the Animal kingdom.

Some microbiologists argue that there are really only three kingdoms: (1) Eukaryota, which contains

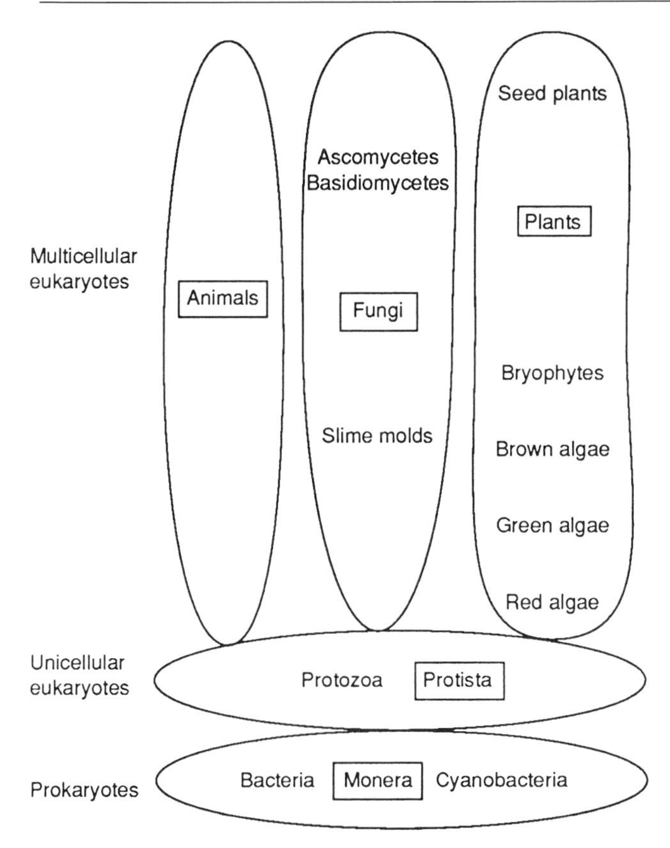

Figure 3 Diagram showing the five-kingdom classification system of earth's living organisms. Note by this scheme that the heterotrophic microorganisms all occupy their own separate kingdoms: The bacteria comprise the kingdom Monera; the protozoa, the Protista; the fungi, the Fungi. [General scheme courtesy of R. Whittaker.]

all eukaryotic organisms (i.e., fungi, protozoa, algae, plants, and animals), (2) Eubacteria, and (3) Archaebacteria (Fig. 4). Evidence for this three-kingdom classification system comes from comparisons of the ribosomal RNA (rRNA) from different types of organisms. All organisms have ribosomes, and the ribosome is a highly conserved structure. Therefore, the sequence of nucleotides in rRNA has been used to compare the similarities among widely different organisms. Ribosomal RNA analyses indicate that two separate groups of prokaryotic organisms exist: the Eubacteria and the Archaebacteria, which, by this analysis, are as distantly related to one another as they are to eukaryotic organisms. If additional evidence supports this three-kingdom system of classification, it may become more widely accepted among biologists.

Although they are biological entities, viruses are not considered organisms because they are not cellular; i.e., they do not have cytoplasm with enzymes and metabolic pathways and they do not carry out reproduction in the same manner as organisms. Viruses will not be treated further here.

III. Heterotrophic Microorganisms

A. Prokaryotic

1. Characteristics of Eubacteria

As already indicated, bacteria are prokaryotic organisms and, based on analyses of rRNA as well as other evidence, bacteriologists separate the bacteria into two different groups: Eubacteria and Archaebacteria. The Eubacteria contains the more common bacteria, those with which most people are familiar. Thus, the bacteria that cause bacterial diseases (e.g., cholera, plague, bacterial pneumonia, whooping cough, tetanus, botulism, scarlet fever, bacterial dysentery, diphtheria) are all Eubacteria. Furthermore, they are also the most common organisms that grow in usual soil and water habitats.

Almost all of the Eubacteria contain a cell wall constituent called peptidoglycan, which is not found in Archaebacteria or anywhere else in the biological world. Peptidoglycan is an unusual polymer that consists of a chain of alternating amino sugars (glucosamine and muramic acid) that is interconnected with short segments of amino acids—a peptide chain. Some of the amino acids in the peptidoglycan are D-amino acids, such as D-alanine and D-glutamic acid. These D-amino acids are found only in Eubacteria, and in eubacteria they are never found in enzymes and other proteins, only in the cell-wall peptidoglycan (some organisms also produce extracellular slime layers with D-glutamic acid). It is of interest to note that the peptidoglycan is the site of action of the antibiotic penicillin (and its derivatives—e.g., ampicillin). Cells that contain peptidoglycan are susceptible to penicillin because they cannot synthesize their cell walls properly if this antibiotic is present in the environment. Since only Eubacteria have peptidoglycan, they are the only organisms that are inhibited by the antibiotic. This is an example of selective toxicity, a concept that explains how antibiotics can be used in human therapy to kill pathogenic bacteria without adversely affecting the patient.

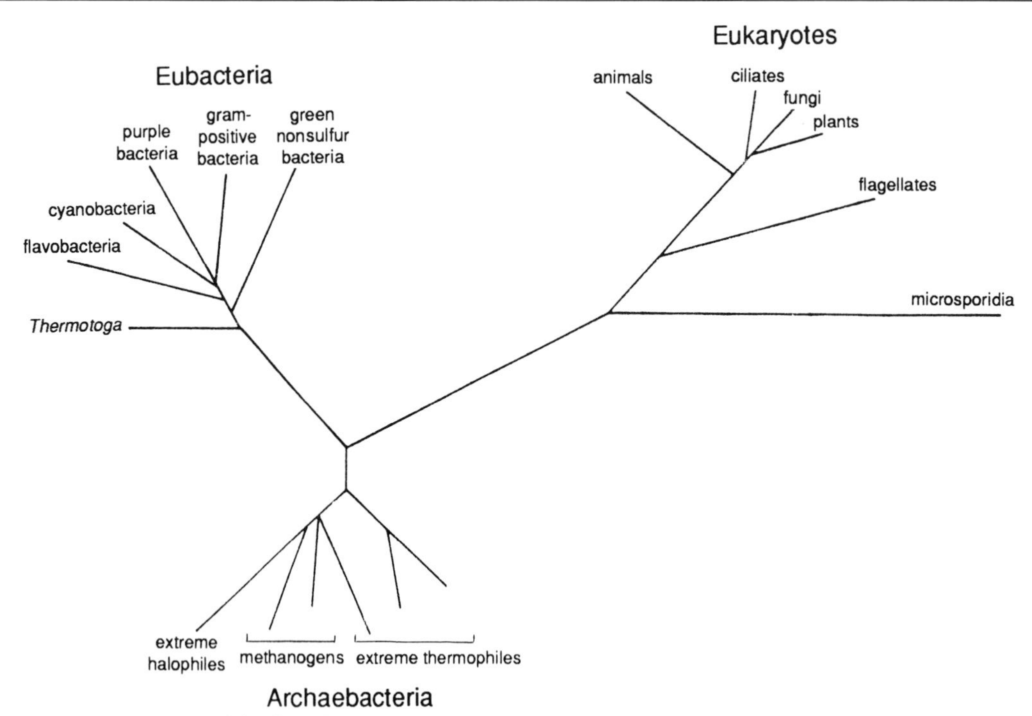

Figure 4 Diagram of the three-kingdom system of classification based on phylogenetic studies using 16S ribosomal RNA information. Note that the Eubacteria are as distant from the Archaebacteria as both groups are from the Eukaryota [Adapted from Woese, C. (1987). *Microbiol. Rev.* **51,** 221–271.]

2. Heterotrophic Eubacteria

Tables III and IV contain listings of the many different groups of heterotrophic Eubacteria. The Eubacteria are separated into two broad groups based on their Gram stain reaction, a simple laboratory staining test that differentiates between the two groups. The basis for the difference in Gram stain is related to the structure of the eubacterial cell wall. In gram-positive bacteria, the major cell component is peptidoglycan. Most gram-negative bacteria also have

Table III Gram-Negative Heterotrophic Eubacteria

Group	Representative genera
1. Spirochetes	*Treponema, Leptospira, Borrelia*
2. Aerobic spirilla	*Spirillum, Aquaspirillum, Campylobacter*
3. Aerobic rods and cocci	*Pseudomonas, Rhizobium, Agrobacterium, Methylomonas, Acetobacter, Legionella, Neisseria*
4. Facultative anaerobes	*Escherichia, Salmonella, Klebsiella, Erwinia, Yersinia, Proteus, Vibrio, Photobacterium, Haemophilus, Zymomonas*
5. Anaerobic rods	*Bacteroides*
6. Dissimilatory sulfate and sulfur-reducing bacteria	*Desulfovibrio, Desulfomonas*
7. Rickettsia and chlamydia	*Rickettsia, Chlamydia*
8. Cell wall-less eubacteria	*Mycoplasma, Spiroplasma*
9. Budding and prosthecate bacteria	*Caulobacter, Hyphomicrobium, Prosthecomicrobium*
10. Planctomyces group	*Planctomyces, Pirellula*
11. Sheathed bacteria	*Sphaerotilus, Leptothrix*
12. Gliding, nonfruiting bacteria	*Cytophaga, Beggiatoa, Simonsiella*
13. Fruiting, gliding bacteria	*Myxococcus*

Table IV Heterotrophic Gram-Positive Eubacteria (Including Actinomycetes)

Group	Representative genera
1. Gram-positive cocci	*Micrococcus, Staphylococcus, Streptococcus, Sarcina, Leuconostoc*
2. Deinococci	*Deinococcus*
3. Nonspore-forming rods	*Lactobacillus, Listeria, Caryophanon*
4. Spore-forming rods	*Bacillus, Clostridium, Desulfotomaculum*
5. Irregular rods	*Arthrobacter, Corynebacterium, Propionibacterium, Cellulomonas, Actinomyces, Bifidobacterium*
6. Acid-fast bacteria	*Mycobacterium,* some *Nocardia*
7. Actinomycetes and streptomycetes	*Frankia, Actinoplanes, Streptomyces*

peptidoglycan, but they have smaller amounts of it and contain an additional major cell-wall layer called the outer membrane, which consists of protein, lipid, and some polysaccharide (see Fig. 1). Because of the simplicity and utility of the test in differentiating between these two bacterial groups, the Gram stain is still commonly used by bacteriologists.

3. Heterotrophic Archaebacteria

Archaebacteria do not produce peptidoglycan, a major difference between them and typical Eubacteria (Table V). They also differ from Eubacteria in the environments in which they live and activities in which they are involved (Table V). The archaebacteria live in extreme environments and carry out unusual metabolic reactions. [*See* ARCHAEBACTERIA (ARCHAEA).]

There are three different groups of Archaebacteria (Table VI). One group, the methanogens, produce methane gas. These are obligate anaerobes that cannot grow in the presence of oxygen. Therefore, they are found deep in aquatic sediments or in special anaerobic environments such as the rumen of cattle or the termite gut. There are two subgroups of the methanogens. One group produces methane gas from hydrogen gas and carbon dioxide. Because this group does not use organic material as a carbon and energy source, these bacteria are classified as autotrophs. The other methanogens are heterotrophic bacteria that use methanol or acetate as carbon sources from which they produce methane gas.

Another archaebacterial group is the extreme halophilic bacteria. These bacteria grow in saturated salt brines. They grow as ordinary heterotrophs and balance the external osmotic pressure by maintaining a high internal concentration of potassium ions. Some of these bacteria can carry out a primitive type of light-driven adenosine triphosphate (ATP) production (photophosphorylation) by virtue of a special pigment they produce called bacteriorhodopsin, which is similar to the rhodopsin found in human eyes. In these bacteria, the pigment is associated with a membrane called the "purple membrane," which is responsible for energy production.

The final archaebacterial group is the extreme thermophiles. Some of these bacteria can grow at temperatures as high as 110–120°C. They grow in

Table V Differences between Eubacteria and Archaebacteria

Characteristic	Eubacteria	Archaebacteria
Peptidoglycan	+[a]	–
Unusual capabilities		
Methane production	–	Methanogens
Growth in saturated salt solutions	–	Extreme halophiles
Growth at 70–120°C	–	Extreme thermophiles

[a] Some Eubacteria do not produce peptidoglycan (e.g., the mycoplasmas lack cell walls entirely and therefore do not produce peptidoglycan).

Table VI Heterotrophic Archaebacteria

Group	Representative genera
Extreme halophiles	*Halobacterium*
Extreme thermophiles[a]	
Sulfur reducers	*Pyrococcus*
Methanogens[b]	*Methanosarcina*

[a] Many extreme thermophiles are autotrophic being able to use H_2S or hydrogen gas as energy sources.

[b] Many methanogens are autotrophs that use CO_2 as their sole source of carbon and hydrogen gas as an energy source.

hot springs fed by volcanic fumaroles, such as those at Yellowstone Park or the deep sea hydrothermal vents. Some of these bacteria are autotrophic sulfur oxidizers, such as the genus *Sulfolobus*. *Sulfolobus* species oxidize sulfur to sulfuric acid and grow at temperatures approaching the boiling point of water. They are called thermoacidophiles because they grow at high temperatures and very low pH values (pH 1–6). This genus is facultatively autotrophic in that it can grow both autotrophically and heterotrophically. As an autotroph, it uses sulfur as an energy source and carbon dioxide as a carbon source; as a heterotroph, it can use complex organic carbon compounds such as yeast extract. Many of the sulfur-oxidizing Eubacteria, including members of the genus *Thiobacillus*, are also facultative autotrophs.

Some of the extreme thermophiles use organic compounds that they oxidize while reducing sulfur compounds. For example, *Pyrococcus furiosus*, which grows at temperatures from 70 to 103°C, oxidizes organic components of yeast extract or peptone anaerobically and reduces elemental sulfur (S^0) to hydrogen sulfide in a type of anaerobic respiration.

B. Eukaryotic

1. Fungi

All fungi are heterotrophic. Most live as saprophytes on the organic remains of plants and animals; however, some are plant or animal parasites or pathogens. All have rigid cell walls composed of cellulose or chitin, or both. They derive their nutrition from soluble organic compounds.

The fungi are classified in two divisions (Table VII). The division Mastigomycota produce motile spores called zoospores. Included here are the chytrids, a group of aquatic fungi, and the plasmodia-forming species. The second division is called the Amastigomycota, which do not produce motile spores. The Amastigomycota are separated into four major subdivisions (Table VII). The simplest types are the Zygomycotina (zygomycetes), which produce zygosopres in their sexual stage. One example of this group is *Rhizopus stolonifer*, also called the "black bread mold" because it is commonly found on stale bread. The Ascomycotina (or ascomycetes) is one of the most common groups. It includes the unicellular fungi (yeasts) as well as a group of macroscopic, mycelial types. The ascomycetes produce their sexual spores in a distinctive structure called an ascus, or sac. The group of morels are ascomycetous mushrooms.

The Basidiomycotina (basidiomycetes) are another very important group of the fungi. These comprise the most common mushroom and toadstool types. Thus, *Agaricus bisporus*, the common edible store mushroom, is a basidiomycete. Basidiomycetes produce their sexual spores on a special sexual structure called a basidium. These basidia are borne in the aerial fructifications (mushrooms) of these organisms. The basidiospores are dropped into the air and may be dispersed long distances by the wind.

The final group of the fungi are the Deutermomycotina (also called deuteromycetes or Fungi Im-

Table VII Major Groups of Fungi

Scientific group (vernacular name)	Representative genera or groups
Division: Mastigomycota	
Class: Chytridiomycetes	chytrids
Class: Hyphochytridiomycetes	*Rhizidiomyces*
Class: Plasmodiophoromycetes	*Plasmodiophora*
Class: Oomycetes	*Saprolegnia*
Division: Amastogomycota	
Subdivision: Zygomycotina (zygomycetes)	*Rhizopus, Mucor, Pilobolus*
Subdivision: Ascomycotina (ascomycetes)	
a. Unicellular (yeasts)	*Saccharomyces, Debaryomyces*
b. Filamentous	*Penicillium, Fusarium, Alternaria, Aspergillus*
Subdivision: Basidiomycotina (basidiomycetes)	*Agaricus, Boletus*
Subdivision: Deuteromycotina (Fungi Imperfecti or deuteromycetes)	*Candida, Trichosporon, Geotrichum Botrytis, Alternaria, Fusarium*

perfecti) a collection of organisms whose sexuality is not yet determined. This group contains many of the common molds such as the genera *Alternaria*, *Penicillium*, *Nigrospora*, and *Fusarium*.

2. Protozoa

The protozoa are complex unicellular organisms. They fall into several phyla based largely on their morphology, type of motility, and life cycles (Table VIII). Important parasites, such as in the genera *Plasmodium* (malaria), *Giardia* (giardiasis), *Trypanosoma* (sleeping sickness), *Trichomonas* (trichomoniasis), *Entamoeba* (amoebic dysentery), and *Naegleria* (encephalitis), have been the most thoroughly studied. However, their role in ecological processes is also of considerable interest. Because they feed like higher animals, i.e., by phagotrophy, they are the simplest organisms that process particulate food internally. They and a large variety of invertebrate animals ingest particulate organic materials (detritus). Bacteria are attached to this organic material and they, too, are involved in its decomposition. Thus, these bacteria, i.e., those that are attached to the detritus, and those that colonize the animals, and the animals themselves all act in a concerted manner in the degradation of the particulate organics, a process referred to by microbial ecologists as the microbial loop.

IV. Metabolism of Organic Compounds

A. Fermentation

The simplest metabolic process involved in the degradation of organic substances is fermentation. In this process, which occurs in the absence of oxygen, a variety of organic compounds can be broken down. For example, lactic acid bacteria ferment sugars to produce lactic acid. This fermentation proceeds via glycolysis using the enzymes of the Emden–Meyerhof pathway. A sugar such as glucose is oxidized anaerobically to pyruvic acid and the chemical bond energy from the process is captured in the synthesis of ATP. During the oxidation, a dehydrogenase enzyme containing the coenzyme nicotinamide adenine dinucleotide (NAD^+) is reduced to $NADH^+H^+$. As this accumulates in the cells, it in turn is used to reduce the pyruvic acid ($C_3H_4O_3$) to lactic acid ($C_3H_6O_3$). Therefore, the ultimate product of the pathway is lactic acid, not pyruvic acid. This fermentation is carried out by the gram-positive lactic acid bacteria such as members of the genera *Streptococcus* and *Lactobacillus* as well as others. As in all fermentations, an organic compound (i.e., a sugar) is oxidized for energy and another one (i.e., pyruvic acid) is reduced.

Other sugar fermentations include the yeast fermentation by *Saccharomyces cerevisiae*, or brewer's yeast, in which ethanol and carbon dioxide are produced, and the mixed acid fermentation by the enteric bacterium *Escherichia coli*. In this latter example, acetic, lactic, and succinic acids as well as carbon dioxide and hydrogen are produced from sugars. There are also bacteria that produce propionic acid (*Propionibacterium* from Swiss cheese) and others that produce butyric acid and butanol as well as others that produce a variety of other products. Amino acids can also be fermented by some bacteria, particularly the gram-positive spore-forming bacteria in the genus *Clostridium*. Even polymeric compounds such as cellulose and chitin can be broken down anaerobically by fermentative processes.

In fermentations, other organic products are usu-

Table VIII Phyla of Protozoa and Representative Genera

Phylum and subphylum (vernacular group)	Representative genera or groups
Sarcomastigophora	
Sarcodina (amoebae)	*Amoeba*, *Entamaeba*, *Naegleria*, *Giardia*, *Dictyostelium*, *Acanthamoeba*, foraminiferans
Mastigophora (flagellates)	*Trypanosoma*, *Trichomonas*
Labyrinthomorpha	*Labryinthula*
Apicomplexa (immotile forms)	*Plasmodium*, *Toxoplasma*
Microspora	
Myxozoa	*Ceratomyxa*
Ciliophora (ciliates)	*Paramecium*, *Stentor*, *Tetrahymena*, *Bursaria*, *Vorticella*

ally formed such as the organic acids mentioned earlier. In addition, gases such as hydrogen and carbon dioxide are frequent products. These substances can be used by other bacteria in the environments in which they are produced. For example, the hydrogen and carbon dioxide are substrates for methanogenic bacteria, which can obtain energy from the oxidation of hydrogen and form methane gas as a product of their metabolism.

B. Respiration

1. Aerobic Respiration

Aerobic respiration occurs in the presence of oxygen, O_2 . This process is carried out by almost all plants and animals. It is also a process used by many aerobic microorganisms that oxidize organic compounds. Many bacteria that respire have the enzymes of the glycolytic pathway and therefore produce pyruvic acid as an intermediate product. If the organisms has the tricarboxylic acid cycle (TCA cycle), then the pyruvic acid can be further metabolized. In one passage through the TCA cycle, pyruvic acid can be completely oxidized to carbon dioxide and water. The carbon dioxide is produced directly from decarboxylation of the intermediate organic acids such as pyruvic acid, α-ketoglutaric acid, and citric acid. Hydrogen ions and electrons are produced during the oxidation. The electrons are passed through an electron transport chain of cytochromes and react with oxygen to produce water. Therefore, a sugar molecule that is respired biologically can be completely oxidized to carbon dioxide and water:

$$C_6H_{12}O_6 + 6\ O_2 \rightarrow 6\ CO_2 + 6\ H_2O$$

Organisms that use sugars as carbon and energy sources may oxidize these materials completely to inorganic products during the sequential processes of glycolylsis and respiration. ATP is produced in respiration from the ATPases that are located in the cell membrane and operate on the gradient of hydrogen ions (protons) established between the interior and exterior of the cell, during respiration.

Biological respiration is analogous to chemical combustion in that the organic substrate is completely oxidized to carbon dioxide and water; however, in biological systems, this process occurs at much lower temperatures and part of the chemical bond energy is captured in the formation of ATP. This ATP can be used in biosynthetic reactions.

2. Anaerobic Respiration

A number of bacteria can grow in the absence of oxygen and carry out a process that resembles aerobic respiration. For example, nitrate is an oxidized inorganic substance (like O_2) that can accept electrons from the cytochrome respiration chain of some bacteria (like *Pseudomonas*). The result is that the nitrate is reduced to N_2 and N_2O. This process is referred to as anaerobic respiration because it occurs in the absence of oxygen; yet organisms that carry it out have a TCA cycle and cytochrome enzymes. When nitrate is reduced to nitrogen gases, the process is called denitrification. Other bacteria can use oxides of iron and manganese as electron acceptors in respirations. These organisms can be found in the surface of sediments where oxygen is depleted.

Another type of anaerobic respiration occurs with sulfate-reducing bacteria such as *Desulfovibrio* and *Desultomaculum*. These bacteria use organic compounds as energy sources and can oxidize them anaerobically in the presence of sulfate. The sulfate is reduced to hydrogen sulfide in this process, which is called sulfate reduction.

A final type of anaerobic respiration occurs with methanogens. Autotrophic methanogens use carbon dioxide as an acceptor of electrons and reduce it to methane gas. [*See* ANAEROBIC RESPIRATION.]

V. Ecological Considerations

A. Eutrophs versus Oligotrophs

Bacteria and fungi are categorized further on the basis of their ability to use organic carbon sources depending on whether they grow in organic-rich or organic-poor environments. The most commonly studied microbes, such as *E. coli*, baker's yeast (*S. cerevisiae*), and the animal and plant pathogens, grow in organic-rich environments that can be readily simulated on artificial media such as nutrient agar. These bacteria are referred to as eutrophic or copiotrophic bacteria. Other bacteria grow in environments that have low concentrations of organic materials, and these are referred to as oligotrophic bacteria. [*See* LOW-NUTRIENT ENVIRONMENTS.]

In general, eutrophic bacteria have rapid growth rates. Some can divide in less than 15 min. They also have low affinities for transport of nutrients into the cell. Oligotrophic bacteria, on the other hand, do not grow rapidly on any medium, rich or poor in nutrients, but have high affinities for their substrates and can therefore use organic compounds that occur in low concentration in the environment. The eutrophs can outcompete the oligotrophs in organic-rich habitats because their transport systems can provide nutrients to the organisms when they occur in high concentration and they can grow more quickly than the oligotrophs. However, in many natural habitats such as the ocean and freshwater lakes and soils, organic carbon sources occur in low concentrations. In these habitats, the oligotrophs have the advantage because their nutrient uptake systems are effective at providing organic carbon to the organisms, and rapid growth is not advantageous because the low concentrations of substrate will not support it. Not only do the oligotrophs grow well in such environments, but they are responsible for maintaining the low concentration of nutrients found in such habitats.

B. Substrates for Heterotrophic Microbes

1. Particulate Organic Material

Among the more refractory compounds found in nature are the polymers, such as cellulose and lignin, which are produced by plants. A number of microorganisms can degrade cellulose. For example, some of the anaerobic ciliate protozoans that reside in the rumen, a stomach compartment of ruminant animals, can degrade cellulose. The basidiomycetes are a group of fungi that are also important in cellulose decomposition. Finally, many bacteria also produce cellulase. This includes both anaerobic bacteria as well as aerobic ones.

Until recently, it was not thought that microorganisms could degrade lignin.However, it is now known that a certain group of fungi, called the white-rot fungi, are lignolytic. The ligninase enzyme is an unusual peroxidase that attacks this complex insoluble material and breaks several different bonds. Energy is not derived from the breakdown of lignin, but the enzyme is used by these fungi to allow the organism to remove the lignin from areas in the plant tissue to allow access to the cellulose, which is degraded by these organisms as an energy source.

Chitin is produced as the tough exoskeleton material of insects, crustaceans, and certain other invertebrate animals and is also the cell-wall component of many fungi. It is similar to cellulose in that it is a hexose polymer whose subunits are linked by a β 1–4 bond. However, the subunit monomer is not glucose, but *N*-acetyl glucosamine. Chitin is degraded by some microorganisms that have a chitinase, including both bacteria and fungi, as well as by some higher animals. More recently, some plants have been shown to produce chitinases in their roots. Microorganisms are know to break down chitin anaerobically or aerobically, depending on the species and habitat.

Many other particulate or insoluble organic materials can be degraded by microorganisms.

2. Dissolved Organic Materials

As a group, microorganisms show great diversity in their metabolic capabilities. Thus, a large number of organic compounds can be degraded by them including polymeric materials such as polysaccharides, lipids, proteins and nucleic acids.

The simplest carbon compound is methane, which is a product of methane-producing archaebacteria. Methane can be broken down by bacteria called methanotrophs. This is a special group of bacteria that live in aerobic environments that receive methane, produced in an anaerobic environment. Thus, these bacteria reside at the sediment (anaerobic)–water (aerobic) interface of many aquatic environments. In shallow ponds or bogs, bubbles of methane gas can be seen coming to the surface of the water and breaking. Therefore, not all of the methane is being oxidized in these habitats and some escapes into the atmosphere. In deeper habitats such as lakes, the bubbles, are not seen because the methane oxidizers completely oxidize the methane. It is interesting to note that this group of methanotrophs are closely related to a group of autotrophic bacteria, the ammonia oxidizing nitrifying bacteria. *Methylomonas* is an example of a genus of methanotrophic bacteria. They are members of the Proteobacteria, a large group of the gram-negative bacteria. [*See* METHANOGENESIS.]

Different types of bacteria can utilize organic acids, sugars, and amino acids as carbon sources for growth. Indeed, even organic compounds that are toxic to higher life forms like methanol, formaldehyde, and phenol can be broken down by some bacteria provided they are exposed to them at low concentrations. It is this principle that is used in the practice of bioremediaton of toxic organic materials. Thus, heterotrophic microorganisms are being used to clean up contaminated sites such as oil spills and toxic waste dumps containing halogenated compounds. [*See* BIOREMEDIATION.]

Bibliography

Balows, A., Trüper, H. G., Dworkin, M., Harder, W., and Schleifer, K. H. (eds.) (1991). "The Prokaryotes. A Handbook on the Biology of Bacteria. Ecophysiology, Isolation, Identification, Applications." 2nd ed. Springer-Verlag, New York.

Holt, J. G. (ed.-in-chief) (1984–1989). "Bergey's Manual of Systematic Bacteriology," Vols. I–IV. Williams and Wilkens, Baltimore, Maryland.

Lee, J. L., Hutner, S. H., and Bovee, E. C. (eds.) (1985). "Illustrated Guide to the Protozoa." Allen Press, Lawrence, Kansas.

Lovely, D. R. (1991). *Microbiol. Rev.* **55,** 259–287.

Moore-Landecker, E. (1990). "Fundamentals of the Fungi," 3rd ed. Prentice-Hall, Englewood Cliffs, New Jersey.

Stanier, R. Y., Ingraham, J. L., Wheelis, M. L., and Painter, P. R. (1986). "The Microbial World." Prentice-Hall, Englewood Cliffs, New Jersey.

Woese, C. (1987). *Microbiol. Rev.* **51,** 221–271.

High-Pressure Habitats

Edward F. DeLong
University of California, Santa Barbara

Glossary

Barophile Organism that grows optimally or preferentially at hydrostatic pressures >1 atmosphere

Barotolerant (baroduric) Having reduced sensitivity to elevated hydrostatic pressure; barotolerant organisms typically have growth rates that are relatively unaffected by increasing hydrostatic pressure

Obligate barophile Organism that has an obligate growth requirement for hydrostatic pressures >1 atmosphere

Pascal SI (International System of Units) unit of pressure; 1 Pascal (Pa) = 1 Newton m^{-2}

Pressure Force per unit area; commonly measured in atmospheres (atm); 1 atm = 1.01325×10^5 Pa = 1.01325 bar

HIGH-PRESSURE HABITATS are environments where ambient pressures are substantially >1 atmosphere (1 atm = 1.01325×10^5 Pa = 1.01325 bar). The deep earth, oceanic abyssal plains and trenches, hydrothermal vents, and cold seeps found at continental margins, are typical habitats of elevated hydrostatic pressure. The high pressures found in these regions have marked influence on phase equilibria, chemical equilibria, and chemical kinetics and, thus, affect a large variety of biological processes. Environmental pressure, in conjunction with temperature, largely defines the physical conditions that allow microbial growth and survival. The unique properties of microorganisms that have evolved in high-pressure habitats reflect their specific adaptation to the physical variable of pressure.

I. High-Pressure Habitats

High-pressure habitats make up a significant fraction of the earth's biosphere. Over 77% of the seafloor lies below 3000 m, and oceanic habitats with hydrostatic pressures >300 atm are a common feature of the marine environment. Habitats of elevated hydrostatic pressure include the deep earth, oceanic abyssal plains and trenches, hydrothermal vents, and cold seeps found at continental margins. Each of these individual environments may vary with respect to conditions of temperature, salinity, oxygen concentration, pH, or nutrient availability. Thus, considerable diversity exists among different high-pressure environments.

A. The Deep Earth

The interstitial spaces within and between the rocks that comprise the earth's upper crust are largely fluid filled, and geological events occurring in the upper crust of the earth take place in a fluid environment. Most subterranean pore fluid is water, and this is the medium that transmits hydrostatic pressure from the ground surface to subterranean regions. In regions of substantial vertical permeability, deep earth pressures are referred to as "normal hydrostatic pressure." This is the pressure generated by a column of water extending from the ground surface to a given depth (about 1 atm/10 m of depth; Fig. 1B).

Depending on the surrounding geological conditions, pressures higher or lower than expected normal hydrostatic pressure can be encountered. For instance, uplifting of solid materials from the deep earth may result in elastic expansion. This expansion leads to an increase in the volume of associated pore spaces in the rock, which causes a decrease in the hydrostatic pressure of the trapped pore water. Abnormally high pressures are most often encountered in the deep earth at depths >3000 m, but they are also found at shallower depths. Water-bearing

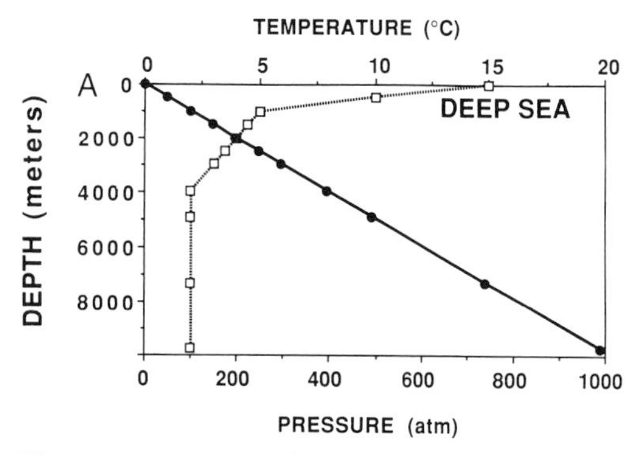

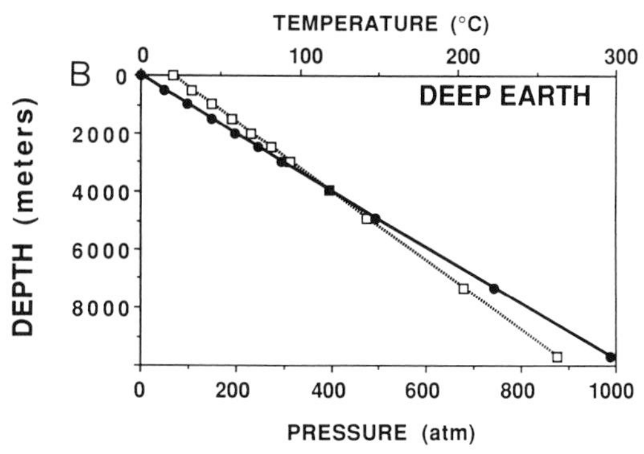

Figure 1 Average conditions of temperature and pressure in the deep sea and deep earth. (A) Average temperature (open squares, broken line) and hydrostatic pressure (closed circles, solid line) found in temperate regions of the earth's oceans, as a function of depth. The increase in pressure is not strictly linear with depth, although nearly so. The *in situ* specific gravity of seawater, and the variation of gravity with latitude and depth, will influence the pressure–depth profile. (B) Average hydrostatic pressure (closed circles, solid line) and temperature (open squares, broken line) in the deep earth, assuming a 2.5°C increase in temperature for every 100 m of depth. 1 atm = 1.01325×10^5 Pa = 1.01325 bar. [From Saunders, P. M., and Fofonoff, N. P. (1976). *Deep-Sea Res.* **23,** 109–111.]

sedimentary muds, which are buried and compacted over time, may give rise to such overpressurized pockets. Pore water that cannot be squeezed out may be trapped in these sedimentary muds and, thus, must sustain part of the weight of overlying rock and sediment. This results in subterranean pressures as much as 250 atm greater than the expected normal hydrostatic pressure.

At shallower depths of 10–20 m, the temperature of the subsurface is approximately equal to the mean annual air temperature. At greater depths (excluding geothermal regions), temperature increases about 2–3°C for every 100 m in depth (Fig. 1B). Subterranean regions of substantial hydrostatic pressure (>300 atm) are consequently warm habitats as well. Below 1000 m, deep-earth temperatures are about 50–55°C. Higher temperatures, favorable for the growth of thermophilic microorganisms, occur at greater depths. [*See* THERMOPHILIC MICROORGANISMS.]

The chemical environment of the deep earth is variable and depends on local geological conditions. Many deep-earth microbial habitats may be high in salinity. Based on available data for shallower groundwaters, most deep subterranean regions are likely to be oligotrophic and largely depleted in organic carbon. Nitrogen and phosphate, if present, may also be present at low concentrations in many deep-earth environments. Some deeper regions may also be depleted in oxygen. In some areas, inorganic carbon availability is another potentially limiting factor for the growth of subsurface autotrophic organisms.

Although the deep earth represents a habitat where pressure-adapted bacteria may thrive, few microbiological studies have been performed at depths >1000 m, where pressure effects on microbial growth become substantial. Largely due to sampling problems, most investigators to date have focused on the characterization of isolates from subterranean depths of up to several hundred meters. Data pertaining to indigenous bacterial populations from depths >500 m are few. Some workers have suggested that, provided there is enough water, pore space, and nutrients, microbial life may exist as deep as 4000 m in the terrestrial subsurface.

B. The Cold, Deep Sea

The cold, deep sea constitutes a major portion of the earth's total biosphere. Nearly 75% of the oceans' total volume lies below 1 km. Deep-sea hydrostatic pressures increase approximately 1 atm for every 10 m increase in depth (Fig. 1A). The average depth of the world's ocean is about 3800 m, which corresponds to a hydrostatic pressure of 380 atm. Thus, a major portion of the marine environment experiences hydrostatic pressures >300 atm. The very deepest oceanic trenches, at tectonic subduction zones, extend to depths as great as 11 km. Thus, ambient hydrostatic pressures in these habitats are >1000 atm.

Other conspicuous deep-sea features include a constant cold temperature of about 2°C (Fig. 1A), darkness, and remoteness from primary productivity. The only appreciable light in abyssal and hadal environments is generated *in situ* by bioluminescent organisms. With few exceptions, most deep-sea communities are predominantly heterotrophic and ultimately depend on the primary productivity of surface waters. The major proportion of oceanic primary productivity is consumed near the surface, and organic material that does sink to greater depths is further degraded during transit to the abyss. Hence, the overall flux of photosynthetically derived material to the deep sea decreases with increasing depth. Excluding hydrothermal vent and cold seep regions, the deep sea is characteristically a nutrient-limited, heterotrophic ecosystem. [*See* HETEROTROPHIC MICROORGANISMS; BIOLUMINESCENCE, BACTERIAL.]

In today's deep sea below 1000 m, a continuous pressure gradient coincides with isothermal temperatures of about 2°C. The origin of most deep-sea bottom waters resides in the Antarctic circumpolar regions, where contemporary glacial conditions create a cold, dense water mass that sinks downward and eventually percolates throughout the deep oceans worldwide. This constant flow of deep, cold Antarctic bottom water has led to the evolution of a largely stenothermal deep-sea fauna, which thrives at near-freezing temperatures. The predominantly psychrophilic nature of indigenous deep-sea bacteria illustrates this point. [*See* LOW-TEMPERATURE ENVIRONMENTS.]

C. Enclosed Deep-Sea Basins

Although high pressure and low temperature characterize the deep regions of today's seas, these conditions did not always prevail. The deep sea of 100 million yr ago is thought to have been about 15°C warmer than contemporary abyssal environments. Ancient seas are not the only examples of warm, deep regions. Present-day, enclosed basins with shallow sills, isolated from Antarctic bottom waters, also attain relatively high temperatures in comparison to the average, cold deep-sea habitat.

The Mediterranean and Sulu Sea basins have temperatures >4°C, at depths as great as 5000 m. Several isolated deeps in the Mediterranean Sea represent unique habitats of moderate temperatures (13.5°C), which coincide with hydrostatic pressures of up to 500 atm. Interestingly, these physical conditions are not unlike the deep sea of 100 million yr ago. Another "warm" basin, the Sulu Sea, extends to depths of about 5000 m, with corresponding bottom temperatures of about 10°C. The shallow sills bordering this sea prevent cold Pacific bottom water from mixing with the deep water of the Sulu Sea basin. The Sulu Sea is also atypical with respect to the relatively low oxygen concentration found at its greatest depths. Enclosed basins such as these represent high-pressure habitats with temperatures significantly higher than the average deep-sea environment.

D. Hydrothermal Vents

Extremely warm, high-pressure habitats are common at midoceanic submarine ridges, where magma rises from the deep earth to form new oceanic crust. In these regions, hydrothermal fluids containing highly reduced chemical species are emitted into the surrounding, oxygenated, cold deep-sea waters. Hydrothermal vents are found at seafloor spreading centers worldwide, including sites along the East Pacific Rise, the Mid-Atlantic Ridge, and the Western Pacific back-arc basins. These submarine hot springs occur over a wide range of depths and hydrostatic pressures. Commonly studied vent systems have hydrostatic pressures ranging from 1 to 300 atm. Some of the deepest hydrothermal vent communities described to date are found at depths of approximately 3700 m, at the Mariana Bac-Arc Basin and also along the Mid-Atlantic Ridge. Other regions of potential hydrothermal activity have been reported as deep as 5800 m. It is possible that geothermal activity at deep trench subduction zones could give rise to high-temperature environments of very high pressures, up to 1100 atm.

Unlike other deep-sea habitats, hydrothermal vent communities do not solely depend on photosynthetically derived surface primary productivity. Instead, microbial oxidation of geothermally produced, highly reduced chemical compounds fuels the chemosynthetic production of new biomass. In many hydrothermal vent communities, hydrogen sulfide serves as the principle reductant emitted from hydrothermal vent fluids. Here, production of new organic carbon is largely due to free-living or symbiotic chemolithotrophic sulfur bacteria. These autotrophic microorganisms utilize the energy liberated by the oxidation of high-energy sulfide compounds, to "fix" CO_2 . The chemosynthetically produced organic carbon supports the anomolously high biomass found in deep-sea hydrothermal vent communities.

In environments of elevated hydrostatic pressures, water remains liquid at temperatures well beyond 100°C. Therefore, organisms inhabiting deep-sea hydrothermal vents may experience extremely high temperatures. Temperatures as great as 350°C have been measured in "black smoker" hydrothermal fluid emissions. Hydrothermal vent habitats are unusual with respect to the very wide range of temperatures found in and around the vent plumes. Temperature gradients spanning from well above 300°C to ambient deep-sea temperatures of 2°C, exist in the vicinity of these deep-sea hot springs.

E. Cold Seeps

At some continental slope regions, fluids laden with methane, hydrogen sulfide, or hydrocarbons emanate from deep submarine cracks and fissures. Relatively dense populations of marine organisms, reminiscent of hydrothermal vent communities, are often found in the vicinity of these seeps. As in hydrothermal vent sites, highly reduced chemical compounds emitted in seep fluids are presumably oxidized by autotrophic bacteria. Bacterial grazing, or nutritional symbioses with chemosynthetic bacteria, are thought to support the abundant biomass found at these sites. Deep-sea seeps are generally isothermal with ambient deep-sea waters, and in this respect they are fundamentally different from hydrothermal vent habitats. A variety of cold water seeps have been found in the Gulf of Mexico, off the West Coast of the United States, and on the coast of Japan.

The geochemical environment of seep communities depends largely on local conditions. Seep fluids in the Gulf of Mexico consist of hypersaline pore waters, originating from Cretaceous limestones. These pore waters contain high concentrations of methane, ammonia, and sulfide, all of which could support the growth of chemoautotrophic bacteria. High hydrocarbon concentrations are found at seeps off the coast of Louisiana, and these petrochemicals fuel the growth of heterotrophic, hydrocarbon oxidizing bacteria. These bacteria presumably provide a prokaryotic food source for the associated seep community. Methane emitted in seep fluids may also support the growth of methanotrophic bacteria, which are found in symbiotic associations with bivalves and possibly other metazoa living near the cold seeps.

Hydrostatic pressures at abyssal seep communities can be substantial. In the Gulf of Mexico, seep communities reach depths of 3200 m, corresponding to hydrostatic pressures of >300 atm. At the Oregon Subduction Zone, seeps occur at depths greater than 2000 m below the surface. The deepest seep communities reported to date, found off the coast of Japan at depths of nearly 6000 m, have hydrostatic pressures around 600 atm. Bacteria that originate from habitats this deep usually show marked pressure adaptation.

II. Effects of High Pressure on Biological Systems

The latter part of the nineteenth century was a period of fundamental development and exploration in the field of oceanography as well as microbiology. One important finding of early oceanographic expeditions was the discovery of an indigenous deep-sea fauna at depths as great as 6000 m. The discovery of life in the deep ocean displaced earlier theories that viewed the deep sea as an azoic desert. These observations prompted early investigators to explore some of the main themes of deep-sea biology, where research interests continue to this day. In addition, the discovery of life in the abyss prompted study of the effects of ecologically relevant hydrostatic pressure on biological systems. In the late nineteenth century, the French workers Regnard and Certes investigated the influence of pressure on microbial growth and fermentation. Their early studies demonstrated that high pressure has significant and variable influence on a variety of biological processes.

Pressure effects on biological systems are complex. The influence of hydrostatic pressure on chemical reaction rates and equilibria, macromolecular structure and assembly, and cell growth and viability can be either neutral, inhibitory, or stimulatory. All pressure effects on biological processes are ultimately based on volume changes. Any process that involves a net volume increase will tend to be inhibited by elevated hydrostatic pressure. Those processes that undergo net negative volume changes tend to be enhanced by elevated pressures. It is difficult to predict *a priori* how pressure will affect any given physiological process.

A. Pressure Effects on Chemical Reactions

Both chemical equilibria and reaction rate are influenced by elevated hydrostatic pressures. The effect of pressure on chemical equilibrium at constant temperature is described by the following equation:

$$(\partial \ln K_{eq}/\partial P)_T = -\Delta V/RT,$$

where K_{eq} is the equilibrium constant for the reaction, P is the pressure (atm), ΔV is the volume difference between the initial and final state of the system, R is the gas constant (cm^3 atm K^{-1} mol^{-1}), and T is the absolute temperature (°K). The volume change, ΔV, is associated with the entire system at equilibrium, including both the solute and the surrounding solvent. For most biological reactions, the solvent is water. Sources of volume changes in biochemical reactions arise from volume changes associated with the reacting molecules as well as from changes in the density of the surrounding solvent. Significantly, volume changes in many reactions are primarily due to the alteration of local water structure (hydration density) around reacting molecules. In particular, reactions that involve a net production of charged species usually have a negative ΔV (Table I), due to the electroconstriction of water molecules around ionized products. As a consequence, elevated pressure tends to shift equilibria toward production of ionized species. This effect is important to consider when designing high-pressure physiological or biochemical experiments. The acid/base equilibria of many buffers (e.g., $H_2PO_4^-$; Table I) are highly perturbed by pressure changes, because they involve net production of ionized species. However, the acid/base equilibria of buffers that do not involve net production of ionized species (e.g., Tris; Table I), are relatively insensitive to pressure variation.

The relative rates of chemical reactions are influenced by both temperature and pressure. The following equation describes the influence of pressure (P) on reaction rate:

$$k_p = k_o{}^{(-P\Delta V\ddagger/RT)},$$

Table I Volume Changes (ΔV) Associated with Acid/Base Ionization

Reaction	ΔV (cm^3 mol^{-1})
$H_2O + H_2PO_4^- \rightarrow HPO_4^{2-} + H_3O^+$	−24.0
$H_2O + CH_3COOH \rightarrow CH_3COO^- + H_3O^+$	−11.0
$H_2O + TrisH^+ \rightarrow Tris + H_3O^+$	+1.0
H_2O + protein − COOH → protein − COO^- + H_3O^+	−11.0
$H_2O + CH_3NH_2 \rightarrow CH_3\text{-}NH_3^+ + OH^-$	−20.0
H_2O + protein − NH_2 → protein − NH_3^+ + OH^-	−18.0

$TrisH^+$, Tris-(hydroxymethyl)aminomethane. [Marquis, R. E., and Matsumura, P. (1978). Microbial life under pressure. *In* "Microbial Life in Extreme Environments" (D. S. Kushner, ed.), pp. 105–157. Academic Press, London. Neuman, R. C., Kauzmann, W., and Zipp, A. (1973). *J. Phys. Chem.* **77,** 2687–2691.]

where k_p and k_o are the rate constants at elevated pressure and 1 atm, respectively, R is the gas constant (cm^3 atm K^{-1} mol^{-1}), T is the absolute temperature (K), and ΔV‡ is the "activation volume" of the reaction. Transition state theory suggests that reactants must first combine favorably to form an activated complex, or transition state, before forming the product. The apparent activation volume, ΔV‡, represents the difference in volume between the reactants and this transition state. The extent of the volume difference between reactants and activated complex (ΔV‡), will largely determine the influence of pressure on transition state formation and, hence, reaction rate.

B. Pressure Effects on Biochemical and Physiological Processes

Individual biochemical reactions vary with respect to both the magnitude and the direction of associated net volume changes (Table I) and, hence, pressure effects. Relatively moderate hydrostatic pressures can affect enzyme structure, regulation, and catalysis. The kinetic properties of enzymes can be significantly altered by very small pressure increases, as low as 50 atm. For example, the substrate-binding affinity of some dehydrogenases appears to be particularly sensitive to shifts in pressure. Ecologically relevant pressures can disrupt the association and assembly of structural biopolymers, such as actin or flagellin. The structure and function of cell membranes is also pressure-sensitive. Increased pressure promotes tighter packing and decreased molecular motion of the fatty acyl chains in membrane phospholipids. Thus, elevated pressure results in increased microviscosity (decreased fluidity) of the membrane lipid bilayer (see Fig. 5A). This pressure-induced decrease in membrane fluidity can influence membrane-associated function. For example, membrane-bound adenosine triphosphatases (ATPases) from shallow-water fish are known to be inhibited by high hydrostatic pressures. This pressure sensitivity is thought to be a consequence of pressure-induced changes in membrane fluidity rather than direct pressure effects on the ATPase itself. These diverse effects on biochemical structure and function indicate the pervasive and complex influence of pressure life processes. [*See* CELL MEMBRANE: STRUCTURE AND FUNCTION.]

The bulk of research on the physiological effects

of high pressure have involved experiments with microorganisms adapted to life at 1 atm. These studies have revealed numerous pressure-sensitive physiological processes. Application of moderate, nonlethal hydrostatic pressure can cause cessation of cell division and filament formation in some marine and terrestrial microorganisms. Motility and flagellar synthesis can also be inhibited by pressures in the vicinity of 200–400 atm. A number of *in vivo* and *in vitro* studies comparing the pressure effects on cellular DNA, RNA, and protein synthesis have indicated that protein synthesis is most sensitive to pressure perturbation. In general, biosynthetic pathways appear more perturbed by pressure than catabolic ones, although there are exceptions to this general trend.

Pressures as low as 200 atm are lethal to some bacteria. For example, the combined effects of low temperature and high pressure, corresponding to a water depth of 1500 m, are lethal for the marine bacterium *Vibrio harveyi*. Cell death results at this habitat temperature and pressure, irrespective of the rate of pressure increase and temperature decrease. These and similar results suggest that, with respect to physical factors, temperature and pressure stratify ocean depths into regions of growth, survival, and death for marine microorganisms. The actual causes of pressure-induced cell death are not well understood.

To fully appreciate the influence of pressure on the ecology and evolution of microorganisms, it is important to consider the combined effects of pressure and other physical and chemical parameters, such as temperature, pH, salinity, and nutrient concentration. For example, organisms adapted for life at 1 atm tend to be most tolerant to elevated hydrostatic pressures at their optimal growth temperatures. At 35°C, *Escherichia coli* grows at pressures approaching 500 atm, but at 10°C growth is inhibited by pressures <100 atm. For microorganisms adapted to 1 atm, there is a general tendency for the growth temperature range to be constricted with increasing growth pressure. Other variables, such as pH, oxygen, and nutrient concentration are also expected to influence the pressure tolerance of any given species.

Because pressure effects on individual biochemical reactions or cellular functions may be either neutral, inhibitory, or stimulatory, the overall influence of pressure on a given pathway or process may be difficult to predict. For optimal function at relevant habitat pressures, a conglomerate of cellular structures and functions must interact cooperatively and synergistically. Thus, it has been suggested that integrated genetic and metabolic regulatory circuits may be especially sensitive to pressure perturbation. However, pressure adaptation is unlikely to be relegated to a single locus. In analogy to the adaptations of thermophiles, a consortium of coordinated modifications most likely characterize a truly pressure-adapted species.

III. Microbial Life in High-Pressure Habitats

Considering the diverse habitats where substantial hydrostatic pressures exist, and the physiological versatility of microbial species, it is not surprising that microorganisms have adapted well to extremes of pressure. Metabolically active, deep-sea microbial populations were hinted at as early as 1895, during the famous oceanographic explorations of the Challenger Expedition. In Volume 5 of the Challenger Reports, devoted to deep-sea deposits, one investigator noted: "The low temperature at the bottom of the ocean and possibly also the pressure retard putrefaction, but it is evidently incorrect to state that putrefaction does not exist in the greater depths, for everywhere there is evidence to the contrary." It was not until the late 1950s, however, that the first evidence of pressure-adapted deep-sea bacteria was reported. Using a most probable numbers dilution technique, it was shown that some mixed bacterial populations from the Philippine Trench appeared capable of growth at pressures of 1000 atm but not 1 atm. These data suggested the existence of barophilic (pressure-loving) deep-sea microbes, which are specifically adapted for growth at high hydrostatic pressures. Today, the term barophile is used to describe organisms that grow optimally at pressures >1 atm. Obligate barophiles are those organisms that have an absolute growth requirement for pressures >1 atm.

Our current understanding of pressure-adapted bacteria is largely restricted to studies of psychrophilic, heterotrophic barophiles. To date, very little evidence for barophilic thermophiles has been reported. The following discussion therefore focuses on pressure-adapted bacteria from the cold, deep sea.

A. Ecology of Indigenous Deep-Sea Bacteria

Most barophilic bacteria isolated from cold deep-sea habitats have been enriched for, and cultivated in, high-nutrient seawater-based media containing high levels (>0.1% weight/volume) of peptone or yeast extract. High-pressure enrichments of deep-sea samples, including sediments, seawater, and the fecal pellets, gut contents, and decaying carcasses of metazoans have all yielded barophilic isolates. Barophilic bacteria isolated to date have been most readily retrieved from high-nutrient environments. They appear to predominate in enteric, epiphytic, or copiotrophic habitats. In particular, the gut contents and fecal pellets of deep-sea invertebrates (amphipods and holothurians) and vertebrates (fish) have been a rich source of barophilic deep-sea bacterial isolates. The barophilic growth response has been demonstrated in bacteria originating from abyssal regions worldwide, including the Arctic, Antarctic, Pacific, and Atlantic oceans (Table II).

Experiments designed to measure the assimilation and degradation of radiolabeled nutrients in deep-sea bacterial populations often detect little or no microbial activity after extended incubation periods at *in situ* temperatures and pressures. Carbon utilization experiments on deep-sea microbial assemblages, incubated at both *in situ* pressures and 1 atm, indicate that heterotrophic barophiles often represent a minor fraction of the microbial biomass in deep seawater and sediment samples. This is apparently due to the presence of a greater number of allochthonous species, which sink from shallower

Table II Origin and Growth Properties of Some Deep-Sea Barophilic Bacteria

Isolate	Reference[a]	Sample origin	Depth of origin (m)	Pressure optima range[b] (atm)	G[c] (hr)
PE31	1	Eastern North Pacific	3584	140–280	5.1
PE36	2	Eastern North Pacific	3584	300–400	6.9
W145	3	South Atlantic	4575	140–410	3.3
27A	4	Central North Atlantic	4900	300–500	6.2
51A	4	Central North Atlantic	4900	400–500	7.1
WHB37	5	Antarctic	5000	300–400	11.1
WHB46	5	Antarctic	5000	300–500	12.5
72	4	Central North Atlantic	5100	100–300	5.5
27	4	Central North Atlantic	5100	100–300	5.9
F1	4	Central North Atlantic	5100	200–400	6.7
MT52	1	Mariana Trench Wall	5672	280–420	8.8
CNPT3	2	Central North Pacific	5782	200–400	8.6
UM40	3	Puerto Rico Trench	5920	270–540	3.3
PT64	2	Philippine Trench	7100	500–600	15.4
AT24	6	Aleution Trench	6350	300–400	12.0
BNL1	7	Puerto Rico Trench	7410	740–920	13.0
MT199	2	Mariana Trench	8961	700–850	35.0
MT41	2	Mariana Trench	10,476	700–950	35.0

[a] References: (1) Yayanos, A. A., Dietz, A. S., and Van Boxtel, R. (1982). *Appl. Environ. Microbiol.* **44,** 1356–1361. (2) Yayanos, A. A. (1986). *Proc. Natl. Acad. Sci. USA* **83,** 9542–9546. (3) Deming, J. W., Hada, H., Colwell, R. R., Leuhrsen, K. R., and Fox, G. E. (1984). *J. Gen. Microbiol.* **130,** 1911–1920. (4) Jannasch, H. W., and Wirsen, C. O. (1984). *Arch. Microbiol.* **139,** 281–288. (5) Weyland, H., and Helmke, E. (1989). Barophilic and psychrophilic bacteria in the Antarctic Ocean. *In* "Recent Advances in Microbial Ecology," (T. Hattori, Y. Ishida, Y. Maruyama, R. Morita, and A. Uchida, eds.), pp. 43–47, Japan Scientific Societies Press, Tokyo, Japan. (6) Yayanos, A. A., and DeLong, E. F. (1987). Deep-sea bacterial fitness to environmental temperatures and pressures. *In* "Current Perspectives in High Pressure Biology," (H. W. Jannasch, R. E. Marquis, and A. M. Zimmerman, eds.), pp. 17–32. Academic Press, London. (7) Deming, J. W., Somers, L. K., Straube, W. L., Swartz, D. G., and Macdonell, M. T. (1988). *Sys. Appl. Microbiol.* **10,** 152–160.

[b] Pressure optima range, in atmospheres, which corresponds to the range of pressures that bracket the optimal growth pressure in complex, nutrient-rich media at 2°C. Data extracted from cited references (see footnote a).

[c] G, approximate generation time, in hours, at the optimal growth pressure at 2°C in complex, nutrient-rich media. Approximate values were estimated from cited references (see footnote a).

waters and accumulate at depth. Most certainly, the overall rates of bacterial activity and productivity decrease with increasing depth in the ocean.

However, the very existence of barophilic bacteria indicates at least the potential for active microbial metabolism and degradation of organic substrates in the deep sea. Doubling times ranging between 3 and 35 hr are typical for barophilic bacteria grown in high-nutrient media at *in situ* temperatures and pressures in the laboratory (Table II). *In situ* estimates of the growth rates of deep-sea microbial assemblages vary widely. In one study, estimated doubling times of mixed microbial populations in some deep-sea sediments ranged from 5.7 to 693 days. In contrast, measurements in deep-sea sediments at 4500 m, which were enriched in natural phytodetritus, yielded bacterial doubling times of 11.5 hr at *in situ* temperatures and pressures. This value is quite close to the measured doubling times of cultured deep-sea barophilic isolates (see Table II).

With due consideration to the physical and trophic contraints imposed by the deep-sea environment, it is likely that many of the ecological roles of surface-adapted bacteria can be extended to deep-sea bacteria. The relative contributions of barophilic bacteria to deep-sea trophic dynamics and nutrient cycling will largely depend on the quality and local concentration of nutrients and perhaps also on the density of metazoan species with which heterotrophic barophiles are often associated. Results of some field studies suggest that deep-sea microbial populations may be responsible for 30–40% of the total *in situ* organic carbon consumption. Natural populations of barophilic flagellates and bacteria have been observed, indicating the potential for functional deep-sea microbial food loops similar to those operating in surface waters. Barophilic bacteria may also be involved in symbiotic associations with mobile scavengers. Animal-associated barophilic deep-sea bacterial isolates have been shown to be a rich source of polyunsaturated fatty acids (PUFAs). These lipids are essential nutrients for most metazoans and were previously thought to be synthesized only in the photic zone. PUFAs are found in large amounts in the lipids of deep-sea vertebrates and invertebrates alike, but they are known to be highly labile and rapidly degraded during transit to the abyss. It is quite possible that barophilic, PUFA-containing bacteria represent a major source of these essential nutrients in deep-sea food webs. They may be an important source of other essential nutrients in the deep sea as well.

B. General Properties of Heterotrophic, Barophilic Bacteria

In the late 1970s, pure cultures of barophilic bacteria were isolated and maintained in the laboratory. The first described barophilic isolate was a gram-negative, spirillum-shaped bacterium recovered from a depth of 5700 m in the Central North Pacific. This isolate, strain CNPT3, was shown to grow over a very wide pressure range (1–800 atm), with optimal growth occurring at pressures >300 atm (Fig. 2C). This deep-sea bacterium was also strictly psychrophilic, with a maximum growth temperature around 15°C. Interestingly, at temperatures greater than about 6°C, this isolate was shown to be an obligate barophile, with an absolute growth requirement for hydrostatic pressures >1 atm. Subsequently, many other barophilic bacteria have been isolated. More recently, barophilic flagellates have been detected in mixed microbial assemblages.

Barophilic bacteria are readily retrieved from deep-sea samples originating from depths greater than about 2000 m. When maintained at cold, deep-sea temperatures, most characterized barophilic bacteria can survive reasonably prolonged periods (hours to days) of decompression. Thus, barophiles can be successfully recovered from deep-sea samples that have been decompressed but kept cold, and the majority of isolates in culture have been obtained in this fashion. Many barophilic bacteria obtained from depths shallower than 5000–6000 m can grow (albeit nonoptimally) at 1 atm and 2°C. Obligate barophiles, which have an absolute growth requirement for pressures >1 atm, have been isolated exclusively from depths exceeding 5000 m.

The three-dimensional growth plots shown in Fig. 2 succinctly summarize some cardinal properties of heterotrophic, psychrophilic deep-sea bacteria. These plots illustrate the growth properties of six bacteria isolated from increasingly greater depths (1957–10,476 m) in the North Pacific Ocean. It is evident from Fig. 2 that extreme psychrophily, coupled with optimal growth at pressures >1 atm, are key features of indigenous deep-sea bacteria. The bacterium SC1 from 1957 m (Fig. 2A) is clearly not barophilic and grows best at 1 atm at all growth temperatures. However, isolates obtained from depths >2000 m display marked barophilic charac-

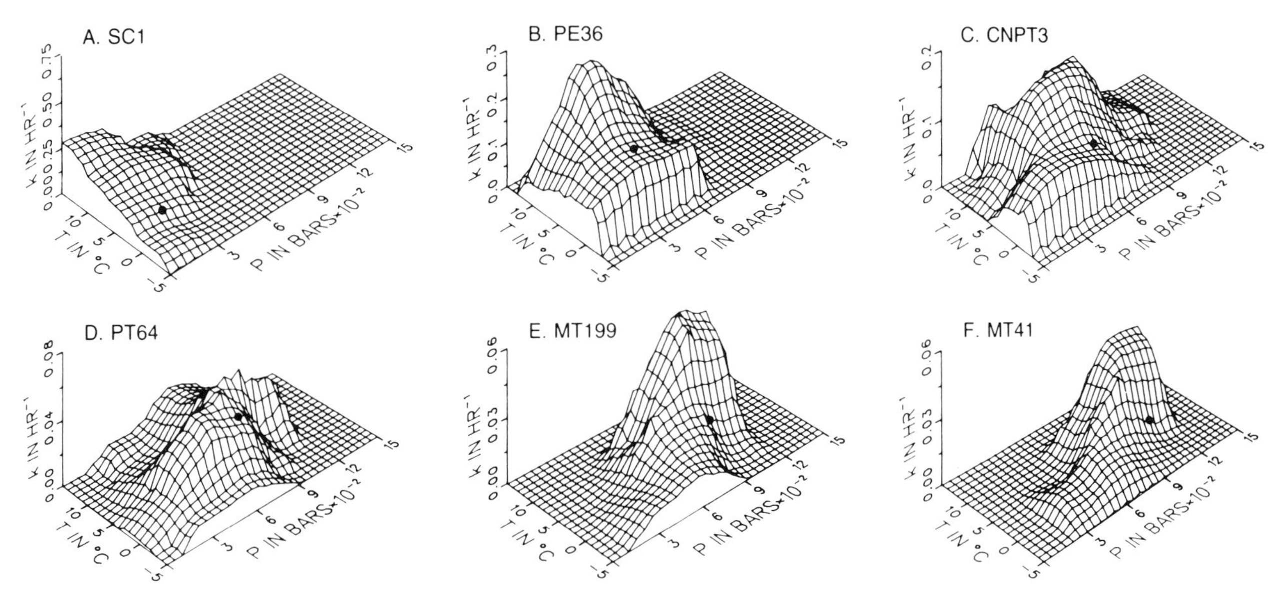

Figure 2 The combined effect of temperature (T) and pressure (P) on the specific growth rate (k) of six deep-sea bacterial isolates (A–F). The black dot corresponds to the temperature and pressure found at the depth of origin of each isolate. 1 atm = 1.01325 $\times 10^5$ Pa = 1.01325 bar. [From Yayanos, A. A. (1986). *Proc. Natl. Acad. Sci. USA* **83**, 9542–9546.]

ter and grow most rapidly at pressures substantially >1 atm (Fig. 2B–F). At suboptimal hydrostatic pressures, temperature optima for these bacteria is about 2°C. At higher growth pressures, temperature optima increase to 8–10°C for most barophiles. The maximum growth pressure of barophilic bacteria increases with increasing depth of origin. Some barophiles capable of growth at 1 atm and 2°C acquire a strict requirement for high growth pressures at higher growth temperatures. Whether an organism is obligately barophilic or not depends on the growth temperature.

When cultivated in rich media at optimal growth pressures and 2–4°C, mean generation times of heterotrophic deep-sea bacteria range from about 3 to 35 hr (Table II). Figure 3A shows the approximate growth pressure optima (estimated from Table II), plotted as a function of the depth of origin, for a number of barophilic isolates. In nutrient-rich media at 2°C, the growth pressure optima of heterotrophic, barophilic bacteria tends to be somewhat less than the pressure corresponding to the isolate's depth of origin. Another general trend can be seen when the mean generation time of barophilic isolates is plotted as a function of their depth of origin (Fig. 3B). The greater the capture depth of the barophilic isolate, the slower the growth rate (greater generation time). The scatter of data points in these plots likely reflects both variability in experimental conditions and physiological differences of the various isolates. It should be emphasized that the relationships illustrated in Fig. 3 are general trends only and are based on comparisons of heterotrophic deep-sea isolates whose phylogenetic relationships and physiological similarities are largely unknown at the present time.

Most barophilic isolates characterized to date are heterotrophic, gram-negative eubacteria. Barophilic bacteria are not easily identified by routine taxonomic methods, because most standard test procedures are designed for screening microorganisms at 1 atm and 20–37°C. However, some biochemical and physiological tests can be adapted for screening microorganisms grown at high pressure. In addition, molecular phylogenetic analyses are also providing insight into the evolutionary relationships of barophilic bacteria. Comparative studies of the 5S ribosomal RNA sequences of two different species of barophilic isolates place these organisms within the gamma subdivision of the proteobacteria. One barophilic species is most clearly affiliated with the genus *Shewanella* and has been assigned the species name *Shewanella benthica*. Based on small subunit RNA sequence analyses, other barophilic bacteria, isolated from both the Atlantic and Pacific oceans, are

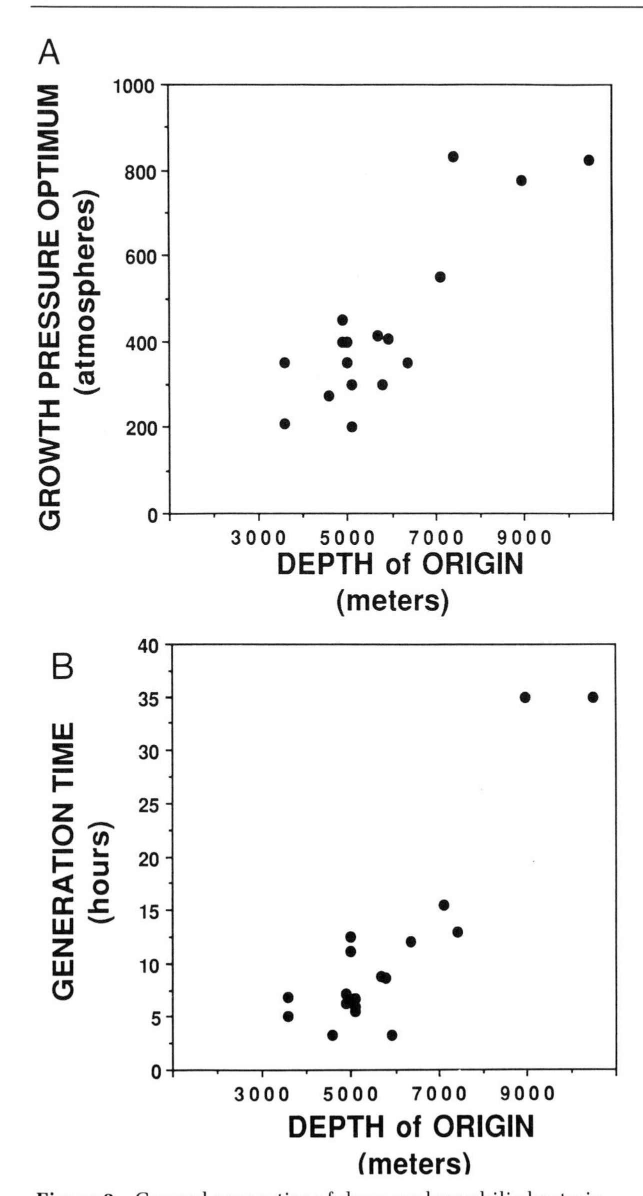

Figure 3 General properties of deep-sea barophilic bacteria that correlate with depth of origin. (A) Approximate growth pressure optimum of barophilic bacterial isolates as a function of depth of origin. (B) Approximate generation time of barophilic bacterial isolates as a function of depth of origin.

also affiliated with the genus *Shewanella*. These data may indicate a specific enrichment of *Shewanella* species in the high-nutrient, high-pressure regimes typically used for isolation of barophilic deep-sea bacteria. One other obligately barophilic strain, isolated from 7410 m in the Puerto Rico Trench, appears more closely related to a nonbarophilic psychrophile, *Vibrio* (*Colwellia* gen. nov.) *psychroerythrus*. These data suggest that the barophilic phenotype may have arisen independently in several different lineages.

Most barophilic bacteria isolated to date are heterotrophic and grow well in seawater-based, high-nutrient media. Some of these barophilic, "gamma" Proteobacteria are facultatively anaerobic and grow fermentatively using glucose. It is worth noting that some of the terrestrial relatives of barophilic bacteria are capable of growing anaerobically using nitrate, iron, or thiosulfate as terminal electron acceptors. One barophile, *S. benthica*, can reduce both iron and thiosulfate and, thus, may be capable of anaerobic respiration using these compounds. It remains to be seen whether or not other barophilic *Shewanella* species have the ability to respire anaerobically using a large suite of different terminal electron acceptors, a property that may have important implications for the ecology of these microorganisms.

C. Understanding the Basis of Pressure Adaptation

To date, few data on the physiological, biochemical, and molecular bases of adaptation to high hydrostatic pressure are available. However, barophilic bacteria, because they are so readily grown and manipulated in the laboratory, are excellent model systems for studying adaptation to high-pressure environments. Ongoing and future studies of barophilic microorganisms should provide a better understanding of the molecular basis of pressure adaptation.

Physiological studies of bacterial isolates from the cold deep-sea have demonstrated their extremely psychrophilic and barophilic nature. Colony-forming ability was used to assay the thermal sensitivity of isolate CNPT3 (see Fig. 2C). At 10°C and 1 atm of hydrostatic pressure, this barophile lost colony-forming ability only very slowly. However, at temperatures >20°C, thermal inactivation was rapid and accompanied by gross changes in cell morphology. Thermal death was not appreciably alleviated by application of elevated pressure during heat inactivation.

In a similar fashion, colony-forming assays were used to measure the kinetics of inactivation caused by decompression of an obligate barophile, MT41 (Fig. 2F). Even when kept at 0°C, this isolate lost colony-forming ability rapidly upon decompression. Decompression-induced inactivation was accompanied by the formation of extracellular membrane vessicles, "ghost" cells, and cell lysis. The specific mechanism of this decompression-induced cell lysis and death is not known at the present time.

When cultivated at nonoptimal pressures approaching 1 atm, barophilic bacteria grow more slowly, have a decreased temperature range for growth, and may display aberrant morphological characteristics. The morphology of the barophilic isolate F1, cultivated in either nutrient-rich media (Fig. 4A, B) or pyruvate media (Fig. 4C, D) is shown in Fig 4. In both nutrient regimes, cells grown at 300 atm (Fig. 4B, D) appeared as uniform ovoid rods, but when cultured at 1 atm (Fig. 4A, C) formed elongated filaments. At higher growth pressures beyond 300 atm, cell morphology was unaffected. Similar phenomena have been observed in several barophilic strains isolated from both the Atlantic and Pacific oceans. It is perhaps relevant to note that some nonbarophilic microorganisms form filaments when cultivated at pressures higher than their optimal growth pressure. So nonoptimal growth pressure, either elevated pressures for surface-dwelling bacteria or decreased growth pressures for the barophiles, apparently interferes with processes important to cell division. Whether or not these pressure-induced morphological responses have a common physiological basis is not clear.

Several studies have investigated the effect of pressure on the membrane lipids of barophilic bac-

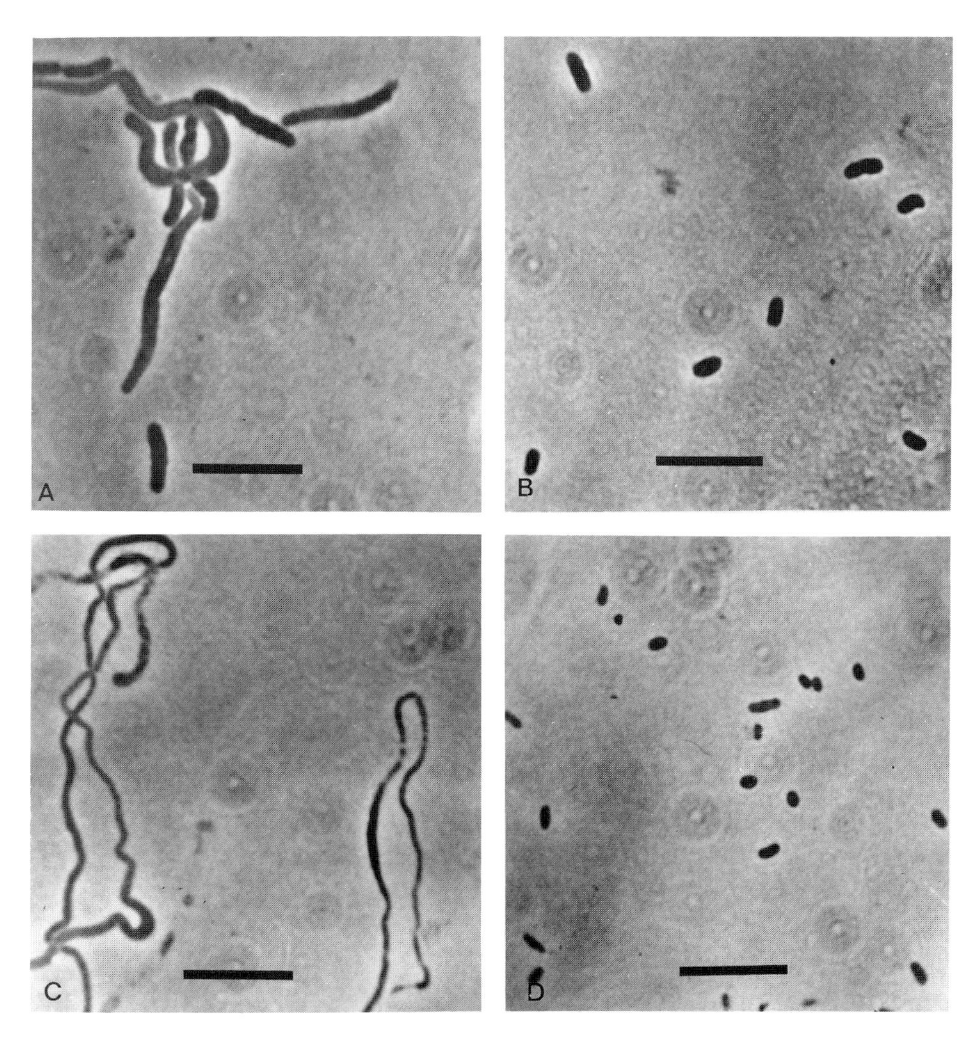

Figure 4 Photomicrographs of the barophilic isolate F1. Cells were grown in either complex nutrient-rich media (A, B) or defined pyruvate media (C, D). Cells were cultivated at 3°C at either 1 atm (A, C) or 300 atm (B, D). Bar = 10 μm. [From Jannasch, H. W., and Wirsen, C. O. (1984). *Arch. Microbiol.* **139,** 281–288.]

teria. Increasing pressure and decreasing temperature are known to physically affect membrane lipids in a very similar fashion (Fig. 5A). Both low temperatures and high pressures act to increase microviscosity (decrease fluidity) in membrane lipids and, thus, can provoke a transition from the fluid, liquid crystalline state to the more solid gel phase (Fig. 5A). Barophilic bacteria appear to have evolved adaptive mechanisms to counteract these temperature- and pressure-induced effects on membrane lipids, analogous to the acclimation responses of cold-adapted species to low temperature. Specifically, several deep-sea bacteria can increase the amounts of unsaturated fatty acids incorporated into their membrane lipids, as a function of increasing growth pressure. Figure 5B shows pressure-induced increase in the ratio of unsaturated fatty acids to saturated fatty acids in the membrane lipids of isolate CNPT3. This increase in fatty acid unsaturation presumably aids in the maintenance of optimal membrane fluidity and function, counteracting the lipid-solidifying effects of high hydrostatic pressure. Studies employing inhibitors of protein and RNA synthesis suggest that the increase of unsaturation results from pressure-induced increases in enzyme activity in isolate CNPT3. Pressure-induced increases in membrane fatty acid unsaturation have been demonstrated in several other barophilic deep-sea bacterial isolates.

Pressure-induced changes in the protein composition of barophilic isolates have also been observed. The barophilic isolate SS9, retrieved from a depth of 2500 m in the Sulu Sea, was shown to have different patterns of protein synthesis at different growth pressures. Specifically, synthesis of an outer membrane protein, designated OmpH, was induced at high growth pressures. A 70-fold increase in OmpH production was observed in cells grown at 280 atm compared with those grown at 1 atm. The gene encoding this protein was cloned and used as a probe to quantitate OmpH gene expression as a function of growth pressure. A 10-fold increase in OmpH RNA transcripts was observed in cells grown at 280 atm, suggesting pressure-sensitive transcriptional regulation of the OmpH protein. At the present time, the function and adaptive significance of OmpH is not known. Available data suggests that barophilic bacteria respond to changes in pressure at both the level of transcription (OmpH expression in SS9) and enzymatic activity (unsaturated fatty acid synthesis in CNPT3).

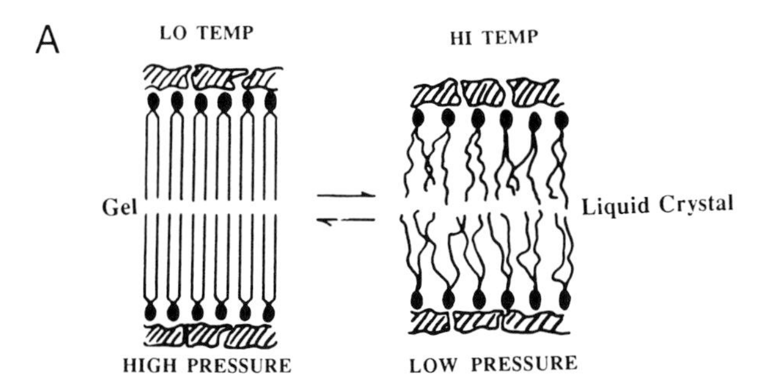

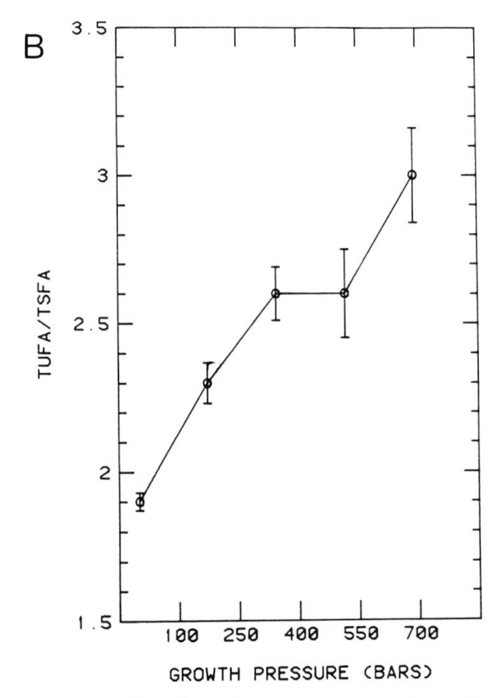

Figure 5 The effect of pressure on membrane lipids. (A) Effect of pressure and temperature on the physical state of membrane lipids. (B) Effect of growth pressure on the ratio of unsaturated fatty acids to saturated fatty acids (TUFA/TSFA) in the membrane lipids of isolate CNPT3. 1 atm = 1.01325×10^5 Pa = 1.01325 bar. [From DeLong, E. F., and Yayanos, A. A. (1985). *Science* **228,** 1101–1103. Copyright 1985 by the AAAS.]

Bibliography

Alongi, D. M. (1990). *Deep-Sea Res.* **37,** 731–746.

Bartlett, D., Wright, M., Yayanos, A. A., and Silverman, M. (1989). *Nature* (London) **342,** 572–574.

DeLong, E. F., and Yayanos, A. A. (1986). *Appl. Environ. Microbiol.* **51,** 730–737.

Deming, J. W., Somers, L. K., Straube, W. L., Swartz, D. G., and Macdonnell, M. T. (1988). *Syst. Appl. Microbiol.* **10,** 152–160.

Jannasch, H. W., and Wirsen, C. O. (1984). *Arch. Microbiol.* **139,** 281–288.

Lochte, K., and Turley, C. M. (1988). *Nature (London)* **333,** 67–69.

Marquis, R. E. (1984). Reversible actions of hydrostatic pressure and compressed gases on microorganisms. *In* "Repairable Lesions in Microorganisms" (A. Hurst and A. Nasim, eds.), pp. 273–301. Academic Press, London.

Weyland, H., and Helmke, E. (1989). Barophilic and psychrophilic bacteria in the Antarctic Ocean. *In* "Recent Advances in Microbial Ecology" (T. Hattori, Y. Ishida, Y. Maruyama, R. Morita, and A. Uchida, eds.), pp. 43–47. Japan Scientific Societies Press, Tokyo, Japan.

Yayanos, A. A. (1986). *Proc. Natl. Acad. Sci. USA* **83,** 9542–9546.

History of Microbiology

Milton Wainwright
University of Sheffield

Joshua Lederberg[1]
The Rockefeller University

Glossary

Antibiotics Antimicrobial agents produced by living organisms

Bacterial genetics Study of genetic elements and hereditary in bacteria

Chemotherapy Systemic use of chemical agents to treat microbial infections

Molecular biology Science concerned with DNA and protein synthesis of living organisms

Monoclonal antibodies Specific antibodies produced by *in vitro* clones of B cells hybridized with cancerous cells

ALTHOUGH MICROORGANISMS were first observed using primitive microscopes as early as the late 1600s, the science of microbiology is barely 150 years old. In this time, major developments have been made in our understanding of microbial physiology, ecology, and systematics. This knowledge has been successfuly applied to broaden our awareness of the nature and etiology of disease, with the result that the majority of the traditional killer diseases have now been conquered. Similar strides have been made in the use of microorganisms in industry, and more recently attempts are being made to apply our knowledge of microbial ecology and physiology to help solve environmental problems. A dramatic development and broadening of the subject of microbiology has taken place since World War II. Microbial genetics, molecular biology, and biotechnology in particular have blossomed. It is to be hoped that these developments are sufficiently opportune to enable us to conquer the latest specter of disease facing us, namely AIDS. Any account of the history of a discipline is, by its very nature, a personal view; hopefully, what follows includes all the major highlights in the development of our science.

The period approximating 1930–1950 was a 'vicennium" of extraordinary transformation of microbiology, just prior to the landmark publication on the structure of DNA by Watson and Crick in 1953. We have important milestones for the vicennium: Jordan and Falk (1928) and "System of Bacteriology" (1930) at its start are magisterial reviews of prior knowledge and thought. Dubos (1945) and Burnet (1945) anticipate the modern era, and Werkman and Wilson (1951) and Gunsalus and Stanier (1962) document its early and continued progress in monographic detail. The *Annual Review of Microbiology*, starting in 1947 (and several other *Annual Reviews*), and *Bacteriological Reviews*, starting in 1937, offer invaluable snapshots of the contemporary state of the art. These works can be consulted for many of the pertinent bibliographic citations, and they will be explicitly repeated here only when important for the argument.

This account will center on the fundamental biology of microbes and give scant attention to continuing advances in the isolation of etiological agents of disease and of vaccines and immunodiagnostic procedures. Most of the agents of common bacterial infections had been characterized by "1930," but

[1] For the period from 1930.

the vicennium was distinguished by important work on the classification of enteric (diarrheal) bacteria and, above all, by the isolation and new study of viruses and rickettsia with methods such as cultivation virus in the chick embryo (Kilbourne, 1987).

I. Observations without Application

Macroscopic manifestations of microbial growth such as bacterial and algal slimes have been recognized since antiquity. However, it was the Dutch microscopist van Leeuwenhoek (Fig. 1) who provided the first observations of bacteria at the microscopic level. van Leeuwenhoek, a draper in Delft, Holland, ground his own lenses to make microscopes with short-focal length lenses giving magnifications of between ×30 and ×266. Descartes had earlier described a similar crude form of microscope, but the quality of his lenses did not allow for magnifications sufficient to see bacteria. van Leeuwenhoek, in contrast, used his homemade microscopes (Fig. 2) to examine microorganisms in rainwater, well water, and seawater as well as water infused with peppercorns. His observations were forwarded to the Royal Society in London on October 1676 and were later published in the Society's *Philosophical Transactions*. In 1683, van Leeuwenhoek contributed a second letter to the Society describing his various microscopical investigations, including novel observations on bacteria present in the scurf of teeth. Published in 1684, these observations include the first drawings of bacteria ever to appear. These drawings are stil extant and clearly show that van Leeuwenhoek observed bacilli, streptococci, and many other characteristic forms of bacteria. van Leeuwenhoek's meticulous drawings also show protozoa such as *Vorticella, Volvox,* and *Euglena.* At about the same time, Huygens also reported observations on a number of free-living protozoa, including species of *Paramecium.* Van Leeuwenhoek also gains credit for describing the first parasitic protozoan, when in 1681 he observed his own fecal stools during a bout of diarrhea and described large populations of what later became known as *Giardia lamblia.*

Figure 1 Anton van Leeuwenhoek (1632–1723).

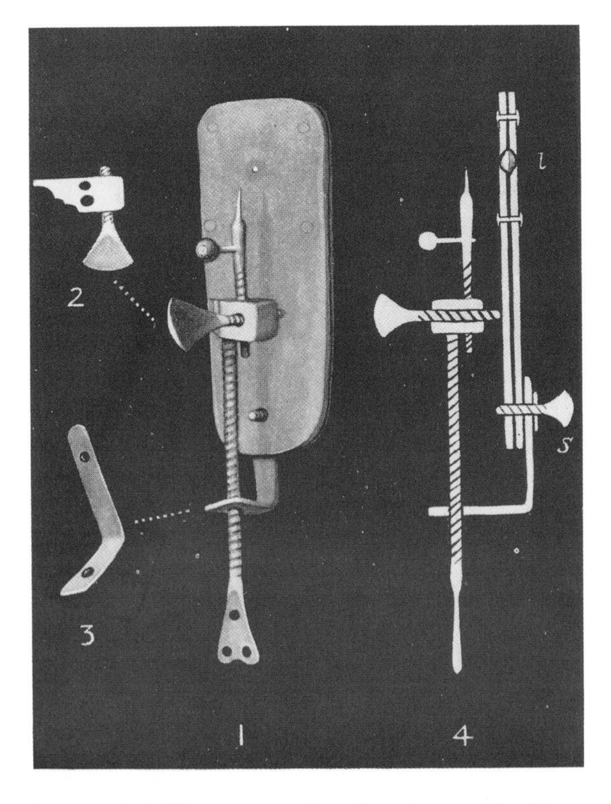

Figure 2 Dobell's reconstruction of van Leeuwenhoek's microscope.

Although van Leeuwenhoek also observed yeast cells in beer, the first illustrations of filamentous microscopic fungi were provided by Robert Hooke, again in a letter to the Royal Society, this time in 1667. However, the most important early work on molds appeared in the following century when the Tuscan botanist, Pietro Antonio Micheli described some 900 species, including important genera such as *Aspergillus* and *Mucor*. It is also worth noting that molds have been used from ancient times to treat infections, an approach termed mold therapy, which was based on folk medicine rather than on any scientific rationale.

Because the connection between microorganisms and fermentation or disease was never made during this period, observations made by the first microscopists had surprisingly little impact on human affairs. Despite this, Cicero and the Renaissance scholar Fracastorius had previously suggested that fevers might be caused by minute animals, collectively described as *contagium vivum*, but it was to be many centuries before the role of microorganisms in disease became recognized, eventually to replace the view that disease resulted from odors or other invisible "miasmas."

II. The Spontaneous Generation Controversy

The view that life arises *de novo* from inanimate objects was widely held from the Middle Ages until remarkably recent times; Van Helmont even provides us with a recipe for the production of mice. The tenacity with which spontaneous generation lingered on is highlighted by the fact that H. Charlton Bastian, one of the concept's chief proponents, died in 1915, still totally convinced of its merit. Although a scientific rationale was apparently provided to account for spontaneous generation by Needham and Buffon as early as 1745, these ideas were quickly dismissed by Spallanzani in the following year. Further developments then had to await the work of Schwann, who in 1837 showed that "air which had been heated then cooled left unchanged a meat broth which had been boiled." Yet by the middle of the seventeenth century, the concept of spontaneous generation held on tenaciously. Then, in 1858, Pouchet published a paper entitled "Proto-organisms . . . Borne Spontaneously in Artificial Air and Oxygen Gas." The French Academy of Sciences was moved by Pouchet's work to offer a prize to anyone who could settle the controversy once and for all. Despite discouragement from his friends who cautioned against becoming embroiled in the controversy, Louis Pasteur (Fig. 3) realized that if microbiology was to advance as a rational science the idea that microorganisms arose spontaneously would need to be experimentally defeated.

Pasteur's studies were published in memoir in 1861 and effortlessly took the prize offered by the Academy. He first of all showed that when air is filtered through cotton wool, large numbers of microorganisms are held back. Pasteur then successfully repeated Schwann's work, but his most famous and successful experiments involved the use of swan-necked flasks, with which he showed that heat-sterilized infusions could be kept sterile in an open flask as long as the open part was tortuous enough to allow any microorganism to settle on the sides of the tubes before reaching the liquid.

It is often assumed that Pasteur's experiments

Figure 3 Louis Pasteur (1822–1895).

immediately brought about the defeat of the theory of spontaneous generation, but this is far from true. Pouchet, for one, remained convinced that Pasteur's experiments did not defeat the concept. The controversy continued over the next quarter of a decade or so. Proponents of Pasteur's views included British scientists such as Huxley, William Roberts, John Tyndall, and the American Jeffries Wyman. The main counter-arguments were provided by the last and most dedicated of the important heterogenesists, H. Charlton Bastian. How this rearguard action by Bastian and others nearly carried the day is remarkable. However, experiments by the mathematician and physicist John Tyndall on the existence of heat-stable forms of certain bacteria (the removal of which involved the process of repeated heating and rest, referred to as tyndalization) finally convinced the scientific establishment of the error of Bastian's arguments. Bastian summed up his views on spontaneous generation in his book *The Evolution of Life* (published as late as 1905), and then died in 1915, still a confirmed believer.

III. Tools of the Trade

The science of microbiology needed two major developments to assure its progress. The first involved improvements in microscopes and associated means by which microorganisms could be better visualized, and the second involved developing methods for culturing microorganisms, thereby ironically liberating the science from total dependence on microscope-based observation.

Compound microscopes first began to appear in Germany at the end of the sixteenth century, and during the following century Robert Hooke developed instruments with magnifications of 3–500×. Although Hooke made major advances in observing microorganisms, he also recognized cellular structure in a variety of life forms. His microscopes, like those of his contemporaries, suffered from chromatic aberration (whereby a ring of colored light prevents accurate focusing on small objects such as bacteria). It was not until the early nineteenth century, when achromatic lenses were introduced by Professor Amici of the University of Medina, that this problem was solved, thereby enabling the light microscope to be developed to its full potential.

The next major development was the introduction of staining procedures, which allowed the fine visualization of microorganisms to occur. The staining of histological specimens was first carried out by the German botanist Ferdinand Cohn in 1849, his work being based on vegetable dyes such as carmine and hematoxylin. By 1877, Robert Koch (Fig. 4) was using methylene blue to stain bacteria, a process in which he developed the standard techniques of preparing dried films, and with the aid of coverslips was preparing permanent preparations. By 1882, Koch had succeeded in staining the tubercle bacillus with methylene blue, employing heat to encourage the stain to penetrate the waxy envelope. Two years later, the Danish pathologist Hans Christian Gram introduced his famous stain, which allowed bacteria to be characterized as gram-positive if they retained the violet dye or gram-negative if they did not. This distinction was later to be correlated with differences in biochemical and morphological characteristics, allowing bacteria to be classified into the two broad groupings still in use today.

Differential staining techniques soon followed, allowing Frederick Loeffler in 1890 to demonstrate the presence of bacterial flagella. During this period, rapid developments occurred in methods for identi-

Figure 4 Robert Koch (1843–1910).

fying bacteria and demonstrating their involvement as causal agents of specific diseases.

The light microscope was eventually developed to its theoretical limits and further progress in microscopy had to await the appearance of the ultraviolet microscope in 1919 (which for the first time allowed certain elementary viruses to be seen). Then, in 1934, the Belgian physicist Marton built the first electron microscope, which achieved magnifications of 2–300,000×, compared to 1200× and 2500× achieved by the light and ultraviolet microscopes, respectively. A further major development in microscope technology came in 1965 with the introduction of the scanning electron microscope.

The first semisynthetic medium designed for cultivating bacteria was introduced in 1860 by Pasteur and consisted of ammonium salts, yeast ash, and candy sugar. Prior to this, meat broths had been used for bacterial growth medium, an approach that persisted well into this century in the laboratories devoted to medical bacteriology. Mycologists, too, tended to rely on undefined media such as potato dextrose agar, although the introduction of Czapek Dox medium eventually provided an ideal defined substrate on which molds could be grown.

In 1872, Ferdinand Cohn developed the idea of the basal medium, to which various additions could be made as required. These early media were always liquid-based and it was not until the introduction first of gelatine and then agar in 1882 that the use of solid media became commonplace. The latter introduction of silica gel media then allowed for rapid advances to be made in the study of chemolithotrophic bacteria such as *Thiobacillus thiooxidans*.

By 1887, a simple and prosaic development revolutionized microbiology when Petri, one of Koch's assistants, introduced the Petri dish. This simple invention provided a far more versatile means of culturing microorganisms than did use of the bulky bell jars employed previously.

From 1898 onward, the Dutch school of microbiologists led by Beijerinck developed the art of enrichment culture, which led to the isolation of both nitrifying and cellulolytic bacteria. Studies on gas gangrene during the first war encouraged McIntosh and Fildes to develop the anaerobic jar. A vast array of selective media were then developed that involved amendments such as tetrathionate broth, tellurite, and crude penicillin. Finally, the introduction of central media supplies after the war liberated the microbiologist and their technicians from the tedium of preparing media in-house. No longer did mycologists, for example, have to spend hours peeling and boiling potatoes when potato dextrose agar was available ready to rehydrate, sterilize, and use.

None of the preceding developments in media preparation would have been useful without the introduction of an efficient means of sterilization. Pasteur's colleague, Chamberland, developed autoclaves—essentially large pressure cookers—in 1884. More recently, gamma rays and ethylene oxide sterilization have allowed for the introduction of factory-sterilized plastics including Petri dishes, another relatively simple development that has, nevertheless, had a marked stimulatory effect on the recent progress of microbiology. [*See* STERILIZATION.]

IV. Microorganisms as Causal Agents of Disease

In 1788, an epidemic of smallpox broke out in the English county of Gloucestershire. Edward Jenner, a country doctor and pupil of the famous anatomist John Hunter, decided to try and prevent his patients from contracting the disease by employing the standard method of inoculation using a mild dose of the infection. Jenner, who had suffered under the blood purgers and inoculists in his youth, was himself immune to smallpox. He aimed to make the traditional inoculation method as rational and reliable as he could. While on his regular rounds, he was surprised to find that patients who had already suffered from cowpox did not react in the normal way to inoculation with smallpox. Although Jenner was aware of the old wives' tale suggesting that cowpox gave protection against the disease, it was not until 1796 nearly a quarter of a century after he had first heard these suggestions, that he decided to act. His first experimental inoculation involved a local boy named James Phipps, who, after receiving cowpox, became immune to smallpox. In June 1798, Jenner presented a paper on his work to the Royal Society, and the effect was remarkable— within a few years, vaccination was commonplace.

Despite Jenner's breakthrough, there was still no convincing explanation to account for the appearance and spread of infections, and by the mid-1800s there was still little that could be done to counter infectious disease. Childbed or puerperal fever was a particularly terrible blight that affected every one of the lying-in hospitals in Europe. During a single month in 1856 in a Paris hospital, 31 recent mothers died of the infection. Vienna of the 1840s had a particularly bad reputation for this disease, despite having one of the most enlightened hospitals in Europe. It was here that the Hungarian doctor, Ignaz Semmelweiss joined the staff of the lying-in clinic of the Vienna General Hospital in 1844. In his first few months of practice, he heard yet another wives' tale, this one associating the high death rate from childbed fever found in the teaching division of the hospital with the high frequency of examination by doctors and their students. Semmelweiss began to collect statistics and soon became aware that the highest rates of infection and mortality occurred in the teaching clinic. This information led him to surmise that the contagion was being transmitted by the doctors and medical students, many of whom examined the wombs of patients without washing their hands, even after coming directly from mortuary duty. Semmelweiss suggested that anyone examining patients should first wash their hands in chlorine water. The results of this simple remedy were phenomenally successful, with mortality rates being reduced from around 11 to 3% within 1 yr. Semmelweiss was slow to write an account of his work, but eventually in 1857 he provided a rambling and highly egotistical survey of his work, which completely failed to make any impression.

Eventually, however, the view that infection was spread by some organic particle did at last become widely accepted, although the exact nature of such particles was unknown. The effect of this ignorance was devastating; during the Crimean War of 1853–1856, for example, a single regiment of the British Army lost 2162 men, with 1713 dying not from wounds or the effects of trauma but from disease. The infamous hospital diseases of erysipelas, pyemia, septicemia, and gangrene made surgical wards nightmares of suffering and death. The causes and mechanisms of disease transmission remained essentially unknown. By 1865, however, Pasteur had concluded that disease must be airborne, a view that galvanized the English surgeon Joseph Lister into action. Lister reasoned that he could reduce mortality due to sepsis by covering wounds with dressings containing chemicals that killed these airborne germs without preventing the entry of air. He knew that carbolic acid had recently been used to sterilize sewage, and with the help of the chemist Anderson he obtained a supply of the sweet-smelling dark liquid that was commonly called German creosote. Lister published his findings in *The Lancet* in 1867. In contrast to Semmelweiss's efforts, Lister's work attracted immediate attention—The age of antiseptic surgery was soon underway.

Once it became realized that microscopic organisms present in the air were responsible for transmitting disease, the next important development was to isolate these organisms and then conclusively demonstrate their role as causal agents of any given disease. Yet, some authorities continued to argue that microorganisms were not the cause of disease, but merely grew on the weakened infection site. In May 1882, Robert Koch dismissed this view when he announced the discovery of the tubercle bacillus; the search for other disease-causing microorganisms then gathered momentum. The introduction by Koch of his famous postulates finally established a means of conclusively demonstrating the involvement of a microorganism as a causal agent of a given disease, and the way lay open to disease prevention and cure.

Major developments were next made in our understanding of immunity. The first rational attempts to produce artificial active immunity was made by Pasteur in 1880 during his work on fowl cholera. By 1882, the Russian biologist Metchnikoff had made the first observations of cellular immunity and coined the term phagocyte. By 1891, Ehrlich had distinguished between active and passive immunity, and 6 years later Kraus published the first account of precipitation reactions when immune sera were added to cell-free filtrates of homologous bacterial cultures.

Nearly 250 million people have been vaccinated against tuberculosis with the bacille Calmette–Guérin (BCG) vaccine, yet its originator, Charles Calmette, remains a largely unknown figure. Calmette, a disciple of Pasteur, was the first Director of the Pasteur Institute in Lille, France, and later became Assistant Director of the Pasteur Institute in Paris. With Guérin, he set about to prepare a protective vaccine against tuberculosis. He spent 13 years developing an attenuated virus, which by not recovering its lost virulence remained both stable and safe. This vaccine, BCG, was first used in 1921, but because of considerable resistance to its use was not widely accepted until after Calmette's death in 1933.

Modern developments in immunology include the work of F. Macfarlane Burnet, who in 1957 published his clonal selection hypothesis.

V. Chemotherapy and Antibiosis

The origin and early development of the concept of chemotherapy is somewhat unusual in that it can be credited to the work of one man, the German chemist Paul Ehrlich. Ehrlich had the vision to apply his knowledge of specific staining of bacteria to the search for chemical compounds that would inhibit the growth of pathogenic bacteria *in vivo*. In 1891, he showed that methylene blue was useful for the treatment of malaria, but because this dye showed no advantage over quinine it was not widely used. By 1902, Ehrlich was concentrating his attention on the organic arsenic compounds, which he hoped would defeat experimental trypanosomiasis in mice. At this point, he and his Japanese bacteriologist assistant Shiga found that atoxyl (sodium arsanilate) was ineffective against mouse trypanosomiasis. This turned out to be a somewhat inexplicable error when the British bacteriologist Thomas was soon to show that atoxyl was in fact extremely effective against trypanosomiasis in mice.

A second equally inexplicable error followed when Schaudinn and Hoffman concluded that *Treponema pallidum* was a protozoan. Ironically, this error proved productive because it pointed to the likelihood that the antiprotozoal agent atoxyl, or a similar compound, might cure syphilis. In 1906, Robert Koch used atoxyl to treat trypanosomiasis in humans. This was the year in which Ehrlich became director of the newly opened George Speyer Institute, which was devoted to chemotherapy research. It was here that the first major chemotherapeutic agent salvarsan was developed. Salvarsan was first discovered in 1907 and was initially found to be inactive against the experimental mouse trypanosomiasis system. Then in 1909, a young Japanese scientist, Hata, joined Ehrlich's laboratory, bringing with him a system that he had developed for the artificial transmission of *T. pallidium* in rabbits. To his evident surprise, Hata found that salvarsan was in fact effective against syphilis in mice; by 1909, the drug was proving spectacularly successful in treating the disease in humans.

Following Ehrlich's death in 1915, research continued into chemotherapy, but little progress was made, with the exception that in 1932 Atebrin became available as the first synthetic drug for prophylactic use against malaria. The next major advance in chemotherapy came in 1935, when Domagk discovered the antibacterial effect of the red dye prontosil. This compound had a dramatic effect on lobar pneumonia in humans, reducing death rates by by two-thirds. In the same year as its discovery, the French scientist Trefouel showed that the active ingredient of prontosil was not the chromophore, but the sulphonamide moeity (sulfanilamide). Sulphonamides were widely used with success to treat bacterial infections from the mid-1930s until the middle of the following decade.

The concept of chemotherapy reached its zenith with the sulphonamides, but such compounds were soon eclipsed by the arrival of first penicillin and then a range of other antibiotics.

Antibiotics (cf. Waksman, MacFarlane, Wilson) had a spectacular beginning with the famous discovery of penicillin by Fleming in 1928, a mold spore having accidentally lodged on agar plates seeded with staphylococci. The story of the discovery of penicillin by Alexander Fleming is probably the best known in the history of medicine, although much that has been written on the subject borders on fairy tale. The important point about Fleming's initial observation, made during the late summer of 1928, was that it represented an extremely rare phenomenon, not merely an example of microbial antagonism, but one of bacterial lysis brought about by mold contaminant. Fleming probably initially thought that he had discovered a fungal variant of lysozyme, a lytic substance that he had previously found in various body fluids. It was this lytic phenomenon that distinguished Fleming's observations from the numerous observations of microbial antagonism that had been reported since Pastuer's time. It is likely that had he observed microbial antagonism, rather than lysis, Fleming would have ignored his observations, regarding them as an example of common phenomenon that was of little interest.

Fleming, however, understood the significance of what he observed. He soon showed that the contaminant produced the antibacterial substance in culture broth, which he called penicillin. Then, with help from various surgeon colleagues, Fleming used crude penicillin-rich filtrates to treat superficial bacterial infections, unfortunately without much success. The first documented cures with penicillin were in fact achieved (using the crude broths) by a former student of Fleming's, Cecil George Paine, who worked at Sheffield University.

In his first famous paper on penicillin, Fleming detailed its properties and antibacterial spectrum and suggested that it, or a similar substance, might find a use in medicine. Unfortunately, neither he nor his colleagues could purify penicillin, an obvious necessary first step for its successful introduction into medicine. It is worth pointing out, however, that Fleming was not alone in being unable to achieve this essential purification step; other attempts such as those made by famous fungal product biochemist Harold Raistrick also proved unsuccessful.

Fleming's notebooks show that despite being unable to purify penicillin, he continued working on crude penicillin throughout the 1930s, during which time he also attempted to isolate other microorganisms capable of producing antibacterial products. Unfortunately, this work was not published and the medical potential of antibioisis remained undeveloped until the discovery of gramicidin in 1939. This substance, which was discovered by Rene Dubos, was unfortunately too toxic for intravenous use; therefore, it was limited to use on a number of superficial infections.

At about the time when gramicidin was being first developed, Florey, Chain, and Heatley managed to purify penicillin and demonstrate its remarkable antibacterial effects when used systematically. The isolation of the antibiotic from the crude culture filtrates was a formidable chemical task, but it was undertaken successfully in the late 1930s by Florey and Chain in England. Industrial production of penicillin soon followed as a joint U.S.–British war project. For this to be feasible required a substantial effort in strain improvement, which was conducted, however, along empirical rather than rational genetic lines (Wilson 1976). This was nevertheless the forerunner of the modern fermentation industry and biotechnology; its antecedents had been the production of butanol and acetone as munitions solvents during World War I and the peacetime production of citric acid by a mold fermentation. Penicillin's introduction into medicine as the first successful antibiotic stimulated the search for similar compounds. A particularly successful antibiotic screening program, devoted to soil actinomycetes, was carried out by Selman Waksman and his students at Rutgers. The first major product of this research, actinomycin, was, like gramicidin, too toxic to be of medical use as an antibiotic, although it was later used as an anticancer agent.

S. Waksman and R. J. Dubos had been studying the biochemical and ecological interrelations of soil microbes. The role of secreted antibiotics in ecological competition provided a rational for seeking these substances. *Tyrothricin* (Dubos, 1939; cf. Crease, 1989) was the first antibiotic to be clinically applicable, but its systemic toxicity limited its application to topical treatment. In Waksman's hands, the same paradigm led to the discovery of streptomycin (1944), which when used in conjunction with periodic acid–Schiff and isoniazid helped to defeat tuberculosis. Thereafter, a continued stream of new antibiotics with untold human benefit. It would be some time before the mode of action of antibiotics would even begin to be understood (cf. Gottlieb and Shaw, 1967) and to allow rational principles to assist in their improvement.

Although Waksman received the Nobel Prize for streptomycin, his triumph was marred by his tardy treatment of the codiscoverer of the antibiotic, Albert Schatz. Schatz, one of Waksman's graduate students, successfully sued Waksman and Rutgers for a share of the royalties for streptomycin. His later attempts to gain a share of the Nobel Prize for his work (he was senior author on the first streptomycin papers and coassignee with Waksman of the streptomycin patents) were, however, unsuccessful.

Streptomycin was soon followed by antibiotics such as chloramphenicol, neomycin, tetracycline, and the first effective antifungal antibiotic, nystatin (discovered by Elizabeth Hazen and Rachel Brown). Penicillinase-resistant penicillins such as methicillin then appeared, followed by semisynthetic penicillins, and finally broad-spectrum compounds like ampicillin.

VI. Microbial Metabolism and Applied Microbiology

Developments in the study of microbial metabolism were, from the outset, closely associated with attempts to use microorganisms for industrial purposes, a trend that continues in modern biotechnology. It is not surprising then to find that the first scientific paper devoted to microbial metabolism (appearing in 1857) can also be regarded as the first citation in applied microbiology or biotechnology. Again Pasteur was responsible for this development, the paper being devoted to an explanation of the causes of the repeated failures of industrial alcohol fermentations. This was an important paper for two reasons: first, because it laid the foundation of the view, later to be amply validated, that microbial activity was responsible for many industrially important fermentations, and, second, because it introduced quantitative treatment of data on microbial growth and metabolism.

Pasteur also addressed problems associated with the microbiology of wine-making. Among the suggestions that he made was a method for improving the keeping qualities of wine by heating it to 68°C for 10 min, followed by rapid cooling, a process subsequently referred to as pasteurization. By 1872, Pasteur's work had been developed to the point where Ferdinand Cohn could suggest that microorganisms play a major role in the biological cycling of the elements responsible for soil fertility and the proper functioning of natural ecosystems. The first fruits of Cohn's theory came in 1888 when the Dutch microbiologist Beijerinck isolated the symbiotic N-fixing bacterium *Rhizobium* from the root nodules of legumes. During these studies, Beijerinck also developed the enrichment technique to the isolation of microorganisms, an approach later to be refined and developed by the Dutch school of microbiologists.

Many of the following breakthroughs in microbial metabolism were associated with studies on soil microbiology and its association with soil fertility. In 1889, Winogradsky described the autotrophic iron and sulfur bacteria, and in the following year the free-living N-fixing *Azotobacter* and the nitrite-oxidizing bacterium *Nitrobacter*. Although many of Winogradsky's so-called pure cultures appear to have been contaminated, his work was nevertheless important because he was the first to appreciate the concept of chemoautotrophy and to relate this growth strategy to the major natural cycles. It was a lack of appreciation of this concept that had hindered the work of others interested in processes such as nitrification. Despite this, the soil chemist Warrington nevertheless did important work on the factors that influence this process in agricultural soils.

Waksman and Joffe isolated and described *T. thiooxidans* in 1922, and over the next quarter of a century, major contributions to the science of microbial physiology came from, among others, Wieland, who in 1900 demonstrated the importance of biological oxidations using microorganisms. Other work of note came from Marjorie Stephenson and J. H. Quastel on enzymes. In 1924, A. J. Kluyver published an important article entitled "Unity and Diversity in the Metabolism of Micro-organisms," a paper that demonstrated the fundamental unity underlaying the apparent diversity of microbial metabolism. By 1930, Karstrom had established the concept of constitutive and adaptive enzymes. By now, microbiology had begun to be a cornerstone of biochemistry and the boundaries between the subjects were soon blurred. In 1941, Lipmann advanced the concept of the high-energy bond, and major developments in theories on the working of enzymes came from Monod's lab.

While most of the early developments in microbial metabolism were centered on bacteria, fungal metabolism, because of its importance to many industrial fermentation (e.g., citric acid production), was by no means neglected. The seminal work in this area came in 1940, when Jackson Foster published his "Chemical Activities of the Fungi." Studies on fungal metabolism obviously gained impetus following the introduction, while the isolation of antibiotics such as streptomycin also gave a boost to the study of a neglected group of organisms—the actinomycetes. It was Selman Waksman who initiated work in these organisms during the early part of this century, a period when the actinomycetes were regarded as fungi rather than bacteria.

VII. Nutrition, Comparative Biochemistry, and Other Aspects of Metabolism

Microbes, first yeast and then bacteria, played an important part in the discovery of vitamins and other growth factors. Growth could be measured in test tubes far more expeditiously and economically than in mice, rats, or humans. Conversely, the realization that microbes shared virtually all of the complex growth factor requirements of animals was an important impetus to "comparative biochemistry," the view that they had a common evolution and a similar underlying architecture. One of the essential amino acids, methionine, was first discovered by Mueller (1922) as a growth factor required by diphtheria bacilli. Mueller joined a school founded by Twort (1911), including Lwoff, Fildes, Knight, and Tatum, that made nutrition a branch of general biochemistry. They perceived that the requirement for a growth factor belied a loss or deficiency of synthetic power; lacking internal synthesis, the organism had to look to the nutrient environment for supply of substance. This also implied that organisms with simple nutrition had to be empowered with complex biosynthetic capability—leaving us humiliated by our species' inferiority to *Escherichia coli,* but that in turn is less capable than the green plant! Besides the practical utility of these findings, they led to a well-founded respect for the complexity of microbial cells.

By "1930," a number of growth factors had been shown to be important in bacterial nutrition, including factors V and X, later shown to be diphosphopyridine-nucleotide and heme, respectively, for hemophilic bacteria; *Mycobacterium phlei* factor, later shown to be vitamin K, for *M. pseudotuberculosis;* and tryptophane for *Salmonella typhi* (Fildes, 1936). Starting with the work of W. H. Peterson, H. Wood, E. Snell, and E. L. Tatum at the University of Wisconsin and B.C.J.G. Knight and P. Fildes in England, a number of bacterial growth factors were identified with B vitamins (extensively cataloged by Johnson and Johnson, 1945). By "1950," most of the known trace growth factors had been identified and associated with nutritional requirements of particular bacteria, as also had been most of the amino acids and a host of other metabolites (Snell, 1951). During the vicennium, most of the vitamins were also identified as co-enzymes, playing a role in the function of specific metabolic enzymes [e.g., thiamin for keto-acid decarboxylases, niacin for dehydrogenases, pyridoxal for transaminases, pantothenate in the citric acid cycle (Schlenk, 1951)]. The 20 canonical amino acids were listed and could be shown to be incorporated into bacterial protein.

A host of other biochemical pathways were also detailed with the help of new methodologies of radioisotopic tracers and chromatography. Of special significance in bacterial metabolism was the demonstration of heterotrophic assimilation of CO_2. This view of CO_2 as an anabolite was contrary to its usual image as a waste product. The specific requirements for CO_2 as a nutrient helped to clear up difficulties in the cultivation of fastidious bacteria and eventually of tissue cells.

By 1941, microbiology and genetics overlapped when G. W. Beadle, Tatum, and coworkers at Stanford University began to use the red bread mold *Neurospora crassa,* an approach in which mutants were employed to help elucidate genetic mechanisms, thereby allowing a number of microbial pathways to be worked out for the first time.

In due course, especially after Beadle and Tatum (1941), the power of synthesis came to be understood as the capability of individual specific genes. This in turn led to concepts and experiments on the genetic underpinnings of metabolism.

The birth of molecular biology followed the work of Watson and Crick in 1853, when microbiology entered into a new phase, allowing it to overlap with many other sciences, leading to the appearance of numerous exciting developments.

The major conceptual theme of change in microbiology during the vicennium was the convergence of the discipline with general biology. As noted by Dubos (1945),

> To the biologist of the nineteenth century, bacteria appeared as the most primitive expression of cellular organization, the very limit of life. Speaking of what he considered "the smallest, and at the same time the simplest and lowest of all living forms," Ferdinand Cohn asserted: "They form the boundary line of life; beyond them, life does not exist, so far at least as our microscopic expedients reach; and these are not small." The minute dimensions of bacteria were considered by many to be incompatible with any significant morphological differentiation; it encouraged the physical chemist to treat the bacterial cell as a simple colloidal system and the biochemist to regard it as a "bag of enzymes."

Still dominated by the medical importance of microbes, the views of microbiologists in "1930" had not evolved much further, although "System" (1930) does have a brief chapter on bacterial cytology and allusion to ongoing controversy over the existence of nuclear structures. Far more attention is given to the Gram stain!

While a few differences in the detail of intermediary metabolism and biosynthetic options have been discovered (e.g., for lysine), it remains true that pathways conveniently noted in bacteria have usually been reliable predictors of the same steps in higher plants and animals. It is possible today to relate this functional conservatism to evolutionary affinity with currently available tools of DNA sequencing.

A. Induced Enzyme Formation, or "Enzymatic Adaptation"

One of the most intriguing phenomena of bacterial physiology is the plasticity of enzyme expression dependent on the chemical environment. For example, *E. coli* grown on a glucose medium exhibits very low levels of β-galactosidase (lactase). When glucose is replaced by lactose, there is a growth delay followed by the abundant production of lactase. Thousands of comparable examples are now known, and the pursuit of the mechanism of this phenomenon has been of outstanding importance in the development of molecular genetics. Anecdotal reports of enzyme adaptation can be traced back to Wortmann (1882, cited *in* Karstrom, 1930); they were collected, together with new experimental observations, by Henning Karstrom for his doctoral dissertation in Virtanen's laboratory in Helsinki. In this turning-point review [Karstrom, 1930, followed by the more accessible Karstrom, 1937; Dubos 1940 (1945)], bacterial enzymes are classified as constitutive or adaptive according to their independence, or otherwise, of the cultural environment. Except for glucose metabolism, most sugar-splitting enzymes are adaptive—resulting in substantial biosynthetic economy for a bacterium or yeast that may only rarely encounter, say, maltose now, or lactose next week. During the vicennium, the work of Stephenson and Yudkin (1963) and Gale (1943) furnished additional clearcut examples of the adaptive response, and Dubos (1945) offers a critical appraisal of the fundamental biological issues. Several theories allowed for the stabilization of preformed enzyme by a substrate, or a Le Chatelier-like principle of mass action, to encourage enzyme synthesis. They shared the presumption that the enzyme molecule itself was the receptor of the inducing substrate. Other hypotheses lent the substrate an instructive role in shaping the specificity of the enzyme. Further progress would depend on the postulation of an enzyme-forming system distinct from the enzyme—and this would emerge under the impetus of genetic studies to be described later. At the very end of the vicennium, Lederberg *et al.* (1951) described a noninducing substrate of lactose, the analog altrose-β-D-galactoside, which pointed to a separation of those specificities. This substrate also allowed the selection of constitutive-lactase formers, showing that lactose was not required for the conformation of the enzyme, but that the latter could be derived directly from the genetic constitution. The debate continued until the mid-1950s (see Lederberg, 1956, p. 51; Monod, 1956); it was mooted by the spectacular progress of the Pasteur Institute group in showing that enzyme induction was the neutralization of an endogenous repressor that inhibited the expression of the lactase gene in the absence of an inducer (Jacob, 1965).

The simultaneous induction of several steps in a metabolic pathway, usually by an early substrate, was exploited to delineate the later steps, notably in the oxidation of aromatic compounds by pseudomonads.

Among technical innovations, one of the most ingenious was the chemostat (Novick and Szilard, 1950). This allowed microbial populations to be maintained for the first time in a well-defined steady state, albeit under limitation for one specific nutrient.

VIII. Microbial Genetics

During the last two decades of the nineteenth century, it was realized that bacterial species were not as stable as had first been thought. Pure line cultures that had been maintained for many generations suddenly underwent dramatic changes in morphology, metabolic properties, and pathogenicity. As more pure cultures were obtained, this variability, or dissociation as it was called, became even more apparent. Then in 1925, R. M. Mellon published a paper describing a primitive from of sexuality in coli-typhoid bacteria. This work had little contemporary impact on the contemporary view that bacteria were anucleate organisms that reproduced without sexuality by binary fission.

Bacterial genetics was substantially nonexistent in 1930. As late as 1942, the eminent British biologist Julian Huxley would suggest of bacteria that "the entire organism appears to function both as soma and germ plasm and evolution must be a matter of alteration in the reaction system as a whole" (Huxley, 1942. Such ideas gave little encouragement to efforts to dissect out individual genes along the Mendelian lines that had been so successful with *Drosophila* and other animals and plants. Some work with fungi had gotten off to a promising start early in the century (Blakeselee, 1902). Authentic but sporadic observations of bacterial mutation (Beijerinck, 1901) were outnumbered by wooly-minded speculations that embraced variations of colony form as manifestations of cellular life cycles among the bacteria (see Dubos, 1945; Lederberg, 1992). These clouds of speculation probably discouraged more serious-minded experimentation.

Mention has already been made of the impact of the work of Beadle and Tatum on mutants in *Neurospora* on our understanding of microbial physiology. However, by initiating the field of biochemical genetics, these studies had even greater impact on the science of genetics. Prior to 1941, genetic research was dominated by work on the fruit fly, *Drosophila melanogaster*. Much was learned from studying morphological mutations in this organism, but efforts to disentangle the biochemical basis of these characteristics resulted only in frustration. Beadle and his microbial biochemist colleague Tatum turned their attention to studying the red bread mold *N. crassa* and soon obtained mutants with nutritional defects such as blocks in the biosynthesis of vitamins like pyridoxine and thiamine. This allows for rapid improvements to be made in genetic analysis, an approach that was subsequently extended by other workers using bacteria. The first fruits of such application came in 1943 when Luria and Delbrück showed by means of their "fluctuation test" that spontaneous mutations occurred in bacteria, to both phage resistance and streptomycin resistance at similar frequencies, as had been observed in other organisms.

The study of bacterial genetics was dramatically advanced during the 1940s following the recognition of antibiotic resistance in pathogenic bacteria. Here was a practical problem, the solution to which provided an obvious impetus to studies aimed at determining its cause. [*See* Antibiotic Resistance.]

Bacteria did of course suffer from the serious methodological constraint of the apparent lack of any recombinational (sexual or crossing) mechanism by which to analyze and reconstitute gene combinations. They would prove, however, to be marvelous material for mutation studies (cf., e.g., Ames, 1975) once the concepts were clarified, for which a major turning point was the work of Luria and Delbrück (1943). In a fashion that reminds one of Gregor Mendel, they studied bacterial mutation by quantitative counts. They used resistance to (bacterio)phage as the marker. Like resistance to antibiotics, or growth on a nutritionally deprived medium, the phage is an environmental agent that makes it easy to count exceptional cells against a preponderant background that can be selectively wiped out. Most importantly, they distinguished between mutational events, which engender resistant clones, and mutant cells, which are counted when you plate a population with the selecting phage.

Luria (1984) in his charming book "A Slot Machine, A Broken Test Tube," recounts how his observation of a jackpot in a gambling den inspired his premonition of the skewed statistics that would govern the numbers of mutants. The fit of experimental numbers to those statistics is subject to great theoretical uncertainty, but they were a corroboration of the clonal model. One of the first articles on bacteria to be published in *Genetics,* the paper promptly attracted broad attention and was widely regarded as having proved "that bacteria have genes." The gist of the demonstration was that mutations to phage resistance agree with a clonal distribution and, thus, render more likely their "preadaptive" occurrence, that is, within the growth of the population rather than at the time of the challenge with the selective agent. It therefore harkens more to Darwin than to Mendel; nevertheless, it was a turning point in geneticists' appreciation of bacteria. The statistical methods, which are helpful in the quantitative estimation of mutation rate, have been improved (Sarkan, 1991).

The themes of nutrition and mutation among microbes had occasional false starts, with observations of strain variability and the "training" of exacting bacteria to dispense with growth factors (Knight, 1936). However, lacking a conceptual framework of "genes in bacteria," these had little fruit prior to the work of Beadle and Tatum (1941) on *Neurospora*. Beadle had begun his research program with Ephrussi on the genes for eye color in *Drosophila* (Burian, 1989). Tatum was engaged to do the biochemical work but found the material almost intractable—When he approached success, he was scooped by Butenandt on the identification of kynurenine as a pigment precursor. Nor was it clear how much closer to the primary gene product this chemistry would bring them. The following account is taken from J. Lederberg's memoir on E. L. Tatum, who was his teacher from 1946 to 1947 (Lederberg, 1990).

> This jarring experience, to have such painstaking work overtaken in so facile a fashion, impelled Beadle and Tatum to seek another organism more tractable than *Drosophila* for biochemical studies of gene action.
>
> In Winter Quarter 1941, Tatum offered a new graduate course in comparative biochem-

istry. In it, he called upon his postdoctorate experience with Kogl in Utrecht, in 1937, and recounting the nutrition of yeasts and fungi, some of which exhibited well-defined blocks in vitamin biosynthesis. Beadle, attending some of these lectures, recalled the elegant work on the segregation of morphological mutant factors in *Neurospora* that he had heard from B. O. Dodge in 1932. The conjunction was that *Neurospora* had an ideal life-cycle for genetic analysis with the immediate manifestation of segregating genes in the string of ascospores. *Neurospora* also proved to be readily cultured on a well defined medium, requiring only biotin as a supplement. By February 1941, the team was X-raying *Neurospora* and seeking mutants with specific biosynthetic defects, namely nutritional requirements for exogenous growth factors.

Harvesting nutritional mutants in microorganisms in those days was painstaking hand labor; it meant examining single-spore cultures isolated from irradiated parents, one by one, for their nutritional properties. No one could have predicted how many thousands of cultures would have to be tested to discover the first mutant: isolate #299 in fact required pyridoxine. Furthermore, the trait segregated in crosses according to simple Mendelian principles, which foretold that it could in due course be mapped onto a specific chromosome of the fungus. Therewith *Neurospora* moved to center stage as an object of genetic experimentation.

In their first paper, they remarked "that there must exist orders of directness of gene control ranging from one-to-one relations to relations of great complexity." The characteristics of mutations affecting metabolic steps spoke to a direct and simple role of genes in the control of enzymes. These were therefore hypothesized to be the primary products of genes. Indeed, insome cases, genes might themselves be enzymes. This was an assertion of what came to be labeled the one-gene : one-enzyme theory, which has become the canonical foundation of modern molecular genetics, albeit with substantial correction and elaboration of detail, especially with regard to the intermediating role of messenger RNA, which could hardly be thought of in 1941. It would be a mistake to focus too sharply on the numerical 1 : 1 assertion; more important was the general assumption of simplicity, and that the details of gene expression could be learned as an outcome of such studies—as indeed they were (see also Horowitz, 1990).

The recruitment of *Neurospora* for what have become classical genetic studies offered further encouragement that bacteria, albeit somewhat more primitive, might be handled in similar fashion. By 1944, Gray and Tatum had produced nutritional mutants in bacteria, including some in a strain that has dominated bacterial genetics ever since, namely *E. coli* strain K-12. These mutants were soon to be put to a most striking use.

In 1944, O. T. Avery and his colleagues concluded that the transforming principle involved in transformation in pneumococci was DNA. This was a major breakthrough, because until then it was thought that the significant part of the nucleoprotein of the chromosome molecule was the protein, the nucleic acid merely acting as a sort of binding agent. The role of DNA was initially puzzling, because it was difficult to see how a polymer that contained only four bases could possibly code for the complex phenotype of even the simplest of organisms. Meanwhile, classic genetic approaches were yielding a wealth of new discoveries. In 1945, Tatum showed that the mutant rate of bacteria could be increased using X-rays, whereas 2 yr later, Tatum and Lederberg demonstrated genetic recombination between two nutritionally defective strains of *E. coli K12*.

The first gene map of *E. coli* K12 appeared, and over the next few years progress was made in explaining the phenomena of conjugation, transduction, and transformation. William Hayes, working at the postgraduate medical school in Hammersmith announced in 1952 his discovery that in conjugation recombination occurred due to the one-way transfer of genetic material, and during the same year Lederberg and Cavalli coined the terms fertility plus (F^+) for donor cells and fertility minus (F^-) for recipient cells. The recognition of these mating types made it clear that conjugation was a primitive form of sexuality, with the recipient F^- cell being the zygote. More advances came when Lederberg, Cavalli, and Lederberg discovered high-frequency recombinant mutants from the F^+ type of *E. coli* K12, a finding that was subsequently confirmed by Hayes. These mutant strains (Hfr) differed from the wild-type F^+ strains, first in transferring various genetic markers at a rate hundreds of times greater than the original strains and, second, in not producing an alteration in the mating type of the recipient cell. However, although the frequency of transfer of the various markers differed, it was the same for any given strain of Hfr.

Between 1955 and 1958, Jacob and Wollman used their famous "interrupted mating experiment" to determine the mechanism of gene transfer in *E. coli* K12. Jacob and Wollman coined the term episome, and in 1963 Cairns confirmed the circular nature of the bacterial chromosome using autoradiography. Bacterial genetics further progressed following the report published in 1961 by Watanabe explaining infectious drug resistance.

A. The Pneumococcus Transformation

What might be regarded as the first major breakthrough in microbial genetics came in 1928 when Griffith published on detailing "transformation" in pneumococci, a study that laid the foundation for later work by Avery and his colleagues. A further development in our understanding of transformation came in 1933, when Alloway showed that rough type 1 cells could be changed into genetically stable smooth type II cells, by growing them in the presence of a cell-free extract of a heat-killed broth culture of smooth type II cells. This work demonstrated the existence of a soluble "transforming agent." [*See* GENETIC TRANSFORMATION, MECHANISMS.]

Apart from cataclysmic happenings in global war, 1944 will also be remembered for the publication of "Studies on the Chemical Nature of the Substance Inducing Transformation of Pneumococcal Types," by Avery, MacLeod, and McCarty.[1] The pneumococcus transformation was stumbled upon by Fred Griffith in London, in 1928, in the course of his studies on the serosystematics of pneumonia. Extracts of one serotype evidently could transform cells of another into the type of the first. In retrospect, it is hard to imagine any interpretation other than the transmission of a gene from one bacterial cell to another, but this interpretation was inevitably dimmed by the poor general understanding of bacterial genetics at that time.

This vagueness was compounded by two outstanding misinterpretations: (1) that the transmissible agent was the polysaccharide itself and (2) that the agent was a "specific mutagen." Concerning the first, it is sometimes overlooked that Griffith understood the distinction well enough. Better than many of his followers, he had at least the germ of a genetic theory: "By S substance I mean that specific protein structure of the virulent pneumococcus which enables it to manufacture a specific soluble carbohydrate." In regard to the second misinterpretation, Dobzhansky wrote that ". . . we are dealing with authentic cases of induction of specific mutations by specific treatments—a feat which geneticists have vainly tried to accomplish in higher organisms." This formally correct attribution, from a most influential source, obfuscates the idea that the agent is the genetic information. Muller had much greater clarity: In his 1946 Pilgrim Trust Lecture to the Royal Society, he remarked,

> . . . in the *Pneumococcus* case the extracted "transforming agent" may really have had its genetic proteins still tightly bound to the polymerized nucleic acid; that is, there were, in effect, still viable bacterial "chromosomes" or parts of chromosomes floating free in the medium used. These might, in my opinion, have penetrated the capsuleless bacteria and in part at least taken root there, perhaps after having undergone a kind of crossing over with the chromosomes of the host. In view of the transfer of only a part of the genetic material at a time, at least in the viruses, a method appears to be provided whereby the gene constitution of these forms can be analyzed, much as in the cross-breeding test on higher organisms. However, unlike what has so far been possible in higher organisms, viable chromosome threads could also be obtained from these lower forms for in vitro observation, chemical analysis, and determination of the genetic effects of treatment.

Other "classical" geneticists had virtually nothing to say about Griffith's work and would have judged themselves incompetent to assess its experimental validity. They began to pay closer attention

[1] It is awkward to have such a nondescript term as "transformation" applied to such an important, specific phenomenon. But when it was first discovered and named, there was no warrant to give it any narrower connotation. Avery had the power of new coinage but was hardly the likely personality.

after 1944, but again had little training in bacterial chemistry to enable them to form critical judgments about the claims presented them.

In Avery's world, however, Griffith was a central figure and his observations could not be ignored. His basic observations were confirmed in Avery's laboratory (see Dubos, 1986), and in due course Avery felt compelled to pursue the chemical extraction and identification of the substance responsible for the transformation. Sixteen years after Griffith, this was achieved, and DNA was thrust into the scientific consciousness as the substance of the gene.

In retrospect, it is difficult to give proper credit to the logical validity of a large range of alternative interpretations and to reconstruct the confusions about what was meant by "gene" and "genetic." Recall that until 1951 the only marker observed in transformation was the capsular polysaccharide, the biosynthesis of which was itself subject to many conjectures [e.g., about the role of starter fragments in self-assembly (discussed by Lederberg, 1956)]. Avery undoubtedly somewhat intimidated by Dobzhansky's authority, was reluctant to put his speculations about the genetic significance of transformation in print; his famous letter to his brother Roy surfaced only years later. There, but not in the paper, he remarks that the . . . [transforming substance is] thereafter reduplicated in the daughter cells and after innumerable transfers [it] can be recovered far in excess of the amount originally used. . . . Sounds like a virus—may be a gene. But with mechanisms I am not now concerned—One step at a time—and the first is, what is the chemical nature of the transforming principle? Someone else can work out the rest (quoted *in* Dubos, 1976). As late as 1948, so distinguished a geneticist as G. W. Beadle still referred to the phenomenon as a "first success in transmuting genes in predetermined ways" (note transmuting, not transmitting!). This obscuration of the pneumococcus transformation became less troublesome with the overall development of bacterial genetics.

Indeed, the controversy raged on the chemical claim that the substance was DNA (and nothing else!). [This story is detailed by Judson (1979) and in McCarty's personal memoir (1987).] Alfred Mirsky, Avery's colleague at the Rockefeller Institute, was a vocal critic of the chemical identification of the transforming agent. Some believe he was quite persuaded that this was an instance of gene transfer, but the more reluctant to concede that the evidence to date settled so important a question as the chemical identity of the gene as pure DNA (versus a complex nucleoprotein). Avery himself had cause to worry—There had been much resistance to his earlier proofs that pneumococcal polysaccharides, free of protein, were immunogenic. Wendell Stanley's first claims that crystalline tobacco mosaic virus was pure protein had to be subject to humiliating correction when ribonucleic acid was also found therein. We should recall that when most biologists of that era used terms such as protein, nucleic acid, or nucleoprotein, it can hardly be assumed that they had today's crisp connotations of defined chemical structure. These issues could only be settled by the few experts who had worked with these materials experimentally—and it was a daunting task to prove that there were too few molecules of any contaminating protein in the "DNA" to account for its genetic specificity. Maclyn McCarty's meticulous work continued to provide ever more persuasive evidence that it was DNA, and the contemporaneous studies of Chargaff showed that DNA was far more complex than Levene had figured it to be and, therefore, capable of the subtlety demanded of a "gene." Rigorous proof about "DNA alone" was really not furnished prior to the production of genetically active synthetic DNA three decades later. By 1952, Hershey and Chase gave evidence from an independent quarter that DNA alone penetrated the phage-infected cell. In the following year, the structural models of DNA as a double helix (Watson and Crick, 1953) lent final plausibility to "DNA alone."

This episode is sometimes painted as unreasonable resistance to a new idea (Stent, 1972). This is hardly a fair assessment of a controversy that was settled within 9 years and that required the emergence of a new class of workers, and conversion of some of the old ones, to deal with new techniques and experimental materials. That controversy continued is appropriate to the spirit of scientific skepticism—more to worry about when challenging new ideas are merely ignored.

All these discoveries, taken together, gave substance to Luria's vision of the virus as a genetic element that is coordinated with the genome of the host, but with pathogenetic consequence that has evolved to suit the needs of the parasite. The host may also co-evolve to reach an equilibrium compatible with the survival of both partners—a general principle in the evolution of pathogenicity (Th. Smith, 1934).

Prospects of cytoplasmic heredity fascinated many workers, even during the working out of the nuclear (Mendelian) basis of microbial biology, perhaps as a carryover of Huxley's idea of the persistent soma. In the course of the discussion, there were angry ripostes as to whether a given entity was really a plasmagene, or perhaps a virus, or perhaps a symbiont. The term and concept "plasmid" was introduced (in 1952) to stress the operational vacuity of those distinctions. A particle could be at the same time a virus (if one focuses on pathology), or symboint, or plasmagene (if one focuses on the genetic role). As a prophage, it may even be integrated into the chromosome, with a potential reappearance later. And it would be impossible to say whether a virus had evolved its pathogenicity, having once been a benign organelle, or vice versa, or both at different evolutionary epochs. One might even revive Altmann's old picture of the mitochondria as originally symbiotic bacteria, an allusion founded merely on the limitations of cytological analysis.

The vicennium worked a transformation—the "biologization" of the microbe. It was an extraordinarily exciting and fertile time, with new phenomena to be found in every culture dish. One could even learn to treasure one's contaminations.

IX. Viruses and Lysogeny: The Plasmid Concept

A. Biology of the Virus

The cardinal discovery for virology was the isolation and crystallization of the tobacco mosaic virus (Stanley, 1935), which sharpened many questions about this boundary of living existence (Pirie, 1937). A more convenient system for virus biology proved, however, to be the viruses attacking bacterial hosts, the (bacterio)phages, especially in the hands of the Delbrück school (Adams, 1959). Their life cycle was worked out in some detail, eventually culminating in two cardinal experiments:

Hershey and Chase (1952): The DNA of the attacking phage particle is sufficient to initiate infection. The DNA (not the entire phage) replicates in the host bacterium and then generates the capsid and assembles itself into mature, infectious phage particles.

Hershey (1946): Different phage genomes can undergo genetic recombination, enabling the construction of linkage maps. These would eventually be constructed in ultimate detail, matching the DNA sequence of the nucleotides.

Viruses were defined by Luria (1953, p.) as "submicroscopic entities, capable of being introduced into specific living cells and of reproducing inside such cells only." He pointed out that this is a methodological rather than taxonomic criterion; such a definition might well embrace a wide range of diverse entities. By 1950, he insisted that the phages exhibited "parasitism at the genetic level," taking over the metabolic direction of the host cell and exploiting a wide repertoire of its genetic capabilities. Whether or not other viruses, in plant and animal cells, would share these attributes remained to be seen (Luria, 1953; Adams, 1959; Hayes, 1964; Galpern, 1988; Burnet, 1945).

B. Lysogeny

Not long after the Twort-d'Herelle discovery of the bacteriophages (1915–1917), bacterial cultures were found that appeared to have established a durable symbiosis with a resident phage. The Delbruck school tended to dismiss these as contaminants, despite persuasive arguments of Burnet and Lush (1936). Lwoff and Gutmann (1950) reentered the controversy and showed that lysogenic *Bacilli* carried a "prophage," a genetic capability of producing the phage. At the same time, Lederberg and Lederberg (1951, 1953) had discovered that *E. coli* K-12 was lysogenic, for a phage they named "lambda," as a parallel (or so they thought) for the kappa particles in *Paramecium*. Crosses of lysogenic with sensitive strains, however, showed that the capacity to produce lambda segregated in close linkage with a chromosomal marker (gal); therefore, they invoked Lwoff's concept and terminology of prophage. However, the working out of that story, and of the phenomena of phage-mediate transduction, belongs to the next era.

X. Virology

The term virus was originally an unspecific term coined by Pasteur to mean any living organism that caused disease. This terminology was used well into the 1930s; thus, the word antivirus was used by Besredka to refer to bacterial filtrates that could apparently cure infections.

The realization that disease could be transmitted by inoculation of cell-free lesions from plant and animal infections led to the introduction of the concept of "filterable virus." Iwanowski's discovery in 1892 of tobacco mosaic disease in plants is usually credited as the first demonstration that a filterable virus could cause disease. Then in 1898, Loeffler and Frosch showed that a filterable virus was apparently the cause of foot and mouth disease. In the same year, S. M. Chapman introduced the use of fertile hens eggs as a means of cultivating viruses. This approach was later to be used by Pyton Rous in his work on the fowl sarcoma that bears his name. By 1915, a new class of virus affecting bacteria but neither plants nor animals, was discovered by F. W. Twort. His observations were extended in 1917 by D. Herrelle, who over the next 13 years published a series of papers on what was initially called the Twort–Herelle phenomenon, but which later became known as bacteriophage.

The development first of the ultraviolet microscope and then of tissue culture techniques in the 1920s added impetus to research on virus structure and cultivation. Maitland's work in 1928 was a major advance in tissue culture techniques, but because of the tedious nature and lack of antibiotics to control bacterial contaminants they were not widely adopted.

By 1931, the potential of the fertile egg for culturing viruses was finally appreciated in the work first of Goodpasture and then of the Australian Macfarlane Burnet. Burnet used this approach to culture the influenza virus, which previously had to be grown in ferrets.

John Enders did much to develop the art of culturing viruses, which finally enabled the development of a range of vaccines. Enders' outstanding contribution to the study of viruses began with his work on mumps when he showed that a virus could be grown in chick embryos and, after successive generations, would loose its ability to cause the disease while retaining the capacity to immunize against it. In this way, the modified virus could be used to prepare vaccines to control the disease. Prior to 1949, for example, the poliomyelitis virus could only be propagated in monkeys. Enders showed that the virus could be grown in culture of nonnervous tissues, and by using this technique Salk developed his famous vaccine, which essentially defeated infantile paralysis. The application of Enders' tissue culture techniques led to the isolation of many other viruses: in 1954, the year when he received the Nobel Prize, Enders himself, for example, succeeded in isolating the measles virus.

The introduction of the electron microscope in 1934 proved a great asset to research on viruses. In 1956, Watson and Crick proposed on theoretical grounds that virus particles must be made up of a nucleic acid core and a surrounding shell comprised of protein subunits, a structure later seen in 1959 under the electron microscope by Horne and Nagington.

Antibiotics aided virus research, allowing for contamination-free studies, so that by 1949, poliomyelitis virus could be grown on nonneural tissues such as minced monkey kidney.

In 1952, the name of the patient Helen Lane became cryptically immortalized when Gay and his colleagues established the famous continuous cell line of HeLa cells, derived from a carcinoma of the patient's cervix uterus. Then, in the following year, Scherer succeeded in growing poliomyelitis virus in these cells.

In 1954, Younger published his technique for growing trypsinized cells in monolayers on glass. This allowed viral infection of cells to be recognized by detecting the cytopathic effect, which allowed for the routine screening for the presence of viruses.

XI. Mycology and Protozoology, Microbiology's Cinderellas

Filamentous fungi and protozoa (i.e., molds and animacules) were observed soon after the earliest microscopes were developed. Studies of these organisms continued largely unnoticed as bacteriology developed. The fact that neither of these groups of microorganisms cause major diseases in the developed world tended to hinder the rapid development of both mycology and protozoology. The principle motivation for studying fungi came from their ability to infect important crop plants. This resulted in a close association between mycology and botany, with the unfortunate result that many microbiologists in the past, as today, regard fungi as lying outside the orbit of their subject.

As early as 1767, Torgioni-Tozetti advanced the view that rust diseases of cereals are caused by microscopic fungi, but experimental proof of the role of fungi as phytopathogens had to await the monograph by Prevost, who in 1807 described experimental smut infections. Prevost also showed that fungal infections could be prevented by soaking seeds in a solution of copper sulfate and thus, he inadvertently became the originator of the pesticide industry. It was Anton de Bary, however, who did the most to develop the science of plant pathology.

During the early part of this century, attention was also focused on the role that fungi play in soil fertility. It soon became evident that while fungi are not as metabolically diverse as soil bacteria, they nevertheless play an important role, principally as agents of decay of organic forms of carbon and nitrogen, in the degradation of leaf litter and humus. Waksman and his colleagues were particularly active in demonstrating the role played by fungi in soils.

Waksman was also one of the first microbiologists to appreciate the industrial importance of molds, he and his group investigating the production of butyric acid and butyl alcohol from starch-rich materials, and then, in 1930, examined lactic acid production by species of *Rhizopus*. The foundation for the development of studies on mold metabolism was laid by Raistrick and his numerous collaborators working at the London School of Hygiene and Tropical Medicine during the 1920s.

Ringworm was the first human disease to be shown to be caused by fungi; described in 1839 by Schoenlein, it was soon followed by the recognition by the Swede F. T. Berg that *Candida albicans* was the causal agent of thrush. Medical mycology was slow to develop, however, and it was not until 1910 tha Sabouraud introduced a medium suitable for the isolation and growth of pathogenic fungi. Systemic mycoses were discovered at the turn of this century, while it was as late as 1934 before Monbreun conclusively demonstrated that histoplasmosis is caused by *Histoplasma capsulatum*. Medical mycology has tended to lag behind other aspects of medical microbiology, although the importance of fungal infections such as pneumocystis pneumonia and candidiasis in the AIDS syndrome is likely to accelerate developments in this area of the subject.

The development of protozoology as a science is almost exclusively devoted to the role of protozoa as agents of disease. Although initially referred to as animacules, by 1764 Wrisberg had introduced the term infusoria, while the first generic name for a protozoan, *Paramecium,* was introduced by Hill in 1752. The term protozoa was first used by the German Goldfuss in 1817. By 1836, Alfred Donne working in Paris had shown that a flagellate was responsible for vaginal discharge in women. It was, however, the colonial expansion of the European powers that provided the stimulus to studies in medical protozoology. The first observations of parasites in the blood of malaria sufferers was made in 1880 by Alphonse Laveran. A long list of diseases were then shown to be caused by protozoa including Texas cattle fever in 1893, Malta fever in 1895, malaria in 1898, sleeping sickness in 1902.

Protozoa have yet to be widely used in industrial microbiology and biotechnology, and their role in the environment has been subject to only limited study; therefore, the history of the development of protozoology in these areas will have to await future developments.

XII. The Modern Period

What then of the landmarks of the recent history of science? Without a doubt, the most obvious development in our science that has taken place since the last war has been the rise in the status of a single organism, the colon bacterium *E. coli*. Using this single organism, scientists such as Niremberg, Holley, Jacob, and Monod have revolutionized our thinking on biology. One practical outcome of this work was the development of an *E. coli* strain by W. Gilbert and others in 1978 that produces human insulin.

A perusal of the list of awards for the Nobel Prize for research in microbiology in the widest sense shows that since 1958 particular recognition has been given to work on genetics, virology, and immunology. Knowledge derived from such studies have had a profound effect on our understanding of the life process, and recent developments in biotechnology have provided real benefits in our lives. [*See* BIOTECHNOLOGY INDUSTRY: A PERSPECTIVE.]

The key technique that has made genetic engineering possible was devised by Herbert Boyer and Stanley Cohen. Boyer working at the University of California collaborated with Cohen of Stanford University to develop a method of splicing genes from a donor into a recipient bacterium. In 1973, they took a gene from the plasmid of one organism and spliced it into a plasmid from another to produce recombinant DNA. When inserted into a recipient bacterium, the foreign genes not only survived but also affected the host in the way it had affected the donor, and was also copied as the cell divides. Boyer later used this approach to insert genes from human proteins into bacteria, and, thus, heralded a biotechnological revolution. A similar revolution was initiated by the production of monoclonal antibodies by Kohler and Milstein (Interestingly, what might be termed ''natural monoclonal antibodies'' had been observed a few years earlier by Joseph Sinkovicks.)

A microbiologist who left science even as late as the mid-1970s to follow other pursuits would now hardly recognize his or her former subject. Studies on the genetics and molecular biology of microorganisms have made particularly rapid progress in the intervening years. We have also seen major improvements in the way we apply microorganisms in biotechnology and, more recently, to address environmental problems. The appearance of AIDS has once and for all shattered our cozy belief that we had all but conquered infectious disease. HIV will undoubtedly not be the last new infectious agent to confront us in the future; if for no other reason than to combat such infections, our science will need to continue to develop at the rapid rate seen in the past few decades.

Bibliography

Amsterdamska, O. (1991). Stabilizing instability: the controversy over cyclogenic theories of bacterial variation during the interwar period. *J. Hist. Biol.* **24,** 191–222.

Brock, T. (1990). "The Emergence of Bacterial Genetics." Cold Spring Harbor Laboratory Press.

Burnet, F. M. (1945). "Virus as Organism. Evolutionary and Ecological Aspects of Some Human Virus Diseases." Harvard University Press, Cambridge, Massachusetts.

Collard, P. (1976). "The Development of Microbiology." Cambridge University Press, Cambridge.

Dubos, R. J. (1945). "The Bacterial Cell in Its Relation to Problems of Virulence, Immunity and Chemotherapy." Harvard University Press, Cambridge, Massachusetts.

Dubos, R. J. (1976). "The Professor, The Institute and DNA." Rockefeller University Press, New York.

Gunsalus, I. C., and Stanier, R. Y. (eds). (1962). "The Bacteria: A Treatise on Structure and Function. Biosynthesis." Academic Press, New York.

Jordan, E. O., and Falk, I. S. (eds). (1928). "The Newer Knowledge of Bacteriology and Immunology." University of Chicago Press, Chicago, Illinois.

Judson, H. F. (1979). "The Eighth Day of Creation-makers of the revolution in biology." Simon and Schuster, New York.

Lechevali, H. A., and Solotorovsky, M. (1965). "Three Centuries of Microbiology." McGraw-Hill, New York.

Lederberg, J. (1987). "Genetic Recombination in Bacteria: A Discovery Account." *Ann. Rev. Genet.* **21,** 23–46.

Lederberg, J. (1990). "Edward Lawrie Tatum. Biographical Memoirs." *Nat. Acad. Sci.* **59,** 357–386.

Lederberg, J. (1991). "The Gene (H. J. Muller 1947)." *Genetics* **129,** 313–316.

Lederberg, J. (1992). "Bacterial Variation since Pasteur; Rummaging in the Attic: Antiquarian Ideas of Transmissible Heredity, 1880–1940." *ASM News* **58(5),** 261–265.

Medical Research Council (Great Britain). (1930). "A System of Bacteriology in Relation to Medicine." His Majesty's Stationery Office, London.

Summers, W. C. (1991). "From Culture as Organism to Organism as Cell: Historical Origins of Bacterial Genetics." *J. Hist. Biol.* **24,** 171–190.

Wainwright, M., and Swan, H. T. (1986). *Medical History* **30,** 42–56.

Wainwright, M. (1990). "Miracle Cure—The Story of Antibiotics." Blackwell, Oxford.

Weatherall, M. (1990). "In Search of a Cure." Oxford University Press, Oxford.

Werkman, C. H., and Wilson P. W. (eds). (1951). "Bacterial Physiology." Academic Press, New York.

Hospital Epidemiology

Robert Latham and William Schaffner
Vanderbilt University School of Medicine

Glossary

Infection rate (incidence) Number of events (infections) occurring per patients at risk over a specified time

Nosocomial infections Infections occurring as direct result of hospitalization, including adverse consequences of procedures and therapy

Prevalence Number of events present in patients at risk at any one point in time

Surveillance Identification of a "case"; requires specific definition of what constitutes a case and a systematic mechanism for finding these patients

Universal precautions Infection control policy and philosophy requiring that all patients (and their secretions) be treated as if they are potentially infectious to health care workers and other patients

HOSPITAL EPIDEMIOLOGY, in today's busy, modern medical institution, refers to that group of medical professionals whose special charge is to prevent and control infections acquired by patients during their stay in that institution. Although infections have been known to complicate hospitalizations throughout medical history, the need for individuals with specific training in infection control and epidemiology has been recognized only in recent years. The development of this "new" discipline has, in large part, been engendered by the complexity of modern medicine. With advancements in treatment modalities for previously hopeless diseases and development of new, but often invasive, techniques for diagnostic and therapeutic purposes, patients admitted to hospitals today are often immunocompromised and subjected to a multitude of tests that increase their chances of developing an infection as a result of their hospital stay. Infection in these patients often causes serious illness and even occasional death. As a result, the need for innovative research to improve ways to detect, prevent, and control these infections has become apparent.

In addition, the 1980s and 1990s have witnessed an increased awareness of the costs involved in delivering care to patients in medical institutions and the financial importance of avoiding prolonged hospital stays. This increased awareness not only occurred among physicians and health care providers, but also among hospital administrators and government officials. Not surprisingly, these latter individuals viewed infection control practitioners as essential elements to an efficient hospital. Moreover, organizations charged with overseeing hospital function have mandated that these practitioners be part of the hospital staff.

Consequently, hospital epidemiology and infection control are enjoying an enormous growth in both number of practitioners and their functions. In this article, we hope to provide the reader with a brief overview of hospital epidemiology: the tools of the trade; the problems encountered; the expanding duties; and, finally, a glimpse into what the future may hold for practitioners of this discipline.

I. Tools of the Hospital Epidemiologist

As the word epidemiologist implies, infection control practitioners are interested in the detection, elimination, and, most of all, the prevention of epidemics. Epidemics are defined as unusual occurrences of disease. This might be an unexpected increase in the number of cases of a particular illness that usually occurs at a very low rate. An epidemic could be the appearance of even one patient with an illness that is new to the area, particularly if the illness has the ability to spread to others. For example, suppose the rate of wound infections among surgical patients at Hospital A is 1% during most months; however, in June, the rate of infections among these patients was found to be 4%. This increase certainly seems unusual and is above the expected month-to-month variation observed at Hospital A. Closer inspection reveals that much of the increase in cases is attributable to an inordinate number of *Staphylococcus aureus* infections among patients of Doctor X. *Staphylococcus aureus* is known to colonize the nostrils without causing symptoms, and cultures of Doctor X reveal *S. aureus* of the same strain causing infection in the patients. This suggested that the physician was disseminating the implicated organism throughout his surroundings, including the operating room and the patients' open wounds (despite routine aseptic precautions) resulting in a subsequent epidemic. Similarly, recognition of even one case of Lassa fever in a patient hospitalized in the United States after recent travel to Africa is regarded as an epidemic and has important infection control implications if further cases among hospital employees and other patients are to be prevented. [*See* EPIDEMIOLOGIC CONCEPTS.]

Unfortunately, even in the best of hospitals with the best of physicians, patients sometimes develop complications as a consequence of their hospitalization. When infections are acquired by patients after they have been admitted to the hospital, such infections are termed nosocomial (a word constructed from Greek roots meaning diseases relating to a hospital). For infection control practitioners to detect the unusual occurrence of disease in the hospital, they must be familiar with the normal incidence of nosocomial infections (baseline rates) in their institution. Surveillance of their patient population for the presence of diseases attributable to being hospitalized is essential to determine baseline rates of infections. An efficient surveillance system requires (1) clear and precise definitions of illnesses to be surveyed, (2) a systematic means of collecting the relevant data, (3) tabulation of the data in a meaningful way, (4) analysis and interpretation of the data, and (5) dissemination of the results to those people who are in a position to be affected by and can effect the results (e.g., surgeons, intensivists, oncologists, nurses).

In most U.S. hospitals, routine surveillance occurs for the most commonly recognized nosocomial infections—urinary tract infections (UTIs), surgical wound infections (SWIs), pneumonias, and bacteremia (see later). Precise definitions for these infections are provided by the Centers for Disease Control (CDC) and are widely used.

The mechanism of data collection in infection control varies from institution to institution, depending on its size and complexity. The key elements in the surveillance mechanism are that it is systematic and consistent. Only with systematic data collection over time can an accurate rate of occurrence for an individual problem be defined. Variability in the intensity of surveillance can alter perceived rates of infection. This is not to say that a surveillance system must be rigid and lack flexibility. As infection control priorities change, practitioners can focus the intensity of surveillance toward a specific problem. In so doing, however, a new baseline is generated and rates of infection cannot be accurately compared with those observed in the past.

Active case finding is the preferred mechanism for surveillance in hospitals. Sources of information are numerous and include the microbiology laboratory, head nurses of patient care areas, employee health records, autopsy information, and discharge summaries. Which of these elements become part of an active surveillance system at a particular institution is determined by personnel resources and the identified infection control problems. Regardless of the makeup of the surveillance system, the infection control practitioner must be frequently visible on the wards of the hospital. This enhances communication with hospital personnel, enabling early detection of potential problems, and serves as a strong reminder for health care workers to follow proper infection control practices as they interact with patients and each other.

In tabulating the collected data on nosocomial infections, attention is directed to where and when the infections occurred in the hospital. To the experienced practitioner, this simple place–time arrangement of the data often will allow identification of an unusual clustering of infections. The role of any analysis is always to present the data in a clear and accurate manner so that people responsible for responding to the information will find it helpful.

The keystone of data analysis is the calculation of infection rates. Rates are defined as the number of infections occurring per patients at risk over a specified period of time. The number of infections (numerator data) is identified by the surveillance system. The actual number of patients at risk (denominator data) is often more difficult to define. Most infection control practitioners use hospital admissions or discharges per unit of time as their denominator in calculating rates. The ease of acquiring this information is a major advantage. For some infections, however, more meaningful rates can be produced with more specific denominator data. For example, an accurate assessment of clean surgical wound infection rates (see later) can be done if knowledge of the number of "clean" operations performed during a certain time are obtainable from operating room logs.

Interpretation of the data is concerned with answering the question, "Is there a problem?" Comparison of rates and assessment for clustering of infections by demographic features is usually sufficient to answer this question. Occasionally, additional information and continued observation are required to clarify the situation when data are not definitive. Finally, the infection control practitioner must ensure that the data are disseminated on a regular basis to hospital personnel in a position to take appropriate action.

In addition to surveillance efforts and epidemiology, infection control practitioners must be familiar with various aspects of microbiology. Similarities in antimicrobial susceptibility patterns among bacterial isolates can be an important clue that a series of infections share a common source and that more investigation is warranted. Knowledge of the antimicrobial resistance among pathogens prevalent in the hospital is critical. Changes in antimicrobial susceptibilities can signal the introduction of new pathogens into the hospital or represent emergence of new genetic resistance determinants that are easily transferred by plasmids among pathogens. In many medical centers with access to research laboratories, new developments in the area of microbiology have been utilized by infection control practitioners to support epidemic investigations. Agarose gel electrophoresis of bacterial plasmids, restriction endonuclease digestion of plasmid and chromosomal DNA, and ribosome typing have been utilized to better characterize pathogens involved in these investigations. As these and other similar techniques become more widely available, infection control practitioners must familiarize themselves with them to know how best to apply this technology to support their efforts.

II. Specific Problems in Hospital Epidemiology

A. Urinary Tract Infections

UTIs account for 30–40% of nosocomial infections in this country. Virtually all of these infections are attributable to instrumentation of the urinary tract during hospitalization, approximately 80% are secondary to the use of a Foley catheter for drainage purposes, and the remainder follow instrumentation for diagnostic or therapeutic purposes. Specific host factors associated with an increased risk of developing a UTI following instrumentation include severe underlying disease, female sex, and older age. In addition to host factors, the risk of developing a UTI in patients with Foley catheters relates strongly to the duration of catheter use. Only a small number of patients (approximately 1%) develop bacteriuria after an in-and-out catheterization. In contrast, about 50% of patients experiencing urinary catheterization for 7–10 days develop bacteriuria.

Gram-negative enteric bacilli account for the majority of these infections, with *Escherichia coli* being the most common isolate identified. In the National Nosocomial Infection Study conducted yearly by the CDC, *E. coli* is reported as the cause of 20–30% of these infections; *Pseudomonas aeruginosa* accounts for 10–15%. A number of other gram-negative bacilli are responsible for up to 30% of the remainder. Enterococci are the most common gram-positive isolate reported (15–20%).

In the past when urinary catheter drainage systems were constructed to allow interruptions and breaks in the system, most infections occurred by inadvertent contamination of the urine during catheter manipulation by hospital personnel. After entry, organisms ascend intraluminally through the catheter to the bladder and establish infection. Because of this mode of infection, closed drainage systems have become the standard used in patients in this country.

Although many catheter-related infections were prevented, adoption of closed drainage systems have not completely eliminated these infections. Colonization of periurethral and rectal areas with uropathogens has been shown to precede the development of UTIs, suggesting that organisms still can gain entry into the bladder by ascending up the exterior surface of the catheter. This mode of infection appears to be particularly important in women. *Escherichia coli* and *Proteus* infections generally appear early during the hospital course, whereas infection with *Pseudomonas* and *Serratia* occur later. This observation further supports the concept of colonization preceding ascension into the bladder since early infections should occur with the host's "normal" flora and those occurring later are caused by organisms establishing colonization of the patient during hospitalization.

Most nosocomial UTIs are relatively benign infections and can be eradicated by simple removal of the Foley catheter. The greatest morbidity and mortality attributable to these infections results from secondary bacteremia. Rates of bacteremia from nosocomial UTIs are estimated to be 1/50–150 catheterized patients. Although the rate of secondary bacteremia is low, UTIs are so common that they account for a large number of septic episodes in an institution. In most series on gram-negative bacteremia, nosocomial UTIs were the recognized cause in 30–40% of patients.

The mainstay of therapy for nosocomial UTIs is removal of the Foley catheter. Because of concern about the possibility of a secondary bacteremia, antibiotics are often utilized when a febrile patient with a UTI is encountered. Elimination of bacteriuria after catheter removal can also be enhanced by appropriate antibiotics. Unfortunately, prolonged antibiotic use while the catheter is in place can lead to superinfection with a bacterial strain that is resistant to the antibiotic being used and should be avoided if possible.

Avoidance of Foley catheter use is the most efficient means to prevent nosocomial UTIs. Unfortunately, in many situations, their use is required; daily assessment of their continued need should subsequently be undertaken and they should be removed as soon as possible. In situations when their use is required, attention should be directed to ensure aseptic technique upon insertion, maintenance of the integrity of the closed drainage system, and keeping the drainage bag in a dependent position.

Numerous other methods for prevention of catheter-associated UTIs have been investigated and include antibiotic- and silver-impregnated catheters, administration of antibiotics to the meatal-catheter junction, H_2O_2 administration into the drainage bag, and bladder irrigations with a number of antibacterial substances. Unfortunately, these efforts have been largely unsuccessful in preventing infections. Prophylactic administration of systemic antibiotics prevents catheter-associated UTIs only when the length of catheterization is brief but leads to superinfection with resistant organisms when catheterization is prolonged. For this reason, use of systemic antibiotics is not routinely recommended.

B. Surgical Wound Infections

Historically, concern about SWIs by physicians formed the basis for modern infection control programs. Even before establishment of the germ theory, Lister demonstrated the benefits of antisepsis in the practice of surgery and Semmelweis identified the lack of antiseptic techniques as a cause of an epidemic of nosocomial puerperal sepsis in Vienna. In modern hospitals, SWIs are second in number to UTIs among nosocomial infections but are foremost among causes of excessive morbidity and mortality. For that reason, SWIs remain a major concern to modern hospital epidemiologists.

The most amazing and least understood phenomenon about SWIs is why all wounds do not become infected. Standard preoperative skin antisepsis is not sufficient to sterilize the surgical field for the entire operation; bacteria are routinely present in wounds at the time of closure. These organisms primarily come from the patient's own skin flora and reside in the deeper areas of the epidermis in hair follicles and sebaceous glands protected from the skin antiseptic. Although the skin antiseptic reduces the quantity of organisms on the surface of the skin, organisms in deeper layers continue to multiply during the operation. At the time of closure, the quantity and type of bacteria with specific virulence factors are major determinants of infection. In the majority of patients, host factors are sufficient to counter the microbial determinants and prevent the development of infection. The nature of the host factors that are mobilized at operation are poorly understood and need to be better defined.

The type of operation is a major determinant of the degree of bacterial contamination of the wound that is anticipated during surgery. For this reason, surgical procedures are classified as follows: clean, no contamination expected; clean-contaminated, nontraumatic wound with minor break in technique or entry of respiratory, gastrointestinal, or genitourinary tract without major spillage into wound; contaminated, a traumatic wound or one into which is spilled infected material or gross contents of the gastrointestinal/genitourinary tract. Not surprisingly, rates of SWIs increase with the degree of contamination.

Although most SWIs are caused by the patients own flora, occasionally organisms colonizing operating room personnel or the operating room environment are implicated as sources of individual infections and, rarely, epidemics of SWIs. Hematogenous seeding of wounds is controversial and, if it occurs, is thought to be extremely unusual.

About 50% of SWIs are caused by gram-positive organisms with *S. aureus* as the most common isolate followed by enterococci and coagulase-negative staphylococci. *Escherichia coli* and pseudomonas account for about 10% each of SWI and are the most common gram-negative bacilli isolated. Infrequent causes include mycoplasmas, rhodococcus, and legionella species and a host of other organisms that, when identified, should alert the infection control practitioner to investigate for source of acquisition.

A number of host factors increase the risk that the patient will develop a SWI. These include the extremes of age, diabetes, steroid therapy, obesity, malnutrition, and concomitant infection elsewhere at the time of surgery. The majority of the host factors presumably are associated with poor immune response of the individual or compromise precise and clean wound closure. Clearly, surgical precision is a factor in development of SWI with wound hematomas or devitalized tissue providing ample substrate for multiplication of organisms.

Surgery in modern hospitals involves a number of techniques to minimize the risk of subsequent SWI. Close attention to the architecture, airflow, traffic of personnel, and housekeeping ensure a clean environment. Before the operation, patient and surgeon perform a carefully orchestrated series of maneuvers to decrease the quantity of organisms on the patient's skin and the surgeon's hands. For many surgical procedures, prophylaxis with antibiotics also diminishes the rate of SWIs. To be effective, antibiotics should have activity versus the likely pathogens, be given so that peak serum levels are achieved when tissues are opened, and be discontinued shortly after the surgical procedure to minimize side effects and colonization with resistant organisms.

C. Nosocomial Pneumonia

Pneumonia accounts for 10–15% of nosocomial infections and, like SWIs, is a cause of major morbidity and mortality in the hospital. In addition, development of pneumonia in the hospital extends the patient's stay an average of 7 days and greatly contributes to the costs incurred.

About 50% of nosocomial pneumonias are caused by gram-negative bacilli including *Pseudomonas, Enterobacter, E. coli,* and *Klebsiella*. In many intensive care units, one gram-negative bacillus emerges as the predominant resident flora and accounts for the majority of infections in that institution. *Staphylococcus aureus* is the most common gram-positive organism isolated from patients with nosocomial pneumonia. With the emergence of methicillin-resistant *S. aureus* as a major cause of nosocomial infections in U.S. hospitals, the proportion of cases attributable to this organism has increased in recent years. The role of anaerobes in development of nosocomial pneumonia remains unclear, largely because of the difficulty in adequately processing respiratory specimens in the laboratory for detection of these organisms. Despite these difficulties, several investigators have identified anaerobes in approximately one-third of nosocomial pneumonias in their studies, most often in association with an aerobic organism. *Legionella* species can be major causes of nosocomial pneumonia and are particularly problematic for infection control practitioners. These organisms can colonize the water systems of hospitals, thus exposing a vast array of patients to the potential of becoming infected. In this setting, these organisms can be very resistant to sterilization efforts. Finally, a number of viral respiratory pathogens are increasingly recognized as causes of nosocomial pneumonia and as infections that predispose to subsequent superinfections with bacterial respiratory pathogens.

Most people aspirate small quantities of oral contents into their respiratory tract at night while they are recumbent and asleep. Pneumonia rarely develops because organisms colonizing the oral mucosa are quantitatively and qualitatively controlled by the cleansing action of eating, mucosal antibody, and proper dental hygiene. In addition, the ciliary action of the tracheobronchial epithelium prevents small aspirated particles from reaching lung parenchymal tissues. Hospitalization is often associated with undermining of these defense mechanisms. Patients are often sedated and immobilized in a recumbent position, thus increasing the chance of aspiration. Moreover, the use of endotracheal tubes and bronchoscopy bypasses these local defense mechanisms and may serve as the means of entry for nosocomial pathogens.

Colonization of the respiratory tract with nosocomial pathogens increases the risk of subsequent pneumonia. Organisms normally colonizing mucosal surfaces resist the adherence for other organisms. Prior antibiotic use eliminates the endogenous flora and allows colonization with resistant pathogens. In intensive care units, colonization of patients with the resident gram-negative bacilli occurs rapidly; about 40% become colonized in 4 days. In addition, gram-negative colonization of the oropharynx is increased in the elderly, patients with severe underlying disease, and patients with neutropenia.

The stomach and its acid environment are normally sterile. Gastric alkalinization by use of H-2 blockers or antacids allows colonization of the stomach with nosocomial organisms and, when these contents are aspirated, increases the risk of nosocomial pneumonia. Sucralfate does not interrupt the stomach acid barrier to colonization and, thus, should be used preferentially when possible.

Recent efforts to decrease the incidence of nosocomial pneumonias have been directed toward prevention of colonization of the oropharynx by use of topical antibiotics. A number of regimens with gram-negative activity have been shown to decrease colonization and infection in patients in intensive care units. Concerns about selection of resistant organisms and cost efficacy of these topical regimens exist and need to be answered before their routine use can be supported.

D. Bacteremias

Although nosocomial bacteremias are infrequent, they are major causes of morbidity and mortality. Gram-negative sepsis produces shock in 30–40% of patients; once it occurs, mortality approaches 50%. Bacteremias can be primary (without any identifiable source) or secondary to an established infection. A secondary source should always be sought in the bacteremic patient. Particular emphasis should be directed toward indwelling intravascular catheters. Approximately 80% of nosocomial bacteremias are associated with intravascular catheters. The major use of these catheters is for the administration of fluids; as many as 25% of hospitalized patients receive these fluids. In addition, other diagnostic and therapeutic measures require access to the vascular system. Many of these procedures (intravenous pyelogram, arteriography, etc.) involve only brief exposure of the vascular structures and are associated with little risk of infection. Unfortunately, many of the technological developments that have enabled proliferation of intensive care units and prolonged care of critically ill patients involve the use of intravascular devices to provide easy vascular access and monitoring of the patient. The incidence of infection of intravascular devices increases directly proportional to the length of time the devices remain in place.

Most episodes of catheter-related sepsis are caused by extrinsic bacterial contamination at the cannulation site with spread along the external surface of the intravascular device. Extrinsic contamination also occurs at junctions in the external line as a result of faulty infection control technique. Intrinsic contamination of the infusate by the manufacturer occurs but is extremely unusual today.

There is an association between the type of contamination and the organism causing the bacteremia. *Staphylococcus aureus* and *Staphylococcus epidermidis* are common skin organisms that gain entrance to the bloodstream by tracking along the catheter. Together, these organisms account for approximately 50% of nosocomial bacteremia. Gram-negative organisms can also gain access along the surface of the catheter. Certain gram-negative bacilli (*Enterobacter, Flavobacterium,* and *Citrobacter*) thrive in liquid environments and occasionally cause bacteremia as a result of inadvertent contamination of infusate that is not changed in a timely manner. *Candida* sepsis is associated with the use of total parenteral nutrition solutions and broad-spectrum antibiotics.

Diagnosis of catheter-related bacteremia is not difficult in the febrile patient without other sources of infection who has redness, tenderness, and/or purulence at the intravenous catheter site. Unfortunately, such a characteristic presentation is unusual. A more common situation is the newly febrile patient in the intensive care unit with multiple vascular lines that appear normal and with numerous other potential sources of infection. When the catheter is clearly infected, the mainstay of therapy is catheter removal with subsequent administration of systemic antibiotics. Identification of bacteremia with staphylococcus or candida warrants consideration of catheter removal even in the absence of signs of local infection. Attention to strict nursing procedures in the care of all lines is important, particularly in the intensive care unit. Peripheral catheters should not be left in place for >72 hr and infusions changed every 24 hr.

Although infections of temporary catheters are

easily treated by catheter removal, replacement of a Hickman or Broviac catheter requires a surgical procedure and is often very difficult in such patients. The management of infections of these catheters is highly dependent on the site of catheter involvement. Cellulitis around the catheter exit site is usually amenable to antimicrobial therapy and does not require catheter removal. When the catheter is infected where it passes beneath the skin (tunnel infections), antibiotics alone rarely are sufficient and the catheter usually must be removed.

E. Miscellaneous Problems

1. *Clostridia difficile* Colitis

In addition to the more traditionally recognized nosocomial infections, diarrhea is increasingly recognized as a nosocomial problem. Numerous causes for outbreaks of diarrhea have been reported in the literature and include viral, bacterial, and a host of noninfectious etiologies. The magnitude of each of these causalities in the endemic setting is largely unclear. *Clostridia difficile* colitis is a major cause of diarrhea in the hospital, complicating the use of systemic antimicrobial therapy. Patients who develop this infection can have *C. difficile* as part of their own flora at admission or can acquire the organism during their hospital stay. Acquisition of these organisms occurs via the contaminated hands of hospital personnel or indirectly through contamination of the hospital environment. Only about one-third of patients who become colonized with *C. difficile* develop diarrhea; asymptomatic individuals can be sources of spread to the environment and other patients.

2. Infections in Immunocompromised Hosts

Infection control for immunocompromised hosts is an area of continued controversy. Despite lack of documented efficacy, expensive laminar air-flow rooms and reverse isolation have become the norm in many cancer and transplant centers. Similarly, the use of prophylactic antibiotics and selective decontamination of the gastrointestinal tract with oral nonabsorbable antibiotics have their proponents and their critics. Infection control practitioners should have input into these activities at their institutions. More importantly, they should routinely provide transplant and cancer specialists with surveillance data about infections in their units and constantly consider ways to improve infection control measures in these very susceptible patients.

Close surveillance of immunosuppressed patients often has additional benefits for the infection control practitioner. In many ways, these patients may serve as the initial group in which a hospital-wide problem becomes apparent. For example, a small cluster of *Legionella pneumophilia* infections among renal transplant patients may be the first clue to widespread colonization of the hospital water system with this organism. [*See* IMMUNE SUPPRESSION.]

3. Antimicrobial Resistance

Since the development of penicillin and the subsequent emergence of penicillin-resistant staphylococci, each new antibiotic introduced into our therapeutic armamentarium has been followed by identification of organisms that have acquired the ability to resist the activity of that antibiotic. Nowhere is this more evident than in the hospital setting, where antimicrobial pressure is great. In this setting, bacteria can exchange genetic resistance determinants throughout the same species and, in some cases, from one bacterial species to another. With continued antibiotic pressure, resistant organisms become the predominant flora in the hospital and specific genetic resistance determinants can persist among hospital flora for extended periods. A single resistant bacterial strain can be widespread throughout the institution, or multiple pathogens can occur, each having a specific geographic locale within the hospital. Infection control practitioners working with microbiologists must monitor the antimicrobial susceptibilities of nosocomial pathogens and inform physicians about changes observed. [*See* ANTIBIOTIC RESISTANCE.]

III. Control Measures

The most important function of the infection control practitioner is to initiate and monitor control measures in an effort to prevent infections in patients. Universal precautions are now utilized by most hospitals in this country to prevent transmission of infection from patient to health care worker and from patient to patient via the hands of health care workers. Universal precautions require that each health care worker assume that all bodily secretions of patients (blood, urine, feces, saliva, etc.) are potentially infectious and take proper precautions to minimize exposure to them. These precautions are often as simple as wearing gloves but occasionally include gowns and masks when soilage of clothes is possible or aerosolized fomites expected. All articles used in the care of one patient should be discarded before administering to another. Handwashing by all health care workers before and after they come in contact with a patient cannot be overemphasized as a critical component of any infection control program.

In outbreak situations, specific control measures must be tailored to the situation. The outbreak investigation often provides clues to the mechanism of transmission for the pathogen. With this in mind, necessary control measures can be directed at the specific problem and less-effective, empirical measures avoided. For example, cohorting patients with methicillin-resistant *S. aureus* may occasionally be effective at diminishing spread or even eliminating the organism from a hospital because the organism is passed from patient to patient by health care workers. Similar control measures would not be effective in preventing acquisition of multiresistant gram-negative bacilli among intubated patients who are inadvertently colonized by contaminated equipment used by respiratory therapy personnel.

IV. Additional Duties

The main concern of infection control practitioners has always been prevention and control of infections in patients admitted to their hospital. Recently, there has been a tremendous increase in efforts to also extend these activities to hospital personnel. This is, in part, related to a large emphasis being placed on all industries by governmental agencies to provide a safe workplace. Perhaps more importantly, a greater awareness of the risks of hospital personnel acquiring infections has occurred in recent years. This awareness has been stimulated greatly by the AIDS epidemic and the associated potential, albeit small, of acquiring a life-threatening illness during the performance of patient care activities. Infection control practitioners armed with their knowledge of infections, disease transmission, and preventive measures are required to have major input into hospital employee health programs. Specific programs for tuberculosis screening, rubella protection, and hepatitis immunization, among others, are now part of virtually all employee health programs in this country. Influenza vaccination is utilized by some in an effort to maintain adequate workforce on site during the "flu" season and prevent transmission of illness to patients from employees.

Efforts to prevent transmission of the human immunodeficiency virus (HIV) to patients and employees has been a major preoccupation of infection control programs over the past several years. A needlestick from an HIV-positive patient is associated with an approximately 0.4% seroconversion rate. The risk of acquiring infection from other types of exposure (blood-to-mucous membrane, blood-to-nonintact skin, etc.) are not as well defined; they are probably less than that observed with needlesticks, but likely not zero. Although efficacy of postexposure prophylaxis with zidovudine is unproven, many experts have recommended its use in these settings. Employee health programs following these recommendations must establish means to expeditiously identify and counsel the patient about the risks and benefits of taking zidovudine. Baseline blood tests need to be obtained, and follow-up with a physician familiar with the use of the drug established.

Growing concerns about the potential dangers of HIV-positive health care workers and the risks of transmitting infection to patients while performing invasive procedures promises to provide another area where infection control practitioners will be critically involved in defining "risky" procedures, counseling the health care worker, monitoring the performance of the individual, and advising hospital officials and medical staff about issues relative to the activities of the individual worker.

In addition to preventing infections among health care workers, close working relationships with occupational health personnel must be established to ensure workers exposed to potentially noxious fumes, chemicals, and other substances are protected and that their workplace is safe. Infection control practitioners are often the first to be notified of spills or leaks causing possible employee exposures and should be knowledgeable about procedures to provide immediate assistance to the worker until appropriate occupational health workers can be contacted.

The infection control practitioner must become involved with virtually all aspects of the hospital. New products introduced may have a cost-saving potential if looked at with only the eyes of the budget officer. If that same item results in increased infections, then it not only is detrimental to patient care but also negates the desired financial benefit. Thus, proposed new products that deal with patient care should be evaluated by the infection control practitioner. Similarly, changes in waste disposal, patient rooms, visiting policies, and numerous other normal hospital functions can inadvertently pose a risk to patient care and should be reviewed by infection control personnel.

This broad involvement with hospital functions by the infection control practitioner even extends their activities to the local community. Many policies such as waste disposal require compliance with safety guidelines issued by local and federal authorities to minimize the risk of contamination of the community environment. Proper management of infected, noninfected, and feared-to-be-infected waste must be ensured by the infection control practitioner.

In a very short time, infection control has evolved from a vague entity to a multi-faceted discipline with involvement in all areas of the hospital and patient care. As discussed earlier, AIDS is making new demands on this discipline. As we approach the twenty-first century, the emphasis in patient care is moving toward ensuring high-quality care for patients. At present, it is largely unclear how quality can be measured, but it is very likely that future infection control personnel with knowledge in epidemiology will be instrumental in defining quality parameters and establishing hospital programs of quality control. [*See* ACQUIRED IMMUNODEFICIENCY SYNDROME (AIDS).]

Bibliography

Bennett, J. V., and Brachman, P. S. (ed.) (1986). "Hospital Infections," 2nd ed. Little, Brown and Company, Boston.

Castle, M., and Ajeman, E. (ed.) (1987). "Hospital Infection Control: Principles and Practice," 2nd ed. Delmar Publishing, Albany, New York.

Martone, W. J., and Garner, J. S. (ed.) (1991). *Am. J. Med.* **91**(suppl. 3B).

Wenzel, R. P. (ed.) (1987). "Prevention and Control of Nosocomial Infections." Williams and Wilkins, Baltimore, Maryland.

Hypha, Fungal

C. H. Dickinson
University of Newcastle

Glossary

Compartments Cell-like lengths of hyphae in higher fungi that are separated by porate septa and functionally dependent on their neighbors, which involves the movement of cytoplasm and water

Septa Regular, cross walls in the distal portion of many hyphae that are clearly visible under the light microscope. The septa, one key feature of the higher fungi, develop by the centripetal extension of the hyphal wall and have several functions for the growth and protection of the fungi.

Spitzenkorper Dark granule in living and actively growing hyphae that is involved in the supply of vesicles to the hyphal apex

Yeasts Fungi that normally grow as single-celled organisms whose increase in biomass is accomplished by either budding or splitting to form new individuals

A HYPHA (plural hyphae) is a tubular, branched, normally colorless thallus, which is employed by most fungi for both vegetative growth and many forms of reproduction. The development of hyphae give fungi a competitive advantage in many situations, where they are growing saprobically or as pathogens of animals and plants (Fig. 1). Hyphae are versatile, plastic, and highly efficient organs that facilitate the exploitation of both solid and liquid resources. They enable fungi to spread across and between resources, although in some species this activity is enhanced by hyphal modifications that increase both the distance that can be covered and the rate of spread. However, for long-distance dispersal or when fungi must survive over long periods, the hyphae may become modified to form reproductive structures, leading to the production of asexual or sexual propagules.

The formation of hyphae is a major feature of the kingdom Fungi, but not all members of the group have adopted this mode of growth. Some fungi are normally found growing as single-celled forms, termed "yeasts" (Fig. 1), although there are close affinities between these forms and their hyphal relatives. Apart from the fungi, the only other group of organisms that produce comparable thalli are certain members of the Actinomycetes (kingdom Monera, Bacteria), notably species of the industrially important genus *Streptomyces*. Many algae also develop tubular thalli, but there are significant differences between the growth and development of these filaments and the tubular thalli formed by the fungi.

Hyphal growth involves extension of the tip of the tube and the formation of branches, which are also usually initiated near the apex. This process of hyphal extension and branching growth from a single propagule, or from two or more fused propagules, gives rise to a colony. Fungal colonies are spherical if there are no substrate or environmental constraints, such as when fungi grow in aerated liquids. It is, however, more common for the colony to be circular, with only limited development in the third dimension, as is seen in the fairy ring fungus (*Marasmius oreades*), whose giant colonies reach several meters in diameter after many years of growth. The term "colony" thus implies a functional entity arising from a defined origin. By contrast, the term "mycelium" describes a collection of hyphae, which might constitute the whole or only a small part of a colony.

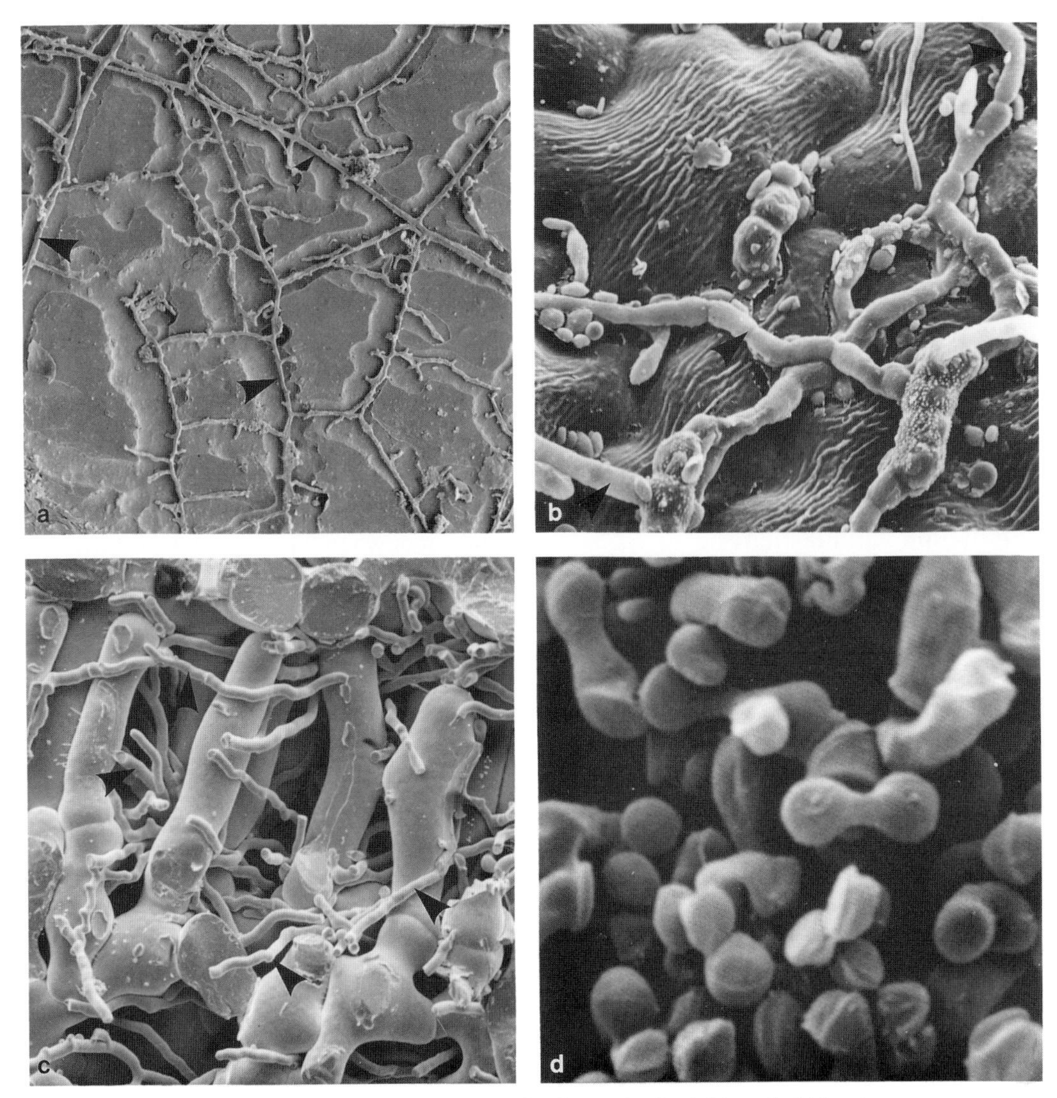

Figure 1 Composite plate showing (a) hyphae on cellophane after burial in soil, (b) hyphae growing in a complex microbial milleu on a leaf surface, (c) hyphae of a rust fungus (*Uromyces*) growing in the intercellular spaces of a broad bean (*Vicia*) leaf, and (d) yeast showing splitting. Hyphae in a, b, and c are shown by arrows.

I. Microscopic Features of Hyphae

Examination of the leading edge of a fungal colony with the light microscope permits the construction of a longitudinal optical section of a hypha (Fig. 2). Fungal hyphae range in diameter from 3 to 30 μm, with a median width of about 5 μm. Their length is less easily defined, being very variable even in the same species. A further complication in this respect is that continuous growth at the tip is often accompanied by an approximately equal rate of senescence of the older portion of the tube. This gives rise to the concept of a functional hyphal length, which for many species is about 200 to 500 μm.

The branching pattern of fungal hyphae is very variable, being determined by the genetic characteristics of each species (Fig. 2) and by the environmental and nutritional conditions. For example, adverse temperatures, suboptimal partial pressures of oxygen, and low concentrations of certain antibiotics can inhibit branch formation in the leading hyphae. Such effects show up clearly when fungi are grown

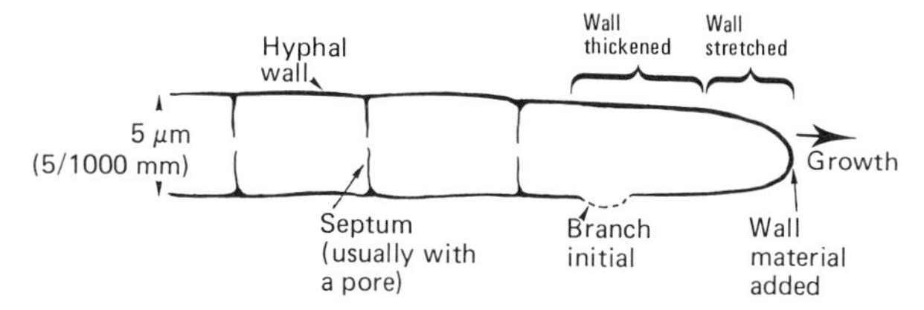

Figure 2 Diagram of the apical portion of a hypha as seen using the light microscope.

Table I Functions of Septa

1. Spatial separation of genetic material
2. Damage limitation when growing in hostile environments
3. Strengthening the tube
4. Seal off living hypha from dead compartments
5. Separate vegetatively active compartments from others undergoing morphogenetic changes associated with reproduction or the production of chlamydospores, etc.

on laboratory media, but it is not yet known exactly how these factors operate in natural environments.

The shape of a hyphal tip approximates to a parabola and is described as a hyphoid curve. Evidence relating this shape to the method of wall formation will be presented later. Behind the apical zone the hypha has parallel walls foming a regular tube. The more distal portion of many hyphae is characterized by presence of regular, cross walls, termed "septa," which are clearly visible under the light microscope. Such regular septa are, however, only found in certain fungi, and they have come to be regarded as one of the key features that helps to distinguish the "higher" fungi (Ascomycotina and Basidiomycotina) from other, mostly aseptate, "lower" fungi (Zygomycotina and Mastigomycotina). Careful study of the cross walls of the higher fungi within the functional length of the hypha shows that they do not form a complete barrier, but rather they remain porate, allowing the movement of cytoplasm and water along the tube. This functional dependence of each part of the hypha on its neighbors leads to the conclusion that the lengths of hyphae separated by pores are not cells, in the sense that this term is used elsewhere in biology, but should rather be termed "compartments." In most of the lower fungi the absence of regular septa leads to the development of an extended, multinucleate hypha that can be regarded as coenocytic (c.f., the use of this term to describe a similar condition in algae such as *Vaucheria*). When cross walls are formed in such coenocytic lower fungi, they form complete barriers, and as such they seal off one part of the hyphal system from the remainder. These cross walls may fulfil several possible functions in different circumstances and different fungi (Table I).

As one traces back along the length of a hypha, the tube becomes increasingly vacuolate and eventually it appears to be almost empty. The main portion of each compartment or the coenocytic hypha becomes dominated by a large vacuole, and the cytoplasm is reduced to a thin layer around the periphery of the tube. Eventually the remaining metabolites, and maybe some wall material, are moved forward from the oldest portion of the hypha, which is then scaled off by the existing septal pore being finally blocked, or by the *de novo* formation of a septum. The dead hypha then makes no further contribution to the colony.

II. Fine Structure of Hyphae

Electron micrographs of fungal hyphae, particularly those taken using the technique of freeze substitution, reveal that the main submicroscopic features present are typical of the other eukaryotic organisms. Thus, the mature hypha contains numerous organelles (e.g., mitochondria, Golgi bodies, ribosomes, microtubules and microfilaments, nuclei, vesicles, and a ground plasm with an extensive endoplasmic reticulum). Where fungal hyphae differ from other eukaryotes is in the distinctive zonation of these structures along the functional length of the hypha.

The extreme tip of the hypha is characterized by the presence of numerous ribosomes and small vesicles and by the absence of mitochondria and nuclei. In the subapical zone there are numerous mitochondria but fewer vesicles and ribosomes. Behind this mitochondria-rich zone is a region characterized by the presence of numerous nuclei and Golgi bodies. These latter are actively producing small vesicles. As will be seen later, the activity in this hinterland supports the continuously extending hyphal tip.

The main somatic hypha is characterized by the presence of elongate vacuoles, and in the higher fungi there are one or more nuclei per compartment, according to the life cycle of the particular species. In the lower fungi the nuclei are distributed along the length of the hypha. The older parts of the hypha are also characterized by the presence of storage granules that consist of lipid and glycogen globules.

The hypha is surrounded by a wall composed of a skeleton of randomly orientated chitin microfibrils, except in the Oomycete fungi (a division of the Mastigomycotina), which have cellulose microfibrils instead of chitin (Table II). These microfibrils are embedded within an amorphous matrix of chitosan and/or other hemicelluloses, and there are also glycoproteins present that may form a distinct layer. It has been suggested that hyphal walls are very complex structures composed of four or more distinct layers, but too little information is presently available to make accurate generalizations on this feature. Outside the hyphal wall there is often a layer of mucilaginous polysaccharide, which may perform a variety of functions, including water regulation and adhesion to surfaces, in a similar manner to the sheath found around many bacterial cells.

III. Hyphal Growth

The process of hyphal extension can be observed in many fungi, and particularly in members of the Zygomycotina, using the light microscope at magnifications of × 100 to × 400. By growing fungi on translucent media, the growth of the hyphal apex can be observed over a period of a few minutes, which must be one of the few instances where growth can be seen to be taking place without any assistance in the form of time-lapse photography or physical measurements.

The hyphal wall appears to play a vital role in regulating the morphology and size of each fungus. This becomes obvious if it is removed, using various enzymes, when the protoplasts assume a spherical shape. Hyphae extend by elongation of the apical section of the tip compartment, with the extension zone being 3 to 30 μm in length in different species. Evidence for this has been obtained from the use of radioisotope-labeled wall precursors and antibodies tagged with fluorescent dyes, as well as by simpler methods where hyphae are marked with Indian ink. As hyphae age the length of this zone of elongation tends to increase, as does the diameter of the tube that is formed. There is no evidence for any further growth in length caused by intersusception of wall material further back in the mature part of the wall. There is, however, some apposition of wall material behind the tip, which results in the wall becoming thickened and thus perhaps more rigid in older compartments.

Considerable progress has been made in understanding the organization of the hyphal apex. The extreme tip is devoid of major organelles, but it contains large numbers of small vesicles, which have been identified as being of several different types. In the higher fungi there is a concentration of these vesicles near the hyphal apex. This concentration appears to correspond to the optically dense granule known as the Spitzenkorper, which can be seen as a dark granule in living and actively growing hyphae with the aid of the light microscope. Electron microscope studies show that this body is involved in the supply of vesicles to the apex.

It has been suggested that hyphal extension is a steady-state process in which there is a continuous secretion of wall polymers and of lytic and plasticiz-

Table II Wall Components in the Main Groups of Fungi[a]

Fungal group	Cellulose	Chitin	Chitosan	Mannan
Mastigomycotina				
Chytridiomycetes	−	+	+	−
Hyphochytridiomycetes	+	+	−	−
Oomycetes	+	−	−	−
Zygomycotina	−	+	+	−
Ascomycotina				
Hemiascomycetes (yeasts)	−	+	?	+
Euascomycetes	−	+	?	−
Basidiomycotina	−	+	?	−

[a]This table is based on the relatively few analyses that have been performed to date, and it must be noted that exceptions do occur to the general statement that is made for each group.

ing enzymes at the apex. These materials are transported to the inner margin of the existing wall in vesicles, which are thought to discharge at the plasma membrane. Here there are enzyme complexes that synthesize chitin (or cellulose as appropriate) and β 1-3 glucan chains. These enzyme complexes have been isolated in a zymogenic form from small cytoplasmic vesicles, which are known as chitosomes. The lytic enzymes are thought to maintain the wall polymers in a plastic state while they are stretched because of hydrostatic pressure exerted along the whole length of the hypha. Eventually, toward the rear of the extension zone, the wall becomes rigid because of cross-bonding between the polymer chains.

A better understanding of the way in which hyphae develop has been obtained from studies on the Spitzenkorper, which is thought to be the body that supplies vesicles to the elongation zone. Using computer modeling procedures, it can be demonstrated that the exact siting and changes in position of this body can determine the shape of the resulting hypha. Hence if the vesicle supply body is in the center of a spherical cell, the spherical shape of the cell will be maintained as it increases in size. If, however, the vesicle supply body is eccentrically sited within a spherical body (which may be equated to a fungal spore), growth will occur from one side of the sphere, and if the body moves to maintain its position with respect to this growth point, a tubular structure (i.e., a hypha) will result.

There has recently been much discussion about the question of polarity in the development of fungal hyphae. This stems from the observations that there are many gradients in fungal hyphae from the tip to the subapical zones. In the aquatic mold *Achlya*, an electric current of 1 $\mu A\ cm^{-2}$ has been shown to flow through the hypha from the tip toward the distal end. The protons enter the hypha at the apex by symport transport with amino acids such as methionine. Protons are expelled from the cytoplasm by pumps located predominantly in the somatic hyphal regions. It is thought that vesicle transport to the apex may be by some form of electrophoretic transport mechanism, but this has not been confirmed experimentally.

More attention has recently been concentrated on the role of calcium in maintaining apical growth. Fluorescent stains reveal a gradient of calcium ions in hyphae, with maximum concentrations occurring at their tips. Calcium uptake only occurs at the hyphal apex. Interactions between calcium and calcium-binding proteins (e.g., calmodulin) may be important in cytoskeleton regulation, and in turn they may control cytoplasmic streaming and transport to the apex, as well as resulting exocytosis.

IV. Branch and Septum Formation

Branches arise shortly behind the hyphal tip or at the anterior end of more distal compartments. In both instances there is thought to be a localized secretion of wall softening enzymes, and this allows the hydrostatic pressure to cause the wall to bulge and hence give rise to a new growing point. In some instances the formation of branches may be coordinated with the development of a septum.

Septa develop by the centripetal extension of the hyphal wall; this process usually takes only a few minutes. The same layering of wall components can be seen in the septum as in the wall itself. As noted before, most septa remain porate in the first instance, and they allow movement of organelles from one compartment to the next. The pores are of two types, with a simple circular opening being characteristic of most Ascomycotina and a more elaborate barrel-shaped opening being produced by the Basidiomycotina. Septa are often associated with ancillary structures in the adjacent cytoplasm. Some of these structures are involved in plugging the pore when the need arises for a complete barrier between one compartment and its neighbor.

V. Functional Aspects of Hyphae

Substantial exchange of materials occurs along the functional length of the hypha as it grows across or through resource materials. Water, oxygen, and nutrients are abstracted from adjacent resources along the whole length of the hypha, with the exception of the extension tip zone. At the same time, carbon dioxide, extracellular enzymes, and metabolic waste materials are excreted from the same portion of the hypha.

VI. Differentiation and Plasticity

Although the typical shape and development of the hypha are well defined, the fungi have a remarkable ability to modify their thallus in a manner that facilitates growth in situations where normal hyphal development would appear to be inappropriate. Such responses are best exemplified by those fungi termed the "ycasts," which normally grow as single-celled organisms whose increase in biomass is accomplished by either budding or splitting to form new individuals. These fungi do not form a coherent taxonomic group and are mostly classified in the Ascomycotina and the Basidiomycotina. Despite this taxonomic diversity, they do, however, have some universal ecological preferences, such as their requirement for free water and for a supply of relatively simple nutrients such as sugars.

Yeastlike cells are produced in some species as a regular feature of an otherwise hyphal-dominated life cycle. It is also well established that many fungi that normally grow in a yeast-like manner can under specific conditions revert to producing a tube-like hypha, and vice versa, there are other hyphal fungi that can be experimentally induced to grow as if they were yeasts. In addition a few species make such transformations fairly commonly in the natural environment, perhaps as a response to changing physiological and biochemical conditions. Some of the factors that can bring about such transformations have been identified (Table III), but there may well be other pressures that influence the form of vogetative development of fungi growing in natural ecosystems.

Other substantial changes in the growth form of fungi involve the development of reproductive structures, ranging from morphologically simple chlamydospores, derived from a single hyphal compartment, to enormous and complex fruiting bodies best exemplified by the mushrooms and toadstools. Chlamydospores and a wide range of conidial spores are formed by processes that essentially involve sections of the hypha becoming spherical and then detached from the neighboring compartments. In the case of chlamydospores, their release is dependent on the adjoining hyphae becoming moribund, but some conidial fungi have evolved complex mechanisms that enable huge numbers of spores to be produced and dispersed in an efficient and rapid manner. This is well exemplified by *Penicillium,* which produces spherical conidia in chains on dedicated hyphal branches that are displayed at the apex of a vertical hypha, termed the "conidiophore."

Table III Polymorphism in *Aureobasidium pullulans*

Growth forms under natural and experimental conditions	Factors known to affect growth and development
Hyaline hyphae	pCO_2
Melanin-pigmented hyphae	pO_2
	Carbon source
Yeastlike cells	Nitrogen source
Swollen hyaline cells	Available divalent cations
Melanin-pigmented chlamydospores	Temperature
	Trace organic nutrients in yeast extract

Other fruiting structures are formed by the production of three-dimensional tissues that develop as a result of a complex series of hyphal branching processes. These tissues appear almost indistinguishable from the parenchyma produced by meristems in green plants. They allow the fungi to develop claborate fruiting bodies, such as are formed by the mushrooms and toadstools. The hyphal masses forming these fruit bodies have an elegant system of responses to a variety of environmental stimuli. Notable among these are their responses to gravity, with the stalk tissue, the cap tissue, and the gill tissue responding in different ways to provide the perfect arrangement for successful spore dispersal.

VII. Ecology of Fungi and Versatility of Hyphae

The longevity of hyphae, and the consequent form of the colony, varies according to the species. Three patterns are evident:

1. The short-lived type in which rapidly growing hyphae exploit the substrate and then the whole colony turns its activity to reproduction. All the food gained is employed to give maximum sporulation (e.g., *Pilobolus*).
2. A longer-lived type in which the whole colony remains more or less viable, but where the outermost hyphae absorb food that is translocated to the center of the colony, where it is used to form reproductive structures that are usually more massive than in type 1 (e.g., some small toadstools such as *Coprinus*).
3. A very long-lived species in which the fungus grows outward from an initial inoculum, using up the substrate and excreting waste materials as it goes. In this case the hyphae are continually extending at the tip and senescing at a somewhat

similar rate from the oldest end outward. This gives rise to a hollow colony in which the main activity, including both vegetative growth and reproduction, is around the perimeter (e.g., the fairy ring fungus *Marasmius oreades*).

The hyphae of many individual fungal species exhibit or possess particular characteristics that clearly fit the individuals concerned for their peculiar ecological niches. Some of these modifications are widespread, as with those species able to synthesize melanin, which impregnates hyphal walls and thus protects them against desiccation and the antagonistic activities of other microbes and animals. Carotinoid pigments accumulate in the cytoplasm of other hyphae where they may function as light receptors (e.g., in *Pilobolus* and *Phycomyces*) or in repair mechanisms after damage caused by uv radiation (e.g., in *Sporobolomyces*). Pathogens of plants have hyphal modifications that facilitate their adhesion to the host's surface together with others will sequentially allow penetration and the establishment of nonlethal physiological links with living host cells. Even more dramatic modifications are seen in some pathogens of animals, such as where hyphal loops are developed that can respond almost instantaneously to contact with nematodes (Fig. 1). The sudden inflation of one or more hyphal compartments then traps the nematode in a lethal embrace.

The hypha may thus be seen to have evolved in many ways that have enabled the fungi to become extremely effective saprobic and pathogenic organisms. Much is already known about hyphal growth, but there are undoubtedly aspects that still require further study, especially where fungi are growing in the natural environment.

Bibliography

Bartnicki-Garcia, S., Hergert, F., and Gierz, G. (1989). *Protoplasma* **153,** 46–57.

Heath, I. R., ed. (1990). "Tip Growth in Plant and Fungal Cells." Academic Press, New York.

Roberson, R., and Fuller, M. S. (1988). *Protoplasma* **146,** 143–149.

Trinci, A. P. J. (1978). *Sci. Prog.* **65,** 75–99.

Trinci, A. P. J. (1984). Regulation of hyphal branching and hyphal orientation. *In* "The Ecology and Physiology of the Fungal Mycelium" D. H. Jennings and A. D. M. Rayner, eds), pp. 23–52. Cambridge University Press.

Wessels, J. G. H. (1988). *Acta Botanica Neerlandica* **37,** 3–16.

I

Identification of Bacteria, Computerized

Stanley T. Williams
Liverpool University

Glossary

Classification Orderly arrangement of individuals into units on the basis of similarity, each unit should be homogenous and different from all others

Frequency matrix Tabulation of the percentage positive character states for a range of taxa

Identification Matching of an unknown against knowns in a classification, using the minimum number of diagnostic characters

Identification coefficients Means of expressing the degree of relatedness of an unknown to a known taxon

Probabilistic identification Determination of the likelihood of an unknown identifying with a known taxon

Taxon General term for any taxonomic group (e.g., strain, species, genus)

BACTERIAL IDENTIFICATION is still largely based on the determination of a wide variety of phenetic characters. These include morphological features, growth requirements, and physiological and biochemical activities. Hence large data banks are constructed for each taxon, most attention being paid to genera and species. Computation is now an essential tool for the construction, evaluation, and application of these data for the identification of unknown strains of medical, industrial, ecological, or scientific importance. The ever-increasing rates of isolation of novel strains and their generation by manipulation *in vitro* require the continued development of rapid and objective computerized systems for identification. The theoretical and practical aspects of some of these systems are discussed later.

I. Principles of Bacterial Identification

The taxonomy of any group of organisms is based on three sequential stages: classification, nomenclature, and identification. The first two stages are the prime concern of professional taxonomists, but the end product of their studies should be an identification system that is of practical value to other workers. Therefore, an identification system is clearly dependent on the accuracy and data content of classification schemes and the predictive value of the name assigned to the defined taxa.

The ideal identification system should contain the minimum number of features required for a correct diagnosis, which is predictive of the other characters of the taxon identified. However, the minimum number of characters required is dependent on both the practical objectives of the exercise and the clarity of the taxa defined in classification. Thus many enterobacteria can be identified using relatively few physiological and biochemical tests, the numerous serotypes of *Salmonella* are recognized by their reactions to specific antisera, and the accurate identification of *Streptomyces* species requires determination of up to 50 diverse characters. However, many new "species" of streptomycetes have been proposed solely on their ability to produce a novel metabolite.

Whatever the aims and scope of an identification system, computation is an established and rapidly developing means of providing a quick assessment of character data. The idea that mathematical models could be used for identification of bacteria was first proposed in the 1960s, and the practical implications for computerized identification of enterobacteria were demonstrated by workers at the Central Public Health Laboratory, United Kingdom, in the 1970s. During the same period, the use of computers for numerical classification was initiated and developed. This concept was subsequently applied to many bacterial groups, and these studies provided large data banks that were ideal for the development of computerized identification schemes. However, many bacterial taxonomists were slow to realize this potential, but these data now form the basis of many probabilistic identification schemes.

A wide and increasing range of computerized systems for the identification of bacteria is now reported in the scientific literature and allied to commercial kits. This reflects both the expansion of techniques used to determine characters for the classification of bacteria and the rapid developments in computer technology. The practical and theoretical impacts of these developments will be assessed.

II. Sources of Data for Bacterial Identification

The range of characters used for both the classification and identification of bacteria is extensive. The sources of information may be categorized as follows:

1. Morphology and physiology (e.g., gross and fine structure, motility, nutrition, and enzyme production)
2. Cell components (e.g., wall composition, lipids, whole cell pyrolysis)
3. Proteins (e.g., sequences, serology, phage typing)
4. Nucleic acids (e.g., DNA/DNA hybridization, DNA and rRNA sequencing, DNA fingerprinting, nucleic acid probes)

Computerized systems are increasingly used for the rapid analysis of data from all these sources. However, most computerized identification systems have been applied to phenetic data obtained from category 1, where the programs are used not just to analyze the data provided but also to select the most diagnostic characters and to evaluate the diagnostic value of the identifications obtained. This article will concentrate on this aspect of computer identification, but other references to the use of computers in the analysis of taxonomic data will be mentioned and also found in other articles. [*See* TAXONOMIC METHODS.]

III. Selection of Diagnostic Characters

The ideal diagnostic character is one that is consistently positive or negative within one taxon, and this differentiates it from most of all selected taxa. This is seldom achieved with one character, but selected groups of characters may approach this ideal. How far this is achieved depends on the consistency of the taxa studied and the objectives of the identification, which in turn influence the principles and methods used to select characters and to identify unknown strains.

Few characters can be regarded as entirely constant. Character variation may be real (e.g., strain variation) or induced by experimental error. Many tests and observations are difficult to standardize completely within or between laboratories. Therefore, any identification system should ideally take account of these sources of variation. The well-known and widely used dichotomous key does not always meet these criteria. The characters used have been selected and weighted carefully and are

Table I Diagnostic Table for *Proteus* Species

Characters	Species			
	P. mirabilis	*P. morganii*	*P. rettgeri*	*P. vulgaris*
Gelatin hydrolysis	(+)	−	−	(+)
Citrate utilization	v	−	(+)	(−)
Indole production	−	+	+	(+)
H_2S production	+	−	−	(+)
Ornithine decarboxylase	+	+	−	−
Mannitol acidified	−	−	+	−

+, 95% positive; (+), 85–95% positive; v, variable 16–84% positive; (−), 5–15% positive; −, 0–4% positive.
[From Sneath, P. H. A. (1978). Identification of microorganisms. *In* "Essays in Microbiology" (J. R. Norris and M. H. Richmond, eds.), pp. 10/1–10/32. J. Wiley & Sons, Chichester.]

based on years of experience of taxonomists. However, such keys are sequential, being based on +/− reactions of single tests at each level of the dichotomy, and hence an aberrant result at any stage can lead to a misidentification. Also, the weighting of characters' importance determines their order in the keys; this is often a highly subjective decision. Nevertheless, in the hands of experienced operators such keys can produce reliable identifications for intensively studied groups (e.g., the enterobacteria) with or without the use of computer analysis. They are also more reliable for identification of higher taxa (e.g., genera).

The diagnostic table (see Table I) is designed to take more account of strain variation and test error. It avoids the all-or-nothing decisions required in dichotomous keys by introducing categories based on test variations within the taxa included. These tables introduce an element of mathematical assessment of variations within taxa, but often their practical value is somewhat limited. Although they present an accurate summary of the reliability of characters within a taxon, they are difficult to access for the identification of unknown strains, either by eye or by computer, although the latter can sometimes be useful.

Therefore the next logical step is to produce a matrix that includes the full range of percentage positive scores of the minimum number of characters needed to differentiate between the taxa of interest. This is termed a "frequency (or probabilistic) identification matrix" (Table II). Such a matrix provides a comprehensive data base, and computer programs play a major part in the construction, assessment, and use of these data banks. As always, the minimum number and type of characters in the matrix depend on the number and distinctiveness of the taxa included and the practical aims of the exercise.

IV. Construction of Frequency Matrices

A. Selection of the Data Base

As stated previously, any identification system should be based on data from a sound classification scheme. Numerical classification schemes use a large number of unweighted characters to define the limits and integrity of related taxa, usually species or related genera, at selected levels of overall similarity (usually >80% for species). This is determined using computer analysis of various coefficients, such as the simple matching coefficient (S_{SM}), which includes both positive and negative matches, and the Jaccard coefficient (S_J), including positive matches only. In addition, numerical classification also provides percentage frequencies of positive character states for all strains within each taxon (or cluster) defined. Such data provide an ideal basis for the construction of a frequency matrix for identification of unknown strains against the defined taxa. For

Table II Frequency Identification Matrix

Characters	Taxa			⟶
	A	B	C	
1	0.80	0.95	0.30	Frequencies of positive character states
2	0.35	0.54	0.99	
3	0.99	0.01	0.65	
4	0.01	0.21	0.84	
5	0.01	0.40	0.72	
↓				

ease of computation, a frequency of 100% is entered as 0.99, and strains that are uniformly negative as 0.01 (see Table II). Ideally the data base should comprise at least 100 characters and at least 20 taxa, but the latter will clearly depend on the scope of the numerical classification and the diversity of the groups studied.

Whatever the size and scope of the classification matrix, by no means all the unweighted characters used will have sufficient diagnostic value for use in construction of an identification matrix. Therefore, the major task is to determine a minimal battery of reliable tests that will distinguish between the taxa.

The first stage is to check the quality of the classification matrix before proceeding further. Two criteria of particular relevance are (1) the homogeneity of the taxa and (2) the degree of separation or minimal overlap between them. These can be assessed using appropriate statistics.

An assessment of the homogeneity of a taxon is provided by calculation of the standard deviation of strains from the center of the cluster of strains. Generally, this will be about 0.2, but if it clearly exceeds the value, the homogeneity and delineation of the taxon are dubious. Calculation of standard deviation has little value if the number of strains in a taxa is less than five. The OVERMAT and OUTLIER programs include a calculation of the standard deviation of taxa.

The heterogeneity of groups defined by numerical classification may be due to the inclusion of so-called aberrant strains, or "outliers." This may occur if the clusters are delineated at low similarity levels or if useful ancillary characters have not been included. Whatever the causes, it is useful to be able to detect and remove such strains before constructing the identification matrix. This can be achieved by the OUTLIER program, which uses as an objective criteria the degree to which the strain data fit a chi-squared distribution. The program uses 1.0 data, and calculations are based on a choice of one of four identification coefficients. Two of these, the -log Willcox likelihood and taxonomic distance squared, were found to be the most useful for detection of aberrant strains.

Another means of evaluating the classification matrix is to assess the degree of overlap between the groups defined. The OVERMAT program is designed to determine overlap between groups in a matrix constructed using positive values for character states. For each pair of groups in the matrix, a disjunction index (W) and a corresponding nominal overlap (V_G) is calculated, the latter ranging from 1.0 for complete overlap to 0 for complete separation. The significance of the calculated overlap is assessed using a noncentral t statistic against a selected critical overlap value (W_o). The latter should not exceed 5%. If pairs of taxa show significant overlap, further characters may be used to affect the separation of their combination into a single taxon.

B. Selection of Characters for the Identification Matrix

Once it can be assumed that the classification matrix is sound, the next step is to select the minimum number of characters from it that are required for the distinction between all the taxa included. One approach is to attempt to select diagnostic characters by eye and/or the basis of preconceived ideas. With a large data matrix, the former is difficult and the latter may be ill-conceived. There is some controversy about the minimum number of tests needed to separate a range of taxa effectively. One guideline is that the number of tests should at least equal the number of taxa. This may apply to relatively small and tightly defined taxa, particularly for genera rather than species. However, for many taxa (e.g., *Bacillus, Clostridium,* Enterobacteriaceae, and *Streptomyces*), the large number of species would necessitate use of an excessive number of characters. Most of these problems can be solved if (1) the aims of the identification exercise are clear and (2) the selection of characters is approached objectively.

The ideal diagnostic character should have a distribution of positive-positive occurrence as near as possible to 50% among the taxa in the matrix. Characters that are either positive or negative for all taxa are clearly of no diagnostic value, as are those that have a frequency of 50% within all or most taxa. In small matrices, the useless characters can be detected by eye. However, this is more difficult with large matrices, and the selection and evaluation of useful characters are usually difficult or impossible to achieve subjectively. However, various separation indices have been devised that can rank characters in order of their diagnostic value. The CHARSEP program incorporates several of these indices and provides a useful means for selection of characters. The use of one index, the variance separation potential (VSP), is illustrated in Table III; this index is based on the variance within taxa multiplied by separation potential. Values greater than 25%

Table III Examples of the Use of the CHARSEP Program to Determine the Most Diagnostic Characters of *Streptomyces* Species

Characters	No. of taxa in which character is predominantly		VSP index (%)
	+ve	−ve	
Good diagnostic characters			
Resistance to phenol (0.1% w/v)	6	8	55.8
Spiral spore chains	7	6	54.9
Degradation of lecithin	13	4	48.6
Antibiosis to *Bacillus subtilis*	4	5	44.5
Poor diagnostic characters			
Production of blue pigment	22	0	0.07
Blue spore mass	22	0	0.20
Spore surface with hairy appendages	21	0	0.32
Use of L-arginine	0	19	1.32
Proteolysis	0	14	3.02

[From Williams, S. T., Goodfellow, M., Wellington, E. M. H., Vickers, J. C., Alderson, G., Sneath, P. H. A., Sackin, M. J., and Mortimer, A. M. (1983). *J. Gen. Microbiol.* **129**, 1815–1830.]

Table IV Examples of the Use of the DIACHAR Program to Evaluate Diagnostic Scores for Characters of *Streptoverticillium* Species

Species	Sum of scores
Streptoverticillium olivoreticuli	20.34
S. salmonis	19.01
S. ladakanum	18.81
S. hachijoense	18.10
S. abikoense	16.64
S. mobaraense	14.19

[From Williams, S. T., Locci, R., Vickers, J., Schofield, G. M., Sneath, P. H. A., and Mortimer, A. M. (1985). *J. Gen. Microbiol.* **131**, 1681–1689.]

indicate acceptable characters. From such data it is possible to exclude most of the characters that have contributed little or nothing to the definition of taxa in the classification and hence are unsuitable for inclusion in an identification matrix. It is also advisable to exclude any characters that have a high diagnostic score but separate the same pairs of taxa. These can be detected by careful visual examination of the classification matrix after the theoretically best characters have been defined by CHARSEP.

A further program (DIACHAR) can be used to rank characters according to their diagnostic potential. This ranks the diagnostic scores of each character for each group in a frequency matrix and also provides the sum of scores for all selected characters in each group (Table IV). The higher the score, the greater the diagnostic value of the selected characters. Sometimes it is desirable to select a few characters that, although of low overall separation potential, are shown by DIACHAR to be diagnostic for a particular taxon.

Using these procedures, useful identification matrices have been constructed for a wide variety of bacterial groups. These include *Bacillus* (30 characters for 44 taxa); Enterobacteriaceae (47 for 100); gram-negatives (83 for 66); *Mycobacterium* (34 for 13); *Streptomyces* (50 for 52), and *Streptoverticillium* (41 for 24). The minimum number of characters clearly varies with the scope of the exercise and the degree of distinctiveness of the taxa. When relatively large numbers of characters are required, the use of miniaturized test systems, particularly for physiological characters, saves much time and labor. These are used increasingly, and some are available as commercial kits. Other approaches to large groups involve the construction of more than one matrix. Thus the large number of *Streptomyces* species have been accommodated in a matrix containing those with many strains, and one with those comprising fewer strains. A system with a primary (supermatrix) to aid choice from a range of secondary (submatrices) has also been applied to the identification of a large number of marine isolates. *Bacillus* species have been split into a matrix for aerobes and one for anaerobes, with some characters common to both.

V. Applications of Identification Matrices

The identification of an unknown bacterium involves determination of its relevant characters and the matching of these with an appropriate data base that defines known taxa. As emphasized previously, the theoretical and practical aims of this exercise will also influence the choice of identification systems. The ideal objective is to assign a name to the unknown that is not only correct but predictive of some or all its natural characters. Computerized identification schemes provide a more flexible system than those of sequential systems (e.g., dichotomous keys).

Computerized identification can be achieved in two ways: numerical codes or probabilistic systems.

A. Numerical Codes

These are usually based or +/− character reactions. They are applied to a relatively small set of characters that have been selected for their good diagnostic value and are applied to clearly defined taxa. Numerical codes require determination of a series of character states and the conversion of the binary results into a code number that is then accessed against the identification data base. Such identification systems are particularly appropriate for the analysis of test results obtained when commercial identification kits are used. An example is the API 20E kit that generates a unique seven-digit number from a battery of 21 tests (Table V). The tests are divided into groups of three, and the results are coded 1, 2, 4 for a positive result for tests in each group. These values are then used to produce a score that reflects the test results, which can be accessed against the identification system. Organisms that generate profile numbers that are not in the identification system can be tested against appropriate computer-assisted probabilistic identification systems. Numerical codes have proved to be convenient and effective, particularly for well-studied groups such as the Enterobacteriaceae. The accuracy and scope of such systems are being extended by development of new data bases, use of new sources of taxonomic information, and improvements in computer technology (see Section VII).

B. Probabilistic Identification

Probabilistic schemes are designed to assess the likelihood of an unknown strain identifying to a known taxon. Therefore such schemes require frequency matrices based on percentage values for diagnostic characters for each taxon and suitable computer programs to access the character data for an unknown strain. In theoretical terms, the taxa are treated as hyperspheres in an attribute space (a-space) in which the dimensions are the characters. The center of the hypersphere (taxon) is defined by the centroid (the most typical representative), and the critical radius encompasses all the members of the taxon. Ideally, each taxon will be distinct from any others if the data matrix has been well constructed. To obtain an identification, the diagnostic characters for an unknown strain are determined and its position in the a-space calculated. If it falls within the hypersphere (taxon) of a known taxon, it is identified. Thus, in essence, probabilistic identification systems allow for an acceptable number of "deviant" characters in both the known taxa and the unknown strains. Therefore, once the most diagnostic characters of an unknown strain have been determined, it is necessary to match this against an appropriate matrix of known taxa (i.e., to determine the likelihood of its identification). Various coefficients for calculating this have been proposed, but the most widely used is the so-called "Willcox probability." This is the likelihood (L) of identification of an unknown (u) against a particular taxon (J), divided by the sum of the likelihood of u against all taxa (q) in the matrix expressed as

$$L_{uJ}/\Sigma q\, L/_{uJ}.$$

Table V Steps in the Use of a System (API 20E) for the Rapid Computerized Identification of a Bacterium

Kit tests							Oxidase test
A B C	D E F	G H I	J K L	M N O	P Q R	S T	
Results							
+ − +	+ + +	− − +	− − −	+ − −	− + −	+ +	−
Values allocated for a positive response							
1 2 4	1 2 4	1 2 4	1 2 4	1 2 4	1 2 4	1 2	4
Cumulative scores for groups of three tests							
5	7	4	0	1	2	3	
Input of scores to computer-based identification matrix							

[Modified from Austin, B., and Priest, F. (1986). "Modern Bacterial Taxonomy." Van Nostrand Reinhold, United Kingdom.]

A maximum identification score (>0.999) will be obtained if the unknown matches to one taxon in the matrix and shows low affinity to all others. However, choice of the minimum acceptable score varies with the group of bacteria and the objectives of the identification (see Section VI).

Other useful identification coefficients are taxonomic distance and the standard error of taxonomic distance. The former expresses the distance of an unknown from the centroid of any taxon with which it is being compared; a low score, ideally less than 1.5, indicates relatedness. The standard error of taxonomic distance assumes that the taxa are in hyperspherical normal clusters. An acceptable score is less than 2.0 to 3.0, and about half the members of a taxon will have negative scores, because they are closer to the centroid than average.

All three of these identification coefficients can be determined for an unknown strain against an identification matrix by the MATIDEN program. Input can be from a coding form printed by the IDEFORM program. An example of the MATIDEN output is given in Table VI.

This combination of identification coefficients can often provide clear-cut evidence of a positive identification (e.g., Table VI), but in some instances a value judgment may still be necessary. The level of scores regarded as acceptable tends to vary to some extent with the group under study and the aims of the investigator. This can be illustrated by some examples of the choice of minimum Willcox probabilities, the most frequently used coefficient. For the well-defined, homogeneous taxa of the Enterobacteriaceae, a threshold score of greater than 0.999 has been used. In contrast, minimum acceptable scores for less clearly defined taxa include 0.85 for *Streptomyces* species and 0.95 for *Bacillus* species.

Table VI Example of the Output Provided by the MATIDEN Program to Identify an Unknown Streptomycete Strain against an Identification Matrix

	1	2	3
S. albidoflavus	0.999	0.276	−1.653
S. chromofuscus	0.671×10^{-1}	0.420	2.362
S. atroolivaceus	0.379×10^{-1}	0.420	3.099

Characters against *S. albidoflavus*—None

Additional characters that assist in separating *S. albidoflavus* from
S. chromofuscus—None
S. atroolivaceus—None

Isolate P571 best identification to *S. albidoflavus*. Scores for coefficients: 1 (Willcox probability), 2 (taxonomic distance), 3 (standard error of taxonomic distance).

[From Williams, S. T., Goodfellow, M., Wellington, E. M. H., Vickers, J. C., Alderson, G., Sneath, P. H. A., Sackin, M. J., and Mortimer, A. M. (1983). *J. Gen. Microbiol.* **129**, 1815–1830.]

VI. Evaluation of Identification Matrices

Once an identification matrix has been constructed, it is important that its diagnostic value is assessed before it is recommended for use by microbiologists who are not necessarily expert taxonomists. Matrices can be evaluated by both theoretical and practical means.

A. Theoretical Evaluation

A convenient way of achieving this is provided by the MOSTTYP program. This accesses the matrix using the selected identification coefficients (Willcox probability, taxonomic distance, and the standard error of taxonomic distance) included in the MATIDEN program. The MOSTTYP program determines the identification scores of the hypothetical median organism of each taxon in the matrix. Therefore it provides the best possible identification scores for each taxon included in the matrix. If any of these are unsatisfactory, practical identification of unknowns against such taxa will inevitably be unreliable. An example of acceptable scores provided by the MOSTTYP program for *Streptoverticillium* species is given in Table VII.

B. Practical Evaluation

If the theoretical evaluation of the identification matrix proves to be satisfactory, the next stage is its practical assessment. The first step in this is to input the diagnostic character states of known taxa to the matrix. Ideally, this should involve the redetermination of the diagnostic characters of a random selection of taxon representatives that have been included in the construction of both the classification and identification matrices. This provides another assessment of experimental error in the determination of character states and its impact on the identification system. Ideally, the selected representatives should then identify closely to their taxon when their identification coefficients are determined

Table VII Identification Scores for the Hypothetical Median Organism of *Streptoverticillium* Species Provided by the MOSTTYP Program

Species	Identification scores		
	Willcox probability	Taxonomic distance	Standard error of taxonomic distance
Streptoverticillium hachijoense	1.000	0.247	−2.953
S. ladakanum	1.000	0.259	−2.914
S. netropsis	0.999	0.220	−3.560
S. griseocarneum	0.999	0.260	−3.189
S. cinnamoneum	0.996	0.249	−3.200

[From Williams, S. T., Locci, R., Vickers, J., Schofield, G. M., Sneath, P. H. A., and Mortimer, A. M. (1985). *J. Gen. Microbiol.* **131**, 1681–1689.]

against the matrix. With a well-constructed matrix, bad identification scores are rare, but when they occur they may reflect the random choice of an atypical representative of a poorly defined taxon rather than experimental error.

The final practical evaluation of a matrix clearly involves assessment of its success in identifying unknown strains. Therefore, the appropriate characters for unknowns are determined, input to the matrix, their identification scores are determined (e.g., by the MATIDEN program) and assessed by the investigator. Success rates vary with the bacterial group under study. Probabilistic identification of gram-negative bacteria has been most effective. Thus, of 933 strains of fermentative gram-negative bacteria and 621 strains of nonfermenters, 98.2% and 91.5% respectively, were identified at a Willcox probability level of greater than 0.999. Of 243 vibrios isolated from freshwater, 71.6% were identified at a level greater than 0.999 and 79.4% at greater than 0.990. When such stringent coefficient levels are applied to gram-positive bacteria, the results are often less impressive. For example, when a probability of greater than 0.999 was applied to coryneform bacteria only 50% of the unknown strains identified; at a level of greater than 0.995, only 42% of streptomycete isolates were identified. However, if less stringent coefficients are applied to take account of the heterogeneity of such groups, a higher rate of useful identifications can be achieved. Thus 73% of 153 streptomycete isolates were identified using a Willcox probability of greater than 0.85, together with other criteria provided by the MATIDEN program. These were (1) low scores for taxonomic distance and its standard error, (2) all first scores significantly better than those for the next best taxa, and (3) a small number of characters of the unknown strain listed as atypical of the taxon to which it identified (see Tables VI and VIII).

To date most probabilistic identification systems

Table VIII Examples of Identification Scores for Unknown Streptomycete Isolates

Isolate number	Identification	Identification scores		
		Willcox probability	Taxonomic distance	Standard error of taxonomic distance
1.	*Streptomyces albidoflavus*	0.999	0.34	−0.04
2.	*S. rochei*	0.992	0.35	0.27
3.	*S. griseoruber*	0.913	0.32	1.90
4.	*S. chromofuscus*	0.907	0.35	0.45
5.	Not identified	0.840	0.41	3.35
6.	Not identified	0.790	0.39	0.78
7.	Not identified	0.640	0.41	1.80
8.	Not identified	0.520	0.40	1.41

[From Williams, S. T., Vickers, J. C., and Goodfellow, M. (1985). Numerical identification of streptomycetes. *In* "Computer-assisted Bacterial Systematics" (M. Goodfellow, D. Jones, and F. G. Priest, eds.), pp. 289–306. Academic Press, London.]

have been tested against and applied to natural or "wild" isolates. However, there is an increasing use of genetic manipulation of such strains for scientific, medical, ecological, and industrial purposes. For a variety of reasons, not least patent laws, it is important to compare manipulated strains with each other and with their wild types. This is still a developing area in bacterial taxonomy, but probabilistic systems can be useful. For example, streptomycete strains that had been manipulated by various means, such as mutagens, plasmid transfer, and genetic recombination, were compared with their parent strains against an identification matrix. Most of the manipulated strains identified to the same species as their parents, indicating that most of the selected diagnostic characters were unchanged and that the identification matrix could accommodate some character state changes.

C. Source of Computer Programs

All the programs mentioned above were developed by Prof. P.H.A. Sneath (Department of Microbiology, Leicester University, United Kingdom.) They are written in BASIC, and full details of the programs can be obtained from the following publications.

Sneath, P.H.A. (1979). MATIDEN program. *Computers Geosci.* **5,** 195–213.
Sneath, P.H.A. (1979). CHARSEP program. *Computers Geosci.* **5,** 349–357.
Sneath, P.H.A. (1980). DIACHAR program. *Computers Geosci.* **6,** 21–26.
Sneath, P.H.A. (1980). MOSTTYP program. *Computers Geosci.* **6,** 27–34.
Sneath, P.H.A. (1980). OVERMAT program. *Computers Geosci.* **6,** 267–278.
Sneath, P.H.A., and Langham, C.D. (1989). OUTLIER program. *Computers Geosci.* **15,** 939–964.
Sneath, P.H.A., and Sackin, M.J. (1979). IDEFORM program. *Computers Grosci.* **5,** 359–367.

VII. Other Applications of Computers to Bacterial Identification

This article has concentrated on the use of computers to construct identification systems as well as to access them. However, computer programs are increasingly used solely to access taxonomic data for the identification of unknown strains. As discussed in Section V, nonprobabilistic, sequential systems (cf., dichotomous keys) based on a relatively limited number of +/− character states can be used to identify well-defined taxa such as some of the enterobacteria. They are based primarily on physiological characters, and a variety of useful programs suitable for the desk-top computers have been devised. Developments in computer technology can also provide a more direct link between the determination of test results and their evaluation. An example is the use of so-called "breathprints," for identification of gram-negative, aerobic bacteria. This relies on a redox dye to detect the increased respiration when a carbon source is oxidized. A range of substrates in a microtiter plate are inoculated with a strain; if a substrate is used, a pigment is formed by the redox dye, indicating a positive reaction. The pattern of these on the plate provides a breathprint, which can then be compared with those of known taxa using a system consisting of a microplate reader and a computer.

During the past decade, the data bases for both the classification and identification of bacteria have been extended and improved by the inclusion of diagnostic characters provided by chemical analysis of cell components. These include cell wall amino acids, membrane lipids, and proteins, which have been particularly useful for the definition and identification of higher taxa such as genera or families. Sometimes the data provided is of an unequivocal (+/−) nature, but in other cases computer analysis is needed to generate a data base for known taxa and to identify unknown strains. A few examples are discussed briefly below.

Analysis of bacterial lipids involves gas chromatography, which results in a printout of a set of peaks that are defined chemically but are difficult to evaluate and compare quantitatively by eye. Various programs have been used to transform the data for principal component analysis and to provide similarity and overlap coefficients for comparison of unknown strains.

The use of polyacrylamide gel electrophoresis (PAGE) to analyze protein patterns of bacteria is well established in bacterial taxonomy. The results are obtained in the form of stained bands on the gels that are difficult to analyze and compare by eye. A variety of computer programs has been devised to facilitate their analysis. Initially, the stained gels are scanned by a densitometer, which can be linked to a microcomputer with an analogue-digital converter. This produces a continuous trace of the densitometer scan, removes background effects, and feeds the data to a file for comparison with other stored traces. The unknown is then compared with known taxa using various similarity coefficients.

Another rapidly developing method of bacterial identification is pyrolysis-mass spectrometry. This involves the thermal degradation of a small sample of cells in an inert atmosphere or vacuum, leading to production of volatile fragments. Under controlled conditions, these are characteristic of a taxon, and they are separated and analyzed in a mass spectrometer. Assessment of the traces obtained requires software (now commercially available) for performing principal components, discriminant, and cluster analysis, to allow known taxa to be distinguished and unknowns identified.

The developments in nucleic acid techniques are having a marked and exciting impact on bacterial taxonomy, where they provide a genetic assessment of taxa, which can be used to supplement or revise the existing phenetic systems. Techniques such as DNA reassociation, DNA–rRNA hybridization, and DNA and RNA sequencing are increasingly used in bacterial classification, whereas nucleic acid probes and DNA fingerprints are of great potential in identification. Computation is increasingly used in the analysis and application of such data. Despite the relative novelty of these sources of taxonomic data, ultimately a convenient and accurate means of comparing data for unknown strains with those of established taxa is still required. Thus, when determining DNA fingerprints it is useful to have a permanent record of fragment size. Precise migration measurements on gels should not be spoiled by inaccurate assessments of fragment size. Fragments have a curvilinear relationship between the mobility of their bands on gels and their molecular sizes. A number of programs have been devised to transfer and assess such measurements.

Thus computation has an established and developing role in all stages and aspects of bacterial identification.

Bibliography

Bochner, B. (1989). *Nature (Lond.)* **339,** 157–158.

Langham, C. D., Williams, S. T., Sneath, P. H. A., and Mortimer, A. M. (1989). *J. Gen. Microbiol.* **135,** 121–133.

Langham, C. D., Sneath, P. H. A., Williams, S. T., and Mortimer, A. M. (1989). *J. Appl. Bacteriol.* **66,** 339–352.

Priest, F. G., and Alexander, B. (1988). *J. Gen. Microbial.* **134,** 3011–3018.

Priest, F. G., and William, S. T. (1992). Computer-assisted identification. *In* "New Bacterial Systematics" (M. Goodfellow, ed.). Academic Press, London.

Immune Suppression

Edwin W. Ades, Diane C. Bosse, and J. Todd Parker
Centers for Disease Control

Glossary

Antibody Globulin protein, referred to as an immunoglobulin, that will interact with the specific antigen that induced its synthesis

Antigen Foreign particle, typically a protein or glycoprotein, that induces a host immune response

B lymphocytes Set of lymphoid cells that produce antibodies

Cell-mediated immunity Immune response mediated by T lymphocytes

Class II major histocompatibility complex Group of glycoproteins that are expressed predominantly on B cells and macrophages; associated with antibody production, mixed lymphocyte reaction stimulation, graft versus host reactions, and immune response genes

Cytokines Protein released from cells that act as chemical messages; one such group, from lymphoid cells, are called interleukins

Mononuclear phagocytes (macrophages) Group of lymphoid cells that phagocytize antigen and process it for use during an immunological response

T lymphocytes Group of lymphoid cells that act as helper, cytotoxic, or suppressor cells during an immunological response

IMMUNE SUPPRESSION is the naturally occurring dampening of the process of nonself recognition known to occur by a repertoire of negative reponses that occur during a normal immune response. This can be caused by pathogenic infections (viral, bacterial, or parasitic), chronic ailments (cancer, cirrhosis, or diabetes), or physical stress (inflammation, burn, or pregnancy). Great strides are being made in identifying the immune system's components and regulatory mechanisms.

I. Immune Response

Our ability to manipulate the immune response has become more of a reality as our understanding of the components of the immune response system improves: specificity (ability to discriminate related antigens), memory (a more rapid and heightened response on secondary exposure to antigen), recruitment (induction of secondary antigen–nonspecific mechanisms of antigen disposal), and regulation (dampening or amplification). This article will address the latter; the downside of regulation or balance.

The repertoire of negative responses that occur during a normal immune response have been associated with immune suppression (Is) genes, which determine whether or not and, if so, to what extent an organism can produce active suppression. The role of the Is genes contrasts with that of the immune response (Ir) genes, which trigger a positive response. Active suppression can mean that the nonresponsiveness due to Is genes differs from the nonresponsiveness due to the absence of Ir genes. "Immunosuppression—the prevention or diminution of the immune response, as by irradiation or by administration of antimetabolites, antilymphocyte serum, or specific antibody; called also immunodepression" ["Dorland's Medical Dictionary," 27th ed. (1988). W. B. Saunders, Philadelphia. p. 823] is the mechanism by which we may down-regulate the repertoire of immune responses.

II. Immune Response to Antigens

A. Overview

The primary goal of a normal immune response to foreign antigens is elimination of the antigen. This normal immune response to a T-dependent antigen involves three cell types: the T cell, the B cell, and the mononuclear phagocyte (macrophage). Although most of the antigen phagocytized by the mononuclear cell is destroyed, a subset of the mononuclear phagocytes (the antigen-presenting cells) "process" the antigen and express parts of it on their cell surface in conjunction with a Class II major histocompatibility complex (MHC)-encoded determinant. The processed antigen–MHC/Ia complex is recognized by a pre-helper T cell, which has membrane receptors for both Ia and the "carrier" portion of the processed antigen. This "antigen-carrier T-helper cell" then interacts with "resting" T-helper cells, which proliferate to make a clone of antigen-specific T-helper cells. The antigen-carrier T-helper cells also interact with "resting" B cells and stimulate them to proliferate and divide into clones of antibody/immunoglobulin-secreting cells (plasma cells). Suppressor T cells are stimulated by the same mechanism and act as regulators of the antigen-specific T-helper cells. Immunoregulation occurs by either direct cell–cell interaction or via soluble factors. [*See* B CELLS; T LYMPHOCYTES.]

B. Feedback

Few biological problems are as intriguing as when antigen is introduced into a host. The existence of complex circuits of T, B, and macrophage cells emphasizes the fact that many cells involved in the immune response are not themselves final effector cells. In general, the normal host immune response proceeds in a stepwise fashion. For each step, a feedback inhibition regulates the previous reaction. This cascade of reactions and counterreactions was described in the early 1970s by a researcher who demonstrated that, for each discrete cell population within the immune system, a discrete regulatory population existed. In short, for a given primary antigenic determinant on a foreign particle, there are antibodies (Ab1) that recognize and bind to the antigen. For a given Ab1, there is a subsequent Ab2 (idiotypic antibody), which binds to Ab1, thereby regulating the available quantity of Ab1. Ab2, in its antigen-binding region, possesses the "internal" image of the primary antigen, thereby controlling the amount of excess Ab1. For any Ab2 (idiotype), there is an Ab3 (anti-idiotype). The Ab3 is virtually identical to Ab1 and has the same antigenic specificity. This loop of idiotypic and anti-idiotypic antibodies provides a simple, although sophisticated, network of regulatory interactions for antibody populations. Recently, it has been shown that the "idiotype network" for B-cell responses, which produces antibodies, can also be described for some T-cell responses.

III. Induced Causes of Suppression

Immune suppression, or the dampening of an immune response, can arise at almost any point associated with the complicated schema of antigen elimination and can be positive, negative, or both. In the case of chronic infections, the continuous attempts at antigen elimination may result in immunopathological symptoms, and immune suppression may be the action necessary to reverse this negative response against the host. Alternatively, immune suppression may prevent the host from responding appropriately to antigenic challenge. Therefore, balance, or the loss thereof, is of primary importance to a completely functional immune system; it can be upset by induced causes (e.g., pathogens, trauma, alcoholism, diabetes) or by "natural" causes (e.g., pregnancy, congenital and acquired immunodeficiencies, age).

A. Pathogenic Suppression

Perhaps primary of the induced causes are the pathogens. The major goal of a pathogen is its survival or, more specifically, its evasion of immunosurveillance. Bacteria and parasites often perform this evasive action by conserving epitopes of mammalian cells in their own surface complexion. Among these conserved epitopes are a group of proteins known as the heat-shock proteins (HSPs). The HSPs have been shown to provide vital functions in the survival of bacteria, yeast, and parasites. They are highly conserved, widely distributed throughout prokaryotes as well as eukaryotes, are induced by alternative transcriptional factors, and play a critical role in the induction of "survival" pathways of the invading pathogen as a response to host stress. Proteins of the HSP 60 and HSP 70 families are found in virtually all species of prokaryotic and eukaryotic species. The HSP 70 family proteins have been associated with a variety of infections, including malaria (*Plasmodium falciparum*), schistosomiasis (*Schistosoma mansoni* and *Schistosoma japonicum*), leprosy (*Mycobacterium leprae*) and tuberculosis (*Mycobacterium* tuberculosis), syphilis (*Treponema pallidum*), Legionnaires' disease (*Legionella pneumophila*), Lyme disease (*Borellia burgdorferi*), and lymphogranuloma venereum (*Chlamydia trachomatous*), whereas HSP 65, an HSP 60 family protein, has been associated with syphilis (*T. pallidum*). The HSPs have also been shown to play a role in the lytic induction of bacteriophage lambda.

Viral immune suppression can be divided into two general groups: viruses that abrogate the immune response and those that interrupt the multiple facets of interconnecting immune-signaling cascade events. The primary mechanism is through the abrogation of the immune response by infection of lymphoid cells, resulting in the impairment of responsiveness to antigenic and mitogenic stimuli. Viruses that infect lymphoid cells include those that infect T-helper cells [e.g., human immunodeficiency virus (HIV)], thereby interrupting the induction of the T-cell-mediated response, which results in repression of the total T-cell population while enhancing the T-suppressor population. One of the most striking clinical features in patients infected with HIV is the development of opportunistic infections and malignancy after the immune system has become severely compromised. Other viruses (e.g., poliovirus, the herpes family of viruses) are believed to activate prostaglandin synthesis, thereby impairing the ability of T cells to respond to the network of cytokines [interleukin-1 and interleukin-2 (IL-1 and IL-2)]. [*See* ACQUIRED IMMUNODEFICIENCY SYNDROME (AIDS).]

B. Nonpathogenic Suppression

Many nonpathogenic-induced causes also upset the balance of the immune system. Trauma, whether caused by physical injury, burn, or surgery, has been associated with various kinds of suppression of the immune system. Burn patients are exquisitely sensitive to, and have a high incidence of, *Pseudomonas* infections. This increased incidence in infection is probably associated with decreased neutrophil function and possibly with the appearance of suppressor T cells, which actively suppress a normal immune response. Defects related to alcoholism and cirrhosis in patients include abnormal pulmonary and systemic defense mechanisms associated with a decrease in granulocyte numbers and defective functions, decreased serum complement levels, abnormal leukocyte chemotaxis, and defects in cell-mediated immunity, which is reflected in a decreased ability to be sensitized to the antigen keyhole limpet hemocyanin. Diabetic patients present a unique immunocompromised profile. Abnormalities in the diabetic's immune system with respect to cell-mediated immunity and neutrophil function have been noted. Infections in these patients create special problems because the infection(s) can lead to difficulty in management of the disease.

IV. Natural Causes of Suppression

A. Pregnancy

Among the "natural" disturbances to the immune system is pregnancy. Pregnancy initiates a unique immune response network, in that the host initiates both a unique immune response and immune suppression toward the same entity. Pregnancy-associated immune suppression is the result of the repression of cell-mediated immunity. A variety of distinct immune modulators have been identified from the serum and amniotic fluid of pregnant women, among these α-fetoprotein (AFP), α_2-macroglobulin (α_2-MAC), and uromodlin. AFP, produced by the fetal liver, is transferred to the maternal serum via the placenta. Micromolar quantities of AFP stimulate the proliferation of suppressor T cells and the inhibition of macrophage surface Ia antigen expression. In plasma concentrations similar to that of AFT, α_2-MAC suppresses CMI *in vitro*. However, neither AFP nor α_2-MAC has been shown to lower host resistance to infection. Uromodulin, a protein isolated from the urine of pregnant women, suppresses the transformation of T lymphocytes at 80-pM quantities without impairing B-lymphocyte response.

B. Congenital and Acquired Immunodeficiencies

Congenital and acquired immunodeficiencies or specific immune defects associated with immune suppression will cause changes to the immune system resulting in immune suppression. Many detailed reviews that summarize the various forms of these defects have been written.

C. Age

The age-associated decline in the immune response occurs in both animals and humans; increased knowledge about it represents an opportunity to restore the response or halt its decline. T-lymphocyte function is most severely affected, manifested by a decline in delayed-type hypersensitivity reactions and delay in allograft rejection. The elderly have decreased antibody responses, poor generation of memory responses, and an increased incidence of autoantibodies and benign monoclonal gammopathies. Such abnormalities of immune function may contribute to the increased incidence of infections in the elderly.

Several explanations have been offered for this immunologic senescence. With age, the thymus gland involutes and has a decreased capacity to promote T-lymphocyte differentiation. Furthermore, hormone secreted by the thymus gland cannot be detected in humans >60 yr of age. Alteration of T-cell subpopulations, a decrease in T-cell proliferation responses, and a marked deficiency of T-lymphocyte precursors have been described in elderly humans. These findings could reflect either an absolute deficiency of inducible precursors or an impaired response of precursors to inductive signals (i.e., these effects occur through cell–cell contact or the elaboration of factors (lymphokines) with biological effects of their own). Many, but not all, of these factors have been given interleukin designations (IL-1 to IL-11), thereby consolidating a confusing array of old names (see Table I). Researchers have shown that T cells of elderly adults and animals both produce less T-cell growth hormone or IL-2 and show a decreased receptivity and decreased binding of normal IL-2 than activated T cells of young individuals. In addition to a qualitative and/or quantitative T-cell deficiency state, overactive inhibitory mechanisms in the elderly may contribute to diminished immune responses. Excessive suppressor cell activity, increased sensitivity to suppression, and augmented autologous anti-idiotypic antibody responses have been described to occur with aging. The sensitivity of lymphocytes to endogenous immunomodulators (e.g., prostaglandin E, histamine, hydrocortisone) changes with aging. Biochemical alterations of aged lymphocytes have been described as well: low activity of 5′-nucleotidase, alterations of lactic dehydrogenase isoenzyme patterns, and low basal levels of cyclic adenosine monophosphate and high levels of cyclic guanosine monophosphate. The progressive incidence of immunosenescence is an important target for therapeutic intervention and will lead to a better understanding of the effector mechanisms that mediate the final process of immune response or its suppression. [*See* INTERLEUKINS.]

V. Manipulation of the Immune System

A. Modulatory Drugs

The past decade has seen a renewed interest in manipulating the immune response both to prevent and to treat infections. The manipulation of the immune system by drugs (e.g., corticosteroids), natural or synthetic products (monoclonal antibodies, cytokines), or derivatives of these products represents an attractive adjunct in the treatment of infections. Many of the experimental foundations for this approach have been laid by the expanding knowledge of endogenous regulators and mediators of lymphomyeloid cooperation and differentiation. Recombinant human interferon and specific monoclonal antibodies have been produced *in vitro*, and both have been used, or will soon be used, in clinical trials against numerous viruses and pathogenic organisms. The role of these modulatory agents may be threefold: (1) enhancement of phagocytosis of infectious agents, (2) restoration of impaired immune functions, and (3) treatment of infection(s) without exerting selective pressure on microbial populations, which is an inherent problem with antibiotic therapy. An increasing number of cytokines (Table I) are being discovered that are capable of modifying immune responses, although their ultimate role in the therapy of infections is still uncertain. Care must be taken, however, because these immunostimulants can exhibit suppressive activities depending on the route and the timing of injection.

B. Cytokines

The direct therapeutic use of cytokines against infections is as yet uncommon; however, a group of cytokines known as colony-stimulating factors

Table I Lymphokines and Cytokines

	Functions
Human interleukin-1 (IL-1) α (17 kDa)	Stimulates thymocyte proliferation, B-cell maturation and proliferation, T-cell activation, fibroblast proliferation
Human IL-1 β (17.5 kDa)	Stimulates thymocyte proliferation, B-cell maturation and proliferation, T-cell activation, fibroblast proliferation
Human IL-2 (15 kDa)	T-cell proliferation and differentiation, including helper and suppressor cells
Human IL-3 (15 kDa)	Supports colony growth of granulocyte, macrophage, megakaryocyte, mast cell, and erythrocyte precursors
Human IL-4 (15 kDa)	Stimulates growth, maturation, and immunoglobin production of B cells; stimulates survival and growth of T cells
Human IL-5 (24 kDa)	Stimulates eosinophil colony formation, B-cell proliferation, immunoglobulin synthesis by B cells
Human IL-6 (20.5 kDa)	Promotes maturation and immunoglobulin production of B cells; stimulates IL-2 production and growth and maturation of T cells
Human IL-7 (17 kDa)	Stimulates maturation of early B and T cells; acts synergistically with IL-2 to promote proliferation of mature T cells
Human endothelial IL-8 (8.5 kDa)	Promotes neutrophil chemotaxis and degranulation; inhibits neutrophil adhesion to cytokine-activated endothelial cells
Human monocyte IL-8 (8 kDa)	Promotes neutrophil chemotaxis and degranulation more avidly than IL-8 (endothelial); inhibits neutrophil adhesion to endothelial cells
IL-9	Development of erythroid progenitors
IL-10 (17 kDa)	Cytokine synthesis inhibitory factor
IL-11 (20 kDa)	Promotes maturation and immunoglobulin production; supports proliferation of plasmacytoma cells
Human interferon(s) (17–25 kDa)	Strong inhibitor of virus proliferation; stimulates macrophages and natural killer cells, induces Class I and Class II major histocompatibility complex
Human tumor necrosis factor (TNF)-α (17 kDa)	Lyses many transformed cells, especially lymphomas and sarcomas; stimulates growth of many normal fibroblasts
Lymphotoxin or TNF-β (20–25 kDa)	Produced only by activated T cells; growth factor for lymphoid cells
Transforming growth factor (TGF)α	Exerts its effects through the epidermal growth factor receptor; inhibits proliferation and differentiation
TGF-β (25 kDa)	Inhibits proliferation and differentiation on lymphoid cells; inhibits and reduces proliferation on nonlymphoid cells
Human colony-stimulating factor–granulocyte-macrophage (GM-CSF) (14 kDa)	Stimulates proliferation and maturation of human granulocytes, macrophages, and eosinophils; enhances activity of neutrophils
G-CSF (19.6 kDa) and M-CSF (70–90 kDa)	Stimulates proliferation and differentiation of progenitors to neutrophils (G), monocytes, or macrophages (M)
Erythropoietin	A growth factor for erythroid precursors

(CSFs; i.e., granulocyte–macrophage-CSF, erythropoietin) have been recently used in cancer patients to decrease the patient's susceptibility to infection during therapy. In general, the immunologic defects associated with cancer therapy fall into three major categories: (1) decrease in neutrophils (neutrophenia), (2) altered immunoglobulin levels, and (3) lowered cell-mediated immunity (CMI). Neutropenia and altered immunoglobulin levels result directly from cellular toxicity due to administration of chemotherapeutic agents and immunosuppressive agents. Decreased CMI has been attributed to alteration of the normal CD4+/CD8+ lymphocyte ratio and enhanced expression of soluble immunomodulatory factors, such as IL-6 and transforming growth factor-β (TGF-β). It is believed that TGF-β aids in neoplasia by inhibiting the growth and induction of immune system cells, thereby inhibiting immune surveillance mechanisms. IL-6 (formerly called B-cell growth factor) has also been implicated in the maintenance of B-cell neoplasias. HSPs have also been implicated in the malignant cells' evasion of the host immunosurveillance. [*See* CYTOKINES IN BACTERIAL AND PARASITIC DISEASES.]

C. Immunosuppressive Drugs

Major advances in the use of organ and bone marrow transplantation as a therapeutic modality have been made possible by immunosuppressive drugs [i.e., exogenous immunosuppression in the form of steroids, azathioprine, or cyclosporin A (CsA)]. In general terms, steroids are the least specific, and their mechanism of action is the least well characterized. Azathioprine, an analog of 6-mercaptopurine, acts by inhibition of purine metabolism so as to block cell division. CsA binds to high-affinity cytoplasmic receptors termed immunophilins (immunosuppressant budding proteins) and blocks the generation/utilization of IL-2 by T cells, thereby preventing reactivity. The actual list of effects mediated by CsA continues to grow. New immunosuppressive agents such as FK-506, rapamycin, 15-deoxyspergualin, anti-CD-3-monoclonal antibodies, and various analogs of CsA are under current investigation. The immunosuppressive agents are necessary in transplantation to prevent a host immune response against the transplanted graft; however, these agents cause a predisposition to infections due to impaired leukocyte function and cell-mediated immunity. This is the most important complication and cause of death in transplant patients; therefore, the balance between adequate immune suppression (to allow acceptance of the graft) and oversuppression leading to infection is critical to these patients. The key to maintaining this balance and predicting the effectiveness of therapy may be in the ability to predict patient responses by *in vitro* tests. Thus far, *in vitro* analysis of patients' blood during therapy often does not agree with *in vivo* analysis, and neither may have predictive value for the final effect therapy may have on the patients.

D. Biological Response Modifiers

The ideal situation would be one in which the specific defects in the immunosuppressed host are known, allowing administration of a specific biological response modifier (BRM). This BRM could either mimic the action of a depressed cytokine (natural biochemical messenger(s) of the immune system) level or antagonize the action of an abnormally high level of another cytokine. Additionally, a BRM could augment the activity of reduced numbers of immunoreactive cells, perhaps by inducing the proliferation of such cells. Until disease-associated specific immunologic defects are identified, we will probably have to rely on currently available agents to augment a less-specific phase of the immune response.

VI. Conclusion

Thus, although great strides have been made in the understanding of the immune system and immune suppression, much is left to discover. As we come to understand how pathogens manipulate the system, how induced effects can alter the system to the system's detriment, and how natural effects such as pregnancy can change the balance for the good of the system, we will come closer to being able to manipulate the immune system for the patient's benefit.

Bibliography

Burdon, R. H. (1986). *Biochem. J.* **240,** 313–324.

Chaouat, G. (1987). *J. Reprod. Immunol.* **10,** 179–188.

Cosimi, A. B. (1981). *In* "Clinical Approach to Infection in the Compromised Host" (R. H. Rubin and L. S. Young, eds.), pp. 607–659. Plenum Press, New York.

Donahue, R. E., et al. (1990). *Blood* **75**(12), 2271–2279.

Foon, K. A. (1989). *Cancer Res.* **49,** 1621–1639.

Go, N. F. (1990). *J. Exp. Med.* **172**(6), 1625–1632.

Green, D. R., and Faist, E. (1988). *Immunol. Today* **9**(9), 253–255.

Kaufman, S. H. E. (1990). *Immunol. Today* **11**(4), 129–142.

Long, E. O., and Jacobson, S. (1989). *Immunol. Today* **10**(2), 45–51.

Makinudan, T., and Kay, M. M. B. (1980). *In* "Advances in Immunology" (H. G. Kunkel and F. J. Dixon, eds.), pp. 207–239. Academic Press, Orlando, Florida.

Oliviera, D. B. G. (1989). *Clin. Exp. Immunol.* **75,** 167–177.

Paul, S. R., et al. (1990). *Proc. Natl. Acad. Sci.* **87,** 7512–7520.

Paul, W. E. (1989). "Fundamental Immunology" (W. E. Paul, ed.). Raven Press, New York.

Pollock, R. E., and Roth, J. A. (1989). *Semin. Surg. Oncol.* **5**(6), 414–419.

Rinaldo, C. R., Jr. (1990). *Annu. Rev. Med.* **41,** 331–338.

Sanchez, E. R., et al. (1985). *J. Biol. Chem.* **260,** 12398–12401.

Schreiber, S. L. (1991). *Science* **251,** 283–287.

Twomey, J. J., Luchi, R. J., and Kouttab, N. M. (1982). *J. Clin. Invest.* **70,** 201–208.

Waldman, T. A., and Broder, S. (1977). *In* "Progress in Clinical Immunology," Vol. III (R. S. Schwartz, ed.), pp. 155–197. Grune and Stratton, Orlando, Florida.

Weinberg, E. D. (1987). *Microb. Pathogen.* **3,** 393–397.

Yonemoto, W., et al. (1982). *In* "Heat Shock from Bacteria to Man" (M. J. Schlesinger, M. Ashburner, and A. Tissieres, eds.), pp. 289–298. Cold Spring Harbor, New York.

Industrial Effluent Processing

N. Kosaric and R. Blaszczyk
The University of Western Ontario

Glossary

Nutrients Organic and inorganic compounds that are necessary for microbial growth

Pollutants All compounds in industrial streams that exceed acceptable health levels

Wastewater Industrial effluents

INDUSTRIAL EFFLUENTS are liquid streams that accompany industrial production. These effluents usually contain compounds undesirable and often detrimental to the natural environment. Processing of industrial effluents is performed to decrease the concentration of contaminants to the level accepted by legislation. An important method for industrial effluent processing is the microbial wastewater treatment. Food industry wastewaters due to their content of nutrients for microbial growth are also most convenient for microbial treatment. Microbial wastewater treatment may be performed under aerobic or anaerobic conditions.

I. Nature and Origin of Industrial Wastewaters

Clean drinking water is one of our most important resources. Total stock of water on earth is estimated to be 1.38×10^{18} m^3 (see Table I); however, 97.2% of this amount is in oceans and only 0.3% could be available as groundwater located down to 1 km.

Mass (industrial) production of goods demands a large quantity of water. Typical rates of water use for various industries are shown in Table II. During industrial processes, a certain amount of substrates, by-products, and final products pass to the water streams and contaminate them. Examples of different types of wastes generated by selected industries are shown in Table III.

The number and concentration of pollutants greatly depends on the technology used. Generally, new technologies generate less wastewaters than the old (sometimes more concentrated). Some wastewater streams may be treated or reused in the plant, discharged as effluent, or sent into other treatment processes. Others may be combined into a single large flow and treated biologically.

The object of any treatment is to remove certain waste materials as specified by governmental regulations. Materials to be removed are as follows:

1. Soluble organics (natural products of animal and plant life and certain synthetic chemicals)
2. Suspended and colloidal solids
3. Heavy metals (e.g., mercury, chromium, lead, cadmium)
4. Acidity and alkalinity (must be neutralized)
5. Oils, greases, and other floating materials
6. Excess inorganic nutrients such as nitrogen and phosphorus
7. Color, turbidity, and odor
8. Priority pollutants (regulated by the U.S. Environmental Protection Agency; the EPA list of organic priority pollutants is shown in Table IV).

II. Properties of Effluents from Selected Industries

A. Definitions

The composition of industrial effluents varies with time and usually cannot be precisely defined. Several overall parameters are used to define quality of industrial effluents.

Table I Total Stocks of Water on Earth

Location	Amount (10^{15} m^3)	Percentage of world supply
Oceans	1350	97.2
Icecaps and glaciers	29	2.09
Groundwater within 1 km	4.2	0.30
Groundwater below 1 km	4.2	0.30
Freshwater lakes	0.125	0.009
Saline lakes and inland seas	0.104	0.007
Soil water	0.067	0.005
Atmosphere	0.013	0.0009
Water in living biomass	0.003	0.0002
Average in stream channels	0.001	0.00007

[From Masters, G. M. (1991). "Introduction to Environmental Engineering and Science." Prentice-Hall, Englewood Cliffs, New Jersey.]

Table II Typical Rates of Water Use for Various Industries

Industry	Range of flow (m^3/100 kg of product)
Cannery	
Green beans	50–70
Peaches and pears	15–20
Other fruits and vegetables	4–35
Chemical	
Ammonia	100–300
Carbon dioxide	60–90
Lactose	600–800
Sulfur	8–10
Food and beverage	
Beer	10–16
Bread	2–4
Meat packing[a]	15–20
Milk products	10–20
Whisky	60–80
Pulp and paper	
Pulp	250–800
Paper	120–160
Textile	
Bleaching[b]	200–300
Dying[b]	30–60

[a] Live weight.
[b] Cotton.
[From Metcalf & Eddy, Inc. (1991). "Wastewater Engineering: Treatment, Disposal, Reuse." McGraw-Hill, New York.]

Table III Examples of Wastes Generated by Selected Industries

Type of wastes	Type of plant
Oxygen-consuming	Breweries, canneries, dairies, distilleries, packinghouses, pulp and paper mills, tanneries, textile mills
With high-suspended solids	Breweries, canneries, coal washeries, coke and gas plant, iron and steel industries, distilleries, packinghouses, pulp and paper mills, tanneries
With high-dissolved solids	Chemical plants, sauerkraut canneries, tanneries, water-softening plants
Oily and greasy	Laundries, metal finishing, oil fields, packinghouses, petroleum refineries, tanneries, wool-scouring mills, iron and steel industries
Colored	Electroplating, pulp and paper mills, tanneries, textile dyehouses
Toxic	Atomic energy plants, chemical plants, coke and oven by-products, petrochemical plants, electroplating, pulp and paper mills, tanneries
High acid	Chemical plants, coal mines, electroplating shops, iron and steel industries, sulfite pulp mills
High alkaline	Chemical plants, laundries, tanneries, textile-finishing mills
High temperature	Bottle-washing plants (dairies, beverage producers), electroplating, laundries, power plant in all industries, textile-finishing mills

1. Biochemical oxygen demand (BOD_5): The amount of oxygen (mg/liter) added to a wastewater to support microbial activity over a period of 5 days.
2. Chemical oxygen demand (COD): The amount of oxygen (mg/liter) needed to chemically oxidize a given wastewater.
3. Total organic carbon (TOC): The amount of carbon detected as CO_2. TOC measurement (using

Table IV EPA List of Organic Priority Pollutants

Compound name
1. Acenaphthene
2. Acrolein
3. Acrylonitrile
4. Benzene
5. Benzidine
6. Carbon tetrachloride (tetrachloromethane)
Chlorinated benzenes (other than dichlorobenzenes)
7. Chlorobenzene
8. 1,2,4-Trichlorobenzene
9. Hexachlorobenzene
Chlorinated ethanes (including 1,2-dichloroethane, 1,1,1-trichloroethane, and hexachloroethane)
10. 1,2-Dichloroethane
11. 1,1,1-Trichloroethane
12. Hexachloroethane
13. 1,1-Dichloroethane
14. 1,1,2-Trichloroethane
15. 1,1,2,2-Tetrachloroethane
16. Chloroethane (ethyl chloride)
Chloroalkyl ethers (chloromethyl, chloroethyl, and mixed ethers)
17. Bis (chloromethyl) ether
18. Bis (2-chloroethyl) ether
19. 2-Chloroethyl vinyl ether (mixed)
Chlorinated napthalene
20. 2-Chloronaphthalene
Chlorinated phenols (other than those listed elsewhere; includes trichlorophenols and chlorinated cresols)
21. 2,4,6-Trichlorophenol
22. *para*-Chloro-*meta*-cresol
23. Chloroform (trichloromethane)
24. 2-Chlorophenol
Dichlorobenzenes
25. 1,2-Dichlorobenzene
26. 1,3-Dichlorobenzene
27. 1,4-Dichlorobenzene
Dichlorobenzidine
28. 3,3′-Dichlorobenzidine
Dichloroethylenes (1,1-dichloroethylene and 1,2-dichloroethylene)
29. 1,1-Dichloroethylene
30. 1,2-*trans*-Dichloroethylene
31. 2,4-Dichloroephenol
Dichloropropane and dichloropropene
32. 1,2-Dichloropropane
33. 1,2-Dichloropropylene (1,2-dichloropropene)
34. 2,4-Dimethylphenol
Dinitrotoluene
35. 2,4-Dinitrotoluene
36. 2,6-Dinitrotoluene
37. 1,2-Diphenylhydrazine
38. Ethylbenzene
39. Fluoranthene
Haloethers (other than those listed elsewhere)
40. 4-Chlorophenyl phenyl ether
41. 4-Bromophenyl phenyl ether
42. Bis (2-chloroisopropyl) ether
43. Bis (2-chloroethoxy) methane
Halomethanes (other than those listed elsewhere)
44. Methylene chloride (dichloromethane)
45. Methyl chloride (chloromethane)
46. Methyl bromide (bromomethane)
47. Bromoform (tribromomethane)
48. Dichlorobromomethane
49. Trichlorofluoromethane
50. Dichlorodifluoromethane
51. Chlorodibromomethane
52. Hexachlorobutadiene
53. Hexachlorocyclopentadiene
54. Isophorone
55. Naphthalene
56. Nitrobenzene
57. 2-Nitrophenol
58. 4-Nitrophenol
59. 2,4-Dinitrophenol
60. 4,6-Dinitro-O-cresol
Nitrosamines
61. N-Nitrosodimethylamine
62. N-Nitrosodiphenylamine
63. N-Nitrosodi-*n*-propylamine
64. Pentachlorophenol
65. Phenol
Phthalate esters
66. Bis (2-ethylhexyl) phthalate
67. Butyl benzyl phthalate
68. Di-*n*-butyl phthalate
69. Di-*n*-octyl phthalate
70. Diethyl phthalate
71. Dimethyl phthalate

Continues

Table IV (*Continued*)

Polynuclear aromatic hydrocarbons (PAH)
72. Benzo(a)anthracene (1,2-benzanthracene)
73. Benzo(a)pyrene (3,4-benzopyrene)
74. 3,4-Benzofluoranthene
75. Benzo(k)fluoranthene (11,12-benzofluoranthene)
76. Chrysene
77. Acenaphthylene
78. Anthracene
79. Benzo(ghi)perylene (1,12-benzoperylene)
80. Fluorene
81. Phenanthrene
82. Dibenzo(a,h)anthracene (1,2,5,6-dibenzanthracene)
83. Ideno (1,2,3-cd)pyrene (2,3-o-phenylenepyrene)
84. Pyrene
85. Tetrachloroethylene
86. Toluene
87. Trichloroethylene
88. Vinyl chloride (chloroethylene)

Pesticides and metabolites
89. Aldrin
90. Dieldrin
91. Chlordane (technical mixture and metabolites)

DDT and metabolites
92. 4-4′-DDT
93. 4,4′-DDE(P,P′-DDX)
94. 4,4′-DDD(p,p′-TDE)

Endosulfan and metabolites
95. Endosulfan-α
96. Endosulfan-β
97. Endosulfan sulfate

Endrin and metabolites
98. Endrin
99. Endrin aldehyde

Heptachlor and metabolites
100. Heptachlor
101. Heptachlor epoxide

Hexachlorocyclohexane (all isomers)
102. BHC-α
103. BHC-β
104. BHC (lindane)-γ
105. BHC-δ

Polychlorinated biphenyls (PCB)
106. PCB-1242 (Arochlor 1242)
107. PCB-1254 (Arochlor 1254)
108. PCB-1221 (Arochlor 1221)
109. PCBO1232 (Arochlor 1232)
110. PCB-1248 (Arochlor 1248)
111. PCB-1260 (Arochlor 1260)
112. PCB-1016 (Arochlor 1016)
113. Toxaphene
114. 2,3,7,8-Tetrachlorodibenzo-p-dioxin (TCDD)

[From Eckenfelder, W. W., Jr. (1989). "Industrial Water Pollution Control." McGraw-Hill, New York.]

an infrared analyzer) lasts only a few minutes and for simple chemical compounds can be related to COD and BOD_5. For real industrial effluents, the TOC : COD and TOC : BOD ratios are estimated experimentally. Oxygen demand and TOC of industrial wastewaters for selected industry all presented in Table V.

4. Total suspended solids (TSS): All organic and inorganic suspended solids (mg/liter).

5. Volatile suspended solids (VSS): Includes only those solids (mg/liter) that can be oxidized to gas at 550°C. At that temperature, most organics are oxidized to CO_2 and H_2O while inorganics remain in ash. Estimate of the components of total (dissolved and suspended) solids in different types of wastewaters is shown in Table VI.

B. Composition and Properties of Effluents from Selected Industries

1. Petroleum Industry Waste

Petroleum refineries utilize from 0.5 to 40 volumes of water per volume of processed petroleum depending on the types of processes used, modernization of the plant, integration with the petrochemical industry, cooling system design, etc. A comparison of average waste flows and loadings from petroleum refineries for old, prevalent, and new technologies is shown in Table VII.

The major function of water in a refinery is for cooling purposes while relatively small quantities are used for boiler feed, direct processing, fire protection, sanitary uses, and others. Origin and

Table V Oxygen Demand and Organic Carbon of Industrial Wastewaters

Waste	BOD_5 (mg/liter)	COD (mg/liter)	TOC (mg/liter)	BOD : TOC	COD : TOC
Chemical[a]	—	4260	640	—	6.65
Chemical[a]	—	2410	370	—	6.60
Chemical[a]	—	2690	420	—	6.40
Chemical	—	576	122	—	4.72
Chemical	24,000	41,300	9500	2.53	4.35
Chemical refinery	—	580	160	—	3.62
Petrochemical	—	3340	900	—	3.32
Chemical	850	1900	580	1.47	3.28
Chemical	700	1400	450	1.55	3.12
Chemical	8000	17,500	5800	1.38	3.02
Chemical	60,700	78,000	26,000	2.34	3.00
Chemical	62,000	143,000	48,140	1.28	2.96
Chemical	—	165,000	58,000	—	2.84
Chemical	9700	15,000	5500	1.76	2.72
Nylon polymer	—	23,400	8800	—	2.70
Petrochemical	—	—	—	—	2.70
Nylon polymer	—	112,600	44,000	—	2.50
Olefin processing	—	321	133	—	2.40
Butadiene processing	—	359	156	—	2.30
Chemical	—	350,000	160,000	—	2.19
Synthetic Rubber	—	192	110	—	1.75

[a] High concentration of sulfides and thiosulfates.
[From Eckenfelder, W. W., Jr. (1989). "Industrial Water Pollution Control." McGraw-Hill, New York.]

nature of pollutants in refinery effluents are shown in Table VIII. Because of its composition, the wastewaters from the petroleum industry have a detrimental effect on soil and natural water reservoirs. Microbial treatment of wastewaters from refinery industry is rather difficult as compared to municipal wastewaters, as refinery wastewaters contain organics that are difficult to degrade microbiologically (e.g., phenols).

2. Pulp and Paper Industry Waste

The pulp and paper industry, representing the fifth largest industry in the United States and the largest single industry in Canada, is expanding at a steady rate of approximately 5% per year (statistics from Canada). The industry is one of the largest water consumers as well as polluters, and the problem of reducing pollution from old pulp and paper mills and design of processes and equipment to combat pollution is one of the most vexing problems confronting this industry. Effluent volumes from the manufacture of pulp and paper products is presented in Table IX and untreated effluent loads in Table X. Waste liquors from pulp and paper industry and, in particular, sulfite waste liquors could be considered as a potential raw material rather than waste. They contain about 50% of the processed wood. By-product recovery from sulfite liquors is practiced in many instances, although not to such large scale that would eliminate the water pollution problem. [*See* Petroleum Microbiology.]

Primarily due to contents of pentoses and hexoses (20–30% of the solids), the waste sulfite liquor is a good fermentation substrate. Both aerobic and anaerobic systems are practiced. In aerobic systems, production of yeast for food is most important and *Candida utilis* and *Monilia murmanica* are most cultivated. The protein content of *Candida* yeast is as high as 47–55%. Under efficient conditions, an overall 45% yield of dried yeast can be expected

Table VI Estimate of the Components of Total (Dissolved and Suspended) Solids in Wastewater

Component	Dry weight (g/capita · day) Range	Typical
Water supply	10–18	14
Domestic wastes		
Feces (solids, 23%)	30–70	40
Ground food wastes	30–80	45
Sinks, baths, laundries, and other sources of domestic wash waters	60–100	80
Toilet (including paper)	15–30	20
Urine (solids, 3.7%)	40–70	50
Water softeners	[a]	[a]
Total for domestic wastewater, excluding water softeners	190–360	250
Industrial wastes	150–400	200[b]
Total domestic and industrial wastes	340–760	450
Nonpoint sources	9–40	18[c]
Storm water	18–40	27
Total for domestic, industrial, nonpoint, and storm water	360–840	860

[a] Variable.
[b] Varies with the type and size of facility.
[c] Varies with the season.
[From Metcalf & Eddy, Inc. (1991). "Wastewater Engineering: Treatment, Disposal, Reuse." McGraw-Hill, New York.]

based on the sugar feed yielding 27% by weight of the assimilated sugar as protein. Anaerobic fermentation comprises production of ethanol by *Saccharomyces cerevisiae* that converts the hexoses to alcohol and CO_2. Up to 95% of the fermentable sugars can be converted to ethanol during a residence time of 15–20 hr at 30–35°C. A pulp mill processing 500 tons/day may produce 10,000 U.S. gal/day (37.5 m^3/day) of 95% alcohol.

3. Food Industry Waste

Waste from food industries is generally characterized by a very high organic content. When released to water streams, it can support an uncontrolled microbial proliferation (e.g., eutrophication, algae growth, sludge deposits). The waste varies considerably in strength and quantity (BOD from 100 to 100,000 mg/liter). A small seasonal plant can create a pollution load equivalent to 15,000–25,000 people while larger plants in North America have a population equivalent to at least one-quarter of a million people. Suspended solids, almost completely absent from some waters, are found in others in concentrations as high as 120,000 mg/liter. The waste may be highly alkaline (pH = 11.0) or highly acidic (pH = 3.5). Mineral nutrients may be absent or present in excess of the optimum ratio for microbial growth, but generally the wastes represent a good nutritional medium for microorganisms, so that biological treatment for this type of waste is recommended and practiced. Wastewater parameters from the canning industry are shown in Table XI, from slaughterhouses in Table XII, from poultry plants in Table XIII, and from the dairy industry in Table XIV.

III. Microbial Treatment of Industrial Effluents (Secondary Treatment)

Primary treatment of industrial effluents implies removal of suspended solids or conditioning of wastewaters for discharge into either a receiving body of water or a secondary treatment facility. Primary treatment usually includes equalization, neutralization, sedimentation, oil separation, and flotation.

Table VII Average Waste Flows and Loadings from Petroleum Refineries for Old, Prevalent, and New Technology

Type of technology	Flow (liter/m^3)		BOD (g/liter)		Phenol (g/liter)		Sulfide (g/liter)
	Average	Range	Average	Range	Average	Range	Average
Old	8000	5400–11,800	1.50	1.15–1.70	0.11	0.11–0.13	0.038
Prevalent	3000	2500–5000	0.38	0.30–0.61	0.04	0.03–0.05	0.011
New	1500	650–1900	0.19	0.08–0.23	0.02	0.004–0.023	0.011

Table VIII Origin and Nature of Pollutants in Refinery Effluents

Process units	Pollutants
Handling of crude oil	Oil, sludge and oily emulsions, sulfur- and nitrogen-containing corrosion inhibitors, inorganic salts, suspended matter
Crude oil distillation	Hydrocarbons, coke, organic acids, inorganic salts, sodium chloride, phenols, sulfur, sour condensate (sulfides and ammonia)
Thermal cracking	Phenols, triphenols, nitrogen derivatives, cyanides, hydrogen sulfide, ammonia
Alkylation, polymerization, isomerization processes	Acid sludge, spent acid, phosphoric, sulfuric, hydrofluoric, and hydrochloric acids, oil, catalyst supports, aluminum or antimony chloride
Refining and reforming processes	Hydrogen sulfide, ammonium sulfide, gums, catalyst supports
Purification and extraction processes	Phenols, glycols, amines, acetonitrile, acids, spent caustic
Sweetening, stripping, filtration	Sulfur compounds (H_2S, mercaptans), nitrogen compounds, sulfonates, acids, alkalies, inorganic salts, copper chloride, suspended matter

A. Microorganisms Employed for Secondary Treatment

The removal of carbonaceous BOD, the coagulation of nonsettleable colloidal solids, and the stabilization of organic matter are accomplished biologically using a variety of microorganisms, principally bacteria. The chemoheterotrophic organisms are of primary importance because of their requirement for organic compound as carbon and energy source.

Table IX Effluent Volumes from the Manufacture of Pulp and Paper Products

Process	m^3/1000 kg of product
Pulp manufacture	
Kraft and soda pulps	60–150
Sulfite pulp	170–250
Semichemical pulp	125–170
Groundwood pulp	15–40
Deinked pulp	85–145
Pulp bleaching	
Kraft and soda pulp	60–250
Sulfite pulp	125–210
Neutral sulfite pulp	170–250
Paper manufacture	
White papers	85–170
Tissues	30–145
Kraft papers	8–40
Paperboard	8–60
Specialty papers	85–420

The principal groups of organisms found in surface water and wastewater are classified as eukaryotes, eubacteria, and archaebacteria. Classification and brief characterization of these microorganisms are shown in Table XV.

Microorganisms convert the organic matter into various gases and into cell tissue. The cell tissue is organic itself and must be removed from wastewaters (see Section IV). Typical composition of bacterial cells is reported in Table XVI. An approximate formula for the organic fraction is $C_5H_7O_2N$. The formulation $C_{60}H_{87}O_{23}N_{12}P$ can be used when phosphorus is also considered.

All elements that are found in microorganisms should be present in wastewaters. Municipal wastewaters typically contain adequate amounts of nutrients (both inorganic and organic) to support biological treatment for the removal of carbonaceous BOD. In industrial wastewaters, nutrients may not be present in sufficient quantities and an addition is necessary for proper growth of bacteria and subsequent degradation of organic waste. An example of such a requirement is when pulp and paper waste liquors, which generally do not contain nitrogen and phosphorus, are treated. [*See* WASTEWATER TREATMENT, MUNICIPAL.]

Organisms that generate energy by enzyme-mediated electron transport from an electron donor to an external electron acceptor are said to have a respiratory metabolism. When molecular oxygen is

Table X Untreated Effluent Loads from Pulp and Paper Manufacture

Effluent	kg/1000 kg of product	
	Suspended solids (range of design values[a])	5-day BOD (range of design values[a])
Pulps		
Unbleached sulfite	10–20	200–350
Bleached sulfite	12–30	220–400
Unbleached kraft and soda	10–15	12–25
Bleached kraft and soda	12–27	22–40
Unbleached groundwood	15–40	8–12
Bleached groundwood	22–42	12–30
Neutral sulfite semichemical	40–90	125–250
Textile fiber	150–250	100–150
Straw	200–250	200–250
Deinked	200–400	30–80
Fine papers		
Bond-mimco	25–50	7–20
Glassine	5–8	7–12
Book or publication papers	25–50	10–25
Tissue papers	15–50	10–15
Coarse papers		
Boxboard	25–35	10–20
Corrugating brand	25–35	12–30
Kraft wrapping	8–12	2–8
Newsprint	10–30	5–10
Insulating board	25–50	75–125
Specialty papers		
Asbestos	150–200	10–20
Roofing felt	25–50	20–30
Cigarette papers	50–400	10–15

[a] Design value depends on yield.

Table XI BOD and Suspended Solids of Cannery Waste

Product	5-day BOD (mg/liter)	Suspended solids (mg/liter)
Apples	1680–5530	300–600
Apricots	200–1020	200–400
Cherries	700–2100	200–600
Cranberries	500–2250	100–250
Peaches	1200–2800	450–750
Pineapples	26	—
Asparagus	16–100	30–180
Beans, baked	925–1440	225
Beans, green wax	160–600	60–150
Beans, kidney	1030–2500	140
Beans, lima, dried	1740–2880	160–600
Beans, lima, fresh	190–450	420
Beets	1580–7600	740–2220
Carrots	520–3030	1830
Corn, cream style	620–2900	300–675
Corn, whole kernel	1120–6300	300–4000
Mushrooms	76–850	50–240
Peas	380–4700	270–400
Potatoes, sweet	1500–5600	400–2500
Potatoes, white	200–2900	990–1180
Pumpkin	1500–6880	785–1960
Sauerkraut	1400–6300	60–630
Spinach	280–730	90–580
Squash	4000–11,000	3000
Tomatoes	180–4000	140–2000

used as the electron acceptor, the process is known as aerobic respiration. Oxidized inorganic compounds such as nitrate and nitrite can function as electron acceptors for some respiratory organisms in absence of molecular oxygen. Processes that use these organisms are often referred to as anoxic. Organisms that depend on the molecular oxygen are called obligate aerobes. Fermentative metabolism does not involve the participation of an external electron acceptor. Organisms that generate energy by fermentation and that can exist only in an environment that is devoid of oxygen are obligate anaerobes.

Facultative anaerobes can grow in either the presence or absence of molecular oxygen.

B. Aerobic Treatment of Industrial Effluents

1. Theory of Aeration

The solubility of oxygen in liquids follows Henry's law. Oxygen saturation for distilled water at standard pressure is 14.6 mg/liter at 0°C and 7.6 mg/liter

Table XII Approximate Range of Flows and Analyses for Slaughterhouses, Packinghouses, and Processing Plants

Operation	Waste flow (m^3/1000 mg live weight slaughtered)	Typical analysis (mg/liter)		
		BOD	Suspended solids	Grease
Slaughterhouse	4–17	2200–650	3000–930	1000–200
Packinghouse	6–30	3000–400	2000–230	1000–200
Processing plant	8–33	800–200	800–200	300–100

	Approximate waste loadings (kg/1000 kg live weight slaughtered)		
	BOD	Suspended solids	Grease
Slaughterhouse	9.2–10.8	12.5–15.4	4.2–3.3
Packinghouse	18.7–11.7	12.5– 0.7	6.3–5.8
Processing plant	6.7	6.7	2.5–3.3

at 30°C. Presence of contamination decreases the saturation level. Oxygen demand for industrial effluents ranges several thousand milligrams of oxygen per liter of effluent (refer to Section II). Such an amount of oxygen must be provided from gas to the liquid phase by aeration systems. An adequate rate of oxygen transfer from gas to liquid is essential hence proper process equipment design is required for good performance of the process. Rate of oxygen transfer can be calculated from the following equation:

$$dC/dt = K_{L}a(C_{WS} - C),$$

where dC/dt = mass of oxygen per unit time per unit volume, $K_{L}a$ = overall oxygen-transfer coefficient (1/time unit), and $(C_{WS} - C)$ = the difference between dissolved oxygen saturation in liquid and actual oxygen concentration in the bulk of the liquid phase (mass of oxygen/volume unit).

For a continuous aerobic treatment process, which is designed for removal of carbonaceous organic matter, the steady-state operating dissolved oxygen level range is between 0.5 and 1.5 mg/liter.

During biological oxidation, several mineral elements are essential for the metabolism of organic matter. The nitrogen requirement is about 4.3 kg N/100 kg $BOD_{removed}$ and the phosphorus requirement is 0.6 kg P/100 kg $BOD_{removed}$. The requirement for other trace nutrients is specified in Table XVII. Usually a concentration of trace elements in a

Table XIII Composition of Poultry Plant Waste

	Range
5-day BOD (mg/liter)	150–2400
COD (mg/liter)	200–3200
Suspended solids (mg/liter)	100–1500
Dissolved solids (mg/liter)	200–2000
Volatile solids (mg/liter)	250–2700
Total solids (mg/liter)	350–3200
Suspended solids (% of total solids)	20–50
Volatile solids (% of total solids)	65–85
Settleable solids (mg/liter)	1–20
Total alkalinity (mg/liter)	40–350
Total nitrogen (mg/liter)	15–300
pH	6.5–9.0

Table XIV Dairy Wastewaters

Product	BOD (kg/100 kg)	Volume (liter/100 kg)
Creamery butter	0.34–1.68	3410–11,300
Cheese	0.45–3.0	10,780–19,300
Condensed and evaporated milk	0.37–0.62	2590–3500
Ice cream[a]	0.15–0.73	5180–1000
Milk	0.05–0.26	1670–4180

[a] Per 375 liters of product.

Table XV Classification of Microorganisms

Group members	Cell structure	Characterization	Representative
Eukaryotes	Eukaryotic[a]	Multicellular with extensive differentiation of cells and tissue	Plants (seed plant ferns, mosses)
		Unicellular or coenocytic, or mycelial; little or no tissue, differentiation	Protists (algae, fungi, protozoa)
Eubacteria	Prokaryotic[b]	Cell chemistry similar to eukaryotes	Most bacteria
Archaebacteria	Prokaryotic[b]	Distinctive cell chemistry	Methanogens, halophiles, thermacidophiles

[a] Contain true nucleus.
[b] Contain no nuclear membrane.
[From Metcalf & Eddy, Inc. (1991). "Wastewater Engineering: Treatment, Disposal, Reuse." McGraw-Hill, New York.]

carrier water is sufficient for microbial growth at aerobic conditions.

2. Kinetics of Aerobic Treatment

The removal of a single (i.e., simple carbonaceous BOD) component typically follows zero-order kinetics to a low residual level, which means that the rate of removal per unit of biomass is more or less constant to a limiting carbon concentration below which the rate becomes concentration-dependent and decreases. The rate constant, K (unit 1/day), may be considered as the fraction of BOD, COD, or TOC removed per day of retention. Table XVIII lists values of K (BOD) for various wastewaters.

In a mixture of substances being removed at different rates, the kinetics can be formulated in accordance with the Monod relationship:

$$(-1/X_v)(dS/dt) = (\mu_m/Y)(S/K_s + S),$$

where S = substrate concentration, μ_m = maximum specific growth of microorganisms, X_v = volatile suspended solids (proportional to the cell mass), K_s = Monod's constant, which is defined as the substrate concentration when the rate is one-half the maximum rate, and Y = biomass yield coefficient.

Table XVI Typical Composition of Bacterial Cells

Element	Percentage of dry mass: Range	Percentage of dry mass: Typical
Carbon	45–55	50
Oxygen	16–22	20
Nitrogen	12–16	14
Hydrogen	7–10	8
Phosphorus	2–5	3
Sulfur	0.8–1.5	1
Potassium	0.8–1.5	1
Sodium	0.5–2.0	1
Calcium	0.4–0.7	0.5
Magnesium	0.4–0.7	0.5
Chlorine	0.4–0.7	0.5
Iron	0.1–0.4	0.2
All others	0.2–0.5	0.3

[From Metcalf & Eddy, Inc. (1991). "Wastewater Engineering: Treatment, Disposal, Reuse." McGraw-Hill, New York.]

Table XVII Trace Nutrient Requirements for Biological Oxidation

	mg/mg BOD
Mn	10×10^{-5}
Cu	14.6×10^{-5}
Zn	16×10^{-5}
Mo	43×10^{-5}
Se	14×10^{-10}
Mg	30×10^{-4}
Co	13×10^{-5}
Ca	62×10^{-4}
Na	5×10^{-5}
K	45×10^{-4}
Fe	12×10^{-3}
CO_3	27×10^{-4}

[From Eckenfelder, W. W., Jr. (1989). "Industrial Water Pollution Control." McGraw-Hill, New York.]

Table XVIII Reaction Rate Coefficients for Organic Wastewaters

Wastewater	K (day^{-1})	Temperature (°C)
Potato processing	36.0	20
Peptone	4.03	22
Sulfite paper mill	5.0	18
Vinyl acetate monomer	5.3	20
Polyester fiber	14.0	21
Formaldehyde, propanol, methanol	19.0	20
Cellulose acetate	2.6	20
ZAO dyes, epoxy, optical brighteners	2.2	18
Petroleum refinery	9.1	20
Vegetable tannery	1.2	20
Organic phosphates	5.0	21
High nitrogen organics	22.2	22
Organic intermediates	20.6	26
	5.8	8
Viscose rayon and nylon	8.2	19
	6.7	11
Soluble fraction of domestic sewage	8.0	20

[From Eckenfelder, W. W., Jr. (1989). "Industrial Water Pollution Control." McGraw-Hill, New York.]

Table XIX Relative Biodegradability of Certain Organic Compounds

Biodegradable Organic Compounds[a]	Compounds generally resistant to biological degradation
Acrylic acid	Ethers
Aliphatic acids	Ethylene chlorohydrin
Aliphatic alcohols (normal, iso, secondary)	Isoprene
Aliphatic aldehydes	Methyl vinyl ketone
Aliphatic esters	Morpholine
Alkyl benzene sulfonates with exception of propylene-based benzaldehyde	Oil
Aromatic amines	Polymeric compounds
Dichlorophenols	Polypropylene benzene sulfonates
Ethanolamines	Selected hydrocarbons
Glycols	Aliphatics
Ketones	Aromatics
Methacrylic acid	Alkyl-aryl groups
Methyl methacrylate	Tertiary aliphatic alcohols
Monochlorophenols	Tertiary aliphatic sulfonates
Nitriles	Trichlorophenols
Phenols	
Primary aliphatic amines	
Styrene	
Vinyl acetate	

[a] Some compounds can be degraded biologically only after extended periods of seed acclimation.

[From Eckenfelder, W. W., Jr. (1989). "Industrial Water Pollution Control." McGraw-Hill, New York.]

A constant mass of microbial cells is synthesized from a given weight of organic matter removed expressed as an oxygen demand. About 0.5 g VSS is produced per gram of BOD_5 from biodegradable organic compounds. Completely biodegradable are organic compounds from the food processing industry. Biodegradability of some organics from other industries is shown in Table XIX.

3. Performance of Aerobic Processes

a. Activated Sludge Process

The activated sludge process is performed in a tank, (i.e., reactor) supplied with liquid feed for microorganisms (wastewaters + nutrients required) and air. During aeration, intensive mixing must be provided. When mixing and aeration is stopped, microorganisms settle and can be separated from the clear liquid. Separation may be performed in the same reactor (then the process is performed in a batch mode) or in another tank, named a settler (then the process may be performed in a continuous mode and concentrated microorganisms, named sludge, may be recycled), dependent on reactor design and operation).

A continuous system for wastewater treatment is shown in Fig. 1. A definition of symbols from this figure is presented in Table XX.

Activated sludge processes are usually fast and highly effective but require a large amount of energy for mixing and aeration. A large amount of sludge (70–120 kg dry solids/1000 m^3) is produced, depending on the process and origin of the wastewaters. This sludge is generally difficult to filter and its disposal is cumbersome.

b. Aerated Lagoons

Aerated lagoons are basins having a depth varying from 4 to 12 ft (1.2 to 3.6 m) in which oxygenation of wastewaters is accomplished by aeration units.) The difference between aerated lagoons and the activated sludge system is that lagoons are flow-through

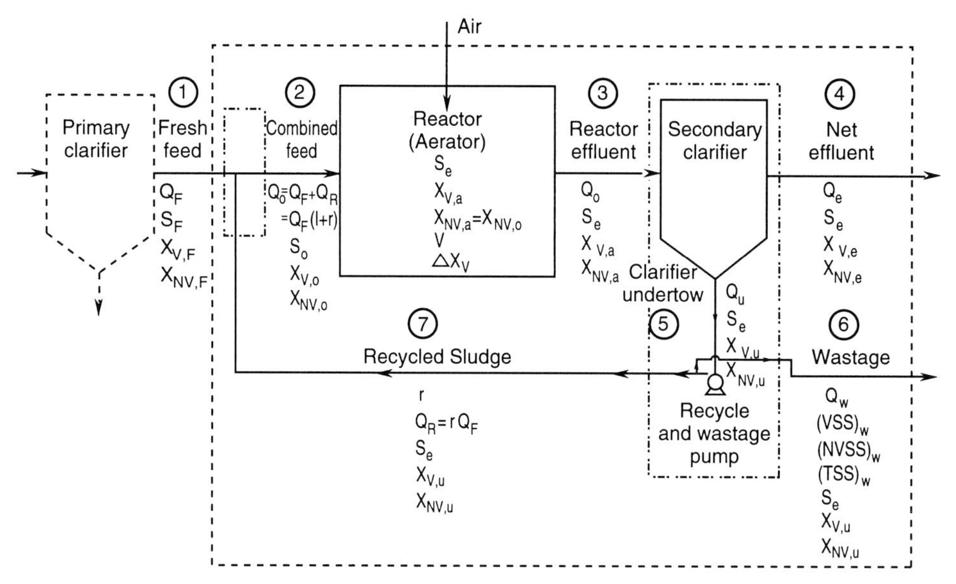

Figure 1 Conventional activated sludge process. See Table XX for a definition of symbols. [From Ramalho, R. S. (1983). "Introduction to Wastewater Treatment Process." Academic Press, New York.]

devices and no recycle of sludge is provided. Solid concentration in the lagoon is a function of wastewater characteristics and retention time. It is usually between 80 and 200 mg TSS/liter, i.e., much lower than that for conventional activated sludge units (2000–3000 mg TSS/liter). Because of this difference, lagoons require much more area for wastewater oxidation than do activated sludge processes. Performance of lagoon systems for wastewaters from several industries is presented in Table XXI.

C. Anaerobic Treatment of Industrial Effluents

1. Characteristics of Anaerobic Processes

Organic materials, in the absence of exogenous electron acceptors such as oxygen, nitrate, and sulfate, can be converted into methane and carbon dioxide through a complex series of microbial interaction. Anaerobic degradation of organic compounds is shown in Fig. 2. In this process, most of the chemical energy in the starting substrate is released as methane and may be recovered.

The aerobic conversion of 1 kg COD requires 2 kW of electricity (for mixing and oxygen supply) and produces 0.5 kg of biomass (dry weight). Anaerobically, 1 kg COD gives rise to 0.35 m^3 biogas (equivalent to about 0.4 liter of liquid fuel) and 0.1 kg of biomass, which can be dewatered if required.

Traditionally, anaerobic digestion was utilized almost exclusively for the stabilization of sewage sludge. The process received little application in the treatment of organic industrial wastes due to several limitations, including the low achievable rates of performance, the inability to withstand hydraulic and organic shockloads, and poor process control. These problems, all inherent to conventional digesters, were associated with difficulties in retaining biomass within the digester and with a very long retention time (up to 50 days). New reactor designs and new methods of bacterial bed preparation allowed this time to be shortened considerably down to a few hours.

Acetate is the most important compound quantitatively produced in the fermentation of organic substrate by the bacterial population with propionate production of secondary consequence. A number of microorganisms can digest soluble organic matter into acetic acid such as *Lactobacillus*, *Escherichia*, *Staphylococcus*, *Micrococcus*, *Bacillus*, *Pseudomonas*, *Desulfovibrio*, *Selemonas*, *Veillonella*, *Sarcina*, *Streptococcus*, *Desulfobacter*, and *Desulfomonas*.

Methanogenic bacteria convert acetate and H_2/CO_2 into methane. These bacteria are very sensitive to oxygen and are obligate anaerobes. Some of

Table XX Definition of Symbols Used in Fig. 1

Key

For suspended solids, double subscripts are utilized (e.g., $X_{V,i}$, $X_{NV,i}$.

The first subscript (V or NV) designates volatile and nonvolatile suspended solids, respectively.

The second subscript (i) refers to the specific stream in question:

F, fresh feed (stream 1)

o, combined feed (stream 2)

a, reactor effluent (stream 3)

e, net effluent (stream 4)

u, underflow from secondary clarifier (stream 5)

Symbols

1. Flow rates

Q_F , fresh feed; m^3 (stream 1)

Q_R , recycle; m^3 (stream 7)

r, recycle ratio; dimensionless ($r = Q_R/Q_F$)

Q_o , combined feed; m^3; $Q_o = Q_F + Q_R = Q_F (1 + r)$ (stream 2)

(volume of combined feed = volume of reactor effluent, i.e., Q_o (stream 2) = Q_o (stream 3)

Q_e , net effluent; m^3 (stream 4)

Q_w , wastage; m^3 (stream 6) (notice that $Q_F = Q_e + Q_w$)

Q_u, clarifier underflow; m^3; $Q_u = Q_w + Q_R = Q_w + Q_F$ (stream 5)

2. Concentrations (mg/liter) of soluble BOD

S_F , soluble BOD of fresh feed

S_o , soluble BOD of combined feed

S_e , soluble BOD of effluent

3. Concentrations (mg/liter) of volatile suspended solids (VSS)

$X_{V,F}$, VSS in fresh feed

$X_{V,o}$, VSS in combined feed

$X_{V,a}$, VSS in reactor; this also is equal to concentration of VSS in reactor effluent (complete mix reactor at steady state)

$X_{V,u}$, VSS in secondary clarifier underflow

$X_{V,e}$, VSS in net effluent

4. Concentrations (mg/liter) of nonvolatile suspended solids (NVSS)

$X_{NV,F}$, NVSS in fresh feed

$X_{NV,o}$, NVSS in combined feed

$X_{NV,a}$, NVSS in reactor ($X_{NV,a} = X_{NV,o}$); this also equals concentration of NVSS in reactor effluent (complete mix reactor at steady state)

$X_{NV,u}$, NVSS in secondary clarifier underflow

$X_{NV,e}$, NVSS in net effluent

5. Wastage

$(VSS)_w$, kg/day of VSS in wastage

$(NVSS)_w$, kg/day of NVSS in wastage

$(TSS)_w$, kg/day of TSS in wastage

6. Reactor volume

V, reactor volume, m^3

7. Sludge production

X_V (kg/day)

[From Ramalho, R. S. (1983). "Introduction to Wastewater Treatment Process." Academic Press, New York.]

Table XXI Performance of Lagoon Systems: Summary of Average Data from Aerobic and Facultative Ponds

Industry	Area/1000 m^2	Depth (m)	Detention (days)	Loading [kg/(m^2 day)]	BOD removal (%)
Meat and poultry	5.3	3.0	0.9	0.0080	80
Canning	27.9	5.8	1.8	0.0157	98
Chemical	125.4	5.0	1.5	0.0175	87
Paper	340	5.0	1.5	0.0116	80
Petroleum	62.7	5.0	1.5	0.0031	76
Wine	28.3	1.5	0.5	0.0245	
Dairy	30.4	5.0	1.5	0.0024	95
Textile	12.5	4.0	1.2	0.0183	45
Sugar	80.9	1.5	0.5	0.0096	67
Rendering	8.9	4.2	1.3	0.0040	76
Hog feeding	2.4	3.0	0.9	0.0396	
Laundry	0.8	3.0	0.9	0.0058	
Miscellaneous	60.7	4.0	1.2	0.0062	95
Potato	10.2	5.0	1.5	0.0123	

[From Eckenfelder, W. W., Jr. (1989). "Industrial Water Pollution Control." McGraw-Hill, New York.]

the notable species that have been classified are *Methanobacterium formicicum, M. bryantic,* and *M. thermoautotrophicum; Methanobrevibacter ruminantium, M. arboriphilus,* and *M. Smithii; Methanococcus vannielli* and *M. voltae; Methanomicrobium mobile; Methanogenium cariaci* and *M. marinsnigri; Methanospirillum hungatei,* and *Methanosarcina barkei.* [*See* METHANOGENESIS.]

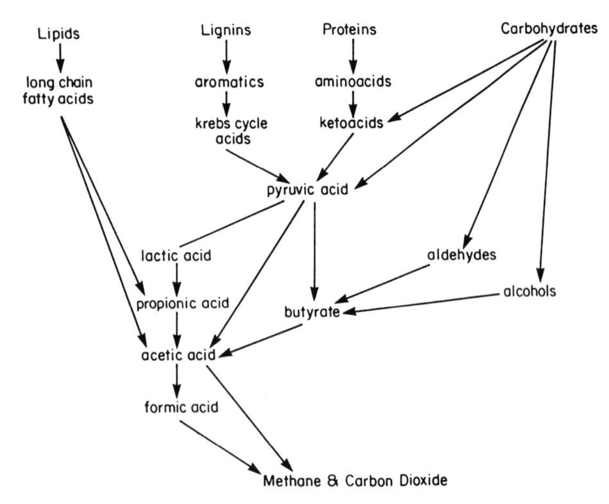

Figure 2 Biological sequences for breakdown of solids to methane and carbon dioxide. [From Kosaric, N., and Blaszczyk, R. (1990). *Adv. Biochem. Eng. Biotechnol.* **42,** 27–62.]

2. Anaerobic Reactors

a. Anaerobic Contact Reactor

In the anaerobic contact reactor system, anaerobic microorganisms are recycled and added to wastewaters and mixed. Because of relatively slow biomass accumulation during anaerobic digestion, microorganisms must be separated by settling, centrifugation, or flotation. Separation is difficult and takes a long time. A schematic of ACR is shown in Fig. 3.

b. Anaerobic Filter

In the anaerobic filter (AF) system, anaerobic microorganisms attach to a solid packing material

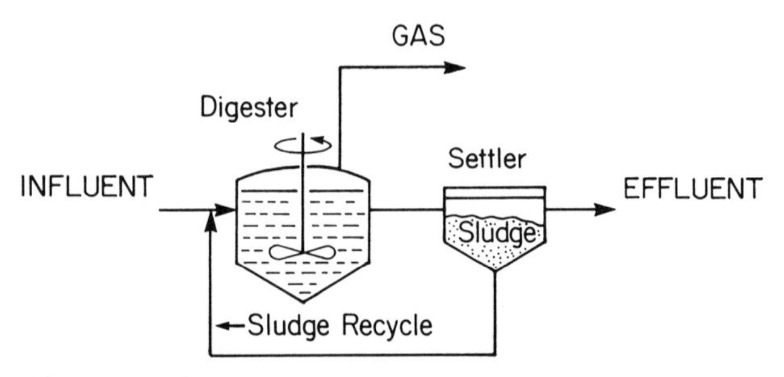

Figure 3 Schematic of an anaerobic contact reactor. [Figs. 3–6 from Kosaric, N., and Blaszczyk, R. (1990). *Adv. Biochem. Eng. Biotechnol.* **42,** 27–62.]

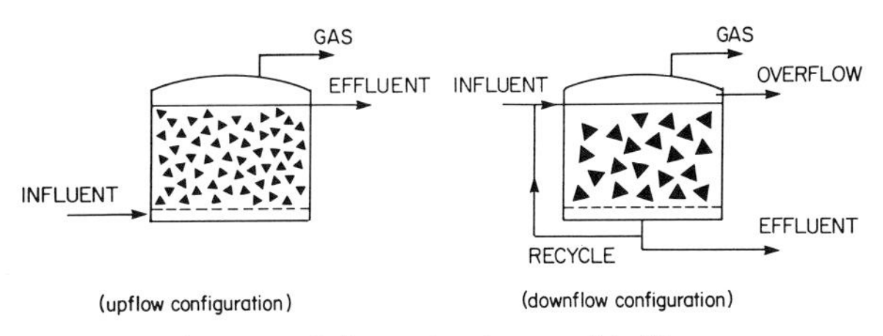

Figure 4 Schematic of anaerobic filters.

within the reactor and remain there for a long time (SRT > 100 d relative to HRT of 0.5–2 d). Process is very effective. The main limitation is plugging of the reactor by accumulated bacteria. Downflow and upflow configuration is recognized. A schematic of upflow and downflow AF is shown in Fig. 4.

c. Upflow Anaerobic Sludge Blanket Reactor

In the upflow anaerobic sludge blanket reactor (UASBR) system, anaerobic microorganisms attach to each other and develop spherical aggregates with diameters of 1–3 mm. No support for microorganisms is necessary. These aggregates settle very well. Their settling velocity is in the range of 20–90 m/hr, which can result in a low retention time of even a few hours. A schematic of a UASBR is shown in Fig. 5.

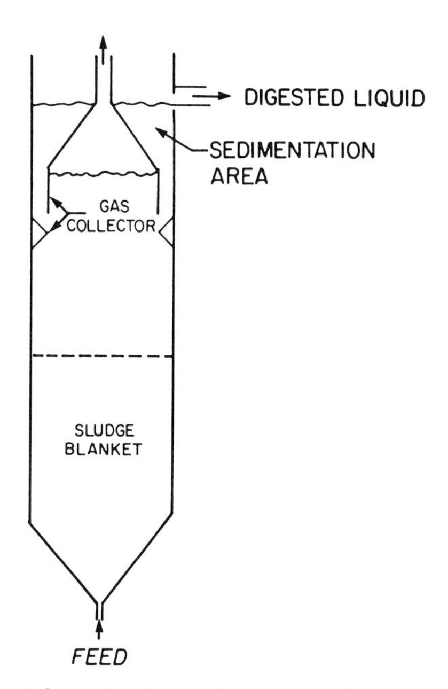

Figure 5 Schematic of an upflow anaerobic sludge blanket reactor.

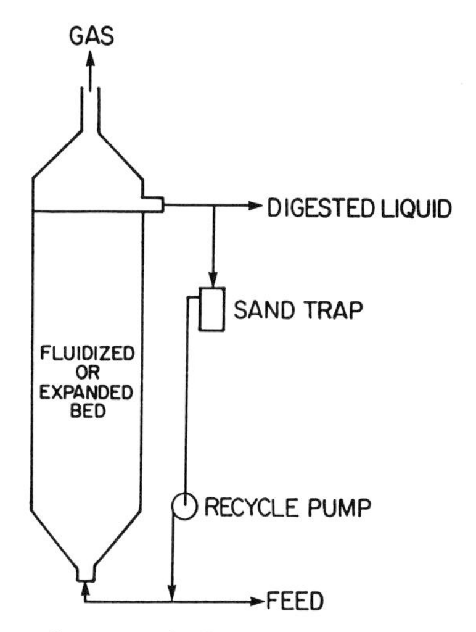

Figure 6 Schematic of a fluidized (expanded) bed reactor.

d. Fluidized or Expanded Bed Reactor

In a fluidized or expanded bed reactor, microorganisms attach to the small particles of the carrier, which may be unporous, such as sand, sepiolite, carbon, and different types of plastics, or porous, such as polyurethane foam and pumice. Settling velocity of particles covered with microorganisms is up to 50 m/hr. A proper superficial liquid rate expands the bed. When the bed expansion is up to 30%, the reactor is named the expanded bed reactor. When expansion is between 30 and 100%, the reactor is named the fluidized bed reactor. A schematic of the fluidized/expanded bed reactor is shown in Fig. 6.

Table XXII Comparison of Anaerobic Digestion Systems

Reactor type	Typical loading rate (kg COD/m Day)	Hydraulic retention time (Days)
CSTR	0.25–3.0	10–60
Contact	0.25–4.0	12–15
UASBR	10–30	0.5–7
CASBER	4.0–5.0	0.5–12
RBC	0.005–0.02	0.4–1
Anaerobic filter	1.0–40	0.2–3
AAFEB	1.0–50	0.2–5
AFB	1.0–100	0.2–5

CSTR, continuous stirrer tank reactor; UASBR, upflow anaerobic sludge blanket reactor; CASBER, carrier assisted sludge bed reactor; RBC, rotation biological contactor; AAFEB, anaerobic attached film expanded bed; AFB, anaerobic fluidized bed. [From Stronach S. M., *et al.* (1986). "Biotechnology Monograph." Springer-Verlag, Berlin.]

Table XXIII Performance of Anaerobic Processes

Wastewater	Process	Loading [kg/(m³ · day)]	HRT (hr)	Temperature (°C)	Removal (%)
Meat packing	Anaerobic contact	3.2 (BOD)	12	30	95
Meat packing		2.5 (BOD)	13.3	35	95
Keiring		0.085 (BOD)	62.4	30	59
Slaughter house		3.5 (BOD)	12.7	35	95.7
Citrus		3.4 (BOD)	32	34	87
Synthetic	Upflow filter	1.0 (COD)	—	25	90
Pharmaceutical		3.5 (COD)	48	35	98
Pharmaceutical		0.56 (COD)	36	35	80
Guar gum		7.4 (COD)	24	37	60
Rendering		2.0 (COD)	36	35	70
Landfill leachate		7.0 (COD)	—	25	89
Paper mill foul condensate		10–15 (COD)	24	35	77
Synthetic	Expanded bed	0.8–4.0 (COD)	0.33–6	10–3	80
Paper mill foul condensate		35–48 (COD)	8.4	35	88
Skimmed milk	USAB	71 (COD)	5.3	30	90
Sauerkraut		8–9 (COD)	—	—	90
Potato		25–45 (COD)	4	35	93
Sugar		22.5 (COD)	6	30	94
Champagne		15 (COD)	6.8	30	91
Sugar beet		10 (COD)	4	35	80
Brewery		95 (COD)	—	—	83
Potato		10 (COD)	—	—	90
Paper mill foul		4–5 (COD)	70	35	87

[From Eckenfelder, W. W., Jr. (1989). "Industrial Water Pollution Control." McGraw-Hill, New York.]

Table XXIV Performance of Anaerobic and Facultative Ponds

Industry	Area/1000 m^2	Depth (m)	Detention (days)	Loading (kg/m^2 day)	BOD removal (%)
	Summary of average data from anaerobic ponds				
Canning	10	1.8	15	0.0435	51
Meat and poultry	4	2.2	16	0.1401	80
Chemical	0.6	1.1	65	0.0060	89
Paper	287	1.8	18.4	0.0386	50
Textile	8.9	1.8	3.5	0.1593	44
Sugar	142	2.1	50	0.0267	61
Wine	15	1.2	8.8		
Rendering	4	1.8	245	0.0178	37
Leather	10.5	1.3	6.2	0.3336	68
Potato	40	1.2	3.9		
	Summary of average data from combined aerobic–anaerobic ponds				
Canning	22	1.5	22	0.0686	91
Meat and poultry	3.2	1.2	43	0.0296	94
Paper	10200	1.7	136	0.0031	94
Leather	18.6	1.2	152	0.0056	92
Miscellaneous industrial wastes	570	1.2	66	0.0142	

[From Eckenfelder, W. W., Jr. (1989). "Industrial Water Pollution Control." McGraw-Hill, New York.]

Ranges of process parameters for the preceding types of reactors are presented in Table XXII. Performance of the anaerobic process for selected industries is shown in Table XXIII.

e. Anaerobic Ponds

The biological process of anaerobic ponds is the same as that occurring in anaerobic digestion reactors but lasts much longer due to limited circulation and diffusion. A large process area is needed. Open ponds cannot be used near populated areas because of the unpleasant odor.

f. Facultative Ponds

The facultative pond is divided by loading stream and thermal stratification into an aerobic surface and an anaerobic bottom. The aerobic surface layer has a diurnal variation, increasing in oxygen content during the daylight hours due to algal photosynthesis and decreasing during the night. Sludge deposited on the bottom undergoes anaerobic decomposition.

Performance of anaerobic and facultative ponds for some industries is presented in Table XXIV.

IV. Sludge Treatment

The sludge separated during industrial effluent treatment contains most of the undesired compounds removed from wastewaters. A special treatment procedure must be involved before disposal of the remaining sludge. A list of possible procedures is presented in Table XXV.

Table XXV Sludge Processing and Disposal Methods

Processing or disposal function	Unit operation, unit process, or treatment method
Preliminary operations	Sludge pumping
	Sludge grinding
	Sludge blending and storage
	Sludge degritting
Thickening	Gravity thickening
	Flotation thickening
	Centrifugation
	Gravity belt thickening
	Rotary drum thickening
Stabilization	Lime stabilization
	Heat treatment
	Anaerobic digestion
	Aerobic digestion
	Composting
Conditioning	Chemical conditioning
	Heat treatment
Disinfection	Pasteurization
	Long-term storage
Dewatering	Vacuum filter
	Centrifuge
	Belt press filter
	Filter press
	Sludge drying beds
	Lagoons
Heat drying	Dryer variations
	Multiple effect evaporator
Thermal reduction	Multiple hearth incineration
	Fluidized bed incineration
	Co-incineration with solid wastes
	Wet air oxidation
	Vertical deep well reactor
Ultimate disposal	Land application
	Distribution and marketing
	Landfill
	Lagooning
	Chemical fixation

[From Metcalf & Eddy, Inc. (1991). "Wastewater Engineering: Treatment, Disposal, Reuse." McGraw-Hill, New York.]

A. Biological Sludge Treatment—Biogasification

Sludge separated after biological treatment of industrial effluent is mainly composed of bacterial cells, which contain a large part of initial BOD from the industrial effluents. The reduction of BOD in the sludge may be performed aerobically or anaerobically.

1. Aerobic Digestion of Sludge

In this process, a mixture of primary digestible sludge from primary treatment and activated sludge from aerobic biological treatment is aerated for an extended period of time. This results in cellular destruction with a decrease of volatile suspended solids and, consequently, reduction of the amount of sludge that is to be disposed. The reduction results from conversion by oxidation of a substantial part of the sludge into volatile products—CO_2, NH_3, and H_2. This oxidation occurs when the substrate in an aerobic system is insufficient for energy maintenance and synthesis (endogenous decay).

2. Anaerobic Digestion of Sludge

Anaerobic sludge digestion is among the oldest forms of biological wastewater treatment. It is performed as the following:

a. Standard-rate digestion in a single-stage process, in which the function of digestion, sludge thickening, and supernatant formation are carried out simultaneously.
b. Single-stage high-rate digestion, in which the sludge is mixed intimately by gas recirculation, mechanical mixers, pumping, or draft tube mixers. When a two-stage high-rate digestion process is employed, the second stage is used for storage and concentration of digested sludge and for the formation of relatively clear supernatant.

During anaerobic digestion of sludge, a gas (65–70% of CH_4, 25–30% of CO_2, and a small amount of N_2, H_2, and H_2S) is produced. Typical biogars values vary from 0.75 to 1.1 m^3/kg of volatile solids destroyed.

3. Composting

The composting process involves complex destruction of organic material coupled with the production of humic acid to produce a stabilized end-product. The microorganisms involved fall into three major categories: bacteria, actinomycetes, and fungi.

Approximately 20–30% of the volatile solids are converted to CO_2 and H_2O. During the process, the compost heats to temperatures in the pasteurization range and enteric pathogenic organisms are destroyed.

Composting may be accomplished under anaerobic and aerobic conditions. Aerobic composting accelerates material decomposition and results in a higher rise in the temperature necessary for pathogenic destruction and minimizes the potential for nuisance odors.

Bibliography

Eckenfelder, W. W., Jr. (1989). "Industrial Water Pollution Control." McGraw-Hill, New York.

Kosaric, N. and Blaszczyk, R. (1990). Microbial aggregates in anaerobic wastewater treatment. *In* "Advances in Biochemical Engineering/Biotechnology," vol. 42, pp. 27–62. Springer Verlag, Berlin Heidelberg.

Metcalf & Eddy, Inc. (1991). "Wastewater Engineering: Treatment, Disposal, Reuse." McGraw-Hill, New York.

Masters, G. M. (1991). "Introduction to Environmental Engineering and Science." Prentice-Hall, Englewood Cliffs, New Jersey.

Ramalho, R. S. (1983). "Introduction to Wastewater Treatment Process." Academic Press, New York.

Vriens, L., van Soest, H., and Verachtert, H. (1990). *Critical Rev. Biotech.*, **10,** 1–46.

Stronach, S. M., Rudd, T., and Lester, J. N. (1986). Anaerobic digestion process in industrial wastewater treatment. *In* "Biotechnology monographs," Vol. 2, (S. Aiba, L. T. Fan, A. Fiechter, and K. Schügerl, eds.) Springer Verlag, Berlin.

Infectious Waste Management

Gerald A. Denys
Methodist Hospital of Indiana

Glossary

Disposal Dilution and dispersal into air or into sewers or containment in a landfill

Guideline Recommendations issued by government agencies or professional organizations

Hazardous waste Material or substance that poses a significant threat to human health or environment, requiring special handling, processing, or disposal; hazardous waste often refers to chemical, radioactive, or mixed wastes

Infectious waste Material capable of producing an infectious disease; a subset of medical waste

Medical waste Material generated as a result of patient diagnosis, treatment, or immunization

Regulations Requirements developed by government at the federal, state, and local levels that are mandatory and enforceable by law

Sharps Hypodermic needles, syringes, disposable pipettes, capillary tubes, microscope slides, coverslips, and broken glass

Standard Performance of activities or quality of products established by professional organizations and have no force of law

Treatment Process that reduces or eliminates the hazardous properties or reduces the amount of waste

Universal precautions All blood and body fluids are considered potentially infectious, and appropriate procedures and barrier precautions should be used to prevent personnel exposure

SINCE 1988, the subject of infectious medical waste has gained considerable public and governmental attention. This has in part been due to the public alarm over identifiable medical waste washed up on our beaches and fears created over the acquired immune deficiency syndrome (AIDS) epidemic. Today's health care industry not only struggles with rising costs but also the tarnished image of polluting the environment. The adoption of universal precautions has resulted in an increase in the quantity of wastes to be managed in many institutions, with little change in the volume of truly infectious material. It has become important to educate employees to identify and segregate infectious waste. Cost of waste disposal continues to rise because of stricter regulations and shortages of available landfill space. Viable treatment and disposal options are also limited or cost-prohibitive.

I. Identification of Infectious Waste

A. Definitions

Medical waste refers to waste generated by a hospital, veterinary clinic, or laboratory, as the result of the diagnosis or treatment of a patient. Infectious waste is usually considered a subset of medical waste; however, there is no consensus on the exact definition. The definition of infectious waste and items included in these definitions varies according to federal, state, and local authorities. A number of synonyms for infectious waste have included biomedical, biological, microbiological, pathological, biohazardous, medical, hospital, and red bag waste. The Environmental Protection Agency (EPA) defines infectious waste as "waste capable of producing an infectious disease." Although the Centers for Disease Control (CDC) definition includes "microbiological waste (i.e., cultures, stocks), blood and blood products, pathological waste, and sharps." The landfill operator often regulates waste by his or her own criteria. If red bags or items resembling medical waste are found in a load of trash, even if previously treated, he or she will not accept it.

Identifying waste as infectious is difficult because there is no way to assess the microbial content of different types of waste. Decisions as to what is infectious must be made by determining the potential risk for infection or injury to individuals who segregate, handle, store, treat, or dispose of medical waste. Because infectious waste requires special treatment, definitions by regulatory agencies can have serious economic implications. Without a national standard, a range of 3% to 90% of hospital waste may be interpreted as infectious.

B. Sources

In the United States, 158 million tons of municipal solid waste are created each year. Medical waste from regulated generators represented only about 1%. Hospitals (86%) are considered the primary generator of regulated medical waste. Infectious waste is only a small subset of this total. Within a hospital, the primary generation points of infectious waste are the surgical and autopsy suites, isolation wards, laboratories, and dialysis units. The quantity of waste produced per patient is dependent on how infectious waste is classified. Other generators of infectious waste are varied and are included in Table I.

C. Types

Although federal agencies such as the CDC and EPA have published general guidelines for categories of infectious waste, some state and local regulators may be more explicit. The CDC designates five categories of waste that could be regulated as infectious medical waste (Table II). The EPA also recommends isolation waste from patients with highly communicable diseases should be managed as infectious in accordance with CDC guidelines.

Table I Generators of Infectious Waste

Health care and related facilities[a]
Academic and industrial research laboratories
Pharmaceutical industry
Veterinary hospital and offices
Funeral homes
Food, drug, and cosmetic industry

[a] Includes hospitals, outpatient clinics, ambulatory surgery centers, medical and diagnostic laboratories, blood centers, dialysis centers, nursing homes and hospices, physician and dental offices, diet or health care clinics, emergency medical health providers, and home health agencies.

Table II Categories of Waste Designated as Infectious by CDC and EPA

Category	CDC	EPA
Microbiological[a]	Yes	Yes
Blood and blood products[b]	Yes	Yes
Pathological[c]	Yes	Yes
Sharps[d]	Yes	Yes
Communicable disease isolation waste	No	Yes
Contaminated animal carcasses[e]	Yes	Yes
Surgery and autopsy waste[f]	No	Optional
Contaminated laboratory waste[g]	No	Optional
Hemodialysis waste[h]	No	Optional
Contaminated equipment	No	Optional

[a] Includes cultures and stocks of infectious agents.
[b] Includes all human blood, serum, plasma, blood products, and items saturated or soaked with blood.
[c] Includes tissues, organs, body parts, and body fluids.
[d] Includes contaminated and unused needles, syringes, scalpel blades, disposable pipettes, capillary tubes, microscope slides, coverslips, and broken glass.
[e] Includes body parts and bedding.
[f] Includes soiled dressing, sponges, drapes, and gloves.
[g] Includes specimen containers, gloves, and laboratory coats.
[h] Includes tubing, filters, sheets, and gloves.

All cultures and stocks of infectious agents from medical, research, and industrial laboratories should be managed as infectious waste. These waste types contain large numbers of microorganisms in high concentrations and presents a great risk to exposure if not treated. Discarded biologicals and waste from vaccine production by pharmaceutical companies for human and veterinary use should also be managed as infectious waste because pathogens may be present. Pathological waste includes tissues, organs, body parts, and body fluids removed during surgery or autopsy. Special handling of this category of waste is indicated because of the infectious potential and for aesthetic reasons. Human blood and blood products should be managed as infectious. Waste in this category includes all human blood, serum, plasma, and blood products. A major concern is the occupational risk of acquiring human immunodeficiency virus (HIV) and hepatitis B virus (HBV) associated with handling human source materials. Contaminated and unused sharps have been known to be an occupational hazard to all handlers of medical waste and require special containers to protect against injury and disease. Sharps include hypodermic needles, syringes, scalpel blades, disposable pipettes, capillary tubes, microscope slides, coverslips, and broken glass. Animal carcasses, body parts, and bedding exposed to an infectious agent and used for medical research are capable of transmitting an infectious disease. These waste types are similar to pathological and blood categories. Other categories of waste that are not designated as infectious but may be considered potentially infectious include wastes from surgery and autopsy that have come in contact with pathological waste (i.e., soiled dressings, sponges, drapes, and gloves). Contaminated laboratory waste (i.e., specimen containers, gloves, and lab coats), hemodialysis waste (i.e., tubing, filters, sheets, and gloves), and contaminated equipment and parts that may come in contact with an infectious agent also are included in this section. All these waste categories should be managed as infectious if grossly contaminated with blood or body fluids.

D. Reduction and Segregation

One problem many medical facilities experience is the high volume of noninfectious waste such as flowers, newspapers, paper towels, and cups that end up in the infectious waste red bags. As much as 50 to 60% of red bag waste is noninfectious material, which could be handled as general trash. A clear definition of infectious waste needs to be established to assure that the waste is segregated at the point of generation. Reducing the amount of waste requiring treatment can also reduce waste handling costs. Source segregation, however, requires educating employees, providing proper containers and frequent monitoring. Reduction in waste volume is also beneficial. Smaller quantities of waste generated will result in less handling and treatment at a cost savings. The increased use of disposable items however, has added to the volume of waste. Methods for reducing waste generation include use of reusable supplies, repackaging, and recycling materials. Waste treatment technologies such as incineration, mechanical/chemical, steam/compaction, and microwave/shredding are methods of reducing the volume of infectious solid waste.

II. Public Health Risks

The probability of developing an infection from medical waste is extremely low. Four requirements must take place for infection to occur. First, a sufficient number of viable pathogens must be present in the waste. Second, the pathogen must exhibit a virulence factor. Infection does not necessarily result in disease. Third, the pathogen must gain entry into the host. Transmission of microorganisms may occur by four routes: breaks in the skin by cuts, scrapes, or puncture wounds; contact onto mucous membranes by splashing of eyes, nose, or mouth; inhalation; and ingestion. The last requirement for infection is a susceptible host for the pathogen to infect. If any of these conditions are not met, disease transmission cannot occur and the risk of infection is eliminated. The only form of medical waste that has been associated with the transmission of an infectious disease is contaminated sharps.

In September 1990, the Agency for Toxic Substances and Disease Registry prepared a report for Congress on the public health risks of medical waste. The report estimated that less than one to four cases of AIDS and 80 to 160 hepatitis B cases per year may occur in health care workers as a result of contact with medical waste sharps. Medical waste-related HIV and HBV infections are a public health concern for selected occupations such as janitorial and laundry workers, nurses, emergency medical personnel, and refuse workers. The general public, however, is not likely to be adversely affected by medical waste from the traditional health care setting. Other findings indicate that the general public may come into contact with medical waste because of the increase in-home health care. The public may also be at risk of HIV and HBV infection from intravenous drug users' related waste. Studies have shown that household waste contains more potentially pathogenic microorganisms than medical waste. There is also no epidemiologic evidence that medical waste treated by chemical, physical, or biological means have caused disease in the community. In addition, untreated medical waste can be disposed of in properly operated sanitary landfills if workers follow proper procedures to prevent contact with waste during handling and disposal operations.

III. Treatment and Disposal Options

A. Traditional Methods

Incineration is the most common method of treating and disposing of medical waste. In this process, combustible material is transformed into noncombustible ash. Of the incinerators currently available, the controlled-air type is most widely used. The principle of a controlled-air incinerator involves two sequential combustion processes, which occur in two separate chambers (Fig. 1). Waste is fed into a primary chamber where the combustion process begins in an oxygen-starved atmosphere. The combustion temperature (1600–1800°F) is regulated by air to volatilize and oxidize the fixed carbon in the waste. The combustion process is completed in the upper chamber where excess air is introduced and the volatile gas–air mixture is burned at a higher temperature (1800+°F). For efficient decontamination and burning of combustible material, the proper temperature must be maintained for an appropriate period of time and mixed with a sufficient amount of oxygen. Although incineration appears to be the ideal way to destroy hazardous components of waste and reduce waste, it has its drawbacks. Extreme variations in waste content, or an improperly operated incinerator can result in the release of viable microorganisms to the environment via stack emissions, ash residue, or wastewater. Air emissions and ash disposal regulations have become more stringent, thus increasing the complexity and cost of owning and operating an incinerator. Community awareness and more stringent permitting and licensing requirements, especially in populous areas, have shut down or limited operations.

Approximately one third of hospitals in the United States treat their microbiological waste using a steam sterilizer or autoclave. An autoclave is an insulated pressure chamber in which saturated steam is used to obtain elevated temperatures. Air is removed from the chamber by either gravity displacement or use of a pre-vacuum cycle. A gravity displacement autoclave is most commonly used for medical waste (Fig. 2). Lighter steam is fed into the chamber to displace heavier air. Commercial autoclaves usually operate at a temperature of 132°C (270°F), pressurized to about 60 to 75 psi and at an operational cycle time of 1 hr. For steam sterilization to be effective, time, temperature, and direct steam contact with the infectious agents are critical. Factors that can influence treatment effectiveness include waste density, physical state and size, and organic content. Because waste composition is so variable, sterilization may not be achieved. Wastes that should not be autoclaved include antineoplastic agents, toxic chemicals, and radioisotopes, which may not be destroyed, or volatile chemicals, which could be vaporized and disseminated by heat.

The sanitary landfill and sanitary sewer have been used in the United States for solid waste and liquid wastes disposal, respectively. Studies have shown that pathogenic microorganisms are significantly reduced in a properly operated landfill. Most states, however, have restricted the use of landfills for disposal of regulated medical waste. Also landfills do

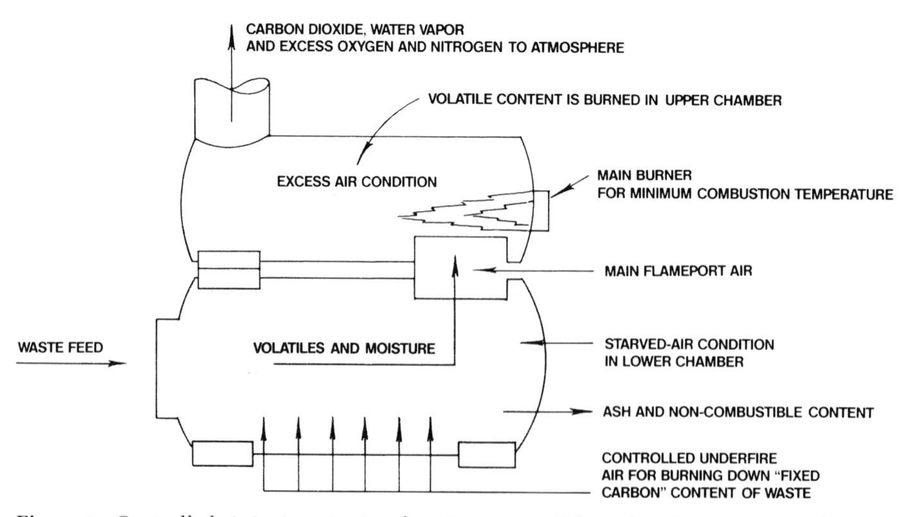

Figure 1 Controlled air incinerator involves two sequential combustion processes taking place in a dual chamber. Combustion process begins in an oxygen-starved atmosphere (lower chamber) and is completed in an air-excess air condition at a higher temperature (upper chamber). (Courtesy of Joy Energy Systems, Inc., Charlotte, North Carolina.)

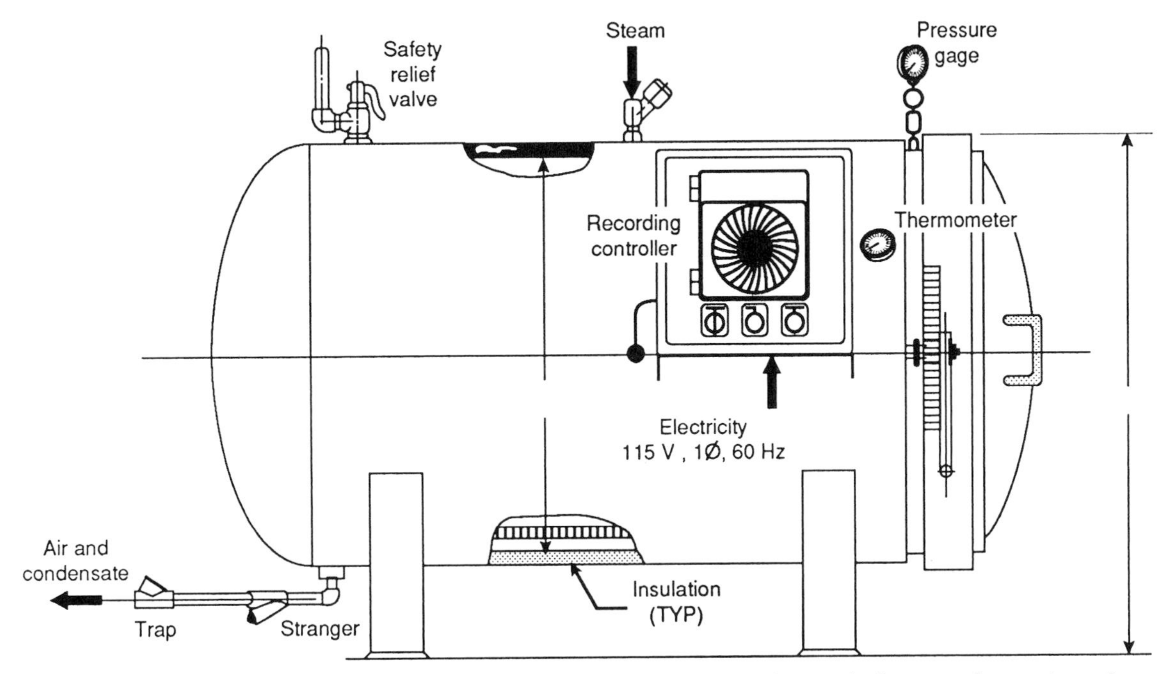

Figure 2 Steam sterilizer or autoclave operates as an insulated pressure chamber in which saturated steam is used to elevate temperature. (Courtesy of The Mark-Costello Company, Carson, California.)

not provide a long-term solution for medical waste disposal because it is estimated that one third of the remaining landfills will reach capacity within the next 5 years. However, landfills are currently used for the disposal of treated and reduced waste such as incinerator ash.

Many hospitals still pour blood and blood products down the drain. This type of waste stream comprises a small portion of sanitary sewer discharges and is diluted by large amounts of residential sewage. Secondary water treatment methods are very effective in lowering the microbial load of sewage. [*See* WASTEWATER TREATMENT, MUNICIPAL.]

Other methods described for decontaminating medical waste include chemical treatment (e.g., sodium hypochlorite and ethylene oxide gas) and radiation. These methods are less popular and considered an adjunct to previously described technologies. [*See* HAZARDOUS WASTE TREATMENT, MICROBIAL TECHNOLOGIES.]

B. Emerging Technologies

Several combined technologies for medical waste treatment and disposal have been developed. These methods have evolved to improve decontamination and reduce waste volume. Such treatment technologies include mechanical/chemical, steam/compaction or steam/shredding, and microwave/shredding. The efficacy of each method is dependent on the contact time, number of microorganisms in the waste to be treated, organic content, volume, and physical state of the waste.

Mechanical/chemical treatment uses a high-speed hammermill to pulverize and mix the waste with a disinfectant solution (Fig. 3). Infectious wastes in bags or containers are placed on an enclosed conveyer, transported to a feed hopper, and sprayed with a bleach solution. The waste and disinfectant pass through a preshredder and then into an ultrahigh speed hammermill. The solid material is captured in a rotary separator, and liquids are either discharged to the sanitary sewer or recirculated through the system. Solids continue to travel up a screw auger conveyer to be deposited in a waste cart. The entire process is maintained under negative pressure, and air is discharged through HEPA filters. Unrecognizable solid material is reduced up to 85% and considered general refuse.

Steam/compaction waste handling systems employ an autoclave combined with a compactor (Fig. 4). Infectious waste is first placed into the sterilizer chamber and steam-treated. After loading, the system is self-operating. After a 55-min cycle time, the chamber liner begins moving treated waste out of

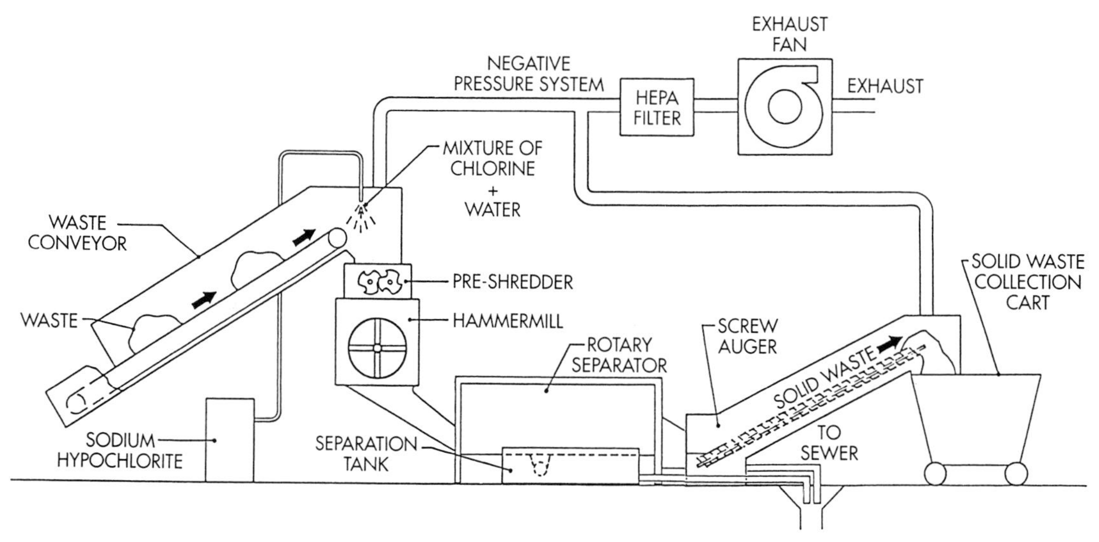

Figure 3 Mechanical/chemical treatment process. Medical Safe TEC. Z-12,500–Theory of Operation. Infectious waste is shredded, pulverized, and decontaminated with sodium hypochlorite solution. (Courtesy of Medical SafeTEC, Inc., Indianapolis, Indiana)

the sterilizer chamber. Treated waste is then dropped into the compaction chamber. A piston compacts the waste into a roll-off container. The compacted waste is reduced 50% but not physically destroyed. Waste is then sent to a landfill or municipal incinerator for final disposal. This system is also designed for the compaction and disposal of general refuse.

Microwave/shredding treatment is a process that shreds wastes and then heats it by microwaving (Fig. 5). Infectious waste is automatically fed into a hopper for shredding and sprayed with steam. The

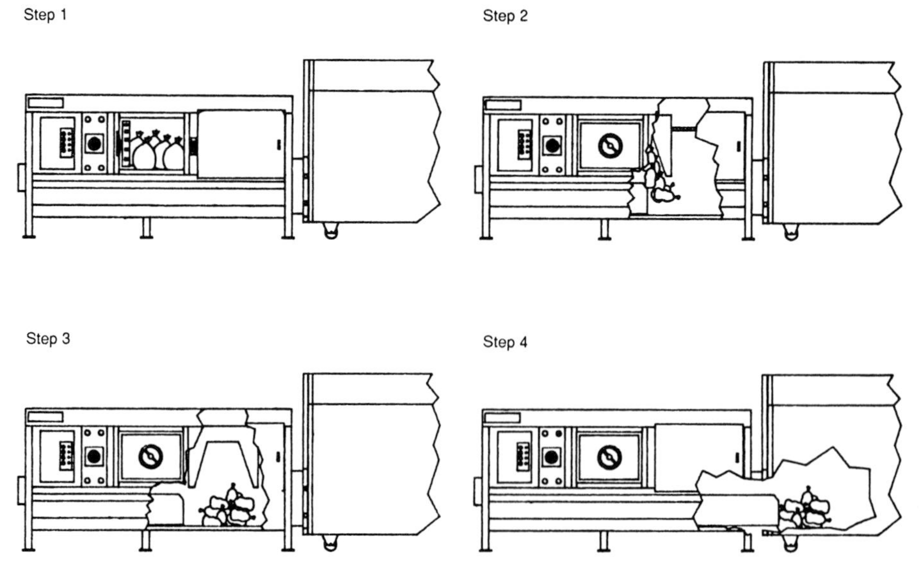

Figure 4 Order of operation for infectious waste in the steam/compaction treatment process. Infectious waste is steam-treated in a sterilizer chamber and then compacted before final disposal. STEP 1, Infectious waste is placed into the sterilizer chamber, door is closed and "start" button depressed; STEP 2, Upon completion of the sterilization cycle, the liner moves the waste out of the sterilizer chamber; STEP 3, the sterilized waste is discharged into the compaction chamber; STEP 4, The waste is then compacted into the roll-off container. (Courtesy of San-I-Pak, Inc., Tracy, California.)

waste moves up a screw conveyer where it is heated at 203°F for 30 min with multiple microwaves. The waste then enters a temperature holding section before discharge into a waste container. Air is discharged through HEPA filters. To ensure proper processing, the waste should contain less than 10% liquid by weight and less than 1% metallic content. The volume reduction of waste is up to 85% before final disposal into a landfill or municipal incinerator.

Other approaches to the treatment of infectious waste are currently being developed and evaluated for both large and small waste generators. These technologies include chemical, irradiation, mechanical, and thermal processes.

C. Selection Considerations

A number of factors should be considered when selecting a treatment method. No single technology is applicable for all types of infectious waste and for all institutions. More than one treatment method may be used for different infectious waste streams. The advantages and disadvantages of onsite versus offsite treatment options should also be considered. Onsite treatment allows one to control costs and reduces the potential for liability of the waste leaving a facility. Sharing a treatment facility may be a attractive alternative. Federal, state, and local regulations may have impact on the treatment selection. Cost considerations are also very important when selecting an infectious waste method. A comparison of infectious waste treatment methods and factors for selection are presented in Table III. Once a treatment method has been selected, it is important to establish a quality assurance program to ensure that it is functioning. Important elements of a quality assurance plan should include indicator tests of equipment operations and treatment effectiveness (i.e., biological indicator tests of stream treatment, tests of incinerator stack gas, and HEPA filter tests of mechanical treatment), criteria to measure effec-

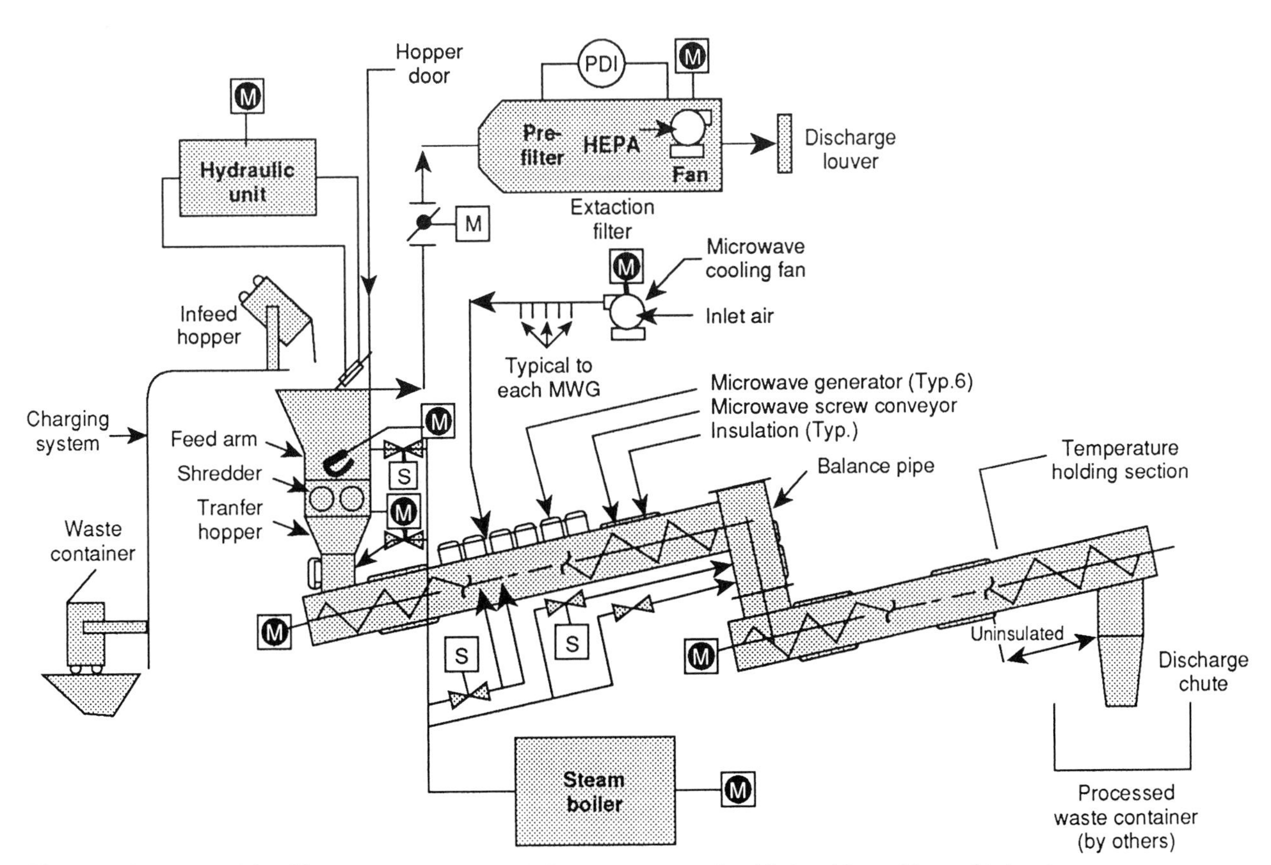

Figure 5 Microwave/shredding treatment process. Infectious waste is shredded and heated by multiple microwaves before final disposal. (Courtesy of ABB Sanitec, Inc., Roseland, New Jersey.)

Table III Comparison of Infectious Waste Treatment Methods

Factor	Type of treatment technology					
	Steam sterilization	Incineration	Mechanical/chemical	Steam/compaction	Microwave/shredding	Comments
Operations						
Applicability	Most infectious wastes*	Almost all infectious wastes	Most infectious wastes*	Most infectious wastes*	Some infectious wastes*, **	* Not chemical/radioactive/pathological ** Not approved for syringes, blood, and body fluids
Equipment operation	Easy	Complex	Moderately complex	Easy	Moderately complex	
Operator requirements	Trained	Highly skilled	Trained	Trained	Trained	
Need for waste separation	To eliminate nontreatable wastes	None	To eliminate nontreatables; for proper feeding	Same for steam sterilization	To eliminate nontreatables, syringes, blood, and body fluids	
Need for load standardization	Yes	Yes*	Yes	Yes	Yes	* Wet waste such as dialysis fluid
Effect of treatment	Appearance of waste unchanged	Waste burned	Waste shredded and ground	Minimum change in appearance	Waste shredded only	
Volume reduction	30%	85–95%	Up to 85%	50%	Up to 85%	Depended on waste type
Occupational hazards	Low	Moderate	Moderate	Low	Low	
QA/QC indicator	*Bacillus stearothermophilus*	*Bacillus subtilis*	FAC*	*B. stearothermophilus*	*B. subtilis*	*Free available chlorine
QA/QC validation	Temperature chart	Temperature chart	Calibration chart	Temperature chart	Temperature chart	
Testing	Easy, inexpensive	Complex, expensive	Moderately, inexpensive	Easy, inexpensive	Easy, inexpensive	

Potential side benefits	None	Energy recovery	None	None	None	
Onsite/offsite location	Both	Both	Both	Both	Both	
Regulatory requirements						
Medical waste tracking regulations	Applicable	Recordkeeping	Not applicable*	Applicable	Not applicable*	* Physical destruction
Applicable environmental regulations	Wastewater	Air emissions, ash disposal, wastewater	Wastewater	Wastewater	None	
Releases to air	Low risk via vent	High risk via emissions	Low risk via HEPA filter	Low risk via vent	Low risk via HEPA filter	
Releases to water	Low risk via drain	Low risk via scrubber water	Moderate risk via wastewater*	Low risk via drain	Low risk (evaporation)	* Low risk (recirculation)
Disposal of residue	To sanitary landfill; potential problem with red bags	Ash may be a hazardous waste; if so, to RCRA-permitted landfill	Liquid: Effluent to sanitary sewer; Solid: Residue to sanitary landfill	Same as for steam sterilization	Residue to sanitary landfill	
Permitting requirements	None	For siting, air emissions	None	None	None	
Costs						
Capital costs	Low	High	Moderate	Low	High	
Labor costs	Low	High	Moderate	Low	Low	
Operating costs	Low	High	Moderate	Low	Moderate	
Maintenance costs	Low	High	Moderate	Low	Moderate	
Downtime	Low	High	Low to moderate	Low	Moderate to high	

[Adapted from Reinhardt, P. A., and Gordon, J. G. (1991). "Infectious and Medical Waste Management." Lewis Publishers, Chelsea, Michigan.]

tiveness (i.e., operation and maintenance schedule), and records of data and review for compliance (i.e., temperature and calibration charts).

IV. Infectious Waste Management Plan

A. Regulations, Guidelines, and Standards

Regulations, guidelines, and standards can influence how infectious waste is managed. To date, the EPA has not issued any specific regulations governing the management of infectious waste. The EPA's only regulatory action under the Resource Conservation and Recovery Act (RCRA) was to implement a 2 year demonstration project to track medical waste from start to finish (Medical Waste Tracking Act of 1988). Findings from this project may result in permanent regulatory actions. Other federal regulations relevant to infectious waste management include rulings by OSHA on hazard communication/right-to-know and the prevention of occupational exposure to bloodborne pathogens. Employers are required to inform workers who handle infectious waste about the risks of exposure to these wastes. Although regulatory requirements may differ, many states regulate, or are about to regulate, the management of infectious waste. Local ordinances also affect infectious waste treatment and disposal methods.

Rather than issuing regulations, various federal, state, and professional organizations have published guidelines on infectious waste management. Guidelines have been published by federal agencies such as the EPA, CDC, OSHA, and the National Institute for Occupational Safety and Health (NIOSH). The National Committee for Clinical Laboratory Standards has also prepared guidelines for protecting laboratory workers from infectious diseases transmitted by blood and tissue.

Standards relevant to infectious waste management have also been established. The Joint Commission on Accreditation of Healthcare Organizations (JCAHO) which accredits hospitals has published Standard #PL.1.10 that pertains to hazardous materials and wastes. The American Society for Testing and Materials (ASTM) develops standards for testing the strength of plastic used to manufacture red bags and defining puncture resistance in sharps containers.

B. Components

A comprehensive infectious waste management plan should be established at each institution. Table IV lists the components which should be included in such a plan. Figure 6 illustrates the complex waste management planning and implementation process.

A waste management system is needed for the management of all wastes. It should be based on regulatory requirements and institutional policy. Written procedures and standard operating practices should be established and personnel responsible for implementing the system identified. Policy and procedures should be presented, training provided, and implementation required.

Waste identified as infectious should be discarded in designated types of containers or red bags to protect waste handlers and minimize waste handling. All sharp objects should be placed directly into impervious, rigid, and puncture-resistant containers. Glass and liquids should be placed in disposable cardboard containers and leak-proof boxes or containers with secured lids. Mixed waste containing infectious and radioactive, or infectious and toxic chemical waste must be segregated and directed to the appropriate treatment procedure based on the severity of the hazardous materials. Waste handling and collection procedures should be established to minimize the potential for exposure. Collection carts should be covered and disinfected after use. When waste is transferred off-site for treatment and disposal, plastic bags should be packed in rigid containers such as plastic barrels or heavy cartons. The waste should be transported in closed and leak-

Table IV Components of the Infectious Waste Management Plan

Waste management system
Identification of infectious waste
Waste discard
Waste handling and collection
Waste storage
Waste Treatment
Disposal of treated waste
Off-site transport for treatment
Contingency planning
Emergency planning
Training
Record keeping
Quality assurance–quality control
Incident–accident analysis
Review procedures and practices

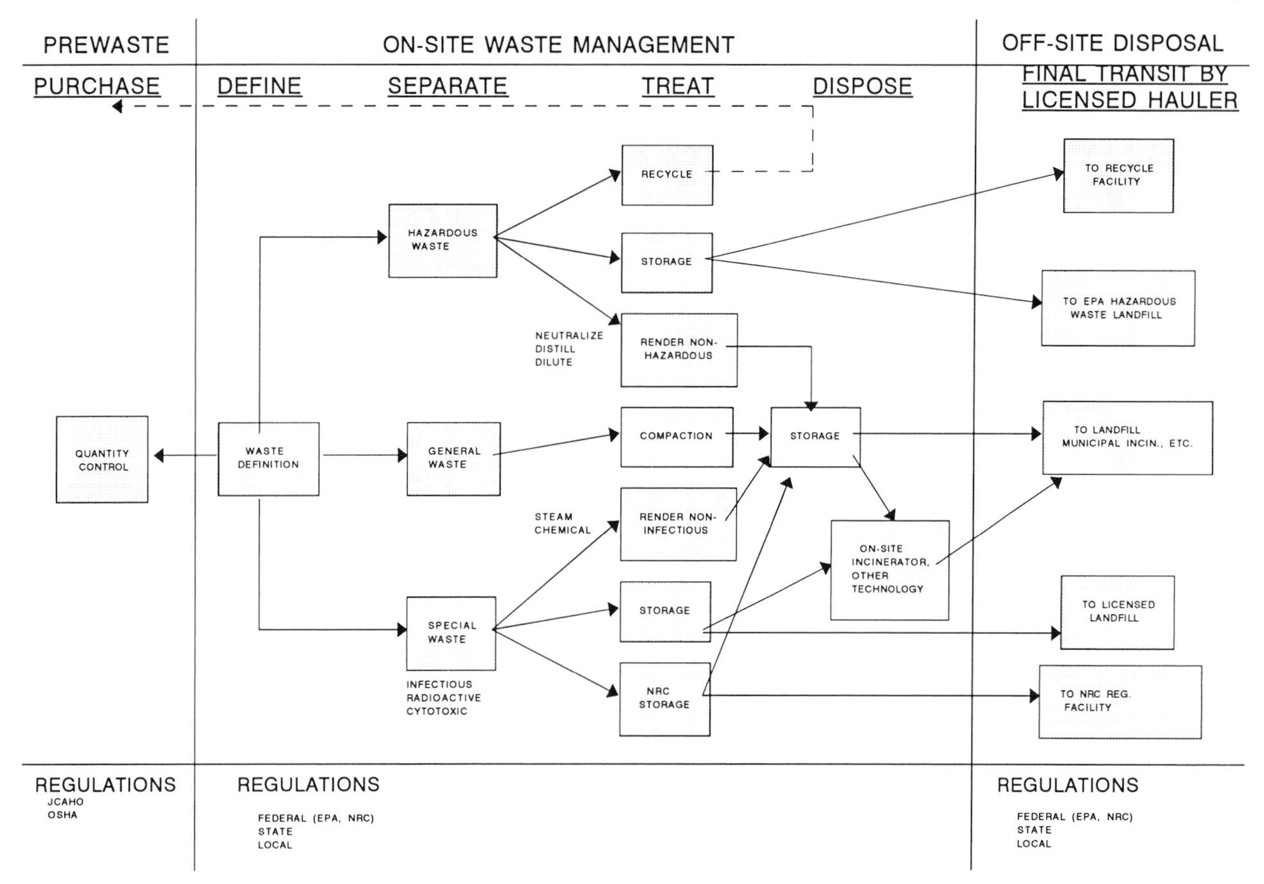

Figure 6 Waste management planning and implementation process.

proof dumpsters or trucks. All infectious waste packages and containers should be marked with the universal biohazard symbol. Treatment of infectious waste within the same day of collection may not be possible; therefore, plans for storage must be made. Factors to consider when storing infectious waste include the integrity of the packaging, storage temperature, storage time, and storage area. The storage area must be kept locked with limited access to authorized personnel, and be kept free from rodents and vermin. Various treatment technologies for infectious waste are available. Even after treatment, certain types of waste such as sharps and pathological waste may require additional treatment according to state and local requirements to render it unrecognizable. Special treatment considerations are needed for mixed infectious, radioactive, and toxic chemical wastes. After the waste is treated, solid residues are usually buried in a landfill or incinerated. Treated liquid wastes are discharged to the sanitary sewer system.

A contingency plan for the treatment of infectious waste should be established when equipment fails or personnel problems arise. Alternatives are waste storage, an exchange agreement with another institution, or off-site treatment by a licensed contractor. Emergency planning should be established in the event of an accidental spill or loss of containment. This plan should include clean-up procedures, use of personal protective equipment, and disposal of spill residue.

Training of all personnel who generate or handle infectious waste is necessary for implementation of the system. Training programs should be based on universal precautions to minimize occupational exposure to potentially infectious material within the health care setting. According to OSHA's hazardous communication/right-to-know regulations and bloodborne pathogens rule, education and training of personnel are required. Educational programs need to be developed and implemented when the infectious waste management plan is first developed, new employees are hired, and whenever infectious management practices are changed.

Recordkeeping is also essential to conform with regulatory requirements and document compliance. Quality assurance and quality control procedures are needed to ensure that the infectious waste management system is working. Examples of quality assurance activities may include spot checks for proper waste collection, transport and storage, documentation of treatment effectiveness, and analysis of accident and incident records. Review of accident or incident reports also help identify occupational issues and deficiencies in the waste management system or practices.

C. Implementation

Implementation is the most difficult part of any infectious waste management plan. Planing requires developing policy and procedures, training, and assigning responsibility. Policy should be determined by administrative operations personnel, generating departments, and service departments and should be based on the user group and legal considerations. Once a draft policy is determined to be operational, it should be presented to appropriate administrative authorities such as an infection control committee, safety committee, or nursing, laboratory, and hospital administration for input.

Development of effective educational material and training programs for all employees are also important. Small group education and training should be given to managers and supervisors with follow-up inservice given to all employees. Waste haulers and treatment personnel should also be educated on policy. The responsibility of infectious and medical waste management has traditionally been assigned to service personnel. Shared responsibility among many departments should be considered. This will require administrative support, coordination of assignments, and communication. Formation of an interdisciplinary waste management committee should also be considered to develop waste management strategies and long-range planning.

There are several advantages to implementing a comprehensive infectious waste management plan. It provides compliance with all regulatory requirements and meets standards required for certification. Other benefits includes minimizing hazard and occupational exposures, minimizing risks and liability, achieving quality assurance, and reducing costs.

Bibliography

Agency for Toxic Substances and Disease Registry. (1990). "The Public Health Implications of Medical Waste: A Report to Congress." U.S. Department of Health and Human Services (PB91-100271), Atlanta, Georgia.

Centers for Disease Control (1988). *Morbid. Mortal. Weekly Rep.* **37,** 377–388.

Centers for Disease Control. (1990). *Morbid. Mortal. Weekly Rep.* **39** (no. RR-1).

Garner, J. S., and Favero, M. S. (1985). "CDC Guidelines for the Prevention and Control of Nosocomial Infections. Guideline for Handwashing and Hospital Environmental Control." U.S. Department of Health and Human Services (PB85-923404), Atlanta, Georgia.

Meaney, J. G., and Cheremisihoff, P. N. (1989). *Pollution Engineering* (October), pp. 93–106.

Reinhardt, P. A., and Gordon, J. G. (1991). "Infectious and Medical Waste Management." Lewis Publishers, Chelsea, Michigan.

Rutala, W. A., Odette, R. L., and Samsa, G. P. (1989). *JAMA* **262,** 1635–1640.

Rutala, W. A., and Weber, D. J. (1991). *N. Engl. J. Med.* **325,** 578–582.

U.S. Department of Labor, Occupational Safety and Health Administration. (1989). *Fed. Regist.* **54,** 23042–23139.

U.S. Environmental Protection Agency. (1986). "Guide for Infectious Waste Management." Washington, D.C. (EPA/530-SW-86-014).

U.S. Environmental Protection Agency. (1989). *Fed. Regist.* **54,** 12326–12395.

Influenza

Edwin D. Kilbourne
Mount Sinai School of Medicine

Glossary

Epitope Single antigenic determinant that together with others defines an antigen
Pandemic Worldwide epidemic
Pleomorphic Of variable shape
Reassortant Virus of mixed genotype produced as the result of coinfection with two distinguishable parental viruses
Strain Isolate or variant of a virus subtype

INFLUENZA is an acute infection of the respiratory tract that occurs in explosive epidemics and is characterized by cough, headache, fever, myalgia, and prostration. It is caused by any of three antigenically distinct types of orthomyxovirus—A, B, or C.

I. History

Because survival of its causative viruses depends on a direct chain of person-to-person transmission and no latent phase of infection has been demonstrated, influenza as a human disease probably originated not earlier than the beginning of communal human existence. Any account of its origin must consider the present-day existence of influenza A viruses in migratory and domestic avian species, swine, and horses, some of which appear to contribute sporadically to the origin of novel viruses capable of producing pandemics in the human population (see Section IV.A). Acute epidemics of febrile incapacitating respiratory disease resembling the disease we now call influenza have been described for at least the past 500 years. The name "influenza" reflects early speculation about its cause and origins, as *influenza de freddo* (influence of the cold; Ital.).

Evaluation of the history of influenza is confounded by difficulties in distinguishing it retrospectively from other acute respiratory infections and by the absence of pathognomonic stigmata of the illness such as the paralysis or characteristic rash that assist in the credible reconstruction of poliomyelitis or smallpox epidemics. Although the record prior to 1889 must be interpreted with caution, it contains clear descriptions of acute wintertime epidemics of febrile respiratory disease affecting all ages. Modern serological studies of human antibody prevalence provide firm evidence indicating that viruses antigenically similar to viruses now circulating were involved in epidemics at the end of the last century. The laboratory isolation from humans of influenza A virus in 1933 (Smith, Andrews, and Laidlaw) and influenza B virus in 1940 (independently by Magill and Francis) has permitted the precise etiological definition of the annual epidemics of influenza that occur throughout the world. Global surveillance by the World Health Organization (WHO) has aided in the definition of the impact of influenza during the past half century. Influenza C virus, first isolated by Taylor in 1950 and associated with epidemic disease by Francis *et al.* in 1953, contributes less to human illness, and its virus has a somewhat ambiguous relationship to the influenza viruses A and B (see Section II.A).

The peculiar historical importance of influenza lies in its capacity to produce catastrophic pandemics at unpredictable intervals as a consequence of the antigenic mutability of its viruses. No other event in history has killed so many people in so short a time span as the notorious pandemic of 1918 at the close of the first world war. Within 10 mo in 1918–1919, the disease killed more Americans (500,000) than have died on the battlefields of all American wars from 1918 to 1963. Worldwide, at least 20,000,000 died, and the enormous morbidity disrupted daily life and community services. No virus was isolated at the time, but retrospective study of serum from persons exposed to infection in 1918 revealed antibody reactive with the swine influenza virus recovered from pigs by R. E. Shope in 1931. (Swine influenza appeared as a new disease in the American Midwest, also in 1918.)

Within the era of modern virology, two unequivocal pandemics have occurred, one in 1957 and the other in 1968 with the introduction of markedly changed influenza A viruses into the population. In 1946–1947, a pervasive but less virulent and less antigenically different viral mutant induced worldwide (i.e., pandemic) disease.

In 1976, an abortive outbreak of serially transmissible disease in military recruits in Fort Dix, New Jersey, resulted from swine influenza virus infection from an unknown source. This revisitation by a virus antigenically similar to the putative 1918 virus occasioned great concern and a controversial national vaccination campaign.

In 1977, yet another unexpected event occurred with the reappearance of a virus of the 1950s in the younger population not previously exposed to immunizing infection in that earlier decade. This H1N1 virus (see Table I) has continued to circulate in the human population coincidently with variants of the H3N2 virus introduced in the 1968 pandemic. Cocirculation of antigenically different subtypes of influenza A viruses had not previously been known to occur, one subtype having been replaced successively by a different subtype in the past. Since 1968 and 1977, progressive mutations of H1N1 and H3N2 viruses have continued, attended by regional epidemics, but new subtype introductions and pandemics have not occurred.

II. The Influenza Viruses

A. Taxonomy and Comparative Virology

The viruses causing influenza have been arbitrarily and alphabetically named in the chronological order of their isolation as influenza A, B, and C viruses. In common usage, these letter designations define antigenically distinct types of influenza virus referred to as, for example, type A influenza virus or influenza A virus (see Table I). Formally, influenza viruses are classified as orthomyxoviruses (orthomyxoviridae) because of the affinity of these viruses for mucus or mucins (Gr., *myxa*). The prefix "ortho-" distinguishes them from the later recognized viruses of superficially similar biologic and structural properties, now designated as the paramyxoviruses (measles, mumps, respiratory syncytial and parainfluenza viruses).

Table I Classification of the Influenza Viruses[a]

Taxonomic status	International name	English vernacular name	Viral envelope antigens[b]
Family	Orthomyxoviridae	Influenza virus group	—
Genus	Influenza virus	Influenza virus	—
Type	A	Influenza A virus	—
Subtypes (human viruses)			H1N1
			H2N2
			H3N2
Strain designation (example)		A/England/1/51 (H1N1)[c]	
		A/Swine/Iowa/15/30 (H1N1)[d]	
Type	B	Influenza B virus	
Subtypes	None		
Strain designation (example)		G/Great Lakes/1/54	—[e]
Genus	Not named		
Type	C		
Subtypes	None		
Strain designation (example)		C/Paris/1/67	

[a] Data from Fourth Report of the International Committee on Taxonomy of Viruses (1982) and World Health Organization Memorandum (1980).
[b] H, hemagglutinin; N, neuraminidase.
[c] Type/place of isolation/strain number/date of isolation (subtype).
[d] Host of origin is given when original isolate is from a nonhuman host.
[e] H and N antigens but no subtypes have been designated.
[From Kilbourne, E. D. (1987). "Influenza." Plenum Medical Books, New York.]

The parainfluenza viruses also are enveloped RNA viruses and have hemagglutinating and neuraminidase activity like the influenza viruses; however, they are distinguished by important differences in replication and genetics. Like influenza viruses, they possess as genomes negative-stranded (i.e., nonmessenger) RNA, but dissimilarly, their RNA is not in divided segments (see Section II.D). Influenza A virus subtypes are defined by their two external glycoprotein antigens (Table I). These two virion proteins, the hemagglutinin (HA) and neuraminidase (NA) appear to have evolved separately and occur in various combinations (e.g., H1N1, H3N1,). The precise definition of strain variants within subtypes is particularly important with influenza viruses because of their rapid evolution and the epidemiologic significance of their progressive antigenic variation. Such strains or variants bear designations reflecting the virus type, the geographic site of their isolation, the isolate number, the year of isolation, and the antigenic subtype [e.g., A/Hong Kong/1/68 (H3N2)]. It should be appreciated that even such detailed nomenclature may be insufficient in scientific communications because it does not identify viral genes other than those coding for HA and NA that may have been transferred to a virus by genetic recombination or gene reassortment. In this case, viruses may be further identified by laboratory appellations that define their precise genotype or biologic phenotypic attributes such as plaque-forming capacity or mouse virulence.

Influenza viruses share with a number of unrelated RNA viruses (e.g., reoviruses, birnaviruses) the property of possessing a divided or segmented genome. Such viruses have a potential evolutionary advantage in that they can reassort genetic information with genetically related viruses, thus providing a biparental contribution similar to that of sexual reproduction. This property may facilitate interspecific infection and extension of host range (see Section II.D).

B. Viral Structure and Composition

1. Virion Structure

By conventional electron microscopy (Fig. 1), influenza virus particles measure 80–120 nm in diameter. Their mass, estimated by scanning transmission electron microscopy, is 174×10^6 daltons. Almost half of the total mass of the virion consists of surface glycoproteins. The relatively small amount of

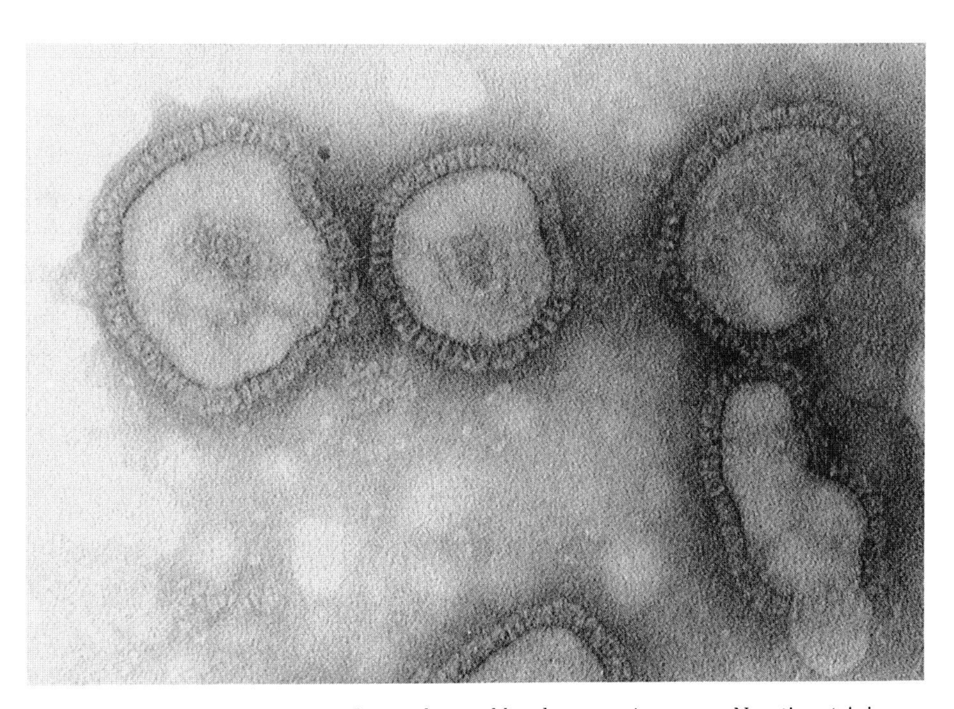

Figure 1 Influenza virus particles as observed by electron microscopy. Negative staining with a 4% sodium phosphotungstate at pH 7.0. Magnification: 120,000x. [Courtesy of A. H. Erickson.]

protein demonstrated within the virus particle suggests that a significant number of particles may contain less than a full complement of ribonucleoprotein (RNP). This finding is in accord with biologic and genetic evidence of genetically defective particles and inefficient replication of influenza virus (see Sections II.C and II.D). Both spherical and filamentous particles of virus are present in influenza virus preparations. Spherical particles predominate in standard laboratory strains and filamentous ones in viruses newly isolated from their natural hosts. Both morphological variants are infectious. The proportion of spherical to filamentous forms produced is an inherent property of the strain that may change with adaptation to a new host through mutation and selection. The spherical particles visualized by electron microscopy appear pleomorphic unless special care is taken in their purification and fixation. This pleomorphism reflects the plasticity of the host-derived lipid bilayer into which the HA and NA surface projections (spikes) are inserted to comprise the envelope surrounding the particle.

Influenza A, B, and C viruses are closely related in structure, chemical composition, and biologic activity. They possess lipid envelopes, derived from the host cells in which they replicate, bearing spikes that comprise virus-coded glycoproteins (Fig. 2). In the case of influenza A and B viruses, the major envelope proteins are the HA and the NA (Fig. 1). Influenza A virus envelopes also contain a third small glycoprotein, M_2; only HA and NA have been identified on the influenza B virus surface, but a third virus-specific glycoprotein, NB, is found in infected cells and might also prove to be a virion component. Within influenza A and B, virions are eight linear single-stranded segments of genomic RNA, each contained in separate helical nucleocapsids or RNPs in which viral nucleoprotein (NP) is in close association with the replicase and transcriptase enzymes PB1, PB2, and PA. The interior of the virion is bounded in all three viral types by a matrix, or M, protein (M1 in influenza A, M in influenza B and influenza C viruses). In influenza A and B viruses, there are 400–500 HA and NA spikes per virion. The proportion of HA to NA spikes is genetically determined and varies from one virus strain to another, but the ratio approximates 5 : 1. Only a single glycoprotein has been found on the surface of influenza C virus, and this virus contains only seven RNA genomic segments.

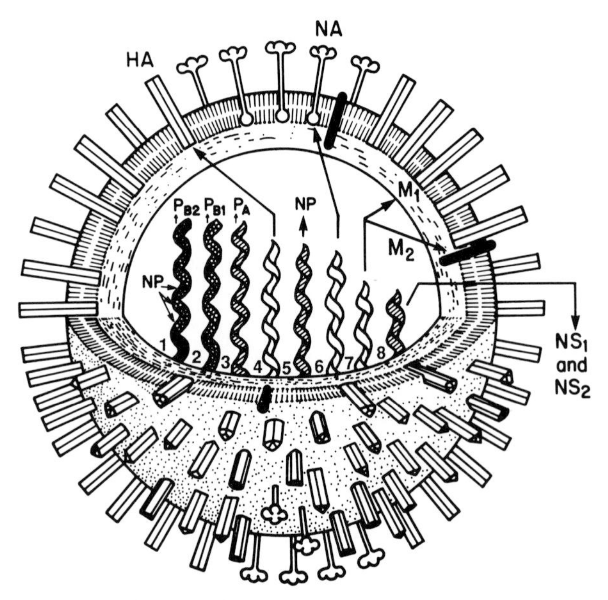

Figure 2 Schematic three-dimensional representation of influenza A virus virion. HA, hemagglutinin; NA, neuraminidase; NP, nucleoprotein. [Courtesy of R. G. Webster.]

2. The Envelope Glycoproteins

The accessibility of influenza virus HA and NA to removal from the virion by proteases and their intrinsic interest as mediators of attachment to and detachment from cells has resulted in such extensive study of their properties that they are among the best defined of all proteins.

a. Hemagglutinin

Influenza virus HA is a trimer of 224,640 molecular weight and is triangular in cross-section (see Fig. 1). Each monomer of the HA contains two disulfide-linked chains, HA (328 amino acids) and HA_2 (221 amino acids). Monomers are associated together hydrophobically as the trimeric virion spikes. As shown schematically in Fig. 3, each monomer possesses a distal globular region joined by a long stalk to a smaller globular region that anchors each monomer in the virion membrane. Anchoring is effected by the carboxy terminus of HA_2. The stem includes all of HA_2 and parts of HA_1. The distal globular portion of HA_1 contains the receptor site through which the virion attaches to cells and also the majority of antigen biotypes by which neutralization of virus by antibody is effected. Mutations of these epitopes enable the influenza virus to escape neutralization and undergo the progressive antigenic evolution that leads to new epidemics.

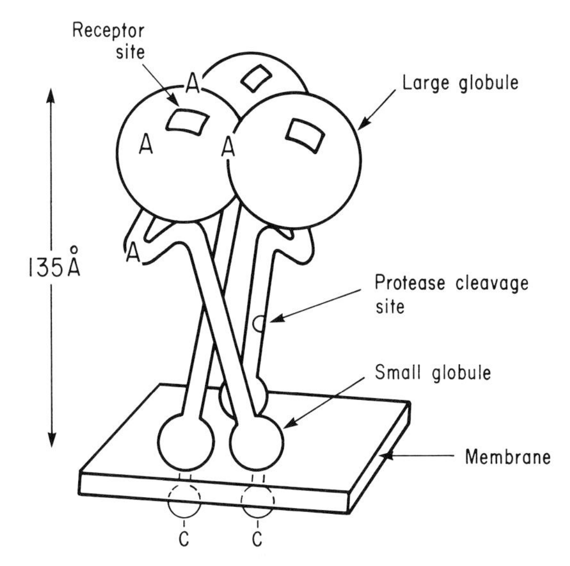

Figure 3 Diagram of the hemagglutinin trimer showing major sites involved in antigenic variation (A) and protrusion of carboxy-terminal residues (C) into the virion interior. [Adopted by D. W. Kingsbury from Wilson, I. A., Skehel, J. J., and Wiley, D. C. (1981). *Nature (London)* **289,** 366–373. Courtesy of Dr. Kingsbury.]

b. Neuraminidase

The influenza virus NA is an exoglycosidase (E. C.3.2.1.18 acylneuraminyl hydrolase) that cleaves the sialic acid acetyl neuraminic acid from α-ketosidic linkages of adjacent sugars. The enzyme in the native form is a 240,000,000 molecular weight tetramer of four identical polypeptide chains. On the virion surface, it presents as mushroom-shaped spikes interspersed with HA trimers. Enzymatic activity resides in the boxlike heads that are attached by narrow stalks to the virion. In contrast to the HA (a class I glycoprotein anchored by a carboxy terminus), the NA stalk is anchored in the virion membrane by an amino-terminal cytoplasmic tail in a manner characteristic of class II glycoproteins. The active site of the enzyme lies in a large pocket in the surface of the boxlike head of the NA. Amino acids at this site are highly conserved. However, most of the antigenic sites identified encircle the hydrolytic site and undergo frequent mutation as in the HA, but probably with less frequency. Antibodies to the NA do not directly neutralize the virus but can diminish multicycle virus replication and favorably modify infection in cell culture and *in vivo*. Antibody inhibits enzymatic activity sterically because of the proximity of the active site to the major antigenic sites.

3. Virion Chemical Composition

In addition to or contained within its structural proteins, the virus contains carbohydrate, lipid, and, of course, RNA. Carbohydrate is contained in the three glycoproteins of the virus and accounts for 5–8% of the mass. Glycosylation sites on the HA are variable in number (four to six) in different strains. The oligosaccharide side chains attached to each glycosylation site are a mixture characteristic for that site of complex (type I) or simple (type II) combinations of oligosaccharides. Differences among viruses in oligosaccharide location and composition are both host cell and virus strain-dependent.

The viral NA contains the same oligosaccharides, glucosamine, mannose, galactose, and fucose found in the HA. Potential glycosylation sites are also variable in number and location in different viral variants. Lipids comprise 20–24% of the virus particle and are located mainly in the virus membrane. Most virion lipids are acquired from the host cell plasma membrane during budding of the particle and therefore reflect the composition of the host cell. Some viral proteins, including HA, contain fatty acids bound by acyl chains. Specificity for their incorporation is at the level of polypeptide acyltransferases.

Genomic RNA constitutes approximately 1% of the viral mass. The total molecular weight of RNAs in influenza A virus is 4.9×10^6 and for influenza B and C viruses 5.3×10^6; the influenza C virus has not been well defined.

C. Viral Replication

The stages, if not all the details, of influenza infection are well defined at the virus cell level, as described later. Virus replication probably proceeds in the same way during the course of influenza, modified, of course, by the host's immune responses.

1. Virus Entry

Infection begins with the attachment of a pocket at the tips of the trimeric HA spikes of influenza virus to the neuraminic acid-containing receptors of host cells. A critical next step is the cleavage of HA_2 from HA_1 by cellular proteases. The HA polypeptide chains remain joined by disulfide bonds, but a new N terminus is created on the HA_2-subunit, the C terminus of which remains imbedded in the viral membrane. This step, in turn, endows the virion with the potential for fusion activity, which is activated at the acid pH of the endosomes within which the virus is carried into the cell. The precise mechanism of fusion of virus to the endosomal membrane is not understood but apparently depends on marked conformational change in the HA induced at the reduced pH in the endosome. Fusion is followed by release of the virion contents into the cell without a requirement for lysosomal enzymes.

2. Replication and Transcription of Viral RNA

Influenza viruses are categorized as negative-stranded viruses, meaning that their eight genomic RNA segments are of negative (i.e., nonmessage) polarity. Thus, the essential initial process is the transcription of input virion RNA (vRNA) into virus specific messenger RNA (mRNA) to enable the production of the viral proteins (PB1, PB2, PA, and NP) that are required for further transcription and also for subsequent synthesis of viral genomic RNA (vRNA). Viral mRNA synthesis is initiated by host cell primers, capped, m^7GpppN^m RNA fragments produced by cellular RNA polymerase II. Elongation of mRNA is terminated short of reaching the 5′ ends of vRNAs, at which point polyadenylation occurs (Fig. 4). Copying of input parental vRNAs to synthesize template RNA needed for production of new vRNAs requires newly synthesized virion proteins. Initiation in this case does not require a primer. Early termination and polyadenylation does not occur as in mRNA synthesis. NP apparently is required in the switch from mRNA to template RNA synthesis.

Synthesis of vRNA destined for incorporation into virus particles similarly is primer-independent and depends on virion protein synthesis.

All RNAs are transcribed and synthesized in the cell nucleus. Early, both mRNA and vRNA (and, of course, template RNA) are produced, but later, synthesis of vRNA predominates.

3. Synthesis of Viral Proteins

Within 1 hr of infection, new viral proteins are detectable in infected cells concomitant with a decrease in host cell protein synthesis. Early in infection, the synthesis of NP and the nonstructural protein NS_1 predominates; later, synthesis of NS_1 declines and synthesis of M_1 and the glycoproteins HA and NA increase. Synthesis of PB1, PB2, and PA continue at a low rate throughout infection. All virion proteins except those coded by RNA 7 and 8 are translated from monocistronic viral mRNAs. RNA segments 7 and 8 contain partially overlapping genes that code for, respectively, M_1 and M_2 and NS_1 and NS_2 proteins. Expression of these proteins is regulated through RNA splicing. A third potential gene product of RNA 7 (M_3) has not been identified.

Proteins of the replication–transcriptase complex (PB1, PB2, PA, and NP) accumulate in the nucleus, although PA has been found in the cytoplasm. NP in association with RNA in RNP nucleocapsids migrates peripherally for incorporation into virus particles. As glycosylated membrane proteins, HA and NA are inserted cotranslationally into the rough endoplasmic reticulum and transported via the Golgi complex to the plasma membrane of the cell. The elongating peptide chains are first glycosylated in rough endoplasmic reticulum vesicles and further glycosylation occurs during peripheral transport in smooth, uncoated vesicles. In cell culture monolayers, the proteins are transported to apical surfaces. Folding of HA monomers and their association into trimers and coalescence of NA monomers into tetramers occur prior to their arrival at the cell surface after passage through the endoplasmic retic-

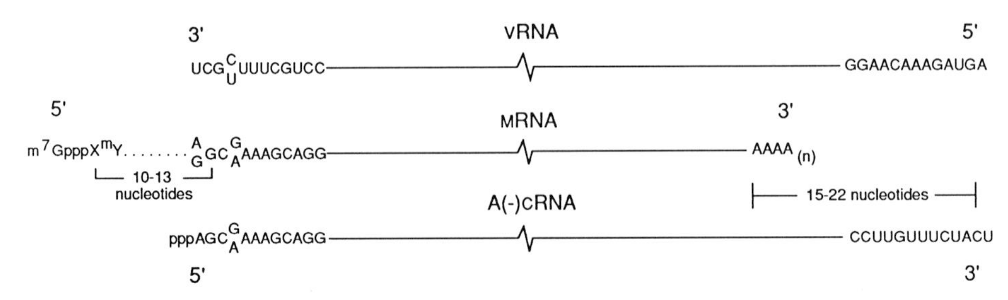

Figure 4 Schematic representation of influenza virus virion RNA (vRNA), messenger RNA (mRNA), and A (−) cRNA (complete transcripts of the vRNA segments that do not contain poly-A). The conserved nucleotides at the ends of each RNA segment are shown. [From Lamb, R. A., and Choppin, P. W. (1983). *Annu. Rev. Biochem.* **52,** 490.]

ulum. The remaining structural proteins, including M_1 and NP, reach the plasma membrane directly by diffusion.

4. Assembly and Release of Virus Particles

Assembly of virus structural proteins is not completely understood but probably involves a transmembrane interaction between the M_1 protein (seen as a dense layer beneath the plasma membrane) and surface glycoprotein spikes discernible projecting from the plasma membrane prior to their budding off in completed virus particles. The underlying RNPs appear only in locations immediately beneath patches of HA and NA. In the final event of viral maturation, particles are pinched off from the cell surface by mechanisms unknown. These particles are free of neuraminic acid, reflecting the presence of active NA on the virion surface. The enzymatic action of NA probably facilitates virus release and prevents the aggregation of released particles. As studied in cell cultures in the laboratory, influenza virus infection is often an abortive or inefficient process that leads to the formation of genetically defective particles of virus. So-called defective interfering particles have deletions principally in the RNA of genes coding for the RNA polymerase complex and interfere with normal viral replication. Defective interfering particles are not known to have any role in the pathogenesis of influenza.

D. Influenza Virus Genetics and Evolution

The genetics of influenza virus is complex and involves high mutation frequency, rapid evolution, and the capacity of the virus to reassort its genes with influenza viruses of the same type. All these processes are directly reflected in the unusual epidemiology of this constantly changing virus.

1. The Viral Genome

The eight vRNA segments of influenza A virus and their 10 gene products are shown in Table II. As described in Section II.C, segments 7 and 8 have overlapping reading frames, which, through spliced mRNAs, each code for two different proteins. The structural relationship of the RNP complexes bearing each of the vRNAs within the virion is not known. It is assumed that each virion contains the appropriate combination of eight RNPs in equimolar amounts, but viral defectiveness and heterozygosity are known to occur.

2. Cloning of Influenza Virus Genes

The cloning of all influenza virus genes in bacterial plasmids or animal virus vectors has greatly facilitated study of gene nucleotide sequences and the replication and function of their coded proteins.

These technical advances have now made directed site-specific mutagenesis possible in defining the molecular basis of changes in viral biologic activity.

3. Mutation and Evolution

RNA viruses in general demonstrate high mutation frequency, in part because they lack proofreading exonucleases to remove misincorporated bases from nascent RNA strands. Mutants appear at a frequency of about 10^{-5}/replication, with an overall genome mutation frequency of 10%.

In influenza viruses, all genes appear to be governed by the same molecular evolutionary clock and have equivalent high rates of silent (nonamino acid changing) substitutions. However, their individual gene products are subject to different selection pressures and therefore evolve at different rates. Notably, the external glycoproteins, HA and NA, evolve most rapidly during their antigenic drift away from the immunologic pressures of the host (Fig. 5). The rate of variation is estimated at 0.6–0.9%/yr by sequential cumulative base changes along a common lineage. In contrast, internal and nonstructural virus proteins evolve more slowly and not in continuum. [*See* EVOLUTION, VIRAL.]

4. Viral Heterogeneity and Variation

Given their rapid rates of mutation and evolution, it is not surprising that influenza viruses as propagated in the laboratory are genetically heterogenous. This fact is readily demonstrated by selection of antigenic or other variants with monoclonal antibodies or by changes in experimental host (host-range mutants). Appreciation of this polymorphism is important in assessing the genetic potential of the virus in genetic reassortment, pathogenesis, and epidemiology.

5. Genetic Reassortment (Recombination)

Within each type (A, B, or C), all strains of influenza virus are genetically compatible. Thus, coincident infection of a cell with any pair of influenza A viruses, for example, has the potential for producing

Table II Genes and Gene Products of Influenza A Virus

Segment	Number of nucleotides[a]	Encoded protein	Molecular weight[b]	Molecules per virion	Structural features	Function
1	2341	P_{B2}	85,700	30–60	Basic protein	Cap recognition of host cell RNA
2	2341	P_{B1}	86,500	30–60	Basic protein	With P_{B2}, P_A, and NP are components of RNA replication–transcriptase complex
3	2233	P_A	82,400	30–60	Acidic protein	
4	1778	HA	61,468	500[c]	Glycoprotein (trimer)	Attachment, fusion, and entry into cells
5	1565	NP	56,101	1000	Coiled filaments	Binds virion RNAs; replication complex
6	1413	NA	50,087	100[c]	Glycoprotein (tetramer)	Enzyme; releases virus from cells
7	1027	M_1	27,801	3000	—	
		M_2	11,010	20–60	Glycoprotein	Virion membrane protein
8	890	NS_1	26,815	—	Nonstructural	Functions unknown
		NS_2	14,216	—	Nonstructural	

[a] In prototype A/PR/8/34 strain.
[b] Deduced from RNA sequence (polypeptide monomers).
[c] Ratio of HA : NA varies among strains.
HA, hemagglutinin; NA, neuraminidase; NP, nucleoprotein.

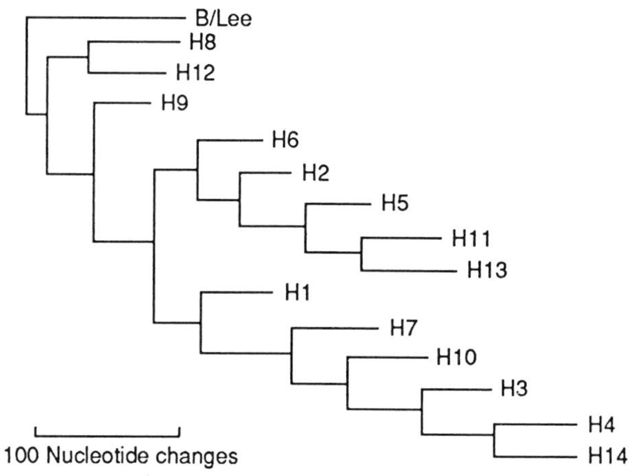

Figure 5 Phylogenetic tree for influenza A virus hemagglutinin (HA) genes of different subtypes, human (H1, H2, and H3) and animal. Nucleotide sequencing corresponding to residues 63–343 were analyzed with a maximum parsimony algorithm. The tree was rooted to the B/Lee/40 HA gene. The number of variables represented is 189; the total tree length is 1344 nucleotide changes. The consistency index (proportion of changes due to forward mutation) is 0.448. Horizontal distance is proportional to the minimum number of nucleotide distances to join nodes and gene sequences. Length of vertical lines is arbitrary. [From Kawaoka, Y., Yamnikova, S., Chambers, T. M., Lvov, D. K., and Webster, R. G. (1990). *Virology* **179**, 759–767.]

reassortant viruses of mixed genotype (Fig. 6). This reassortment occurs with a high frequency inconsistent with the rate of classical DNA recombination and, in fact, reflects the actual reassortment of whole genes in variable combinations. This event instantly creates new viruses of multiparental origin and new virulence and epidemiologic potential. It is not clear whether or not reassortment of RNAs is completely random (i.e., without linkage), but many different gene combinations are readily demonstrated following mixed infection. During mixed infection, viral pseudotypes may be found that bear a single parental genome but that have acquired mixed or heterologous envelope proteins. Such particles breed true to genotype and probably have no significance in natural infection. This phenomenon of phenotypic mixing is not unique to influenza viruses. Genetic reassortment has been demonstrated to occur in animals and humans and probably has an important role in the epidemiology of the virus, particularly in the origin of pandemics. [*See* EPIDEMIOLOGIC CONCEPTS.]

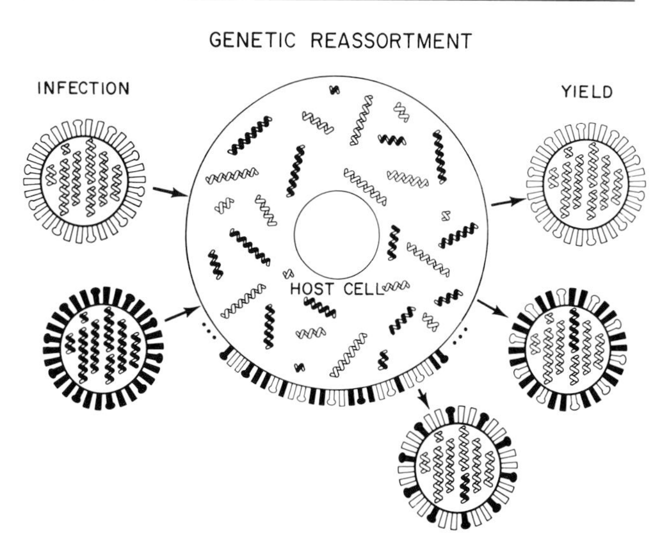

Figure 6 Intratypic genetic reassortment of influenza viruses, illustrating diagramatically 3 of 256 possible reassortant progeny from infection with two parental viruses. Note phenotypically mixed envelopes of two progeny. [From Kilbourne, E. D. (1987) "Influenza," Plenum Medical Books, New York.]

III. Influenza

A. Influenza in Humans

1. Clinical Description

The clinical expression of influenza is variable, ranging from asympt omatic to fatal infection. Typically, onset of illness is rapid and prostrating, and almost invariably attended by cough, malaise, headache, and myalgia. Coryza, sore throat, and, less commonly, substernal pain also indicate that the primary site of infection is the respiratory tract. Other common symptoms include extraocular muscle pain, chilliness, and nasal and conjunctival burning and discharge. Unlike other viral respiratory infections, fever and systemic symptoms predominate. Recovery usually occurs in <1 wk, coincident with decreased shedding of virus (Fig. 7). The simultaneous occurrence of other cases in the community is an aid to diagnosis. In young children, fever and rhinitis predominate as symptoms, and vomiting and diarrhea may be a significant part of the presenting illness. The severity of influenza is principally host-determined in relation to age, physiologic state, or prior immunization by natural infection or vaccination. The intrinsic virulence of influenza virus probably changes little despite its high mutation rate. The exceptional morbidity and mortality seen in pan-

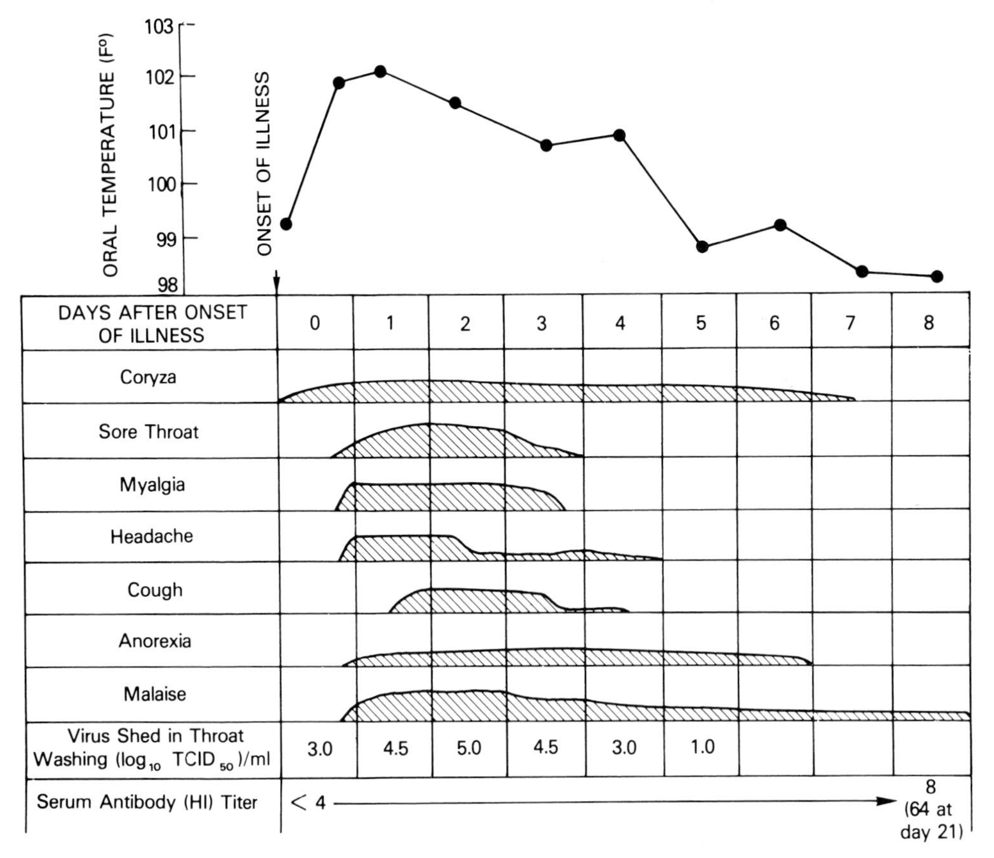

Figure 7 Clinical characteristics of a naturally occurring case of influenza caused by infection of a normal 28-yr-old man by A/Victoria (H3N2) influenza virus. The height of each shaded curve is proportional to the magnitude of the corresponding symptom. HI, hemagglutination inhibition; TCID, tissue culture infective dose. [From Dolin, R. (1976). *Am. Acad. Fam. Physician* **14,** 74.]

demics is almost certainly a consequence of the antigenic novelty of the new pandemic virus that enables it to elude the immunity established to its antigenically different predecessors.

2. Immunity to Influenza

Despite the recurrent nature of influenza in the community, single infections can result in solid homologous immunity. Immunity is attained as a composite derived from successive infections with the same or related viruses within a given type or subtype. Therefore, the frequency and severity of disease in interpandemic periods tends to decrease with increasing age. Reasons for higher mortality in the elderly (see later) are not entirely clear but probably reflect compromised cardiopulmonary function as well as a greater susceptibility to secondary bacterial pneumonia.

Immunity in influenza is mediated principally by antibody to the HA and NA antigens. Although immunity is most directly correlated with serum antibody levels, secretory antibody in the respiratory tract probably makes a significant contribution to resistance. In addition, cellular immunity to HA and NA and possibly to the internal antigen NP are probably important in the modulation of infection.

3. Cytopathology and Pathogenesis of Influenza

a. Cytopathic Effects of Influenza Virus

Present evidence suggests that all cells in which influenza virus infection occurs are destroyed and, thus, that prolonged persistence of the virus genome is precluded and the occurrence of temperate or latent infection is unlikely. Abortive replication occurs in lymphocytes, but cells so infected do not proliferate or produce IgG.

In productive infections, cytopathic effects include rounding of cells, cytoplasmic retraction, granularity and vacuolization, nucleolar enlargement, and clumping of chromatin. Changes are evident by electron microscopy at 4 hr after infection with the acquisition of cytoplasmic dense spots and the formation of fibrillar and electron-dense structures in the nucleus. Also, early in infection aggregation of ribosomes is seen in the cytoplasm. Late changes are those characteristic of cytocidal viral infections in general, rupture of cytoplasmic nuclear and nucleolar membranes, and cytoplasmic vacuolization.

As cytolytic viruses, influenza viruses produce plaques in a variety of cell lines, provided that a protease such as trypsin is added to ensure cleavage and activation of the HA protein. Because inactivated virus (as in vaccines) has some toxic effects, it would appear that cell damage from virus can occur in the absence of viral replication. If this is so, toxicity manifest by, for example, cell damage or pyrogenic effects has not been incontestably associated with any one viral component. A variety of *in vitro* effects of infective virus on leukocytes (principally polymorphonuclear leukocytes) have been reported but in most instances do not occur after inactivation of viral infectivity. In view of the fact that the viral receptor, neuraminic acid, is ubiquitously distributed among cells, the usual restriction of influenza virus replication to respiratory tract cells is remarkable. As discussed later, this organ specificity may reflect differences in the availability of suitable proteases for the cleavage of HA that is essential to productive infection.

b. Pathogenesis and Pathology

Influenza is initiated by virus transmitted from one human respiratory tract to another. The primary site of viral deposition in unclear. Whether in natural infection virus initially infects cells of the upper tract and then descends to involve the trachea, bronchi, and sometimes the lung, or infection is an ascending process after primary alveolar implantation by aerosol, is uncertain. Both modes of infection may occur. In any case, it is clear that the principal target of infection is the ciliated epithelium of the middle respiratory tract. This epithelium rapidly undergoes necrosis and desquamation. Shedding of virus from the nasopharynx ceases coincidently with recovery from illness after 4–7 days. Cellular regeneration starts early in infection, even before desquamation begins.

Involvement of the lung parenchyma in primary influenza virus pneumonia is characteristic. Lungs are heavy, hemorrhagic, and edematous. Cellular infiltration is limited in degree, alveoli are denuded of epithelium, and intra-alveolar hemorrhage is frequent. The presence of acellular hyaline membranes lining alveoli is characteristic of influenza pneumonia.

Despite the severity of systemic symptoms, viremia has only rarely been demonstrated in influenza and is not an important component of the pathogenesis of typical illness. In severe disease, infection of nonrespiratory organs can occur that must result from blood-borne dissemination of virus.

c. Pathophysiology of Infection

Pulmonary function is impaired during and after uncomplicated influenza. Dysfunction includes airway hyperreactivity, restrictive ventilatory defects, and increased alveolar–arterial oxygen tension gradients, even in patients with normal chest X-rays.

4. Unusual Manifestations and Complications

a. Pneumonia

Some minor signs of transient pulmonary involvement (rales, roughened breath sounds) may be found in 1–5% of carefully examined patients whose chest X-rays are negative. In addition, influenza may be complicated by primary influenza viral (diffuse) pneumonia, secondary bacterial pneumonia (focal or lobar) following influenza, or a simultaneous combination of the two. Common secondary bacterial pathogens are pneumococci, staphylococci, and group A streptococci. Although primary influenza viral pneumonia can occur without obvious underlying disease, it is most common in patients with antecedent compromised cardiopulmonary function, especially in those with rheumatic valvular heart disease. Women in the final trimester of pregnancy are also at greater risk. Most patients who die following influenza die with/of bacterial pneumonia.

b. Rare Complications

Acute myopathy is a rare complication of influenza and is of three types: (1) acute transient crural myopathy in children with influenza B manifest as incapacitating leg pain, (2) acute myopathy with rhabdomyolysis and associated renal dysfunction in adults with influenza A, and (3) myopathy in the elderly without obvious symptoms of influenza. Nephropathy may occur secondary to myopathy and myoglobinemia resulting from muscle destruction. Carditis and encephalitis are equally rare complications. Postinfluenza asthenia and depression are considered to be common consequences of influenza, although a causal link to influenza has not been established.

c. Reyes Syndrome

This serious, often fatal illness, related to impaired liver function, has been linked to influenza, principally influenza B, but is not specific for influenza having been associated also with other viral infections such as varicella. Administration of aspirin appears to be a cofactor in the pathogenesis of this condition and should be avoided.

5. Therapy of Influenza

The usual treatment for influenza is supportive and nonspecific including bed rest, light diet, and maintenance of adequate hydration. Symptomatic treatment with nonsalicylate analgesics and antipyretics can be given as well as cough-suppressant drugs in moderation.

Amantadine hydrochloride at an oral dosage of 200 mg/day can modify the clinical course of influenza A if given within 48 hr of onset. The drug is specific for influenza A virus infections, and its effect is mediated through effects on the viral M_2 protein.

6. Laboratory Diagnosis of Influenza

Specific diagnosis of influenza depends on isolation of virus from nasopharyngeal washings or secretions, by demonstration of specific antibody response, or, less commonly, by fluorescent antibody staining of exfoliated epithelial cells of the respiratory tract. Virus can be recovered in aneuploid continuous cell lines such as Madin-Darby canine kidney cells, if trypsin is provided as an exogenous protease, or in chick embryos by allantoic sac inoculation.

The standard technique for the measurement of serum antibody is the hemagglutination-inhibition (HI) test, in which the agglutination of human or chick erythrocytes by virus is inhibited by specific antibody. Because of the prevalence of influenza virus infection, demonstration of antibody in a single serum specimen is not sufficient for diagnosis; rather, an increase in antibody titer with paired acute and convalescent sera must be shown. ELISA with viral antigens permits the serologic diagnosis of influenza, even in laboratories not equipped to propagate the virus. [*See* ELISA TECHNOLOGY.]

B. Influenza in Animals

Although many mammalian and avian species are sporadically infected with influenza A viruses, serially propagated epizootics are most often observed in domestic chickens, turkeys, swine, and horses. The seal is the only wild mammal in which epizootics have been seen. Infection in domesticated and wild ducks appears to be mainly asymptomatic.

Representative strains of all human influenza A virus subtypes, i.e., H1N1, H2N2, and H3N2, have been recovered from animals. In addition to these, viruses representing a total of 14 different HA and 9 different NA subtypes have been found, mostly in migrating birds.

HA and NA antigens are found in avian viruses in different combinations, strongly suggesting that genetic reassortment among influenza A virus subtypes is not unusual in nature. Transmission of viruses among species including humans has been demonstrated; it may explain the emergence of human pandemic viruses (see Section IV).

In mammals, infection is confined to the respiratory tract, but in birds infection is often generalized, with respiratory, intestinal, and even central nervous systems involved. Certain avian infections are highly malignant and are known as fowl plague. (Fowl plague virus was the first influenza virus to be isolated.) Influenza in domestic fowl can be devastating, requiring destruction of flocks with serious economic consequences.

IV. The Epidemiology of Influenza

The survival of influenza viruses in humans requires the continual evolution of antigenic variants that can infect populations immunized by preceding variants (see Fig. 8). Viral gene sequences indicate that mutation in both the HA and NA antigens of influenza A viruses are cumulative and sequential. Influenza C virus mutations are essentially noncumulative and divergent, but change in influenza B viruses involves both the sequential and divergent mutations. These studies of the molecular basis of viral evolution are in accord with the differing epidemiology of the three types of influenza in which the frequency and severity of disease declines progressively with influenza viruses A, B, and C.

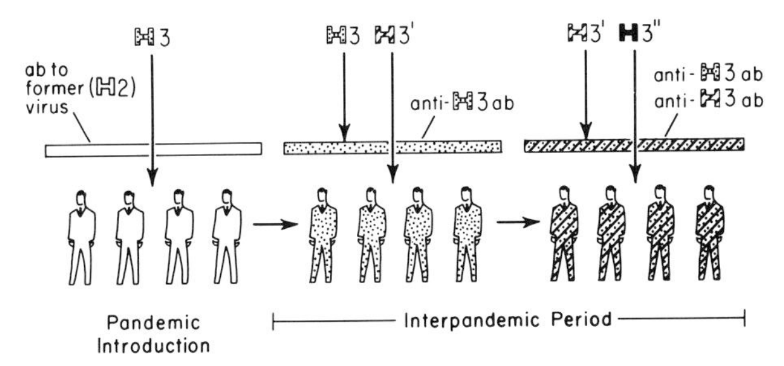

Figure 8 Selection of antigenic mutants as a function of population antibody. New pandemic viral subtype H3 transcends the barrier of antibody to unrelated previously prevalent virus H2 and readily infects the population. When a critical percentage of the population has been infected with H3, survival of H3 is impeded and antigenically changed mutant H3′ and later H3″ have survival advantage (minor antigenic variation or antigenic "drift"). [From Kilbourne, E. D. (1975). The Epidemiology of Influenza. In "The Influenza Viruses and Influenza," (E. D. Kilbourne, ed.), p. 522. Academic Press, New York. Reproduced with permission.]

A. Influenza A

Influenza A, associated with minor antigenic change in the virus, occurs every winter in the form of regional epidemics. At unpredictable and variable intervals (10–30 yr), global pandemics occur in which the virus has been found to have undergone major mutations in one or both external antigens. The origin and outcome of epidemic (interpandemic) and pandemic influenza will be considered separately.

1. Pandemic Influenza

Pandemic viruses, new to the human population (see Section I), probably originate as genetic reassortants of human and animal viruses that provide novel surface antigens alien to previous human experience. Therefore, people of all ages are susceptible to infection, and the pandemic virus spreads so rapidly that it may appear to have originated simultaneously in many different sites. Historical evidence, however, points to the Far East as a source of pandemics, a fact that may reflect the high population density in countries such as China and the proximity to humans of pigs and ducks that carry animal influenza viruses.

By definition, pandemics involve all people and, because of universal susceptibility, move rapidly to produce high morbidity within a relatively brief (<1–2 yr) period. Although the case fatality rate of influenza ordinarily is low (<0.01%), relatively large numbers of people die in pandemics because morbidity rates (the incidence of illness) are so high. The pandemic of 1918 may have been uniquely virulent in its high mortality rates in young adults, in addition to the very young and very old in whom mortality usually occurs. The course of disease in the typical uncomplicated case of pandemic disease is indistinguishable from that of interpandemic influenza. In both pandemic and epidemic influenza, mortality is principally a consequence of secondary bacterial pneumonia.

It is notable that in the pandemics of 1957 and 1968 the pandemic viruses drove out the previously circulating viral subtypes. The "juvenile" pandemic with H1N1 virus in 1977 did not do so, however, and H1N1 and H3N2 viruses continue to circulate at this writing. Based on serologic incidence, recycling of major subtype HA antigens has occurred (Fig. 9), suggesting that the number of HA subtypes capable of infecting humans may be limited.

2. Interpandemic (Epidemic) Influenza

Following pandemic introduction of a new viral subtype, mutual adaptations of virus and humans occur; as specific immunity to the new virus develops in response to infection, progressive selection of antigenic viral variants takes place (Fig. 8). As a consequence, regional epidemics of influenza continue to occur, varying in severity and frequency principally in relation to the magnitude of viral antigenic change. Although interpandemic influenza is more sporadic and less striking in its impact than influenza in pandemic form, it imposes a huge burden of morbidity and mortality every year. The death rate in the United States alone may reach 20,000, even in a nonpandemic winter. The striking effect of influenza on community life and health services provides methods for its ascertainment, as shown in Fig. 10, which shows the impact on school absenteeism, hospital emergency room visits, previous admissions, and mortality of a single epidemic.

3. Seasonal Factors in Influenza

The wintertime prevalence of influenza is not well understood but may reflect both indoor crowding during inclement weather and a demonstrated increased survival of the virus in aerosols at the low relative humidity characteristic of indoor winter environments.

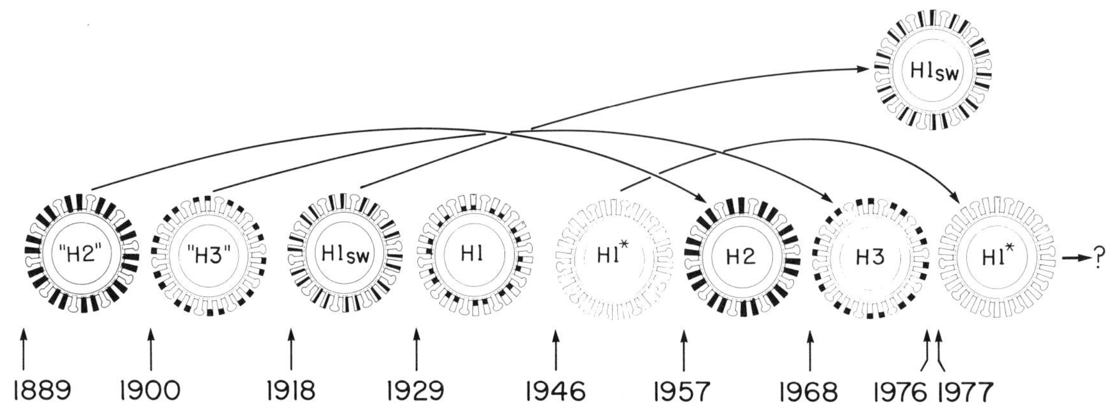

Figure 9 Influenza pandemics of the past century (major antigenic variation). Periods of prevalance of major hemagglutinin subtypes of influenza A viruses. The years of introduction of the new subtypes are indicated by arrows. The designations H1, H2, etc., refer to the hemagglutinin subtypes of the viruses without reference to their neuraminidase antigens. Note that the H1 intrasubtypic variants—$H1_{sw}$ (swine influenza virus), H1, and H1*—were associated with significant antigenic change and with major epidemics and are therefore represented. Identification of viral subtypes prior to 1933 (marking the first isolation of a human influenza virus) is deduced from studies of human antibodies. $H1_{sw}$ appeared abortively in the human population in 1976. H3 and H1* subtypes continue to circulate concordantly. [From Kilbourne, E. D. (1987). "Influenza." Plenum Medical Books, New York.]

B. Influenza B

The epidemiology of influenza B resembles that of influenza A on a reduced scale except that it does not occur in pandemics. This does not imply that influenza B does not kill. Like influenza A, it may pave the way for bacterial pneumonia or produce primary influenza virus pneumonia. Perhaps related to its slower antigenic evolution, influenza B affects principally children, adolescents, and young adults. Epidemics tend to occur in 3-yr rather than 2-yr cycles.

C. Influenza C

Infection with influenza C virus occurs primarily in children; well-defined epidemics have not been seen in the general population. The virus appears to survive without the necessity for sequential antigenic drift.

V. Prevention and Control

As a contagious, rapidly spreading disease, influenza knows no national boundaries. For this reason and because of its pandemic potential, influenza is under worldwide surveillance by WHO-affiliated laboratories. Prompt identification and antigenic analysis of new strains is an essential guide for the annual or biannual revision of vaccines needed to protect against this constantly changing virus. The WHO network also functions to pool and collate information on the incidence and progression of the disease. Despite these efforts, it must be said that epidemics of influenza remain essentially uncontrolled.

A. Vaccination

1. Vaccines

Presently licensed vaccines contain inactivated purified virus or viral components. The vaccines are trivalent, including representative strains of the two prevalent influenza A subtypes, H3N2 and H1N1, and a single strain of influenza B virus. Vaccines are standardized by regulatory agencies to contain a uniform concentration of HA antigen of each strain. Vaccine is given by parenteral injection. Immunity is attained coincident with demonstrable increase in HI antibody 2–3 wk after administration.

Attenuated live virus vaccines have been used with some success in the past, notably in the Soviet Union, and at present are under active development but not yet licensed in the United States. Promising experimental vaccines include temperature-sensitive and cold-adapted mutants.

Modern vaccines are relatively nonreactogenic, except in very young children in whom febrile reactions are common. In school children, fever may

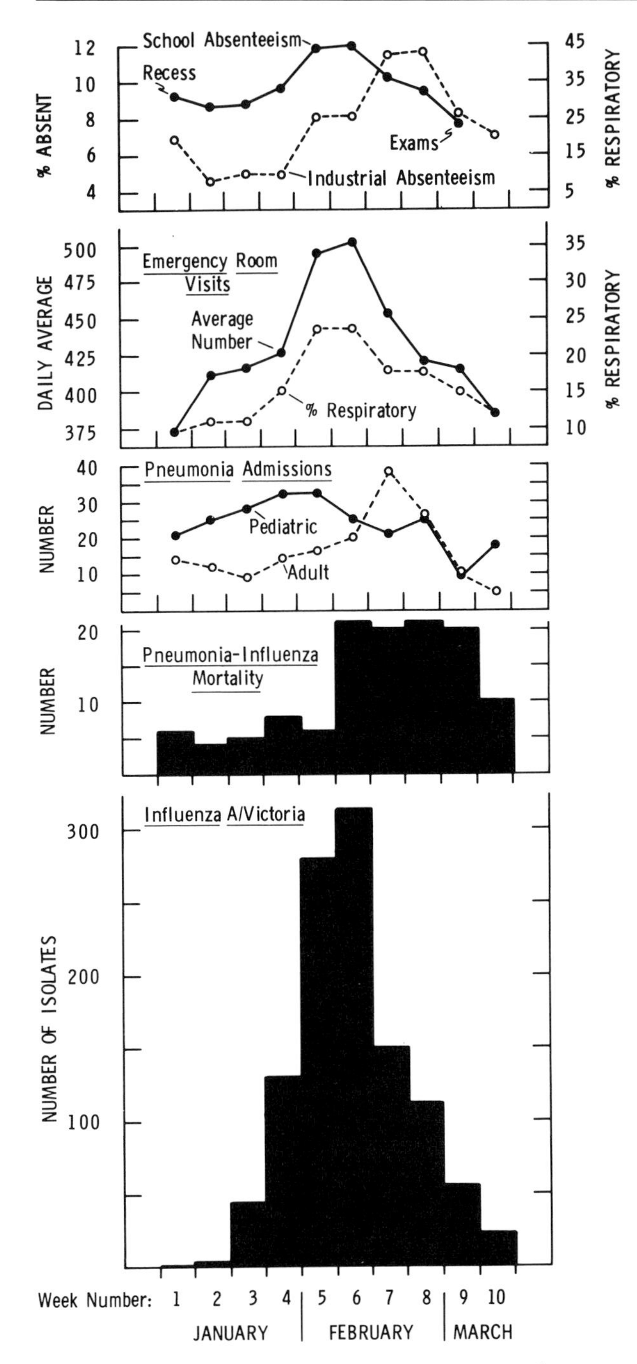

Figure 10 Correlation of the nonvirologic indices of epidemic influenza with the number of isolates of influenza A/Victoria virus according to week, Houston, 1976. Industrial absenteeism is indicated by percentage with respiratory complaints. [Reprinted with permission, from Glezen and Couch (1978). *New Engl. J. Med.* **298,** 589.]

occur in as many as 10%, but 1% of adults become febrile. Local erythema, pain, tenderness, and itching may occur at the site of injection, but these symptoms rarely interfere with normal activity. Neurological complications such as Guillain-Barre syndrome are extremely rare and are not specific for influenza vaccine.

2. Groups Targeted for Vaccination

Inactivated influenza vaccine is 75–90% effective against challenge with homologous virus, but immunity usually lasts <1 yr. Furthermore, universal immunization in childhood is impractical because of vaccine toxicity in that age group. Therefore, vaccine is recommended principally for those at special risk of serious consequences of illness. These include (1) adults and children with chronic cardiovascular or pulmonary disease, diabetes, or other metabolic diseases, (2) residents of nursing homes or other chronic care facilities, and (3) medical personnel—for their own protection and that of their patients.

In addition, those >65 yr of age, persons involved in activities essential to the community, or travelers or anyone who does not wish to become ill or incapacitated should be immunized.

Even vaccination limited to these categories is presently inadequate in the United States, probably reaching only 25% of high-risk groups.

3. Chemoprophylaxis

Influenza A can be prevented by oral administration of amantadine (L-adamantanamine hydrochloride) or the related drug rimantadine. With a recommended dosage for amantadine of 100–200 mg/day, infection is not entirely precluded, because some virus shedding and antibody response occur.

Because these drugs produce significant side effects related to action on the central nervous system, they should be regarded as a supplement to immunization and restricted to short-term administration during epidemics to high-risk subjects who have escaped vaccination or to those allergic to influenza vaccine.

B. Prospects for Ultimate Control

Influenza A virus has attributes both of an obligate human parasite and (in pandemics) a zoonosis. Therefore, interpersonal transmission is susceptible to containment if adequate (i.e., universal) vaccination can be carried out in early childhood and appropriately spaced booster injections are administered thereafter. In addition, animals identified as hosts most likely to contribute new antigens to pandemic viruses could be closely monitored, immunized, or otherwise controlled.

Bibliography

Air, G. M., Laver, W. G., and Webster, R. G. (1987). Antigenic variation in influenza viruses. *In* "Antigenic Variation: Molecular and Genetic Mechanisms of Relapsing Disease" (J. M. Cruse and R. E. Lewis, Jr., eds.) pp. 20–59. Karger, Basel.

Crosby, A. W. (1976). "Epidemic and Peace, 1918." Greenwood Press, Westport, Connecticut.

Kilbourne, E. D. (1987). "Influenza." Plenum Medical Books, New York.

Klenk, H.-D., and Rott, R. (1988). *Adv. Virus Res.* **34,** 247–281.

Krug, R. M. (ed). (1989). "The Influenza Viruses." Plenum Press, New York.

Murphy, B. R., and Webster, R. G. (1990). Orthomyxoviruses. *In* "Virology," 2nd ed. (B. N. Fields and D. M. Knipe, eds.), pp. 1091–1152. Raven Press, New York.

Wiley, D. C., and Skehel, J. J. (1987). *Annu. Rev. Biochem.* **56,** 365–394.

Insecticides, Microbial

Allan A. Yousten
Virginia Polytechnic Institute and State University

Brian A. Federici
University of California, Riverside

Donald W. Roberts
Boyce-Thompson Institute for Plant Research

Glossary

Biological control Use of natural enemies, including predators, parasites, and diseases, to control insect pests of agricultural, medical, or veterinary importance

***cry* genes** Genes found in naturally occurring isolates of bacterium, *Bacillus thuringiensis;* they encode proteins that are toxic to insects; the genes are frequently located on large plasmids that may be transmissible by conjugation between bacteria

Delta endotoxin A protein toxin produced by the bacterium *Bacillus thuringiensis* that is lethal when ingested by insects of a specific order

Granulosis virus Type of enveloped, double-stranded DNA virus reported only from insects of the order Lepidoptera (moths, butterflies) that replicates initially in the nucleus and later in the cytoplasm of infected cells; virions occluded individually in small occlusion bodies called granules

Integrated pest management Practice for managing insect pest populations whereby chemical, biological, and cultural techniques are integrated to achieve effective, environmentally safe pest control; emphasis on the reduction or elimination of synthetic chemical insecticides

Nuclear polyhedrosis virus Type of enveloped, double-stranded DNA virus reported from insects and other invertebrate organisms; replicates in nuclei of infected cells and occluded in large protein crystals known as polyhedra

Parasporal inclusion body Aggregate of one or more proteins formed inside a bacterial cell at the time of sporulation; sometimes called crystals, may be toxic to insects that ingest them, and found in several different species that are pathogenic for insects

Recombinant viral insecticide Insect virus that has been genetically engineered using recombinant DNA technology to improve insecticidal properties

MICROBIAL INSECTICIDES are composed of microorganisms that produce disease and ultimately death in insects. The microorganisms are introduced into the environment where they may persist and limit insect populations for extended periods of time. More commonly, however, the microorganisms or the microorganisms and an associated toxin are delivered into the insect-infested area where they will produce disease only for limited periods. Even though these microorganisms may be detectable in the area for a long time, they are incapable of maintaining themselves in sufficient numbers to provide effective insect control. When used in this way, they are similar to synthetic organic chemical insecticides; however, they differ in important ways.

A significant advantage of microbial insecticides is safety, which results from the high degree of specificity for the animals they affect. Another advantage is that relatively little resistance has been found among insects exposed to certain microbial insecticides for many years, which has allowed their use in situations where synthetic chemicals have become ineffective. In addition, microbial insecticides may often be brought onto the market more rapidly and at a lower cost than synthetic chemicals.

Although specificity is an advantage when considered in terms of safety, it is also a disadvantage in some agricultural situations where several different insect pests are present simultaneously. Another disadvantage is that some microbial insecticides are more affected by environmental factors, such as humidity and sunlight, than are synthetic chemicals. Knowledge of insect ecology and of the environment are important to the successful use of microbial insecticides. The entomopathogenic microorganisms that are likely to make the most important contributions to agriculture and to public health in future years are bacteria, viruses, and fungi.

I. Bacteria

A. Introduction

A large number of bacterial species have been isolated from insects, and some of these have proven to be pathogenic; however, considerations of safety and effectiveness have resulted in only a very few being developed as insecticides. The following species have already proven effective and are commercially available or are close to that level of development.

B. *Bacillus popilliae* and *Bacillus lentimorbus*

Bacillus popilliae and *B. lentimorbus* are the etiological agents of milky disease in larvae of the family Scarabaeidae (Coleoptera). The Japanese beetle is among the best known of these insects and is the target of commercially available insecticides containing spores of the bacteria. These facultative, nutritionally fastidious, spore-forming bacilli have a gram-positive cell wall profile, although the gram stain is reported to be negative during vegetative cell growth. There have been a number of similar bacilli isolated from related larvae displaying milky disease, but their relationship to *B. popilliae* and *B. lentimorbus* is unclear. Some consider them all (including *B. lentimorbus*) to be subspecies of *B. popilliae*. *Bacillus popilliae* produces a proteinaceous parasporal body at the time of sporulation whereas *B. lentimorbus* does not. There is no evidence that the parasporal body of *B. popilliae* plays any significant role in the course of the disease. These bacteria metabolize a number of sugars including trehalose, which is present in high concentrations in the hemolymph of the host insect. *Bacillus popilliae*, the subject of most metabolic studies, appears to lack a complete tricarboxylic acid cycle. This characteristic and the fact that the bacteria are catalase negative have been suggested to be related to the inability of the bacteria to sporulate well *in vitro*.

The course of infection in the Japanese beetle larva is initiated following ingestion of spores. In the first phase, spores germinate in the larval gut and penetrate the epithelium. In the second phase, lasting about 1 week or less, vegetative cells proliferate in the hemolymph. During the third phase of about 1 week, vegetative growth continues, but some cells sporulate. In the fourth and final phase, large numbers of spores are present in the hemolymph, and death of the larva occurs 2 to 3 weeks following initial infection. The dying larva displays a milky white color giving the condition the name, "milky disease." The hemolymph of the dead larva contains very high numbers of spores, reported to be up to 5×10^{10}/ml of hemolymph. With the disintegration of the cadaver, the spores are released into the soil to be consumed by additional larvae as they feed on plant roots.

Insecticides containing the spores of *B. popilliae* are produced by collecting infected larvae from beneath sod or by collecting uninfected larvae and infecting them in the laboratory. The spores collected from dead or dying larvae are blended with inert carriers and sold as a powder designed to be introduced into soil in areas where protection is sought from Japanese beetles. A U.S. government-sponsored program using powder produced in this way distributed spores widely in the northeast United States in the 1940s and 1950s. Following its registration in 1948, the powder has been produced commercially. Although the bacteria grow readily in ordinary bacteriological media, only a small percentage sporulate. This has frustrated attempts to produce the spores *in vitro*. A patent issued in 1989 described a process for obtaining *B. popilliae* spores *in vitro*, but it now appears that this process has been unsuccessful.

C. *Bacillus thuringiensis*

Bacillus thuringiensis is the most successful and most widely used microbial insecticide. This gram-positive, spore-forming, facultative bacterium is related to *B. cereus*. Comparisons of *B. thuringiensis* and *B. cereus* using a large number of phenotypic tests were unable to distinguish between the two species; however, *B. thuringiensis* has been retained as a distinct species because it produces several related proteins that are toxic when ingested by insects. Within the species, the strains have been grouped by serotyping of the flagellar H antigens. At this writing, at least 30 serovarieties have been identified. In nature it has been isolated from diseased insects, from the soil, and from the leaf surface of many plants. This bacterium is easily grown and sporulates well under laboratory conditions, and uses a variety of carbohydrates as carbon and energy sources.

The pathogenicity of *B. thuringiensis* is caused by the toxic proteins that it produces. These toxins, referred to as "delta endotoxins," are synthesized by the cells beginning at about stage III of sporulation. They aggregate within the cells as parasporal inclusion bodies or "crystals." The inclusions are readily visible by phase contrast microscopy, and their presence adjacent to an elliptic spore in an unswollen sporangium allows a tentative identification of the bacterial species. Furthermore, there is a general correlation between the shape of the inclusions and the group of insects affected by the toxin. Bypyramidal inclusions (Fig. 1) contain protein(s) toxic to lepidopteran larvae; irregular, composite inclusions (Fig. 2) contain proteins toxic to dipteran larvae; and square or rhomboidal inclusions (Fig. 3) contain proteins toxic to coleopteran larvae. *Bacillus thuringiensis* strains can be grouped according to pathotypes, referring to the orders of insects that they kill. A few strains have been isolated with parasporal inclusions for which no susceptible insect has been identified.

The nucleotide sequence of many of the *cry* (crystal protein) genes has been determined and have been arranged into four major groups (*cry*I, *cry*II, *cry*III, and *cry*IV) based on the nucleotide sequence similarity of the genes and the host range of the toxins. The *cry* toxin genes of all four groups have been found to be located on plasmids of various sizes. These plasmids may be transmissible by conjugation between strains of *B. thuringiensis* or closely related bacteria. The *cry*I genes encode proteins of 131–139 kDa, which have toxicity for Lepidoptera. At least 20 related but distinct nucleotide sequences have been identified among the *cry*I genes. One to three of these genes are found in the

Figure 1 Carbon replica of *B. thuringiensis* parasporal inclusions (crystals) containing lepidopteran-active toxin.

majority of *B. thuringiensis* serovarieties and were the earliest described and brought into use as insecticides. The second group, the *cry*II genes, encode toxins killing both Diptera and Lepidoptera (*cry*IIA) or Lepidoptera alone (*cry*IIB). The molecular mass of the *cry*II-encoded proteins is about 71 kDa. The *cry*III genes encode proteins of 73 kDa that are toxic to coleopterans. The *cry*IV genes encode dipteran-toxic proteins that are heterogeneous in mass, ranging from 72 to 135 kDa. These toxins are found in a composite inclusion body that also contains an unrelated cytolytic toxin (28 kDa) that may function synergistically with the *cry*IV proteins. The large protoxins encoded by *cry*I, *cry*IVA, and *cry*IVB have the toxic segment localized in the N-terminal half of the molecule, the C-terminal half not being essential to toxicity. Also, several of the toxins have a region of hydrophobicity near the N-terminal end that may be involved in interaction of the toxin with the membrane of gut epithelial cells.

Any *B. thuringiensis* isolate may possess one or more of the *cry* genes. Because the toxins ecoded by these genes differ in their toxicity for any one insect,

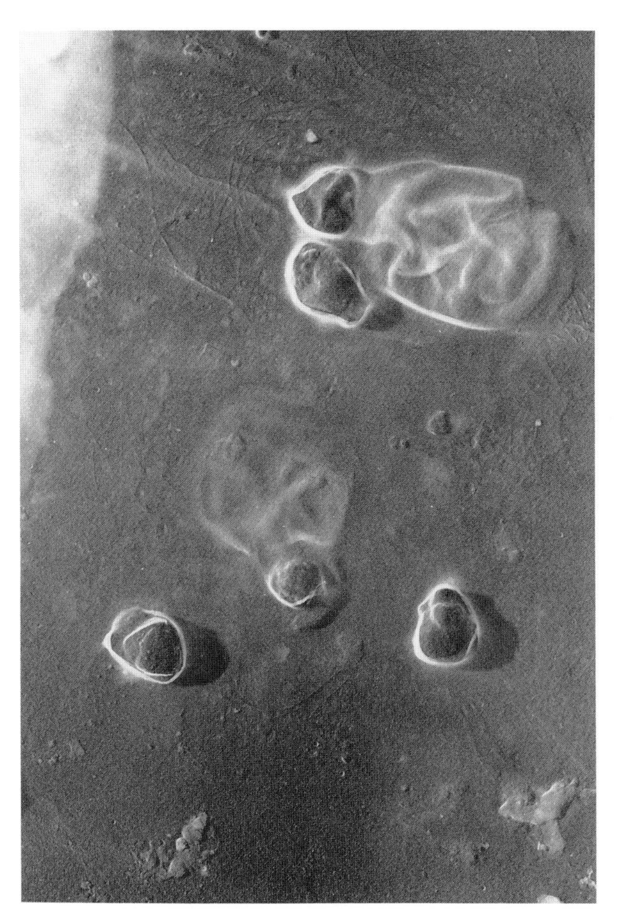

Figure 2 *B. thuringiensis* parasporal inclusions containing dipteran-active toxins. In thin sections, these inclusions reveal discrete segments held together in a single composite body.

different isolates possess different levels of toxicity for insects. This information has opened up the possibility of combining different toxin genes in a single strain to modify the host range of the bacteria included in an insecticide. It may also be possible to use site directed mutagenesis to change the amino acid sequence of a toxin to increase toxicity or to vary the spectrum of activity of the toxin.

Following ingestion by larvae, inclusions dissolve in the alkaline gut. For many of the proteins, this is followed by proteolytic cleavage converting a protoxin into an active toxin. The mode of action is controversial but may involve the introduction of pores into the membrane of susceptible cells, which is followed by a colloid-osmotic lysis, disintegration of the gut, and death of the animal.

Bacillus thuringiensis is produced on a large scale by industrial fermentation techniques. At the completion of growth, sporulation, and associated toxin production, the cells are recovered by centrifugation or filtration. Subsequent formulation depends on the target insect (i.e., whether the toxin is directed at agricultural pests on plants, forest insects, or mosquito or blackfly larvae in water). The toxin must be applied repeatedly to the area for it to be protected, much like a chemical insecticide. This is because the bacteria lack the ability to reproduce themselves in numbers adequate to provide effective insect control. Despite world-wide use of *B. thuringiensis* in agriculture for about 30 years and heavy use of a dipteran-active strain in an onchocerciasis control program in Africa during the 1980s, there have been only a few isolated reports of the development of resistance to these toxins. The high degree of specificity of the toxins provides safety to animals including beneficial insects found in the same habitats as the target pests.

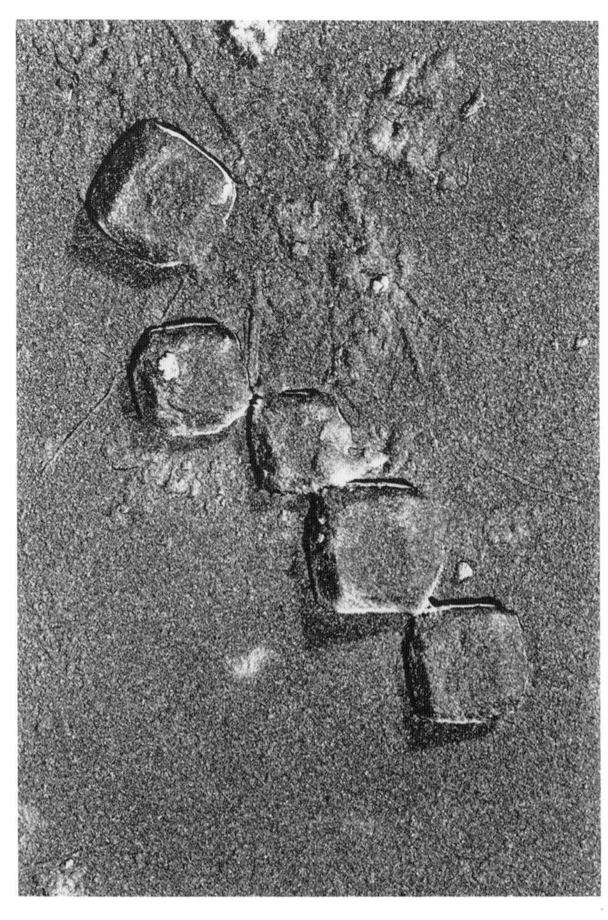

Figure 3 *B. thuringiensis* parasporal inclusions containing coleopteran-active toxin.

D. *Bacillus sphaericus*

These aerobic, gram-positive bacilli form round spores that swell the sporangium. They are common inhabitants of soil and are relatively inert in their metabolism, failing to metabolize glucose or several other sugars because of missing enzymes for sugar transport and catabolism. They are readily grown in media containing proteinaceous substrates. The majority of strains in culture collections have no pathogenicity for insects, and the first pathogenic isolate was described in 1964. DNA homology studies have shown that the pathogenic strains are distinct from the type strain of the species and from most nonpathogenic strains; however, they are not easily distinguished by the usual phenotypic tests used in taxonomic studies. The pathogenic strains are grouped by serotyping and by bacteriophage typing. Pathogenicity is limited to mosquito larvae and is caused by production of a toxin that accumulates within the bacterial cell as a parasporal inclusion at the time of sporulation. In this respect it is similar to *B. thuringiensis,* however, the toxin produced by *B. sphaericus* is not related to any of the protein toxins present in *B. thuringiensis*. Unlike the *cry*IV-encoded toxins of *B. thuringiensis,* that of *B. sphaericus* has no effect on blackflies.

The toxin of *B. sphaericus* is unusual in that it is a binary toxin. Both proteins, one of 41.9 kDa and one of 51.4 kDa, are required to kill larvae, and both occur in the parasporal inclusion and are released into the alkaline larval gut. Both are cleaved by gut proteases to smaller forms that have increased toxicity to cultured mosquito cells. The genes for both the 41.9 and the 51.4 kDa proteins have been cloned and sequenced from several isolates, and only small variations in amino acid sequence have been found between toxins from different strains. The mode of action of these toxins is unknown, but the pathology resembles that produced by the *cry*IV-encoded toxins of *B. thuringiensis*. In addition to the highly toxic strains producing the 41.9 and 51.4 binary toxin, there are strains that are much less toxic and that produce a 100-kDa protein toxin not located in a parasporal inclusion.

Although not presently produced commercially, *B. sphaericus* is of interest as a potential mosquito larvicide because it is reported to persist in the aquatic environment for a longer time than *B. thuringiensis*. The reasons for this are somewhat unclear, but it decreases the need for frequent applications of the bacteria.

E. Future Developments

Screening programs continually uncover new bacterial insect pathogens. For example, *Serratia entomophila,* pathogenic for New Zealand grass grubs; strains of *Bacillus,* pathogenic for nematodes and snails; and *Clostridium bifermentans,* pathogenic for mosquito larvae, have been described in recent years. The vast numbers of anaerobic bacteria have been largely neglected in the almost single-minded emphasis on the aerobic sporeformers. Genetic manipulation of *B. thuringiensis* toxins is already producing strains having better insecticidal performance. The genes for these toxins have been inserted into *Pseudomonas* to provide persistence in the field; into *Clavibacter xyli,* a corn endophyte, to control corn borer; into *Rhizobium* to provide the toxin in root systems; into cyanobacteria to control mosquitoes; and into plants to provide protection for a variety of crops.

II. Viruses

A. Introduction

Viruses are obligate intracellular parasites that are known from all types of organisms. Biochemically, they consist of a DNA or RNA genome surrounded by a layer of protein subunits collectively known as the capsid. Some of the more complex types of viruses also contain lipid in the form of an envelope that typically is derived from membranes of the host cell.

As they do in other types of organisms, viruses cause important diseases in insects, and it has been known for more than 50 years that some types of viruses attacking insects periodically cause spectacular declines in insect populations. The lethal effects that these viruses have on insect populations suggested that viruses could be used as "natural enemies" to control insect pests, and over the past few decades there has been a considerable effort to develop insect viruses as either classical biological control agents or viral insecticides. A classical biological control agent is a living organism or virus that, when introduced into a pest population, results in permanent establishment of the agent accompanied by reduction of the population on a long-term basis to a point where it no longer is a significant economic pest. A viral insecticide, on the other hand, does not lead to long-term control, but rather must be applied periodically, much like a synthetic chemical insecticide, though not as frequently.

Though the use of insect viruses as classical biological control agents provides an ideal solution to pest control problems, this tactic, unfortunately, is not effective in most cases. A noteworthy exception is control of the European spruce sawfly by a nuclear polyhedrosis virus in Canada. This pest was introduced into Canada during the early part of the century and was a major forest pest by the 1930s. Around 1935, a sawfly nuclear polyhedrosis virus was introduced along with parasitic wasps to control this pest, and within 5 years the sawfly was effectively controlled by these natural enemies. Subsequent studies showed that the virus was the principal factor responsible for this control. In most cases, however, viruses are used as insecticides.

The viruses used as insecticides have a very narrow host spectrum and typically are only active against the target insect pest and a few closely related species, making them much safer environmentally than most of the broad spectrum chemical insecticides. Though this is certainly an advantage, viruses also have disadvantages, a key one being that as obligate parasites mass production for commercial use requires that they be grown in living insect hosts. In this section the properties of the major types of viruses that attack insects will be reviewed, with the discussion focusing on the nuclear polyhedrosis viruses, the virus type that has been most widely used as microbial insecticide and that shows the most promise for more widespread use in the future.

B. Major Types of Viruses

1. Occluded and Nonoccluded Viruses

The viruses that attack insects are divided into two broad nontaxonomic categories, the occluded and nonoccluded viruses. The occluded viruses are so named because after formation in infected cells the mature virus particles (virions) are occluded within a protein matrix, forming protective paracrystalline bodies that are generically referred to as either inclusion or occlusion bodies. In the nonoccluded viruses, the virions occur freely or occasionally form paracrystalline arrays of virions that are also known as inclusion bodies; these, however, have no occlusion body protein interspersed among the virions.

The occluded viruses of insects include the cytoplasmic polyhedrosis viruses (cytoplasmic RNA viruses), entomopoxviruses (cytoplasmic DNA viruses), nuclear polyhedrosis viruses (nuclear DNA viruses), and granulosis viruses (nuclear/cytoplasmic DNA viruses), but the overwhelming majority of those developed or under development as microbial insecticides belong to the latter two groups, both of which belong to the insect virus family Baculoviridae. The baculoviruses are large DNA viruses that form enveloped rod-shaped virus particles that are packaged by the virus into the occlusion bodies already noted and illustrated in Figs. 4a and 4b. Literally thousands of nuclear polyhedrosis viruses (NPVs) and granulosis viruses (GVs) have been discovered, but only those that attack important lepidopteran (caterpillar) or hymenopteran (sawfly) pests are used or considered for use as insecticides.

2. General and Molecular Biology

The NPVs and GVs cause infectious diseases of insects and gain entry to their hosts primarily by being eaten. In the case of lepidopterans, insect larvae ingest occlusion bodies while feeding, and the virus invades and initially colonizes the host's stomach. Once established, the virus spreads to most other tissues where it infects and grows within the cell nuclei (NPVs) or in nuclei and the cytoplasm (GVs), eventually killing the host. The time between infection and death varies but depends mostly on the amount of virus consumed and the stage of larval development at the time of infection. Young larvae that consume several polyhedra, for example, can be killed within 48–72 hr, but older larvae that feed on the same amount of virus may not die for 1 week or longer. The initial transmission and infection process is similar for the NPVs in sawfly larvae, except that the infections are restricted to stomach cells.

Most NPVs and GVs are very specific in their host range, attacking only one or several closely related species. A few, such as the NPV isolated from the alfalfa looper, *Autographa californica*, have a broad host spectrum that extends to over 40 lepidopteran species. In comparison to chemicals, this is still a very narrow spectrum of activity and is one of the reasons that these viruses are so safe for nontarget organisms.

With respect to their molecular biology, the genome of the NPVs and GVs is a large, circular, double-stranded DNA molecule with a size of approximately 130 kilobase pairs. It is estimated that the genome of each virus is capable of encoding

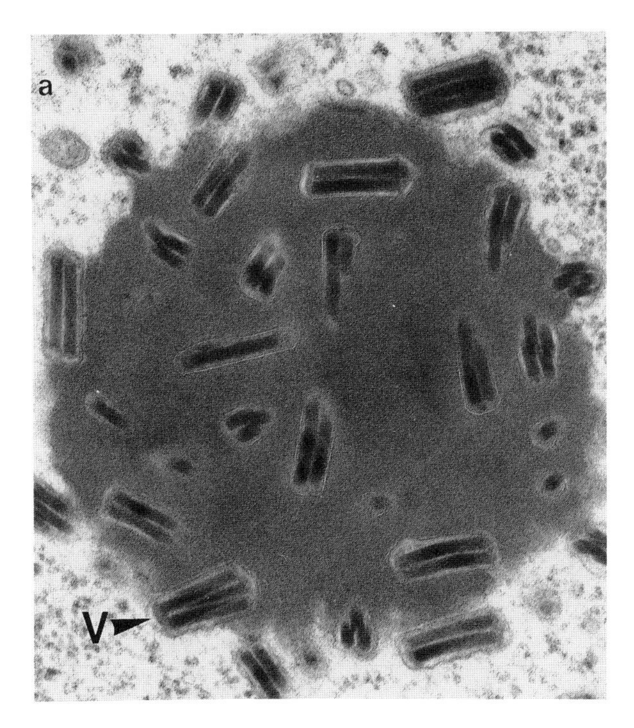

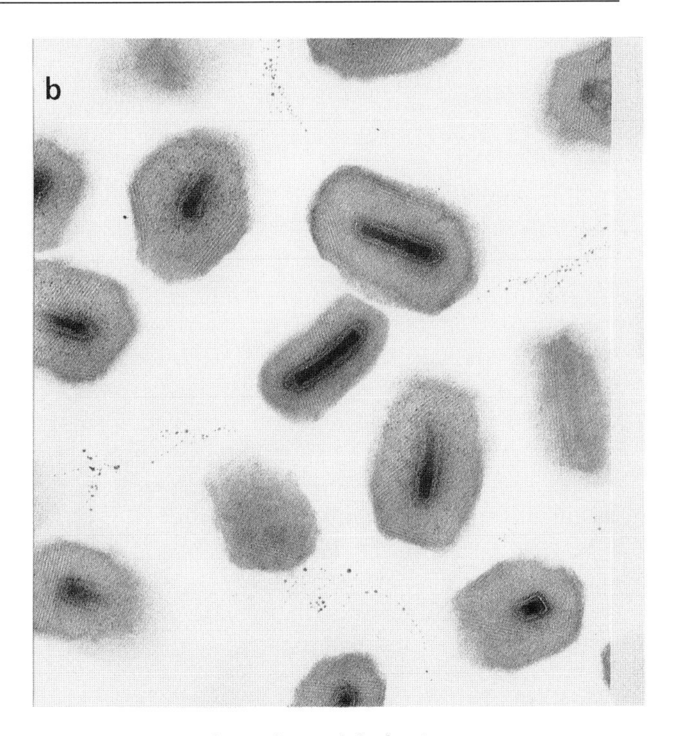

Figure 4 Electron micrographs illustrating the internal structure of occlusion bodies of a nuclear polyhedrosis virus (a) and a granulosis virus (b). In the nuclear polyhedrosis viruses, many virions (V) are occluded in each occlusion body whereas in the granulosis viruses, there is only one virion per occlusion body. The nuclear polyhedrosis viruses and granulosis viruses are the most common types of insect viruses used as insecticides. Magnification is approximately 52,000×.

about 50 proteins, making these viruses among the largest and most complex known. During viral replication, the synthesis of these proteins is divided into four major sequential phases. Virions are assembled and occluded in occlusion bodies during the last phase, which is also the time period during which the occlusion body protein, a very highly expressed protein of 29 kDa, is also synthesized. In the NPVs, numerous virions are occluded within each occlusion body, and, as these are typically polyhedral in shape, they are often referred to as polyhedra. In the GVs, each virion is occluded in a single small occlusion body, and these, being granular in shape, are referred to as granules. In the case of NPVs, hundreds of polyhedra form within each infected nucleus and, in the GVs, thousands of granules form per cell.

C. Conventional Viral Insecticides

The major insect pests, be they of agricultural or medical importance, fall into two broad groups with respect to their feeding—they are either chewing or sucking insects. Sucking insects feed directly in host tissues and are not good targets for viral insecticides because, unlike fungi, they cannot penetrate through insect cuticle. Thus, the NPVs and GVs, which must be eaten to be effective, can only be used from a practical standpoint against insects with chewing mouthparts. Fortunately, many of the most important agricultural insect pests, caterpillars for example, are chewing insects. In addition, many NPVs and GVs have been isolated from these important insect pests.

1. Target Pests: Key Examples

The primary targets for viral insecticides are the numerous caterpillar (cutworms, loopers, armyworms, and bollworms) and sawfly pests that attack crops, such as vegetables, cotton, maize, sorghum, rice, fruit crops, and forests throughout the world. Some particularly important target pests include species of *Heliothis*, such as *H. zea*, *H. virescens*, and *H. armigera*, that are major pests of grain crops and cotton in many regions of the world; species of *Spodoptera*, such as *S. littoralis* and

S. litura, major pests of cotton in the Middle East and Asia, and *S. exigua* and *S. frugiperda,* major pests of vegetable crops in many areas of the world; and species of *Mamestra* that attack vegetable crops in Europe. With respect to forests, the major targets include sawfly larvae of the genus *Neodiprion* in North America and Europe, and caterpillar pests such as larvae of the gypsy moth and Douglas fir tussock moth in the United States.

For many of the above pests, viral insecticides serve as useful methods of control when used either alone or in conjunction with reduced levels of chemical insecticides in integrated pest management programs. As in the case of chemical insecticides, the use of an insect virus as a control agent in many countries requires that it be registered as an insecticide. The following are examples of viral insecticides registered by the U.S. Environmental Protection Agency; the NPV of *Heliothis* spp. developed for use in cotton (commercial name, Elcar), the NPV of the Douglas fir tussock moth for use in forests (TM-Biocontrol 1), the NPV of the gypsy moth for use in hardwood forests (Gypcheck), and the NPV of *Neodiprion sertifer* for use against the pine sawfly. The GV of the codling moth, an important pest of apples, is currrently in the process of being registered. A similar range of viruses have been registered in Europe.

Though viruses are registered and used in industrialized countries, their use in less developed nations has actually been much more extensive. For example, a variety of different NPVs and GVs, especially those infective for *Spodoptera* and *Heliothis* spp. are widely used for control of vegetable and field crop pests in many parts of the world. The reasons for this are that registration is relatively simple or not required, labor to produce the virus is inexpensive, and the viruses are cheaper than the chemical insecticides available locally. The viruses in most of these countries are produced by farmers' cooperatives or "cottage industries." Usage appears to be increasing, which testifies to the efficacy of these viruses, at least by the standards used within these countries.

2. Production and Use

As obligate intracellular parasites, viruses can only be grown in living cells. Thus, NPVs and GVs must be produced in their larval hosts. To do this, large cultures of the target pest are reared, usually in large environmental chambers, either on an artificial diet or on a natural food source. When the larvae are slightly beyond the middle phase of their growth, they are fed an amount of virus sufficient to kill them before they pupate. This ensures maximum yield of virus. Just prior to or shortly after death, the virus-infected larvae are collected and formulated into an insecticidal powder or liquid concentrate. These insecticide powders or liquids are then applied to crops or forests using equipment and methods similar to those used to apply chemical insecticides. Another possibility for production is to grow these viruses *in vitro* in large cell cultures, but cell culture technology has not yet made this commercially feasible.

3. Limitations

Due largely to the slow speed-of-kill, narrow host spectrum, and necessity of using larvae for mass production, relatively few viruses have been registered as insecticides in the United States or Europe. One reason for this is that viruses have had to compete with available chemical insecticides, which in general have been cheaper; however, the increasing costs of chemical insecticides and the development of resistance to these in many insect populations, has resulted in increased interest in developing more viruses as insecticides. In addition, the development of recombinant DNA techniques offers the potential for increasing the host range of insect viruses in a controlled manner and the possibility for increasing the speed-of-kill.

D. Improved Recombinant Viral Insecticides

As noted above, two commercial limitations of viral insecticides are that they have a very narrow host range and if not used against young larvae, which is often not practical, they can take from 5–10 days to kill the target pest, a period that can result in significant economic loss. To overcome these limitations, techniques of recombinant DNA technology are being used to genetically engineer more efficacious viral insecticides. The basic objective here is to find a gene for an insecticidal protein that can be introduced into the viral genome under the control of a strong promoter and that will result in an expanded host range, but one still limited to important insect pests and an increased speed-of-kill. Several candidate genes have been identified, including those for insect peptide hormones, regulatory enzymes, and insecticidal proteins from mites, spiders, and scorpions. Several recombinant viruses that express these proteins have been constructed, and it has been shown that the speed-of-kill has been reduced to a period of several days in late instar larvae, as opposed to a week or more for the wild-type virus. These results indicate that the goal of producing a genetically engineered viral insecticide that is environmentally safe and commercially feasible should be a reality within the next few years. [*See* RECOMBINANT DNA, BASIC PROCEDURES.]

E. Conclusions

The successful use of viruses in developed countries in the future will continue to depend on the regulatory environment, their efficacy, and their economic competitiveness. As long as other cheaper materials, such as chemicals and *Bacillus thuringiensis*, are available and effective, they will be used. The viruses likely to be used in the future will be conventional viruses targeted against major pests for which chemicals and bacterial insecticides are not available and recombinant viruses that kill target insects more quickly and have a broader host range.

III. Fungi

A. Introduction

The entomopathogenic fungi fill an important niche in microbial control of insect pests. Virtually all insect orders are susceptible to fungal diseases. Fungi are particularly important in the control of pests that feed by sucking plant juices because these insects have no means of ingesting pathogens, and they are important for Coleoptera control because viral and bacterial diseases are unknown for many of the coleopteran pests. There are approximately 700 species of entomopathogenic fungi in almost 100 genera. Accordingly, there is the potential for developing microbial control programs with fungi for virtually all pest insect species. Only a very small percentage of possible fungus/insect combinations have been tested for their potential as microbial control systems.

From the 1880s through the early 1900s, the spectacular epizootics caused by entomopathogenic fungi led to studies of their potential use for pest control. Interest in fungi as pest control agents waned, however, as synthetic chemical insecticides were used more frequently. More recently, because of the myriad difficulties that have been gradually encountered in the use of chemical insecticides, the field of biological control has been undergoing a renaissance. In particular, the knowledge of entomopathogenic fungi is increasing rapidly.

B. Diversity of Entomopathogenic Fungi

Insect-infecting fungi are found in virtually all taxonomic groups except the higher Basidiomycetes and the dematiaceous Hyphomycetes. Entomogenous fungal species are also diverse in their degree of virulence; they range from obligate pathogens, through facultative pathogens attacking only weakened hosts, to commensal or symbiotic fungi. Research aimed at pest control usually targets obligate pathogens of economically important pests. Taxonomically more primitive fungi with motile spores have been one focus for mosquito control efforts in aquatic habitats; however, the majority of entomogenous fungi with control potential are in the Order Entomophthorales (Class Zygomycetes) or the Class Hyphomycetes. The Entomophthorales are generally characterized by species with heightened host specificity and great epizootic potential. Many species in the Hyphomycetes have broader host ranges and are generally easier to grow *in vitro*. As a result, almost all species that have been registered for use, are being actively used, or both are in the Hyphomycetes.

In addition to having a wide spectrum of hosts, entomopathogenic fungi often have wide geographic ranges. Considerable genetic diversity can be found among various isolates of single species from different hosts and localities. In addition to naturally occurring fungal isolates, there is the potential to modify characteristics of entomopathogenic fungi through genetic manipulation techniques. Three of the most common hyphomycetous entomopathogenic fungi have been transformed in this way.

C. Pathogenicity

In virtually all cases, the infective unit for an entomopathogenic fungal species is a spore, usually a conidium. The course of disease development is initiated by the spore adhering to the cuticle of an insect. The spore germinates and the resulting germ tube either penetrates the cuticle directly or produces an appressorium from which an infection peg grows into the insect cuticle.

An understanding of the mechanisms for triggers, receptors, and the biochemical machinery for appressorium initiation and their relationship to the expression of virulence by insect pathogens has been slow to form; however, the development of an *in vitro* system whereby differentiation of *Metarhizium anisopliae* is induced when conidia germinate on a flat hydrophobic surface in a sparce nutritional environment has facilitated meaningful examinations of the biological significance and function of appressoria. A model for appressorium formation has been proposed in which a localized induction signal disrupts the apical Ca^{2+} gradient required for localized exocytosis and maintenance of polar germ-tube growth. This initiates appressorium formation by turgor pressure against an expanded area of the cell wall produced by randomly dispersed exocytosis of new wall material over the entire cell surface. The fungus penetrates to the hemocoel where it normally grows in a yeastlike phase called hyphal bodies or blastospores. A fungus may produce toxins in the hemocoel that aid in overcoming the immune response of the host or in causing other disruptions of host physiology. The fungus eventually invades virtually all internal organs of the insect after which it penetrates to the outside of the cuticle and produces new conidia. In dry conditions, the fungus may lie dormant within the dead host for long periods of time rather than emerging to the exterior and producing conidia. Some Entomophthorales produce heavy-walled resting spores that remain within cadavers and serve as the overwintering stage of these fungi. [*See* Hypha, Fungal.]

A full understanding of the factors affecting the ability of the fungus to enter, kill, and subsequently sporulate on its host is paramount in the development of fungi as control agents. Factors that interfere in any of the critical steps could render the pathogen less efficient, if not ineffective. Only recently have the complex interactions occurring between the host integument and fungal pathogen been explored in detail. Features affecting these interactions include molting, host defense mechanisms, fungal strain, presence of nutrients on host integument, and humidity.

A possible indirect use of entomopathogenic fungi for insect control would be the detection, isolation, characterization, and, finally, commercial development of toxins produced in culture by these fungi. The secondary metabolites of entomopathogenic fungi include a number of compounds toxic to insects. The known toxic compounds are quite diverse chemically but a number of them (e.g., destruxins and bassianolide) are depsipeptides.

D. Current Approaches to Field Use

1. Permanent Introduction

Frequently called classical biological control, permanent introduction entails establishment of a fungal species in a area with host populations where the pathogen does not occur. This method for control is clearly the least labor-intensive and least costly over the long-term because it involves a limited number of releases; its results are aimed at naturally occurring long-term control.

The pathogen, *Entomophaga maimaiga,* has been used successfully in both classical biological control and inoculative release efforts against the gypsy moth. *Entomophaga maimaiga* was introduced from Japan to northeastern North America in 1910–1911 to control gypsy moth populations that had been introduced from France four decades earlier. This fungus was not believed to have become established until in 1989 and 1990 it caused extensive mortality in gypsy moth populations in 10 northeastern states. *Entomophaga maimaiga* is now distributed in those areas where the gypsy moth has been established for some time. Therefore, at present, efforts are under way to introduce this fungus to the leading edge of the ever-increasing moth distribution.

Although classical biological control introductions are frequently attempted with parasitic insects, colonization attempts of exotic fungi to control exotic pests have seldom been reported in the literature. Obviously this tactic is underemployed in biological control programs.

2. Inoculative Augmentation

Inoculative augmentation involves releasing a pathogen in the field with the expectation that it will cycle in the host population to provide effective control. Frequently, inoculative releases are repeated during a season (mycoinsecticide), and it is not expected that effective populations of the pathogen will carry through to the next year. This method has been used with fungi most frequently for pest control on an annual basis.

The coleopterous family Scarabaeidae contains many species whose larval stages are devastating to roots and whose adults may or may not cause foliar damage to agricultural and horticultural plants. The group includes Japanese beetles, European cockchafers, and rhinoceros beetles. The larval stages, which may last 1 to 3 years, usually remain within the root zone. This makes application of fungi, as well as other types of control, difficult. A novel, long-term approach to control of the cockchafer is under development in Switzerland where the insect has a 3 year life cycle and the population is synchronous. Adults are present only once every 3 years. Adults aggregate at the edges of pastures, where they previously developed through the immature stages, and feed and mate in the surrounding trees. *Beauveria brongniartii* blastospores are sprayed on these border trees. The adult female, after feeding, returns to the pasture where she burrows several centimeters into the ground to oviposit. Many fungus-exposed adults die in the field either underground or on the surface. The fungus sporulates on the cadavers, thereby introducing foci of large numbers of conidia throughout the field. It has been noted that, after the second generation following application, the fungus maintains the cockchafer population at nondamaging levels.

Spittle bugs are serious pests in sugar cane and improved pasture lands in Brazil. Conidia of *Metarhizium anisopliae* have been used for control of these insects for more than 10 years. Production of the fungus is by small companies or grower cooperatives. The spittle bug populations in sugar cane fields normally are reduced by about 40% following fungus application. Another attractive feature of fungus use, in contrast to chemical insecticides, is that hymenopterous parasites used for control of a lepidopterous borer of cane are not affected. Since 1986 approximately 100,000 hectares of cane have been treated annually.

Perhaps the largest program entailing fungi for insect control is that of the People's Republic of China to treat pine forests with *Beauvaria bassiana* conidia for control of pine moth larvae. At least 1 million hectares are involved and applications are usually needed at 3-year intervals.

3. Conservation or Environmental Manipulation

This involves manipulation of the host environment to enhance activity of the fungal pathogen. Such manipulations have proven very successful in systems that are well understood, and use of this approach is clearly a goal of integrated pest management. This approach would include such tactics as selection of chemical pesticides or timing of chemical pesticide application to cause minimal damage to the entomopathogenic fungi. Because of moisture requirements of most fungi infecting insects, environmental manipulations generally center around maintaining the pathogen in a moist environment where pests are also abundant. An elegant example of environmental manipulation of fungal entomopathogens is a system developed for control of the alfalfa weevil. This is based on cutting alfalfa early and leaving it in windrows for several days. The adult weevils aggregate in the windrows, which provide a moist and warm microclimate for transmission and development of *Erynia* spp. Simulation modeling of this system, with the addition of early season sampling to regulate insecticide application, has led to altered alfalfa weevil control recommendations and projected increased net profits to growers.

4. Compatibility in Integrated Pest Management

The restricted host ranges of fungi offer control with limited threat to nontarget organisms, including compatibility with parasites and parasitoids and other pathogens. The most successful examples of integrated use are found in greenhouse pest control where *Verticillium lecanii* and *Aschersonia aleyrodis* can be used with predatory mites and beneficial wasps for effective control of aphids and whiteflies, respectively. Also, *Metarhizium* is used in Brazil for spittle bug control in cane fields simultaneously with release of *Apanteles* wasps parasitic on sugar cane borers.

E. Mass Production and Formulation

Although some potentially useful entomopathogenic fungi are difficult to mass produce, the great majority grow profusely on very simple, inexpensive media. The most common procedure for large-scale production is to grow the fungus on plant materials such as bran, cracked barley, rice, or even peat soil, until sporulation. The resulting conidia, with or without plant substrate, are introduced into the field, usually by means of standard pesticide application equipment.

A recent breakthrough has been the development of a technique for producing dry, viable mycelium that can be stored under refrigeration. When introduced into fields, particles of dry mycelium rapidly produce infectious conidia following rehydration from dew, rain, or soil moisture. This method can be used with many species of fungi and has been successfully field tested.

Formulation is frequently very simple or nonexistent. Nevertheless, development of proper formulations is crucially needed. These would be designed to improve shelf life, persistence in the field, range of conditions under which the fungus is effective, spread in the insect habitat, and adherence to substrates where insects reside.

F. Biotic and Abiotic Limitations

As with other biological control agents, fungi are limited by an array of biotic and abiotic factors. Biotic limitations are poorly understood. These include microbial antagonists on the host integument, leaf surface or in the soil; host feeding behavior, physiological condition, including humoral and cellular immune reactions, and age; and fungal strain.

Abiotic factors include inactivation by sunlight, desiccation, certain insecticides or fungicides, and temperature and humidity thresholds for germination and growth. Traditionally, because of a germination requirement for free water or a relative humidity of over 90%, relative macrohumidity has been considered as the most serious constraint on the use of fungi for insect control; however, studies indicate that fungal infections can occur at relatively low macrohumidities. Microhumidities at the surface of the host integument or on the foliage may be sufficient for spore gemination and host penetration. Incompatibility with other components of pest control such as insecticides and fungicides can also be a serious limitation.

G. Future Research Approaches with Fungi

Although entomopathogenic fungi have obvious high potential for pest control, this potential has seldom been properly exploited. With the increased interest in use of fungi in other biotechnologies, major improvements in mass cultivation, storage, and formulation technology can be expected. These advances, coupled with increased understanding of pathogenesis with emphasis on epizootiology, host specificity, virulence and use of recombinant DNA technology, fungi should play an increasingly important role in integrated pest management.

Bibliography

Baumann, P., Clark, M., Baumann, L., and Broadwell, A. (1991). *Microbiol. Rev.* **55,** 425–436.

Blissard, G. W., and Rohrmann, G. F. (1990). *Ann. Rev. Entomol.* **35,** 127–155.

Burges, H. D. (ed.) (1981). "Microbial Control of Pests and Plant Diseases 1970–1980." Academic Press, London, 949, pp.

Granados, R., and Federici, B. (1986). "The Biology of Baculoviruses," Vol. I and II. CRC Press, Boca Raton, Florida.

Hofte, H., and Whiteley, H. R. (1989). *Microbiol. Rev.* **53,** 242–255.

Payne, C. (1988). *Phil. Trans. R. Soc. Lond. B.* **318,** 225–248.

Roberts, D. W. (1989). *Mem. Inst. Oswaldo Cruz, Rio de Janeiro.* **84,** 89–100.

Roberts, D. W., Fuxa, J., Gaugler, R., Goettel, M., Jaques, R., and Maddox, J. (1991). Use of pathogens in insect control. *In* "Handbook of Pest Management in Agriculture" (D. Pimentel, ed.), 2nd ed., pp. 243–278. CRC Press. Boca Raton, Florida.

Samson, R., Evans, H., and Latge, J. P. (1988). "Atlas of Entomopathogenic Fungi." Springer-Verlag, The Netherlands. 187 pp.

Stahly, D. P., Andrews, R. E., and Yousten, A. A. (1991). The Genus *Bacillus:* Insect Pathogens. *In* "The Procaryotes" (A. Balows, H. Truper, M. Dworkin, W. Harder, and K. Scheifer, eds.), 2nd ed., pp. 1697–1745. Springer-Verlag, New York.

Interferon

Charles E. Samuel
University of California, Santa Barbara

Glossary

Antiviral state Physiologic condition of the host cell characterized by a reduced ability of viruses to multiply

Induced protein Protein whose production is increased in response to a stimulus such as treatment of cells with interferon

Interferon Cellular protein synthesized in response to viral infection or antigen stimulation, which functions to regulate the expression of cellular genes, the products of which may affect several biologic processes including virus replication, cell growth, and the immune response

Transcription Process by which a complementary single-stranded RNA (messenger RNA) is synthesized from a template gene, which, for cells, is double-stranded DNA

Translation Process by which a polypeptide chain composed of amino acids is synthesized from the genetic information encoded in a messenger RNA (mRNA) molecule, using cellular ribosomes, transfer RNA, and associated enzymes and factors to decipher the nucleotide code of the mRNA

Virus Small infectious agent composed of a nucleic acid genome (RNA or DNA) surrounded by a protein capsid coat and, in some cases, also by a lipid envelope; they require a host cell, minimally for energy production and protein production, in order to multiply

INTERFERONS (IFNs) are a family of regulatory proteins that can affect a number of functions in animal cells, including the ability of viruses to multiply. The name "interferon" derives from the property by which the class of proteins was discovered—interferons were identified as secreted factors produced by influenza virus-infected cells that can transfer a virus-resistance state to uninfected cells and, thus, "interfere" with subsequent virus replication. This fundamental observation concerning viral interference, made in London, England, in 1957 by the late Alick Isaacs and Jean Lindenmann, led to a wealth of knowledge about animal cell molecular and cellular biology through the subsequent study of the IFN system.

The IFN system, as shown by Fig. 1, includes the cells that produce IFN and the cells that respond to IFN. Considerable progress has been made in understanding the molecular biology of the human IFN system. The genes encoding the IFNs, their receptors, and the proteins that mediate many of their biologic effects have been molecularly cloned and characterized. The availability of complete complementary DNA (cDNA) clones of components of the IFN systems contributed significantly to elucidating both the biology and the biochemistry of the antiviral actions of IFNs.

I. Interferon Genes, Proteins, and Receptors

A. Type I (α, β, B)

The family of IFN proteins may be grouped into two types, I and II, on the basis of their physical and biological properties. Type I IFNs are induced by virus infection. Type I IFNs constitute a multigene famly; they include the alpha (α) or leukocyte (α_I) IFNs, the beta (β) or fibroblast IFNs, and the omega ω or trophoblast (α_{II}) IFNs.

The type I human IFN genes, proteins, and receptors have been extensively studied. There are more than 20 human IFN-α genes, but only one IFN-β gene. All of the type I IFN genes lack introns. They

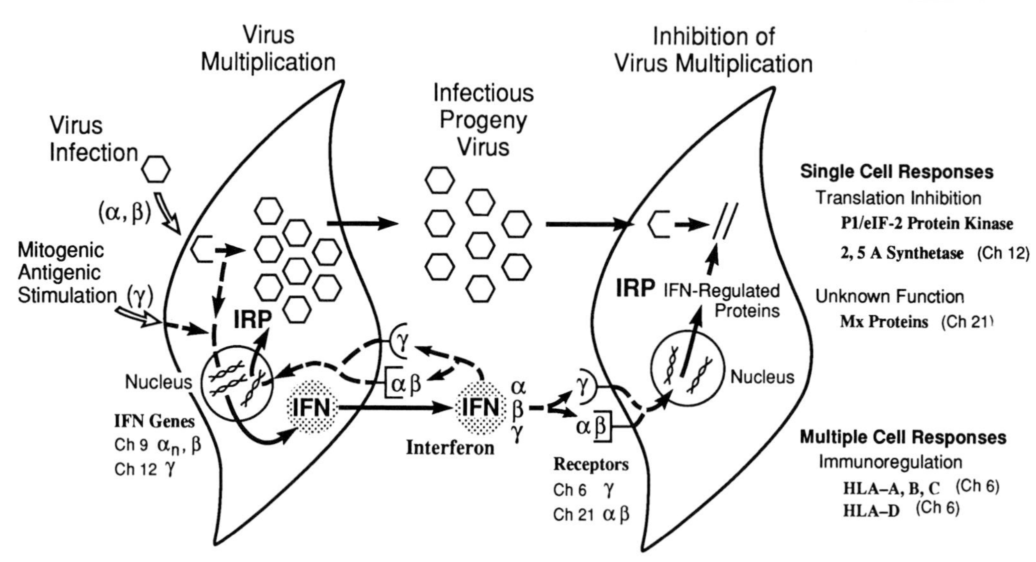

Figure 1 Schematic summary diagram of the interferon (IFN) system based on information obtained from the study of human cells. The illustration pictures virus particles as open hexagons and IFN proteins as shaded circles. The infected (stimulated) cell shown on the left is depicted as the IFN-producing cell in response to induction with virus particles (α- and β-IFNs) or mitogens and antigens (γ IFN). The IFN-treated cell shown on the right is responding to the presence of IFN by producing new proteins that block virus replication. Some of the IFN-regulated proteins (IRP) responsible for the inhibition of virus replication within single cells (P1/eIF-2α protein kinase; 2′,5′-oligoadenylate synthetase; protein Mx) or within whole animals via multiple cell responses [histocompatibility antigens (HLA)] are listed on the right. The chromosome (Ch) assignment of the IFN genes, IFN receptors, and IFN-regulated proteins is for human cells.

are closely linked, mapping to the short arm of human chromosome 9.

The human IFN-α proteins, with one exception, are all 166 amino acids in length. Most α IFNs are not glycosylated, and they function as monomers. The single β IFN protein, similar to the α IFN proteins, is likewise 166 amino acids. However, unlike the α IFNs, the β IFN is N-glycosylated and functions as a dimer. The IFN-ω proteins are larger, 172 amino acids, and they are N-glycosylated.

IFNs exert their actions through type-specific cell-surface receptors, which are also often species-specific. The human IFN-α receptor, encoded on chromosome 21, has been molecularly cloned. Mouse cells normally do not bind or respond biologically to human α IFN. However, transfected mouse cells expressing the cDNA for the human IFN-α receptor display two biochemical and biological activities expected of human IFN-α receptor-positive cells: They possess binding sites for human IFN-α, and they are sensitive to the antiviral activity of human IFN-α. Competitive binding studies indicate that the human type I IFN-α and IFN-β subspecies share a common receptor. However, transfected mouse cells expressing the human IFN-α receptor exhibit a poor sensitivity to human IFN-β. A converse functional dissociation between human IFN-α and IFN-β has been observed with human cell variants that lack IFN-α binding sites, yet are still sensitive to IFN-β.

B. Type II (γ)

The type II IFN, known as gamma (γ) or immune IFN, is induced by mitogenic or antigenic stimulation of T lymphocytes and natural killer cells. In contrast to the type I IFN genes, the single human type II γ IFN gene possesses three introns and maps to the long arm of chromosome 12. The type II IFN-γ protein is smaller than the type I IFN proteins; IFN-γ is 146 amino acids in size. IFN-γ is N-glycosylated and functions as a tetramer.

The subunit of the human IFN-γ cell-surface receptor to which the IFN-γ protein binds has also been molecularly cloned; it is encoded on human chromosome 6. However, for cells to become sensitive to the antiviral activity of human IFN-γ, additional species-specific accessory factor(s) or receptor subunits are required along with the subunit encoded on human chromosome 6. At least one of the accessory proteins required for antiviral activity is encoded by human chromosome 21.

The IFN receptor-mediated signal transduction pathways responsible for induction of the synthesis of the interferon-regulated protein (IRP) products have not yet been clearly resolved. Several second-messenger signal transduction pathways have been implicated in IFN action, including G proteins and protein kinase A, calcium and phospholipid-dependent protein kinase C, altered arachidonic acid metabolism, and altered Na^+/H^+ exchange and cytoplasmic alkalinization.

II. Biological Effects of Interferons

IFNs possess many biologic activities. IFNs can modulate diverse cellular functions ranging from cell growth and differentiation and the immune response to affecting the efficiency of virus multiplication. Many of the biological effects of IFNs can be dissociated and are due to different molecular mechanisms. Furthermore, the actions of the type I and type II IFNs are often synergistic, consistent with the notion that the IFN-induced proteins may act either singularly or in combination with one another by different molecular mechanisms.

At the biologic level, the antiviral effects of IFN may be viewed to be virus-type nonspecific. That is, treatment of cells with one IFN type or subspecies often leads to the generation of an antiviral state effective against many different RNA and DNA animal viruses. However, at the biochemical level, the antiviral action of IFN is often virus-type selective. That is, the apparent molecular mechanism primarily responsible for the inhibition of virus replication may differ considerably between virus types and even between host cells. The relative sensitivity to IFN of different animal viruses varies tremendously. All three of the fundamental components involved in the IFN-mediated antiviral response may play a role in determining the relative effectiveness of the response: the species of IFN administered, the kind of cell treated with IFN, and the type of virus used to challenge the IFN-treated host cell. Furthermore, the relative sensitivity to IFN observed for a particular IFN–cell–virus combination is the result of an equilibrium among the many agonists and antagonists that contribute to the overall response.

III. Mechanisms of Interferon Action

The antiviral activity of IFNs, the property that led to their discovery by Isaacs and Lindenmann, remains one of the most widely studied aspects of IFN research. Much of our knowledge about the biochemical basis of IFN action derives from antiviral studies. Both in cell culture and in animals, IFNs induce the synthesis of proteins known as IRPs, which cause the inhibition of virus multiplication.

Among the more than two dozen IRPs that have been identified are four IFN-induced proteins that have been especially well characterized: the P1/eIF-2α protein kinase, the 2′,5′-oligoadenylate synthetase, protein Mx, and the major histocompatibility antigens.

A. P1/eIF-2α Protein Kinase

A protein kinase, designated the P1/eIF-2α protein kinase, is induced following treatment of most species of animal cells with type I IFNs. The IFN-induced human P1/eIF-2α protein kinase is a cyclic adenosine monophosphate (cAMP)-independent, serine–threonine protein kinase that depends on RNA for activation. The P1/eIF-2α kinase, when activated by an RNA-dependent autophosphorylation, catalyzes the phosphorylation of protein synthesis initiation factor eIF-2α. Phosphorylation of eIF-2α causes an inhibition of protein synthesis at the initiation step of translation.

Considerable evidence supports the notion that the P1/eIF-2α protein kinase plays a central role in the establishment of an antiviral state effective against a number of different kinds of animal viruses. Inhibition of viral protein synthesis is a principal cause of the IFN-induced inhibition of virus multiplication. This inhibition apparently is due to an alteration of a component of the translation machinery other than the messenger RNA (mRNA) template; the IFN-induced inhibition of viral mRNA translation correlates with the induction and activation of the P1/eIF-2α protein kinase.

A wide array of different viruses have developed effective but varied measures to modulate the activity of the kinase. Several viruses have effective mechanisms to prevent the function of the IFN-induced, RNA-dependent P1/eIF-2 protein kinase, including adenovirus, human immunodeficiency virus (HIV), influenza virus, poliovirus, reovirus, and vaccinia virus. This broadly based ability of both DNA and RNA viruses to inhibit the P1/eIF-2α protein kinase may occur by several different mechanisms—for example, by production of an RNA that blocks activation (adenovirus VA RNA; HIV TAR RNA) or by production of a protein that either binds the activator RNA (reovirus $\sigma 3$, vaccinia virus SKIF) or causes the degradation of the kinase (poliovirus).

The ability of viruses to block the function of the RNA-dependent P1/eIF-2α protein kinase is balanced against their ability to activate the kinase. Virus-specific RNAs that are capable of activating the kinase are synthesized by many viruses including adenovirus, reovirus, HIV, encephalomyocarditis (EMC) virus, mengovirus, and poliovirus. The balance between virus-mediated activation and inhibition of the P1/eIF-2α protein kinase may reflect an important role of the kinase in translational control in animal cells even in the absence of treatment with exogenous IFN.

B. 2′,5′-Oligoadenylate Synthetase

The 2′,5′-oligoadenylate synthetase is one of the three enzymes that collectively constitute the 2′,5′-oligoadenylate (2-5A) synthetase-nuclease pathway. The other two enzymes are a 2′,5′-oligoadenylate-dependent endoribonuclease and a 2′,5′-phosphodiesterase. The 2-5A synthetase is an IFN-induced protein that, like the P1/eIF-2α protein kinase, depends on RNA for its activation. The activated 2-5A synthetase catalyzes the synthesis from adenosine triphosphate (ATP) of a unique family of oligonucleotide products of the general structure ppp $(A2'p5')_nA$, abbreviated 2-5A. The only biochemical function so far identified for 2-5A is as an activator of an endoribonuclease, designated RNase L or F, which is present in a latent form in most types of animal cells, both untreated and IFN-treated. The 2-5A-activated RNase degrades both viral and cellular single-stranded RNA. The 2′,5′-phosphodiesterase, likewise typically present in both untreated and IFN-treated cells, catalyzes the degradation of 2-5A to yield AMP plus ATP.

Different forms of the human synthetase have been identified by immunological, biochemical, and molecular genetic criteria. Two of the forms (40 kDa and 46 kDa) of the enzyme are derived from a single gene by differential splicing, yielding 2-5A synthetase enzymes that differ at their C-terminal regions. The gene for the small (40–46 kDa) forms of human 2-5A synthetase has been mapped to chromosome 12.

Among the many different families of animal viruses examined, only in the case of picornaviruses (Mengo virus, EMC virus) does a strong correlation seem to exist between the activation of the IFN-induced 2-5A synthetase and resistance to viral infection. The 2-5A synthetase is not sufficient to establish an antiviral state against either vesicular stomatitis (VS) virus or herpes simplex virus (HSV-2). The possible antiviral activities, and normal cellular functions of the different forms of 2-5A synthetase have not yet been unequivocally resolved. The different forms of 2-5A synthetase are localized to different subcellular sites and display different requirements for activation by double-stranded RNA, different IFN dose responses, and different kinetics of synthesis. It is therefore possible that the different enzyme forms are involved in different biological activities modulated by IFN.

C. Mx Proteins

Cellular resistance against orthoMyxovirus infection in mice is due in large part to a trait inherited by a single dominant allele, designated Mx1. While the Mx proteins play a major role in controlling influenza pathogenicity in mice, all other mammalian species so far examined that are normally prone to influenza virus infection, including humans, rats, pigs, and horses, also respond to IFN treatment by synthesizing one or more Mx-related proteins. Both type I IFNs, α and β, induce the Mx proteins; type II γ IFN, by contrast, is a poor inducer of the Mx proteins. Human cells treated with either α or β IFN synthesize two proteins, MxA and MxB, that exhibit high homology to the mouse Mx protein. The human *MxA* and *MxB* genes encoding the human Mx proteins both map to chromosome 21. However, in contrast to the murine Mx1 protein, which is found in the nucleus in IFN-treated Mx^+ mouse cells, the human Mx proteins are found predominantly, if not exclusively, in the cytoplasm of IFN-treated human cells.

A combination of genetic and biochemical evidence strongly suggests that Mx proteins possess intrinsic antiviral activity. Studies with cells in culture transfected with cDNA copies of *Mx* genes, and with *Mx*-transgenic mice, indicate that the expression of the Mx protein is sufficient to establish an antiviral state that is often selective against influenza virus. Picornaviruses (Mengo virus and EMC virus) and a togavirus (Semliki forest virus) are not inhibited by Mx, but some Mx proteins do inhibit the rhabdovirus, VS virus. However, the molecular mechanism by which the Mx proteins act to inhibit virus multiplication is not yet resolved. Because the Mx proteins from different species may each display unique antiviral activity and subcellular localization, the principal molecular mechanism may not be identical for all Mx-like proteins. Indeed, the biochemical activities of the cellular Mx proteins appear quite varied and range from identification as a histocompatibility antigen to a component of microtubule-based motility or protein sorting pathways.

D. Major Histocompatibility Antigens

The expression of the human major histocompatibility antigens (HLAs) is induced by IFN treatment of human cells. Both type I (α and β) and type II (γ) IFNs induce HLA expression. Furthermore, both class I (HLA-A, -B, and -C) and class II (HLA-D) HLA expression is induced by IFN treatment. Expression of the highly polymorphic HLAs is essential for immunocompetent cells to present foreign antigens, including peptides derived from viral proteins, to T cells during the course of generation of specific immune responses. IFN-induced expression of histocompatibility antigens may represent an important contribution to the antiviral actions of IFN within the whole animal at the cell–cell level, perhaps by enhancement of the antigen-specific lytic effect of cytotoxic T lymphocytes.

IV. Clinical Studies of Interferons

Three classes of the human IFN, α, β, and γ, display significant clinical activity. IFN-α and IFN-γ have been approved by the U.S. Food and Drug Administration: IFN-α for hairy cell leukemia and genital warts and IFN-γ for chronic granulosis disease. The IFNs also show clinical activity against other diseases, including hepatitis B and C and chronic myelogenous leukemia (IFN-α) and multiple sclerosis (IFN-β). It is now clear that IFNs are indeed important therapeutic agents. The present availability of large quantities of these important proteins, produced by biotechnology from recombinant DNA clones, should facilitate the further clinical development of IFN as a therapeutic agent for the treatment of cancers, viral diseases, and other clinical disorders.

Bibliography

Borden, E. C. (1988). *Pharmac. Ther.* **37,** 213–229.

DeMaeyer, E., and DeMaeyer-Guignard, J. (1988). "Interferons and Other Regulatory Cytokines." John Wiley & Sons, New York.

Fields, B. N., and Knipe, D. M. (1990). "Virology," 2nd ed. Raven Press, New York.

Isaacs, A, and Lindenmann, J. (1957). *Proc. R. Soc. Lond. B.* **147,** 258–267.

Pestka, S., Langer, J. A., Zoon, K. C., and Samuel, C. E. (1987). *Annu. Rev. Biochem.* **56,** 727–777.

Samuel, C. E. (1988). *Prog. Nucleic Acid Res. Mol. Biol.* **35,** 27–72.

Samuel, C. E. (1991). *Virology* **183,** 1–11.

Weissmann, C., and Weber, H. (1986). *Prog. Nucleic Acid Res. Mol. Biol.* **33,** 251–300.

Interleukins

Michael L. Misfeldt
University of Missouri-Columbia School of Medicine

Glossary

Class I/II antigens Classes of molecules encoded within the major histocompatibility complex (MHC)

Cytokines Factors produced by cells that affect other cells

Effector cells Term that in context means those lymphocytes or phagocytes that produce the end effect

Interferons Heterogeneous group of molecules elaborated by infected or stimulated host cells that protect noninfected cells from viral infection

Interleukins Factors released from leukocytes or other cells that have defined biologic activity

Lymphokines Soluble products of lymphocytes, other than antibodies, that are involved in signaling between cells of the immune system

INTERLEUKINS are defined as proteins produced by leukocytes that can affect leukocytes or other target cells. An interleukin designation is assigned to any protein that fulfills this definition. However, additional proteins fit the classification as interleukins but were named before this classification. These proteins as well as interleukins are termed cytokines because they function as intercellular signals that regulate local and/or systemic inflammatory responses. Cytokines that are produced by lymphocytes are termed lymphokines, whereas peptides produced by monocytes or macrophages are given the term monokines. Thus, the terms cytokines, lymphokines, and interleukins may be used interchangeably to designate those peptide molecules that modulate host responses to foreign antigens or host injury by regulating the growth, mobility, and differentiation of leukocytes and other cells.

I. General Properties of Interleukins/Cytokines

Interleukins (ILs) are defined as proteins produced by leukocytes that can affect leukocytes or other target cells. These proteins can either act as short-range messengers between cells interacting in an immune response or as hormones circulating in the blood that can affect cells at distant sites. The term "interleukin" was introduced in 1979 and used to designate such proteins. An IL designation is assigned to any protein that fits the general criteria of a protein produced by leukocytes that affects other cells and for which the gene encoding that protein has been cloned and the amino acid sequence of the product has been determined. A number of well-known proteins that fit the criteria for classification as ILs, the interferons (IFNs), transforming growth factors (TGFs), and tumor necrosis factors (TNFs), however, have not been assigned an IL designation even though they clearly fit the criteria because they were named before the introduction of the IL designation. Yet, these proteins, like ILs, function as cytokines. Therefore, all these protein molecules will be discussed in this review. The characteristics of these proteins are listed in Table I.

Our understanding of ILs has been greatly facilitated by a number of major technological developments that have occurred in recent years, including the ability to isolate single lymphocytes and grow them into clones of identical cells. In addition, developments in monoclonal antibody and molecular cloning techniques have enhanced our abilities to isolate and purify large quantities of homogeneous protein material for biochemical analysis. In addition, the development of recombinant DNA technol-

Encyclopedia of Microbiology, Volume 2

Table I Characteristics of Interleukins/Cytokines

Cytokine	Molecular weight	Principal cell sources
IL-1	17,500	Macrophages and others
IL-2	15,500	T lymphocytes
IL-3	14,000–28,000	T lymphocytes
IL-4	20,000	$CD4^+$ T lymphocytes
IL-5	18,000	$CD4^+$ T lymphocytes
IL-6	22,000–30,000	Fibroblasts/monocytes/T lymphocytes
IL-7	25,000	Stromal cells
IL-8	8800	Monocytes and others
IL-10	16,000–21,000	$CD4^+$ T lymphocytes
TNFα, β	17,000–18,000	Macrophages/$CD4^+$ T lymphocytes and others
IFNα, β, γ	18,000–25,000	Leukocytes/fibroblasts/T lymphocytes
TGFβ	25,000	Platelets/activated T lymphocytes and B lymphocytes

ogy has enabled researchers to produce ILs in sufficient quantities, free from contamination with other biologically active molecules that has enabled researchers to functionally characterize these protein molecules. Thus, because of these advances, there has been a tremendous explosion in not only the number of proteins that have been given an IL designation but also in our understanding of the biological activities that are associated with a given IL.

With the recent availability of purified recombinant ILs/cytokines, experimental data has now been obtained that indicates that cytokines function in a pleiotropic manner. Certain cytokines may also possess overlapping or redundant biological activities. As cytokines represent antigen nonspecific molecules, their specificity depends on the cells that produce them and the target cells that respond to them. The target cell response is dependent on the expression of specific cytokine receptors. In addition, target cell lineage and stage of differentiation may affect the target cell response. Finally, differences in processing rates between receptors for different cytokines on a single target cell type may also affect target cell responses. Because it is unlikely that each and every cytokine would be produced in response to an antigenic challenge, the specific target cells stimulated by antigen determines which cytokines will be produced and the effector mechanisms activated. Therefore, the specific target cells present and the specific cytokines produced at the site of initial antigenic challenge will determine the type of immune response generated.

II. Interleukin/Cytokine Receptors

Although cytokine receptors are specific for a given cytokine (Table II), cytokine receptors can be grouped into families based on certain shared features. One group of receptors has been designated the hemopoetin receptor family, based on the fact that most ligands act on cells of the immune system. Members of this family include the receptors for IL-2 (IL-2Rβ or p75), IL-3, IL-4, IL-6, and IL-7. These receptors are integral membrane glycoproteins with an orientation of the N termini outside the plasma membrane and a single hydrophobic membrane-spanning region. The major region of structural homology is at the N termini, and this region contains four conserved cysteines that form disulfide bonds and contribute to the tertiary structure of the receptor. A Trp-Ser-X-Trp-Ser motif has also been observed in this receptor family and is located close to the transmembrane domain. Thus, there are similarities between the extracellular regions of this family of receptors. However, the cytoplasmic domains of these receptors are for the most part unrelated and more than likely trigger unique intracellular signals that determine the specific cellular response.

The second receptor family is the TNF receptor family. These receptors are also integral membrane glycoproteins that are structurally related at the primary sequence level. The observed homology lies within the N-terminal extracellular domain. Recently, there have been reports that indicate that a soluble version of the TNF receptor may exist. This soluble protein, which has been shown to bind both forms of TNF, TNFα and TNFβ, could function as an important immunoregulatory molecule.

The final family of receptors is the immunoglobulin superfamily receptor family. These cell surface molecules are composed of a series of independently folded domains, each of which is held together by a centrally located disulfide bond. Of the human cytokine receptors, only the IL-1 and IL-6 receptors contain this immunoglobulinlike sequence motif. Like the other receptor families, the similarities to

Table II Characteristics of Cytokine Receptors

Receptor	Molecular size	Expression levels (receptors/cell)	Cellular distribution
IL-1α	82,000	10^2–10^4	Lymphocytes, fibroblasts
IL-2	IL-2Rα/55,000 IL-2Rβ/75,000	10^3–10^5	Activated B cells, activated T cells
IL-3	120,000–140,000	10^2–10^3	Cells of myelomonocytic and B-cell lineage
IL-4	140,000	10^2–10^3	B cells, T cells
IL-5	46,500	10^3–10^4	Activated B cells, eosinophils
IL-6	80,000	10^2–10^4	T cells, B cells, fibroblasts
IL-7	70,000–75,000	10^2–10^4	T cells, pre-B cells, fibroblasts
IL-8	65,000	10^3–10^4	Granulocytes, macrophages
TNF	80,000	10^2–10^4	Cells of many lineages
IFNγ	80,000	10^2–10^4	Cells of many lineages
TGFβ	260,000 70,000 p73 53,000 p53	10^2–10^4	Cells of many lineages

other family members is confined to the extracellular ligand-binding region of the receptor molecules.

Although a number of cytokine receptors have been cloned and shown to share certain features, certain cytokine receptors still remain to be characterized. After all the cytokine receptors have been characterized, the major question to be answered will be to determine the nature of the signals transmitted across the cell membrane on cytokine binding to its specific receptor. Once this has been elucidated, researchers should have an increased understanding of the mechanisms by which cytokines trigger specific target cell responses.

III. Interleukins

A. Interleukin 1

Interleukin 1 (IL-1) is a polypeptide that is produced primarily by macrophages after infection, injury, or antigenic challenge. Interleukin 1 was initially described in the 1940s as endogenous pyrogen (EP) for its ability to induce fever. Gery and Waksman described a protein, lymphocyte-activating factor (LAF), that augmented T-lymphocyte responses to T-cell mitogens, which was indistinguishable from EP. The term interleukin 1 (IL-1) now includes the initially described EP and LAF activity, as well as mononuclear cell factor, catabolin, osteoclast activating factor, hemopoietin 1, and other activities.

Two biochemically distinct but related molecules have been isolated IL-1α and IL-1β. Nonallelic genes code for these two molecules, which share small stretches of amino acid homology (approximately 25% homology for human IL-1). Interleukin 1β mRNA predominates over IL-1α mRNA, and this prevalence has been observed in the proportion of the IL-1 forms measured in the circulation and in body fluids. The IL-1α mRNA is about 2.3 kb in length, whereas the IL-1β mRNA measures between 1.4 and 1.8 kb. Both IL-1α and IL-1β are initially synthesized as precursor polypeptides that are subsequently processed to mature IL-1. The sequence similarity of the IL-1α molecule or the IL-1β molecule between different species is about 60–70%. However, the sequence identity between IL-1α and IL-1β within a single species is only approximately 25%. Despite this sequence variability, the IL-1α and IL-1β precursor proteins display a number of shared features. The bioactive region of both molecules can be localized to the C-terminal half of the molecule. Interleukin 1α is initially synthesized as a precursor molecule of 268–271 amino acids in length, whereas IL-1β is translated as a precursor of 266–269 amino acids. However, neither of the two precursor molecules has a classical N-terminal hydrophobic signal peptide, and therefore the exact mechanisms by which IL-1 is secreted are unknown. Both precursor molecules are cleaved, resulting in biologically active molecules of 17–19 kDa.

Interleukin 1 lacks a distinct signal peptide, and as such, a considerable amount of IL-1 has been shown to remain cell-associated, either intracellularly or as part of the cell membrane. A 31-kDa and a 22-kDa form of IL-1 has been found associated with the cell that could represent a membrane form of IL-1. In fact, membrane-bound IL-1 has been reported to be biologically active and may account for the immunostimulatory activities of IL-1 in local tissues. Experiments have indicated that most membrane-bound IL-1 exists in the form of IL-α, and the IL-1β molecule is the predominant form found secreted into the extracellular fluids.

Interleukin 1 is predominantly produced by activated macrophages, but it is also synthesized by synovial fibroblasts, keratinocytes, Langerhans cells, B lymphocytes, astrocytes, microglial cells of the brain, and vascular endothelial cells.

A wide variety of biological activities has been ascribed to the molecule now known as IL-1 (Tables III and IV). The IL-1 molecule has numerous nonimmunologic effects, including inducing fever, increasing slow-wave sleep, increasing sodium excretion, inducing neutrophilia, and increasing leukocyte adherance, as well as many other biological activities (Table III). In addition to the nonimmunologic effects, IL-1 also possesses many immunologic effects, including inducing T-cell activation, enhancing antibody production, inducing cytokine synthesis, and inducing cytotoxicity, and other activities (Table IV). Studies using recombinant IL-1 have confirmed that all these multiple biological activities can be attributed to IL-1.

Table III Nonimmunologic Effects of Interluekin 1

Effect
Effect on central nervous system
Induces fever
Increases ACTH
Increases neuropeptides
Increases slow-wave sleep
Effect on metabolism
Increases acute-phase proteins
Increases sodium excretion
Induces hypozincemia and hypoferremia
Decreases cytochrome p450 enzyme
Hematologic effect
Increases nonspecific resistance
Increases tumor killing
Induces neutrophilia
Induces lymphopenia
Effect on vascular cells
Increases leukocyte adherance
Increases prostaglandin (PGI and PGE_2) synthesis
Increases platelet activating factor
Increases cardiac heart rate and cardiac output

Table IV Immunological Effects of Interleukin 1

Cell type	Effect
T lymphocytes	T-cell activation IL-2 production Increases IL-2 receptor expression Induction of IFNγ, IL-3, and other lymphokines
B lymphocytes	Enhances antibody production Synergizes with B-cell growth and differentiation factors
Natural killer cells	Synergizes with IL-2 and IFNγ for cell lysis Induction of cytokine synthesis
Macrophages	Induce prostaglandin (PGE_2) synthesis Induction of cytotoxicity Induce synthesis of cytokines IL-6, GM-CSF, IL-8

B. Interleukin 2

Interleukin 2 (IL-2) is a T-cell-derived lymphokine that was initially called T-cell growth factor (TCGF) because of its ability to stimulate T-cell proliferation and sustain T-cell growth in culture. Interleukin 2 has also been described as a growth and differentiation factor for B cells and a growth factor for natural killer (NK) cells.

Interleukin 2 is encoded by a single, nonallelic gene on chromosome 4 in humans. The translation product of the IL-2 gene is a polypeptide of 153 amino acids that includes a signal peptide of 20 amino acid residues. Differences in *O*-glycosylation result in mature molecules that range from 15 to 17 kDa. However, no biological activity has been ascribed to the carbohydrate moiety as recombinant IL-2 produced in *Escherichia coli* has the identical activity as the lymphocyte-derived molecule.

Interleukin 2 exerts its biological effects by binding to a cell surface receptor, and as such, it closely mimics other hormone/receptor systems. The IL-2 receptor is a complex of two distinct polypeptide chains. The α chain, also known as Tac or p55, binds the IL-2 molecule with an affinity of 1×10^{-7} to 1×10^{-8} *M*. The larger β-chain, known as p75, binds the IL-2 molecule with an intermediate affinity of 1×10^{-9} *M*. The high-affinity IL-2 receptor, which is a heterodimer of the p55 and p75 chains, is capable of binding IL-2 with an affinity of 1×10^{-11} *M*. The initial finding that the p55 chain, TAC, contained a short intracytoplasmic tail of 13 amino acids suggested that p 55 may be unable to deliver any intracellular signal. Once p75 was cloned and it was determined that it contained a long intracytoplasmic tail of 286 amino acids, with a serine-rich region, an acidic region, and a proline-rich region, it is now believed that the p75 chain is essential for signal transduction.

The purification of IL-2 and the subsequent cloning of the gene represented a major breakthrough in cytokine research, as it represents the central molecule of the immune response and is essential for the proliferation and differentiation of antigen-stimulated T cells and B cells as well as affecting NK and lymphokine-activated killer (LAK) cells. [*See* T LYMPHOCYTES; B CELLS.]

C. Interleukin 3

Interleukin 3 (IL-3) is a hemopoietic growth factor produced by T lymphocytes. Interleukin 3 has been reported to possess a number of biological activities, and as such, IL-3 was initially known by a number of alternative names (Table V). The initial isloation of IL-3 revealed that IL-3 was a glycoprotein with a molecular mass ranging from 28 to 32.5 kDa, the differences attributed to carbohydrates. A cDNA for IL-3 was synthesized and shown to code for a protein with a molecular mass of 16 kDa containing 166 amino acids, which included a signal sequence of 20 residues at the N-terminal end. The synthetic molecule was shown to contain the biological activity, thus indicating that the carbohydrate was not essential for biological activity. Interleukin 3 interacts with a receptor on target cells that shares homology with other cytokine receptors such as IL-4, IL-6, and the IL-2Rβ chain. The major activity of IL-3 is to promote the production and differentiation of multiple blood types including granulocytes, monocytes, erythrocytes, and megakaryocytes and to stimulate the production of mucosal mast cells from bone marrow stem cells. It has also been shown to stimulate lymphocytes at an early differentiation stage that is associated with the acquisition of the enzyme 20α-hydroxysteroid dehydrogenase (20α SDH).

Table V Alternative Names for Interleukin 3

Name	Activity
Multicolony stimulating factor (M-CSF)	Growth of hemopoietic stem cells
P-cell stimulating factor	Growth of mast cells from spleen cultures
WEHI-3B factor	Long-term growth of bone marrow cultures
Mast-cell growth factor	Proliferation of mast cells in cultures of spleen and bone marrow
Histamine-producing cell-stimulating factor	Growth and maturation of mast cells

D. Interleukin 4

Interleukin 4 (IL-4) is a pleiotropic cytokine derived from T cells, which was initially described for its ability to enhance DNA synthesis in resting mouse B lymphocytes stimulated with anti-IgM antibodies. Thus, it was designated B-cell growth factor (BCGF). However, after the observations that this molecule could also induce the expression of MHC class II molecules and enhance B-lymphocyte responsiveness to anti-IgM antibodies, the molecule was given the designation B-cell stimulatory factor-1 (BSF-1). These names have since been replaced with the designation IL-4 to describe its pleiotropic actions on a wide range of target cells. In the murine system, IL-4 is predominantly produced by the murine T-cell subset T_{H2} (Table VI). In the human system, IL-4 has been reported to be produced by $CD4^+$ T lymphocytes. Interleukin 4 has been cloned and shown to code for a protein of 153 amino acids in humans and a 140-amino-acid protein in mice. Re-

Table VI Patterns of Lymphokine Production by Murine $CD4^+$ T-Cell Subsets

T_{H1}
Interleukin 2
Interferon γ (IFNγ)
Lymphotoxin (TNFβ)
T_{H2}
Interleukin 4
Interleukin 5
Interleukin 6
Interleukin 10
Both T_H types
Interleukin 3
Interleukin 8
GM-CSF
TNFα

moval of the signal peptide yields a secreted protein of 129 amino acids in humans and 120 amino acids in the mouse. However, as is the case with other ILs, IL-4 in a deglycosylated form retains full biologic activity.

The IL-4 gene is found in a cytokine complex located on chromosome 11 in the mouse and on chromosome 5 in humans. The IL-4 gene is linked to the IL-5 gene. The genes for IL-3 and GM-CSF are also found at a similar location.

The initial description of IL-4 functional activity was its ability to regulate B-cell growth and the expression of membrane antigens. Mouse B cells exposed to low concentrations ($\leq 5\ \mu g/ml$) of anti-IgM antibody failed to undergo DNA synthesis unless IL-4 was present. Preculturing resting B lymphocytes with IL-4 enhanced the B-cell response to either anti-IgM or to the B-cell mitogen lipopolysaccharide (LPS). In addition, IL-4 has been observed to increase or induce the expression of MHC class II molecules and the low-affinity receptor for IgE, Fc_{ε} RII (CD23).

Interleukin 4 has also been shown to function as a switch factor for IgE and IgG1. Murine B cells stimulated with LPS and IL-4 produce significant amounts of IgE and IgG1, whereas treatment with only LPS yielded virtually no IgE and modest amounts of IgG1. This effect of IL-4 on IgE and IgG1 production has also been observed when the B-cell "costimulant" is an activated T cell rather than LPS.

The effect of IL-4 is also observed *in vivo* as a monoclonal anti-IL4 antibody, or a monoclonal antibody to the IL-4 receptor will block the IgE response to helminthic infection or to *in vivo* polyclonal B cell activation. Thus, IL-4 can induce the expression of different classes of Ig both *in vivo* and *in vitro*. In addition, IL-4 can inhibit the production of IgG2a by B cells that have been treated with LPS and IFNγ. Thus, the balance of IL-4 and IFNγ produced during an immune response may be a major determining factor in the generation of an immune response.

Interleukin 4 has also been shown to exert an effect on T lymphocytes. IL-4 has been shown to act as an autocrine growth factor for a specific subset of long-term cultured murine T cells, the T_{H2} cells. Furthermore, IL-4 has also been reported to enhance the proliferation of cytotoxic T-cell (CTL) precursors and their differentiation into CTL effectors.

Interleukin 4 mediates its activity by binding to specific receptors on target cells. The responding target cells express low numbers of receptors per cell (~400), and these receptors have an affinity of $\sim 10^{-10}\ M$. On activation of B lymphocytes, either with anti-Ig antibodies of LPS, or activation of T lymphocytes with T-cell mitogens, one can observe an increase in IL-4 receptor expression up to approximately 2000 receptors per cell.

The exact characterization of the IL-4 receptor has not been achieved. However, a cDNA clone has been isolated that codes for a 120 kDa form of the IL-4 receptor. This cDNA clone has been expressed in cos-7 cells and results in IL-4 membrane receptors, which can bind IL-4 as efficiently as the naturally occurring receptors on mouse cells. In addition to the 120 kDa form of the IL-4 receptor, there has also been a report that cDNA has been isolated that appears to represent an alternatively spliced form of the receptor and encodes the extracellular domain of the IL-4 receptor followed by a 114-nucleotide insert that codes for the addition of six amino acids followed by a stop codon. This cDNA does not encode the transmembrane or cystolic region of the IL-4 receptor, and the inferred amino acid sequence of this molecule suggests that it may represent a soluble form of the IL-4 receptor with a molecular mass of 40 kDa. Although no evidence has been obtained that indicates that this p40 form of the IL-4 receptor can be secreted by normal cells, it suggests that a soluble form of the IL-4 receptor may have important immunoregulatory potential.

E. Interleukin 5

Interleukin 5 (IL-5) is a product of T lymphocytes. In the mouse, IL-5 is produced exclusively by the T_{H2} population as is IL-4. IL-5 was initially described for its ability to function as a T-cell replacing factor (TRF). In addition, IL-5 was also shown to possess B-cell growth factor activity (BCGF-II), which was distinct from other BCGFs, and eosinophil differentiation factor (EDF) activity. Based on gene cloning studies, the gene for IL-5 was shown to code for a 14-kDa protein molecule that contained N-linked glycosylation sites. These glycosylation sites help to explain the experimental findings that the mature IL-5 molecule has a molecular mass of 20 kDa. The gene for IL-5 has been shown to be closely linked with the genes coding for IL-3, IL-4, and GM-CSF. Interleukin 5 has been observed to preferentially stimulate an isotype switch in B cells expressing surface IgM (sIgM) to IgA expressing cells.

F. Interleukin 6

Interleukin 6 (IL-6) is another cytokine that displays multifunctional activity. As is the case with other ILs, IL-6 had been given a number of alternative names based on previously described biological activities. These names are listed in Table VII. Some of these names included β_2-interferon (IFN-β_2), B-cell stimulating factor 2 (BSF-2), 26-kDa protein, hybridoma/plasmacytoma growth factor (HPGF or IL-HP1), hepatocyte stimulating factor (HSF), and T-cell activation factor (TAF). However, molecular cloning studies that provided recombinant IL-6 revealed that a single protein product of a single gene possessed all these activities, and that protein was IL-6. Complementary DNA for human IL-6 codes for a single polypeptide chain of 212 amino acids, which includes a signal sequence of 28 hydrophobic residues that are cleaved before secretion. Thus, mature secreted human IL-6 has a molecular mass of 26 kDa. Murine and human IL-6 have 42% homology at the amino acid level, which may explain why IL-6 is not species-specific. Interleukin 6 is produced by a number of cells, including monocytes, endothelial cells, epithelial cells, and T lymphocytes and B lymphocytes. In addition to its effects on B cells and T cells, IL-6 also functions as part of the first defense mechanism against infection by stimulating the production of proteins with direct or indirect antibacterial effects.

G. Interleukin 7

Interleukin 7 (IL-7) represents a 25-kDa protein that was originally described as a product of stromal cells derived from bone marrow. Interleukin 7 was reported to promote the *in vitro* growth of B-cell precursors but had no effect on mature B cells. However, more recently it has been shown that IL-7 can also support the proliferation of murine thymocytes, human thymocytes, and T lymphocytes isolated from murine lymph nodes and spleen by functioning as a co-stimulatory molecule.

Complementary DNA for murine IL-7 codes for a protein of approximately 15 kDa that has two N-linked glycosylation sites. Glycosylation of the molecule can increase the molecular mass to 25 kDa. Interleukin 7 contains six cysteine residues within the molecule that are involved in intrachain disulfide bonds. Reduction of the disulfide bonds with 2-mercaptoethanol (2-me) results in the loss of biological activity. Murine and human IL-7 show approximately 60% homology at the amino acid level and, like other cytokines such as IL-6, do not function in a species-specific manner.

Table VII Activities Associated with Interleukin 6

Previous name	Biological activity
Interferon β_2 (IFN-β_2)	Antiviral activity produced by virus-infected fibroblasts
BSF-2/BCDF	Increased antibody production
HPGF/IL-HP1	Supported growth of hybridomas and plasmacytoma
HSF	Stimulated the production of acute-phase proteins
TAF	Costimulates PHA-treated T lymphocytes

H. Interleukin 8

Interleukin 8 (IL-8) is a member of the small cytokine family or scy family. This family not only includes IL-8 but also macrophage inflammatory proteins (MIP-1, MIP-2) and monocyte chemoattractant protein (MCP-1). Interleukin 8 was initially described as a neutrophil activating protein (NAP-1). Interleukin 8 has been shown to induce the adherence of neutrophils to endothelial cells and subendothelial matrix proteins. More recent studies have revealed that IL-8 increases the binding activity of cell surface CD11/CD18 complexes, thus promoting leukocyte adhesion. Interleukin 8 has also been shown to be chemotactic for T lymphocytes. Interleukin 8, as well as other members of the scy family, is produced by a variety of cell types including endothelial cells, fibroblasts, hepatocytes, activated macrophages, and T cells. Interleukin 8 has been reported to have a molecular mass ranging between 6.5 and 10 kDa, and cDNA has been shown to code for a protein of 8 kDa. Because IL-8 has only recently been described, its full range of biological activities remains to be determined.

I. Interleukin 10

Interleukin 10 (IL-10) was originally known as cytokine synthesis inhibitory factor (CSIF). Cytokine synthesis inhibitory factor was shown to be an acid-labile 35- to 40-kDa homodimer that is produced by murine T_{H2} cells and not by T_{H1} cells. A murine IL-10 cDNA clone has been isolated and shown to code for a molecule of 178 amino acids, and the first 18 residues appear to be a hydrophobic leader sequence characteristic of a secreted protein. Two potential N-linked carbohydrate attachment sites have been identified that may account for the heterogeneity of the monomers. A human IL-10 cDNA clone has also been isolated and shown to code for a single biologically active 18-kDa polypeptide that lacked significant carbohydrates, indicating that carbohydrates are not required for biological activity.

The mechanism of action of IL-10 is to inhibit the synthesis of most or all the cytokines producted by T_{H1} cells. This inhibition of cytokine synthesis by T_{H1} cells has also been reported to occur in CTL clones. As with T_{H1} cells, IL-10 inhibited the synthesis of cytokines by CTL clones but had no effect on the proliferation of these CTL clones.

IV. Other Soluble Mediators/Cytokines

A. Tumor Necrosis Factor

Tumor necrosis factor was originally described as an activity in supernatants of endotoxin-treated macrophages that was capable of destroying tumor tissue in a tumor-bearing animal. Although TNF was so named because of its tumor necrotizing ability, TNF has many other biological effects of which a large number of activities overlap with those associated with IL-1 (Table VIII). In addition to TNF produced by monocytes and macrophages, supernatants from antigen-stimulated lymphocytes were found to contain an activity that could kill syngeneic lymphocytes *in vitro* and thus was named *lymphotoxin* (LT). Recent cloning of the genes for both the macrophage TNF activity and the lymphocyte LT activity has revealed a high degree of homology for these two proteins, and it has been shown that the two molecules bind to the same receptor. Thus, based on this information, it has been proposed that the macrophage-derived TNF now be called TNFα and lymphotoxin would be called TNFβ.

Table VIII Activities Associated with TNF

Target Cell	Effects
Monocytes/macrophages	Coactivates to cytocidal stage Induces prostaglandins and other cytokines (IL-1, IL-6, IL-8)
Endothelial cells and vascular smooth muscle	Increases adhesiveness by up-regulating I-CAM
Adiopocytes	Increases lipolysis Decreases lipoprotein lipase levels
Neutrophils	Activates their functions

Both of the genes coding for TNFα and TNFβ have been cloned. The degree of homology between TNFα and TNFβ is 46% at the nucleotide level and 28% at the amino acid level. The precursor form of the TNFα molecule contains 236 amino acids and before or during secretion the 70 N-terminal amino acids are removed to yield a mature form of TNFα of 17.3 kDa, which contains 157 amino acids. Tumor necrosis factor β initially contains 204 amino acids in its precursor form, which undergoes cleavage yielding a 18.6-kDa, 171-amino-acid mature form of TNFβ.

Like other IL/cytokine molecules, TNFs exert their biological activity by binding to a cell surface receptor. Studies have indicated that there is considerable variation in the number of TNF receptors expressed on the surface of cells. Numbers ranging from 1000 to 10,000 receptors per cell have been reported, depending on the cell type analyzed. However, all the receptors are of high affinity, with binding affinities in the order of $2–6 \times 10^{-10}$ *M*.

Tumor necrosis factor has a range of activities that closely proximate those described for IL-1. Tumor necrosis factor, like IL-1, can also have both beneficial and harmful effects *in vivo*. Tumor necrosis factor has been described to be a powerful inflammatory agent that has been implicated in a number of disease processes, including septic shock and certain autoimmune diseases.

B. Interferons

Interferons (IFNs) are defined as proteins that exert nonspecific antiviral activity. Interferons can be classified as one of two major types, based on the nature of the inducing agent and the producing cell. Classical or type I IFNs are secreted by all nucleated cells after viral infection. Type I IFNs are subdivided into two subtypes, IFNα and IFNβ, based on structural differences as detected by specific antibodies. Type I IFNs are the predominant types of IFN produced by virus-infected leukocytes and fibroblasts. Another IFN type, type II, or now commonly known as IFNγ, is the product of lymphocytes activated by specific antigens or mitogens. Both type I and type II IFNs have been cloned and revealed little, if any, structural homology between the two types. However, IFNα and IFNβ, both type I, display a high degree of structural similarity with an amino acid homology of approximately 29%. In addition, IFNα has been shown to consist of multiple subspecies (approximately 13 or 14 nonallelic subtypes) that display about 80% homology in their amino acid sequences. Interferon α and IFNβ both contain about 165 amino acids in their secreted mature form, which is derived after cleavage of their signal peptide, which consists of about 20 amino acids. In contrast, the mature IFNγ is a 17-kDa protein of 146 amino acids that contains an excess of basic amino acid residues, which may contribute to the distinguishing property that IFNγ is labile at pH 2.0.

The IFNs mediate their effects through interaction with specific receptors like other cytokines. Interferon-γ receptors that are specific for the type II IFN have been found ranging from levels of 2,400 receptors per human fibroblast to 10,000 receptors per HeLa cell. These receptors possess a high-affinity binding capacity (3×10^{-10} *M*), which is characteristic of many hormonelike receptor–ligand interactions.

Interferon γ exerts various biological effects on a wide variety of cells within the immune system (Table IX). Interferon γ can act synergistically with other cytokines, such as TNFα and TNFβ, to augment antiviral immunity. Interferon γ has been reported to possess growth-inhibitory properties, as well as the ability to destroy cultured, transformed cells such as HeLa cells. Interferon γ can also promote cell differentiation as indicated by its ability to increase the cytotoxicity of NK cells and specific cytolytic T lymphocytes. One of the other important biological activities of IFNγ is its ability to induce an increase in cell surface expression of both MHC class I and class II antigens by increasing gene expression. In addition to its ability to up-regulate MHC gene expression, IFNγ can also increase the expression of the high-affinity receptors for the Fc region of IgG (FcγR) found on mature neutrophils and monocytes. Interferon γ has also been reported to have an effect on antibody production. In mice, IFNγ added to polyclonally activated B cells can increase the synthesis of IgG2a antibodies and suppress the production of IgE antibodies. Finally, IFNγ has been shown to function as a macrophage activating factor (MAF), which stimulates macrophages into effector cells that possess increased phagocytic capabilities, increased cytolytic activity for tumor cells and intracellular bacteria, and increased respiration and production of H_2O_2. Therefore, IFNγ represents a lymphokine that possesses potent immunomodulatory activities in addition to its antiviral activity and, as such, may have important immunotherapeutic value. [*See* INTERFERON.]

Table IX Biological Properties of IFNγ

Action of IFNγ	Target cell
Inhibits viral replication	All nucleated cells
Increases MHC expression	Macrophages
Increases phagocytosis	Macrophages
Increases cytotoxicity	Macrophages, natural killer cells, cytotoxic T cells
Increases antibody production of an IgG2a isotype	B lymphocytes
Decreases antibody production of an IgE isotype	B lymphocytes

C. Transforming Growth Factor β

The TGFβs are multifunctional molecules that regulate inflammation, tissue repair, and immune function. Biologically active TGFβ is a 25-kDa molecule that represents a dimer of two disulfide-linked 112-amino-acid polypeptides. These 112-amino-acid polypeptides can be any one of three isomeric forms (i.e., $TGF\beta_1$, $TGF\beta_2$, and $TGT\beta_3$). Homodimers and heterodimers have been found, indicating that a number of different biologically active TGFβs exist. Transforming growth factor β_1 cDNA has been cloned and shown to code for a 391-amino-acid precursor peptide in humans and a 390-amino-acid precursor peptide in mice. The mature TGF contains 112 amino acids, which represents the C terminus of the precursor form, which is generated by proteolytic cleavage. The 112-amino-acid mature forms of human and mouse TGTβ differ by only one amino acid. Human $TGF\beta_2$ and $TGF\beta_3$ have also been cloned and shown to code for precursor molecules of 414 and 412 amino acids, respectively. The homology between $TGF\beta_2$ and $TGF\beta_1$ is about 71%, and the homology between $TGF\beta_3$ and $TGF\beta_1$ and $TGF\beta_2$ is about 80%.

Transforming growth factor β, which is produced by both activated T and B lymphocytes, has been reported to inhibit IL-2-dependent T-cell proliferation and IL-2 receptor up-regulation and IL-2-dependent B-cell proliferation and differentiation. Furthermore, TGFβ also transforms certain normal cells into neoplastic cells, indicating that TGFβ possesses diverse biological activities.

V. Summary

Since the discovery of IL-2 in 1976, more than 20 different molecules have been discovered that are immunologically active. These cytokines have been shown to have profound effects both *in vitro* and *in vivo*. These cytokines are essential for many of the proliferative and differentiative activites of immune cells. Although some cytokines have limited function, most cytokines are pleiotropic. Cytokines also overlap in their biological activities. Finally, cytokines can also act in a synergistic manner as well as in an antagonistic manner, and therefore, the exact cytokines produced on antigenic challenge may affect the outcome of the host's immune response.

As each cytokine becomes better defined relative to its numerous biological activities, our understanding of the complexity of the immune system will be greatly enhanced. Therefore, in the very near future, it may be possible to use individual cytokines or combinations of cytokines as immunotherapy. Cytokines may be used for such disease states as atherosclerosis; autoimmunity; infections, including viral infections by the human immunodeficiency virus (HIV), the causative agent of AIDS; and cancer. [*See* CYTOKINES IN BACTERIAL AND PARASITIC DISEASES.]

Bibliography

Dawson, M. M. (1991). *In* "Lymphokines and Interleukins" (M. M. Dawson, ed.). CRC Press, Boca Raton, Florida.

Dinarello, C. A. (1988). *FASEB J.* **2,** 108–115.

Dinarello, C. A., and Savage, N. (1989). *Crit. Rev. Immunol.* **9,** 1–20.

Dower, S. K., Smith, C. A., and Park, L. S. (1990). *J. Clin. Immunol.* **10,** 289–299.

Kelso, A. (1989). *Curr. Opinion Immunol.* **2,** 215–225.

Mosmann, T. R., and Moore, K. W. (1991). *Immunol. Today* **12,** A49–A53.

Mosmann, T. R., Schumacher, J. H., Street, N. F., Budd, R., O'Carra, A., Fong, T. A. T., Bond, M. W., Moore, K. W. M., Sher, A., and Fiorentino, D. F. (1991). *Immunol. Rev.* **123,** 209–229.

Paul, W. E. (1991). *Blood* **77,** 1859–1870.

Pennica, D., Shalaby, M. R., and Palladino, M. A., Jr. (1987). Tumor necrosis factors alpha and beta. *In* "Recombinant Lymphokines and Their Receptors" (S. Gillis, ed.), pp. 301–317. Marcel Dekker, New York.

Sherry, B., and Cerami, A. (1991). *Curr. Opinion Immunol.* **3,** 56–60.

Ion Transport

Simon Silver
University of Illinois College of Medicine

Mark Walderhaug
Food and Drug Administration

Glossary

Chromosomal-governed Protein is encoded by a gene on the bacterial chromosome

Efflux Outwardly oriented transport, either as exchange (with no net concentration change) or leading to a net reduction in the cellular concentration

Influx Initial process of unidirectional cellular uptake; not to be confused with accumulation, which refers to net concentrations

Nutrient ion Inorganic cation or anion required for normal cell growth

Plasmid-governed Protein is encoded by a gene on the bacterial plasmid rather than the cell chromosome

Toxic ion Inorganic cation or anion that inhibits cellular growth (note that an ion may be both nutrient at low concentrations and toxic at high concentrations)

Transport Movement of a substance across a biological membrane via a specific membrane protein

MICROBIAL CELLS accumulate needed inorganic cations and anions by highly specific membrane transport systems, each of which consists of one or a few proteins. For every needed ion there is one or more separate transport systems. Regulation occurs at the level of physiological function and at the level of genetic control of the amounts of proteins synthesized. The same (or additional) transport systems that function for net uptake also function during cation/cation or anion/anion exchange and during conditions of net efflux of ions (to control intracellular ion concentrations, pH, osmotic turgor, and related cellular processes). In addition to the large number of chromosomally determined ion transport systems, genes for resistances to toxic heavy metal cations and oxyanions occur on plasmids. These genes often code for energy-dependent ion efflux transporters that assure low intracellular concentrations of toxic ions.

I. Introduction

Microbial cells are usually exposed to inorganic compounds as cations and anions in solution rather than as solid precipitates or elemental compounds. For transport into the cells, ions cannot passively diffuse across biological membranes. Highly specific systems consisting of membrane-embedded proteins are almost always responsible for transport (Fig. 1). This conclusion is reasonable for major inorganic nutrients such as K^+, Mg^{2+}, SO_4^{2-}, or PO_4^{3-}. Frequently, bacterial cells have parallel transport systems: constitutively synthesized relatively nonspecific systems for peaceful times of relative nutrient abundance and inducibly synthesized highly specific systems for stressful times of nutrient starvation or toxic ion assault (Fig. 2). Even such readily diffusable compounds as ammonia and carbon dioxide are transported across microbial membranes by specific transport proteins.

In this article, we discuss the entire Periodic Table of essential inorganic nutrients, both "macro" and "micro" needs. In addition, the transport systems for abundant cations and anions that are not essen-

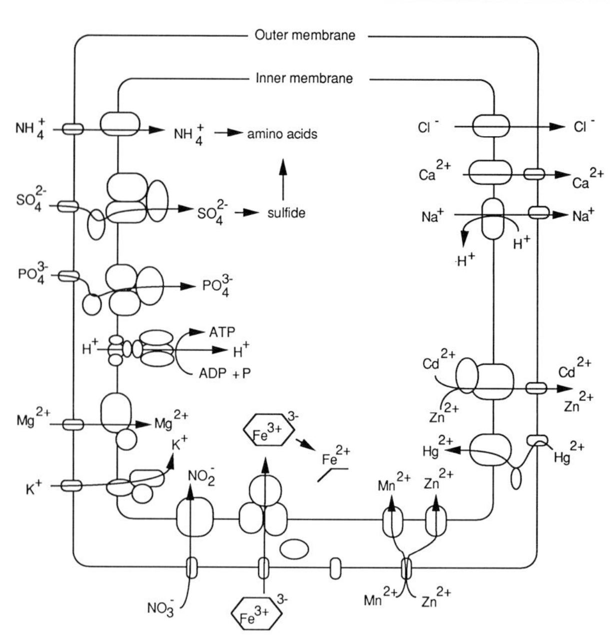

Figure 1 Summary of some bacterial membrane transport systems for inorganic cations and anions. Oval shapes represent individual polypeptide components and their positions in selected examples.

tial for microbial nutrition (Na^+, Ca^{2+}, and perhaps Cl^-) are described. These systems are outwardly oriented (i.e., for efflux) and function to maintain low intracellular concentrations and transmembrane ion gradients that can be coupled to other bioenergetic and cellular functions. Protons (H^+) form a most important transmembrane gradient, which establishes the membrane potential and pH gradient that is coupled to many other transport processes,

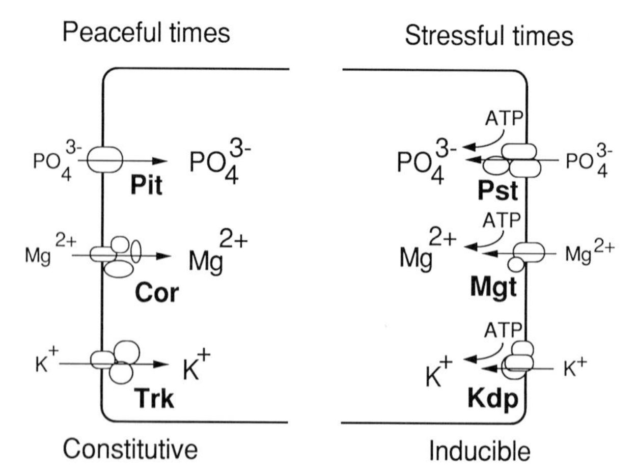

Figure 2 Parallel transport systems for phosphate, magnesium, and potassium constitutively synthesized for times of global peace (or abundance) and inducibly synthesized in times of stress (or starvation).

and to the synthesis of ATP by oxidative- and photophosphorylation (Harold, 1986). Proton-coupled energetic processes are discussed elsewhere in this volume, so we will not discuss proton transport. Following Peter Mitchell's well-established chemiosmotic concepts (Harold, 1986; Rosen and Silver, 1987), net uptake of cations will often be driven by the membrane potential (internal charge negative) and movement is an electrogenic uniport process. Net uptake of anions must be driven as cotransport with protons (the cellular outside is generally acid relative the inside pH) so the net uptake is charge-neutral or even net-positive, or as an antiport process with the outward movement of another anion providing the cellular energy.

Because microbial cells have encountered toxic cations and anions since the origin of life, membrane transport systems for most toxic ions have also evolved. Whereas the transport systems for nutrients are oriented inward (for uptake), the transport systems for toxic cations are generally outwardly oriented to provide mechanisms of resistance. Some ions, such as copper and zinc, are required as micronutrients at low concentrations and are toxic at high concentrations. For these cations, separate uptake and efflux systems are found. The genes for membrane proteins for uptake systems (and their regulation) generally occur on the bacterial chromosomes, because they are needed in all cells. The genes for toxic metal ion resistances are generally found on plasmids (and sometimes transposons), because they are only needed during toxic onslaught. Efflux transport systems for cations must be energized by ATP or by antiport coupling with uptake of another cation, frequently H^+, Na^+, and Ca^{2+}.

Despite the growth of understanding of ion transport in microbial cells, little is known except for a few well-studied microbial species. This article discusses the best characterized transporters, generally from *Escherichia coli* or a few other species. As more organisms are characterized in terms of ion transport, additional mechanisms and new patterns may be found.

II. Chromosomal-Governed Transport of Nutrient Ions

A. Potassium

Potassium is the major intracellular cation for all cell types: plants, animals, and microbes. Frequently the intracellular concentrations are kept at 0.1 to

0.5 *M* K^+, even when the extracellular concentrations drop to starvation (sub-micromolar) concentrations. Maintaining intracellular K^+ requires parallel potassium transport systems: first a relatively high-rate, low-affinity system (for times of K^+ abundance) and second a high-affinity, high-specificity, low-rate system for times of K^+ starvation (Fig. 2).

In *E. coli,* the high-rate system is called Trk (for *tr*ansport of *K*$^+$) (Fig. 3). Trk is synthesized constitutively and shows no sign of gene level regulation; however, the ATP requirement for Trk may be for regulatory purposes rather than for energy. The number of separate K^+ transport systems thought to occur in *E. coli* has gone up and down over the years; Fig. 3 summarizes recent genetic evidence. The four genes *trkA, trkE, trkG,* and *trkH* are unlinked on the *E. coli* chromosome, but their products appear to form two separate Trk systems with distinct inner membrane proteins, TrkG and TrkH. The Trk systems appear to share the membrane-associated TrkA and TrkE proteins. TrkA may be regulated by ATP; the Trk systems do not appear to couple ATP hydrolysis as the energy form driving K^+ transport. The TrkA protein is a peripheral membrane protein, anchored to the cytoplasmic membrane by TrkE, TrkG, and/or TrkH. TrkA protein is found in the cytoplasm under conditions of

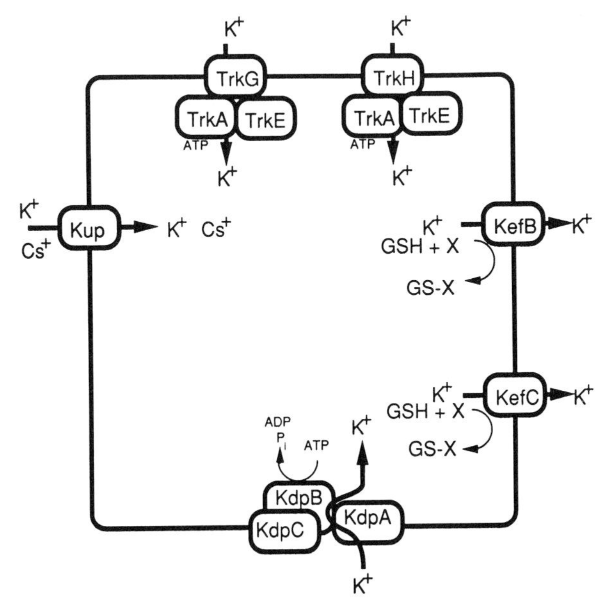

Figure 3 Pathways for potassium uptake and efflux in *E. coli*. Three-component TrkG and TrkH and one-component Kup systems are constitutively synthesized, whereas the three-component Kdp system is synthesized under conditions of osmotic upshock or potassium starvation. Trk is regulated by ATP, whereas Kdp is an ATPase.

overexpression and by itself does not function in K^+ transport. There is a third constitutively synthesized K^+ uptake system, Kup (which was previously called TrkD). The Kup system has a relatively low rate and low alkaline earth ion specificity. However, Kup is the sole pathway in *E. coli* for the uptake of Cs^+. TrkG, TrkH, and Kup are K^+ uptake pathways that were formerly thought to be a single system. There are two pathways for K^+ efflux by the KefB and KefC proteins (Fig. 3) (formerly called TrkB and TrkC). Efflux by KefB and KefC is regulated physiologically (effectively gated) by an undefined product of glutathione (Fig. 3). The Kef efflux pathways for K^+ are not genetically or physically connected with the uptake pathways, but they function in a coordinated fashion to maintain K^+ homeostasis.

Synthesis of the high-affinity (less than 0.2 μM K^+ affinity constant) Kdp (*K*$^+$ *dep*endent growth) system (Fig. 3) is induced under conditions of potassium starvation or in rapid response to a decrease in membrane turgor (i.e., osmotic upshock). The additional *kdpD* and *kdpE* gene products regulate Kdp function, the first by sensing the osmotic pressure across the cell membrane and the second by activating (after activation itself via phosphorylation from KdpD) the DNA transcriptional initiation site of *kdpABC*. Then the three components of the membrane ATPase (KdpA, KdpB, and KdpC) are synthesized. All three components are membrane proteins, and KdpA appears to be associated with the initial recognition of K^+. However, the KdpB protein is the membrane-embedded ATPase subunit, homologous in sequence and in function to the more familiar animal Na^+/K^+ and sarcoplasmic reticulum Ca^{2+} ATPases. [*See* ATPases and Ion Currents.]

Although K^+ transport has not been studied in detail in other bacterial types, it has always been found when sought. The two types of systems [(1) high-rate, low-affinity, constitutively synthesized and (2) lower-rate, high-affinity, inducibly synthesized] probably occur in most other bacteria as well.

Rubidium is frequently an alternative substrate for K^+ uptake systems, and Rb^+ can replace K^+ for growth in some bacteria (not in *E. coli*). Cs^+ is generally a poorer K^+ substitute than Rb^+, and Cs^+ will not replace K^+ for physiological functions. Because of the abundance of Na^+ in usual microbial environments, the K^+ transport systems discriminate strikingly against Na^+ (and the sodium analog Li^+).

B. Magnesium

Magnesium is the second most abundant intracellular inorganic cation, but Mg^{2+} has roles quite different from K^+. Mg^{2+} functions primarily as a catalytic cofactor for many enzymes. Most intracellular Mg^{2+} is not osmotically free, but is bound (in kinetically exchangeable forms) with cellular polyanions, such as nucleic acids and lipids. Thus, Mg^{2+} does not contribute to osmotic processes. There are three known Mg^{2+} transport systems in *Salmonella typhimurium,* the best studied microbe. Two of these systems are inducibly regulated. The MgtA and MgtB proteins are members of the E_1E_2 class of ATPases (recently called P ATPases) described above for Kdp). The MgtA system may consist of two polypeptides, but this is not currently clear. The MgtB ATPase (which is 75% identical at the amino acid level with MgtA) may consist of a single polypeptide. The third *S. typhimurium* Mg^{2+} transport system is called Cor (for *co*balt *r*esistance, as Co^{2+} is an alternative substrate for this system). The Cor system is predominate (with the highest V_{max} for Mg^{2+} transport) in most gram-negative bacteria under most growth conditions (Fig. 2). Cor is synthesized constitutively. Mg^{2+}, Co^{2+}, and Ni^{2+} are the substrates for the Cor system. Cor consists of four polypeptides, determined by the *corA, corB, corC,* and *corD* genes. CorA appears to be the primary inner membrane protein responsible for uptake via the Cor system. CorB, CorC, and CorD function only at high extracellular Mg^{2+} concentrations and may mediate Cor-specific $Mg^{2+}{}_{in}/Mg^{2+}{}_{out}$ exchange. Mutants missing one or two of these proteins show reduced Mg^{2+} efflux, and mutants lacking all three proteins completely lose Mg^{2+} efflux.

C. Iron

Iron is the best known and perhaps the most important inorganic micronutrient. Almost all bacterial cells contain Fe^{2+} (incorporated in haem proteins, such as cytochromes, and in nonhaem iron proteins). The only known exception is lactic acid bacteria, which contain no haem proteins and substitute Mn^{2+} for some intracellular roles normally played by Fe^{2+}. Bacteria have highly specific iron transport systems. In the best-understood organism, *E. coli,* there are five known Fe^{3+} transport systems and another for transport of Fe^{2+} under anaerobic conditions (when Fe^{2+} is stable). The five Fe^{3+} transport systems are named for the organic iron chelates involved. Such iron transport chelates are called siderophores, from the Greek words for "iron bearing." The five systems for *E. coli* are (1) the enterochelin system, which uses a trimeric catechol (Fig. 4) synthesized by *E. coli;* (2) the aerobactin system, which uses a hydroxamic acid chelate consisting of two modified lysine residues coupled to a citric acid residue (Fig. 4); (3) the citrate-Fe^{3+} transport system, which depends on exogenously (i.e., environmentally) available citrate (Fig. 5); (4) the ferric hydroxamate uptake (Fhu) system, which takes up Fe^{3+} chelated with hydroxamate siderophores synthesized by other bacteria (or even fungal cells); and (5) a low-affinity system, which functions at high concentrations of Fe^{3+}, apparently without the need for a siderophore transport cofactor. The genes governing aerobactin uptake may be on the chromosome [as in *Klebsiella* (formerly called *Aerobacter*) *aerogenes*] or on a plasmid (as in the aerobactin system that functions as a virulence factor on the ColV-class plasmids in enteric bacteria). The third siderophore shown in Fig. 4 is Desferal, the only clinically used siderophore. Desferal is the mesylate salt of desferrioxamine B and is kept in hospitals for treatment of iron and aluminum poisoning. Desferrioxamine B is made by the filamentous gram-positive bacterium *Streptomyces pilosus* and is also a hydroxamate siderophore, like aerobactin.

Each of the five pathways for Fe^{3+} uptake in *E. coli* (and the comparable systems in other bacteria) involves a series of gene products for (1) regulation, (2) transport, and (3) energy coupling. We describe the iron dicitrate system (Fig. 5) in detail as just one example. Three regulatory proteins are involved. FecR and FecI couple in order to sense the presence of citrate in the periplasmic space and to transmit the signal to the DNA operator/promoter region of the *fec* genes, which activates synthesis of the transport proteins. The molecular mechanism of this signal transduction is not known. Both extracellular citrate (sensed by FecR) and the absence of intracellular Fe^{2+} (sensed by the global iron-regulatory protein Fur, which is also used as a repressor protein for the alternative iron transport systems) are needed for transcription of the *fec* genes. The *fec* operon system consists of five specific proteins. The first, FecA, is a large outer membrane protein (Fig. 5) and a member of the TonB class of energy-dependent outer membrane proteins. The inner membrane protein TonB stretches across the periplasmic space

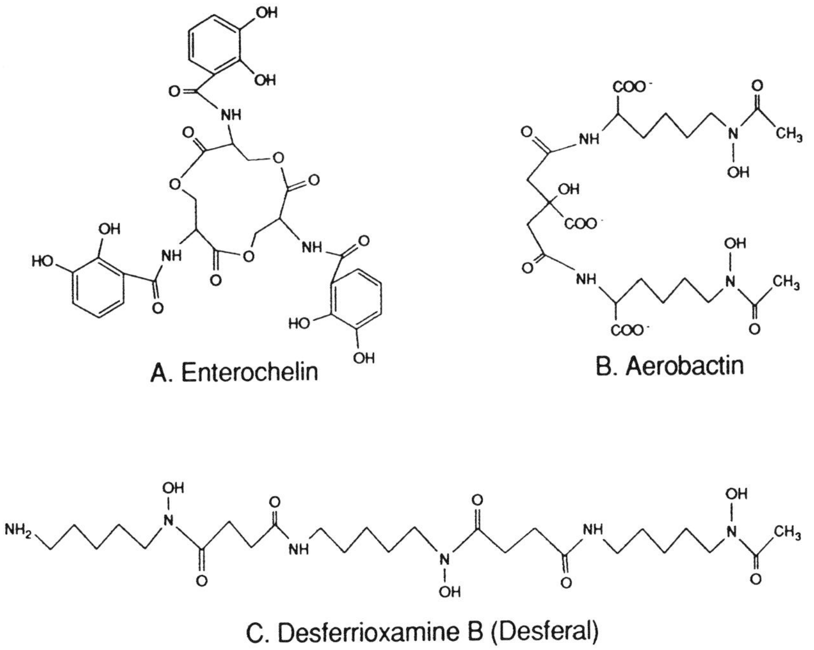

A. Enterochelin

B. Aerobactin

C. Desferrioxamine B (Desferal)

Figure 4 Typical iron-transport chelates (siderophores): (A) the catechol enterochelin (also called enterobactin); (B) the hydroxamic acid aerobactin; and (C) the hydroxamic acid desferrioxamine B (Desferal), which is used clinically for iron poisoning.

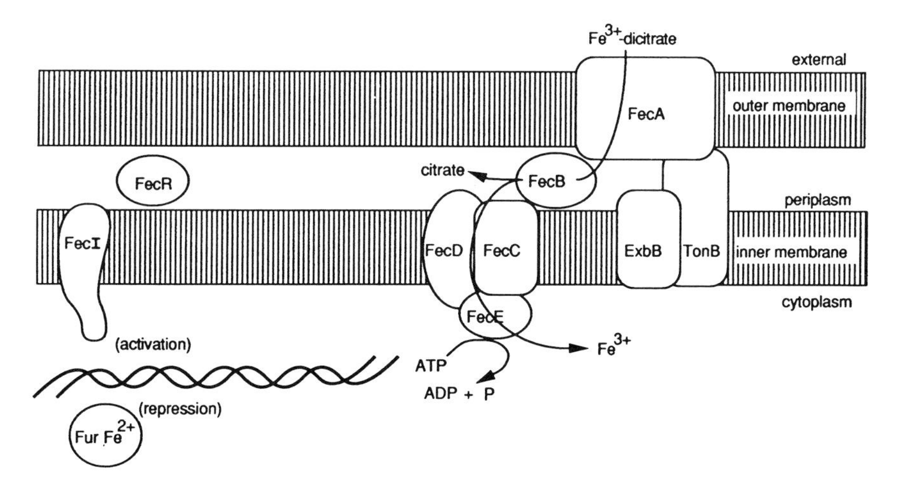

Figure 5 Iron citrate transport system of *E. coli* and its regulation. [Adapted from Silver, S., and Walderhaug, M. (1992). *Microbiol. Rev.* **56**, 195–228.] Iron dicitrate crosses the outer membrane protein FecA, with energy provided by the inner membrane TonB/ExbB complex. Fe^{3+} binds to the periplasmic FecB protein. Fe^{3+} is transported through the FecC, FecD, FecE complex, leaving citrate behind; energy-coupling is from ATP hydrolysis. FecR and FecI are a two-component regulatory system that senses extracellular citrate (FecR) and positively activates synthesis of the Fec system. Intracellular Fur protein senses internal Fe^{2+} and negatively regulates (i.e., represses) synthesis of the same system.

to make physical contact with a TonB box of conserved amino acid residues on the FecA protein and to transmit the energized state of the inner membrane to open the outer membrane channel for iron dicitrate. There is no evidence that energy expenditure is required directly for transport of iron dicitrate through FecA. TonB is stabilized by and dependent on the ExbB protein, as is also the case for several other members of the TonB family. Once in the periplasmic space between the outer and inner membrane, the iron dicitrate associates with the Fe^{3+}-dicitrate binding protein FecB (Fig. 5). Citrate is stripped off and released extracellularly, and the Fe^{3+} associates inner membrane proteins FecC and FecD, which form the inner membrane Fe^{3+} channel. In this aspect, the Fe^{3+}-citrate system differs from most siderophore-dependent transport pathways with which the Fe^{3+} along with the siderophore enters the cell. The inner membrane protein pair FecC and FecD are homologous in sequence and presumably have functional regions similar to those of other TonB-dependent systems. These systems all have inner membrane protein pairs and a membrane-associated ATP-binding protein (FecE in this case), which couples ATP hydrolysis to the transport of Fe^{3+} into the cell. Intracellular Fe^{3+} is rapidly reduced to Fe^{2+} and incorporated either into haem groups (by the enzyme ferrochelatase) or into nonhaem iron proteins. If one considers the example of iron dicitrate transport in *E. coli* and how much is known about its details (Fig. 5) and the molecular mechanisms involved, one gains respect for the exactness of microbial ion transport mechanisms and their regulation, especially if one generalizes this to the other iron-siderophore transport systems in *E. coli* and other bacteria, and then to non-iron transport systems.

D. Manganese

Much less is known about microbial Mn^{2+} transport than about K^+, Mg^{2+}, or Fe^{3+} transport. Bacteria have highly specific transport systems for accumulating needed Mn^{2+} (Fig. 1) in addition to the frequent uptake of Mn^{2+} by Mg^{2+} transport systems. Because Mn^{2+} is needed for specific purposes (e.g., superoxide dismutase and *Bacillus* sporulation), a general divalent cation transport system alone does not provide sufficient flexibility for a frequently magnesium-rich but manganese-starved world. The energy source driving Mn^{2+} uptake appears to be the membrane potential in some cases, because bacterial membrane ghosts of both *E. coli* and *Bacillus subtilis* can carry out Mn^{2+} transport against a concentration gradient in the absence of ATP. The genes or proteins involved in Mn^{2+} transport have not been identified, although there have been attempts. In one study, selection of Mn^{2+}-resistant mutants resulted in the isolation of Fur^- mutants (because of interactions of Mn^{2+} with the Fur protein, in the absence of excess Fe^{2+}). In a second study, a Cd^{2+}-resistant mutant of *B. subtilis* was isolated that appears to have lost the recognition for Cd^{2+} by the chromosomally determined Mn^{2+} carrier. Mn^{2+} transport in this mutant strain was, however, unimpaired.

Mn^{2+} transport in *Bacillus* is tightly regulated; regulation in other organisms has not been investigated. Under conditions of Mn^{2+} starvation, *B. subtilis* cells increase the V_{max} for Mn^{2+} uptake by 50-fold, without a change in affinity (K_m). The increased uptake is specific for Mn^{2+}. When micromolar concentrations of Mn^{2+} are added to Mn^{2+}-starved cells, the V_{max} for Mn^{2+} uptake drops rapidly, indicative of inactivation of transport function. The molecular and genetic details of Mn^{2+} transport and its regulation have not been addressed.

E. Phosphate

Phosphate is the major intracellular inorganic oxyanion and is an inorganic nutrient usually accumulated by bacteria in an inorganic form. In contrast, sulfur may be taken up preferentially as organosulfur compounds (e.g., cysteine) rather than as inorganic sulfate.

Bacterial cells frequently have parallel PO_4^{3-} transport systems (Fig. 2) for times of relative PO_4^{3-} abundance and for times of phosphate-limited growth. In many environmental settings, phosphate availability often limits growth of bacterial cells. *E. coli*, which once again is the best studied organism, has two phosphate transport systems (Fig. 2), Pit (for *Pi t*ransport) and Pst (for *p*hosphate-*s*pecific *t*ransport). The Pit system consists of a single membrane protein that carries out accumulation of phosphate driven by the transmembrane pH gradient. Pit is constitutively synthesized and has the same K_m for PO_4^{3-} (the nutrient substrate) as for AsO_4^{3-} (a toxic analog). Mutational inactivation of Pit induces synthesis of the Pst system. This switch affords the cell a degree of arsenate resistance, because the K_m for PO_4^{3-} for Pst is 100 times lower than the K_i for AsO_4^{3-}. The Pst system (Fig. 6) consists of five proteins and its synthesis is induced by phosphate starvation or arsenate toxicity. Regulation of the Pst system (part of the Pho regulon of phosphorus-related gene products) is complex and includes (1) sensing of external phosphate, (2) transmission of that signal into a phosphorylated regulatory membrane protein (PhoR), (3) followed by transphosphorylation from PhoR to a cytoplasmic regulatory protein (PhoB), that then (4) binds to the specific Pho box nucleotide sequence upstream of the *pst* genes, (5) turning on mRNA transcription.

The protein components of the Pst phosphate transport system are shown in Fig. 6A. First, there is the relatively specific anion channel protein (PhoE) through the outer membrane. No energy is involved, however, in passage of PO_4^{3-} across the outer membrane. Periplasmic phosphate binds to the PstS phosphate-binding protein. The crystal structure of this protein is solved, so the molecular details of how hydrogen bonding determines the specificity of PstS for PO_4^{3-} are understood. PO_4^{3-} is delivered from PstS to the PstA and PstC inner membrane proteins (that are related to the Fe^{3+} transport inner membrane proteins shown in Fig. 5 and are homologous one to the other). Transport across the inner membrane is probably energized by ATP hydrolysis through the membrane-associated PstB protein (Fig. 6A), which belongs to the same class of ATP-binding membrane-associated proteins as FecE (Fig. 5). Thus we see patterns emerging for ion transport mechanisms. Not all properties of every transport system are unique. Specificity is needed for only a few components (e.g., the periplasmic binding proteins and inner membrane transport proteins). Generally all components are distinct, but they fall into families of sequence-related (and presumedly evolutionarily related and functionally related) proteins.

F. Sulfate

Sulfate transport is basically similar to phosphate transport. After transport, however, sulfate is reduced to sulfide and incorporated into amino acids (Fig. 1) and other sulfur-containing small molecules. The primary sulfate transport systems in *E. coli* and *Salmonella* are summarized in Fig. 6B. The protein components are similar to those of the Pst system (Fig. 6A). Initially SO_4^{2-} must pass through an outer membrane porin protein (which has not been identified). There are two periplasmic binding proteins: Sbp, which binds SO_4^{2-} specifically, and CysP (for its role in cysteine biosynthesis), which binds both SO_4^{2-} and $S_2O_3^{2-}$ (Fig. 6B). These two binding proteins are less than 50% identical in their amino acid sequences. Both proteins, however, feed SO_4^{2-} to the inner membrane proteins CysT and CysW, which function in SO_4^{2-} transport similarly to the phosphate proteins PstA and PstC. CysA is the membrane-associated ATP-binding protein that provides energy for sulfate transport.

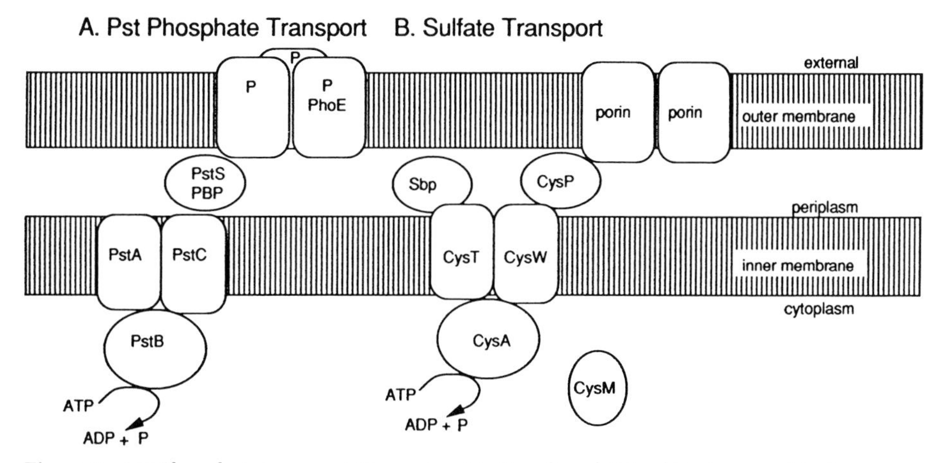

Figure 6 (A) Phosphate transport (Pst system) in *E. coli* and (B) sulfate transport in *S. typhimurium* [adapted from Silver, S., and Walderhaug, M. (1992). *Microbiol. Rev.* **56**, 195–228.] See text for description of components.

G. Trace Nutrient Transport Systems

For the sake of completion, we will list additional known (or potential) trace nutrient cations and oxyanions even when the evidence is indirect. Among the cations that are required for some intracellular enzymes are Co^{2+}, Cu^{2+}, Ni^{2+}, and Zn^{2+}. All of these cations may be transported by less specific divalent cation (Mg^{2+}) systems. Oxyanions of V, Cr, Se, Mo, and As may be nutrients in trace amounts. Si is essential for diatoms and higher organisms, but is not a nutrient for bacteria.

Na^+ and Ca^{2+} are not trace elements, but they are needed for only a few bacteria and under specific conditions. Na^+ and Ca^{2+} appear to have primarily extracellular roles. Na^+ is used especially by bacteria that grow in high-salinity environments as a cotransport substrate that is carried into the cell along with amino acids or sugars, much as by the cells of eukaryotes. The uptake of these organic nutrients is driven by the transmembrane Na^+ gradient. Na^+ is then effluxed (by either a Na^+/H^+ antiporter or a Na^+ ATPase) to maintain the sodium gradient (internal low/external high) that can be coupled to metabolic energy. There are also examples of specific Na^+-effluxing membrane decarboxylase enzymes generating the Na^+ gradient in some bacteria. Intracellular Na^+ does not occur in most bacteria at concentrations that would activate enzymes. Ca^{2+} frequently activates bacterial exoenzymes, such as nucleases and proteases. There are no microbial specific Ca^{2+}-activated intracellular enzymes. Unlike eukaryotic cells, bacteria do not use Ca^{2+} fluxes as signaling mechanisms. Massive amounts of Ca^{2+} are accumulated in bacterial spores (as a complex of Ca^{2+}-dipicolinic acid, which can accumulate up to 20% of spore dry weight). Ca^{2+}-dipicolinate is used to inactivate metabolism in the spore cytoplasm and to provide the heat resistance characteristic of bacterial spores. In spores, the function of Ca^{2+} is structural or physical, but not physiological.

III. Plasmid-Governed Toxic Ion Resistances Involving Transport

A. Mercury

The first heavy metal resistance system to be thoroughly studied and understood was that for mercury. The basic resistance mechanism involves enzymatic detoxification rather than transport. The enzyme organomercurial lyase cleaves the mercury-carbon bond in organomercurials, such as phenylmercury and methylmercury, releasing inorganic Hg^{2+} and a small organic compound. Inorganic mercuric reductase then reduces Hg^{2+} to Hg^0, which is relatively nontoxic and rapidly volatilizes when stirred under aerobic conditions. The mercury resistance systems have been reviewed in detail by Silver and Misra (1988), Misra (1992), and Silver and Walderhaug (1992). Here we will discuss only the Hg^{2+} transport system that is always a component of the multigene mercury resistance systems. [*See* MERCURY CYCLE.]

Because the membrane surface of bacterial cells is susceptible to mercury poisoning and because detoxification by Hg^{2+} reductase must be intracellular (because high-energy NADPH is a cofactor), a Hg^{2+} transport system is encoded as part of the resistance mechanism. For gram-negative bacteria, MerP is the first specific component of a dithiol "bucket brigade," which functions so that Hg^{2+} is passed on from cysteine pair to cysteine pair (Fig. 7). MerP is

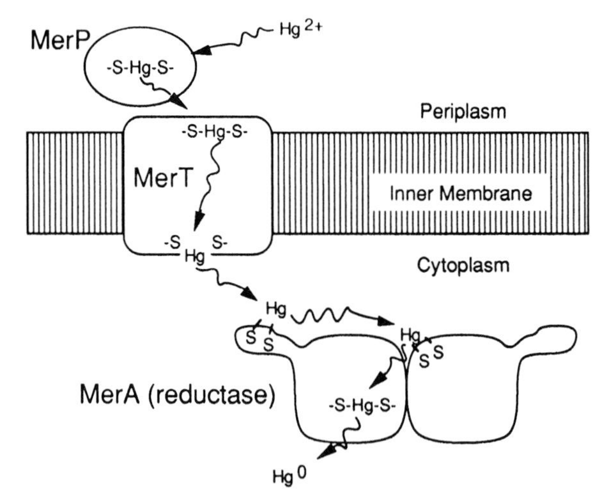

Figure 7 Mercury transport component of the mercury detoxification system ("bucket brigade"). MerP binds extracellular Hg^{2+} in a di-cysteine complex and passes the Hg^{2+} on to the first of two cysteine pairs in the inner membrane protein MerT. The second cysteine pair of MerT (at the cytoplasmic membrane surface) passes Hg^{2+} on to the N-terminal cysteine pair of mercuric reductase (the *merA* gene product). Hg^{2+} is then passed to the C-terminal cysteine pair (possibly of the other subunit) and from there associates with the redox-active cysteine pair of the active site of the opposite subunit, where Hg^{2+} is reduced to Hg^0. Hg^0 dissociates from the enzyme and passively diffuses from the cell.

the smallest known periplasmic binding protein, having only 70 amino acids after cleavage of the transmembrane leader sequence. There is a single (and required) cysteine pair in MerP. Next, Hg^{2+} is passed on to the first of two pairs of cysteine residues in MerT, an inner membrane transport protein. Transport has been directly associated with the activity of MerT and MerP, but it is not known whether metabolic energy is involved. Hg^{2+} is next transferred across the membrane to the second cysteine pair in MerT (Fig. 7). From this cysteine pair on the inner surface of MerT, Hg^{2+} is probably passed directly to the cysteine pair at the N terminus of MerA, the mercuric reductase enzyme. The region of this N-terminal cysteine pair is conserved among versions of this enzyme in all gram-negative bacteria. The Hg^{2+} is never free in the cytoplasm but is passed on from thiol pair to thiol pair. From the N-terminal cysteine pair, the Hg^{2+} is passed to the C-terminal cysteine pair, which is conserved in all mercuric reductases. The N-terminal cysteine pair functions only in transport and is not needed for enzyme activity; the C-terminal cysteine pair is required for enzymatic function. The N-terminal 100 amino acids (which contain the first cysteine pair) are missing from the recent X-ray crystallography structure of mercuric reductase (because these residues lacked fixed positions but were flexible within the crystal). Finally, the Hg^{2+} is coordinated between the C-terminal cysteine pair on one monomer subunit and the active site cysteine pair on the second subunit. Reduction occurs with electrons passed from NADPH via FAD. Metallic Hg^0 is released from the enzyme and passively diffuses from the cell without the need for protein carriers. The overall bucket brigade process (Fig. 7) means that Hg^{2+} is never free to inhibit cellular processes.

B. Cadmium

There are several known systems (different mechanisms) for cadmium resistance in bacteria. The two best understood systems, *cadA* of gram-positive bacteria such as *Staphylococcus* and *czc* (for *c*admium, *z*inc, and *c*obalt resistances) in the gram-negative soil microbe *Alcaligenes*, both maintain low intracellular Cd^{2+} concentrations by energy-dependent pumping from the cell. Otherwise the two systems are quite different. The CadA efflux system is an ATPase of the widespread (in bacteria and in eukaryotes) E_1E_2 or P class, which includes the Kdp K^+ ATPase described above and the Na^+/K^+ and Ca^{2+} ATPases of animal membranes. These ATPases have a number of shared properties (Fig. 8), including a substrate-recognition region. In CadA, this region includes two cysteine residues and is closely homologous to the Hg^{2+}-binding MerP protein and the N-terminal Hg^{2+}-binding region of

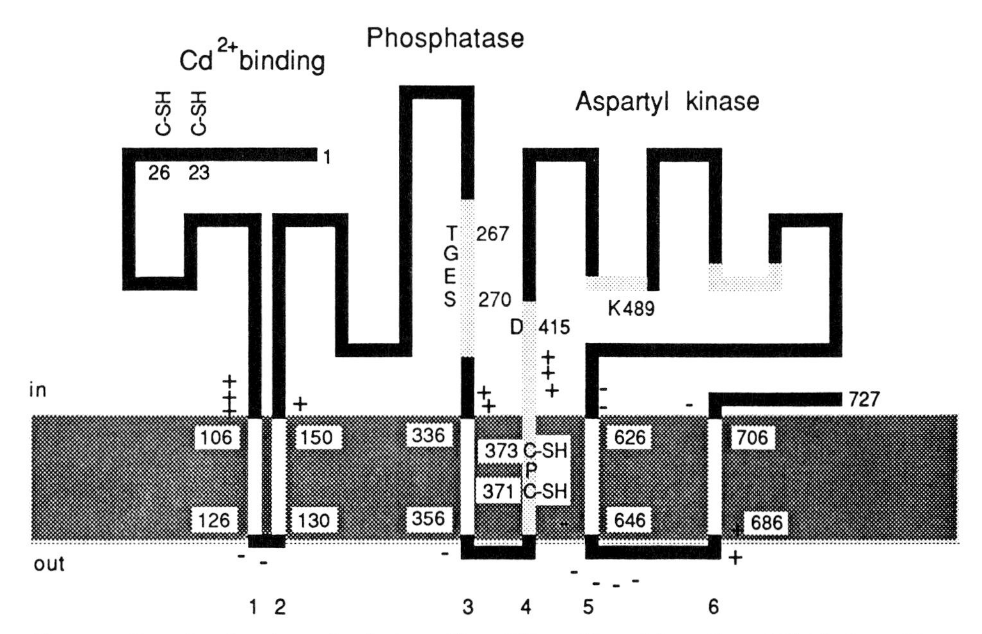

Figure 8 Functional regions of cadmium efflux ATPase. [From Silver, S., and Walderhaug, M. (1992). *Microbiol. Rev.* **56**, 195–228.] See text for details.

mercuric reductase. After CadA is anchored in the membrane by a hydrophobic hairpin region (Fig. 8), there is a 190-amino acid domain that functions as a phosphatase and includes a conserved T-G-E-S tetrapeptide common to these enzymes. After another transmembrane hairpin, which appears to contain the transport channel (and includes another cysteine pair and a conserved proline residue), there is a 250-amino acid cytoplasmic domain. This domain contains a conserved heptapeptide (in KdpB and CadA polypeptides of bacteria, in plant H^+ ATPases, and in animal Na^+/K^+ and Ca^{2+} ATPases) including aspartate-415, which is phosphorylated during the ATPase cycle. The primary characteristic of E_1E_2 ATPases that differentiates them from other ATPases (such as the H^+ translocating ATPase of photo- and oxidative-phosphorylation) is the presence of this covalently phosphorylated intermediate state. Lysine-489 position (Fig. 8) is thought to be involved in ATP binding to the protein. After a third transmembrane hairpin, the CadA ATPase ends; some eukaryote E_1E_2 ATPases continue for another 200 amino acid residues. The CadA Cd^{2+} efflux ATPase consists of two polypeptides: CadA (Fig. 8) and a smaller, highly charged protein called CadC. For comparison, Kdp consists of three polypeptides (KdpA, KdpB, and KdpC), but all three are hydrophobic membrane proteins. Expression of the CadA system is regulated by the presence of toxic divalent cations.

In contrast, the Czc system of *Alcaligenes* consists of three polypeptides but probably is not an ATPase. It is not known how Czc energizes Cd^{2+}, Zn^{2+}, and Co^{2+} efflux. The large (1063 amino acids) central membrane protein CzcA appears to associate with the smaller CzcB and CzcC proteins, which in turn affect the substrate specificity range of the divalent cation efflux system. Expression of Czc is also regulated by toxic divalent cations.

C. Arsenic

The mechanism of arsenic resistance, which is fundamentally the same in gram-negative and gram-positive bacteria, consists of an efflux system that uses an unusual membrane ATPase. Experts refer to the arsenic ATPase as an "orphan," because it does not belong to known classes of ATPases, such as the E_1E_2 ATPases or the F_0F_1 ATPases of mitochondria, chloroplasts, and bacteria. Alternatively, we hope that the arsenic ATPase will be the patriarch of a new clan of interesting membrane-transport enzymes. The Ars (*ars*enic resistance) system confers resistances to (and therefore must pump out) arsenate (AsO_4^{3-}), arsenite (AsO_2^-), and antimonate (SbO_2^-). In gram-negative bacteria, the Ars ATPase consists of three polypeptides (ArsA, ArsB, and ArsC) and its regulation involves a pair of additional polypeptides (ArsD and ArsR). ArsR is a repressor protein that binds the operator site and prevents initiation of *ars* mRNA synthesis. Induction with oxyanions or mutational inactivation of ArsR turns on mRNA synthesis. ArsD appears to be a secondary regulator that lowers Ars ATPase synthesis. Of the three ATPase components, ArsB is the integral inner membrane protein that must contain the oxyanion transport channel. ArsB contains 430 amino acids and differs from the CadA polypeptide in that ArsB lacks large nonmembrane regions. ArsA is the ATPase subunit and is attached to the membrane somewhat like FecE (Fig. 5) or PstB and CysA (Fig. 6). There is no overall sequence homology, however, between ArsA and the other ATPase subunits, except for the short ATP-binding motif. ArsA appears to have originated by a tandem duplication and gene fusion, because the first half of ArsA is closely homologous in sequence to the second half. ArsA sits on the ArsB portion of the membrane as a dimer, so there are four ATP-binding sites per complex. The molecular mechanism of ATP energy coupling is not known. Together ArsA and ArsB confer energy-dependent efflux of AsO_2^- and SbO_2^-. The small hydrophilic (but presumedly membrane-associated) ArsC protein is needed in addition to ArsA and ArsB for arsenate efflux and resistance.

D. Chromium

For chromium resistance determined by plasmids of *Pseudomonas* and *Alcaligenes,* a transport effect determines resistance. There is reduced uptake by the induced resistant cells. However, unlike the systems described above, there is no evidence for chromate efflux as contrasted with diminished chromate uptake. Further work is needed to establish the mechanism. The chromate resistance determinant has been sequenced in two cases. There is a large (about 400 amino acids) central *chrA* protein that is hydrophobic and presumedly constitutes a chromate membrane channel. Other than *chrA*, there are two additional potential genes whose roles are not known.

E. Copper

The two best-studied bacterial copper resistance systems appear to have fundamentally different mechanisms of resistance, but both involve membrane transport proteins (Fig. 9). The copper resistance system of the plant pathogenic bacterium *Pseudomonas syringae* consists of four plasmid genes [and the corresponding proteins (Fig. 9)]. CopA and CopC are periplasmic copper-binding proteins. During growth on Cu^{2+} salts, the cells become bright blue as they accumulate copper, while the cell-free medium loses the color of the copper salt. How the outer membrane protein CopB and the inner membrane protein CopD cooperate with the periplasmic proteins to localize (and sequester) copper in the periplasm is not known. Whether chromosomal genes are involved in addition to the plasmid genes is not known but likely.

Both chromosomal and plasmid gene products are involved in *E. coli* copper transport, accumulation, and efflux (Fig. 9). Copper appears to be transported into the cellular interior by the parallel membrane transport systems CutA and CutB, encoded by chromosomal genes. In the cytoplasm, Cu^{2+} is highly toxic (if free) because it catalyzes oxidative damage. Therefore, intracellular Cu^{2+} is sequestered by the two chromosomal gene products CutE and CutF and by the additional PcoC protein, which is determined by a resistance plasmid gene (Fig. 9). When copper overload occurs, *E. coli* cells probably efflux Cu^{2+} by a system consisting of two chromosomally encoded proteins, CutC and CutD, plus (if the resistance plasmid is present) two plasmid-encoded proteins, PcoB and PcoA. The models for copper transport, homeostasis, and efflux in Fig. 9 are tentative; they are based on new and incomplete evidence. However, that we can produce such explicit models marks real progress in our understanding of the transport and cellular regulation of a critical trace nutrient such as Cu^{2+}.

F. Other Inorganic Cations and Anions

There are additional toxic metal and oxyanion resistance systems that we will not discuss for lack of understanding and space. Thallium ion resistance, for example, was recently reported. Since Tl^{+} is a toxic analog of K^{+} (and sometimes a substrate for potassium transport systems), we anticipate transport is responsible for this resistance.

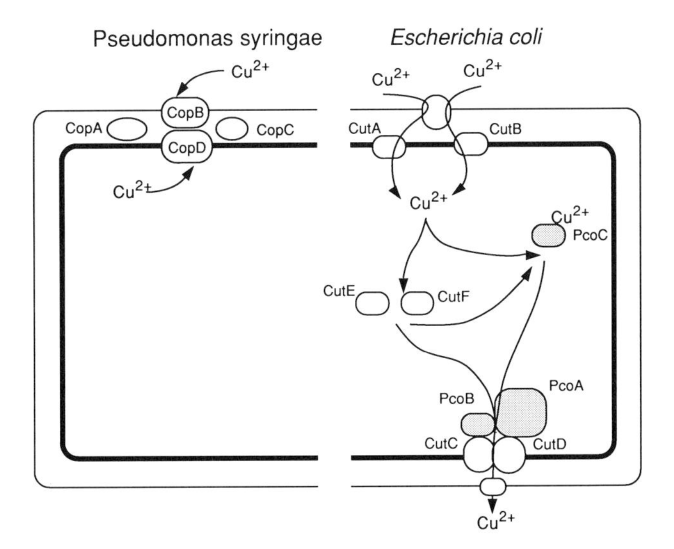

Figure 9 Models of plasmid-resistance systems for copper in (A) *Pseudomonas syringae* and (B) *E. coli*. In *Pseudomonas*, Cu^{2+} is thought to be sequestered extracellularly on CupB and CupC proteins in the periplasmic space. In *E. coli*, Cu^{2+} enters the cell through parallel chromosomally encoded uptake systems, is sequestered in the cytoplasm by the plasmid-encoded PcoC protein, and is effluxed from the cell by a 4-component complex consisting of chromosome encoded (CutC and CutD) and plasmid-encoded (PcoA and PcoB) proteins.

IV. Conclusions

For every inorganic cation or anion (generally oxyanion) that bacterial cells may be exposed to, there are separate, genetically determined, and carefully governed membrane transport systems to assure that the needs for growth are met and to avoid toxic metal overload. This new perspective on ion transport in bacteria departs from older perspectives in several ways. First, there is the frequent involvement of primary ATPases. The chemiosmotic picture of ion transport focused on secondary transport systems coupled to the membrane potential and proton gradient. Whereas transport systems operating under conditions of substrate abundance (Figs. 1 and 2) are frequently chemiosmotic in energy coupling, those operating under conditions of starvation or stress are frequently ATPases. The chromosomal Pst, Kdp, Fec, and Mgt uptake systems, as well as plasmid resistance systems, function in a similar fashion. A second refinement on earlier ideas is the occurrence of physically and genetically separate uptake and efflux transporters. Regulation of ion content at the physiological level is not simply a matter of running transporters in reverse.

We now know that regulation occurs at the gene level as well as at the level of physiological control. There are proteins that sense both intracellular (e.g., Fur for Fe^{2+}) and extracellular (e.g., FecR for Fe^{3+}-dicitrate and KdpD) ion concentrations and transfer that information to DNA to initiate synthesis of mRNA for the appropriate genes.

For bacterial ion transporters in general, two families of ATPases are well represented. The E_1E_2 ATPases are always (or P) cation translocating membrane enzymes. The cation differs from system to system (including H^+, Ca^{2+}, Na^+, K^+, Mg^{2+}, and Cd^{2+}), and the direction of transport may be inward or outward. In some cases (the Na^+/K^+ ATPase of animals and what may be the CadA Cd^{2+}/H^+ antiporter), the same complex of one to three polypeptides appears to carry out coupled movement of cations in both directions. The second frequently found family of ATPases consists of a periplasmic substrate-binding protein, two inner membrane channel-forming proteins, and a peripheral membrane-associated ATP-binding protein. These transport systems accumulate both cations (iron, Fig. 5) and anions (phosphate and sulfate; Fig. 6). Related systems for uptake of amino acids and sugars are well known.

We are basing these generalities on a few well-studied systems in a few bacterial species. In time, additional studies will show whether the known patterns are widely used and whether additional patterns (and families of transporters) exist.

Bibliography

Harold, F. M. (1986). "The Vital Force: A Study of Bioenergetics." W. H. Freeman, New York.

Hughes, M. N., and Poole, R. K. (1989). "Metals and Microorganisms." Chapman and Hall, London.

Misra, T. K. (1992). Heavy metals, bacterial resistances. *In* "Encyclopedia of Microbiology" (J. Lederberg, ed.). pp. 361–369. Academic Press, San Diego.

Rosen, B. P., and Silver, S., eds. (1987). "Ion Transport in Prokaryotes." Academic Press, San Diego.

Silver, S. (1978). Transport of cations and anions. *In* "Bacterial Transport" (B. P. Rosen, ed.), pp. 221–324. Marcel Dekker Publ., New York.

Silver, S., and Misra, T. K. (1988). *Annu. Rev. Microbiol.* **42,** 717–743.

Silver, S., and Perry, R. D. (1982). Bacterial inorganic cation and anion transport systems: A bug's eye view of the Periodic Table. *In* "Membranes and Transport"(A. N. Martonosi, ed.), vol. 2, pp. 115–121. Plenum Press, New York.

Silver, S., and Walderhaug, M. (1992). *Microbiol. Rev.* **56**, 195–228.

Isolation

Jennie C. Hunter-Cevera and Angela Belt
The Biotic Network

Glossary

Actinomycetes Diverse group of bacteria that share the common property of a "filamentous" and often branched growth habit

Autochtonous Indigenous population (i.e., group of microorganisms that are part of the natural biota)

Biota Population of microorganisms associated with a particular niche or habitat found within a specific ecosystem

Ecosystem System formed by the interaction of a community of organisms with their environment

Enrichment Producing a special *in vitro* environment where one group or type of microorganisms is increased in numbers and favored over another group

Eubacteria True bacteria (i.e., no filaments produced during its growth cycle)

Habitat Place where a given animal or plant naturally lives or grows, such as warm seas, mountain tops, and fresh waters

Niche Specific role or position of an organism within a ecosystem

Selection In biology, any process, whether natural or artificial, by which certain organisms or characteristics are permitted or favored to survive and reproduce in, or as if in, preference to others

MANY METHODOLOGIES AND TECHNIQUES are used to isolate microorganisms found in nature. Although the same principles apply to isolation of human and animal pathogens, clinical techniques are more thoroughly described elsewhere. No one knows exactly how many microorganisms really exist in nature. There are numerical estimates of almost every group of microorganisms existing in our biosphere; however, these numbers are representative of the successful isolation techniques employed. The numbers and types of microorganisms that one isolates from any given sample depends on two factors: (1) the limiting biophysical parameters within an ecosystem and (2) the media and method used to isolate any general or specific group of microbes. The ability to isolate microorganisms from a specific niche within a defined habitat or ecosystem is determined by the procedures employed by the researcher more than by the conditions and microbes existing in nature.

I. Collection and Handling of Samples

A. Questions to Consider

A number of important questions need to be answered before starting experiments to isolate microbes. In the tradition of the investigative reporter, one could query what, who, how, where, and when. For example, What? What is the purpose of this experiment? Are you trying to isolate a microbe that can utilize or degrade a particular substrate? Are you interested in the effects of seasonal fluctuation on a given population within the ecosystem being examined? Are you interested in quantitating the numbers and types of microbes present in a specific sample type? Are you trying to determine the function or role of the microbes present as part of the natural biota? Are you trying to isolate as many as possible isolates of one given species or genus for taxonomic characterization and definition? Who? What microorganisms do you want to isolate: anaerobes, aerobes, autotrophs, heterotrophs? Where? Where do you collect the samples: forests, meadows, marine waters, gardens? Specific substrates: plants, soils, waters, rocks, insects? When? When do you collect the sample? Are the numbers and/or types isolated influenced by seasonal changes? How? How do you isolate the microorganism: direct or indirect examination of the sample, selective or enrichment techniques?

These examples illustrate the necessary thought processes one goes through before designing and implementing isolation experiments. It should also help direct where one would sample for the desired microorganisms of interest.

The methods listed and described in this article are by no means set in stone. As a scientist involved in the isolation of microorganisms from nature, one should be as creative and innovative as possible in designing isolation methods and media. It is not the diversity of microbes present in a sample that limits the success in recovery of many different isolates but, rather, one's own imagination and application of these ideas.

B. Where and What to Sample

To isolate the resident biota from a particular ecosystem, especially those species occurring in unique microenvironmental niches, considerable attention should be paid to sampling. Samples should be representative of a site (e.g., a particular soil type and its horizons, leaf litter and detritus, rhizosphere plane and zone, marine sand, sediment and muds, plant surfaces, parts or water column). Seasonal and temporal aspects of collecting should be considered, because a true autochtonous population may occur transiently (e.g., actinomycete numbers decrease after a heavy rainfall).

Once samples are collected, they should be examined immediately upon return to the laboratory or stored at 5°C overnight in an area separate from actual plating, screening, and culture collection facilities to minimize the chances of mite infestations. [*See* CULTURE COLLECTIONS: METHODS AND DISTRIBUTION.]

C. Documentation

It is extremely important to document the sample type, i.e., describe the environment from which it was collected, time, and date. Also for record keeping, it is beneficial to give each sample a number.

D. Permits and Regulations

One should check with county, state, and/or federal government agencies to determine what regulations and permits are needed for handling and containment of soil, plant, and water samples. The U.S. Department of Agriculture strictly enforces the movement and transfer of soils and plant materials and certain microbes within states and across state lines as well as the importation of foreign samples.

E. Collection Materials

Certain aseptic supplies and tools are needed to collect samples. Sterile containers (centrifuge bottles) or bags (NASCO Whirlpak bags) to carry the sample are useful as are sterile tools (gloves, spatulas spoons, scissors, scapels) for the actual sampling process. The type of container employed depends on what is being collected. For example, if collecting water samples, the containers ned to be leak-proof. [*See* SPECIMEN COLLECTION AND TRANSPORT.]

II. Isolation Approaches

A. Media Design and Incubation

There are probably as many media recipes as there are cookbooks. Media design is very important because this combination of ingredients will determine which colonies initially appear on plates or grow in broth from the sample being examined. When designing media, certain parameters should be considered such as ecological parameters of the sample niche, broth and/or solid media, and synthetic and/or natural extracts. These parameters are discussed in Section II.A.1.

Incubation times can vary depending on the group or type of microorganisms being isolated. For example, fungi can take as long as 10 days or more to appear on isolation agar plates, whereas soil bacteria usually appear within 1–5 days. The temperature of incubation is another factor to consider when designing isolation experiments. Typical incubation temperatures used for isolating microorganisms from nature range between 20° and 28°C. Water-saturated filter paper disks attached (by adsorption) to the lids of the petri dishes are used to prevent drying out of agar plates that are incubated for long periods of time or at higher temperatures. Once isolated in pure form, a temperature profile (15–40°C) is recommended to determine the optimal temperature for growth and maintenance.

1. Ecological Parameters

Ecological parameters to consider include low or high temperature and pH, ionic strength, redox potential, salinity, light, nitrogen sources, surface : volume ratios, and even substrate concentration (e.g., in regard to fungi, using media with low C:N ratios yields more discrete countable colonies, resulting in more effective isolation and identification). Most of these parameters are measured in the field or in the laboratory. The isolation media are then modified to suit the ecosystem of the habitat being examined. For example, seawater is added to isolation agars at different concentrations to match the salt gradient in a salt marsh; pH is lowered to suit the different horizons present in a forest; and temperature of incubation is raised for thermophiles (55–70°C) and lowered for psychrophiles (4–15°C). [*See* ECOLOGY, MICROBIAL.]

2. Broth and Solid Media

Most microorganisms can be isolated on media that thas been solidified with 1–2% agar or gelatin; however, a few species of microorganisms are inhibited on solid media, and broth techniques must be employed for initial isolation and maintenance. Such examples can be found within all the five major groups of microorganisms discussed in this article: bacteria, algae, actinomycetes, fungi, and protozoa.

3. Synthetic versus Natural

Although less suitable for the initial isolation of strains from nature, commercially prepared media are readily available for the isolation of clinical strains of microorganisms. However, some of the more selective media (e.g., red violet bile, potato dextrose) are suitable when either diluted quarter or half strength and augmented with the addition of natural extracts added to aid the artificial environment on or in which the microorganisms initially grow. The extracts, which stimulate the natural environment, are added to these commercially prepared media to a concentration of 5–15% of total liquid volume. However, if making "lean" isolation agars from "scratch," it is recommended that extracts be added in the concentration of 10–50% of total liquid volume.

Infusions and extracts are usually made by gently heating materials (rocks, plants, detritus, bark, etc.) collected within the ecosystem being examined. The tea bag technique (placing the material inside cheesecloth, tieing it with a string, placing the bag in water, and then gently heating this material for 20–30 min) works well for making most natural extracts. This extract is decanted and filter-sterilized or autoclaved for 1 hr to ensure killing of spores.

4. Chemically Defined or Complex Media

Media employed for isolation can be either chemically defined or complex. Chemically defined media require that each ingredient represents a single simple compound, such as glucose or glycerol for a carbon source and ammonia or arginine for a nitrogen source. In addition, cofactors or growth supplements in the form of individual minerals such as magnesium and cobalt are added as well as vitamins such as biotin and thiamine.

Complex media represent just the opposite of defined media; i.e., the sources of carbon, nitrogen, and mineral/vitamin supplements are mixed or coming from more than one source. Complex media can consist of two or more simple sugars or a complex carbohydrate for a carbon source, such as cellulose or corn steep liquor. For a nitrogen source, either yeast extract or peptones are commonly used. There are small amounts of "nutritive substances" in yeast extract, such as minerals, that aid in the growth and replication of microorganisms.

B. Selective and Enrichment Isolation

The mixed biota of soil, water, plants, etc., are a group of competing species or metabolic types. Changing the physical and chemical environment will promote dominance of one or a few particular types. Antibiotics, chemical agents, and physical treatments are used to select and enrich for specific groups of microorganisms. Table I lists some of the many available methods and chemical agents used to isolate specific groups of microorganisms from nature.

C. Isolation Techniques

There are basic methods and techniques that apply to the isolation of algae, actinomycetes, bacteria, fungi, and protoza. Summaries or these techniques follow.

1. Dilution-Spread Plate

An example of the dilution-spread plate method is as follows. Mix 5 g (wet weight) soil with 99 ml sterile distilled water, phosphate buffer, or quarter-strength soil infusion or extract in a 250-ml flask and shake at 150 rpm for 25 min at 26°C. Using sterile 1-ml-wide mouth pipettes or a pipette aid with eppendorff tips, serially dilute 1 ml of the soil–liquid suspension in 9 ml of an appropriate diluent to 10^{-10} if necessary. Some soil composition types will affect turbidity observations; i.e., a large clay content may tend to make the suspension more turbid, yet microbial numbers may be low. To ensure equal mixing of the sample, always shake or vortex for 30 sec before dispensing the suspension. Spread-plate 0.1 ml volumes of three appropriate dilutions (based on turbidity) onto three to seven different isolation agars with the aid of a sterile (alcohol-flame) bent glass rod (75° angle) and a turn table (Fisher). The dilution-spread plate technique is also used to sample plants, rocks, waters, etc.

2. Dilution-Pour Plate

The dilution-pour plate technique uses the same dilution procedure as in the dilution-spread plate method with the following difference. Dispense 20 ml molten (45°C) isolation agar into individual sterile test tubes and place in a 45°C water bath. Add 1 ml of a dilution sample to each tube of medium. Each tube can be used to prepare one plate. Use at least the final three dilutions for isolation plates. Work quickly to minimize sample exposure to 45°C. Roll the melted medium tube gently between one's hands to mix contents and pour into a 100-mm plastic petri dish. Allow agar to solidify before moving plates to an incubation chamber.

3. Stamping

The stamping technique is more often used to isolate actinomycetes from soils. This technique has also been applied successfully for the isolation of fungi from plant debris and soils.

Air-dry soil and plant samples in sterile petri dishes under a laminar flow hood for 3–10 days. Then, gently ground the samples to a fine powder with a sterile pestle and "stamp" them onto isolation agars. Directly press a small circular sponge (Dispo culture plug, 16 mm; Scientific Products) into the dried powder and then remove. Shake any excess powder off. Then, inoculate a stack consisting of 9–12 plates of alternating different isolation agars by successive "stamping" of the agar surface 13 times (10 stamps on the outside perimeter and 3 in the middle) with the sponge to achieve a diminutive effect.

4. Rolling and Particle Implants

Chop plant parts (flowers, leaves, upper stem, middle, base, roots) aseptically with scissors or scalpels, preferable in a laminar-flow hood. Then, either implant the pieces into the agar surface or gently roll them over the surface of the agar in a streaking manner. The plant parts may, before implanting, also be dried for 2–7 days in sterile petri plates to encourage the development of spore-formers.

5. Filtration and Filter Imprints

Filtering techniques apply to isolation of microorganisms from water samples. Filter 50 ml of a water sample through a 0.22-μm disposable filter. When the sediment has settled on the filter membrane, add 1 ml sterile diluent with a flat, wide-mouth 2-ml sterile pipette and then gently scrape the membrane with the side of the pipette or a rubber policeman. Transfer the diluent suspension to 9 ml of a corresponding diluent and vortex for 5 min. Serially dilute and sample and spread-plate in the manner described in Section II.C.1.

III. Specific Microbial Groups

A. Algae

The algae make up a large group and, depending on the classification scheme used, consist of at least 6 divisions, about 15 classes, and more than 22,000 named species. Algae are very diverse.

Physically, their forms include unicellular motile and nonmotile forms, filamentous and colonial types, and complexly organized macroscopic forms with a thallus consisting of differentiated and specialized structures.

The habitats they occupy are equally diverse: terrestrial forms, benthic or deep-sea forms, marine and freshwater species, those that thrive in thermal hotsprings and others that proliferate in the freezing temperatures of Antarctica. Some algae are free-living, whereas others are symbiotic with fungi, animals, or higher plants.

An understanding of the ecology and similarities as well as differences in physiological requirements of the various classes of algae will help when preparing a comprehensive isolation scheme. The basic strategy using the enrichment/inhibition approach, as outlined in Table I, can be applied to selectively encourage the growth of one algal species over another and for the separation of the algae culture from contaminating bacteria. For example, diatoms, blue-green algae, and bacteria can often be eliminated from green algae cultures by selective use of antibiotics. Motile unicellular forms can be isolated by manipulating the physical environment. In addition, thermophilic hotspring forms will require very different environmental conditions than marine deep-water algae.

Algae can be isolated by the direct method, where target cells are physically isolated from the sample for further culturing and purification. These methods involve examining the sample under a dissecting microscope and selecting algal cells using a micromanipulator or capillary pipette. A limitation with this method is the size of the algae being isolated. Very small algae, which are too small to be seen well at this magnification, are not suited to this method. The size of the capillary opening will limit the usefulness of this technique for some larger algae; however, many filaments and reproductive cells can be isolated with the method. A modification is the glass hook method, where a hook is made on the end of Pasteur-type pipette by passing through a flame and is then used to tease and separate filamentous forms.

Another method, which relies on the physical separation of sample components (vs. nutritional separation), is separation by centrifugation and successive washings. With this method, cells can be partially purified from contaminants in the sample prior to placing them into media for further isolation.

Another physical method involves phototactic migration. Motile algae are placed in one side of a glass container (e.g., a baking dish, pipette with a light source placed at the opposite end). The algae will then migrate through the media toward the light source (hopefully), leaving nonmotile, nonphototactic cells behind.

With the implant method, pieces of sample (e.g., soils, rocks, algae bloom) are placed directly on an agar surface and embedded slightly. Filamentous forms grow out from the point of implantation.

The streak plate method is the same as that used for bacterial or yeast isolations, where a sample is cross-streaked onto suitable agar media with a sterile loop to achieve single colonies and then incubated under appropriate conditions.

Similarly, the spray plate method utilizes agar plates; however, the sample is sprayed onto the surface using a Pasteur-type pipette attached to a sterile air source.

The spread plate method can be useful for isolation of single colonies of filamentous and unicellular forms and is performed as described in Section II.C.1 using media and growth conditions suitable for algae.

The pour plate method, also adapted from bacteriological work, requires placing the sample or dilutions of the sample into cooled molten agar, mixing gently, pouring into petri dishes, and incubating in light.

With the broth method, dilutions of sample are placed in appropriate liquid media and incubated with or without shaking.

The hanging drop method involves placing a drop of the sample on a coverslip and inverting over a ringed or depression slide. Motile cells should swim up to the coverslip, with the sample source settling in the bottom of the hanging drop. The drop of sample material can then be removed and the motile cells washed off the surface of the coverslip into suitable media.

Fewer methods have been developed for isolating and growing larger macroscopic forms of algae or seaweeds and are reported elsewhere.

B. Bacteria

Over 300 genera are listed in "Bergey's Manual." They are classified by cell wall types (based on retaining certain dyes), preference for oxygen or an inorganic element for a terminal electron donor, fermentative or nonfermentative, spore- or non-spore-formers, motility with or without flagella, oxidase-positive or -negative, etc. One can only begin to imagine the tremendous amount of isolation techniques developed to isolate all 300 genera. It has also been said that the bacteria listed in "Bergy's Manual" only represent 10% of what really exist in nature. Some genera require specific enrichment or selection techniques for isolation in pure culture. Much literature is available on the selective isolation of each of these genera; however, some of the genera associated with samples from specific ecosystems can be isolated via direct examination methods, i.e., use of dilution spread plate techniques on agars that are developed to isolate a wide range of organisms. The incorporation of natural extracts and environmental biophysical parameters into media, as well as the diluent employed in plating out the sample, can affect the number and types of bacteria isolated by direct examination. Described below are but a few of the many genera found in nature and some of the isolation techniques used by microbiologists.

1. Aerobic Organotrophic

Gliding bacteria are motile but do not have flagella. Media containing dead bacterial cells, cellulose, or other polymers are used to isolate these polymorphic bacteria.

Budding or stalked bacteria can be isolated from freshwater by use of selective agars of low nutrition concentration and enrichment techniques such as peptone enrichments, antibiotics that inhibit peptidoglycan synthesis, and hanging glass slide cultures.

Bacteria associated with plant material such as members of the Pseudomonaceae are frequently isolated by use of the dilution spread plate technique or by rolling a plant part across an agar plate surface.

Aerobic spore-formers such as members of the genus *Bacillus* often appear as the "unwanted colonies" on dilution-spread plates. The spores produced are very resistant to desiccation and heat.

2. Aerobic Chemolithotrophic

Bacteria in the aerobic chemolithotrophic group can only be isolated on media containing no organic components and compounds that are reduced inorganic substances such as sulfur, nitrite, or ammonia. These inorganic compounds are then oxidized to produce energy for the cells. *Thiobacillus,* a sulfur-oxidizing bacterium that is in part responsible for acid mine drainage problems, requires low pH for growth.

Continuous culture enrichments prove to be an extremely efficient method for the isolation of *Nitrosomonas,* an NH_4^+-oxidizing bacteria. A high degree of enrichment can be achieved at a dilution rate of 0.04 hr^{-1} and pure cultures are obtained within 6–7 wk. [*See* CONTINUOUS CULTURE.]

3. Facultative Anaerobes

Facultative anaerobes are numerous and occur in all the major taxonomic groups of bacteria. Some of them are pigmented such as *Chromobacterium* (purple) and *Flavobacterium* (yellow, orange, red, or brown). These can easily be distinguished on media favoring the development of pigment, for example, containing a mineral salts solution.

Members of the Enterobacteriaceae ferment sugars and produce acid and gas. Use of dyes that indicate a change in pH are incorporated into selective media.

4. Anaerobes

For enumeration and isolation of anaerobic bacteria from samples such as sewage digester fluid or anaerobic muds, the anaerobic roll tube method (agar-coated test tubes) combined with mixed gas phase (95% N_2 and 5% CO_2) is often used.

Photosynthetic bacteria require light and some can use sulfide as the sole electron donor for photosynthesis. *Desulphovibrio* also uses sulfates as the terminal electron acceptor and, when isolated on selective media containing sulfates and iron salts, the cultures turn black.

Methane-producing bacteria, such as *Methanococcus,* produce methane by reducing CO_2 via the fermentation of acetate or methanol or by the oxidation of hydrogen or formate. They are isolated by use of enrichment broths.

Anaerobic spore-formers such as *Clostridium* and *Desulfotomaculum* can be isolated using reduced media; i.e., all the oxygen is replaced by a mixture of nitrogen, hydrogen, or carbon dioxide. The dilution-spread plate technique is one method frequently used to isolate *Clostridium,* and the isolation is done inside an anaerobic chamber.

C. Actinomycetes

Actinomycetes are bacteria that at some time in their life cycle exhibit filamentous growth, spores, and/or coccoid elements. They are regarded as "industrially important bacteria" because they are well known for their ability to produce antibiotics and enzymes. In nature, they are responsible for the "earthy" odor associated with soil; geosmin is a neutral oil produced by streptomyces as a secondary metabolite. Actinomycetes are also involved in the mineralization of soil (e.g., cellulose) and help hold together soil particles with their fine hyphae.

They are widespread in nature, mainly as aerobic, saprophytic mesophiles, although a few are pathogenic for plants, animals, and humans. They can be isolated from soils, plants, waters, and even lake bottoms. Their numbers range between 1.0×10^5 and 1.0×10^8/g dry temperate zone soil; this should be qualified in that the plate count method emphasizes the sporing ability of an organism rather than the cellular mass.

The most abundant actinomycetes isolated from soils using standard plate dilution techniques are *Streptomyces* spp. followed by *Nocardia* spp. and *Micromonospora* spp. Soil is diluted between 10^{-2} and 10^{-7} in sterile saline (0.85% NaCl) and 0.1-ml aliquots are spread-plated onto actinomycete media such as water, collodial chitin, arginine—glycerol salts, and starch casein media. To retard fungi from growing on the isolation plates, antifungal agents such as cyclohexamide and nystatin are filter-sterilized and added to media filtration to a final concentration of at least 50 μg/ml each.

Use of the stamping method described in Section II.C.3 will facilitate a greater number of diverse species. In addition, an increase in the number of one species can be obtained through the use of selective isolation techniques such as prior heating of soils (70°C, 10 min) to obtain *Thermomonospora,* baiting with wet pollen grains to isolate *Actinoplanes,* or adding chitin to increase the numbers of *Micromonospora*. For additional enrichment agents, see Table I.

D. Fungi

Estimates of the numbers of species of fungi range from 65,000 to 100,000 or more with 700–1500 new species being described yearly. An understanding of the ecology and similarities as well as differences in physiological requirements of the various classes of fungi will help when preparing a comprehensive isolation scheme. The basic strategy of using the enrichment/inhibition approach, as outlined in Table I, can be applied to selectively encourage the growth of one fungus over another. For example, lignicolous fungi metabolize lignin-containing substrates such as wood. Coprophilous fungi occur on dung. Some pathogenic fungi have specific host requirements.

1. "Lower" Fungi

"Lower" fungi can be isolated from nearly any substrate. In general, their growth is more rapid than that of higher fungi. To prevent overgrowth of the plates by any one type of fungus, compounds that allow growth but tend to inhibit the growth rate and colony size are useful (Table I). Bacteria, which have a faster generation time than fungi, must be excluded by adding selective antibiotics or changing the physical conditions, for example, by lowering pH or excessive drying of a soil sample. A number of methods are useful to isolate from soil, waters, and plant materials. Described below are some of those most commonly and successfully used.

Direct isolation involves sieving, decanting, observing the sample with a microscope, and physically isolating the target spores or mycelial fragments with a needle or micromanipulator. This procedure is time-consuming and can be tedious but may be best for some situations (e.g., isolating spores of endomycorrhizal fungi from soil or sand).

The implant method works well for soil or plant samples. The sample is plated directly onto the agar surface, pressing in slightly. The fungus will spread out from the point of implantation. Positioning the implant at the edge of the plate will provide a longer linear distance for the growing front. When seeking to isolate fungi that are internally located in the plant sample (e.g., from mycorrhizal roots or internal plant pathogens), surface sterilization with 20% bleach, 30% hydrogen peroxide, or mercuric chloride (1:1) to eliminate competitors is advised. These sterilants are all caustic or toxic and should be handled with care using appropriate laboratory safety measures.

The dilution plate method described in Section II.C.1 may also be used. One gram of soil is added to each dilution tube of 9.0 ml diluent, agitated for 15 min, serially diluted, and spread onto agar surfaces. Plant parts are added to diluent, with the addition of a mild detergent such as Tween, agitated, filtered, and processed as described earlier. When sampling from waters, consider the potential microbial richness of the sample (clear water vs. sewage or polluted) when determining the appropriate dilution needed to yield countable colonies.

With the maceration method, the previous steps are preceded by first macerating the plant sample. One should be cautious using this last technique because maceration can cause deleterious effects on the organisms in the sample, whether by physical injury or through the release of inhibitory compounds.

The Warcup method is similar to the pour plate method used to isolate bacteria. The sample is moistened with a little sterile water, placed in a petri dish, and covered with warm (45°C) molten agar.

The stamping method using foam plugs, as described in Section II.C.3, can also be used for soil

Table I Physical and Chemical Treatments for Enrichment/Selection of Specific Groups and/or Genera

Treatment	Inhibits:	Enriches/selects for:
Germanium	Diatoms	Other algae
Penicillin	Bacteria	Pure algal cultures
Triphenylmethane dyes	Gram-positive bacteria	Gram-negative bacteria
Malachite green		
Brilliant green		
Crystal violet		
Basic fushin		
Phenylethyl alcohol	Gram-negative bacteria	Gram-positive bacteria
N_2 sole nitrogen		Nitrogen fixers
NH_4, NO_2/inorganic media		Nitrifying bacteria
Elemental sulfur	Heterotrophs	*Thiobacilli*
Thallium		*Mycoplasma*, Enteroccoceae
Low temperature (0–5°C)	Thermophiles, mesophiles	Psychrophiles, psychrotrophs
High temperature (>55°C)	Psychrophiles	Thermophiles
Heat shock at 70–80°C/10 min	Non-spore-formers	Heat-resistant spore-formers
Air-dry sample/grind		Spore-formers
Acidic pH	Bacteria	Fungi, low pH-tolerant bacteria
Alkaline pH	Fungi	Bacteria, high pH-tolerant fungi
NaCl/seawater		Halophiles
Selenite		*Salmonella*
Cycloheximide/nystatin	Lower fungi	Actinomycetes
Wet pollen grains		Actinoplanaceae
Chitin/cellulose		*Micromonospora*
60°C for 40 min plus novobiocin		*Micromonospora*
75°C for 6–18 hr plus novobiocin		*Thermomonspora*
120°C/ 1 hr		*Microbispora*
		Streptosporangium
100°C/ 1 hr		*Microbispora*
		Microtetrasopra
Tetracycline		*Nocardia*
55°C/6 min		*Rhodococcus*
100°C/15 min		*Actinomadura*
Rose bengal	Bacteria	Fungi
Pimaricin	Most soil, fungi	*Phytophthora*,
		Pythium
Vancomycin/colistin/nystatin	Fungi	Bacteria, actinomycetes
Ortho-phenyphenol	Microfungi	Wood-decaying fungi
Benlate	Lower fungi	Higher fungi
Pentachloronitrobenzene	Lower fungi	Higher fungi
2,6-Dichloro-4-nitroaniline	Lower fungi	Higher fungi

samples and often results in the isolation of more numbers and types of spore-formers. Using a similar process, the balloon method involves rolling an inflated balloon across the surface of the plant sample and then pressing the balloon surface onto agar.

2. "Higher" Fungi

The term "higher" fungi, as used here, refers to the basidiomycetes and ascomycetes that form macroscopic soprocarps or mushrooms. The mycelia of most of these fungi are generally nondescript, making it difficult to differentiate species from one another. Consequently, most attempts to isolate these involves the fruiting structures or spores.

As a rule, the growth rate of most higher fungi is much slower than that of the lower fungi. Therefore, in addition to using previously described methods to exclude competing bacteria, the lower fungi, which will otherwise tend to outgrow and overrun the plate, also must be excluded by selective use of fungicides that do not inhibit the higher fungi (see Table I).

To obtain tissue culture isolates, collect fresh, firm sporocarps. Brush the surface clean (a surface sterilant may be used but is seldom necessary) and asceptically tear open the mushroom, exposing the interior. Using a sterile needle, forceps, or tweezers, remove a piece of tissue from the interior of the pileus or stipe and place it on an agar using the implant method described earlier. The mycelia produced from the tissue culture method will be identical to the source (usually dikaryotic).

Mushroom cultures can also be isolated from spores. By isolating hyphae from a single germinating spore, before fusion with compatible hyphae, monokaryotic mycelia can be obtained. The least contamination-prone way to collect spores is by making a spore print. This can be done directly onto agar or another sterile surface as follows. Carefully slice a wedge of pileus and attach it to the inner surface of a perti dish lid, with the spore surface pointing downward. Vaseline or another gel can be used to adhere the upper surface to the glass. Place the petri dish lid over a petri dish bottom containing agar or sterile liquid or filter paper. Allow th spores to drop (this may take several hours), replace the lid with a sterile one, and then store in the cold or incubate. This method can also be used with slants of media.

The spore print method works only for fungi that forcibly discharge their spores. A combination maceration–sucrose gradient centrifugation method may be required to collect spores from other fungi. Macerate cleaned sporecarps briefly in a high-speed blender. Transfer and separate spores by centrifugation in a 20–40–60% sucrose gradient. Most spores will collect between the 20–40% layer. Repeated centrifugations for further separation from contaminants may be required. Alternatively, one may macerate the sporocarps and separate spores directly onto media by serial dilution.

Two considerations when isolating from spores are viability and dormancy. The spores of some higher fungi may have a limited period of viability that can be from days to years. This may depend on environmental conditions, intra- and interspecies variability, and even how the spores were collected and from what part of the hymenium.

Viability can be tested by using a combination of vital stains (e.g., tetrazolium salts to measure succinate dehydrogenase activity) and nuclear staining to monitor the condition of the spores.

Viable spores may remain dormant for a number of reasons. The environmental conditions may need modification (more or less moisture, temperature, pH). Some spores are inhibited by the close proximity of other spores and will require further dilutions. Conversely, with some species the presence of foreign mycelia or spores can stimulate germination. The *Rhodotorula* streaking method sometimes works to break dormancy of spores that require an external stimulation to germinate. With this method, spore solutions are plated onto agar and incubated overnight. Half the plate is dusted with activated charcoal and inoculated at the interface with *Rhodotorula glutinous*. Plates are wrapped with parafilm, inverted, and incubated at 25°C. The plates should be examined periodically for germination over the next few weeks.

3. Yeasts

Yeasts may be isolated using the plating methods previously described.

To prevent growth of contaminating bacteria, broad-spectrum antibiotics, inactive against yeasts may be incorporated into the media. Alternatively, the pH can be lowered to 3.5–4.0. Some inorganic acids are inhibitory toward yeasts and should be avoided.

Other enrichment methods are necessary to exclude the lower fungi or molds from yeast isolations. Exclusion of air by using the pour plate method or by adding a layer of paraffin has been used to select fermentative yeast species over other competing fungi.

One can also place the sample in acidified liquid media on a shaker for several days. The shaking action limits sporulation of the filamentous fungi, favoring the multiplication of both fermentative and nonfermentative yeast cells, which then may be separated by plating.

High osmophilic tolerant or obligate yeasts can be enriched for by growing under higher osmotic conditions (e.g., in 30–50% glucose media). Osmotically sensitive strains do better on sterile plant parts incorporated into water agar.

E. Protozoa

Over 300 species of protozoa have been isolated from plant litter and soils. The protozoa population is considered cosmopolitan. However, species may occur with different frequencies under differing soil conditions and are often restricted in distribution within a particular soil. Many species are considered to be transient and are directly influenced by the amount of moisture in the sample. Amoebae and flagellates are, under most soil conditions, the more prolific protozoa averaging about 10^6/g in moist forest litters.

The choice of culture medium depends on the study objectives. The easiest technique involves pouring 15–20 ml of liquid over an agar (plain or enriched) surface in a petri dish. Liquid soil extracts appear to be the most suitable media. Colpoda and small flagellates appear within 1–2 days. Small naked amoebae colonize within 3–5 days later, and finally, after 10–15 days, other ciliates and testacea appear. The dilution-culture technique has been used extensively for determining protozoan abundance.

Bibliography

Aaronson, S. (1989). Enrichment culture. *In* "Practical Handbook of Microbiology" (W. M. O'Leary, ed.), pp. 337–347. CRC Press, Baton Roca, Florida.

Goodfellow, M., and Williams, E. (1986). *Biotechnol. Gen. Eng. Rev.* **4,** 213–262.

Gray, T. R. G. (1990). Soil bacteria. *In* "Soil Biology Guide" (D. L. Dindal, ed.), pp. 15–31. John Wiley and Sons, New York.

Holt, J. (1989). "Bergy's Manual of Systematic Bacteriology." Williams and Wilkins, Baltimore, Maryland. 2648 pp.

Hunter-Cevera, J. C., and Eveleigh, D. E. (1990). Actinomycetes. *In* "Soil Biology Guide" (D. L. Dindal, ed.), pp. 33–47. John Wiley and Sons, New York.

Hunter-Cevera, J. C., Fonda, M. E., and Belt, A. (1986). Isolation of cultures. *In* "Manual of Industrial Microbiology and Biotechnology" (A. Demain and N. Solomon, eds.), pp. 3–23. American Society for Microbiology, Washington, D.C.

Lousier, J. D., and Bamforth, S. S. (1990). Soil protozoa. *In* "Soil Biology Guide" (D. L. Dindal, ed.), pp. 97–136. John Wiley and Sons, New York.

Stein, J. R. (ed.) (1979). "Handbook of Phycological Methods: Culture Methods and Growth Requirements." Cambridge University Press, Cambridge, England. 448 pp.

Trupei, H. G., et al. (1992). "The Procaryotes: A Handbook on the Biology of Bacteria; Ecophysiology, Isolation, Identification, Applications." Springer-Verlag, New York. 4126 pp.

Tsao, P. H. (1970). *Annu. Rev. Phytopathol.* 6, 157–186.

L

Laboratory Safety and Regulations

Edward L. Gershey and Robert C. Klein[1]
The Rockefeller University

Glossary

Hazard communication standard This right-to-know standard obliges employers to inform their workers of the presence of and hazards from hazardous materials in the workplace. To be effective, this information program should include both written hazard communications as well as systematic training in the safe handling of those materials.

Hazardous materials As defined by the Environmental Protection Agency, these materials can be divided into five broad categories: acute and chronic health, fire, reactive or pressure release hazards.

Laboratory standard This standard took effect in January, 1991 and recognizes the fundamental differences between laboratory and industrial settings in their potential for and nature of worker exposures to hazardous materials. This standard applies exclusively to workplaces where the laboratory use of hazardous chemicals occurs.

Medical waste Also known as "infections" waste, the EPA and OSHA define infectious waste in broad terms. This waste can include animal waste, carcasses, and bedding; blood and blood products; cultures and stocks; all isolation waste and laboratory waste with infectious agent contact; pathological and surgical wastes; and sharps, including needles, syringes, scalpels, and glass. Definitions often vary among federal, state, and local authorities.

OSHA Occupational Safety & Health Administration was established in 1970 as part of the Occupational Safety & Health Act; it exists to promulgate and enforce regulations and standards for protecting workers as well as conduct work-site inspections and issue citations and violations to prevent work-related injury and illness

Permissible exposure limits (PEL) Specific numerical standards for exposure to airborne chemical concentrations

Superfund Superfund Amendments and Reauthorization Act of 1986 extended the Comprehensive Environmental Response, Compensation, and Liability Act of 1980; it extends the hazard communication program to the entire community and establishes requirements for the planning for and reporting of incidents involving emergency response, leaks and spills, and emissions.

[1] *Present affiliation:* Doucet and Mainka, P. C.

MICROBIOLOGY has become an interdisciplinary enterprise where health and safety issues are of increasing concern. Careful analysis of accidents and injuries, the presence of hazardous materials, and the recognition of health risks from chronic and acute exposures to harmful substances have produced many regulations. These regulations serve to heighten our awareness of potential hazards and establish a basis for programmatically assuring a safer and healthier workplace and environment. Unfortunately, in as much as the regulatory process is a political one, some regulations have more of their basis in public perception than in true risk avoidance. Key terms such as radioactivity, toxicity, and infectivity have very divisive effects on our society and have led to actions that heavily impact all laboratory operations, sometimes without significantly reducing risks to workers or the public. Today, successful science requires not only organizational skills and intellectual insights, but also personal, personnel, public, and environmental protection.

Ultimately, each individual is responsible for his or her own safety. To many, this means using common sense. To the extent that one recognizes that a potential hazard exists, the appropriate response may indeed be intuitive. However, gathering experience unwittingly through unsafe laboratory practices is risky. A well-designed safety program will aid in identifying potential hazards and share the common wisdom necessary for reducing the risks of laboratory work. The regulatory process addresses the complex interactions among the workplace, the environment, and society. These interactions frequently require professional assistance to properly gauge the risks, potential environmental and public health impacts, and community relations. Ideally, regulations should be promulgated in proportion to the actual risk of an activity, but in reality they often reflect the risks as perceived by the public. Whether or not regulations are truly commensurate with the hazard is a function of how successful, or unsuccessful, we scientists are in communicating the risks and benefits of our work.

While an individual laboratory worker might not be held personally responsible for compliance with regulations, an awareness of them will facilitate compliance, and vigilance will (hopefully) assure that someone takes responsibility.

I. Regulatory Agencies and Advisory Bodies

The following abridged lists of major regulatory agencies and advisory bodies are useful outlines for following and understanding the interactions that guide laboratory operations.

A. Federal Level

U.S. Occupational Safety and Health Administration (OSHA)
U.S. Department of Labor (DOL)
U.S. Environmental Protection Agency (EPA)
U.S. Department of Transportation (DOT)
U.S. Department of Agriculture (USDA)
U.S. Nuclear Regulatory Commission (NRC)
U.S. Public Health Service, Department of Health and Human Services
  National Institute for Occupational Safety and Health (NIOSH)
  National Institutes of Health (NIH)
  Centers for Disease Control (CDC)
  Recombinant Advisory Committee (RAC)
  Food and Drug Administration (FDA)
National Fire Protection Association (NFPA)
Building Officials and Code Administrators International, Inc. (BOCA)
American Conference of Governmental Industrial Hygienists (ACGIH)
American Society of Heating, Refrigerating and Air-Conditioning Engineers, Inc. (ASHRAE)
Joint Commission on Accreditation of Healthcare Organizations (JCAHO)
American Association for Accreditation of Laboratory Animal Care (AAALAC)
College of American Pathologists Standards for Accreditation of Medical Laboratories
National Safety Council

B. State and Local Levels

Departments of:

Buildings
Environmental Conservation
Fire Protection and Prevention
Health
Labor
Transportation

Various regulatory initiatives that began in the manufacturing sector now extend into research, diagnostic, and development laboratories as well. In the United States, federal legislation and enforcement is frequently superceded and extended by individual states and municipalities that have enacted their own statutes and established local enforcement plans. These local agency regulations must be at least as stringent as their federal counterparts. For illustration, as well as the authors' familiarity, New York State and New York City regulations, among the most restrictive in the country, will be used as examples.

II. Occupational Safety and Health

The most basic and wide-reaching federal safety legislation is the Occupational Safety and Health (OSH) Act of 1970 (Public Law 91-596). This act established the U.S. Occupational Safety and Health Administration (OSHA) to promulgate and enforce regulations and standards for protecting workers as well as conduct work-site inspections and issue citations and violations to prevent work-related injury and illness. Under the "general duty clause" of the Act, OSHA requires that "each employer shall furnish to each of his employees employment and a place of employment which are free from recognized hazards that are causing or are likely to cause death or serious physical harm to his employees." The Act also established the National Institute for Occupational Safety and Health (NIOSH) to develop and periodically revise recommendations for limiting exposure to potentially hazardous substances or conditions in the workplace.

Enforcement of OSHA regulations is achieved by the issuance of citations. These must be either complied with or contested. Failure to follow either of these responses will lead to violations and possible civil and/or criminal penalties. The OSH Act also established a review commission and a specific administrative procedure for handling appeals. For industry, enforcement has sometimes been accompanied by much publicity, large fines (hundreds of thousands to millions of dollars), and criminal penalties; however, the average fine assessed for serious violations has been on the order of several hundred dollars.

The OSH Act not only promulgates many specific regulations, but it also serves as a model for the exercise of regulatory authority. The OSHA principals of hazard identification, assessment, and control of exposure, written compliance programs, worker "right-to-know" training, and documentation will influence every workplace, especially complex laboratory work environments, well into the twenty-first century.

In 1989, the most frequently served (alleged) OSHA violations were (in decreasing order):

1910.1200(e)(1)	Hazard Communication	written hazard communication program
1903.2(a)1	OSHA Notice	posting describing OSHA to employees
1904.2(a)	Recordkeeping, OSHA 200 Log	annual posting of illness and injuries
1910.1200(h)	Hazard Communication	employee information and training
1910.1200(g)	Hazard Communication	Material Safety Data Sheets
1910.215(b)(9)	Abrasive Wheel Machinery	safety guarding
1910.212(a)(1)	Machine, General Requirements	safety guarding
1926.404(b)(1)	Wiring Design, Protection	construction
1926.404(f)(6)	Wiring Design, Protection	construction
1926.21(b)(2)	Safety Training, Education	construction
1910.20(g)(2)	Employee Medical Records	access to information
1910.132(a)	Personal Protective Equipment	provision and maintenance

Of this group, the second most frequently cited is the easiest to comply with—posting the OSHA Notice "in a conspicuous place or places where notices to employees are customarily posted." The three categories of violations for the Hazard Communication Standard comprise nearly 10% of the total number of violations issued. Specifically, they refer to the need for a written hazard communication program that informs employees of the potential hazards of the workplace and trains them to recognize and minimize exposure to hazardous materials. This includes having Material Safety Data Sheets (MSDSs) on hand for all hazardous materials and training workers on how to read and interpret them. In practice, this last point often requires the services of an industrial hygienist or other professional familiar with exposure and toxicological data. While the points related to safety guarding and construction were promulgated for machine shops and major construction projects, the same principles are applicable to common laboratory equipment (e.g., unguarded vacuum pumps, unenclosed centrifuges) and more clearly to the construction or renovation of laboratory spaces.

OSHA's enforcement policies are most severe for "willful" and "serious" violations. A violation is considered willful if an employer can be shown to have had prior knowledge of a dangerous condition but failed to take corrective action. Where OSHA considers the condition(s) to be in flagrant violation of the law, it may propose penalties for individual rather than groups of similar violations. Serious violations are issued whenever exposures exceed their standard as well as other conditions that are likely to cause death or serious physical harm. Under the federal Budget Summit agreement (30 September 1990), the maximum fine for each willful violation may be increased from the original $10,000 outlined in the OSH Act to as much as $70,000. At present, serious and willful violations may be accompanied by criminal penalties, including imprisonment. The instance-by-instance penalty scheme has been most frequently applied for inadequate recordkeeping procedures. The degree to which violations and fines are levied is heavily influenced by an employer's history of prior violations and whether or not these violations undermine the effectiveness of any existing safety and health program. Thus, even in the absence of serious illness, death, or high rates of worker injuries, the repetitive practice of simply paying fines without taking corrective action may constitute bad faith in complying with regulatory requirements.

Only those OSHA standards most applicable to laboratory work are discussed in this section. Interested readers are encouraged to consider a fuller review of OSHA standards. For many of the OSHA sections, the titles are self-explanatory, but some are not. In cases where OSHA is not the lead agency, identification of the lead agencies and their relevant regulations will be included along with a discussion of the OSHA standard. While realization that these standards exist may be reassuring, it is essential that someone or some group at your workplace not only remain aware of the regulations but also be responsible for (and actively pursue) regulatory interpretation and code compliance.

A. Ventilation (29 CFR Part 1910.94)

Ventilation is one of the most important physical characteristics of the laboratory environment, yet it is probably one of the least understood and least scrutinized. Poor ventilation has been connected with "sick building syndrome"; outbreaks of *Legionella* and *Streptococci;* exposure to pollen, mold, mites, fungi, chemicals, and other allergens; and general exacerbation of stress and discomfort. A properly functioning ventilation system (including a fume hood or other method of directly capturing hazardous vapors, gases, and dusts) plays a key role in health, safety, and fire prevention. While OSHA generally gives specific ventilation requirements in other standards, this standard draws heavily on the American Society of Heating, Refrigerating and Air-Conditioning Engineers, Inc. (ASHRAE) standard (1). ASHRAE defines acceptable indoor air quality as "air in which there are no known contaminants at harmful concentrations and with which a substantial majority (80%) of the people exposed do not express dissatisfaction." Although laboratory environments differ from residential and office settings, ASHRAE unfortunately does not devote much attention to them. For example, variations in humidity and temperature that might not be appreciably noticeable to the occupants may in fact be damaging to laboratory equipment and experiments. Filters and water condensates (common in laboratory ventilation systems) support the accumulation and growth of spores and other microorganisms and are a common source of air contamination. The frequently used medium efficiency filters, capable of removing up to 95% of the large, nonrespirable particles from outdoor air, are designed to protect the ventilation equipment, not to ensure clean air for occupants or experimental conditions. For these reasons and because of the potential for accumulations of toxic vapors or particulates, air in laboratories should not be recycled and instead should be exchanged with fresh air at a rate higher than in nonlaboratory areas.

While not explicitly discussed in the standard, fume hoods and ducted extractors not general room ventilation, must be used to control hazardous materials at their source. Although fume hoods can be effective at capturing low-density vapors in a still environment with face velocities as slow as 30–50 feet per minute (fpm), a higher flow rate (on the order of 100–150 fpm) is preferable because movements and pedestrian traffic near the hood face will cause turbulence and affect air flow into the hood. Unless highly engineered from the design stage and scrupulously built during the construction phase, fume hoods should be provided with dampers or blast gates to allow fine-tuning of the system. Fume hood exhaust ducts should not be "ganged" in series with other kinds of exhaust lines nor should fume hood systems share an exhaust fan with general room ventilation systems. Workers must be instructed to work in fume hoods in such a manner as to minimize turbulence of the air flowing into the hood, i.e., keeping storage to a minimum within the hood, raising bulky items off of the hood working surface, maintaining the sash at the level indicated for good face velocity, and notifying maintenance personnel at the first sign of operating problems. Fume hoods must be inspected at least annually by an engineer or industrial hygienist. For high-hazard operations (e.g., radioiodinations, hydrogen fluoride work, perchloric acid), hoods should be monitored more frequently or even continuously with solid state pressure or flow sensing instruments.

In addition to source capture and control of hazardous materials, supplying sufficient room-air changes of fresh air can further minimize the possibility of exposures to hazardous materials. To achieve good mixing, fresh air should be supplied and exhausted from opposite ends of a room, bearing in mind that air flows like a fluid and avoiding conditions that promote turbulence. Supply air grilles should be located to minimize drafts on nearby equipment such as biological safety cabinets and fume hoods. New York City Building Code 27-777.1(b) recommends six air changes per hour or 1 cubic foot per minute (cfm) per square foot of occupied area, whichever is greater. ASHRAE recommends supplying 20 cfm per person and 30% more fresh air than in nonlaboratory settings. In its new "Occupational Exposure to Hazardous Chemicals in Laboratories" standard (laboratory standard), OSHA recommends 4–12 room-air changes per hour. Although high fresh air flows require balancing the supply and exhaust ventilation systems and consume more energy, they result in a safer laboratory environment. Chemical storage rooms require special attention because organic vapors are often heavier than air and may accumulate on the storage room floor if air is not exhausted at the floor as well as the ceiling level. Although used mainly for storage and unattended procedures, environmentally controlled rooms (i.e., hot and cold rooms) are usually designed as closed boxes without ventilation. These rooms are, however, frequently accessed by researchers and should therefore be viewed as extensions of the laboratory. Particular attention should be given to the kinds of materials stored and used in such confined spaces. As yet, compliance with the ASHRAE standard is voluntary, and no ventilation standards exist for protection against indoor air pollution, especially in laboratories. Additional regulatory standards would be helpful to ensure that, despite the high cost of installation and maintenance, ventilation does not receive the low priority it is frequently assigned. Additional information on ventilation is available from ASHRAE (2) and the American Conference of Governmental Industrial Hygienists (ACGIH) (3).

B. Noise (29 CFR Part 1910.95)

Although noise levels above OSHA-prescribed limits are rarely encountered in laboratories, certain pieces of equipment can present problems. OSHA's noise standard follows a two-tiered approach. It established a permissible exposure level of 90 decibels (on the A-weighted scale, dBA) for an 8-hr period and incrementally shorter permissible exposures for higher levels of sound (up to a maximum of 115 dBA for no more than 15 min of exposure). However, before this level is reached, the employer must implement a program at levels ≥85 dBA. This "hearing conservation program" must identify both the source(s) and affected individuals and ensure that the 90-dBA permissible exposure level is not exceeded. Employees receiving noise exposures ≥85 dBA must also be so notified and receive hearing protection, training, and audiometric testing (baseline and annual follow-up). For some guidance in recognizing noise levels, a whisper is approximately 30 dBA, office noise averages 55–65 dBA, a vacuum cleaner generates 70 dBA at a 10-ft distance, heavy urban traffic can reach 75 dBA, and portable power tools can exceed 90 dBA. Levels much above 65 dBA are generally recognized as annoying, while those in the 85–90-dBA range require shouting at close distance to be understood. Prolonged exposures >85 dBA can cause diminished hearing capacity, especially in the mid- to high-frequency range. More attention is being given to the nonauditory health effects of noise, but few standards exist. In laboratories, noise from fans, fume hoods, and ventilation ducting are often found to be the most disturbing and should be addressed at the laboratory design stage. Cell disruptors and sonicators are common sources of noise in the laboratory and hearing protection devices are advisable during their use. Other sources include microcentrifuges, vacuum pumps, and cooling fans located on electronic equipment, especially when many units are used in the same room.

C. Ionizing Radiation (29 CFR Part 1910.96)

Various national and international advisory agencies are actively involved in the continuous reassessment of radiation protection standards and include the National Council on Radiation Protection and Measurements, the International Commission on Radiological Protection, the International Atomic Energy Agency, and the United Nations Scientific Committee on the Effects of Atomic Radiation. OSHA's regulations for exposure to ionizing radiation principally yield to standards set by the Nuclear Regulatory Commission (NRC). The NRC's regulations are found in Title 10 of the Code of Federal Regulations; those that most directly impact on laboratory operations are as follows:

Part 19 Notices, Instructions, Reports to Workers
Part 20 Standards for Protection against Radiation
Part 30 Rules of General Applicability to Domestic Licensing of Byproduct Material
Part 31 General Domestic Licenses for Byproduct Material
Part 33 Specific Domestic Licenses of Broad Scope for Byproduct Materials
Part 35 Human Uses of Byproduct Material
Part 61 Management and Disposal of Low-Level Wastes

The NRC has formed agreements with over half of the states with the state agencies superceding rulemaking and enforcement authority. In addition to the NRC, the FDA's National Center for Devices and Radiological Health, the Environmental Protection Agency's (EPA) Office of Radiation Programs, the Department of Interior's Mining Enforcement and Safety Administration, and the Department of Transportation have jurisdiction over different and sometimes overlapping aspects of control of radioactive materials. In many cases, the authority of the federal agencies is shared by various states and local agencies as well.

Radioactive materials are probably the most highly regulated of all hazardous materials. (Disposal of this material is discussed later.) Before receiving, possessing, or using radioactive materials, an institution or laboratory must obtain a license from the NRC or one of the NRC's agreement agencies. The institution or laboratory must have a designated Radiation Safety Officer and often an oversight committee. An adequate radiation protection program includes mechanisms for oversight and documentation of the requisition, receipt, storage, use, and disposal of all radioactive materials. The program must also provide for as well as document a monitoring system that includes personnel monitoring, contamination surveys of laboratory areas, and maintenance and calibration of survey and X-ray equipment. Monitoring is guided by the principal of assuring that any radiation exposures are kept below maximum permissible limits and "as low as reasonably achievable" (ALARA). Hazard communication and training are essential ingredients for providing protection and minimizing exposures (4, 5).

D. Nonionizing Radiation (29 CFR Part 1910.97)

While the hazards of ionizing radiation have been long recognized and highly regulated, the safety of nonionizing radiation has only recently been questioned. Nonionizing radiation includes all regions of the electromagnetic radiation spectrum with wavelengths longer than the far ultraviolet, i.e., the longer ultraviolet and visible lights, infrared, and microwave, and radio, TV, and power transmission radiations. The OSHA standard, covering electromagnetic energy of frequencies from 10 MHz to 100 GHz, recommends a limit of 10 mW/cm^2 averaged over any possible 0.1-hr period and references the American National Standards Institute standards C95.1-1966 and C95.2-1966. Of more relevance may be the threshold limit values (TLVs) suggested by the ACGIH (6) for power density (mW/cm^2), electric fields (V^2/m^2), and magnetic fields (A^2/m^2). Also worth considering are the ACGIH's TLVs for exposures to ultraviolet light, infrared radiation, and static magnetic fields.

E. Respiratory Protection (29 CFR Part 1910.134)

Respiratory protection in a laboratory is best addressed through source control and ventilation, specifically fume hoods, dedicated exhausts, biological safety cabinets, containment rooms, and general room ventilation. However, allergies, particularly noxious chemicals, and pathogens may still require the use of respirators for certain operations or until an engineered solution is implemented. Two types of respirators exist. The first, fitted with cartridges specific for particulates or various chemical vapors or gases, depends on the physical action of breathing to power the filtration of contaminated air. A leak-free facial fit and the wearer's physical condition are more important for this type of respirator than for the second, which actually supplies clean breathing air under positive pressure. As one might expect, the latter type of respirator provides better protection and should be used for emergencies and highly hazardous materials. Respirator selection is not simple and should be contemplated only in consultation with an industrial hygienist or occupational physician.

Although wearing a respirator is not common in laboratories, their use is highly regulated by OSHA, requiring a written procedure for evaluating the hazardous conditions present as well as written operating instructions covering their routine and emergency use, cleaning, and maintenance. Also, anyone using a respirator must undergo periodic medical evaluation, which may involve stress testing, to determine whether or not he or she is "physically able to perform work and use the equipment." In addition to fit testing, the wearer must receive documented training in the maintenance and use of this equipment—all performed by a knowledgeable professional. Where possible, records of each episode of respirator use should be kept on file with the employee's medical records.

F. Fire Protection (29 CFR Part 1910.155-163)

While the fire protection standard includes many details on fire fighting and fire suppression equipment, a few of these details are worth emphasizing for safe (and cost-effectively insured) laboratory operations. First, a functional communication system must be audible throughout all areas of the workplace. Although a majority of institutions still use a gong-based fire alarm system, a loudspeaker-based public address system permitting both verbal and tonal communication is the most flexible. For example, with such a system, specific instructions besides just those to evacuate can be given to select areas. A written evacuation plan must also be on file. Fire extinguishers must be inspected annually and records thereof maintained. This standard also has specific requirements for initial and refresher training of fire brigades, including medical evaluation.

Most fire protection standards for laboratories are ultimately based on the National Fire Codes (7) established by the National Fire Protection Association (NFPA). An extensive master index aids in identifying the many laboratory operations covered by the Code. However, the local agencies and fire departments most commonly set and enforce fire protection standards. For example, in New York City, the Department of Buildings and the New York City Fire Department's Bureau of Fire Prevention have this authority. Laboratory design, construction, and renovations require a Certificate of Occupancy and a permit from the Department of Buildings (given only with approval of the Fire Department). In addition, the New York City Fire Department, under Fire Prevention Directive 1-66 (revised), also regulates the storage and use of chemicals in college, university, hospital, research, and commercial laboratories. According to this Directive, a permit to operate any laboratory containing chemicals must be obtained for each and every laboratory room. The Directive sets forth chemical storage limits and conditions and requirements for placarding, providing information (MSDSs), labeling, and recordkeeping. Furthermore, a person with a degree in chemistry or its equivalent must also obtain a Certificate of Fitness for general supervision of the laboratory with regard to the safety provisions of the Directive. Additional certificates are required to operate special equipment (e.g., oxygen torches).

Smoke detectors are essential for detecting and limiting the spread of slow-starting laboratory fires. They should be installed in all laboratories and wired to a central alarm system wherever possible. Where fire extinguishers are required, either dry chemical or carbon dioxide units are suitable for most laboratory areas. A major disadvantage of dry chemical extinguishers, however, is that they leave behind a residue that may irreversibly damage sensitive equipment. To prevent the spread of a fire, laboratory doors should always be kept closed. However, the need to constantly grapple with door knobs while transporting delicate or hazardous materials gives rise to the common (but illegal) practice of holding doors open with chocks or strings. One generally accepted solution is to install smoke and heat-sensitive, fire-rated door closures.

G. Hazardous Materials (29 CFR Part 1910.101–120)

Subpart H of the hazardous materials standard provides definitions and handling guidelines for work with compressed gases (specifically acetylene, hydrogen, oxygen, and nitrous oxide), flammable and combustible liquids, vapors, gases, anhydrous ammonia, and hazardous wastes.

H. Air Contaminants (29 CFR Part 1910.1000)

Subpart Z (revised 1989) of Toxic and Hazardous Substances covers exposures to various air contaminants. While recognizing that the safest practice is to keep airborne chemical concentrations as low as reasonably achievable, this section contains specific numerical standards for exposure. These standards, known as permissible exposure limits (PELs), carry the force of law. Most of these standards have been adopted from maximum exposure levels known as threshold limit values, or TLVs, recommended by the ACGIH (8) for selected chemicals. Initially, employers are required to use any combination of engineering, work practice, and respiratory protection control measures to reduce exposures. The values listed for individual chemicals generally do not consider effects that may be additive or synergistic with other airborne chemicals in the workplace or that may be influenced by the hypersusceptibility of certain individuals due to age, genetic factors, personal habits, medication, previous exposure(s), or preexisting medical conditions. However, there are methods for calculating the PEL for mixtures. TLVs and PELs represent airborne concentrations of contaminants that are designed to protect most workers under continuous occupational exposure (8 hr/day, 40 hr/wk, 50 wk/yr) for the duration of their working life. PELs and TLVs are time-weighted averages (TWA) based on exposure averaging over the course of an 8-hr working day. Short-term exposure limits (STEL) address acute exposures. In practice, the STEL assumes a 15-min or shorter period of exposure, occurring no more frequently than four times per day with 1-hr between exposures. In addition to the TWAs and STELs, for certain chemicals there are also "ceiling limits," concentrations that must not be exceeded during any part of an exposure, even if the TWA remains below the PEL or TLV. Despite the adjective "permissible," OSHA does not intend for PELs to be reached because "action levels" (levels at which steps must be taken to reduce exposures) are specified for many compounds and are generally set at one-half of the PEL.

I. OSHA-Listed Hazardous Chemicals

The safest method of reducing worker exposures to hazardous materials is by substituting the offending compound with less dangerous materials. Where substitution is not feasible, other methods of isolation or automation are useful to remove physically the potential for contact between worker and toxicant. When all other methods have been exhausted, respirators and other protective equipment (e.g., gloves, faceshields, chemically resistant clothing) come into play. Of the several hundred health and safety standards promulgated by OSHA, only 20-odd toxic and hazardous chemicals are specifically listed with individual standards (under Subpart Z; i.e., 29 CFR 1910 Parts 1001–1101). For each of these chemicals, OSHA gives general information about their chemical and physical properties, health hazards, treatment and protection data, exposure limits, methods for their control, medical surveillance, special requirements, regulated areas, written compliance program, training, and recordkeeping. While this list includes the only carcinogens for which OSHA has promulgated its own standards, the Laboratory Standard (1910.1450, discussed later) now recognizes over 100 compounds that must be treated as carcinogens (see Table I). More importantly, by adoption of categoric recommendations from objective, third-party scientific advisory agencies, OSHA has introduced a more efficient and flexible mechanism for keeping worker protection standards current than their original approach, which required legislative approval on a compound-by-compound basis.

The first step toward managing the risks from hazardous agents is to compile an inventory system to account for their ordering, receipt, distribution, and rate of use. Unlike production facilities, laboratories tend to use relatively small amounts of a very wide

Table I Substances Specifically Regulated by OSHA under the Laboratory Standard (29 CRF 1910.1450)[a]

	Substance	IARC group[b]
	A-α-C(2-amino-9*H*-pyrido[2,3,*b*]indole)	2B
	Acetaldehyde	2B
	Acetamide	2B
•	2-Acetylaminofluorene	—
	Acrylamide	2B
•	Acrylonitrile	2A
	Adriamycin	2A
	AF-2[2-(2-furyl)-3-(5-nitro-2-furyl)acrylamide]	2B
	Aflatoxins	1
	para-Aminoazobenzene	2B
	ortho-Aminoazotoluene	2B
•*	4-Aminobiphenyl	1
	2-Amino-5-(5-nitro-2-furyl)-1,3,4-thiadiazole	2B
	Amitrole	2B
*	Analgesic mixtures containing phenacetin	1
	Androgenic steroids	2A
	ortho-Anisidine	2B
	Aramite™	2B
•*	Arsenic and arsenic compounds	1
•*	Asbestos	1
	Auramine, technical grade	2B
	Azaserine	2B
*	Azathioprine	1
•*	Benzene	1
•*	Benzidine	1
	Benzidine-based dyes	2A
	Benzo[*a*]pyrene	2A
	Benzo[*b*]fluoranthene	2B
	Benzo[*f*]fluoranthene	2B
	Benzo[*k*]fluoranthene	2B
	Benzyl violet 4B	2B
	Beryllium compounds	2A
	Betel quid with tobacco	1
	Bis)chloroethylnaphthyl)amine	1
	Bischloroethyl nitrosourea	2A
•*	Bis(chloromethyl) ether	1
	Bleomycins	2B
	Bracken fern	2B
	1,3-Butadiene	2B
*	1,4-Butanediol dimethanesulfonate ("Myleran")	1
	Butylated hydroxyanisole	2B
	β-Butyrolactone	2B
	Cadmium compounds	2A
	Carbon-black extracts	2B
	Carbon tetrachloride	2B
	Carrageenan, degraded	2B
*	Chlorambucil	1
	Chloramphenicol	2B
	Chlordecone ("Kepone")	2B
	α-Chlorinated toluenes	2B
	1-(2-Chloroethyl)-3-cyclohexyl-1-nitrosourea	2A
	1-(2-Chloroethyl)-3-(methylcyclohexyl)-1-nitrosourea	1
	Chloroform	2B

Continues

Continued

	Chlorophenols	2B
	Chlorophenoxy herbicides	2B
	4-Chloro-*ortho*-phenylenediamine	2B
	para-Chloro-*ortho*-toluidine	2B
*	Chromium VI compounds	1
	Cisplatin	2A
	Citrus Red No. 2	2B
•	Coal tar pitches	1
•	Coal tars	1
•	Cotton dusts	—
	Creosotes	2A
	para-Cresidine	2B
	Cycasin	2B
*	Cyclophosphamide	1
	Dacarbazine	2B
	Daunomycin	2B
	N,N′-Diacetylbenzidine	2B
	2,4-Diaminoanisole	2B
	4,4′-Diaminodiphenyl ether	2B
	2,4-Diaminotoluene	2B
	Dibenz[*a,f*]acridine	2B
	Dibenz[*a,h*]acridine	2B
	Dibenz[*a,h*]anthracene	2A
	7H-Dibenzo[c,g]carbazole	2B
	Dibenzo[*a,e*]pyrene	2B
	Dibenzo[*a,h*]pyrene	2B
	Dibenzo[*a,i*]pyrene	2B
	Dibenzo[*a,l*]pyrene	2B
•	1,2-Dibromo-3-chloropropane	2B
	para-Dichlorobenzene	2B
•	3,3′-Dichlorobenzidine	2B
	3,3′-Dichloro-4,4′-diaminodiphenyl ether	2B
	Dichloro-diphenyl-trichloroethane	2B
	1,2-Dichloroethane	2B
	Dichloromethane	2B
	1,3-Dichloropropene, technical grade	2B
	Diepoxybutane	2B
	Di(2-ethylhexyl)phthalate	2B
	1,2-Diethylhydrazine	2B
*	Diethylstilboestral	1
	Diethyl sulphate	2A
	Diglycidyl resorcinol ether	2B
	Dihydrosafrole	2B
	3,3′-Dimethoxybenzidine (*ortho*-dianisidine)	2B
•	*para*-Dimethylaminoazobenzene	2B
	trans-2[(Dimethylamino)methylimino]-5-(2-(5-nitro-2-furyl)vinyl-1,3,4-oxadiazole	2B
	3,3′-Dimethylbenzidine (*ortho*-tolidine)	2B
	1,1′-Dimethylhydrazine	2B
	1,2-Dimethylhydrazine	2B
	Dimethylcarbamoyl chloride	2A
	Dimethyl sulphate	2A
	1,4-Dioxane	2B
	Epichlorohydrin	2A
	Erionite	1
	Ethyl acrylate	2B
	Ethylene dibromide	2A

Continues

Continued

•	Ethyleneimine	—
•	Ethylene oxide	2A
	Ethylene thiourea	2B
	Ethyl methanesulphonate	2B
	N-Ethyl-*N*-nitrosourea	2A
•	Formaldehyde	2A
	2-(2-Formylhydrazino)-4-(5-nitro-2-furyl)thiazole	2B
	Furniture- and cabinet-making	1
	Glu-P-1 (2-amino-6-methyldipyrido[1,2-α:3′,2′-*d*]imidazole)	2B
	Glu-P-2 (2-aminodipyrido[1,2-α:3′,2′-*d*]imidazole)	2B
	Glycidaldehyde	2B
	Griseofulvin	2B
	Hematite mining, underground, with exposure to radon	1
	Hexachlorobenzene	2B
	Hexachlorocyclohexanes	2B
	Hexamethylphosphoramide	2B
	Hydrazine	2B
	Indeno[1,2,3-*cd*]pyrene	2B
	IQ (2-amino-3-methylimidazo[4,5-*f*]quinoline)	2B
	Iron–dextran complex	2B
	Lasiocarpine	2B
•	Lead compounds, inorganic	2B
	MeA-α-C(2-amino-3-methyl-9H-pyrido[2,3-*b*]indole)	2B
	Medroxyprogesterone acetate	2B
*	Melphalan	1
	Merphalan	2B
	5-Methoxypsoralen	2A
*	8-Methoxypsoralen and ultraviolet light	1
	2-Methylaziridine	2B
	Methylazoxymethanol and its acetate	2B
•	Methyl chloromethyl ether	1
	5-Methylchrysene	2B
	4,4′-Methylene bis(2-chloroaniline)	2A
	4,4′-Methylene bis(2-methylaniline)	2B
	4,4′-Methylenedianiline	2B
	Methyl methanesulphonate	2B
	2-Methyl-1-nitroanthraquinone	2B
	N-Methyl-*N*-nitrosourethane	2B
	N-Methyl-*N*′-nitro-*N*-nitrosoguanidine	2A
	N-Methyl-*N*-nitrosourea	2A
	Methylthiouracil	2B
	Metronidazole	2B
	Mirex	2B
	Mitomycin C	2B
	Monocrotaline	2B
	MOPP and combined chemotherapy preparations	1
	5-(Morpholinomethyl)-3-[(5-nitrofurfurylidene)amino]-2-oxazolidinone	2B
*	Mustard gas	1
	Nafenopin	2B
•	1-Naphthylamine	3
•*	2-Naphthylamine	1
	Nickel compounds	1
	Niridazole	2B
	5-Nitroacenaphthene	2B
•	4-Nitrobiphenyl	3

Continues

Table 1 *Continued*

	Nitrofen, technical grade	2B
	1-[(5-Nitrofurfurylidene)amino]-2-imidazolidonone	2B
	N-[4]-(5-Nitro-2-furyl)-2-thiazolyl]acetamide	2B
	Nitrogen mustard	2A
	Nitrogen mustard *N*-oxide	2B
	2-Nitropropane	2B
	N-Nitrosodiethylamine	2A
•	*N*-Nitrosodimethylamine	2A
	N-Nitrosodi-*n*-butylamine	2B
	N-Nitrosodi-ethanolamine	2B
	N-Nitrosodi-*n*-propylamine	2B
	3-(*N*-Nitrosomethylamino)propionitrile	2B
	4-(*N*-Nitrosomethylamino)-1-(3-pyridyl)-1-butanone	2B
	N-Nitrosomethylethylamine	2B
	N-Nitrosomethylvinylamine	2B
	N-Nitrosomorpholine	2B
	N-Nitrosonornicotine	2B
	N-Nitrosopiperidine	2B
	N-Nitrosopyrrolidine	2B
	N-Nitrososarcosine	2B
	Oestrogen replacement therapy	1
	Oestrogens, nonsteroidal	1
	Oestrogens, steroidal	1
	Oil Orange SS	2B
	Oral contraceptives, combined	1
	Oral contraceptives, sequential	1
	Panfuran S (containing dihydroxymethylfuratrizine)	2B
	Phenacetin and analgesics	2A
	Phenazopyridine hydrochloride	2B
	Phenobarbital	2B
	Phenoxybenzamine hydrochloride	2B
	Phenytoin	2B
	Polybrominated biphenyls	2B
	Polychlorinated biphenyls	2A
	Ponceau 3R	2B
	Ponceau MX	2B
	Potassium bromate	2B
	Procarbazine hydrochloride	2A
	Progestins	2B
	1,3-Propane sulfone	2B
•	*β*-Propiolactone	2B
	Propylene oxide	2A
	Propylthiouracil	2B
	Saccharin	2B
	Safrole	2B
	Shale oils	1
	Silica, crystalline	2A
	Sodium *ortho*-phenylphenate	2B
	Soots	1
	Sterigmatocystin	2B
	Streptozotocin	2B
	Styrene	2B
	Styrene oxide	2A
	Sulfallate	2B
	Talc containing asbestiform fibers	1
	2,3,7,8-Tetrachlorodibenzo-*para*-dioxin	2B
	Tetrachloroethylene	2B
	Thioacetamide	2B

Continues

Continued

	4,4′-Thiodianiline	2B
	Thiourea	2B
*	Thorium dioxide	—
	Tobacco products, smokeless	1
	Tobacco smoke	1
	Toluene diisocyanates	2B
	ortho-Toluidine	2B
	Toxaphene, polychlorinated camphenes	2B
	Treosulphan	1
	Tris(1-aziridinyl)phosphine sulphide (Thiotepa)	2A
	Tris(2,3-dibromopropyl) phosphate	2A
	Trp-P-1 (3-Amino-1,4-dimethyl-5H-pyrido[4,3-*b*]indole)	2B
	Trp-P-2 (3-amino-1-methyl-5H-pyrido[4,3-*b*]indole)	2B
	Trypan blue	2B
	Uracil mustard	2B
	Urethane	2B
	Vinyl bromide	2A
•*	Vinyl chloride	1

[a] International Agency for Research on Cancer (IARC)-designated carcinogens and suspect carcinogens; [From the IARC (1987). "IARC Monographs on the Evaluation of Carcinogenic Risks to Humans: Overall Evaluations of Carcinogenicity." Supplement 7. Lyons, France.]; compounds regulated by OSHA [•, Subpart Z—Toxic and Hazardous Substances (29 CFR 1910 Subpart Z) as of 19 January 1989]; and compounds considered "known carcinogens" by the National Toxicology Program [*, (1989). "Fifth Annual Report on Carcinogens." Report NTP 89-239.

[b] Carcinogen groups: 1, known carcinogenicity; 2A, probable; 2B, possible; 3, not classifiable due to insufficient or conflicting data.

variety of reagents, making their tracking even more important. Through an inventory, it is possible to identify those chemicals and personnel for whom hazard information, training, and exposure monitoring may be required.

A discussion of two carcinogenic chemicals not unfamiliar to many microbiologists, ethylene oxide and formaldehyde, should illustrate Subpart Z standards. Since exposure limits for these two chemicals exist, the onus is on the laboratory to prove that PELs were not or could not be reached. Sampling can be performed relatively easily with passive dosimeters similar to radiation film badges but will require subsequent laboratory analysis. More aggressive air sampling can provide more representative results but should only be designed and conducted in consultation with an industrial hygienist.

1. Ethylene Oxide (29 CFR Part 1910.1047)

Ethylene oxide (C_2H_4O) exists as a gas at room temperature and is commonly used as a cold sterilant, especially for surgical equipment, dressings, and other items that would be damaged by high-temperature, high-pressure autoclaving. Unfortunately, ethylene oxide is also a potent and well-documented carcinogen. Whenever possible, presterilized, disposable materials should be substituted for those requiring ethylene oxide treatment. Where used, OSHA requires that a Compliance Program be prepared with a written emergency plan that includes a means of alerting employees and a leak detection schedule. If the PEL (1 ppm) has been exceeded, medical surveillance must be conducted for 30 days. If exposures exceed either the action level (0.5 ppm) or STEL, or if an affected individual(s) requests and the doctor agrees, blood, fertility, or pregnancy testing must be conducted. Records of monitoring, medical surveillance, and employment must be kept for 30 yr postemployment. As part of the hazard communication standard, ethylene oxide can be used only in an area designated by warning signs, i.e., "Danger, Ethylene Oxide, Cancer and Reproduction Hazard, Authorized Persons Only, Respirators and Protective Clothing May Be Required in Area." All containers must be labeled and MSDS information must be on hand. If exposure is at the action level or above the STEL, training must be conducted annually and work practices must be posted. Sterilizers must be equipped with local exhaust ventilation to remove any residual gas at its source prior to opening the unit upon completion of a sterilization cycle. The door, gaskets, cylinder, vacuum hoses, filters, and valves should be tested biweekly for leaks. Air monitoring should be conducted in sterilizer areas to establish a baseline and then representative sampling should be done. The results of this monitoring must be reported to employees. Of course, given appropriate engineering controls, equipment maintenance, and work practices, exposures should never reach PELs. Note that ethylene oxide is flammable, heavier than air, does not require oxygen to burn, and readily permeates rubber and leather. The EPA also regulates ethylene oxide as a pesticide, fungicide, and antimicrobial.

2. Formaldehyde (29 CFR Part 1910.1048)

Unlike ethylene oxide, which is used in fixed and hopefully adequately ventilated locations, formaldehyde (CH_2O) is used in many areas and for many procedures. Many laboratories fail to recognize the potential for exposure during routine operations with formaldehyde (e.g., perfusion of animals, specimen preservation, storage). A first step toward managing health hazards from this suspect carcinogen is to identify where and how it is used and to assess the potential for exposing workers. Protective measures should rely heavily on engineering controls (such as local ventilation) and training workers on the established work procedures. As for other OSHA-listed chemicals, copies of the MSDS, information on health effects, and written work and emergency procedures must be on hand. The exposure limits for formaldehyde are 1 ppm PEL and 2 ppm STEL(15 min). Although monitoring need not be initiated unless there is a likelihood that these limits are exceeded, the employer must be able to show that such levels are unlikely. In many instances, this is not easily accomplished without monitoring the exposure of representative employees. If these levels are exceeded, employees must be notified in writing. Areas where levels exceed the PEL or the STEL must be posted with warning signs and access limited to those persons who have been trained (and annually retrained) to recognize the hazards of formaldehyde. Engineering controls and work practices must be adopted to keep exposures below the PEL. Respirators should be used only when it is not feasible to lower exposures by other means or while engineering controls are being implemented. In the event of an overexposure, a series of very explicit steps must be taken, including a standardized questionnaire covering medical and work histories and a medical examination. Records of exposure determinations, monitoring, and individual medical evaluations must be kept for 30 yr postemployment.

J. "Right-to-Know" (29 CFR Part 1910.1200)

The Hazard Communication ("Right-to-Know") standard obliges employers to inform their workers of the presence of and hazard from hazardous materials in the workplace. The cornerstone of this standard is the MSDS. These must be available to all employees and must contain the following:

- The identity of the compound as written on the container as well as any chemical or trivial names
- Physical and chemical characteristics
- Life safety and health hazards (especially carcinogenicity)
- Maximum permissible exposure levels and primary routes of entry into the body
- Safe handling and use precautions
- Control measures
- Emergency and first-aid procedures

The EPA has recently proposed that if a company has any information on how a chemical might adversely affect the environment, that information should be added to the existing MSDS. A major weakness of MSDSs is that they are frequently incomplete, they are often written by the manufacturer simply to comply with a regulation, and, because manufacturers cannot be sure of how their products may be used, information relevant to particular applications is not included. Lastly, because of the high rate of functional illiteracy, many workers with poor reading skills may remain unaware of potential dangers. For these reasons, simply providing an MSDS to a laboratory worker does not fulfill the employer's obligation to inform his or her employees of the hazards inherent in that workplace.

Although the EPA is generally recognized as the major identifier of hazardous chemicals, there are other useful listings of hazardous chemicals. Note that the absence alone of a given substance from these extensive listing cannot be taken to mean that it is not hazardous. If a hazardous chemical is present in a mixture at $>1\%$, the entire mixture must be treated as hazardous. Where questions exist about the hazardous characteristics of a chemical, other supportive data should be consulted. For rapid access, several computerized databases at the National Library of Medicine are available through the Medical Literature Analysis & Retrieval System (MEDLARS). These include the following:

CCRIS	Chemical Carcinogenesis Research Information System
ChemID	Chemical Identification
CHEMLINE	Chemical Dictionary Online
DART	Developmental and Reproductive Toxicology
EMIC	Environmental Mutagen Information Center
HSDB	Hazardous Substances Data Bank
IRIS	Integrated Risk Information System
RTECS	Registry of Toxic Effects of Chemical Substances
TOXLINE	Toxicology Information Online and Toxicology Literature from Special Sources
TRI	Toxic Release Inventory

To be effective, the communication of hazards to employees should include, at the least, a written hazard communication program containing a list of all hazardous chemicals in the workplace, with each chemical referenced to its MSDS, as well as systematic training in the safe handling of those materials. Training must include methods to detect the release of hazardous chemicals, the health hazards they present, protective measures to take in the event of an exposure, and explanation of the MSDS and other information. Although no record of training is officially required, the burden of proof that employees have received training rests with the employer. Records on workers exposed to any OSHA regulated substance must include employee name, address, social security number, name of the chemical(s), and Chemical Abstracts Service (CAS) number(s) of the materials to which the employee is routinely exposed. Chemical storage containers must be labeled in English with the chemical name and appropriate hazard warning. Portable containers used by employees for immediate work with the chemicals are exempt from these requirements. Laboratories or operations where employees handle only sealed containers of chemicals need only comply with labeling standards and limited MSDS and employee information requirements.

K. Laboratory Standard (29 CFR Part 1910.1450)

The Occupational Exposure to Hazardous Chemicals in Laboratories ("Laboratory Standard") standard took effect on 31 January 1991. It recognizes the fundamental differences between laboratory and industrial settings in their potential for, and nature of, worker exposures. This standard applies exclusively to workplaces where the "laboratory use" of hazardous chemicals occurs. Laboratory use means that "(i) Chemical manipulations are carried out on a 'laboratory scale'; (ii) Multiple chemical procedures or chemicals are used; (iii) The procedures involved are not part of a production process . . . [and] (iv) Protective laboratory practices and equipment are available and in common use to minimize the potential for employee exposure to hazardous chemicals." Laboratory scale refers to volumes of chemicals that can be manipulated by one person. For laboratory environments, this standard has, in the main, superceded the Subpart Z standards with a set of performance-based standards, "permitting a greater degree of flexibility to laboratories in developing and implementing employee safety and health programs." Specific procedures are not prescribed, allowing laboratories to tailor their plans to meet their particular circumstances. Key to this standard is the development and implementation of a written Chemical Hygiene Plan, the effectiveness of which must be reviewed at least annually. The Chemical Hygiene Plan consists of (at least) the following:

- Assignment of a chemical hygiene officer
- A description of standard operating procedures, including health and safety features
- Engineering, personal protective equipment and administrative steps to reduce worker exposures
- A program to assure that engineering controls (e.g., fume hoods, biological safety cabinets) are working properly and regularly maintained
- The circumstances under which a particular laboratory operation will require prior approval from the supervisor before commencement
- Designation of special work areas for handling OSHA-specified "select carcinogens"

- A description of employee training programs and a means for effective hazard communication
- Procedures for medical consultation and exams
- Decontamination procedures
- Hazardous waste disposal
- Monitoring for chemical exposures
- Recordkeeping

Employers are required to monitor employee exposures to hazardous substances whenever there is "reason to believe that exposures routinely exceed the action level or PEL," at which point the requirements typical of general industry standards (e.g., periodic monitoring, posting, recordkeeping) apply. Standard operating procedures must include the designation of specific areas for work with carcinogens and substances with high acute toxicity and must describe the containment devices to be used as well as procedures for decontamination and waste removal. "Select carcinogens" (see Table I for listing) include those substances specifically regulated by OSHA under Subpart Z (Part 1910.1000–1450), carcinogens (Group 1) and suspect carcinogens (Groups 2A and 2B) designated by the International Agency for Research on Cancer, and compounds considered "known carcinogens" or "reasonably anticipated to be carcinogens" by the National Toxicology Program. The training program must provide information on standards and hazard communication requirements similar to those described earlier. Specifically, the employer must ensure that employees are appraised of the hazards of chemicals present in their work area at the time of initial assignment and prior to reassignment to new exposure situations. Training must include methods for detecting the presence or release of hazardous materials, information on physical and health hazards of chemicals in the work area, and the use of worker-initiated protective measures such as personal protective equipment and emergency procedures. The information provided to employees should include applicable details of the employer's written Chemical Hygiene Plan (e.g., standard operating protocols), PELs, symptoms associated with overexposure, and the availability of reference materials including MSDSs. Employees must be retrained at a frequency determined to be adequate by the employer. Aside from the recordkeeping, a major undertaking for many laboratories will be the requirement for designating special work areas for many common, albeit highly hazardous chemicals. A likely outcome of this requirement will be the designation of entire laboratories as controlled areas.

Other agencies regulate various aspects of laboratory operations. The National Institutes of Health (NIH) (9) provide general information for protection when handling carcinogens and OSHA (10) provides guidelines for work with cytotoxic material. The Food and Drug Administration's "Good Laboratory Practices for Nonclinical Laboratory Studies" (21 CFR Part 58) carries explicit requirements for the conduct and documentation of experiments under their aegis. This emphasizes quality assurance with regard to laboratory and animal facilities, equipment maintenance, laboratory protocols, purity of standards, controls and calibrations, and maintaining laboratory records. It is the responsibility of a designated quality assurance officer to verify that investigators adhered strictly to protocols and, if deviations are necessary, that they have been adequately justified and documented.

L. Bloodborne Pathogens

Although entitled "Bloodborne Pathogens" (29 CFR Part 1910.1030), this standard applies more generally to so-called infectious materials, including blood and blood products, body fluids (i.e., saliva, semen, vaginal secretions, and cerebrospinal, synovial, pleural, pericardial, peritoneal, and amniotic fluids), and contaminated sharps, pathological wastes, microbiological wastes, unfixed human tissue or organ, and human immunodeficiency virus- (HIV) or hepatitis B virus- (HBV) containing materials, including animals. A written infection control plan is required to identify and document tasks and procedures where occupational exposures may take place, describing the steps taken to minimize or eliminate employee exposure. Chief among these are the "universal precautions"—for example, wearing gloves and other appropriate personal protective clothing and eye protection, if necessary; washing hands; not eating, drinking, storing food, smoking, applying cosmetics, handling contact lenses, or mouth pipetting in the laboratory; minimizing aerosol generation; and disposing of needles and syringes safely. Housekeeping details include cleaning and disinfecting all surfaces and equipment on a regular, often daily basis and also immediately after a spill, using mechanical means to collect potentially contaminated broken glassware, using labeled, leakproof primary and secondary containers for the storage and transportation of infectious wastes, collecting and disposing of waste and laundry properly, and assuring that custodial and laundry workers wear gloves and other appropriate personal protective equipment. HIV and HBV research laboratories and production facilities are specifically required to do the following:

- Mark work areas with biohazard signs and limit access thereof
- Have a sink in the room for handwashing
- Wear personal protective clothing and gloves in the work area and remove them before leaving
- Protect vacuum lines with filters and traps
- Keep doors closed
- Affix biohazard warning labels to refrigerators, freezers, and other containers holding blood and other potentially infectious materials
- Conduct all work with potentially infectious materials in a biological safety cabinet that is certified at least annually (no work conducted in open vessels on open bench)
- Minimize the generation of infectious aerosols by enclosing potential sources
- Dispose of needles (not resheathed) in puncture-proof containers
- Transport wastes for autoclaving in puncture- and leak-proof containers
- Decontaminate all infectious waste
- Report all spills or accidents
- Prepare a biosafety manual describing, among other things, the laboratory's standard operating procedures
- Advise all personnel of risks and assure they read and follow procedures

In addition to the requirements outlined above, production facilities must also do the following:

- Separate their work areas from access corridors by passage through a dual set of self-closing doors
- Have interior room surfaces that are water-resistant for cleaning
- Seal all penetrations between the facility and rooms adjacent, above, and below
- Have a sink installed near the work area exit that is operable without the use of one's hands (i.e., foot treadle)
- Run under negative air pressure relative to adjacent areas and exhaust room air directly outdoors

HBV vaccination or antibody titer testing must be offered free-of-charge to all employees who have occupational exposures to blood or other potentially HBV-infected materials. Medical surveillance is mandatory and includes postexposure follow-up by a licensed physician, documentation of the route(s) of exposure and the status of source patient(s), collection and testing for HIV or HBV of blood from source patient(s) (with permission) and exposed employee, and follow-up testing and counseling of the exposed employee. Laboratory personnel must also receive a copy of the standard and an explanation of its contents as well as training on precautionary measures, modes of transmission, and the prevention of HBV, HIV, and other bloodborne infectious diseases. Pregnant personnel must be counseled, as early as possible, regarding possible risks to the fetus from these agents. Lastly, this standard requires the documentation of medical records and training sessions.

In addition to this standard, the Department of Health and Human Services has issued guidelines for health-care workers (11). This reference is a useful compilation of NIOSH recommendations, OSHA regulations, Centers for Disease Control (CDC) guidelines, and information provided by the Joint Commission on Accreditation of Health-Care Organizations (JCAHO), the NFPA, EPA, and other regulatory and advisory groups. Also, the National Institute of Health's Safety Division issued specific guidelines for working with HIV in 1988, which have been endorsed by the Department of Labor (12, 13) These recommendations are based on guidelines established by the NIH and CDC (discussed later).

III. Health and Human Services

A. Biosafety (NIH/CDC)

Although the NIH/CDC guidelines (14) are not, strictly speaking, regulations, they have been accepted by many granting agencies, accrediting bodies, and the scientific community at large. These guidelines center on the concept of containment as afforded by facility design, safety equipment (e.g., biological safety cabinets, personal protective items), work practices, and training. Four different levels of containment are prescribed according to the level of hazard inherent in a particular practice and are summarized in Table II. These guidelines provide information and a biohazard rating of some of the viruses found in biomedical laboratories; the National Cancer Institute has also provided guidance documents (15). A list of infectious agents based on the CDC classification appears in Table III.

B. Recombinant DNA (NIH)

In addition to the biosafety guidelines discussed earlier, the NIH has also established guidelines for recombinant DNA (16). These guidelines specify containment levels and containment principles similar to those already described and are applicable to any recombinant DNA research conducted at an institution receiving funding from the NIH. Many other granting agencies require adherence to these guidelines as well. For public relations and possible legal reasons, many institutions have found it prudent to conform to these guidelines. Compliance includes establishing an Institutional Biosafety Committee (IBC), consisting of five or more members, at least two of whom are not affiliated with the institution and who represent the interests of the surrounding community with respect to health and environmental protection. If research is conducted at BL3 or BL4 (biosafety levels), a Biological Safety Officer must also sit on the committee. The IBC roster must be registered with the NIH's Office of Recombinant DNA Activities and updated annually. The committee's function is to review proposals for compliance with NIH guidelines. The registration document may also be required to contain a description of the sources of DNA, the nature of the inserted sequences, the hosts and vectors to be used, whether or not a foreign gene will be expressed and, if so, what protein(s) will be produced, and the containment conditions specified in the guidelines. In practice, many IBCs ask the investigator to cite the paragraphs of the guidelines he or she thinks are relevant and to include an abstract of the project. The IBC review should include assessment of the required containment levels, facilities, procedures, practices, and training. Only projects explicitly covered by the guidelines may be approved without redress to the NIH. Besides this initial review, the committee must also periodically review all recombinant DNA work conducted at the institution and adopt emergency plans for accidental spills and personnel contamination. [*See* RECOMBINANT DNA, BASIC PROCEDURES.]

The vast majority of laboratory experiments are exempt from the NIH guidelines if the recombinant DNA molecules (1) are not in organisms or viruses, (2) consist entirely of DNA segments from a single, nonchromosomal or viral DNA source, (3) consist entirely of DNA from a prokaryotic or eukaryotic host, (4) consist entirely of plasmids (excluding viruses) when propagated in that host (or a closely related strain of the same species), or (5) were transferred to another host by well-established physiological means (prokaryotic DNA only).

In addition, a list of exempt DNA classes in Appendix C of the guidelines is updated periodically by the NIH. Three groups of experiments are not exempt and, due to an increased recognition of their associated risks, require either notification of the IBC at their commencement, approval of the IBC before commencement, or NIH and IBC approval before commencement. The most stringently regulated experiments are those that involve the deliberate formation of recombinant DNAs containing genes for the biosynthesis of molecules toxic to vertebrates, releases to the environment, transfer of drug resistance to disease agents, or transfer of recombinant DNA or derived RNA to human subjects. Experiments requiring IBC approval before commencement involve human or animal pathogens that require BL2 containment, experiments involving whole animals or plants, or work in volumes >10 liters. Experiments requiring notice of the IBC at project commencement can be performed at BL1 levels and include recombinant DNA molecules containing no more than two-thirds of the genome of any eukaryotic virus (lacking helper virus).

C. Animals (NIH)

Animal use and care is stringently regulated by several federal regulations. The Animal Welfare Act of 1969 (as amended in 1970 and 1976; Public Laws 89-544, 91-579, and 94-279, respectively) provides broad protection to animals under the authority of the U.S. Department of Agriculture (USDA). It also regulates the transport, purchase and sale, care, handling, and use of animals for research and other purposes. The Endangered Species Act of 1973 (PL 93-205) designates the Department of the Interior (*via* the Fish and Wildlife Service) to protect both endangered and rare species as well as the habitats needed to sustain these organisms. For research and

Table II Containment Levels

Biosafety level (BL)[a]	Hazard	Typical organisms	Practices and techniques	Safety equipment
1	No known hazard		Standard microbiological practices: limit access; daily decon of work surfaces; minimize aerosols; transport wastes in leak-proof, covered containers; wear lab coats; no mouth pipetting, eating, drinking, or applying cosmetics; wash hands before leaving lab; inactivate infectious materials	Room with door and sink
2	Moderate hazard	Most parasites, fungi, and bacteria and viruses not mentioned in the higher BLs	Above plus: warning signs describing agent; educate about hazards; prepare biosafety manual; wear gloves; special handling for needles and syringes	Above plus: biosafety cabinet; autoclave capabilities available
3	Serious/lethal	*Yersinia pestis*, *Brucella* spp., *Mycobacterium tuberculosis*, Semliki Forest virus, Rift Valley fever	Above plus: special clothing; only organisms directly related to work permitted in laboratory; perform all work in biological safety cabinet; decon all wastes	Above plus: double self-closing doors; room under negative pressure, exhaust to outdoors; protect vacuum lines with filters/traps; autoclave in lab
4	Life-threatening	Marburg, Ebola, Lassa, Junin, Machupo viruses	Most stringent practices and facilities	Maximum containment; change room with exit shower; class III biosafety cabinet

[a] BL designations have replaced the former P designations (e.g., P1, P2, P3).

[Adapted from U.S. Department of Health and Human Services (1988). "Biosafety in Microbiological and Biomedical Laboratories," 2nd ed. HHS Publication No. (CDC) 88-8395. U.S. Government Printing Office, Washington, D.C.]

Table III The Center for Disease Control's Classification of Infectious Agents

CLASS 1 AGENTS

All bacterial, aparasitic, fungal, viral, rickettsial, and chlamydial agents not included in higher classes

CLASS 2 AGENTS

Bacterial

Acinetobacter calcoaceticus
Actinobacillus (all species)
Aeromonas hydrophila
Arizona hinshawii (all serotypes)
Bacillus anthracis
Bordetella (all species)
Borrelia recurrentis
B. vincenti
Campylobacter fetus
C. jejuni
Chlamydia psittaci
C. trachomatis
Clostridium botulinum
Cl. chauvoei
Cl. haemolyticum
Cl. histolyticum
Cl. novyi
Cl. septicum
Cl. tetani
Corynebacterium diphtheriae
C. equi
C. haemolyticum
C. pseuotuberculosis
C. pyogenes
C. renale
Edwardsiella tarda
Erysipelothrix insidiosa
Escherichia coli (all enteropathogenic, enterotoxigenic, enteroinvasive, and strains bearing K1 antigen)
Haemophilus ducreyi
H. influenzae
Klebsiella (all species)
Legionella pneumophila
Leptospira interrogans (all serotypes)
Listeria (all species)
Moraxella (all species)
Mycobacteria (all species except those listed in Class 3)
Mycoplasma (all species except *M. mycoides* and *M. agalactiae*, which are in Class 5)
Neisseria gonorrhoeae
N. meningitidis
Pasteurella (all species except those listed in Class 3)
Salmonella (all species and all serotypes)
Shigella (all species and all serotypes)
Sphaerophorus necrophorus
Staphylococcus aureus
Streptobacillus moniliformis
Streptococcus pneumoniae
S. pyogenes
Treponema carateum
T. pallidum
T. pertenue
Vibrio cholerae
V. parahemolyticus
Yersinia entercolitica

Fungal Agents

Actinomycetes (including *Nocardia* species, *Actinomyces* species, and *Arachnia propionica*)
Blastomyces dermatitidis
Cryptococcus neoformans
Paracoccidioides braziliensis

Parasitic Agents

Endamoeba histolytica
Leishmania sp.
Naegleria gruberi
Schistosoma mansoni
Toxocaro canis
Toxoplasma qondii
Trichinella spiralis
Trypanosoma cruzi

Viral, Rickettsial, and Chlamydial Agents

Adenoviruses (all types)
Cache Valley virus
Corona viruses
Coxsackie A and B viruses
Cytomegaloviruses
Echoviruses (all types)
Encephalomyocarditis virus
Flanders virus
Hart Park virus
Hepatitis (associated antigen material)
Herpes viruses [except *Herpesvirus simiae* (Monkey B virus), which is in Class 4]
Influenza (all types except A/PR8/34, which is in Class 1)
Langat virus
Lymphogranuloma venereum agent
Measles virus
Mumps virus
Parainfluenza virus (all types except Parainfluenza virus, 3 SF4 strain, which is in Class 1)
Polioviruses (all types, wild and attenuated)
Poxviruses (all types except alastrim, smallpox, and white pox, which are Class 5 and monkey pox, which, depending on experiments, is in Class 3 or 4)
Rabies virus (all strains except rabies street virus which should be in Class 3)
Reoviruses (all types)
Respiratory syncytial virus
Rhinoviruses (all types)
Rubella virus
Simian viruses [all types except *Herpesvirus simiae* (Monkey B virus) and Marburg virus, which are in Class 4]
Sindbis virus
Tensaw virus
Turlock virus
Vaccinia virus
Varicella virus

Continues

Continued

Vesicular stomatitis virus
Vole rickettsia
Yellow fever virus, 17D Vaccine strain

CLASS 3 AGENTS

Bacterial Agents

Bartonella (all species)
Brucella (all species)
Francisella tularensis
Mycobacterium avium
  M. bovis
  M. tuberculosis
Pasteurella multocide Type B (buffalo and other foreign virulent strains)
Pseudomonas mallei
  P. pseudomallei
Yersinia pestis

Fungal Agents

Coccidioides immitis
Histoplasma capsulatum
  H. capsulatum var. *duboisii*

Parasitic Agents

None

Viral, Rickettsial, and Chlamydial Agents

Arboviruses (all strains except those in Classes 2 and 4)
Arboviruses indigenous to the United States are in Class 3 except those listed in Class 2; West Nile and Semliki Forest viruses may be classified up or down depending on the conditions of use and geographical location of the laboratory)
Dengue virus, when used for transmission or animal inoculation experiments
Lymphocytic choriomeningitis virus (LCM)
Monkey pox, when used *in vitro*
Rickettsia (all species except Vole rickettsia when used for transmission or animal inoculation experiments)
Yellow fever virus (wild, when used *in vitro*)

CLASS 4 AGENTS

Bacterial Agents

None

Fungal Agents

None

Parasitic Agents

None

Viral, Rickettsial, and Chlamydial Agents

Ebola fever virus

Hemorrhagic fever agents (including Crimean hemorrhagic fever (Congo), Junin, and Machupo viruses and others as yet undefined)
Herpesvirus simiae (Monkey B virus)
Lassa virus
Marburg virus
Monkey pox when used for transmission or animal inoculation experiments
Tick-borne encephalitis virus complex (including Russian spring–summer encephalitis, Kyasnur forest disease, Omsk hemorrhagic fever, and Central European encephalitis viruses)
Venezuelan equine encephalitis virus (epidemic strains, when used for transmission or animal inoculation experiments)
Yellow fever virus (wild, when used for transmission or animal inoculation)

CLASS 5 AGENTS

Animal Disease Organisms That Are Forbidden Entry into the United States by Law

Foot and mouth disease virus

Animal Disease Organisms and Vectors That Are Forbidden Entry into the United States by USDA Policy

African horse sickness virus
African swine fever virus
Besnoitia besnoiti
Borna disease virus
Bovine infectious petechial fever
Camel pox virus
Ephemeral fever virus
Fowl plague virus
Goat pox virus
Hog cholera virus
Louping ill virus
Lumpy skin disease virus
Mycoplasma agalactiae (contagious agalactia of sheep)
  M. mycoides (contagious bovine pleuropneumonia)
Nairobi sheep disease virus
Newcastle disease virus (Asiatic strains)
Rickettsia ruminatium (heart water)
Rift valley fever virus
Rhinderpest virus
Sheep pox virus
Swine vesicular disease virus
Teschen disease virus
Trypanosoma evansi
  T. vivax (Nagana)
Theileria annulata
  T. bovis
  T. hirci
  T. lawrencei
  T. parva (East Coast fever)
Vesicular exanthema virus
Wesselsbron disease virus
Zyonema

CLASSIFICATION OF ONCOGENIC VIRUSES ON THE BASIS OF POTENTIAL HAZARD

Low-Risk Oncogenic Viruses

Rous sarcoma
Simian Virus 40 (SV40)
CELO

Continues

Continued

Ad7-SV40	Adenovirus
Polyoma	Shope fibroma
Bovine papilloma	Shope papilloma
Rat mammary tumor	
Avian leukosis	
Murine leukemia	**Moderate-Risk Oncogenic**
Murine sarcoma	**Viruses**
Mouse mammary tumor	Ad2-SV40
Rat leukemia	Feline Leukemia (FelV)
Hamster leukemia	HV Saimiri
Bovine leukemia	Epstein-Barr (EBV)
Dog sarcoma	SSV-l
Mason-Pfizer monkey virus	GalV
Marek's	HV ateles
Guinea pig herpes	Yaba
Lucke (Frog)	Feline sarcoma (FeSV)

diagnostic needs, it is the Public Health Service and specifically NIH funding that requires that institutions follow the NIH's guidelines (17) as a basis for developing and implementing an institutional animal care and use program. These guidelines require the establishment of an animal care and use committee consisting of a scientist with experience in animal research from the institution, a certified veterinarian, and an interested person not affiliated with the institution. Additional members may be required by other regulations and policies in many states. The program should be directed by a veterinarian who is trained or experienced in laboratory animal science and medicine. If the program is directed by a qualified professional who is not a veterinarian, a veterinarian should still be associated with the program. The committee is responsible for evaluating the animal care and use program. It must review the use of animals and submit an annual report on the status of the program. The NIH guidelines cover the physical plant, animal husbandry, veterinary care, and research aspects of animal work. Some basic considerations important for the conduct of animal experimentation include the selection of an appropriate animal model, avoidance or minimization of pain and stress, surgical procedures, euthanasia, chemical and physical restraints, and alteration of life support systems. In addition to these guidelines, the NIH/CDC document (18) mentioned earlier presents a containment strategy for vertebrates that parallels that described in the section on Biosafety.

IV. Community Right-to-Know

Besides extending the Comprehensive Environmental Response, Compensation, and Liability Act of 1980, more frequently known as Superfund, the Superfund Amendment and Reauthorization Act of 1986 (SARA) extended the hazard communication program to the community. Title III of the SARA mandated a program subtitled the "Emergency Planning and Community Right-to-Know Act." This program established requirements for the planning for, and reporting of, incidents involving emergency response, leaks and spills, and emissions. The governor of each state must have an emergency response commission, which in turn generally designates committees to oversee local planning districts. Each local committee must have a district plan for emergency response actions in the event of the release of extremely hazardous substances. Facilities must maintain an inventory of the chemicals used and, if specified amounts of certain chemicals are exceeded, must report this fact to state and district agencies as well as their local fire department. Chemicals and their "reportable quantities" are derived from the EPA's 300-plus item "extremely hazardous substances list" (40 CFR Part 355). Depending on the chemical, the EPA's threshold reporting quantity can range from 1 to 10,000 pounds. Hazardous materials are divided into five broad categories: acute and chronic health, fire, reactive or pressure release hazards—essentially the same classifications used by the EPA under the Resource, Conservation, and Recovery Act of 1976 (RCRA) and OSHA under the Hazard Communication standard. In addition to legislated planning, the public is ensured access to this information. Section 313 requires the filing of annual reports of emissions of hazardous chemicals and Section 304 requires reporting unauthorized leaks or spills of these substances. While most of the threshold reporting limits are beyond the contents of individual laboratories, the law treats different laboratories and buildings within an institution collectively. Without an interactive requisition and inventory system, it is, in many instances, difficult to know whether or not these thresholds have been exceeded. More significantly, without such recordkeeping it is difficult to prove that such quantities are not, or have not been, on hand. In most cases, institutions and laboratories will be below the threshold reporting limits except in more local jurisdictions where state (e.g., California) and even cities (e.g., New York City) have adopted considerably lower reporting limits.

V. Toxic Substances Control

Two major federal laws govern the manufacture, sale, and use of toxic substances. The Toxic Substances Control Act of 1976 (TSCA) requires the EPA to collect data on the effects of chemicals on human health and the environment while the Hazardous Substances Act governs the interstate transportation of such substances. As the lead agency, EPA has the authority to restrict the production, use, and (indirectly) the shipment of certain chemical substances and mixtures. At present, the regulated TSCA inventory includes all substances used in commerce and numbers well over 60,000 compounds. The Hazardous Substances Act directs the Consumer Products Safety Commission to evaluate consumer products according to their hazard. The Commission must classify and list such substances along with those already designated as hazardous by other agencies. The Commssion has the authority to ban substances from commercial use and to require that all products be labeled with the appropriate cautionary information. The TSCA and the Hazardous Substances Control Act have primary impact on manufacturers and importers of chemicals and only a minor direct impact on laboratories; new chemical substances being used in small quantities for research and development are exempt from these regulations, although they are regulated under OSHA's "laboratory standard."

VI. Chemical Waste and Emissions

A. Solids

The primary goals of the Resource, Conservation, and Recovery Act of 1976 (RCRA) are to protect human health and the environment from hazardous waste by reducing the amount of material generated, to conserve energy and natural resources, and to manage the disposal of hazardous waste properly. RCRA defines hazardous wastes and makes all who generate them responsible from "cradle to grave," i.e., from the time waste is generated or arrives on-site to the time it is disposed of, including any impacts that may arise from it in the future. In response to RCRA, the EPA established the hazardous waste classification criteria of corrosivity, ignitability, reactivity, and toxicity and developed procedures to petition for the delisting of certain nonhazardous wastes. Although many laboratory chemicals do not meet EPA criteria for consideration as hazardous waste (a determination for which the generator is responsible), they may need to be dealt with as hazardous due to uncertainties about their chemical, physical, and health hazards. This Act requires generators to maintain written procedures, waste classifications, emergency and standard operating procedures, job-specific duties, health and safety information, and employee training. The National Research Council (19) has written an excellent reference that describes the impacts of RCRA on laboratories and provides practical suggestions for the disposal of hazardous chemicals.

All generators, transporters, and treatment, storage, and disposal facilities (TSDFs) must obtain an EPA identification number and manifest their waste from the point of generation to final disposal. Generators include institutions, universities, independent laboratories, and all other facilities that generate or accumulate 100 kg or more of hazardous waste in a calendar month. Initially, the EPA exempted "small quantity generators," those producing <1000 kg of waste per month, from some of the provisions. However, the Hazardous and Solid Waste Amendments of 1984 (HSWA) lowered the exemption limit to 100 kg/mo and state enactments continue to lower this amount. The EPA estimates that reducing the ceiling for exemption from 1000 to 100 kg/mo changed the number of regulated generators from 15,000 to >100,000. In 1986, the EPA established somewhat abbreviated requirements for facilities generating between 100 and 1000 kg of hazardous waste per month.

Shipments of hazardous waste must be accompanied by a Uniform Hazardous Waste Manifest (EPA Form 8700-22). Many states have designed their own manifest forms, but they must include the same information requested on the EPA manifest, namely the names, addresses, telephone numbers and EPA ID numbers of all parties concerned, descriptions of each unit of hazardous waste including DOT designations, hazard class, UN/NA identification numbers, and quantity of all containers including any special handling instructions. A signed certification by the generator attesting to the validity of the information must be provided. Since 1985, generators have also been required to certify that they are actively implementing an on-site hazardous waste reduction program and have selected the safest available treatment and disposal methods for their wastes.

While small generators must certify that waste minimization programs have been implemented, they must obtain a TSDF permit to treat wastes for recycling or volume reduction, including the common practice of neutralizing acids. Without a TSDF permit, waste may not be stored on-site for >90 days. Small-quantity generators (100–1000 kg/mo) may hold waste on-site for up to 180 days (270 days, if the waste will be shipped 200 mi or more) without a special permit. Reportedly, <200 commercial facilities have TSDF permits. Meeting the requirements for obtaining this permit is difficult and costly and provides strong disincentives for small generators to treat and dispose of their wastes on-site.

Some wastes may require treatment prior to disposal. For example, the HSWA established a phase-out period for the land burial of all untreated liquid hazardous wastes, to be fully implemented by 1992. The land burial of noncontainerized liquid hazardous wastes was prohibited in 1985 unless it could be demonstrated that leakage would not affect drinking water. Burial of cyanides, heavy metals, halogenated chemicals, and wastes with a pH of 2 or lower was banned in 1988. After 1990, when the last third of the hazardous wastes were evaluated, land disposal of untreated hazardous wastes, except for those not requiring treatment to protect the public health and environment, was banned.

B. Clean Air and Clean Water Acts

The Clean Air and Clean Water Acts are products of the environmental movement of the 1960s and 1970s and have had dramatic effects in improving the quality of air and water over the last 20 years. These acts provided a comprehensive national framework to protect air quality by limiting emissions from both stationary and mobile sources and protect water quality by limiting the discharge of pollutants. The EPA's National Ambient Air Quality Standards constitute the fundamental guidelines used to measure air quality and set maximum concentrations of specific pollutants that may not be exceeded. Standards have been issued for both "industrial" discharges and drinking water (at the tap) but, unfortunately, ambient water quality standards similar to those for air have not been developed.

C. Vapors and Gases

Ambient air quality, including the impacts from laboratories (e.g., fume hood emissions), is regulated by the federal Clean Air Act Amendments of 1970 and 1977 (reauthorized in 1990) with the EPA as the lead agency. While the law specifies limits for carbon monoxide, lead, nitrogen dioxide, ozone, particulates, and sulfur oxides, the EPA also established National Standards for Hazardous Air Pollutants. Charged with regulating every pollutant likely to pose a risk to public health without regard to cost, the EPA has only ruled on eight chemicals or groups of chemicals (asbestos, benzene, beryllium, coke oven emissions, inorganic arsenic, mercury, radionuclides, and vinyl chloride) in the last 20 years. With the 1990 reauthorization of the Clean Air Act, the EPA is now required to regulate the sources of pollution rather than individual pollutants and must consider technology-based standards (e.g., maximum achievable control technology). This new approach promises to be costly but very effective.

The original Clean Air Act required each state to submit an implementation plan to the EPA describing ambient air quality and strategies for achieving and maintaining the standards. Some states have passed more stringent regulations, especially for emissions from readily identifiable sources such as automobiles (e.g., California) and incinerators (e.g., New York). New incinerator standards will have a major financial (as well as public health) impact on hospitals and biomedical institutions that rely on incineration, either on- or off-site, for the disposal of hazardous chemical and biologicals. A reduction in on-site incineration will increase disposal and transportation costs for medical and certain other hazardous wastes and increase the potential for contact between the public and these materials. In some municipalities, a permit is required for laboratory construction and renovations and, in other areas, an environmental impact review must precede actual work. For laboratories, the principal source of air emissions is fume hoods. These emissions, although regulated, generally have a trivial impact on ambient air quality. However, for some compounds (e.g., certain radionuclides, carcinogens) emission levels may exceed standards, requiring treatment such as filtration or adsorption prior to release.

D. Liquids

The Clean Water Act of 1977 required the EPA to establish ambient water quality criteria, dictating the kinds and amounts of materials that may be released into navigable waters. States may apply to the EPA for authorization to administer various aspects of the National Pollutant Discharge Elimination System (a permitting procedure) and public pretreatment programs. The EPA has developed a list of 126 priority pollutants, based on factors including biochemical oxygen demand, total suspended solids, pH, fecal coliform, oil, grease, and so-called nonconventional effluent standards. The EPA considers toxicity, environmental persistence, impact on native organisms, and the extent to which controls are technologically or economically feasible. Historically, toxic and hazardous materials have also been released to municipal sewage systems, a disposal method that depends greatly on volume dilution to minimize hazards. While it is clear that corrosive, flammable, water reactive, highly toxic, and water-immiscible materials should not be poured down the drain, many investigators still do. Regardless of how small, no amount of certain chemicals (e.g., pyridine, carbon disulfide, azides) may be tolerable in laboratory plumbing because of odor, toxicity, or flammability. Ultimately, the impact of releases of chemicals and otherwise potentially environmentally harmful materials depends less on their physical and chemical properties than on whether or not they are biodegradable and whether or not the degradation products are toxic to waste water treatment plant organisms or the environment into which they are subsequently released. Even if the EPA does not prohibit the sewage disposal of specific compounds, the laboratory or institution may be held accountable for the impacts of their discharges. In a similar manner, the Safe Drinking Water Act addresses groundwater protection and has been used with varying degrees of success to limit industrial development over primary drinking water aquifers.

VII. Low-Level Radioactive Waste

Low-level radioactive waste (LLRW) is a highly visible target for regulation, most recently under the Low-Level Radioactive Waste Policy Act (Public Law 96-573), Low-Level Radioactive Waste Policy Amendments Act (Public Law 99-240), and their various state reenactments. LLRW must be segregated according to form and isotope to reduce effectively the volume of waste that must be shipped off-site for disposal (20, 21). Wastes contaminated with isotopes of short half-life (88 days or less) can be held for decay, whereas those with longer lives should be shipped to one of the three commercial LLRW facilities for shallow land burial. LLRW must be packaged, manifested, and transported according to the instructions provided by certified haulers, who, in turn, should be in compliance with the various local, state, and federal transportation and site-specific disposal guidelines.

Despite the carefully monitored and highly defined ways in which radioactive waste may be disposed of, a particular method may not be legal if LLRW contains other wastes. LLRW that also contains hazardous chemicals is called "mixed waste." While mixed wastes generated during biomedical research consist primarily of organic solvents used in the extraction and purification of radiolabeled biomolecules and contain low levels of ^{3}H and ^{14}C, (22) this waste stream is variable and difficult to manage. Currently, there are no commercial disposal outlets for mixed wastes nor may they be stored on-site legally for more than 180 days. Mixed waste may not be treated to separate the radioactivity or render the chemicals nonhazardous without an RCRA Part B permit (discussed earlier). The NRC has stated that radioactive medical waste will be subject to all existing NRC regulations as well as the EPA's new medical waste regulations (23). Of course, minimizing the generation of these kinds of waste should be the first and most important responsibility of all researchers. But for the remaining fraction that will continue to be generated, deregulation of certain mixed wastes by the NRC and radioactive medical wastes by the EPA and/or NRC would be the safest, most cost-effective, and practical method for dealing with these materials.

VIII. Medical Waste

Medical waste constitutes <1% of all municipal solid waste yet a national climate of increasingly stringent regulations for all hazardous materials, the widespread fear of AIDS, and the media coverage of the washing up of syringes and needles on beaches in the Northeast in the late 1980s led to a strong public outcry for stricter control of medical waste. This outcry resulted in passage of the federal Medical Waste Tracking Act of 1988 (40 CFR Part 259). The Act authorized the EPA to establish a pilot program in 10 states, beginning in 1989 in New York, New Jersey, Connecticut, Rhode Island, and Puerto Rico. The Act also directed the Agency for Toxic Substances and Disease Registry to report on the "potential for infection or injury from the segregation, handling, storage, treatment, or disposal of medical wastes . . . estimate the number of people injured or infected annually by sharps . . . or other means related [to medical waste] . . . including what percentage of disease, such as AIDS and hepatitis B, cases nationally may be traceable to medical wastes." The report (24) concluded that the general public is not likely to be adversely affected by medical waste and that, in fact, no spread of disease from medical waste has been reported nor does medical waste contain any greater quantity or different types of microbiological agents than ordinary residential waste and domestic sewage.

The major impact on laboratories from these new regulations stems from the increasingly broad definition of what is considered "infectious" waste. While EPA and OSHA are currently the lead agencies in regulating biomedical waste, the definition of biomedical waste varies from agency to agency and state to state. Where differences in federal, state, and local regulations exist, generators must comply with the most stringent. Table IV highlights some of the differences among federal, New York State, and New York City laws.

Under the Tracking Act, medical waste generators are responsible for developing a plan to identify, segregate, package, monitor, treat, and dispose of these materials. By strict definition, to be considered infectious, the possibility of the presence of an infectious agent in the waste must exist. However, because of the public's inability to distin-

Table IV Regulated Categories of Biomedical Waste[a]

	New York City	New York State	Federal
Animal waste/carcasses/bedding	Yes	Yes	Yes
Blood/blood products	Yes	Yes	Yes
Cultures and stocks	Yes	Yes	Yes
In-home health-care waste	No	No	No
Isolation waste	Yes	Yes	Yes
Laboratory wastes			
All materials from laboratory	Yes	No	No
Only materials with infectious agent contact	Yes	Yes	Yes
Pathological wastes	Yes	Yes	Yes
Sharps: needles, syringes, scalpels, glass	Yes	Yes	Yes
Surgical waste, including gloves, gowns, and masks	Yes	No	No
Small quantities (<50 lb/mo of any of the above)	Yes	Yes	No

[a] Material must be inventoried, treated, and destroyed on-site or manifested and removed by a licensed medical waste transporter for disposal at a licensed medical waste disposal facility. Small-quantity generators are required only to maintain records of amounts treated and shipped.

guish between infectious and noninfectious waste and their misperception of the associated risks, many facilities manage their entire medical waste stream as if it were infectious. Any facility that produces or transports 50 lb or more of regulated waste per month must comply with the Tracking Act. Generators must segregate their regulated waste into three categories: sharps, fluids (any amount of fluid >20 ml), and solids. These wastes must be stored in such a way as to protect the public. In practice, this means using puncture- and leak-proof containers as well as refrigerating any waste that might decompose. These packages must be manifested, marked as "infectious waste" or "medical waste," and have the shipping date, name, permit number, and address of both the shipper and any transporter(s) indelibly marked on both the outer rigid container and any inner containers. If a generator ships <50 lb/mo off-site, the transporter and generator need not use the manifest but the generator must maintain a written log of the materials either treated on-site or shipped for disposal. A small-quantity generator may apply for a permit to use a nonpermitted transporter for shipping medical waste off-site. A multiple-copy tracking manifest, including proof of destruction, and heavy liability insurance have become a standard part of the transportation and disposal operation.

Laboratory workers should be aware of several details of the federal medical waste law:

1. Medical waste must be collected in *marked, puncture-proof containers for sharps* and segregated as soon as practicable. Wastes may not be kept on patient floors >24 hr unless they are in rooms used exclusively for waste storage; wastes may only be stored on premises for up to 30 days.
2. *Treated, regulated medical waste* is that treated (e.g., autoclaved or chemically inactivated with bleach) to reduce substantially or eliminate its potential for causing disease but has not yet been destroyed.
3. *Destroyed, regulated medical waste* is that ruined, torn apart, or mutilated through processes such as thermal treatment, melting, shredding, grinding, tearing, or breaking so that it is no longer recognizable as medical waste. Autoclaved wastes must be either destroyed on-site or removed by a certified hauler to a medical waste disposal facility. Note that treatment residue and ash from the incineration of medical waste is considered ordinary solid waste unless it contains other materials that are hazardous.
4. A generator who *treats* and *destroys or disposes of medical waste on-site* (e.g., incineration, burial, or sewer disposal as permitted) is not subject to tracking requirements for that waste but *must maintain records* of the quantity (by weight) for at least 3 years.

While private carters charge about $0.20/lb for regular trash, the 1990 cost for infectious waste disposal is between $0.50 and $1.00/lb. Medical waste disposal costs have risen 10-fold, but the Act specifically forbids passing this increase on to patients. The EPA is authorized to issue injunctions and to impose civil and criminal penalties ranging from $25,000/day for each violation of the Tracking Act and 2 years imprisonment to $250,000 and 15 years for violations that place someone at risk of serious injury or death; those convicted of knowing endangerment may be fined up to $1 million. States may perform independent inspections and enforce additional provisions. Although several fines have been levied for violations of medical waste regulations, subsequent investigations of most of these incidents of improper disposal have pointed to deliberate actions or the results of municipal wastewater treatment plant malfunction. Given the liabilities associated with mishandling of laboratory wastes, one of the unanticipated results of this legislation will be a reduction in recycling. [*See* INFECTIOUS WASTE MANAGEMENT.]

IX. Shipment of Hazardous Materials

A dozen agencies share the responsibility for ensuring that hazardous materials are transported as safely as possible, issuing regulations for the classification, design, and performance of containers, procedures for handling and transporting, and guidelines for documenting and labeling hazardous materials. The U.S. Department of Transportation (DOT) (49 CFR Parts 100–199) is the lead agency and governs all shipments of hazardous materials (49 CFR Part 172.101) between states and to foreign countries as well as all modes of transport, except those made through the U.S. Postal Service (39 CFR Part 123). The DOT must coordinate its regulations with the agencies governing the vehicles or vessels themselves. These agencies include the following:

- Office of Hazardous Materials Transportation—serves as the DOT's liaison with the EPA (49 CFR Parts 171-177)
- Bureau of Motor Carrier Safety—motor carrier standards, regulates vehicles (49 CFR Parts 350-399)
- Federal Highway Administration—inspection and enforcement of highway transportation, depots, and trans-shipment points (49 CFR Part 177)
- National Highway Traffic Safety Administration—regulation of vehicles
- Federal Railroad Administration—rail transportation, holdings in depots or freight yards (49 CFR Part 174)
- Federal Aviation Administration—domestic and foreign air carriers operating in the United States and airport cargo holding areas (49 CFR Part 175)
- U.S. Coast Guard—bulk shipments by ship or barge, port areas; all domestic and foreign ships in U.S. waters (49 CFR Part 176)
- U.S. Public Health Service—restricts the importation of etiologic agents and vectors of human diseases (42 CFR Part 71.156)

Other agencies have jurisdiction over other aspects of hazardous materials transportation. For example, the NRC and the U.S. Departments of Defense and Energy all regulate shipments of radioactive materials and both the U.S. Department of Agriculture and the Public Health Service control shipments of biological materials.

While most state and local regulations concerned with the transport of hazardous materials deal only with emergency response and enforcement and fall within the guidelines of federal regulations, some states and localities have established additional regulations. Most frequently, these regulations deal with route restrictions and shipment notifications or registration and special permits for shippers and carriers of hazardous materials. Anyone working with or supervising those working with hazardous materials must be aware that transport of any quantity of these materials is highly regulated. These materials must be properly identified, packaged, labeled, and accompanied by appropriate shipping papers before being transported over any public conveyance, whether that transport is by air, rail, water, or road and whether by commercial carriers, private methods, or the U.S. Postal Service.

X. Inspections

In the process of complying with the various regulations described in this article, it is important to keep sight of employer and employee rights in dealing with regulatory agencies and not to forfeit those rights out of ignorance. For example, OSHA has the right to enter the premises where work is performed by an employee "without delay and at reasonable times" and to question employers, owners, operators, agents, and employees. However, Fourth Amendment protection against unreasonable searches allows the institution to request that OSHA (and other regulatory agencies) obtain a warrant before an unscheduled inspection. Obtaining a warrant requires showing probable cause (e.g., evidence of an existing violation or employee complaints and a reasonable administrative protocol for conducting the inspection). Both employer and employee have the right to be informed of the purpose of the inspection. The need for a warrant can be excepted if the conditions are observable by the public, the observed condition is in obvious and plain view of inspectors while on the premises, or if the employer consents to the inspection. Private interviews with employees are permitted on the employer's premises. The Hazard Communication ("Right-to-Know") standard guarantees OSHA as well as employees or their representatives the opportunity to examine records of exposure to toxic substances and harmful physical agents, medical records and analyses thereof, all of which must be kept for 30 years postemployment. However, access need not be immediate but in a "reasonable time, place and manner," generally not more than 15 days after the request. Employees may also access others' exposure records where other employees had "similar job duties or working conditions that were related" to their own past, present, or future assignments. OSHA may access personal medical records without the employee's written consent. Representatives for the employer and employees may, and should, accompany the inspector during an inspection, observing testing procedures and seeking information as to the details of the complaint. Moreover, the employer is entitled to a copy of any written complaint but with the name of the employee(s) deleted. Both employer and employee may attend opening and closing conferences where the inspector will describe the alleged violation, note applicable standards, and advise of the right to contest any citations, which must then be filed within 15 days.

The best defense is to be prepared. Establish on paper and through practice a well-developed and annunciated health and safety program and that you meet or exceed standards. Document these efforts with letters, standard operating protocols, worksheets, change orders, monitoring records, photos, or contracts. At the very least, laboratory safety programs should be designed to meet the spirit, if not always the letter, of the law. If the tone of this paragraph sounds defensive, bear in mind that regulatory agencies tend to proceed as if you are guilty unless you can substantiate your knowledge of the regulations by documenting the actions you have taken to comply.

XI. Conclusions

No one article could contain all there is to know about regulations, even those limited to laboratory operations. It is important that someone at an institution remain aware of not only the current regulations and opportunities for their review, but also of proposed or pending regulations. Working to ensure that the regulations with which one will have to comply are reasonable and effective is easier than explaining later why one did not comply. However, even well-intentioned regulations may not be practical or effective. Sometimes, federal regulations are reinterpreted at state and local levels where uncertainty and insufficient technical expertise dictate the approach that "stricter is safer."

The active participation of scientists in the political and educational process can help assure that truly dangerous practices and work areas are regulated and that the regulations protect the health and safety of laboratory workers, the public, and the environment. Because these interactions are complex, it is likely that participation will require an increasing commitment to sustain interest in and understanding of the broader issues at hand. Insufficient commitment to public education on the part of the scientific community contributes to a general pattern of scientific illiteracy, public distrust, and subsequent political responses that result in overreaction and overregulation. Unfortunately, the public and political quest for zero risk has led to extensive regulations, even in the absence of demonstrable hazard—the treatment and disposal of LLRW and medical waste are cases in point.

Regulations are an important factor for protecting workers and the environment and it is clear that the initial cost for protection is far lower than that for remediation and compensation after an accident or release of toxic materials. While compliance alone does not guarantee safety, it should be satisfying to know that personal risks and environmental impacts have been reduced to a level as low as reasonably achievable. Where health and safety issues are involved, it is, as the catchphrase goes, better to be proactive than reactive.

Bibliography

1. American Society of Heating, Refrigerating and Air-Conditioning Engineers, Inc. (1989). "Ventilation for Acceptable Indoor Air Quality." Standard 62-1989. ASHRAE, Atlanta.
2. American Society of Heating, Refrigerating and Air-Conditioning Engineers, Inc. (1989). "Ventilation for Acceptable Indoor Air Quality." Standard 62-1989. ASHRAE, Atlanta.
3. American Conference of Governmental Industrial Hygienists (1988). "Industrial Ventilation," 20th ed. ACGIH, Cincinnati.
4. Cember, H. (1983). "Introduction to Health Physics," 2nd ed. Pergamon Press, New York.
5. Shapiro, J. (1990). "Radiation Protection: A Guide for Scientists and Physicians," 3rd. ed. Harvard University Press, Cambridge.
6. American Conference of Governmental Industrial Hygienists (1990). "Threshold Limit Values and Biological Exposure Indices for 1989-1990." ACGIH, Cincinnati, Ohio.
7. National Fire Protection Association (1990). "National Fire Codes." NFPA, Quincy, MA.
8. American Conference of Governmental Industrial Hygienists (1990), "Threshold Limit Values and Biological Exposure Indices for 1989-1990." ACGIH, Cincinnati, Ohio.
9. U.S. Department of Health and Human Services (1981). "Guidelines for the Laboratory Use of Chemical Carcinogens." NIH Publication No. 81-2385. U. S. Government Printing Office, Washington, D. C.
10. American Journal of Hospital Pharmacy (1986). OSHA work-practice guidelines for personnel dealing with cytotoxic (antineoplastic) drugs; Appendix 7, **43,** 1193–1204.
11. U. S. Department of Health and Human Services (1988). "Guidelines for Protecting the Safety and Health of Health Care Workers." NIOSH Publication No. 88-119. U. S. Government Printing Office, Washington, D. C.
12. National Institutes of Health (1988). "Working Safely with HIV in the Research Laboratory: Biosafety Level 2/3." U. S. Government Printing Office, Bethesda, Maryland.
13. U. S. Department of Labor (1987). Guidelines on AIDS and Hepatitis B. *Fed. Regist.* **52,** 41818.
14. U. S. Department of Health and Human Services (1988).

"Biosafety in Microbiological and Biomedical Laboratories," 2nd. ed. NIH Publication No. 88-8395. U. S. Government Printing Office, Washington, D. C.
15. U. S. Department of Health, Education and Welfare (1974). "Biological Safety Manual for Research Involving Oncogenic Viruses." DHEW Publication No. 76-1165. U. S. Government Printing Office, Washington, D. C.
16. U. S. Department of Health and Human Services (1986). Guidelines for research involving recombinant DNA molecules. *Fed. Regist.* **51,** 16958-16985.
17. U. S. Department of Health and Human Services (1985). "Guide for the Care and Use of Laboratory Animals," revised. NIH Publication No. 85-23. U. S. Government Printing Office, Washington, D. C.
18. U. S. Department of Health and Human Services (1988). "Biosafety in Microbiological and Biomedical Laboratories," 2nd. ed. NIH Publication No. 88-8395. U. S. Government Printing Office, Washington, D. C.
19. National Research Council (1983). "Prudent Practices for Disposal of Chemicals from Laboratories." National Academy Press, Washington, D. C.
20. Party, E., and Gershey, E. L. (1989). Recommendations for radioactive waste reduction at biomedical/academic institutions. *Health Phys.* **56,** 571-572.
21. Gershey, E. L., Klein, R. C., Party, E., and Wilkerson, A. (1990). "Low-Level Radioactive Waste: From Cradle to Grave." Van Nostrand Reinhold, New York.
22. Linins, I., Klein, R. C., and Gershey, E. L. (1991). Management of mixed wastes from biomedical research. *Health Phys.* **61,** 421-426.
23. Nuclear Regulatory Commission (1989). Information Notice 89-85. NRC, Washington, D.C.
24. Agency for Toxic Substances and Disease Registry (1990). "The Public Health Implications of Medical Waste: A Report to Congress." HHS Publication No. PB91-100271. U. S. Department of Health and Human Services, Atlanta.

Leprosy

Thomas P. Gillis and Robert C. Hastings
Gillis W. Long Hansen's Disease Center

Glossary

Lepromatous leprosy (LL) Disseminated form of leprosy with host lacking cellular immunity to *Mycobacterium leprae;* generally large numbers of bacilli in lesions

Lepromin Name given to various preparations of killed *M. leprae* used for skin testing; original preparation by Mitsuda was autoclaved *M. leprae* from human tissues

Mitsuda reaction Hypersensitivity granuloma in response to antigens of *M. leprae* monitored at 3 weeks after injection

T-cell anergy Absence of T-cell response to antigen caused by suppression or lack of responder T cells

Tuberculoid leprosy (TT) Localized form of leprosy with host exhibiting strong cellular immunity to *M. leprae;* few or no bacilli in lesions

LEPROSY (HANSEN'S DISEASE) is a chronic granulomatous infection caused by *Mycobacterium leprae*. Global estimates for leprosy are between 10 and 12 million cases, with only about one half receiving treatment. Predilection of *M. leprae* for peripheral nerves can cause nerve damage resulting in sensory and motor paralysis and deformities. Clinical leprosy exhibits symptoms and underlying pathology of a spectral nature. At one extreme of the spectrum is the localized form of the disease, tuberculoid leprosy (TT), in which few or no bacilli are demonstrable. At the opposite pole of the spectrum is the disseminated form of the disease, lepromatous leprosy (LL), in which large amounts of *M. leprae* are found in the skin. Patients whose disease is somewhere between the two polar forms are referred to as borderline leprosy patients. The causative agent of leprosy, *M. leprae,* has never been grown on artificial media, which has slowed research efforts to understand the basic metabolic, biochemical, and pathogenic nature of this obligate intracellular pathogen of humans.

I. History and Epidemiology

Accurate descriptions of clinical leprosy exist in Chinese and Indian writings dating to 600 B.C. There is no convincing evidence that leprosy existed in Asia Minor or Europe until after the return of troops of Alexander the Great of Macedonia from the India campaign in about 327 B.C. Thus there was no disease we would recognize as leprosy in lands occupied by the Jewish people at the time the Old Testament was written. This, plus the inaccuracies of the descriptions of the disease, makes it clear that the Old Testament descriptions of leprosy did not refer to the modern disease called leprosy.

By 250 B.C. modern leprosy was recognized by Alexandrian and Greek physicians. After the Crusades the disease was introduced into Europe. At the time of the Great Plague there was a curious, and not yet adequately explained, abrupt fall both in the number of leprosy hospitals and apparently in the number of leprosy patients. The last places in Europe in which the disease remained in substantial numbers were in Scandinavia. In the leprosy hospital in Bergen, G. H. Armauer Hansen described the leprosy bacillus in 1874. Current World Health Organization (WHO) estimates put the total number of registered leprosy patients in Europe at 13,400.

The disease was and remains endemic in China, India, and sub-Sahara Africa. Leprosy is now disappearing from the mainland of Japan, and in China control of the disease is improving. Many new cases continue to appear in India. Most known leprosy cases are in Southeast Asia; current WHO estimates are 2,556,000, 68,700, 22,100, and 121,500 for India,

China, The Philippines, and Indonesia, respectively.

There is no convincing evidence that leprosy existed in the New World before the voyages of Columbus. Current estimates are 337,800 known cases in the Americas with the vast majority in Brazil (257,000 cases). In the United States, leprosy is predominantly a disease of immigrants. Most leprosy cases are found in the population centers of the United States; however, the disease is still indigenous in Louisiana, California, Hawaii, and Texas.

II. Clinical Leprosy

The cardinal signs of leprosy are (1) one or more hypopigmented skin lesions, (2) loss of sensation, (3) peripheral nerve enlargement, and (4) finding acid-fast bacilli in the skin. A wide variety of other signs or symptoms may occur in leprosy patients. Commonly nasal stuffiness, eye inflammation, unnoticed burns heralding extremity anesthesia, and a variety of generalized skin lesions are presenting signs or symptoms of the disseminated form of the disease. Localized disease is typically a ring worm–type skin lesion that is anesthetic.

Inherent to an understanding of clinical leprosy is an understanding of the graded nature of the host resistance to the infection. In areas exhibiting the highest endemicity for leprosy, the prevalence of the disease rarely exceeds 5% of the population. It is reasonable to assume that in these highly endemic areas virtually the entire population is exposed to leprosy bacilli more or less continuously. Even under these conditions, 95% of these populations do not develop overt signs or symptoms of the disease. It is likely that everyone in the population is infected with *M. leprae,* and yet overt disease develops in only 1 in 20. It is thought that a somewhat higher proportion may develop an early leprosy characterized by a nondescript macular skin lesion with at most minor diminution in sensation and with a nonspecific chronic dermatitis histopathology. This early disease is called indeterminate leprosy. If followed for a year, 75% of indeterminate leprosy cases will self-heal. At some time or other perhaps as much as 20% of a highly endemic population may have had indeterminate leprosy, of whom three fourths self-healed, leaving 5% of the population who proceeded to develop overt leprosy from their indeterminate disease.

The characteristics of the overt leprosy that evolves from indeterminate leprosy all follow from the degree of cell-mediated immunity (CMI) expressed by the patient toward *M. leprae* on the one hand and by the duration of untreated disease on the other. A far-advanced patient with disseminated disease will bear enormous burdens of leprosy bacilli, whereas a patient diagnosed early in the course of his or her disseminated disease will have much lower burdens of bacilli and fewer chances for a variety of complications.

The classic means of assessing CMI in leprosy is to test the ability of an individual to develop an epithelioid cell granuloma in his or her skin in response to an intradermal injection of killed leprosy bacilli. This test is called the Mitsuda lepromin test. Traditionally the bacilli have been obtained from humans with the disease. In recent years the bacillary count in the reagent has been standardized, and the bacilli are produced from experimentally infected nine-banded armadillos. Lepromin reactivity peaks macroscopically at about 3 weeks after injection. Microscopically a positive lepromin or a positive Mitsuda mimics the histopathological lesion of localized leprosy. A negative lepromin fails to become indurated and histopathologically reproduces the lesion of disseminated leprosy. Clinical experience has shown that a large majority of individuals who have never been exposed to the leprosy bacillus are usually Mitsuda-positive, indicating that lepromin testing serves as a microvaccination. Thus, unlike the tuberculin test, the lepromin does not have diagnostic value, but it is valuable as prognostic test and is used widely for that purpose in many endemic areas. It is widely accepted that if a healthy individual is tested with lepromin three times and all three tests are negative, then that individual is not only susceptible to leprosy but is susceptible to the disseminated type of leprosy.

A. Tuberculoid Leprosy

The localized form of leprosy is called tuberculoid, so named because the skin lesions it causes have an histologic appearance that is similar to that seen in tuberculous lesions. This is the localized form of the disease because the patient has CMI directed against antigens of the leprosy bacillus. Additionally the patient has delayed-type hypersensitivity (DTH) directed toward antigens of *M. leprae*. The patient's CMI, characterized by a positive lepromin test, keeps the disease localized and provides a favorable prognosis for self-healing in many cases and rapid resolution of the bacterial disease in all cases after beginning appropriate chemotherapy. Only one or a few skin lesions are seen, and they are sharply demarcated from surrounding normal skin. Bacilli may or may not be demonstrated in routine skin biopsies. The characteristic histopathological picture is one of well-formed tubercles with central accumulations of epithelioid cells and occasional Langhans' giant cells. The diagnostic features are selective granulomatous inflammation, destruction of dermal nerve fibers, and the finding of an occasional acid-fast bacillus. Characteristically the so-called polar tuberculoid lesions impinge on the basal layer of the epidermis, whereas in those cases with slightly less competent CMI and consequently with slightly more disseminated disease the lesions do not involve the basal layer of the epidermis. These slightly more disseminated tuberculoid cases are more common than the so-called polar tuberculoid cases and are termed borderline tuberculoid (BT) patients. Polar tuberculoid leprosy is termed TT disease.

As might be expected, the preferential localization of the leprosy bacilli in peripheral nerves, both nerve trunks and dermal nerve fibers, together with competent CMI and a potentially powerful DTH to antigens of the leprosy bacillus leads to an intense inflammation of nerves containing *M. leprae*. Fortunately the degree of dissemination of the bacilli is limited, and therefore the extent of the peripheral nerve involvement is relatively limited. However, the extent of destruction of the nerves that are involved is frequently total. For this reason the degree of anesthesia in the skin lesions is nearly total, and if peripheral nerve trunks are affected there is often nearly total sensory and motor loss in the area of distribution of the nerve trunk. Early diagnosis and early treatment stop the spread of bacilli and hence the spread of bacillary antigens, thus lessening the chances for new nerve damage. However, even during effective antibacterial treatment, DTH mechanisms may destroy peripheral nerves already containing bacilli and hence bacillary antigens.

B. Lepromatous Leprosy

The disseminated form of leprosy is called LL. The patient lacks CMI against *M. leprae* and lacks DTH against any antigen of the leprosy bacillus. The nature of this failure to develop CMI against any of the antigens of the leprosy bacillus is not known. Lepromatous patients who have no localizing features to their disease whatsoever, the so-called polar lepromatous (LLp) cases, are lepromin-negative when they have active disease. These patients were presumably lepromin-negative before they contracted their disease, and definitely remain lepromin-negative for the rest of their lives even though effective chemotherapy will essentially eliminate all traces of leprosy bacilli from their bodies. Nerve involvement occurs incidentally and not selectively. The macrophage accumulation may occupy 90% of the dermis. There is frequently epidermal atrophy, and there is characteristically a zone of the dermis between the macrophage layer and the epidermis that is free of involvement, the so-called subepidermal free zone. There may be up to 10^{13} leprosy bacilli in a far-advanced LL patient. Under optimum chemotherapy the numbers of bacilli fall 0.5 to 1.0 logs per year. The bacilli cease to be detectable with routine acid-fast stains at a total body load of approximately 10^{7} organisms.

In the era before treatment was available, patients with LLp died of obstructive laryngeal involvement, secondary bacterial infections caused by open ulcers from both traumatic injuries of insensitive extremities and from progressive bacillary disease, and renal amyloidosis. In the absence of complications, with effective treatment nasal stuffiness promptly disappears, and there is a halt in the progression of skin anesthesia and noticeable improvement in the skin lesions.

C. Borderline Leprosy

Patients whose disease is somewhere between polar tuberculoid and LLp in its characteristics are said to have borderline leprosy. As mentioned earlier, if their disease is near tuberculoid it is termed BT. Similarly cases with borderline leprosy near lepromatous are said to have borderline lepromatous (BL) leprosy. Patients in the middle of the borderline spectrum are rare. They are said to have midborderline (BB) leprosy. Movements of a given borderline patient in classification over time toward the lepromatous type of disease are called downgrading reactions. Similarly, movements of a given borderline case over time toward tuberculoid are called reversal reactions or upgrading reactions. Many patients on effective chemotherapy tend to upgrade as their bacillary loads and hence antigenic loads decline, allowing their CMI to recover.

D. Reactions

In approximately 50% of leprosy patients under treatment, acute inflammatory reactions develop that are immunologically mediated responses to antigens of *M. leprae*. Reactions are classified according to the type of immune response that underlies the clinical episode.

Type 1 reactions, or upgrading (reversal) reactions occur in individuals who have significant T-cell immune responses to *M. leprae;* therefore, they occur only in individuals whose underlying disease classification is TT, BT, BB, or BL. Because effective T-cell immunity is absent in LL patients, no type 1 reactions can develop. Type 1 reactions are most common in BT disease. The typical clinical presentation is acute edema and acute erythema of a pre-existing macular or plaque-like skin lesion, frequently with acute inflammation of one or more nerve trunks in the region of the skin lesion(s). The histopathology of a reversal-reaction lesion is edema superimposed on a typical BT granuloma. The extreme of the DTH reaction is caseation necrosis, and it occurs almost exclusively in nerves containing antigens of *M. leprae*. Necrosis with abscess formation is in nerve trunks with irreparable loss of function. The eventual outcome of the type 1 reaction depends on the balance between the relative effectiveness of CMI and the proliferation of *M. leprae*. With enhanced effective immunity the patient may move in classification toward tuberculoid (upgrading/reversal) reaction and clear all bacilli. If there is a reduction in effective CMI and enhanced numbers of leprosy bacilli, the patient moves in classification toward lepromatous (downgrading reaction). At the time of the acute episode, one can only diagnose type 1 reaction. At a later time, after the acute episode, one can then determine if the patient's disease classification has changed, and if so, whether he or she has experienced an upgrading (reversal) reaction or a downgrading reaction. The management of type 1 reactions is high-dosage glucocorticoids to reverse, if possible, and to prevent further deterioration in peripheral nerve function. It is important to continue specific antibacterial chemotherapy without interruption because steroids in the absence of chemotherapy strongly favor downgrading.

Type 2 reactions are traditionally thought to be clinical manifestations of antigen–antibody complexes forming in or near blood vessel walls, fixing complement there and creating acute inflammation resulting in vasculitis. Clinically the skin lesions resemble those seen with erythema nodosum, thus the common name for type 2 reactions is erythema nodosum leprosum (ENL). Because patients with good CMI do not form large amounts of antibodies against the leprosy bacillus and patients with poor CMI do, ENL is seen in BL and LL patients and never in the tuberculoid forms of the disease. Typically an LL patient develops a crop of tender, erythematous, nodular skin lesions with fever. The overall distribution of the ENL lesions follows a temperature-linked pattern similar to the underlying bacterial densities. For this reason ENL is thought to represent a multifocal Arthus-type hypersensitivity rather than a circulating immune complex disorder. ENL can affect any tissue containing *M. leprae* and its antigens. Thus ENL-type inflammation can occur in the eyes, nose, lymphatics, joints, testicles, and most importantly, the peripheral nerves that are infected by the bacteria.

Histopathologically, ENL is characterized frequently by a vasculitis and always by a neutrophilic infiltrate superimposed on a lepromatous or near lepromatous infiltrate. Because bacterial antigens continue to be present for long periods of time, ENL frequently becomes chronic with repeated crops of skin lesions and fever lasting from a few months to many years. Management of ENL includes high-dose glucocorticoids for acute iritis and acute peripheral neuritis to reverse, if possible, acute losses of motor function, blindness and acute sensory losses of critical areas, particularly the palms, soles, and cornea. The anti-inflammatory action of high doses of steroids is evident within 6 hr; therefore they are the agents of choice in an ENL patient with significant acute peripheral neuritis because no other drug works this rapidly. If peripheral nerve function is stable, the drug of choice for ENL is thalidomide (100 mg 4 times daily until the reaction subsides and then 100 mg every other day for maintenance). Because of its universal teratogenicity, thalidomide is given to fertile women with extreme care or not at all, depending on the regulatory and ethical environment. Thalidomide usually begins to reduce fever by the first day of treatment, usually no further ENL lesions develop and existing ones resolve, and usually within 4 days all signs and symptoms of ENL have disappeared. Overall more than 95% of ENL patients respond to thalidomide. Large doses of clofazimine (300 mg daily) also control ENL. ENL is usually brought under control within 30 days. Clofazimine can be given in these doses safely for 6 months, but beyond this period of time there is an increasing risk of toxicity to the small intestine caused by drug accumulation in the bowel wall and blockage of mesenteric lymphatics with resultant lymphedema and impaired absorbtion and bowel motility. This small bowel involvement is most marked in the distal ileum, is the dose-limiting toxicity of clofazimine, and can be fatal.

III. Treatment

The treatment of leprosy from the point of view of public health and long-term cost-effectiveness has as its major goal the reduction of future cases by treating the current infected population to the extent possible with the available resources. The most cost-effective means of delivering treatment for this purpose is through primary health care. The objective of treatment is achieved when the patient is no longer infectious. When resources are limited the public health approach to treatment may offer the best, if not the only, opportunity to impact the overall burden of leprosy in a population. Of necessity such an approach must be simple, cheap, rigid, and brief. Such an approach was advised by the WHO in 1982. The WHO regimen for paucibacillary leprosy (indeterminate, BT, and TT) consists of daily dapsone (100 mg) and monthly rifampin (600 mg) for 6 months. In patients with large numbers of bacilli, so-called multibacillary leprosy (BB, BL, and LL), treatment should consist of daily dapsone (100 mg), daily clofazimine (50 mg), monthly rifampin (600 mg), and monthly clofazimine (300 mg), with the monthly doses given under supervision. This treatment is to be given until the patient becomes negative for bacilli or at least for 2 years.

The other extreme in the treatment of leprosy is based on individualization of care for each leprosy patient. This approach assumes adequate resources to deliver health care in general and comprehensive health care for leprosy patients in particular. The objective of this approach is prevention of deformity and disability in the individual patient. Because this approach assumes delivery of care under the personal guidance of a physician and assumes the availability of a variety of medical and surgical specialties, it is not cost-effective in terms of reducing the human reservoir of *M. leprae*. Clearly, it will reduce the chances for disability and deformity in the individual patient, however. The cost-effectiveness of preventing these deformities and disabilities depends on the balance between available resources and the value one places on deformities and disabilities of that individual. The relatively intense, and therefore costly, medical attention overshadows costs for drugs. Consequently, the individualized treatment regimens can be relatively expensive, relatively complicated, and can be administered indefinitely if felt advisable.

A more individualized drug regimen has been used for more than 20 years in the United States. Because of an increasing number of patients developing secondary dapsone-resistant disease in the 1960s and because of regularly monitoring newly diagnosed patients for drug sensitivities, it was felt advisable and safe to treat LL patients with rifampin (600 mg daily) and dapsone (100 mg daily), or clofazimine (100 mg daily) for dapsone-resistant patients for 3 years and then to discontinue the rifampin but to continue the dapsone (100 mg daily) or clofazimine (100 mg daily) in dapsone-resistant patients for life. This individualized approach was possible because the patient was followed as needed for the rest of his or her life for all relevant rehabilitation needs regarding extremity sensory losses, motor losses, monitoring for eye complications, etc.

The case is much different in immunologically competent TT patients. These patients tend to self-heal in the absence of any chemotherapy. As a practical matter, dapsone-resistant TT simply is not a significant problem. For this reason, an individualized treatment of TT would consist of dapsone monotherapy (100 mg daily) until a period of time after the patient's disease activity has disappeared. There is no need for continuing chemotherapy to prevent relapse because the absence of disease activity indicates the bacilli have been cleared, and continuing effective CMI will prevent disease from developing in the event that reinfection occurs.

Borderline patients followed on an individual basis will have treatment tailored to their immune status and the degree of disease progression in each case. In general, BL patients with bacterial loads and immunologic status approaching LL will be treated more like LL patients. BT patients will be treated somewhat longer but probably still with dapsone monotherapy like TT patients.

IV. Microbiology

A. Classification

Mycobacterium leprae is a microaerophilic, nonmotile, nonsporeforming, acid-fast staining actinomycete that usually forms slightly curved or straight rods. The organism is generally observed as pleomorphic rods with an average width of 0.25 to

0.30 μm and length of 2.1 μm. *M. leprae* stains gram-positive, but standard differential staining of the bacilli is accomplished by acid-fast staining with carbolfuchsin.

Electron microscopic studies of *M. leprae* reveal essential uniformity in ultrastructure with other members of the genus. The more prominent structures include cytoplasmic inclusions, mesosomes, and a highly complex lipid and carbohydrate-rich cell wall comprised of fibrous and band structures external to the cytoplasmic membrane.

B. Metabolism

Because *M. leprae* cannot be grown on artificial media, it has been necessary to study the metabolic characteristics of the bacilli derived from infected tissues of either patients with leprosy or, more recently, with bacilli obtained from experimentally infected athymic nude mice or armadillos. Limitations on both the availability of bacilli for study and the problems inherent in purifying and differentiating *M. leprae* enzymatic activity from that of host cells, has slowed the pace of obtaining a thorough understanding of the metabolic capabilities of *M. leprae*.

For energy generation, *M. leprae* oxidizes glucose primarily through the Embden-Meyerhof pathway with some activity present in the hexose monophosphate pathway. As with other mycobacteria, glycerol is also oxidized through these pathways. A major difference between *M. leprae* and other mycobacteria, however, is its ability to produce very high levels of the first enzyme of the hexose monophosphate pathway, glucose 6-phosphate dehydrogenase, suggesting a unique characteristic of energy production for *M. leprae*. [*See* MYCOBACTERIA.]

Demonstration of all enzymes of the tricarboxylic acid (TCA) cycle in *M. leprae*, plus the ability of intact bacilli to oxidize important intermediates in the TCA cycle to carbon dioxide and the demonstration of two potential regulatory mechanisms for the TCA cycle, suggests the presence and activity of this key metabolic pathway. Although not extensively characterized, an electron transport system would appear to be present in *M. leprae* because of detectable reduced nicotinamide adenine dinucleotide oxidase activity. Short-term maintenance of intracellular adenosine triphosphate (ATP) pools by *M. leprae in vitro* and demonstrated generation of ATP via *M. leprae*-derived adenylate kinase provide evidence that *M. leprae* can generate its own ATP. [*See* ATPASES AND ION CURRENTS.]

Demonstration of enzymes active in amino acid formation and degradation (e.g., glutamate decarboxylase, gamma-glutamyl transpeptidase, and 3,4-dihydroxyphenylalanine oxidase) provides evidence for amino acid metabolism in *M. leprae*. *M. leprae* can also incorporate amino acids into protein after uptake from the media; however, mechanisms for such activity are not understood. Nucleic acid metabolism in *M. leprae* is poorly characterized, but it is known that *in vivo*-grown *M. leprae* does not synthesize purines *de novo*. In addition, *M. leprae* exhibits mechanisms for uptake of purines and pyrimidines and has enzymes capable of interconverting purine bases. Very little is known regarding lipid metabolism except that acetate and palmitate can be incorporated into the phenolic glycolipid I molecule of *M. leprae*, depending on the culture conditions, and that *M. leprae* can rapidly oxidize palmitate to carbon dioxide.

C. Cell Wall Chemistry

Mycobacterial cell walls have been and continue to be a primary focus for biochemical and immunological research because of their chemical complexity and adjuvant-like properties. *M. leprae* shares this rich array of complex lipooligosaccharides, carbohydrates, glycolipids, glycopeptidolipids, and proteins with other mycobacteria but has some unique features that distinguish the species. For example, although sharing the meso-diaminopimelic acid substitutions in the peptidoglycan with other mycobacteria, *M. leprae* contains the unusual substitution of glycine for alanine in the tetrapeptide portion of the peptidoglycan. Covalently linked and external to the peptidoglycan is an arabinogalactan-mycolic acid complex constituting almost three fourths of the cell wall mass. Linear arrays of D-arabinose-D-galactose polymers with appendages of D-arabinose make up the arabinogalactan layer. Mycolic acids are esterified to terminal arabinose moieties of the arabinogalactan, contributing significantly to the hydrophobicity of *M. leprae* and other mycobacteria. In *M. leprae* these long-chain fatty acids form two groups (the alpha and beta mycolates) and can be used in differentiating *M. leprae* from other mycobacteria. [*See* CELL WALLS OF BACTERIA.]

Although the above components constitute the bulk of the cell wall of *M. leprae*, speculation as to the chemical nature of the so-called capsular-like material observed as an electron transparent zone in thin sections has focused on a different cell wall–

associated glycolipid, phenolic glycolipid-I (PGL-I). PGL-I is unique to *M. leprae* and has been found in large quantities in infected tissues; the sugar moiety of the molecule provides a strong immunogenic stimulus for immunoglobulin M antibody in LL. The basic chemical structure of PGL-I is a trisaccharide moiety composed of 3,6-di-*O*-methyl-beta-D-glucose (1 → 4), 2,3-di-*O*-methyl-alpha-L-rhamnose (1 → 2), 3-*O*-methyl-alpha-L-rhamnose linked to a phthiocerol lipid core through a phenolic group. The terminal sugar (3,6-di-*O*-methyl-beta-D-glucose) represents the immunodominant region of the molecule, and synthetic constructs of this moiety have been used to develop serologic tests to detect antibodies to PGL-I. Another group of antigenic cell wall–associated polysaccharides of *M. leprae* are the arabinose- and mannose-containing phosphorylated lipooligosaccharides, referred to as lipoarabinomannans. These molecules are highly immunogenic in *M. leprae* infections but are common immunogens within the genus, limiting their usefulness as diagnostic tools in serology.

Cell wall proteins of *M. leprae* fall into two basic categories: those loosely associated with the cell wall and removed by detergent washing and those tightly bound to the cell wall, resisting detergent solubilization. The best studied example from the former group is the 65-kDa protein or the *Escherichia coli* GroEL homolog. Implied in its sequence homology with GroEL is its function as a heat-shock protein. Immunoelectron microscopy with monoclonal antibodies to the 65-kDa protein localized it in both the cell wall and cytoplasmic compartments. This type of diverse localization of the 65-kDa protein supports the suggested role of "molecular chaperon," possibly transporting other proteins through the cytoplasmic membrane or associating with newly translated proteins undergoing molecular folding in various compartments of the cell.

Exemplifying the firmly bound proteins of the cell wall of *M. leprae* are the low-molecular-weight proteins of the so-called cell wall protein complex. These molecules have not been fully characterized chemically but do demonstrate a strong immunogenic potential, particularly toward T cells, and have been shown to induce protective immunity in mice.

D. Genetics

1. General Characteristics

Although phenotypic characteristics place the leprosy bacillus in the genus *Mycobacterium*, genotypic analyses have been less supportive of this taxonomic grouping. For example, the guanine-plus-cytosine (G+C) content and size of the *M. leprae* genome are significantly different from other mycobacteria. G+C for *M. leprae* is approximately 56%, with most other mycobacteria exhibiting greater than 60% G+C. The genome of *M. leprae* (2.2×10^9 daltons) is smaller than that of most other mycobacteria, which range in size from 2.8×10^9 to 4.5×10^9 daltons. Further supporting the evidence that *M. leprae* is widely divergent from other mycobacteria is the low degree of DNA homology between most species of mycobacteria and *M. leprae*, as judged by DNA binding studies using total genomic DNA. In one study, maximum binding of *M. leprae* DNA to heterologous sample DNA barely exceeded 10% of *M. leprae* : *M. leprae* DNA hybridization. In contrast to these types of analyses, taxonomic relationships between *M. leprae* and other mycobacteria as judged by comparative nucleic acid sequence analysis of 16S rRNA places the leprosy bacillus in the slowly growing subgroup of mycobacteria.

Intraspecies comparisons of individual *M. leprae* isolates have shown an incredibly high degree of homogeneity at the DNA level, as judged by restriction fragment length polymorphism (RFLP) analysis. Numerous human and animal isolates from different continents demonstrated this characteristic. One interpretation of these results is that unique evolutionary pressures acting on *M. leprae*, as a result of intracellular parasitism, are similar in human and animal populations, producing highly adapted organisms exhibiting minimal variation between infectious strains.

2. Recombinant DNA

The establishment of recombinant DNA libraries containing the entire *M. leprae* genome have revolutionized leprosy research in areas as diverse as immunology, biochemistry, and microbiology. For example, major contributions have been seen in identifying, cloning, and expressing major protein antigens of *M. leprae*. Various proteins have been tested for immunologic reactivity as well as identified functionally, albeit indirectly, via sequence homology with other prokaryotic and eukaryotic gene sequences. Two particularly well-studied proteins of *M. leprae* are the 65- and 70-kDa proteins. Both proteins share a high degree of homology with other prokaryotic heat shock proteins of similar size. [*See* RECOMBINANT DNA, BASIC PROCEDURES.]

Other genes cloned from *M. leprae* recombinant libraries have been shown to complement known enzymatic pathways in auxotrophic mutants of *E. coli*. For example, the gene for glt A (citrate synthase) and aro B (dehydroquinate synthetase) have been identified using this approach. This approach, although functional, has had limited success, possibly because of problems inherent in mycobacterial expression in *E. coli*. However, it is likely that future developments in recombinant vectors, designed to shuttle mycobacterial DNA between cultivable mycobacteria and *E. coli*, will overcome these problems and provide the tools for further study of the genetics of mycobacteria and particularly that of *M. leprae*. A spin-off of this type of research has been the development of recombinant bacteria possessing genes that express either defined antigenic proteins for potential vaccine delivery systems or segments of genomic DNA from pathogenic mycobacteria to be used in studying the nature of virulence. Both approaches hold great promise and should advance our understanding of protective immunity and the basis for pathogenesis in leprosy.

V. Immunology

A. Host–Parasite Interactions

The initiation, duration, and resolution of infectious diseases are determined by the properties of the infecting microorganism and the response of the infected host. Clinical symptoms and the course of leprosy are the by-product of complex interactions between two biological entities striving to maintain optimal conditions for survival.

Thus far, *M. leprae* has been grown only in living hosts, and because it is found primarily inside intact host cells, it is referred to as an obligate intracellular pathogen. The term "opportunistic pathogen" has not been used to describe *M. leprae*, primarily because leprosy patients are not immunocompromised individuals with increased incidence of opportunistic infections and cancer.

A parasite's ability to inflict damage on its host is referred to as virulence or pathogenic potential. The most successful pathogens minimize host tissue damage maintaining an optimal environment for proliferation increasing the likelihood of infecting other hosts. *M. leprae* appears to be highly efficient as a human pathogen because its infectivity is suspected to be relatively high but its virulence as measured by toxicity or mortality is very low.

B. Serology in Leprosy

It is well established that antibodies produced in response to an *M. leprae* infection are not protective. This is exemplified in the types of antibody responses observed in the polar forms of the disease, lepromatous and tuberculoid. In LL, copious amounts of anti-*M. leprae* immunoglobulins are seen in the absence of CMI. These antibodies recognize cross-reactive and species-specific proteins, carbohydrates, and glycolipid antigens. In contrast, patients with TT demonstrate strong manifestations of CMI reactivity (positive Mitsuda response) with minimal antibody response to *M. leprae*.

The discovery of the immunoreactive species-specific glycolipid, PGL-I, ushered in a decade of serologic studies around 1980, with the goal of developing a test for early diagnosis of leprosy. It was soon discovered that most patients with LL and a very low percentage of patients with TT produced antibodies to PGL-I. Contact studies have demonstrated only minimal increased risk of developing leprosy for individuals testing positive for PGL-I antibody. This kind of information, coupled with our current understanding of the dynamics of leprosy (i.e., very low prevalence and incidence even in many endemic areas) and the fact that approximately one half of new cases arise from the "noncontact" population, argues against the widespread use of serologic testing for diagnosis of leprosy.

C. Cell-Mediated Immunity in Leprosy

It is generally accepted that CMI constitutes the bulk of protective immunity against most intracellular pathogens, including *M. leprae*. This response in leprosy can be conceptualized as an interaction between T cells, antigen-presenting cells, and parasitized host cells, primarily macrophages and Schwann cells. Destruction and elimination of many intracellular bacteria require, at a minimum, antigen-reactive T cells that provide lymphokine-mediated activation of effector cells, resulting in the eventual death of the parasite. The initial T-cell signals are produced in response to major histocompatibility complex (MHC)-related antigen recognition by antigen-specific T cells. The magnitude and type of the resulting T-cell response play a major role in determining the outcome of the infection. Understanding the basic underlying immunologic mechanisms for protective immunity and pathogenesis in leprosy has focused research efforts in four major areas: (1) T-cell responses to *M. leprae* antigens, (2) T-cell anergy in LL, (3) genetic control of CMI in leprosy, and (4) the role of the macrophage in host resistance to *M. leprae*.

1. T-Cell Responses to *M. leprae* Antigens

During the past decade, immunochemical analysis of *M. leprae* strengthened our understanding of the antigenic nature of the bacillus and the immunologic response to *M. leprae* during infection. The major objectives in antigenic analysis of *M. leprae* have been to identify markers specific for the leprosy bacillus and to identify antigens capable of stimulating protective immunity in hopes of developing a vaccine to prevent disease. The former goal has been accomplished through the discovery of both protein epitopes and PGL-I described earlier. The latter goal has remained more elusive, primarily because of the lack of appropriate animal models for studying protective immunity and a minimal understanding of what constitutes protective immunity in leprosy.

Human and murine T-cell responses have been demonstrated to most of the proteins located from *M. leprae*. These include proteins with molecular masses of 7, 14, 16, 17, 18, 22, 30/31, 35, 36, 38, 65, and 70 kDa. Some of these proteins (e.g., 70 and 65 kDa) were selected as recombinant proteins using monoclonal antibodies raised to *M. leprae*, whereas others (e.g., 7, 17, and 28 kDa) were purified in native form from extracts of *M. leprae*. In most cases these proteins have been shown to stimulate antibody responses as well as lymphoproliferative responses in either leprosy patients, Bacille Calmette-Guerin (BCG)-vaccinated individuals, or contacts of leprosy patients.

Vaccine studies in mice using subcellular fractions of the bacillus have focused attention on the highly immunoreactive cell wall as the fraction containing protective epitopes. These fractions are devoid of all major carbohydrate and lipid components and contain primarily proteins and minor amounts of peptidoglycan. Thus far, the only purified protein demonstrating protection in mice is the 35-kDa protein, which was purified from extracts containing cell walls and given along with a potent adjuvant.

Developments in DNA cloning, protein expression, and peptide synthesis of *M. leprae* antigens have reduced the long-standing problem in leprosy immunochemistry of insufficient materials for study. Ironically, now researchers are inundated with highly purified antigenic moieties of *M. leprae* but find the methods for testing relevance to protective immunity underdeveloped.

2. T-Cell Anergy in Lepromatous Leprosy

It is well established that helper T-cell responses to *M. leprae* antigens are present in TT but absent in LL. This has been demonstrated both *in vitro* by lymphoproliferative responses of LL and TT patients and *in vivo* using the lepromin skin test. Comparative immunohistologic staining of LL and TT skin lesions also suggests a major difference in T-cell responses as reflected in the CD4/CD8 ratios of the respective granulomas. For example, although both TT and LL patients exhibit normal CD4/CD8 ratios (2 : 1) in peripheral blood, only TT patients mimic this ratio in skin lesions. In addition, the cells in the tuberculoid lesions are arranged in a distinct architecture, with CD4 cells in the center of epithelioid granulomas and CD8 cells on the periphery. Skin lesions from LL patients show mostly macrophages, with far fewer lymphocytes and a markedly lower ratio (0.5 : 1) of CD4/CD8 T cells with a random distribution of T-cell subsets.

A fundamental cause–effect question remains unanswered in leprosy. Is the specific T-cell anergy in LL the result of active suppression of the immune response induced by *M. leprae* antigens or is there a pre-existing condition of the individual who develops TT or LL on infection? Evidence is available to suggest that at least one *M. leprae* antigen, PGL-I, is capable of stimulating CD8 suppressor T cells in some LL patients as well exhibiting antiproliferative responses *in vitro*. Moreover, CD8-positive clones, obtained from LL lesions, have been shown to mediate MHC class II antigen-restricted suppression of CD4-positive clones after interacting with lepromin. [*See* T LYMPHOCYTES.]

3. Genetics in Leprosy

Genetic studies have yet to uncover direct linkages between human leukocyte antigen (HLA) genes and susceptibility to leprosy; however, associations have been reported between the incidence of different forms of leprosy and particular HLA haplotypes, suggesting a role for HLA selection in determining outcome of disease once infected with *M. leprae*. For example, individuals with the HLA-DR2 or DR3 haplotype may be predisposed to the TT form of leprosy, whereas the incidence of DR2-DQW1 may be increased in individuals with LL.

4. Macrophages and Leprosy

The macrophage is the central element in the host's defense against *M. leprae*. Immunologic dogma places the macrophage in both the afferent and efferent limbs of the immune response. Antigen processing, presentation, and monokine secretion are three major functions of the macrophage in the afferent stage of the developing immune response. In LL, all three activities have been shown to be altered in already infected macrophages, suggesting potentially important sites for disrupting the normal development of the immune response to the bacilli. For example, infected macrophages have been shown to be defective in their ability to present *M. leprae* antigens to sensitized T cells, and studies have also shown that macrophages may be defective in producing IL-1, an important monokine for T-cell division.

Normal efferent functions of macrophages culminate in killing intracellular *M. leprae*. This process is generally referred to as microbicidal activity and is a result of T-cell-mediated macrophage activation. While this appears to occur in TT, the infected and highly parasitized macrophage is conspicuously inept at killing and eliminating *M. leprae* in LL. Experimental data on macrophages from LL patients are contradictory, showing normal levels of oxidative metabolism in some patients and deficiencies in others, the latter result suggesting a concomitant deficiency in microbicidal capacity.

Experimental data in *M. leprae*–infected mice suggest that heavily infected macrophages become unresponsive to activation signals (e.g., gamma interferon) and that this unresponsiveness is paralleled by an increase in macrophage production of prostaglandin E_2, a molecule with potent immunosuppressive characteristics. These data suggest that activation of heavily infected macrophages to kill and clear *M. leprae* is unlikely. Rather, activation of newly immigrating mononuclear phagocytes may be necessary to destroy and remove *M. leprae* in skin lesions. Human studies administering lymphokines (e.g., IL-2, IFN-γ) either locally or systemically underscore the effect of local immunologically mediated change in LL lesions. For example, administration of IL-2 to LL patients induced local upgrading reactions accompanied by mononuclear cell infiltration and evidence of bacterial clearance.

5. Vaccines

Early vaccine trials in leprosy relied on cross-reactive immunity induced with the BCG strain of *Mycobacterium bovis*. Previous studies using this live, attenuated vaccine for protection in tuberculosis had shown variable levels of protection in different tuberculosis endemic populations. Protective efficacy of BCG in leprosy was not unlike that observed in tuberculosis vaccine studies, with levels of protection varying from 20% to 80%. For example, in Uganda, BCG vaccination reduced significantly the incidence of TT in children, but data available for measuring efficacy in LL was insufficient to observe an effect. In Burma and New Guinea, less impressive protective effects (34% and 46%, respectively) were reported, and marked variation between groups within both trials were apparent, suggesting other uncontrolled variables operative within populations affecting the outcome of vaccine efficacy.

A newer approach for a whole bacteria vaccine in leprosy has extended the above approach by adding heat-killed *M. leprae* to the standard BCG vaccine. This type of immunostimulatory mixture is also being tested for its ability to reconstitute immunologic reactivity in anergic patients as a form of immunotherapy.

Finally, recombinant vaccines expressing various *M. leprae* protein antigens are under development. Viral and bacterial vectors are being studied; however, it is unclear at this time which antigens are appropriate for stimulating protective immunity and what type(s) of vector(s) will be optimal for delivery of the appropriate stimulus to the immune system.

Bibliography

Bryceson, A., and Pfaltzgraff, R. E., eds. (1990). "Leprosy," 3rd ed. Churchill Livingstone, London.

Gaylord, H., and Brennan, P. J. (1986). Leprosy: Antigens and host parasite interactions. *In* "Parasite Antigens: Toward New Strategies for Vaccines" (T. W. Pearson, ed.), pp. 49–89. Marcel Dekker, New York.

Hastings, R. C., ed. (1985). "Leprosy." Churchill Livingstone, London.

Hastings, R. C., Gillis, T. P., Krahenbuhl, J. L., and Franzblau, S. G. (1988). *Clin. Microbiol. Rev.* **1**(3), 330–348.

Ratledge, C., Stanford, J., and Grange, J. M., eds. (1982, 1983, 1989). "The Biology of Mycobacteria," vols. I, II, and III. Academic Press, London.

Rees, R. J. W., ed. (1988). "Tuberculosis and Leprosy." Churchill Livingstone, London.

Leucine/Lrp Regulon

Elaine Newman and Rongtuan Lin
Concordia University

Richard D'Ari
Institut Jacques Monod

Glossary

CAP Cyclic AMP receptor protein, also known as catabolite gene activator protein; a DNA-binding protein that regulates transcription of many genes involved in carbohydrate metabolism and is known to bend DNA

C-1 units Single carbon units, carried on tetrahydrofolic acid and used in biosynthesis of many cell components

Fnr Transcriptional regulator of the expression of anaerobic respiratory function in *Escherichia coli*

lrp Gene that codes for Lrp, previously known as *rbl, oppI, ihb,* and *livR*

Lrp Leucine-responsive regulatory protein; a DNA-binding protein that regulates transcription of genes of the leucine regulon

LysR Positive regulatory factor controlling expression of *lysA;* prototype for a large class of prokaryote transcription factors

TFIID TATA-box recognition factor; a eukaryote general transcription factor

A SURPRISING NUMBER of *Escherichia coli* enzymes have been shown to be induced by the addition of L-leucine to the medium in which the cells are grown. Recently this response has been shown to be mediated by an interaction of leucine and a DNA-binding protein known as leucine-responsive regulatory protein or Lrp, encoded by the *lrp* gene. The group of genes regulated in this way is known as the leucine/Lrp regulon.

I. Leucine/Lrp Regulon: A New Global Response

Escherichia coli responds rapidly to environmental changes, synthesizing new proteins, ceasing to synthesize others, and even inactivating some existing proteins. The transcriptional patterns change quickly, accomplishing a remarkable adjustment brought about by monitoring and reacting to environmental signals. From a relatively small number of key signals, the organism deduces the state of the world that surrounds it at any instant, usually very successfully.

These environmental responses require coordinated changes in the cell's machinery, which it brings about by coordinated changes in the rate of transcription of a number of genes. In a typical system, a single protein, called a global regulator, controls the transcription rate of several genes. Changes in that protein will then bring about a response of all these coregulated genes.

Examples of well-studied global responses include those triggered by starvation (for nitrogen, carbon, or phosphate), by DNA damage (the SOS regulon), by changes in temperature (the heat shock regulon), and by changes in oxygen tension (*oxyR* regulon). In all these cases, a change in the environment affects a general regulator, which then alters transcription rates.

A recently discovered global response in *E. coli* is the leucine regulon, regulated by the leucine-responsive regulatory protein (Lrp), which controls transcription from a number of genes (Table I), usually—but not always—in response to exogenous L-leucine. Because the regulon is not limited to regulation by leucine as was originally thought, we are now giving it a new name—the leucine/Lrp regulon—at least until further information suggests a more descriptive name.

Although only recently recognized as a general regulator of metabolism, Lrp has been known for some time under a variety of names. As *ilvIH*-binding protein, it was described as the regulator of synthesis from *ilvIH*. As LivR, it regulated the transport of branched-chain amino acids. As OppI, it controlled oligopeptide permease. As R*bl* (*r*egulation *b*y *l*eucine), it was described as the regulator of the leucine regulon. All these are now considered to be different names for a single gene product, Lrp. There is, therefore, a great deal more information available about Lrp than one would expect from the recent date at which it was realized that Lrp governs a large regulon.

The leucine/Lrp regulon is of interest because of the diversity and large number of functions it comprises, the surprising number of different mechanisms it uses at different promoters, and the fact that it responds to an amino acid that is otherwise considered to be rather metabolically inert in *E. coli*.

II. Genes Known to Be Regulated by Lrp

L-Leucine has been known to regulate the synthesis of many enzymes. In one of the earliest examples, Pardee and Prestidge in 1955 showed that leucine induced an L-serine deaminase in *E. coli*, whereas L-serine did not, with the paradoxical effect that the extent of serine degradation depended on the availability of L-leucine.

Table I Identified Genes Known To Be Regulated by Lrp

		Effect of		
Gene[a]	Activity encoded	LRP[b]	Leucine[c]	LRP + leucine[d]
Biosynthesis				
ilvIH	Acetolactate synthetase III	A	decr	
?	Leucine biosynthesis	A	NT[e]	
serA	Phosphoglycerate dehydrogenase	A	decr	
gcv	Glycine cleavage enzymes	A	none	
lysU	Lysyl-tRNA synthetase II	R	incr	
Uptake				
oppA–D	Oligopeptide permease	R	incr	
livJK	Branched chain amino acid transport	None	none	R
Degradation				
sdaA	L-Serine deaminase I	R	incr	
tdh	Threonine dehydrogenase	R	incr	
Regulation				
lrp	Leucine-responsive protein	R	none	
?	Gene regulating *sdaB*	None	none	A

[a] Genes identified as Lrp targets are listed according to their putative metabolic categories.
[b] The effect of Lrp on gene transcription as judged by the effects of an *lrp* insertion mutation is listed as A (activator) or R (repressor).
[c] The effect of leucine is given as decr (decrease) and incr (increase).
[d] When neither Lrp nor leucine alone has an effect, the effect of both together is given as A or R.
[e] NT, not testable in absence of L-leucine.

One would of course expect L-leucine to be involved in regulating its own biosynthesis. However, it is also involved in regulating synthesis from many other genes, both increasing and decreasing gene expression. It is this general regulatory role that is mediated through its interaction with Lrp. In the following sections we list the enzymes known to be regulated by Lrp, classifying them by whether they are decreased, increased, or unaffected by leucine.

III. Enzyme Activities Decreased by L-Leucine

1. Acetohydroxy acid synthetase III (*ilvIH*) is one of three isoenzymes synthesized by *E. coli* that catalyze the first step in L-isoleucine and L-valine biosynthesis.
2. Phosphoglycerate dehydrogenase (*serA*) catalyzes the NAD-linked dehydrogenation of phosphoglycerate to phosphohydroxypyruvate, the first step in L-serine biosynthesis.
3. The *livJK* gene product catalyzes the transport of branched-chain amino acids into the cell. Synthesis of this permease is decreased in the presence of L-leucine.

IV. Enzyme Activities Increased by L-Leucine

1. L-Serine deaminase (*sdaA*) converts L-serine to pyruvate.
2. L-Threonine dehydrogenase (*tdh*) converts threonine to γ-amino-α-ketobutyric acid, the first step in the synthesis of glycine from L-threonine.
3. Lysyl-tRNA synthetase II (*lysU*) activates L-lysine and condenses it with lysyl-tRNA, forming lysyl-tRNA for protein synthesis. L-lysine is the only amino acid for which two aminoacyl-tRNA synthetases are known.
4. *oppA–D* and *tppA,B* code for two different systems for tripeptide uptake by *E. coli*. *oppA–D* have been shown to be regulated by L-leucine and Lrp; *tppB* is induced by L-leucine, but regulation by Lrp has not yet been tested.
5. Synthesis of one of the serine transport systems is also induced by L-leucine.

V. Enzymes Controlled by Lrp but Apparently Not in Conjunction with L-Leucine

1. Lrp controls synthesis from genes involved in L-leucine biosynthesis, and mapping in the leucine biosynthetic operon at 2 min. Regulation of the leucine operon is complex, and both leucine and Lrp affect it, but independently. Leucine limitation increases expression from these genes. However, this effect does not depend on Lrp.
2. The glycine-cleavage system, which converts glycine to C1-THF, carbon dioxide, and ammonia using a series of four enzymes, is totally dependent on Lrp for synthesis, and L-leucine has no effect on transcription.
3. Lrp itself is autoregulated, that is, Lrp protein decreases transcription from *lrp*; again L-leucine has no effect.

VI. Leucine Regulon Includes a Large Number of Genes

The number of genes that are regulated by Lrp has been estimated at about 30 by a comparison of two-dimensional gels of proteins extracted from an *lrp* mutant and its parent. A more sensitive estimate based on isolation of *lrp*-regulated λp*lac*Mu insertions (discussed subsequently) suggests that the number may be much higher, and that the extent of regulation can vary enormously from 2–50 fold.

The amount of Lrp was estimated by Western blot analysis at about 0.1% of the total protein of cells grown in minimal medium, or 3000 Lrp dimers per cell. This is similar to the number of catabolite gene activator protein (CAP) molecules per cell, but much higher than that of nonglobal regulators such as LacI, DsdC, and AraC, which are present at less than 100 molecules per cell.

VII. Possible Metabolic Rationale for the Leucine Regulon

For several *E. coli* global response regulons, the metabolic function is clear, because the products of the co-regulated genes are obviously related in their function. Examples of this are those operons involved in heat shock, DNA repair, detoxification of peroxides, phosphate starvation, and nitrogen starvation.

The metabolic rationale of the leucine regulon is not nearly so clear. We know that L-leucine is a signal to the transcription system, but not what is signaled. As can be seen in Table I, the reactions controlled by the regulon do not fit into an obvious metabolic pattern. Lrp controls expression of genes involved in amino acid biosynthesis (*serA, ilvIH,* and *gcv*), amino acid and peptide transport (*livJ, livK, oppABCDF,* and the gene coding for a serine transporter), amino acid degradation (*sdaA* and *tdh*), and amino acyl–tRNA synthesis (*lysU*).

A consideration of the type of regulation of each operon on this list suggests that the regulon may be involved in integrating biosynthetic and degradative activities of the cell. Thus Lrp in the presence of L-leucine would decrease expression of genes involved in the degradation of exogenous compounds and increase expression of genes involved in the biosynthesis of compounds starting from simple intermediates such as pyruvate. Lrp would then represent a mechanism for controlling a switch from rich intestinal-type environments to poorer environments of the laboratory and outside world.

Autoregulation of synthesis of Lrp and similarly regulated repressors provides a mechanism to maintain a nearly constant level of gene product. In Luria broth (LB), an Lrp-independent mechanism leads to a large decrease in Lrp production, suggesting that Lrp is less important in rich medium or works only on high affinity promoters. The full effects of Lrp would then be seen only in minimal medium, that is, Lrp is needed to organize metabolism in a way that depends on biosynthesis and not on degradation. Although this seems likely, it should be noted that the enzymes and genes studied thus far were chosen according to the interests of the investigators. Identifying the many other genes of the regulon may change this picture.

VIII. Lrp Gene as Major Regulator of Single-Carbon Metabolism

Whatever its other metabolic roles may be, the Lrp has a profound effect on one-carbon (C-1) metabolism. In *E. coli* grown in glucose-minimal medium, C-1 units are derived from serine, either by serine hydroxymethyltransferase (SHMT, the *glyA* gene product), with the concomitant synthesis of glycine, or by glycine cleavage by the *gcv* enzymes. Lrp is absolutely required for *gcv* transcription. Consequently, in an *lrp* mutant, C-1 metabolism is greatly altered and SHMT is responsible for all synthesis of both glycine and C-1. The *lrp* mutant's absolute dependence on SHMT is indicated by the fact that an *lrp glyA* mutant cannot grow in glucose-minimal medium with glycine unless the endproducts of C-1 metabolism are provided, whereas the parent *lrp*$^+$ *glyA* strain can derive all its C-1 from glycine.

IX. Surprising Variety in the Mechanisms by Which Lrp Interacts with Its Target Promoters

It is well established that a number of important *E. coli* regulatory proteins, CAP for example, activate transcription of some operons and decrease transcription of others. In the case of CAP, both activation and repression require that CAP interact with its effector, cyclic AMP.

Lrp, like CAP, activates expression of some operons and decreases transcription of others. What is unusual for Lrp is the complexity of its mechanisms. Six patterns of interaction between L-leucine, Lrp, and promoter regions of DNA have been described (Table II).

Lrp increases or decreases gene expression using a mechanism sometimes involving L-leucine and sometimes not. When L-leucine is involved, it may be required for Lrp to act or it may be antagonistic to it. For both increased or decreased expression, there are three types of interaction known—Lrp acts alone, Lrp action requires L-leucine, or Lrp action is reversed by L-leucine.

Table II Classes of Interactions at Lrp-Regulated Promoters

Effect of lrp	Prototype
Increases expression	
1. leucine reduces effect	*ilvIH*
2. leucine is needed for effect	*sdaB* regulatory gene?
3. leucine has no effect	*gcv*
Decreases expression	
4. leucine reduces effect	*sdaA*
5. leucine is needed for effect	*livJK*
6. leucine has no effect	*lrp*

For example, in the case most studied *in vitro*, Lrp activates transcription from the *ilvIH* promoter and L-leucine lowers expression by interfering with the action of Lrp. On the other hand, Lrp decreases expression from the *sdaA* promoter and L-leucine induces expression of this operon, presumably by interfering with the repressive action of Lrp.

The most surprising finding is that, in some cases, Lrp has a very large effect on transcription but L-leucine has little or no effect (Classes 3,6). This is true of the *gcv* operon (Class 3) which, as described earlier, is not transcribed to a physiologically significant level in *lrp* mutants. It is also true of the *lrp* gene itself, for which transcription is reduced by Lrp (Class 6). L-Leucine does not significantly affect transcription of either *gcv* or *lrp*. Most screens used to identify genes of the leucine regulon were not designed to find Class 3 and 6. Now that their existence is recognized, more genes in these classes should be identified.

X. Characteristics of the Lrp Protein

The Lrp protein is a homodimer, with a monomer molecular mass of 19,000. It is encoded by the gene now known as *lrp*, which has been sequenced in two laboratories—once as *oppI* and once as *lrp*—producing identical sequences. The protein has been purified to over 96% purity; its 38 N-terminal amino acids have been determined, and agree perfectly with the sequence predicted from the nucleic acid sequence.

The amino acid sequence and the protein sequence have been examined by computer analysis. The *lrp* sequence bears very little resemblance to any of the known bacterial global regulators. In particular, it is quite different from the LysR family, from CAP, and from Fnr. The amino acid sequence of Lrp shows 25% identity with AsnC, an activator protein that activates transcription of the neighboring *asnA* gene coding for one of the two asparagine synthetases. This similarity extends throughout the amino acid sequence. It is difficult to know whether this similarity has any physiological significance, since the effects of AsnC that have been identified so far indicate it to be a local regulator rather than a global one.

Because of prior work with Lrp under different names, a considerable number of mutations affecting Lrp have been isolated, so that structure/function information should be available soon.

XI. *In Vitro* Studies on Lrp Interactions with Its Promoters

In gel retardation studies using purified Lrp, binding has been shown upstream of *ilvIH, lysU, sdaA, serA, gcv,* and *glyA*. Lrp, studied as the *ilvIH*-binding protein, binds to two sites upstream of *ilvIH,* from −190 to −260 and from −100 to −40 relative to the start of transcription. The fact that this binding was specifically decreased by L-leucine suggests that these are the physiologically relevant sites. The *ilvIH* promoter is currently under intensive study.

The structure of the DNA upstream of *lysU* and *sdaA* (Class 4, Table II) seems to be somewhat different from that upstream of *serA* (Class 1, Table II), and all are different from that of *ilvIH*. In the first three genes, Lrp protects a long area against DNase I—88, 75, and 94 bp, respectively. For *sdaA*, two binding sites have been differentiated, a high affinity region and a much lower affinity site near the start of the coding region.

In gel retardation studies, a reaction between Lrp and L-leucine can be demonstrated. Lrp is removed from *sdaA* or *lysU* DNA by 8–16 m*M* leucine and by somewhat higher concentrations of alanine and isoleucine. Binding at *serA* is complex. At low Lrp concentration, only one retarded band is seen. At higher protein concentrations, that band disappears and four retarded bands are seen. Addition of L-leucine at high protein concentrations removes protein entirely from some molecules and changes the pattern of retarded bands. The two most retarded bands are lost, and a new band slightly less retarded than these two is seen.

XII. Is There a Concensus Sequence for Lrp Binding?

The upstream region of a considerable number of Lrp-regulated genes has already been sequenced, so one would expect to be able to identify a consensus sequence for Lrp binding. Such a sequence has recently been suggested: TTTATTCtNaAT. However, this is an extremely widespread sequence in both procaryotic and eucaryotic genes, and falls in the coding region of other *E. coli* genes such as *fhlA* and *fiaA*. No other consensus sequence has been identified so far.

Although this sequence may not be a consensus site for Lrp binding, it could indicate that Lrp function is based on DNA flexibility, and that this putative binding site is one of several readily bent sequences upstream of Lrp target genes. Looping or coiling of upstream DNA is suggested by the repeating areas of DNA hypersensitive to DNaseI seen on all footprints studied so far. It seems, then, that Lrp might act as a bending protein as part of its mechanism for regulating specific operons.

XIII. Possible Role of Lrp in Chromosome Folding

The first Lrp-regulated genes map in scattered fashion through the *E. coli* genome. Those studied so far show binding to an unusually long stretch of DNA—some 85 bp compared with the 26–50 protected against DNase I by CAP at its various sites and the 37–44 protected by AraC and integration host factor (IHF). This may indicate that Lrp loops the DNA extensively or that several molecules of Lrp bind at each operator. It may also mean that, although Lrp binds at this long stretch when tested alone, in the cell it interacts with a variety of other proteins binding in the same area.

Many regulatory proteins bring about substantial bending of DNA. CAP has been shown to bend DNA by 90° in crystals of CAP with a 30-bp fragment of DNA. If Lrp binds at 100 promoters, and bends the DNA at each site, one would expect this to have a considerable effect on the overall folding of DNA in the cell. This could be true of any of the global regulators, including CAP. If these regulators bend DNA in a concerted fashion, then the chromosomal structure of DNA in *E. coli* might be partially determined by the binding of several general regulators.

In keeping with the suggestion that TFIID and the dnaB gene product both form structures for the accretion of the other proteins of an initiation complex, we suggest that Lrp, CAP, and similar proteins, in addition to their other more specific functions, are organizers of DNA structure and determine the specificity for folding DNA in the cell. Then nonspecific proteins like HU fill in the spaces made accessible by the DNA organizers. Further, the DNA folding pattern may vary according to the environment if binding of the organizers also varies.

This implies that, during evolution, *E. coli* has taken advantage of proteins attached at intervals on the DNA and used them for efficient packaging of DNA in the cell. This use of proteins for purposes other than those for which the protein first evolved is similar to the use of the arginine repressor, coded by *argR*, as the regulator of arginine biosynthesis and as part of the *cer* recombination system of colE1.

Bibliography

Lin, R. T., D'Ari, R., and Newman, E. B. (1990). *J. Bacteriol.* **172,** 4529–4535.

Platko, J. V., Willins, D. A., and Calvo, J. M. (1990). *J. Bacteriol.* **172,** 4563–4570.

Schultz, S. C., Shields, G. C., and Steitz, T. A. (1991). *Science* **253,** 1001–1007.

Stirling, C. J., Szatmari, G., Stewart, G., Smith, M., and Sharratt, D. (1988). *EMBO J.* **7,** 4389–4395.

Willins, D. A., Ryan, C. W., Platko, J. V., and Calvo, J. M. (1991). *J. Biol. Chem.* **266,** 10768–19774.

Low-Nutrient Environments

Richard Y. Morita
Oregon State University

Glossary

Energy charge Equals [(ATP) + 1/2(ADP)]/[(ATP) + (ADP) + (AMP)]; also called adenylate energy charge

Starvation-survival Physiological state of the microbial cell that results from insufficient nutrients (energy) for cell growth (size increase) and reproduction

Syntrophy Ecological relationship in which organisms provide nourishment for each other

LOW-NUTRIENT ENVIRONMENTS, termed oligotrophic environments, mainly lack organic matter for the growth of heterotrophic bacteria. The lack of other nutrients such as phosphate or nitrogen (ammonium, nitrate, or nitrate) can also limit heterotrophic microbial growth in such ecosystems. Because organic matter is the energy source for growth of heterotrophic bacteria, this article will concentrate on the lack of bioavailable organic matter (energy) in ecosystems as it affects microbes.

I. Oligotrophic Environments

Most ecosystems have oligotrophic components. Even in eutrophic ecosystems, some physiological types of bacteria that are in dire need of specific nutrients and/or energy will be present. Oligotrophic environments result from situations where plant life is limited; thereby, production of organic matter for heterotrophic bacteria is also limited. However, most of the oligotrophic nature of ecosystems is due to the presence of various physiological types of bacteria decomposing the present organic matter to a point where little or no organic matter is left. If sufficient organic matter (energy) is present, the rate of decomposition is rapid, provided other conditions are favorable. [*See* Organic Matter, Decomposition.]

A. Levels of Organic Matter in Soil and Aquatic Systems

Each ecosystem has its own level of organic matter, but average values are 25 mg organic C per gram for a good loam and 10 mg organic C per liter for freshwater systems and the ocean. Nutrient broth contains 5000 mg of peptone and 3000 mg of yeast extract per liter. Thus, it can be seen that laboratory media contain organic matter far in excess of that found in most systems. Most of the organic C in nutrient broth represents a readily available energy source, whereas most of the organic C found in natural ecosystems are not available to the indigenous microorganism.

In some ecosystems, the main energy source for the production of microbial cells is based mainly on sulfur and methane. As a result, the first trophic level in these ecosystems is the chemolithotrophic bacteria.

B. Measurement of Activity

Thus far, there has not been any accurate and dependable technique to measure microbial activity or microbial productivity in natural systems. The use of radioactive material added in small amounts to determine heterotrophic activity of bacteria in an ecosystem has been employed, but the more proper name should be heterotrophic potential, mainly because a specific, readily utilizable substrate has been added. In some instances, the added amount of substrate may be more than the amount already present initially, or just a little more of the substrate may greatly increase the system's activity. Incorporation of a specific substrate (e.g., thymidine incorporation, leucine incorporation) has been employed extensively, but again we are adding an extra energy source, Thymidine can be degraded to furnish energy, but not all bacteria will take up thymidine. Some evidence indicates that an added amount of labile organic matter provides the cells with sufficient energy to enhance the accessibility of dissolved recalcitrant organic C. Perturbation of soil samples will also increase activity. Thus, any method that employs the addition of an energy-yielding compound, especially labile ones, should be viewed as suspect as a measure of microbial activity or productivity.

C. Availability of Energy

Most of the organic matter found in ecosystems is recalcitrant. Infrared analysis of the carbon dioxide formed by the complete combustion of organic C is used as a measure of the total organic C in samples. Organic C is measured in samples from various environments, but this measurement does not indicate how much of the organic C is available for the indigenous microbes. However, a correlation probably exists between the total organic C in environmental samples and the amount available for microorganisms. The best measure of bioavailable energy in any environmental samples is still the biochemical oxygen demand, formerly called the biological oxygen demand, because it measures the amount of energy available to the indigenous microbes present, as indicated by the amount of oxygen utilized. Although one can chemically measure the amount of various labile compounds in environmental samples, it does not necessarily represent the amount that is available to microorganisms in the samples.

Chlorophyll-driven photosynthesis is the main process for the synthesis of organic matter on earth. One of the main products of photosynthesis is cellulose. Not all heterotrophic bacteria produce cellulase or can utilize other products formed during photosynthesis. Thus, syntrophy is the main mechanism by which most bacteria in nature receive the nutrients they need for growth and reproduction. As a result, within any ecosystem, there are active and inactive microbial forms. Growth of microbes, at best, is sporadic in nature and nutrient availability is the main factor limiting growth. One of the best examples of this sporadic growth is the growth of the endolithic micoorganisms in the Ross Desert of Antarctica, where active metabolism is limited to a few hundred hours each year. Natural growth is slow under such conditions. [*See* CELLULASES.]

There are many reasons why the organic matter in soil is not available for the indigenous microorganisms. However, the main limitation may be the presence of liquid water. The more labile compounds resulting from plant debris are readily utilized, leaving the more recalcitrant compounds such as humic substances, tannins, and lignins. Much of the decomposition of plant material above the surface of the soil occurs in the litter zone. Many of the more labile compounds are rendered unavailable, being complexed to lignins, phenolics, etc., while others may be adsorbed to clay minerals. Tannins and phenolic compounds are known to inhibit many enzymes, including cellulases, proteases,and lipases. The litter zone should not be considered a part of the soil system. Various fractions of plant debris in soil (e.g., phenols) can remain in soil for many years. Soils help protect organic matter from microbial decomposition. This is important because organic matter aids the water-holding capacity and workability of the soil. Thus, soil has been termed "grossly oligotrophic." The term oligotrophic is not used by soil microbiologists in the United States, but the term is applicable.

The same mechanisms active in making organic matter unavailable to the microorganisms in soil apply to the aquatic environment. In addition, a large amount of organic matter enters the nearshore environment due to freshwater runoff, which carries much recalcitrant organic C with it. Approximately 0.4% of the organic matter synthesized in the photic zone of the ocean enters the deep sea as dissolved organic C, and the age of this organic matter has been estimated to be 3,400 yr. It has been estimated that only 10–30% of the organic C in the ocean is available for biological decomposition; the remainder is regarded as recalcitrant. The oxygen utilization rate in the deep ocean, by all organisms, has been calculated to be 0.004 (Pacific Ocean) and 0.002 (Atlantic Ocean) ml/liter/yr. When this is coupled with the residence time of water masses, which may be over a millennium, one realizes the slow metabolic rates taking place in the deep sea. If sufficient organic matter were present in the deep sea, then it would become anoxic like the Black Sea. Fortunately, the recalcitrant nature of the organic matter to the microorganisms in the deep sea leaves energy for all the other deep-sea organisms, a blessing in disguise. In some ecosystems where specific amino acids have been measured, the data show that only a small fraction of the total amount of amino acids are available to the microbes. As mentioned previously, what you can measure chemically does not necessarily mean that all of it is bioavailable to the microorganisms.

When attempts were made to grow marine microbes in a chemostat employing an organic C concentration equal to the marine waters, the microbes were washed out of the chemostat. This is another indication of the oligotrophic nature of the marine environment.

II. Oligotrophic, Low-Nutrient, and Copiotrophic Bacteria

Oligotrophic bacteria are those organisms that grow in low nutrient media with 1–15 mg C/liter or 10–50 mg C/liter, depending on whose value is employed. Oligotrophic bacteria have been divided into two categories: facultative and obligate. The facultative oligotrophs are capable of being adapted to grow at higher concentrations than the definition permits, whereas the obligate oligotrophs cannot be adapted to grow at higher organic concentrations. Although there are indications that obligate oligotrophic bacteria exist, there is no unequivocal proof of their existence. Some investigators prefer that the term "oligotrophic" would refer to the environment rather than to a specific group of bacteria. However, the Japanese investigators in this field have modified it to designate those bacteria capable of growing on medium containing <1 mg of organic C per liter. To prove that an obligate oligotroph exists, the organism must be tested against a variety of and combinations of substrates, and each of these under different pH values and Eh values. Various combinations of all these factors must be checked out. As it now stands, only one medium or substrate is usually employed and only at a given pH value and temperature under aerobic conditions. The concept that no one medium can satisfy the nutrient requirements for all bacteria seems to have been forgotten. Thus, proving that an oligotroph is an obligate oligotroph involves much work and time. Although it has been demonstrated that low nutrient media produces more colonies on an agar plate than does rich media, it should be recognized that this method entails a form of resuscitation for starved bacteria present. Thus, the statement that a larger number of colonies on an agar plate represents the large number of oligotrophs in a system must be seriously questioned.

Copiotrophic bacteria are those that grow readily in media that are 100 times richer in nutrients than the environment from which the organism was isolated. Because most ecosystems on earth are oligotrophic, most bacteria are capable of better growth at an organic concentration greater than the environment from which they were isolated. Just where the cutoff should be in terms of the organic concentration becomes a matter of debate, especially in light of the fact that few investigators will measure the organic C in samples from which they make their isolations. If organic C is measured, should one use the total organic C or the bioavailable organic C to the organism to determine whether or not it is a copiotroph? In addition, one must also prove that the organism is capable of growing and multiplying in the natural sample without the addition of added nutrients. Also, the "bottle," or "surface," effect must be ruled out when a determination of growth in a natural sample is made. If isolated on a medium and then placed in the filter-sterilized natural sample, fragmentation could result when the cells are placed in a starvation situation, which could also give erroneous results. Researchers have demonstrated that increased cell numbers (without biomass increase) can result also when artificial seawater is supplemented with an amino acid concentration equal to that found in nearshore waters, or even 10 times the natural amino acid concentration.

The definition of low nutrient bacteria remains vague. The low nutrient bacteria were isolated from seawater agar plates to which no added energy source was introduced. However, there is organic matter in regular Difco agar as well as in other types of agar.

It is known that bacteria isolated from lake water can grow in the laboratory without the addition of organic matter. In this case, the amount of organic matter and ammonium that is absorbed from the laboratory air is sufficient to permit the organisms to grow.

III. Starvation-Survival in Oligotrophic Environments

Because there is a lack of energy in most ecosystems, the normal state of most of the bacteria present is the starvation mode. A continually active microbial biomass in soil has never been shown, and there is a growing realization that many of the microorganisms isolated from soil by conventional means were present in a dormant state. Considerable attention has been paid to these so-called active microorganisms. The calculation (approximation) in soil (nonagricultural) on the turnover times of microbial biomass-C ranges widely, depending on the soil in question. Nevertheless, calculations between 5.5 times per year (generation time of 66.36 days) and 0.4 times per year (generation time of 912.5 days) are common. Extrapolation of the data indicate that maintenance energy demand surpasses values known for average C input to soils, if maintenance values are derived from pure culture energy of maintenance studies.

If insufficient energy is the general rule in ecosystems, then microbes in oligotrophic environments must be in a starvation mode, and there must be mechanisms whereby nonspore-forming microbes can survive these conditions. Lack of other nutrients besides energy may lead to radically different starvation states. Recognizing that microbes have evolved ca. 2.3 billion yr longer than other forms, undoubtedly physiological mechanisms have been developed by the microbes to survive this situation. Unfortunately, microbiologists do not deal with the Darwin's "survival of the species" nor "survival of the fittest." The term "starvation-survival" has been defined as a physiological state resulting from an insufficient amount of energy for growth (increase in cellular biomass) and reproduction. Because growth in nature depends largely on syntrophy, there may be instances where the amount of energy-yielding nutrients are insufficient for the cell to reproduce or even for the cell to increase in biomass. Within the past decade, there has been interest among microbiologists to study starvation-survival. However, it should be noted that these studies generally entail growth of microorganisms in nutrient-rich laboratory media (gluttonous bacteria) before the onset of the physiological process involved when the cells are placed in a starvation menstruum. Although attempts have been made to grow microbes in a chemostat employing natural seawater concentrations of organic matter, they have failed because the cells are washed out of the chemostat due to their slow rate of division. A good approximation to a chemostat with the dilution rate of 200 days would be a batch culture stored for 100 days and studied for a month. When grown in the laboratory, harvested by centrifugation, washed twice, and then placed in a menstruum containing no energy source, cells will undergo the starvation-survival process. There appears to be four patterns of starvation-survival when cells are placed in a starvation-survival menstruum, as illustrated in Fig. 1. Curves A, C, and D of Fig. 1 were conducted on freshly isolated bacteria from the marine environment, whereas curve B illustrates the pattern of starvation-survival of *Nitrosomonas cryotolerans*, a chemolithotrophic bacterium. Organisms that fit the pattern illustrated by curve C have been studied the most.

Figure 2 gives the total cells (acridine orange direct count [AODC]), viable cells (CFUs), and optical density of Ant-300 cells grown at various dilution rates ($D = 0.015$, doubling time = 46.2 hr; $D = 0.057$, doubling time = 12.2 hr; $D = 0.170$, doubling time = 4.1 hr) and batch culture with starvation time. These cells were harvested by centrifugation, washed several times, and then placed in artificial seawater containing no energy source. The concentration of cells for the starvation-survival process was 3×10^7 cells/ml, and after the end of the third stage in all cases, the viable cell count was ca. 10^5/ml. This latter value represents approximately 0.3% of the initial inoculum. The viable counts fall off after the number of cells has increased to its peak level, whereas the total cells remain level after its peak level (Fig. 2). However, the optical density decreases with time, an indication that the cells have become smaller. No cryptic growth could be demonstrated in this organism during the starvation process. Batch-grown cells die off faster in stage 2 compared with cells grown in chemostat cul-

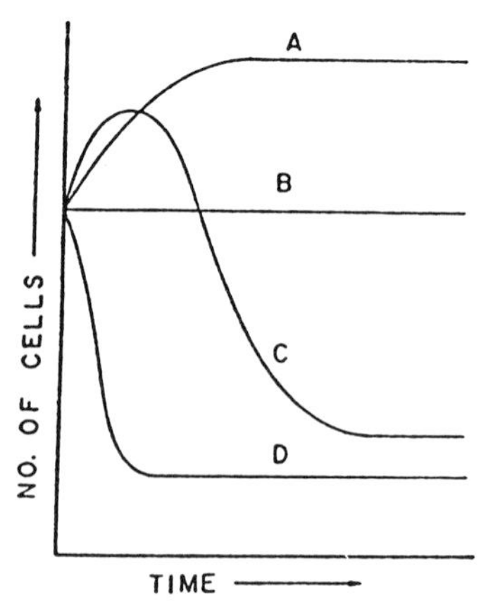

Figure 1 Starvation-survival patterns of microorganisms. [From Morita, R. Y. (1985). "Bacteria in Their Natural Environments: The Effect of Nutrient Conditions." pp. 111–130. Academic Press, Orlando.]

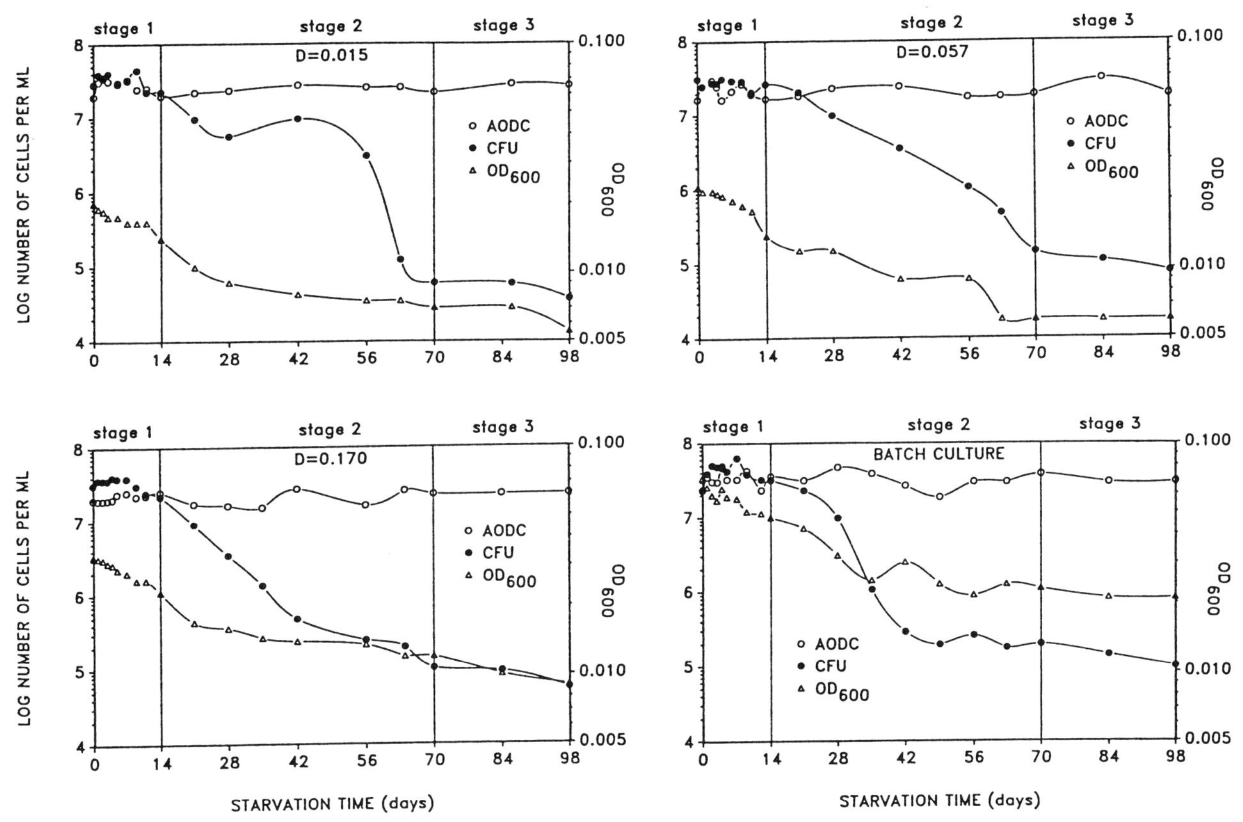

Figure 2 Total cells (AODC), viable cells (CFU), and optical density (OD_{600}) of ANT-300 cells with starvation time for cells from different dilution rates and batch culture. [From Moyer, C. L., and Morita, R. Y. (1989). *Appl. Environ. Microbiol.* **55,** 1122–1127.]

tures; the die-off is delayed in $D = 0.015$ cells. However, from an ecological viewpoint, only one cell per environment needs to survive, and when conditions (including nutrients) become right, the cell will grow, multiply, and take its place in the ecosystem.

Arbitrarily, three stages of starvation-survival were made. Naturally, each species will have its own periods at which the three stages delineate.

A. Degrees of Starvation-Survival

The interval of time between a microorganism having a sufficient supply of energy and until it again experiences the same situation may be a matter of hours, days, months, or even years. Because growth is sporadic in nature and relies on syntrophy, the supply of right substrate(s) and its concentration(s) will determine whether or not the starved cell will grow and multiply. If the amount of energy for growth is insufficient, many will still utilize it because the mechanism for obtaining energy is present. Cells of Ant-300 starved for 72 days have been shown to degrade a concentration of C^{14}-U-glutamate at 10^{-12} *M;* none of the radioactivity is incorporated into the cell. Studies have shown that the ability to utilize or degrade substrates remains intact in starved cells and that the longer the starvation period, the longer the lag period when the cells are placed in laboratory media. Cells apparently still contain a sufficient amount of adenosine triphosphate (ATP) which is necessary to allow active transport of a substrate when it is encountered in the environment. In *N. cryotolerans,* when starved for ammonium, the ATP level remained the same for 2 wk and then decreased and stabilized after 4 wk, whereas the electron transport system remained unchanged, and the energy charge, which was initially 0.68, stabilized at 0.50. In soils, a high energy charge can be found.

Thus, one can encounter in nature recently starved to starved organisms that have been found in ancient rock, deep strata of the geological material, deep aquifers, and salt crystals that were laid down centuries ago. Living bacteria have been isolated from these materials, and the bacteria, in all probability, experienced a dearth of nutrients before they became embedded or lodged in these ancient materials. If *Escherichia coli, Enterobacter agglomerans, Klebsiella pneumoniae,* and *Pseudomonas cepacia* are placed in sterile well water at a concentration of approximately 10^7 cells/ml, viable counts of 10^4–10^5 can be obtained after incubation at room temperature after 624 days. The oldest pure culture survival studies in the absence of energy have been performed with 14 various strains of *Pseudomonas syringae* subsp. *syringae*. These organisms were cultured on an agar surface, then with a loop transferred to sterile distilled water, capped, and placed at 10°C for 24 yr. Initial inoculum was varied from 10^7 to 10^9 cells/ml, and after 24 yr the viable count was found to be 10^5–10^6, with two exceptions where the obtained viable count was 0 and 10^2. In soil that was stored 54 yr, sulfur oxidizers could still be isolated, and there are other reports on the longevity of microbial cultures.

B. Physiological Processes Associated with Starvation-Survival

The following changes that take place when bacteria are starved have been performed with numerous species. Each species probably undergoes its own mechanism for survival due to the lack of energy. As yet, no exhaustive study has been made of any one species of bacteria.

Generally, cell size decreases with increasing time of starvation. (However, in *N. cryotolerans,* cell size does not decrease with starvation but remains the same as when the starvation process was initiated.) The resulting cells have been termed ultramicrocells, dwarf cells, minicells, picoplankton, and nanoplankton. The first term has been used the most. Ultramicrocells have been found in the environment; some are capable of passing through a 0.2-μm membrane filter and subsequently have been found to grow on media. Most of the bacteria (60–80%) in the ocean and soil have been found to be between 0.4 and 0.8 μm. The various genera that have been isolated from material passing through <1.0-μm membrane filters are definitely not the genral size indicated in Bergey's Manual.

The size of starved and nonstarved cells of Ant-300 grown in a chemostat under various dilution rates is given in Table I. Note that batch-grown cells represent "gluttonous" bacteria, whereas $D = 0.015$ bacteria do not represent the starved bacteria in nature but come as close as experimentation allows. Note that cell size differs between starved and unstarved cells grown at different dilution rates compared with the batch-cultured cells. In the chemostat, multiplication takes place even when the cells are small, which brings up the question as to how large a cell must be before division of the cell takes place.

Resuscitation procedures employed before undertaking viable counts of natural samples are extremely rare. In all probability, many of these nonculturable organisms do have the ability to metabolize and are dead only by the traditional microbial definition of death. Nonviability in an environmental sample may be due to the use of the wrong medium, pH, Eh, temperature, etc. Because of the oligotrophic nature of most ecosystems, the normal state of bacteria is to some degree of the starved state.

When Ant-300 cells (a marine psychrophilic vibrio) are grown in laboratory culture under optimal conditions, they may have more than one nuclear body. On the other hand, in the starved state, the amount of DNA, RNA, and proteins is much smaller than in the nonstarved state (Fig. 3 and Table I). Note that the DNA is much smaller in $D = 0.5$ cells than in the batch-grown (gluttonous) cells (Table I). Figure 3 is based in AODCs, mainly because the cells are not lysed and the acridine orange is an

Table I Effect of Growth Rate on DNA Content in Relation to Average Cell Volume for Unstarved and Starved Ant-300 Cells

Culture	DNA/cell (fg ± SEM)	Cell volume (μm^3)	Estimated nucleoid volume (μm^3)	Nucleoid volume/cell volume
Unstarved				
Batch	23.66 ± 0.01	5.94	0.33	0.06
$D = 0.170\ hr^{-1}$	24.69 ± 0.01	1.16	0.35	0.30
$D = 0.057\ hr^{-1}$	20.12 ± 0.06	0.59	0.28	0.48
$D = 0.015\ hr^{-1}$	15.33 ± 0.27	0.48	0.21	0.45
Starved[a]				
Batch	1.23 ± 0.17	0.28	0.017	0.06
$D = 0.170\ hr^{-1}$	1.03 ± 0.35	0.19	0.014	0.08
$D = 0.057\ hr^{-1}$	1.45 ± 0.02	0.18	0.020	0.11
$D = 0.015\ hr^{-1}$	1.28 ± 0.02	0.05	0.018	0.40

[a] All cell samples were taken from stage 3 of starvation-survival.
[From Moyer, C. L., and Morita, R. Y. (1989). Effect of growth rate and starvation-survival on cellular DNA, RNA, and protein of a psychrophilic marine bacterium. *Appl. Environ. Microbial.* **55**, 2710–2716.]

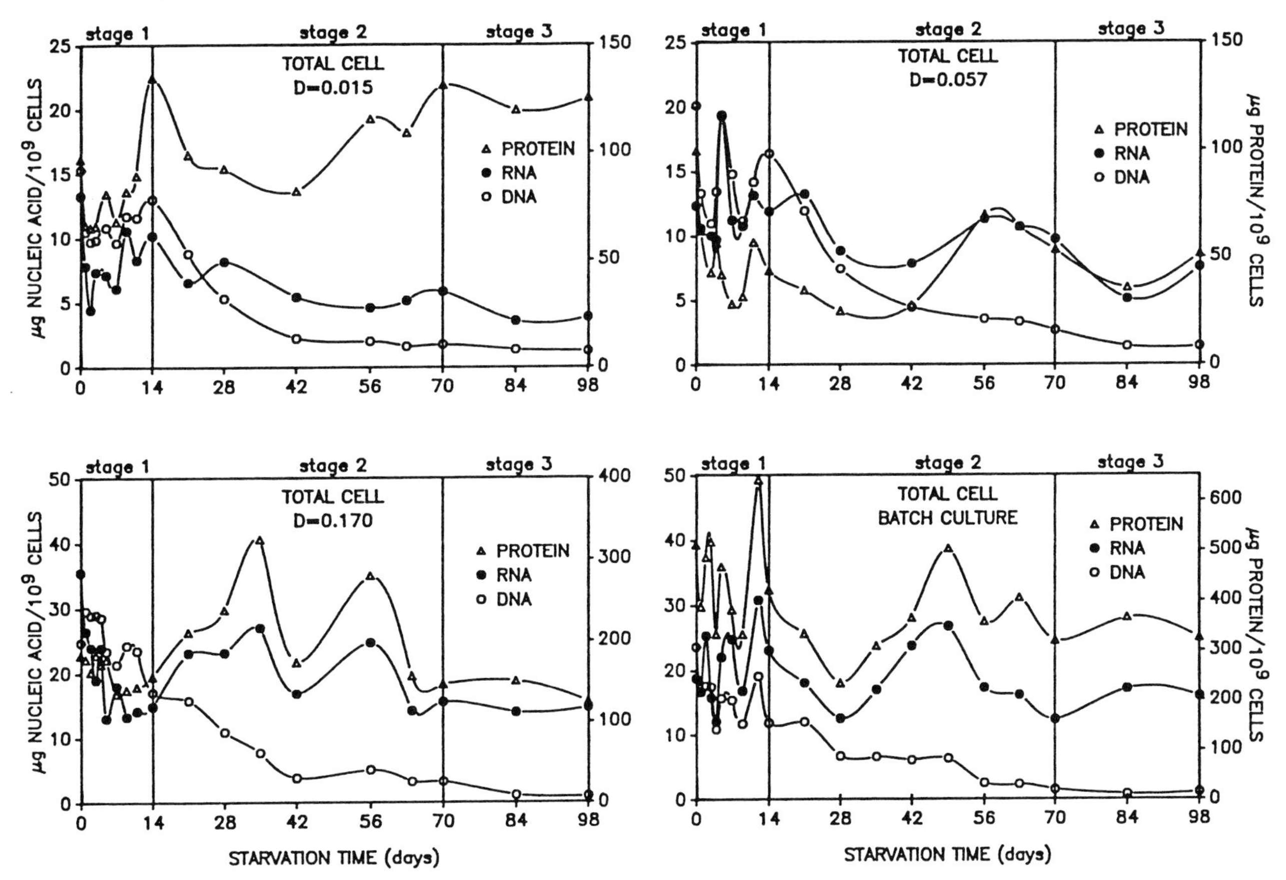

Figure 3 DNA, RNA, and protein per total cell, with starvation times and stages for cells grown under different growth rate conditions. [From Moyer, C. L., and Morita, R. Y. (1989). *Appl. Environ. Microbiol.* **55**, 2710–2716.]

intercalator of DNA. The determination of the nucleoid volume (Table I) is important because many investigators equate the microbial biomass in marine environments to the amount of DNA that can be measured. In addition, it also shows that starved cells have a smaller amount of DNA per cell.

In nature, the rate of microbial metabolic activity is determined mainly by the concentration of organic matter present in the ecosystem, provided that water in the liquid state, inorganic ions, etc., are satisfactory. The rate can be very low in the deep sea, as evidenced by the fact that oxygen consumption has been estimated to be 0.002–0.004 ml/liter/yr for all the organism present. This indicates that the rate of metabolic activity must be in tune with the amount of organic matter present. If there were a fast rate of metabolism of the heterotrophic bacteria in the deep sea, all of the organic matter would be utilized quickly, leaving none for the higher forms of life present. Recognizing that most of the organic matter in this situation is recalcitrant, this low metabolic rate is a blessing in disguise. On the other hand, the microbes in ancient rocks and salt crystals do not have access to a readily utilizable energy source and, thus, are deprived of energy for long periods of time.

In Figs. 2 and 3, starvation-survival was arbitrarily divided into three stages. Stage 1 represents a period when the number and amount of proteins, DNA, and RNA greatly fluctuate. Employing gluttonous cells, a rapid oxygen uptake takes place. The ratios of DNA : RNA, DNA : protein, and RNA : protein fluctuate. Also, endogenous energy sources rapidly decrease. However, these studies are made with gluttonous bacteria, whereas in nature these reserve endogenous energy sources would probably not exist. A great decrease in the phospholipid content of the cells also occurs. Stage 1 is when a drastic reorganization of the cells takes place. During stage 1, chemotaxis appears after 48 hr of starvation and lasts for approximately 7 days. This permits the organism to seek out nutrients during this period because the primary function of chemotaxis is to permit the organisms to find energy. In addition, cellular proteins are degraded and some "starvation-specific" proteins synthesized. During this stage, the formation of fibrillar structures appears in some organisms, permitting increased surface adhesion. In addition, certain marine bacteria become more hydrophobic during the starvation period. Adhesion aids the microbe in its initial stages of starvation because surfaces are known to attract organic matter (bottle, or surface, effect).

Stage 2 is a period when the number of viable cells slowly decreases. Certain cellular proteins continue to be degraded, while others are not. New proteins are also synthesized during this period, labeled by some researchers as the "starvation" protein, and are thought to be important in the viability of the cells during the starvation-survival process. This is a "fine-tuning" period of the cells in preparation for stage 3.

Stage 3 is when the cells are in a state of metabolic arrest—a situation somewhat analogous to bacterial spore. Because no substrates exist, no metabolism takes place. There apparently is an undetectable energy of maintenance. The concept of energy of maintenance was developed in light of the data for a clone of gluttonous bacteria. As mentioned previously, the data concerning the turnover time for bacteria in soil would indicate that the concept of energy of maintenance is not valid and probably applies to a clone of bacteria *in situ* and not for naturally occurring bacteria in oligotrophic environments, where survival periods may be very long. In this metabolic arrest state, the organism can survive long periods of time, perhaps giving the organisms some "biological" control over time.

VI. Conclusion

Research involving microorganisms in the oligotrophic environment is still in its infancy. The proof of the existence of obligate oligotrophic bacteria must be established experimentally. Studies dealing with bacteria in the same physiological state as those found in oligotrophic environments must be initiated to rule out studies done with the gluttonous bacteria. Because of the oligotrophic nature of most ecosystems and the unavailability of energy sources, most bacteria in nature are in some state of starvation-survival. Thus, we must combine Koch's "feast or famine" concept with Poindexter's "fast or famine" concept. Definitely, new techniques must be developed for a better understanding of the oligotrophic environment. The ability to survive during periods of nutrient deprivation must also be taken into consideration when evolutionary processes are addressed.

Acknowledgment

Published as technical paper No. 9450, Oregon Agricultural Experiment Station.

Bibliography

Iscobellis, N. S., and DeVay, J. E. (1986). *Appl. Environ. Microbiol.* **52,** 388–389.

Johnstone, B. H., and Jones, R. D. (1988) *Mar. Ecol. Prog. Ser.* **49,** 295–303.

Kjelleberg, S., Hermansson, H., Marden, P., and Jones, G. W. (1987). *Annu. Rev. Microbiol.* **41,** 25–50.

Morita, R. Y. (1982). *Adv. Microbiol. Ecol.* **6,** 171–198.

Morita, R. Y. (1985). Starvation and miniaturization of heterotrophic bacteria, with special reference on maintenance of the starved state. *In* "Bacteria in their Natural Environments: The Effect of Nutrient Conditions" (M. Fletcher and G. Floodgate, eds.), pp. 111–130. Academic Press, Orlando.

Morita, R. Y. (1988). *Can. J. Microbiol.* **34,** 436–441.

Moyer, C. L., and Morita, R. Y. (1989). *Appl. Environ. Microbiol.* **55,** 1122–1127.

Moyer, C. L., and Morita, R. Y. (1989). *Appl. Environ. Microbiol.* **55,** 2710–2716.

Roszak, D. B., and Colwell, R. R. (1987). *Microbiol. Rev.* **51,** 365–379.

Shilo, M. (ed.) (1978). "Strategies of Microbial Life in Extreme Environments" Verlag Chemie, Weinheim.

Low-Temperature Environments

Richard Y. Morita
Oregon State University

Glossary

Endolithotropic Living inside rock, usually limestone or sandstone

Thermocline In the stratification of warm surface water over cold, deeper water, the transition zone of rapid temperature decline between the two layers

Upwelling Transport of water from the deep ocean to the surface, replacing the surface water that has moved offshore

LOW-TEMPERATURE ENVIRONMENTS (sometimes referred to as the psychrosphere) dominate the biosphere. Yet, in terms of ecological and physiological studies, low-temperature research lags behind studies in the mesophilic and thermophilic range.

I. Low-Temperature Environments

A. Arctic and Antarctic Environments

Fourteen percent of the earth's surface is in the polar regions. Snow and ice, including the continental ice sheet, are one of the major features of the many polar environments. Within the polar regions, the temperature can fluctuate greatly due to solar radiation. This also occurs in temperate regions. For instance, on the rock surfaces in the Horlick Mountains (4° from the geographical South Pole), the temperature has been recorded to rise from −15° to 28.8°C within 3 hr due to solar radiation, even though the air temperature was around 0°C. Freezing and melting of snow and ice play major roles in the activities of the microorganisms in certain cold environments. Although both polar regions are cold and lack sunlight during the winter, their differences lie in their land masses, land topographical features, large-scale water transport patterns, and the magnitude of nutrient supply. The latter is due to the major river inputs in the Arctic, which are lacking in the Antarctic.

Both polar regions have different types of ecosystems such as the tundra, deserts, sea–ice fronts, snow, glaciers, continental ice sheets, mountains, lakes (some ice-covered year-round), rivers, ice-bubble habitats, and meltpools. In some ice-covered lakes, water below the ice may be warmer. Each has its own microbial community structure. In addition to the cold environment, the microbial population must contend with conditions of high salinity and dehydration created by water freezing, different light intensities including seasonal darkness, etc. The opportunity for microbial growth occurs mainly during the period of snow and ice melt because the frozen state offers little opportunity for growth. Melting of the snow packs is enhanced by wind-blown sediment, hence the melting snow becomes increasingly "dirty" and the melting enhanced due to the absorption of radiant energy. Snow appears to have a rather high nitrogen content.

B. Oceans

The oceans, which represent 71% of the earth's surface and 90% by volume, is 5°C or colder. The portion in the lower latitudes is below the thermocline (discontinuity layer), and, in the polar regions, this thermocline may be at the surface of the ocean. This thermocline is an imperfect barrier that permits only a small amount of organic matter, produced by photosynthetic organisms in the photo zone, from sink-ing into the cold, deep sea. Unlike the land masses, this environment is rather constant in that the temperature and salinity does not vary greatly. In the deep sea, there is no light; however, where the discontinuity layer comes to, or near, the surface in the polar regions, the light intensity varies depending on the season and climatic conditions [*See* MARINE HABITATS, BACTERIA.]

Even in nearshore environments, the water temperature may be low, especially when an upwelling situation exists.

C. Upper Atmosphere

At altitudes >10,000 m, the temperature is <10°C, while at altitudes >3000 m, the temperature is consistently <5°C. With increasing higher altitudes, the temperature progressively decreases, and temperatures below −40°C have been recorded.

D. High Mountains

Where snow or ice remains year-round, low temperatures are always present, except on surfaces that receive solar radiation. At other portions of the mountains where snow and ice melt, the temperature is cold part of the year. The same holds true for the mountain lakes, and if a thermocline exists in these freshwater bodies, then a permanent cold temperature exists.

A cold permanent environment dominates the biosphere. If we add the winter season, this low-temperature environment becomes a major feature of the earth's environment. Thus, for all pragmatic purposes, we can divide the cold environments into two categories: psychrophilic (permanently cold) and psychrotrophic (seasonally cold or temperature fluxes into the mesophilic range) environments.

II. Low-Temperature Organisms

A. Psychrotrophs

Psychrotrophs are defined as microorganisms capable of growing at 5°C and below. Although growth of a psychrotroph may be slow at 5°C, this definition permits the maximal and optimal temperature for growth to occur in the mesophilic temperature range. Most microbes described in the microbiology of foods and dairy products fit this category. For example, mesophilic bacteria such as *Bacillus megatherium* and *Bacillus subtilis* and some *Arthrobacter* and *Corynebacterium* can grow at <5°C. Even the pathogen *Yersinia pestis* has been reported to grow at −2° as well as at 40°C. The question that arises is whether or not these organisms growing at low temperatures on rich media are capable of growth at similar temperatures in their respective environments. Psychrotrophs are widely distributed in nature and are present in ecosystems where there is a flux of temperature, especially above the maximal temperature for growth of psychrophiles. However, it should also be noted that their abundance appears to be greater than true psychrophiles in permanent cold environments. For example, in the surface soils of the Antarctic as well as in the water below the thermocline, psychrotrophs will dominate. In the former environment, this is probably due to the soil receiving sufficient solar radiation to cause thermal death of psychrophiles during a short period of the austral summer months. In addition, psychrotrophs have adapted to the environment because microorganisms do have this flexibility. Nevertheless, when one considers the energy flow in cold regions, photosynthesis only occurs periodically depending on the latitude. The photosynthetic material is the main energy source for the heterotrophs except for the photosynthetic organisms and chemolithotrophs.

Snow algae can be found where snow is melting and where the snow surface is red, green, or yellow. Most of these snow algae are psychrotrophs and receive their energy and increased temperature from the sun. However, algae can also be found in the ice.

B. Psychrophiles

Psychrophiles have been defined as organisms having an optimal temperature for growth at 15°C or lower, a maximal temperature for growth at about 20°C, and a minimal temperature for growth at 0°C or lower. For organisms higher than bacteria, cryophiles have been known to exist. Cryophilic yeasts occur in the extreme environments of the Ross Desert of Antarctica. However, the existence of true cold-loving bacteria was not documented without question until 1964. Because most bacteria isolated prior to 1862 did not meet the true definition of psychrophile (with one exception, which was not documented to any great degree), various investigators employed other terms to describe the cold-loving bacteria isolated. Terms such as Glaciale Bakterien, rhigophile, psychrotolerant, psychrocartericus, psychrobe, thermophobic bacteria, cryophile, facultative psychrophile, obligate psychrophile, and psychrotrophic bacteria have been used. Because of this situation, many organisms described in the literature as psychrophiles are not truly psychrophilic and should be classified as psychrotrophic. Naturally, one should recognize that a continuum of upper and lower temperatures exists within the microbial world, but the preceding definition recognizes that one existing thermal group of microorganisms are as physiologically unique as the thermophiles.

The isolation of psychrophiles is not a difficult process. The prerequisites are that all material be kept cold and that the source material for the bacteria be from an environment that is permanently cold. The sample from which psychrophiles are to be isolated must never be allowed to warm to room temperature and must come from a psychrophilic environment. Their abnormal thermosensitivity is probably one of the reasons why many investigators working in the Arctic and Antarctic were not able to isolate psychrophiles. All barophiles (pressure-loving bacteria) are true psychrophiles. The abnormal thermal sensitivity of psychrophiles is probably the main reason why investigators prior to 1964 could not isolate these organisms.

The distribution of marine psychrophiles is always associated with permanent cold ecosystems, but the number of psychrophiles in relation to the psychrotrophs is low, even in cold environments, according to the older literature. More recent studies in the African upswelling regions and the Antarctic Ocean demonstrate that more colonies are formed from the same samples when the agar plates are incubated at 2°C than at 20°C. If both a psychrotroph and a psychrophile are introduced into the same culture tube, capable of growing in the medium employed, and incubated at 5°C, the psychrophile, naturally, will outgrow the psychrotroph. Thus, in the Arctic and Antarctic as well as in the waters below the thermocline, psychrotrophs will dominate in numbers, mainly because of the lack of utilizable energy. Because a variety of different ecosystems exist in low-temperature environments, no definite statement can be made as to the distribution of microorganisms. The ability to isolate organisms from any ecosystem does not indicate that the specific microorganism is active under the ecosystem's environmental factors. The distribution of psychrophilic an psychrotrophic bacteria in different ecosystems in nonpolar soil samples is tabulated in Table I, in Arctic and other cold north soil samples in Table II, and in Antarctic soil samples in Table III.

The cell yield for *Vibrio marinus* MP-1 in laboratory medium has been recorded to be 1.3×10^{12} cells/ml at 15°C in 24 hr and 9×10^{9} cells/ml at 3°C in 24 hr, whereas the best cell yields for a psychrotrophic bacteria, *Pseudomonas fluorescens* (considered a psychrophilic strain in the literature) was recorded to be 9×10^{8} cells/ml at 20°C in 72 hr and 3.4×10^{3} at 0°C in 72 hr. Thus, low temperatures are not detrimental to large cell yields, provided that the water is in a liquid state. In growing psychrophiles, there is a long lag time, but if they are kept in the log growth phase, high cell yields result. The minimal temperature for growth for the psychrophiles thus far isolated has not been determined, but growth definitely occurs <0°C. The lowest temperature recorded for the growth of bacteria is at −11°C, whereas the lowest temperature recorded for an enzyme reaction is −25°C.

C. Occurrence of Other Thermo-Groups in Low-Temperature Environments

Due to natural conditions (wind, rain, etc.), the transport of other thermal groups into the low-temperature environment occurs. Thus, the presence of mesophiles in low-temperature environments can easily be demonstrated. The presence of thermophiles has also been demonstrated in low-temperature environments of the ocean as well as in permafrost. However, their activities are probably nil because the temperature is below the minimal growth temperature, which, in turn, dictates the ability to take up substrate. Because the temperature is low, survival of these organisms occurs.

III. Physiological and Biochemical Characteristics

Most microorganisms isolated from the environment have the ability to grow better at temperatures 10–20°C higher than the temperature from which they were taken. Thus, this fact is reflected in the physiological and biochemical characteristics of the organism.

A. Temperature–Growth Profiles

Although the temperature–growth curves of psychrophilic bacteria have been recorded, generally the minimum temperature for growth is not recorded due to the difficulty of keeping media in the liquid state at temperatures below freezing. The recording of −11°C was observed by colony formation on an

Table I Incidence of Psychrophilic and Psychrotrophic Bacteria in Nonpolar Soil Samples

Location	Soil types	Incubation temperature (°C)	Bacterial counts (g^{-1} dry wt soil)
Temperate soil		4.5	7×10^5
Temperate (France)	Garden	4	4.3×10^7 g^{-1} wet wt soil
French Alps	Glaciers	4	$0.6–3.6 \times 10^6$ g^{-1} wet wt soil
1. Vegetable gardens (13)[a]		3.5	2.4×10^5
2. Salt marsh pastures (4)			1.0×10^5
3. Sand dune pastures (7)			6.0×10^5
4. Lowland pastures (23)			39.8×10^5
5. Lapland pastures (17)			107.0×10^5
6. Rhizosphere (10)			230.0×10^5
7. Lowland raised bogs (3)			16.0×10^5
8. Heathmoors (5)			13.0×10^5
9. Arable soils (8)			5.9×10^5
Lapland (near glacier)	Forest soil	2	Below ice: $2.8–3.9 \times 10^6$ No ice: $1.5–2 \times 10^6$
Louisiana	Garden, footpath, lawn	0	Summer: <10 Winter: <100
Eastern Canada	Frozen soil	3	Surface to 2 cm: 6.1×10^6 6 cm: 6.5×10^6 10 cm: 1.3×10^6
Washington State	Garden soil Cultivated soil Uncultivated soil	0	$0.92–1700 \times 10^3$ $23–810 \times 10^3$ $420–3100 \times 10^3$
Temperate soil (near area receiving cannery waste water)		2	$1.5–7.4 \times 10^5$ g^{-1} wet wt soil

[a] Total number of samples in parentheses.
[From Baross, A. J., and Morita, R. Y. (1978). "Microbial Life in Extreme Environments." Academic Press, London.]

agar surface instead of liquid culture. Psychrophiles do have a long lag period but can be kept in the log growth phase.

Sufficient temperature-growth curves, such as those in Fig. 1, exist in the literature to substantiate the definition of psychrophiles given earlier. However, one should use the various thermal group definitions for utilitarian purposes because there is a continuum of organisms from the low-temperature range to the high-temperature range.

B. Thermosensitivity

Cells held above the maximum growth temperature lose their viability. Naturally, this is a time-dependent process. Why cells lose their viability still remains unanswered. Some of the protein in cell-free extracts held at temperatures at and above their maximum growth temperatures will undergo denaturation. Although at temperatures 2° above the maximum growth temperature, macromolecular synthesis (DNA, RNA, and protein) takes place in cells, at 5°C above the maximum growth temperature, it does not. Leakage of intracellular materials [amino acids, proteins (including some enzymes), DNA, and RNA] does take place when cells are exposed to temperatures above the maximal growth temperature. When leakage of large quantities of their intracellular materials occur, 90% of the cells are dead. Small quantities of intracellular materials do leak from the cells before death of the cell occurs but the question is how much leakage of intracellular materials must take place for death of the cell to occur? It is also recognized that some enzymes of

Table II Incidence of Psychrophilic and Psychrotrophic Bacteria in Arctic and Other Cold Northern Soil Samples

Location	Soil type	Incubation temperature (°C)	Bacterial counts (g^{-1} dry soil)		
Arctic coast					
Point Lay Alaska	Peat	2			51×10^4
	Loam	2			30×10^4
Wainwright	Peat	2			38×10^4
	Loam	2			4×10^4
Barrow	Peat	2			10×10^4
	Loam	2			7×10^4
Cape Simpson	Peat	2			14×10^4
	Loam	2			5×10^4
Pitt Point	Peat	2			17×10^4
	Clay	2			9×10^4
Barter Island	Loam	2			$13–26 \times 10^4$
Alaskan Arctic	Peat to permafrost layer	2		Surface:	9.6×10^3
				15 cm:	9.3×10^3
				30 cm:	5.4×10^2
				>30 cm:	<1
Inuvik, North West Territories	Uncultivated	2		Surface:	$0.7–1500 \times 10^4$
	Cultivated	2		Surface:	$4.9–280 \times 10^4$
Alaska tundra, Napaskiak	Boggy soil	3–5		Sept. range:	$1.2–150 \times 10^3$
				Average:	35×10^3
				June range:	$0.1–1300 \times 10^3$
				Average:	150×10^3
MacKenzie Valley, North West Territories		4		Actinomycetes	Other bacteria
	Organic soils		2.5–30 cm:	71×10^4	$<10^3$
			30–35 cm:	33	$<10^3$
	Subarctic gleyed acid, brown wooded		0–23 cm:	$5–10 \times 10^5$	17×10^3
			23–70 cm:	3×10^4	2×10^2
	Subarctic brown wooded (cultivated)		0–10 cm:	5.6×10^6	$<10^3$
			10–70 cm:	$4–50 \times 10^4$	$<10^3$
	Subarctic brown wooded (uncultivated)		0–40 cm:	$80–212 \times 10^4$	$<10^3–60 \times 10^3$
			>40 cm:	1×10^4	$<10^3$
Baffin Island, North West Territories		10		Counts g^{-1} dry wt soil Actinomycetes	Other bacteria
	Sandy soil			$14–17.8 \times 10^3$	$8–159 \times 10^3$
	Gravel			$1–11.7 \times 10^3$	$10–25 \times 10^3$
	Wet soil or rock edges			0.8×10^3	11.2×10^3

[From Baross, J. A., and Morita, R. Y. (1978). "Microbial Life in Extreme Environments." Academic Press, London.]

Table III Incidence of Psychrophilic and Psychrotrophic Bacteria in Antarctic Soil Samples

Location	Soil type	Incubation temperature (°C)	Bacterial counts (g^{-1} dry wt soil)
Signy Island	Peat	10	Surface to 1–2 cm: 486×10^3
			6–7 cm: 1160×10^3
			11–12 cm: 2200×10^3
Victoria Land (dry valley)	Dry soil	2	0–2 cm: <10
			15 cm: 8×10^3
	Permafrost		25 cm: 1.8×10^4
	Moist soil (near pond)	2	$1.2–14 \times 10^5$ (range of three samples)
Ross Island			
Cape Royal	Dump (human contamination)	2	2.8×10^5
	Soil 1		0
	Soil 2		5.2×10^3
McMurdo Sound			
	Areas inhabited by humans	2	$0.18–63 \times 10^6$
	Soil 3		1.1×10^3
	Soil 4		73×10^3
	Soil 5		38
Dry Valley			1
Cape Evans (inhabited by humans)			6.8×10^5
Wright Valley	Mummified seal carcasses	2	$0.1–1 \times 10^3$
Dry Valleys	1. Taylor Valley soil	2	$0.2–6 \times 10^3$
	2. Wright Valley soil		$0.3–9 \times 10^3$
	3. Marble Point soil		$0.46–25 \times 10^3$
	4. Strand Moraines		$0.053–27 \times 10^3$
Macquarie Island	1. Soils	10	
	a. Gravel		4.3×10^5
	b. Basalt		2.9×10^6
	2. Rhizosphere		
	a. Soil		1.67×10^6
			14.2×10^6
	b. Root		4.81×10^6 g^{-1} root
			2870×10^8 g^{-1} root
	Algal soil crests	2	$1.6–2.6 \times 10^5$
			$0.96–1.2 \times 10^5$
Wheeler Dry Valley	Sand	2	Surface to 2 cm: $0.2–150 \times 10^3$
			2–15 cm: $1–100 \times 10^3$
			18–33 cm: $0.5–4.8 \times 10^3$
			60 cm: 100
Matterhorn Valley	Sand	2	Surface to 2 cm: $<10–370$
			2–10 cm: $<10–20$
	Loam or sandy loam	2	Surface to 2 cm: 10–180
			2–10 cm: 3.5×10^4
Coalsack Bluff	Arid	2	Surface: $<10–13{,}000$
Antarctic interior	Desert soil		2–10 cm: 0–2700
Farthest south soil sample	Desert soil	2	5–25
Victoria Valley	Sand	2	0–2 cm: 2.8×10^4
			2–15 cm: 2.6×10^3
			15–25 cm: 5.6×10^4
			25–30 cm: 2.7×10^7
	Sandy loam	2	0–2 cm: <10
			2–15 cm: 30
			15–25 cm: 0
			25–30 cm: 5

(continues)

Continued

	Sand (miscellaneous samples)		0–25 cm:	<10–730
Dry Valley		5	Surface to 2 cm:	<10–2.5 × 10^5
South Victoria Land dry valleys		2	Surface:	10–10^4
			Subsurface:	10–1.7 × 10^5
Taylor Dry Valley	Saline soil	2	0% NaCl[a]:	5.1 × 10^3
			5% NaCl:	1.2 × 10^3
			15% NaCl:	<40
	Nine soil samples	0	<100–20,000	
	Animal feces (four samples)	0	0.32–6000 × 10^4	

[a] Percent NaCl concentration in the isolation medium.
[From Baross, J. A., and Morita, R. Y. (1978). "Microbial Life in Extreme Environments." Academic Press, London.]

the cells (see Section II.C) are also abnormally thermal labile and it may also bring about death of the cell. Leakage of protein, RNA, and DNA from the cell takes place when the cell is exposed to a few degrees above its maximal growth temperature. Between the maximal and optimal growth temperature, intracellular proteins are denatured, which may explain the difference between the optimal and maximal growth temperature of organisms. Destruction of the membrane also appears to take place. Is death due to a single event or a combination of many events? Although death of psychrophiles due to temperatures above the maximum growth temperatures has been attributed to the inability of glutamic acid, histidine, and proline to acylate to the t-RNA, these studies were done at 5°C above the maximum growth temperature of the organisms and not 1 or 2°C above the maximum growth temperature where protein the macromolecular synthesis does take place. There are probably too many subtle changes that take place in the cell at 1 and 2°C above the maximum growth temperature.

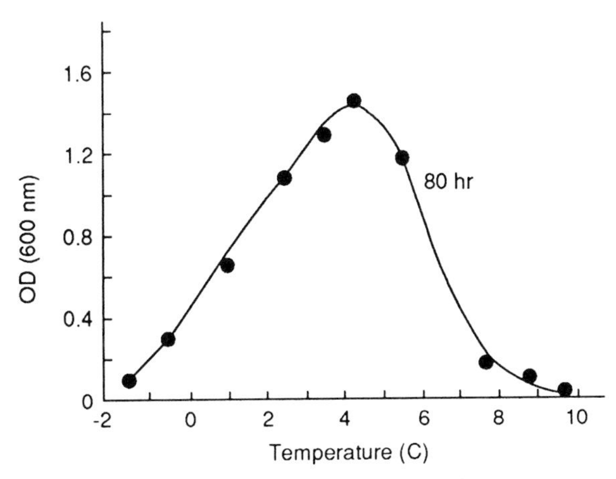

Figure 1 Growth of an Antarctic psychrophile (AP-2-24), designated tentatively as a *Vibrio* sp. Incubation period was 80 hr in Lib-X medium employing a temperature gradient incubator. [From Morita, R. Y. (1975). *Bacteriol. Rev.* **39,** 144–157.]

C. "Psychrophilic" Enzymes

Although only a few enzymes from psychrophilic bacteria have been studied in relation to their activity in the low temperature range, they show the characteristic maximum, minimum, and optimum temperatures that are quite a bit lower than their counterpart from mesophilic bacteria. These enzymes were obtained from *Vibrio marinus* MP-1, which has an optimum growth temperature at 15°C and a maximum growth temperature of 20 °C. They could be classified as "psychrophilic" due to their abnormal thermolability; however, not all the enzymes isolated from psychrophilic bacteria fall into the psychrophilic range. In addition, a specific psychrophilic enzyme from one psychrophile is not necessarily psychrophilic in another psychrophile. It is possible that we have not, as yet, isolated a psychrophilic in which all the enzymes are truly psychrophilic. It is truly unfortunate that further studies on psychrophilic enzymes have not been performed on psychrophiles that have lower cardinal temperatures than *V. marinus* MP-1.

Malic dehydrogenase obtained from *V. marinus* MP-1 has been the most studied enzyme from a psychrophile, but lactic dehydrogenase, succinic dehydrogenase, hexokinase, aldolase, and phosphoglucose isomerase from the same organism also possessed psychrophilic characteristics. Partially purified malic dehydrogenase (Fig. 2) exhibits psychrophilic characteristics, but the enzyme is unstable without the presence of ammonium sulfate, which protects the enzyme from thermal inactiva-

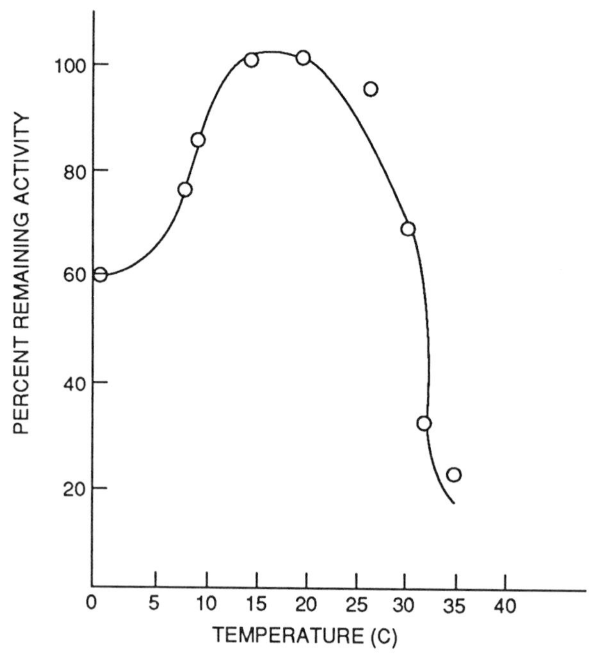

Figure 2 The effect of 30-min exposure of partially purified malic dehydrogenase on increasing temperature. [From Langridge, P. A., and Morita, R. Y. (1966). *J. Bacteriol.* **92**, 418–423.]

tion. Thus, the curve in Fig. 2 should be skewed to the lower temperatures. Psychrophilic aldolase, on the other hand, appears to have the same amino acid composition (within experimental error) as aldolase from thermophilic organisms, so that the primary structure is more or less identical.

IV. Microbial Activities in Low-Temperature Environments

As in all oligotrophic environments, the amount of energy available for heterotrophic bacteria is limited, especially when water is in the frozen state. Photosynthesized material is the bases of the organic matter, but photosynthesis does not occur below −6°C. This becomes especially true for all the cold environments on earth; therefore, the growth of psychrophilic and psychrotrophic bacteria is limited. For instance, for yeast from Ross Desert soil to multiply, glucose must be added to the soil. Microbes have been isolated from the ice in both polar regions, but it is doubtful that growth takes place due to the frozen nature of the water, which does not permit the organism to have access to the limited amount of energy present The ecosystem selects which microogranisms will grow, and the ecosystem is not a constant one, especially when season of the year dictates the environmental temperature.

A. Polar Regions

Early expeditions in the 1900s in the Antarctic concentrated on elucidating and classifying the microbes from seawater, soil, air, and animals. Microbial activity in the waters off Antarctica is comparable with that of the other parts of the ocean; however, within the Antarctic continent, many geological features are unusual, and all the geological features that have been investigated have shown the presence of microbes. The data on the occurrence and activities of bacteria in arctic, alpine, and seasonally ice-covered lakes are summarized in Table IV and in three antarctic lakes in Table V. Apparently, there is a greater predominance of cyanobacteria in the Antarctic relative to the Arctic benthic and soil ecosystems.

There is very little liquid water in the Antarctic continent, but some lakes located near the ocean can undergo partial to complete thawing during the Antarctic summer. Some of the inland lakes are permanently frozen to the bottom, whereas other lakes have brackish water at the bottom, such as lakes Vanda and Bonney. Where there is brackish water at the bottom of such lakes, microbiological and photosynthetic activity have been observed to occur just below the ice layer. In lakes that are permanently frozen, fish and zooplankton are absent.

In the cold deserts of Antarctica, microbial community developments are mainly restricted to three types of habitats: endolithic communities inside rocks, freshwater communities in transient pools, streams, and rivers, and ice-covered lakes. All organisms in these ecosystems have their own distinctive strategies for survival. One of the main factors for the limited growth of bacteria and other microorganisms in the dry valleys of the Antarctic is the lack of liquid water. Without the water, metabolic processes cannot take place and makes nutrients unavailable to the cells. The endolithic communities in the Beacon sandstone in the Ross Desert only grow for a short period each year when liquid water is available. The endolithic communities consist of diverse combinations of cyanobacteria, green algae, filamentous fungi, yeasts, lichens, and associated heterotrophic bacteria. The desert varnish formed on the out portion of the sandstone help to keep the material from desiccation.

B. Polar Oceans

The range of glutamic acid uptake ranged from 1 to 180 ng/liter/hr and 1 to 90 ng/liter/hr in the Arctic and Antarctic marine waters, respectively. These measurements were made employing identical techniques from the same laboratory. These ranges only indicate that the activity varies with the location (water mass) from which the samples were taken. There is also a range in the viable counts of bacteria (10^1–10^4), depending on the water mass sampled.

C. Other Environments

Psychrotrophic bacteria have been reported in permanently cold caves, but the investigations on these environments are few. The presence of various thermal groups in permanently cold caves in the Arctic, Lapland, the Pyrenees, the Alps, and Romania, where the temperatures range from −0.8 to 5°C and are stable throughout the year, has been investigated. Plate counts were invariably higher when incubated at 20° than at 2°C. Most of the organisms isolated belonged to the genera *Arthrobacter, Pseudomonas,* and *Flavobacterium,* with the *Arthrobacter* predominating. *Arthrobacter* is considered to be part of permanent indigenous miroflora of these caves. None of the cave isolates produced a pigment which may have been lost due to the fact that no sunlight was present. Many of the psychrophiles isolated from the soil of these caves resembled *Arthrobacter glaciales.*

Only psychrotrophs were found in the subterranean Karst caves (10–12°C) of Indiana, which were formed during the retreat of the Wisconsin ice from Indiana 20,000 yr ago. It was anticipated *a priori* that a permanent microbial organism could develop with optimum temperatures that coincided with the cave temperatures. This time period may be too short for the organisms to evolve to a "psychrophilic" from other thermal groups due to the long generation time under low-temperature conditions as well as the lack of ample supplies of energy. Furthermore, the cave temperatures are 5–10°C warmer than the permanently cold environments from which true psychrophiles are usually isolated.

Numerous bacteria can be found in the troposphere (ground level to 10,000 m). Bacteria can be isolates at 3000 m above the polar regions, where the temperature is −20°C, as well as in the stratosphere (27,000 m, −40°C). Bacteria isolated from an altitude of 7000 m were shown to be typical soil forms. Likewise, bacteria near or over the ocean may be marine bacteria. Bacteria counts can exceed $500/m^3$. This suggests that many of the bacteria isolated from the upper atmosphere are capable of surviving the freezing temperature and are introduced into the atmosphere by natural process, mainly strong winds. Nevertheless, it has been reported that many bacteria isolated from air samples >3000 m had the ability to grow at 0°C. The incidence of psychrophiles in the atmosphere has not been investigated to any degree but it is possible that, over both polar regions, psychrophiles may be active in the atmosphere, utilizing organic matter formed by gas bubbles in the ocean and then being airborne.

A "biological zone" has been proposed for the atmosphere where bacteria grow in water droplets containing organic matter as well as to carry on nitrogen fixation. Bacterial growth could occur in clouds, thereby enriching the rain with organic matter. There are relatively high levels of cobalamin, biotin and niacin in raindrops, presumably bacteria in origin.

In the older literature, psychrophiles (actually psychrotrophs) were reported in various types of food. True psychrophiles are probably not involved in the spoilage process of frozen food, mainly due to their thermosensitivity. The only food in which they might be responsible for spoilage is the prawns caught in the Antarctic, where the air temperature is also cold and the prawns are stored in ice. Generally, spoilage takes place when the frozen food is thawed and left for some time at room temperature before being used or refrozen. However, spoilage can take place if the food is held at refrigeration temperature (e.g., 5°C). Spoilage of dairy products is the result of psychrotrophs that have the ability to grow at refrigeration temperature. [*See* REFRIGERATED FOODS.]

V. Conclusion

A myriad of ecosystems exist within low-temperature environments, and each has its own microbial community that can vastly differ from one ecosystem to the other. Energy flow from photosynthesis material (organic matter) is limited, depending on the amount of solar radiation the system

Table IV Incidence and Activity of Bacteria in Some Arctic, Alpine, and Seasonally Ice-Covered Lakes

Lake	Number of bacteria	V_{max} μg liter^{-1} hr^{-1}	Turnover time (hr)	Phytoplankton productivity (mg C m^{-2} day^{-1})
Char, arctic	0.1–2 × 10^8 liter^{-1}	1–8 × 10^{-3} (glucose)	43–1700	25
Meretta, arctic	2–80 × 10^8 liter^{-1}	0.1–7.5 × 10^{-1} (glucose)	5–175	125
Vorderer Finstertaler (8 mo ice and snow cover)	2–8 × 10^8 liter^{-1}	Ratio of bacteria to phytoplankton Winter: 3 : 1 Summer: 0.03 : 1		1.8
Lake Lotsjon, Sweden	Winter temperature 0.4–0.6	(Acetate) Winter: 3.3–8.7 × 10^{-1} Summer: 80 × 10	 4–22 0.5	
Lapland lakes (6) (ice- and snow-covered most of year)		1–2 × 10^{-3} (glucose)	—	50
Erken (ice-covered)	Estimate to be 50 × 10^6 liter^{-1} during winter (<4°C)	(water beneath the ice) Winter samples 2.3–4.5 × 10^{-2} (glucose) 1.9–5.4 × 10^{-2} (acetate)	 60–100 200–430	500
Wingra, Wisconsin (winter ice cover)	Bacterial No./g sediment 4°C counts: 0.7–44 × 10^5 20°C counts: 0.7–70 × 10^5		Optimum uptake of [^{14}C-] glucose by winter population is 25°C	
Marion Lake, Canada	Bacteria cm^{-3} sediment	V_{max} μg C g^{-1} hr^{-1} sediment		31

Marion Lake, Canada	Bacteria cm^{-3} sediment	Glycine	Acetate	Glucose
Winter (<3°C)	5 × 10^4	0.1–0.5	3–5	1–3
Summer (>10°C)	2 × 10^6	2–3	40–50	10–20

West Blue Lake, Manitoba (ice surface during winter)	Organisms ml^{-1} on different substrates[a]				18–67 (summer only)
	a. Succinic	24.6	2.8×10^{-2}	33	
	b. Pyruvic	21.7	7.7×10^{-2}	300	
	c. Fumaric	13.1	4.8×10^{-2}	50	
	d. Malic	12.0	3.6×10^{-2}	8	
	e. Lactic	6.0	1.2×10^{-1}	303	
	f. Acetic	5.8	3.1×10^{-2}	416	
	g. Citric	2.4	8.0×10^{-3}	100	
	h. Glycollic	2.1	9.0×10^{-3}	950	
	i. Complete medium	92.4			
Upper Klamath Lake (naturally eutrophic; ice surface during winter)	February samples only, 5°C				No data
	a. Glutamate	4–6		10–15	
	b. Aspartate	4–5		5–15	
	c. Asparagine	5		20–40	
	d. Lysine	2		10–15	
	e. Proline	>5		25–135	
	f. Alanine	2–3		20–60	
	Bacterial counts ml^{-1} at 2°C				
Dolomite Lake	65				
Shell Lake	290				
Gravel Pit Lake	85				
Hospital Lake	110				
Duck Lake	83				
Twin Lake	93				
Boot Lake	60				
Hidden Lake	240				

[a] Average of four different sample times; complete medium is a *Cytophaga* medium.

[From Baross, J. A., and Morita, R. Y. (1978). "Microbial Life in Extreme Environments." Academic Press, London.]

Table V Incidence and Activity of Microorganisms in Three Antarctic Lakes

Lake and properties	Primary productivity (mg C m^{-2} day^{-1})	Number of bacteria	Isolation temperature (°C)	Microbial activity
Bonney				
Depth: 32 m	31	High concentration just below zone of photosynthetic activity (20 m), 1.2×10^5 ml^{-1} to $<10^3$ ml^{-1} at 30 m	Direct counts	
Temperature: perennially ice-covered: −2° to 7°C, high temperature at 15 m				
Salinity: 0.25% at 11 m; 21.7% at 30 m; mostly NaCl and $MgCl_2$		Spring: 0–400 ml^{-1} Summer: 3–460 ml^{-1} Decrease with depth	0°C plate counts	
		Distinct vertical populations at 4, 6, 8–9, 12, and 15–16 m; 10^2–10^3 ml^{-1}	4°C plate counts	[^{14}C]acetate max. activity 15×10^3 cpm at 5 m, decrease with depth
Vanda				
Depth: 60 m	29	No zone of high bacterial densities noted, summer 0.4–1 $\times 10^3$ ml^{-1} blue-green algae		[^{14}C]glucose surface: no counts; 10 m 180 cpm
Temperature: perennially ice-covered; 0°C at surface to 24.2°C at 60 m (solar trap)				
Salinity: 12%, 10% $CaCl_2$		Summer surface water, 210 ml^{-1}; 30 m is 5 ml^{-1}; 40–50 m is 175–180 ml, bottom 0 ml^{-1}	0°C plate counts	
Don Juan				
Depth: 11 cm	No photosynthesis	*Bacillus*, *Micrococcus*, *Corynebacterium*, and a yeast species	Representative isolates grew at both 0° and 25°C; grew well in nutrient medium made with Don Juan water, particularly at low temperatures	
Temperature: no ice cover; −24° to −3°C				
Salinity: 47.4% solutes, most of which is $CaCl_2$				
		Apparent toxicity of pond water; 10^{-2} dilution gave 3 organisms, whereas 10^{-3} dilution gave 90. Only microorganism recovered was *Arthobacter parvulus*	Plate counts at 20°C	

[From Baross, J. A., and Morita, R. Y. (1978). "Microbial Life in Extreme Environments." Academic Press, London.]

receives. Although low temperature is one of the factors in psychrophilic environments, this is not necessarily true for psychrotrophic environments where wide fluctuations of temperature can occur. Nevertheless, the organism's ability to survive the cold temperature (as well as desiccation, lack of energy at times, etc.) is one of the main factors in their evolution. The ability to survive permits the organism to grow well when conditions are more optimal. Much more research is needed on the various ecosystems within the low-temperature environments.

Acknowledgment

Published as technical paper No. 9451, Oregon Agricultural Experiment Station.

Bibliography

Atlas, R. M., and Griffiths, R. P. (1984). Bacterial populations of the Beaufort Sea. *In* "The Alaskan Beaufort Seas Ecosystems and Environments." Academic Press, Orlando.

Baross, J. A., and Morita, R. Y. (1978). Life at low temperatures: Ecological aspects. *In* "Microbial Life in Extreme Environments" (D. J. Kushner, ed.), pp. 9–71. Academic Press, London.

Herbert, R. A. (1986). The ecology and physiology of psychrophilic microorganisms. *In* "Microbes in Extreme Environments" (R. A. Herbert and G. A. Codd, eds.), pp. 1–23. Academic Press, Orlando.

Langridge, P., and Morita, R. Y. *J. Bacteriol.* **92,** 418–423.

Morita, R. Y. (1975). *Bacteriol. Rev.* **39,** 144–167.

Gregory, P. H. (1973). "The Microbiology of the Atmosphere." Halsted Press, New York.

Vincent, W. F. (1988). Microbial ecosystems of Antarctica. Cambridge University Press, Cambridge.

Wynn-Williams, D. D. (1990). *Adv. Microb. Ecol.* **11,** 71–146.

Lyme Disease

Leonard H. Sigal
University of Medicine and Dentistry of New Jersey–Robert Wood Johnson Medical School

Glossary

Blood–brain barrier Variety of cells and membranes that prevent the direct continuity of serum and cerebrospinal fluid as well as the direct transfer of large molecules between the two fluids

Borrelia burgdorferi Causative agent of Lyme disease

Enzyme-linked immunosorbent assay Technique to detect antibodies in fluids; a compound or cell can be placed on the inner wall of a plastic well and antibodies binding to it detected by the addition of an anti-human immunoglobulin that has been conjugated to an enzyme; with the addition of a compound that changes color upon being digested by the enzyme, color change in the well can be detected as a measure of the specific antibody present in the fluid being tested

Erythema chronicum migrans Pathognomonic skin lesion of early Lyme disease; usually starts as an erythematous macule (flat lesion) or papule (small, raised lesion), which then expands, often with clearing of the center, so that the resulting lesion is an expanding erythematous ring

Immunoblot (Western blot) Technique to detect antibodies against specific components of an organism or mixture of proteins; the proteins are separated by electrophoresis and transferred to a solid support, which is then incubated with the test serum; detection of bound antibodies can use enzyme-linked immunosorbent assay (ELISA) technology; alternatively, the detection system may make use of a radioactive label rather than the enzyme used for ELISA

Indirect immunofluorescence Technique to detect antibodies in a fluid directed against an organism or tissue; the organism or tissue is placed on a glass slide and antibodies are detected by the addition of a fluoresceinated anti-human immunoglobulin

Mitogen Compound that nonspecifically stimulates all B or T cells; mitogens that affect only B cells or only T cells have been described

Oligoclonal bands Cerebrospinal fluid can be subjected to electrophoresis and immunoglobulin detected; if only a few electrophoretically defined immunoglobulin clones are found (oligoclonal bands), one can say that local production is likely, and implicating local infection or a local immune response; were the immunoglobulin present to be the result of serum leakage across the blood–brain barrier, one would expect no discrete bands to be seen, rather a broad smear of serum antibodies

Polymerase chain reaction Technique to detect small amounts of a specific gene, by making repeated copies of the gene using a polymerase and specific synthetic substrate

LYME DISEASE (LD) is a multisystem inflammatory disease caused by the recently described organism *Borrelia burgdorferi*. Lyme disease was identified as a new disease after epidemiologic investigation of an outbreak of juvenile rheumatoid arthritis in three small communities in Connecticut in 1975; however, the history of LD is over 100 years old. The skin rash of LD, erythema chronicum migrans (ECM), was first described in Sweden in 1909.

In the following decades, the syndrome of lymphocytic meningitis, painful radiculitis, and cranial nerve palsy was described and, in 1951, related to preceding ECM and bite by the tick *Ixodes ricinus*. This tick-borne meningopolyneuritis was later called Bannwarth syndrome and thought to be due to an infection, likely a spirochete. The cutaneous lesions acrodermatitis chronica atrophicans and lymphadenosis benigna cutis are often preceded by ECM and are now known to be due to the same pathogen, *B. burgdorferi*. In the past 17 years, LD has been recognized across North America (and in two areas of Mexico) and Europe and in certain regions of Africa, Asia, and Australia. The disease is spread by Ixodid ticks (*I. dammini, I. scapularis*, and *I. pacificus* in the United States and Canada; *I. ricinus* in Europe; *I. persulcatus* in Asia; and, possibly, *I. hyocyclus* in Australia); other ticks may also spread the disease. LD is now known to have many clinical manifestations and may mimic many other conditions. Laboratory diagnostic testing is often helpful and, once diagnosed, LD usually responds to antibiotic therapy. The complex microbiology and chemistry of *B. burgdorferi* helps to explain the specific immune responses *in vivo* and the immunomodulatory effects described *in vitro*, which may explain some of the manifestations of the disease.

I. Clinical Description of Lyme Disease

A. Early, Localized Disease

Within 1 mo of the tick bite (recalled by about 30% of patients) the pathognomonic skin rash of LD, ECM, occurs in 50–70% of patients at the site of the bite. Usually beginning as an erythematous asymptomatic macule or papule, ECM usually expands, often with central clearing. The lesion occasionally will burn, sting, or hurt, and it may be indurated; vesicular, necrotic, or urticarial lesions have been reported rarely. If untreated, ECM usually fades spontaneously, although occasionally the untreated rash may persist for many months. In about 50%, multiple lesions are noted; secondary lesions are thought to be due to hematogenous dissemination of infection. *Borrelia burgdorferi* has been grown from the expanding erythematous ring of the ECM lesion. Many patients are totally asymptomatic and unaware of the rash. Early LD is often associated with nonspecific symptoms, described as a "viruslike syndrome," including fever, lethargy, malaise, and fatigue. Some patients describe arthralgias and myalgias. Complaints related to focal pathology also occur, including pain on neck flexion, malar rash, conjunctivitis, erythematous throat, and temporomandibular (and other) joint pains may occur. Regional or generalized lymphadenopathy (usually of the nodes draining the area with the ECM lesion) is seen in about one-third of patients. Rarely hepatosplenomegaly, right upper quadrant tenderness, or muscle tenderness is noted.

B. Early, Disseminated Disease

1. Neurologic Features

Approximately 10–15% of untreated patients with early LD develop the neurologic features of early disseminated LD (Bannwarth's syndrome), usually 2–3 mo after ECM but occasionally as long as 9 mo later. Proper therapy of early localized disease usually prevents progression; no studies to determine the proportion of patients who progress despite treatment have been done, but it is likely to be a rare occurrence. A lymphocytic meningitis, with headache, mild neck stiffness, and photophobia, may occur, but Kernig's and Brudzinski's signs (clinical markers of meningeal irritation) are absent. Meningitis, cranial neuropathy, and peripheral neuropathy, when found together, constitute Bannwarth's syndrome. The organism has been grown from samples of cerebrospinal fluid (CSF) and oligoclonal bands in the CSF with anti-*B. burgdorferi* reactivity, and concentrations of both cellular (T and B cell) and humoral specific immunity in the CSF suggest that a local immune response is mounted in meningitis and radiculitis.

Fever may be mild or absent, but fatigue and malaise are common. Mild encephalopathy (difficulty with concentration and memory and emotional lability) may be prominent. Any of the cranial nerves can be affected, but facial nerve palsy is the most common, often bilateral. Facial palsies are often self-limited, usually lasting less than 2 mo. Radiculoneuropathy may affect the limbs or trunk.

2. Cardiac Features

Approximately 8–10% of patients with untreated early LD develop carditis, which may be an isolated finding or occur with neurologic disease (see earlier). Lyme carditis causes inflammation and dysfunction of the cardiac conduction system (fluctuating degree of atrioventricular conduction defect and occasional tachyarrhythmias) and the heart muscle (myopericarditis and mild congestive heart failure). Cardiac LD can occur coincident with ECM and neurologic, and/or articular features of LD. Patients may complain of lightheadedness, syncope, palpitations, shortness of breath, or chest pain. *Borrelia burgdorferi,* or a form of an organism similar to it, has been seen on endomyocardial biopsies and at postmortem and grown from the myocardium of a patient with chronic dilated cardiomyopathy. Conduction defect may be the first and only manifestation of LD. Carditis is usually reversible, with or without antibiotic therapy, but there have been a few fatal cases and a small number where the conduction defect was permanent despite adequate antibiotic therapy.

C. Late Disease

1. Arthritis

In a study of 55 patients done prior to the routine use of antibiotic therapy for early LD, 44 (80%) experienced articular complaints over the course of a 6-year follow-up. Ten patients (18% of the group) had ECM with, or followed in a few weeks to months by, arthralgias. A second group of 28 patients (51% of the group) had ECM followed days to years later by intermittent true arthritis. One-half of these patients had preceding arthralgias. The arthritis usually affected one or only a few joints and was usually asymmetric. The knee was the most frequent joint affected, in 27 of the 28 patients. Some patients had a migratory and intermittent arthritis. This pattern of arthritis can resemble juvenile rheumatoid arthritis; it was the investigation of a cluster of newly reported cases of juvenile rheumatoid arthritis that brought LD to the attention of the medical world in 1977. Finally, six patients (11%) experienced chronic Lyme arthritis, occurring months to years after ECM. One to three joints were affected; in four patients the knee was involved.

2. Tertiary Neuroborreliosis

The neurologic features of early, disseminated LD were identified over 70 years ago. In the last decade, a syndrome of encephalomyelopathy, polyneuritis, and mental and/or psychiatric changes has been ascribed to chronic LD, occasionally as the first feature of previously latent infection. This syndrome is known as tertiary neuroborreliosis, a term meant to suggest a possible analogy with neurosyphilis, another chronic spirochetal infection of the central nervous system infection. The late neurologic features of LD occur months to years after the onset of LD, sometimes as the only manifestation of LD, occasionally coeval with Lyme arthritis. The central nervous system is usually affected, in contrast to the predominantly peripheral nervous system features of earlier neurologic damage. Lymphocytic pleocytosis is rare in tertiary neuroborreliosis.

D. Pregnancy and Lyme Disease

The effect of *B. burgdorferi* infection on pregnancy represents an area of major concern. The initial reports of LD-complicated pregnancies suggest that this infection might cause adverse outcomes (including congenital neurologic damage, intrauterine death, premature labor, and perinatal death due to congenital cardiac anomalies). Subsequently reported experiences, in clinical series and prospective analyses of pregnancies complicated by LD, have been far more encouraging. This is not to say that LD has absolutely no effect on the unborn child; however, if there is an adverse effect, it seems to be quite uncommon.

E. Ophthalmologic Features

Ophthalmologic damage associated with other features of LD, or as the sole manifestation of *B. burgdorferi* infection, has been reported. Cases of conjunctivitis, corneal infiltrations, and acute visual impairment or loss have been described, the latter due to presumed *B. Burgdorferi*-induced vitritis, optic neuritis, retinal vasculitis, neuroretinitis, choroiditis, iridocyclitis, and panophthalmitis. In addition, cranial neuropathies can affect the intrinsic musculature of the eye. The presence of extraocular problems compatible with LD often suggests the diagnosis. These case reports of LD-related eye disease may represent a true association or may be mere clinical coincidences. The proposed ocular features of LD may represent recrudescence of the organism from a privileged site within the eye.

II. Laboratory Diagnostic Testing in Lyme Disease

LD remains a clinical diagnosis, with laboratory testing used as an adjunct only. A number of techniques have been employed in LD testing.

A. Indirect Immunofluorescence

Indirect immunofluorescence (IFA) is performed by placing the organism on a glass slide and then applying the serum to be tested. After the excess serum is removed, an anti-human immunoglobulin (known as a secondary) antibody, typically produced in goat or sheep and with a fluorescent molecule added to it, is added. Detection of adherent patient antibody is then detected by exposing the slide to the appropriate ultraviolet light and viewing the fluorescence. The usefulness of this technique is hampered by frequent false-positive testing and the fact that large numbers of samples cannot be done easily. IFA has largely been replaced by enzyme-linked immunosorbent assay (ELISA).

B. Enzyme-Linked Immunosorbent Assay

1. Standard Enzyme-Linked Immunosorbent Assay

ELISA is done by coating plastic microtiter plates with whole or modified organism (a number of different *B. burgdorferi*-derived preparations are used, none yet proved to be superior) and then adding the patient serum to be tested. Once the excess has been washed away, a secondary antibody is added; the difference between IFA and ELISA is that the secondary has an enzyme conjugated to it. Thus, after the excess secondary antibody has been washed away, the ELISA well contains a sandwich with the plastic wall of the plate as the base, followed by the organism, the adherent serum antibody, the secondary antibody, and finally the enzyme; the amount of enzyme in the sandwich is proportional to the amount of patient antibody binding to the organism on the well wall. Addition of a chemical that changes color when it is digested by the enzyme (known as the chromagen) is the detection step, and automated ELISA readers allow colorimetric reading of whole plates.

The results of ELISA can then be expressed in a number of ways. First, the optical density (OD, which is a measure of the amount of color generated) can be reported and compared with an "upper limit of normal," generated from a statistical analysis of the OD readings derived from normal sera; by these means, each laboratory can set its own upper limit of normal. Second, the OD may be expressed as a percentage of the OD generated from an established positive. Third, an established positive serum can be serially diluted and run in the assay in parallel with the patient sample, generating a standard curve. The OD of the patient's sample is then compared with the standard curve and the result is expressed as the equivalent of the dilution of the standard serum giving the same OD. Finally, the patient serum may be serially diluted and run in the assay, with the result expressed as the last dilution, or titer, at which the sample's OD exceeds that of the normal sera used. Each technique has its adherents, and none is proven superior.

The advantage of ELISA is that it can be "scaled up" to do large numbers of samples and the specificity can be increased by defining a more stringent upper limit of normal (although this will sacrifice some sensitivity in the assay). ELISA is still prone to false-positivity but less so than IFA. In LD, treponemal infections (e.g., syphilis, pinta, yaws, gingivitis due to *Treponema denticola*), other spirochetal and nonspirochetal infections, and infections causing polyclonal B cell activation (e.g., Epstein-Barr virus) can cause sufficient cross-reacting antibody to produce a false-positive ELISA. [*See* ELISA TECHNOLOGY.]

2. Antigen-Capture Enzyme-Linked Immunosorbent Assay

A modification of the standard ELISA starts by coating the well with an isotype-specific anti-human immunoglobulin, which "captures" the serum antibody added next. The subsequent step adds a preparation of organism, followed by addition of an anti-organism serum made in an animal not of the species generating the original anti-human immunoglobulin. The organism is bound only by the patient's antibodies that have anti-*B. burgdorferi* activity. The secondary antibody has an enzyme conjugated to it, and the addition of chromagen and color measurement complete the antigen-capture ELISA. This technique is very useful in detecting small amounts of antibody (e.g., in CSF).

3. Concentration of Reactivity at Sites of Inflammation

It is very important to establish that arthritis or neurologic disease is due to LD, because this is a potentially treatable infection; the alternative is that a patient has serum antibodies and a coincident, but unrelated, clinical problem. Thus, testing CSF or synovial fluid and comparing the amount of antibody with that present in serum may be crucial in patient care. Antigen-capture ELISA may be a way to do this kind of testing with great ease; i.e., the capture of equal amounts of serum antibody from normal and possible LD CSF may allow one to determine that there is evidence of LD. Alternatively, the OD derived from CSF and serum done in parallel can be subjected to compensation for the different amounts of albumen and total amount of immunoglobulin of the isotype being tested (serum has more immunoglobulin than CSF and the amount of albumen in the CSF is a measure of the breakdown of the blood–brain barrier and, thus, a marker of the passive leakage of antibody from serum into CSF).

Another marker of intrathecal synthesis of specific antibody is the finding of oligoclonal bands, as detected by immunoelectrophoresis, indicating that there are discrete families of antibody present in the CSF. These may be studied to see if they have anti-*B. burgdorferi* activity.

C. Immunoblot (Western Blot)

IFA and ELISA determine the presence of antibodies against whole or modified organism. By means of sodium dodecyl sulphate–polyacrylamide gel electrophoresis (SDS-PAGE), one can separate a mixture of proteins into discrete bands of separate molecular weight, each representing a single constituent. The proteins thus spread out in the gel can then be transferred onto a nitrocellulose or nylon filter paper and then incubated with the serum being tested. Antibodies in the patient's serum can be detected by the application of ELISA technology, as already described; alternatively, one can use a radioactive marker on the secondary antibody and then expose the filter paper to X-ray film and read the bands that develop.

The advantage of immunoblotting is that one can see which specific proteins are bound by the serum. It has become apparent that antibodies to certain proteins are seen widely in the normal population; for example, antibodies against *B. burgdorferi*'s flagellin are found in non-LD patients, representing antibodies against the flagellins of other microorganisms in the normal gut flora or causing urinary tract infections. Patterns of reactivity compatible with early or late infection can be read in immunoblot, and positive ELISA results can thereby be corroborated.

D. T-Cell Proliferative Responses

In 1986, the first report of the proliferation of T cells from patients with LD on *in vitro* exposure to *B. burgdorferi* appeared. Despite early enthusiasm for T-cell testing as a diagnostic test, there seems little use for this assay, due to the frequency of false-positive and negative results. Concentration of antigen-specific T cells has been reported in LD CSF and synovial fluid, suggesting that comparison of levels of T-cell proliferation in inflammatory fluids with peripheral blood might be a useful test for LD.

E. Polymerase Chain Reaction

Polymerase chain reaction (PCR) allows for rapid amplification of specific messenger RNA or DNA sequences, so that rare sequences or small numbers of organisms can be identified or used for further study (e.g., sequencing, cloning). The potential for PCR in diagnostic testing is great, in that PCR can amplify a very small number (theoretically between one and five per sample) of *B. burgdorferi* or other genes, thereby allowing the detection of the organism in biologic materials.

The known genetic material of the organism is searched for a sequence that is absolutely unique, unshared by any other known gene. Oligonucleotides that are complementary to the sequences on both strands of the DNA flanking the unique area are manufactured. The biologic material to be tested is then mixed with a vast excess of these "primers" and of adenosine, guanosine, cytidine, and ribosylthymine triphosphates, and the mixture is heated to denature the DNA to be detected into single strands. With cooling, most of the single strands will anneal with the primers. A DNA polymerase (often the thermostable Taq polymerase) is then added, which then uses the single strand as a template, starting at the annealed primer. Thus, the single strand becomes a double strand of DNA, which upon denaturing gives two strands. By repeating this cycle 30 times, one can generate over 1,000,000,000 copies of the DNA, easily visualized on agarose gel electrophoresis.

Although the technique is relatively new, PCR will probably be very useful in LD and other infectious diseases. However, the very sensitivity of the assay predisposes to false-positive results, due to a poorly controlled assay or a sample that has been inadvertently contaminated. PCR can demonstrate that DNA of the organism is present; it does not distinguish between DNA derived from living and from dead organisms. Therefore, a positive PCR does not equate with active infection. [*See* POLYMERASE CHAIN REACTION (PCR).]

F. Antigen Detection Assays

Detection of organism-derived antigen in biologic fluids represents another potential diagnostic test. This technique was applied to urine previously, by concentrating the urine on a filter paper and probing it with monoclonal antibodies to certain *B. burgdorferi* proteins. An assay using this technique was available commercially and then removed from the market; many LD researchers thought the assay was not accurate or useful. Others are now refining the assay.

III. Therapy

A. Early, Localized Disease

If untreated, ECM spontaneously resolves in a median of 28 days (range, 1 day to 14 mo). The rash usually disappears within a few days of the start of antibiotic therapy. Current oral regimens for early, localized disease include 3–4 wk of doxycycline (100 mg twice a day) and amoxicillin (500 mg three or four times daily); some clinicians add probenecid 500 mg three times daily. Other drugs have been tested for early LD, including azithromycin and cefuroxime axetil.

A relatively mild Jarisch–Herxheimer reaction is noted in about 15% of patients. The Jarisch–Herxheimer reaction was first described in syphilis and consists of a remarkable worsening of clinical signs and symptoms of disease 1–2 days after the institution of spirocheticidal therapy. This worsening is self-limited and lasts less than 1 day. The severity of the initial illness and the presence of multiple ECM lesions correlate with progression to later manifestations of LD. Even mild or inapparent early, localized LD can progress to later manifestations of LD, so even the mildest cases should be treated.

B. Early, Disseminated Disease

Intravenous therapy for neurologic disease is indicated, with the exception of isolated facial palsy, which may be treated with oral antibiotics if CSF analysis does not reveal inflammatory changes. If a pleocytosis, elevated protein, or concentration of specific anti-*B. burgdorferi* antibodies is found, intravenous therapy for facial palsy is indicated. LD-associated meningitis and neuropathy may resolve spontaneously; intravenous therapy is designed to speed the resolution of damage and prevent progression to later LD. The addition of corticosteroids may speed recovery from painful radiculitis, although no evidence indicates that corticosteroid therapy decreases the duration or severity of the facial palsy. No studies have addressed the issue of duration of intravenous therapy. The current recommendation is 2–4 wk of a third-generation cephalosporin, cefotaxime (3 g twice a day, or 2 g thrice daily) or ceftriaxone (1 g twice a day, or 2 g once a day); these two agents are equivalent and both are superior to penicillin.

The initial experience with Lyme carditis suggested that treatment with antibiotics was not needed, because the heart block usually resolved spontaneously or after therapy with anti-inflammatory agents, corticosteroids [usually in high doses (40–60 mg/day)] or aspirin. Most recent cases have been treated with antibiotics, but no studies have been done to establish an optimum regimen. Our recommendation is that patients be admitted to the hospital for monitoring and treated with intravenous antibiotics, as described already. Indications for temporary pacemaker placement are the same as those for patients with other forms of heart block. The heart block of Lyme carditis is usually self-limited, with resolution usually beginning in a few days. However, other examples of Lyme carditis have not resolved spontaneously, or even after adequate therapy.

C. Late Disease

Lyme arthritis can often be cured with intravenous antibiotic therapy, most often using the cephalosporin regimens described in Section III.B. Hydroxychloroquine and synovectomy have been used successfully in patients whose arthritis has not responded to antibiotic therapy; these modalities are also used in rheumatoid arthritis and other chronic inflammatory arthropathies. Therapy with oral amoxicillin plus probenecid or doxycycline has been reported as effective in Lyme arthritis, but these oral regimens may not have sufficient penetration into the neuraxis to assure the sterilization of clinically inapparent central nervous system foci of infection. We currently recommend intravenous therapy for Lyme arthritis; oral therapy of Lyme arthritis requires that neurologic disease is not clinically evident and that a CSF analysis is normal.

No studies have investigated the optimal therapy for late, neurologic LD. The anecdotal experience of many researchers in Europe and the United States suggests that intravenous antibiotics, as for the other late features of LD, may be effective. Typically resolution of clinical manifestations is slow, with some patients having residual neurologic dysfunction that appears to be permanent.

D. Pregnancy and Lyme Disease

Current evidence suggests that pregnancy complicated by LD should *not* be terminated. The patient should be treated with the antibiotic regimen indicated for her clinical manifestations of LD, although some physicians always treat with intravenous antibiotics in this circumstance.

IV. The Organism—*Borrelia burgdorferi*

Borreliae are spirochetes, with the typical structure of spirochetes, including helical shape, an outer cell membrane surrounding the protoplasmic cylinder (which consists of the cytoplasm, the inner cell membrane, and a peptidoglycan layer), and flagella (found in the periplasmic space, between the protoplasmic cylinder and the outer cell membrane; the flagella are inserted at the ends of the protoplasmic cylinder). *Borrelia burgdorferi* is somewhat longer and narrower than most *Borrelia* (20–30 μm in length, 0.2–0.3 μm in width). Division of *B. burgdorferi* into two genomic species has been proposed, based on patterns of ribosomal RNA gene-restriction patterns. The ability to culture *B. burgdorferi* in modified Kelly [(also known as BSK-II (Barbour–Stoenner–Kelly)] medium has allowed detailed study of the organism. [*See* SPIROCHETES.]

The outer surfacee membrane consists of 45–62% protein, 23–50% lipid, and 3–4% carbohydrate. Outer surface proteins (osp) A and B (31 and 34 kDa, respectively) are major constituents of this layer; ospA and ospB are found on the same plasmid and are cotranscribed, which is fairly unusual for gram-negative bacterial membrane proteins. Other noteworthy proteins of this organism include flagellin (41 kDa and the immunodominant protein, especially early in infection), heat-shock proteins (60 and 66 kDa, both analogs of the *Escherichia coli* GroEL heat-shock protein, and 72 kDa, analog of the *E. coli* dnaK protein), and a penicillin-binding protein (94 kDa). The function of proteins with molecular weights of approximately 88, 83, 75, 73, 39, 29, 25, 20, and 18 are unknown; these weights may vary, presumably due to intracellular modification.

Bacteriophages and plasmids have been identified within the organism; between four and seven plasmids were found in each isolate tested, and the plasmid profile varies greatly. Virulence factors may be encoded on some of these plasmids, as changes in infectivity and loss of plasmids occur in parallel during *in vitro* culture. At least one linear plasmid with covalently closed ends has been identified, a form of DNA seen in some animal viruses but not previously in a prokaryotic organism. The telomeric structure of these linear plasmids is similar to that of eukaryotic viruses, suggesting that the novel linear plasmids of *Borrelia* may result from horizontal genetic transfer across kindgoms. [*See* BACTERIOPHAGES; PLASMIDS.]

V. *In Vivo* and *in Vitro* Immunologic Changes Induced by *Borrelia burgdorferi*

The causative agent of LD is capable of inducing specific and nonspecific immune reactions, affecting the function of T and B cells and macrophages.

Immune regulation, as measured *in vitro* by suppressor cell assays, is modified by incubation of normal cells with *B. burgdorferi*. Peripheral blood mononuclear cells from patients with LD have deficient suppressor cell activity (coincident with a T lymphopenia), both of which tend to normalize as disease resolves. The organism can stimulate normal T cells to produce gamma-interferon and interleukin-6 (IL6); the organism can also stimulate glial cells to produce IL6 *in vitro*. In addition to these nonspecific changes, antigen-specific T cells develop, which can be detected by lymphocyte proliferation assays. Antigen-specific T cells can be found in the blood but are concentrated in the CSF and synovial fluid of patients with Lyme meningitis or arthritis, respectively. As noted earlier, this concentration of specific immune reactivity may be used for diagnostic purposes. [*See* T LYMPHOCYTES.]

Borrelia burgdorferi contains a B cell mitogen (a mitogen is a substance that nonspecifically stimulates B or T cells). The presence of a true lipopolysaccharide in this organism has been disputed. Nonetheless, the organism contains a substance capable of polyclonal B-cell activation, with measurable *in vitro* production of immunoglobulin. [*See* B CELLS.]

Specific anti-*B. burgdorferi* IgM antibodies can be detected as early as 2 wk after the tick bite, although in some cases the antibody response may be delayed as long as 6–8 wk. Another peculiarity of the immune response in LD is that early, even inadequate, antibiotic therapy may permanently suppress the antibody response. IgG antibodies are usually detectable less than 1 wk after the IgM, and elevated levels of IgG may persist for long periods. Immune complexes, presumably containing specific antibody and borrelial antigens (the former has been demonstrated experimentally) can be detected in serum and inflammatory fluids, although a pathogenetic role for these complexes has not been established. Detection of relative concentration of specific antibodies in CSF and synovial fluid can be used for diagnostic purposes. Some of the antibodies to flagellin can also bind *in vitro* to human axons (indirect immunofluorescence on sections of normal human nerve) and to human and rat neuroblastoma cell lines, suggesting that the immunopathogenesis of Lyme neurologic disease may include autoimmune phenomena predicated on molecular mimicry.

Decreased natural killer cell activity has been documented in patients with ECM and in those with active arthritis but was normal in convalescent patients.

Human mononuclear phagocytes make IL1 upon exposure to the organism, and studies with a murine monocytelike cell line, RAW264, suggest that messenger RNA for tumor necrosis factor is increased after *in vitro* stimulation of the cells with *B. burgdorferi*. The organism is taken up by monocytes and may survive in the intracellular compartment. Intracellular killing of the organism within polymorphonuclear cells and monocytes has been documented, invoking oxidative mechanisms. Phagocytosis is increased by the addition of immune serum, and immune serum increases the intensity of the respiratory burst. Certain components of the organism, including the flagellin, are chemotactic factors for phagocytes, and a host-derived neutrophil chemoattractant factor has been identified in Lyme arthritis synovial fluid.

VI. Vaccine Development

Clinical experience with LD suggests that seropositivity from preceding exposure to LD does not generally protect humans from a second episode of LD. However, there is reason to believe that the second episode of LD in patients with preceding severe manifestations of LD may be milder than the initial disease. Of interest is the observation that patients with previous chronic LD previously cured with antibiotic therapy may be protected from developing subsequent disease. Thus, there is reason to believe that a vaccine might be effective in LD.

Mice inoculated with *B. burgdorferi* ospA, recombinant or purified from the organism, can be protected from infection. Monoclonal antibodies to ospA prevent infection in severe combined immunodeficiency mice. Other preparations of the organism (e.g., thimerosal inactivated organism) have also been shown effective as vaccines under experimental conditions. A dog vaccine has been marketed, although the efficacy of this preparation has not been demonstrated.

OspA is highly conserved across the spectrum of *B. burgdorferi* isolates and currently is the best studied candidate antigen for a vaccine, because it may be useful in all the geographic areas where LD has been described. A major concern is that antibodies to ospA appear only late in natural infection, which may limit its usefulness as a human vaccine active against early disease.

Bibliography

Barbour, A. G., and Hayes, S. F. (1986). *Microbiol. Rev.* **50,** 381–396.

Benach, J. L., and Bosler, E. M. (1988). *Ann. N. Y. Acad. Sci.* **539,** 1–513.

Rahn, D. W., and Malawista, S. E. (1991). *Ann. Intern. Med.* **114,** 472–481.

Sigal, L. H. (1989). *Semin. Arthritis Rheum.* **18,** 151–167.

Sigal, L. H. (1991). *Arthritis Rheum.* **34,** 367–370.

Sigal, L. H. (1992). *DRUGS* (in press).

Stanek, G. M., Flamm, H., Barbour, A. G., and Burgdorfer, W. (1986). *Zentralbl. Bakteriol. Hyg. A* **263,** 1–495.

Steere, A. C. (1989). *New Engl. J. Med.* **321,** 586–596.

Steere, A. C., Malawista, S. E., Craft, J. E., Fischer, D. K., and Garcia-Blanco, M. (1984). *Yale J. Biol. Med.* **57,** 445–705.

Szczepanski, A., and Benach, J. L. (1991). *Microbiol. Rev.* **55,** 21–34.

ISBN 0-12-226892-X